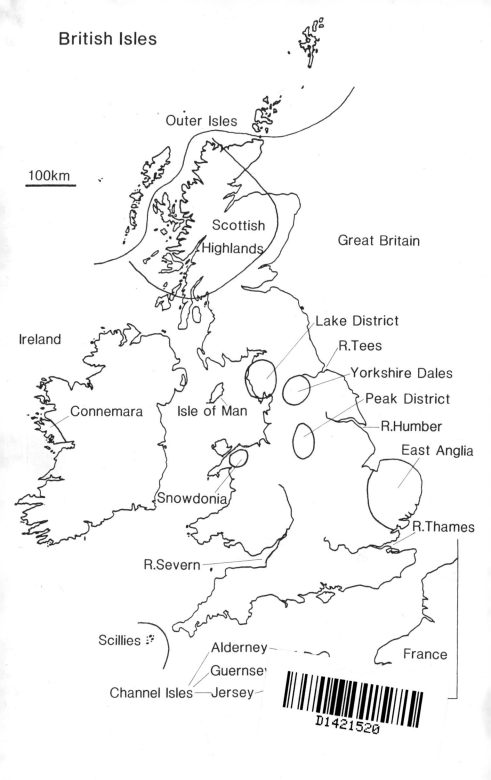

British Isles

Outer Isles

100km

Scottish Highlands

Great Britain

Ireland

Lake District

R.Tees

Yorkshire Dales

Peak District

R.Humber

East Anglia

Connemara

Isle of Man

Snowdonia

R.Thames

R.Severn

Scillies

Alderney

Guernsey

Channel Isles — Jersey

France

D1421520

NEW FLORA *of the* BRITISH ISLES

CLIVE STACE

with illustrations mainly
by
Hilli Thompson

CAMBRIDGE
UNIVERSITY PRESS

Published by the Press Syndicate of the University of Cambridge
The Pitt Building, Trumpington Street, Cambridge CB2 1RP
40 West 20th Street, New York, NY 10011-4211, USA
10 Stamford Road, Oakleigh, Victoria, 3166, Australia

First published 1991
Reprinted 1992

Printed in Great Britain by
St Edmundsbury Press Ltd, Bury St Edmunds, Suffolk

British Library cataloguing in publication data

Stace, Clive A.
New flora of the British Isles.
I. Title
581.0941

ISBN 0 521 42793 2

C O N T E N T S

PREFACE

In writing this book I have attempted to produce exactly the kind of Flora that for twenty to thirty years I have wanted for my own use. Such a Flora would be as complete, up-to-date and user-friendly as possible, would be selectively illustrated, and would be available at a reasonable price.

No doubt many people will find camera-ready copy of the sort utilized here less attractive than traditional type-setting, but the costs of the latter would have increased the price of the book very steeply without, in my opinion, increasing its utility in any way. Others might well decry the consistent use of English names (albeit completely subsidiary to the scientific Latin names), but I strongly believe that the study of wild plants by many more people with very diverse backgrounds is important if we are to convince the politicians that we must effectively conserve our native plant genetic diversity. For the same reason I have used fewer technical terms and fewer abbreviations of them than is usual in Floras; indeed, with hindsight I believe I should have used even fewer.

None of the above, however, is to be seen in any way as compromising or diminishing the need for absolute accuracy. When it is necessary to use a greater magnification than a hand-lens, or to cut sections of an organ, in order to see the diagnostic features, I have never pretended otherwise. The lack of a means of magnifying objects above x20 in good illumination, or of the ability to measure accurately to within 0.1mm, not only prevents one from obtaining certain data but, more seriously, is a frequent cause of misinterpretation or mismeasurement of plants. The remedy is obvious, and no more expensive than are the essential tools of a photographer, ornithologist or golfer.

The Flora is designed to enable field-botanists and those working with herbarium specimens to identify plants that are found **in the wild** in the British Isles. This is, I believe, a new criterion; more usually the origin and performance of plants are given higher significance when deciding where to draw the line between those to be included and those to be excluded. However, when one encounters a plant it is often not possible to know whether it is native or alien, or whether it has arrived accidentally or been planted, and one cannot know whether it will still be there next year! Hence a pragmatic approach has been adopted. The list of species included, as well as the data provided for each of them (especially their nomenclature and distribution), are as up-to-date as the information I possessed in October 1990 would allow.

The decision to write this Flora was made in 1983, and the actual preparation of the text took almost seven years. This decision was made due to the collapse of plans by a group of taxonomists to write a multi-volume definitive or critical Flora of Great Britain and Ireland, which had been discussed over the period December 1973 to January 1985. In the early 1980s it became clear that this project would founder (just as a similar one did in the 1930s), and it is regrettably the case that the British Isles still lack a truly complete Flora. Much information (especially concerning aliens) used in writing the present book was originally obtained by me for the abandoned floristic project in the late 1970s. However, this Flora in no way replaces the latter.

In the Acknowledgements I have tried to convey my indebtedness to all those who have helped me over the past 40 or more years and during the writing of this book. Despite all their help I am aware that many imperfections remain, and doubtless errors will be uncovered as well. I should be very grateful to hear of those encountered by readers. If this Flora helps others to achieve anything approaching the degree of enjoyment and satisfaction that I have gained from the study of our wild plants, then it will have succeeded in its main aim.

Ullesthorpe, Leicestershire   CLIVE A. STACE
March 1991

ACKNOWLEDGEMENTS

This book certainly could not have been written without the assistance so readily given me by over 200 friends and correspondents over the past twelve years, quite apart from the help and encouragement I received for many years prior to that. Rather than make lame reference to people 'too numerous to mention', I have attempted to list all of those who have directly helped me. Their participation has ranged from essentially one-word answers to questions such as 'Is the _Phyteuma_ _spicatum_ naturalized in your area blue- or yellow-flowered?', to detailed advice on taxonomic problems, the provision of specimens, or careful proof-reading or factual checking of draft accounts. Others, mentioned elsewhere, have assisted in more specific ways, such as by preparing drawings.

As well as thanking those who have directly helped in the preparation of this Flora, I wish to pay tribute to a number of people who have been instrumental in guiding me along the road to becoming a botanist. From the age of about 8 my parents actively encouraged my interest in natural history and helped to develop contacts with the local museum and societies in Tunbridge Wells, Kent. In my teens I received a tremendous amount of tuition and stimulation from Aline Grasemann (Tunbridge Wells Natural History Society) and Dik Shaw (my biology master). Our 'regional expert', Francis Rose, was also very influential, especially after the formation of the Kent Field Club in 1955. My first post-card from Miss Grasemann (dated 13th September 1953) reads: "If you are not doing anything else next Saturday what about coming over to the Dykes from Tonbridge Castle and seeing the Orange Balsam? And if you want to come and look anything up in 'Clapham' afterwards do.' 'Clapham' (Clapham, Tutin & Warburg 1952) was to be a Christmas present from her three months later!

In all my years at University, as both student and lecturer, I have been fortunate in working with many very clever and helpful people. I should especially mention my postgraduate supervisor, Arthur Exell, and one of my professors at Manchester, David Valentine. The other great source of inspiration to me has been the Botanical Society of the British Isles. In the fifties and sixties those of its members who were of particular help to me, both on excursions and by post, were Joan and Peter Hall, Douglas Kent, Ted Lousley, David McClintock and Ted Bangerter.

Finally, my wife Margaret and sons Richard and Martin

deserve special thanks. To someone who is totally addicted
to field botany encouragement is both unnecessary and
inappropriate, but their understanding and support have been
crucial.
Of all those who have assisted me during the writing of this
book, a few demand particular mention. This project has been
very much a joint project with my wife and it has absorbed
much of her spare time over the past five years. She played
a major role in the planning of the format and in proof-
reading, and carried out all the inputting and the
preparation of camera-ready copy. Douglas Kent, who has been
preparing a new checklist of vascular plants of the British
Isles in parallel with my work, has given me the benefit of
his vast knowledge of alien as well as native plants and of
nomenclatural matters. He has corrected many nomenclatural
errors in my drafts. Hilli Thompson has skilfully prepared
the great majority of the line-drawings, patiently taking
account of all my demanding and pernickety criticisms and
requests for alterations. Peter Hall has painstakingly
proof-read all the text, ensuring that there are many fewer
mistakes and inconsistencies than would otherwise have been
the case, and has helped in numerous places by drawing on his
(and his wife Joan's) long field experience. Many people
have advised me on alien plants, particularly on those that
should and those that should not be treated by me, but by far
the most help has been received from David McClintock and
Eric Clement. Their knowledge of British alien plants is
unparalleled, and they shared it freely with me. I am more
indebted than I can adequately express to all the above for
their generous co-operation.
The long list of others to whom I offer my sincere thanks
follows: Kenneth Adams, John Akeroyd, Abdul-Karim Al-Bermani,
David Allen, Mark Atkinson, John Bailey, Peter Ball, G.H.
Ballantyne, Bernard Baum, E.P. Beattie, Stan Beasley, Peter
Benoit, E.M. Booth, Humphrey Bowen, Paul Bowman, John Bowra,
Eileen Bray, Anne Brewis, Mary Briggs, Dick Brummitt, Elaine
Bullard, John Burlison, Rodney Burton, Andrew Byfield, Mary
Caddick, Douglas Chalk, Arthur Chater, Eric Chicken, Tony
Church, Peter Clough, P.R. Colegate, Ann Conolly, Tom Cope,
Pam Copson, Roderick Corner, Eva Crackles, Diana Crichton,
Gigi Crompton, David Curry, Tom Curtis, Kery Dalby, Dick
David, John Day, David Dow, Ursula Duncan, Trevor Elkington,
Gwynn Ellis, Trevor Evans, Lynne Farrell, F. Fincher, Bryan
Fowler, Christopher Fraser-Jenkins, Jeanette Fryer, Gill
Gent, Joan Gibbons, Vera Gordon, Richard Gornall, Gordon
Graham, Florence Gravestock, Peter Green, Eric Greenwood,
Adrian Grenfell, Paul Hackney, G. Haldimann, Joan Hall,
Geoffrey Halliday, Gordon Hanson, Ray Harley, Gerald
Harrison, John Harron, Clare Harvey, Chris Haworth, Stan
Heyward, Sonia Holland, Kathleen Hollick, Florence Houseman,
Enid Hyde, Bertil Hylmoe, Ruth Ingram, Martin Ingrouille,
Charlie Jarvis, Charles Jeffrey, Clive Jermy, John Jobling,

Bengt Jonsell, Stephen Jury, Joachim Kadereit, John Kelsey, Archie Kenneth, Michel Kerguelen, Mary Kertland, Mohammed Khalaf, Peter Knipe, Doreen Lambert, Jacques Lambinon, Meredith Lane, David Lang, Peter Langley, Ailsa Lee, Alan Leslie, Frances Le Sueur, Richard Libbey, David Long, David Mabberley, Hugh McAllister, David McCosh, Len Margetts, Mary Martin, John Mason, Brian Mathew, C.T.F. Medd, Desmond Meikle, Ronald Melville, Guy Messenger, J.M. Milner, Alan Mitchell, M. Morris, Mike Mullin, A.R.G. Mundell, Michael Nelhams, Charles Nelson, Alan Newton, Hans den Nijs, Tycho Norlindh, Elizabeth Norman, John Palmer, Richard Palmer, Richard Pankhurst, Rosemary Parslow, James Partridge, Ron Payne, Franklyn Perring, Ted Phenna, Eric Philp, Ann Powell, Chris Preston, Tony Primavesi, Cecil Prime, Jim Ratter, John Richards, Dick Roberts, A.W. Robson, R.G.B. Roe, Francis Rose, Krzysztof Rostanski, Fred Rumsey, Jocelyn Russell, Alison Rutherford, Patience Ryan, Bruno Ryves, Margaret Sanderson, Maura Scannell, H.D. Schotsman, Walter Scott, Peter Sell, Ted Shaw, B. Shepard, Alan Silverside, B. Simpson, David Simpson, W.A. Sledge, Philip Smith, Sven Snogerup, Keith Spurgin, Rod Stern, Olga Stewart, Allan Stirling, Lawrence Storer, David Streeter, Barbara Sturdy, George Swan, Joan Swanborough, Eric Swann, Pierre Taschereau, Michael Taylor, Nigel Taylor, David Tennant, Richard Thomas, Stephanie Thomson, Goeran Thor, John Trist, Ian Trueman, Bill Tucker, Maureen Turner, Tom Tutin, David Valentine, Ted Wallace, Geoffrey Watts, David Webb, Mary McCallum Webster, Sarah Webster, David Welch, Derek Wells, Terry Wells, Christopher Westall, Ann Westcott, Mike Wilkinson, Arthur Willis, Chris Wilmot-Dear, Stan Woodell, Brian Wurzell, Goronwy Wynne, Peter Yeo, Jerzy Zygmunt.

INTRODUCTION

The following paragraphs are intended to explain the contents
and arrangement of the Flora and the reasons for the various
conventions adopted and decisions taken.

TAXONOMIC SCOPE

All vascular plants (pteridophytes, gymnosperms and
angiosperms) are included, as is traditional in British
Floras. These are placed in classes (Lycopodiopsida, etc.),
subclasses (Magnoliidae, etc.), families (Lauraceae, etc.),
genera, species and subspecies. In addition the Synopsis
lists within the angiosperms superorders (Magnoliiflorae,
etc.) and orders (Laurales, etc.). Below the family level,
subfamilies or tribes (both in the case of Asteraceae and
Poaceae) are defined only for those families with 20 or more
genera. Below the genus level, subgenera or sections are
defined only for those genera with at least 20 species.
 Apomictic microspecies are covered in full in most genera,
but not for the three notorious genera Rubus, Taraxacum and
Hieracium, for which specialist accounts already exist or are
in preparation by experts. In these genera a separation into
relatively easily recognized groups of microspecies (here
called sections) is provided instead. A full account of
these genera would have greatly exceeded my own abilities and
the scope of one volume.
 The coverage of alien taxa has been as thorough and
consistent as possible. Many more aliens are included than
in any other British Flora, yet a considerable number of
aliens traditionally to be found in other Floras have been
omitted. To merit inclusion, an alien must be either
naturalized (i.e. permanent and competing with other
vegetation, or self-perpetuating) or, if a casual, frequently
recurrent so that it can be found in most years. All this
applies as much to garden-escapes or throw-outs as to
unintentionally introduced plants. Rarity, or the
requirement of a highly specialized habitat, have not been
taken into consideration (any more than is the case with
natives). Cultivated species have been included if they are
field-crops or forestry-crops or, in the case of trees only,
ornamentals planted on a large scale. Exclusively garden
plants, however abundant, whether crops or ornamentals, have
not been covered, but most of the commoner taxa are included
anyway because of their occurrence as escapes or throw-outs.
Also exluded are non-tree ornamentals planted en masse on new
roadsides or in parks, etc. The aim of this re-vamped and

expanded set of criteria is to include **all taxa that the plant-hunter might reasonably be able to find 'in the wild' in any one year.** Any such plant, whether native, accidentally introduced or planted, affects wild habitats and is part of the ecosystem, and botanists and others might be expected to need or want to identify it. Ornamental trees (but not shrubs or herbs) have been included because they are long-lived and frequently persist decades after all other signs of planting have disappeared from the area, so that the finder could not be expected to know that they were once planted. Doubtless, some additions to and removals from the list finally adopted are justified, but the selection of taxa is as judicious as it is largely due to the enormous help I have received from many correspondents (but especially Eric Clement, Douglas Kent and David McClintock), who have made alien plants their special study and have generously given me the benefit of their advice.

As well as the taxa treated 'fully' (i.e. keyed out, and provided with a numbered entry), other taxa that narrowly miss qualification, or fall far short of qualification despite their frequent inclusion in other Floras, or have been erroneously included in the past, are also briefly mentioned. These are covered under families with a number followed by a letter (e.g. 11A), or under the headings 'Other genera' or 'Other spp.' that follow the keys to genera or species.

All interspecific and intergeneric hybrids are included, but the level of treatment varies. Hybrids that have attained distributions no longer tied to those of their parents (i.e. those that occur at least sometimes in the absence of both parents) are treated exactly like species, except that the multiplication sign is inserted between the generic name and the specific epithet (e.g. _Salix_ x _rubens_), and the parental formula is given (e.g. _S. alba_ x _S. fragilis_). This has been normal procedure for some genera (e.g. _Circaea_, _Mentha_) in the past, but a consistent application of the criterion has resulted in many more such taxa being similarly treated. Other hybrids are placed in their appropriate systematic position, but are not keyed and are not provided with their own number; they always occur with at least one parent and their identity can usually be deduced because of this. They are provided with as much information as their situation appears to warrant. The only exceptions to the above are that all the hybrids in the uniquely promiscuous genus _Euphrasia_ are not listed, and some highly fertile hybrids that can occur in the absence of both parents (e.g. _Geum_, _Hyacinthoides_) are not treated as separate entities, since they form a spectrum of variation linking that of their parents.

GEOGRAPHICAL SCOPE
This Flora deals with the British Isles, comprising Great

Britain (England, Scotland and Wales), Ireland (Northern
Ireland and Eire), the Isle of Man, and the Channel Isles
(Bailiwicks of Jersey and Guernsey). The Bailiwick of
Guernsey includes Guernsey, Alderney, Sark, Herm and various
lesser islands. The above are always referred to in their
correct, strict senses, rather than loosely, except that a
distinction between Great Britain and the Isle of Man is made
only where necessitated by particular patterns of
distribution. The United Kingdom (Great Britain and Northern
Ireland) is not referred to in the text.
The smallest geographical unit utilized is usually the
vice-county. There are 111 of these in Great Britain and 40
in Ireland, with the Isle of Man and Channel Isles
representing two others. All 153 are mapped and listed on
the end-papers. The Scillies are part of W Cornwall, but
they have a distinctive flora and are therefore frequently
referred to separately.

CLASSIFICATION AND NOMENCLATURE
I have attempted to use the most up-to-date and accurate
classification and nomenclature, but in the interests of
stability I have adopted names that represent departures from
normal usage in the British literature only where the
evidence is unequivocal or overwhelming. The writing of this
book coincided with the preparation by Douglas Kent of a new
standard list of vascular plants of the British Isles
(replacing that of Dandy (1958)). We have collaborated
closely in our respective works and have adopted the same
classification, sequence and nomenclature; I have benefited
greatly from my colleague's expertise and from the results of
his research, which he has generously put at my disposal.
Where a recent authoritative revision of a group is
available I have generally followed it, even where I am
inclined to disagree with some aspects. The family sequence
and circumscription followed is that of Cronquist (1981) for
the angiosperms, and Derrick et al. (1987) for the
pteridophytes. Many sources have been followed for generic
and specific sequences, but in general the less specialized
taxa precede the more specialized ones.
A limited list of synonyms is provided. These are the
accepted names used in the pre-1970 Dandy (1958), Clapham et
al. (1962, but not 1952), and Tutin et al. (1964-1980), and
those used in the post-1970 British literature and in Greuter
et al. (1984-) and Walters et al. (1984-). This selection
should enable the reader to equate the names used in this
work with those in virtually any standard modern reference.
The abbreviations of authors used are those in Meikle
(1980).
English names are given for all the taxa. I am quite
convinced that the provision of English names is important in
increasing the numbers of people with an interest in and
knowledge of wild plants. Despite the fact that the Latin

names are scientifically more meaningful and in my view
always preferable, English names, if consistently and
logically applied, can be no less accurate and their use by
those who find them easier to remember should not be too
strongly disparaged. In all but a few cases I have used the
names adopted in Dony et al. (1986), but about 1000 species
that I have included are not listed in that work. Many
sources have been consulted in order to find suitable English
names for these other species; the American, Australasian,
South African and horticultural literature was especially
helpful. However, about 300 names have been, of necessity,
coined anew. My list of names not found in Dony et al.
(1986) has been scrutinized by Franklyn Perring, who has
kindly commented upon them, and in most cases his suggestions
were adopted.

DESCRIPTIONS
The descriptions of all the taxa are brief diagnoses
providing what I consider to be the most important
characters, and they have been made as consistent and
comparable as possible. The data provided in the family
descriptions cover all the genera treated as well as 'Other
genera', and the data in the generic descriptions likewise
cover 'Other spp.' as well as the fully treated species.
However, it is important to note that no account of variation
outside the British Isles is taken; indeed, it is
specifically excluded, and the reader must beware of using
the descriptions as definitions of the taxa on a world-wide
or even European scale.
    In order to compile a description of a genus, the generic
diagnosis should be read in conjunction with the family
diagnosis. Species and subspecies descriptions should
likewise be supplemented by the family and the generic
diagnoses and the key to species (where provided). For
reasons stated later on, however, it could be misleading to
use the family and generic keys in order to compile
descriptions.
    No generic description is given if there is only one genus
included in the family, subfamily or tribe, as in these cases
the description of the latter would be the same as that of
the former.
    In compiling the descriptions I have naturally made use of a
very wide range of literature. I have attempted to avoid the
repetition of errors thereby encountered by checking most of
the measurements and other characters on actual specimens. I
have examined material of virtually all the species covered
in this work, most of it in the fresh state, over the past
six or seven years. I have grown (or allowed to grow, or
failed to prevent from growing) about a quarter of the
species in my own garden; these, and others in other gardens
(including the University of Leicester Botanic Garden) and
locally in the wild, I have been able to observe closely over

the changing seasons.

Many measurements given, especially those describing plant heights or lengths, should be prefixed 'normally', 'usually' or 'mostly'. It is often misleading to give ranges including the extremes that have been encountered (e.g. a grass species 2–153cm high); usually the normal range is much more useful. More exceptional measurements are often given in brackets, e.g. 3–6(9)cm, but even these do not always represent the extremes. In the case of trees, however, the maximum heights known in the British Isles are given, taken from Mitchell (1982). Measurements given without qualification are lengths; those separated by a multiplication sign (e.g. 3–6 x 1–2cm) are lengths and widths respectively.

Certain conventions in terminology will become apparent after usage, especially if the Glossary is consulted. For example, 'above' and 'below' are used only to imply the upper and lower parts of a plant; upper and lower surfaces of an organ are referred to as 'upperside' and 'lowerside', or sometimes more specifically as adaxial and abaxial sides. The term 'leaf', unless otherwise stated, refers to the leaf-blade, excluding the petiole; this fact is especially important in the case of leaf length/breadth ratios.

IDENTIFICATION KEYS

The primary means of identification is the keys to families, genera, species and subspecies. The great majority of these are dichotomous keys. In order to save space no line-gap is left between couplets, but alternate couplets are slightly indented to effect visual separation. Despite this appearance all the dichotomous keys are of the bracketed version, which I consider to be generally superior to (i.e. easier to use than) the indented type followed by some Floras.

In constructing the keys I have attempted to avoid as many as possible of the pitfalls that I have personally encountered over the years. Keys are a vital part of a Flora, yet are one of the most difficult aspects to master and they provide a frequent barrier for the beginner. Long keys are particularly daunting, so I have subdivided keys wherever necessary by providing a general key to a series (A, B, etc) of supplementary keys. Hence few keys contain more than 20 couplets and very few more than 30.

In a small number of genera multi-access keys are used instead of dichotomous keys, i.e. where I consider them to be more reliable (e.g. Epilobium, Cotoneaster). These are usually cases where some important characters are difficult to observe or are likely to be misinterpreted, so that it is hazardous to rely upon them in isolation (as often encountered in a couplet). In other cases 'difficult' characters are allowed for by providing two or more routes in a key. For example the (superior) ovaries of Rosa are liable to be wrongly scored as inferior, the (five) leaflets of

Lotus are often mistaken for three plus two stipules, and the (white) petals of _Berteroa_ often fade to yellow when dried. In these and in many other cases both alternatives are allowed for. A consequence of this is that the 'information' given in a number of keys to families and genera is sometimes taxonomically inaccurate. These keys are provided solely for the purposes of identification, and should not be used to compile descriptions of taxa. The keys to species and subspecies, however, should be free of any such misleading data, and can be considered as part of the description of the species and subspecies. Some notoriously difficult characters (e.g. aerial stems present/absent in _Viola_; inflorescences axillary/terminal in _Trifolium_; structure of the throat of the corolla-tube in Boraginaceae) have been deliberately more or less ignored in the keys.

No species key is provided if the genus includes only two species, and no subspecies key if a species contains only two subspecies; in both cases the two taxa are immediately adjacent and no key is needed. The keys to families include all the families not fully treated (e.g. 46A. Basellaceae) and take full account of all the genera mentioned in certain families under 'Other genera'; likewise, the keys to genera take account of all the species mentioned in certain genera under 'Other spp.'.

I have assumed that the reader is familiar with the use of dichotomous and multi-access keys, but I provide here some hints that I have found very valuable in the past. The keys are intended for use both indoors and in the field, and with both fresh and dried matrial. However, certain characters are not suited to field observation and, where special dissection or high magnification is absolutely necessary, no pretence is made that less satisfactory characters will suffice. The use of insufficient magnification is a frequent cause of misidentification. _Before_ starting on the keys it is important to examine in detail the structure of the flowers, making sure that the number, shape and arrangement of the various parts are fully ascertained. If the flowers are not all bisexual then the distribution of the sexes must be understood. The structure of the gynoecium usually presents the greatest problems; sectioning with a razor-blade vertically and transversely is often required. If fresh material is being collected, observations on underground parts, woodiness of stems and range of leaf-shape should be recorded. If possible, flowers (and fruits) of varying ages should be gathered. Mistakes are often made in distinguishing between a compound leaf (no buds in axils of leaflets) and a group of simple leaves (with buds, often very rudimentary, in axils).

In general, flowers are needed for identification by means of the keys, but there are some exceptions such as near the start of the General Key and in Keys A and B of the Key to Families of Angiosperms. Apart from non-flowering material,

it is usually not possible to key out a range of
abnormalities such as extreme horticultural variants (e.g.
flore pleno or otherwise with more floral parts than usual,
extremely dissected leaves or petals, unusual colour
variants), abnormally tall or dwarfed plants, monstrosities
such as many-headed Plantago and leafy-stemmed Taraxacum,
plants with petaloid or leafy bracts, gall-induced
variations, and various odd mutants (e.g. Fraxinus leaves
with 1 leaflet). In the wild such plants usually occur with
normal ones.

Finally, four tips. Firstly, before using a key to genera
read carefully the family description and any notes that
follow it, and before using a key to species read carefully
the genus description and any subsequent notes. These
descriptions and notes always contain useful data and
sometimes vital ones, since special terms and conventions
(e.g. 'spikelet length' in Festuca is not actually the total
spikelet length) are often defined. Secondly, read the whole
of both alternatives of each couplet before attempting to
choose between them. Thirdly, if there is genuine doubt
about which alternative to choose, follow both, as one will
usually soon show itself to be unsuitable. Fourthly, if a
nonsensical answer is obtained, check back to ensure that the
frequent error of choosing the correct alternative but
following the wrong subsequent route has not been committed.

ILLUSTRATIONS

Some sort of illustration is provided for over half the taxa
treated. The page on which each occurs is indicated in bold
in the right-hand margin of the text. The purpose is not to
picture a representative sample of the taxa, but (1) to
provide drawings of (mostly alien) species for which ready
sources are not available in the literature; and (2) to
illustrate diagnostic parts (e.g. seeds, leaves, flowers) of
more critical groups of taxa on a comparative basis.

The illustrations are either line drawings or photographs.
The former have mostly been executed by Hilli Thompson, to
whom I am greatly indebted for the tremendous trouble she has
taken to capture accurately the minute detail of the
specimens. However, the choice of subject-matter, the supply
of material, and the checking of accuracy, was carried out by
me, and if there are faults in those respects they are my
responsibility. The few drawings not made by Hilli Thompson
are all attributed and acknowledged in the appropriate
caption; other artists who made drawings especially for this
Flora were Kery Dalby (183), Jerzy Zygmunt (533), Fred Rumsey
(749), Dick Roberts (1171) and Sue Ogden (1186-9).

The photographs, most of them taken via the light microscope
or scanning electron microscope, have been prepared by me or
by various colleagues in the School of Biological Sciences,
University of Leicester, on my behalf, except in the few
cases specifically acknowledged. I am extremely grateful for

the help in this respect given me by Abdul-Karim Al-Bermani, John Bailey, Jenny Haywood, George McTurk, Ian Riddell and Andrew Scott.

CONSERVATION AND RARITY

By far the greatest threat to our wild flora is the destruction of habitats, still continuing at a most alarming rate in the name of everything from 'economic development' to 'leisure activity'. When populations of plants (or animals) are decimated they become highly vulnerable to secondary pressures, of which collecting is one. There can be little objection to the accumulation of a reference collection of plants, providing uncommon species are excluded and populations of even common ones are not significantly reduced. Indeed, a collection of accurately determined plants is the best way of learning them and of enabling identification of extra species encountered later. Often, however, only a small part of the plant (e.g. a basal leaf or a single flower) is needed for diagnostic purposes, and rarely are underground parts essential. Usually, even where they are, they can be adequately substituted by notes made in the field. It should be noted that it is actually illegal in Great Britain to uproot any wild plant, even common weeds, without the land-owner's permission, and there are more specific regulations governing Nature Reserves and very rare species (see below).

Since it is only botanists who have a good knowledge of our wild flora, it is vital that they consider themselves under a special obligation to protect it by example and by persuasion.

Rare species are referred to in this Flora under three categories, marked by **R**, **RR** and **RRR** in the right-hand margin; no plants in any of these categories should be collected, damaged or disturbed (e.g. by trampling, or by 'arranging' the immediate surroundings during photography). Even species not so marked are frequently rare in some areas (e.g. montane species in the south, southern species in Scotland, or Continental species in Ireland); where they are rare they should be respected as much as species that are rare throughout the British Isles. The 'R' signs of rarity are given for only native species; aliens, subspecies (with few exceptions) and hybrids are not treated. The precise meanings of these signs are as follows:

R – Rare, found in not more than 100 different 10 x 10km grid-squares in the British Isles (there are over 3500 of these grid squares in total). Such species are being monitored by the Scarce Plant Project of the Institute of Terrestrial Ecology, and I am grateful to Alison Stewart for help in compiling the list for my purposes. The above scheme does not cover Ireland, but I have extended it to do this. Sometimes the abundance of a species in Ireland has removed a plant from the list.

RR - Very rare, found in not more than 15 different 10 x 10km
grid-squares in the British Isles. Such plants are listed
in the British Red Data Book, but the current edition
(Perring & Farrell 1983) is now slightly out of date and I
have attempted to update the list by additions or removals.
The Irish Red Data Book (Curtis & McGough 1988) is also
useful in this respect, but most of the species in it are
not rare in a British context; sometimes the reverse is
true. Lynne Farrell, of the Nature Conservancy Council,
has helped to interpret the Red Data Book entries.

RRR - Very rare and endangered or vulnerable. These species
are all listed in the Schedule of Protected Plants of the
Wildlife and Countryside Act 1981, updated after the first
quinquennial review to include 93 species. A second
quinquennial review, due to be completed in 1991, will add
a few more. This Act covers only Great Britain. Further
Acts cover Eire and Northern Ireland, and species not found
in Great Britain are also given the **RRR** status where
applicable. Under the British Act it is an offence to
pick, remove or destroy any part (including seeds) of the
species in the Schedule, to attempt to do so, or to trade
in these species.

Endemic or extinct taxa are indicated in the text.

BIBLIOGRAPHY

Hundreds of books and thousands of articles in journals have been used in writing this book. In addition to those listed below, which were the ones most frequently used, special mention must be made of the numerous local Floras together covering most of the counties of the British Isles. These are packed with valuable information and were freely consulted, especially those dealing with rich areas or with regions at the extremities of the British Isles.

Bailey, L.H. & Bailey, E.Z. (1976). Hortus third (revised by staff of L.H. Bailey Hortorium). MacMillan, New York.

Bean, W.J. (1970-88). Trees and shrubs hardy in the British Isles, 8th ed. (revised by Clarke, D.L.), 1-4 + Supplement. John Murray, London.

Clapham, A.R., Tutin, T.G. & Warburg, E.F. (1952). Flora of the British Isles. Cambridge University Press, Cambridge. 2nd ed. (1962); 3rd ed. (by Clapham, A.R., Tutin, T.G. & Moore, D.M.) (1987).

Clapham, A.R., Tutin, T.G. & Warburg, E.F. (1959). Excursion Flora of the British Isles. Cambridge University Press, Cambridge. 2nd ed. (1968); 3rd ed. (1981).

Clayton, W.D. & Renvoize, S.A. (1986). Genera graminum. H.M.S.O., London.

Cronquist, A. (1981). An integrated system of classification of flowering plants. Columbia University Press, New York.

Curtis, T.G.F. & McGough, H.N. (1988). The Irish Red Data Book, 1. Vascular plants. Stationery Office, Dublin.

Dandy, J.E. (1958). List of British vascular plants. British Museum, London.

Dandy, J.E. (1969). Watsonian vice-counties of Great Britain. Ray Society, London.

Derrick, L.N., Jermy, A.C. & Paul, A.M. (1987). Checklist of European pteridophytes. Sommerfeltia, 6. Oslo.

Dony, J.G., Jury, S.L. & Perring, F.H. (1986). English names of wild flowers, 2nd ed. Botanical Society of the British Isles, London.

Edees, E.S. & Newton, A.N. (1988). Brambles of the British Isles. Ray Society, London.

Ellis, R.G. (1983). Flowering plants of Wales. National Museum of Wales, Cardiff.

Greuter, W., Burdet, H.M. & Long, G. (1984-). Med-Checklist, 1-. Conservatoire et jardin botaniques, Geneva.

Hubbard, C.E. (1954). Grasses. Penguin Books,

Harmondsworth. 2nd ed. (1968); 3rd ed. (revised by Hubbard, J.C.E.) (1984).

Jermy, A.C., Arnold, H.R., Farrell, L. & Perring, F.H. (1978). Atlas of ferns of the British Isles. Botanical Society of the British Isles, London.

Jermy, A.C., Chater, A.O. & David, R.W. (1982). Sedges of the British Isles. Botanical Society of the British Isles, London.

Lousley, J.E. & Kent, D.H. (1981). Docks and knotweeds of the British Isles. Botanical Society of the British Isles, London.

Mabberley, D.J. (1987). The plant-book. Cambridge University Press, Cambridge.

Meikle, R.D. (1980). Draft index of author abbreviations compiled at The Herbarium, Royal Botanic Gardens, Kew. Royal Botanic Gardens, Kew.

Meikle, R.D. (1984). Willows and poplars of Great Britain and Ireland. Botanical Society of the British Isles, London.

Mitchell, A. (1982). The trees of Britain and northern Europe. Collins, London.

Page, C.N. (1982). The ferns of Britain and Ireland. Cambridge University Press, Cambridge.

Perring, F.H. (1968). Critical supplement to the Atlas of the British flora. Thomas Nelson, London.

Perring, F.H. & Farrell, L. (1983). British Red Data Books, 1. Vascular plants. Royal Society for Nature Conservation, Lincoln.

Perring, F.H. & Walters, S.M. (1962). Atlas of the British flora. Thomas Nelson, London.

Rich, T.C.G. (1991). Crucifers of Great Britain and Ireland. Botanical Society of the British Isles, London.

Scannell, M.J.P. & Synnott, D.M. (1987). Census catalogue of the flora of Ireland, 2nd ed. Stationery Office, Dublin.

Stace, C.A. (1975). Hybridization and the flora of the British Isles. Academic Press, London.

Stearn, W.T. (1983). Botanical Latin, 3rd ed. Thomas Nelson, London.

Tutin, T.G. et al. (1964-80). Flora Europaea, 1-5. Cambridge University Press, Cambridge.

Tutin, T.G. (1980). Umbellifers of the British Isles. Botanical Society of the British Isles, London.

Walters, S.M. et al. (1984-). The European garden flora, 1-. Cambridge University Press, Cambridge.

Willis, J.C. (1973). A dictionary of the flowering plants and ferns, 8th ed. (revised by Shaw, H.K.A.). Cambridge University Press, Cambridge.

L Y C O P O D I O P S I D A
1. Lycopodiaceae
2. Selaginellaceae
3. Isoetaceae

E Q U I S E T O P S I D A
4. Equisetaceae

P T E R O P S I D A
5. Ophioglossaceae
6. Osmundaceae
7. Adiantaceae
8. Pteridaceae
9. Marsileaceae
10. Hymenophyllaceae
11. Polypodiaceae
11A. Cyatheaceae
12. Dicksoniaceae
13. Dennstaedtiaceae
14. Thelypteridaceae
15. Aspleniaceae
16. Woodsiaceae
16A. Davalliaceae
17. Dryopteridaceae
18. Blechnaceae
19. Azollaceae

P I N O P S I D A
20. Pinaceae
21. Taxodiaceae
22. Cupressaceae
23. Araucariaceae
24. Taxaceae

M A G N O L I O P S I D A

MAGNOLIIDAE (Dicotyledons)
MAGNOLIIFLORAE
Laurales
25. Lauraceae
Aristolochiales
26. Aristolochiaceae
Nymphaeales
27. Nymphaeaceae
28. Ceratophyllaceae
Ranunculales
29. Ranunculaceae
30. Berberidaceae
Papaverales
31. Papaveraceae
32. Fumariaceae
HAMAMELIFLORAE
Hamamelidales
33. Platanaceae
Urticales
34. Ulmaceae
35. Cannabaceae
36. Moraceae
37. Urticaceae
Juglandales
38. Juglandaceae
Myricales
39. Myricaceae
Fagales
40. Fagaceae
41. Betulaceae
CARYOPHYLLIFLORAE
Caryophyllales
42. Phytolaccaceae
42A. Nyctaginaceae

101. Anacardiaceae
102. Simaroubaceae
102A. Rutaceae
Geraniales
103. Oxalidaceae
104. Geraniaceae
105. Limnanthaceae
106. Tropaeolaceae
107. Balsaminaceae
Apiales
108. Araliaceae
109. Apiaceae
ASTERIFLORAE
Gentianales
110. Gentianaceae
111. Apocynaceae
Solanales
112. Solanaceae
113. Convolvulaceae
114. Cuscutaceae
115. Menyanthaceae
116. Polemoniaceae
117. Hydrophyllaceae
Lamiales
118. Boraginaceae
119. Verbenaceae
120. Lamiaceae
Callitrichales
121. Hippuridaceae
122. Callitrichaceae
Plantaginales
123. Plantaginaceae
Scrophulariales
124. Buddlejaceae
125. Oleaceae
126. Scrophulariaceae
127. Orobanchaceae
128. Gesneriaceae
129. Acanthaceae
130. Lentibulariaceae
Campanulales
131. Campanulaceae
Rubiales
132. Rubiaceae
Dipsacales
133. Caprifoliaceae
134. Adoxaceae
135. Valerianaceae
136. Dipsacaceae
Asterales
137. Asteraceae

LILIIDAE (Monocotyledons)
ALISMATIFLORAE
Alismatales
138. Butomaceae
139. Alismataceae
Hydrocharitales
140. Hydrocharitaceae
Najadales
141. Aponogetonaceae
142. Scheuchzeriaceae
143. Juncaginaceae
144. Potamogetonaceae
145. Ruppiaceae
146. Najadaceae
147. Zannichelliaceae
148. Zosteraceae
ARECIFLORAE
Arecales
148A. Arecaceae
Arales
149. Araceae
150. Lemnaceae
COMMELINIFLORAE
Commelinales
151. Commelinaceae
Eriocaulales
152. Eriocaulaceae
Juncales
153. Juncaceae
Cyperales
154. Cyperaceae
155. Poaceae
Typhales
156. Sparganiaceae
157. Typhaceae
ZINGIBERIFLORAE
Bromeliales
158. Bromeliaceae
LILIIFLORAE
Liliales
159. Pontederiaceae
160. Liliaceae
161. Iridaceae
162. Agavaceae
163. Dioscoreaceae
Orchidales
164. Orchidaceae

HOW TO USE THIS BOOK

It is strongly recommended that, before consulting the information in this book, the contents page, the introductory chapter and this page (including the reverse of it) be read carefully.

In order to look up a family, genus or species, the Index at the end of the book should be used; both Latin and English names are indexed.

In order to identify a plant it is necessary first to decide whether it is a pteridophyte, gymnosperm or angiosperm. Many Floras purport to do this by means of keys, but in reality the questions posed (e.g. plant reproducing by spores; ovules enclosed in a carpel) amount to the same as the decision called for here. In practice it is best to become familiar with the range of form and structure found in the relatively few pteridophytes and gymnosperms, all other vascular plants being angiosperms. In the case of pteridophytes, the few that do not have divided fern-like leaves can easily be learnt, and in the case of gymnosperms, all have simple narrow leaves (except (Araucaria) and woody female cones (except Taxus and Juniperus). It is especially crucial to distinguish between superficially similar but unrelated plants that provide pitfalls for the unwary. Well-known examples are mosses and Lycopodiopsida; Equisetum and Hippuris; Lemna and Azolla; Isoetes and Littorella; Pilularia and Juncus; and Alnus and conifers. If flowers, spore-bearing sporangia or woody cones are evident, then the task is an easy one, If not, familiarity and experience will soon prevent one from falling into traps such as the above. According to the decision, follow the generic or family keys starting on pages 1, 46 and 63, the positions of which are marked by black-edged pages. These will lead to a family or genus which will provide further keys as necessary enabling one to arrive at the genus, species and subspecies. Where relevant, keys to genera are given under each family, to species under each genus, and to subspecies under each species. Before using the keys the appropriate part of the introductory chapter should be studied.

The following spoof entry indicates the arrangement of the information given in each species account:

1. **Accepted Latin Name** Author(s) (Synonyms Authors) – English Name. Brief description to give habit and comparative diagnostic features, not always repeating those in species key. Status; most characteristic habitats; distribution in BI; area of most likely origin if not native.

Illustrations are numbered according to the page on which they appear, not in a sequence from 1 onwards. References to illustrations are given in the right-hand margin adjacent to the relevant taxon by sole means of a bold number.

Rarity and conservation status are similarly referred to in the margin by means of **R, RR** and **RRR.** For the precise

meaning of these symbols, see pages xx–xxi.
A glossary is placed after the systematic accounts (marked
by a black-edged page).
Signs and abbreviations are listed below.
Maps and a ruler are provided on the end covers.

SIGNS AND ABBREVIATIONS

BI	– British Isles
CI	– Channel Isles
Br	– Great Britain
En	– England
Ir	– Ireland
Sc	– Scotland
Wa	– Wales

N, E, S, W, NE, etc.	– points of compass
C, M, MW, etc.	– central, Mid-, Mid-West, etc.
Leics, W Kent, etc.	– (vice-counties) see end papers

Jan, Feb, Mar, etc.   – months of year

agg.	– aggregate
auct.	– of various authors but not the original one
c.	– about
excl.	– excluding
FIG	– Figure (number following is the page number)
incl.	– including
intrd	– introduced
natd	– naturalized
nom. illeg.	– illegitimate name
nom. nud.	– name invalid since without description
nom. inval.	– name invalid for some other reason
R.	– River
sp., spp.	– species (singular and plural)
ssp., sspp.	– subspecies (singular and plural)
var., vars	– variety, varieties

$\pm$	– more or less
$>$, $<$	– more than, less than
$\geq$	– over and including; at least; not less than
$\leq$	– up to and including; at most; not more than
0	– absent
x	– times (2x, etc. = twice, etc.); or indicating a hybrid
2n	– sporophytic chromosome number

KEYS TO GENERA OF PTERIDOPHYTES (LYCOPODIOPSIDA, EQUISETOPSIDA, PTEROPSIDA)

General key

1 Leaves scale-like, in whorls fused into sheath at
  each node; stems jointed   **4/1. EQUISETUM**
1 Leaves not in a fused whorl at each node; stems
  not jointed   2
  2 Plants free-floating on water, with 2-lobed
    leaves on short stems   **19/1. AZOLLA**
  2 Plant rooted on solid substratum   3
3 Leaves simple but lobed almost to midrib or just
  pinnate, when mature covered on underside by dense
  reddish-brown scales   **15/3. CETERACH**
3 Leaves variously divided, not covered on lowerside
  by reddish-brown scales   4
  4 Leaves simple, not lobed or lobed <1/2-way to
    midrib   Key A
  4 Leaves compound, or simple and lobed >1/2-way
    to midrib (rarely a few ± simple)   5
5 Sporangia borne on leaves or parts of leaves or
  special branches distinctly different from
  vegetative leaves   Key B
5 Sporangia borne on normal foliage leaves   6
  6 Sori on margins of leaves either in protruding
    indusia or at least partly covered by indusium-
    like folded-over leaf-margin   Key C
  6 Sori on underside of leaves, sometimes near
    margin but then not covered by folded-over
    leaf-margin   Key D

Key A -Leaves simple, not lobed or lobed <1/2-way to midrib
1 Stem a rhizome or stolon, or very short and leaves
  single to tufted from ground; leaves usually >1cm   2
1 Stem elongated and aerial; leaves <1cm   6
  2 Leaves filiform, <5mm wide   3
  2 Leaves linear to ovate-elliptic, >5mm wide   5
3 Plant rhizomatous; leaves borne singly (often close
  together) and rolled in flat spiral when young
                                               9/1. PILULARIA
3 Plant with very short corm-like stem; leaves 1-2 or
  in a rosette, not rolled in flat spiral when young   4
  4 Leaves borne in rosette, with sporangia at base
    on upperside   **3/1. ISOETES**
  4 Leaves 1-2; sporangia borne on spike-like
    special branches   **5/1. OPHIOGLOSSUM**

5 Leaves cordate at base, with sporangia borne in
   linear sori on lowerside   **15/1. PHYLLITIS**
5 Leaves cuneate at base; sporangia borne on spike-
   like special branches   **5/1. OPHIOGLOSSUM**
   6 Leaves distinctly serrate along most of margin
      (x10 lens), the youngest ones with minute ligule
      near base on upperside; heterosporous
                                          2/1. SELAGINELLA
   6 Leaves entire, serrate only at base, or obscurely
      serrate along margin, without ligule; homosporous   7
7 Stems all ascending to erect, dividing into equal
   branches; sporangium-bearing leaves not in
   differentiated cones   **1/1. HUPERZIA**
7 Main stems procumbent, with shorter branches;
   sporangium-bearing leaves in differentiated cones   8
   8 Branches flattened, with leaves in 2 alternating,
      opposite pairs   **1/4 DIPHASIASTRUM**
   8 Branches not flattened, with leaves borne in
      whorls, alternately or spirally   9
9 Sterile and sporangium-bearing leaves similar,
   without either hair-points or scarious margins
                                          1/2. LYCOPODIELLA
9 Either sterile leaves with hair-points or sporangium-
   bearing leaves with scarious, toothed margins
                                          1/3. LYCOPODIUM

Key B - Leaves compound, or simple but lobed >1/2-way to
        midrib; sporangia borne on leaves or branches
        that are different from foliage leaves
1 Leaves simple and deeply lobed or 1-pinnate, the
   lobes or leaflets not or scarcely lobed   2
1 Leaves >2-pinnate or 1-pinnate with deeply lobed
   leaflets   5
   2 Stalk from ground bearing 1 pinnate vegetative
      branch and 1 sporangium-bearing branch
                                          5/2. BOTRYCHIUM
   2 Stalks from ground either a vegetative leaf or a
      sporangium-bearing leaf   3
3 Sorus-bearing pinnae with distinct flat, green
   central region, the sori clearly marginal **8/1. PTERIS**
3 Sorus-bearing pinnae without green flat region,
   or if with one then sori clearly on its lowerside   4
   4 Sterile leaves triangular-ovate in outline, <2x
      as long as wide   **16/2. ONOCLEA**
   4 Sterile leaves oblanceolate to lanceolate in
      outline, >3x as long as wide   **18/1. BLECHNUM**
5 Stalks from ground each bearing very different
   vegetative and fertile branches   **6/1. OSMUNDA**
5 Stalks from ground either a vegetative leaf or a
   sporangium-bearing leaf   6
   6 Sterile leaves >2-pinnate, finely divided, ±
      parsley-like   7

    6 Sterile leaves regularly 2-pinnate, or 1-pinnate
       with deeply lobed pinnae   8
 7 Perennial with densely scaly rhizome; sori near
    leaf margin which is folded over to cover it
                                           7/1. CRYPTOGRAMMA
 7 Annual with very short sparsely scaly rhizome; sori
    on leaf lowerside, not covered   **7/2. ANOGRAMMA**
    8 Lowest pinna on each side bearing another pinna
       near its base   **8/1. PTERIS**
    8 Lowest pinna on each side ± like upper ones,
       not bearing another pinna   9
 9 Leaves borne singly spaced out along rhizome; fertile
    leaves green on upperside   **14/1. THELYPTERIS**
 9 Leaves borne in tufts from apices of branches of
    rhizome; fertile leaves brown at maturity
                                          16/1. MATTEUCCIA

Key C - Leaves compound, or simple but lobed >1/2 way to
        midrib; sporangia borne on edge of normal vegetative
        leaves
 1 Sori a continuous line round margins of pinnules   2
 1 Sori few-many discrete groups of sporangia, some-
    times close together   3
    2 Leaves 1-2-pinnate, tufted, ≤75cm excl. petiole;
       rhizomes short, scaly   **8/1. PTERIS**
    2 Leaves 3-pinnate, borne singly, ≤2(5)m excl.
       petiole; rhizomes long, pubescent   **13/1. PTERIDIUM**
 3 Rhizome trunk-like, >20cm thick, covered with old
    leaf-bases; some leaves >1m incl. petioles
                                           12/1. DICKSONIA
 3 Rhizome horizontal, <1cm thick, not covered with
    leaf-bases; leaves <50cm incl. petioles   4
    4 Ultimate leaf-segments >5mm wide; indusia formed
       from folded-under flap of pinnule   **7/3. ADIANTUM**
    4 Ultimate leaf-segments <5mm wide; indusia formed
       from tubular or 2-valved protruding outgrowth
       from pinnule   5
 5 Distal part of petiole winged; rhizomes pubescent;
    mature indusia with protruding bristle, tubular
                                         10/2. TRICHOMANES
 5 Petiole not winged; rhizomes glabrous; indusia without
    protruding bristle, of 2 valves   **10/1. HYMENOPHYLLUM**

Key D - Leaves compound, or simple but lobed >1/2-way to
        midrib; sporangia borne on lowerside of normal
        vegetative leaves
 1 Leaves simple, or 1-pinnate with the pinnae entire
    to toothed <1/2 way to midrib   2
 1 Leaves 1-pinnate with the pinnae divided >1/2 way to
    midrib, or 2-more-pinnate   6
    2 Sori narrowly elliptic to linear   **15/2. ASPLENIUM**
    2 Sori circular to very broadly elliptic   3

3 Indusium 0   4
3 Indusium present   5
   4 Leaves regularly pinnate or nearly so
                             11/1. POLYPODIUM
   4 Leaves (on 1 plant) very variably and irregularly
       pinnately lobed   **11/2. PHYMATODES**
5 Pinnae <1.5cm wide, with sori in one row either side
   of midrib   **17/1. POLYSTICHUM**
5 Pinnae >1.5cm wide, with sori distributed $\pm$ evenly
   all over lowerside   **17/2. CYRTOMIUM**
   6 Sori linear to oblong or C- to V-shaped, >1.5 x
       as long as wide   7
   6 Sori orbicular to broadly elliptic-oblong, <1.5x
       as long as wide   8
7 Sori linear to oblong, with the margin next to
   midrib straight   **15/2. ASPLENIUM**
7 Sori oblong to C- or V-shaped, with the margin next
   to midrib curved or bent   **16/3. ATHYRIUM**
   8 Leaves borne singly spaced out along rhizome   9
   8 Leaves borne in tufts from apices of branches of
       rhizome   12
9 Leaves 2-pinnate, or 1-pinnate with the pinnae
   deeply lobed   10
9 Leaves 3-pinnate, at least at base   11
  10 Pinnae all $\pm$ parallel, the longest ones several
     removed from the basal one; indusium present
                            14/1. THELYPTERIS
  10 Lowest pair of pinnae bent back away from plane
     of others, the longest one the basal or the next
     to basal; indusium 0   **14/2. PHEGOPTERIS**
11 Indusium 0   **16/4. GYMNOCARPIUM**
11 Indusium present, persistent   **16/5. CYSTOPTERIS**
  12 Indusium consisting of ring of hairs or narrow
     scales arching over sorus when young; petiole
     with joint c.1/3 way from base   **16/6. WOODSIA**
  12 Indusium 0, vestigial or well developed and
     membranous; petiole not jointed   13
13 Indusium a flap-like hood; leaves slender, with few
   or 0 scales on petiole   **16/5. CYSTOPTERIS**
13 Indusium 0, vestigial or reniform or peltate;
   leaves often large and with many scales on petiole   14
  14 Pinnules with teeth contracted into very fine
     acuminate apices; indusium peltate **17/1. POLYSTICHUM**
  14 Pinnules untoothed or with rounded to acute teeth;
     indusium 0, vestigial or reniform   15
15 Sori in rows on pinnules distinctly nearer margin
   than midrib; fresh fronds with faint lemon scent
   when crushed   **14/3. OREOPTERIS**
15 Sori either rather scattered on pinnules or in rows
   no nearer margin than midrib; fresh fronds without
   lemon scent   16

16 Indusium reniform, very obvious; widespread
                                17/3. DRYOPTERIS
16 Indusium 0 or vestigial; mountains of Sc
                                16/3. ATHYRIUM

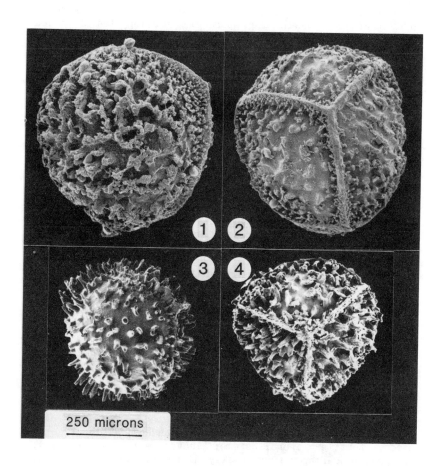

250 microns

FIG 5 – **Isoetes** megaspores. 1–2, **I. lacustris,** outer and inner faces. 3–4, I. **echinospora,** outer and inner faces. Courtesy of A.C. Jermy and Natural History Museum, London.

LYCOPODIOPSIDA - CLUBMOSSES AND QUILLWORTS
(Lepidophyta, Lycopodineae, Lycopsida)

Herbaceous plants with simple or sparingly branched stems and simple leaves with 1 vein. Sporangia homosporous or heterosporous, borne singly in leaf axils or on upperside of leaf near its base, the sporangium-bearing leaves often aggregated into cones. Gametophyte of homosporous species free-living, subterranean, mycorrhizal and saprophytic; gametophytes of heterosporous species much reduced and retained within spore, which lies on the ground.

1. LYCOPODIACEAE - Clubmoss family

Stems elongated, not, little or considerably branched, bearing roots and leaves without ligules. Homosporous; sporangia in leaf-axils, the sporangium-bearing leaves often differentiated into cones.
Moss-like plants whose leaves have true midribs and stomata.

1. HUPERZIA Bernh. - Fir Clubmoss
Stems all ascending to erect, dividing into equal, non-flattened branches; leaves spirally arranged, often with bud-like outgrowths in their axils (these effect vegetative propagation); sporangium-bearing leaves not differentiated into cones, similar to sterile leaves.

1. **H. selago** (L.) Bernh. ex Schrank & C. Martius (Lycopodium selago L.) - Fir Clubmoss. Stems to 25cm; leaves 4-8mm, patent to erecto-patent, lanceolate to narrowly ovate, entire or nearly so. Native; heaths, moors, grassy or rocky places on mountains; common in NW Br S to Wa, rather scattered in Ir, rare and very scattered in lowland Br, formerly locally frequent there.

2. LYCOPODIELLA Holub - Marsh Clubmoss
Stems procumbent, with non-flattened branches, giving rise to erect, fertile lateral stems; leaves spirally arranged; sporangium-bearing leaves weakly differentiated into apical cones.

1. **L. inundata** (L.) Holub (Lepidotis inundata (L.) P.   R
Beauv., Lycopodium inundatum L.) - Marsh Clubmoss.
Procumbent stems dying back quickly behind, to c.20cm; erect stems to 8(10)cm; leaves 4-6mm, erecto-patent, linear to

narrowly ovate, entire; sporangium-bearing leaves broader at base. Native; wet heaths, often on bare peaty soil, sometimes submerged; scattered almost throughout Br and Ir, formerly much more common.

3. LYCOPODIUM L. - Clubmosses
Stems procumbent, with non-flattened branches, with erect sterile and fertile lateral stems; leaves spirally arranged or in whorls; sporangium-bearing leaves well differentiated with apical cones.

1. **L. clavatum** L. - Stag's-horn Clubmoss. Procumbent stems to 1m or more; erect stems to 25cm; leaves 3-5mm, erect to erecto-patent, linear-lanceolate, with long white apical point, minutely toothed; sporangium-bearing leaves ovate to broadly ovate with long white apical point and scarious toothed margin; cones (1-3) borne at apex of distinct peduncle 1.5-20cm with very sparse leaves. Native; heaths, moors, mountains, mostly in grassy places; formerly throughout Br and Ir, now absent from much of lowlands.
2. **L. annotinum** L. - Interrupted Clubmoss. Procumbent stems R
to 60cm; erect stems to 25cm; leaves 4-10mm, patent to erecto-patent, linear-lanceolate, acute, + entire; sporangium-bearing leaves ovate, acuminate, with scarious toothed margin; cones (1) borne at apex of leafy stems (distinct peduncle 0). Native; moors and mountains on thin soil over rocks, often among Calluna; local in C & N (+ entirely mainland) Sc, extinct in S Sc, N En and N Wa except Westmorland (1 site).

4. DIPHASIASTRUM Holub - Alpine Clubmosses
Stems procumbent, often + subterranean, with flattened erect branches arising in fan-like groups; leaves in alternating opposite pairs; sporangium-bearing leaves well differentiated into apical cones.

1. **D. alpinum** (L.) Holub (D. complanatum ssp. alpinum (L.) Jermy, Diphasium alpinum (L.) Rothm., Lycopodium alpinum L.) - Alpine Clubmoss. Procumbent stems to 50(100)cm; erect branches to 10cm, slightly flattened, glaucous; leaves on erect branches and upperside of procumbent stems 2-4 x c.1mm, entire, appressed, sessile; ventral leaves petiolate, c.0.5mm wide, with >1mm free from stem; lateral leaves fused to stem for c.1/2 their length; cones at apices of normal leafy shoots. Native; moors and mountains among grass and Calluna, often very exposed; locally common in N & W Br S to Derbys and S Wa (formerly to S Devon), N, E & W Ir.
2. **D. complanatum** (L.) Holub (Diphasium complanatum (L.) RR
Rothm., Lycopodium complanatum L.) - Issler's Clubmoss. Differs from D. alpinum in more robust; erect branches strongly flattened, scarcely glaucous; ventral leaves sessile, c.1mm wide, with <1mm free from stem; lateral leaves

fused to stem for c.2/3 their length; cones at apices of sparsely-leafed peduncles. Native; heaths and lowland moors; formerly very sparsely scattered in C & N Sc and W En, extinct except in few sites in S Aberdeen and W Sutherland. Our plant is ssp. **issleri** (Rouy) Jermy (D. issleri (Rouy) Holub, Diphasium issleri (Rouy) Holub), probably derived from hybrids with D. alpinum.

2. SELAGINELLACEAE - Lesser Clubmoss family

Stems elongated, little or considerably branched, bearing roots on end of special leaf-less branches or on small corm-like swelling at base of stem; leaves serrate, with microscopic outgrowths (ligules) on upperside near base. Heterosporous; sporangia in leaf-axils, the sporangium-bearing leaves in ill- to well-defined cones with megasporangia at base and microsporangia at apex.
Distinguished from Lycopodiaceae in presence of ligule, heterospory, and roots being borne on specialized leaf-less stem-like outgrowths or small corm-like swellings.

1. SELAGINELLA P. Beauv. - Lesser Clubmosses
  1. S. selaginoides (L.) P. Beauv. - Lesser Clubmoss. Stems decumbent to procumbent, to 15cm, the branches not flattened, bearing erect fertile branches to 6(10)cm with terminal rather ill-defined cones; roots borne from small corm-like swelling at base of stem; leaves all of 1 sort, 1-3mm, those in cones similar but larger. Native; damp places among moss and short grass on mountains; locally common in BI S to Co Limerick, Merioneth and SE Yorks, formerly to W Cork, Derbys and S Lincs.
  2. S. kraussiana (Kunze) A. Braun - Kraus's Clubmoss. Stems procumbent, to 1m, the branches dorsiventrally flattened, bearing well-defined cones increasing in length with age and apparently not terminal; roots borne at ends of special leaf-less branches; leaves of 2 sorts, 2 rows on upperside of stems c.1-2mm, 2 rows on sides of stems c.2-4mm; cones with 4 closely overlapping rows of leaves. Intrd; grown as ground cover in mild damp regions, + natd in shrubberies and damp shady places; scattered in S & W Br, E, W & S Ir and CI N to Herts, Man and W Mayo, Co Durham; Africa.

3. ISOETACEAE - Quillwort family

Stems short and corm-like, bearing roots at base and a rosette of long, erect, + subulate leaves with minute ligule on upperside near base. Heterosporous; sporangia + embedded in leaf-base below ligule; megasporangia produced each year on older leaves, microsporangia on younger ones, the youngest leaves not bearing sporangia.

Similar only to certain angiosperms (notably Lobelia, Littorella, Subularia, Juncus); in absence of sporangia the leaves with 4 air-cavities seen in transverse section (only 1 in I. histrix) and the peculiar corm-like 2-3-lobed stem are diagnostic.

1. ISOETES L. - Quillworts

1 Plant only seasonally submerged, with leaves Oct-Jun; leaf-bases dark, shiny, horny, persistent
3. I. histrix
1 Plant submerged for all or most of year, with leaves Jan-Dec; leaf-bases not dark, shiny and horny, not persistent   2
    2 Megaspores 530-700 microns across, with blunt, anastomosing tubercles on outer face; leaves stiff, remaining apart when plant removed from water   **1. I. lacustris**
    2 Megaspores 440-550 microns across, with acute spines on all faces; leaves flaccid, falling together when plant removed from water
2. I. echinospora

Other spp. - The amphidiploid derivative of I. lacustris x I. echinospora (**I. brochonii** Mot.) might also occur, but needs careful research, as does the relation to it of the Irish **I. morei** Moore.

1. I. lacustris L. - Quillwort. Leaves 8-25(40)cm x 2-5mm,   5
usually ± erect, ± stiff, with 4 longitudinal, septate air canals, parallel-sided for most of length then tapered to acute, often asymmetric point; megaspores 530-700 microns across, with blunt anastomosing tubercles; 2n=110. Native; in clear upland lakes, mostly on stony substrata, down to 6m depth; locally frequent in Ir and N & W Br, but absent in En except S Devon and Lake District.
1 x 2. I. lacustris x I. echinospora occurs by a reservoir in Cumberland with both parents; it is intermediate and sterile; 2n=66.
2. I. echinospora Durieu (I. setacea auct. non Lam.) -   5
Spring Quillwort. Leaves 4-15cm x 2-3mm, usually spreading   R
to ± erect, ± flaccid, with 4 longitudinal, septate air canals, tapered to very acute apex from low down; megaspores 440-550 microns across, with acute spines; 2n=22. Native; in similar places to I. lacustris but rarely with it, mostly on silty substrata; similar distribution to I. lacustris, but rare in N, commoner in SW En E to Dorset, only in W Ir.
3. I. histrix Bory - Land Quillwort. Leaves 1-4(10)cm x   RR
c.1mm, variously spreading, with 1 longitudinal non-septate air canal, tapered to very acute apex from low down; megaspores 400-560 microns across, with blunt tubercles; 2n=20. Native; sandy or peaty hollows on cliff-tops near

sea, where water lies in Winter; extremely local in Guernsey,
Alderney and Lizard Peninsula (W Cornwall).

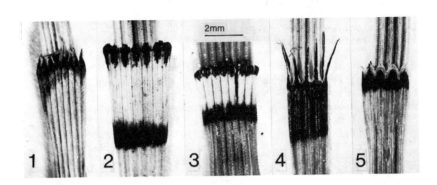

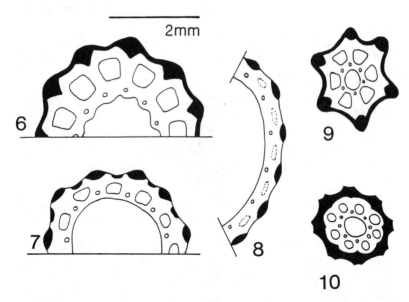

FIG 10 – **Equisetum.** 1-5, leaf-sheaths. 1, **E. ramosissimum.** 2,
E. x moorei. 3, **E. hyemale.** 4, **E. x trachyodon.** 5, **E.
variegatum.** 6-10, leaf-internode sections. 6, **E. arvense.** 7,
E. x litorale. 8, **E. fluviatile.** 9, **E. palustre.** 10, **E.
variegatum.**

E Q U I S E T O P S I D A - Horsetails
(Calamophyta, Equisetinae, Sphenopsida)

Herbaceous rhizomatous perennials; aerial stems elongated, jointed, simple or bearing whorls of branches at nodes; leaves simple, with 1 vein, borne in whorls and fused into sheath round stem. Sporangia homosporous, borne in clusters under peltate specialized branches which are packed into well defined terminal cones. Gametophyte free living, green and photosynthetic.

4. EQUISETACEAE - Horsetail family

The jointed ridged aerial stems without branches or with whorls of branches at each node, and with whorls of leaves forming a fused sheath at each node, are unmistakable. The cones are terminal either on the normal green vegetative stems or on special unbranched brownish or whitish stems produced earlier than the vegetative ones. The common name Mare's-tail is often used but is a misapplication; true Mare's-tail is Hippuris (Hippuridaceae), whose whorled 'branches' at each node are actually leaves.

1. EQUISETUM L. - Horsetails

1 Stems brown or whitish, simple, with cone at apex   2
1 Stems green, simple or branched, with or without
  cone at apex   5
  2 Leaf-sheaths with (15)20-30(40) teeth; cones
    (2)4-8cm   **12. E. telmateia**
  2 Leaf-sheaths with 3-20 teeth; cones 1-4cm   3
3 Leaf-sheaths with teeth united into 3-6 obtuse lobes
  at least at some nodes   **10. E. sylvaticum**
3 Leaf-sheaths with (3)6-20 separate (sometimes
  slightly adherent) acute teeth   4
  4 Base of stem and leaf-sheaths usually tinged green;
    green branches very soon produced   **9. E. pratense**
  4 Green colour absent from stem and leaf-sheaths;
    branches normally not produced   **8. E. arvense**
5 Leaf-sheaths normally with conspicuous black bands
  near top and bottom; teeth falling off before or as
  soon as shoots fully expanded   6
5 Leaf-sheaths with 0-1 conspicuous black bands, or
  black ± all over; teeth present on fully expanded
  shoots   7

6 Leaf-sheaths c. as long as wide, without teeth
   from very early on; stems perennial   **1. E. hyemale**
6 Leaf-sheaths distinctly longer (usually c.1.5-2x)
   than wide, with teeth until shoots fully expanded;
   stems wholly or largely dying down in winter
                                             2. E. x moorei
7 Stems perennial with previous year's cones
   persisting; cones obtuse to apiculate at apex   8
7 Stems annual; previous year's cones not persisting;
   cones rounded at apex   10
   8 Teeth of leaf-sheaths with broad scarious margins
      each much wider than black centre, obtuse at
      maturity; stem-ridges 4-10   **5. E. variegatum**
   8 Teeth of leaf-sheaths at least near tip with
      narrow scarious margins no wider than black
      centre, tapering to fine point; stem-ridges 8-20   9
9 Stems usually ± well branched; spores fertile;
   central hollow of stem mostly ≥1/2 as wide as stem
                                             4. E. ramosissimum
9 Stems not or sparsely branched; spores sterile;
   central hollow of stem mostly <1/2 as wide as stem
                                             3. E. x trachyodon
   10 Stem internodes white, often ≥1cm wide, with
      18-40 ridges; stems with whorls of branches ±
      to top   **12. E. telmateia**
   10 Stem internodes green, mostly <1cm wide, with
      4-30 ridges, if >18 then at least top part of
      stem without whorls of branches   11
11 Branches regularly branched again; teeth of leaf-
   sheaths united into 3-6 lobes (fewer than stem-
   ridges)   **10. E. sylvaticum**
11 Branches 0, or present but not or sparsely and
   irregularly branched again; teeth of leaf-sheaths
   not fused, as many as stem-ridges   12
   12 Stem-internodes with central hollow >3/4 as wide
      as stem, with 10-30 ridges (usually >20 in
      stems >8mm wide)   **6. E. fluviatile**
   12 Stem-internodes with central hollow <3/4 (usually
      c.1/2 or less) as wide as stem, with 4-20 ridges
      (stems rarely >8mm wide)   13
13 Stem with peripheral hollows c. same size as central
   hollow; stem-internodes with 4-9(12) ridges
                                             11. E. palustre
13 Stem with peripheral hollows <1/2 size of central
   hollow; stem-internodes with 6-20 ridges   14
   14 Internodes of branches mostly 3-angled, the
      lowest shorter than adjacent leaf-sheath on main
      stem   **9. E. pratense**
   14 Internodes of branches mostly 4(or more)-angled,
      the lowest as long as to longer than adjacent
      leaf-sheath on main stem   15
15 Cones always produced on green stems; stem-internodes

with central hollow c.1/2 as wide as stem
                                               7. E. x litorale
15 Cones only exceptionally produced on green stems;
stem-internodes with central hollow distinctly <1/2
as wide as stem   **8. E. arvense**

1. **E. hyemale** L. - Rough Horsetail. Stems to 1m, evergreen,   10
simple, rough to touch, with 10-30 2-angled ridges; cones on
normal vegetative shoots, apiculate at apex.   Native; in
ditches and on river- or stream-banks, often in dense
vegetation; scattered through most of Br and Ir, decreasing.
2. **E. x moorei** Newman (E. hyemale x E. ramosissimum) -   10
Moore's Horsetail.   Stems to 60cm, deciduous to semi-
evergreen, simple, slightly rough to touch, with 10-15
2-angled ridges; cones on normal vegetative shoots,
apiculate, with sterile spores. Native; dunes and banks by
sea in Co Wexford and Co Wicklow in absence of E.
ramosissimum and usually of E. hyemale.
3. **E. x trachyodon** A. Braun (E. hyemale x E. variegatum) -   10
Mackay's Horsetail.   Stems to 1m, evergreen, simple to
sparsely branched, slightly rough to touch, with 6-14 ridges;
cones on normal vegetative shoots, apiculate, with sterile
spores.   Native; sandy lake-shores and river-banks and damp
places in dunes; very scattered in Ir and Sc, Cheshire, often
far from 1 or both parents.
4. **E. ramosissimum** Desf. - Branched Horsetail. Stems to 1m,   10
± evergreen, usually well branched in lower part, slightly   RRR
rough to touch, with 7-20 rounded ridges; cones on normal
vegetative shoots, obtuse to apiculate.   Probably intrd; in
long grass by river at 1 site in S Lincs since 1947, in rough
grass near sea at 1 site in N Somerset since c.1963; Europe.
5. **E. variegatum** Schleicher - Variegated Horsetail. Stems   10
to 40(80)cm (often <20), evergreen, usually rough to touch,   R
simple or branched at base, with 4-10 2-angled ridges; cones
on normal vegetative shoots, bluntly apiculate. Native; dune
slacks, river-banks, lake-shores, wet stony mountain sites;
scattered in Ir and W & N Br, very rare in C & S En.
6. **E. fluviatile** L. - Water Horsetail.   Stems to 1.5m,   10
deciduous, smooth to touch, simple or with whorls of branches
in middle region, with 10-30 very low ridges; cones on
vegetative but shorter and less branched shoots, rounded at
apex. Native; ponds, ditches, marshes, backwaters, in or by
water; common throughout BI.
6 x 11. **E. fluviatile x E. palustre = E. x dycei** C. Page
occurs in Outer Hebrides and W & SW Ir; it is intermediate,
resembling a weak plant of E. x litorale but with fewer
whorled branches.
7. **E. x litorale** Kuehl. ex Rupr. (E. fluviatile x E.   10
arvense) - Shore Horsetail. Stems to 1m, deciduous, smooth
to touch, with dense whorls of branches except at base and
extreme apex, with 6-20 low ridges; cones on normal
vegetative shoots, rounded at apex. Native; wet places by

rivers and lakes, in ditches and dune-slacks; scattered
throughout Br and Ir, probably overlooked, sometimes in
absence of 1 or both parents.
8. E. arvense L. - Field Horsetail. Vegetative stems to **10**
80cm, deciduous, smooth to touch, with dense whorls of long
branches, with 6-20 rounded ridges; cones on special,
unbranched brown stems to 20(30)cm appearing before
vegetative stems, rounded at apex. Native; grassy places,
damp places, dune-slacks and rough, waste and cultivated
ground, often a very pervasive weed; abundant throughout BI.
8 x 11. E. arvense x E. palustre = E. x rothmaleri C. Page
was found in N Ebudes in 1972 and Herts in 1987; it is intermediate between the parents (cones on normal vegetative
shoots) and sterile; endemic.
9. E. pratense Ehrh. - Shady Horsetail. Vegetative stems to   **R**
50cm, deciduous, rough to touch, with whorls of thin branches
often slightly swept to 1 side, with (8)12-20 rounded ridges;
cones on shorter, pale shoots at first unbranched but then
resembling others, rounded at apex. Native; banks of rivers
and streams, often on open soil in shade, and flushed grassy
areas, mostly upland; local in BI S to Westmorland and
Fermanagh. Often not coning.
9 x 10. E. pratense x E. sylvaticum = E. x mildeanum Rothm.
occurs in 3 sites in M & E Perth; it is intermediate in all
respects and no cones have been found.
10. E. sylvaticum L. - Wood Horsetail. Vegetative stems to
50(80)cm, deciduous, slightly rough to touch, with whorls of
thin delicate branched branches, with 10-18 rather flat-
topped ridges; cones on shorter, pale shoots at first
unbranched but then resembling others, rounded at apex.
Native; damp woods, hedgerows and stream-banks in lowlands,
open moorland in uplands; throughout Br and Ir, common in N &
W, rare in most of C & E En.
10 x 12. E. sylvaticum x E. telmateia = E. x bowmanii C.
Page was found in S Hants in 1986; it is intermediate in all
characters (cone-bearing stems with only sparse green
branches) and has sterile spores; endemic.
11. E. palustre L. - Marsh Horsetail. Stems to 60cm, **10**
deciduous, smooth to touch, simple or with usually rather
sparse whorled branches, with 4-9(12) rounded ridges; cones
on vegetative (but often shorter and less branched) shoots,
rounded at apex. Native; all kinds of wet or marshy ground;
very common throughout BI. Unbranched stems usually differ
from E. variegatum in rounded, smooth stem-ridges; branched
stems differ from E. arvense in branches with teeth appressed
(not spreading) and lowest internode shorter (not longer)
than main stem sheaths.
11 x 12. E. palustre x E. telmateia = E. x font-queri Rothm.
occurs in scattered places in W Br from Worcs to N Ebudes; it
is intermediate in all characters (but cones on vegetative
shoots) and has sterile spores.
12. E. telmateia Ehrh. - Great Horsetail. Vegetative stems

to 1.5(2)m, deciduous, smooth to touch, white, with very
dense whorls of branches, with 18-40 very low ridges; cones
on special, unbranched, brown stems to 25(30)cm appearing
before vegetative stems, rounded at apex. Native; damp shady
places, woods and wayside banks; throughout BI, but rare in
C, E & N Sc.

FIG 15 – **Hymenophyllum.** 1-2, indusia and leaf-apices of **H.
wilsonii.** 3-4, indusium and leaf-apices of **H. tunbrigense.**

PTEROPSIDA - FERNS
(Filicopsida, Filicineae)

Herbaceous plants with simple or variously branched, frequently subterranean stems (or rhizomes) and usually much divided (rarely simple) leaves with several to many veins, the veins usually much branched and often anastomosing. Sporangia homosporous or rarely heterosporous, often grouped in specialized regions (<u>sori</u>) on lowerside of leaves or on specialized leaves or parts of leaves, never in cones, often covered by special flaps of tissue (<u>indusia</u>). Gametophyte of homosporous species free living, either non-green, subterranean and mycorrhizal, or green and photosynthetic; gametophytes of heterosporous species much reduced and retained within spore, which lies on or under water. All except <u>Azolla</u>, <u>Ophioglossum</u> and <u>Botrychium</u> have the young leaves flatly spirally coiled in 1 plane. Only <u>Azolla</u> and <u>Pilularia</u> are heterosporous and only these 2 have simple non-branching veins in leaves. Only <u>Phyllitis</u>, <u>Pilularia</u> and <u>Ophioglossum</u> have undivided leaves; in <u>Azolla</u> they are only 2-lobed. Primary divisions of a leaf are <u>pinnae</u>, ultimate divisions of a leaf at least 2-pinnate are <u>pinnules</u>. In most ferns the stalk arising from ground level appears to be a stem, bearing leaves, but in fact it is the petiole of a usually much divided leaf. <u>Pteridium</u> and <u>Osmunda</u> are perhaps the spp. most likely to be misinterpreted. In all our ferns the true stem is a rhizome, subterranean or on the surface, but often it is oblique (as in <u>Dryopteris</u> etc.) and partly above ground, and in <u>Dicksonia</u> it becomes the 'trunk'.

5. OPHIOGLOSSACEAE - Adder's-tongue family

Rhizome short or corm-like, without scales; leaves borne singly, with erect stem-like petiole and sterile blade often plus 1 fertile blade; sterile blade simple and entire or 1-pinnate, not spirally coiled when young; fertile blade a simple spike or a panicle, the spike or panicle-branches bearing sporangia in a row either side of axis; homosporous; gametophytes non-green, subterranean.
The leaves, divided into 2 parts, are unique.

1. OPHIOGLOSSUM L. - Adder's-tongues
Sterile blade simple, entire; fertile blade a simple spike of sunken sporangia.

1 Sterile blade rarely >2cm, linear to narrowly
   elliptic, the vein-islets without free vein endings;
   spores ripe Jan-Mar   **3. O. lusitanicum**
1 Sterile blade rarely <2cm, oblong-elliptic to
   broadly so, the vein-islets with minute free vein-
   endings within them; spores ripe Apr-Aug   2
2 Sterile blade mostly 3-3.5cm; sporangia 6-14
   either side of spike   **2. O. azoricum**
2 Sterile blade mostly 4-15cm; sporangia 10-40
   either side of spike   **1. O. vulgatum**

1. O. vulgatum L. - Adder's-tongue. Leaves to 30(45)cm;
sterile blade (3)4-15(30)cm, rounded to cuneate at base;
fertile blade 1.5-5(7)cm; spores 26-41 microns across.
Native; grassland, dune-slacks, ditches, open woods, mostly
in lowlands; frequent throughout most of BI.
2. O. azoricum C. Presl (O. vulgatum ssp. ambiguum (Cosson &   R
Germ.) E. Warb.) - Small Adder's-tongue. Leaves to 10cm;
sterile blade (1.5)3-3.5cm, strongly narrowed to ± stalked at
base; fertile blade 0.8-2cm; spores 38-47 microns across.
Native; barish or grassy places on sandy or peaty damp soils
near sea; very scattered round coasts of BI, but not in E Ir,
W Sc or E En S of Cheviot. Possibly derived from O. vulgatum
x O. lusitanicum.
3. O. lusitanicum L. - Least Adder's-tongue. Leaves to 2cm;   RRR
sterile blade 0.6-3cm; fertile blade 0.3-1.5cm, with 3-8
sporangia on either side; spores 23-32 microns across.
Native; very short turf by sea; local in Guernsey and
Scillies.

2. BOTRYCHIUM Sw. - Moonwort
Sterile blade pinnate; fertile blade a panicle of axes with
sessile but not sunken sporangia.

1. B. lunaria (L.) Sw. - Moonwort. Leaves to 30cm; sterile
blade 2-12cm, 1-pinnate, with asymmetric fan-shaped pinnules;
fertile blade 1-5cm.   Native; dry grassland, mostly in
uplands; throughout Br and Ir, especially N & W Br.

6. OSMUNDACEAE - Royal Fern family

Rhizomes rather short, thick, not scaly, suberect to
ascending, eventually forming solid, raised clumps; leaves in
tufts at apices of rhizome-branches, spirally coiled when
young, with stout ± erect petiole and large 2-pinnate blade,
the lower parts sterile, the upper parts often fertile, not
green, covered in masses of sporangia; homosporous;
gametophytes green, surface-living.
  The Royal Fern, with its large leaves sterile below and
fertile above, is unmistakable.

1. OSMUNDA L. - <u>Royal Fern</u>
  1. O. regalis L. - <u>Royal Fern</u>. Leaves (incl. petiole)
(0.3)0.6-2(4)m, with 5-15 pairs of pinnae; pinnules oblong-
lanceolate, 2-8 x 1-1.5m, crenate to subentire. Native;
fens, bogs, wet woods and heaths on peaty soil, also grown in
gardens and natd in ditches, woods and hedgerows; throughout
BI, common in parts of W Ir and W Br, absent from large areas
of E Br.

7. ADIANTACEAE - <u>Maidenhair Fern family</u>
(Cryptogrammaceae, Gymnogrammaceae)

Rhizomes short to long, scaly; leaves tufted at apices of
rhizome-branches or close together along rhizomes, spirally
coiled when young, of 1 sort (bearing sporangia on normal
pinnules) or of 2 sorts (fertile and sterile), 1-4-pinnate;
sori linear along veins with no indusium and flat leaf-
margin, or oblong to suborbicular near vein-endings and
protected by reflexed leaf-margin; homosporous; gametophytes
green, surface-living.
Not recognisable as a family on superficial characters, but
each of the 3 genera is very distinctive.

1. CRYPTOGRAMMA R. Br. - <u>Parsley Fern</u>
Rhizome short; leaves tufted at apices of rhizome-branches,
of 2 sorts (fertile and sterile), 2-4-pinnate; sori oblong,
becoming contiguous when mature, protected by continuous
reflexed leaf-margin.

  1. C. crispa (L.) R. Br. ex Hook. - <u>Parsley Fern</u>. Leaves
(incl. petiole) to 30cm, erect, the sterile ones 2-3-pinnate
with pinnules c.5-10 x 3-7mm, the fertile ones 3-4-pinnate
with narrower pinnules. Native; rocky places on acid soils
on mountains; locally frequent in Wa, Sc, En S to SW Yorks,
very local in C & N Ir, N & S Devon.

2. ANOGRAMMA Link - <u>Jersey Fern</u>
Plant annual; rhizome very short; leaves tufted, of 2 sorts
(fertile and sterile), 1-3-pinnate; sori linear along veins,
without indusium; leaf-margin flat.

  1. A. leptophylla (L.) Link - <u>Jersey Fern</u>. Leaves (incl. **RR**
petiole) to 10cm, erect, the sterile ones 1-2-pinnate with
pinnules c.3-8 x 2-6mm, the fertile ones 2-3-pinnate with
narrower pinnules. Native; damp shady hedgebanks; frequent
in Jersey, 1 site in Guernsey.

3. ADIANTUM L. - <u>Maidenhair Fern</u>
Rhizome rather short; leaves produced singly but close
together, all of 1 sort, 1-3-pinnate; sori suborbicular at
tips of pinnules, covered by discrete reflexed flaps of

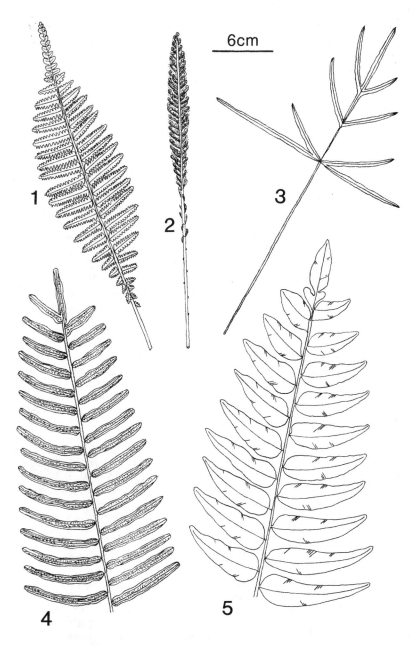

FIG 19 — **Pteropsida.** 1–2, **Matteuccia struthiopteris.** 1, sterile. 2, fertile. 3, **Pteris cretica,** fertile. 4–5, **Blechnum cordatum.** 4, fertile. 5, sterile.

pinnule.

1. A. capillus-veneris L. - <u>Maidenhair</u> <u>Fern</u>. Leaves (incl.   R
petiole) to 25(45)cm, erect to pendent, with very fine
blackish petiole, rhachis and branches; pinnules obtrullate
to obtriangular, often ± lobed at apex, c.0.6-3cm x 0.6-3cm.
Native; limestone cliffs, grykes and rock crevices near sea,
also grown in gardens and natd in similar places and on walls
and bridges, always in moist sheltered spots; W Br from
Cornwall to Man and Westmorland, W Ir from W Cork to W
Donegal, CI, natd in scattered places in En.

8. PTERIDACEAE - <u>Ribbon</u> <u>Fern</u> <u>family</u>

Rhizomes short, scaly; leaves tufted at apices of rhizome-
branches, spirally coiled when young, ± of 1 sort but those
bearing sori with narrower divisions, 1-2-pinnate, the basal
pinna on each side bearing another pinna near its base; sori
linear along margins of leaves, protected by reflexed leaf-
margin; homosporous; gametophyte green, surface-living.
The pinnate (or 2-pinnate) leaves with long narrow pinnae,
the lowest of which on each side give rise to an extra pair,
are diagnostic.

1. PTERIS L. - <u>Ribbon</u> <u>Ferns</u>

Other spp. - P. vittata L., from tropics, differs in having
≥10 pinnae with cordate bases; it was formerly natd on a hot
colliery tip in W Gloucs and has existed on the wall of a
garden in Oxon since before 1924. **P. serrulata** Forsskaol,
from W Mediterranean, differs in its 2-pinnate leaves; it was
formerly natd in a basement enclosure in W Gloucs.

1. P. cretica L. - <u>Ribbon</u> <u>Fern</u>. Leaves (incl. petiole) to   **19**
75cm, with blade ovate-triangular in outline with 1-5 pairs
of finely toothed oblong-linear pinnae c.7-16 x 0.7-2cm and
cuneate at base.   Intrd; much grown as a pot-plant or in
gardens, natd on walls, old buildings and rock-faces in very
sheltered places; scattered in S & C En; S Europe.

9. MARSILEACEAE - <u>Pillwort</u> <u>family</u>

Rhizomes long, thin, pubescent, not scaly; leaves borne
singly but often close together, spirally coiled when young,
of 2 sorts (sterile and fertile), the sterile ones filiform
(?petioles only represented), the fertile ones small,
globose, shortly stalked, completely enclosing the sporangia;
heterosporous.
<u>Pilularia</u> could be mistaken only for an angiosperm, but the
spirally coiled young leaves and globose sporocarps

distinguish it.

1. PILULARIA L. - Pillwort
1. **P. globulifera** L. - Pillwort. Leaves erect, 3-8(15)cm x   R
c.0.5mm; fertile leaves c.3mm across, on stalks c.1mm.
Native; on silty mud by lakes, ponds and reservoirs,
submerged for at least part of year; scattered throughout
most of BI but much less common than formerly, now frequent
only in CW Ir, CS En and SC Wa.

10. HYMENOPHYLLACEAE - Filmy-fern family

Rhizomes thin, glabrous or pubescent but not scaly, surface-
running; leaves borne singly along rhizome, spirally coiled
when young, of 1 sort, 1-3-pinnate, very thin, membranous and
translucent; sori in tubular or valve-like indusia at vein-
ends, protruding from edge of leaf; homosporous; gametophyte
green, surface-living.
The very thin leaves with protruding marginal sori in
tubular or valve-like indusia are unique.

1. HYMENOPHYLLUM Smith - Filmy-ferns
Rhizomes filiform, glabrous; petiole not winged; leaves 1-2-
pinnate; indusium of 2 valves, without protruding bristle.

1. **H. tunbrigense** (L.) Smith - Tunbridge Filmy-fern. Leaves   15
(incl. petiole) 2-5(10)cm, the blade usually elliptic-oblong
in outline, the veins ending just short of leaf-margin (x10
lens), with cells mostly c.1.5x as long as wide; indusium
valves conspicuously dentate; spores 40-50 microns across.
Native; shaded, damp rock-faces and tree-trunks; local in Ir
and W Br from Cornwall to N Ebudes, E Sussex.
2. **H. wilsonii** Hook. - Wilson's Filmy-fern. Leaves (incl.   15
petiole) 3-10(20)cm, the blade usually narrowly elliptic-
lanceolate in outline, the veins mostly reaching leaf-margin
(x10 lens), with cells mostly >2x as long as wide; indusium
valves entire; spores 60-75 microns across. Native; similar
places to H. tunbrigense, often with it; similar distribution
in Ir and W Br but commoner and N to Shetland, not E Sussex.

2. TRICHOMANES L. - Killarney Fern
Rhizomes thin, pubescent; petiole winged at least distally;
leaves 2-3-pinnate; indusium tubular, with protruding bristle
when mature.

1. **T. speciosum** Willd. - Killarney Fern. Leaves (incl. RRR
petiole) 7-45cm, the blade usually triangular-ovate in
outline, the veins ending just short of leaf-margin (x10
lens). Native; very sheltered, damp rock-faces, often near
waterfalls or at cave entrances; extremely local in SW, NW &
CN En, Wa, SW Sc and Ir (mainly SW), formerly commoner and

still surviving as gametophytes in parts of N & SW En and N
Wa where sporophytes no longer develop.

11. POLYPODIACEAE - Polypody family

Rhizomes extended, scaly; leaves borne singly along rhizome,
spirally coiled when young, of 1 sort, 1-pinnate to simple
but very deeply pinnately lobed, the lobes/pinnae linear,
entire to shallowly serrate; sori orbicular to elliptic, on
leaf lowerside; indusium 0; homosporous; gametophyte green,
surface-living.
The only ferns with pinnate or deeply pinnately lobed leaves
with ± parallel-sided pinnae/lobes and orbicular to elliptic
sori without indusia.

1. POLYPODIUM L. - Polypodies
Leaves pinnate to very deeply pinnately lobed right to base
of blade; sori in row on either side of midribs of pinnae,
not sunken.
Microscopic examination is necessary for certain
identification except with extreme or very typical examples.
In the hybrids all or most of the spores are empty and
shrivelled; plants with all or most spores full and turgid
can be separated as in the key. Ten sporangia per plant
should be measured to obtain mean figures.

1 Leaf-blades mostly ≤2x as long as wide; pinnae
   usually narrowly acute, often markedly serrate;
   sporangia mixed with hairs (paraphyses) which
   are ≥0.5mm   **3. P. cambricum**
1 Leaf-blades >2x (up to c.6x) as long as wide; pinnae
   rounded to acute, usually subentire; sporangia
   without paraphyses   2
   2 Sori usually orbicular; mature leaves parallel-
      sided in proximal 1/3 to 1/2; annulus at yellow
      sporangium stage dark orange-brown, with mean of
      10-14 thick-walled cells   **1. P. vulgare**
   2 Sori usually broadly elliptic; mature leaves
      scarcely parallel-sided; annulus at yellow
      sporangium stage pale buff to golden-brown, with
      mean of 7-9 thick-walled cells   **2. P. interjectum**

1. P. vulgare L. - Polypody. Leaves (incl. petiole) to
25cm, usually narrowly oblong, mostly 3-6x as long as wide
with 12-30 pinnae on each side; rhizome-scales mostly <6mm,
triangular to narrowly so; thickened cells of annulus 7-17
(mean per plant 10-14); 2n=148 (tetraploid). Native; on
rocks, walls, tree-trunks and banks, often on acid soils;
frequent to common ± throughout Br and Ir, the most exposure-
and acid-tolerant sp.
   1 x 2. P. vulgare x P. interjectum = P. x mantoniae Rothm. &

U. Schneider occurs in scattered localities throughout Br and Ir (the commonest hybrid); it has a mean of 9-10 thickened annulus cells, is intermediate in other characters, has no paraphyses, and is sterile; 2n=185 (pentaploid).

1 x 3. **P. vulgare** x **P. cambricum** = **P. x font-queri** Rothm. occurs in a few scattered places in Br N to S Sc; it has a mean of 11-14 thickened annulus cells, is intermediate in other characters, has no paraphyses, and is sterile; 2n=111 (triploid).

2. **P. interjectum** Shivas (P. vulgare ssp. prionodes (Asch.) Rothm.) - Intermediate Polypody. Leaves (incl. petiole) to 40cm, usually narrowly oblong-lanceolate, mostly 2-4x as long as wide with 12-30 pinnae on each side; rhizome-scales mostly 3.5-11mm, triangular-acuminate to narrowly so; thickened cells of annulus 4-13 (mean per plant 7-9); 2n=222 (hexaploid). Native; similar places to P. vulgare but more calcicole and usually in more shaded places; similar distribution to P. vulgare, but commoner in S (and in CI) and rare in N (not in Orkney or Shetland).

2 x 3. **P. interjectum** x **P. cambricum** = **P. shivasiae** Rothm. occurs in very scattered places in Ir, W Br and Guernsey; it has a mean of 5-11 thickened annulus cells, is intermediate in other characters, has paraphyses in some but not all plants, and is sterile; 2n=148 (tetraploid).

3. **P. cambricum** L. (P. australe Fee, P. vulgare ssp.   R serrulatum F. Schultz ex Arcang.) - Southern Polypody. Leaves (incl. petiole) to 40cm, usually triangular-ovate to oblong-ovate, mostly 1.25-2x as long as wide with 9-22 pinnae on each side; rhizome-scales mostly 5-16mm, lanceolate; thickened cells of annulus 4-19 (mean per plant 5-10); 2n=74 (diploid). Native; mostly on base-rich rocks, sometimes on tree-trunks, in moist places; scattered in Ir, W Br N to C Sc, scattered in S En to W Kent, Guernsey.

2. PHYMATODES C. Presl - Kangaroo Fern

Leaves pinnately lobed to varying extent on 1 plant, but base of blade not lobed and gradually cuneate; sori in row on either side of main midrib and of midribs of main lobes, sunken.

1. **P. diversifolia** (Willd.) Pichi Serm. (Microsorum   24 diversifolium (Willd.) Copel.) - Kangaroo Fern. Leaves (incl. petiole) to 60cm but usually much less; blade varying from entire to pinnately lobed nearly to midrib, the lobing on 1 leaf very uneven. Intrd; grown in gardens, natd on shady walls and damp places in woods; Guernsey and Scillies; New Zealand and Australia.

11A. CYATHEACEAE

Cyathea dealbata (Forster) Sw. is a tree-fern from New

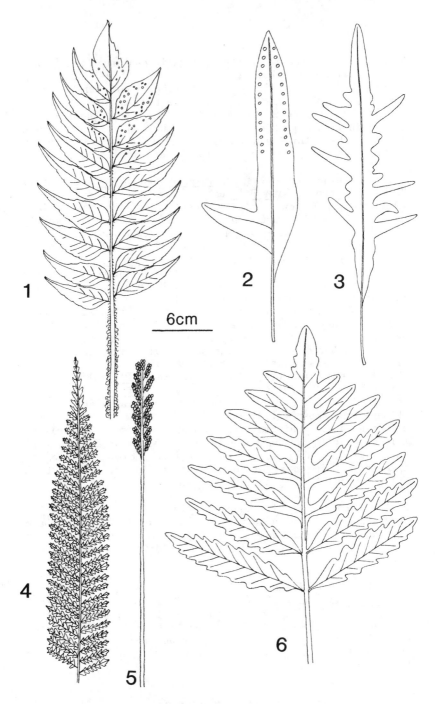

FIG 24 – Pteropsida. 1, **Cyrtomium falcatum**. 2-3, **Phymatodes diversifolia**. 4, **Dicksonia antarctica** (small piece of 1 pinna). 5-6, **Onoclea sensibilis**. 5, fertile. 6, sterile.

CYATHEACEAE   25

Zealand with huge leaves on a thick trunk, differing from
Dicksonia in possessing dense scales on rhizome (trunk) and
petioles; it is self-sown in wild gardens in S Kerry and
might become natd in the wild.

12. DICKSONIACEAE - Tree-fern family

Rhizomes very thick, suberect, trunk-like, with dense hairs
but no scales; leaves without scales, borne in dense tuft at
rhizome apex, spirally coiled when young, of 1 sort,
3-pinnate, rather Dryopteris-like but very large; sori
marginal on pinnules with deeply lobed margins, protected by
a 2-valved indusium becoming hard when mature; homosporous;
gametophyte green, surface-living.
The only natd tree-ferns (see also Cyatheaceae, with which
it is often united).

1. DICKSONIA L'Her. - Australian Tree-fern
  1. D. antarctica Labill. - Australian Tree-fern. Rhizome 24
eventually to 15m but often <1m; leaves up to 200 x 60cm.
Intrd; grown in gardens in SW, natd in woods and shady places
in S Kerry, Scillies and E & W Cornwall; S & E Australia.

13. DENNSTAEDTIACEAE - Bracken family
(Hypolepidaceae pro parte)

Rhizomes very extensive, deeply subterranean, pubescent but
without scales; leaves borne singly, without scales, spirally
coiled when young, of 1 sort, (2-)3-pinnate, with strong,
long, stem-like petiole and rhachis, the lower pinnae often
remote; sori marginal, ± continuous, enfolded by reflexed
margin of pinnule; homosporous; gametophyte green, surface-
living.
Bracken is unmistakable in appearance and in the scent of
its crushed fresh leaves.

1. PTERIDIUM Gled. ex Scop. - Bracken
  1. P. aquilinum (L.) Kuhn - Bracken. Leaves (incl. petiole)
stiffly erect, to 3(5)m, pubescent below and on lowerside,
the blade usually rather leathery and shiny on upperside.
Native; woods, heaths and moors, often dominant over large
areas, usually on acid dry soils, rarely on calcareous ones;
abundant throughout BI. Very variable, often variously
divided into sspp. or vars. but a world-wide study is
essential before a useful classification can be adopted.
Ssp. **atlanticum** C. Page has the young leaves densely white-
pubescent, without the reddish-brown hairs of ssp. **aquilinum,**
and showing a very gradual acropetal unfolding; it occurs
often on limestone in W Sc. Ssp. **latiusculum** sensu C. Page
is a small wiry plant with early ± simultaneous leaf-

unfolding, steeply inclined pinna orientation (pinna divisions in ± same plane as leaf-rhachis), and abundant cinnamon as well as some white hairs; it occurs in C Sc, especially in native Pinus woods. However, all intermediates occur and are fertile, and similar plants are found elsewhere in Br; it is probably not the same as the N American Pteris latiuscula Desv. on which it was based.

14. THELYPTERIDACEAE - Marsh Fern family

Rhizomes long and slender or short and thick, with (sometimes scarce) scales; leaves borne singly or in dense terminal tufts, with sparse or very sparse scales, spirally coiled when young, of 1 or 2 sorts, almost 2-pinnate, with entire to crenate or sinuate pinnules; sori in a row on pinnule lowerside usually near to margin, with indusium 0 or small, thin and soon withering; homosporous; gametophyte green, surface-living.
Recognizable by the submarginal ± round sori with 0 or flimsy and quickly withering indusia on somewhat Dryopteris-like leaves.

1. THELYPTERIS Schmidel - Marsh Fern
Rhizomes long and slender, bearing leaves singly at intervals or sometimes in sparse tufts; leaves ± of 2 sorts, straight, the fertile usually longer and with narrower pinnules recurved at margins; lowest pair of pinnae usually ≥1/2 as long as longest pair and parallel with it; petiole c. as long as blade or longer.

1. **T. palustris** Schott (T. thelypteroides Michaux ssp. glabra Holub) - Marsh Fern. Leaves to 80cm (sterile) or 1.5m (fertile), the blade lanceolate to narrowly elliptic in outline, subglabrous, pale green. Native; marshes and fens, usually shaded among taller herbs or shrubs; scattered in Br and Ir N to C Sc, frequent only in E Anglia, decreasing.

2. PHEGOPTERIS (C. Presl) Fee - Beech Fern
Rhizomes long and slender, bearing leaves singly at intervals; leaves of 1 sort, with blade bent sharply away from plane of petiole; lowest pair of pinnae ± as long as longest pair, reflexed; petiole as long as blade or longer.

1. **P. connectilis** (Michaux) Watt (Thelypteris phegopteris (L.) Slosson) - Beech Fern. Leaves to 50cm, the blade triangular to ovate-triangular in outline with strongly acuminate pinnae, minutely pubescent, pale green. Native; damp woods, shady rocky places and banks, on acid soils; rather common in W & N Br, absent from most of C, S & E En, scattered in Ir.

3. OREOPTERIS Holub - <u>Lemon-scented</u> <u>Fern</u>

Rhizomes short and stout, bearing leaves in tuft at apex; leaves of 1 sort, straight; lowest pair of pinnae extremely short, much <1/2 as long as longest; petiole much shorter than blade.

1. O. limbosperma (Bellardi ex All.) Holub (<u>Thelypteris</u> <u>oreopteris</u> (Ehrh.) Slosson, <u>T.</u> <u>limbosperma</u> (Bellardi ex All.) H.P. Fuchs) - <u>Lemon-scented</u> <u>Fern</u>. Leaves to 1.2m, the blade oblanceolate in outline, with pinnae gradually reducing in length basally to very short basal ones, with numerous minute glands on lowerside giving lemon scent when fresh leaf is bruised, pale green. Native; damp shady places and woods on acid soil; similar distribution to <u>Phegopteris</u> but more common, and also frequent in SE & SW En.

15. ASPLENIACEAE - <u>Spleenwort family</u>

Rhizomes short or very short, with scales; leaves borne in tufts at end of rhizomes, with scales (often sparse), spirally coiled when young, of 1 sort, simple and entire to 3-pinnate; sori oblong to long and linear, on leaf lowerside, with indusium the same shape as sorus and attached to side of it, or without indusium but whole leaf lowerside covered with scales; homosporous; gametophyte green, surface-living.
 Distinguished by the elongated (often very long) sori either with a straight-sided indusium or with indusium 0 but whole leaf lowerside covered with scales; most spp. are small and/ or have relatively little-divided leaf-blades.

1. PHYLLITIS Hill - <u>Hart's-tongue</u>

Leaves simple, entire, with few or 0 scales on lowerside; sori linear, long, strictly parallel with each other and with lateral veins, each apparent sorus actually 2 lying adjacent, the pair with a linear indusium on each side with facing openings.

1. P. scolopendrium (L.) Newman (<u>Asplenium</u> <u>scolopendrium</u> L.) - <u>Hart's-tongue</u>. Leaves (incl. petiole) to 60cm but often much less; blade linear-oblong, cordate at base, up to 7cm wide; petiole <1/2 as long as blade. Native; shady moist rocky places, banks, walls and woods; common throughout BI except very scattered in N Sc.

1 x 2. PHYLLITIS x ASPLENIUM = X ASPLENOPHYLLITIS Alston

The 3 combinations that have been found in BI are all very rare and highly sterile, and all show intermediacy between <u>Phyllitis</u> and the <u>Asplenium</u> parent. The blades are pinnate below and simple above, the latter character separating them from the somewhat similar <u>Asplenium</u> <u>marinum</u>. In most plants there is a mixture of single sori (as in <u>Asplenium</u>) and paired sori (as in **Phyllitis**).

1 Pinnae in lower part of blade with conspicuous gaps
   between, not or scarcely longer than wide
                                              3. X A. confluens
1 Pinnae in lower part of blade without or with very
   small gaps between, often overlapping, usually c.2x
   as long as wide   2
   2 Longest pinna (2)3-4(8) from bottom; sori
      2-3(4)mm, usually confined to outer 1/2 of blade
      and near to pinna-margin   **2. X A. microdon**
   2 Longest pinna the lowest or next to lowest; sori
      4-6mm, usually confined to inner 1/2 of blade
      and near to pinna-midrib   **1. X A. jacksonii**

1. X A. jacksonii Alston (Asplenium x jacksonii (Alston)
Lawalree; P. scolopendrium x A. adiantum-nigrum). Leaves
(incl. petiole) up to 16(20)cm. Native; found in W Cornwall,
N Devon and CI in 19th Century; endemic.
2. X A. microdon (T. Moore) Alston (Asplenium x microdon (T.
Moore) Lovis & Vida; P. scolopendrium x A. obovatum). Leaves
(incl. petiole) up to 20(30)cm. Native; found in W Cornwall
and Guernsey in 19th Century, rediscovered on hedgebanks in
Guernsey in 1965; endemic.
3. X A. confluens (T. Moore ex E. Lowe) Alston (Asplenium x
confluens (T. Moore ex E. Lowe) Lawalree; P. scolopendrium x
A. trichomanes). Leaves (incl. petiole) up to 10(20)cm.
Native; found in NE Yorks, Westmorland and N Kerry in 19th
Century, refound in S Kerry in 1982.

2. ASPLENIUM L. - Spleenworts
Leaves 1-3-pinnate or sparsely and irregularly divided into
linear segments, with few or 0 scales on lowerside; sori
oblong to linear, usually parallel with lateral veins of
pinnae or pinnules, single and each with 1 indusium.
  Hybrids are quite numerous but all are very rare and
scarcely contribute to identification problems. They are all
sterile and ± obviously intermediate.

1 Leaves irregularly and sparsely divided into linear
   segments   **8. A. septentrionale**
1 Leaves 1-3-pinnate; pinnules or pinnae not linear   2
   2 Leaves 1-pinnate   3
   2 Leaves 2-3-pinnate at least in part   5
3 Distal part of rhachis green-winged; larger pinnae
   >12mm, >2x as long as wide   **4. A. marinum**
3 Rhachis not winged; pinnae <12mm, <2x as long as wide 4
   4 Rhachis blackish   **5. A. trichomanes**
   4 Rhachis green   **6. A. trichomanes-ramosum**
5 Longest pinnae at c. middle of blade, the basal
   pair slightly to considerably shorter; petioles
   usually much shorter than blade   **3. A. obovatum**

5 Longest pinnae clearly the basal ones; petioles
  nearly as long as to longer than blade   6
  6 Petioles green; blade 1-2-pinnate; indusia
    with fringed margins   **7. A. ruta-muraria**
  6 Petioles reddish-brown to blackish; blade
    (2-)3-pinnate; indusia subentire   7
7 Blade and pinnae acute to very shortly acuminate;
  spores (32)39-45(52) microns across; widespread
                      1. A. adiantum-nigrum
7 Blade and pinnae very long-acuminate; spores
  (25)30-35(39) microns across; S Ir only **2. A. onopteris**

Other spp. - There is no good evidence for the existence of
either **A. fontanum** (L.) Bernh. or **A. cuneifolium** Viv. in the
wild in BI as sometimes claimed. Specimens of the former
were probably in gardens or temporarily escaped just outside,
and records of the latter from serpentine rocks in Sc were
errors for A. adiantum-nigrum with unusually wide, blunt
ultimate leaf-segments.

1. **A. adiantum-nigrum** L. (A. cuneifolium auct. non Viv.) - 30
Black Spleenwort. Leaves 2-3-pinnate, triangular-ovate in
outline, to 50cm incl. petiole c. as long as blade; pinnules
ovate to lanceolate, with obtusely cuneate base, with small
dark scales (x10 lens) on lowerside; sori linear to linear-
oblong. Native; rocky places in woods, banks and open sites,
walls; common throughout most of BI.
1 x 2. **A. adiantum-nigrum** x **A. onopteris** = **A. x ticinense** D.
Meyer occurs in W & M Cork and Co. Kilkenny.
1 x 3. **A. adiantum-nigrum** x **A. obovatum** = **A. x sarniense**
Sleep occurs in Guernsey; endemic.
1 x 8. **A. adiantum-nigrum** x **A. septentrionale** = **A. x contrei**
Calle, Lovis & Reichst. was found in Caerns in 19th Century.
2. **A. onopteris** L. (A. adiantum-nigrum ssp. onopteris (L.) 30
Heufler) - Irish Spleenwort. Leaves 3-pinnate, triangular in RR
outline, to 50cm incl. petiole usually longer than blade;
pinnules lanceolate to linear-lanceolate, with acutely
cuneate base, without scales; sori linear to linear-oblong.
Native; dry banks and rock-faces mostly on limestone and near
sea; very local in N & S Kerry, W & M Cork and Co Kilkenny,
formerly elsewhere in S Ir.
3. **A. obovatum** Viv. (A. billotii F. Schultz) - Lanceolate 30
Spleenwort. Leaves 2-pinnate at least at base, oblong- R
lanceolate in outline, to 30cm incl. petiole usually much
shorter than blade; pinnules ovate to suborbicular; sori
oblong. Native; rocks, hedgebanks and walls, especially near
sea; common in CI and SW En, frequent in W Wa, very scattered
in Ir and W Br N to W Sutherland, formerly W Kent and E
Sussex. Our plant is ssp. **lanceolatum** (Fiori) Pinto da
Silva.
4. **A. marinum** L. - Sea Spleenwort. Leaves 1-pinnate,
oblong-lanceolate in outline, to 40cm (but often much less)

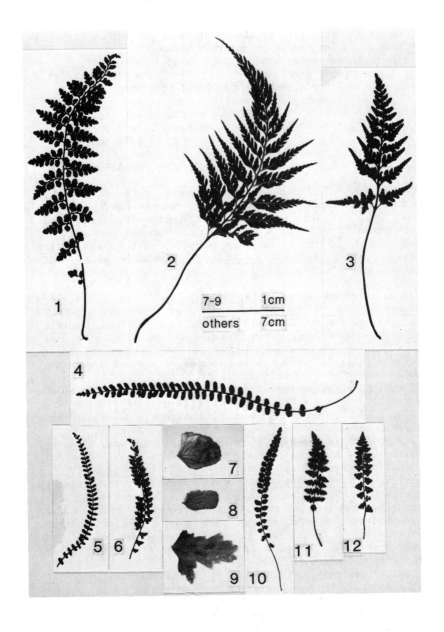

FIG 30 – Pteropsida. 1–10, Asplenium. 1, **A. obovatum**. 2, **A. onopteris**. 3, **A. adiantum–nigrum**. 4–6, **A. trichomanes**. 4, ssp. **quadrivalens**. 5, ssp. **trichomanes**. 6, ssp. **pachyrachis**. 7–9, pinnae of **A. trichomanes**. 7, ssp. **quadrivalens**. 8, ssp. **trichomanes**. 9, ssp. **pachyrachis**. 10, **A. trichomanes–ramosum**. 11–12, Woodsia. 11, **W. ilvensis**. 12, **W. alpina**.

incl. petiole much shorter than blade; pinnae oblong to
ovate-oblong or narrowly so; sori linear. Native; walls,
cliffs and rock-crevices close to sea (often sea-sprayed);
frequent on coasts of BI except E & S coast of En from S
Hants to SE Yorks, formerly E Sussex.
   5. A. trichomanes L. - Maidenhair Spleenwort. Leaves 1-
pinnate, linear in outline, to 20(40)cm incl. petiole much
<1/2 as long as blade; sori linear to oblong-linear. Native;
the 3 sspp. are often distinguishable only with difficulty.
 1 Pinnae oblong-triangular due to expanded basal
   bulges, usually attached to petiole in centre of
   base, usually conspicuously serrate, minutely
   pubescent on lowerside   **c. ssp. pachyrachis**
 1 Pinnae suborbicular to oblong, without expanded basal
   bulges or with 1 on proximal side only, usually
   attached to petiole at proximal corner, usually
   subentire, + glabrous   2
 2 Pinnae suborbicular to rhombic, often asymmetric
   due to basal bulge on proximal side only, to 8mm,
   flat to concave on upperside; sori <2mm, <6(9)
   per pinna; rhizome-scales <3.5mm; calcifuge
                                        a. ssp. trichomanes
 2 Pinnae oblong, usually + symmetric, to 12mm,
   convex to flat on upperside; sori <3mm, <9(12)
   per pinna; rhizome-scales <5mm; calcicole
                                       b. ssp. quadrivalens
   a. Ssp. trichomanes. Smaller and more delicate than ssp.   30
quadrivalens, with thinner petiole and rhachis, and thicker,
subsessile pinnae; spores (23)29-36(42) microns; 2n=72.
Non-calcareous rocks and walls; scattered in Sc, local in Wa
and Lake District, Co Down.
   b. Ssp. quadrivalens D. Meyer. Less delicate and longer   30
than ssp. trichomanes, with thicker petiole and rhachis, and
thinner, distinctly stalked pinnae; spores (27)34-43(50)
microns; 2n=144. Calcareous or neutral rocks or walls (incl.
mortar in walls in acidic regions); common over most of BI.
Hybridises with ssp. trichomanes (=nothossp. x **lusaticum** (D.
Meyer) Lawalree) and ssp. pachyrachis (=nothossp. x **staufferi**
Lovis & Reichst.).
   c. Ssp. pachyrachis (Christ) Lovis & Reichst. Differs from   30
ssp. quadrivalens in leaves smaller and more delicate,
usually appressed to rock, with rigid, fragile petiole and
rhachis; and see key (couplet 1). Limestone rocks and walls;
W Gloucs, Mons and Herefs, probably elsewhere on limestone in
W Br.
   5 x 7. A. trichomanes x A. ruta-muraria = A. x clermontiae
Syme was found in Co Down and possibly Westmorland in 19th
Century.
   5 x 8. A. trichomanes x A. septentrionale = A. x
alternifolium Wulfen is known in Merioneth, Caerns and
Cumberland and there are a few other records from N & W Br.
   6. A. trichomanes-ramosum L. (A. viride Hudson) - Green   30

Spleenwort. Differs from A. trichomanes in leaves to
15(20)cm incl. petiole; rhachis green (not blackish); and
asymmetrically ovate pinnae with serrate margins. Native;
base-rich (mainly limestone) upland rock-crevices; rather
local in W & N Br S to S Wa and Derbys, formerly to Warks.
  7. A. ruta-muraria L. - Wall-rue. Leaves rather irregularly
and sparingly 1-2(3)-pinnate, triangular-ovate in outline, to
8(15)cm incl. petiole longer than blade; pinnules trullate,
rhombic or obtriangular, minutely serrate, often long-
stalked; sori linear, becoming merged. Native; rocks and all
kinds of walls with base-rich substratum, incl. mortar in
acid areas; common throughout most of BI.
  7 x 8. A. ruta-muraria x A. septentrionale = A. x murbeckii
Doerfler is known in Cumberland, with old records in C Sc.
  8. A. septentrionale (L.) Hoffm. - Forked Spleenwort.   R
Leaves irregularly and sparsely divided into linear to very
narrowly elliptic often subdivided segments, to 8(15)cm incl.
petiole usually longer than blade; sori linear. Native; in
rock-crevices in acid areas, mostly upland; very local in W &
N Br from Cards to S Aberdeen, W Galway, formerly S Devon to
W Ross.

3. CETERACH Willd. - Rustyback
Leaves simple but lobed ± to base or pinnate, often former
distally and latter proximally, densely covered on lowerside
with scales; sori linear, without indusium, becoming merged.

  1. C. officinarum Willd. (Asplenium ceterach L.)   -
Rustyback. Leaves linear-oblong in outline, to 15(20)cm
incl. petiole <1/2 as long as blade; scales on lowerside
whitish at first, then reddish-brown; lobes/pinnae widest at
base which is decurrent on to rhachis at both ends. Native;
base-rich crevices and mortar cracks in walls; common in SW
Br N to NW En, and in Ir and CI, scattered E to E En and N to
Easterness.

16. WOODSIACEAE - Lady-fern family
(Athyriaceae)

Rhizomes short (to long), with scales; leaves borne in tufts
at end of rhizomes, or sometimes singly at intervals, with
scales, spirally coiled when young, of 1 or 2 sorts (fertile
and sterile), 1-4-pinnate; sori variable, with or without
indusium; homosporous; gametophyte green, surface-living.
  The family in our flora most defying definition except by
reference to vascular architecture; the 6 genera differ
widely in superficial characters.

1. MATTEUCCIA Tod. - Ostrich Fern
Rhizomes short; leaves borne in apical tufts, of 2 sorts;
sterile leaves 2-pinnate or nearly so; fertile leaves

shorter, with longer petiole, 1-pinnate, not green; sori in
1-2 contiguous rows on each pinna, protected by tightly
inrolled leaf-margin.

1. M. struthiopteris (L.) Tod. - Ostrich Fern. Sterile **19**
leaves up to 1.5m incl. petiole much <1/2 as long as blade,
1-pinnate with deeply lobed pinnae or just 2-pinnate, with
pinnae gradually reducing in length basally to very short
basal ones (superficially closely resembling Oreopteris
limbosperma); fertile leaves up to 60cm, with pinnae very
narrow due to margin inrolling. Intrd; grown in gardens,
natd in shady places; scattered in S & C Sc, Westmorland and
N Ir; Europe.

2. ONOCLEA L. - Sensitive Fern
Rhizomes long; leaves borne singly, of 2 sorts; sterile
leaves 1-pinnate or nearly so with winged rhachis and entire
to lobed pinnae; fertile leaves shorter, 2-pinnate or nearly
so, with non-green pinnae; sori in compact groups protected
by tightly inrolled leaf-margin.

1. O. sensibilis L. - Sensitive Fern. Sterile leaves up to **24**
1m incl. petiole much longer than blade, 1-pinnate at base,
simple and deeply lobed above, the lower pinnae lobed, the
lowest pinna the longest; fertile leaves up to 80cm, with
pinnae very narrow due to inrolled margins. Intrd; grown in
gardens, natd in shady places; scattered in CW and SW Sc and
NW En, Jersey; E N America and E Asia.

3. ATHYRIUM Roth - Lady-ferns
Rhizomes short; leaves borne in apical tufts, of 1 sort,
2-pinnate with deeply lobed to crenate pinnules; sori on
lowerside of pinnules, not protected by inrolled leaf-margin,
with 0, inconspicuous or well-developed indusium.

1 Sori oblong to curved with well-developed J- or
   C-shaped indusium   **1. A. filix-femina**
1 Sori orbicular, with 0 or vestigial indusium   2
   2 Leaves oblanceolate to narrowly elliptic, up to
      30(70)cm incl. petiole c.1/8 to 1/4 as long as
      blade; sori commoner near base of blade
                                            2. A. distentifolium
   2 Leaves irregularly narrowly oblong-elliptic, up to
      20(30)cm incl. petiole <1/8 as long as blade;
      sori commoner near apex of blade   **3. A. flexile**

1. A. filix-femina (L.) Roth - Lady-fern. Leaves to **34**
1(1.5)m, suberect, straight, with petiole usually 1/4 to 1/3
as long as blade; blade sometimes + 3-pinnate; indusium
well-developd, J- or C-shaped. Native; damp woods, shady
hedgebanks and rocky places, mountain screes, marshes; common
throughout most of BI.

FIG 34 - Pteropsida. 1–3, **Athyrium**. 1, **A. filix–femina**. 2, **A. distentifolium**. 3, **A. flexile**. 4–6, **Cystopteris**. 4, **C. fragilis**. 5, **C. dickieana**. 6, **C. montana**. 7–8, **Gymnocarpium**. 7, **G. robertianum**. 8, **G. dryopteris**.

2. **A. distentifolium** Tausch ex Opiz (A. alpestre (Hoppe) **34**
Ryl. ex T. Moore non Clairv.) – Alpine Lady-fern. Differs **R**
from small plants of A. filix-femina only as in key (couplet
1). Native; acid gullies, boulder-slopes and scree, sometimes with A. filix-femina, rarely below 600m; local in C & N
Sc.
3. **A. flexile** (Newman) Druce (A. alpestre var. flexile **34**
(Newman) Milde, A. distentifolium var. flexile (Newman) **R**
Jermy) – Newman's Lady-fern. Differs from A. distentifolium
in leaves bent at or just above base of blade to produce
spreading, often ± flexuous (not suberect, straight) leaves;
and see key (couplet 2). Native; damp acid rocky places at
1040-1140m, often with A. distentifolium; very local in C Sc;
endemic. Perhaps only a var. of A. distentifolium.

4. GYMNOCARPIUM Newman – Oak Ferns
Rhizomes long; leaves borne singly, of 1 sort, 2-3-pinnate,
with petiole much longer than blade, with ± triangular blade;
sori on underside of pinnules rather near margin, orbicular
to elliptic, without indusium.

1. **G. dryopteris** (L.) Newman (Thelypteris dryopteris (L.) **34**
Slosson) – Oak Fern. Leaves to 40cm (incl. blackish
petiole); blade yellowish- to mid-green, non-glandular; basal
pinnae each nearly as large as rest of blade, these 3 units
rolled separately as 3 'balls' in the young leaf. Native;
damp woods and shady rocks, banks and ravines, often in ±
acid humus-rich soil and often with Phegopteris connectilis;
frequent in N & W Br S to Severn-Humber estuaries, rare and
scattered SE of that line and in N Ir.
2. **G. robertianum** (Hoffm.) Newman (Thelypteris robertiana **34**
(Hoffm.) Slosson) – Limestone Fern. Leaves to 50cm (incl. **R**
greenish-brown petiole); blade dull green, with small glands
that are also on rhachis and top of petiole; basal pinnae
each c.1/2 as large as rest of blade, each pinnule and pinna
rolled separately and these in turn rolled into 1 'ball' in
the young leaf. Native; open or partly shaded scree-slopes
or rocky places on limestone; local in En and Wa, rare and
very scattered in Sc and W Ir.

5. CYSTOPTERIS Bernh. – Bladder-ferns
Rhizomes long with leaves borne singly, or short with leaves
in a terminal tuft; leaves of 1 sort, 2-3(4)-pinnate; sori
orbicular, on lowerside of pinnules, not protected by
inrolled leaf-margin, with flap-like indusium that becomes
reflexed to expose sporangia.

1 Rhizomes elongated, bearing leaves singly; leaves
   triangular-ovate, the lowest pinna the longest
                                          3. C. montana
1 Rhizome short, bearing terminal tuft of leaves;
   leaves narrowly oblong to lanceolate, the longest

pinnae near the middle of leaf   2
2 Spores rugose; adjacent pinnules and pinnae
  strongly overlapping   **2. C. dickieana**
2 Spores spinose; adjacent pinnules and pinnae
  scarcely overlapping   **1. C. fragilis**

1. C. fragilis (L.) Bernh. (C. regia (L.) Desv.) - Brittle   34
Bladder-fern. Leaves tufted, to 25(45)cm (incl. petiole
shorter than blade), lanceolate to oblanceolate or narrowly
oblong, 2-pinnate with toothed to deeply lobed pinnules or
3(-4)-pinnate. Native; shady rocks and walls or rocky woods
on basic soils, incl. mortar in acid areas; common N & W Br,
very scattered in S & E, frequent but scattered in Ir.
2. C. dickieana Sim - Dickie's Bladder-fern. Leaves to   34
20(25)cm (incl. petiole shorter than blade), 2(-3)-pinnate;   RRR
differs from C. fragilis as in key. Native; basic rocks in
sea-caves; Kincardines, formerly inland in M Perth.
3. C. montana (Lam.) Desv. - Mountain Bladder-fern. Leaves 34
borne singly, to 45cm (incl. petiole longer than blade),   R
triangular-ovate, 3(-4)-pinnate. Native; shady wet basic
rock-ledges, gullies and scree above 700m; local in C Sc,
formerly Westmorland.

6. WOODSIA R.Br. - Woodsias
Rhizomes short, with leaves borne in a terminal tuft; leaves
of 1 sort, 1-pinnate with deeply lobed pinnae or 2-pinnate on
proximal pinnae; petiole with a visible joint c.1/3 way up
(an eventual abscission point); sori orbicular, on lowerside
of pinnules, not protected by inrolled leaf-margin, with an
indusium consisting of basal ring of hairs or narrow scales.

1. W. ilvensis (L.) R.Br. - Oblong Woodsia. Leaves to   30
10(15)cm (incl. petiole shorter than blade), pubescent, with RRR
scales (c.2-3mm) on leaf lowerside; pinnae 7-15 on each side,
the longest ones oblong or ovate-oblong and c.1.5-2x as long
as wide with 3-8 lobes on each side; spores 42-50 microns.
Native; crevices in mostly neutral rocks from 360m to 720m;
very local in Caerns, Cumberland, Dumfriess and Angus,
decreasing and formerly more widespread.
2. W. alpina (Bolton) Gray - Alpine Woodsia. Leaves to   30
8(15)cm, rather sparsely pubescent, with scales (c.1-2mm)   RRR
only on petiole and on rhachis of blade and pinnae; pinnae
5-10 on each side, the longest ones triangular-ovate and
c.1-1.5x as long as wide, with 1-4 lobes on each side; spores
50-57 microns. Native; similar places to W. ilvensis but on
more basic rocks from 580m to 920m; very local in Caerns and
C Sc, decreasing.

16A. DAVALLIACEAE

DAVALLIA canariensis (L.) Smith, Hare's-foot Fern, from SW

Europe and Macaronesia, was formerly natd on a wall in Guernsey; it has a long densely silky-scaly rhizome producing 3(-4)-pinnate leaves to 70cm (inc. petiole c. as long as blade) with sori on lowerside each covered by indusium attached at base and sides and opening towards leaf-margin.

17. DRYOPTERIDACEAE – Buckler-fern family
(Aspidiaceae pro parte)

Rhizomes short, densely scaly; leaves borne in tufts at end of rhizome or its branches, spirally coiled when young, with scales, usually of 1 sort, 1-3(4)-pinnate; sori orbicular, on leaf lowerside, covered by peltate or reniform indusium; homosporous; gametophyte green, surface-living.
Familiar ferns with usually 2-3-pinnate densely tufted leaves (1-pinnate in some spp.) with peltate or reniform indusia attached at 1 point only.

1. POLYSTICHUM Roth – Shield-ferns
Leaves 1-2-pinnate, of 1 sort; sori in a row down each side of pinna or pinnules or sometimes of main lobes of pinnules; indusium peltate, attached in centre.

1 Leaves 1-pinnate with shallowly toothed (not lobed)
  pinnae, even when producing sori   **3. P. lonchitis**
1 Leaves 1-pinnate with deeply lobed pinnae to 2-(3)-
  pinnate   2
  2 Lowest pinnae nearly as long as longest pinnae;
    proximal pinnules on each pinna shortly stalked,
    with a blade ± right-angled at base   **1. P. setiferum**
  2 Lowest pinnae c.1/2 as long as longest pinnae;
    proximal pinnules on each pinna sessile or
    pinnules not even differentiated, with a blade
    acute-angled at base   **2. P. aculeatum**

Other spp. – P. munitum (Kaulf.) C. Presl (Western Sword-fern), from W N America, was formerly natd on a hedgebank in Surrey; it is 1-pinnate like P. lonchitis but the pinnae are linear and the basal ones are nearly as long as the longest.

1. P. setiferum (Forsskaol) Moore ex Woynar – Soft Shield-   **41**
fern. Leaves to 1.5m (incl. petiole much shorter than blade), 2(-3)-pinnate, rather soft; pinnules shortly stalked, their blades right-angled or nearly so at base with the edge nearer the leaf rhachis ± at right-angles to the pinna rhachis. Native; woods and hedge-banks in moist places; frequent in CI, S & W Br and Ir, becoming rarer in NE and absent from most of N & E Sc.
1 x 2. P. setiferum x P. aculeatum = P. x bicknellii (Christ) Hahne is scattered throughout BI N to C Sc and probably frequent though overlooked; it is intermediate and

sterile.
1 x 3. P. setiferum x P. lonchitis = P. x lonchitiforme
(Hal.) Bech. occurs at Glende, Co Leitrim, with the parents
and P. x illyricum; it is intermediate and sterile.
2. P. aculeatum (L.) Roth - Hard Shield-fern. Leaves to 1m **41**
(incl. petiole much shorter than blade), 1-2-pinnate, rather
hard in texture; pinnules sessile, their blades acute-angled
at base with the edge nearer the leaf rhachis at an acute
angle to the pinna rhachis. Native; similar places to P.
setiferum but more upland and northern; frequent in Br and
Ir, common in N & W Br but less common than P. setiferum in
SW En and S Wa. Juvenile plants are often very similar to
mature ones of P. lonchitis, but lack sori.
2 x 3. P. aculeatum x P. lonchitis = P. x illyricum (Borbas)
Hahne occurs with the parents in Co Leitrim, W Ross and W
Sutherland; it is intermediate and sterile.
3. P. lonchitis (L.) Roth - Holly-fern. Leaves to 30(60)cm **41**
(incl. petiole much shorter than blade), 1-pinnate, hard in
texture; lowest pinnae much shorter than longest ones; pinnae
usually with 1 basal lobe but otherwise only shortly toothed,
asymmetrically and narrowly triangular-ovate. Native; basic
rock-crevices, scree and ravines mostly above 600m (but down
to 150m); local in N En, C & N Sc and W Ir, Caerns,
Dumfriess, natd on bridge in Northants.

2. CYRTOMIUM C. Presl (Phanerophlebia C. Presl) - House
Holly-fern
Leaves 1-pinnate, of 1 sort; sori scattered all over lower-
side of pinnae; indusium peltate, attached in centre.

1. C. falcatum (L.f.) C. Presl (Phanerophlebia falcata **24**
(L.f.) Copel., Polystichum falcatum (L.f.) Diels) - House **41**
Holly-fern. Leaves to 1.2m (incl. petiole shorter than
blade), very hard in texture; pinnae usually with 1 basal
lobe but otherwise entire to very shortly obtusely toothed,
asymmetrically triangular-ovate. Intrd; grown in
conservatories etc., natd on and by walls, among maritime
rocks and in other shady places; scattered in W Br from
Scillies to WC Sc, W Cork, CI; E Asia.

3. DRYOPTERIS Adans. - Buckler-ferns
Leaves 2-3(4)-pinnate, usually of 1 sort; sori in a row down
each side of pinnules or sometimes of main lobes of pinnules;
indusium reniform, attached at the notch.
A difficult genus, especially due to the extent of
hybridization. Most of the hybrids are sterile, but some
produce good spores with an unreduced chromosome number that
develop into gametophytes that reproduce apogamously (grow
directly into new sporophytes without fertilization). Such
taxa, e.g. D. affinis and D. remota, are best treated as
spp.; D. x complexa is theoretically in the same category,
but its 2 parents are already so close that recognition of a

third (intermediate) species of very sporadic occurrence is
unrealistic. The apogamous taxa can hybridize with sexual
taxa by producing male gametes that fertilize the female
gametes of the sexual taxa, but they do not themselves
produce female gametes. Such hybrids are usually also
apogamous (D. affinis and its hybrids demonstrate this).
Hybrids between sexual spp. are sterile.

1 Leaves 1-pinnate with deeply lobed pinnae, to
  2-pinnate with pinnules lobed to c.1/2 way to midrib   2
1 Leaves 2-pinnate with pinnules lobed nearly to
  midrib, to 3(-4)-pinnate   7
  2 Leaves ± of 2 sorts, the fertile longer and
    more erect, lanceolate–oblong, parallel–sided
    for most of length; pinnae with <15 pinnules/
    lobes each side; pinnules with mucronate teeth
                                            7. D. cristata
  2 Leaves of 1 sort, lanceolate–elliptic, scarcely
    parallel–sided; pinnae with >15 pinnules each
    side; pinnules with acute, obtuse or 0 teeth   3
3 Leaf dull green, ± mealy due to many minute glands
  on both surfaces; lowest pinna with proximal 3–4
  pinnules on both sides ± same size   6. D. submontana
3 Leaf clear green (of various shades), not or
  slightly glandular; lowest pinna with pinnules
  successively smaller from base distally, or just
  the 2 proximal ± same size   4
  4 Pinnules evenly lobed nearly 1/2 way to midrib;
    petiole c.1/3 to 1/2 as long as blade; extremely
    rare   4. D. remota
  4 Pinnules entire to lobed distinctly <1/2 way to
    midrib; petiole usually ≤1/3 as long as blade;
    common   5
5 Pinnules parallel–sided for most of length, broadly
  rounded to ± truncate (but often toothed) at apex;
  pinnae with dark blotch where they join rhachis;
  petioles with dense golden scales   3. D. affinis
5 Pinnules distinctly tapering, rounded to ± obtuse
  at apex; pinnae without dark blotches at base;
  petioles with sparse to ± dense greyish– or pale–
  brown scales   6
  6 Rhizome not or little branched, hence leaf-tufts
    scattered; pinnules with erect or convergent
    teeth at apex; sori c.1.5mm across, along ±
    whole length of pinnules, with non–glandular
    indusia with edges not tucked under; widespread
                                           2. D. filix–mas
  6 Rhizome well branched, hence leaf-tufts in
    groups; pinnules with teeth spreading fan-wise
    at apex; sori c.1mm across, ± confined to
    proximal 1/2 of pinnules with ± glandular
    indusia with edge well tucked under before

sporangial dehiscence; mountains   **1. D. oreades**
7 Leaf dull green, $\pm$ mealy due to many minute
  glands on both surfaces; lowest pinna with proximal
  3-4 pinnules on both sides $\pm$ same size   **6. D. submontana**
7 Leaf clear green (of various shades), not or
  slightly glandular on upperside; lowest pinna with
  pinnules successively smaller from base distally   8
  8 Petioles with dense golden scales at least in
    proximal 1/2; pinnules evenly lobed $\leq$5/6 way to
    midrib; extremely rare   **4. D. remota**
  8 Petioles variously clothed but not with dense
    golden scales; leaves 2-3(4)-pinnate, if only
    2-pinnate then at least lowest pinnules on
    lowest pinnae lobed $\pm$ to base; common   9
9 Pinnules distinctly concave on upperside, with
  numerous minute glands on lowerside and sometimes
  upperside   **5. D. aemula**
9 Pinnules flat or convex on upperside, glands 0 or
  rare on lowerside (except sometimes on indusia) and
  0 on upperside   10
  10 Scales on petiole usually with very distinct dark
     centre (sometimes uniformly pale in small upland
     plants); pinnules usually convex on upperside;
     indusia glandular   **9. D. dilatata**
  10 Scales on petiole uniformly pale or dark brown,
     or somewhat darker in centre (the dark suffusing
     outwards, not in a distinct zone); pinnules
     usually flat; indusia glandular or not (if
     glandular then scales not pale)   11
11 Leaf-blades ovate-triangular, with all pinnae $\pm$
   in 1 plane; petioles with scales mid- to dark-brown
   or with darker centres; lowest pinnae with lowest
   pinnule on basal side usually $\geq$1/2 as long as
   its pinna   **10. D. expansa**
11 Leaf-blades narrowly triangular-lanceolate, with
   lower pinnae twisted into $\pm$ horizontal plane;
   petioles with pale brown scales; lowest pinnae with
   lowest pinnule on basal side <1/2 as long as its
   pinna   **8. D. carthusiana**

1. D. oreades Fomin (_D. abbreviata_ auct. non DC.) – <u>Mountain</u> **41**
<u>Male-fern</u>. Leaves to 0.5(-1.2)m (incl. petiole 1/8 to 1/4 as
long as blade), 1-pinnate with deeply divided pinnae to
2-pinnate, with crenate lobes/pinnules; petiole with rather
numerous dull pale-brown scales; blade dull mid-green,
without dark blotch at base of pinna, with lowest pinnae <1/2
as long as longest; 2n=82.   Native; rocky places on
mountains, in open or slight shade, scree-slopes; frequent
above 240m in Wa, N En S to MW Yorks, Sc, S Kerry, Co Down.
   1 x 2. D. oreades x D. filix-mas = D. x mantoniae Fraser-
Jenkins & Corley is scattered in N Wa, Lake District and Sc.
   1 x 5. D. oreades x D. aemula = D. x pseudoabbreviata Jermy

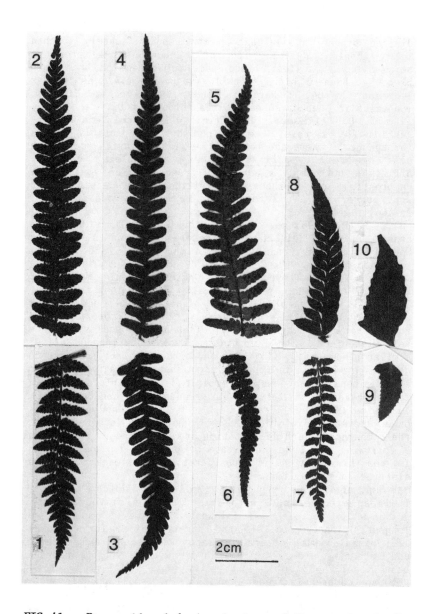

FIG 41 - Pteropsida. 1-6, basal pinna of **Dryopteris**. 1, **D. remota**. 2, **D. filix-mas**. 3-5, **D. affinis**. 3, ssp. **affinis**. 4, ssp. **borreri**. 5, ssp. **cambrensis**. 6, **D. oreades**. 7-9, basal pinna of **Polystichum**. 7, **P. setiferum**. 8, **P. aculeatum**. 9, **P. lonchitis**. 10, basal pinna of **Phanerophlebia falcata**.

was found on Mull (M Ebudes) in 1967; endemic.
2. D. filix-mas (L.) Schott - <u>Male-fern</u>.   Leaves to 1.2m **41**
(incl. petiole 1/4 to 1/3 as blade), 1-pinnate with deeply
divided pinnae to 2-pinnate, with usually acutely toothed
lobes/pinnules; petiole with rather numerous pale brown to
straw-coloured scales; blade pale mid-green, without dark
blotch at base of pinna, with lowest pinnae c.4/5 as long as
longest; 2n=164.   Native; woods, hedge-banks, ditches,
mountains in open or shade; common throughout BI.
   2 x 3. D. filix-mas x D. affinis = D. x complexa Fraser-
Jenkins (<u>D.</u> x <u>tavelii</u> auct. non Rothm.) probably occurs
throughout BI where the parents co-exist. Hybrids occur with
all 3 sspp. of <u>D. affinis</u>: nothossp. **x complexa** (with ssp.
<u>affinis</u>); nothossp. **x contorta** Fraser-Jenkins (with ssp.
<u>cambrensis</u>); and nothossp. **x critica** Fraser-Jenkins (with
ssp. <u>borreri</u>).   All 7 taxa occur in 1 small wood in Leics.
2n=164 or 205.
   2 x 8. D. filix-mas x D. carthusiana = D. x brathaica
Fraser-Jenkins & Reichst. (<u>D.</u> x <u>remota</u> auct. non (A. Braun ex
Doell) Druce) was found in Westmorland in 1859 and is still
in cultivation.
   3. D. affinis (Lowe) Fraser-Jenkins - <u>Scaly</u> <u>Male-fern</u>.
Leaves to 1.5m (incl. petiole usually <1/4 as long as blade),
1-pinnate with deeply divided pinnae to 2-pinnate, with
subentire to acutely or obtusely toothed lobes/pinnules;
petiole with very dense golden-brown scales; blade
yellowish-green, with dark blotch at base of pinna. Native;
similar places to <u>D. filix-mas</u> and often with it; frequent to
common throughout BI (but less so than <u>D. filix-mas</u>).   <u>D.</u>
<u>affinis</u> consists of apogamous diploids or triploids derived
from hybridization between <u>D. oreades</u>, the non-British <u>D.</u>
<u>caucasica</u> (A. Braun) Fraser-Jenkins & Corley, and at least 1
other ancestral diploid. 3-5 sspp. have been recognized, but
their characters and delimitation are still under study. The
3 following are best understood at present, but recognition
is possible only after considerable experience; relative
distributions are unknown.
   a. Ssp. affinis (<u>D.</u> <u>pseudomas</u> (Wollaston) Holub & Pouzar). **41**
The most extreme ssp., with shiny leaves with very densely
golden-scaly petioles, with lowest pinnae c.1/2 as long as
longest; pinnae parallel- and straight-sided for proximal
1/2; pinnules with subtruncate apex with short obtuse teeth
and subentire sides, and the lowest with a slight, rounded
basal lobe; 2n=82.
   b. Ssp. cambrensis Fraser-Jenkins (ssp. <u>stilluppensis</u> auct. **41**
non (Sabr.) Fraser-Jenkins).   Intermediate between ssp.
<u>affinis</u> and <u>D. oreades</u>, with rather shiny leaves with densely
reddish-golden-scaly petioles, with lowest pinnae usually
<1/2 as long as longest; pinnae tapering from base to apex;
pinnules with very broadly rounded apex with obtuse teeth and
obtusely toothed often revolute sides, and the lowest with a
substantial rounded basal lobe often overlapping leaf

rhachis; 2n=123. Probably mostly in N & W.

c. Ssp. borreri (Newman) Fraser-Jenkins (ssp. <u>stilluppensis</u> **41**
(Sabr.) Fraser-Jenkins, ssp. <u>robusta</u> Oberholzer & Tavel ex
Fraser-Jenkins, <u>D. borreri</u> (Newman) Newman ex Oberholzer &
Tavel, <u>D. tavelii</u> Rothm., <u>D. woynarii</u> auct. non Rothm.).
Closest to <u>D. filix-mas</u>, with scarcely shiny leaves with
moderately densely light-golden-scaly petioles, with lowest
pinnae >1/2 as long as longest; pinnae parallel-sided for
proximal 1/2 but uneven due to various-lengthed pinnules;
pinnules with rounded apex often with a large tooth on each
'shoulder' and well-toothed sides, and the lowest with a
large pointed basal lobe; 2n=123. Probably the most
widespread ssp.

4. D. remota (A. Braun ex Doell) Druce (<u>D. woynarii</u> Rothm.) **41**
- <u>Scaly Buckler-fern</u>. Leaves to 75cm (incl. petiole c.1/3 to **RR**
1/2 as long as blade), 2-pinnate with pinnules lobed c.1/3 to
5/6 way to midrib; petiole with dense golden-brown scales
often with darker centre; blade narrowly ovate, with lowest
pinnae c.4/5 as long as longest; 2n=123. Native; formerly in
woods in N Kerry and SE Galway, possibly Dunbarton; extinct.
Apogamous hybrid derivative of <u>D. affinis</u> x <u>D. expansa</u>,
although <u>D. expansa</u> is not known in Ir.

5. D. aemula (Aiton) Kuntze - <u>Hay-scented Buckler-fern</u>.
Leaves to 75cm (incl. petiole >1/2 as long as blade), 3(-4)-
pinnate; petiole with few pale scales; blade triangular-
ovate, with lowest pinnae the longest; 2n=82. Native; moist
shady places in woods, ravines and hedge-banks; local in Ir
and W Br from W Cornwll to Outer Hebrides and Orkney, very
scattered in E Br except frequent in acid parts of Kent and
Sussex.

6. D. submontana (Fraser-Jenkins & Jermy) Fraser-Jenkins (<u>D.</u> **R**
<u>villarii</u> (Bellardi) Woynar ex Schinz & Thell. ssp. <u>submontana</u>
Fraser-Jenkins & Jermy) - <u>Rigid Buckler-fern</u>. Leaves to 75cm
(incl. petiole >1/2 as long as blade), 2-pinnate with
pinnules lobed 1/3 to 3/4 way to midrib; petiole with rather
sparse pale brown scales; blade narrowly ovate- to
lanceolate-triangular, with lowest pinnae the longest;
2n=164. Native; in limestone crevices, grykes and scree;
very locally frequent in NW En, 1 site each in Derbys, Denbs
and formerly Caerns.

7. D. cristata (L.) A. Gray - <u>Crested Buckler-fern</u>. Leaves **RR**
to 60(100)cm (incl. petiole 1/3 to 1/2 as long as blade),
1-pinnate with very deeply divided pinnae to just 2-pinnate,
with mucronate-toothed lobes/pinnules; petiole with sparse,
pale brown scales; blade lanceolate-oblong, with lowest
pinnae nearly as long as longest; 2n=164. Native; wet
heaths, dune-slacks, marshes and fens, often with <u>Thelypteris</u>
<u>palustris</u>; very local and decreasing in Surrey, Berks, E
Suffolk, E & W Norfolk and Renfrews, formerly elsewhere
scattered in En.

7 x 8. D. cristata x D. carthusiana = D. x uliginosa (A.
Braun ex Doell) Kuntze ex Druce occurs in E & W Norfolk,

formerly in other sites of D. cristata.

8. D. carthusiana (Villars) H.P. Fuchs (D. lanceolato-cristata (Hoffm.) Alston, D. spinulosa Kuntze) – Narrow Buckler-fern. Leaves to 80(100)cm (incl. petiole c. as long as blade), 2-3-pinnate; petiole with sparse, pale brown scales; blade narrowly ovate-oblong, with lowest pinnae ± as long as longest; 2n=164. Native; damp or wet woods, marshes, fens and wet heaths; frequent throughout most of Br and Ir.

8 x 9. D. carthusiana x D. dilatata = D. x deweveri (J. Jansen) Wachter occurs frequently with the parents scattered over Br and Ir.

8 x 10. D. carthusiana x D. expansa = D. x sarvelae Fraser-Jenkins & Jermy occurs in Westerness and Kintyre, discovered in 1978.

9. D. dilatata (Hoffm.) A. Gray (D. austriaca Woynar ex Schinz & Thell. non Jacq.) – Broad Buckler-fern. Leaves to 1(1.5)m (incl. petiole 1/4 to 2/3 as long as blade), 3(-4)-pinnate; petiole with numerous scales with dark centres and paler edges; blade ovate to triangular-ovate, with lowest pinnae the longest; 2n=164. Native; woods, hedge-banks, ditches, shady places on heaths and mountains; common throughout BI.

9 x 10. D. dilatata x D. expansa = D. x ambroseae Fraser-Jenkins & Jermy occurs in N & W Br S to Caerns.

10. D. expansa (C. Presl) Fraser-Jenkins & Jermy (D. assimilis S. Walker) – Northern Buckler-fern. Leaves to 80(100)cm (incl. petiole c. as long as blade), 3-4-pinnate; petiole with fairly numerous pale- to reddish-brown scales often with darker centres; blade triangular-ovate, with lowest pinna usually the longest; 2n=82. Native; cool, often damp places in woods and mountain crevices and scree; locally frequent in Sc, Wa and En S to Westmorland.

18. BLECHNACEAE – Hard-fern family

Rhizomes long or short, scaly; leaves coriaceous, borne in tufts at ends of rhizome branches, spirally coiled when young, sparsely scaly, of 2 sorts (of 1 sort in Woodwardia), 1-pinnate with entire pinnae (with deeply acutely lobed pinnae in Woodwardia); sori linear, continuous either side of pinna midrib (stout, discrete, in row either side of pinna-lobe midribs in Woodwardia), covered by indusium same shape as sorus and opening towards midrib; homosporous; gametophyte green, surface-living.

The sori having openings towards the midrib and the leaves being coriaceous and 1-pinnate (or nearly 2-pinnate in Woodwardia) are diagnostic, as are the 1-pinnate separate sterile and fertile leaves of Blechnum.

Other genera – WOODWARDIA radicans (L.) Smith, from SW Europe, regenerates in gardens in S Kerry and W Cornwall and might escape into the wild; it differs from <u>Blechnum</u> as above, with leaves to 2m that root at tips from a scaly bud.

1. BLECHNUM L. - <u>Hard-ferns</u>

Other spp. - B. penna-marina (Poiret) Kuhn, from extreme S S America and Australasia, has been natd for a short time in a few places and might well be again; it resembles a very small <u>B. spicant</u>, with leaves ≤20 x 1.5cm and a slender creeping rhizome.

1. **B. spicant** (L.) Roth - <u>Hard-fern</u>. Leaves to 50(75) x 4cm, the fertile more erect and slightly longer than the sterile; pinnae entire, the sterile up to 2 x 0.5cm, the fertile as long but much narrower, attached to rhachis over whole width of their base. Native; woods, heaths, moors, grassy and rocky slopes on acid soils, often in rather dry places; common throughout most of BI but absent from much of C & CE En.

2. **B. cordatum** (Desv.) Hieron. (<u>B.</u> <u>chilense</u> (Kaulf.) Mett.) **19** - <u>Chilean</u> <u>Hard-fern</u>. Leaves to 100 x 20cm, the fertile scattered among the sterile; pinnae sub-entire, the sterile up to 10 x 2.5cm, the fertile as long but much narrower, greatly narrowed at base and attached to rhachis over short width. Intrd; grown in gardens, natd in shady places and by streams in SW En, W Sc and SW Ir; S S America. The identity of our plant is not certain; possibly >1 sp. is present.

19. AZOLLACEAE - <u>Water</u> <u>Fern</u> <u>family</u>

Stems very slender, branching, floating on water, with hanging, simple roots, without scales; leaves on 2 opposite sides of stem, small, 2-lobed, not spirally coiled when young, of 1 sort; sori borne on lower lobe of 1st leaf of each branch; heterosporous.
The only floating fern.

1. AZOLLA Lam. - <u>Water</u> <u>Fern</u>
1. **A. filiculoides** Lam. - <u>Water</u> <u>Fern</u>. Stems 1-5(10)cm; leaves up to 2.5 x 1.5mm in surface view; plant green in early part of season but becoming bright red in Autumn. Intrd; natd on ponds and in canals and dykes, often covering water surface like a <u>Lemna</u> (and often with it); fairly frequent but often sporadic in C & S Br N to MW Yorks, Man, E & S Ir, CI; tropical America.

P I N O P S I D A - CONIFERS, GYMNOSPERMS
(Coniferopsida, Coniferae, Gymnospermae, Taxopsida)

Trees or shrubs with simple, usually evergreen leaves. Male sporangia borne on sporophylls arranged in male cones. Female sporangia borne in naked ovules either borne singly and terminally or borne on cone-scales in female cones. Female gametophyte greatly reduced and retained in ovule. Fertilized ovule (seed) retained on sporophyte until ripe.

VEGETATIVE KEY TO GENERA OF PINOPSIDA

1 Some leaves at least 1cm wide   **23/1. ARAUCARIA**
1 All leaves <1cm wide   2
   2 Leaves in opposite pairs or in threes at each node   3
   2 Leaves 1 at each node, borne spirally but often apparently 2-ranked   6
3 At least some leaves in threes, not appressed to stem (some other cultivated Cupressaceae key out here)   **22/5. JUNIPERUS**
3 Leaves in opposite pairs, usually ± appressed to stem (some cultivated *Juniperus* spp. key out here)   4
   4 Leaves obtuse; ultimate branchlets not flattened, spreading in 3 dimensions   **22/1. CUPRESSUS**
   4 Leaves acute to acuminate; ultimate branchlets flattened, mainly or wholly spreading in 1 plane   5
5 Foliage with a resinous or oily scent when crushed
     22/2 & 3. X CUPRESSOCYPARIS & CHAMAECYPARIS
5 Foliage with a sweet aromatic scent when crushed
     22/4. THUJA
   6 Leaves long and needle-like, borne in groups of 2-5 on very dwarf short-shoots   **20/7. PINUS**
   6 Leaves all borne singly on long-shoots or most borne in clusters of >10 on short-shoots   7
7 Most leaves borne in dense clusters on short-shoots   8
7 All leaves borne singly on long-shoots   9
   8 Leaves deciduous, dorsiventrally flattened
     20/5. LARIX
   8 Leaves evergreen, 3-5-angled in section   **20/6. CEDRUS**
9 Leaves tapered from broad base to narrowly acute apex, not dorsiventrally flattened   10
9 Leaves narrowed at base and apex, flattened or not   11
   10 Trunk with thick spongy outer bark; leaves wider than thick   **21/1. SEQUOIADENDRON**
   10 Trunk with thin stringy outer bark; leaves thicker

                than wide   **21/2. CRYPTOMERIA**
11 Winter vegetative buds green; leaves and wood
   without resin-ducts   **24/1. TAXUS**
11 Winter vegetative buds brown; leaves and wood
   normally with resin-ducts   12
   12 Leaves with distinct slender, short, green
      petiole ± appressed to twig; resin-duct 1 per
      leaf   **20/3. TSUGA**
   12 Leaves sessile, though often much narrowed at base
      and sometimes borne on brown, petiole-like peg;
      resin-ducts usually >1 per leaf, sometimes 0-1   13
13 Leaves borne on distinct brown, petiole-like pegs
   remaining on twig when leaf falls   **20/4. PICEA**
13 Leaves sessile on twigs or on slightly raised cushions
                                                          14
   14 Winter buds conical, shiny, sharply pointed
                                                  20/2. PSEUDOTSUGA
   14 Winter buds rounded at apex   **20/1. ABIES**

KEY TO FAMILIES OF PINOPSIDA

1 Some leaves at least 1cm wide   **23. ARAUCARIACEAE**
1 All leaves <1cm wide   2
   2 Leaves in opposite pairs or in 3s at each node
                                        22. CUPRESSACEAE
   2 Leaves 1 at each node and borne spirally (but
      often apparently 2-ranked), or tightly clustered
      in groups of 2-many on short-shoots   3
3 Vegetative buds with green bud-scales; usually
   dioecious; ovules solitary, surrounded by red
   succulent upgrowth at seed maturity   **24. TAXACEAE**
3 Vegetative buds with brown bud-scales or proper bud-
   scales 0; monoecious; ovules borne in cones that
   become woody at seed maturity   4
   4 Vegetative buds with brown bud-scales; bracts and
      cone-scales distinct   **20. PINACEAE**
   4 Vegetative buds without proper bud-scales;
      bracts and cone-scales completely fused, their
      distinction difficult   **21. TAXODIACEAE**

20. PINACEAE - Pine family

Evergreen or deciduous resiniferous trees; vegetative buds
with brown bud-scales; leaves borne spirally, ± entire,
linear to needle-like, borne singly on long-shoots or in
clusters of 2-many on short-shoots. Monoecious; male
sporangia 2 per sporophyll; female cones with woody, spirally
arranged cone-scales bearing 2 ovules; seeds winged and
becoming detached from cone-scale at maturity; cone-scales
each with a distinct bract below it.
 Distinguished from other coniferous families by the

combination of spirally borne, very narrow leaves, scaly
buds, and ovules borne in cones in which each cone-scale has
a distinct bract below it.
Some genera bear short lateral stems (short-shoots) of very
limited length, in addition to the normal extension shoots
(long-shoots). 'Twigs' refers to stems of the previous 1-2
years' growth.

1 Leaves borne in groups of 2-5 on very dwarf short-
   shoots borne in axils of small scale-leaves   **7. PINUS**
1 Leaves all borne singly on long-shoots, or most
   borne in dense clusters on short-shoots   2
   2 Most leaves borne in dense clusters on short-
      shoots   3
   2 All leaves borne singly on long-shoots   4
3 Leaves deciduous, flattened; female cones <5cm,
   eventually falling whole   **5. LARIX**
3 Leaves evergreen, 3-5-angled in section; female
   cones >5cm, eventually disintegrating on tree   **6. CEDRUS**
   4 Leaves with a distinct slender, short, green
      petiole appressed to twig; female cones <3cm
                                                      3. TSUGA
   4 Leaves sessile though usually narrowed at base
      and sometimes borne on a brown petiole-like peg;
      female cones >3cm   5
5 Leaves borne on distinct brown, petiole-like pegs;
   female cones pendent, falling whole; bracts not
   protruding beyond cone-scales   **4. PICEA**
5 Leaves sessile or borne on slightly raised cushions   6
   6 Winter buds conical, sharply pointed; female cones
      pendent, falling whole; bracts 3-lobed, protruding
      beyond cone-scales but not reflexed   **2. PSEUDOTSUGA**
   6 Winter buds rounded to obtuse; female cones erect,
      disintegrating on tree; bracts not protruding
      beyond cone-scales, or protruding, 1-lobed and
      reflexed   **1. ABIES**

1. ABIES Miller - <u>Firs</u>
Evergreen; leaves borne only on long-shoots, single, sessile,
falling to leave disc-like scars, with 2 resin-ducts, dorsiventrally flattened; female cones erect, the scales deciduous
at maturity.

1 Leaves grey-green on upperside, appressed to twig near
   base and then widely divergent; female cones mostly
   >15cm, with exserted bracts   **3. A. procera**
1 Leaves dark or bright green on upperside, equally
   divergent from twig along length; female cones mostly
   <15cm   2
   2 Twigs conspicuously pubescent (hairs >0.25mm);
      buds not resinous; female cones with exserted
      bracts   **1. A. alba**

2 Twigs minutely pubescent (hairs <0.25mm); buds
   resinous; female cones with included bracts
                                        2. A. grandis

Other spp. - A. nordmanniana (Steven) Spach (Caucasian Fir)
from the Caucasus is nowadays more often planted than A. alba
and has been used in some forestry trials. It differs from
all the above in its very sparsely pubescent to ± glabrous
twigs, from A. procera in its dark or bright green leaf
uppersides, and from the other spp. in the leaves on the
upperside of the twigs short and forward pointing, not
leaving a parting. **A. cephalonica** Loudon (Greek Fir), from
Greece, differs from all other spp. in its very rigid leaves
projecting all around the stem; it is planted on a very small
scale and self-sown seedlings appear rarely in S En.

1. A. alba Miller - European Silver-Fir. Tree to 48m; twigs
conspicuously pubescent; leaves 15-30mm, those on upperside
of twigs laterally spreading and leaving a distinct parting,
dark green on upperside, with 2 whitish stripes on lowerside;
female cones 10-15(20)cm. Intrd; formerly much planted for
timber, especially in N and W, and also in parks, often
self-sown; mountains of C Europe.
2. A. grandis (Douglas ex D. Don) Lindley - Giant Fir. Tree   57
to 60m (tallest tree of any sp. in BI); twigs minutely
pubescent; leaves 20-60mm, those on upperside of twigs
laterally spreading and leaving a distinct parting, bright
green on upperside, with 2 whitish stripes on lowerside;
female cones 5-10cm. Intrd; increasingly used in forestry,
especially in N and W, occasionally self-sown; W N America.
3. A. procera Rehder - Noble Fir. Tree to 47m; twigs   51
densely reddish pubescent; leaves 10-35mm, those on upperside
of twigs short, forward and upward pointing, not leaving a
parting, grey-green with paler bands near apex on upperside,
rather similar on lowerside; female cones 12-25cm. Intrd;
used on a small scale in forestry mostly in W and in parks in
N and W, occasionally self-sown; W N America.

2. PSEUDOTSUGA Carriere - Douglas Fir
Evergreen; vegetative buds sharply pointed; leaves borne only
on long-shoots, sessile, single, falling to leave slightly
raised cushions, with 2 resin-ducts, dorsiventrally
flattened; female cones pendent, falling whole, with
exserted, forward-pointing, 3-toothed bracts.

1. P. menziesii (Mirbel) Franco (P. taxifolia Britton) -   51
Douglas Fir. Tree to 58m; twigs sparsely pubescent; leaves
20-35mm, those on upperside of twigs laterally spreading and
leaving a distinct parting, light to dark green on upperside,
with 2 whitish to pale green stripes on lowerside; female
cones 5-10cm. Intrd; very widely planted for timber, and
also in parks, occasionally self-sown; W N America.

3. TSUGA (Antoine) Carriere – <u>Hemlock–spruces</u>
Evergreen; vegetative buds subacute to rounded; leaves borne
only on long–shoots, single, distinctly petiolate, falling to
leave short brown pegs, with 1 resin–duct, dorsiventrally
flattened; female cones pendent, falling whole, with included
bracts.

Other spp. – **T. canadensis** (L.) Carriere (<u>Eastern Hemlock–
spruce</u>) from E and C N America is frequent in parks and has
been used in small–scale forest plots. It differs from <u>T.
heterophylla</u> in its leaves tapering from near base to apex
and with narrower white stripes on lowerside.

1. T. heterophylla (Raf.) Sarg. – <u>Western Hemlock–spruce</u>. 51
Tree to 46m; twigs densely pubescent; leaves 6–20mm, ±
parallel–sided, dark green on upperside, with 2 very broad
whitish stripes on lowerside, those laterally spreading
distinctly longer than those on upperside of twigs; female
cones 1.5–2.5cm. Intrd; frequent in plantations, often mixed
with hardwoods, often self–sown; W N America.

4. PICEA A. Dietr. – <u>Spruces</u>
Evergreen; vegetative buds acute to rounded; leaves borne
only on long–shoots, single, sessile, falling singly to leave
distinct brown pegs, with 0–2 resin–ducts, dorsiventrally
flattened or 4–angled in section; female cones pendent,
falling whole, with minute bracts. Spp. with flattened
leaves have the leaf twisted at the base so that the
morphologically true lower surface is uppermost.

1 Leaves dorsiventrally flattened   2
1 Leaves about as thick as wide or thicker than wide,
  4–angled in section   3
  2 Twigs glabrous; leaves 15–25mm, sharply pointed at
    apex; female cones 6–10cm   **1. P. sitchensis**
  2 Twigs pubescent; leaves 8–18mm, obtuse to rounded
    but shortly mucronate; female cones 3–6cm
                                       2. P. omorika
3 Leaves bright to dark green with faint stripes, with
  a resinous smell when crushed; female cones 10–20cm
                                          3. P. abies
3 Leaves bluish–green with rather conspicuous stripes,
  with an unpleasant smell when crushed; female cones
  up to 7.5cm   4
  4 Twigs glabrous; leaves 8–18mm   **4. P. glauca**
  4 Twigs pubescent; leaves 15–25mm   **5. P. engelmannii**

1. P. sitchensis (Bong.) Carriere – <u>Sitka Spruce</u>. Tree 57
to 55m; twigs glabrous; leaves 15–25mm, flattened, dark green
on true upperside, with 2 broad whitish stripes on true
lowerside; female cones 6–10cm. Intrd; abundant forest tree
in W and also in parks, often self–sown; W N America.

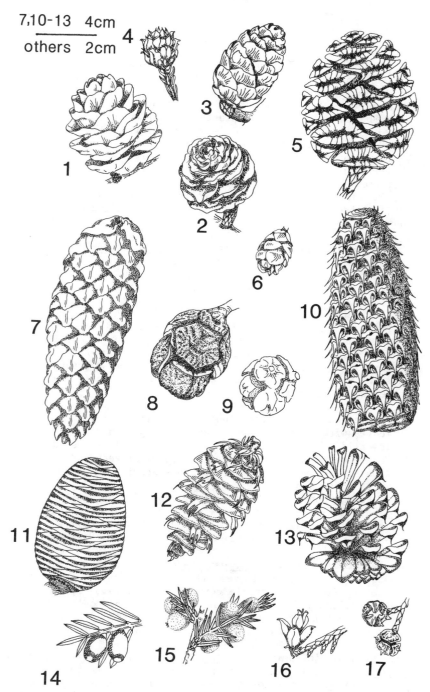

7,10-13 4cm
others 2cm

FIG 51 – Female cones of Pinopsida. 1, **Larix x marschlinsii**.
2, L. kaempferi. 3, L. decidua. 4, **Cryptomeria**. 5,
Sequoiadendron. 6, **Tsuga**. 7, Picea abies. 8, Cupressus. 9, X
Cupressocyparis. 10, Abies procera. 11, Cedrus libani. 12,
Pseudotsuga. 13, Pinus contorta. 14, Taxus. 15, Juniperus.
16, **Thuja**. 17, Chamaecyparis lawsoniana.

2. P. omorika (Pancic) Purkyne – <u>Serbian</u> <u>Spruce</u>. Tree to 31m; twigs pubescent; leaves 8–18mm, flattened, yellowish- or bluish-green on true upperside; with 2 broad whitish stripes on true lowerside; female cones 3–6cm. Intrd; planted on a small scale in plantations in W; Jugoslavia.

3. P. abies (L.) Karsten – <u>Norway</u> <u>Spruce</u>. Tree to 46m; 51
twigs usually glabrous, sometimes pubescent; leaves 10–25mm, 57
4-angled, bright to dark green; female cones 10–20cm. Intrd; abundant forest tree and in shelter-belts and parks, occasionally self-sown; Europe and W Asia.

4. P. glauca (Moench) Voss – <u>White</u> <u>Spruce</u>. Tree to 31m; twigs glabrous; leaves 8–18mm, 4-angled, bluish-green; female cones 2.5–6cm. Intrd; used in shelter-belts and very small-scale forestry in W; N N America.

5. P. engelmannii Parry ex Engelm. – <u>Engelmann</u> <u>Spruce</u>. Tree to 27m; twigs pubescent; leaves 15–25mm, 4-angled, bluish- green; female cones 3.5–7.5cm. Intrd; used in shelter-belts and very small-scale forestry in W; W N America.

5. LARIX Miller – <u>Larches</u>
Deciduous; leaves borne singly on leading long-shoots and in dense clusters on lateral short-shoots, sessile, falling singly, with 2 resin-ducts, dorsiventrally flattened; male cones borne singly, <2cm; female cones erect, eventually falling whole, with bracts exserted at anthesis but ± concealed at maturity.

1 Female cone-scales erect, not with recurved tips;
   leaves with inconspicuous greenish stripes on
   lowerside   **1. L. decidua**
1 Female cone-scales with patent or somewhat recurved
   tips; leaves with rather conspicuous greyish or
   whitish stripes on lowerside   2
   2 Female cones ovoid, usually c.1.25–1.5 x as long
      as wide   **2. L. x marschlinsii**
   2 Female cones broadly ovoid to subglobose, usually
      1–1.25 x as long as wide   **3. L. kaempferi**

1. L. decidua Miller – <u>European</u> <u>Larch</u>. Tree to 46m; young 51
shoots not glaucous; leaves with inconspicuous greenish stripes on lowerside; female cones (1.5)2–3.5(4.5)cm, ovoid, usually c.1.25–1.5 x as long as wide, the cone-scales often slightly wavy but not recurved at tip. Intrd; much planted for forestry and in parks, commonly self-sown; mountains of C Europe.

2. L. x marschlinsii Coaz (<u>L.</u> x <u>henryana</u> Rehder, <u>L.</u> x 51
<u>eurolepis</u> A. Henry nom. illeg.; <u>L.</u> <u>decidua</u> x <u>L.</u> <u>kaempferi</u>) – <u>Hybrid</u> <u>Larch</u>. Tree to 30m, fertile, intermediate between parents in most characters, but female cones more similar to those of <u>L.</u> <u>decidua</u> in shape and cone-scales more similar to those of <u>L.</u> <u>kaempferi</u> in tip recurving. First noticed at Dunkeld, E Perths, in 1904 among progeny from seed collected

from L. kaempferi growing in mixed stands; now planted for
forestry more than either parent, occasionally self-sown and
often originating anew.
3. L. kaempferi (Lindley) Carriere (<u>L. leptolepis</u> (Siebold &   **51**
Zucc.) Gordon) - <u>Japanese</u> <u>Larch</u>. Tree to 37m; young shoots
glaucous; leaves with conspicuous whitish stripes on lowerside; female cones 1.5-3.5cm, broadly ovoid to subglobose,
usually c.1-1.25 x as long as wide, the cone-scales
distinctly recurved at tip. Intrd; much planted for forestry
and for land reclamation schemes, sometimes self-sown; Japan.

6. CEDRUS Trew - <u>Cedars</u>
Evergreen; leaves borne singly on leading long-shoots and in
dense clusters on lateral short-shoots, sessile, falling
singly, with 2 resin-ducts, 3-5-angled in section; male cones
borne singly, >2cm; female cones erect, disintegrating on
tree, with minute bracts.

1 Tip of tree pendent; leaves on short-shoots mostly
   3-3.5cm   **1. C. deodara**
1 Tip of tree stiffly curved to 1 side or erect; leaves
   on short-shoots mostly 1.5-2.5cm   2
  2 Leaves mostly 1.5-2.5cm, abruptly tapered at tip;
     twigs usually glabrous; female cones mostly
     7-10cm   **2. C. libani**
  2 Leaves mostly 1.5-2cm, rather gradually tapered to
     c.0.5mm translucent tip; twigs very shortly
     pubescent; female cones mostly 5-8cm **3. C. atlantica**

1. C. deodara (Roxb. ex D. Don) Don - <u>Deodar</u>. Tree to 37m   **57**
with pendent tip; leaves bright green, mostly 3-3.5cm on
short-shoots, rather gradually tapered to translucent tip <u><</u>
0.4mm; female cones rather rare, 8-12cm. Intrd; formerly
used in shelter-belts and small-scale forestry, persisting
and occasionally self-sown; Afghanistan to W Himalayas.
2. C. libani A. Rich. - <u>Cedar-of-Lebanon</u>. Tree to 40m with   **51**
tip stiffly curved to 1 side or erect; leaves dark green to   **57**
glaucous, mostly 1.5-2.5cm on short-shoots, abruptly tapered
to scarcely translucent tip c.0.2mm; female cones mostly
7-10cm. Intrd; commonly planted in parks and avenues for
ornament; Lebanon to Turkey.
3. C. atlantica (Endl.) Carriere (<u>C. libani</u> ssp. <u>atlantica</u>   **57**
(Endl.) Battand. & Trabut) - <u>Atlas</u> <u>Cedar</u>. Tree to 39m with
tip stiffly curved to 1 side or erect; leaves dark green to
glaucous (more often so than in <u>C. libani</u>); differing from <u>C.</u>
<u>libani</u> in less well developed low horizontal branches and in
characters in key, of which the longer (c.0.5mm) translucent
leaf-tip is probably most reliable. Intrd; widely planted in
parks, etc., mainly as cv. 'Glauca', but less common than <u>C.</u>
<u>libani</u>, self-sown in Surrey; Morocco.

7. PINUS L. - Pines
Evergreen; leaves borne in groups of 2, 3 or 5 on very dwarf
short-shoots, sessile, with 2-many resin-ducts, semicircular
to variously angled in section; male cones borne in clusters;
female cones eventually falling whole, with minute bracts.
On very young plants long leaves are borne singly on long-
shoots, but otherwise long-shoots bear only scale-leaves.

```
1 Leaves in pairs                                               2
1 Leaves in threes or fives                                     7
  2 Leaves mostly >10cm                                         3
  2 Leaves mostly <10cm                                         4
3 Bud-scales recurved at apex; female cones 8-22cm
                                                  5. P. pinaster
3 Bud-scales appressed at apex; female cones 3-9cm
                                                    2. P. nigra
  4 Upper part of trunk pale orange-red; cone-scales
    + without prickle on outer face; leaves glaucous,
    twisted                                     1. P. sylvestris
  4 Upper part of trunk grey-brown to red-brown; cone-
    scales usually with distinct prickle on outer face;
    leaves green, twisted or not                                5
5 Leaves distinctly twisted; cone-scales with rather
  long slender prickle                            4. P. contorta
5 Leaves scarcely twisted; cone-scales with very short
  stout prickle                                                 6
  6 Buds obtuse to subacute; leaves <8cm; female
    cones <5cm; resin-ducts just below leaf surface
                                                     3. P. mugo
  6 Buds acuminate; leaves often >8cm; female cones
    often >5cm; resin-ducts deep-seated in leaf
                                                    2. P. nigra
7 Leaves in threes                                              8
7 Leaves in fives                                               9
  8 Leaves 10-15cm, bright green, rarely >1mm wide
                                                   6. P. radiata
  8 Leaves (10)15-25cm, dull green, <2mm wide
                                                 7. P. ponderosa
9 Twigs glabrous; tip of female cone-scales distinctly
  thickened                                         8. P. peuce
9 Twigs with minute hairs at base of short-shoots; tip
  of female cone-scales scarcely thickened      9. P. strobus
```

Other spp. - Of the many other spp. grown for ornament in
parks, etc., the 5-leaved **P. wallichiana** A. B. Jackson (P.
chylla Lodd. nom. nud.) (Bhutan Pine) from the Himalayas is
perhaps the commonest. It differs from P. peuce and P.
strobus in its glabrous twigs, longer, less rigid leaves
(8-20cm), and longer female cones (15-25cm).

1. P. sylvestris L. - Scots Pine. Tree to 36m with trunk   57
pale orange-red above; leaves in pairs, 2-8cm, glaucous,

twisted; female cones 2.5–7.5cm. Native; forming pure or mixed woodland in Highlands of Sc S to M Perth; very widely planted in BI for timber and ornament and commonly natd, especially on sandy soils. Native plants retain their pyramidal shape until late in life and have short leaves and cones; they are often distinguished as ssp. **scotica** (P.K. Schott) E. Warb.

2. **P. nigra** Arnold – see sspp. for English names. Tree to 42m with trunk dark grey above; leaves in pairs, 8–18cm, not glaucous, scarcely twisted; female cones 3–9cm. Intrd.

a. **Ssp. nigra** – <u>Austrian</u> <u>Pine</u>. Crown wide with long side branches; leaves dark green, 8–12cm, rather stiff. Commonly planted in shelter-belts and for ornament, often self-sown; C and SE Europe.

b. **Ssp. laricio** Maire (<u>P.</u> <u>nigra</u> var. <u>maritima</u> (Aiton) Melville) – <u>Corsican</u> <u>Pine</u>. Crown columnar with short side branches; leaves bright green, 10–18cm, rather flexible. Commonly planted in shelter-belts and for ornament, especially near sea in E, and also for forestry on sandy soils in S, often self-sown; Corsica, Sicily and S Italy.

3. **P. mugo** Turra – <u>Dwarf</u> <u>Mountain-pine</u>. Shrub to 4m with irregularly spreading branches; leaves in pairs, 3–8cm, bright green, ± straight; female cones 2–5cm. Intrd; very hardy and planted in upland areas as windbreak, especially for young plantations; C SE Europe.

4. **P. contorta** Douglas ex Loudon – <u>Lodgepole</u> <u>Pine</u>. Tree to   **51** 25m with trunk dark red-brown above; leaves in pairs, 3–8.5cm, bright green, twisted; female cones 2–6cm, with characteristic long prickles on ends of cone-scales. Intrd; now one of the most frequently planted forestry conifers but rarely used for other purposes, self-sown in Cards and Scillies; W N America.

5. **P. pinaster** Aiton – <u>Maritime</u> <u>Pine</u>. Tree to 35m with trunk dark red-brown above; leaves in pairs, (10)15–20(25)cm, dark to greyish green, sometimes twisted; female cones 8–22cm. Intrd; used in small-scale forestry and shelter-belts, especially on sandy soil or by sea in S, often self-sown; Mediterranean and SW Europe. Most trees grown in BI are referable to ssp. <u>pinaster</u> (ssp. <u>atlantica</u> Hug.-del-Vill.), but distinctness of sspp. among cultivated plants is dubious.

6. **P. radiata** D. Don – <u>Monterey</u> <u>Pine</u>. Tree to 41m; leaves in threes, 10–25cm, bright green, very slender; female cones 7–14cm, very asymmetrical, with small non-persistent point at apex of cone-scales. Intrd; commonly planted in W Wa, SW En and CI in parks, field borders and small plantations, especially near sea, self-sown in Cornwall; California.

7. **P. ponderosa** Douglas ex Lawson & P. Lawson – <u>Western</u> <u>Yellow-pine</u>. Tree to 40m; leaves in threes, (10)15–25cm, deep yellow-green, stout; female cones 7–15cm, ± symmetrical, with strong persistent point at apex of cone-scales. Intrd; frequently planted in parks and occasionally used in experimental forestry plantations in W; W N America.

8. P. peuce Griseb. - <u>Macedonian Pine</u>. Tree to 30m; twigs glabrous; leaves in fives, 7-12cm; female cones 8-15cm, with tips of cone-scales distinctly thickened. Intrd; occasionally grown in trial plantations in uplands in W, and in parks; Balkans.

9. P. strobus L. - <u>Weymouth Pine</u>. Tree to 40m; twigs with tuft of hairs at base of short-shoots; leaves in fives, 5-14cm; female cones 8-20cm, with tips of cone-scales scarcely thickened. Intrd; occasionally grown in trial plantations in W, and in parks, occasionally self-sown; C & E N America.

21. TAXODIACEAE - <u>Redwood family</u>

Evergreen or deciduous resiniferous trees; vegetative buds without proper bud-scales; leaves borne spirally, entire, linear, needle-like or scale-like, borne singly on long-shoots. Monoecious; male sporangia 2-8 per sporophyll; female cones with woody, spirally arranged cone-scales bearing 2-12 ovules; seeds winged or not, becoming detached from cone-scale at maturity; cone-scales and bracts wholly or partially fused.
Distinguished from Pinaceae in the indistinct bracts fused to the cone-scales; the usually >2 male and female sporangia; and the buds lacking proper bud-scales.

1 Trunk with thick, spongy outer bark; leaves wider than
   thick; female cones >4cm, without recurved points at
   apex of scale   **1. SEQUOIADENDRON**
1 Trunk with thin, stringy outer bark; leaves thicker
   than wide; female cones <3cm, with recurved points at
   apex of scales   **2. CRYPTOMERIA**

Other genera - 3 spp. in other genera, differing from the above in their dorsiventrally flattened leaves spreading in 1 plane on opposite sides of the twigs, are frequently grown as specimen trees in parks, etc. **SEQUOIA** Endl. is evergreen with a thick spongy bark and female cones like those of <u>Sequoiadendron</u> but <3cm; **S. sempervirens** (Lambert) Endl. (<u>Coastal Redwood</u>), from Californian coast, is the only sp. **TAXODIUM** Rich. and **METASEQUOIA** Miki have a stringy bark and the leaves and ultimate lateral twigs are deciduous; **T. distichum** (L.) Rich. (<u>Swamp Cypress</u>) from SE United States has apparently alternate leaves and **M. glyptostroboides** Hu & W. C. Cheng (<u>Dawn Redwood</u>) from China has apparently opposite leaves.

1. SEQUOIADENDRON Buchholz - <u>Wellingtonia</u>
Evergreen; trunk with thick, spongy, red-brown outer bark; leaves tapering from decurrent base to pointed apex; male cones solitary at apex of twigs; female cones without recurved point at apex of scale.

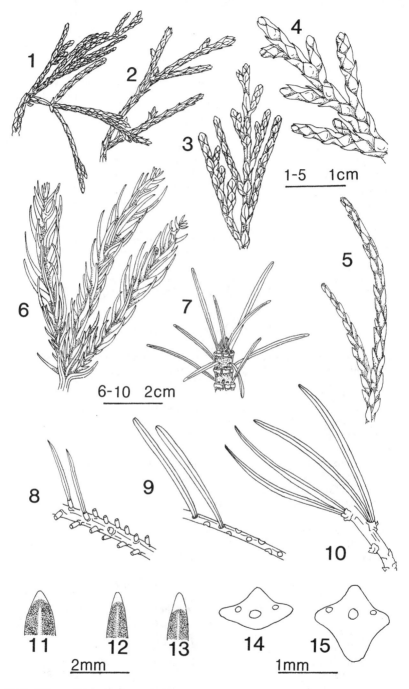

FIG 57 – Pinopsida. 1–10, leafy shoots. 1, **Cupressus** 2, **Chamaecyparis lawsoniana**. 3, X **Cupressocyparis**. 4, **Thuja**. 5, **Sequoiadendron**. 6, **Cryptomeria**. 7, **Cedrus atlantica**. 8, **Picea sitchensis**. 9, **Abies grandis**. 10, **Pinus sylvestris**. 11–13, leaf-apices of **Cedrus**. 11, **C. libani**. 12, **C. deodara**. 13, **C. atlantica**. 14–15, leaf-sections of **Picea**. 14, **P. sitchensis**. 15, **P. abies**.

1. S. giganteum (Lindley) Buchholz (<u>Sequoia wellingtonia</u>   51
Seemann) - <u>Wellingtonia</u>. Tree to 50m; leaves with free   57
portion 3-7mm, appressed; female cones 4.5-8cm. Intrd;
commonly planted in parks and along drives, and sometimes in
larger groups; Sierra Nevada, California.

2. CRYPTOMERIA D. Don - <u>Japanese Red-cedar</u>
Evergreen; trunk with thin, stringy, red-brown outer bark;
leaves (except juvenile foliage) tapering from decurrent base
to pointed apex; male cones lateral, in groups behind apex of
twigs; female cones with recurved points at apex of scale.

1. C. japonica (L.f.) D. Don - <u>Japanese Red-cedar</u>. Tree   51
to 37m but usually much less; leaves with free portion   57
6-15mm, ± appressed; female cones 12-30cm. Many cultivars
grown in parks and gardens, some retaining juvenile foliage
(leaves longer, not decurrent, spreading). Intrd; grown in
rather small-scale forestry plots in Wa; China and Japan.

22. CUPRESSACEAE - Juniper family

Evergreen resiniferous trees or shrubs; vegetative buds
without proper bud-scales; leaves opposite or whorled,
needle-like or scale-like. Monoecious or dioecious; male
sporangia 3-5 per sporophyll; female cones with woody or
fleshy cone-scales arranged in opposite pairs or in threes
and bearing 1-many ovules; seeds winged or not, becoming
detached from woody cones or remaining attached to succulent
cones at maturity.
Distinguished from other conifers by the opposite or whorled
leaves and female cone-scales. Juvenile foliage is
needle-like and patent, with leaves usually in whorls of 3
but sometimes opposite. Mature foliage is scale-like and
appressed with opposite leaves. Except in <u>Juniperus</u> and
certain cultivars of other genera the juvenile foliage is
lost at a very young stage. In all other genera 'leaves'
refers to those of mature foliage.

1 Female cone berry-like at maturity, dispersed whole;
    leaves usually borne in threes   **5. JUNIPERUS**
1 Female cone dry and woody at maturity, the scales
    opening to release the seeds; leaves usually
    opposite   2
    2 Most female cones >2cm; leaves obtuse;
        ultimate branchlets not flattened, spreading in 3
        dimensions   **1. CUPRESSUS**
    2 Female cones <2cm; leaves acute to acuminate;
        ultimate branchlets flattened, mainly or wholly
        spreading in 1 plane   3
3 Female cones elongated, with flattened scales; leaves
    with sweet aromatic smell when crushed   **4. THUJA**

3 Female cones ± globose, with peltate scales; leaves
   with resinous or oily smell when crushed   4
   4 Female cones <12mm   **3. CHAMAECYPARIS**
   4 Female cones mostly 15-20mm   **2. X CUPRESSOCYPARIS**

1. CUPRESSUS L. - Cypresses
Twigs spreading in 3 dimensions, not flattened in 1 plane;
juvenile foliage lost at very young stage; mature foliage
with leaves opposite, scale-like, obtuse, appressed to twig;
monoecious; female cones ripening in second year, ± globose,
with 4-7 decussate pairs of peltate, woody cone-scales each
with 8-20 narrowly winged seeds.

1. C. **macrocarpa** Hartweg ex Gordon - Monterey Cypress. Tree   51
to 37m; leaves dark green, 1-2mm; female cones 20-35mm, with   57
stout conical protuberance on each cone-scale.   Intrd;
commonly planted in parks, village greens, roadsides, etc.,
especially near coast in SW, self-sown in Scillies and
Jersey; California.

2. X CUPRESSOCYPARIS Dallimore (Cupressus x Chamaecyparis) -
Leyland Cypress
(See Chamaecyparis key)
Variously intermediate between the parental genera; more
similar to Chamaecyparis in vegetative characters but near
half-way in cone characters; ultimate branchlets flattened,
mainly spreading in 1 plane but some leaving it; female cones
usually with 4 pairs of cone-scales and c.5-6 seeds per
scale.

1. X C. **leylandii** (A. B. Jackson & Dallimore) Dallimore   51
(Cu. macrocarpa x Ch. nootkatensis) - Leyland Cypress.   57
Tree to 34m; foliage like that of Ch. nootkatensis but
ultimate branchlets patent to erect, not pendent, and not
spreading entirely in 1 plane; female cones 15-20mm, with a
strong conical protuberance on each cone-scale.   Arose in
1888 in a female cone of Ch. nootkatensis growing near Cu.
macrocarpa at Welshpool, Monts; now very commonly planted as
a fast-growing screen in parks, around playing fields,
factories, etc., and beside roads.

3. CHAMAECYPARIS Spach - Cypresses
Twigs spreading in 1 plane, flattened; juvenile foliage
usually lost at very young stage; mature foliage with leaves
opposite, ± scale-like, acute to acuminate, ± appressed
to twig; monoecious; female cones usually ripening in first
year, ± globose, with 3-6 decussate pairs of peltate, woody
cone-scales each with 2-5 narrowly winged seeds.

Key to Chamaecyparis and X Cupressocyparis
1 Female cones with small prickle on each cone-scale;
   ultimate branchlets with white markings beneath   2

1 Female cones with prominent central conical spine
   on each cone-scale; ultimate branchlets without white
   markings beneath   3
   2 Leaves usually acuminate, with inconspicuous
      glands on back;   female cones c.5-6mm
                                            2. Ch. pisifera
   2 Leaves usually acute or narrowly acute, the dorsal
      row with a conspicuous translucent gland on back;
      female cones c.7-9mm   **1. Ch. lawsoniana**
3 Ultimate branchlets pendent; female cones 8-12mm
                                         3. Ch. nootkatensis
3 Ultimate branchlets patent to erect; female cones
   mostly 15-20mm   **X Cu. leylandii**

  1. C. lawsoniana (A. Murray) Parl. - Lawson's Cypress. Tree   51
to 41m, extremely variable in habit and colour with numerous   57
cultivars; branchlets marked with white beneath; leaves ≤
c.2mm, acute to narrowly acute; male cones pinkish-red.
Intrd; very widely planted in groves and as windbreaks and
screens, and mixed or alone in forestry plantations, commonly
self-sown; W United States.
  2. C. pisifera (Siebold & Zucc.) Siebold & Zucc. - Sawara
Cypress. Tree to 26m, with many variable cultivars, several
retaining juvenile foliage; differing from C. lawsoniana in
its acuminate leaves with less conspicuous dorsal gland and
smaller female cones with smaller apical spine on
cone-scales.   Intrd; commmonly planted in parks and
occasionally self-sown; Japan.
  3. C. nootkatensis (Lambert) Spach - Nootka Cypress. Tree
to 30m; differing from other 2 spp. in lack of white markings
on lowerside of branchlets, leaves ≤ c.3mm, and larger cones
which do not open until the second year and have a prominent
conical spine on end of cone-scales.   Intrd; frequent in
parks and sometimes found as experimental forestry plantings;
W N America.

4. THUJA L. - Red-cedars
Differing from Chamaecyparis in the elongated female cones
with usually 10-12 flattened cone-scales each with a recurved
apical spike and 2-3 seeds.

  1. T. plicata Donn ex D. Don - Western Red-cedar. Tree to   51
42m; vegetatively similar to Chamaecyparis lawsoniana but   57
leaves longer (≤ 6mm) and with a sweet aromatic smell when
crushed; male cones blackish, becoming yellowish when ripe;
female cones 10-12mm. Intrd; very commonly planted in parks
and as windbreaks, etc., and usually mixed in forestry
plantations, frequently self-sown; W N America.

5. JUNIPERUS L. - Junipers
Twigs spreading in 3 dimensions, not flattened in 1 plane,
juvenile foliage with erect to patent leaves usually borne in

whorls of 3; adult foliage with appressed scale-like opposite leaves; usually dioecious; female cones berry-like, with succulent ± fused cone-scales; seeds not winged, 1 per cone-scale, retained within cone at dispersal.

Other spp. - Several spp. and a great many cultivars are grown in parks and on roadsides as ground-cover or bushes in shrubberies and rockeries, etc. Most of these spp. (notably **J. chinensis** L. (<u>Chinese Juniper</u>) from China and Japan) bear both juvenile (3-ranked) and adult (opposite) foliage, but often the former is eventually lost or, in many cultivars, the latter is never developed. In some cultivated spp. the juvenile leaves are borne in opposite pairs.

1. J. communis L. - <u>Common Juniper</u>. Tree or shrub to 51 16m, never attaining adult foliage; leaves 4-20mm, linear, with a single broad white band on upperside; female cones 5-10mm, ± globose. Native; little planted.
a. Ssp. communis. Spreading shrub to erect tree; leaves mostly 8-20 x 1-1.5mm, ± patent, gradually tapered to sharp point so that branchlets are prickly to touch. Very local but often common throughout Br and Ir, on both limestone and acid soils.
b. Ssp. alpina Celak. (ssp. <u>nana</u> (Willd.) Syme). Procumbent matted shrub, leaves mostly 4-12 x c.1.5mm, erecto-patent, abruptly tapered to short sometimes blunt point so that branchlets are scarcely sharp to touch. NW Wa, NW En, W Ir and C & NW Sc, on rocks and moorland mostly in upland areas.
There are frequent intermediates between these sspp.

23. ARAUCARIACEAE - <u>Monkey-puzzle family</u>

Evergreen resiniferous trees; vegetative buds without proper buds-scales; leaves borne spirally, broad and flat. Normally dioecious; male sporangia 5-15 per sporophyll; female cones with woody, spirally arranged cone-scales bearing 1 ovule; seeds not winged, remaining attached to cone-scale which falls at maturity.
Differs from other conifers in the broad many-veined leaves.

1. ARAUCARIA Juss. - <u>Monkey-puzzle</u>
1. A. araucana (Molina) K. Koch - <u>Monkey-puzzle</u>. Tree 62 to 29m; younger branches densely clothed in erecto-patent leaves; leaves 2.5-5cm, leathery, lanceolate-triangular, at least some >1cm wide at base, narrowed to sharp apical spine; male cones c.10-12cm; female cones c.15-20 x 12-15cm. Intrd; commonly planted in rows or groups by drives, wood-borders, etc., very rarely self-sown; Chile and W Argentina.

24. TAXACEAE - <u>Yew family</u>

Evergreen non-resiniferous trees or shrubs; vegetative buds
with green bud-scales; leaves borne spirally, linear.
Normally dioecious; male sporangia 4-9 per sporophyll; ovules
borne singly (not in cones), terminal on very short lateral
branches; seed surrounded except at apex by succulent
upgrowth (aril), falling with it at maturity.
Distinguished from other conifers by the lack of resin and
the single ovules with a fleshy aril after fertilization.

1. TAXUS L. - <u>Yew</u>
  1. T. baccata L. - <u>Yew</u>. Large bush or spreading tree to   51
28m, often with multiple trunks; leaves 10-30 (45) x 2-3mm,
dark green with two pale stripes on lowerside; aril c.1cm,
red. Native; very local but often common throughout En, Wa
and Ir, but very rare in Sc; on well-drained limestone and
also locally on acid sandstone, sometimes dominant; also very
widely grown and often self-sown. Numerous cultivars, with
varied habit and leaf and aril colour, are grown.

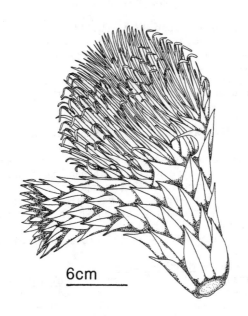

6cm

FIG 62 - Female cone of **Araucaria araucana**.

MAGNOLIOPSIDA - ANGIOSPERMS, FLOWERING PLANTS
(Angiospermopsida, Angiospermae, Anthophyta)

Trees, shrubs, climbers, herbs or variously reduced plants of extremely varied growth-form. Male sporangia borne in specialized organs (stamens) grouped 1-many in male or bisexual flowers. Female sporangia borne in ovules enclosed 1-many together in carpels; carpels grouped 1-many in female or bisexual flowers. Female gametophyte greatly reduced and retained in ovule. Fertilized ovule (seed) retained within carpel on sporophyte until ripe, then dispersed separately or within carpel (fruit).

KEY TO FAMILIES OF MAGNOLIOPSIDA
(Dicotyledons and Monocotyledons)

Before using these keys the section on Identification keys in the Introduction should be read carefully. In particular, it is essential to work out fully the structure of the flowers <u>before</u> starting on the keys. The keys have been made as user-friendly as possible, but it must be admitted that pitfalls still abound. Only after considerable experience should the supplementary keys (A, B, etc.) be used directly; it is advisable to start always with the General key. In either case it is of course of paramount importance to arrive at the correct supplementary key; hence the General key should be used very carefully and a good working knowledge of it be built up.

The distinctions between 2 perianth whorls that are similar as opposed to different, and between sepaloid and petaloid perianth segments, are subjective. Wherever the answer is considered equivocal both alternatives are allowed for here. In some cases, however, the inner and outer whorls of perianth segments are quite different, but a lens is needed to discover this due to their small size (e.g. <u>Empetrum</u>, <u>Ruscus</u>) and careless observation will produce the wrong answer.

Because of the variation shown, many families are keyed out in several different positions in up to 5 different supplementary keys. In all cases where only 1 genus from a family containing >1 genus is referred to, that genus is stated.

<u>General key</u>
1 Plants consisting of floating or submerged ±

undifferentiated pad-like fronds $\leq$10(15)mm, sometimes with narrow stalk-like part at 1 end, with
or without roots dangling in water (rarely stranded
temporarily on mud) (beware <u>Azolla</u>)   **150. LEMNACEAE**
1 If plants free-floating then with clearly
  differentiated stems and leaves   2
    2 Aquatics with some leaves or parts of leaves
      modified as small bladders to catch minute
      animals (<u>Utricularia</u>)   **130. LENTIBULARIACEAE**
    2 Leaves never modified as small bladders   3
3 Aquatic or mud plants with at least some leaves in
  whorls of $\geq$3, the leaves linear or $\pm$ so or
  divided into linear segments   <u>Key A</u>
3 If aquatic or on mud then leaves not whorled and/
  or not linear or with linear segments   4
    4 Woody plant parasitic on aerial parts of trees,
      the roots buried in living host branches
                          88. VISCACEAE
    4 If growing on aerial parts of trees then merely
      epiphytic, with roots not buried in living host
      branches   5
5 Trees with unbranched stem and terminal rosettes of
  huge pinnate or palmate leaves; or seedlings with
  leaves ribbed alternately on each surface
                    148A. ARECACEAE
5 If trees then not with single terminal rosette of
  compound leaves; if seedlings then not with leaves
  ribbed alternately on each surface   6
    6 Plant consisting of 1-few rosettes of many linear
      simple leaves usually $\geq$1m, either borne on
      ground or at tips of woody branches   **162. AGAVACEAE**
    6 If leaves all in 1 or few rosettes then much
      <1m and often not linear   7
7 Plant consisting of dense hemispherical mass often
  $\geq$1m across, with narrow pineapple-like leaves with
  strongly spiny margins   **158. BROMELIACEAE**
7 If plant in a dense hemispherical spiny mass then
  a growth-form of plant without pineapple-like leaves 8
    8 All inflorescences entirely replaced by
      vegetative propagules (bulbils or plantlets)   <u>Key B</u>
    8 At least some inflorescences bearing flowers
      (or fruits)   9
9 Perianth 0, or of 1 whorl, or of $\geq$2 whorls of or
  a spiral of similar segments   10
9 Perianth of 2 (rarely more) distinct whorls or
  rarely a spiral, the inner and outer differing
  markedly in shape/size/colour   12
    10 Perianth $\pm$ corolla-like, usually white or
      distinctly (often brightly) coloured   <u>Key C</u>
    10 Perianth $\pm$ calyx-like or bract-like (greenish
      to brownish, or scarious or much reduced) or 0   11
11 Trees or shrubs, sometimes very short or

procumbent but then with woody stems producing
growth in subsequent years   Key D
11 Herbs, with non-woody (or only basally woody)
   stems dying to ground level after 1(-few) year(s) Key E
   12 Flowers all male or all female, the male or
      female parts 0 or extremely vestigial   Key F
   12 At least some flowers bisexual or both male
      and female flowers present   13
13 Petals fused at base for varying distances to apex   14
13 Petals free, or rarely ± fused just at basal point,
   or rarely ± fused near apex but free at base   15
   14 Ovary superior (hypogynous or perigynous)   Key G
   14 Ovary inferior (epigynous) or partly so   Key H
15 Ovary inferior (epigynous) or partly so   Key I
15 Ovary superior (hypogynous or perigynous)   16
   16 Carpels and styles free, or carpels fused
      just at base   Key J
   16 Carpels and/or styles fused wholly or for
      greater part, or carpel 1   Key K

Key A - Aquatic or mud plants with whorled leaves, the leaves
        either linear or divided into linear segments (beware
        Equisetum, with whorled lateral branches)
1 Leaves simple, sometimes prominently toothed   2
1 Leaves divided (forked or pinnate)   6
  2 Tip of stems erect and emergent at flowering;
    leaves in whorls of 4-12   3
  2 Stems always submerged (unless stranded by
    drought); leaves opposite or in whorls of 3-6(8)   4
3 Leaves entire, 6-12 per whorl   121. HIPPURIDACEAE
3 Leaves with few to many pricklets on margins, 4-8
  per whorl (Galium)   132. RUBIACEAE
  4 Leaves with stipules free from leaf-base
                                    147. ZANNICHELLIACEAE
  4 Leaves without stipules; often with sheathing leaf-
    base or with 2 minute scales at base in axil   5
5 Leaves with distinctly widened base shortly
  sheathing stem   146. NAJADACEAE
5 Leaves wider near base than near apex but narrowed
  at extreme base, sometimes slightly clasping stem
  but not sheathing it   140. HYDROCHARITACEAE
  6 Leaves forked 1-4 times   28. CERATOPHYLLACEAE
  6 Leaves 1-2-pinnate   7
7 Leaves with flat segments; flowers conspicuous,
  >15mm across (Hottonia)   70. PRIMULACEAE
7 Leaves with filiform segments; flowers inconspicuous,
  <6mm across (Myriophyllum)   80. HALORAGACEAE

Key B - Plants with all inflorescences entirely replaced by
        vegetative propagules
1 Leaves reniform, palmately lobed (Saxifraga)
                                        75. SAXIFRAGACEAE

1 Leaves linear to lanceolate, entire   2
  2 Inflorescences proliferating by producing small
    plantlets in place of flowers   3
  2 Inflorescences replaced by axillary solitary or
    terminal clusters of small solid structures
    ('bulbils')   4
3 Stems solid; leaves unifacial, flattened-cylindrical
  (Juncus)   **153. JUNCACEAE**
3 Stems hollow; leaves bifacial, flat or inrolled or
  infolded   **155. POACEAE**
  4 Bulbils sessile, forming $\pm$ globose compact
    terminal head or $\pm$ globose mass at ground level
    (Allium, Gagea)   **160. LILIACEAE**
  4 Bulbils on slender stalks, solitary in leaf-axils
    (Lysimachia)   **70. PRIMULACEAE**

Key C - Perianth of 1 whorl, or of ≥ 2 whorls or a spiral of $\pm$
    similar segments, $\pm$ corolla-like (usually white or
    distinctly coloured) (beware taxa with distinct inner
    and   outer   perianth   whorls   but   1   or   other   soon
    falling, e.g. Papaveraceae)
1 Flowers bisexual or monoecious   2
1 Flowers dioecious   29
  2 Ovary inferior (flowers epigynous)   3
  2 Ovary superior (flowers hypogynous to perigynous)   13
3 Stamens ≥ 6 4
3 Stamens ≤ 5 5
  4 Leaves ovate or broadly so, cordate at base; ovary
    6-celled; styles/stigmas 6   **26. ARISTOLOCHIACEAE**
  4 Leaves narrow, gradually tapered to base; ovary
    3-celled; styles/stigmas 1 or 3   **160. LILIACEAE**
5 Stamens 1-3   6
5 Stamens 4-5 (some may drop very early)   8
  6 Perianth with 5 segments; ovule 1 (or 0) per cell
                                           135. VALERIANACEAE
  6 Perianth with 6 segments; ovules numerous per cell   7
7 Style obvious, with 3 obvious branches or 3 separate
  stigmas; stamens 3   **161. IRIDACEAE**
7 Styles 0; stamens 1 or 2   **164. ORCHIDACEAE**
  8 Leaves simple, in whorls of ≥ 4 **132. RUBIACEAE**
  8 Leaves not in whorls of ≥ 4 9
9 Petals free, inserted on top of ovary   10
9 Petals fused into tube at least proximally   11
  10 Stamens 4; carpels and styles 1 (Sanguisorba)
                                                  76. ROSACEAE
  10 Stamens 5; carpels and styles 2   **109. APIACEAE**
11 Flowers borne in open raceme-like cymes **87. SANTALACEAE**
11 Flowers borne in dense capitula   12
  12 Stamens 4, free, $\pm$ exserted; ovary and fruit
    surrounded by epicalyx   **136. DIPSACACEAE**
  12 Stamens 5, their anthers fused into tube round

style; ovary and fruit not surrounded by epicalyx
                                    137. ASTERACEAE
13 Tree or shrub or woody climber   14
13 Herb   18
  14 Tepals 10; stamens 5   **97. STAPHYLEACEAE**
  14 Tepals 3-8; stamens 6-many or flowers female   15
15 Climber or scrambler   16
15 Strongly self-supporting shrub   17
  16 Leaves simple; stipules scarious, fused round
    stem; stamens 6-9   **48. POLYGONACEAE**
  16 Leaves usually ternate or pinnate; stipules 0;
    stamens >10 (Clematis)   **29. RANUNCULACEAE**
17 Tepals free, creamy- to greenish-yellow; flowers
dioecious; leaves evergreen, sweetly scented when
crushed   **25. LAURACEAE**
17 Tepals fused into tube (or arising from tubular
hypanthium), purple; flowers bisexual; leaves
deciduous, not sweetly scented   **83. THYMELAEACEAE**
  18 Carpels >2, free or + so   19
  18 Carpels 1, or >1 and fused   23
19 Carpels each with 1-2 ovules   20
19 Carpels each with several to many ovules   21
  20 Flowers in racemes; fruit succulent
                       42. PHYTOLACCACEAE
  20 Flowers solitary or in cymes; fruit not succulent
                       29. RANUNCULACEAE
21 Tepals 1(-2), white; inflorescence a forked spike
just above water-surface   **141. APONOGETONACEAE**
21 Tepals >5, usually coloured; inflorescence not a
forked spike   22
  22 Leaves simple, linear, without petiole;
    inflorescence an umbel   **138. BUTOMACEAE**
  22 Leaves compound and/or with well developed
    petiole; inflorescence not an umbel
                       29. RANUNCULACEAE
23 Stamens >10   **29. RANUNCULACEAE**
23 Stamens 3-9   24
  24 Ovary 1-3-celled, each cell with 2-many ovules   25
  24 Ovary 1-celled (or 3-celled with 2 cells +
    aborted and empty), with 1 ovule   26
25 Ovary 1-celled; tepals and stamens 5 (Glaux)
                        70. PRIMULACEAE
25 Ovary 3-celled; tepals and stamens 4 or 6 **160.LILIACEAE**
  26 Leaves pinnate (Sanguisorba)   **76. ROSACEAE**
  26 Leaves simple, usually entire   27
27 Perianth >15mm across, with tube >2cm
                     42A. NYCTAGINACEAE
27 Perianth <10mm across, with tube <1cm (often 0)   28
  28 Emergent aquatic with blue flowers; tepals and
    stamens always 6   **159. PONTEDERIACEAE**
  28 Flowers never blue; tepals mostly 5; stamens
    (4-)8(9)   **48. POLYGONACEAE**

29 Flowers in dense capitula closely surrounded by
   ≥1 row of bracts   **137. ASTERACEAE**
29 Flowers not in dense capitula   30
   30 Tepals 4 or 6   31
   30 Tepals 5   32
31 Tree or shrub; tepals 4; leaves elliptic   **25. LAURACEAE**
31 Herb; tepals 6; leaves reduced to scales, replaced
   by cylindrical cladodes (<u>Asparagus</u>)   **160. LILIACEAE**
   32 Leaves simple, entire or ± so   **48. POLYGONACEAE**
   32 At least leaves on stem pinnate or deeply
      pinnately lobed   33
33 Tepals free; basal leaves pinnate (<u>Trinia</u>)
                                             109. APIACEAE
33 Tepals fused into tube proximally; basal leaves
   simple and ± entire (<u>Valeriana</u>)   **135. VALERIANACEAE**

<u>Key</u> <u>D</u> - Perianth 0 or of 1 or more whorls or a spiral of ±
        similar segments, ± calyx-like or bract-like (usually
        greenish or brownish or scarious); plants with woody
        stems
1 Leaves pinnate or ternate   2
1 Leaves simple, often deeply lobed   7
   2 Stems spiny (<u>Aralia</u>)   **108. ARALIACEAE**
   2 Stems not spiny   3
3 Shrubs <2m; flowers in dense capitula   4
3 Trees >2m; flowers various   5
   4 Stems ± erect; flowers in clusters of ±
     campanulate capitula (<u>Artemisia</u>)   **137. ASTERACEAE**
   4 Woody stems procumbent; flowers in spherical
     capitula solitary at apex of erect herbaceous
     stem (<u>Acaena</u>)   **76. ROSACEAE**
5 Leaves alternate or spiral; fruit a nut, sometimes
   with wing ± surrounding it; monoecious   **38. JUGLANDACEAE**
5 Leaves opposite or ± so; fruit 1 or 2 achenes each
   with elongated wing on 1 side; often dioecious or
   ± so   6
   6 Stamens c.8; fruit of 2 achenes each with
     elongated wing   **100. ACERACEAE**
   6 Stamens 2; fruit of 1 achene with elongated
     wing   **125. OLEACEAE**
7 Leaves opposite or ± so   8
7 Leaves alternate or spiral   12
   8 Plants dioecious; male and female flowers in
     elongated dense catkins (<u>Salix</u>)   **62. SALICACEAE**
   8 Plants dioecious or not; flowers sometimes in
     pendent panicles (if so not dioecious) but not in
     compact catkins   9
9 Leaves entire   10
9 Leaves palmately lobed or serrate   11
   10 Most leaves >5cm; tepals fused into tube
      proximally (<u>Coprosma</u>)   **132. RUBIACEAE**
   10 Leaves all <3cm; tepals 0 or ± free   **91. BUXACEAE**

11 Leaves palmately lobed, the lobes entire to serrate;
   fruit of 2 winged achenes   **100. ACERACEAE**
11 Leaves not lobed, serrate; fruit a black berry
                                            93. RHAMNACEAE
   12 Stems scrambling, creeping or climbing   13
   12 Plant a self-supporting shrub or tree   14
13 Leaves evergreen, palmately veined and often
   palmately lobed; fruit a black or rarely yellow
   berry (Hedera)   **108. ARALIACEAE**
13 Leaves deciduous, pinnately veined, not lobed;
   fruit and achenes surrounded by white succulent
   tepals (Muehlenbeckia)   **48. POLYGONACEAE**
   14 Flowers borne on inside of hollow receptacles
      that become succulent in fruit (Ficus) **36. MORACEAE**
   14 Flowers not borne on inside of hollow receptacles 15
15 At least male flowers (if dioecious then female
   flowers also) in pendent or rigid catkins or in
   pendent tassels or globular heads   16
15 Flowers not in catkins, if in tight groups then
   not pendent   22
   16 Male and female flowers in separate, spherical,
      pendent capitula; leaves with petiole hollow
      at base and forming cap over axillary bud
                                            33. PLATANACEAE
   16 Flowers not in spherical pendent capitula;
      base of petiole not concealing axillary bud   17
17 Leaves densely mealy; male and female flowers in
   same catkin (Atriplex)   **44. CHENOPODIACEAE**
17 Leaves glabrous to pubescent, not mealy; male and
   female flowers in separate catkins or heads
   (or dioecious)   18
   18 Leaves dotted with translucent glands, with
      strong aromatic scent when crushed   **39. MYRICACEAE**
   18 Leaves not gland-dotted   19
19 Fresh stems and leaves with latex; fruits red to
   black, succulent (Morus)   **36. MORACEAE**
19 Latex absent; fruits not succulent   20
   20 Ovary 1-celled, with many ovules; fruit a capsule
      with many plumed seeds; or flowers all male
                                            62. SALICACEAE
   20 Ovary 2-6-celled, each cell with 1-2 ovules;
      fruit a nut, sometimes borne in husk formed
      from enlarged scales   21
21 Ovary 3- or 6-celled, with 3-9 styles   **40. FAGACEAE**
21 Ovary 2-celled, with 2 styles   **41. BETULACEAE**
   22 Leaves peltate, palmately lobed (Ricinus)
                                            92. EUPHORBIACEAE
   22 Leaves not peltate, not palmately lobed   23
23 Leaves with dense ± sessile scales at least
   on lowerside, appearing mealy   **79. ELAEAGNACEAE**
23 Leaves without scales, glabrous to pubescent   24

24 Leaves <20 x 2mm, succulent (<u>Suaeda</u>)
                                      44. CHENOPODIACEAE
24 At least most leaves >20mm and >2mm wide, not
   succulent   25
25 Flowers in umbels of umbels, epigynous (<u>Bupleurum</u>)
                                      109. APIACEAE
25 Flowers not in umbels of umbels, hypogynous or
   perigynous   26
  26 Stamens 8-12, or flowers all female with 1-celled
     ovary; leaves exstipulate   27
  26 Stamens 4-5, or flowers all female with 2-4-
     celled ovary; leaves stipulate when young   28
27 Dioecious; tepals longer than perianth-tube and
   falling from it after flowering   **25. LAURACEAE**
27 Flowers bisexual; tepals shorter than perianth-tube
   and not falling separately from it   **83. THYMELAEACEAE**
  28 Flowers all bisexual; ovary 1-celled; fruit a
     winged achene   **34. ULMACEAE**
  28 Flowers bisexual to dioecious; ovary 2-4-celled
     or 0; fruit a berry   **93. RHAMNACEAE**

<u>Key E</u> - Perianth 0 or of 1 or more whorls or a spiral of ±
          similar segments, ± calyx-like or bract-like (usually
          greenish or brownish or scarious); herbs
1 Flowers numerous in dense capitula closely
   surrounded by sepal-like bracts   **137. ASTERACEAE**
1 Flowers not in dense capitula, or if so then not
   closely surrounded by sepal-like bracts   2
  2 Leaves in whorls of ≥4 (<u>Rubia</u>)   **132. RUBIACEAE**
  2 Leaves opposite or in whorls of 3   3
3 Leaves at least partly opposite or whorled; aquatic
   or marsh plants with floating, procumbent or very
   weakly ascending stems   4
3 Leaves all alternate or all basal, or if some
   or all opposite then plant not aquatic, or if so
   then stems self-supporting   12
  4 Leaves fused in opposite pairs, forming
     succulent sheath round stem   **44. CHENOPODIACEAE**
  4 Leaves not fused in succulent sheath round stem   5
5 Tepals 0; stamens 1-4 (or flowers female)   6
5 Tepals 4-6 on at least some flowers; stamens 4-12
   (or flowers female)   8
  6 Flowers bisexual; stamens 4; fruits on stalks
     >10mm; only upper leaves opposite   **145. RUPPIACEAE**
  6 Monoecious; male flowers with 1-2 stamens; fruits
     on stalks ≤10mm; ± all leaves opposite or in 3s   7
7 Stigmas linear, 2 per ovary; ovary developing into
   4 nutlets   **122. CALLITRICHACEAE**
7 Stigmas peltate, 1 per carpel; 1-4(more) carpels per
   flower developing into nutlet   **147. ZANNICHELLIACEAE**
  8 Tepals 5-6 (or 0 in female flowers)   9
  8 Tepals 4   10

9 Flowers monoecious, hypogynous; tepals 5 in male
  flowers, 0 in female flowers; female flowers and
  fruit with 2 prominent basal bracteoles
                                          44. CHENOPODIACEAE
9 Flowers bisexual, perigynous; tepals 6; flowers
  without bracteoles   **82. LYTHRACEAE**
  10 Flowers in terminal flat-topped cymes; stamens 8
     (Chrysosplenium)   **75. SAXIFRAGACEAE**
  10 Flowers solitary or in spikes in leaf-axils;
     stamens 4   11
11 Flowers in long-stalked axillary spikes
                                     144. POTAMOGETONACEAE
11 Flowers solitary and sessile in leaf-axils
   (Ludwigia)   **85. ONAGRACEAE**
  12 Flowers greatly reduced, arranged in units
     largely composed of leafy or membranous scaly
     bracts, with perianth 0 or represented by
     bristles or minute scales, aerial; leaves linear
     grass-like, sheathing the stem proximally   13
  12 Flowers with obvious structure, mostly with
     perianth, if greatly reduced with 0 or obscure
     perianth then not arranged in units as above
     and often subaquatic or on water surface; leaves
     various   14
13 Flowers with bract above as well as below (if not
   then stems hollow); stems usually with hollow
   internodes, circular or rarely compressed or +
   quadrangular in section; leaf-sheaths usually with
   free overlapping margins   **155. POACEAE**
13 Flowers never with bract above; stems usually with
   solid internodes, often + triangular in section;
   leaf-sheaths usually cylindrical, with fused
   margins   **154. CYPERACEAE**
  14 Aquatic or marsh plants with linear leaves   15
  14 If leaves linear then plants not in water or
     marshes; if aquatic then leaves not linear   26
15 Leaves all basal; inflorescence a tight capitate
   mass on long leafless stem   **152. ERIOCAULACEAE**
15 If leaves all basal, then inflorescence not a single
   terminal tight capitate mass   16
  16 Flowers very small, many tightly packed in dense
     spherical or elongated conspicuous clusters   17
  16 Flowers not many together in dense clusters   19
17 Flowers bisexual,; fresh leaves with strong spicy
   scent when crushed (Acorus)   **149. ARACEAE**
17 Flowers unisexual, the male and female in clearly
   separated parts of inflorescence; leaves without
   spicy scent   18
  18 Flowers in globose heads   **156. SPARGANIACEAE**
  18 Flowers in cylindrical spikes   **157. TYPHACEAE**
19 Leaves very thin, ribbon- or thread-like, mostly
   subaquatic   20

19 Leaves thicker, not ribbon- or thread-like   22
20 Flowers bisexual, borne in stalked spikes
                                    144. POTAMOGETONACEAE
20 Flowers dioecious or monoecious, borne in stalked
   or sessile spathes   21
21 Flowers dioecious, in short- or long-stalked
   spathes; tepals 3; fresh-water (Vallisneria)
                                    140. HYDROCHARITACEAE
21 Flowers monoecious, in sessile spathes; tepals 0;
   marine   **148. ZOSTERACEAE**
22 Tepals 4-5, or 0   23
22 Tepals 6   24
23 Tepals 5 (or 0 in female flowers); leaves alternate
                                    44. CHENOPODIACEAE
23 Tepals 4; leaves all basal (Subularia) **63. BRASSICACEAE**
24 Flowers in branched cymes, sometimes compact
                                    153. JUNCACEAE
24 Flowers in a simple terminal raceme   25
25 Leaves on stems 0 or few and near base, without
   pore at apex; flowers numerous, without bracts
                                    143. JUNCAGINACEAE
25 Leaves several on stems, with prominent pore at
   apex; flowers <12, with bracts   **142. SCHEUCHZERIACEAE**
26 Flowers small, in dense spike on axis, the axis
   sometimes extended distally as sterile
   projection, with large spathe at base often
   partly or wholly obscuring flowers   **149. ARACEAE**
26 If flowers in single dense spike then without
   large spathe at base   27
27 Stems entirely rhizomatous, producing large
   (usually >1m across) leaves and huge (usually >50cm)
   elongated dense panicles   **81. GUNNERACEAE**
27 If stems entirely rhizomatous then leaves <10cm
   and flowers solitary or in whorls or umbels   28
28 Inflorescence consisting of units arranged in
   umbels, each unit consisting of several male
   flowers (each of 1 stamen) and 1 female flower
   (of 1 stalked ovary) all surrounded by 4 or 5
   conspicuous glands; plants with copious white
   latex (Euphorbia)   **92. EUPHORBIACEAE**
28 Inflorescence not consisting of units as above;
   plants without copious white latex   29
29 Leaves opposite; stems procumbent to weakly
   ascending   30
29 If leaves opposite then stems self-supporting and
   ± erect   32
30 Tepals 5, free   **47. CARYOPHYLLACEAE**
30 Tepals 3-4, fused   31
31 Leaves <15mm; perianth 4-lobed, greenish
   (Nertera)   **132. RUBIACEAE**
31 Leaves >20mm; perianth 3-lobed, brownish-purple
   (Asarum)   **26. ARISTOLOCHIACEAE**

32 Plants aquatic or in wet bogs; leaves simple,
   entire or ± so   **144. POTAMOGETONACEAE**
32 Plants usually on dry ground, if in marshes or
   bogs then leaves not simple and entire   33
33 Perianth of 6 lobes or segments, 3 in outer and 3 in
   inner whorl   34
33 Perianth lobes or segments 2-5, or 4-5 in each of
   2 whorls, rarely 6 and then not in 2 whorls of 3   37
   34 Leaves or leaf-like organs linear, without
      basal lobes   35
   34 Leaves broader than linear, or if not then with
      basal lobes   36
35 Dioecious; leaves reduced to scales, their normal
   function replaced by clusters of 4-10(more)
   cladodes (<u>Asparagus</u>)   **160. LILIACEAE**
35 Flowers bisexual; leaves reduced to scales or not,
   but not replaced by clusters of cladodes **153. JUNCACEAE**
   36 Twining climber; ovary inferior; fruit a red
      berry   **163. DIOSCOREACEAE**
   36 Not climbing; ovary superior, fruit an achene
      (<u>Rumex</u>)   **48. POLYGONACEAE**
37 Flowers epigynous, semi-epigynous, or perigynous
   with deeply concave hypanthium   38
37 Flowers hypogynous, or perigynous with flat to
   saucer-shaped hypanthium, or flowers all male
   (dioecious)   43
   38 Ovary with 3-8 cells; stigmas 3-8 (<u>Tetragonia</u>)
                                       43. AIZOACEAE
   38 Ovary with 1-2 cells; styles/stigmas 1-2   39
39 Tepals 4, or 4 plus 4 epicalyx segments beneath;
   stamens 1-4, or ≥8   40
39 Tepals 5; stamens 5   41
   40 Stamens 1-4, or ≥10; fruit 1-many achenes
                                       76. ROSACEAE
   40 Stamens 8; fruit a capsule (<u>Chrysosplenium</u>)
                                    75. SAXIFRAGACEAE
41 Ovary 2-celled; fruit a 2-celled schizocarp
                                      109. APIACEAE
41 Ovary 1-celled; fruit a 1-celled achene   42
   42 Leaves linear or ± so   **87. SANTALACEAE**
   42 Leaves ovate to lanceolate or deltate (<u>Beta</u>)
                                44. CHENOPODIACEAE
43 Tepals 2; stamens 12-18, conspicuous (<u>Macleaya</u>)
                                   31. PAPAVERACEAE
43 Tepals 3-5, or 5 with 5 epicalyx segments beneath,
   or sometimes 0 in female flowers, rarely 2 and then
   stamens also 2   44
   44 Leaves opposite; flowers often dioecious   45
   44 Leaves alternate; flowers rarely dioecious   49
45 Leaves deeply palmately lobed to ± palmate
                                   35. CANNABACEAE
45 Leaves simple, at most toothed   46

46 Leaves >1.5cm; stigmas branched or conspicuously
   papillose; at least male flowers in axillary
   spikes (catkins)   47
46 Leaves <1.5cm; stigmas ± smooth or minutely
   papillose; flowers not in spikes   48
47 Leaves with strong stinging hairs; tepals 4;
   fruit an achene (Urtica)   **37. URTICACEAE**
47 Leaves without stinging hairs; tepals 3; fruit a
   2-celled capsule (Mercurialis)   **92. EUPHORBIACEAE**
   48 Tepals 3; leaves conspicuously stipulate
      (Koenigia)   **48. POLYGONACEAE**
   48 Tepals 4-5; leaves exstipulate   **47. CARYOPHYLLACEAE**
49 Gynoecium composed of 1 carpel with 1 ovule,
   producing a 1-seeded fruit, or all flowers male   50
49 Gynoecium composed of 2-many (often fused) carpels,
   with many ovules in total, producing a many-seeded
   fruit or many 1-seeeded fruits   54
   50 Leaves palmately lobed to base or ± so
      (Cannabis)   **35. CANNABACEAE**
   50 Leaves simple, entire or pinnately lobed but not
      ± to base   51
51 Leaves with well-developed stipules fused into short
   tube round stem   **48. POLYGONACEAE**
51 Leaves exstipulate   52
   52 Tepals scarious, with 3-5 often similar
      bracteoles just below; fruit often dehiscent
                                             45. AMARANTHACEAE
   52 Tepals herbaceous or 0; flowers often without
      bracteoles; fruit always indehiscent   53
53 Tepals 3-5 or 0; styles 2-3, with smooth to
   papillate stigmas   **44. CHENOPODIACEAE**
53 Tepals 4; style 1; stigma much branched   **37. URTICACEAE**
   54 Ovary of 2 carpels completely fused; tepals 4
                                             63. BRASSICACEAE
   54 Ovary of 2-many carpels, free or <1/2 fused, if
      >1/2 fused then 5   55
55 Each carpel with several to many ovules, producing
   several to many seeds   56
55 Each carpel with 1-2 ovules, producing 1 seed   57
   56 Leaves palmate or ± so; flowers >1cm across
      (Helleborus)   **29. RANUNCULACEAE**
   56 Leaves 2-ternate to 2-pinnate; flowers <1cm
      across (Astilbe)   **75. SAXIFRAGACEAE**
57 Carpels fused, elongated distally into sterile column
   ending in 1 style with 5 stigmas   **104. GERANIACEAE**
57 Carpels free, each with 1 stigma   58
   58 Tepals 5, with 5 epicalyx segments just beneath
      (Sibbaldia)   **76. ROSACEAE**
   58 Tepals 4-5, without epicalyx   **29. RANUNCULACEAE**

Key F - Perianth of 2 (rarely more) distinct whorls or rarely
        a spiral, the inner and outer differing markedly in

shape/size/colour; flowers all male or all female
1 Tree or shrub, sometimes procumbent but with
  distinctly woody stems   2
1 Herb   11
  2 Leaves pinnate; tree with white flowers
    (Fraxinus)   **125. OLEACEAE**
  2 Leaves simple, or if pinnate then shrub with
    yellow flowers   3
3 Leaves all scale-like, functionally replaced by
  leaf-like stem outgrowth bearing flowers on their
  faces (Ruscus)   **160. LILIACEAE**
3 Leaves photosynthetic, not bearing flowers on their
  faces   4
  4 Stems procumbent; leaves Erica-like, with
    revolute margins   **65. EMPETRACEAE**
  4 Stems ascending, spreading or erect; leaves not
    Erica-like   5
5 Leaves stipulate, either pinnate or simple and
  entire   **76. ROSACEAE**
5 Leaves exstipulate, or if stipulate then simple
  and serrate   6
  6 Leaves palmately lobed (Ribes)   **73. GROSSULARIACEAE**
  6 Leaves entire to serrate or shallowly spinose-
    pinnately lobed   7
7 Leaves gland-dotted, <2cm   **69A. MYRSINACEAE**
7 Leaves not gland-dotted, most or all >2cm   8
  8 Corolla white; at least some leaves usually with
    spine-tipped teeth   **90. AQUIFOLIACEAE**
  8 Corolla purple or greenish-yellow; leaves not
    spiny   9
9 Flowers ± hypogynous, with large nectar-secreting
  disc at base   **89. CELASTRACEAE**
9 Flowers clearly epigynous or perigynous, the inferior
  ovary or hypanthium distinct even in male flowers   10
  10 Leaves closely serrate, usually stipulate
    (Rhamnus)   **93. RHAMNACEAE**
  10 Leaves entire or remotely serrate, exstipulate
                         86. CORNACEAE
11 Climbing plant bearing tendrils (Bryonia)
                          61. CUCURBITACEAE
11 Plant not climbing, without tendrils   12
  12 Aquatic or marsh plants with all leaves in
    basal rosette(s)   13
  12 If aquatic or marsh plants then all leaves not
    in basal rosettes   14
13 Petals 3, free, conspicuous, white
                     140. HYDROCHARITACEAE
13 Petals 4, fused, inconspicuous, ± scarious (female
  flowers present low down in leaf-rosette but easily
  missed) (Littorella)   **123. PLANTAGINACEAE**
  14 Sepals 3; petals 3   15
  14 Sepals and petals not both 3   16

15 Leaves reduced to scales, replaced by bunches of
   $\pm$ cylindrical cladodes (Asparagus)   **160. LILIACEAE**
15 Leaves flat, with 2 basal lobes (Rumex)
                                                 48. POLYGONACEAE
   16 Flowers in dense capitula closely surrounded
      by $\geq$1 row of bracts   **137. ASTERACEAE**
   16 Flowers not in dense capitula though sometimes
      crowded   17
17 At least some leaves ternate to pinnate   18
17 All leaves simple (sometimes lobed)   20
   18 Basal leaves simple; stem-leaves 1-pinnate
      (Valeriana)   **135. VALERIANACEAE**
   18 All leaves compound, at least some 2-3-pinnate
      or -ternate   19
19 Stamens $\leq$10; carpels usually 2; plant <1m (Astilbe)
                                                 75. SAXIFRAGACEAE
19 Stamens >10; carpels usually 3; plant usually
   $\geq$1m (Aruncus)   **76. ROSACEAE**
   20 Leaves petiolate, truncate to cordate at base   21
   20 Leaves sessile or $\pm$ so, rounded to cuneate
      at base   22
21 Leaves not lobed, entire; tepals 5, all $\pm$ white;
   plant >(25)100cm (Fallopia)   **48. POLYGONACEAE**
21 Leaves lobed, serrate; sepals 5, green; petals 5,
   white; plant <25cm (Rubus)   **76. ROSACEAE**
   22 Leaves alternate; petals 4; stamens 8 or ovaries
      4 (Sedum)   **74. CRASSULACEAE**
   22 Leaves opposite; petals 5; stamens usually 10 or
      ovary 1   **47. CARYOPHYLLACEAE**

Key G - Perianth of 2 (rarely more) distinct whorls or rarely
        a spiral, the inner and outer differing markedly in
        shape/size/colour; petals fused at base for varying
        distances to apex; ovary present in at least some
        flowers, superior
1 Stems twining, not green; leaves reduced to small
  scales, not green   **114. CUSCUTACEAE**
1 If stems twining then leaves expanded and green   2
   2 Plant wholly lacking green colour, yellow to
     brown, sometimes red- or purple-tinged
                                                 127. OROBANCHACEAE
   2 Plant with obvious green colouring   3
3 Leaves linear to subulate, all in basal rosette under
  water or on mud; flowers inconspicuous, solitary
  on pedicels or densely clustered on leafless scapes   4
3 Leaves not both linear to subulate and all in basal
  rosette   6
   4 Flowers bisexual, solitary; stamens inluded in
     corolla (Limosella)   **126. SCROPHULARIACEAE**
   4 Flowers in dense heads or spikes on scapes, or if
     solitary then unisexual and stamens exserted
     from corolla   5

5   Tepals 4, densely fringed at apex; flowers in
    capitate clusters, unisexual   **152. ERIOCAULACEAE**
5   Tepals >4, not fringed at apex; flowers either
    bisexual and in spikes, or unisexual and solitary
                                          123. PLANTAGINACEAE
    6   Basal leaves peltate, succulent, glabrous;
        inflorescence a terminal raceme (Umbilicus)
                                          74. CRASSULACEAE
    6   Leaves not peltate and succulent   7
7   Flowers pea-like, zygomorphic; petals 5, 1 upper, 2
    lateral, and 2 lower fused to form keel; stamens 10
                                          78. FABACEAE
7   Flowers not pea-like with 5 petals and 10 stamens   8
    8   Sepals 5 (3 outer small, 2 inner large); petals
        3; stamens 8   **96. POLYGALACEAE**
    8   Not with the combination sepals 5, petals 3,
        stamens 8   9
9   Stamens >10   10
9   Stamens <10   12
    10 Carpels 5-many, usually ± fused; leaves
        stipulate   **54. MALVACEAE**
    10 Carpel 1; leaves exstipulate   11
11 Tree or shrub; flowers numerous in dense clusters,
    each with perianth <1cm   **77. MIMOSACEAE**
11 Herb; flowers in elongated raceme or sometimes few,
    each with perianth >1cm (Consolida)   **29. RANUNCULACEAE**
    12 Stamens 2   13
    12 Stamens >2   16
13 Leaves all in a basal rosette; flowers solitary on
    erect pedicels (Pinguicula)   **130. LENTIBULARIACEAE**
13 Plant not with all leaves in a basal rosette and
    flowers solitary on erect pedicels   14
    14 Ovary 4-celled, each cell with 1 ovule
                                          120. LAMIACEAE
    14 Ovary 2-celled, each cell with 2-many ovules   15
15 Perianth actinomorphic; fruit a berry or capsule
    with <4 seeds   **125. OLEACEAE**
15 Perianth slightly to strongly zygomorphic; fruit a
    capsule with usually >4 seeds   **126. SCROPHULARIACEAE**
    16 Ovary 4-celled with 1 ovule per cell; fruit a
        cluster of 1-seeded nutlets; plant bisexual   17
    16 Ovary not 4-celled with 1 ovule per cell, or if
        so then fruit a berry and plant dioecious   19
17 Leaves alternate; flowers usually in cymes spirally
    coiled when young; stems not square in section
                                          118. BORAGINACEAE
17 Leaves opposite; flowers not in spirally coiled
    cymes; stems usually ± square in section   18
    18 Ovary scarcely lobed at flowering, with terminal
        style and capitate stigma   **119. VERBENACEAE**
    18 Ovary deeply lobed at flowering, with usually
        basal style and (1-)2 linear stigmas   **120. LAMIACEAE**

19 Sepals 2; petals fused only at base   **46. PORTULACACEAE**
19 Sepals usually >2, if 2 then petals fused for >1/2
   of length   20
   20 Tree or shrub (sometimes very dwarf)   21
   20 Herb; if stems woody then climbing or trailing   31
21 Leaves opposite or whorled   22
21 Leaves alternate or spiral   26
   22 Stamens 5-10   23
   22 Stamens 4   24
23 Stamens 5; leaves mostly >1cm wide; flowers >15mm
   across   **111. APOCYNACEAE**
23 Stamens 8 or 10, or if 5 then leaves all <5mm wide
   and flowers <10mm across   **66. ERICACEAE**
   24 Flowers dull, brownish (Plantago)
                                       123. PLANTAGINACEAE
   24 Flowers white or brightly coloured   25
25 Flowers >2.5cm, in loose inflorescences
                                       126. SCROPHULARIACEAE
25 Flowers <2.5cm, in dense inflorescences
                                       124. BUDDLEJACEAE
   26 Dioecious, either stamens or ovary rudimentary   27
   26 Flowers bisexual, with functional stamens and
      ovary   28
27 Leaves gland-dotted, <2cm, serrate; fruit purple
                                       69A. MYRSINACEAE
27 Leaves not gland-dotted, most >2cm, entire or with
   spiny teeth; fruit red to yellow   **90. AQUIFOLIACEAE**
   28 Stamens 4   **124. BUDDLEJACEAE**
   28 Stamens 5-10   29
29 Plant a dense cushion <10cm with solitary flowers
   on erect stalks <5cm; stamens 5, with anthers
   opening by slits   **69. DIAPENSIACEAE**
29 Plant rarely a dense cushion, if so then stamens 8
   or 10 with anthers opening by apical pores   30
   30 Stamens 5; flowers actinomorphic   **112. SOLANACEAE**
   30 Stamens 8 or 10, or if 5 then flowers slightly
      zygomorphic   **66. ERICACEAE**
31 Inner perianth-segments (corolla-lobes) and stamens
   on same radius   32
31 All or most inner perianth-segments (corolla-lobes)
   and stamens on alternating radii   33
   32 Styles 5; stigmas linear   **49. PLUMBAGINACEAE**
   32 Styles 1; stigma capitate   **70. PRIMULACEAE**
33 Aquatic or bog plant with showy corollas densely
   fringed distally with hairs or narrow serrations
                                       115. MENYANTHACEAE
33 If aquatic or marsh plant then corollas not fringed
   or fringed at base of lobes   34
   34 Leaves all in basal rosette; corolla violet with
      yellow centre and yellow anthers   **128. GESNERIACEAE**
   34 If leaves all in basal rosette then corolla not
      violet with yellow centre   35

35 Flowers in compact spikes; perianth scarious and
   brownish; stamens long-exserted (<u>Plantago</u>)
                                        123. PLANTAGINACEAE
35 If flowers in compact spikes then not with both
   scarious perianth and long-exserted stamens   36
   36 Flowers in axils of spiny bracts; sepals 4, 2
   large and 2 small; corolla with 3-lobed lower
   lip and 0 upper lip   **129. ACANTHACEAE**
   36 If bracts spiny then sepals not 2 large and 2
   small and corolla not with large lower and 0
   upper lip   37
37 Leaves opposite   38
37 Leaves alternate   41
   38 Ovary 1-celled   39
   38 Ovary 2-celled, or ovaries 2 and each 1-celled   40
39 Ovule 1   **42A. NYCTAGINACEAE**
39 Ovules many   **110. GENTIANACEAE**
   40 Ovaries 2, with common style expanded into ring
   below stigma   **111. APOCYNACEAE**
   40 Ovary 1, with styles not expanded into ring
   below stigma   **126. SCROPHULARIACEAE**
41 Stamens normally 6-8; inner or outer tepals with keel
   on abaxial side, otherwise similar to others; ovary
   1-celled, with 1 ovule (<u>Fallopia</u>)   **48. POLYGONACEAE**
41 Stamens 3-5; outer tepals sepaloid, inner petaloid;
   ovary 1-5-celled, with total of $\geq$4 ovules   42
   42 Ovary 3-celled; stigmas 3 on 1 style
                                         116. POLEMONIACEAE
   42 Ovary 1-5-celled, if 3-celled then stigmas 1-2
   per style   43
43 Stamens 3 or 4, at least in most flowers
                                      126. SCROPHULARIACEAE
43 Stamens 5, at least in most flowers   44
   44 Styles 2, or style 1 but divided distally into 2
   branches/long stigmas   45
   44 Style 1, with 1-2 short stigmas at apex   46
45 Ovary 2-lobed, each lobe with 1 style; corolla
   1.5-2.5mm (<u>Dichondra</u>)   **113. CONVOLVULACEAE**
45 Ovary not lobed, with 1 style divided distally;
   corolla 6-10mm   **117. HYDROPHYLLACEAE**
   46 Ovary 1-2(3)-celled, with total of 4(-6) ovules;
   stems trailing or climbing   **113. CONVOLVULACEAE**
   46 Ovary 2(-5)-celled, each cell with many ovules;
   stems usually not trailing or climbing   47
47 Flowers distinctly zygomorphic; corolla divided
   >1/2 way to base; anthers not forming close cone
   round style (<u>Verbascum</u>)   **126. SCROPHULARIACEAE**
47 Flowers actinomorphic or slightly zygomorphic;
   corolla divided <1/2 way to base or if >1/2 way to
   base then flowers actinomorphic and anthers forming
   close cone round style   **112. SOLANACEAE**

<u>Key H</u> - Perianth of 2 (rarely more) distinct whorls or rarely
    a spiral, the inner and outer differing markedly in
    shape/size/colour; petals fused at base for varying
    distances to apex; ovary present in at least some
    flowers, inferior or partly so
1 Stamens 8 to numerous (or 4-5 but appearing 8-10
   because filaments split to base with each 1/2
   bearing one 1/2-anther)   2
1 Stamens <u>5</u> (sometimes alternating with staminodes)   4
   2 Tree (<u>Eucalyptus</u>)   84. MYRTACEAE
   2 Herb or shrub <2m   3
3 Flowers 5 in 1 terminal cluster, greenish; leaves
   ternate; herb   134. ADOXACEAE
3 Flowers 1-several, axillary, white to red; leaves
   simple; shrub (<u>Vaccinium</u>)   66. ERICACEAE
   4 Flowers in dense heads surrounded by row(s) of
      sepal-like bracts   5
   4 Flowers not in dense heads, or if so then with
      only 2 bracts at base   8
5 Ovary 2-celled; sepals 4-5, fused proximally   6
5 Ovary 1-celled; calyx represented by a cup, or by
   often >5 narrow teeth, bristles, scales or hairs   7
   6 Leaves alternate; stamens borne on receptacle;
      ovules numerous   131. CAMPANULACEAE
   6 Leaves whorled; stamens borne on corolla-tube;
      ovules 1 per cell (<u>Sherardia</u>)   132. RUBIACEAE
7 Anthers usually 5, fused into tube round style;
   ovary and fruit not enclosed in epicalyx
                                              137. ASTERACEAE
7 Anthers 4, not fused; ovary and fruit enclosed in
   tubular epicalyx   136. DIPSACACEAE
   8 Stamens 1-3 (vestigial if flowers female)   9
   8 Stamens 4-5 (0 or vestigial if flowers female)   11
9 Flowers numerous, + closely packed in flat or
   domed inflorescences, <1cm across   135. VALERIANACEAE
9 Flowers rather few in spikes or racemes, or
   solitary, >2cm across   10
   10 Stamens 3, with obvious filaments and anthers
                                              161. IRIDACEAE
   10 Stamens 2, with sessile anthers (<u>Cypripedium</u>)
                                              164. ORCHIDACEAE
11 Leaves opposite or in whorls   12
11 Leaves alternate or spiral   13
   12 Leaves >4 per node, or 2 per node and each with
      stipules   132. RUBIACEAE
   12 Leaves 2 per node, exstipulate   133. CAPRIFOLIACEAE
13 Flowers unisexual; stamens 3 (0 in female flowers),
   2 with 2 pollen-sacs and 1 with 1   61. CUCURBITACEAE
13 Flowers bisexual; stamens 5, all with 4 pollen-sacs   14
   14 Stamens opposite corolla-lobes; stigma 1
      (<u>Samolus</u>)   70. PRIMULACEAE
   14 Stamens alternating with corolla-lobes; stigmas

2-5   **131. CAMPANULACEAE**

Key I – Perianth of 2 (rarely more) distinct whorls or rarely
        a spiral, the inner and outer differing markedly in
        shape/size/colour; petals free, or rarely fused just
        at basal point, or near apex but free at base; ovary
        present in at least some flowers, inferior or partly
        so.
1 Petals >8   2
1 Petals <6   3
    2 Aquatic with floating flowers and leaves
      (Nymphaea)   **27. NYMPHAEACEAE**
    2 Terrestrial plant; leaves very succulent
                                               43. AIZOACEAE
3 Sepals and petals each 3   4
3 Sepals and petals each 2 or 4-6   7
    4 Flowers zygomorphic; stamens 1-2   **164. ORCHIDACEAE**
    4 Flowers actinomorphic; stamens 3-12 or 0 in
      female flowers   5
5 Outer whorl of tepals sepaloid; stamens 9-12 or 0 in
  female flowers; plant aquatic   **140. HYDROCHARITACEAE**
5 Both whorls of tepals petaloid; stamens 3-6;
  plant terrestrial   6
    6 Stamens 6 (Galanthus)   **160. LILIACEAE**
    6 Stamens 3 (Libertia)   **161. IRIDACEAE**
7 Tree or shrub   8
7 Herb, rarely woody at base   15
    8 Stamens 4-5 or 0 in female flowers   9
    8 Stamens 8-many   11
9 Flowers in true umbels, arising at 1 point, sometimes the umbels further aggregated   **108. ARALIACEAE**
9 Flowers solitary or in racemes or panicles, sometimes in corymbose umbel-like clusters but not
  arising at 1 point   10
    10 Origin of sepals separated from ovary wall by
       saucer- to cup-shaped hypanthium **73. GROSSULARIACEAE**
    10 Flowers without hypanthium (sepals arising direct
       from ovary wall)   **86. CORNACEAE**
11 Leaves alternate   12
11 Leaves opposite or whorled   13
    12 Fruit a woody capsule; style 1 (Leptospermum)
                                             84. MYRTACEAE
    12 Fruit surrounded by succulent or pithy
       hypanthium; styles (1)2-many   **76. ROSACEAE**
13 Fruit a capsule; styles 2-4(6), free or united
   proximally   **72. HYDRANGEACEAE**
13 Fruit a berry; style 1   14
    14 Hypanthium 5-16mm beyond ovary apex; flowers
       pendent on long pedicels (Fuchsia)   **85. ONAGRACEAE**
    14 Hypanthium not or scarcely extending beyond
       ovary apex; flowers not pendent (Amomyrtus)
                                             84. MYRTACEAE

15 Leaves simple, all or most >1m across; inflorescence
   an elongated very dense panicle >50cm   **81. GUNNERACEAE**
15 If leaves >50cm then compound; if inflorescence
   >30cm then not dense and elongated   16
   16 Flowers in umbels   17
   16 Flowers not in umbels, but sometimes corymbose 19
17 Styles 2; fruit a dry 2-celled schizocarp **109. APIACEAE**
17 Styles 1 or 5; fruit a succulent drupe or berry   18
   18 Plant with a single terminal umbel; fruit red;
      leaves simple (Cornus)   **86. CORNACEAE**
   18 Plant with umbels in large panicles; fruit dark
      purple to black; leaves 1-2-pinnate (Aralia)
                           108. ARALIACEAE
19 Sepals 5   20
19 Sepals 2 or 4   21
   20 Petals bright yellow; leaves pinnate   **76. ROSACEAE**
   20 Petals not bright yellow, or if so then leaves
      simple   **75. SAXIFRAGACEAE**
21 Sepals 2; petals yellow, <1cm; fruit a capsule
   opening transversely (Portulaca)   **46. PORTULACACEAE**
21 If sepals 2 then petals white; if petals yellow
   then >1cm; if fruit a capsule then not opening
   transversely   22
   22 Fruit an achene without bristles, with
      persistent sepals; leaves subsessile (Haloragis)
                        80. HALORAGACEAE
   22 Fruit a many-seeded capsule, or if a 1-2-celled
      achene then with many hooked bristles and without
      persistent sepals, and leaves long-petiolate
                      85. ONAGRACEAE

Key J - Perianth of 2 (rarely more) distinct whorls or rarely
      a spiral, the inner and outer differing markedly in
      shape/size/colour; petals free or rarely fused just
      at basal point, or near apex but free at base; >1
      ovary present in at least some flowers, superior,
      free or fused just at extreme base
1 Tree with flowers borne in pendent unisexual globose
   clusters   **33. PLATANACEAE**
1 Herb or shrub; flowers not in pendent unisexual
   globose clusters   2
   2 Sepals and petals each 3   3
   2 Sepals and petals each >3   5
3 Carpels 3; stamens 3; flowers <3mm across (Crassula)
                    74. CRASSULACEAE
3 Carpels >6; stamens >6 or 0 in female flowers;
   flowers >5mm across   4
   4 Carpels with numerous ovules, forming follicles
      in fruit; sepals purple-tinged green; leaves
      linear   **138. BUTOMACEAE**
   4 Carpels with 1 ovule forming achenes in fruit,
      or if with >1 ovules forming follicles in fruit

then sepals green and leaves ovate with cordate
    base   **139. ALISMATACEAE**
5 Flowers >8cm across; petals bright red   **50. PAEONIACEAE**
5 If flowers >5cm across then petals not bright red   6
  6 Leaves either opposite or very succulent
                                             74. CRASSULACEAE
  6 Leaves all basal, alternate or spiral, not or
    only slightly succulent   7
7 Shrub   **76. ROSACEAE**
7 Herb, sometimes with ± woody surface rhizome   8
  8 Flowers perigynous; origin of sepals (often of
    stamens and petals too) separated from base of
    ovary by obvious flat or saucer- to bowl-shaped
    hypanthium   9
  8 Flowers hypogynous or ± so, the stamens, petals
    and sepals free and arising directly at base of
    carpels   10
9 Carpels 2, each with many ovules; fruit a pair of
  follicles; leaves simple and unlobed; petals bright
  pink (Bergenia)   **75. SAXIFRAGACEAE**
9 Carpels 3-many, or if 2 then each with 1 ovule;
  fruit 1-many achenes; plant not with both simple
  unlobed leaves and pink petals   **76. ROSACEAE**
  10 Carpels with 1 ovule; fruit an achene   11
  10 Carpels with few to many ovules; fruit a follicle 12
11 Flowers zygomorphic; at least some petals very
   deeply divided (Sesamoides)   **64. RESEDACEAE**
11 Flowers actinomorphic; no petals deeply divided
                                            29. RANUNCULACEAE
   12 Carpels 5-many   **29. RANUNCULACEAE**
   12 Carpels 2-4   13
13 Stamens 5-10 (or 0 in female flowers); carpels
   rarely >2   **75. SAXIFRAGACEAE**
13 Stamens >10 (or 0 in female flowers); carpels
   usually >2   14
   14 Leaves palmate or deeply palmately lobed
                                            29. RANUNCULACEAE
   14 Leaves 2-3-pinnate (Aruncus)   **76. ROSACEAE**

Key K - Perianth of 2 (rarely more) distinct whorls or rarely
        a spiral, the inner and outer differing markedly in
        shape/size/colour; petals free, or rarely fused just
        at basal point, or near apex but free at base; ovary
        present in at least some flowers, superior, of 1
        carpel or of >1 wholly or mostly fused carpels
1 Plant yellowish-brown, lacking green pigment in
  all parts   **68. MONOTROPACEAE**
1 Plant with at least leaves or stems at least partly
  green   2
  2 Leaves modified to form tubular 'pitchers' with
    small blade ('hood') at entrance, all in basal
    rosette   **55. SARRACENIACEAE**

2   Leaves not modified as 'pitchers'   3
3 Leaves covered on upperside with very sticky
    glandular hairs, reddish, all basal   **56. DROSERACEAE**
3 Leaves not all basal and covered with very sticky
    glandular hairs   4
    4   Flowers zygomorphic   5
    4   Flowers actinomorphic   19
5 At least 1 sepal or petal with conspicuous basal
    spur or pouch   6
5 Flowers without basal spur(s)   10
    6   Spur(s) formed from petal(s); ovary 1-celled   7
    6   Spur(s) formed from sepal(s); ovary 3-5-celled   8
7 Leaves pinnate or ternate, exstipulate; sepals 2;
    petals 4   **32. FUMARIACEAE**
7 Leaves simple, stipulate (the stipules sometimes
    deeply divided or pinnate); sepals 5; petals 5
                                                  58. VIOLACEAE
    8   Sepals 3; petals 5 but apparently 3 due to
        fusion of 2 pairs of laterals   **107. BALSAMINACEAE**
    8   Sepals 5; petals 5   9
9 Upper sepal with spur fused to pedicel; ovary
    5-celled, with distal sterile beak (<u>Pelargonium</u>)
                                                  104. GERANIACEAE
9 Upper 1-3 sepals with free spurs; ovary 3-celled,
    without distal beak   **106. TROPAEOLACEAE**
    10 Stamens 8 or 10, all or all but 1 with filaments
        fused into tube   11
    10 Stamens 3-many, free   12
11 Flowers with sepal uppermost (on top-line);
    stamens 8; anthers opening by pores   **96. POLYGALACEAE**
11 Flowers with petal uppermost (on top-line);
    stamens 10; anthers opening by slits   **78. FABACEAE**
    12 Tree, or shrub >1m high; ovary 3-celled   13
    12 Herb, or shrub <1m high; ovary 1-2-celled   14
13 Leaves palmate, opposite; flowers white to red
                                              99. HIPPOCASTANACEAE
13 Leaves 1-2-pinnate, alternate; flowers yellow
                                              98. SAPINDACEAE
    14 Stamens 3; petals 3, filiform, brownish
        (<u>Tolmiea</u>)   **75. SAXIFRAGACEAE**
    14 Stamens 5-many; petals >3, not filiform, white
        or pink   15
15 Ovary 2-celled; petals 4, the lower 2 and upper 2
    forming 2 different pairs   16
15 Ovary 1-celled; petals 4-6(8), if only 4 then upper
    one, lateral two and lower one of 3 different forms   17
    16 Leaves palmate; capsule >2cm   **62A. CAPPARACEAE**
    16 Leaves simple to pinnate; capsule <2cm
                                              63. BRASSICACEAE
17 Petals 4-6(8), at least some deeply lobed;
    stamens 7-many (<u>Reseda</u>)   **64. RESEDACEAE**
17 Petals 5, or apparently 4 due to fusion of lower 2,

```
    entire to very shallowly lobed or toothed          18
    18 Leaves paripinnate              77A. CAESALPINIACEAE
    18 Leaves ternate (Thermopsis)              78. FABACEAE
19 Stamens >12, >2x as many as petals                 20
19 Stamens 1-12, <2x as many as petals                28
    20 Petals >9                                       21
    20 Petals <6                                       22
21 Aquatic plant with leaves and flowers at or just
    above water surface (Nuphar)        27. NYMPHAEACEAE
21 Terrestrial plant with succulent leaves   43. AIZOACEAE
    22 Stamens with filaments fused into tube round
        styles                           54. MALVACEAE
    22 Stamens free or + united into bundles, but not
        forming tube round styles                     23
23 Ovary with 1-2(5) ovules per cell                  24
23 Ovary with several to many ovules per cell         25
    24 Ovary 5-celled; inflorescence stalks fused to
        narrowly oblong papery bract     53. TILIACEAE
    24 Ovary 1-celled; inflorescence stalks not fused
        to long papery bract             76. ROSACEAE
25 Leaves simple, entire                              26
25 Leaves simple and toothed or lobed, or compound    27
    26 Leaves with translucent and/or coloured sessile
        glands; styles 3 or 5           52. CLUSIACEAE
    26 Leaves without sessile glands, style 0 or 1
                                         57. CISTACEAE
27 Sepals 2(-3), sepaloid, usually caducous
                                       31. PAPAVERACEAE
27 Sepals (3)4-5, petaloid, not caducous  29. RANUNCULACEAE
    28 Perianth in 3-6 whorls each of 2-4 segments
                                       30. BERBERIDACEAE
    28 Perianth in 2 whorls each of >2 segments       29
29 Stems herbaceous, sometimes woody just at base     30
29 Stems wholly or mostly woody                       54
    30 Sepals 2; petals 5                              31
    30 Sepals >3; petals as many as sepals or fewer   32
31 Stems twining or sprawling, >1m; underground tubers
    present; capsule indehiscent       46A. BASELLACEAE
31 Stems decumbent to erect, <50cm; underground tubers
    0; capsule dehiscent               46. PORTULACACEAE
    32 Sepals 3; petals 2-3                            33
    32 Sepals >3; petals >3                            35
33 Ovary 1-celled, with 1 ovule; sepals and petals
    both sepaloid or petaloid (Rumex)   48. POLYGONACEAE
33 Ovary 3-celled, each cell with 2-many ovules; petals
    petaloid; sepals sepaloid                         34
    34 Leaves opposite, stipulate, <1cm; flowers <5mm
        across                           51. ELATINACEAE
    34 Leaves alternate, with sheathing base but
        exstipulate, >1cm; flowers >10mm across
                                       151. COMMELINACEAE
35 Ovary 1-celled, at least apically, with 1-many
```

ovules   36
35 Ovary 2–10–celled throughout with 1–many ovules
   per cell   42
   36 Stamens 5, alternating with 5 conspicuous deeply
      divided staminodes (Parnassia)   75. SAXIFRAGACEAE
   36 Deeply divided staminodes 0   37
37 Ovary with 1 ovule   38
37 Ovary with few to many ovules   40
   38 Leaves exstipulate; flowers showy, usually pink
      or blue, >4mm across   49. PLUMBAGINACEAE
   38 Leaves stipulate; flowers inconspicuous, green,
      yellowish-green or white   39
39 Leaves simple, entire   47. CARYOPHYLLACEAE
39 Leaves compound or conspicuously lobed   76. ROSACEAE
   40 Style 1, >1mm, with 1 or 5 stigmas at apex
                                                67. PYROLACEAE
   40 Styles +0 or 2–5, free, if 1 then either
      <1mm or divided into 3 distally   41
41 Ovary with ovules on parietal placentas; styles
   fused proximally; leaves exstipulate   60. FRANKENIACEAE
41 Ovary with ovules on free–central placenta; styles
   free, or 0, or rarely fused proximally and then
   leaves stipulate   47. CARYOPHYLLACEAE
   42 Flowers perigynous, with tubular or cup– or
      bowl–shaped hypanthium bearing sepals and petals
      at apex   43
   42 Flowers hypogynous, with sepals and petals borne
      at base of ovary   45
43 Hypanthium becoming hard and protective at fruiting;
   carpels 2, each with 1 ovule   76. ROSACEAE
43 Hypanthium not becoming hard and protective;
   carpels 2(–4), each with many ovules   44
   44 Sepals and petals each usually 6, with epicalyx
      segments outside; leaves + all on stem, at least
      the lower opposite or in whorls of 3   82. LYTHRACEAE
   44 Sepals and petals each 5, without epicalyx
      segments; leaves all or nearly all basal
      (Rodgersia)   75. SAXIFRAGACEAE
45 Stem–leaves all in single whorl of 3–8, with 1
   flower above (Paris)   160. LILIACEAE
45 Stem–leaves (if present) not all in 1 whorl   46
   46 Petals and sepals each 4, at least in most
      flowers   47
   46 Petals and sepals each 5 in all flowers   50
47 Stamens 4 or 6   48
47 Stamens 8   49
   48 Stamens 4; capsule dehiscing by 8 valves
      (Radiola)   95. LINACEAE
   48 Stamens (4–)6; capsule dehiscing by 2 valves or
      breaking transversely or indehiscent
                                             63. BRASSICACEAE
49 Leaves simple, entire, stipulate   51. ELATINACEAE

49 Leaves 2-3-pinnately lobed, exstipulate (<u>Ruta</u>)
            102A. RUTACEAE
 50 Fruit a 5-celled schizocarp with 1 seed per
  cell, elongated distally into sterile column 51
 50 Fruit a 5- to many-seeded capsule, without a
  sterile column   52
51 Leaves stipulate; petals white to red, blue or
 purple   **104. GERANIACEAE**
51 Leaves exstipulate; petals yellow **105. LIMNANTHACEAE**
 52 Ovary 2-celled; styles 0 or 2 (<u>Saxifraga</u>)
           75. SAXIFRAGACEAE
 52 Ovary 5-celled; styles 5   53
53 Capsule opening by 5 valves; leaves ternate or
 palmate   **103. OXALIDACEAE**
53 Capsule opening by 10 valves; leaves simple,
 entire   **95. LINACEAE**
 54 Sepals 3; petals 2-3   55
 54 Sepals and petals each >3   57
55 Woody climber (<u>Fallopia</u>) **48. POLYGONACEAE**
55 Procumbent to erect shrub   56
 56 Leaves <3mm wide; flowers in small axillary
  clusters   **65. EMPETRACEAE**
 56 Leaves replaced by leaf-like flat cladodes >3mm
  wide, bearing 1-2 flowers on their surface
  (<u>Ruscus</u>)   **160. LILIACEAE**
57 Stamens 2x as many as petals   58
57 Stamens <2x as many as petals at least on most
 flowers   62
 58 Woody stems ± procumbent; leaves >6cm, simple,
  serrate (<u>Bergenia</u>) **75. SAXIFRAGACEAE**
 58 Woody stems erect or ± so; leaves usually
  compound or lobed   59
59 Leaves simple, serrate; ovules many in each cell
         64A. CLETHRACEAE
59 Leaves compound, or simple and lobed; ovules 1-2
 per cell   60
 60 Ovary 2-celled; leaves opposite **100. ACERACEAE**
 60 Ovary 4-6-celled; leaves alternate or opposite 61
61 Fruit a 4(-5)-celled capsule, not winged; leaves
 ternate, or pinnate or pinnately lobed with segments
 <1cm wide; shrub   **102A. RUTACEAE**
61 Fruit a group of 1-6 winged achenes; leaves pinnate
 with leaflets mostly >1cm wide; tree **102. SIMAROUBACEAE**
 62 Leaves opposite   63
 62 Leaves alternate   67
63 Stems procumbent; ovary 1-celled **60. FRANKENIACEAE**
63 Stems erect; ovary >1-celled   64
 64 Leaves simple and not lobed   65
 64 Leaves pinnate   66
65 Fruit a pinkish-red succulent capsule with 4-5
 orange seeds; leaves exstipulate **89. CELASTRACEAE**
65 Fruit a black berry with 2-4 non-orange seeds;

leaves stipulate at least when young   **93. RHAMNACEAE**
66 Sepals 5, free; stamens 5; ovules many in each
    cell; fruit an inflated capsule   **97. STAPHYLEACEAE**
66 Sepals 4, fused into tube proximally; stamens 2;
    ovules 2 in each cell; fruit a winged achene
    (Fraxinus)   **125. OLEACEAE**
67 Petals and sepals each 4; stamens 6; ovules many
    in each cell   **63. BRASSICACEAE**
67 Petals and sepals each 5, or if 4 then stamens 4
    and ovules 1-2 per cell   68
    68 Leaves <5mm, ± scale-like   **59. TAMARICACEAE**
    68 Leaves >5mm, not scale-like   69
69 Leaves with strongly revolute margins, densely
    rusty-tomentose on lowerside; stamens mostly 6-8
    (Ledum)   **66. ERICACEAE**
69 Leaves with margins not or scarcely revolute,
    not rusty-tomentose (sometimes white-tomentose) on
    lowerside; stamens 4-5   70
    70 Stems trailing or climbing; leaves palmate or
        palmately lobed   **94. VITACEAE**
    70 Stems self-supporting; leaves pinnate, pinnately
        lobed or unlobed   71
71 Petals purple; ovary 1-celled with many ovules;
    fruit a capsule   **71. PITTOSPORACEAE**
71 Petals not purple; ovary 1-4-celled, each cell with
    1-2 ovules; fruit a berry or drupe   72
    72 Ovary 1-celled with 1 ovule   **101. ANACARDIACEAE**
    72 Ovary 2-4-celled, each cell with 1-2 ovules   73
73 Leaves exstipulate, at least some with strong
    marginal spines; petals white; fruit a red to
    yellow drupe   **90. AQUIFOLIACEAE**
73 Leaves stipulate, without spines (stems often spiny);
    petals greenish; fruit a black berry   **93. RHAMNACEAE**

MAGNOLIIDAE - DICOTYLEDONS
(DICOTYLEDONIDAE)

Often trees or shrubs; commonly with secondary thickening from a permanent vascular cambium; vascular bundles usually in a ring in the stem; primary root commonly persisting; leaves usually with pinnate or palmate major venation and reticulate minor venation; flower parts mostly in fours or fives; pollen grains mostly radially symmetrical, commonly with 3 pores and/or furrows; cotyledons normally 2; endosperm typically nuclear or cellular. Numerous exceptions to all the above occur.

25. LAURACEAE - Bay family

Trees or shrubs; leaves aromatic, simple, evergreen, entire, usually petiolate, alternate, exstipulate. Flowers solitary or in few-flowered clusters in leaf-axils, dioecious, perigynous, actinomorphic; perianth of 1 whorl of 4 $\pm$ free lobes; male flowers with 8-12 stamens; female flowers with 2-4 staminodes and a 1-celled ovary with 1 ovule; style 1; stigma capitate; fruit 1-seeded berry.
Easily recognized by the aromatic leaves and dioecious flowers with 4-lobed perianth.

1. LAURUS L. - Bay
1. L. nobilis L. - Bay. Shrub or tree exceptionally to 18m; leaves 5-10cm, laurel-like, glabrous, usually acute, with distinctive aromatic smell; flowers c.1cm across, cream. Intrd; widely planted in small numbers for culinary use, sometimes persisting in wild places in S, natd (probably bird-sown) in scrub and on cliffs near sea in SW Ir, S Wa, SW En and Jersey; Mediterranean.

26. ARISTOLOCHIACEAE - Birthwort family

Perennial herbs; leaves simple, entire, strongly cordate, exstipulate. Flowers solitary and terminal or 1-8 in leaf-axils, bisexual, epigynous, actinomorphic or zygomorphic; perianth of 1 whorl, tubular with 1 or 3 terminal lobes; stamens 6-12 in 1-2 whorls, arising from base or side of stylar column; ovary 6-celled, each cell with many ovules; styles 6, united into a column with 6-lobed stigma; fruit a capsule.

2 distinctive genera easily recognized by their cordate
leaves and weird flowers.

1 Aerial stem $\leq$ 10cm, densely pubescent; petiole
  much longer than lamina; flower solitary, terminal;
  perianth 3-lobed   **1. ASARUM**
1 Aerial stem >10cm, $\pm$ glabrous; petiole shorter
  than lamina or $\pm$ absent; flowers 1-8, axillary;
  perianth 1-lobed   **2. ARISTOLOCHIA**

1. ASARUM L. - <u>Asarabacca</u>
Rhizomatous herb with aerial stems bearing scales, usually 2
apparently opposite foliage leaves, and 1 terminal flower;
flowers actinomorphic; perianth campanulate, 3-lobed; stamens
12, in 2 whorls, arising from base of stylar column.

1. A. europaeum L. - <u>Asarabacca</u>. Rhizome on soil surface;
stems, petioles and pedicels pubescent; flowers 12-15mm, with
brownish-purple perianth, close to soil on short pedicel.
Intrd (sometimes claimed native); in woods, increasingly rare
and very scattered over Br N to C Sc; Europe.

2. ARISTOLOCHIA L. - <u>Birthworts</u>
Aerial stems not or little branched, bearing numerous
alternate leaves and 1-8 flowers in each leaf-axil; flowers
zygomorphic; perianth with rounded swollen base, narrowly
tubular upper part and 1-lobed apex; stamens 6, in 1 whorl,
arising from side of stylar column.

1. A. clematitis L. - <u>Birthwort</u>. Rhizomatous; $\pm$ glabrous; 91
stems to 1m; petioles longer than pedicels, c. 1/2 as long as
lamina; flowers 2-3.5cm x <5mm, with yellowish-brown lobe,
(1)2-8 in each leaf-axil. Intrd; formerly grown medicinally
and persisting in increasingly few places; rough ground, very
scattered over En and Wa, 1 or 2 records in S Sc; Europe.
2. A. rotunda L. - <u>Smearwort</u>. With an underground tuber; $\pm$ 91
glabrous; stems to 60cm; petioles absent or very short, much
shorter than pedicel; flowers 2.5-5cm x <5mm, with dark brown
lobe, 1 in each leaf-axil. Intrd; natd in Surrey since at
least 1918 (and formerly Kent) on chalky slopes; S Europe.

27. NYMPHAEACEAE - <u>Water-lily</u> family

Aquatic perennial herbs with stout rhizomes; leaves
alternate, simple, entire, mostly floating on water surface,
with deep basal sinus, with long petiole, stipulate or not.
Flowers solitary on long pedicels in leaf-axils, borne at or
above water surface, bisexual, hypogynous to $\pm$ epigynous,
actinomorphic; perianth of (3)4-6(7) free sepals and 9-33
free spirally-arranged petals; stamens 37-200, spirally
arranged, at least the outer with broad petaloid filaments;

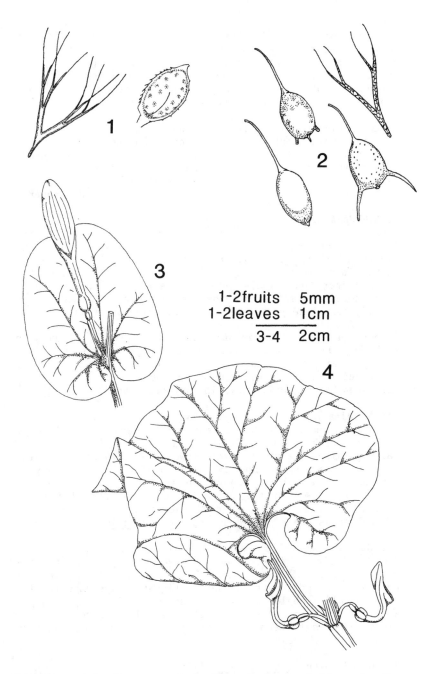

FIG 91 – 1-2, Leaves and fruits of **Ceratophyllum**. 1, **C. submersum**. 2, **C. demersum** (3 fruit types). 3-4, Flowering nodes of **Aristolochia**. 3, **A. rotunda**. 4, **A. clematitis**.

ovary globose to bottle-shaped, 8-many-celled, each cell with
many ovules; style ± absent; stigma a very broad rayed disc
entire or lobed at margin; fruit rather spongy, often
described as a berry-like capsule, dehiscing irregularly.
2 unmistakeable genera, with large long-petioled leaves and
flowers on or near water surface.

1 Leaf-veins forming a reticulum near leaf-margin;
   petals white, outermost much longer than the
   usually 4 sepals   **1. NYMPHAEA**
1 Leaf-veins forking near leaf-margin, not re-joining;
   petals yellow, shorter than the usually 5 sepals
                                                       2. NUPHAR

1. NYMPHAEA L. - White Water-lilies
Mature leaves rarely submerged, if so similar to floating
leaves; leaf-veins forming a reticulum near leaf-margin;
sepals (3)4(5), green to reddish-brown externally; petals
white, inserted at a range of levels on side of ovary, the
outer longer than sepals; stamens inserted on side of ovary
above petals; ovary subglobose; stigmatic rays projecting as
curved, horn-like processes; fruit ripening under water.

Other spp. - Various exotic spp. and cultivars, often with
pink or yellow flowers, are planted in ponds and lakes and
may persist or spread after gardening activities cease.
Their identity needs investigating; most are referable to **N.
marliacea** Latour-Marl., covering various hybrids of N. alba.

1. N. alba L. - White Water-lily. Petals 12-33, the outer
2-8.5cm; stamens 46-125; stigmatic rays (and carpels) 9-25.
Native; in lakes, ponds, dykes and slow-flowing rivers.
a. Ssp. alba. Leaves 9-30cm; flowers 9-20cm across, opening
wide; pollen-grains usually with projections of varying
lengths; stamens borne almost to top of ovary and leaving
scars to top of fruit; stigmatic rays usually >14; fruit
usually obovoid. Throughout BI except CI, but absent from
several areas and replaced by ssp. occidentalis in parts of N
and W.
b. Ssp. occidentalis (Ostenf.) N. Hylander. Leaves 9-13cm;
flowers 5-12cm across, usually never opening wide; pollen-
grains usually with projections of ± uniform length; stamens
not borne on upper part of ovary; stigmatic rays usually <16;
fruit usually subglobose, without stamen-scars in upper part.
W Ir and N & W Sc, especially Hebrides and Shetland.
   Intermediates between the sspp. occur and are not confined
to areas where their ranges meet.

2. NUPHAR Smith - Yellow Water-lilies
Mature leaves usually floating and submerged, the latter
thinner and with undulate margins; leaf-veins forking and not
rejoining near leaf-margin; sepals 5(-7), yellowish-green;

petals yellow, shorter than sepals; petals and stamens inserted at base of ovary; ovary bottle-shaped; stigmatic rays not projecting or projecting as flat, obtuse bulges; fruit ripening above water.

1 Leaf-blades erect, held above water surface; petioles ± terete **4. N. advena**
1 Leaf-blades horizontal, floating on water surface; petioles trigonous or dorsiventrally compressed   2
  2 Leaves with 23-28 lateral veins on each side; stigmatic disc 10-15mm across, circular or slightly crenate at margin, with 9-24 rays; flowers 3-6cm across **1. N. lutea**
  2 Leaves with 11-22 lateral veins on each side; stigmatic disc 6-11mm across, crenate to lobed at margin, with 7-14 rays; flowers 1.5-4cm across   3
3 Leaves with 15-22 lateral veins on each side; stigmatic disc 7.5-11mm across, crenate at margin, with 9-14 rays; stamens 60-100; pollen <25% fertile **2. N. x spenneriana**
3 Leaves with 11-18 lateral veins on each side; stigmatic disc 6-8.5mm across, distinctly lobed at margin, with 7-12 rays; stamens 37-65; pollen >90% fertile **3. N. pumila**

1. N. lutea (L.) Smith - <u>Yellow</u> <u>Water-lily</u>. Floating leaves to 40 x 30cm, with 23-28 lateral veins on each side, with a very narrow basal sinus; petiole trigonous; stigmatic disc 10-15mm across, circular or slightly crenate at margin. Native; in lakes, ponds, dykes and rivers; frequent throughout BI except CI and extreme N Sc.
2. N. x spenneriana Gaudin (<u>N. lutea</u> x <u>N. pumila</u>) - <u>Hybrid</u> <u>Water-lily</u>. Intermediate between the parents; floating leaves to 18 x 14cm, with 15-22 lateral veins on each side; stigmatic disc 7.5-11mm across, crenate at margin; pollen <20% fertile, seed fertility c.20%. Native; scattered in extreme N En and S & C Sc largely outside area of <u>N. pumila</u>, and 1 record in Merioneth; introgression of <u>N. lutea</u> into <u>N. pumila</u> occurs sometimes.
3. N. pumila (Timm) DC. - <u>Least</u> <u>Water-lily</u>. Floating leaves   R to 17 x 12.5cm, with 11-18 lateral veins on each side, with a broader sinus than in <u>N. lutea</u>; petiole dorsiventrally compressed, ± keeled beneath; stigmatic disc 6-8.5mm across, distinctly lobed at margin. Native; in ponds and lochs in highland areas but often at low altitude; local in N & C Sc, 1 locality in Salop, planted and natd in Surrey.
4. N. advena (Aiton) Aiton f. - <u>Spatter-dock</u>. Leaves erect, held above water, rarely floating, to 40 x 30cm, with 20-38 lateral veins on each side, with narrowly to broadly triangular basal sinus; petiole ± terete; stigmatic disc (6-)12-20mm across, ± circular at margin; flowers sometimes red-tinged. Intrd; planted as ornament and ± natd in 2

places in Surrey, perhaps overlooked elsewhere; E N America.

28. CERATOPHYLLACEAE - Hornwort family

Submerged aquatic perennial herbs; leaves in whorls of
(3)6-8(12), regularly forked into linear segments,
exstipulate. Flowers monoecious, subaquatic, sessile,
solitary in leaf-axils, male and female at different nodes,
actinomorphic, hypogynous,; perianth of numerous free green,
narrow segments in 1 whorl; male flowers wth 10-25 stamens;
female flowers with 1-celled ovary with 1 ovule; style 1;
stigma minutely bifid; fruit an achene.
Distinguished from other subaquatic plants by the whorled,
bifid leaves.

1. CERATOPHYLLUM L. - Hornworts
1. C. demersum L. - Rigid Hornwort. Stems branched, up to   **91**
1m; leaves dark green, rigid, forked 1-2x, with conspicuously
denticulate segments; fruit 4-5mm, rather rarely produced,
smooth or slightly warty, apiculate to long-spined at apex,
often also with 2 long spines at base or these missing (var.
inerme Gay ex R.-Smith) or reduced to tubercles (var.
apiculatum (Cham.) Asch.). Native; in ponds, ditches and
slow rivers; scattered over En and Wa, rare and very local in
Sc, Ir and CI.
2. C. submersum L. - Soft Hornwort. Leaves lighter green,   **91**
less rigid, forked 3-4x, with rather sparsely denticulate   **R**
segments; fruit 4-5mm, rather commonly produced (Aug-Oct),
conspicuously warty, apiculate to obtuse at apex, without
basal spines. Native; in ponds and ditches, mostly near sea;
very local in C & SE En and S Wa.

29. RANUNCULACEAE - Buttercup family

Herbaceous annuals or perennials, sometimes woody climbers;
leaves borne spirally or sometimes opposite or whorled,
simple or variously compound, usually petiolate, usually
exstipulate. Flowers variously arranged, bisexual, hypogynous, usually actinomorphic, sometimes zygomorphic;
perianth of 1-2 whorls of free segments, the outer (sepals)
often petaloid, of various colours, the inner (petals or
honey-leaves) bearing nectaries and often reduced or absent,
sometimes (when petals O) a whorl of sepal-like bracts
outside the petaloid sepals; stamens usually numerous, rarely
as few as sepals; carpels 1-many, sometimes partially or
rarely fully fused, if many usually spirally arranged, with
1-many ovules; fruit usually an achene or follicle, rarely a
berry or capsule.
Very variable in floral morphology, but most genera have
spirally arranged leaves without stipules and produce a head

of achenes or follicles from each flower, which is always
hypogynous and often with the sepals more conspicuous than
the petals.

1 Woody climber; leaves opposite; perianth of 4
   segments in 1 whorl   **12. CLEMATIS**
1 Herbaceous, not climbing; leaves alternate or whorled;
   perianth rarely of 4 segments, often in 2 whorls   2
   2 Ovary with many ovules; fruit a follicle, capsule
      or berry   3
   2 Ovary with 1 ovule; fruit an achene   11
3 Flowers with 1 carpel   4
3 Flowers with (2)3 or more carpels   5
   4 Flowers actinomorphic, whitish; sepals not spurred;
      fruit a berry   **8. ACTAEA**
   4 Flowers zygomorphic, usually blue, sometimes white
      or pink; the upper sepal with a conspicuous spur;
      fruit a follicle   **7. CONSOLIDA**
5 At least 1 of the petals or sepals conspicuously
   hooded or spurred   6
5 Petals and sepals not spurred or hooded   7
   6 Flowers zygomorphic, the upper sepal hooded,
      only the 2 upper petals spurred   **6. ACONITUM**
   6 Flowers actinomorphic, the sepals not hooded, each
      of the 5 petals spurred   **16. AQUILEGIA**
7 Flowers white to reddish, bluish or greenish,
   sometimes tinged with purple   8
7 Flowers yellow   9
   8 Annuals; sepals bluish; follicles fused up to apex;
      leaves divided into fine linear entire segments
                                                  5. NIGELLA
   8 Perennials; sepals green, white, violet or purple;
      follicles free or fused at base only; leaves
      divided into wider, toothed segments   **3. HELLEBORUS**
9 Stem-leaves 3, in a whorl just below flower **4. ERANTHIS**
9 Stem-leaves 1-many, not whorled and usually not
   just below flower   10
   10 Leaves deeply lobed; perianth of 2 whorls, the
      inner consisting of small nectaries   **2.TROLLIUS**
   10 Leaves simple, finely toothed; perianth of 1
      whorl   **1. CALTHA**
11 Perianth in a single whorl   12
11 Perianth apparently in two whorls   14
   12 Stamens longer than perianth; flowering stems with
      alternate leaves and many flowers   **17. THALICTRUM**
   12 Stamens shorter than perianth; flowering stems
      with 1-few flowers and whorled leaves or bracts   13
13 Styles scarcely elongating in fruit; whole plant
   glabrous to shortly pubescent   **9. ANEMONE**
13 Styles greatly elongating in fruit and becoming
   feathery; whole plant with long hairs   **11. PULSATILLA**
   14 Outer whorl of apparent perianth of 3 segments;

inner whorl of perianth segments blue or white
                                                        10. HEPATICA
14 Outer whorl of apparent perianth usually of 5
    segments, if of 3 then inner whorl yellow   15
15 Flowers solitary on leafless stem; leaves all
    linear, in a basal rosette; sepals spurred **15. MYOSURUS**
15 Flowers on usually branched, leafy stems; lowest
    leaves not linear; sepals not spurred   16
16 Petals red with a blackish basal blotch   **14. ADONIS**
16 Petals yellow or white, without a dark basal
    blotch   **13. RANUNCULUS**

1. CALTHA L. - Marsh-marigold
Herbaceous perennials; leaves spirally arranged, simple,
exstipulate; flowers solitary or in few-flowered cymes,
without a whorl of bracts below, actinomorphic; perianth of 1
whorl of 5-8(10) petaloid sepals; stamens numerous; carpels
5-15, free, spirally arranged; fruit a follicle.

1. **C. palustris** L. (C. radicans T. F. Forster) - **97**
Marsh-marigold.   Rhizomatous,   glabrous;   basal   leaves
long-petioled, cordate, denticulate; flowers 1-5cm across,
saucer-shaped; sepals golden yellow.   Native; in marshes and
ditches and beside ponds and streams.   Plants with procumbent
stems   rooting   at   nodes   and   little-branched,   eventually
turning up to produce usually 1 small flower, are found in N
Wa, N En, Sc and Ir in upland areas, and are best recognised
as var. **radicans** (T. F. Forster) Hook. (ssp. minor auct. non
Miller).

2. TROLLIUS L. - Globeflower
Herbaceous perennials; leaves borne spirally, palmate or
deeply palmately lobed, exstipulate; flowers solitary or in
few flowered cymes, without a whorl of bracts below, actinomorphic; perianth of 2 whorls; sepals 5-15, petaloid; petals
5-15, in the form of narrow, small nectaries; stamens
numerous; carpels numerous, free, spirally arranged; fruit a
follicle.

1. **T. europaeus** L. - Globeflower. Aerial stems to 70cm, **97**
glabrous, erect, little branched, arising from stout stock;
basal leaves long-petioled, palmate, with 3-5 lobed leaflets;
flowers 2.5-5cm across, $\pm$ globose; sepals pale yellow,
hiding petals.   Native; damp places in grassland or woods,
often upland; local in Wa, N En, Sc and NW Ir.

3. HELLEBORUS L. - Hellebores
Herbaceous perennials; leaves spirally arranged or all basal,
palmate with long, toothed leaflets, the lateral ones joined
at base, exstipulate; flowers in few- to many-flowered cymes,
or solitary, without whorl of bracts below, actinomorphic,
appearing very early in season; perianth of 2 whorls; sepals

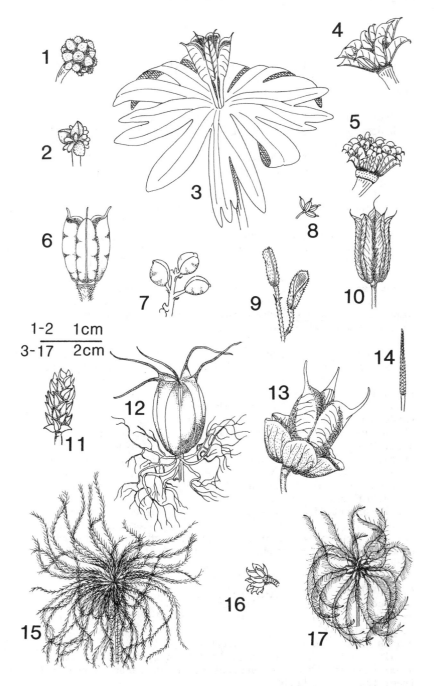

Fig 97 – Fruits of **Ranunculaceae**. 1, **Ranunculus ficaria** ssp.
ficaria. 2, **R.f.** ssp. **bulbilifer**. 3, **Eranthis**. 4, **Caltha**. 5,
Trollius. 6, **Aconitum napellus**. 7, **Actaea**. 8, **Thalictrum
minus**. 9, **Consolida**. 10, **Aquilegia vulgaris**. 11, **Adonis**. 12,
Nigella. 13, **Helleborus foetidus**. 14, **Myosurus**. 15,
Pulsatilla. 16, **Anemone nemorosa**. 17, **Clematis vitalba**.

5, usually not brightly coloured; petals 5-12, in the form of
small tubular nectaries; stamens numerous; carpels 2-5,
usually slightly fused at base; fruit a follicle.

1 Stems lasting from 1 Spring to next, many-flowered;
  bracts ± entire; leaves all on stem; fresh plant
  stinking when crushed   **1. H. foetidus**
1 Stems lasting from Winter to late Spring, 2-4-flowered;
  bracts deeply divided; leaves all basal; fresh plant
  not stinking when crushed   2
    2 Follicles fused for c.1/4 their length, sessile;
      flowers mostly 3-5cm across, pale green
                                                  2. H. viridis
    2 Follicles free to base, shortly stalked; flowers
      mostly 5-7cm across, yellowish-green to purplish
                                                  3. H. orientalis

Other spp. - Several European spp. are cultivated and may
self-sow in 'wild gardens' or persist for a while, e.g. **H.
niger** L. (Christmas-rose) with white sepals and **H. atrorubens**
Waldst. & Kit. with violet sepals.

1. H. foetidus L. - Stinking Hellebore. Not rhizomatous;   97
stems to 80cm, erect; flowers 1-3cm across, deeply cup-   R
shaped; sepals yellowish-green usually tinged purplish.
Probably native; woods and scrub on calcareous soils; very
local in En and Wa N to Yorks and Lancs, also common in
gardens and natd outside native range in Br.
2. H. viridis L. - Green Hellebore. Rhizomatous; stems to
40cm, erect; flowers 3-5cm across, saucer-shaped; sepals
green. Native; woods and scrub on calcareous soils; very
local in En and Wa N to Yorks and Lancs, also grown in
gardens and natd in Br and Ir. The British plant is ssp.
occidentalis (Reuter) Schiffner.
3. H. orientalis Lam. - Lenten-rose. Rhizomatous; stems to
60cm, erect; flowers 5-7cm across, saucer-shaped; sepals
yellowish-green to purplish. Intrd; much grown in gardens,
sometimes natd in woods and parks; Surrey and E Gloucs;
Turkey.

4. ERANTHIS Salisb. - Winter Aconite
Herbaceous perennials, with underground tubers; leaves all
basal except bracts, palmate or deeply palmately lobed,
exstipulate; flowers solitary, with whorl of 3 leaf-like
bracts just below, actinomorphic; perianth of 2 whorls;
sepals usually 6, petaloid; petals usually 6, in the form of
small tubular nectaries; stamens numerous; carpels usually 6,
free; fruit a follicle.

1. E. hyemalis (L.) Salisb. - Winter Aconite. Aerial stems   97
to 15cm, erect, lasting from winter to late spring; flowers
2-3cm across; sepals bright yellow. Intrd; common in gardens

and often becoming well natd in woods, parks and roadsides; scattered in Br N to C Sc; S Europe.

5. NIGELLA L. – <u>Love-in-a-mist</u>
Annuals; leaves spirally arranged, pinnate, exstipulate; flowers solitary or in few-flowered cymes, with whorl of bracts below, actinomorphic; perianth of 2 whorls; sepals 5, petaloid; petals 5, in the form of clawed nectaries; stamens numerous; carpels usually 5, fused up to apex; fruit a capsule.

1. N. damascena L. – <u>Love-in-a-mist</u>. Stems up to 50cm, **97** simple or little branched, glabrous; leaves finely divided with linear segments, at least 3 in a cluster just below each terminal flower; sepals blue; capsule strongly inflated, $\pm$ globose. Intrd; commonly grown in gardens and often persisting on waste ground and rubbish tips; En and Wa; S Europe.

6. ACONITUM L. – <u>Monk's-hoods</u>
Herbaceous perennials; leaves spirally arranged, palmate or deeply palmately lobed, exstipulate; flowers in terminal racemes, each in axil of small bract, zygomorphic; perianth of 2 whorls; sepals 5, petaloid, the upper one forming an elongated hood; petals 2-10, in the form of nectaries, the 2 upper large and enclosed in the sepal-hood, the others very small or absent; stamens numerous; carpels 3(-5), $\pm$ fused at base; fruit a follicle.

1 Flowers yellow or cream; upper sepal >2x as high as
  wide; leaves not divided to base   **3. A. vulparia**
1 Flowers blue or blue and white; upper sepal <2x as
  high as wide; at least lower leaves divided to base   2
  2 Pedicels pubescent to densely pubescent; flowers
    usually blue; upper sepal $\pm$ as high as wide,
    gradually tapered into forward-projecting spur
                                   1. A. napellus
  2 Pedicels glabrous to sparsely pubescent; flowers
    blue and white or blue; upper sepal distinctly
    higher than wide, abruptly narrowed into forward-
    projecting spur   **2. A. x cammarum**

Other spp. – All records of **A. variegatum** L. and many of **A. compactum** (Reichb.) Gayer are errors for <u>A. x cammarum</u>.

1. A. napellus L. (<u>A. anglicum</u> Stapf) – <u>Monk's-hood</u>. Stems **97** to 1.5m, erect; leaves divided to base, with narrow deeply **101** divided segments; pedicels conspicuously appressed- **R** pubescent; flowers blue to violet; upper sepals $\pm$ as wide as high; pollen full; seeds fertile. Native; very local in shady places by streams; probably only in SW En and S Wa, but commonly cultivated and natd sparsely over much of Br in native-type habitats as well as waste places. Cultivated

plants are very variable in habit, leaf-lobing, flower colour
and flowering time; they may be referred to >4 sspp. but
distinctions between them break down in garden material. A.
anglicum comes under ssp. napellus and true A. compactum
under ssp. vulgare (DC.) Rouy & Fouc.; the latter has less
tapering apices to leaf-lobes, and a more compact inflorescence with more appressed pedicels.
2. A. x cammarum L. (A. napellus x A. variegatum L.) - 101
Hybrid Monk's-hood. Stems to 1.5m, erect; leaves as in A.
napellus; pedicels glabrous to sparsely appressed-pubescent;
flowers blue to violet or variegated with white; upper sepal
distinctly higher than wide; pollen empty; seeds sterile.
Intrd; grown in gardens and natd in damp shady places;
frequent in Sc, rare in Wa and En. Most natd plants are cv.
'Bicolor', with variegated flowers, of garden origin.
3. A. vulparia Reichb. - Wolf's-bane. Stems to 1m, erect; 101
leaves not divided to base and with broader and less divided
segments than in A. napellus; pedicels conspicuously
appressed-pubescent; flowers yellow; upper sepals c.3x as
high as wide. Intrd; grown in gardens and occasionally natd
by streams and in woods; C En to N Sc; C Europe.

7. CONSOLIDA (DC.) Gray (Delphinium L. pro parte) - Larkspurs
Annuals; leaves spirally arranged, palmate with finely
divided segments, exstipulate; flowers in terminal racemes,
each in axil of bract and with bracteoles on pedicel,
zygomorphic; perianth of 2 whorls; sepals 5, petaloid, the
upper one long-spurred; petals 4, elaborately shaped, the
upper two fused and with a nectariferous spur enclosed in the
sepal-spur; stamens numerous; carpel 1; fruit a follicle.

Other spp. - C. orientalis (Gay) Schroedinger (Delphinium
orientale Gay) (Eastern Larkspur) differs from C. ajacis in
its upper bracteoles overlapping the base of the flower (not
so in C. ajacis) and its <12mm sepal-spur. C. regalis Gray
(Delphinium consolida L.) (Forking Larkspur) differs from the
other 2 spp. in having glabrous follicles, and branches
arising at wide angles. Both come from Europe and formerly
occurred as casuals.

1. C. ajacis (L.) Schur (C. ambigua auct. non (L.) P. Ball & 97
Heyw., Delphinium ambiguum auct. non L.) - Larkspur. Stems 101
to 1m, simple or with branches arising at rather narrow
angles; at least lower bracts deeply divided; flowers blue,
pink or white; sepal-spur 12-18mm; follicle pubescent.
Intrd; much grown in gardens and a common escape in Br;
formerly a corn-field weed natd in E Anglia; Mediterranean.

8. ACTAEA L. - Baneberry
Herbaceous perennials; leaves borne spirally, ternate or
pinnate, exstipulate; flowers in terminal racemes, each in
axil of a small bract, actinomorphic; perianth usually of 2

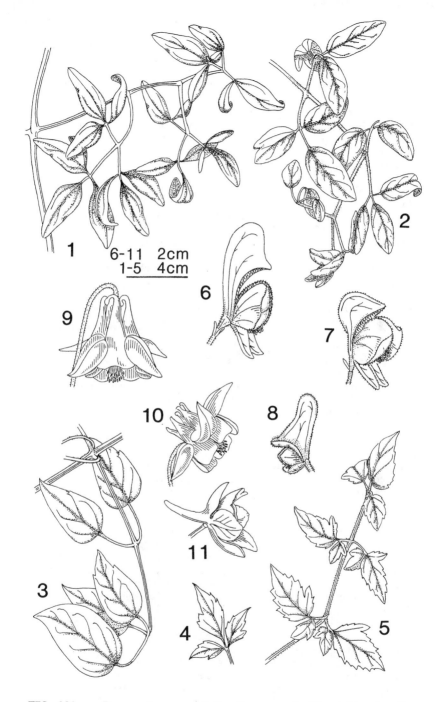

FIG 101 – Ranunculaceae. 1-5, leaves of **Clematis**. 1, C. flammula. 2, **C. viticella**. 3, **C. vitalba**. 4, **C. montana**. 5, C. tangutica. 6-8, flowers of **Aconitum**. 6, **A. x cammarum**. 7, **A. napellus**. 8, **A. vulparia**. 9-10, flowers of **Aquilegia**. 9, **A. vulgaris**. 10, **A. pyrenaica**. 11, flower of **Consolida** ajacis.

whorls; sepals (3)4(5), petaloid; petals 4-6 or 0, petaloid,
without nectar; stamens numerous; carpel 1; fruit a berry.

1. A. spicata L. - <u>Baneberry</u>. Stems to 60cm, simple or   **97**
little branched; leaves with broad leaflets; flowers small   **R**
and crowded, white; berry 10-13mm, black. Native, limestone
pavements and sparse woodland on limestone; local in Yorks,
Westmorland and Lancs.

9. ANEMONE L. - <u>Anemones</u>
Herbaceous perennials; leaves all basal, palmate, palmately
lobed or ternate, exstipulate; flowers solitary or few with a
whorl of 3 leaf-like bracts some way below, or in terminal
few-flowered cymes, actinomorphic; perianth of 1 whorl of
5-20 petaloid sepals; stamens numerous; carpels numerous,
free, spirally arranged; fruit an achene, the style remaining
shorter than fertile portion.

1   Stem >30cm, usually branched with several flowers;
    sepals densely sericeous on lowerside; autumn
    flowering   **4. A. x hybrida**
1   Stem <30cm, simple, with 1(-3) flowers; sepals
    glabrous to sparsely pubescent on lowerside; spring
    flowering   2
2   Sepals yellow, mostly 5   **3. A. ranunculoides**
2   Sepals white to pink or blue, mostly 6 or more   3
3   Sepals (5)6-7(9), mostly white or pinkish
                                          1. A. nemorosa
3   Sepals (8)10-15(18), mostly blue   **2. A. apennina**

Other spp. - **A. blanda** Schott & Kotschy, from SE Europe,
is grown in gardens and sometimes persists; it differs from
<u>A. apennina</u> mainly in its nodding head of achenes and up to
20 glabrous sepals.

1. A. nemorosa L. - <u>Wood</u> Anemone. Stems 5-30cm, erect, with **97**
whorl of 3 leaf-like bracts 1-6cm below the solitary flower;
sepals (5)6-7(9), glabrous or sparsely pubescent near base on
lowerside, usually white, often variously tinged with pink or
purple and sometimes pale blue; head of achenes nodding.
Native; woodland, hedgerows, and open grassland in wetter
districts; throughout BI but local in S Ir.
2. A. apennina L. - <u>Blue</u> Anemone. Differs from <u>A. nemorosa</u>
in its pedicel to 12cm; sepals (8)10-15(18), narrower,
usually blue (sometimes pink or white), sparsely pubescent
near base on lowerside; anthers paler yellow; head of achenes
erect. Intrd; commonly cultivated and frequently persisting
as a throwout or escape in woodland, hedgerows and rough
ground; scattered in En and Wa; S Europe.
3. A. ranunculoides L. - <u>Yellow</u> Anemone. Differs from <u>A.
nemorosa</u> in its fewer (sometimes 0) basal leaves with ±
sessile or incomplete (not stalked) main leaf-segments, and

1(3) flowers with usually 5 yellow sepals. Intrd; grown in
gardens and sometimes persisting as a throwout or escape in
shady places; very scattered in En and Sc; Europe.
4. A. x hybrida Paxton (<u>A.</u> x <u>japonica</u> auct.; <u>A. hupehensis</u>
(Lemoine) Lemoine x <u>A. vitifolia</u> Buch.-Ham. ex DC.) -
<u>Japanese</u> <u>Anemone</u>. Stems 40-150cm, erect, rather sparsely
branched above with whorls of + leaf-like bracts at each node
and 1 flower terminating each branch; sepals up to c.30, the
inner white to purple, densely sericeous on lowerside.
Intrd; much cultivated and persisting on old garden sites or
as an escape or throwout; rare throughout BI; garden origin.

10. HEPATICA Miller - <u>Liverleaf</u>
Herbaceous perennials; leaves all basal, distinctively
3-lobed, exstipulate; flowers solitary, actinomorphic;
perianth apparently of 2 whorls, but inner whorl of 6-10
coloured segments is calyx and outer whorl of 3 sepal-like
segments is bracts; stamens numerous; carpels numerous, free,
spirally arranged; fruit an achene, the style remaining
shorter than fertile portion.

1. H. nobilis Schreber - <u>Liverleaf</u>. Leaves on petioles to
15cm, cordate at base, with entire lobes; flowers on
peduncles to 15cm arising from leaf-rosette; flowers 15-25mm
across, blue, rarely white. Intrd; grown in shady places in
wild gardens, natd in E Gloucs and SE Yorks; Europe.

11. PULSATILLA Miller (<u>Anemone</u> subg. <u>Pulsatilla</u> (Miller)
Thome) - <u>Pasqueflower</u>
Herbaceous perennials; leaves all basal, 2-pinnate,
exstipulate; flowers solitary, with a whorl of 3 almost
leaf-like bracts just below, actinomorphic; perianth of 1
whorl of 6 petaloid sepals; stamens numerous, the outer
sterile and nectariferous; carpels numerous, free, spirally
arranged; fruit an achene, the style becoming greatly
elongated and feathery.

1. P. vulgaris Miller (<u>Anemone</u> <u>pulsatilla</u> L.) - <u>Pasque-</u>   97
<u>flower</u>. Stems to 30cm, simple, with + spreading hairs;   R
sepals 6, 2-5cm, deep violet-purple, densely sericeous on
lowerside; style becoming 3-5cm in fruit. Native; dry
calcareous grassland; very local in C & E En, from W Gloucs
to Cambs and N Lincs, formerly further N.

12. CLEMATIS L. - <u>Traveller's-joys</u>
Woody climbers; leaves opposite, pinnate or ternate or rarely
simple, exstipulate, the petioles and rhachis twining round
supports; flowers in axillary cymes, often with 2 small
opposite bracteoles below, actinomorphic; perianth of 1 whorl
of usually 4 petaloid sepals; stamens numerous; carpels
numerous, free, spirally arranged; fruit an achene, the style
becoming greatly elongated and feathery or not so.

1 Leaves ternate, the primary divisions serrate
                                        4. C. montana
1 Leaves pinnate, the primary divisions divided further
   or not   2
   2 Flowers yellow   **3. C. tangutica**
   2 Flowers white to blue or purple   3
3 Flowers blue to purple; styles glabrous, not
   elongating in fruit   **5. C. viticella**
3 Flowers white or cream; styles pubescent,
   elongating in fruit   4
   4 Leaves pinnate; sepals pubescent on both surfaces
                                        1. C. vitalba
   4 Leaves 2-pinnate; sepals pubescent only on lowerside
                                        2. C. flammula

Other spp. - **C. cirrhosa** L., from S. Europe, grows
outside gardens on walls in Guernsey; it is an evergreen with
simple, lobed or ternate leaves, nodding whitish flowers, and
fused bracteoles.

1. C. vitalba L. - Traveller's-joy. Rampant deciduous **97**
climber to 30m; leaves pinnate; leaflets rounded to cordate **101**
at base; flowers in dense clusters; sepals c.1cm, creamish-
or greenish-white; style becoming long and feathery in fruit.
Native; hedgerows, scrub and woodland on base-rich soils;
locally abundant in En and Wa N to SW Yorks, also natd in
scattered localities N to C Sc and over most of Ir and CI.
2. C. flammula L. - Virgin's-bower. Sprawling deciduous **101**
climber to 6m; leaves 2-pinnate; leaflets rounded to cuneate
at base; flowers in dense clusters; sepals c.1cm, white;
style becoming long and feathery in fruit.   Intrd; natd on
cliffs and dunes by sea; E Kent, W Cornwall and Caerns;
Mediterranean.
3. C. tangutica (Maxim.) Korsh. - Orange-peel Clematis. **101**
Deciduous climber to 4m; leaves pinnate with primary
divisions mostly ternate; leaflets cuneate to cordate at
base; flowers usually 1 in each leaf-axil; sepals 3-5cm,
yellow; style becoming long and feathery in fruit.   Intrd;
natd on sand in W Cornwall; China.
4. C. montana Buch.-Ham. ex DC. - Himalayan Clematis. **101**
Vigorous deciduous climber to 6m; leaves ternate; leaflets
cuneate to rounded at base; flowers in groups of 1-6; sepals
1.5-2.5cm, white to pink or bluish; style becoming long and
feathery in fruit.   Intrd; much grown in gardens and rarely
escaping or persisting over hedges and walls; very scattered
in Br N to Mid-W Yorks; Afganistan to Taiwan.
5. C. viticella L. - Purple Clematis. Deciduous climber **101**
to 4m; leaves pinnate with primary divisions ternate;
leaflets cordate to broadly cuneate at base; flowers 1-3 in
each leaf-axil; sepals 1.5-3cm, blue to purple; style
glabrous, elongating little in fruit.   Intrd; natd in hedges;
very scattered, Dorset and Surrey to Salop; S Europe.

13. RANUNCULUS L. - Buttercups

Herbaceous perennials or annuals, some aquatic; leaves spirally arranged, simple to much divided, stipulate or not; flowers usually in cymes, sometimes solitary and leaf-opposed, without a whorl of bracts below, actinomorphic; perianth of 2 whorls; sepals 3 or 5, sepaloid; petals 5 or fewer by reduction (to 0) or 7-12, petaloid, usually with a small nectar-secreting pit (nectar-pit) on inner face; stamens usually numerous, sometimes 5-10; carpels numerous, free, spirally arranged; fruit an achene, the style remaining shorter than fertile portion.

General key
1 Sepals 3; petals 7-12; many roots modified as whitish, swollen tubers with rounded apices   **18. R. ficaria**
1 Sepals 5; petals usually 5, sometimes <5 or many; root tubers rarely present, if so then with finely tapering apices   2
  2 Petals yellow   Key A
  2 Petals white, often yellow at base   3
3 Plant erect, terrestrial; largest leaves basal; achenes not transversely ridged   **17. R. aconitifolius**
3 Plant rarely erect, often aquatic; basal leaves 0; achenes transversely ridged   Key B

Key A - Sepals 5; petals yellow, usually 5
1 Leaves entire or toothed at margin, unlobed   2
1 At least some leaves divided at least 1/4 way to base 7
  2 Flowers 2-5cm across; achenes c.2.5mm **12. R. lingua**
  2 Flowers <2(2.5)cm across; achenes 1-2(2.3)mm   3
3 Lower leaves cordate at base   4
3 All leaves cuneate at base   5
  4 Stems erect; achenes tuberculate
                                16. R. ophioglossifolius
  4 Stems procumbent to decumbent; achenes smooth
                                13. R. flammula
5 Stems erect to ± procumbent, usually rooting only at lower nodes; beak c.1/8 to 1/10 as long as rest of achene; widest petals usually >4mm across
                                13. R. flammula
5 Stems procumbent, rooting at all or most nodes; beak c.1/3 to 1/6 as long as rest of achene; widest petals usually <4mm across   6
  6 Beak c.1/5 to 1/6 as long as rest of achene; basal leaves usually >1.5mm wide   **14. R. x levenensis**
  6 Beak c.1/3 to 1/4 as long as rest of achene; basal leaves usually <1.2mm wide   **15. R. reptans**
7 Sepals strongly reflexed at anthesis   8
7 Sepals not reflexed at anthesis   13
  8 Plant perennial with swollen stem-base
                                3. R. bulbosus
  8 Plant annual, without swollen stem-base   9

9 Achenes c.1mm, on elongated receptacle, + smooth
   on sides   11. R. sceleratus
9 Achenes >2mm, on + spherical receptacle, with
   tubercles or spines on sides   10
   10 Achenes 5-8mm including beak of 2-3mm; longest
      spines on sides of achene at least 1mm
                                          6. R. muricatus
   10 Achenes <5mm including beak of <1.5mm; longest
      tubercles on sides of achene <1mm   11
11 Flowers <8mm across; receptacle glabrous
                                          7. R. parviflorus
11 Flowers >9mm across; receptacle pubescent   12
   12 Achenes with tubercles confined to edge of faces,
      close to border; beak c.0.5-0.75mm   4. R. sardous
   12 Achenes with faces covered with tubercles; beak
      c.0.75-1mm   5. R. marginatus
13 Plant with procumbent stems rooting at nodes
                                          2. R. repens
13 Stems usually not procumbent, not rooting at nodes 14
   14 Plant annual; receptacle pubescent; achenes with
      conspicuous spines   8. R. arvensis
   14 Plant perennial; receptacle glabrous; achenes
      glabrous or pubescent but without spines   15
15 Some roots swollen into fusiform tubers; achenes
   borne in an elongated head   9. R. paludosus
15 All roots thin; achenes borne in a + spherical head 16
   16 Basal leaves with a reniform outline, glabrous to
      sparsely appressed-pubescent; achenes pubescent
                                          10. R. auricomus
   16 Basal leaves with a polygonal or polygonal-rounded
      outline, conspicuously pubescent; achenes glabrous
                                          1. R. acris

Key B - Sepals 5; petals white, 5; basal leaves 0
1 Laminar (floating or aerial) leaves the only ones
   present   2
1 Capillary (normally submerged) leaves present   6
   2 Receptacle pubescent; leaves divided usually
      >1/2 way into 3(-5) main lobes   3
   2 Receptacle glabrous; leaves divided <1/2 way into
      3-5(7) main lobes   5
3 Petals >5.5mm, contiguous at anthesis; achenes
   narrowly winged and borne on ovoid receptacle when
   completely mature   23. R. baudotii
3 Petals <6mm, not contiguous at anthesis; achenes
   not winged, borne on + spherical receptacle   4
   4 Leaves usually 3-lobed, rarely 5-lobed; petals to
      4.5mm, c.1.5x as long as sepals; pedicels strongly
      reflexed when fruit ripe   22. R. tripartitus
   4 Leaves often 5-lobed; petals to 6mm, c.2x as long
      as sepals; pedicels remaining erect when fruit
      ripe   21. R. x novae-forestae

5 Leaf-lobes broadest at their base; leaf-sinuses very
   open, obtuse to subacute; petals <4.5mm, little longer
   than sepals   **19. R. hederaceus**
5 Leaf-lobes broadest above their base; leaf-sinuses
   narrowly acute; petals >4.5mm, 2-3x as long as
   sepals   **20. R. omiophyllus**
   6 Plant with both laminar and capillary leaves   7
   6 Plant with only capillary leaves   13
7 Petals <6mm, not contiguous at anthesis   8
7 Petals 5-20mm, contiguous at anthesis   9
   8 Laminar leaves usually 3-lobed, rarely 5-lobed;
      petals < 4.5mm, c.1.5x as long as sepals;
      pedicels strongly reflexed when fruit ripe
                                          22. R. tripartitus
   8 Laminar leaves often 5-lobed; petals < 6mm,
      c.2x as long as sepals; pedicels remaining erect
      when fruit ripe   **21. R. x novae-forestae**
9 Achenes glabrous when immature, narrowly winged and
   borne on ovoid receptacle when completely mature
                                          23. R. baudotii
9 Achenes usually pubescent when immature, not winged
   at maturity, borne on + spherical receptacle   10
   10 Pedicel in fruit rarely >50mm, shorter than
      petiole of opposed laminar leaf; petals <10mm,
      with circular nectar-pit   **25. R. aquatilis**
   10 Pedicel in fruit usually >50mm, usually longer
      than petiole of opposed laminar leaf; petals
      usually >10mm, with pear-shaped nectar-pit   11
11 Leaves intermediate between laminar and capillary
   usually common, mostly as capillary leaves with
   some slightly flattened segments; mature achenes
   not formed   **27. x kelchoensis**
11 Leaves intermediate between laminar and capillary
   rarely present; fertile achenes regularly produced   12
   12 Capillary leaves rigid or flaccid, shorter than
      adjacent stem internode, with markedly divergent
      segments   **26. R. peltatus**
   12 Capillary leaves flaccid, usually longer than
      adjacent stem internode, with + parallel
      segments   **28. R. penicillatus**
13 Leaves rigid, circular in outline, the segments all
   lying in 1 plane   **31. R. circinatus**
13 Leaves with segments not lying in 1 plane   14
   14 Achenes narrowly winged and borne on ovoid
      receptacle when completely mature   **23. R. baudotii**
   14 Achenes not winged at maturity, borne on +
      spherical receptacle   15
15 Petals usually <6mm, with lunate nectar-pit, not
   contiguous at anthesis   **24. R. trichophyllus**
15 Petals >5mm, with circular or pear-shaped (rarely
   lunate) nectar-pit, contiguous at anthesis   16
   16 Receptacle pilose to glabrous; mature leaves

       longer than adjacent internode, rarely >4x
       forked, with ± parallel segments   **30. R. fluitans**
    16 Receptacle densely pubescent; mature leaves shorter
       to longer than adjacent internode, usually some
       >4x forked, with divergent to ± parallel segments   17
    17 Pedicel in fruit rarely >50mm; petals <10mm, with
       circular nectar-pit   **25. R. aquatilis**
    17 Pedicel in fruit usually >50mm; petals usually
       >10mm, with pear-shaped (rarely lunate) nectar-pit   18
    18 Mature achenes not formed; pedicels not
       elongating   **29. R. x bachii**
    18 Fertile achenes regularly produced from each
       flower; pedicels elongating in fruit   19
    19 Ultimate segments of well-developed leaves <100
                                               26. R. peltatus
    19 Ultimate segments of well-developed leaves >100,
       often >200   **28. R. penicillatus**

Subgenus **1** – RANUNCULUS (spp. 1-17).   Petals yellow
except in R. aconitifolius (white), normally 5; sepals 5;
achenes not transversely ridged; roots rarely modified as
tubers and then not with rounded apices.

 1. R. acris L. – Meadow Buttercup.   Erect perennial to 1m; **110**
basal leaves deeply palmately lobed, pubescent; flowers 15-
25mm across; sepals not reflexed; achenes 2-3.5mm, glabrous,
smooth, with short hooked beak.   Native; grassland, especially damp and calcareous.   Very variable; 3 vars may be
usefully recognised. Var. **villosus** (Drabble) Coles is common
in undisturbed areas of N Sc (including the islands) and W &
C Ir; it differs from var. acris in having little-branched
and few-flowered stems, leaves with relatively broad lobes,
and early-formed basal leaves with hairs mostly >1.2mm (not
mostly <1.2mm), but intermediates occur in N Sc.   Var.
pumilus Wahlenb. (ssp. borealis auct.) has similar stems but
early-formed basal leaves are glabrous and the leaves are
relatively shallowly divided; it occurs in Cairngorms, C Sc.
 2. R. repens L. – Creeping Buttercup.   Perennial with strong **110**
creeping stems rooting at nodes and ± erect flowering stems
to 60cm; basal leaves triangular-ovate in outline, with 3
main segments, the middle one long-stalked and borne above
the 2 laterals, usually pubescent; flowers 20-30mm across;
sepals not reflexed; achenes 2.5-3.8mm, glabrous, smooth,
with short curved beak.   Native; wet grassland, woods,
streamsides, marshes and duneslacks, and as a weed; abundant
throughout BI.
 3. R. bulbosus L. – Bulbous Buttercup.   Erect perennial to **110**
40cm with swollen corm-like stem-base; basal leaves ovate in
outline, with 3 main lobes, the middle one sessile or
stalked, pubescent; flowers 15-30mm across; sepals strongly
reflexed at anthesis; achenes 2-4mm, glabrous, finely pitted,
with short hooked beak.   Native; dry grassland and fixed

dunes; common in most of BI but absent from parts of Sc and
Ir.   Ssp. **bulbifer** (Jordan) P. Fourn. is continuously
connected to ssp. bulbosus and probably not worthy of
recognition. A variant described from maritime dunes in CI
and W BI as var. **dunensis** Druce has large flowers (25-30mm
across), rather short stout stems, and stems and upper leaves
with dense, long, white, patent hairs, but similar plants
predominate all over Ir and parts of BI and do not merit ssp.
status.
    4. R. sardous Crantz - Hairy Buttercup. Erect annual to 110
40cm; leaves similar to those of R. bulbosus; flowers 12-25mm
across, paler yellow than in R. bulbosus; sepals strongly
reflexed at anthesis; achenes 2.5-4mm, glabrous, smooth apart
from few tubercles just inside border, with short curved
beak.   Probably native; grassland and cultivated land;
frequent near S & E coasts of En and in CI, very local
(formerly commoner) elsewhere in Br N to C Sc.
    5. R. marginatus Urv. - St Martin's Buttercup. Erect annual 110
to 40cm; lower leaves with 3 broad incomplete lobes, usually
sparsely pubescent; flowers 12-25mm across; sepals strongly
reflexed at anthesis; achenes 3-5mm, glabrous, densely
tuberculate on sides, with curved beak c.1mm.   Intrd; + natd
as weed of cultivated ground in extreme SW En, especially
Scillies, rare casual elsewhere; E Mediterranean. British
examples are referable to var. **trachycarpus** (Fischer & Meyer)
Azn.; the type has smooth achenes, and might occur here.
    6. R. muricatus L. - Rough-fruited Buttercup. Erect annual 110
to 40cm; lower leaves usually 3-lobed to <1/2 way, glabrous
to very sparsely pubescent; flowers 6-16mm across; sepals
strongly reflexed at anthesis; achenes 5-8mm, glabrous, with
spines at least 1mm on sides, wih curved beak 2-3mm. Intrd;
natd as weed of cultivated ground in SW En and rare casual
elsewhere; S Europe.
    7. R. parviflorus L. - Small-flowered Buttercup. Spreading 110
decumbent to erect annual to 40cm;; lower leaves 3-5-lobed   R
usually to c.1/2 way, pubescent; flowers 3-6mm across; sepals
strongly reflexed at anthesis; achenes 2.5-3.5mm, glabrous,
with short tubercles bearing minute hooked spines on sides,
with short hooked beak.   Native; open ground of all sorts,
especially near coast; frequent near coast in SW En, S Wa and
CI, very local elsewhere N to Yorks and SE Ir, formerly
commoner.
    8. R. arvensis L. - Corn Buttercup. Erect annual to 60cm; 110
lowest leaves very shallowly lobed, middle ones very deeply   R
ternately or pinnately lobed, sparsely appressed-pubescent;
flowers 4-12mm across; sepals not reflexed; achenes 6-8mm,
glabrous, with spines >1mm on sides, with slightly curved
beak 3-4mm. Native; weed of cultivated ground, especially
cornfields; formerly frequent in En and Wa except W, but now
much rarer, very scattered in E Sc.
    9. R. paludosus Poiret - Jersey Buttercup. Erect perennial 110
to 40cm, some roots developed as fusiform tubers; inner basal RR

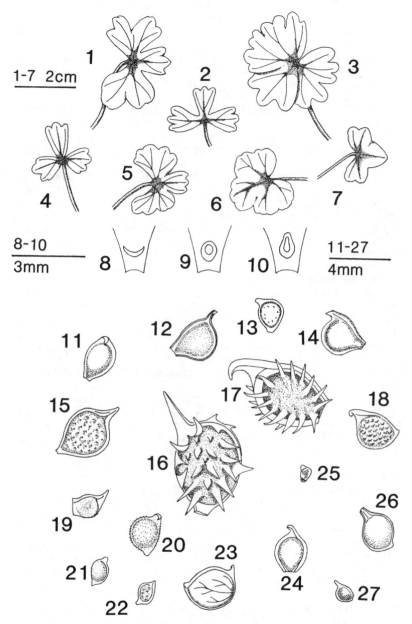

FIG 110 - **Ranunculus.** 1-7, leaves. 1, **R. peltatus.** 2, **R. baudotii.** 3, **R. aquatilis.** 4, **R. tripartitus.** 5, **R. penicillatus.** 6, **R. omiophyllus.** 7, **R. hederaceus,** 8-10, nectar-pits at base of petals. 8, lunate. 9, circular. 10, pear-shaped. 11-27, achenes. 11, **R. acris.** 12, **R. repens.** 13, **R. sardous.** 14, **R. bulbosus.** 15, **R. marginatus.** 16, **R. arvensis.** 17, **R. muricatus.** 18, **R. parviflorus.** 19, **R. paludosus.** 20, **R. auricomus.** 21, **R. flammula.** 22, **R. ophioglossifolius.** 23, **R. aconitifolius.** 24, **R. lingua.** 25, **R. sceleratus.** 26, **R. ficaria.** 27, **R. aquatilis.**

leaves narrowly and deeply ternately divided, pubescent; stem-leaves 1-2, small; flowers 20-30mm across; sepals not reflexed; achenes 2.5-3mm, slightly pubescent, + smooth, with + straight or minutely hooked beak c.1mm, borne on much elongated receptacle. Native; dry grassy places; Jersey.

10. R. auricomus L. - Goldilocks Buttercup. Erect perennial **110** to 40cm; basal leaves reniform in outline, very variably 3-lobed, glabrous to sparsely appressed-pubescent; stem-leaves few, deeply divided; flowers with variably developed (sometimes 0) petals, 15-25mm across when complete; sepals not reflexed; achenes 3-4mm, very shortly pubescent, smooth, with curved or hooked beak >1mm. Native; woods and hedgebanks; frequent throughout En but very scattered in Sc, Wa and Ir. Apomictic; several hundred agamospecies have been described from the Continent. Our plants are probably different from any of these, and probably well over 100 could be recognized, but they have not yet been worked out.

11. R. sceleratus L. - Celery-leaved Buttercup. Erect **110** annual to 60cm; lower leaves deeply 3-lobed, + glabrous, shiny; flowers 5-10mm across; sepals strongly reflexed at anthesis; achenes c.1mm, glabrous, smooth, with vestigial beak. Native; in marshy fields, ditches, ponds and streamsides; frequent throughout En and CI but more scattered and mostly coastal in Sc, Wa and Ir.

12. R. lingua L. - Greater Spearwort. Erect strongly **110** stoloniferous perennial to 120cm; lowest leaves ovate, cordate, + entire, but often withered at flowering; stem-leaves lanceolate to oblanceolate, narrowly cuneate, shallowly toothed, + glabrous; flowers 20-50mm across; pedicels terete; sepals not reflexed; achenes c.2.5mm, glabrous, minutely pitted, very narrowly winged, with short curved beak. Native; marshes and pondsides; scattered through BI except N Sc; commonly grown and often + natd.

13. R. flammula L. - Lesser Spearwort. Erect to procumbent **110** perennial to 50cm; stem-leaves lanceolate to oblanceolate or linear, narrowly cuneate, + glabrous; flowers 7-20(25)mm across; pedicels furrowed; sepals not reflexed, achenes 1-2(2.3)mm, glabrous, minutely pitted, not winged, plus beak c.1/8 to 1/10 as long. Native; all kinds of wet places; throughout BI.

1 Lowest leaves with lamina cordate at base
                                            b. ssp. minimus
1 Lowest leaves with lamina cuneate at base or 0   2
  2 Lowest leaves persistent, with well developed
    lamina   **a. ssp. flammula**
  2 Lowest leaves caducous, with lamina 0 or much
    reduced   **c. ssp. scoticus**

  a. Ssp. flammula. Erect to procumbent, often rooting at lower nodes; lowest leaves oblong-lanceolate, cuneate, + entire; achenes 1.1-1.6x as long as wide. Range of sp.

  b. Ssp. minimus (A. Bennett) Padm. Decumbent to procumbent, rooting only at lower nodes; lowest leaves broadly ovate,

cordate, $\pm$ entire; flowers >15mm across; achenes 1-1.4x as long as wide. N Sc (mainland and N & W Isles) and Clare.
c. Ssp. scoticus (E. Marshall) Clapham. Erect; lowest leaves consisting of long petiole with reduced or no lamina, caducous; flowers usually 1, mostly 10-15mm across; achenes as in ssp. flammula. NW Sc and NW Ir.
14. R. x levenensis Druce ex Gornall (R. flammula x R. reptans) - Loch Leven Spearwort. Differs from R. flammula ssp. flammula in its procumbent stems rooting at most nodes; flowers c.5-12mm across, solitary on upturned flowering stem; achenes 1-1.5(1.9)mm plus beak c.1/5 to 1/6 as long; and see key. Native; barish lake shores on pebbly or silty substrata; local in Br from Lake District to N Aberdeen, mostly in absence of R. reptans but usually with R. flammula, with which it backcrosses.
15. R. reptans L. - Creeping Spearwort. Differs from R. **RR** flammula and R. x levenensis in its procumbent stems rooting at $\pm$ all nodes; solitary flowers c.5(-10)mm across; achenes 1-1.5mm plus beak c.1/3 to 1/4 as long; and see key. Native; same habitats and localities as R. x levenensis and giving rise to it, now extinct except for occasional non-persistent reintroductions (probably by geese).
16. R. ophioglossifolius Villars - Adder's-tongue Spearwort. **110** Erect annual to 40cm; lowest leaves ovate, cordate, glabrous, **RRR** $\pm$ entire; stem-leaves broadly to narrowly elliptic, shallowly and distantly toothed; flowers 5-9mm across; sepals not reflexed; achenes c.1.5mm, glabrous, with small tubercles on sides, with very short beak. Native; marshy ground; Gloucs, formerly Dorset, S Hants and Jersey.
17. R. aconitifolius L. - Aconite-leaved Buttercup. Erect **110** perennial to 60cm; lowest leaves deeply palmately 3-7-lobed; flowers 10-25mm across; sepals not reflexed; petals white; achenes 2-4mm, glabrous, smooth apart from raised veins, with short hooked beak. Intrd; grown in gardens, often as flore pleno, and natd in a few damp places and by streams north-wards from Yorks; Europe.

Subgenus 2 - FICARIA (Schaeffer) L. Benson (sp. 18). Petals yellow, 7-12; sepals 3; achenes not transversely ridged; many roots modified as tubers with rounded apices.

18. R. ficaria L. - Lesser Celandine. Glabrous ascending **110** perennial to 25cm; basal leaves ovate, cordate, with long petiole; flowers 10-30mm across; sepals not reflexed; achenes c.2.5mm, shortly pubescent, smooth, with very short beak. Native; in damp meadows, woods and hedgebanks, and beside streams; common throughout BI.
a. Ssp. ficaria (var. fertilis Clapham nom. nud.). Tubers **97** not formed in leaf-axils; full head of ripe achenes usually produced; diploid (2n=16); mean no. chloroplasts per stomatal guard-cell c.13-17.
b. Ssp. bulbilifer Lambinon (ssp. bulbifer Lawalree nom. **97**

illeg., var. bulbifera Marsden-Jones nom. illeg.). Tubers formed in leaf-axils after anthesis; few (rarely >6) ripe achenes produced in each head; tetraploid (2n=32); mean no. chloroplasts per stomatal guard-cell c.24-28. Other differences between the sspp. are less reliable. Ssp. ficaria tends to have larger flowers with wider petals and entire (not shallowly lobed) leaves. Both sspp. occur throughout BI, but locally one may be absent or much less common than the other. Ssp. bulbilifer is less common in Ir and W Br, but in E Br is less tolerant of open conditions. Triploids (2n=24), recognized by extremely small flowers, total sterility, usually 0 (rarely few) axillary tubers, and c.20-22 chloroplasts per stomatal guard-cell, occur widely, usually close to diploids or tetraploids or both, and are probably hybrids. The presence of other taxa in BI, e.g. ssp. **ficariiformis** (F. Schultz) Rouy & Fouc., requires elucidation.

Subgenus 3 - BATRACHIUM (DC.) A. Gray (spp. 19-31). Petals white, normally 5; sepals 5; achenes transversely ridged; roots not modified as tubers; stems with broadly lobed (laminar) normally floating leaves, very finely divided (capillary) normally submerged leaves, or both (hetero-phyllous); transitional leaves rare except in hybrids.

19. R. hederaceus L. - Ivy-leaved Crowfoot. Procumbent **110** annual or perennial; leaves all laminar, divided <1/2 way into 3-5(7) lobes widest at base; petals (1.2)2.5-3.5(4.3)mm, not contiguous at anthesis, with lunate nectar-pit; sepals not reflexed; receptacle glabrous; immature achenes glabrous. Native; on mud and in shallow water; frequent throughout BI.

20. R. omiophyllus Ten. (R. lenormandii F. Schultz) - **110** Round-leaved Crowfoot. Procumbent annual or perennial; leaves all laminar, divided <1/2 way into 3-5(7) lobes widest above base; petals (4)5-6(7)mm, with lunate nectar-pit; sepals reflexed; receptacle glabrous; immature achenes glabrous. Native; on wet mud and in shallow ponds and streams; frequent in Wa, W & S En, SW Sc and S Ir.

20 x 26. R. omiophyllus x R. peltatus = R. x hiltonii Groves & J. Groves was known in E Sussex between 1896 and 1926 in a stream with both parents. It was ± intermediate, with the lower leaves finely divided but only sub-capillary, but fertile with large pollen grains. It might have been an amphidiploid. Recently a sterile hybrid has been found in S Hants and there is an old record for E Cornwall. Endemic.

21. R. x novae-forestae S. Webster (R. lutarius auct. non (Revel) Bouvet; R.omiophyllus x R. tripartitus) - New Forest Crowfoot. Like R. tripartitus but laminar leaves often 5-lobed with shallower sinuses; petals <6mm; pedicels often erect when fruit ripe; achenes <c.60% fertile. In and by pools in New Forest, S Hants, formerly also Glam.

22. R. tripartitus DC. - Three-lobed Crowfoot. Procumbent **110**

or subaquatic annual or perennial; heterophyllous or leaves **R**
all laminar; laminar leaves divided >1/2 way into 3(-5) lobes
widest well above base; petals 1.25-4.5mm, not contiguous at
anthesis, with lunate nectar-pit; sepals reflexed, blue-
tipped; receptacle pubescent; immature achenes glabrous.
Native; wet mud, ditches and ponds; extreme S & W En and Wa
from E Sussex to Cheshire, formerly Ir (W Cork).
  22 x 25. R. tripartitus x R. aquatilis is found in W Cornwall usually near both parents. It is sterile, weak and non-
persistent, with many transitional leaves, and ± intermediate. Endemic.
  23. R. baudotii Godron - Brackish Water-crowfoot. 110
Procumbent or subaquatic annual or perennial; heterophyllous, **R**
or with capillary leaves only, or rarely with laminar leaves
only; laminar leaves divided >1/2 way into 3(-5) lobes;
capillary leaves with rigid divergent segments; petals
5.5-10mm, contiguous at anthesis, with lunate nectar-pit;
sepals reflexed, usually blue-tipped (sometimes so in other
spp.); receptacle pubescent, becoming ovoid in fruit;
immature achenes glabrous, becoming very narrowly winged at
full maturity.   Native; ditches and ponds near sea, often
brackish; scattered round coasts of BI.
  23 x 24. R. baudotii x R. trichophyllus = R. x segretii
A. Felix is found in coastal regions near the parents N to
Flints and S Lincs.   It is sterile and intermediate, but
closer to R. trichophyllus, and develops transitional
leaves.
  23 x 25. R. baudotii x R. aquatilis = R. x lambertii A.
Felix is found with both parents in W Cornwall.   It is
sterile and intermediate, with transitional leaves.
  23 x 26. R. baudotii x R. peltatus has probably been found
recently in S Devon.   It resembles R. baudotii more closely
but is sterile and has elongated nectar-pits. Endemic.
  24. R. trichophyllus Chaix - Thread-leaved Water-crowfoot.
Tufted or subaquatic annual or perennial; leaves all
capillary with flaccid or rigid divergent segments; petals
3.5-6(6.5)mm, not contiguous at anthesis, with lunate nectar-
pit; sepals not reflexed; receptacle pubescent; immature
achenes pubescent.   Native; ponds, ditches, canals and slow
rivers; scattered but often common throughout BI.
  24 x 25. R. trichophyllus x R. aquatilis = R. x lutzii
A. Felix is scattered through En N to Derbys with both
parents.   It is intermediate and sterile, with transitional
leaves.
  24 x 26. R. trichophyllus x R. peltatus is known from Warks
and perhaps elsewhere in En with both parents. It is intermediate and sterile.
  24 x 31. R. trichophyllus x R. circinatus occurs with the
parents in W Suffolk. It is sterile and ± intermediate.
  25. R. aquatilis L. - Common Water-crowfoot.   Tufted or 110
subaquatic annual or perennial; heterophyllous or with
capillary leaves only; laminar leaves divided slightly >1/2

way into (3-)5(-7) lobes, with acute basal sinus; capillary
leaves with flaccid or rigid divergent segments; petals
5-10mm, contiguous at anthesis, with circular nectar-pit;
sepals not reflexed; receptacle pubescent; immature achenes
pubescent. Native; ponds, ditches, canals and slow rivers;
frequent throughout most of BI, commonest sp. of subgenus.
   25 x 26. R. aquatilis x R. peltatus = R. x virzionensis A.
Felix has been found in Warks and there are other unconfirmed
records from En.
   26. R. peltatus Schrank - Pond Water-crowfoot. Tufted or   110
subaquatic annual or perennial; heterophyllous or with
capillary leaves only; laminar leaves divided slightly >1/2
way into (3-)5(-7) lobes, with obtuse basal sinus; capillary
leaves with flaccid or rigid divergent segments; petals
(9)12-15(20)mm, contiguous at anthesis, with pear-shaped
nectar-pit; sepals not reflexed; receptacle pubescent;
immature achenes pubescent. Native; ponds, ditches, canals
and slow rivers; frequent throughout most of BI.
   27. R. x kelchoensis S. Webster (R. peltatus x R. fluitans)
- Kelso Water-crowfoot. Intermediate between the parents but
receptacle pubescent; laminar leaves like those of R.
peltatus; transitional leaves usually produced; very robust;
sterile. Native; slow rivers; scattered in Br N to Berwicks,
especially in EC En and SE Sc, Co Antrim, often in absence of
both parents. Might have given rise to R. penicillatus var.
penicillatus.
   28. R. penicillatus (Dumort.) Bab. - Stream Water-crowfoot. 110
Subaquatic perennial, heterophyllous or with flaccid
capillary leaves only; petals (5)10-15(20)mm, contiguous at
anthesis, with pear-shaped nectar-pit; sepals not reflexed;
receptacle pubescent; immature achenes pubescent. Native;
rivers, usually swift-flowing.
   a. Ssp. penicillatus. Heterophyllous; laminar leaves like
those of R. peltatus; capillary leaves longer than adjacent
stem internode, with flaccid ± parallel segments. Ir, Wa and
W En.
   b. Ssp. pseudofluitans (Syme) S. Webster (var. calcareus
(Butcher) C. Cook, var. vertumnus C. Cook, R. pseudofluitans
(Syme) Newbould ex Baker & Foggitt, R. peltatus ssp.
pseudofluitans (Syme) C. Cook nom. inval.). Leaves all
capillary, shorter to longer than adjacent stem internode,
usually c.6-8x forked with flaccid or rigid, ± parallel or
divergent segments. En, Wa, N Ir and S Sc. Vegetatively
variable.
   R. penicillatus is often very difficult to distinguish from
R. x bachii, R. peltatus x R. fluitans, R. peltatus or R.
fluitans.
   29. R. x bachii Wirtgen (R. fluitans x R. trichophyllus & R.
fluitans x R. aquatilis; R. x bachii probably strictly refers
to former) - Wirtgen's Water-crowfoot. Very robust
perennials closely resembling R. penicillatus ssp.
pseudofluitans (i.e. with capillary leaves only), but highly

sterile. The 2 combinations are indistinguishable morphologically, but former has 2n=24, latter 2n=40. Native;
rivers, usually swift-flowing; very scattered in En, Wa and S
Sc, often replacing R. fluitans. Might have given rise to R.
penicillatus ssp. pseudofluitans.
30. R. fluitans Lam. - River Water-crowfoot. Subaquatic
perennial; leaves all capillary, rarely <8cm, rarely >4x
forked, usually longer than adjacent stem internode, with
flaccid + parallel segments; petals 7-13mm, contiguous at
anthesis, with pear-shaped nectar-pit; sepals not reflexed;
receptacle pilose to glabrous; immature achenes sparsely
pubescent to glabrous. Native; mostly in larger rivers of
moderate flow-rate; scattered in En, especially in N, Wa and
S Sc, and in Antrim, Ir; decreasing.
30 x 31. R. fluitans x R. circinatus occurs in River Black-
adder, Berwicks, with R. circinatus. It is + intermediate
but has a pubescent receptacle.
31. R. circinatus Sibth. - Fan-leaved Water-crowfoot. Subaquatic annual to perennial; leaves all capillary, with
short, rigid, divergent segments lying in 1 plane; petals
4-10(12)mm, barely contiguous at anthesis, with lunate
nectar-pit; sepals not reflexed; receptacle pubescent;
immature achenes pubescent. Native; in ponds, canals and
slow-flowing rivers, usually base-rich; sparsely scattered in
En, Wa, Ir and S Sc.

14. ADONIS L. - Pheasant's-eye
Herbaceous annual; leaves spirally arranged, pinnate,
stipulate; flowers + solitary, without a whorl of bracts
below, actinomorphic; perianth of 2 whorls; sepals 5,
sepaloid; petals 5-8, petaloid, not nectariferous; stamens
numerous; carpels numerous, free, spirally arranged; fruit an
achene, the style remaining shorter than fertile portion.

1. A. annua L. - Pheasant's-eye. Stems simple to branched,   **97**
to 40cm; leaves much divided, with narrow segments; flowers
15-25mm across; petals bright scarlet with dark basal spot;
receptacle elongating as fruits ripen. Intrd; weed of
cultivated and waste ground, formerly locally natd in cornfields in S En but now almost extinct; a rare casual in S En,
Wa and CI, formerly S Ir; S Europe.

15. MYOSURUS L. - Mousetail
Herbaceous annual; leaves all basal, simple, exstipulate;
flowers solitary, without a whorl of bracts below, actinomorphic; perianth of 2 whorls; sepals 5(-7), + sepaloid,
each with a small basal spur; petals 5(-7), in the form of
tubular nectaries; stamens 5-10; carpels numerous, free,
spirally arranged; fruit an achene, the style remaining
shorter than fertile portion.

1. M. minimus L. - Mousetail. Leaves 1-8cm, linear, in a   **97**

basal rosette; flowers solitary on bare scapes to 10cm;   **R**
petals 3-4mm, greenish, inconspicuous; receptacle becoming
much elongated, finally c.2-7cm. Probably native; damp
arable ground; En N to SE Yorks, CI, rare and declining.

16. AQUILEGIA L. - Columbines
Herbaceous perennials; leaves spirally arranged, 2-ternate,
exstipulate; flowers in cymes, without a whorl of bracts
below, actinomorphic; perianth of 2 whorls; sepals 5,
petaloid; petals 5, petaloid, each with a long, backwards-
directed nectariferous spur; stamens numerous, the inner c.10
flat and sterile; carpels 5(-10), free; fruit a follicle.

Other spp. - Some garden escapes passing as A. vulgaris
might actually represent related spp. or hybrids; study of
these is needed.

1. A. vulgaris L. - Columbine. Stems to 1m, branched,   **97**
usually with several well developed leaves and flowers;   **101**
leaves pubescent on lowerside; flowers usually blue,
sometimes white or pink to purple; sepals 15-30mm; petal-spur
15-22mm, strongly hooked at end; follicles 15-20mm. Native;
woods, fens and damp calcareous grassland and scree; local in
BI N to C Sc; also much grown and a frequent escape, becoming
natd in some areas; native populations are usually all
violet-blue-flowered.
2. A. pyrenaica DC. - Pyrenean Columbine. Stems to 25cm, **101**
often simple, usually with 0-1 well developed leaves and 1-3
flowers; leaves glabrous to sparsely pubescent on lowerside;
flowers blue; sepals 20-35mm; petal-spur 10-16mm, gently
curved; follicles 12-17mm. Intrd; planted on rock-ledges at
c.900m, Caenlochan Glen, Angus, known since 1895; Pyrenees.

17. THALICTRUM L. - Meadow-rues
Herbaceous perennials; leaves spirally arranged, pinnate to
ternate, stipulate; flowers in racemes of compound inflorescences without a whorl of bracts below, actinomorphic;
perianth of 1 whorl of 4 small but $\pm$ petaloid sepals; stamens
numerous, more conspicuous than sepals; carpels 2-15, free;
fruit an achene, the style remaining shorter than fertile
portion.

1 Inflorescence a simple raceme; plant rarely >15cm
                                             5. T. alpinum
1 Inflorescence branched with >1 flower per branch;
   plant usually >15cm   2
   2 Filaments thickened, wider than anthers, white to
      lilac or pink   **1. T. aquilegiifolium**
   2 Filaments thin, narrower than anthers, yellowish   3
3 Sepals pink to lilac, c. as long as stamens
                                            2. T. delavayi
3 Sepals yellow, much shorter than stamens   4

4 Inflorescence diffuse; stamens ± pendent;
   leaflets not or little longer than wide; achenes
   with 8-10 ribs   **4. T. minus**
4 Inflorescence ± dense; stamens held stiffly
   erect to patent; leaflets much longer than wide;
   achenes with 6 ribs   **3. T. flavum**

Other spp. - 1 plant of **T. lucidum** L., from E Europe, grows on a river-bank in Berwicks; it differs from T. flavum in its achenes with 8-10 ribs and lanceolate to linear (not obovate to oblong) leaflets of upper leaves.

1. T. aquilegiifolium L. - French Meadow-rue. Scarcely rhizomatous; stems to 1m, erect, usually simple; leaves 2-4-ternate; infloresence compound; flowers in dense clusters, whitish to lilac or pink, with erect to patent stamens with filaments coloured and wider than anthers. Intrd; grown in gardens and ± natd in grassy places as a throwout in a few places in En and Sc; Europe.

2. T. delavayi Franchet - Chinese Meadow-rue. Scarcely rhizomatous; stems to 1m, erect, branched; leaves 3-5-pinnate; inflorescence compound; flowers in ± dense clusters; sepals 6-15mm, pink to lilac; filaments yellowish, narrow. Intrd; natd in grassy area in Cambs; China.

3. T. flavum L. - Common Meadow-rue. Strongly rhizomatous; stems to 1.2m, erect, simple or little-branched; leaves 2-3-pinnate; inflorescence compound; flowers in dense clusters, bright yellow, with erect to patent stamens with narrow yellowish filaments. Native; fens, streamsides and wet meadows; scattered and declining through Br N to S Sc, mostly in E En, very scattered in Ir.

4. T. minus L. - Lesser Meadow-rue. Scarcely to moderately **97** rhizomatous; stems to 1.2m, erect or spreading, often zigzag, simple or branched; leaves 3-4-ternate to -pinnate; inflorescence compound; flowers in diffuse panicles, pale yellow, with ± pendent stamens with narrow yellowish filaments. Native; in varied, usually calcareous habitats such as dunes, limestone cliffs and pavement, grassy banks and hedgerows, scrubland, and lake-sides; scattered in Br and Ir, locally common but absent from large areas including C & SE En; grown in gardens and a frequent persister or throwout outside native range. A very variable and little understood sp.; up to 8 spp. or sspp. have been segregated in Br, based mainly on characters of fruit, habit and indumentum, but until properly investigated they are not worth recognizing.

5. T. alpinum L. - Alpine Meadow-rue. Rhizomatous; stems rarely >15cm, erect, very thin, simple; leaves 2-pinnate to -ternate; inflorescence simple; flowers well spaced, pale yellow, with dangling stamens with very thin filaments. Native; grassy and rocky places on mountains; W & C Sc, very local in N Wa, NW Ir and extreme N En; formerly MW Yorks.

30. BERBERIDACEAE - Barberry family

Shrubs with yellow wood or rarely herbaceous perennials; leaves alternate, simple to pinnate or ternate, entire to toothed, exstipulate; petioles present or 0. Flowers axillary or terminal, actinomorphic, bisexual, hypogynous, solitary or in usually racemose inflorescences; perianth usually of 4-6 whorls of 2-3 free segments each, 1-2 outer whorls sepaloid, the rest petaloid, usually yellow; stamens 4 or 6 in 2 whorls; ovary 1-celled, with 1-many ovules; style short or 0; stigma capitate; fruit a capsule or berry.

Usually distinguishable by its shrubby often spiny habit, yellow wood, perianth of several whorls, and 1-celled ovary.

1 Stems with spines; leaves simple	**1. BERBERIS**
1 Stems without spines; leaves pinnate	**2. MAHONIA**

Other genera - **EPIMEDIUM** L. differs in being herbaceous with 2-ternate leaves, perianth of 6 whorls of 2 (apparently 3 whorls of 4) segments, and the fruit a capsule. **E. alpinum** L. (Barren-wort) from S Europe is grown in gardens; it formerly escaped in woodland etc. and still occasionally persists in neglected gardens and estates. It has 1 2-ternate leaf on the flowering stem and compound glandular-pubescent inflorescences. Most modern records of garden throw-outs or relics probably refer to other taxa; notably **E. pinnatum** Fischer ssp. **colchicum** Boiss. from the Caucasus, with leafless flowering stems and simple glabrous or sparsely glandular-pubescent inflorescences; and **E. x versicolor** Morren (E. grandiflorum Morren x E. pinnatum), with leafless to biternate-leaved flowering stems, and mostly simple (often branched at base), ± glandular-pubescent inflorescences. The last taxon appears to be the commonest garden plant, often with red flowers.

1. BERBERIS L. - Barberries

Shrubs; stems with (1)3(-7)-partite spines bearing short-shoots in their axils; leaves deciduous or evergreen, simple; flowers in axillary racemes, fascicles or panicles, or solitary; perianth of 4-5 whorls each of 3 segments, various shades of yellow; stamens 6; fruit a few-seeded, red to purple-black, often bloomed berry.

1 Leaves deciduous to semi-evergreen, entire to
  serrate, the apex and teeth without spines or with
  weak spines much <1mm; fruit red   2
1 Leaves evergreen, entire to spinose-toothed, but apex
  on most or all leaves with pungent spine ≥1mm;
  fruit bluish-black   5
  2 Leaves 2.5-6cm, with >10 teeth on each side;
    flowers in pendent racemes   **1. B. vulgaris**
  2 Leaves 1-3cm, entire or with <10 teeth on each

side; flowers in 1-many-flowered fascicles or
panicles   3
3 Leaves with several teeth on each side; flowers
numerous in dense panicles   **5. B. aggregata**
3 Leaves usually entire; flowers in loose fascicles
of 1-6   4
  4 Twigs minutely pubescent; spines mostly 3-partite;
  leaves oblanceolate   **4. B. wilsoniae**
  4 Twigs glabrous; spines mostly simple; leaves
  obovate   **2. B. thunbergii**
5 Flowers in fascicles without or with very short
common peduncle   6
5 Flowers in racemes with common peduncle   9
  6 Leaves >3cm, with >3 spinose teeth on each margin   7
  6 Leaves <3cm, with <3 spinose teeth on each margin   8
7 Leaves narrowly elliptic; twigs ± terete; fruit with
nearly sessile stigma   **6. B. gagnepainii**
7 Leaves linear to narrowly lanceolate; twigs with
raised ridges; fruit with distinct style   **7. B. julianae**
  8 Leaves elliptic to obovate, with flat margins
                                             8. B. buxifolia
  8 Leaves narrowly elliptic, with revolute margins
                                           10. B. x stenophylla
9 Spines <1cm, 3-7-partite; leaves 1-3cm; flowers
orange   **9. B. darwinii**
9 Spines 1-3cm, 1-3-partite; leaves 2.5-8cm; flowers
yellow   **3. B. glaucocarpa**

Other spp. - Many spp., hybrids and cultivars are grown in
gardens and several are mass planted in parks, whence
occasional bird-sown bushes (often not coming true from seed)
on walls and in hedges etc. can originate.

1. B. vulgaris L. - <u>Barberry</u>. Deciduous shrub to 3m; spines 121
mostly 3-partite; leaves 2.5-6cm, elliptic to obovate, with
numerous small subspinose teeth; flowers in pendent racemes
30-50mm, yellow; fruit red.   Probably intrd; long natd in
hedges and rough ground throughout most of BI, but very
scattered; Europe.
  2. B. thunbergii DC. - <u>Thunberg's Barberry</u>. Deciduous shrub 121
to 2m; spines mostly simple; leaves 1-3cm, obovate, entire,
green to purple; flowers in short sparse fascicles to 20mm,
yellow suffused with red; fruit bright red.   Intrd; very
commonly mass planted in parks and by roads; occasionally
bird-sown in En; Japan.   A wide range of cultivars is grown
in parks etc.
  3. B. glaucocarpa Stapf - <u>Great Barberry</u>.   Semi-evergreen 121
shrub to 3m; spines mostly 3-partite; leaves 2.5-6(8)cm,
oblanceolate to obovate, entire to sparsely spinose-toothed;
flowers in stiff but often pendent racemes 20-70mm, yellow;
fruit bluish-black, with dense white bloom.   Intrd; planted
for hedging and ornament; long natd in hedges in S Somerset

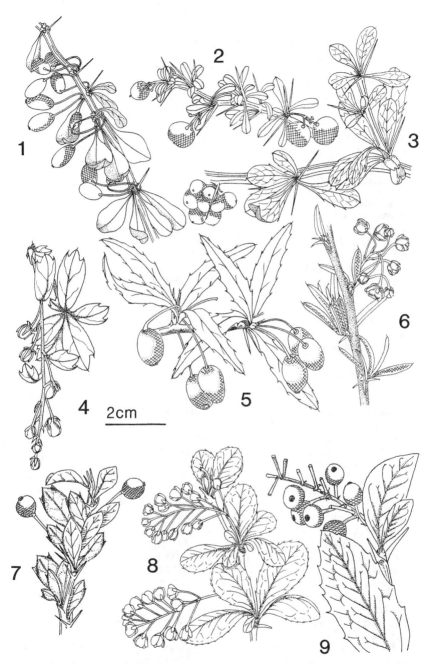

FIG 121 – Shoots of **Berberis**. 1, **B. thunbergii**. 2, **B. wilsoniae**. 3, **B. aggregata**. 4, **B. darwinii**. 5, **B. gagnepainii**. 6, **B. x stenophylla**. 7, **B. buxifolia**. 8, **B. vulgaris**. 9, **B. glaucocarpa**.

and occasionally bird-sown elsewhere in S En; W Himalayas.

4. B. wilsoniae Hemsley - <u>Mrs Wilson's Barberry</u>. Semi- **121**
evergreen shrub to 1m; spines mostly 3-partite; leaves
1-2.5cm, oblanceolate, usually entire; flowers in short
sparse fascicles to 10mm, yellow; fruit pinkish red. Intrd;
commonly cultivated; sometimes bird-sown in a range of
habitats in En, Wa and Ir; W China. Plants differing in
having some leaves with a few spinose teeth and more densely
fascicled or racemose inflorescences are commonly cultivated
and may become natd; they are probably <u>B. aggregata</u> x <u>B.
wilsoniae</u>.

5. B. aggregata C. Schneider - <u>Clustered Barberry</u>. **121**
Deciduous shrub to 2m; spines 3-partite; leaves 1-2.5cm,
oblong to obovate, with rather sparse spinose teeth; flowers
numerous in dense stiff panicles 10-35mm, yellow; fruit pale
red. Intrd; frequently cultivated; natd on chalk and walls
and in hedges; scattered in Br N to S Ebudes; W China.

6. B. gagnepainii C. Schneider - <u>Gagnepain's Barberry</u>. **121**
Evergreen shrub to 2m; spines 3-partite; leaves 3-11cm,
linear to narrowly lanceolate, with numerous spinose teeth;
flowers 3-9 in rather stiff fascicles <3cm, yellow; fruit
bluish-black with conspicuous bloom. Intrd; frequently
planted in gardens, parks and roadsides; natd on walls,
hedges and riverbanks; scattered in Br N to MW Yorks; W
China.

7. B. julianae C. Schneider - <u>Chinese Barberry</u>. Evergreen
shrub to 2m; differs from <u>B. gagnepainii</u> in leaves 3-6cm,
shiny (not matt or nearly so) on upperside; ovules 1-2 (not
(3)4-15); and see key. Intrd; frequently planted in parks
and gardens; natd on walls and roadsides in W Kent and
Surrey; C China.

8. B. buxifolia Lam. - <u>Box-leaved Barberry</u>. Evergreen shrub **121**
to 2m; spines (1)3(-5)-partite; leaves 1-2.5cm, dull,
elliptic to obovate, entire or with 1-few spinose on each
side near apex; flowers 1-2 on pedicels to 20mm, orange-
yellow; fruit dark purple. Intrd; frequently cultivated;
natd on commonland in W Norfolk, occasionally elsewhere in En
and Sc; S America.

9. B. darwinii Hook. - <u>Darwin's Barberry</u>. Evergreen shrub **121**
to 3m; spines 3-7-partite; leaves 1-3cm, glossy on upperside,
obovate or less often oblong or oblanceolate, usually with 1
spinose tooth on each side near apex and often more further
back; flowers in pendent racemes 35-60mm, orange; fruit
bluish-purple. Intrd; commonly cultivated and often mass
planted; occasionally bird-sown in scattered places throughout BI; Chile and Argentina.

10. B. x stenophylla Lindley (<u>B. darwinii</u> x <u>B. empetrifolia</u> **121**
Lam.) - <u>Hedge Barberry</u>. Evergreen shrub to 3m; spines mostly
3-partite; leaves 1.5-2.5cm, narrowly elliptic with revolute
margins, scarcely glossy, entire, with spinose tip; flowers
in loose fascicles or short racemes to 30mm, golden yellow;
fruit bluish-black.   Arose c.1860 near Sheffield; now

abundantly mass planted in parks and by roads and as a hedge; occasionally self-sown in Br and Ir but often coming up closer in characters to B. darwinii.

2. MAHONIA Nutt. - Oregon-grapes

Shrubs; stems without spines; leaves evergreen, pinnate; flowers in axillary fasciculate racemes, yellow, structure as in Berberis; fruit a blue-black, bloomed berry with few seeds.

Other spp. - Some natd Mahonia determined as M. aquifolium might be **M. x wagneri** (Jouin) Rehder (M. aquifolium x M. pinnata (Lagasca) Fedde), with 7-13 leaflets >2x as long as wide.

1. **M. aquifolium** (Pursh) Nutt. - Oregon-grape. Somewhat stoloniferous shrub with ascending stems to 1.5m; leaflets 5-9, 3-8cm, c.2x as long as wide, glossy on upperside, not papillose on lowerside, with c.5-15 spinulose teeth on each side. Intrd; commonly cultivated and often mass planted for ground and game cover; natd in scrub, woodland and hedges etc. throughout Br N to C Sc; W N America.

2. **M. x decumbens** Stace (M. aquifolium x M. repens (Lindley) Don - Newmarket Oregon-grape. Sprawling, strongly stoloniferous shrub to 50cm, with decumbent stems; leaflets 5-7, <2x as long as wide, dull on upperside, minutely papillose on lowerside, with c.8-22 spinulose teeth on each side. Intrd; frequently grown for ground cover and natd in woodland in Wilts, Cambs and Man. M. repens and hybrid occur within the range of M. aquifolium in N America, but crossing also occurs in cultivation. M. repens might also occur wild in Br.

31. PAPAVERACEAE - Poppy family

Herbaceous annuals or perennials usually with white or yellow latex; leaves spiral, shallowly pinnately lobed to 2-pinnate, exstipulate; petioles present or 0. Flowers terminal or axillary, actinomorphic, bisexual, hypogynous or rarely perigynous, solitary or in umbellate or paniculate inflorescences; sepals 2(-3), free or fused, normally caducous; petals 4(-6, or 0), free, often large and brightly coloured and crumpled at first, normally caducous; stamens numerous; ovary of 2-many fused carpels, 1(-2)-celled, with (2-)many ovules; style 1, short or 0; stigma usually large and ± capitate or peltate, lobed or rayed or ± divided; fruit a capsule, dehiscent or not.

The 2 sepals, 4 showy petals, and distinctive latex usually distinguish this family, but Macleaya lacks petals.

1 Petals 0; flowers numerous in large panicles
                                               7. MACLEAYA

1 Petals conspicuous; flowers solitary or in <10-
   flowered inflorescences   2
   2 Sap watery; sepals fused, shed as a hood as flower
      opens; receptacle raised above base of ovary
      (flowers perigynous)   **6. ESCHSCHOLZIA**
   2 Sap a white to orange latex; sepals often adherent
      but not fused; flowers hypogynous   3
3 Capsule >10x as long as wide; stigma with 2 lobes   4
3 Capsule <6x as long as wide; stigma with >3 lobes or
   rays   5
   4 Flowers >3cm across, 1-2 per leaf-axil; capsule
      >10cm, 2-celled   **4. GLAUCIUM**
   4 Flowers <3cm across, mostly in umbels of >3;
      capsule <6cm, 1-celled   **5. CHELIDONIUM**
5 Lobes and teeth of leaves each ending in a long
   weak spine   **3. ARGEMONE**
5 Leaves not spiny   6
   6 Petals white to red or mauve; style absent; stigma
      a 4-20-rayed disk   **1. PAPAVER**
   6 Petals yellow; style present; stigma 4-6-lobed
                                                      2. MECONOPSIS

Other genera - A single sp. of **ROEMERIA** Medikus, **R. hybrida**
(L.) DC. (Violet Horned-poppy), from Europe, with violet
petals and a long linear capsule, was formerly natd in E
Anglia but is now apparently extinct even as a casual.

1. PAPAVER L. - Poppies
Annuals or perennials with white latex unless otherwise
stated; leaves glabrous to pubescent; flowers solitary;
petals red or mauve to white; capsule 1-celled, with 4-20
incomplete placentae projecting inwards, opening by pores
just below the persistent stigma; stigma sessile, a 4-20-
rayed flat disk.

1 Tufted perennials with basal rosette of leaves at
   flowering   2
1 Annuals usually without basal leaves at flowering   3
   2 Petals <45mm; capsule narrowly obovoid to clavate,
      widest immediately below stigma   **2. P. atlanticum**
   2 Petals >45mm; capsule obovoid, widest c.3/4 from
      base to apex   **1. P. orientale**
3 Stem-leaves strongly glaucous, clasping stem at base,
   toothed or lobed but not to near midrib
                                                   3. P. somniferum
3 Stem-leaves green to slightly glaucous, not clasping
   stem, mostly divided as far as midrib   4
   4 Capsule with long stiff hairs; filaments dilated
      distally   5
   4 Capsule glabrous; filaments not dilated   6
5 Capsule <1.5cm, c. as long as wide   **6. P. hybridum**

5 Capsule >1.5cm, at least some >2x as long as wide
                                                    7. P. argemone
   6 Capsule <2x as long as wide   **4. P. rhoeas**
   6 Capsule >2x as long as wide   **5. P. dubium**

Other spp. - Records of **P. lateritium** K. Koch (Armenian
Poppy), from Armenia, were errors for **P. atlanticum**.

1. P. orientale L. - Oriental Poppy. Tufted perennial to **128**
1m; pedicel with conspicuous appressed hairs; sepals usually
3; petals usually 6, 45-80mm, pale pink to orange-red, often
with dark basal blotch; anthers violet; capsule <40mm,
subglobose to obovoid, glabrous; stigma 7-16(20)-rayed, c. as
wide as capsule. Intrd; common in gardens; natd in various
habitats on well-drained soils in En and Sc; SW Asia. Some
natd plants probably belong to **P. bracteatum** Lindley, which
differs in the presence of leaf-like bracts immediately under
the flower, pedicels 3-4mm (not 2-3mm) wide, and petals with
the dark blotch longer than wide (not wider than long or 0).
The 2 spp. are sympatric and intermediates are common; they
might be conspecific.
2. P. atlanticum (Ball) Cosson - Atlas Poppy. Tufted **128**
perennial to 60cm; pedicel with appressed to erecto-patent
hairs; petals 20-40mm, orange-red; anthers yellow; capsule
<25mm, narrowly obovoid to clavate, glabrous; stigma
(5-)6(-8)-rayed, c. as wide as capsule. Intrd; grown in
gardens and natd on walls, roadsides and rough ground
sparsely throughout Br; Morocco.
3. P. somniferum L. - Opium Poppy. Erect strongly glaucous **128**
annual to 50(100)cm; pedicel glabrous to sparsely bristly;
petals 25-50mm, white to deep mauve, sometimes red or
variegated; anthers yellowish or dark purplish; capsule
<90mm, globose to obovoid, glabrous; stigma 5-12(18)-rayed,
c. as wide as capsule.
   a. Ssp. somniferum (ssp. hortense (Hussenot) Syme).
Glabrous or very sparsely bristly; leaves rather shallowly
lobed; stigma-rays never overlapping; very variable,
especially in petal colour, size and shape and capsule size
and shape. Intrd; cultivated in many parts of world for
latex (opium), seeds and ornament; common casual over most of
BI, grown for ornament and on small scale for seed (Blue
Poppy).
   b. Ssp. setigerum (DC.) Arcang. (P. setigerum DC.). Stems,
leaves and sepals conspicuously bristly; leaves rather deeply
lobed; capsule relatively narrow, with overlapping stigma-
rays. Intrd; rare casual on tips etc. in Br; S Europe.
Recent work suggests this might be a distinct sp.
4. P. rhoeas L. (P. strigosum (Boenn.) Schur, P. commutatum **128**
Fischer & Meyer) - Common Poppy. Erect annual to 60(80)cm;
latex usually white, sometimes yellow; pedicels usually with
patent hairs but sometimes with appressed hairs distally;
petals 30-45mm, usually bright scarlet, often with black

blotch at base, sometimes white, pink, mauvish or variegated; anthers bluish-black; capsule <20mm, obovoid to subglobose, glabrous; stigma (5)8-12(18)-rayed, at least as wide as capsule. Native; arable ground, roadsides and waste places; throughout BI and often common, but rare in most of N and W; grown in gardens (Shirley Poppy) and often escaping.

4 x 5. P. rhoeas x P. dubium = P. x hungaricum Borbas (P. x expectatum Fedde) is often recorded but usually in error for abnormal plants of P. rhoeas. There are a few scattered records of the hybrid from S En; it is intermediate in capsule shape and, like some plants of P. rhoeas, sterile.

5. P. dubium L. - Long-headed Poppy. Erect annual to 60cm; 128 pedicels with appressed hairs distally; petals 15-35mm, pink to red (usually paler than in P. rhoeas), sometimes with dark blotch at base; anthers bluish-black; capsule <25mm, narrowly obovoid to clavate, glabrous; stigma (4)7-9(12)-rayed, slightly less wide than capsule; seeds bluish-black. Probably native; similar places to P. rhoeas.

a. Ssp. dubium. Latex white or cream, becoming brown to black when dry; upper leaves with ultimate lobes often >1.5mm wide; anthers brownish- to bluish-black. Most of BI, rarer than P. rhoeas in S & E but more widespread in W & N.

b. Ssp. lecoqii (Lamotte) Syme (P. lecoqii Lamotte). Latex yellow or quickly turning yellow on exposure to air, becoming reddish when dry; upper leaves with ultimate lobes rarely >1.5mm wide; anthers often yellow. Very sparsely scattered in En, Wa and Ir, absent from most of Sc, frequent only on chalk in S En.

6. P. hybridum L. - Rough Poppy. Erect annual to 50cm; 128 pedicels with appressed hairs distally; petals 10-25mm,   R crimson, with dark blotch at base; anthers blue; capsule <15mm, subglobose to broadly obovoid or ellipsoidal, densely appressed-bristly; stigma 4-8-rayed, much less wide than capsule. Probably native; arable fields and waste places; much less common than formerly and now rare and ± confined to E & S En and CI.

7. P. argemone L. - Prickly Poppy. Erect or ascending 128 annual to 45cm; pedicels with appressed hairs; petals 15-   R 25mm, pale scarlet, sometimes with dark blotch at base; anthers blue; capsule <25mm, narrowly obovoid, sparsely ± appressed-bristly; stigma 4-6-rayed, less wide than capsule. Probably native; arable fields and waste places on light soils; much less common than formerly and now rather rare and largely confined to En, Wa and CI.

2. MECONOPSIS Viguier - Welsh Poppy
Perennials with yellow latex; leaves nearly glabrous; flowers solitary; petals yellow; capsule 1-celled, with 4-6 incomplete placentae projecting inwards, opening by elongated pores at apex; style short but distinct; stigma ± capitate, 4-6-lobed.

1. M. cambrica (L.) Viguier - <u>Welsh</u> <u>Poppy</u>. Stems to 60cm, **128**
erect, very sparsely pubescent; leaves pinnate with pinnately   **R**
lobed leaflets; flowers 50-80mm across; anthers yellow;
capsule 20-40mm, narrowly obovoid. Native; shady places
among rocks or under trees especially in hilly country; Wa,
SW En and scattered parts of Ir, but also grown in gardens
and extensively natd mainly in N En and Sc.

3. ARGEMONE L. - <u>Mexican Poppy</u>
Annuals with yellow latex; leaves with a weak spine at tip of
lobes and teeth; flowers solitary, with usually 2 leaf-like
bracts just below; petals 4-6, yellow; capsule 1-celled, with
4-6 incomplete placentae projecting inwards, opening by 4-6
elongated pores at apex; style short but distinct; stigma ±
capitate, 4-6-lobed.

1. A. mexicana L. - <u>Mexican Poppy</u>. Stems much branched, **128**
weakly spiny, spreading, to 90cm but usually much less;
leaves deeply pinnately lobed; flowers 40-60mm across;
capsule 25-45mm, spinose, ellipsoidal. Intrd; wool-alien,
grain-alien and cultivated plant escaping as a casual; very
scattered in S & C En; C America.

4. GLAUCIUM Miller - <u>Horned-poppies</u>
Annuals to perennials with yellow latex; leaves glaucous, the
lower pubescent; flowers solitary; petals yellow or red;
capsule 2-celled, opening from above ± along its length by
2 valves and leaving the seeds embedded in septum; style ±
0; stigma ± capitate, 2-lobed.

Other spp. - **G. corniculatum** (L.) Rudolph (<u>Red</u> <u>Horned-</u>
<u>poppy</u>), an annual from S Europe, has red petals and pubescent
stems; it formerly occurred as a casual and was occasionally
natd.

1. G. flavum Crantz - <u>Yellow</u> <u>Horned-poppy</u>. Biennial to **128**
perennial; stems glabrous, much branched, spreading, to 90cm;
lower leaves deeply pinnately lobed, the upper shallowly
lobed, much less hairy and clasping stem at base; flowers
6-9cm across; petals yellow; capsule 15-30cm, linear,
glabrous. Native; on maritime shingle and less often other
substrata; common on coasts of En, Wa and CI, local in Ir and
very local in S Sc.

5. CHELIDONIUM L. - <u>Greater Celandine</u>
Perennials with orange latex; leaves nearly glabrous; flowers
in umbels of (2-)3-6; petals yellow; capsule 1-celled,
opening from below along whole length by 2 valves; style
short but distinct; stigma ± capitate, 2-lobed.

1. C. majus L. - <u>Greater</u> <u>Celandine</u>. Stems sparsely **128**
pubescent, ± spreading, to 90cm; leaves pinnate with broad,

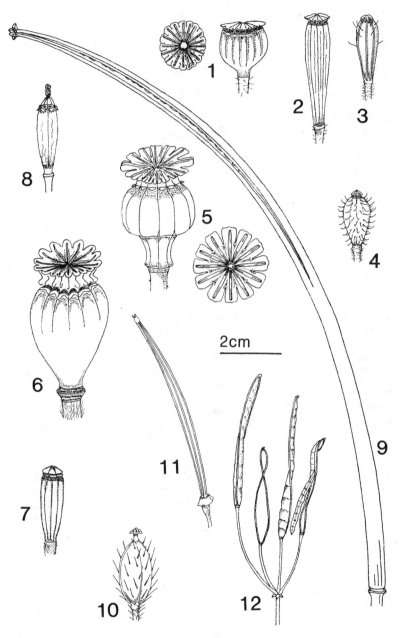

FIG 128 – Fruits of **Papaveraceae**. 1, **Papaver rhoeas**. 2, **P. dubium**. 3, **P. argemone**. 4, **P. hybridum**. 5, **P. somniferum**. 6, **P. orientale**. 7, **P. atlanticum**. 8, **Meconopsis**. 9, **Glaucium**. 10, **Argemone**. 11, **Eschscholzia**. 12, **Chelidonium**.

lobed leaflets; flowers 15-25mm across; capsule 3-5cm, linear, glabrous. Possibly native; hedgerows, walls and other marginal habitats, often near habitation (formerly cultivated); throughout BI but rare in N Sc.

6. ESCHSCHOLZIA Cham. - Californian Poppy
Annuals to perennials with watery sap; leaves glabrous, glaucous; flowers solitary; petals yellow to orange; capsule 1-celled, opening from below along its whole length by 2 valves; style very short; stigma deeply 4-6-lobed.

1. E. californica Cham.- Californian Poppy. Rarely perenn- 128 ating in BI; stems glabrous, little branched, to 60cm, erect to spreading; leaves compoundly pinnate with linear leaflets; flowers 2-12cm across; capsule 7-10cm, linear, glabrous. Intrd; commonly grown as summer bedding plant and often found as casual on tips and roadsides; natd and perennating on dunes, walls and cliff-tops in Guernsey, and in quarries and by railways in Kent; SW N America.

7. MACLEAYA R. Br. - Plume-poppies
Perennials with orange latex; leaves glabrous or sparsely pubescent, white-glaucous on lowerside; flowers small, in crowded, large, terminal panicles; petals 0; capsule 1-celled, opening from above by 2 valves; style very short; stigma deeply 2-lobed.
Perhaps not distinct from **Bocconia** L., which has priority.

1. M. x kewensis Turrill (M. cordata (Willd.) R. Br. x M. microcarpa (Maxim.) Fedde) - Hybrid Plume-poppy. Rhizomatous; stems sparsely pubescent to glabrous, glaucous, erect, to 2.5m; leaves petiolate, cordate at base, pinnately lobed; stamens 12-18, pinkish-buff; ovules 2-4; capsule not developed. Intrd; grown in gardens and found as escape and throwout, long persisting if undisturbed, in En, Ir and S Sc; garden origin from E Asian parents.

32. FUMARIACEAE - Fumitory family

Herbaceous annuals or perennials with watery sap; leaves spiral or all basal, pinnate or ternate, exstipulate; petioles present. Flowers in simple or compound racemes, zygomorphic, bisexual, hypogynous; sepals 2, free, small, caducous; petals 4, free or adherent, white to pink or yellow, upper or upper and lower with basal spur; stamens 2, tripartite; ovary 1-celled, with 1-many ovules; style 1; stigma ± 2-lobed; fruit an achene or capsule.
The distinctive flowers are unique, but the family is linked to Papaveraceae by intermediates and is probably best united with it.

1 Flowers with 2 spurred petals (the upper and lower)
                                                    1. DICENTRA
1 Flowers with 1 spurred petal (the upper)   2
   2 Fruit a 1-seeded achene   **5. FUMARIA**
   2 Fruit a dehiscent capsule, usually with >1 seed   3
3 Annual; leaves with tendrils; flowers <8mm
                                                    4. CERATOCAPNOS
3 Perennial,; leaves without tendrils; flowers >8mm   4
   4 Stems branched, arising from + cylindrical stock;
      flowers cream to yellow   **3. PSEUDOFUMARIA**
   4 Stems simple, arising from + globose tuber; flowers
      white to purple   **2. CORYDALIS**

1. DICENTRA Bernh. - <u>Bleeding-hearts</u>
Perennials, with short branched rhizome; aerial stems with
terminal inflorescence; leaves all or mostly basal; flowers
pink; upper and lower petals spurred at base; fruit a
capsule; seeds several, with an aril.

Other spp. - **D. spectabilis** (L.) Lemaire, from E Asia, has
been recorded either in error for <u>D. formosa</u> or as a very
rare and impermanent casual; it has simple racemes and leaves
on all the stems.

1. D. formosa (Andrews) Walp. - <u>Bleeding-heart</u>. Leaves all
basal, 2-4-ternate; flowering stems to 30cm, leafless;
flowers 15-20mm, pink, in rather dense compound racemes.
Intrd; commonly grown in gardens; natd in shady places
especially by streams mainly in W & N Br; W N America. Some
natd <u>Dicentra</u> might be **D. eximia** (Ker Gawler) Torrey (<u>Turkey-</u>
<u>corn</u>), from E N America, or <u>D. formosa</u> x <u>D. eximia</u>. The
latter sp. differs in its corolla with a more narrowed upper
part, with more prominent inner petals, and separating to
below the middle. The hybrid can be triploid and sterile, or
diploid and fertile.

2. CORYDALIS DC. - <u>Corydalises</u>
Perennials; stems usually simple; inflorescence 1, terminal;
leaves present on aerial stem; flowers white to purple or
yellow; upper petal spurred at base; fruit a capsule; seeds
several, with aril.

1 Flowers yellow; leaves 2-4-pinnate, fern-like;
   subterranean tuber 0   **3. C. cheilanthifolia**
1 Flowers white to purple; leaves 2-3-ternate, not
   fern-like; stems arising from subterranean tuber   2
   2 Bracts narrowly lobed; stem with large scale just
      below lowest leaf   **1. C. solida**
   2 Bracts + entire; stem without large scale
                                             2. C. cava

1. **C. solida** (L.) Clairv. (<u>C. bulbosa</u> (L.) DC., nom. illeg.) - <u>Bird-in-a-bush</u>. Stems ± erect, to 20cm, arising from ± globose solid tuber; flowers 15-25(30)mm incl. basal spur c. 1/2 total length. Intrd; grown in gardens and natd in woods and hedges; scattered over En and Wa; Europe. A variant with entire bracts occurs but is not recorded from BI.

2. **C. cava** (L.) Schweigger & Koerte (<u>C. bulbosa</u> auct. non (L.) DC.) - <u>Hollow-root</u>. Differs from <u>C. solida</u> in hollow tuber; and see key. Intrd; in similar places to <u>C. solida</u>; En and Wa but much less common; Europe.

3. **C. cheilanthifolia** Hemsley - <u>Fern-leaved Corydalis</u>. Stems erect, to 25cm, usually shorter than basal leaves; flowers 10-16mm incl. basal spur c.1/3 total length. Intrd; grown in gardens, natd on walls; Surrey; China.

3. PSEUDOFUMARIA Medikus - <u>Corydalises</u>
Perennials, arising from ± cylindrical stock; stems branched; inflorescences several, terminal and leaf-opposed; flowers cream to yellow; upper petal spurred at base; fruit a capsule; seeds several, with aril.

1. **P. lutea** (L.) Borkh. (<u>Corydalis lutea</u> (L.) DC.) - <u>Yellow Corydalis</u>. Stems erect to spreading or hanging, to 30cm; leaves 2-3-pinnate or -ternate; petioles ridged but not winged; flowers 12-18mm including basal spur 2-4mm, yellow; seeds shiny. Intrd; commonly cultivated and natd on walls and in stony places over most of BI; S Alps.

2. **P. alba** (Miller) Liden (<u>Corydalis ochroleuca</u> Koch, <u>P. ochroleuca</u> (Koch) Holub) - <u>Pale Corydalis</u>. Differs from <u>P. lutea</u> in its very narrowly winged petioles; flowers 10-18mm including basal spur 1-2mm, cream with yellow tip; seeds matt. Intrd; similar places to <u>P. lutea</u> but much rarer and only in S En; CS Europe.

4. CERATOCAPNOS Durieu - <u>Climbing Corydalis</u>
Annuals; stems branched, climbing by means of leaf-tendrils; inflorescences several, leaf-opposed; flowers pale cream; upper petal spurred at base; fruit a capsule; seeds (1)2-3, with aril.

1. **C. claviculata** (L.) Liden (<u>Corydalis claviculata</u> (L.) DC.) - <u>Climbing Corydalis</u>. Stems to 75cm, scrambling; leaves 2-pinnate to -ternate, ending in branched tendril; flowers 4-6mm including basal spur to 1mm. Native; woods and other shady places often on rocks; scattered over most of BI.

5. FUMARIA L. - <u>Fumitories</u>
Annuals to 1m but often much less; stems much-branched, scrambling, thin; leaves all cauline, 2-4-pinnate; inflor-escences leaf-opposed racemes; flowers white to purple; upper petal spurred at base; upper and lateral petals darker coloured at tip; fruit a 1-seeded achene; seeds without aril.

All spp. are similar in appearance and are distinguished by
inflorescence, flower and fruit characters. The upper petal
has a dorsal ridge and lateral margins which may be bent upwards to ± conceal the ridge or spread laterally to reveal
it. The lower petal appears parallel-sided to strongly
spathulate in top or bottom view according to the relative
expansion of the margins distally. Flower colours given
ignore the very dark petal tips. Flower length is measured
from end of basal spur to tip of longest petal on fresh
material; dried material has smaller floral parts (especially
sepals). Fruit shape and measurements refer to those seen in
widest profile on mature, dried fruits. It is essential to
base determinations on well-grown, non-shaded material in
early or mid flowering season. Late or shade-grown plants
may be very atypical, with short (often cleistogamous) paler
petals, longer narrower sepals, relatively long bracts and
less or more recurved fruiting pedicels.

1 Flowers ≥9mm; sepals (2)3–6.5mm; lower petal
   usually ± parallel-sided, rarely subspathulate   2
1 Flowers ≤9mm; sepals <4mm; lower petal distinctly
   spathulate   9
   2 Lower petal with broad margins; fruit 2.75–3 x
      2.75–3mm; flowers 12–14mm   **2. F. occidentalis**
   2 Lower petal with narrow margins; fruit ≤2.75 x
      2.5mm; flowers usually <12mm   3
3 Fruiting pedicels rigidly recurved to patent; sepals
   4–6(6.5)mm   4
3 Fruiting pedicels erecto-patent, often not rigid;
   sepals (2)3–5mm   5
   4 Petals usually creamy-white, sometimes pink to
      red; upper petal with spreading margins not
      concealing dorsal ridge   **1. F. capreolata**
   4 Petals purple; upper petal with erect margins ±
      concealing tip of dorsal ridge   **6. F. purpurea**
5 Lower petal subspathulate   6
5 Lower petal ± parallel-sided   7
   6 Flowers 9–10mm   **5. F. muralis**
   6 Flowers 10–11mm   **F. x painteri**
7 Sepals 2–3 x 1–2mm, with sharp forward-pointing teeth
   all round margin; fruit distinctly rugose when dry
                                               3. F. bastardii
7 Sepals 3–5 x 1.5–3mm, with outward-pointing teeth in
   basal part of margin only or subentire; fruit smooth
   to rugulose when dry   8
   8 Flowers 9–11(12)mm; sepals usually dentate near
      base; racemes c. as long as peduncles **5. F. muralis**
   8 Flowers (10)11–13mm; sepals subentire to
      denticulate near base; racemes distinctly longer
      than peduncles   **4. F. reuteri**
9 Flowers ≥6mm; sepals ≥1.5mm   10
9 Flowers 5–6mm; sepals 0.5–1.5mm   13

10 Bracts at least as long as pedicels, often longer
                                                    8. F. densiflora
10 Bracts shorter than pedicels   11
11 Fruits truncate to retuse at apex, distinctly wider
   than long   **7. F. officinalis**
11 Fruits rounded to subacute at apex, narrower than to
   as wide as long   12
12 Sepals 2-3(3.5) x 1-2mm, with sharp forward-
   pointing teeth all round margin; fruit distinctly
   rugose when dry   **3. F. bastardii**
12 Sepals (2.7)3-5 x 1.5-3mm, with outward-pointing
   teeth in basal part of margin only; fruit smooth
   to rugulose when dry   **5. F. muralis**
13 Corolla white to pale pink; bracts at least as long
   as fruiting pedicels   **9. F. parviflora**
13 Corolla pink; bracts shorter than fruiting pedicels
                                                    10. F. vaillantii

1. F. capreolata L. - White Ramping-fumitory. Flowers 10-   R
13(14)mm, usually creamy-white suffused with pink, rarely
reddish; upper petal with narrow spreading margins; sepals
4-6 x (2)2.5-3(4)mm, dentate mostly near base; fruit 2-2.5 x
2.5mm, orbicular, ± smooth when dry. Native; arable and
waste ground and hedges. The 2 sspp. are rather doubtfully
distinct.
a. Ssp. capreolata. Bracts usually shorter than fruiting   **134**
pedicels; fruit c.2 x 2mm, with rounded to very obtuse apex.
CI; the Continental ssp.
b. Ssp. babingtonii (Pugsley) Sell. Bracts ± equalling
fruiting pedicels; fruit c.2.5 x 2.5mm, with truncate apex.
Scattered over much of Br and Ir, mainly near coasts, but
absent from most of Sc and E En; endemic.
2. F. occidentalis Pugsley - Western Ramping-fumitory.   **134**
Flowers 12-14mm, whitish at first, later pink; upper and   **RR**
lower petals with broad margins; sepals 4-5.5 x 2-3(3.5)mm,
dentate mostly near base; fruit 2.75-3 x 2.75-3mm, orbicular,
minutely retuse at apex, rugose when dry. Native; arable
land and waste places; N Hants, Cornwall and Scillies;
endemic.
3. F. bastardii Boreau - Tall Ramping-fumitory. Flowers   **134**
(8)9-11(12)mm, very pale pink to pink, in raceme longer than   **R**
its peduncle; sepals 2-3(3.5) x 1-2mm, serrate all round
margin; fruit 2-2.4 x 2-2.4mm, broadly ovate, with truncate
base much wider than pedicel apex when dried, obtuse at apex,
rugose when dry. Native; arable and waste ground and hedgebanks; scattered mainly over W part of BI and often common in
Ir, W Wa and SW En. Sometimes (especially var. **hibernica**
Pugsley) appears to merge into F. muralis.
3 x 5. F. bastardii x F. muralis was once recorded as a
sterile plant in Guernsey (CI), but in view of the frequent
cohabitation and the closeness of the parents it might be
overlooked.

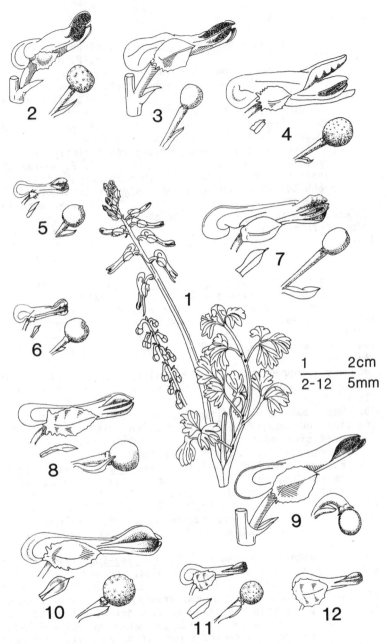

FIG 134 – **Fumaria**. 1, flowering node and fruiting raceme of **F. capreolata**. 2–12, flowers, bracts and fruits. 2, **F. officinalis** ssp. **officinalis**. 3, **F. muralis** ssp. **boraei**. 4, **F. bastardii**. 5, **F. parviflora**. 6, **F. vaillantii**. 7, **F. reuteri**. 8, **F. purpurea**. 9, **F. capreolata** ssp. **capreolata**. 10, **F. occidentalis**. 11–12, **F. densiflora**.

4. F. reuteri Boiss. (F. martinii Clavaud). - Martin's **134**
Ramping-fumitory. Flowers (10)11-13mm, pink, in raceme much **RRR**
longer than its peduncle; sepals 3-5 x 1.5-2.5(3)mm, +
entire; fruit 2.2-2.7 x 1.75-2.5mm, broadly elliptic, obtuse
at apex, + smooth when dry. Native; cultivated ground; very
rare in S En and Guernsey (seen recently only in W Cornwall
and Wight). Our plant is ssp. **martinii** (Clavaud) A. Soler,
perhaps not even subspecifically distinct.
5. F. muralis Sonder ex Koch - Common Ramping-fumitory. **134**
Flowers (8)9-11(12)mm, pink, in raceme c. as long as its
peduncle; sepals (2.7)3-5 x 1.5-3mm, dentate mostly near
base; fruit (1.75)2-2.5 x c.2mm, orbicular to broadly obovate
or ovate, with truncate base c. as wide as pedicel apex when
dried, smooth to rugulose when dry. Native; arable and waste
ground and hedge-banks; scattered over most of BI and the
commonest Fumaria in Ir, Wa, W En and CI. Very variable.
Traditionally 3 sspp. have been recognized: ssp. **boraei**
(Jordan) Pugsley (F. boraei Jordan), the common taxon; ssp.
muralis, very scattered in S & W Br but not recorded for many
years; and ssp. **neglecta** Pugsley, recorded intermittently
from a few fields in E & W Cornwall. They differ in minor
characters of plant and inflorescence robustness, corolla
size and colour, sepal size and serration, and fruit size and
shape, but several variants are difficult to place and many
others occur on the Continent, so that ssp. recognition is
not feasible. The large no. of variants makes separation
from F. bastardii often very difficult.
5 x 7. F. muralis x F. officinalis = F. x painteri Pugsley
was collected as a fertile plant in 1896 and 1907 in two
places in Salop, but not seen since; sterile hybrids have
been recorded from a few scattered places in Wa, S En and
Guernsey (CI), but none recently; endemic and enigmatic. It
has flowers 10-11mm in raceme longer than its peduncle;
sepals 3-3.5 x c.1.5mm, serrate to dentate mainly at base;
fruit not formed or c.2.5 x 2.5mm, squarish-orbicular, with
truncate to subemarginate apex, rugulose when dry.
6. F. purpurea Pugsley - Purple Ramping-fumitory. Flowers **134**
10-13mm, pinkish-purple; upper petal with narrow erect **R**
margins; sepals (4.5)5-6.5 x 2-3mm, + entire to denticulate;
fruit c.2.5 x 2.5mm, squarish-orbicular with truncate apex, +
smooth or rugulose when dry. Native; arable and waste ground
and hedges; rare and very sparsely scattered over BI; endemic.
7. F. officinalis L. - Common Fumitory. Flowers (6)7-
8(9)mm, pink, in raceme longer than its peduncle; sepals
1.5-3.5 x 1-1.5mm, irregularly serrate; fruit 2-2.5 x 2.25-
3mm, broadly transversely elliptic, truncate to emarginate at
apex, rugose when dry. Native; cultivated and waste ground.
a. Ssp. officinalis. Well-formed racemes >20-flowered; **134**
sepals 2-3.5 x 1-1.5mm; 2n=32. All over BI, the commonest
Fumaria in Sc and E & C En, but rare in many parts of the W.
b. Ssp. wirtgenii (Koch) Arcang. Racemes <20-flowered;
sepals 1.5-2 x 0.75-1mm; 2n=48. Perhaps commoner than ssp.

officinalis on light soils in E En, rare elsewhere in BI.
7 x 8. F. officinalis x F. densiflora has been found as a
sterile plant in 4 localities in SE En, only once recently;
endemic.
7 x 9. F. officinalis x F. parviflora was recorded in 1910
in Surrey; endemic.
8. F. densiflora DC. (F. micrantha Lagasca) - Dense-flowered 134
Fumitory. Flowers 6-7mm, pink, densely packed in raceme much R
longer than its peduncle; sepals (2)2.5-3.5 x (1)2-3mm, subentire to irregularly toothed; fruit 2-2.5 x 2-2.5mm,
orbicular, + rounded at apex, rugose when dry. Native;
arable land; frequent on well-drained soils in SE En, very
sparsely scattered and rare over rest of Br and Ir.
9. F. parviflora Lam. - Fine-leaved Fumitory. Flowers 5- 134
6mm, white or very pale pink, in almost sessile raceme; R
sepals 0.5-0.75(1) x 0.5-0.8mm, irregularly serrate; fruit
1.7-2.3 x 1.75-2.5mm, broadly ovate to orbicular, rounded to
subacute at apex, verrucose-rugose when dry. Native; arable
land usually on chalk; E, SE & SC En, very rare in E Sc.
10. F. vaillantii Lois. - Few-flowered Fumitory. Flowers 134
5-6mm, pale pink, in lax raceme much longer than its R
peduncle; sepals 0.5-0.75-1(1.5) x 0.25-0.5mm, irregularly
serrate; fruit 2-2.25 x 2-2.25mm, + orbicular, rounded to
truncate or slightly emarginate at apex, verrucose-rugose
when dry. Native; arable land usually on chalk; SE & SC En.

33. PLATANACEAE - Plane family

Trees; leaves simple, deciduous, palmately lobed, petiolate,
alternate, stipulate when juvenile. Flowers monoecious, in
stalked, dense spherical clusters 2-several together on
pendent unisexual stalks, hypogynous, actinomorphic; perianth
small, fused or not, of 1-2 whorls of usually 3-4 segments;
male flowers with usually 3-4 stamens; female flowers usually
with 3-4 staminodes and 5-8 free carpels each with 1(2)
ovules; style 1; stigma linear; fruit an achene with long
hairs at base.
An unmistakable tree.

1. PLATANUS L. - Planes
1. P. x hispanica Miller ex Muenchh. (P. x hybrida Brot.; P.
occidentalis L. x P. orientalis L.) - London Plane. Tree to
44m; bark with conspicuous large peeling plates; leaves with
hollow petiole-base concealing axillary bud, with sharply
pointed leaf-lobes; fruits in spherical pendent clusters
2-3.5cm across, breaking up in Spring. Intrd; abundantly
planted as street and park tree, especially in S En, and
often producing seedlings; of uncertain horticultural origin.

34. ULMACEAE - Elm family

Trees; leaves simple, deciduous, serrate, usually asymmetrical at base, petiolate, alternate, stipulate when young.
Flowers in small axillary clusters produced before leaves,
hypogynous, actinomorphic, bisexual; perianth inconspicuous,
campanulate, 4-5-lobed; stamens 4-5; ovary 1-celled, with 1
ovule; styles 2; stigma linear; fruit an achene with 2 wide
wings extending beyond both base and apex.
Unmistakable flowers, fruits and asymmetric leaf-bases.

1. ULMUS L. - Elms

An extremely difficult genus, having been interpreted in
widely different ways, with 2-7 spp. and 1-c.12 interspecific
hybrid combinations recognised. The complex hybrid origin of
many taxa postulated by R. Melville may be correct in many
instances, but does not form the basis of a practical classification. The 2-species concept of R.H. Richens is not
sufficiently discriminating to be of taxonomic value. This
account treats the 4 most distinctive taxa as spp. and is a
reasonable compromise between the 2 extreme opinions. Only
binary hybrids are mentioned; the existence of ternary and
quaternary hybrids is highly contentious, as is the history
of the various taxa in Br, but hybridization is (or has been)
undoubtedly very widespread and frequent. Dutch elm disease
has killed most trees in much of S and C Br, including the
Midlands where the greatest variation occurs. Over large
areas the elm population exists solely or largely as hedgerow
suckers, and is now unidentifiable.
Only leaves from the middle of short-shoots in high summer
should be used for identification; leaves from long-shoots,
suckers, epicormic shoots or Lammas shoots must be avoided.
Leaf lengths below are measured from the base of the longer
side of the lamina to its apex. Tree outlines refer to
mature, solitary specimens. All taxa except U. glabra and
some U. x vegeta produce abundant suckers.

1 Rust-coloured hairs abundant on buds; leaves >7cm,
   very rough on upperside, with >12 pairs of lateral
   veins; petiole <3mm, most of it overlapped by base
   of long-side of lamina   **1. U. glabra**
1 Rust-coloured hairs 0 or present on buds; leaves
   usually <7cm, if >7cm smooth on upperside, with c.5-
   18 pairs of lateral veins; petiole usually >5mm,
   not or partly overlapped by base of lamina   2
   2 Leaves usually >7cm, with length x width >28(cm);
      rust-coloured hairs often present on buds   3
   2 Leaves usually <7cm, with length x width <28(cm);
      rust-coloured hairs 0 on buds (except in U. glabra
      hybrids)   4
3 Leaves almost 2x as long as wide, acuminate at apex,
   with 12-18 pairs of lateral veins; tree outline

broadly obovate to orbicular, with long branches from
low down   **2. U. x vegeta**
3 Leaves distinctly <2x as long as wide, acute to shortly
  acuminate at apex, with 10-14 pairs of lateral veins;
  tree outline obovate, narrow below but with spreading
  branches above   **3. U. x hollandica**
  4 Leaf width/length ratio >0.75; leaves usually
    rough on upperside; tree outline obovate to oblong,
    with strong branches at all levels   **4. U. procera**
  4 Leaf width/length ratio <0.75; leaves usually
    smooth on upperside, if rough then tree outline
    very narrow with no strong branches   5
5 Tree outline narrow but irregular, with leading shoot
  arching or pendent; strong branches 0, all partly
  pendent; short-shoots mostly continuing growth as
  long-shoots; leaves usually rough on upperside
                                                     6. U. plotii
5 Tree outline various but never as last, with erect
  leading shoot; some strong branches usually present,
  only the lower or 0 pendent; short-shoots rarely
  continuing growth; leaves smooth on upperside
                                                     5. U. minor

1. U. glabra Hudson - Wych Elm. Tree to 37m; outline ± 139
orbicular; trunk dividing low down into many long spreading
branches; leaves 8-16cm, very rough on upperside, asymmetric
at base, with 12-18 pairs of lateral veins; petiole <3mm,
most of it overlapped by base of long-side of lamina.
Native; woods and hedgerows, sometimes dominant, especially
on limestone; throughout BI, but much commoner in N & W and
not native in much of SE. 2 ill-marked sspp. are sometimes
recognized: ssp. **glabra**, with broadly obovate leaves, more
southern in distribution; and ssp. montana N. Hylander, with
narrowly obovate leaves and typical of the N & W.
  1 x 4.   U. glabra x U. procera has been recorded from
several areas, most reliably Essex and Lincs, but is
certainly rare and perhaps extremely so.
  1 x 5.   U. glabra x U. minor is abundant wherever the 2
parents meet, especially in C En, and also elsewhere due to
introductions. Plants extremely varied according to the U.
minor parent(s) involved and to the degree of back-crossing.
Characters of U. glabra detectable are rust-coloured hairs on
the buds, numerous (>12) pairs of lateral veins, cuspidate
leaf-apex, and coarse, forward-directed serration; the leaves
usually have a smooth upperside. Some of these hybrids might
involve U. plotii as well in their ancestry. See 2 and 3.
  1 x 6. U. glabra x U. plotii = ?U. x elegantissima Horw. is
common within the range of U. plotii. It can be told from U.
glabra x U. minor by the erect habit often with pendent
branches and the small leaves with blunt serrations.
  2. U. x vegeta (Loudon) Ley (U. x hollandica var. vegeta 139
(Loudon) Rehder; U. glabra x U. minor) - Huntingdon Elm.

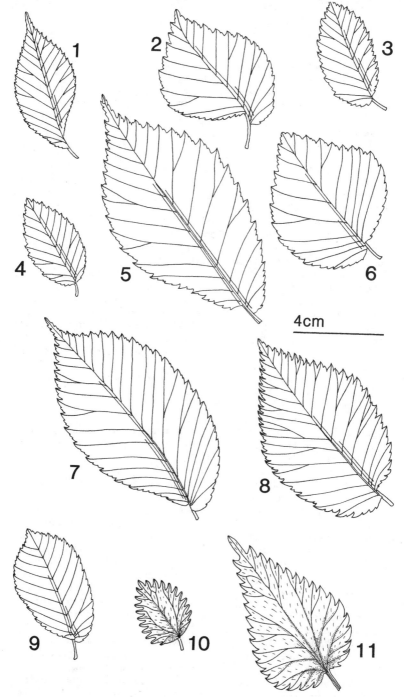

FIG 139 - 1-9, Leaves of **Ulmus**. 1, **U. carpinifolia**. 2, **U. coritana**. 3, **U. minor** ssp. **sarniensis**. 4, **U. plotii**. 5, **U. x vegeta**. 6, **U. procera**. 7, **U. glabra**. 8, **U. x hollandica**. 9, **U. minor** ssp. **angustifolia**. 10-11, Leaves of **Urtica**. 10, **U. urens**. 11, **U. dioica**.

Tree to 32m; outline broadly obovate to + orbicular;
branches long, straight, spreading fan-like to form a broad
crown; leaves similar to those of U. glabra but smooth on
upperside and with a petiole >5mm. Native; hedgerows and
copses; C En and E Anglia; also very widely planted in groups
and avenues.
3. **U. x hollandica** Miller (?U. glabra x U. minor, or U. **139**
glabra x U. minor x U. plotii) - Dutch Elm. Tree to 32m;
outline obovate; branches long, crooked and spreading above
but few low down; leaves relatively shorter and broader and
more shortly acuminate than in U. x vegeta, but with obvious
evidence of U. glabra characters. Native; hedgerows; C & S
En and CI; very widely planted in BI. Plants from CI, with
more asymmetric leaves and longer petioles, probably have U.
minor ssp. sarniensis in their parentage and have been
distinguished as var. **insularum** Richens.
4. **U. procera** Salisb. (U. minor var. vulgaris (Aiton) **139**
Richens) - English Elm. Tree to 33m; outline obovate to
oblong; branches strong at all levels from a wide, tall
trunk; leaves 5-9cm, often + suborbicular, rough on upperside, asymmetric at base, with 10-12 pairs of lateral veins;
petiole c.5-8mm. Probably native; hedgerows; C & S Br,
introduced in Sc, Ir and CI. Rarely reproducing by seed and
rarely forming hybrids.
4 x 5. **U. procera x U. minor** is occasionally found as
isolated individuals within the range of various variants of
U. minor. The influence of U. procera can be seen in the
very broad leaves with a somewhat rough upperside.
4 x 6. **U. procera x U. plotii** is rarely and perhaps doubtfully recorded from within the range of U. plotii, and very
sporadically elsewhere (e.g. N Wa) presumably planted.
5. **U. minor** Miller - see sspp. for English names. Tree to
31m; outline extremely various; leaves extremely various in
size and shape but usually <10cm, + symmetric to strongly
asymmetric at base, smooth on upperside. Native at least in
CI and probably in Br; hedgerows and copses. 2 + uniform,
geographically isolated biotypes can be recognized as sspp.,
but the type ssp. remains one of the most polymorphic taxa in
the British flora. A number of the variants might have U.
glabra and/or U. plotii in their ancestry.
1 Outline + spreading, with major branches <1/2 way
    up tree   **a. ssp. minor**
1 Outline very narrow, with no major branches in lower
    1/2 of tree   2
  2 Trunk persisting to tree apex; branches numerous,
      slightly ascending   **c. ssp. sarniensis**
  2 Trunk ending short of tree apex; branches few, the
      lowest steeply ascending   **b. ssp. angustifolia**
  a. Ssp. minor (U. carpinifolia Suckow, U. coritana Melville, **139**
U. diversifolia Melville) - Small-leaved Elm. Outline
variously spreading; trunk usually dividing <1/2 way up tree.
Kent and Hants to S Wa and C En. The 3 synonyms given

represent widely divergent biotypes, but there are many
others as well and complete intergradation occurs. The
isolated biotype in Hants (U. stricta var. goodyeri Melville)
may merit separate subspecific recognition. The following
are distinctive in their extreme form: **U. carpinifolia** has
narrow leaves with the margin of the long-side ± straight for
the lower 1/3-1/2; **U. coritana** has broad leaves often similar
to those of U. procera in shape but with a very unequal base
and very large patches of hairs around the main vein axils on
lowerside.

 b. Ssp. angustifolia (Weston) Stace (U. angustifolia **139**
(Weston) Weston, U. stricta (Aiton) Lindley, U. minor var.
cornubiensis (Weston) Richens) - Cornish Elm. Outline
narrow; lower branches few and steeply ascending; leaves
4-8cm, narrowly to broadly ovate or obovate, weakly
asymmetric to subequal at base, with 8-12 pairs of lateral
veins; petiole 4-7mm. Cornwall, Devon and Dorset.
 c. Ssp. sarniensis (C. Schneider) Stace (U. sarniensis (C. **139**
Schneider) Bancr., U. minor var. sarniensis (C. Schneider)
Richens) - Jersey Elm. Outline narrow; trunk extending to
apex of tree; branches slightly ascending, progressively
shorter towards tree apex; leaves similar to those of ssp.
angustifolia. Native; Guernsey, probably intrd into rest of
CI and Br, especially as a common roadside tree.
 5 x 6. U. minor x U. plotii = ?U. x viminalis Lodd. occurs
in E Anglia and C En with the parents and is planted elsewhere. The influence of U. plotii is most often shown by the
elongating short-shoots.
 6. U. plotii Druce (U. minor var. lockii (Druce) Richens) - **139**
Plot's Elm. Outline narrow; leading shoot arching or
pendent; branches all weak, pendent at ends; short-shoots
mostly continuing growth as long-shoots; leaves 3-7cm,
elliptic, ± rough on upperside, subequal at base, with 7-10
pairs of lateral veins; petiole 3-6mm, slender. Native;
hedgerows; C En; endemic.

35. CANNABACEAE - Hop family

Annual or perennial herbs; leaves palmately lobed to ±
palmate, petiolate, alternate or opposite, stipulate.
Flowers small and inconspicuous, in varied inflorescences,
usually dioecious, hypogynous, actinomorphic; perianth small
to ± absent, of 1 whorl of 5 completely or partly fused
segments; male flowers with 5 stamens; female flowers with
1-celled ovary with 1 ovule; styles 2; fruit an achene
subtended by persistent bract.
Both representatives are very well known and distinctive.

1 Erect annual; at least upper leaves alternate
                   1. CANNABIS

1 Rampant perennial climber; leaves all opposite
                                                    2. HUMULUS

1. CANNABIS L. - Hemp
Upper leaves alternate, lower sometimes opposite; inflorescences axillary clusters towards stem apex, the male
looser, the female ripening to form irregular groups of
achenes and bracts.

1. C. sativa L. - Hemp. Erect unbranched or little-branched
annual to 2.5m; leaves palmately divided to base or almost so
into 3-9 lanceolate lobes. Intrd; formerly grown for fibre
and still illicitly on small scale for drug; imported in
mixed bird-seed and a common casual on tips, and in parks and
farms where birds are fed; much of BI, especially urban areas
in S; S & W Asia.

2. HUMULUS L. - Hop
Leaves all opposite; male inflorescences loose spreading
axillary panicles; female inflorescences small dense capitate
clusters ripening to cone-like fruiting heads, in rather
loose panicles.

1. H. lupulus L. - Hop. Scrambling perennial climber to 8m;
leaves palmately divided c.3/4 to base into 3-5 ovate lobes;
fruiting heads commonly 3-5cm, up to c.10cm in cultivars.
Native; hedgerows, scrub and fen-carr; S Br and CI;
cultivated especially in SE En and SW Midlands for brewing
industry and very widely natd almost throughout BI.

36. MORACEAE - Mulberry family

Trees or shrubs with latex, leaves simple, deciduous, often
palmately lobed, petiolate, alternate, stipulate when young.
Flowers small and inconspicuous, crowded into dense heads or
into hollow receptacles (figs), monoecious, hypogynous,
actinomorphic; perianth reduced, of 1 whorl of 4-5 free
segments; male flowers with 4-5 stamens; female flowers with
1-celled ovary with 1 ovule; styles 1-2; stigma linear or
capitate; fruiting head a mass of drupes surrounded by fleshy
perianth or fleshy receptacle.
  Both representatives are very well known and distinctive.

1 Fruiting head raspberry-like in appearance; stipule-
     scar on 1 side of stem; latex watery   **1. MORUS**
1 Fruiting head ± pear-shaped; stipule-scar completely
     encircling stem; latex milky   **2. FICUS**

1. MORUS L. - Mulberries
Leaves mostly unlobed; stipules separate, their scars not
encircling stem; fruiting head raspberry-like.

1. M. nigra L. - Black Mulberry. Upright tree to 14m; leaves 6-20cm, usually simple, sometimes palmately lobed, cordate at base, pubescent; fruiting heads raspberry-like, sessile, red when mature. Intrd; waste ground and walls, bird-sown from cultivated trees; natd by Thames in Middlesex, frequently grown in parks and seedlings rarely found in scattered localities; C Asia.

2. FICUS L. - Fig
Latex milky; leaves mosty deeply palmately lobed; stipules fused, their scars encircling stem; fruiting head pear-shaped.

1. F. carica L. - Fig. Spreading shrub or small tree to 10m; leaves 10-20cm, cordate at base, with 3-5 obtuse to rounded lobes, sparsely and roughly hairy; fruiting head the characteristic 'fig', green to blackish at maturity. Intrd; waste ground and walls, especially by rivers; natd in S En, S Wa and CI, and rarely as far N as S Sc, bird-sown or from imported fruit; SW Asia.

37. URTICACEAE - Nettle family

Annual or perennial herbs; leaves simple, opposite or alternate, stipulate or not. Flowers axillary, small and inconspicuous, solitary or in crowded inflorescences, monoecious or dioecious, hypogynous, actinomorphic; perianth of 1 whorl of 4 often partly fused greenish to brownish segments; male flowers with 4 stamens; female flowers with 1-celled ovary with 1 ovule; style 1; stigma much branched; fruit an achene.
The 3 genera appear very different vegetatively, but are characterized by their inconspicuous, unisexual flowers with 4 perianth segments, 4 stamens, 1-celled superior ovary with 1 ovule, 1 style and densely branched stigma.

1 Leaves opposite, usually toothed and with stinging
  hairs, stipulate; stems erect   **1. URTICA**
1 Leaves alternate, entire, without stinging hairs,
  exstipulate; stems procumbent to decumbent   2
  2 Stems decumbent but not rooting at nodes; leaves
    mostly >10mm; flowers crowded   **2. PARIETARIA**
  2 Stems procumbent and rooting at nodes; leaves rarely
    >6mm; flowers solitary   **3. SOLEIROLIA**

1. URTICA L. - Nettles
Annual or perennial with erect stems; leaves opposite, stipulate, toothed and normally with stinging hairs; flowers in dense axillary usually elongate inflorescences, monoecious or dioecious; perianth of free segments, 2 inner longer than 2 outer and enclosing fruit.

Other spp. - **U. incisa** Poiret is a rare wool-alien from Australia and New Zealand. It is a monoecious perennial, glabrous apart from rather sparse stinging hairs, vegetatively rather similar to some subglabrous variants of U. dioica but lacking rhizomes; usually lower inflorescences are female and upper male. **U. pilulifera** L. (Roman Nettle), from S Europe, used to occur as a casual.

1. **U. dioica** L. - Common Nettle. Strongly rhizomatous **139** usually dioecious perennial to 1.5m; leaves and stems usually with abundant stinging hairs and more numerous smaller nonstinging hairs; terminal leaf-tooth longer than adjacent laterals. Very variable, especially in leaf-shape and pubescence; stingless, subglabrous and monoecious variants are known. Native; in many habitats, especially woodland, fens, cultivated ground and where animals defecate; abundant throughout BI.
2. **U. urens** L. - Small Nettle. Monoecious annual to 60cm; **139** inflorescences each with many female and few male flowers; leaves and stems with usually abundant stinging hairs but otherwise glabrous to sparsely hairy; terminal leaf-tooth about as long as adjacent laterals. Probably native; cultivated and waste ground; frequent throughout BI but commoner in E.

2. **PARIETARIA** L. - Pellitory-of-the-wall
Perennial with decumbent stems; leaves alternate, entire, exstipulate; flowers in dense axillary short inflorescences, monoecious; perianth of equal segments fused at least at base and enclosing fruit.

1. **P. judaica** L. (P. diffusa Mert. & Koch) - Pellitory-of-the-wall. Stems to 80cm; much-branched; leaves softly hairy, ovate to elliptic, <5cm. Native; on walls, rocks, cliffs and steep hedgebanks; frequent in BI, except rare in Sc and absent from N.

3. **SOLEIROLIA** Gaudich. (Helxine Req. non L.) - Mind-your-own-business
Perennial with procumbent stems rooting at nodes; leaves alternate, entire, exstipulate; flowers solitary in leaf-axils, monoecious; perianth of equal segments fused at least at base and enclosing fruit.

1. **S. soleirolii** (Req.) Dandy (Helxine soleirolii Req.) - Mind-your-own-business. Stems to 20cm, very slender; leaves hairy, suborbicular, <6mm. Intrd; natd on damp shady walls and banks; frequent in S En and CI, very scattered elsewhere and very rare in Sc; W Mediterranean islands.

38. JUGLANDACEAE - Walnut family

Trees; leaves pinnate, deciduous, alternate, exstipulate.
Flowers monoecious, the male in pendent catkins, the female
in pendent catkins or 1-few in stiff clusters, epigynous, ±
actinomorphic, in axil of a bract and with 2 bracteoles;
perianth small and inconspicuous, 1-whorled, 1-5-lobed; male
flowers with 3-many stamens; female flowers with 1-celled
ovary with 1 ovule, styles 2; stigmas branched; fruit a drupe
or winged nut.
Easily recognized trees with pinnate leaves and distinctive
flowers and fruits.

1 Leaflets entire, aromatic when crushed;fruits 1-few
   in rigid clusters, with green husk and hard-shelled
   'nut' inside   **1. JUGLANS**
1 Leaflets serrate, not aromatic; fruits several-many in
   pendent clusters, with broad suborbicular wing around
   nut   **2. PTEROCARYA**

1. JUGLANS L. - Walnut
Leaves with 3-9 entire, aromatic, ovate lealets; female
flowers 1-few in stiff clusters; fruits drupes, with green
outer husk and hard-shelled edible 'nut' inside.

1. J. regia L. - Walnut. Tree to 24m, not suckering at
base; winter buds sessile, glabrous, with bud-scales; fruits
ellipsoid, mostly >3cm.   Intrd; commonly planted and often
surviving in wild places, but self-sown only in warmer parts;
most of Br; SE Europe and Asia.

2. PTEROCARYA Kunth - Wingnuts
Leaves with (7)15-27(41) serrate, non-aromatic, lanceolate
leaflets; female flowers in pendent catkins; fruits nuts with
broad suborbicular wings derived from bracteoles.

1. P. fraxinifolia (Poiret) Spach - Caucasian Wingnut. Tree
to 35m, usually forming thickets due to suckers if left
undisturbed; winter buds stalked, pubescent, without bud-
scales; fruits suborbicular, c.2cm across, broadly winged.
Intrd; planted on field borders and embankments and occasionally running wild; SE En and SW Sc; Caucasus.

39. MYRICACEAE - Bog-myrtle family

Shrubs; leaves simple, deciduous, alternate, petiolate, exstipulate, strongly aromatic. Flowers normally dioecious,
sometimes monoecious, in stiff catkins, hypogynous, actinomorphic, in axil of a bract, the female with 2 bracteoles;
perianth 0; male flowers with usually 4 stamens; female
flowers with 1-celled ovary with 1 ovule; styles 2; stigmas

linear; fruit a drupe or narrowly winged nut.
Easily recognized by the aromatic foliage and distinctive
catkins.

1. MYRICA L. - Bog-myrtles
1. M. gale L. - Bog-myrtle. Erect shrub to 1.5(2)m; current
season's twigs very sparsely pubescent; leaves 2-6cm, oblanceolate to narrowly so, serrate towards apex, very sparsely
pubescent; catkins appearing before leaves, on twigs of
previous season that do not continue growth; fruit a very
narrowly 2-winged nut. Native; wet moorland and heathland,
bogs and fens; throughout most of Br and Ir but mostly in NW,
there often abundant.
2. M. cerifera L. (M. caroliniensis auct. non Miller) -
Bayberry. Shrub to 3m; current season's twigs densely
pubescent; leaves 3-10cm, oblanceolate to narrowly obovate or
elliptic, usually serrate or dentate towards apex; sparsely
pubescent; catkins appearing before leaves, on twigs of
previous season that continue forward growth and leaf
production; fruit a globose drupe c.4-5mm, with waxy bloom.
Intrd; planted for cover and winter effect and natd on
heathland in New Forest, Hants; E N America.

40. FAGACEAE - Beech family

Trees; leaves simple, deciduous or evergreen, alternate,
petiolate, stipulate at least when young. Flowers
monoecious, inconspicuous, the male in pendent or stiff
catkins or 1-numerous in heads, the female in small groups of
1-few surrounded by numerous small scales, epigynous, actinomorphic; perianth 1-whorled, 4-7-lobed; male flowers with 4-
40 stamens; female flowers with 3- or 6-celled ovary with 2
ovules per cell; styles 3-9; stigmas linear; fruit a nut,
1-3(6) surrounded by a cupule formed from the fused scales.
 Catkin-bearing trees distinguished from other such families
by the 3- or 6-celled ovary with 3-9 styles.

1 Male flowers in elongated catkins; nuts terete or
  with rounded corners; winter buds obtuse to rounded
  at apex   2
1 Male flowers 1-many in short heads; nuts triquetrous;
  winter buds acute at apex   3
  2 Male flowers in stiff catkins; cupule strongly
    spiny, completely enclosing 1-3(6) nuts during
    development   **3. CASTANEA**
  2 Male flowers in pendent catkins; cupule not
    spiny, enclosing only lower part of 1 nut
                                                          4. QUERCUS
3 Male flowers 1-3 in stiff clusters; nuts usually 3 per
  cupule   **2. NOTHOFAGUS**
3 Male flowers numerous in pendent tassels; nuts usually

2 per cupule   **1. FAGUS**

1. FAGUS L. - <u>Beech</u>
Leaves deciduous; male flowers numerous in pendent heads,
with 8-16 stamens; female flowers usually 2 together in
erect, pedunculate clusters; nuts triquetrous, 1-2 per
cupule.

1. F. sylvatica L. - <u>Beech</u>. Tree to 42m; winter buds long **148**
and slender, finely pointed; leaves 4-9cm, ovate to elliptic,
entire, long-ciliate on margin; fruiting cupule with stiff
pointed subulate scales. Native; well drained soils, often
forming pure woods on chalk and soft limestone and sometimes
acid sandstone; SE Wa and S En; also very widely planted and
natd over all BI.

2. NOTHOFAGUS Blume - <u>Southern Beeches</u>
Leaves deciduous or evergreen, often slightly asymmetric at
base; male flowers 1-3 in stiff clusters, with 8-40 stamens;
female flowers 1-3 in ± sessile clusters; nuts triquetrous,
usually 3 per cupule.

Other spp. - Of the several spp. now being tried for
forestry, **N. dombeyi** (Mirbel) Blume (<u>Coigue</u>), from Chile and
W Argentina, with evergreen leaves up to 4cm, is now found in
a few plantations in SW En and Wa.

1. N. obliqua (Mirbel) Blume - <u>Roble</u>. Deciduous tree to **148**
30m; winter buds c.4mm; leaves coarsely and doubly serrate,
± glabrous, with 7-11 pairs of lateral veins; cupules with
short stiff scales. Intrd; now widely planted for forestry
in W En, W Sc and Wa, often producing self-sown offspring;
Chile and W Argentina.
1 x 2. N. obliqua x N. nervosa occurs spontaneously in mixed
plantations mainly in SW En; endemic.
2. N. nervosa (Philippi) Dmitri & Milano (<u>N. procera</u> **148**
Oersted) - <u>Rauli</u>. Deciduous tree to 26m; winter buds c.10mm;
leaves 4-8cm, ovate-oblong, acute to rounded at apex, finely
serrate to subentire or crenate, pubescent on veins on
lowerside, with 14-18 pairs of lateral veins; cupules with
deeply laciniate scales. Intrd; now widely planted and
self-sowing as for <u>N. obliqua</u>; Chile and W Argentina.

3. CASTANEA Miller - <u>Sweet Chestnut</u>
Leaves deciduous; flowers in long rather stiff insect-
pollinated catkins, mostly male but female at base, the male
with 10-20 stamens, the female usually in groups of 3; nuts
with 2-4 rounded angles, 1-3(more) per cupule.

1. C. sativa Miller - <u>Sweet Chestnut</u>. Tree to 35m; leaves **148**
10-30cm, oblong-lanceolate, regularly coarsely serrate with
finely pointed teeth, ± glabrous at maturity; fruiting

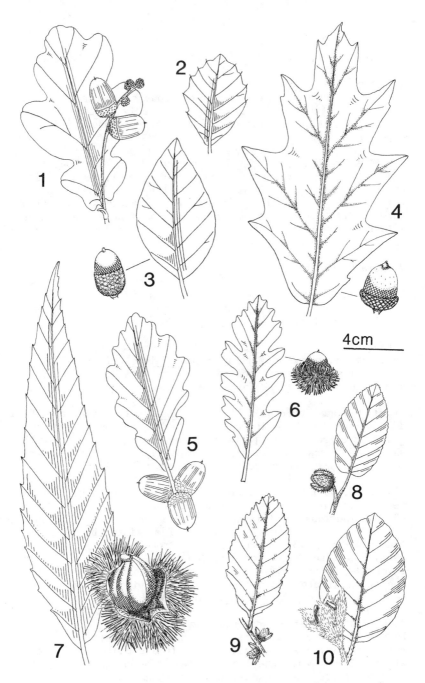

FIG 148 – Leaves and fruits of **Fagaceae**. 1, **Quercus robur**. 2-3, **Q. ilex**. 4, **Q. rubra**. 5, **Q. petraea**. 6, **Q. cerris**. 7, **Castanea sativa**. 8, **Nothofagus nervosa**. 9, **N. obliqua**. 10, **Fagus sylvatica**.

cupule densely covered with pungent spines. Intrd; planted
throughout BI, especially as coppiced woodland in SE En;
setting seed and natd ± only in S En and CI; S Europe.

4. QUERCUS L. - Oaks

Leaves deciduous or evergreen; male flowers in pendent
catkins, with 4-12 stamens; female flowers 1-few in stiff
pedunculate or sessile clusters; nuts (acorns) terete, 1 per
cupule.

1 Fruit cupule with erecto-patent to reflexed scales;
　at least terminal buds surrounded by persistent
　stipules　　　　　　　　　　　　　　　　　　　　　　2
1 Fruit cupule with short appressed scales; buds not
　surrounded by persistent stipules　　　　　　　　　　4
　2 Leaves toothed <1/3 to midrib, the teeth usually
　　<6mm deep　　　　　　　　　　　　**1. Q. castaneaefolia**
　2 Leaves lobed >1/3 to midrib, the lobes usually
　　>6mm deep　　　　　　　　　　　　　　　　　　　　3
3 Tree without green leaves in Winter; leaf-lobes
　obtuse to acute or shortly apiculate　　　　**2. Q. cerris**
3 Tree usually with some green leaves in Winter; leaf-
　lobes mucronate to aristate　　　　　**3. Q. x pseudosuber**
　4 Leaves evergreen, coriaceous, grey-tomentose on
　　lowerside even at maturity　　　　　　　　**4. Q. ilex**
　4 Leaves deciduous, not coriaceous, glabrous to
　　sparsely or patchily pubescent on lowerside
　　at maturity　　　　　　　　　　　　　　　　　　5
5 Leaves with acuminate to aristate leaf-lobes; nut
　with shell tomentose inside　　　　　　　　**8. Q. rubra**
5 Leaves with obtuse to rounded leaf-lobes; nut with
　shell glabrous inside　　　　　　　　　　　　　　6
　6 Tree usually with some green leaves in Winter;
　　leaves densely tomentose when young but becoming
　　glabrous or ± so at maturity, toothed <1/3 to
　　midrib　　　　　　　　　　　　　　**5. Q. canariensis**
　6 Tree without green leaves in Winter; leaves
　　glabrous to finely pubescent on veins at maturity,
　　lobed >1/3 to midrib　　　　　　　　　　　　　7
7 Petiole <1cm; leaf-base cordate with distinct
　auricles; leaf glabrous or with simple hairs on
　lowerside; peduncle 2-9cm, glabrous　　　　**7. Q. robur**
7 Petiole >1cm; leaf-base cuneate to cordate but
　without auricles; leaf with simple hairs along midrib
　and also with some stellate hairs on lowerside;
　peduncle 0-2(4)cm, with clustered hairs　　**6. Q. petraea**

Other spp. - Many other spp. are grown for ornament and some
may appear 'wild' in abandoned woodland or parkland. Of
these, **Q. coccinea** Muenchh. (Scarlet Oak), from E N America,
differs from Q. rubra in its shiny (not matt) leaves which
are scarcely longer than wide and more deeply lobed (at least
2/3 way to midrib); and **Q. x turneri** Willd. (Q. ilex x Q.

robur) (Turner's Oak) is semi-evergreen and differs from Q. x
pseudosuber in its non-mucronate leaf-lobes and stalked,
clustered fruits with short-tipped scales.

1. Q. castaneaefolia C. Meyer - Chestnut-leaved Oak. Tree
to 32m; leaves deciduous, cuneate at base, toothed <1/3 way
to midrib with subacute to mucronate teeth, usually pubescent
on lowerside; petiole 1-3cm; peduncle <2cm; cupule with
reflexed scales up to 1cm. Intrd; infrequently planted,
saplings natd in scrub in W Gloucs; Caucasus.

2. Q. cerris L. - Turkey Oak. Tree to 40m; leaves decid- **148**
uous, cuneate to subcordate at base, lobed to c.1/2 way or
more to midrib with acute to subobtuse lobes, usually
pubescent on lowerside; petiole 1-2.5cm; peduncle <2cm;
cupule with patent to reflexed scales up to 1cm. Intrd;
commonly grown for ornament, often natd on acid sands in S
Br; BI N to C Sc; S Europe.

2 x 7. Q. cerris x Q. robur has been detected in several
places in S En, having originated naturally near both
parents, whose leaf characters it possesses in various
combinations.

3. Q. x pseudosuber Santi (Q. x hispanica auct. non Lam.; Q.
cerris x Q. suber L.) - Lucombe Oak. Tree to 35m; leaves
semi-evergreen, cuneate to sub-cordate at base, lobed
<1/2-way to midrib, with mucronate to aristate lobes, usually
pubescent on lowerside; petiole 0.5-2cm; peduncle <2cm;
cupule with erecto-patent to reflexed scales up to 1cm.
Intrd; rather frequently planted in parks etc., self-sown
(and then widely segregating) in Surrey; garden origin 1762.

4. Q. ilex L. - Evergreen Oak. Tree to 25m; leaves ever- **148**
green, cuneate at base, entire or variously sharply serrate
(especially on juvenile shoots), grey-tomentose on lowerside;
petiole 0.5-1.5cm; peduncle <2cm; cupule with small appressed
scales. Intrd; much planted for ornament, and often for
shelter in E En; self-sown in S & C En, Wa, S Ir and CI;
Mediterranean and SW Europe.

5. Q. canariensis Willd. - Algerian Oak. Tree to 31m;
leaves semi-evergreen, rounded at base, toothed <1/3 way to
midrib with subacute to rounded teeth, usually ± glabrous at
maturity; petiole 0.8-3cm; peduncle <2cm; cupule with
appressed scales up to 1cm. Intrd; infrequently planted,
saplings natd in woodland in Surrey; W Mediterranean.

6. Q. petraea (Mattuschka) Liebl. - Sessile Oak. Tree to **148**
42m, of less spreading habit and with a taller, straighter
trunk than Q. robur; leaves deciduous, usually ± elliptic,
cuneate to cordate with no or weak auricles at base, lobed ±
as in Q. robur but usually with 5-8 pairs of less deep lobes,
with frequent simple and some stellate hairs on lowerside;
petiole 13-25mm; peduncle 0-2(4)cm; cupule with small
appressed scales. Native; almost throughout Br and Ir and
often abundant and forming dense woodland, epecially on
shallow, sandy, acid soils and in N and W, often over 300m

altitude, but absent or only intrd in some areas.
6 x 7. Q. petraea x Q. robur = Q. x rosacea Bechst. occurs throughout BI in areas where 1 or both parents occur, occasionally being commoner than either. It combines in various ways the leaf and fruit characters of the 2 spp., and is fertile.
7. Q. robur L. - Pedunculate Oak. Tree to 37m; leaves **148** deciduous, usually obovate, cordate and with well-marked auricles at base, lobed <1/2 way to midrib with 3-6 pairs of rounded lobes, glabrous or with few simple hairs on lower-side; petiole (0)2-3(7)mm; peduncle 2-9cm; cupule with small appressed scales. Native; almost throughout BI and often abundant and forming dense woodland, on a wide range of soils but especially deep rich ones, rarely over 300m altitude, not native in some areas in N and W and on acid shallow soils.
8. Q. rubra L. (Q. borealis Michaux f.) - Red Oak. Tree to **148** 34m; leaves deciduous, often turning red in Autumn, cuneate at base, lobed c.1/2 way to midrib with acuminate to aristate lobes, ± glabrous; petiole 2-5cm; peduncle <2cm; cupule with small appressed scales. Intrd; much planted for ornament and sometimes for forestry or screening, especially on shallow sandy soils, often self-sowing; En, Wa and CI; E N America.

41. BETULACEAE - Birch Family
(Corylaceae)

Trees or shrubs; leaves simple, deciduous, alternate, petiolate, stipulate when young. Flowers monoecious, inconspicuous, epigynous, actinomorphic; perianth 0 or 1-whorled and very small; male flowers numerous in mostly pendent catkins, 1 or 3 per bract, with 0, 2 or 4 bracteoles per group, with 2-14 stamens; female flowers in small groups on erect or pendent catkins, 2-3 per bract, with 2 or 4 bracteoles per group, with 2-celled ovary with 1 ovule per cell; styles 2; stigmas linear; fruit a nut, winged or not. Distinguished from other monoecious catkin-bearing trees by the simple leaves and 2-celled ovary with 2 styles. Corylaceae (Corylus and Carpinus) are often separated off.

1 Fruits winged, in compact cone-like structure formed
　from dried-out bracts; male flowers 3 per bract　　　　　　　2
1 Fruits not winged, each with enlarged lobed or
　laciniate bracts at base; male flowers 1 per bract　　　　　3
　2 Bracts of fruiting cones falling from axis with
　　the fruits, distinctly 3-lobed; male catkins
　　opening with the leaves; stamens 2, the lobes of
　　each well separated　　　　　　　　　　　　**1. BETULA**
　2 Bracts of fruiting cones persistent, falling with
　　the whole cone, obscurely 5-lobed; male catkins
　　opening before the leaves; stamens 4, the lobes
　　of each slightly separated　　　　　　　　**2. ALNUS**

3 Winter buds rounded at apex; fruits 1-several in
   short clusters, each surrounded by cupule of
   laciniate bracts   **4. CORYLUS**
3 Winter buds acute; fruits several in pendent catkins;
   each with large 3-lobed bract   **3. CARPINUS**

1. BETULA L. - Birches
Trees or shrubs; male flowers 3 per bract, with 2 bracteoles
per group, with minute perianth and 2 stamens; female flowers
in stiff erect catkin, 3 per bract, with 2 bracteoles per
group, without perianth; fruits winged, in compact cone-like
structure which disintegrates from axis at maturity to
release fruits and strongly 3-lobed bracts.

1 Leaves <2cm, + orbicular, obtuse to truncate at apex;
   petiole <5mm; male catkins not exposed in winter,
   erect at least until anthesis, <1cm   **3. B. nana**
1 Leaves >2cm, + ovate, subacute to acuminate at
   apex; petiole >5mm; male catkins exposed in winter,
   pendent, >2cm   2
   2 Leaves acuminate at apex, distinctly doubly
      serrate with prominent primary teeth, truncate to
      broadly cuneate at base, glabrous   **1. B. pendula**
   2 Leaves acute to subacute at apex, rather evenly
      or irregularly serrate without obvious primary
      teeth, rounded to cuneate at base, glabrous to
      pubescent   **2. B. pubescens**

Other spp. - Several spp. are grown in parks and on roadside
verges, especially the N N American **B. papyrifera** Marshall
(Paper Birch) and the Himalayan **B. jacquemontii** Spach
(Jacquemont's Birch). Both have rather large doubly serrate
leaves and whiter barks than the native spp.; the former has
glabrous and the latter pubescent shoots.

1. B. pendula Roth - Silver Birch. Tree to 30m; bark 153
silver-white above, erupting into irregular blackish fissures
below; twigs often pendent, glabrous; leaves ovate, doubly
serrate, acuminate, glabrous; each wing of fruit >2x as wide
as body, extending well beyond stigmas at apex of body;
2n=28. Native; forming woods on light, mostly acid soils,
especially heathland, usually in drier places than B.
pubescens; throughout almost all Br and Ir, commoner in S.
1 x 2. B. pendula x B. pubescens = B. x aurata Borkh. occurs
in many areas, often in absence of 1 or both parents. It
varies in characters between the 2 parents, in chromosome
number (2n=42 or 56), and in fertility (fully fertile to
highly sterile). Its frequency is hard to determine due to
uncertainty of parental limits. Most 'intermediates' have
2n=56 and could be hybrids or B. pubescens. B. pendula and
B. pubescens are most conveniently separated by the Atkinson
Discriminant Function: Positive values = B. pendula; negative 153

Atkinson Discriminant Function = 12LTF + 2DFT − 2LTW − 23
(means of 5 short-shoot leaves)

LTF = Leaf Tooth Factor − number of teeth projecting beyond line connecting tips of main teeth at ends of 3rd and 4th lateral veins, subtracted from total number of teeth between these 2 main teeth. **LTW** = Leaf Tip Width − width in mm of leaf 1/4 distance from apex to base. **DFT** = Distance to First Tooth − distance in mm from apex of petiole to first tooth.

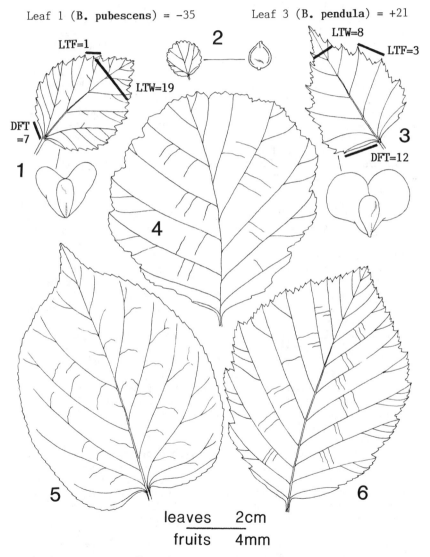

FIG 153 − 1-3, Leaves and fruits of **Betula**, showing Atkinson Discriminant Function. 1, **B. pubescens**. 2, **B. nana**. 3, **B. pendula**. 4-6, Leaves of **Alnus**. 4, **A. glutinosa**. 5, **A. cordata**. 6, **A. incana**.

values = B. pubescens; 93% certainty of correct answer, most
errors being B. pubescens with slightly positive values.
Experimental hybrids are usually closer to B. pubescens in
diagnostic characters and many would be determined as that in
the field.
   2. B. pubescens Ehrh. - Downy Birch. Tree to 24m; bark **153**
variously brown, grey or white but rarely white with strongly
contrasting black fissures below; twigs usually not pendent,
pubescent to glabrous; leaves ovate, singly or irregularly to
somewhat doubly serrate, acute or subacute to slightly
acuminate, pubescent to glabrous; each wing of fruit c.1-1.5x
as wide as body, not extending beyond stigmas at apex of
body; 2n=56. Native; in similar places to B. pendula but
favouring wetter and more peaty soils, especially upland;
throughout almost all Br and Ir, commoner in N.
   a. Ssp. pubescens. Usually a tree; young twigs and petioles
usually pubescent, sometimes glabrous; leaf-laminas mostly
>3cm; each wing of fruit wider than body. Throughout BI but
+ confined to lowland in N.
   b. Ssp. tortuosa (Ledeb.) Nyman (ssp. carpatica (Willd.)
Asch. & Graebner, ssp. odorata sensu E. Warb.). Usually
shrubby; young twigs and petioles pubescent or + glabrous;
leaf-laminas mostly <3cm; each wing of fruit not or little
wider than body. Upland areas of N Br, replacing ssp.
pubescens. A rather ill-defined taxon, but worth recognising
due to its distinctive distribution. Small-leaved variants
of it might have arisen by introgression from B. nana.
   2 x 3. B. pubescens x B. nana = B. x intermedia Thomas ex
Gaudin occurs within the range of B. nana in Sc. It varies
considerably between the parents, especially in leaf characters, but seems only slightly fertile, though introgression
might occur.
   3. B. nana L. - Dwarf Birch. Shrub to 1m, with procumbent **153**
to ascending stems; twigs stiff, pubescent; leaves +   **R**
orbicular, + rounded at both ends, deeply and regularly
crenate, glabrous at maturity; each wing of fruit much
narrower than body and not extending beyond it; 2n=28.
Native; upland moors and bogs on peat; S Northumb, and mainland Sc from Perths and Argyll northwards.

2. ALNUS Miller - Alders
Trees; male flowers 3 per bract, with 4 bracteoles per group,
with minute perianth and 4 stamens; female flowers in small
stalked groups, 2 per bract, with 4 bracteoles per group,
without perianth; fruits narrowly winged, in compact cone-
like structure from which only fruits are released at
maturity; fruits bracts persistent, obscurely 5-lobed.

1 Leaves rounded to cordate at base, regularly crenate-
   dentate; female flower-groups 1-3 on common stalk
                                            3. A. cordata
1 Leaves truncate to cuneate at base, irregularly

(often doubly) serrate; female flower-groups 3-8 on
common stalk   2
2 Leaves broadly obtuse to retuse at apex, sticky
  when young, not or scarcely paler on lowerside,
  with 4-8 pairs of lateral veins   **1. A. glutinosa**
2 Leaves subacuminate to subacute at apex, not sticky
  when young, distinctly paler on lowerside, with
  7-15 pairs of lateral veins   **2. A. incana**

Other spp. - A number of spp. may be planted for ornament
and encountered on roadsides and in parks; commonest is **A.
viridis** (Chaix) DC. (Green Alder), from Europe, with sessile
buds and catkins appearing with the leaves.

1. A. glutinosa (L.) Gaertner - Alder. Tree to 29m; bark 153
dark brown, much fissured; leaves obovate or elliptical to
sub-orbicular, cuneate to rounded or ± truncate at base,
broadly obtuse to retuse at apex, irregularly and usually
shallowly 2-serrate; flowering Feb-Mar; female 'cones'
8-28mm, 3-6 each pedunculate on a common stalk. Native; damp
woods and by lakes and rivers; throughout BI.
1 x 2. A. glutinosa x A. incana = A. x pubescens Tausch
occurs frequently among plantings of either or both parents
on roadsides and in copses, originating from seed collected
from either parent. It combines the leaf and fruit characters of the parents in various combinations.
2. A. incana (L.) Moench - Grey Alder. Tree to 23m; bark 153
grey, ± smooth; leaves ovate, cuneate at base, subacute to
subacuminate at apex, sharply 2-serrate; flowering (Dec) Jan-
Feb; female 'cones' 10-18mm, 3-8 each sessile or shortly
pedunculate on a common stalk. Intrd; planted for shelter
and ornament, especially on poor wet soils in N, over much of
BI; often proliferating from suckers and occasionally self-
sown; Europe.
3. A. cordata (Lois.) Duby - Italian Alder. Tree to 28m; 153
bark greyish, rather smooth; leaves ovate, cordate to rounded
at base, obtuse to shortly acuminate at apex, regularly
crenate-dentate; flowering Feb-Mar; female 'cones' 15-30mm,
1-3 each pedunculate on a common stalk. Intrd; rather
frequently roadside or mass planted for ornament over much of
BI, especially in S, very rarely self-sown; Italy and
Corsica.

3. CARPINUS L. - Hornbeam
Trees; male flowers 1 per bract, without bracteoles, without
perianth, with c.10 stamens; female flowers in pendent
catkin, 2 per bract, with 2 bracteoles per group, with minute
irregularly toothed perianth; fruits not winged, in conspicuous pendent catkin, each nut subtended by much enlarged 3-
lobed bract.

1. C. betulus L. - Hornbeam. Tree to 32m; leaves ovate,

rounded to cordate at base, acute to acuminate at apex, sharply 2-serrate, with 7-15 pairs of lateral veins, glabrous except on veins on lowerside; male catkins opening just before leaves (Apr); nuts subtended by 3-lobed bract to 5cm. Native; forming woods and copses on clay soils in SE En extending to Mons and Cambs; much planted on roadsides and as hedging over rest of BI, often as a fastigiate cultivar.

4. CORYLUS L. - <u>Hazels</u>
Small trees or large shrubs; male flowers 1 per bract, with 2 bracteoles (fused to bract) per group, without perianth, with c.4 stamens; female flowers in small sessile, bud-like group, 2 per bract, with 2 bracteoles per group, with minute irregularly lobed perianth; fruits not winged, large and edible, each nut surrounded by much enlarged, laciniate, fused bracts.

1. C. avellana L. - <u>Hazel</u>. Several-stemmed shrub to 6(12)m; leaves suborbicular, usually cordate at base, obtuse to cuspidate at apex, sharply 2-serrate, softly pubescent; male catkins opening well before leaves (Jan-Mar); nuts 1-few in cluster, usually <2cm, globose to ovoid, each surrounded by girdle of fused, laciniate bracts shorter than to slightly longer than nut. Native; hedgerows, scrub and woodland; whole of BI.
1 x 2. C. avellana x C. maxima, with intermediate fruits, has been reported from Suffolk.
2. C. maxima Miller - <u>Filbert</u>. Shrub or sometimes small tree to 10m; differs from <u>C. avellana</u> in the ovoid nuts usually >2cm, with the fused surrounding bracts much longer than nut and contracted beyond its apex. Intrd; grown for its nuts ('Kentish Cobs') in orchards in Kent and in gardens elsewhere, sometimes found as a relic and rarely self-sown; SE Europe and SW Asia.

42. PHYTOLACCACEAE - <u>Pokeweed family</u>

Herbaceous perennials; leaves alternate, simple, entire, petiolate, exstipulate. Flowers many in leaf-opposed racemes, bisexual, hypogynous, actinomorphic; perianth of 1 whorl of 5 free segments, usually petaloid; stamens 7-16; carpels 7-10 in 1 whorl, $\pm$ free, each with 1 ovule and 1 short style; fruit succulent, berry-like, 7-10-lobed from close adnation of the carpels.
Distinctive in habit and in flower and fruit structure.

1. PHYTOLACCA L. - <u>Pokeweed</u>

Other spp. - 3 other Asian spp., **P. latbenia** (Buch.-Ham.) H. Walter, **P. clavigera** W. Smith and **P. esculenta** Van Houtte, differing from <u>P. acinosa</u> in flower colour, stamen number,

and pubescence, are grown in gardens and may also appear
bird-sown outside. The distinctness of some from P. acinosa
remains to be confirmed. P. americana L. (American Pokeweed),
from N America, with narrower leaves, longer usually arching
racemes and fused carpels, is the sp. usually recorded, but
perhaps always in error for an Asian sp.

1. P. acinosa Roxb. - Indian Pokeweed. Stems branching,
erect, 1.5m; leaves (ovate-)elliptic, <20cm; racemes erect,
many-flowered, to as long as leaves; perianth c.3mm, whitish-
green to red; stamens 7-12; fruit c.4mm, blackish, with red
juice. Intrd; grown in gardens and bird-sown usually as
individual plants outside; occasional in S En, S Wa and CI;
China and Japan.

42A. NYCTAGINACEAE

MIRABILIS jalapa L. (Marvel-of-Peru), from Tropical America,
is grown in gardens in S En and CI and is occasionally found
on rubbish-tips. It is an erect, branched perennial to 1m,
with opposite, simple, entire, exstipulate leaves and a
trumpet-shaped, variously coloured (usually red) perianth
consisting of a narrow tube c.3cm and a limb of 5 spreading
lobes c.25mm across.

43. AIZOACEAE - Dew-plant family
(Tetragoniaceae)

Annuals or slightly to moderately woody perennials, usually
glabrous; leaves opposite or alternate, simple, usually
entire, usually thick and succulent, exstipulate; petioles
present or 0. Flowers solitary or in cymes, actinomorphic,
usually bisexual, epigynous to perigynous; sepals 4-5(6),
free or fused at base to form short tube above ovary, usually
succulent; 'petals' (actually petaloid staminodes) 0 or
numerous and ± free and in 1-several rows, usually linear,
often brightly coloured; stamens 3-numerous, usually in
several rows; ovary with c.3-20 cells, with 1-many variously
arranged ovules in each cell; styles as many as carpels, free
or fused at base; stigmas minute or linear; fruit a hard,
dehiscent or indehiscent capsule with 3-many seeds, or
succulent with many seeds.
Most spp. are easily recognized by their succulent leaves
('ice-plants') and colourful, many-petalled daisy-like
flowers. There remain many taxonomic problems at both
generic and specific levels. Identification is much easier
in the fresh state.

1 Leaves >15mm wide, strongly flattened, abruptly
   petiolate   2

1 Leaves <15mm wide, usually <2x as wide as thick,
   sessile or gradually petiolate   3
   2 Leaves alternate; petals 0   9. TETRAGONIA
   2 Leaves opposite; petals present   1. APTENIA
3 Fruit succulent; seeds embedded in mucilage;
   stigmas c.8-20   8. CARPOBROTUS
3 Fruit woody, without copious mucilage; stigmas <8   4
   4 Leaves c. as long as wide, triquetrous, sparsely
      but conspicuously toothed on all 3 angles
                                          4. OSCULARIA
   4 Leaves >2x as long as wide, triquetrous to
      terete, entire or minutely toothed   5
5 Leaves widest and thickest near apex, triquetrous,
   with a conspicuous (c.0.5mm wide) translucent
   shallowly and irregularly toothed border on each
   angle   7. EREPSIA
5 Leaves not or scarcely wider or thicker near apex,
   terete to trigonous or rarely triquetrous; with a
   very narrow ± entire translucent border   6
   6 Main stems long-procumbent, often rooting at nodes,
      mat-forming   7
   6 Stems upright to ascending or rarely ± procumbent,
      stiff and strongly woody below, forming upright
      or spreading shrub   8
7 Leaves covered with rounded whitish papillae; young
   stems with white, patent or reflexed hairs
                                        6. DROSANTHEMUM
7 Leaves not papillose; stems not hairy   5. DISPHYMA
   8 All leaves <3cm long and/or <5mm thick; styles
      stout (c.0.4mm wide at mid length); stigmas tuft-
      like; flowers (20) 0-50mm across   3. LAMPRANTHUS
   8 Some leaves >3cm long and >5mm thick; styles
      filiform (c.0.1mm wide at mid length); stigmas
      tapered; flowers 18-30mm across   2. RUSCHIA

1. APTENIA N.E. Br. - Heart-leaf Ice-plant
Perennial; leaves flattened, succulent, entire, petiolate;
flowers axillary and terminal, solitary; sepals 4; petals
numerous; stigmas 4; capsule woody, opening by 4 valves, with
axile placentation.

   1. A. cordifolia (L.f.) Schwantes - Heart-leaf Ice-plant. 160
Stems woody below, freely branched, spreading, to 60cm;
leaves ovate, cordate to cuneate at base, acute to obtuse at
apex, papillose; flowers (5)10-18mm across, purplish-red.
Intrd; grown as pot-plant or summer bedding plant and
escaping on walls and dry ground in Scillies and CI; S
Africa.

2. RUSCHIA Schwantes - Purple Dew-plant
Dwarf shrubs, strongly woody below; leaves trigonous to ±
terete, entire, sessile, opposite pairs united into common

sheath; flowers terminal, usually 3-4 together; sepals 5;
petals numerous; stigmas 4-5; capsule woody, opening by 4-5
wingless valves, with parietal placentation.

1. R. caroli (L. Bolus) Schwantes - Purple Dew-plant. Stems **160**
± erect to ± procumbent, to 80cm; leaves 1.5-7cm, 2-8mm wide
and thick, acute at apex, glaucous with green dots; flowers
18-30mm across, purplish-red. Intrd; grown for ornament and
escaping on sea-cliffs in Scillies; S Africa.

3. LAMPRANTHUS N.E. Br. - Dew-plants
Dwarf shrubs, strongly woody below; leaves triquetrous to ±
terete, entire or nearly so, sessile, opposite pairs shortly
united; flowers terminal, 1-many together; sepals 5; petals
numerous; stigmas 5; capsule woody, opening by 5 winged
valves, with parietal placentation.
A large critical S African genus.

Other spp. - **L. conspicuus** (Haw.) N.E. Br., with leaves
5-7cm and purplish flowers c.5cm across, has been reported
from Scillies, and **L. scaber** (L.) N.E. Br., with leaves 2-3cm
and with roughly papillose edges and purplish flowers c.3cm
across, has been reported from Jersey.

1. L. falciformis (Haw.) N.E. Br. - Sickle-leaved Dew-plant. **160**
Erect, bushy, to 30cm; leaves (6)10-15(20) x 1.5-5mm,
falcate, sharply to bluntly triquetrous, markedly mucronate
at apex, conspicuously dotted; flowers 3.5-4.5cm across, pale
pink. Intrd; cultivated near sea and natd on walls, hedgebanks, cliffs and quarries; Pembs, Scillies, ?CI; S Africa.
2. L. roseus (Willd.) Schwantes (?L. multiradiatus (Jacq.) **160**
N.E. Br.) - Rosy Dew-plant. Erect bushy shrub to 60cm,
similar to L. falciformis; differing in leaves (10)20-40mm,
not or slightly falcate, acute or acuminate to obtuse or
obscurely mucronate at apex, less conspicuously dotted;
flowers (2)3-5cm across. Intrd; in similar places to L.
falciformis; W Cornwall, E & W Cork, CI; S Africa.

4. OSCULARIA Schwantes - Deltoid-leaved Dew-plant
Perennial; leaves sharply triquetrous, with conspicuous
distant teeth on all 3 angles, opposite pairs shortly united;
flowers terminal, 1-3 together; sepals 5; petals numerous;
stigmas 5; capsule woody, opening by 5 narrowly winged
valves, with parietal placentation.

1. O. deltoides (L.) Schwantes (Lampranthus deltoides (L.) **160**
Glen) - Deltoid-leaved Dew-plant. Stems woody below, with
spreading branches to 50cm; leaves 6-18mm, scarcely less wide
and thick, narrowed towards base, acute to truncate at apex,
glaucous; flowers 10-20mm across, pink. Intrd; grown on
walls and banks and escaping or persisting; Scillies, very
rare in Guernsey; S Africa.

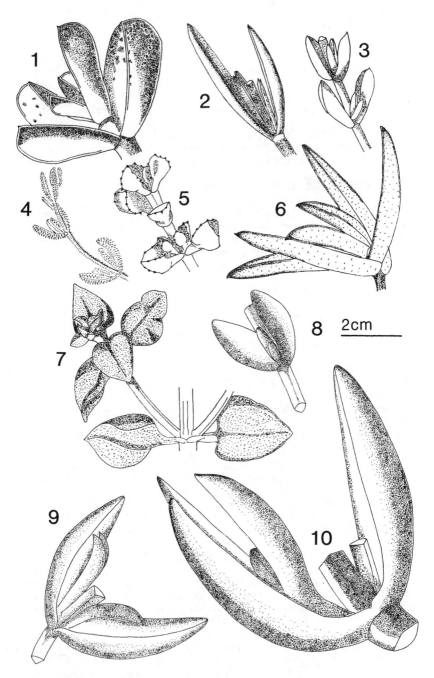

FIG 160 – Aizoaceae. 1, Erepsia. 2, Lampranthus roseus. 3, L. falciformis. 4, Drosanthemum. 5, Oscularia. 6, Ruschia. 7, Aptenia. 8, Disphyma. 9, Carpobrotus glaucescens. 10, C. acinaciformis.

5. DISPHYMA N.E. Br. - <u>Dew-Plants</u>
Perennial, with procumbent, fleshy, somewhat woody main
stems; leaves trigonous to + terete, entire, smooth,
sessile, opposite pairs very shortly united; flowers terminal
on short, + erect, lateral branches, mostly solitary;
sepals 5; petals numerous; stigmas 5; capsule rather spongy,
opening by 5 winged valves, with parietal placentation.

Other spp. - D. australe (Sol. ex G. Forster) J. Black, from
Australia and New Zealand, has been reported from Scillies,
but is only doubtfully distinct from D. crassifolium.

1. D. crassifolium (L.) L. Bolus - <u>Purple Dew-plant</u>. Main **160**
stems to 1m, rooting at nodes, well branched; leaves 12-40 x
c.5mm, dark green, with translucent dots, obtuse to mucronate
at apex; flowers 2.5-4(5)cm across, reddish-purple. Intrd;
grown for ornament near sea and natd on walls, cliffs and
sandy places; W Cornwall, Scillies, E Sussex, Anglesey, CI; S
Africa.

6. DROSANTHEMUM Schwantes - <u>Pale Dew-plant</u>
Perennial, with thin, procumbent, woody main stems; leaves
+ terete, entire, covered in rounded papillae, sessile,
opposite pairs not united; flowers terminal on short, +
erect, lateral branches, mostly solitary; sepals 5; petals
numerous; stigmas mostly 5; capsule rather woody, opening by
mostly 5 winged valves, with parietal placentation.

Other spp. - D. candens (Haw.) Schwantes, from S Africa,
with pure white flowers, has been reported from SW En, but
needs confirming; it might not be a distinct sp.

1. D. floribundum (Haw.) Schwantes - <u>Pale Dew-plant</u>. Main **160**
stems to 80cm, scarcely rooting at nodes, well branched;
younger stems densely pubescent; leaves 6-20 x 2-4mm, covered
with whitish papillae, obtuse at apex; flowers 12-25mm
across, pinkish-mauve. Intrd; grown for ornament near sea
and well natd on walls, rocks and cliffs; W Cornwall,
Scillies, CI; S Africa.

7. EREPSIA N.E. Br. - <u>Lesser Sea-fig</u>
Perennial; leaves sharply triquetrous, with translucent,
shallowly and irregularly toothed edges (especially the
abaxial), opposite pairs shortly united; flowers terminal,
1-3 together; sepals 5; petals numerous; stigmas 5-6; capsule
woody to spongy, opening by 5-6 narrowly winged valves, with
parietal placentation.

1. E. heteropetala (Haw.) Schwantes - <u>Lesser Sea-fig</u>. Stems **160**
woody below, with erect to ascending branches to 30cm; leaves
15-40 x 5-10mm, 6-15mm thick, widest near apiculate apex,
green to reddish; flowers inconspicuous, 10-15mm across,

reddish. Intrd; garden escape or throw-out, spreading vegetatively and from seed; in quarry in Scillies; S Africa. Vegetatively somewhat resembles Carpobrotus.

8. CARPOBROTUS N.E. Br. - Hottentot-figs
Perennial; leaves triquetrous, with translucent edges, entire or minutely toothed, opposite pairs shortly united; flowers terminal, solitary; sepals 5; petals numerous; stigmas c.8-20; fruit succulent, indehiscent, the seeds embedded in mucilage, with parietal placentation.
A critical genus; the identity of many plants natd in BI is still far from clear.

1 Petals yellow, becoming pinkish as they wither
                                                      2. C. edulis
1 Petals pink to purple from first   2
    2 Leaves thickest close to apex (scimitar-shaped),
      distinctly narrower than thick; flowers 8-10
      (12)cm across   **1. C. acinaciformis**
    2 Leaves ± equally thick for most of length, about
      as wide as thick; flowers <8.5(10)cm across   3
3 Flowers 7-8.5(10)cm across; petals often slightly
  paler but not white or yellow at base; ripe fruit
  little or not longer than wide   **2. C. edulis**
3 Flowers 3.5-6(8)cm across; petals ± white or
  yellow at base; ripe fruit usually distinctly longer
  than wide   **3. C. glaucescens**

Other spp. - Some plants might be referable to **C. aequilaterus** (Haw.) N.E. Br., from Australia, or **C. chilensis** (Molina) N.E. Br, from Chile and W USA; these might be conspecific. Plants in SW En, slightly smaller than C. edulis and with yellow-based petals (**C. edulis** var. **chrysophthalmus** Preston & Sell), might belong here. The genus badly needs revising.

1. **C. acinaciformis** (L.) L. Bolus - Sally-my-handsome. 160
Stems procumbent, woody, angled, to 2m; leaves 4-10cm, thickest near apex with strongly curved abaxial angle, much thicker than wide; flowers 8-10(12)cm across; petals pinkish-purple. Intrd; on rocks, cliffs and sand near sea; Devon, Cornwall and Scillies; S Africa. Much confused with purple-flowered plants of C. edulis.
    2. **C. edulis** (L.) N.E. Br. - Hottentot-fig. Stems procumbent, woody, angled, to 3m; leaves 4-9cm, equally thick for most of length, c. as thick as wide; flowers 7-8.5(10)cm across; petals yellow (fading pinkish) or pinkish-purple (var. **rubescens** Druce). Intrd; on rocks, cliffs and sand near sea; CI, Scillies to N Wa and E Kent, very local in W Lancs, Man and S & E Ir; S Africa.
    3. **C. glaucescens** (Haw.) Schwantes - Angular Sea-fig. 160
Similar to C. edulis but smaller; flowers 3.5-6cm across;

petals pinkish-purple, white or very pale pink at base.
Intrd; rocks and cliffs by sea; CI, E Suffolk, Wigtowns; E
Australia.

9. TETRAGONIA L. - New Zealand Spinach
Annual; leaves alternate, flat, ± succulent, entire,
abruptly narrowed to long petiole; flowers axillary, mostly
solitary; sepals 4-5; petals 0; stigmas 3-8; fruit ± woody,
indehiscent, ridged, with 1 seed in each of 3-8 cells.

1. T. tetragonoides (Pallas) Kuntze - New Zealand Spinach.
Stems much branched, procumbent to ascending, to 1m; leaves
to 10cm, ovate to rhombic, cuneate to hastate at base, acute
to obtuse at apex, papillose; flowers very inconspicuous,
yellow-green.   Intrd; cultivated as a leaf-vegetable and
found on rubbish-tips in S En; Australia and Japan to S
America.

44. CHENOPODIACEAE - Goosefoot family

Herbaceous annuals or perennials or shrubs; leaves usually
alternate, rarely opposite, sometimes succulent, simple,
exstipulate, petiolate or sessile.   Flowers small and
greenish, sometimes with bracteoles, usually borne in cymes,
the cymes axillary or in panicles, rarely solitary, bisexual
or unisexual, hypogynous or semi-inferior, actinomorphic;
perianth herbaceous, of 1 whorl of 3-5 free or partly fused
tepals, sometimes 0; stamens as many as tepals or fewer;
ovary 1-celled, with 1 ovule; styles 2-3; stigma linear or
feathery; fruit an achene.
A family of mainly dull-coloured weedy plants usually
recognizable by the 1-whorled herbaceous perianth, 1-celled
superior or semi-inferior ovary with 1 ovule, and 2-3 styles.

1 Leaves fused in opposite pairs, forming succulent
  sheath round stem, with 0 or very short free part   2
1 Leaves not fused to form succulent sheath round stem,
  usually alternate, with distinct free lamina   3
  2 Annual, easily uprooted in entirety   **7. SALICORNIA**
  2 Perennial, with procumbent rhizomes at or just
    below soil surface giving rise to aerial stems
                                          6. SARCOCORNIA
3 Leaves <5mm wide, succulent, entire   4
3 Leaves not both <5mm wide and succulent, usually
  neither, often lobed or toothed at margin   5
  4 Leaves ending in a spine, plant bristly   **9. SALSOLA**
  4 Leaves acute to obtuse but without a spine;
    plant glabrous   **8. SUAEDA**
5 Flowers bisexual, or bisexual and female; fruits
  surrounded by persistent tepals   6
5 Flowers unisexual, the female at least mostly without

          perianth; fruits surrounded by 2 enlarged bracteoles   8
     6 Tepals at fruiting with very short transverse
        wing or tubercle abaxially   **2. BASSIA**
     6 Tepals at fruiting without transverse tubercle or
        wing abaxially, but often with longitudinal keel   7
7 Ovary semi-inferior; receptacle becoming swollen at
   fruiting   **5. BETA**
7 Ovary superior; receptacle not becoming swollen at
   fruiting   **1. CHENOPODIUM**
     8 Stigmas 2; bracteoles almost free to ± completely
        fused, if fused >1/2 way then leaves mealy-white
        and cuneate at base   **4. ATRIPLEX**
     8 Stigmas 4-5; bracteoles ± completely fused; leaves
        green, at least some ± truncate at base   **3. SPINACIA**

Other genera - **AXYRIS** L. would key out as _Atriplex_ but
female flowers have 3 tepals, the achenes have a terminal
wing and the leaves are densely stellate-pubescent. **A.
amaranthoides** L. (_Russian Pigweed_), from Russia, is an
annual to 80cm and occasionally occurs in waste places.

1. CHENOPODIUM L. (_Blitum_ L.) - _Goosefoots_
Annual or perennial herbs; leaves flattened, often mealy,
entire, toothed or lobed; bracteoles 0; flowers bisexual or
some female; tepals mostly 4-5, persistent and surrounding
fruit, with or without abaxial longitudinal keel.
  Vegetatively extremely plastic, especially in habit and leaf
shape.   Testa sculpturing is important and sometimes
essential for identification. It can be examined under >x20
magnification after removal of the pericarp, which may be
effected either by rubbing in the hand or sometimes only
after boiling and dissection. The orientation of the seed
is also important: either 'vertical', with long axis parallel
to the length of the flower; or 'horizontal', with it at
right angles to the length of the flower. 5 spp. (6, 17, 19,
20, 25) were formerly commoner than now and are included only
for comparison, since they are often represented in herbaria.
C. album is extremely variable and may very closely
resemble typical variants of spp. 21, 22 and 24-27, which are
distinguished only with difficulty and by a combination of
characters.   They often occur mixed with C. album and are
often frosted before seeding. Spp. 21, 24, 26 and 27, and
perhaps 22 and 25, are doubtfully distinct from C. album.

1 Stems glandular-pubescent, at least towards apex;
   plant aromatic   2
1 Stems glabrous or mealy, not glandular-pubescent;
   plant not aromatic, sometimes stinking   6
     2 Tepals fused >1/2 way, net-veined on outside
                                        2. C. multifidum
     2 Tepals not fused or fused <1/2 way, not net-veined
        on outside   3

3 Flower clusters in axillary racemes or panicles; at
   least some seeds horizontal; larger leaves rarely <3cm
                                        1. C. ambrosioides
3 Flower clusters sessile and solitary in leaf axils;
   seeds vertical; larger leaves rarely >3cm   4
   4 Tepals rounded abaxially, not keeled, not meeting
      at margins and only partially concealing fruit
                                            3. C. pumilio
   4 Tepals prominently keeled abaxially, + concealing
      the fruit   5
5 Tepals + truncate and variably toothed at apex in
   side view; keel entire   **4. C. carinatum**
5 Tepals with long beak at apex; keel deeply laciniate
   along most of length   **5. C. cristatum**
   6 Stems woody at least below; inflorescence
      branchlets ending in bare weakly spinose points
                                         7. C.nitrariaceum
   6 Stems herbaceous; branchlets not bare and spinose
      at tips   7
7 Flowers in a spike of globose sessile axillary heads
   >5mm across; perianth turning red and fleshy at
   fruiting   **6. C. capitatum**
7 Flowers in racemes or panicles of heads usually <5mm
   across; perianth not turning red and fleshy at
   fruiting   8
   8 Rhizomatous perennial; stigmas 0.8-1.5mm
                                      8. C. bonus-henricus
   8 Annual; stigmas <0.8mm   9
9 Fruiting perianths mostly longer than wide, with
   vertical or oblique seeds; inflorescence glabrous   10
9 Fruiting perianths wider than long, with horizontal
   seeds; inflorescence mealy or glabrous   12
   10 Leaves green on upperside, mealy-grey on
      lowerside   **9. C. glaucum**
   10 Leaves green (to reddish) on both surfaces   11
11 Tepals of lateral fruits in each cluster fused
   <1/2 way; leaves usually strongly toothed or lobed
                                           10. C. rubrum
11 Tepals of lateral fruits fused to near apex; leaves
   usually entire to sparsely lobed or toothed
                                      11. C. chenopodioides
   12 Leaf-blades weakly cordate at junction with
      petiole, at least on some leaves   **14. C. hybridum**
   12 Leaf-blades cuneate at junction with petiole   13
13 Seeds distinctly acutely keeled at edges; tepals
   minutely denticulate   **16. C. murale**
13 Seeds with subacute to rounded unkeeled edges;
   tepals entire   14
   14 Leaves entire or at most with 1 obscure tooth or
      lobe on each side   15
   14 At least lower leaves distinctly toothed and/or
      lobed   18

15 Leaves green (or reddish) on both surfaces, not
   mealy; stems tetraquetrous   **12. C. polyspermum**
15 Leaves mealy-grey at least on lowerside; stems ±
   terete to ridged   16
   16 Leaves ovate-trullate, <2.5cm; tepals rounded
      abaxially; plant stinking like rotten fish
                                                      13. C. vulvaria
   16 Leaves linear to triangular-ovate, rarely ovate-
      trullate, the largest usually >2.5cm; tepals
      keeled abaxially; plant not stinking   17
17 Leaves linear to linear-oblong, densely mealy-grey on
   lowerside, distinctly mucronate at apex, mostly with
   only 1(-2) pairs of lateral veins visible; petiole
   <1cm   **17. C. desiccatum**
17 Leaves usually oblong to ovate, trullate or
   triangular, less densely mealy-grey on lowerside,
   rarely mucronate at apex, mostly with >2 pairs of
   lateral veins visible; longest petioles usually >1cm 24
   18 Stems, leaves and flowers not or very sparsely
      mealy; leaves trullate to triangular   **15. C. urbicum**
   18 At least flowers and small branchlets conspicuously
      mealy; leaves various   19
19 Testa with conspicuous regular pits delimited by
   pronounced reticulum of ridges   20
19 Testa irregularly pitted to almost featureless, often
   with radial and/or tangential furrows, sometimes with
   regular reticulum of slightly raised ridges but with
   flat (not concave) areas within   22
   20 Lower leaves rarely distinctly 3-lobed; tepals
      with strong wing-like keel along whole length;
      testa with almost isodiametric honeycomb-like
      pitting   **20. C. berlandieri**
   20 Lower leaves usually distinctly 3-lobed; tepals
      weakly keeled or strongly so in distal part only;
      testa with radially-elongated pitting   21
21 Plant not stinking like rotten fish; lower leaves
   with elongated central lobe 2-3x as long as side
   lobes; seeds <1.4mm across   **18. C. ficifolium**
21 Plant stinking like rotten fish; lower leaves with
   short central lobe little longer than side lobes;
   seeds >1.4mm across   **19. C. hircinum**
   22 Plant to 2m; young shoots usually extensively
      coloured reddish-purple; larger leaves <14cm, ±
      always some >6cm, ovate-trullate to ovate-
      triangular   **27. C. giganteum**
   22 Plant usually less robust, if >1.5m leaves <6cm;
      plant green or variously red-tinged or -striped,
      but not with young shoots extensively reddened   23
23 Seeds (1.3)1.5-2mm in longest diameter, rarely <1.5mm
   in shortest diameter; perianth often blackish-green,
   rather sparsely mealy; leaves trullate   **21. C. bushianum**
23 Seeds <1.5mm in longest diameter, <1.5mm in shortest

diameter; perianth usually green to grey, variously
mealy; leaves various   24
24 Leaves not or little longer than wide, ovate-
   trullate, not lobed or with 2 basal shallow lobes;
   tepals often fused to c.1/2 way; infloresence
   densely mealy   **22. C. opulifolium**
24 Leaves distinctly longer than wide; tepals
   usually fused to <1/2 way; inflorescence variously
   mealy   25
25 Leaves narrowly oblong, with ± parallel sides, often
   obtuse, with shallow teeth but not lobed; seeds with
   ratio of longest to shortest diameters usually >1.15
                                                24. C. strictum
25 Leaves various but not with parallel sides, usually
   acute; seeds with ratio of longest to shortest
   diameters usually <1.15   26
   26 Seeds with subacute edges as seen in narrowest
      profile   **23. C. album**
   26 Seeds with obtuse to rounded edges as seen in
      narrowest profile   27
27 Plant often >1m, frequently tinged with red; seeds
   usually <1.25mm in longest diameter but flowering
   very late, usually frosted before seeding; tepals
   slightly keeled   **26. C. probstii**
27 Plant <1m, not tinged with red (except rarely in
   leaf-axils); seeds often >1.25mm in longest diameter,
   usually produced well before frosts; tepals
   usually well keeled   **25. C. suecicum**

Other spp. - About 40 additional spp. have been reliably
recorded from Br, especially as grain- and wool-aliens, but
all are rarer than the above except 6, 17, 19, 20 and 25. **C.
quinoa** Willd. (Quinoa), from S America, is becoming
cultivated in gardens as a grain-crop and is likely to occur
on tips; it has large seeds as in C. bushianum, but very
dense inflorescences showing the conspicuously pale straw-
coloured pericarps and perianths.

Section 1 - AMBRINA Hook. f. (sp. 1). Annuals with glandular
hairs, not mealy; flowers in dense axillary clusters arranged
in terminal leafy panicles; tepals fused <1/2 way, rounded
abaxially; stamens (3-)5; seeds <1mm across, vertical and
horizontal.

1. C. ambrosioides L. - Mexican-tea. Strongly aromatic;
stems upright, to 1m; leaves usually lanceolate, entire to
deeply dentate. Intrd; casual or rarely natd in waste places
and tips, from wool, soya-bean waste, bird-seed and other
sources; occasional in S Br; warm parts of America.

Section 2 - ROUBIEVA (Moq.) Volkens (sp. 2). Annuals with
glandular hairs, not mealy; flowers in sessile axillary

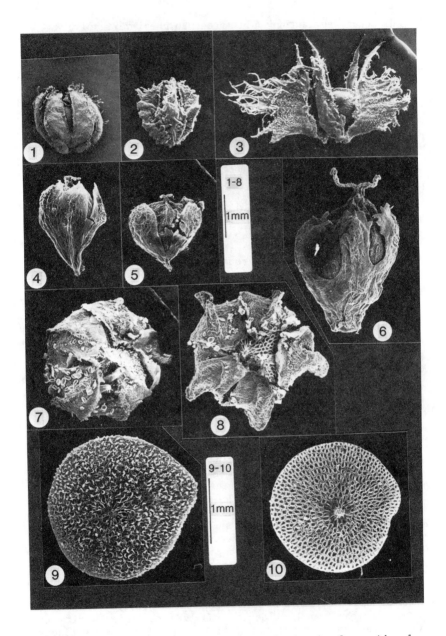

FIG 168 – **Chenopodium**. 1–6, fruiting perianths from side. 1, **C. pumilio**. 2, **C. carinatum**. 3, **C. cristatum**. 4, **C. chenopodioides**. 5, **C. rubrum**. 6, **C. bonus–henricus**. 7–8, fruiting perianths from apex. 7, **C. album**. 8, **C. berlandieri**. 9–10, fruits from apex. 9, **C. album**. 10, **C. berlandieri**.

clusters; tepals fused >1/2 way, rounded abaxially; stamens
5; seeds mostly <1mm, vertical.

2. C. multifidum L. - <u>Scented</u> <u>Goosefoot</u>. Aromatic; stems
procumbent to erect, to 50cm; leaves narrowly oblong, with
narrow deep regular lobes. Intrd; casual on tips and in
fields as wool-alien; rather rare in S Br; S America.

<u>Section</u> <u>3</u> - <u>ORTHOSPORUM</u> R. Br. (spp. 3-5). Annuals with
glandular hairs, not mealy; flowers in sessile axillary
clusters; tepals fused <1/2 way, rounded or keeled abaxially;
stamen 1; seeds <1mm, vertical.

3. C. pumilio R. Br. <u>Clammy Goosefoot</u>. Aromatic; stems **168**
procumbent to ascending, to 50cm; leaves elliptic-oblong,
with shallow to medium rounded lobes; tepals narrow, not
fully concealing fruit, rounded and without keel abaxially.
Intrd; casual or rarely persisting on tips and in fields as
wool-alien; occasional in S Br; Australia.

4. C. carinatum R.Br. - <u>Keeled</u> <u>Goosefoot</u>. Similar to <u>C.</u> **168**
<u>pumilio</u> but tepals widest near middle, shallowly to
conspicuously toothed distally, with ± entire prominent keel.
Intrd; casual on tips and in fields as wool-alien; rather
rare in S En; Australia.

4 x 5. C. carinatum x C. cristatum = C. x bontei Aellen
occurs rarely in S En as a wool-alien; it has intermediate
tepals.

5. C. cristatum (F. Muell.) F. Muell. - <u>Crested Goosefoot</u>. **168**
Similar to <u>C. carinatum</u> but tepals with prominent, deeply
laciniate keel. Intrd; casual on tips and in fields as
wool-alien; rather rare in S En; Australia.

<u>Section</u> <u>4</u> - <u>BLITUM</u> (L.) Benth. & Hook. f. (sect. <u>Morocarpus</u>
Graebner) (sp.6). Annuals, ± glabrous; flowers in sessile
axillary clusters, the uppermost not subtended by bract;
tepals fused <1/2 way, becoming red and succulent in fruit;
stamen often 1; seeds c.1mm, mostly vertical.

6. C. capitatum (L.) Asch. (<u>Blitum</u> <u>capitatum</u> L.) -
<u>Strawberry-blite</u>. Stems erect, to 50cm; leaves triangular-
ovate, usually with prominent lobe on each side near base,
sometimes with more distal lobes. Intrd; casual or sometimes
persisting in fields and waste places; sporadic in Br and Ir;
widespread in warm areas.

<u>Section</u> <u>5</u> - <u>RHAGODIOIDES</u> Benth. (sp. 7). Divaricately
branching, ± mealy shrub; ultimate branches bare and ±
spinose; flowers in small clusters in terminal spicate to
subpaniculate inflorescences; tepals fused <1/2 way, rounded
abaxially; stamens 5; seeds c.1mm, vertical.

7. C. nitrariaceum (F. Muell.) F. Muell. ex Benth. - <u>Nitre</u>

Goosefoot. Straggly shrub to 1m; leaves entire, linear to oblanceolate or narrowly oblong. Intrd; casual in fields and on tips as wool-alien; rather rare in En; Australia.

Section 6. - AGATHOPHYTON (Moq.) Benth. & Hook. f. (sp. 8). Rhizomatous, sparsely mealy herbaceous perennials; flowers in small clusters arranged in terminal spicate or subpaniculate inflorescences; tepals fused <1/2 way, rounded or slightly keeled abaxially; stamens 4-5; stigmas >0.8mm; seeds >1mm, mostly vertical.

8. C. bonus-henricus L. - Good-King-Henry. Stems erect, to **168** 50cm; leaves triangular with prominent basal lobes, otherwise entire or sinuate or sparsely and shallowly lobed. Native; roadsides, pastures and by farm buildings in nitrogen-rich places; scattered and locally common over most of BI.

Section 7 - CHENOPODIUM (sect. Pseudoblitum Benth. & Hook. f.) (spp. 9-10). Annuals, glabrous or mealy; flowers in clusters arranged in terminal leafy panicles; tepals fused <1/2 way, slightly or not keeled abaxially; stamens 2-5; seeds mostly <1mm, vertical and horizontal.

9. C. glaucum L. - Oak-leaved Goosefoot. Stems much **173** branched, procumbent to erect, to 50cm, + glabrous; leaves **RR** narrowly elliptic to elliptic, shallowly but rather regularly lobed, green on upperside, mealy-grey on lowerside; plant rarely red-tinged. Possibly native; waste places on rich soils, often near sea, and casual on tips and in dockland; very local in Br, mostly in S & E En.

10. C. rubrum L. - Red Goosefoot. Stems much to little **168** branched, procumbent to erect, to 80cm; leaves ovate to **173** triangular or elliptic, variably but often strongly lobed; plant often red-tinged, especially at fruiting. Native; cultivated and waste ground, often near sea, and on tips; frequent or common in much of En, local or rare elsewhere.

Section 8 - DEGENIA Aellen (sp. 11). Annuals, + glabrous; flowers in clusters arranged in terminal bracteate spikes or panicles; tepals of lateral flowers fused >1/2 way at fruiting, slightly keeled abaxially; stamens 1-5; seeds <1mm, vertical and horizontal.

11. C. chenopodioides (L.) Aellen (C. botryodes Smith) - **168** Saltmarsh Goosefoot. Resembles a small (to 30cm), usually **173** much branched and procumbent to ascending C. rubrum, but **R** leaves usually triangular and with only the basal lobes well developed; diagnosed by perianth (see key). Native; by dykes and in barish pastures near sea; local in SE En and CI.

Section 9 - LEPROPHYLLUM Dumort. (sect. Chenopodium auct.) (spp. 12-27). Annuals, glabrous to densely mealy; flowers in

clusters arranged in leafy or leafless spikes or panicles;
tepals fused <1/2 way, rounded to keeled abaxially; stamens
5; seeds mostly >1mm, horizontal.

12. C. polyspermum L. - Many-seeded Goosefoot. Plant ± **173**
glabrous; stems usually much branched, decumbent to
ascending, to 1m; leaves ± glabrous, ovate to elliptic, ±
entire; tepals rounded abaxially; testa with raised radial,
sinuate striations. Native; waste and cultivated ground;
common in CI and C & S En, local or rare and not native in
Wa, S Sc and Ir.

13. C. vulvaria L. - Stinking Goosefoot. Plant stinking of **173**
rotten fish, mealy-grey; stems much branched from base, **RRR**
mostly ascending, to 40cm; leaves ovate-trullate, ± entire;
tepals rounded abaxially; testa with faint radial furrows.
Probably native; barish places near sea; rare in S En and CI,
formerly frequent but very much reduced; now most often as
casual in waste places in C & S BI.

14. C. hybridum L. - Maple-leaved Goosefoot. Plant glabrous **173**
to very slightly mealy; stems erect, to 1m; leaves ovate-
triangular, cordate at base, with few acute lobes; tepals
rounded or slightly keeled abaxially; testa covered with
small, regular, deep pits. Probably intrd; waste and arable
ground; rare in Br, mainly S and usually casual; Europe.

15. C. urbicum L. - Upright Goosefoot. Plant ± glabrous; **173**
stems erect, to 1m; leaves ovate-trullate to -triangular,
with variable mostly acute lobes; tepals ± rounded abaxially;
testa with rather faint reticulate furrows. Probably intrd;
waste and cultivated ground, usually from grain; now much
rarer than formerly; rare and very scattered in En and Wa and
usually only casual; Europe. Distinguished from C. rubrum by
the horizontal seeds with black (not brown) testa.

16. C. murale L. - Nettle-leaved Goosefoot. Similar to C. **173**
urbicum but inflorescence much looser, more branched and
somewhat mealy; seeds with acute keel around margin; tepals
minutely denticulate; testa covered with minute rounded pits.
Probably native near sea in S En; waste and cultivated
ground; rarely established, very scattered casual over much
of BI, locally frequent in SE En and CI.

17. C. desiccatum Nelson (C. pratericola Rydb.). Plant **173**
mealy; stems well branched, ascending to erect, to 1m; leaves
linear to linear-oblong, ± entire; tepals strongly keeled;
testa with reticulum of furrows. Intrd; tips and waste
ground; rather rare casual, mainly from grain, S Br; N
America. C. pratericola is possibly a distinct sp.

18. C. ficifolium Smith - Fig-leaved Goosefoot. Plant **173**
mealy; stems ascending to erect, to 1m; leaves distinctly
3-lobed, the central lobe narrow and 2-3x as long as
laterals; tepals rather weakly keeled abaxially; testa with
deep conspicuous pits ± arranged in radial rows and
slightly radially elongated. Native; waste and arable
ground; CI, S & E En and S Wa, casual in rest of Br.

19. C. hircinum Schrader - <u>Foetid</u> <u>Goosefoot</u>. Plant stinking **173** of rotten fish, mealy; stems erect to ascending, to 1m; leaves distinctly 3-lobed, the central lobe the widest but little longer than laterals; tepals keeled abaxially, usually strongly so towards apex; testa ± as in <u>C. ficifolium</u> but pits strongly radially elongated. Intrd; casual on tips and waste ground from wool and bird-seed, formerly fairly frequent but now rare; En and Wa; S America.

20. C. berlandieri Moq. (<u>C.</u> <u>zschackei</u> Murr) - <u>Pitseed</u> **168** <u>Goosefoot</u>. Plant mealy; stems erect, to 1.5m; leaves **173** trullate-ovate, variously toothed and lobed, sometimes the lower ± 3-lobed; tepals with strong wing-like keel abaxially; testa with deep, conspicuous ± honeycomb-like pits. Intrd; tips and waste ground, formerly fairly frequent as grain-alien but now very rare; En and S Wa; N America. Often very closely resembles <u>C. album</u> but has quite different testa markings.

21. C. bushianum Aellen - <u>Soya-bean</u> <u>Goosefoot</u>. Like <u>C.</u> **173** <u>album</u> but only moderately mealy; leaves triangular to trullate, usually very shallowly toothed; tepals with moderate keel abaxially; 2n=36. Intrd; tips and waste ground, mainly from soya-bean waste; S En; N America.

22. C. opulifolium Schrader ex Koch & Ziz - <u>Grey</u> <u>Goosefoot</u>. **173** Like <u>C. album</u> but densely mealy; stems never reddish-purple; leaves ovate-trullate, entire to shallowly lobed or toothed; tepals slightly keeled abaxially; 2n=54. Intrd; tips and waste ground from many sources, often grain-alien; scattered in Br, mainly in S; Europe.

23. C. album L. (<u>C.</u> <u>reticulatum</u> Aellen, <u>C.</u> <u>album</u> ssp. **168** <u>reticulatum</u> (Aellen) Beauge ex Greuter & Burdet) - <u>Fat-hen</u>. **173** Plant variably mealy; stems erect to ascending, often reddish- purple suffused or in axils, to 1.5m; leaves lanceolate to ovate, trullate or triangular, ± entire to shallowly lobed or toothed; tepals with moderate to indistinct keel abaxially; testa ± smooth to weakly radially furrowed or rarely (var. **reticulatum** (Aellen) Uotila) with **173** prominent reticulate ridges; 2n=54. Native; waste and cultivated ground; throughout BI. Extremely variable, especially in height, branching, leaf-shape, mealiness and even testa markings.

<u>C. album</u> has been recorded as hybridising with a number of other spp., notably <u>C. ficifolium</u> (=**C. x zahnii** Murr), <u>C.</u> <u>berlandieri</u> (=**C. x variabile** Aellen), <u>C. opulifolium</u> (=**C. x preissmannii** Murr) and <u>C. suecicum</u> (=**C. x fursajewii** Aellen & Iljin). These 4 hybrids have been reported from several parts of Br, but many if not all plants were probably 1 or other putative parent, the variation of which makes the determination of hybrids extremely hazardous.

24. C. strictum Roth (<u>C.</u> <u>striatum</u> (Krasan) Murr - <u>Striped</u> **173** <u>Goosefoot</u>. Like <u>C. album</u> but stems usually red-striped; leaves narrowly oblong with ± parallel sides, shallowly toothed but not lobed; tepals scarcely keeled; 2n=36. Intrd;

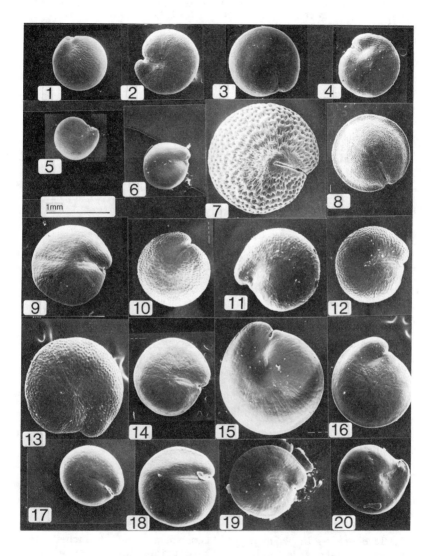

FIG 173 – Seeds of **Chenopodium**. 1, **C. glaucum**. 2, **C. polyspermum**. 3, **C. vulvaria**. 4, **C. urbicum**. 5, **C. rubrum**. 6, **C. chenopodioides**. 7. **C. hybridum**. 8, **C. murale**. 9, **C. desiccatum**. 10, **C. ficifolium**. 11, **C. hircinum**. 12, **C. berlandieri**. 13, **C. bushianum**. 14, **C. opulifolium**. 15, **C. album**. 16, **C. album** var. **reticulatum**. 17, **C. strictum**. 18, **C. suecicum**. 19, **C. probstii**. 20, **C. giganteum**.

tips and waste places mainly from grain and wool; rather rare
in Br, mainly S; Europe.
25. C. suecicum Murr - <u>Swedish</u> <u>Goosefoot</u>. Like <u>C</u>. album but **173**
stems rarely with reddish-purple; leaves ovate-trullate,
usually sharply toothed; tepals moderately to prominently
keeled abaxially; seeds with rather more strongly furrowed
testa; 2n=18. Intrd; tips and waste places; formerly
frequent, now rare, extremely scattered in Br; N Europe.
26. C. probstii Aellen - <u>Probst's</u> <u>Goosefoot</u>. Like <u>C</u>. album **173**
but often to 2m; stems and leaves usually tinged reddish-
purple; lower leaves usually with distinct basal lobe on each
side, sharply toothed; tepals slightly keeled abaxially;
2n=54. Intrd; tips and waste places from wool, bird-seed,
soya-bean and other sources; scattered in Br; N America.
27. C. giganteum D. Don (<u>C</u>. album ssp. <u>amaranticolor</u> Coste & **173**
A. Reynier) - <u>Tree</u> <u>Spinach</u>. Like <u>C</u>. album but usually much
larger and extensively reddish-purple on stems and leaves;
stems to 2m; leaves ovate-trullate to ovate-triangular,
irregularly toothed but scarcely lobed, <14cm; tepals with
indistinct keel abaxially; 2n=54. Intrd; tips and waste
places mainly from wool; scattered in En; India.

2. BASSIA All. (<u>Kochia</u> Roth) - <u>Summer-cypress</u>
Annual herb; leaves flat, not mealy, entire, at least the
lower pubescent; bracteoles 0; flowers bisexual or some
female; tepals 5, developing small transverse wing or
tubercle abaxially at fruiting.

1. B. scoparia (L.) Voss (<u>Kochia</u> <u>scoparia</u> (L.) Schrader) -
<u>Summer-cypress</u>. Stem erect, much branched, forming dense
ovoid bushy plant to 1m; leaves linear to lanceolate,
sessile, < 5cm but usually <3cm; whole plant often becoming
purplish-red in autumn. Intrd; much grown in gardens and
also contaminant in seed and wool; tips and waste places;
throughout much of BI but mostly S; temperate Asia.

3. SPINACIA L. - <u>Spinach</u>
Usually annual herb; leaves flat, not mealy, glabrous;
bracteoles present with female flowers, developing abaxial
spine at fruiting; flowers unisexual, male in dense spikes or
panicles, female axillary; tepals 4-5 in male flowers, 0 in
female flowers.

1. S. oleracea L. - <u>Spinach</u>. Stems erect, well branched, to
1m; leaves rather variable, ovate to triangular, often with 1
basal acute lobe each side; petiole distinct. Intrd; grown
as vegetable (summer spinach) but less so now than formerly;
rather rare on tips and in waste places in En; Asia.

4. ATRIPLEX L. (<u>Halimione</u> Aellen) - <u>Oraches</u>
Annual herbs or perennial shrubs; leaves flattened, often
mealy, entire, toothed or lobed; flowers mostly unisexual,

the male with 5 tepals, the female usually with 0 tepals but
2 bracteoles enlarging and partly concealing fruit.
Vegetatively extremely plastic, especially in habit, mealiness and leaf-shape.   Young plants often closely resemble
Chenopodium spp., but in fruit the genus is very distinct.   A
problem in identification of annual spp. is that lower
leaves, which are often lost before fruit ripens, and also
ripe   fruit   and   surrounding   bracteoles,   are   important
diagnostic characters.   However, populations usually contain
a range of individuals of different ages.

1 Shrubs   2
1 Annual herbs   3
   2 Lower leaves opposite; bracteoles fused to >1/2 way
                                   13. A. portulacoides
   2 All leaves alternate; bracteoles fused only at base
                                   12. A. halimus
3 Bracteoles with small apical lobe much exceeded by 2
   adjacent laterals, fused + to apex   **14. A. pedunculata**
3 Bracteoles without 3 such apical lobes, not fused to
   c.1/2 way   4
   4 Bracteoles orbicular to broadly elliptic, entire,
      papery, present with only some female flowers
                                 1. A. hortensis
   4 Bracteoles not orbicular to broadly elliptic,
      angled or toothed, herbaceous to + woody, present
      with all female flowers   5
5 Bracteoles hardened (cartilaginous) in basal part at
   fruiting; ultimate venation of leaves thick, dark
   green against lighter background (fresh material)   6
5 Bracteoles not hardened at fruiting, remaining
   herbaceous or becoming spongy; ultimate venation of
   leaves very thin, not green   7
   6 Bracteoles 2-5mm, with 3-9 acute teeth in distal
      1/2, not tuberculate abaxially; wool-alien
                                 10. A. suberecta
   6 Bracteoles 6-7mm, with irregular mostly obtuse
      teeth around middle, usually tuberculate abaxially;
      coastal beaches   **11. A. laciniata**
7 Lower leaves linear to linear-lanceolate, entire to
   toothed but without distinct basal lobes; coastal only
                                  8. A. littoralis
7 Lower leaves lanceolate to triangular or trullate,
   with distinct basal lobes; coastal or inland   8
   8 Bracteoles fused for >1/3 their length, often
      to c.1/2 way   9
   8 Bracteoles fused at base only, for <1/4 way   11
9 Lower leaves lanceolate to trullate, acutely cuneate
   at base, with forwardly directed basal lobes;
   bracteoles herbaceous at base; coastal and inland
                                 9. A. patula
9 Lower leaves triangular to trullate, truncate to

obtusely cuneate at base, with laterally or forwardly
directed basal lobes; bracteoles thickened and
spongy at base; coastal only   10
10 Bracteoles sessile, 4-10mm, not foliaceous
      distally   **4. A. glabriuscula**
10 Some bracteoles stalked, up to 20mm and foliaceous
      distally   **5. A. x taschereaui**
11 Some bracteoles >10mm and foliaceous distally, with
      stalks >5mm   12
11 Bracteoles all <10mm, rarely foliaceous distally,
      with stalks <5mm   13
12 Bracteoles to 25mm, strongly foliaceous distally,
      united only at base, with stalks to 25(30)mm
                                                  6. A. longipes
12 Bracteoles to 20mm, sometimes foliaceous distally,
      the smaller ones often united nearly to 1/2way,
      with stalks to 10mm   **5. A. x taschereaui**
13 Lower leaves trullate, cuneate at base; lower littoral
      zone of coasts only   **7. A. praecox**
13 Lower leaves triangular, + truncate at base; coastal
      and inland   14
14 Bracteoles 2-6mm, sessile   **2. A. prostrata**
14 Bracteoles 3.5-9mm, always some with stalks to 1mm
      and often to 5mm   **3. A. x gustafssoniana**

Other spp. - About 20 additional spp. have been reliably
recorded from Br, especially as wool-aliens, but all are
rarer than the above. **A. muelleri** Benth., from Australia,
has been much confused with A. suberecta, but differs in
having leaves truncate to emarginate and mucronate at apex
and bracteoles obovate rather than rhombic. **A. sagittata**
Borkh. (A. nitens Schk. nom. illeg.), from E Europe,
differing from A. hortensis in its leaves white on lowerside
and oblong-cordate bracteoles, is now only a rare casual.

1. **A. hortensis** L. - Garden Orache. Erect annual to 2m; 177
leaves often >10cm, triangular, + truncate at base, basal
lobes present or 0, entire or toothed, green or more often
purplish-red; bracteoles orbicular to broadly elliptic, 5-
15mm, entire, fused only at base, papery, + sessile. Intrd;
sometimes grown for ornament (rarely as leaf-vegetable) and
escaping, also a bird-seed alien, occasional on tips and in
waste places in much of BI, especially S; ?Asia.
2. **A. prostrata** Boucher ex DC. (A. hastata auct. non L.) - 177
Spear-leaved Orache. Erect to procumbent annual to 1m; lower
leaved triangular, + truncate with laterally directed lobes
at base, usually shallowly toothed, green to strongly mealy;
bracteoles triangular, 2-6mm, herbaceous to spongy at base,
fused only at base, entire to dentate, sessile; seeds with
radicle positioned basally or sub-basally and directed
laterally or obliquely. Native; waste places and cultivated
ground, often in saline habitats inland and by sea;

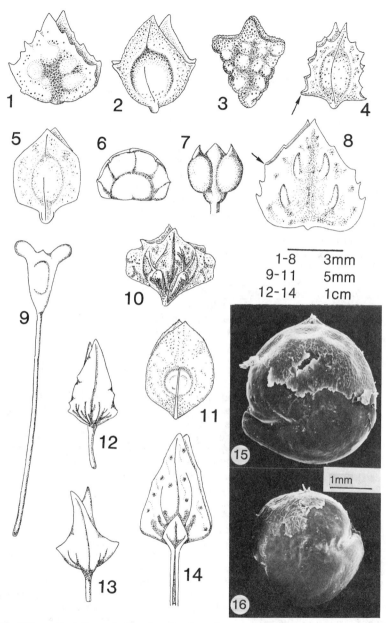

1-8 3mm
9-11 5mm
12-14 1cm

1mm

FIG 177 – **Atriplex.** 1–14, bracteoles (arrows indicate limit of fusion). 1, **A. littoralis.** 2, **A. patula.** 3, **A. portulac-oides.** 4, **A. prostrata.** 5, **A. praecox.** 6, **A. halimus.** 7, **A. suberecta.** 8, **A. glabriuscula.** 9, **A. pedunculata.** 10, **A. laciniata.** 11, **A. hortensis.** 12, **A. x taschereaui.** 13, **A. x gustafssoniana.** 14, **A. longipes.** 15–16, fruits with lower part of pericarp removed. 15, **A. prostrata.** 16, **A. glabriuscula.**

throughout BI.
2 x 4. A. prostrata x A. glabriuscula occurs rarely on
beaches in En and Sc with both parents, but has been much
over-recorded.   It is fertile and has bracteoles spongy and
fused like those of A. glabriuscula, but smaller and in
denser inflorescences.
2 x 8. A. prostrata x A. littoralis = A. x hulmeana
Taschereau occurs locally on coasts in E Anglia and N En with
both parents.   It is fertile and has narrowly trullate,
succulent lower leaves and spongy bracteoles.
3. A. x gustafssoniana Taschereau (A. longipes ssp. 177
kattegatensis Turesson; A. prostrata x A. longipes) -
Kattegat Orache).   Variously intermediate and fertile; var.
kattegattensis (Turesson) Taschereau has bracteoles 3.5-5mm
with variously fused margins and stalks to only 1mm; other
nothomorphs have bracteoles to 9mm, some with stalks to 5mm.
Native; coastal estuarine and sometimes inland saline areas,
often without 1 or both parents, commoner than A. longipes
and as common as A. glabriuscula; scattered round coasts of
Br, but var. kattegatensis only in N Sc.
4. A. glabriuscula Edmondston - Babington's Orache.   Pro- 177
cumbent to rarely erect annual; differs from A. prostrata in
more mealy stems and leaves, in 4-10mm bracteoles more spongy
at base and with margins fused to c.1/2 way, and in seeds
with radicle positioned laterally and directed apically.
Native; sandy or shingly beaches all round coasts of BI, but
rare in many places.
4 x 7. A. glabriuscula x A. praecox occurs in N Sc with both
parents; most plants resemble A. praecox in leaf-shape but A.
glabriuscula in bracteole characters; fertile.
5. A. x taschereaui Stace (A. glabriuscula x A. longipes) - 177
Taschereau's Orache.   Vegetatively close to A. glabriuscula
but some bracteoles to 20mm and foliaceous at apex and with
stalks to 10mm; fertile.   Native; exposed coastal beaches in
Sc, N En and Man, usually with A. glabriuscula but often not
with A. longipes and commoner than it.
6. A. longipes Drejer - Long-stalked Orache.   Erect to 177
procumbent annual to 90cm; lower leaves narrowly triangular   R
or trullate, cuneate with laterally or forwardly directed
lobes at base, entire to shallowly toothed, not mealy;
smaller bracteoles triangular to trullate, ± entire, fused
only at base, herbaceous, 5-10mm, sessile or with stalks to
1mm; larger bracteoles to 25mm with stalks to 25(30)mm,
foliaceous at apex; seeds with basal, laterally directed
radicle.   Native; in taller saltmarsh vegetation; very local
on coasts of En and SW Sc.
7. A. praecox Huelph. - Early Orache.   Procumbent to erect 177
annual to 10(15)cm; lower leaves ovate or lanceolate to   R
trullate, cuneate with laterally or forwardly directed lobes
at base, ± entire, often red-tinged, not mealy; bracteoles
triangular to ovate, ± entire, 3-5mm, herbaceous, united
only at base, sessile or with stalks to 1.5mm; seeds with

sub-basal laterally or obliquely directed radicle. Native; margins of sea inlets just above <u>Fucus</u> zone; SW & N Sc.

8. A. littoralis L. - <u>Grass-leaved</u> <u>Orache</u>. Usually erect 177 annual to 1.5m; lower leaves linear to linear-lanceolate, entire to toothed, without basal lobes, not mealy; bracteoles triangular to trullate with slender acute apex, entire to toothed, 3-6mm, spongy at base, fused only at base, sessile; seeds with sub-basal laterally to obliquely directed radicle. Native; saline open or colonized, usually sandy places near sea, rarely inland as casual; round most coasts of BI, commoner in E.

8 x 9. A. littoralis x A. patula occurs in disturbed ground with both parents in Midlothian. It is largely sterile (triploid) with lower leaves like those of <u>A. patula</u> but bracteoles spongy as in <u>A. littoralis</u> yet united as in <u>A. patula</u>.

9. A. patula L. - <u>Common</u> <u>Orache</u>. Erect to procumbent 177 annuals to 1m; lower leaves lanceolate to trullate, acutely cuneate with forwardly directed lobes at base, entire to toothed, not mealy; bracteoles triangular to trullate, entire to toothed, 3-7(20)mm, herbaceous, fused to c.1/2 way, sessile or with stalks to 4mm; seeds with sub-basal laterally or obliquely directed radicle; 2n=36 (all other spp. except <u>A. portulacoides</u> 2n=18). Native; disturbed and waste ground of all types; throughout BI.

10. A. suberecta Verd. - <u>Australian</u> <u>Orache</u>. Sprawling 177 annual to 60cm; lower leaves trullate- to triangular-ovate, acute to rounded and mucronate at apex, cuneate to truncate at base, coarsely sinuate-toothed, mealy on lowerside; bracteoles 2-5mm, ± rhombic, with 3-9 acute teeth in distal part, hardened at base, fused to ≥1/2 way, sessile or shortly stalked. Intrd; rather frequent wool-alien; Australia.

11. A. laciniata L. - <u>Frosted</u> <u>Orache</u>. Usually decumbent 177 annual to 30(50)cm; lower leaves trullate, coarsely toothed, cuneate with ± distinct lobes at base, strongly whitish-mealy; bracteoles 6-7(10)mm, broadly rhombic, entire to shortly toothed, hardened at base, fused to c.1/2 way, sessile or short-stalked. Native; lower parts of sandy beaches, often on strand-line; most coasts of BI.

12. A. halimus L. - <u>Shrubby</u> <u>Orache</u>. Well-branched erect 177 shrub to 2.5m; lower leaves oblong to elliptic, cuneate at base, lobes and teeth 0, white-mealy; bracteoles orbicular to ovate or reniform, 1.5-3mm, entire to dentate, hardened at base, fused only at base, ± sessile. Intrd; planted as wind-break by sea, natd and spreading vegetatively in CI, rare and not spreading S En; S. Europe.

13. A. portulacoides L. (<u>Halimione</u> <u>portulacoides</u> (L.) 177 Aellen) - <u>Sea-purslane</u>. Well-branched sprawling shrub to 1m; lower leaves oblong to elliptic, cuneate at base, lobes and teeth 0, whitish-mealy; bracteoles rhombic to obtrullate, 2.5-5mm, with 3 large lobes near apex, somewhat cartilaginous at base, fused to >1/2 way, ± sessile. Native; in saline mud

and sand, usually fringing pools or dykes and often flooded
at high tide, rarely on sea-cliffs; coasts of Br N to S Sc, E
Ir and CI.

14. A. pedunculata L. (Halimione pedunculata (L.) Aellen). - **177**
Pedunculate Sea-purslane. Erect annual to 30cm; lower leaves **RR**
elliptic to oblong, entire, cuneate at base, whitish-mealy;
bract-eoles obtriangular, 2-6mm, with 2 large lateral and 1
small apical lobe, $\pm$ herbaceous, fused almost to apex, with
stalks to 3cm. Native; drier, barish parts of salt-marshes;
from Kent to Lincs, extinct since 1938, refound S Essex 1987.

5. BETA L. - Beets
Annual to perennial herbs; roots often swollen; leaves
flattened, not mealy, usually $\pm$ entire; bracteoles 0; flowers
bisexual; tepals 5, persistent; ovary semi-inferior.

1. B. vulgaris L. - Beet. Whole plant often red-coloured or
-tinged. Stems erect to decumbent, little- to much-
branched, to 1.5m; leaves ovate to lanceolate or deltate,
cordate to cuneate at base, often slightly succulent;
tepals green or purplish-red, incurved in fruit; stigmas
usually 2.
1 Usually sprawling perennials; lower leaves mostly
   <10cm; lower bracts mostly 10-35mm; maritime
                                        a. ssp. maritima
1 Usually erect annuals or biennials; lower leaves
   mostly >10cm; lower bracts mostly 2-20mm; cultivated   2
   2 Grown for large foliage; roots not to moderately
     swollen   **b. ssp. cicla**
   2 Grown for greatly swollen roots   **c. ssp. vulgaris**
   a. Ssp. maritima (L.) Arcang. - Sea Beet. Usually much-
branched sprawling perennial; root not strongly swollen;
lower leaves usually <10cm. Native; shores and waste ground
near sea; round coasts of BI except most of N & C Sc.
   b. Ssp. cicla (L.) Arcang. - Foliage Beet. Usually little-
branched, erect annual to biennial; root mostly slightly
swollen; lower leaves usually >20cm. Intrd; cultivated for
its foliage and a common casual or relic. Includes var.
cicla L. (Spinach Beet) and var. **flavescens** (Lam.) Lam.
(Swiss Chard).
   c. Ssp. vulgaris - Root Beet. Usually little-branched,
erect annual to biennial; root strongly swollen; lower leaves
usually >10cm. Intrd; cultivated for its roots and a common
casual or relic. Includes Beetroot, Sugar Beet, Fodder Beet
and Mangel-wurzel.
   2. B. trigyna Waldst. & Kit. - Caucasian Beet. Erect,
usually little-branched perennial to 1m; leaves ovate,
usually cordate at base; tepals whitish-yellow, erect in
fruit; stigmas 3. Intrd; grown for ornament and persisting
or relict in waste places and on tips; S En; SE Europe.

6. SARCOCORNIA A. J. Scott - <u>Perennial Glasswort</u>
Dwarf subshrubs; leaves fused in opposite pairs, forming
succulent sheath round stem which appears composed of short
segments; flower ± immersed in row of segments at ends of
main stem ('terminal spike') and branches in 2 opposite
groups in each segment, each group with 3 flowers in a ±
straight transverse row, the centre one completely separating
and c. as large as the laterals.
Variously included in <u>Salicornia</u> or <u>Arthrocnemon</u> Moq.

1. S. perennis (Miller) A. J. Scott (<u>Salicornia</u> <u>perennis</u> **183**
Miller, <u>Arthrocnemon</u> <u>perenne</u> (Miller) Moss) - <u>Perennial</u>   **R**
<u>Glasswort</u>. Aerial stems to 30cm, some fertile, some not,
erect to decumbent, usually little-branched, becoming
yellowish to reddish, arising from thin extensive rhizomes;
terminal spike 10-40mm; fertile segments 3-4mm, 3-4.5mm wide
at narrowest point; anthers c.1.5mm. Native; mostly middle
and upper parts of salt-marshes; scattered in En and Wa N to
Yorks, frequent only in SE En, Wexford.

7. SALICORNIA L. - <u>Glassworts</u>
Like <u>Sarcocornia</u> but annuals and with 1-3 flowers in each
group, the central 1 of each group of 3 extending much
further apically than the laterals, so that the group is
triangular in shape.
An extremely difficult genus, the problems arising mainly
from great phenotypic plasticity and the inbreeding nature of
the plants, which tend to form numerous distinctive local
populations. At least 20-30 'sorts' can be distinguished in
SE En; possibly only 3 spp. (<u>S. pusilla</u>; <u>S. europaea</u> agg.;
<u>S. procumbens</u> agg.) should be recognized. Identification of
segregates within the latter 2 aggregates should be attempted
only on several fresh well-grown plants from unshaded
populations developing ripe fruit (Sep-Oct).

1 Flowers 1 per group; fertile segments disarticulating
   when fruit ripe, <2 x 2mm   **1. S. pusilla**
1 Flowers mostly 3 per group; fertile segments not
   disarticulating, >2 x 2mm   2
   2 Anthers 0.2-0.5(0.6)mm; stamens 1(-2); central
      flower distinctly larger (c.2x) than 2 laterals;
      fertile segments with distinctly convex sides;
      seeds 1-1.7mm   3
   2 Anthers (0.5)0.6-0.9mm; stamens (1-)2; all 3
      flowers about same size; fertile segments with
      straight or sometimes slightly convex or concave
      sides; seeds (1.3)1.5-2.3mm   5
3 Apical edge of fertile segments with scarious border
   0.1-0.2mm wide; plants deep shiny-green becoming
   reddish-purple   **2. S. ramosissima**
3 Apical edge of fertile segments with scarious border
   <0.1mm wide; plants lighter duller green becoming

yellowish-green sometimes suffused with pinkish-
purple   4
4 Plant glaucous, matt, not reddening or only
   slightly around flowers; branches simple, the
   lowest usually <1/2 as long as main stem,
   curving upwards distally   **4. S. obscura**
4 Plant clear green, not matt, usually reddening;
   branches usually branched, the lowest usually >1/2
   as long as main stem, + straight   **3. S. europaea**
5 Lower fertile segments <3(3.5)mm, <3.5(4)mm
   wide at narrowest point; plant becoming brownish-
   purple or -orange   **5. S. nitens**
5 Lower fertile segments 3-6mm, 3-6mm wide at
   narrowest point; plant becoming pale green to
   yellowish, sometimes tinged purple   6
6 Terminal spike + cylindrical, of 6-15(22) fertile
   segments; plant becoming yellowish-green to bright
   yellow   **6. S. fragilis**
6 Terminal spike usually tapering, of 12-30 fertile
   segments; plant becoming dull green, dull yellow
   or yellowish-brown   **7. S. dolichostachya**

1. S. pusilla J. Woods - <u>One-flowered</u> <u>Glasswort</u>. Erect to **183**
procumbent, simple to much-branched, to 25cm, becoming   **R**
orangy- or purplish-pink; branches + straight; terminal spike
<10mm; lower fertile segments 1-1.5mm, 1-1.5mm wide at
narrowest point; 2n=18. Native; drier parts of salt-marshes;
S Br and S Ir.
1 x 2. S. pusilla x S. ramosissima has been found in S En
close to both parents; it has 1-3 flowers per group but
resembles S. pusilla more closely in segment-shape and colour
and is fertile; endemic.
2-4. S. europaea L. agg. Anthers 0.2-0.5(0.6)mm; stamens
1(-2); central flower distinctly larger than 2 laterals;
fertile segments with distinctly convex sides; seeds 1-1.7mm;
2n=18 (diploid).
2. S. ramosissima J. Woods - <u>Purple</u> <u>Glasswort</u>. Erect to **183**
procumbent, simple to much-branched, to 40cm, usually
becoming dark purple; branches + straight; terminal spike
(5)10-30(40)mm; lower fertile segments 1.9-3.5mm, 2-4mm wide
at narrowest point. Native; mostly middle and upper parts of
salt-marshes; round coasts of BI. Particularly variable in
habit and colour; recent work suggests that this sp. is not
distinct from S. europaea.
3. S. europaea L. - <u>Common</u> <u>Glasswort</u>. Usually erect, much- **183**
branched, to 35cm, usually becoming yellowish-green suffused
with pink or red; branches + straight; terminal spike
10-50(60)mm; lower fertile segments 2.5-4mm, 3-4.5mm wide at
narrowest point.   Native; at all levels in salt-marshes;
round coasts of BI.
4. S. obscura P. Ball & Tutin - <u>Glaucous</u> <u>Glasswort</u>. Usually **183**
erect and little-branched, to 40cm, glaucous, becoming dull

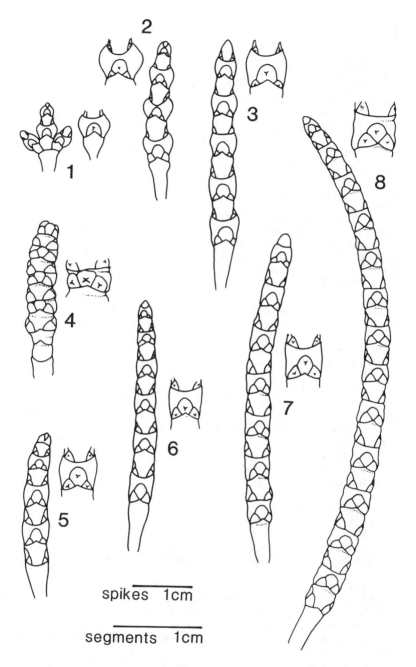

FIG 183 – Fruiting terminal spikes and fruiting segments of *Salicornia* and *Sarcocornia*. 1, *Salicornia pusilla*. 2, *S. ramosissima*. 3, *S. europaea*. 4, *Sarcocornia perennis*. 5, *Salicornia obscura*. 6, *S. nitens*. 7, *S. fragilis*. 8, *S. dolichostachya*. Drawings by D.H. Dalby.

yellowish-green; branches curving upwards distally; terminal
spike 10-40(45)mm; lower fertile segments 2.5-4.5mm,
2.8-4(5)mm wide at narrowest point. Native; on bare mud in
salt-pans and beside channels; S & E coasts of En from S
Lincs to S Hants.

5-7. S. procumbens Smith agg.   Anthers (0.5)0.6-0.9mm;
stamens (1-)2; all 3 flowers c. same size; fertile segments
with straight or slightly convex or concave sides; seeds
(1.3)1.5-2.3mm; 2n=36 (tetraploid).

5. S. nitens P. Ball & Tutin - <u>Shiny Glasswort</u>. Usually **183**
erect and little-branched, to 25cm, becoming light brownish-
purple to brownish-orange with diffuse red tinge; terminal
spike 12-40mm; lower fertile segments (1.8)2-3(3.5)mm,
1.8-3.5mm wide at narrowest point. Native; in middle and
upper parts of salt-marshes; scattered in S Br and S Ir,
Cheviot. The French **S. emerici** Duval-Jouve might be the same
and the name has priority.

6. S. fragilis P. Ball & Tutin (?<u>S. procumbens</u> Smith, <u>S.</u> **183**
<u>lutescens</u> P. Ball & Tutin) - <u>Yellow Glasswort</u>. Usually
erect, little- to much-branched, to 40cm, becoming yellowish-
green to bright yellow, sometimes suffused with red or
purple; terminal spike (15)25-80(100)mm; lower fertile
segments ± cylindrical, 3-5 x 3-6mm. Native; on lower parts
of salt-marshes and along dykes and runnels; most coasts of
Br and Ir.

7. S. dolichostachya Moss - <u>Long-spiked Glasswort</u>. Erect to **183**
procumbent, usually much-branched, to 45cm, becoming dull
green, dull yellow or yellowish-brown, sometimes suffused
with purple, terminal spike (25)50-120(200)mm; lower fertile
segments ± cylindrical, 3-6 x 3-6mm. Native; on lower parts
of salt-marshes and along dykes and runnels; most coasts of
Br and Ir, but rare in SE En. **S. oliveri** Moss might be the
same and the name has priority.

8. SUAEDA Forsskaol ex J. Gmelin - <u>Sea-blites</u>
Annual herbs or perennial shrubs; leaves linear, ± flat on
upperside, rounded on lowerside, succulent, entire, acute to
obtuse; bracteoles 2-3, minute, scarious; flowers bisexual
and female; tepals 5, persistent and partly concealing fruit.

1. S. vera Forsskaol ex J. Gmelin (<u>S. fruticosa</u> auct.) -   **R**
<u>Shrubby Sea-blite</u>. Evergreen branching shrub to 1.2m; leaves
5-18 x 0.8-1.5mm, obtuse, rounded at base; stigmas 3; seeds
vertical, smooth, shining. Native; sand and shingle beaches
and dry upper parts of salt-marshes; coasts of SE En from
Dorset to Lincs.

2. S. maritima (L.) Dumort. - <u>Annual Sea-blite</u>. Simple to
much-branched annual herb to 30(75)cm, often ± woody at
base; leaves 3-25 x 1-2(4)mm, acute to subacute, not contracted or rounded at base; stigmas usually 2; seeds
horizontal, shining but minutely reticulate. Native; middle
and lower parts of salt-marshes, often with <u>Salicornia</u> spp.;

round coasts of BI. Very variable in habit and seed-size.

9. SALSOLA L. - <u>Saltworts</u>
Annual somewhat woody herbs; leaves linear or linear-triangular, subterete to ± flattened, succulent, entire, spine-tipped; bracteoles 2, ± leaf-like; flowers bisexual; tepals 5, persistent and concealing fruit, usually developing a horizontal wing; seeds horizontal.

1. S. kali L. - see sspp. for English names. Stems erect to straggly, to 50(100)cm; leaves widest at base, tapered to spine-tipped apex, the lower 10-40(70) x 1-2mm.
  a. Ssp. kali - <u>Prickly</u> Saltwort. Stems mostly straggly, to 50cm, usually hispid; leaves to 40 x 2mm, usually hispid; tepals stiff, with distinct midrib and apical spine, winged in fruit. Native; natural and disturbed maritime sandy places; round coasts of BI.
  b. Ssp. ruthenica (Iljin) Soo (<u>S.</u> <u>pestifer</u> Nelson) - <u>Spineless</u> Saltwort. Stems mostly erect, to 1m, usually ± glabrous; leaves to 70 x c.1mm, usually glabrous; tepals soft, with obscure midrib and usually no apical spine, often only ridged in fruit. Intrd; casual in waste places in Br from wool waste and probably other sources, persistent in S Essex on ash-tips; Europe.

45. AMARANTHACEAE - <u>Pigweed family</u>

Herbaceous annuals or perennials; leaves alternate, simple, entire, exstipulate, the lower petiolate. Flowers small and usually brownish, with 3-5 bracteoles, usually borne in cymes, the cymes axillary or densely clustered at stem apex, mostly monoecious, sometimes dioecious, hypogynous, actinomorphic; perianth scarious, of 1 whorl of usually 2-5 free tepals, sometimes 0; stamens as many as tepals; ovary 1-celled, with 1 ovule; styles 2-3; stigma linear; fruit an achene or usually a 1-seeded capsule.
Distinguished from Chenopodiaceae by the brownish, scarious perianth and often dehiscent fruit. A difficult wholly alien genus which should be collected late in the year or grown on to obtain fruit. It is important to distinguish between bracteoles and tepals (which can be similar), and male and female flowers. In key and descriptions 'tepals' refers to those of female flowers only.

Other genera - **CELOSIA** L. differs in having several-seeded capsules. **C. argentea** L. (<u>Cockscomb</u>), from the Tropics, with gaudy wax-like inflorescences, is probably the world's ugliest plant; it is much grown as a park bedding-plant and occasionally occurs on tips etc.

1. AMARANTHUS L. - <u>Amaranths</u>

1 Flowering stems leafy to apex, the flowers borne
  in axillary clusters   2
1 Flowering stems leafless towards apex, the flowers
  borne in dense spike-like terminal panicles (often
  also in axillary clusters further back)   7
  2 Bracteoles c.2x as long as perianth, with ±
    spiny tip   **9. A. albus**
  2 Bracteoles shorter than perianth, not with
    spiny tip   3
3 Tepals 4-5   4
3 Tepals 3   5
  4 Tepals subequal, obovate to spathulate; fruit
    indehiscent   **14. A. standleyanus**
  4 Tepals unequal, ovate to elliptic; fruit
    transversely dehiscent   **10. A. blitoides**
5 Tepals shorter than fruit, with apical point <0.5mm
  and ± straight   **13. A. graecizans**
5 Tepals longer than fruit, with apical point >0.5mm
  and usually bent outwards or hooked   6
  6 Tepals with fine, straight to outwardly curved
    (rarely >90 degrees) apical point   **11. A. thunbergii**
  6 Tepals with stiffly hooked (mostly 180-360
    degrees) apical point   **12. A. capensis**
7 Plant wholly male   **7. A. palmeri**
7 Plant with female (and usually male) flowers   8
  8 Tepals (2-)3, c.1/2 as long as fruit; fruit
    inflated (seed much smaller than cavity),
    indehiscent; often perennial   **8. A. deflexus**
  8 Tepals (3-)5, nearly as long as to longer than
    fruit; fruit not inflated, dehiscent or not;
    annual   9
9 Fruits indehiscent or irregularly dehiscent   10
9 Fruits transversely dehiscent   11
  10 Tepals tapered to very acute apex; bracteoles
    longer than perianth   **6. A. bouchonii**
  10 Tepals rounded to retuse at apex, obovate to
    spathulate; bracteoles shorter than perianth
                                              14. A. standleyanus
11 Tepals all tapered to acute apex   12
11 At least the inner tepals obovate to spathulate,
  rounded to retuse at apex though often with mucro   13
  12 Longest bracteoles of female flowers c.2x as
    long as perianth   **4. A. hybridus**
  12 Longest bracteoles of female flowers 1-1.5x as
    long as perianth   **5. A. cruentus**
13 Inflorescence long, pendent or trailing, often red;
  tepals obovate or broadly spathulate to elliptic,
  strongly imbricate   **1. A. caudatus**
13 Inflorescence erect or weakly pendent, rarely red;
  tepals narrowly oblong to spathulate, not or weakly

imbricate   14
14 Tepals with midrib ending below apex (though apex
   often mucronate); stems at base of flowering
   region pubescent to densely so   **3. A.retroflexus**
14 Tepals with midrib extending beyond apex into
   mucro; stems at base of flowering region glabrous
   to sparsely pubescent   15
15 All flowers female; tepals dissimilar, the outer
   longer and tapering to acute apex   **7. A. palmeri**
15 Male and female flowers mixed; tepals all similar
                                                2. A. quitensis

Other spp. - About 22 additional spp. have been reliably
recorded from Br as rare casuals. **A. blitum** L. (A. lividus
L.) used to be quite frequent but almost all recent records
except a few from CI and Surrey are errors; it is similar to
A. deflexus but a glabrous annual (not with stems hairy near
apex) with retuse (not acute to obtuse) leaves, and fruits
little longer than tepals. Of additional spp. recently found
with soya-bean waste a dioecious American taxon with long
narrow, terminal, interrupted, leafless inflorescences is
most common; it might be **A. arenicola** I. M. Johnston.

1. A. caudatus L. - Love-lies-bleeding. Arching annual to **188**
80cm; main inflorescence terminal, very long, thick, pendent,
usually red, sometimes green, yellow or white; tepals 5,
obovate or broadly spathulate to elliptic, strongly
imbricate, fruit dehiscent.   Intrd; commonly grown and
frequent on tips and in waste places in Br and CI, mainly S;
S America.
2. A. quitensis Kunth - Mucronate Amaranth. Annual to 1m, **188**
usually ± erect; main inflorescence terminal, large, with
many side-branches, green; tepals 5, narrowly oblong to
spathulate, with midrib continued into mucro; fruit
dehiscent.   Intrd; infrequent casual from wool, soya-bean,
bird-seed and other sources; Br, mainly S; S America.
Flowers very late and never sets seed.   Probably the wild
progenitor of A. caudatus and perhaps not specifically
distinct.
3. A. retroflexus L. - Common Amaranth. Annual to 1m, **188**
usually ± erect; main inflorescence terminal, thick, with few
side branches, very rarely red; tepals 5, narrowly oblong to
spathulate, sometimes with mucro but midrib usually ending
short of apex; fruit dehiscent.   Intrd; frequent casual in
waste places and on tips from many sources, including wool,
bird-seed and soya-bean waste, sometimes natd; scattered in
BI, mainly S; N America.
3 x 4. A. retroflexus x A. hybridus = A. x ozanonii (Thell.) **188**
C. Schuster & Goldschmidt (A. x adulterinus Thell.) has been
found as a casual in a few places in S En from 1959 onwards;
it has intermediate tepals and is partially fertile.
4. A. hybridus L. (A. hypochondriacus L., A. patulus **188**

FIG 188 – Fruiting perianths of **Amaranthus**. 1, **A. graecizans**. 2, **A. albus**. 3, **A. hybridus**. 4, **A. x ozanonii**. 5, **A. retroflexus**. 6, **A. quitensis**. 7, **A. blitoides**. 8, **A. deflexus**. 9, **A. thunbergii**. 10, **A. bouchonii**. 11, **A. caudatus**. 12, **A. cruentus**. 13, **A. standleyanus**. 14, **A. palmeri**. 15, **A. capensis**.

Bertol., **A. powellii** S. Watson) - <u>Green</u> Amaranth. Annual to
1m, usually + erect; main inflorescence as in **A. retroflexus**
but usually less dense and more branched; bracteoles c.1.5-2x
as long as perianth; tepals (3-)5, lanceolate, tapering to
acute apex; fruit dehiscent. Intrd; frequent casual from
many sources, especially wool, bird-seed and soya-bean waste,
very rarely becoming natd; Br and CI; America. A. powellii,
with bracteoles 5-5.5mm (not 3-5mm) and tepals 3-5 (not 5),
is sometimes kept separate; it has been found in soya-bean
waste in S En.
   5. A. cruentus L. (A. hybridus ssp. cruentus (L.) Thell., **188**
ssp. incurvatus (Timeroy ex Gren. & Godron) Brenan, **A.**
paniculatus L.) - <u>Purple</u> Amaranth. Like A. hybridus but
inflorescence often red and bracteoles c.1-1.5x as long as
perianth. Intrd; fairly frequent casual from several
sources, including soya-bean waste, very rarely becoming
natd; S En; America. Probably the domesticated derivative of
A. hybridus, and perhaps not specifically distinct.
   6. A. bouchonii Thell. (A. hybridus ssp. bouchonii (Thell.) **188**
O. Bolos & Vigo) - <u>Indehiscent</u> Amaranth. Like A. hybridus
but fruit indehiscent. Intrd; casual or sometimes natd in E
Anglia, very rare but probably overlooked elsewhere in Br.
Probably a sporadic mutant of A. hybridus (especially of A.
powellii) and possibly not specifically distinct.
   7. A. palmeri S. Watson - <u>Dioecious</u> Amaranth. Dioecious **188**
annual to 1m, usually + erect; inflorescence a short terminal
spike-like panicle with many axillary clusters below; tepals
5, dissimilar, the outer tapering-acute, the inner +
spathulate and obtuse to rounded or retuse; fruit dehiscent.
Intrd; infrequent casual from grain, soya-bean waste and
other sources; S En; N America.
   8. A. deflexus L. <u>Perennial Pigweed</u>. Procumbent to **188**
ascending perennial herb (only annual where casual) to 40cm;
inflorescence mostly terminal, green to reddish-brown; tepals
2-3, c.1/2 as long as fruit; fruit indehiscent. Intrd;
infrequent casual from several sources in S & C Br and CI,
established as garden weed in CI; S America.
   9. A. albus L. - <u>White Pigweed</u>. Erect to procumbent annual **188**
with stiff whitish stems to 60cm; inflorescences axillary;
tepals 3, shorter than fruit, linear-lanceolate, acute; fruit
dehiscent. Intrd; frequent casual from many sources,
especially wool and bird-seed, occasionally persisting; Br,
mainly S; N America.
   10. A. blitoides S. Watson - <u>Prostrate Pigweed</u>. Procumbent **188**
to decumbent annual to 50cm; inflorescences axillary; tepals
4-5, unequal, the longer c. as long as fruit, oblong-ovate,
acute; fruit dehiscent. Intrd; infrequent casual from
several sources, mainly bird-seed and wool; Br, mainly S; N
America.
   11. A. thunbergii Moq. - <u>Thunberg's Pigweed</u>. Procumbent to **188**
ascending annual to 50cm; inflorescences axillary; tepals 3,
longer than fruit, with long fine usually outwardly curved

apex; fruit dehiscent. Intrd; frequent wool-alien in En and
Sc; S Africa. The leaves often have dark blotch on upper
surface.
12. A. capensis Thell. (_A. dinteri_ auct. non Schinz) - Cape 188
Pigweed. Procumbent to decumbent annual to 40cm; inflorescences axillary; tepals 3, longer than fruit, with fine,
rigidly hooked apices; fruit dehiscent. Intrd; infrequent
casual, mostly as wool-alien; En; S Africa. English material
is referable to ssp. **uncinatus** (Thell.) Brenan (_A. dinteri_
var. uncinatus Thell.).
13. A. graecizans L. - Short-tepalled Pigweed. Procumbent 188
to decumbent annual to 70cm; inflorescences axillary; tepals
3, shorter than fruit, narrowly ovate to elliptic, mucronate;
fruit dehiscent. Intrd; infrequent casual from several
sources including bird-seed and wool; S Br; Mediterranean and
Africa. Ssp. **sylvestris** (Villars) Brenan, with broader
leaves than the type, also occurs in Br, but is not worth
ssp. rank.
14. A. standleyanus L. Parodi ex Covas - Indehiscent 188
Pigweed. Erect to decumbent annual to 70cm; inflorescences
all axillary or some condensed into leafless apical spike-
like panicle; tepals 5, c. as long as fruit, obovate to
spathulate, mucronate; fruit indehiscent. Intrd; infrequent
casual from several sources; S Br; S America.

46. PORTULACACEAE - Blinks family

Herbaceous annuals to perennials; leaves mostly basal or
alternate to opposite, simple, somewhat succulent, entire,
stipulate or not, petiolate or not. Flowers solitary or in
terminal or axillary cymes, bisexual, hypogynous or semi-
inferior, actinomorphic or nearly so; sepals 2, free or
united below; petals 4-6, free or fused proximally; stamens
3-c.12; ovary 1-celled, with 1-many ovules; styles 1-3(6);
stigmas 3-6, linear; fruit a 1-many seeded capsule.
Distinguished from Caryophyllaceae by the 2 (not 5) sepals.

1 Petals yellow; stamens 6-c.12; capsule opening
  transversely; seeds >5   **1. PORTULACA**
1 Petals white to pink; stamens 3-5; capsule opening
  vertically; seeds 1-3   2
  2 Stem-leaves 1 pair, opposite   **2. CLAYTONIA**
  2 Stem-leaves several pairs, opposite or alternate
                                          3. MONTIA

Other genera - CALANDRINIA Kunth has alternate leaves,
flowers with 5 red petals, and many-seeded capsules opening
vertically. **C. ciliata** (Ruiz Lopez & Pavon) DC. (Red-maids),
from W N America, occurs occasionally on sandy soils mainly
in S En and CI; its mode of arrival is uncertain.

1. PORTULACA L. - Common Purslane
Stems with several pairs of opposite to subopposite (or some
alternate) leaves with usually bristle-like stipules; flowers
sessile, in groups of 1-3 with group of leaves just below;
sepals and petals falling before fruiting; stamens 6-c.12;
ovary semi-inferior; fruit a many-seeded capsule, opening
transversely.

1. P. oleracea L. - Common Purslane. Procumbent to erect, +
succulent annual to 50cm; petals (4-)5(6), c.4-8mm, yellow,
free or joined at extreme base. Intrd; natd weed of arable
ground in CI and Scillies, rare casual in En perhaps from
bird-seed; Mediterranean. There seem to be no records in BI
of the pot-herb ssp. **sativa** (Haw.) Celak.

2. CLAYTONIA L. - Purslanes
Stems with 1 pair of opposite leaves; stipules 0; flowers
stalked, several in terminal cyme; sepals persistent to after
seed dispersal; petals 5, free; stamens 5; ovary superior;
fruit a 1-seeded capsule, opening vertically.

1. C. perfoliata Donn ex Willd. (Montia perfoliata (Donn ex
Willd.) Howell) - Springbeauty. Annual to 30cm; stem-leaves
fused to form cup-like structure at base of inflorescence;
petals <5mm, white, entire or slightly notched. Intrd; weed
of cultivated and waste ground; scattered throughout BI and
CI, locally common but rare in W Br and in Ir; W N America.
2. C. sibirica L. (Montia sibirica (L.) Howell) - Pink
Purslane. Annual to 40cm; stem-leaves sessile but not fused;
petals >5mm, pink or sometimes white, deeply notched. Intrd;
barish damp shady places; scattered through Br and CI,
commoner in N & W, very rare in Ir; W N America.

3. MONTIA L. - Blinks
Stems with several alternate or opposite leaves; stipules 0;
flowers stalked, in terminal or axillary groups of 1-3 or
several in terminal cyme; sepals persistent to after seed
dispersal; petals 5, fused at base or free; stamens 3-5;
ovary superior; fruit a usually 3-seeded capsule, opening
vertically.

1. M. fontana L. - Blinks. Annual to perennial; stems **192**
branched, erect to procumbent or floating, 1-20(50)cm; stem-
leaves opposite; basal leaves 0; petals <2mm, white; seeds
black to dark brown, smooth or with projecting cells
('tubercles'). Native; many kinds of damp places, from
streams to seasonally damp hollows. The following 4 taxa
might be worth only varietal status; intermediates occur but
are rare.
1 Seeds smooth on faces and margin, very shiny
                                              a. ssp. fontana
1 Seeds tuberculate at least on margin, dull or

somewhat shiny   2
 2 Seeds with broad rounded tubercles on faces and
    margin, dull   **d. ssp. minor**
 2 Seeds smooth near centre of faces, somewhat shiny   3
 3 Seeds with ≥ 3 rows of narrow long-pointed tubercles
    along margin   **c. ssp. amporitana**
 3 Seeds with usually only 1-4 rows of broad short-
    pointed tubercles along margin   **b. ssp. variabilis**

a. Ssp. fontana. Locally common in N En, W Wa, Sc and Ir;
the commonest ssp. in NW.

b. Ssp. variabilis Walters. Scattered over most of Br and
Ir, but rare in S and E En.

c. Ssp. amporitana Sennen (ssp. <u>intermedia</u> (Beeby) Walters).
Scattered over most of BI, locally common in SW, rare in Sc.

d. Ssp. minor Hayw. (ssp. <u>chondrosperma</u> (Fenzl) Walters).
Scattered over most of BI, rare in N, locally common in S & C
Br.

2. M. parvifolia (Mocino) E. Greene - <u>Small-leaved</u> <u>Blinks</u>.
Stoloniferous perennial; flowering stems erect to ascending,
unbranched, to 20cm; stem-leaves alternate; basal-leaves
long-petiolate; petals >6mm, pink; seeds black, minutely
rough. Intrd; wet banks and rocks by river; natd by R. Cart,
Lanarks; W N America.

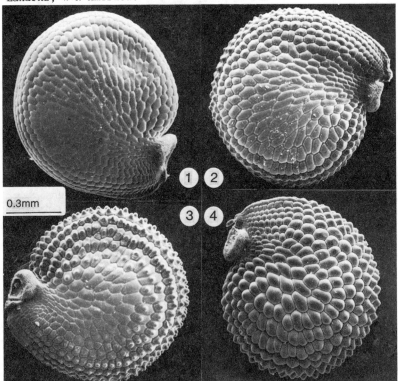

FIG 192 - Seeds of **Montia fontana**. 1, ssp. **fontana**. 2, ssp.
variabilis. 3, ssp. **amporitana**. 4, ssp. **minor**.

46A. BASELLACEAE

ANREDERA cordifolia (Ten.) Steenis (Boussingaultia cordifolia Ten. non (Moq.) Volkens, B. baselloides Hook. non Kunth) (Madeira-vine), from S America, is a twining or sprawling herbaceous perennial with tuber-bearing rhizomes, ovate-cordate leaves, and flowers in branching racemes, 1 plant of which has been established on waste ground in Guernsey for a few years. The flowers are small (c.2mm) and whitish, with 2 sepals and 5 petals; it differs from Portulacaceae in the solitary ovule and seed and indehiscent fruit.

47. CARYOPHYLLACEAE - Pink family
(Illecebraceae)

Herbaceous annuals to perennials, sometimes woody at base; leaves opposite or sometimes whorled or alternate, simple, usually entire, exstipulate or with scarious stipules. Flowers variously arranged but usually in cymes, bisexual or sometimes dioecious or variable, hypogynous or sometimes perigynous, actinomorphic or nearly so; perianth of 2 whorls or petals absent; sepals 4-5, free, green; petals usually 4-5, free, white or coloured, sometimes greenish or 0; stamens 2x as many a sepals, sometimes fewer; carpels fused, with 1 cell with many or sometimes 1-few ovules, with usually free-central placentation; styles 2-5, sometimes very short; stigmas linear to capitate; fruit a many-seeded capsule, sometimes a berry or 1-seeded achene.
Mostly recognisable by the opposite exstipulate leaves, 4-5 sepals and petals, 8 or 10 stamens, and many-seeded capsule with free-central placentation, but exceptions to all these occur. The leaves are usually narrow and the petals white and often bifid.

General key
1 Stipules present, at least partly scarious   Key A
1 Stipules 0   2
    2 Sepals free or joined only at extreme base   Key B
    2 Sepals joined to form distinct calyx-tube   Key C

Key A - Stipules present, at least partly scarious
1 Leaves alternate   **12. CORRIGIOLA**
1 At least lower leaves opposite or whorled   2
    2 Petals conspicuous, about as wide as sepals or
        wider   3
    2 Petals inconspicuous and much narrower than sepals 4
3 Styles 5; capsule opening by 5 valves; petals white
                                             16. SPERGULA
3 Styles 3; capsule opening by 3 valves; petals usually
    pink, sometimes white   **17. SPERGULARIA**
    4 Stigmas 3; fruit with >1 seed   **15. POLYCARPON**

4 Stigmas 2, fruit with 1 seed   5
5 Sepals conspicuous, white, spongy; fruit a dehiscent
   capsule   **14. ILLECEBRUM**
5 Sepals inconspicuous, greenish to brownish, thin;
   fruit an indehiscent achene   **13. HERNIARIA**

Key B - Stipules 0; sepals free or joined only at extreme
       base
1 Flowers perigynous; styles 2; fruit an indehiscent
   achene   **11. SCLERANTHUS**
1 Flowers hypogynous; styles >2; fruit with >1 seed,
   usually dehiscent   2
   2 Capsule-teeth or -valves as many as styles (but
      sometimes slightly bifid), or flowers male   3
   2 Capsule-teeth or -valves 2x as many as styles   6
3 Leaves and capsules succulent; seeds at least 3mm;
   maritime   **3. HONCKENYA**
3 Plant not succulent; seeds <3mm   4
   4 Petals bifid >3/4 to base   **8. MYOSOTON**
   4 Petals entire or slightly emarginate, or 0   5
5 Most flowers with 3 styles and most capsules with
   3 teeth, or flowers male   **4. MINUARTIA**
5 Styles 4-5; capsule with 4-5 valves   **10. SAGINA**
   6 Petals 0   7
   6 Petals present   8
7 Styles 3   **5. STELLARIA**
7 Styles 5   **7. CERASTIUM**
   8 Petals irregularly toothed or jagged   **6. HOLOSTEUM**
   8 Petals entire to deeply bifid   9
9 Petals bifid >1/4 to base   10
9 Petals entire to slightly emarginate   12
   10 Styles 4-5   **7. CERASTIUM**
   10 Styles 3   11
11 Petals divided <1/2 way; alpine   **7. CERASTIUM**
11 Petals divided >1/2 way   **5. STELLARIA**
   12 Styles 4   **9. MOENCHIA**
   12 At least most flowers with 3 styles   13
13 Seeds smooth, shiny, with small, lobed oil-body
   near hilum   **2. MOEHRINGIA**
13 Seeds tuberculate or papillose, without oil-body
                                                   1. ARENARIA

Key C - Stipules 0; sepals joined to form distinct calyx-tube
1 Styles 2 on all flowers   2
1 Styles 3-5 (rarely 2 on few flowers), or flowers male   5
   2 Calyx-tube scarious at joints between lobes
                                                   24. PETRORHAGIA
   2 Calyx-tube herbaceous all round   3
3 Calyx-tube with 5 longitudinal wings   **23. VACCARIA**
3 Calyx-tube without wings   4
   4 Calyx-tube with 2-6 sepal-like bracteoles appressed

    around base; scales absent from base of petal-limb
                                 25. DIANTHUS
4 Bracteoles absent; 2 small scales present on inner
    face of base of petal-limb   **22. SAPONARIA**
5 Calyx-teeth longer than -tube, extending beyond
    petals   **19. AGROSTEMMA**
5 Calyx-teeth shorter than -tube, falling short of petal
    tips   6
  6 Fruit a black berry   **21. CUCUBALUS**
  6 Fruit a capsule   7
7 Styles (2)3(-5), or flowers male; capsule with 2x as
    many teeth as styles   **20. SILENE**
7 Styles 5; capsule with 5 teeth   **18. LYCHNIS**

Other genera - Of the several grown in gardens, **GYPSOPHILA** L., separable from Petrorhagia by its 2 styles and campanulate (not tubular) calyx-tube, is occasionally found as garden escapes. **G. paniculata** L., a tall perennial with calyx <2.5mm, and **G. elegans** M. Bieb., an annual with calyx >2.5mm, are the commonest spp. Other spp. are rare casuals. **TELEPHIUM** L. has alternate leaves and is related to Corrigiola, from which it differs in its woody stems and many-seeded dehiscent fruit; both may be better placed in Molluginaceae. **T. imperati** L., from the Mediterranean, is a rare casual, formerly more common. **PARONYCHIA** Miller differs from Herniaria in its terminal (not axillary) inflorescence and large silvery stipules. **P. polygonifolia** (Villars) DC., from S Europe, has a similar status to T. imperati.

SUBFAMILY 1 - ALSINOIDEAE (genera 1-11). Leaves opposite, exstipulate; sepals free or joined only at extreme base; petals mostly 4-5, small to medium; carpophore 0; fruit a capsule or achene.

1. ARENARIA L. - Sandworts
Annuals to perennials; sepals 5; petals 5, white, entire or ± so; stamens 10; styles 3(-5); fruit a capsule opening by 6(10) teeth; seeds without oil-body.

1 Petals shorter than sepals   **1. A. serpyllifolia**
1 Petals longer than sepals   2
  2 Leaves all petiolate, almost all borne on
    procumbent stems rooting at nodes **4. A. balearica**
  2 Upper leaves sessile, borne on ± erect stems;
    stems not rooting   3
3 Leaves slightly succulent, with obscure midrib, the
  margins glabrous or pubescent only in lower 1/3;
  sepals glabrous or with a few basal hairs
                    2. A. norvegica
3 Leaves not succulent, with distinct midrib, the
  margins pubescent for >1/3 from base; sepals with
  hairs on lower part of margin and back **3. A. ciliata**

Other spp. - A. montana L. (Mountain Sandwort), from SW Europe, has been locally established as a garden escape.

1. A. serpyllifolia L. - Thyme-leaved Sandwort. Decumbent **206** to erect or stiffly spreading annual or biennial to 30cm; flowers numerous; leaves broadly ovate to ovate-lanceolate; petals shorter than sepals; styles 3. Native; open ground on well-drained soils, especially sand and limestone.

1 Capsule conical, straight-sided, with flexible walls
  after seed dispersal; pedicel c.0.3mm thick, longer
  than sepals; sepals 2-3mm   **c. ssp. leptoclados**
1 Capsule flask-shaped, with 'neck', with brittle walls
  after seed dispersal; pedicel c.0.5mm thick; sepals
  3-4.5mm   2
  2 Capsule mostly <2mm wide; pedicels longer than
    sepals; seeds mostly <0.6mm   **a. ssp. serpyllifolia**
  2 Capsule mostly >2mm wide; pedicels often shorter
    than sepals; seeds mostly >0.6mm   **b. ssp. lloydii**

a. Ssp. serpyllifolia. Relatively robust with diffuse inflorescence at fruiting; capsule c.3-3.5 x 1.5-2mm; seeds c.0.5-0.6mm; 2n=40. Throughout BI.

b. Ssp. lloydii (Jordan) Bonnier (ssp. macrocarpa F. Perring & Sell). Relatively robust but with dense inflorescence at fruiting; capsule c.3-3.5 x 2-2.5mm; seeds c.0.55- 0.65mm; 2n=40. Dune ecotype, perhaps only a var. of ssp. serpyllifolia; scattered round coasts of BI, mostly in SW.

c. Ssp. leptoclados (Reichb.) Nyman (A. leptoclados (Reichb.) Guss.). Relatively slender with diffuse inflorescence at fruiting; capsule c.2.5-3 x 1.2-1.5mm; seeds c.0.4-0.5mm; 2n=20. Throughout BI, probably less common than ssp. serpyllifolia, especially in N.

2. A. norvegica Gunnerus - see sspp. for English names. **RRR** Annual to perennial well branched below with ascending stems to 6cm; flowers 1-2(4) per stem; petals 4-5.5mm; styles 3-5.

a. Ssp. norvegica - Arctic Sandwort. Perennial with many non-flowering shoots; leaves obovate; flowers c.9-10mm across; styles 3-5. Native; base-rich scree and river shingle; extremely local in NW Sc and W Ir.

b. Ssp. anglica Halliday - English Sandwort. Annual or biennial with few non-flowering shoots; leaves narrowly elliptic to ovate-lanceolate; flowers c.11-23mm across; styles 3. Native; bare limestone in MW Yorks; endemic.

3. A. ciliata L. - Fringed Sandwort. Similar to A. **RRR** norvegica ssp. norvegica but differs in key characters, and in usually oblanceolate leaves; petals 5-7.5mm; and flowers 12-16mm across. Native; limestone cliffs in Sligo. The Irish plant is ssp. **hibernica** Ostenf. & O. Dahl (var. hibernica (Ostenf. & O. Dahl) Druce); endemic.

4. A. balearica L. - Mossy Sandwort. Procumbent much-branched perennial rooting at nodes, with solitary flowers on long up-turned pedicels, ascending to 5cm; leaves ovate-elliptic to suborbicular; petals c.2x as long as sepals;

styles 3. Intrd; well natd garden plant in damp rocky places and on paths and walls, scattered over most of BI; W Mediterranean islands.

2. MOEHRINGIA L. - Three-nerved Sandwort
Mostly annuals; sepals 5; petals 5, white, entire; stamens 10; styles 3(-4); fruit a capsule opening by 6(-8) teeth; seeds with distinctive oil-body.

1. M. trinervia (L.) Clairv. - Three-nerved Sandwort. Stems decumbent to erect, diffusely branched, to 40cm; leaves ovate, 3-5-veined; petals shorter than sepals. Native; shady places in woods and hedgebanks; throughout most of BI.

3. HONCKENYA Ehrh. - Sea Sandwort
Subdioecious ± succulent perennials; sepals 5; petals 5, greenish-white, entire; stamens 10 in male flowers; styles 3(-5) in female flowers; fruit a globose capsule opening by 3 valves.

1. H. peploides (L.) Ehrh. - Sea Sandwort. Decumbent to erect flowering stems to 25cm arising from extensive stolons or rhizomes; leaves ovate, succulent; petals c. as long as sepals in male flowers, much shorter in female flowers. Native; bare maritime sand and sandy shingle; common round coasts of BI.

4. MINUARTIA L. (Cherleria L.) - Sandworts
Annuals or perennials; sepals usually 5; petals usually 5 or 0, white, entire; stamens 10 or fewer; styles 3(-5); fruit a capsule opening by 3 teeth.

1 Slender annual without non-flowering shoots
                                          5. M. hybrida
1 Mat- or cushion-forming perennials with non-flowering
     shoots   2
   2 Petals 0 or minute; nectaries 10, conspicuous at
      base of sepals   **6. M. sedoides**
   2 Petals at least 1/2 as long as sepals; nectaries
      vestigeal   3
3 Leaves indistinctly 1-veined; pedicels glabrous
                                          4. M. stricta
3 Leaves distinctly 3-veined; pedicels glandular-hairy   4
   4 Petals shorter than sepals   **3. M. rubella**
   4 Petals as long as or longer than sepals   5
5 Sepals 3-veined; usually laxly tufted   **2. M. verna**
5 Sepals 5(-7)-veined; usually densely tufted; SW
   Ireland   **1. M. recurva**

Other spp. - **M. rubra** (Scop.) McNeill is one of several European alpine plants said to have been collected in Sc over 100 years ago but never refound; most if not all are spurious records.

1. M. recurva (All.) Schinz & Thell. – <u>Recurved</u> <u>Sandwort</u>. **RRR**
Densely tufted perennial, woody below, to 5cm; leaves 4-8mm,
linear, recurved; petals longer than sepals; styles 3.
Native; non-calcareous rocks above 500m in W Cork and S
Kerry, discovered 1964.
2. M. verna (L.) Hiern – <u>Spring</u> <u>Sandwort</u>. Laxly tufted to   **R**
loosely-carpeting perennial, scarcely woody below, to 15cm;
leaves 6-20mm, linear, ± straight; petals usually longer than
sepals; styles 3. Native; base-rich rocky places and sparse
grassland, often on lead-mine spoil; locally abundant in N
En, N Wa and W Ir, extremely scattered in Sc, N Ir and SW En.
3. M. rubella (Wahlenb.) Hiern – <u>Mountain</u> <u>Sandwort</u>. Tufted   **RR**
perennial, slightly woody below, to 6cm; leaves 4-8mm,
linear, mostly ± straight; petals c.2/3 as long as sepals;
styles 3-4(5). Native; base-rich rocks on mountains; very
local in N & C Sc.
4. M. stricta (Sw.) Hiern) – <u>Teesdale</u> <u>Sandwort</u>. Laxly **RRR**
tufted perennial, slightly woody below, to 10cm; leaves
6-12mm, linear, mostly ± straight; petals usually slightly
shorter than sepals; styles 3. Native; calcareous flushes in
upper Teesdale, Durham, above 450m.
5. M. hybrida (Villars) Schischkin – <u>Fine-leaved</u> <u>Sandwort</u>.   **R**
Slender erect annual to 20cm; leaves 5-15mm, linear, straight
or recurved; petals distinctly shorter than sepals; styles 3.
Native; dry bare stony places, walls and arable land;
scattered in En, Ir, E Wa and CI, common only in E Anglia.
6. M. sedoides (L.) Hiern (<u>Cherleria</u> <u>sedoides</u> L.) – <u>Cyphel</u>.   **R**
Dense yellow-green moss-like cushion, woody below, to 8cm;
leaves 3.5-6mm, linear-lanceolate, crowded; petals 0 or
minute; styles 3. Native; rocky ledges and slopes, often
damp, mostly on mountains but down to sea-level; N & C Sc.

5. STELLARIA L. – <u>Stitchworts</u>
Annuals or perennials; sepals 5; petals 5, sometimes fewer or
0, bifid, white; stamens 10 or fewer; styles 3; fruit a
capsule opening by 6 teeth.

1 At least lower leaves petiolate; stems terete to
   weakly and smoothly ridged   2
1 All leaves sessile; stems tetraquetrous to strongly
   ridged   5
   2 Petals >1.5x as long as sepals; stems ± equally
      pubescent all round   **1. S. nemorum**
   2 Petals <1.5x as long as sepals; stems with 1(2)
      lines of hairs down each internode   3
3 Sepals mostly 5-6.5mm; stamens >8; seeds mostly
   >1.3mm   **4. S. neglecta**
3 Sepals mostly <5mm; stamens mostly <8; seeds mostly
   <1.3mm   4
   4 Stamens 3-8; sepals mostly >3mm; seeds mostly

>0.8mm; petals usually present   **2. S. media**
4 Stamens 1-3; sepals mostly <3mm; seeds mostly
  <0.8mm; petals usually absent   **3. S. pallida**
5 Bracts entirely herbaceous; petals bifid c.1/2 way to
  base   **5. S. holostea**
5 Bracts entirely scarious or with wide scarious
  margins; petals bifid much >1/2 way to base   6
  6 Petals distinctly shorter than sepals; leaves
    mostly <15mm   **8. S. uliginosa**
  6 Petals as long as to longer than sepals; leaves
    mostly >15mm   7
7 Bracts and outer sepals ciliate; flowers c.5-12mm
  across   **7. S. graminea**
7 Bracts and sepals glabrous; flowers c.12-18mm
  across   **6. S. palustris**

1. S. nemorum L. - <u>Wood</u> Stitchwort. Stoloniferous perennial 212
with decumbent to ascending stems to 60cm; stems pubescent;
lower leaves long-petiolate, ovate, cordate; petals mostly
c.2x as long as sepals. Native; damp woods and shady stream-
sides.

a. Ssp. nemorum. Bracts gradually decreasing in size at
each node of inflorescence, those at 2nd node at least 1/3 as
long as those at 1st; seeds with rounded hemispherical
marginal papillae. Scattered in N & W Br, locally common in
N En.

b. Ssp. montana (Pierrat) Berher (ssp. <u>glochidisperma</u>
Murb.). Bracts abruptly shorter (at most 1/3 at 2nd node)
after 1st node of inflorescence; seeds with cylindrical
marginal papillae. Scattered in Wa.

2. S. media (L.) Villars - <u>Common</u> Chickweed. Sprawling
much-branched annual (often over-wintering); stems procumbent, to 50cm, with 1 line of hairs; lower leaves
petiolate, ovate to elliptic, sometimes ± cordate at base;
sepals 2.7-5.2mm, pubescent or glandular-pubescent; petals
1.1-3.1mm, rarely 0 or minute; stamens 3-5(8); seeds 0.8-
1.4mm. Native; ubiquitous weed of cultivated and open
ground. Stamen number and seed size are the best characters
to separate <u>S. media</u>, <u>S. pallida</u> and <u>S. neglecta</u>.

3. S. pallida (Dumort.) Pire - <u>Lesser</u> Chickweed. Sprawling,
often yellowish-green, much-branched, short-lived annual;
like <u>S. media</u> but sepals 2.1- 3.6mm, glabrous to pubescent;
petals 0 or <1mm; stamens 1-2(3); seeds 0.6-0.9mm. Native;
coastal dunes and shingle, inland on bare sandy soil; locally
frequent in En, Wa and CI, rare and scattered in Sc and Ir.

4. S. neglecta Weihe - <u>Greater</u> Chickweed. Usually ascending
annual to perennial to 80cm; like <u>S. media</u> but sepals 3.5-
5mm, glandular-pubescent; petals 2.5-4mm; stamens 10; seeds
1.1-1.4mm. Native; shady, usually damp places; scattered
through most of Br, mainly in S, very rare in W Ir.

5. S. holostea L. - <u>Greater</u> Stitchwort. Scrambling to
ascending perennial to 60cm; stems with rough angles,

glabrous or pubescent above; leaves lanceolate, sessile;
bracts herbaceous; petals much longer than sepals. Native;
woods and shady hedgerows; common throughout most of BI.
  6. S. palustris Retz. - <u>Marsh</u> <u>Stitchwort</u>. Ascending to ±
erect perennial to 60cm; stems with smooth angles, glabrous;
leaves linear-lanceolate, sessile; bracts scarious except for
green midline; petals slightly to much longer than sepals.
Native; usually base-rich marshes and fens; scattered in Br S
from C Sc, C Ir.
  7. S. graminea L. - <u>Lesser</u> <u>Stitchwort</u>. Decumbent to
ascending perennial to 80cm; stems with smooth angles,
glabrous; leaves lanceolate to linear-lanceolate, sessile;
bracts scarious; petals c. as long as to longer than sepals.
Native; grassy, often dry places; throughout BI.
  8. S. uliginosa Murray (<u>S.</u> <u>alsine</u> Grimm nom. inval.) - <u>Bog</u>
<u>Stitchwort</u>. Decumbent to ascending perennial to 40cm; stems
with smooth angles, glabrous; leaves ovate-lanceolate to
elliptic, sessile; bracts scarious except for green midline;
petals shorter than sepals. Native; streamsides, ditches,
wet tracks, depressions, often on acid soils; throughout BI.

6. HOLOSTEUM L. - <u>Jagged</u> Chickweed
Annuals; sepals 5; petals 5, irregularly toothed, white;
stamens usually 3-5; styles 3; fruit a capsule opening by 6
teeth.

  1. H. umbellatum L. - <u>Jagged</u> Chickweed. Stems ± erect, **RR**
little-branched, to 20cm, glandular-pubescent; leaves
oblanceolate to elliptic, the lowest shortly petiolate;
flowers in umbel on different-lengthed pedicels; petals
longer than sepals. Possibly native; arable fields, banks
and walls on well-drained soils; E Anglia and SE En, but
extinct since 1930.

7. CERASTIUM L. - <u>Mouse-ears</u>
Annuals or perennials; sepals 4-5; petals 4-5, sometimes 0,
retuse or emarginate to bifid to 1/2 way, white; stamens 4, 5
or 10; styles 3-5(6); fruit a capsule opening by 2x as many
teeth as styles.

1 Perennial, with usually ± procumbent non-flowering
    shoots as well as more erect flowering shoots   2
1 Annual, with all shoots producing flowers   8
    2 Styles mostly 3; capsule-teeth mostly 6
                                         1. C. cerastoides
    2 Styles 5; capsule-teeth 10   3
3 Petals <u><</u>1.7(2)x as long as sepals   **7. C. fontanum**
3 Petals <u>></u>(1.7)2x as long as sepals   4
    4 Leaves and upper parts of stems with predominantly
        densely matted or long (>1mm) shaggy hairs   5
    4 Leaves and stems with predominantly short (<1mm)
        ± straight hairs; long shaggy hairs less frequent

```
     or 0                                                    6
5  Leaves linear to narrowly elliptic, densely tomentose
   with matted hairs; most stems with >4 flowers;
   lowland                                      3. C. tomentosum
5  At least lower leaves elliptic to ovate or obovate,
   with many long shaggy hairs; stems with 1-4 flowers;
   arctic-alpine                                   4. C. alpinum
     6  Leaves linear to narrowly oblong; flowering stems
        with short leafy shoots in some leaf axils;
        lowland                                   2. C. arvense
     6  Leaves narrowly to broadly elliptic; flowering
        stems without conspicuous axillary shoots; arctic-
        alpine                                               7
7  Leaves narrowly elliptic to elliptic, either sparsely
   pubescent or with both glandular and non-glandular
   hairs; not Shetland                           5. C. arcticum
7  Leaves elliptic to broadly elliptic, densely
   glandular-pubescent; Shetland                6. C. nigrescens
     8  Sepals with long eglandular hairs projecting
        beyond sepal apex                                    9
     8  Sepals without eglandular hairs projecting beyond
        sepal apex                                          10
9  Inflorescence compact at fruiting; fruiting pedicels
   mostly shorter than sepals; glandular hairs abundant
   on sepals                                    8. C. glomeratum
9  Inflorescence lax at fruiting; fruiting pedicels
   mostly longer than sepals; glandular hairs 0
                                              9. C. brachypetalum
     10 Bracts completely herbaceous; petals and stamens
        usually 4; capsule-teeth usually 8     10. C. diffusum
     10 Bracts with scarious tips; petals and stamens
        usually 5; capsule-teeth usually 10                 11
11 Uppermost bracts scarious for at least apical 1/3;
   petals distinctly shorter (c.2/3 as long) than
   sepals                                     12. C. semidecandrum
11 Uppermost bracts scarious for at most apical 1/4;
   petals c. as long as sepals                 11. C. pumilum
```

1. **C. cerastoides** (L.) Britton – <u>Starwort</u> <u>Mouse-ear</u>. Mat-   R
forming perennial with ascending to decumbent vegetative and
flowering shoots to 15cm, almost glabrous apart from short
glandular hairs on pedicels and sepals; petals nearly 2x as
long as sepals; styles 3(-6). Native; moist rocky places on
mountains; N Sc S to Perths.
2. **C. arvense** L. – <u>Field</u> <u>Mouse-ear</u>. Loose mat-forming
perennial with ascending to decumbent vegetative and
flowering shoots to 30cm, shortly pubescent (often sparsely
so) with mostly eglandular but some glandular hairs; petals
c.2x as long as sepals; styles 5. Native; dry grassland;
most of Br but rare in W, local in Ir.
2 x 3. C. arvense x C. tomentosum = C. x maueri Schulze (<u>C.
decalvans</u> auct. non Schlosser Klek. & Vukot.) occurs in E and

SE En and C Sc near both parents; it is intermediate in
pubescence and produces some good seed. Probably overlooked;
likely to spread away from its parents.

2 x 7. C. arvense x C. fontanum = C. x pseudoalpinum Murr
has occurred in 3 places in Co Durham and S Lincs in rough
grassland with the parents; it resembles C. arvense but with
smaller flowers, wider leaves and denser pubescence, and is
sterile.

3. C. tomentosum L. - Snow-in-summer. Rampant mat-forming
perennial with strong ascending to decumbent vegetative and
flowering shoots to 40cm, densely whitish-tomentose with
matted elglandular hairs; petals c.2x as long as sepals;
styles 5. Intrd; commonly cultivated in rock gardens; natd
widely in BI in dry places; Italy and Sicily. Part of a
difficult complex of taxa, others of which, especially **C.
biebersteinii** DC. from Crimea and **C. decalvans** Schlosser
Klek. & Vukot. from Balkans, have been recorded in error. C.
tomentosum is extremely variable in Italy and Sicily and
several of the variants occur in BI.

4. C. alpinum L. - Alpine Mouse-ear. Tufted or matted   R
perennial with decumbent to ascending vegetative and
flowering shoots to 20cm, pubescent with many long and short,
soft, whitish, eglandular hairs and few or 0 glandular hairs;
petals c.2x as long as sepals; styles 5. Native; rock
outcrops and ledges on mountains; mainland C & N Sc, also
extremely local in S Sc, NW En and N Wa. The British plant
is ssp. **lanatum** (Lam.) Gremli.

4 x 5. C. alpinum x C. arcticum (C. x blyttii auct. non
Baenitz) occurs in N Wa and Sc near its parents; it is intermediate in pubescence and bract-form and is largely sterile.

4 x 7. C. alpinum x C. fontanum = C. x symei Druce occurs on
mountains in Sc with both parents; it is intermediate in
pubescence and flower-size and -number and is sterile.

5. C. arcticum Lange (C. nigrescens ssp. arcticum (Lange)   R
Lusby) - Arctic Mouse-ear. Tufted or matted perennial with
ascending to decumbent vegetative and flowering shoots to
15cm, pubescent with +0 to frequent short hairs, 0 to
frequent (var. **alpinopilosum** Hulten) long wavy hairs, and few
to many glandular hairs; leaves elliptic to narrowly
elliptic; sepals acute to subacute; petals c.2x as long as
sepals; styles 5. Native; rocky ledges and outcrops on
mountains; N Wa, C & N mainland Sc. Var. alpinopilosum,
which might represent introgression from C. alpinum, is much
the commoner var. Plants from 1 area of the Cairngorms, S
Aberdeen, almost entirely lack non-glandular hairs and are
closer to var. **arcticum**, but differ from it and might
represent a new taxon.

5 x 7. C. arcticum x C. fontanum = C. x richardsonii Druce
has been found with the parents in N Wa and N Sc; it is
intermediate but with petals as long as those of C. arcticum
and is largely sterile.

6. C. nigrescens (H. Watson) Edmondston ex H. Watson (C.   RR

arcticum ssp. **edmondstonii** (Edmondston) A. & D. Love) -
Shetland Mouse-ear. Differs from C. arcticum in leaves
elliptic to broadly elliptic, usually dark green and purplish
tinged, with abundant glandular and sparse to frequent short
non-glandular hairs; sepals mostly obtuse.   Native; serpentine rocks on Unst, Shetland; endemic.
   7. C. fontanum Baumg. - Common Mouse-ear. Tufted or matted
perennial with decumbent vegetative and ascending flowering
shoots to 50cm, sparsely to densely pubescent with few or 0
glandular hairs; upper bracts usually with narrow scarious
margins; petals mostly slightly shorter to slightly longer
than sepals, rarely nearly 2x as long; styles 5.   Native;
grassland, and open waste and cultivated ground.   Very
variable; other taxa than the 3 following occur, but all
might be better placed in 1 ssp.
1 Seeds 0.8-1mm, with tubercles c.0.1mm across at base;
   petals c.1.3-1.7x as long as sepals   **c. ssp. scoticum**
1 Seeds 0.4-0.9mm, with tubercles c.0.05mm across at
   base; petals c.0.8-1.2(2)x as long as sepals   2
   2 Lower stem internodes with hairs all round; leaves
      pubescent on both sides   **a. ssp. vulgare**
   2 Lower stem internodes glabrous or with 1-2 lines
      of hairs; leaves ± glabrous or very sparsely
      pubescent with most hairs on margin or lowerside
      midrib   ·   **b. ssp. holosteoides**
   a. Ssp. vulgare (Hartman) Greuter & Burdet (ssp. triviale
(Murb.) Jalas). See key for characters. Common through BI.
   b. Ssp. holosteoides (Fries) Salman, Ommering & Voogd (ssp.
glabrescens (G. Meyer) Salman, Ommering & Voogd, C.
holosteoides Fries, incl. var. glabrescens (G. Meyer) N.
Hylander). Differs from ssp. vulgare in less pubescent stems
and leaves, mostly narrower leaves (length/width mostly >4)
and mostly slightly smaller calyx (4-7mm) and capsules (7-
11mm). Mostly wet places; probably throughout BI but much
commoner in N.
   c. Ssp. scoticum Jalas & Sell. Differs from other sspp. in
larger seeds, usually longer petals, and more prominently
keeled sepals; pubescence usually ± intermediate between that
of other 2 sspp. Rocky and grassy places; very local in C &
N mainland Sc; endemic.
   8. C. glomeratum Thuill. - Sticky Mouse-ear. Erect or
ascending annual to 45cm, with glandular and abundant
eglandular hairs; bracts entirely herbaceous; inflorescence
compact at fruiting; flowers 5-merous; petals c. as long as
sepals.   Native; open places in natural and artificial
habitats; common throughout BI.
   9. C. brachypetalum Pers. - Grey Mouse-ear. Erect annual to **RR**
30cm, with 0 glandular and abundant patent eglandular hairs;
bracts entirely herbaceous; inflorescence diffuse at
fruiting; flowers 5-merous; petals c.1/2 as long as sepals.
Possibly native; chalk grassland and barish places on banks;
Beds (discovered 1947), Northants (1973) and W Kent (1978).

10. C. diffusum Pers. (<u>C.</u> <u>atrovirens</u> Bab.) - <u>Sea</u> <u>Mouse-ear</u>.
Decumbent to erect annual to 30cm, with abundant glandular
and some eglandular hairs; bracts entirely herbaceous;
infloresence diffuse at fruiting; flowers 4-(5-)merous;
petals c.3/4 as long as sepals. Native; dry, open sandy
places, common round coasts of BI and very local inland.

11. C. pumilum Curtis - <u>Dwarf</u> <u>Mouse-ear</u>. Erect to ascending   R
annual to 12cm, with abundant glandular and non-glandular
hairs; bracts with scarious tips <u><</u>1/4 total length;
inflorescence diffuse at fruiting; flowers 5-merous; petals
c. as long as sepals. Native; calcareous bare ground or open
grassland; very scattered, En N to Leics and Wa N to Caerns.

12. C. semidecandrum L. - <u>Little</u> <u>Mouse-ear</u>. Erect to
ascending annual to 20cm, with abundant glandular and non-
glandular hairs; bracts with scarious tips <u>></u>1/3 total
length; inflorescence diffuse at fruiting; flowers 5-merous;
petals c.2/3 as long as sepals. Native; dry open places on
sandy or calcareous soils, especially dunes; frequent
throughout most of BI but rare in N Sc and C Ir.

8. MYOSOTON Moench - <u>Water</u> <u>Chickweed</u>
Perennials; sepals 5; petals 5, white, bifid almost to base;
stamens 10; styles 5; fruit a capsule opening by 5 slightly
bifid teeth.

1. M. aquaticum (L.) Moench (<u>Stellaria</u> <u>aquatica</u> (L.) Scop.)
- <u>Water</u> <u>Chickweed</u>. Plant straggling, decumbent to ascending,
to 1m, glabrous below, with abundant glandular hairs above;
leaves ovate, cordate at base; petals longer than sepals.
Native; marshes, ditches and banks of water-courses; fairly
common in En and Wa N to Durham, formerly S Sc.

9. MOENCHIA Ehrh. - <u>Upright</u> <u>Chickweed</u>
Annuals; sepals 4; petals 4, entire, white; stamens 4; styles
4; fruit a capsule opening by 8 teeth.

1. M. erecta (L.) Gaertner, Meyer & Scherb. - <u>Upright</u>   R
<u>Chickweed</u>. Erect to procumbent, glabrous, + glaucous annual
to 12cm; leaves linear; petals shorter than sepals. Native;
barish places on sandy or gravelly ground; local in En and Wa
N to Norfolk and Caerns, formerly to Cheviot, CI.

10. SAGINA L. - <u>Pearlworts</u>
Annuals or perennials; leaves linear; sepals 4-5; petals 4-5,
sometimes 0, entire, white or whitish; stamens 4, 5, 8 or 10;
styles 4-5; fruit a capsule opening by 4-5 valves.

1 Annual, with all shoots producing flowers   2
1 Perennial, tufted or mat-forming, with short non-
     flowering shoots   3
   2 Leaves obtuse, sometimes minutely mucronate
        with point <0.1mm   **9. S. maritima**

2 Leaves abruptly protracted into fine point >0.1mm
                                                      8. S. apetala
3 Petals 0 or <1/2 as long as sepals   4
3 Petals 3/4-2x as long as sepals   5
  4 Plant a densely compact small cushion with rigidly
    recurved leaves hiding the stems; seed rarely
    produced   **7. S. boydii**
  4 Stems usually partly visible; leaves not or weakly
    recurved; seed abundantly produced   **6. S. procumbens**
5 Leaves at uppermost nodes usually >3x shorter than
  those at lower nodes; petals c.2x as long as sepals
                                                     1. S. nodosa
5 Leaves at uppermost nodes usually <2x shorter than
  those at lower nodes; petals <1.25x as long as sepals 6
  6 Sepals with glandular hairs   **3. S. subulata**
  6 Sepals glabrous   7
7 Plant compactly tufted, with basal leaf-rosette only
  in first year   **2. S. nivalis**
7 Plant a large tuft or mat-forming, with conspicuous
  basal leaf-rosette   8
  8 Sepals 4(5); stamens 4(5)   **6. S. procumbens**
  8 Sepals (4)5; stamens >5, usually 10   9
9 Leaves with apical point usually >0.2mm; ripe capsule
  2.5-3mm   **3. S. subulata**
9 Leaves with apical point 0-0.2mm; ripe capsules 3-4mm 10
  10 Capsules mostly 3.5-4mm, >1.5x as long as
     sepals, fully fertile; stems not or little rooting
     at nodes   **4. S. saginoides**
  10 Capsules mostly 3-3.5mm or not developing, <1.5x
     as long as sepals, usually with few seeds; stems
     usually rooting at nodes   **5. S. x normaniana**

1. S. nodosa (L.) Fenzl - <u>Knotted</u> <u>Pearlwort</u>. Diffuse
perennial with many procumbent to ascending stems to 15cm;
leaves with point <0.1mm; sepals 5, glabrous or glandular-
pubescent; petals c.2x as long as sepals; stamens 10; capsule
c.4mm.   Native; damp, rather open sandy and peaty soil;
throughout BI but only locally common.
2. S. nivalis (Lindblad) Fries (<u>S.</u> <u>intermedia</u> Fenzl) - <u>Snow</u>   **RR**
<u>Pearlwort</u>. Compact tufted perennial with numerous erect to
decumbent stems to 3cm; leaves with point <0.1mm; sepals 4-5,
glabrous; petals slightly shorter than sepals; stamens 8-10;
capsule 2.5-3mm. Native; bare ground near mountain tops;
very rare, C Highlands of Sc.
3. S. subulata (Sw.) C. Presl (<u>S.</u> <u>glabra</u> auct. non (Willd.) **206**
Fenzl, <u>S.</u> <u>pilifera</u> auct. non (DC.) Fenzl) - <u>Heath</u> <u>Pearlwort</u>.
Matted perennial with decumbent stems to 10cm; leaves with
point 0.3-0.6mm, rarely less; sepals 5, glandular-pubescent
or sometimes glabrous (var. **glabrata** Gillot); petals
slightly shorter than to as long as sepals; stamens 10;
capsule 2.5-3mm. Native; dry open ground on sand or gravel;
scattered throughout most of BI but absent from most of C & E

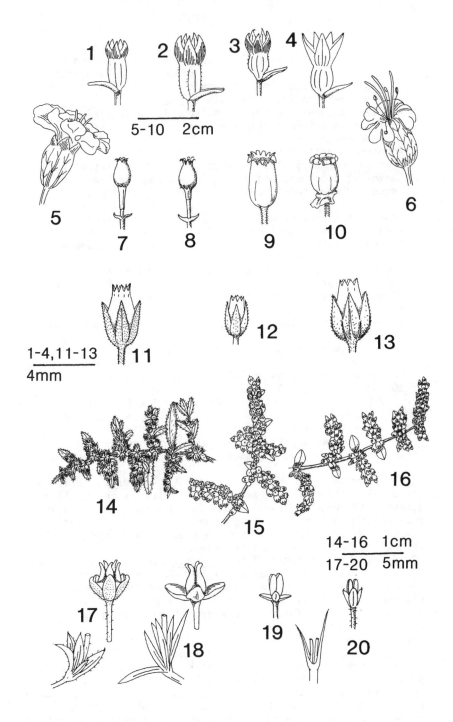

1
2
3 4
5-10 2cm
5
6
7 8 9 10
1-4,11-13
4mm
11 12 13
14
15 16
14-16 1cm
17-20 5mm
17 18 19 20

FIG 206 — see caption opposite

SAGINA   207

Ir and C & E En. Lawn-weeds mostly in S En once known as S.
glabra or S. pilifera seem to be S. subulata var. glabrata.
3 x 6. S. subulata x S. procumbens = S. x micrantha Boreau
ex E. Martin has been found with the parents in Merioneth,
Kintyre and Shetland; it has the habit of S. procumbens but
the pubescence of S. subulata and has low fertility.
   4. S. saginoides (L.) Karsten - Alpine Pearlwort. Matted   R
perennial with decumbent stems to 10cm not or scarcely
rooting at nodes; leaves with point <0.2mm; sepals 5,
glabrous; petals slightly shorter than to as long as sepals;
stamens 10; capsules mostly 3.5-4mm. Native; barish ground
and rock ledges on mountains; C & N mainland Sc.
   5. S. x normaniana Lagerh. (S. saginoides ssp. scotica   RR
(Druce) Clapham; S. saginoides x S. procumbens) - Scottish
Pearlwort. Intermediate between its parents and partially
fertile; differs from S. procumbens in stamen number and
capsule length (see key) and from S. saginoides mainly in
stems rooting at nodes. Native; usually with but sometimes
without parents in range of S. saginoides.
   6. S. procumbens L. - Procumbent Pearlwort. Matted 206
perennial with decumbent stem to 20cm much rooting at nodes;
leaves with point <0.2mm; sepals 4(-5), glabrous; petals 0 or
minute, rarely ± as long as sepals; stamens 4(-5); capsule
2-3mm. Native; paths, lawns, ditchsides and short turf;
common throughout BI.
   7. S. boydii Buch.-White - Boyd's Pearlwort. Densely tufted   RR
perennial with erect stems to 2cm; leaves rigidly recurved,
with point <0.1mm; sepals 4-5, glabrous; petals 0 or minute;
stamens 5-10; capsule extremely rarely developing.
Enigmatic; presumed to have been collected in S Aberdeen in
1878; not seen wild since but still in cultivation; endemic.
Comes true from seed.
   8. S. apetala Ard. - Annual Pearlwort. Diffusely branched
± erect annual to 15cm; leaves with point 0.1-0.4mm; sepals
usually 4, glabrous or glandular-pubescent; petals 0 or
minute; stamens usually 4; capsule 1.6-2.5mm. Native.
   a. Ssp. apetala (S. ciliata Fries, S. filicaulis Jordan). 206
Sepals erect to erecto-patent in fruit, at least the outer
subacute; most seeds >1/3mm. Dry bare ground on heaths and
paths; scattered through most of BI. **S. filicaulis**, more
slender, with smaller parts and glandular rather than

FIG 206 - 1-4, Fruiting perianths of **Scleranthus. 1, S.
perennis** ssp. **prostratus.** 2, ssp. **perennis.** 3, **S. annuus** ssp.
polycarpos. 4, ssp. **annuus.** 5-6, Flowers of **Silene. 5, S.
vulgaris.** 6, **S. maritima.** 7-10, Capsules of **Silene. 7, S.
italica.** 8, **S. nutans.** 9, **S.latifolia.** 10, **S. dioica.** 11-13,
Capsules of **Arenaria serpyllifolia.** 11, ssp. **serpyllifolia.**
12, ssp. **leptoclados.** 13, ssp. **lloydii.** 14-16, Shoots of
Herniaria. 14, **H. hirsuta.** 15, **H. ciliolata.** 16, **H. glabra.**
17-20. Leaves and capsules of **Sagina.** 17, **S. subulata.** 18, **S.
procumbens.** 19, **S. apetala** ssp. **erecta.** 20. ssp. **apetala.**

glabrous pedicels and sepals, may merit equal rank.
  b. Ssp. erecta F. Herm. (<u>S.</u> <u>apetala</u> auct. non Ard.).
Sepals patent in fruit, subobtuse; most seeds <1/3mm. Paths,
walls and bare cultivated ground; common through most of BI
but rare in N Sc.
Possibly these 2 sspp. should be recognized as 2 or 3 vars.
  9. S. maritima G. Don - <u>Sea Pearlwort</u>. Diffusely branched
± erect annual to 15cm; leaves with point <0.1mm; sepals 4,
obtuse, glabrous, erecto-patent in fruit; petals 0 or minute;
stamens 4; capsules 2-2.8mm. Native; damp sand-dunes, rocky
places and cliff-ledges on barish soil; round coasts of BI
and rarely slightly inland on mountains in Sc.

11. SCLERANTHUS L. - <u>Knawels</u>
Annuals to perennials; leaves linear, fused at base in
opposite pairs; flowers greenish, inconspicuous; sepals
usually 5; petals 0; fertile stamens (2)5-10; styles 2; fruit
an achene.

  1. S. perennis L. - <u>Perennial Knawel</u>. Biennial to perennial **206**
to 20cm with woody basal parts and some sterile shoots at **RRR**
flowering time; sepals subacute to obtuse, with white border
c.0.3-0.5mm wide (nearly as wide as green central part), the
tips parallel or incurved over ripe achene; fertile stamens
10. Native.
  a. Ssp. perennis. Stems ascending to erect; achene (incl.
sepals) (3)3.5-4.5mm. Doloritic rocks in Rads.
  b. Ssp. prostratus Sell. Stems procumbent to ± ascending;
achene (incl. sepals) 2-3(3.5)mm. Very local and decreasing
on sandy heaths in Norfolk and Suffolk; endemic.
  2. S. annuus L. - <u>Annual Knawel</u>. Decumbent to erect annual **206**
or biennial to 20cm, without sterile shoots at flowering
time; sepals acute to subacute, with whitish border c.0.1mm
wide (1/2 as wide as green central part); fertile stamens
2-10. Native; dry open sandy ground.
  a. Ssp. annuus. Achene (incl. sepals) 3.2-4.5(5.5)mm, with
divergent sepals when ripe. Scattered over most of BI but
rare in Ir and N Sc.
  b. Ssp. polycarpos (L.) Bonnier & Layens (<u>S.</u> <u>polycarpos</u> L.).
Achene (incl. sepals) 2.2-3.0(3.8)mm, with parallel to
convergent sepals when ripe. Frequent on sandy ground in
Suffolk.

SUBFAMILY 2 - **PARONYCHIOIDEAE** (genera 12-17). Leaves
opposite, alternate or whorled, stipulate; sepals 5, free;
petals 5, often very small; carpophore 0; fruit a capsule or
achene.

12. CORRIGIOLA L. - <u>Strapwort</u>
Leaves alternate; flowers slightly perigynous; petals ±
equalling sepals, white, sometimes red-tipped, entire;
stamens 5; stigmas 3, sessile, capitate; fruit an achene.

This and Telephium might be better removed to Molluginaceae.

Other spp. - C. telephiifolia Pourret was natd for some years recently at Gloucester Docks, E Gloucs, but has now gone; it is a perennial with +0 bracts in the inflorescence but often difficult to distinguish.

1. C. litoralis L. - Strapwort. Glabrous + glaucous **RRR** annual with 1-many decumbent stems to 25cm; leaves linear to oblanceolate, obtuse; inflorescence branches with leaf-like bracts. Native; on sand and gravel by ponds; S Devon, formerly W Cornwall and Dorset, infrequent casual elsewhere mainly by railways in En.

13. HERNIARIA L. - Ruptureworts
Leaves opposite; petals shorter than sepals, filiform; stamens 5; stigmas 2, capitate, sessile or on very short styles or common style; fruit an achene.

1 Calyx pubescent; leaves with many hairs on surfaces
                                        3. H. hirsuta
1 Calyx glabrous or ciliate; leaves with 0 hairs on
  surfaces but often ciliate   2
  2 Fruit acute to subacute, distinctly exceeding
    sepals; seeds c.0.5-0.6mm   **1. H. glabra**
  2 Fruit obtuse to subobtuse, scarcely or not
    exceeding sepals; seeds c.0.7-0.8mm   **2. H. ciliolata**

1. H. glabra L. - Smooth Rupturewort. Annual to perennial **206** with numerous procumbent branched stems to 30cm; stems **RR** usually hairy all round, not or slightly woody at base; 2n=18. Native; dry sandy ground; E En, extinct or rare casual elsewhere in Br.
2. H. ciliolata Meld. - Fringed Rupturewort. Evergreen **206** perennial; differs from H. glabra in stems strongly woody **RR** below; and see key. Inflorescence shape, leaf pubescence and sepal length and pubescence, often used to separate H. ciliolata and H. glabra, seem unreliable. Native; maritime sandy and rocky places.
a. Ssp. ciliolata. Stems hairy on 1 (usually upper) side; leaves ovate to obovate; 2n=72. W Cornwall, Guernsey and Alderney; endemic.
b. Ssp. subciliata (Bab.) Chaudhri (H. ciliolata var. angustifolia (Pugsley) Meld.). Stems hairy all round; leaves narrowly elliptic; 2n=108. Jersey; endemic.
3. H. hirsuta L. (H. cinerea DC.) - Hairy Rupturewort. **206** Annual similar to H. glabra but whole plant surface with patent hairs; 2n=36. Intrd; open waste ground; occasional casual from several sources, sometimes natd; En; Europe. Extreme H. cinerea is distinct but all intermediates occur.

14. ILLECEBRUM L. - <u>Coral-necklace</u>
Leaves opposite; flowers hypogynous; petals much shorter than
sepals, filiform; stamens 5; stigmas 2, sessile, capitate;
fruit a 1-sided capsule opening by 5 valves adherent at apex.

1. I. verticillatum L. - <u>Coral-necklace</u>. Glabrous annual   **R**
with procumbent to decumbent stems to 20cm, often reddish;
leaves obovate, obtuse; flowers densely clustered in leaf-
axils; sepals white, thick, fleshy, hooded, with finely
pointed apex. Native; damp sandy open ground; rare in
extreme S En, rarely natd or casual elsewhere.

15. POLYCARPON L. - <u>Four-leaved Allseed</u>
Leaves opposite or 4-whorled; flowers hypogynous; petals much
shorter than sepals, often emarginate; stamens (1)3-5;
stigmas 3, on separate styles; fruit a capsule opening by 3
valves.

1. P. tetraphyllum (L.) L. (<u>P. diphyllum</u> Cav.) - <u>Four-leaved</u>   **RR**
<u>Allseed</u>. Almost glabrous annual with erect to ascending
much-branched stems to 25cm; leaves obovate, obtuse to
rounded; flowers numerous, small and greenish. Native; open
sandy and waste ground near sea; CI, Dorset, S Devon,
Cornwall and Scillies, casual elsewhere. Plants with leaves
only paired, contracted inflorescences, and stamens 1-3 (not
3-5) are distinguishable as var. **diphyllum** DC.

16. SPERGULA L. - <u>Spurreys</u>
Leaves opposite but often appearing whorled due to axillary
leaf clusters; flowers hypogynous; petals c. as long as
sepals or slightly longer, white, entire; stamens 5-10;
stigmas 5, on separate styles; fruit a capsule opening by 5
valves.

1. S. arvensis L. - <u>Corn Spurrey</u>. Glandular-pubescent **212**
annual with erect to ascending branched stems to 40(60)cm;
leaves linear, furrowed on lowerside; seeds little flattened,
without wing or with circum-equatorial wing <1/10 as wide as
actual seed. Native; calcifuge on usually sandy cultivated
ground or rarely in short maritime turf; throughout BI. Var.
arvensis has seeds with 0 or extremely narrow wing and the
surface covered by clavate papillae; it is usually sparsely
glandular and commoner in SE. Var. **sativa** (Boenn.) Mert. &
Koch has seeds with 0 papillae and a wider wing; it is
usually densely glandular and commoner in NW. Other alien
vars might occur but have not been studied. Var. **nana** E.F.
Linton (ssp. **nana** (E.F. Linton) D. McClint.), a dwarf variant
of var. <u>arvensis</u> hardly worthy of separation, occurs in CI.
  2. S. morisonii Boreau - <u>Pearlwort Spurrey</u>. Annual with **212**
slenderer but more rigid stems to 30cm, ± glabrous to
sparsely glandular-pubescent; leaves linear, not furrowed on
lowerside; seeds strongly flattened, with circum-equatorial

wing >1/4 as wide as actual seed.   Intrd; natd on sandy or
peaty arable land; E Sussex and Kildare; C Europe.

17. SPERGULARIA (Pers.) J.S. & C. Presl – <u>Sea-spurreys</u>
Leaves opposite; flowers hypogynous; petals shorter to longer
than sepals, (white to) pink, entire; stamens (0)5-10;
stigmas 3, on separate styles; fruit a capsule opening by 3
valves.

1	At least some seeds with distinct circum-equatorial wing	2
1	All seeds unwinged	3
2	Flowers (7)10-12(13)mm across; stamens (0-)10, if fewer than 10 the remainder represented as staminodes	**2. S. media**
2	Flowers (4)5-8mm across; stamens (0)2-7(10); staminodes 0(-3)	**3. S. marina**
3	Flowers mostly >8mm across; sepals mostly >4mm; capsule mostly >5mm	4
3	Flowers mostly <8mm across; sepals mostly <4mm; capsule mostly <5mm	5
4	Plant glabrous to glandular-pubescent only in inflorescence, not woody below	**2. S. media**
4	Plant glandular-pubescent over most of stem, woody below	**1. S. rupicola**
5	Seeds >0.6mm; stipules on young shoots fused for >1/3 their length	**3. S. marina**
5	Seeds <0.6mm; stipules on young shoots fused for <1/4 their length	6
6	Upper bracts not much different from stem-leaves; most pedicels much longer than capsules	**4. S. rubra**
6	Upper bracts much shorter than stem-leaves or represented + only by stipules; most pedicels shorter than capsules	**5. S. bocconei**

1. S. rupicola Lebel ex Le Jolis – <u>Rock Sea-spurrey</u>. 212
Perennial with + woody base; stems to 35cm, decumbent,
glandular-pubescent to well below inflorescence; sepals
4-4.5mm, about as long as petals; capsule 4.5-7mm; seeds not
winged.   Native; walls and rocky maritime places; locally
common on coasts of CI, Ir, Wa, W En and SW Sc.
   1 x 3. S. rupicola x S. marina has been found in mixed
populations of the parents in Co Dublin, E Cornwall, Dorset
and Monts; it is a completely sterile vigorous floriferous
perennial intermediate in morphological details.
   2. S. media (L.) C. Presl (<u>S. marginata</u> Kittel nom. illeg., 212
<u>S. maritima</u> (All.) Chiov.) – <u>Greater Sea-spurrey</u>.   Perennial
with non-woody base; stems to 40cm, decumbent to ascending,
glabrous or glandular-pubescent only in inflorescence; sepals
(3.5)4-6mm, usually slightly shorter than petals; capsule
(4)7-9mm; seeds 0.7-1mm (excl. wing), all or many (very
rarely 0) winged.   Native; sandy and muddy maritime places;

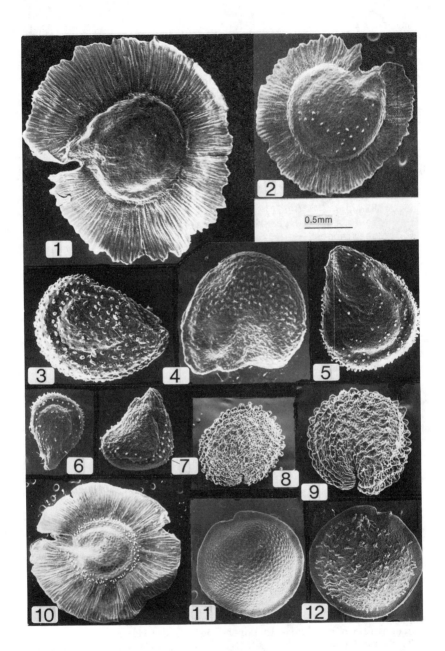

FIG 212 – Seeds of **Caryophyllaceae**. 1, **Spergularia media** (winged). 2, **S. marina** (winged). 3, **S. rupicola**. 4, **S. media** (unwinged). 5, **S. marina** (unwinged). 6, **S. bocconei**. 7, **S. rubra**. 8, **Stellaria nemorum** ssp. **montana**. 9, **S. nemorum** ssp. **nemorum**. 10, **Spergula morisonii**. 11, **S. arvensis** var. **sativa**. 12, **S. arvensis** var. **arvensis**.

common round coasts of Br and Ir, very rarely inland.

3. S. marina (L.) Griseb. (_S. salina_ J.S. & C. Presl) – **212**
Lesser Sea-spurrey. Annual or sometimes perennial; stems
decumbent to procumbent, to 35cm, usually glandular-pubescent
in inflorescence and often some way below; sepals 2.5–4mm,
usually longer than petals; capsule 3–6mm; seeds 0.6–0.8mm
(excl. wing), wingless, winged, or mixed. Native; sandy and
muddy maritime places and inland saline areas; common round
coasts of BI and locally frequent inland.

4. S. rubra (L.) J. S. & C. Presl – _Sand_ Spurrey. Annual or **212**
biennial; stems to 25cm, decumbent, glandular-pubescent in
inflorescence and usually some way below; sepals 3–4(5)mm,
usually longer than petals; capsule 3.5–5mm; seeds not
winged. Native; calcifuge of sandy or gravelly ground;
throughout most of BI and often common.

5. S. bocconei (Scheele) Graebner – Greek Sea-spurrey. **212**
Annual; stems to 20cm, decumbent; densely glandular-pubescent   **RR**
in inflorescence and some way below; sepals 2–4mm, longer
than petals; capsule 2–4mm; seeds not winged. Possibly
native; dry sandy and waste places by sea; frequent in CI,
rare and probably intrd in Cornwall, rare casual elsewhere in
S En and S Wa.

SUBFAMILY 3 – CARYOPHYLLOIDEAE (Silenoideae) (genera 18–25).
Leaves opposite, exstipulate; sepals fused to form calyx-
tube and –lobes; petals 5, mostly medium to large, usually
differentiated into a narrow proximal claw within and widened
distal limb above calyx-tube; stamens 10; petals, stamens and
ovary usually elevated above receptacle on sterile axis
(carpophore); fruit a capsule or berry.

18. LYCHNIS L. (Viscaria Roehl., Steris Adans.) – Catchflies
Bracteoles not forming epicalyx; petal-limb entire or
2–4-lobed, with 2 scales at base, purple or rarely white;
styles 5; fruit a capsule opening by 5 teeth. Sometimes
subdivided, but perhaps best amalgamated with Silene.

1 Leaves and stems with dense white woolly hairs
                                                1. L. coronaria
1 Leaves and stems glabrous to pubescent but not covered
   with long hairs   2
   2 Petal-limb divided much >1/2 way into 4 narrow
      lobes   **2. L. flos-cuculi**
   2 Petal-limb divided up to c.1/2 way into 2 lobes   3
3 Petal-limb divided much <1/2 way, with conspicuous
   scales at base; carpophore >3mm at fruiting
                                                3. L. viscaria
3 Petal-limb divided c.1/2way, with minute knob-like
   scales at base; carpophore <2mm at fruiting
                                                4. L. alpina

Other spp. – **L. chalcedonica** L. (Maltese-Cross), from

Russia, is commonly grown in gardens and occasionally escapes; it has tall erect stems topped by a dense cluster of flowers with scarlet petal-limbs divided c.1/3 to base.

1. L. coronaria (L.) Murray (<u>Silene</u> <u>coronaria</u> (L.) Clairv.) – <u>Rose Campion</u>. Erect branched perennial to 1m, covered with dense white woolly hairs; flowers on long individual stalks; petal-limb purple, entire to emarginate. Intrd; commonly grown in gardens, frequently escaping and sometimes ± natd in waste places; S & C Br, rare casual elsewhere; SE Europe.

2. L. flos-cuculi L. (<u>Silene</u> <u>flos-cuculi</u> (L.) Clairv.) – <u>Ragged-Robin</u>. Erect simple or branched perennial to 75cm, glabrous to sparsely pubescent; flowers loosely clustered in open inflorescence; petal-limb pale purple, deeply 4-lobed. Native; marshy fields and other damp places; throughout BI.

3. L. viscaria L. (<u>Viscaria</u> <u>vulgaris</u> Bernh., <u>Steris</u> <u>viscaria</u> **R** (L.) Raf.) – <u>Sticky</u> <u>Catchfly</u>. Erect simple perennial to 60cm, glabrous to sparsely pubescent; stems very sticky below each node; flowers loosely clustered in open inflorescence; petal-limb purple, emarginate to shortly 2-lobed. Native; cliffs and rocky places; extremely local in Wa and Sc; grown in gardens and rarely escaping.

4. L. alpina L. (<u>Viscaria</u> <u>alpina</u> (L.) Don, <u>Steris</u> <u>alpina</u> **RR** (L.) Sourk. nom. inval.) – <u>Alpine</u> <u>Catchfly</u>. Erect simple perennial to 20cm, glabrous or very sparsely pubescent; stems not sticky; flowers ± densely clustered in compact head; petal-limb pale purple, 2-lobed. Native; mountain serpentine or heavy-metalliferous rocks; extremely rare in Cumberland and Angus.

19. AGROSTEMMA L. – <u>Corncockle</u>
Bracteoles not forming epicalyx; petal limb ± rounded to retuse, without scales, purple or rarely white; carpophore 0; styles 5; fruit a capsule opening by 5 teeth.

1. A. githago L. – <u>Corncockle</u>. Erect simple to slightly branched annual to 1m, with abundant appressed hairs; flowers on long individual stalks; petal-limb purple with dark streaks, exceeded by calyx-lobe. Intrd; cultivated and waste ground, formerly common in cornfields; formerly scattered over most of BI but now very rare and only casual.

20. SILENE L. (<u>Melandrium</u> Roehl.) – <u>Campions</u>
Bracteoles not forming epicalyx; petal-limb entire to 2-lobed, with or without scales at base, white to red or purple or yellow; styles 3 or 5; fruit a capsule opening by 2x as many teeth as styles. Often subdivided or amalgamated with related genera.

1 Calyx <6mm; flowers arranged in whorl-like groups at each inflorescence node   **3. S. otites**

1 Calyx usually >6mm, if <6mm then flowers solitary   2
  2 Plant with only male flowers   3
  2 Plant with at least some female or bisexual
    flowers   6
3 Calyx 20-veined, strongly inflated   11
3 Calyx 10-veined, not or scarcely inflated   4
  4 Dwarf cushion-plant with 1 flower per stem
                                            6. S. acaulis
  4 Stems branched, with >1 flower   5
5 Upper part of inflorescence sticky; calyx-teeth with
  broad scarious margins   14
5 Upper part of inflorescence not or scarcely sticky;
  calyx-teeth with 0 or narrow scarious margins   8
  6 Styles 5; capsule-teeth 10   7
  6 Styles 3; capsule-teeth 6   9
7 Flowers all bisexual; carpophore >5mm 11. S. coeli-rosa
7 Flowers dioecious (sometimes smut-infected stamens
  present in female flowers); carpophore <2mm   8
  8 Corolla white; capsule-teeth erect   9. S. latifolia
  8 Corolla red or pink (rarely white, see text);
    capsule-teeth revolute   10. S. dioica
9 Calyx 20-30-veined, strongly inflated at fruiting   10
9 Calyx 10-veined, usually not or scarcely inflated at
  fruiting   12
  10 Calyx 30-veined, cross-connecting veins not
     apparent; annual   14. S. conica
  10 Calyx 20-veined, wih conspicuous cross-connecting
     veins; perennial   11
11 Bracts largely herbaceous; capsule with patent to
   revolute teeth   5. S. uniflora
11 Bracts scarious; capsule with erect to erecto-patent
   teeth   4. S. vulgaris
  12 Perennial, with non-flowering shoots at flowering
     time   13
  12 Annual, without non-flowering shoots   15
13 Dwarf cushion-plant with 1 flower per stem
                                            6. S. acaulis
13 Stems branched, with >1 flower   14
  14 Carpophore about as long as capsule; petal-limb
     with small knob-like scales at base   1. S. italica
  14 Carpophore <1/2 as long as capsule; petal-limb
     with conspicuous acute scales at base   2. S. nutans
15 Stems, leaves and calyx glabrous   7. S. armeria
15 Stems, leaves and calyx pubescent   16
  16 Calyx >20mm   8. S. noctiflora
  16 Calyx <18mm   17
17 Calyx >12mm, with short sparse hairs   12. S. pendula
17 Calyx <12mm, with long ± patent hairs   13. S. gallica

Other spp. - Many spp. are grown as ornamentals or occur as
impurities in bird-seed and appear occasionally on tips and

waste ground.   The European annuals **S. dichotoma** Ehrh.
(Forked Catchfly), **S. csereii** Baumg., **S. cretica** L., and **S. muscipula** L. formerly occurred frequently as casuals but are now rare.   S. dichotoma resembles S. gallica but has horizontal flowers with whitish petals and calyx 8-15mm. S. csereii resembles S. vulgaris but the calyx has 10 long and 10 short veins (not 20 long veins).   S. muscipula and S. cretica resemble S. armeria but have acute calyx-teeth and all the leaves tapered to a narrow base; the carpophore is pubescent and glabrous respectively.   **S. fimbriata** (Adams ex Fried. Weber & Mohr) Sims (S. multifida (Fried. Weber & Mohr) Rohrb.)   and **S. schafta** Gmelin ex Hohen. are garden perennials from Caucasus.   **S. fimbriata** was formerly natd in W Sc; it has erect stems, a strongly inflated 10-veined calyx and much-divided white petals.   S. schafta is a rare escape formerly natd in S Hants; it has several ± simple decumbent stems to 15cm, each with few flowers, a narrow calyx >20mm, and broadly obovate dentate red petals. **S. conoidea** L., from S Europe, differs from S. conica in being less pubescent and larger in most parts (calyx >15mm, capsule >12mm, seeds >1mm); it is a rare casual and plants reported to be natd in CI were actually variants of S. conica. **S. alpestris** Jacq., from E Alps, was planted on Ben Lawers, Mid Perth, and has persisted since at least 1974; it is a perennial with a rather woody base, narrow leaves and white flowers 1-1.5cm across.

1. S. italica (L.) Pers. - Italian Catchfly. Pubescent 206
perennial with ascending to erect stems to 70cm; stem-leaves linear to oblanceolate; flowers in lax dichasia, sometimes unisexual; calyx 14-21mm; petals whitish, yellowish or pinkish, deeply bifid.   Intrd; roadside banks and quarries; natd since 1863 near Greenhithe, W Kent, a rare casual elsewhere in Br; Europe.

2. S. nutans L. - Nottingham Catchfly. Pubescent perennial 206
with erect stems to 80cm; similar to S. italica but calyx **R**
9-12mm, and see key.   Native; dry grassy or bare places; very scattered but locally common in Br N to MW Yorks, E Sc, CI. British plants are all ssp. **smithiana** (Moss) Jeanm. & Bocq., but this might not be distinct.

3. S. otites (L.) Wibel - Spanish Catchfly. Shortly **RR**
pubescent perennial with erect stems to 80cm; stem-leaves linear to oblanceolate; flowers numerous in whorl-like clusters at each node, ± dioecious; calyx 4-6mm; petals pale yellowish-green, entire.   Native; dry grassy heathland; E Anglia, rare casual elsewhere.

4. S. vulgaris Garcke - Bladder Campion. Glabrous to 206
pubescent perennial with decumbent to erect stems to 80cm; flowers in lax dichasia, often unisexual; calyx 10-18mm, strongly inflated, narrowed at mouth; petals deeply bifid. Grassy places, open and rough ground.

a. Ssp. vulgaris. Underground stolons absent; lower stem-leaves elliptic to ovate-lanceolate; calyx broadly ellipsoid to subglobose; petals white, capsule 6-9(11)mm; seeds 1-1.6mm. Native; throughout most of BI but rare in N & W Sc and N Ir.

b. Ssp. macrocarpa Turrill (_S. linearis_ auct. non Sweet). Underground stolons present; lower stem-leaves narrowly lanceolate to lanceolate; calyx oblong to ellipsoid; petals pinkish to greenish; capsule 11-13mm; seeds 1.6-2.1mm. Intrd; natd at Plymouth, S Devon, since 1921; Mediterranean.

4 x 5. S. vulgaris x S. uniflora occurs when the parents meet, which is rather rarely, in scattered places in Br. Hybrids are completely fertile and have intermediate characters, but precise identification is difficult except in mixed populations.

5. S. uniflora Roth (_S. maritima_ With., _S. vulgaris_ ssp. **206** maritima (With.) A. & D. Love) - Sea Campion. Glabrous or sparsely pubescent perennial with procumbent to ascending stems to 30cm; similar to _S. vulgaris_ but flowers slightly larger and rarely >4 per stem; stem-leaves linear to narrowly lanceolate; calyx often not narrowed at apex; and see key. Native; rocky and shingly coasts, and cliffs, lake-shores and streamsides in mountains; round most coasts of BI and rather rare in mountains of N En, Wa and Sc.

6. S. acaulis (L.) Jacq. - Moss Campion. Sparsely pubescent, densely tufted perennial with erect 1-flowered stems to 10cm; leaves linear; flowers often ± dioecious; calyx 5-9mm; petals pink, emarginate. Native; mountain rock-ledges and scree; N Wa, Lake District, C & NW Sc, rare in N Ir.

7. S. armeria L. - Sweet-William Catchfly. Erect glabrous annual to 40cm; stem-leaves lanceolate to narrowly ovate, cordate, clasping the stem; flowers in lax to dense corymbose cymes; calyx 12-15mm, with obtuse teeth; petals pink, emarginate. Intrd; infrequent garden escape on tips and in waste places; scattered in En and Guernsey; Europe.

8. S. noctiflora L. - Night-flowering Catchfly. Erect **R** glandular-pubescent annual to 50cm; stem-leaves ovate to lanceolate, narrowed to stalk-like base clasping the stem; flowers in sparse cymes; calyx 20-30mm, with long narrow teeth; petals whitish-yellow, deeply bifid. Native; sandy arable soils; scattered through En, Wa, E Sc and E Ir, but rare except in E En; also an infrequent casual.

9. S. latifolia Poiret (_S. alba_ (Miller) E. H. Krause nom. **206** illeg., _S. pratensis_ (Rafn) Godron & Gren.) - White Campion. Erect, pubescent, ± glandular, rarely glabrous (annual to) perennial to 1m; stem-leaves similar to those of _S. noctiflora_; flowers dioecious, in open dichasia, more numerous on male plants; calyx 15-22mm and 10-veined (male), 20-30mm and 20-veined (female), with narrow teeth; petals white, deeply bifid. Native; banks, roadsides, waste and

cultivated ground, mostly on light soils in the open;
throughout most of BI but rare or absent from much of W.
British plants are ssp. **alba** (Miller) Greuter & Burdet, but
ssp. **latifolia** (S. alba ssp. divaricata (Reichb.) Walters, S.
pratensis ssp. divaricata (Reichb.) McNeill & H. Prent., S.
macrocarpa (Boiss. & Reuter) E. H. Krause) from S Europe,
differing in its more finely tapered calyx-teeth and
divergent to recurved capsule-teeth, occurs as a rare casual.

9 x 10. S. latifolia x S. dioica = S. x hampeana Meusel & K.
Werner (S. x intermedia (Schur) Philp nom. illeg. non (Lange)
Bocq.) occurs commonly whenever the parents meet over most of
BI, especially in lowland areas; it is highly fertile and
intermediate in characters, notably the pale pink petals.

10. S. dioica (L.) Clairv. - Red Campion. Erect pubescent, **206**
sometimes glandular, rarely glabrous perennial to 1m; very
like S. latifolia but capsule-teeth revolute and petals red
or pink (rarely white). Native; woods and hedgerows, usually
in shady places in lowlands but on open cliffs and scree-
slopes in mountains; throughout BI but very scattered in S
Ir. White-flowered plants can usually be told from S.
latifolia by the lack of anthocyanin in stem and leaves as
well, but pale pink-flowered plants seem sometimes identical
with certain hybrids. Some plants in Shetland have large
deep red flowers and strong densely pubescent stems and have
been recognized as ssp. zetlandica (Compton) Clapham nom.
inval. (Melandrium dioicum var. zetlandicum Compton).
However they represent only extreme variants with one
combination of characters, and are not worthy of ssp. rank.

11. S. coeli-rosa (L.) Godron - Rose-of-heaven. Erect
glabrous annual to 50cm; leaves linear; flowers few per stem;
calyx 15-28mm, clavate; petals bright pink, bifid,
conspicuous. Intrd; grown in gardens and an occasional
escape on tips and in waste places; SW Europe.

12. S. pendula L. - Nodding Catchfly. Pubescent ascending
annual to 45cm; leaves obovate to elliptic; flowers very few
per stem; calyx 13-18mm, ± inflated; petals bright pink,
emarginate to shallowly bifid. Intrd; grown in gardens,
readily self-sowing and an occasional escape; S Europe.

13. S. gallica L. - Small-flowered Catchfly. Pubescent **R**
erect annual to 45cm; leaves oblanceolate (lower) to linear
(upper); flowers ± erect in monochasia; calyx 7-10mm with
long teeth; petals yellowish-white to pink or red-blotched,
emarginate to retuse. Native; waste places, cultivated land
and open sandy ground; CI and very locally in parts of S Br,
scattered N to C Sc and in Ir but probably intrd.

14. S. conica L. - Sand Catchfly. Densely glandular- **R**
pubescent decumbent to erect annual to 35cm; leaves linear to
linear-lanceolate; flowers rather few per stem; calyx 10-
15mm, becoming strongly ovoid, with long fine teeth; petals
pink, emarginate to shallowly bifid. Native; sandy places,
especially maritime dunes; very local in parts of E & S Br,

CI, rather frequent casual elsewhere.

21. CUCUBALUS L. - Berry Catchfly
Bracteoles not forming an epicalyx; petal-limb bifid, with 2
scales at base, white; styles 3; fruit a berry.

1. C. baccifer L. - Berry Catchfly. Shortly pubescent
perennial with much-branched sprawling stems to 1m; leaves
ovate, acuminate, with short petiole; flowers in open
dichasia; calyx 8-15mm, the petals longer; berry 6-8mm
across, black, not concealed by calyx. Intrd; rough grassy
places; rare casual, natd in W Norfolk; Europe.

22. SAPONARIA L. - Soapworts
Bracteoles not forming an epicalyx; petal-limb ± entire, with
2 scales at base, pink (to white); styles 2; fruit a capsule
opening by 4 teeth.

1. S. officinalis L. - Soapwort. Glabrous perennial with
erect or ascending stems to 90cm; leaves ovate to elliptic,
the largest >5cm; flowers in rather compact corymbose cymes,
c.2.5cm across; calyx glabrous, 15-20mm. Probably intrd;
grown in gardens and a common escape and throwout on waste
ground, roadsides and grassy places, often flore pleno, often
well natd and perhaps native in SW En; throughout most of BI
but rare in N Sc; Europe.
2. S. ocymoides L. - Rock Soapwort. Pubescent perennial
with procumbent to decumbent stems to 50cm; leaves narrowly
elliptic to oblanceolate, <3cm; flowers in rather lax
dichasia, c.1cm across; calyx pubescent, 7-12mm. Intrd;
grown in rock-gardens and sometimes escaping on walls and
stony banks, natd in a few places; En and Wa; S Europe.

23. VACCARIA Wolf - Cowherb
Bracteoles not forming an epicalyx; petal-limb entire or
shortly bifid, without scales, pink; styles 2; fruit a
capsule opening by 4 teeth.

1. V. hispanica (Miller) Rauschert (V. pyramidata Medikus) -
Cowherb. Glabrous glaucous erect annual to 60cm; leaves
lanceolate to ovate, clasping the stem; flowers in lax
dichasia; calyx becoming inflated in fruit, with 5 acute
angles or wings. Intrd; casual mainly from bird-seed on tips
and waysides and in parks, occasionally natd; Br and CI; S &
C Europe.

24. PETRORHAGIA (Ser. ex DC.) Link (Kohlrauschia Kunth) -
Pinks
Bracteoles close to base of calyx, forming epicalyx;
calyx-tube scarious at joints between lobes; petal-limb
emarginate, without scales, pink; styles 2; fruit a capsule

opening by 4 teeth.

1 Mat-forming perennials; flowers solitary; epicalyx
   of (2-)4(5) whitish bracts; calyx <8mm   **3. P. saxifraga**
1 Erect annuals; flowers in heads of (1)3-11; epicalyx
   of several pairs of brown bracts; calyx >8mm   **2**
   2 Seeds tuberculate on surface   **1. P. nanteuilii**
   2 Seeds reticulate on surface   **2. P. prolifera**

1. P. nanteuilii (Burnat) P. Ball & Heyw. (<u>Kohlrauschia</u> **RRR**
<u>nanteuilii</u> (Burnat) P. Ball & Heyw., <u>K. prolifera</u> auct. non
(L.) Kunth) - <u>Childing Pink</u>. Stems to 50cm, glabrous or
puberulous; flowers in compact ovoid heads, opening 1 at a
time; leaf-sheaths c. as long as wide; calyx 10-13mm; seeds
tuberculate; 2n=60. Native; dry grassy places; rare in SC
En, S Wa and CI, casual rarely natd elsewhere.
2. P. prolifera (L.) P. Ball & Heyw. (<u>Kohlrauschia prolifera</u>   **RR**
(L.) Kunth) - <u>Proliferous Pink</u>. Differs from <u>P. nanteuilii</u>
in its reticulate seed-surface; stem usually more densely
pubescent and leaf-sheaths up to 2x as long as wide; 2n=30.
Possibly native; dry banks; Beds, possibly elsewhere in En.
3. P. saxifraga (L.) Link (<u>Kohlrauschia saxifraga</u> (L.)
Dandy) - <u>Tunic-flower</u>. Mat-forming perennial with decumbent
to ascending, glabrous to pubescent stems to 35cm; calyx
3-6mm; seeds ± tuberculate; 2n=60. Intrd; waste places; rare
casual in Br, natd near Tenby, Pembs; Europe.

25. DIANTHUS L. - <u>Pinks</u>
Bracteoles close to base of calyx, forming epicalyx; calyx-
tube not scarious at joints between lobes; petal-limb
variously divided, without scales, usually pink to red;
styles 2; fruit a capsule opening by 4 teeth.

1 Flowers mostly in compact cymose clusters of usually
   >3 surrounded by involucre of leaf-like bracts;
   epicalyx nearly as long as to longer than calyx   **2**
1 Flowers solitary or in lax cymes, sometimes 2-3 close
   together but not surrounded by involucre of leaf-
   like bracts; epicalyx <3/4 as long as calyx   **3**
   2 Annual to biennial, without sterile shoots at
      flowering time; inflorescence pubescent
                                              7. D. armeria
   2 Perennial, with sterile shoots at flowering time;
      inflorescence glabrous   **6. D. barbatus**
3 Petal-limb divided >1/4 way to base into narrow lobes **4**
3 Petal-limb divided <1/4 way to base into narrow or
   triangular teeth   **5**
   4 Leaves on sterile shoots mostly >2cm, narrowly
      acute; inner bracteoles of epicalyx broadly
      obovate, shortly apiculate at apex   **3. D. plumarius**
   4 Leaves on sterile shoots mostly <1.5cm, obtuse to

        subacute; inner bracteoles of epicalyx oblong to
        obovate, apiculate to cuspidate at apex
                                                4. D. gallicus
5 Stems puberulous below; flowers scentless; inner
   bracteoles of epicalyx cuspidate   **5. D. deltoides**
5 Stems glabrous; flowers scented; inner bracteoles
   of epicalyx apiculate   6
   6 Flowers <30mm across; calyx <20mm; stems rarely
      >20cm   **1. D. gratianopolitanus**
   6 Flowers >30mm across; calyx >20mm; stems rarely
      <20cm   **2. D. caryophyllus**

Other spp. - **D. carthusianorum** L., from Europe, is occasionally found as a garden escape; it is a perennial with the
flowers packed into tight clusters, epicalyx c.1/2 as long as
calyx, and usually dark red petals. **D. chinensis** L. (Rainbow
Pink) and **D. superbus** L. are grown in gardens and there are
unconfirmed records of them as escapes.

1. D. gratianopolitanus Villars - Cheddar Pink. Mat- **RRR**
forming, glabrous, glaucous perennial with ± erect flowering
stems to 20cm; flowers <30mm across, usually solitary,
fragrant, pale to deep pink; calyx <20mm. Native; limestone
rock crevices and cliff-ledges; Cheddar Gorge, N Somerset,
rare escape sometimes planted or natd elsewhere.
1 x 2. D. gratianopolitanus x D. caryophyllus,
1 x 2 x 3. D. gratianopolitanus x D. caryophyllus x D.
plumarius, and
1 x 3. D. gratianopolitanus x D. plumarius
are the parentages of a number of escaped garden Pinks, some
of which are occasionally natd on walls etc. in S & C Br.
Characters of the 3 spp. occur in varying combinations.
2. D. caryophyllus L. - Clove Pink. Tufted, glabrous,
glaucous perennial with ± erect flowering stems to 60cm;
flowers >30mm across, 1-5 in lax cymes, fragrant, usually
pink; calyx >20mm. Intrd; grown in gardens and occasionally
natd on old walls; S & C Br, ephemeral outcast elsewhere; S
Europe.
2 x 3. D. caryophyllus x D. plumarius occurs in the same
manner as the above 3 hybrids.
3. D. plumarius L. - Pink. Tufted or mat-forming, glabrous,
glaucous perennial with ± erect flowering stems to 30cm;
flowers 20-40mm across, 1-5 in lax cymes, fragrant, white to
deep pink; calyx >20cm. Intrd; commonly grown in gardens and
sometimes natd on old walls or banks; similar distribution to
D. caryophyllus but commoner; SE Europe.
4. D. gallicus Pers. - Jersey Pink. Mat-forming glaucous
perennial, puberulous below, with ± erect flowering stems to
25cm; flowers 15-25mm across, 1-3 together, fragrant, pink;
calyx 20-25mm. Intrd; grassy coastal dunes; natd in Jersey
since 1892; W coast of Europe.

5. D. deltoides L. - <u>Maiden</u> <u>Pink</u>. Loosely tufted, **R**
puberulous, green to glaucous perennial with ± erect
flowering stems to 45cm; flowers 12-20mm across, 1(-3)
together, scentless, white to pink, usually spotted; calyx
12-18mm. Native; dry grassland; scattered and decreasing in
Br N to C Sc, CI.

6. D. barbatus L. - <u>Sweet-William</u>. Mat-forming, ± glabrous,
green perennial with erect flowering stems to 50cm; flowers
<20mm across, in dense clusters at stem apex, ± scentless,
pink to red, often spotted; calyx 12-18mm. Intrd; commonly
grown in gardens, often escaping or as a throw-out on tips
and waste places; Br, mainly S, rarely persisting; S Europe.

7. D. armeria L. - <u>Deptford</u> <u>Pink</u>. Erect, pubescent, green   **R**
annual or biennial to 60cm; flowers <15mm across, in ±
dense clusters at stem apex, scentless, bright pink, often
spotted; calyx 15-20mm. Native; dry grassy places; very
scattered and decreasing in S Br, CI, introduced further N.

48. POLYGONACEAE - Knotweed family

Herbaceous annuals to perennials or woody climbers; leaves
alternate (rarely subopposite), simple, usually entire, with
fused often scarious stipules sheathing stem. Flowers in
simple or branched racemes, sometimes few or solitary,
bisexual to dioecious or variable, hypogynous, actinomorphic;
perianth of 1-2 usually ± similar whorls, of 3-6 tepals
(2-3 per whorl) free or fused below, greenish, brownish,
white or pink, persistent in fruit; stamens (3)6-9; ovary
1-celled, with 1 basal ovule; stigmas 2-3, sessile or on
styles, capitate to finely divided; fruit an achene.
Usually easily recognized by the distinctive stipules and
tepals and by the single basal ovule.

```
1 Tepals 3; tiny annual with mostly subopposite
  leaves                                          2. KOENIGIA
1 Tepals mostly > 4; annual to perennial with
  alternate leaves                                          2
  2 Tepals 4; leaves reniform                      9. OXYRIA
  2 Tepals mostly 5-6; leaves very rarely reniform        3
3 Tepals 6, in 2 whorls of 3                               4
3 Tepals mostly 5                                          5
  4 Stamens 6; inner 3 tepals enlarging and enclosing
    achene; achene not winged; leaves pinnately
    veined                                         8. RUMEX
  4 Stamens 9; tepals remaining small in fruit;
    achene 3-winged; leaves palmately veined       7. RHEUM
5 Stems woody, twining                                     6
5 Stems mostly herbaceous though often shrub-like          7
  6 Leaves <2cm, rounded at base         6. MUEHLENBECKIA
  6 Leaves >2cm, truncate to cordate at base   5. FALLOPIA
```

7 Outer 3 tepals with longitudinal wing or strong keel
                                               5. FALLOPIA
7 Tepals not winged, rarely weakly keeled   8
   8 Inflorescences ≤6-flowered, all axillary **4. POLYGONUM**
   8 Inflorescences mostly >6-flowered, some or all
      terminal   9
9 Achene >2x as long as perianth, not winged; leaves
   sagittate; filaments flattened or winged **3. FAGOPYRUM**
9 Achene usually <2x as long as perianth, if >2x as
   long then strongly winged; leaves very rarely
   sagittate; filaments not winged or flat   **1. PERSICARIA**

1. PERSICARIA Miller (Aconogonum Reichb., Bistorta Scop.,
Polygonum sects. Aconogonon Meissner, Persicaria (L.)
Meissner, Echinocaulon Meissner, Cephalophilon Meissner,
Bistorta (L.) D. Don) - Knotweeds
Annuals to perennials, rhizomatous, stoloniferous or neither;
inflorescence many-flowered, terminal and axillary, spike-
like to subcapitate or paniculate; tepals mostly 5, not
winged, petaloid, not enlarging in fruit; stamens 8(or 4-7 by
reduction); style 1 and divided into 2 or 3, or 3; stigmas
small, capitate to clavate; achene lenticular (biconvex to
biconcave), trigonous, triquetrous or 3-winged.

1 Stigmas 3 in all flowers; stamens 8; perennials   2
1 Stigmas 2 in all or most flowers; stamens often <8 in
   at least some flowers; annuals (perennial in P.
   amphibia)   9
   2 Inflorescence unbranched; stamens exserted   3
   2 Inflorescence a branched panicle; stamens included 5
3 Lower part of inflorescence with bulbils instead of
   flowers; lower leaves cuneate at base   **8. P. vivipara**
3 Bulbils 0; lower leaves truncate to cordate at base 4
   4 Petioles of basal and lower stem-leaves winged
      above; flowers usually pale pink; stems unbranched
                                            6. P. bistorta
   4 Petioles unwinged; flowers usually red; stems
      usually branched   **7. P. amplexicaulis**
5 Tepals <2.5mm at flowering   6
5 Tepals >2.5mm at flowering   7
   6 Leaves tomentose on lowerside, with densely
      matted, twisted hairs; achene with 3 wings;
      perianth shrivelling in fruit   **4. P. weyrichii**
   6 Leaves glabrous to densely pubescent on lower-
      side, with ± straight hairs; achene not winged;
      perianth becoming fleshy in fruit and ≤3mm
                                            5. P. mollis
7 Tepals fused to c.1/4 to 1/2 way, usually pinkish
                                         2. P. campanulata
7 Tepals free ± to base, white   8
   8 Styles (+ stigmas) <0.5mm; lower leaves cuneate at

base, <3.5cm wide   **1. P. alpina**
8 Styles (+ stigmas) >0.5mm; lower leaves truncate
  to cordate at base, >3.5cm wide   **3. P. wallichii**
9 Stems with recurved prickles   **17. P. sagittata**
9 Stems without prickles, usually glabrous   10
  10 Flowers in + globose heads; styles 3
                                           10. P. nepalensis
  10 Flowers in cylindrical or tapering inflorescences;
     styles 2(-3)   11
11 Perennial with strong rhizomes or stolons, often
   aquatic; stamens exserted at anthesis   **9. P. amphibia**
11 Annual, often rooting at lower nodes; stamens
   included   12
   12 Sessile or stalked glands present on perianth and/
      or peduncle   13
   12 Glands 0 on perianth and peduncle   15
13 Glands on peduncle clearly stalked   **13. P. pensylvanica**
13 Glands on peduncle + sessile or 0   14
   14 Inflorescence dense, the flowers crowded; leaves
      without peppery taste   **12. P. lapathifolia**
   14 Inflorescence lax, the flowers mostly separated;
      leaves with sharp peppery taste (usually strong,
      sometimes faint) when fresh   **14. P. hydropiper**
15 Inflorescence dense, the flowers crowded; leaves
   often with dark blotch   **11. P. maculosa**
15 Inflorescence lax, the flowers mostly separated;
   leaves never with dark blotch   16
   16 Lower leaves usually <5x as long as wide,
      12-30mm wide; achene 2.5-3.5mm   **15. P. laxiflora**
   16 Leaves usually >5x as long as wide, 2-15mm wide;
      achene 2-2.5mm   **16. P. minor**

Other spp. - **P. capitata** (Buch.-Ham. ex D. Don) Gross
(Polygonum capitatum Buch.-Ham. ex D. Don), from Himalayas,
is a perennial similar to P. nepalensis but differs in its
conspicuously glandular-pubescent stems and stipules,
abruptly acute leaves with short petiole with curious lobe at
base, and no tuft of white hairs at base of stipules (present
in P. nepalensis); it is an infrequent garden or pot-plant
escape not hardy in Br. Records of **Polygonum lichiangense** W.
Smith are errors.

1. P. alpina (All.) Gross (Polygonum alpinum All.) - Alpine
Knotweed. Stems erect, to 1m; leaves <15 x 3.5cm, lanceolate, sparsely pubescent; perianth 2.5-3.5mm; styles <0.5mm;
achene 4-5mm, slightly exceeding perianth, trigonous. Intrd;
garden escape natd on river shingle; by R. Dee near Ballater,
S Aberdeen, less permanent elsewhere in Sc; Europe. A plant
natd by a road near Huddersfield, SW Yorks, might be a
variant of this sp. or a related sp.; it differs in wider
leaves and achene with 3 thick wings c.2x as long as

perianth.

2. P. campanulata (Hook. f.) Ronse Decraene (<u>Polygonum campanulatum</u> Hook. f.) - <u>Lesser</u> <u>Knotweed</u>. Stems ascending, to 1m; leaves <u>≤</u>15 x 7cm, lanceolate to ovate, tomentose on lowerside; perianth 3.5-5mm; heterostylous, with styles c.0.5mm or c.1mm; achene 3-4mm, included in perianth. Intrd; garden escape natd in damp shady places; sparsely scattered over BI; Himalayas.

3. P. wallichii Greuter & Burdet (<u>P.</u> <u>polystachya</u> (Wallich ex Meissner) Gross non Opiz, <u>Polygonum</u> <u>polystachyum</u> Wallich ex Meissner) - <u>Himalayan</u> <u>Knotweed</u>. Stems erect, to 1.5m; leaves <u>≤</u>20 x 8cm, lanceolate to narrowly ovate, almost glabrous to densely pubescent on lowerside; perianth (2.5)3-4mm; heterostylous, with styles c.0.6-0.8 or 0.8-1.2mm; achene 1.5-3mm, trigonous, included in perianth. Intrd; garden escape natd in grassy places and roadsides; scattered over Br and Ir; Himalayas. Densely pubescent plants are best separated from <u>P.</u> <u>mollis</u> by the unequal tepals.

4. P. weyrichii (F. Schmidt ex Maxim.) Ronse Decraene (<u>Polygonum</u> <u>weyrichii</u> F. Schmidt ex Maxim.) - <u>Chinese</u> <u>Knotweed</u>. Stems erect, to 1.5m; leaves <u>≤</u>20 x 6cm, lanceolate to narrowly ovate, densely tomentose on lowerside; perianth 1.5-2mm; styles <0.5mm; achene 4-6mm, broadly 3-winged, >2x as long as perianth. Intrd; garden escape natd in rough ground; Wastwater, Cumberland; E Asia.

5. P. mollis (D. Don) Gross (<u>Polygonum</u> <u>molle</u> D. Don, <u>P.</u> <u>rude</u> Meissner, <u>P.</u> <u>paniculatum</u> Blume) - <u>Soft</u> <u>Knotweed</u>. Stems erect, to 1.5m; leaves <u>≤</u>25 x 8cm, lanceolate to narrowly ovate, almost glabrous to densely pubescent on lowerside; perianth 1.5-2.2mm; styles <0.5mm; achene 2-2.5mm, trigonous, included in perianth which becomes succulent and blackish. Intrd; garden escape natd in rough ground; near Coylet, Argyll and Tunbridge Wells, W Kent; Himalayas.

6. P. bistorta (L.) Samp. (<u>Polygonum</u> <u>bistorta</u> L.) - <u>Common</u> <u>Bistort</u>. Stems erect, simple, to 80(100)cm; leaves truncate to cordate at base, the lower with winged petioles; flowers pink; inflorescences dense, 1 per stem. Native; grassy places; throughout most of Br and Ir but very common only in NW En and intrd in Ir and much of S Br.

7. P. amplexicaulis (D. Don) Ronse Decraene (<u>Polygonum</u> <u>amplexicaule</u> D. Don) - <u>Red</u> <u>Bistort</u>. Stems erect, usually branched, to 1m; leaves cordate at base, the upper clasping the stem; petioles not winged; flowers red; inflorescences dense, usually >1 per stem. Intrd; cultivated in gardens, natd in grassy and rough places; C & W Ir, CI and very scattered in Br; Himalayas.

8. P vivipara (L.) Ronse Decraene (<u>Polygonum</u> <u>viviparum</u> L.) - <u>Alpine</u> <u>Bistort</u>. Stems erect to ascending, simple, to 30cm; leaves cuneate at base; petioles not winged; flowers pink; inflorescences dense, 1 per stem, lower part (rarely <u>+</u> all) occupied by dark purple bulbils. Native; grassland and rock-ledges on mountains; Sc and N En, very rare in N Wa and SW Ir.

9. P. amphibia (L.) Gray (Polygonum amphibium L.) – Amphibious Bistort. Rhizomatous perennial of very varied habit, in water with glabrous floating leaves, on dry land with ± erect sparsely pubescent stems to 60cm, intermediate habits common; lower leaves cordate to truncate at base; inflorescence cylindrical, dense, obtuse; peduncle glandular or not; perianth eglandular; achene 2-3mm, shiny, lenticular. Native; in water, wet places, river banks and a weed on rough ground; throughout BI.

10. P. nepalensis (Meissner) Gross (Polygonum nepalense Meissner) – Nepal Persicaria. Decumbent, glabrous to sparsely pubescent annual to 40cm; leaves tapering-acute, the lower narrowed to winged petiole which is often expanded and cordate at base; inflorescence ± globose, with leafy bract at base; peduncles with stalked glands; perianth eglandular, usually mauve; achene 1.5-2mm, lenticular, dull.   Intrd; cultivated and waste ground, rare casual in Br, natd in rough ground in S Somerset and Dorset; Himalayas.

11. P. maculosa Gray (Polygonum persicaria L.) – Redshank. Decumbent to erect ± glabrous annual to 80cm; leaves lanceolate; inflorescence cylindrical, dense, obtuse; peduncle and perianth eglandular; achene 2-3.2mm, shiny, lenticular or (c.10-60%) trigonous.   Native; waste, cultivated and open ground; throughout BI.

11 x 12. P. maculosa x P. lapathifolia = P. x lenticularis (Hy) Sojak (Polygonum x lenticulare Hy);

11 x 14. P. maculosa x P. hydropiper = P. x intercedens (G. Beck) Sojak (Polygonum x intercedens G. Beck);

11 x 15. P. maculosa x P. laxiflora = P. x condensata (F. Schultz) Sojak (Polygonum x condensatum (F. Schultz) F. Schultz; and

11 x 16. P. maculosa x P. minor = P. x brauniana (F. Schultz) Sojak (Polygonum x braunianum F. Schultz) have been recorded occasionally in S & C Br with apparently intermediate characters and varying sterility. They are all rare, probably over-recorded, and need checking.

12. P. lapathifolia (L.) Gray (Polygonum lapathifolium L., P. nodosum Pers.) – Pale Persicaria. Decumbent to erect ± glabrous to tomentose-leaved annual to 1m; leaves lanceolate; inflorescence cylindrical, dense, obtuse; peduncle and perianth with glands with stalks no longer than heads; achene 2-3.3mm, shiny, lenticular or (<1%) trigonous.   Native; waste, cultivated and open, especially damp ground; throughout BI, but very scattered in N Sc. Very variable in habit, flower colour (greenish to deep pink) and pubescence.

12 x 14. P. lapathifolia x P. hydropiper = P. x figertii (G. Beck) Sojak (Polygonum x metschii G. Beck) has been recorded from Surrey, Cambs and Hunts, but needs checking.

13. P. pensylvanica (L.) M. Gomez (Polygonum pensylvanicum

L.) - Pinkweed. Usually erect annual to 1m, glabrous below
but glandular-pubescent above; leaves lanceolate; inflorescence cylindrial, dense, obtuse; peduncle with stalked
glands; perianth eglandular; achene 2.5-3.5mm, shiny,
lenticular or trigonous. Intrd; on tips and waste ground,
frequent with soya-bean waste; S En; E N America.
   14. P. hydropiper (L.) Spach (Polygonum hydropiper L.) -
Water-pepper. Decumbent to ± erect, ± glabrous annual to
75cm; leaves lanceolate to narrowly so; inflorescence
tapering, rather lax, often nodding; peduncle glandular or
not; perianth with many (rarely few) sessile glands; achene
2.5-3.8mm, dull, lenticular or trigonous. Native; damp
places and shallow water, often shaded; throughout BI but
very scattered in N Sc.
   14 x 15. P. hydropiper x P. laxiflora = P. x hybrida (Chaub.
ex St. Amans) Sojak (Polygonum x oleraceum Schur) has been
reported from Surrey, W Gloucs and Co Cavan, but needs
confirmation.
   14 x 16. P. hydropiper x P. minor = P. x subglandulosa
(Borbas) Sojak (Polygonum x subglandulosum Borbas) has been
reported from En N to Yorks, but needs confirmation.
   15. P. laxiflora (Weihe) Opiz (Polygonum mite Schrank, non   R
Persicaria mitis Gilib.) - Tasteless Water-pepper. Annual
resembling P. hydropiper but without the sharp taste; glands
0 or very sparse on perianth and peduncle; bristles at tip of
stipules >3mm (not <3mm); achene 2.5-3.5mm, ± shiny. Native;
similar places to P. hydropiper; rare and very scattered over
En, Wa and NE Ir, and over-recorded.
   15 x 16. P. laxiflora x P. minor = P. x wilmsii (G. Beck)
Sojak (Polygonum x wilmsii G. Beck) has been recorded from
Tyrone, Co Antrim, Berks and Oxon, but needs confirming.
   16. P. minor (Hudson) Opiz (Polygonum minus Hudson) - Small   R
Water-pepper. Decumbent to ascending, ± glabrous annual to
40cm; leaves linear to narrowly elliptic; inflorescence
tapering, rather lax; peduncle and perianth eglandular;
achene 2-2.5mm, shiny, lenticular. Native; damp fields,
ditches and pond-sides; very scattered and rather rare over
most of Br and Ir.
   17. P. sagittata (L.) Gross ex Nakai (Polygonum sagittatum
L.) - American Tear-thumb. Sprawling annual to 1m, glabrous
but with recurved prickles on stems, petioles and midribs;
leaves sagittate; inflorescence rather sparse, ± capitate,
peduncle and perianth eglandular; achene 2-4mm, shiny,
trigonous. Intrd; natd by streams near Castle Cove, S Kerry,
since 1889, now nearly extinct; E N America.

2. KOENIGIA L. - Iceland-purslane
Annuals; inflorescence of 1-few flowers, terminal and
axillary; tepals 3, not winged, not enlarging in fruit;
stamens 3; styles 2; stigmas capitate; achene trigonous.
Superficially like Montia or Peplis, but quite different in

all details.

1. K. islandica L. - <u>Iceland-purslane</u>. Stems to 6cm, erect, **RR**
branched, usually reddish; leaves subopposite, <5mm; flowers
very inconspicuous. Native; damp stony and gravelly ground
>500m; Skye and Mull, W Sc, discovered 1934.

3. FAGOPYRUM Miller - <u>Buckwheats</u>
Annuals or herbaceous perennials; flowers in terminal and
axillary panicles; tepals 5, petaloid, not winged or keeled,
not enlarging in fruit; stamens 8; styles 3, long; stigmas
capitate, small; achene triquetrous, far exserted.

Other spp. - **F. tataricum** (L.) Gaertner (<u>Green</u> <u>Buckwheat</u>) is
a casual from Asia occasionally occurring as an impurity in
F. esculentum seed and rarely sown for gamebirds; it differs
from the latter in its short (c.2mm) greenish-white tepals
and achenes with undulate margins.

1. F. esculentum Moench - <u>Buckwheat</u>. Erect, very sparsely
pubescent, little-branched annual to 60cm; leaves sagittate;
perianth 2.5-4mm; achene 5-7mm, with straight margins.
Intrd; casual on tips and waste ground, formerly widely
cultivated but now only rarely so; formerly common, now
infrequent, over most of BI; Asia.
2. F. dibotrys (D. Don) H. Hara - <u>Tall</u> <u>Buckwheat</u>. Perennial
with herbaceous, erect little-branched stems to 1(2)m; very
like F. esculentum but perianth 2.5-3.5mm; achene 6-8mm.
Intrd; rare garden plant natd by road; Pembs; Asia.

4. POLYGONUM L. (<u>Polygonum</u> sects. <u>Polygonum</u>, <u>Avicularia</u>
Meissner) - <u>Knotgrasses</u>
Annuals or perennials, with strong tap-root; leaves small,
narrowed at base; inflorescences <6-flowered, axillary;
tepals 5, ± petaloid, not or slightly keeled, not enlarging
in fruit; stamens 8; stigmas 3, capitate, small, almost
sessile; achene trigonous.
For correct identification plants must possess ripe achenes
but not be so old that all lower leaves are gone.

1 Uppermost bracts not exceeding flowers, usually
  partly scarious   2
1 All bracts exceeding flowers, leaf-like   3
    2 Achene c.3mm, shiny; tepals erect, green to apex
      in midline; stems often erect   **7. P. patulum**
    2 Achene c. 2mm, ± dull; tepals divergent and
      wholly pink at apex; stems procumbent to ascending
                                             8. P.   arenarium
3 Stems strongly woody below; stipules at upper nodes
  at least as long as internodes, with 6-12 branched
  veins   **1. P. maritimum**

3 Stems not or slightly woody below; stipules at upper
  nodes much shorter than internodes, with wholly or
  mostly unbranched veins   4
  4 Achene shiny, slightly to much longer than
    perianth   **2. P. oxyspermum**
  4 Achene dull, shorter than to slightly longer than
    perianth   5
5 Leaves of main and lateral stems similar in size;
  tepals fused for >1/3; achene 1.5–2.5mm, with 2
  convex and 1 concave sides   **3. P. arenastrum**
5 -Leaves of lateral stems much smaller than those of
  main stem; tepals fused for <1/4; achene 2.5–4.5mm,
  with 3 concave sides   6
  6 Leaves <4mm wide; tepals narrowly oblong, gaping
    near apex to reveal achene   **6. P.rurivagum**
  6 Larger leaves >5mm wide; tepals oblong–obovate,
    overlapping almost to apex   7
7 Leaves ovate–lanceolate to narrowly elliptic, with
  petioles + included within stipules; achene <3.5mm
                                                 4. P. aviculare
7 Larger leaves obovate to narrowly so, with petioles
  well exserted from stipules; achene >3mm   **5. P. boreale**

Other spp. – **P. cognatum** Meissner (<u>Indian Knotgrass</u>), a
perennial from Asia differing from <u>P. maritimum</u> in its
shorter achene (c.3mm) and perianth fused for c.1/2 (not
<1/4), is now a rare grain-casual but was formerly natd
near a few docks and breweries in S En.

1. P. maritimum L. – <u>Sea Knotgrass</u>. Procumbent glaucous **RRR**
perennial; stems woody below, to 50cm; stipules conspicuous,
silvery, longer than upper internodes; achene 4–4.5mm, as
long as or slightly longer than perianth, shiny. Native; low
down on sandy beaches; very rare and sporadic in CI, E & W
Cornwall and Waterford, extinct in other SW En sites.
2. P. oxyspermum Meyer & Bunge ex Ledeb. (<u>P. raii</u> Bab.) –   **R**
<u>Ray's Knotgrass</u>. Procumbent annual or sometimes perennial;
stems sometimes + woody below, to 1m; stipules shorter than
internodes, brownish with silvery tips; achene 3.5–5.5mm,
slightly to much longer than perianth, shiny. Native; low
down on sandy beaches; scattered and decreasing round coasts
of BI, absent from most of E Br. Our plant is ssp. **raii**
(Bab.) D. Webb & Chater.
3. P. arenastrum Boreau – <u>Equal-leaved Knotgrass</u>. Usually
procumbent annual to 30(50)cm; leaves of main and lateral
stems similar in size; achene 1.5–2.5mm, shorter than to very
slightly longer than perianth, dull. Native; all sorts of
open ground; common throughout BI but less so than <u>P.</u>
<u>aviculare</u> except in N Sc.
4. P. aviculare L. – <u>Knotgrass</u>. Procumbent to scrambling
heterophyllous annual to 2m; achene 2.5–3.5mm, shorter than

to very slightly longer than perianth, dull. Native; all
sorts of open ground; commonest sp. of genus throughout BI
except in N Sc.
  5. P. boreale (Lange) Small - Northern Knotgrass. **R**
Procumbent to scrambling heterophyllous annual to 1m; achene
3-4.5mm, shorter to very slightly longer than perianth, dull.
Native; similar places to P. aviculare; Shetland and Orkney
(commonest sp. of genus), Outer Hebrides, M Ebudes and
Sutherland.
  6. P. rurivagum Jordan ex Boreau - Cornfield Knotgrass. **R**
Usually ± erect slender heterophyllous annual to 30cm; achene
2.5-3.5mm, usually very slightly longer than perianth, dull.
Native; cornfields and other arable land; rare and decreasing
in SE En, extremely rare and scattered elsewhere in Br N to C
Sc. Possibly best amalgamated with P. aviculare.
  7. P. patulum M. Bieb. - Red-knotgrass.   Usually ± erect
annual to 1m; bracts not leaf-like towards stem apex; achene
c.3mm, shorter than perianth, shiny. Intrd; rather frequent
grain- and wool-alien on tips, waste places and shoddy-
fields; scattered in Br; S & C Europe. Much confused with P.
arenarium but apparently separable on characters in key;
relative distributions in Br unknown.
  8. P. arenarium Waldst. & Kit. - Lesser Red-knotgrass.
Procumbent or scrambling annual to 50cm; similar to P.
patulum but see key. Intrd; status in Br similar to that of
P. patulum; Mediterranean region. Plants in Br belong to
ssp. **pulchellum** (Lois.) Thell.

5. FALLOPIA Adans. (Bilderdykia Dumort., Reynoutria Houtt.,
Polygonum sects Tiniaria Meissner, Pleuropterus (Turcz.)
Benth.) - Knotweeds
Annuals to robust herbaceous or woody perennials; inflioresc-
ences terminal and axillary, simple to paniculate; tepals 5,
± petaloid, the outer 3 keeled or winged and enlarging to
protect fruit; stamens 8; styles 3; stigmas ± capitate or
much divided; achene trigonous.

1 Rhizomatous herbaceous perennial; stigmas finely
    divided, flowers functionally dioecious   2
1 Twining; annual or woody perennial; rhizomes 0;
    stigmas capitate; flowers all bisexual   3
    2 Leaves rarely >12cm, truncate at base, cuspidate
                                                    1. F. japonica
    2 Leaves often >12cm, cordate to cordate-truncate
        at base, acute to ± acuminate at apex
                                                2. F. sachalinensis
3 Woody perennial; the larger inflorescences well
    branched   **3. F. baldschuanica**
3 Annual; inflorescences with 1 main axis   4
    4 Fruiting pedicels 1-3mm; achene 4-5mm, dull
                                                4. F. convolvulus

4 Fruiting pedicels 3–8mm; achene 2.5–3mm, shiny
                                  5. F. dumetorum

1. F. japonica (Houtt.) Ronse Decraene (<u>Reynoutria japonica</u>
Houtt., <u>Polygonum cuspidatum</u> Siebold & Zucc.) – <u>Japanese
Knotweed</u>. Stems erect to arching, to 2m, often forming dense
thickets; leaves broadly ovate, <u><</u>12(18) x 10(13)cm; inflorescences <u><</u>15cm. Intrd; waste places, tips and by roads,
railways and rivers; frequent to common over BI, originally a
garden escape 1st found in wild in 1886; Japan. Almost all
plants in BI are + female octoploids (2n=88). Almost all
seed set is hybrid. Var. **compacta** (Hook. f.) J. Bailey (<u>R.
japonica</u> var. <u>compacta</u> (Hook. f.) Buchheim) is a dwarf (<1m)
tetraploid (2n=44) with thick leaves + as wide as long with +
undulate margins and usually red-tinged inflorescences; both
sexes are still cultivated and natd plants are scattered
throughout Br.
1 x 2. F. japonica x F. sachalinensis = F. x bohemica
(Chrtek & Chrtkova) J. Bailey (<u>Reynoutria</u> x <u>bohemica</u> Chrtek &
Chrtkova) occurs usually with 1 or both parents, but
sometimes without either due to independent vegetative
dispersal, in scattered sites in En, Wa and Ir; it has
intermediate leaf-shape and -size. Most plants are hexaploid
(2n=66) but some with 2n=44 are probably derived from <u>F.
japonica</u> var. <u>compacta</u>.
1 x 3. F. japonica x F. baldschuanica is the parentage of
much seed produced by <u>F. japonica</u> near plants of <u>F.
baldschuanica</u>. The seed is viable but only 1 wild hybrid has
been found (waste ground by railway, Middlesex, 1987); it has
woody stems with rhizomes and scarcely climbing stems, and
intermediate leaf-shape, perianth and stigmas.
2. F. sachalinensis (F. Schmidt ex Maxim.) Ronse Decraene
(<u>Reynoutria</u> <u>sachalinensis</u> (F. Schmidt ex Maxim.) Nakai,
<u>Polygonum</u> <u>sachalinense</u> F. Schmidt ex Maxim.) – <u>Giant
Knotweed</u>. Similar to <u>F. japonica</u> but often taller (to 3m);
leaves ovate-oblong, <u><</u>38 x 28cm; inflorescences often
shorter (<10cm) and denser; 2n=44. Intrd; in similar places
to <u>F. japonica</u> but rarer; scattered over Br and Ir, locally
common, 1st recorded in wild in 1896. Most plants in BI are
+ female but + male plants are not rare.
3. F. baldschuanica (Regel) Holub (<u>F. aubertii</u> (L. Henry)
Holub, <u>Polygonum aubertii</u> L. Henry, <u>P. baldschuanicum</u> Regel,
<u>Bilderdykia</u> <u>aubertii</u> (L. Henry) Mold., <u>B. baldschuanica</u>
(Regel) D. Webb) – <u>Russian-vine</u>. Stems woody below, twining
and scrambling for many m; leaves ovate-triangular, cordate,
obtuse to acuminate; fruiting pedicels <u><</u>8mm; outer tepals
broadly winged in fruit; achenes 4–5mm, shiny. Intrd;
commonly cultivated and a persistent throw-out or relic in
waste scrubby places or hedges; scattered over most of BI but
rarely well natd; C Asia. <u>F. aubertii</u> perhaps differs in its
smaller achenes and flowers, more papillose inflorescence-
branches and more undulate leaf-margins; it appears to be

rare in cultivation but is very doubtfully specifically
distinct, and other differences claimed are not constant.
4. F. convolvulus (L.) A. Love (Polygonum convolvulus L.) -
Black-bindweed. Annual with trailing or climbing stems to
1(1.5)m; leaves ovate-triangular, cordate to sagittate,
obtuse to acuminate; fruiting pedicels 1-3mm; outer tepals
keeled to narrowly winged (var. **subalatum** (Lej. & Courtois)
Kent) in fruit; achenes 4-5mm, dull. Native; waste and
arable ground; common in most of BI.
5. F. dumetorum (L.) Holub (Polygonum dumetorum L.) - Copse-   RR
bindweed. Very like F. convolvulus but stems climbing to 2m;
leaves more narrowly acuminate; fruiting pedicels 3-8mm;
outer tepals broadly winged in fruit; achenes 2.5-3mm, shiny.
Native; in hedges and thickets; rare and very scattered in S
En, Caerns. Probably not distinct from the American **F.
scandens** (L.) Holub, which has priority.

6. MUEHLENBECKIA Meissner - Wireplant
Woody sprawling or climbing perennials; inflorescences short
axillary or terminal racemes, dioecious; tepals 5, fused for
>1/4 from base, enlarging and becoming white and succulent
in fruit, not keeled or winged; stamens 8, styles 3, stigmas
much divided; achene trigonous.

1. M. complexa (Cunn.) Meissner - Wireplant. Stems to
several m; leaves <2cm, oblong to suborbicular, with distinct
petiole, deciduous; flowers in autumn. Intrd; garden escape
natd on cliffs, walls and rough ground and in hedges; CI,
Scillies, extreme SW En.

7. RHEUM L. - Rhubarbs
Tall rhizomatous herbaceous perennials; leaves large, mostly
basal, palmately veined; flowers in large terminal and
axillary panicles; tepals 6, + petaloid, not winged or
keeled, not enlarging in fruit; stamens usually 9; anthers
versatile; stigmas 3, subsessile, capitate, papillate; achene
triquetrous, with 3 broad membranous wings.

1. R. x hybridum Murray (R. x cultorum Thorsrud & Reis. nom.
nud., R. rhaponticum auct. non L., R. rhabarbarum auct. non
L.) - Rhubarb. Basal leaves often 1m, glabrous, cordate,
very shallowly lobed with entire, obtuse to rounded lobes and
thick, edible petioles; flowering stems to 1.5m, glabrous,
leafless; flowers cream; achenes 6-12mm with wing 2-3mm wide.
Intrd; commonly grown vegetable, on field-scale especially in
N En, often persisting as relic or throwout; scattered
throughout BI; garden origin, probably from Siberian parents.
2. R. palmatum L. - Ornamental Rhubarb. Habit similar to
that of R. x hybridum but leaves sparsely pubescent,
distinctly lobed with acute, dentate lobes; flowering stems
pubescent; flowers reddish. Intrd; grown in gardens as
ornament and formerly medicinally, occasionally found as

outcast in grassy places; very scattered in En; NE Asia.

8. RUMEX L. - Docks
Usually herbaceous perennials (rarely annual or biennial),
sometimes ± rhizomatous; inflorescences terminal and axillary
racemes or panicles with whorled flowers; tepals 6, ±
sepaloid, not keeled or winged, the inner usually enlarging
in fruit and often with swollen tubercle on face; stamens 6;
anthers basifixed; styles 3; stigmas deeply divided; achene
triquetrous.
'Tepals' refer to the inner 3 at fruiting. The keys deal
only with species, not hybrids. All the hybrids are to some
degree sterile, most highly so, with undeveloped achenes;
they are not rare, but mostly occur as single or few plants
and almost always with 1 or both parents.

General Key
1 Lower leaves sagittate to hastate, acid-tasting;
  flowers mostly unisexual   2
1 Leaves not sagittate or hastate, not acid-tasting;
  flowers bisexual   4
    2 All leaves with distinct petiole, most c. as
      wide as long   **2. R. scutatus**
    2 Upper leaves sessile, most distinctly longer
      than wide   3
3 Upper leaves clasping stem; basal lobes pointed ±
  basally (sagittate); tepals becoming much longer
  than achene   **3. R. acetosa**
3 Upper leaves not clasping stem; basal lobes mostly
  pointed laterally or ± forward (hastate); tepals
  not or scarcely longer than achene   **1. R. acetosella**
    4 All tepals lacking swollen tubercle   Key A
    4 At least 1 tepal with distinct swollen tubercle
      on outer face   5
5 Tepals with 1-several teeth (each >0.5mm) on
  each side   Key B
5 Tepals entire to crenate   Key C

Key A - All tepals lacking swollen tubercle
1 Tepals with long hooked teeth   **17. R. brownii**
1 Tepals entire or nearly so   2
    2 Tepals distinctly longer than wide   **7. R. aquaticus**
    2 Tepals c. as long as wide   3
3 Plant rhizomatous; lower leaves <1.5x as long as wide
                                               6. R. pseudoalpinus
3 Plant not rhizomatous; leaves mostly >2x as long as
  wide   **8. R. longifolius**

Key B - At least one tepal with distinct swollen tubercle;
        all tepals with 1-several teeth
1 Tepals mostly <4mm   2
1 Tepals mostly >4mm   3

  2 Tepals mostly <3mm, with teeth c. as long (>2mm);
     tubercle acute distally; anthers 0.4-0.6mm
                                23. R. maritimus
  2 Tepals mostly >3mm, with teeth much shorter (<2mm);
     tubercle obtuse distally; anthers 0.9-1.3mm
                                22. R. palustris
3 Tepals broadly ovate to suborbicular, >5mm wide
   (excl. teeth)   **11. R. cristatus**
3 Tepals ovate-triangular, <4mm wide (excl. teeth)   4
  4 Tubercles on tepals coarsely warty   5
  4 Tubercles on tepals ± smooth   6
5 Perennial; leaves usually constricted just below
   middle (violin-shaped), rounded to cordate at base;
   at least some branches arising at >60 degrees
                                    18. R.pulcher
5 Annual to biennial; leaves rarely violin-shaped,
   rounded to cuneate at base; branches arising at <60
   degrees   **21. R. obovatus**
  6 Annual with basal leaves to 12cm; usually all 3
     tepals with well-developed tubercle
                              20. R. dentatus
  6 Perennial with basal leaves to 40cm; usually only
     1 tepal with well-developed tubercle
                          19. R. obtusifolius

Key C - At least 1 tepal with distinct swollen tubercle; all
        tepals entire to crenate
1 Part of pedicel above joint shorter than tepals   2
1 Part of pedicel above joint c. as long as or longer
   than tepals   3
  2 Tepals entire, with ± smooth tubercles; branches
     few, arising at <45 degrees from main stem
                                   5. R. frutescens
  2 Tepals with some short teeth, with warty
     tubercles; branches numerous, arising at 45
     degrees from main stem   **18. R. pulcher**
3 Tepals mostly <5mm   4
3 Tepals mostly >5mm   9
  4 Tepals <3mm   5
  4 Tepals mostly >3mm   6
5 All 3 tepals with well-developed oblong tubercle
                              14. R. conglomeratus
5 1 tepal with well-developed ± globose tubercle, other
   2 with 0 or rudimentary tubercle   **15. R. sanguineus**
  6 Lower leaves ovate-oblong, strongly cordate at
     base; tepals often with some short teeth
                             19. R. obtusifolius
  6 Lower leaves narrowly oblong, narrowly elliptic or
     lanceolate, cuneate to subcordate at base; tepals
     entire   7
7 Tepals not or scarcely wider than tubercles
                                 16. R. rupestris

7 Tepals much wider than tubercles   8
  8 Lower leaves tightly undulate; stems with short +
    erect branches, erect   13. R. crispus
  8 Leaves not undulate; stems with long branches
    flowering later than main stem, often procumbent
    at base   4. R. salicifolius
9 Tepals nearly as wide as long to wider, + rounded at
  apex, only 1 with well-developed tubercle   10
9 Tepals distinctly longer than wide, tapered to
  rounded to obtuse apex, usually all 3 with + well-
  developed tubercle   12
  10 Basal leaves <1.5x as long as wide, deeply cordate
     at base; petiole longer than leaf   9. R. confertus
  10 Basal leaves >2x as long as wide, cuneate to
     cordate at base; petiole shorter than leaf   11
11 Lower leaves + cordate at base; lateral veins
   arising at >60 degrees from midrib   11. R. cristatus
11 Lower leaves cuneate to truncate at base; lateral
   veins arising at <60 degrees from midrib
                                                       12. R. patientia
  12 Tubercles >3mm; robust waterside plant with basal
     leaves >60cm   10. R. hydrolapathum
  12 Tubercles <3mm; basal leaves <50cm   13
13 Basal leaves ovate-oblong, strongly cordate at base
                                                       19. R. obtusifolius
13 Basal leaves narrowly oblong, narrowly elliptic or
   lanceolate, cuneate to subcordate at base
                                                       13. R. crispus

Other spp. - c. 24 other spp. have been found as rare
casuals in BI. Of these **R. stenophyllus** Ledeb., from C & E
Europe, was formerly natd at Avonmouth Docks, W Gloucs; it
would key out as R. obtusifolius (tepals toothed and with
tubercles) but in habit and leaf-shape more closely resembles
R. crispus.

<u>Subgenus 1</u> - ACETOSELLA Raf. (sp. 1). Dioecious; some roots
horizontal and producing aerial shoots; leaves sagittate,
acid-tasting; tepals remaining shorter than achene, without
tubercles.

1. R. acetosella L. - <u>Sheep's Sorrel</u>. Stems procumbent to **241**
erect; leaves linear to oblong-lanceolate, with narrow,
laterally or forward-directed basal lobes; achene 1-1.5mm.
Native; heathy open ground, short grassland and cultivated
land, mostly on acid sandy soils; throughout BI. Distribution of the 2 sspp. in BI has not been worked out, but ssp.
pyrenaicus is probably mainly in S and absent from N. Extent
of overlap is unknown; both occur in C En and probably
intermediates exist, as in areas of overlap in C Europe.
  a. Ssp. acetosella (R. tenuifolius (Wallr.) A. Loeve).
Tepals forming a loose cover round ripe achene (easily rubbed

off by rolling achenes between finger and thumb). Small
plants with narrowly linear leaves and occurring on very dry
sands are worth no more than varietal rank as var.
tenuifolius Wallr.
 b. Ssp. pyrenaicus (Pourret) Akeroyd (ssp. angiocarpus auct.
non (Murb.) Murb., R. angiocarpus auct. non Murb.). Tepals
tightly adherent to ripe achene (not able to be rubbed off).

Subgenus 2 - ACETOSA Raf. (spp. 2-3). Dioecious; aerial
stems arising from short rhizomes; leaves sagittate to
hastate, acid-tasting; tepals enlarging to much longer than
achene, each with or without tubercle.

 2. R. scutatus L. - French Sorrel. Stems much-branched, 241
erect to spreading, ± woody below, to 50cm; leaves broadly
ovate, ± hastate, with very wide out-turned basal lobes;
tepals 5-8mm, orbicular, cordate, without tubercles. Intrd;
surviving on banks, old walls and rough ground in very
scattered places in Br; C & S Europe.
 3. R. acetosa L. - Common Sorrel. Stems not or little- 241
branched, usually erect, herbaceous, to 1m; leaves ovate or
obovate to narrowly triangular, sagittate, with very acute
backward-directed basal lobes; tepals 2.5-4mm, suborbicular,
± cordate, each with small tubercle near base.
1 Plant papillose-puberulent (hairs ≤0.3mm) on all
    vegetative parts   **b. ssp. hibernicus**
1 Plant usually ± glabrous, with papillae ±
    confined to basal margin of leaves   2
    2 Inflorescence with well-branched branches
                                                 d. ssp. ambiguus
    2 Inflorescence with simple branches   3
3 Leaves thick, succulent; basal leaves c.2x as long
    as wide; stem-leaves (1)2-4; coastal   **c. ssp. biformis**
3 Leaves thin, not succulent; basal leaves usually
    2-4x as long as wide; stem-leaves often >4;
    widespread   **a. ssp. acetosa**
 a. Ssp. acetosa. Plant to 60(100)cm; leaves rarely
succulent; inflorescence usually with several simple
branches. Native; in wide range of grassy places; common
throughout BI.
 b. Ssp. hibernicus (Rech. f.) Akeroyd (R. hibernicus Rech.
f.). Plant to 30(50)cm; leaves succulent; inflorescence with
few simple branches. Native; coastal dunes; NW, W & S Ir, N
Sc, SW En, SW Wa; endemic.
 c. Ssp. biformis (Lange) Valdes-Berm. & Castroviejo. Plant
to 20(30)cm; leaves succulent; inflorescence with few simple
branches. Native; sea-cliffs; W Cornwall, Clare, but perhaps
overlooked.
 d. Ssp. ambiguus (Gren.) A. Loeve (R. rugosus Campdera).
Plant to 1.2m; leaves large, thin; inflorescence with well
branched branches. Intrd; grown as vegetable, natd in Herts
and E Suffolk; origin unknown.

<u>Subgenus</u> 3 - <u>RUMEX</u> (spp. 4-23). Bisexual; aerial stems
usually erect, arising from tap-root or sometimes rhizomes;
leaves cordate to cuneate, larger than in other subgenera;
inner tepals enlarging to much longer than achene, 0, 1 or 3
with tubercle.
'Leaves' refers to lower stem-leaves and basal leaves;
'tepals' refers to inner tepals at fruiting. Hybrids are
frequent in mixed populations; they mostly possess
predictably intermediate characters.

4. R. salicifolius Weinm. (<u>R. triangulivalvis</u> (Danser) Rech. **241**
f.) - <u>Willow-leaved Dock</u>. Decumbent to erect perennial to
50(100)cm, often with long branches from near base; leaves
lanceolate to linear-lanceolate; inflorescence branched,
open; tepals 3-4mm, ovate-triangular, ± entire, all with
narrow warty tubercle. Intrd; natd on waste land by docks,
railways, canals and tips, originating with grain; very
scattered over Br and Ir, especially SE En, often only
casual; N America. Our plant is ssp. **triangulivalvis** Danser.
5. R. frutescens Thouars - <u>Argentine Dock</u>. Rhizomatous **241**
perennial with acending to erect unbranched stems to 30cm;
leaves obovate to oblanceolate, leathery; inflorescence
dense, not or shortly branched; tepals 4-5mm, narrowly
ovate-triangular, entire, all with large ± smooth tubercle.
Intrd; natd on coastal dunes in SW En and S Wa, casual near
docks in Sc, Wa and W En; S America.
5 x 14. R. frutescens x R. conglomeratus = R. x wrightii
Lousley was found in 1952 at Braunton Burrows, N Devon, with
<u>R. frutescens</u>; endemic.
6. R. pseudoalpinus Hoefft (<u>R. alpinus</u> L. 1759 non 1753) - **241**
<u>Monk's-rhubarb</u>. Rhizomatous perennial with erect stems to
70(100)cm; leaves broadly ovate, cordate; inflorescence
dense, with erect branches; tepals 5-6mm, broadly ovate-
triangular, entire, without tubercles. Intrd; natd relic of
old cultivation in grassy places by roads, streams and old
buildings; scattered in Sc and En S to Staffs; Europe.
7. R. aquaticus L. - <u>Scottish Dock</u>. Erect perennial to 2m; **241**
leaves triangular-ovate, cordate; inflorescence rather open, **RR**
with erecto-patent branches; tepals 5-8mm, ovate-triangular,
entire, without tubercles. Native; seasonally flooded ground
by Loch Lomond, Stirlings and Dunbarton.
7 x 13. R. aquaticus x R. x crispus = R. x conspersus
Hartman was found with both parents in Dunbarton in 1976.
7 x 15. R. aquaticus x R. sanguineus = R. x dumulosus
Hausskn. was found near both parents in Stirlings in 1989.
7 x 19. R. aquaticus x R. obtusifolius = R. x platyphyllos
Aresch. (<u>R. x schmidtii</u> Hausskn.) occurs with both parents
in Stirlings and Dunbarton.
8. R. longifolius DC. - <u>Northern Dock</u>. Erect perennial to **241**
1.2m; leaves lanceolate to narrowly ovate; inflorescence
dense, with short erect branches; tepals 4-5.5mm, suborbicular, cordate, entire, without tubercles. Native; damp

open and grassy ground often by water; Sc and En S to S
Lancs, Staffs.
8 x 13. R. longifolius x R. crispus = R. x propinquus
Aresch. is frequent in Sc wherever the parents meet.
8 x 19. R. longifolius x R. obtusifolius = R. x hybridus
Kindb. (R. x <u>arnottii</u> Druce) is frequent in Sc wherever the
parents meet (often near R. x propinquus too), and sparse in
N En.
9. R. confertus Willd. - <u>Russian</u> <u>Dock</u>. Erect perennial to **241**
1.2m; leaves broadly ovate, cordate; inflorescence fairly
dense, with erect branches; tepals 6-9mm, suborbicular,
cordate, ± entire, 1 with small tubercle. Intrd; natd on
roadside in E Kent, formerly Surrey and Oxon; E Europe.
9 x 13. R. confertus x R. crispus = R. x skofitzii Blocki
occurred in Surrey in 1954-5 with R. confertus.
9 x 19. R. confertus x R. obtusifolius = R. x borbasii
Blocki occurred in Surrey in 1954 with R. confertus.
10. R. hydrolapathum Hudson - <u>Water</u> <u>Dock</u>. Erect perennial **241**
to 2m; leaves lanceolate to broadly so, cuneate; inflores-
cence large, rather open, with many erecto-patent branches;
tepals 5-8mm, ovate-triangular, ± entire, all with
elongated smooth tubercle. Native; by lakes, rivers, canals,
ditches and marshes; scattered through BI N to Banffs.
10 x 13. R. hydrolapathum x R. crispus = R. x schreberi
Hausskn. has been found in E Suffolk, Notts, Caerns and Down
near both parents.
10 x 14. R. hydrolapathum x R. conglomeratus = R. x digeneus
G. Beck has been found with both parents in S En.
10 x 19. R. hydrolapathum x R. obtusifolius = R. x
lingulatus Jungner (R. x weberi Fisch.-Benzon) has been found
several times in S En and S Wa.
11. R. cristatus DC. - <u>Greek</u> <u>Dock</u>. Erect perennial to 2m; **241**
leaves broadly lanceolate, cordate; inflorescence dense but
with long erecto-patent branches; tepals 5-8mm, broadly ovate
to suborbicular, cordate, denticulate to dentate with teeth
to 1mm, 1 with large rounded smooth tubercle. Intrd; natd on
waste ground in SE En and S Wa, casual elsewhere in S En; CS
Europe. Perhaps only a ssp. of R. patientia.
11 x 13. R. cristatus x R. crispus = R. x dimidiatus
Hausskn. has occurred with the parents near R. Thames in W
Kent, S Essex and Middlesex.
11 x 19. R. cristatus x R. obtusifolius = R. x lousleyi Kent
occurs with the parents in SE En and SE Wa; endemic.
12. R. patientia L. - <u>Patience</u> <u>Dock</u>. Erect perennial to 2m; **241**
leaves broadly lanceolate, cuneate to truncate at base;
inflorescence ± as in R. cristatus; tepals 5-8mm, broadly
ovate to suborbicular, cordate, ± entire, 1 with small
rounded smooth tubercle. Intrd; natd in a few waste places
by docks and breweries, scattered casual elsewhere in En;
Europe. Several sspp. based on tepal shape are probably not
more than vars; the common one in BI is ssp. **orientalis**
Danser.

12 x 13. R. patientia x R. crispus = R. x confusus Simonkai has been found with the parents in Middlesex, S Essex and W Gloucs.

12 x 14. R. patientia x R. conglomeratus was found in W Kent in 1978; endemic.

12 x 19. R. patientia x R. obtusifolius = R. x erubescens Simonkai has occurred with the parents in several places in S & C En.

13. R. crispus L. - <u>Curled Dock</u>. Erect perennial to 1(2)m; **241** leaves narrowly oblong, narrowly elliptic or lanceolate, cuneate to subcordate at base, tightly undulate; inflorescence open to dense, with short, often few, erect to erecto-patent branches; tepals 3-6mm, ovate-triangular, ± entire, 1-3 with variously developed tubercles. Native.

1 Achene 1.3-2.5mm; tubercles usually <2.5mm, unequal, often only 1 developed; mostly inland   **a. ssp. crispus**
1 Achene 2.5-3.5mm; tubercles ≤3.5mm, usually sub-equal; maritime   **2**
  2 Stems usually <1m; inflorescence dense in fruit; on shingle, dunes and saltmarshes   **b. ssp. littoreus**
  2 Stems often >1m; inflorescence lax in fruit; on estuarine mud   **c. ssp. uliginosus**

a. Ssp. crispus. Waste, rough, cultivated and marshy ground; abundant throughout BI.

b. Ssp. littoreus (J. Hardy) Akeroyd. Maritime shingle, dunes and saltmarshes; scattered round coasts of BI.

c. Ssp. uliginosus (Le Gall) Akeroyd. Tidal estuarine mud; locally common in S Ir and S Br.

13 x 14. R. crispus x R. conglomeratus = R. x schulzei Hausskn. is fairly common with the parents in S & C Br.

13 x 15. R. crispus x R. sanguineus = R. x sagorskii Hausskn. is fairly common with the parents in S & C Br.

13 x 16. R. crispus x R. rupestris occurs in Scillies and Glam with both parents.

13 x 18. R. crispus x R. pulcher = R. x pseudopulcher Hausskn. occurs with the parents in S En, S Wa and CI.

13 x 19. R. crispus x R. obtusifolius = R. x pratensis Mert. **241** & Koch (<u>R.</u> x <u>acutus</u> auct. non L.) occurs ± commonly through BI where parents meet. By far the commonest hybrid in the genus, and usually easily recognized by intermediate leaves and tepals and low fertility. This is the most fertile <u>Rumex</u> hybrid; some seeds are formed and back-crossing occurs.

13 x 22. R. crispus x R. palustris = R. x heteranthos Borbas (<u>R.</u> x <u>areschougii</u> G. Beck) occurs with the parents by lakes and reservoirs in N Essex and Leics.

14. R. conglomeratus Murray - <u>Clustered Dock</u>. Erect **241** (biennial to) perennial to 60(100)cm; leaves ± oblong, broadly cuneate to subcordate; inflorescence very diffuse, with long branches spreading at >30 degrees; tepals 2-3mm, oblong to narrowly ovate, entire, all 3 with oblong tubercle. Native; damp places, grassy or bare, especially by ponds and rivers; throughout BI but sparse in upland areas.

14 x 15. R. conglomeratus x R. sanguineus = R. x ruhmeri
Hausskn. is second most fertile <u>Rumex</u> hybrid; it is recorded
only from S En but probably is frequent over much of Br.
Similarity of parents makes hybrid identification difficult.
14 x 18. R. conglomeratus x R. pulcher = R. x muretii
Hausskn. is frequent with the parents in S En and S Wa.
**14 x 19. R. conglomeratus x R. obtusifolius = R. x
abortivus** Ruhmer occurs frequently with the parents
scattered throughout Br.
14 x 22. R. conglomeratus x R. palustris = R. x wirtgenii G.
Beck occurs with the parents in C & SE En.
14 x 23. R. conglomeratus x R. maritimus = R. x knafii
Celak. occurs with the parents scattered in En.
15. R. sanguineus L. - <u>Wood Dock</u>. Erect perennial to 60 241
(100)cm; leaves and inflorescence as in R. conglomeratus but
branches arising at <30(45) degrees and tubercle usually 1, $\pm$
globose.   Native; damp shady places, mostly in woods or
hedgerows or by water; common throughout most of BI but rare
in N Sc. The common plant is var. **viridis** (Sibth.) Koch;
var. **sanguineus** has blood-red leaf-veins and is a rare
garden escape or casual.
15 x 18. R. sanguineus x R. pulcher = R. x mixtus Lambert
occurs with the parents in S En.
15 x 19. R. sanguineus x R. obtusifolius = R. x dufftii
Hausskn. occurs with the parents scattered over Br and CI.
16. R. rupestris Le Gall - <u>Shore Dock</u>. Erect perennial to 241
50(70)cm; leaves thick, narrowly oblong to lanceolate,   RR
undulate, cuneate to subcordate; inflorescence diffuse, with
long branches spreading at 25-50 degrees; tepals 3-4mm,
oblong to narrowly ovate, entire, all with large tubercle.
Native; damp places on sand or rocks by sea; coast of S Wa,
SW En and CI.
16 x 18. R. rupestris x R. pulcher = R. x trimenii Camus has
occurred with R. rupestris in Cornwall.
17. R. brownii Campdera - <u>Hooked Dock</u>. Rhizomatous 241
perennial with erect stems to 60cm; leaves lanceolate to
ovate, tightly undulate, cordate to cuneate; inflorescence
very diffuse, with few long branches; tepals 2-3.5mm, ovate,
with long hooked teeth, without tubercles. Intrd; wool-alien
scattered in En and Sc and sometimes persisting; Australia.
18. R. pulcher L. - <u>Fiddle Dock</u>. Erect to spreading 241
perennial to 40(50)cm; leaves oblong-obovate, usually

FIG 241 - Fruiting tepals of **Rumex**. 1, **R. obovatus**. 2, **R.
longifolius**. 3, **R. pseudoalpinus**. 4, **R. scutatus**. 5, **R.
aquaticus**. 6, **R. confertus**. 7, **R. cristatus**. 8, **R.
frutescens**. 9, **R. patientia**. 10, **R. pulcher**. 11, **R. dentatus**.
12, **R. hydrolapathum**. 13, **R. crispus**. 14, **R. x pratensis**. 15,
R. obtusifolius var. **obtusifolius**. 16, var. **microcarpus**. 17,
R. salicifolius. 18, **R. brownii**. 19, **R. acetosella**. 20, **R.
acetosa**. 21, **R. maritimus**. 22, **R. rupestris**. 23, **R.
conglomeratus**. 24, **R. sanguineus**. 25, **R. palustris**.

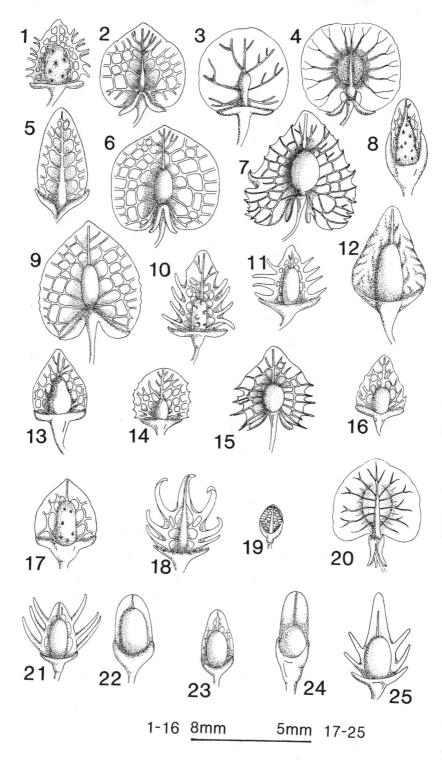

1-16 8mm   5mm 17-25

FIG 241 — see caption opposite

strongly constricted just above cordate to rounded base;
inflorescence very diffuse, with long branches spreading at
45-90 degrees; tepals 4-5.5mm, narrowly to broadly ovate,
usually with well-developed teeth but sometimes with few
short ones and rarely entire, (1-)3 with elongate warty
tubercle. Native; dry grassy places; frequent in CI and S Br
N to N Lincs and Anglesey, rare casual further N and in Ir.
Several sspp. based on tepal-shape are probably not more than
vars; only ssp. **pulcher** is native.

18 x 19. R. pulcher x R. obtusifolius = R. ogulinensis
Borbas occurs with R. pulcher in S En.

19. R. obtusifolius L. - Broad-leaved Dock. Erect perennial 241
to 1(1.2)m; leaves ovate-oblong, cordate; inflorescence
rather open, with erecto-patent branches; tepals 3-6mm,
triangular- to oblong-ovate, with variable teeth as in R.
pulcher, 1(-3) with smooth tubercle. Native; grassland, by
roads and rivers, waste and cultivated ground; abundant
throughout BI. 3 taxa based on tepal-shape are best treated
as vars: var. **obtusifolius** has large tepals with strong
teeth and 1 tubercle; var. **microcarpus** Dierb. (ssp. 241
sylvestris (Wallr.) Celak.) has small tepals with 0 or few
short teeth and 3 tubercles; var. **transiens** (Simonkai) Kubat
(ssp. transiens (Simonkai) Rech.f.) is intermediate. Only
the 1st is native.

19 x 22. R. obtusifolius x R. palustris = R. x steinii A.
Becker occurs with R. palustris in SE En.

19 x 23. R. obtusifolius x R. maritimus = R. x callianthemus
Danser occurs with R. maritimus in SE En.

20. R. dentatus L. - Aegean Dock. Erect annual to 70cm; 241
leaves oblong-lanceolate, rounded to cordate; inflorescence
very diffuse, wih few long erecto-patent branches; tepals
4-6mm, triangular-ovate, with mostly long teeth, (1-)3 with
smooth tubercle. Intrd; rather rare wool-alien, very
scattered in Br. Plants in Br are ssp. **halacsyi** (Rech.)
Rech. f.; SE Europe, Asia.

21. R. obovatus Danser - Obovate-leaved Dock. Erect annual 241
or biennial to 40(70)cm; leaves mostly obovate, rounded to
broadly cuneate at base; inflorescence rather diffuse but
very leafy, with few erecto-patent branches; tepals 4-5mm, ±
as in R. dentatus but all with warty tubercle. Intrd; grain-
alien near mills, wharves and warehouses; very scattered in
Br; Argentina and Paraguay.

22. R. palustris Smith - Marsh Dock. Erect biennial to 241
perennial to 60(100)cm; leaves lanceolate to narrowly   R
elliptic, cuneate; inflorescence diffuse, pale brown in
fruit, with many long, widely spreading then incurved
branches; tepals 3-4mm, narrowly ovate-triangular, with few
long rather rigid teeth, all with large tubercle. Native;
edges of ponds, ditches, gravel-pits and in marshy fields,
usually flooded at times; very local in S & E En N to Yorks,
Mons, rare casual elsewhere. Possibly hybridises with R.
maritimus.

23. R. maritimus L. - <u>Golden</u> <u>Dock</u>. Erect annual to **241** perennial to 40(100)cm, turning golden at fruiting; similar **R** to <u>R. palustris</u> but branches shorter and straighter and tepals 2.5-3mm, with longer more flexible teeth and narrower tubercle. Native; similar places to <u>R. palustris</u> but rarely with it, more widespread but equally scarce; scattered in Br N to S Sc, very scattered in Ir.

9. OXYRIA Hill - <u>Mountain Sorrel</u>
Herbaceous perennials; inflorescence a terminal panicle; tepals 4, sepaloid, not keeled or winged, the inner enlarging in fruit but without tubercles; stamens 6; anthers versatile; styles 2; stigmas deeply divided; achene biconvex.

1. O. digyna (L.) Hill - <u>Mountain Sorrel</u>. Stems to 30cm, erect, little branched, with 0(-2) leaves; basal leaves long-stalked, reniform; achene 3-4mm, broadly winged. Native; damp rocky place on mountains; N Wa, NW En, Sc, rare in Ir.

49. PLUMBAGINACEAE - <u>Thrift family</u>

Perennial herbs; leaves all basal, narrowed to base, simple, entire, exstipulate. Flowers in branched cymes or hemi-spherical heads, bisexual, hypogynous, actinomorphic, 5-merous; calyx tubular below, with free lobes above, scarious at least above; corolla of 5 pink to blue petals fused at base; stamens 5, borne on base of corolla; ovary 1-celled, with 1 basal ovule; styles 5; stigmas linear; fruit a 1-seeded capsule.
Two unmistakable, usually coastal genera.

1 Flowers in dense hemispherical heads with tubular
   sheath of fused scarious bracts beneath   **2. ARMERIA**
1 Flowers in branching cymes, the ultimate units of
   1-5 flowers with 3 scale-like bracts   **1. LIMONIUM**

1. LIMONIUM Miller - <u>Sea-lavenders</u>
Aerial stems branched, the ultimate branches consisting of small clusters (spikelets) of 1-5 flowers with 3 bracts (outer, middle, inner) below, the spikelets aggregated into spikes occupying ends of branches; flowers blue to purple or lilac; styles glabrous.
Plants produce pollen of 1 or 2 sorts (A, coarsely reticu- **245** late; B; finely reticulate) and possess stigmas of 1 of 2 sorts (Cob, with rounded papillae; Papillate, with prominent papillae). Species may be dimorphic (A/Cob and B/Papillate) and self-incompatible; monomorphic (A/Papillate) and self-compatible; or monomorphic and apomictic.

1 Leaves distinctly pinnately veined   2
1 All obvious veins (1-9) arising separately from

petiole-like base; vein-branches from midrib 0 or
indistinct   4
2 Leaves rounded to emarginate at apex, dying off
  before autumn; cliffs   **6. L. hyblaeum**
2 Leaves acute to obtuse, and mucronate, at apex,
  dying in autumn or winter; salt-marshes   3
3 Spikes mostly 1-2cm, with >4 spikelets/cm; outer
  bract 1.7-3mm; anthers yellow   **1. L. vulgare**
3 Longest spikes 2-5cm, with <3 spikelets/cm;
  outer bract 3-4mm; anthers reddish-brown   **2. L. humile**
4 Stems with numerous well-branched non-flowering
  lateral branches below; outer bract scarious
  except on midline   **3. L. bellidifolium**
4 Stems with 0 or few or little-branched non-
  flowering lateral branches below; outer bract
  herbaceous for most of width   5
5 Leaves obovate-spathulate, 11-25mm wide, with 5-7(9)
  obvious veins; inflorescence widest and ± flat at
  top; Channel Isles only   6
5 Leaves linear-oblong to oblanceolate-spathulate,
  rarely obovate-spathulate, 5-15(25)mm wide, with
  1-3(5) obvious veins; inflorescence widest below top
  and tapered above; widespread   **7-15. L. binervosum agg.**
6 Outer bract (1.8)1.9-2.4(2.9)mm; calyx (3.8)4-
  4.6(5.5)mm   **4. L. auriculae-ursifolium**
6 Outer bract (2.6)3-4(4.2)mm; calyx (4.1)4.8-
  5.5(6.3)mm   **5. L. normannicum**

Other spp. - **L. latifolium** (Smith) Kuntze, 1 of the common
gardeners' 'Statice', from SE Europe, has leaves mostly
25-60cm with pinnate venation and outer bracts completely
hyaline; it was reported in 1990 as natd in N Essex.

1. L. vulgare Miller - Common Sea-lavender. Stems erect to
ascending, to 40(60)cm; leaves dying in autumn or winter, up
to 20(30)cm, elliptic to oblanceolate, strongly pinnately
veined; spikes 1-2cm, dense, with 5-8 spikelets in lowest cm,
the lowest 2 spikelets 1.5-3mm apart; dimorphic, self-
incompatible. Native; muddy salt-marshes; locally common
around coasts of CI and Br N to C Sc.
1 x 2. L. vulgare x L. humile = L. x neumanii Salmon occurs
in S & E En and NW Wa with both parents. It is intermediate
in spike and spikelet characters and partially fertile;
introgression may occur.
2. L. humile Miller - Lax-flowered Sea-lavender. Like L.   R
vulgare but spikes lax, the longer 2-5cm, with 2-3 spikelets
in lowest cm, the lowest 2 spikelets 4-10mm apart; monomorphic, self-compatible. Native; similar distribution to L.
vulgare in Br N to SW Sc but rarer, frequent on coasts of Ir.
3. L. bellidifolium (Gouan) Dumort. - Matted Sea-lavender. RR
Stems decumbent, to 30cm, the lowest branches much-branched
and sterile; leaves dying by flowering, up to 4cm, oblanc-

eolate, 1-3(5)-veined; spikes mostly <1cm, very dense, with
<10 spikelets; dimorphic, self-incompatible. Native; drier
parts of salt-marshes; coasts of N Norfolk and Lincs.

4. L. auriculae-ursifolium (Pourret) Druce – <u>Broad-leaved</u> **247**
<u>Sea-lavender</u>. Stems erect, to 30(45)cm; leaves evergreen, **RR**
to 12cm, obovate-spathulate, with 5-7(9) obvious veins;
inflorescence obtrullate in side view; spikes <2cm, very
dense, with <13 spikelets, with 6-8 spikelets in lowest cm;
monomorphic (B/Papillate), apomictic. Native; on rocks by
sea; Plemont Point and formerly Rouge Nez, Jersey.

5. L. normannicum Ingrouille – <u>Alderney Sea-lavender</u>. Like **247**
<u>L. auriculae-ursifolium</u> but stems to 20(25)cm; inflorescence **RR**
+ obtriangular in side view; spikes even denser, <1.5cm,
with larger parts (see key); monomorphic (A/Cob), apomictic.
Native; on maritime rocks and dunes; Alderney and Jersey (St
Ouens and formerly Ronez Point).

6. L. hyblaeum Brullo (<u>L. companyonis</u> auct. non (Gren. & **245**
Billot) Kuntze) – <u>Rottingdean Sea-Lavender</u>. Stems erect, to
20(30)cm; leaves dying off by autumn, to 5cm, broadly

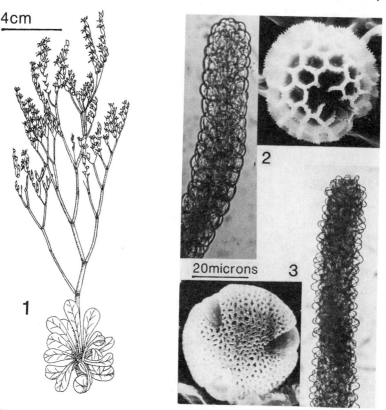

4cm

20microns

FIG 245 – Limonium. 1, **L. hyblaeum**. 2-3, pollen- and stigma-
types. 2, cob stigma and 'A' pollen. 3, papillate stigma and
'B' pollen. Photographs courtesy of M.J. Ingrouille.

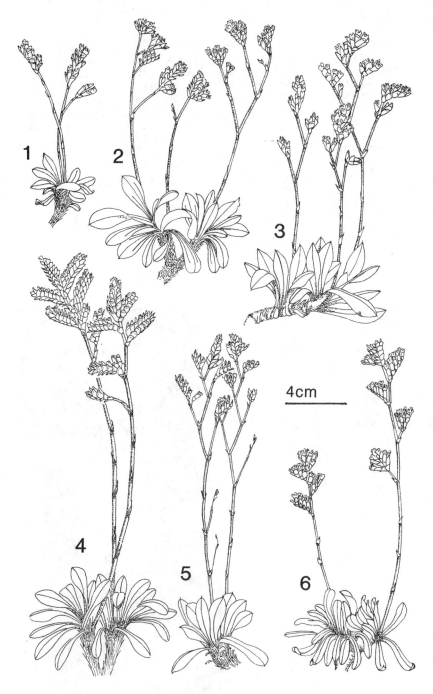

FIG 246 – Limonium binervosum agg. 1, L. parvum. 2, L. britannicum. 3, L. paradoxum. 4, L. recurvum. 5, L. loganicum. 6, L. transwallianum.

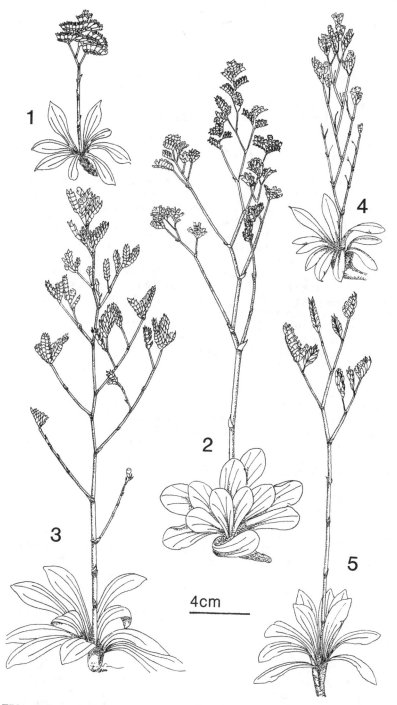

FIG 247 - 1-2, **Limonium auriculae–ursifolium** agg. 1, **L. normannicum**. 2, **L. auriculae–ursifolium**. 3-5. **L. binervosum** agg. 3, **L. procerum**. 4, **L. binervosum**. 5, **L. dodartiforme**.

obovate-spathulate, conspicuously pinnately veined; spikes to
6cm, lax, with 1-2 spikelets in lowest cm, the lowest 2
spikelets 5-10mm apart; monomorphic (A/Cob) and apomictic.
Intrd; well natd on cliffs at Rottingdean and garden escape
elsewhere in E Sussex; Sicily.

7-15. L. binervosum agg. - <u>Rock Sea-lavender</u>. Stems usually
erect, to 30(70)cm, variable in size and branching; leaves
evergreen, to 10(15)cm but often much less, narrowly obovate
to oblanceolate or spathulate, with 1-3(5) obvious veins;
spikes to 3(4.5)cm, lax to dense; monomorphic (A or intermediate or no pollen/Cob, or rarely no pollen/ Papillate),
apomictic. Native; maritime rocks, dunes and salt-marshes;
coasts of BI N to N Lincs, Wigtowns and Donegal.

Spp. in <u>L. binervosum</u> agg. are difficult to distinguish.
Several plants from a population must be examined, as extreme
individuals often cannot be identified; geographical location
is an important aid. Sspp. are even less easy to define, but
if the locality is known the key should enable determination.
Specialist literature should be consulted for more details.

<u>Multi-access key to spp. and sspp. of L. binervosum agg.</u>

Whole of stem strongly tuberculate (rough)	A
Stem not tuberculate (smooth) or tuberculate only above	B
± all stems branched from low down	C
Some stems unbranched in lower half	D
Flowers borne on upper 2/3 of stem	E
Flowers borne on upper 1/3(1/2) of stem	F
Mean spike length <15mm	G
Mean spike length >15mm or spikes with <5 spikelets	H
Spikes dense, with 6-10 spikelets in lowest cm	I
Spikes lax, with 3-5 spikelets in lowest cm, or with <5 spikelets	J
Mean length outer bract >3mm	K
Mean length outer bract <3mm	L
Mean length inner bract <5mm	M
Mean length inner bract >5mm	N
Mean width petal <1.5mm	O
Mean width petal >1.5mm	P

ACF(GH)ILMP		**15. L. recurvum**
G	Portland (tip), Dorset	**a. ssp. recurvum**
(GH)	Portland (other than tip), Dorset; Kerry	
		b. ssp. portlandicum
(GH)	Donegal; Cumberland; Wigtowns	
		c. ssp. humile
H	Clare	**d. ssp. pseudotranswallianum**
ACFGJLMP	Cornwall	**12. L. loganicum**
BC(EF)H(IJ)KNP	Dorset	**14. L. dodartiforme**
BCEGILMO	Pembs	**13. L. transwallianum**
BCF(GH)(IJ)KMP	Pembs	**8. L. paradoxum**
BCF(GH)(IJ)LMP		**10. L. britannicum**
G I	Cornwall(N)	**a. ssp. britannicum**
G I	Cornwall(S); Devon(S)	**b. ssp. coombense**
G I	Devon(N); Pembs	**c. ssp. transcanalis**

H J	Anglesey to Lancs	d. ssp. celticum
BCFHJL(MN)P		7. L. binervosum
M	Norfolk; Lincs	c. ssp. anglicum
M	Essex	d. ssp. saxonicum
N	Kent (S E); Sussex	a. ssp. binervosum
N	Kent (N E)	b. ssp. cantianum
N	Devon	e. ssp. mutatum
N	Channel Isles	f. ssp. sarniense
BDFGILMP	Pembs	11. L. parvum
BDFHI(KL)NP		9. L. procerum

Leaves mostly >8mm wide; branches and
stems ± straight; E Sussex and
widespread in W   a. ssp. procerum
Leaves mostly >8mm wide; branches and
stems wavy; S Devon   b. ssp. devoniense
Leaves mostly <8mm wide; branches and
stems ± straight; Pembs   c. ssp. cambrense

7. L. binervosum (G. E. Smith) Salmon. Stems relatively 247
tall, well-branched from low down, smooth at least below; R
spikes rather lax, with 2-5(6) spikelets in lowest cm; the
only sp. in E En, and also in NW France.

a. Ssp. binervosum. Chalk cliffs and salt-marshes; E Sussex
and S E Kent.

b. Ssp. cantianum Ingrouille. Chalk cliffs and salt-
marshes; N E Kent.

c. Ssp. anglicum Ingrouille. Salt-marshes; E Norfolk to S N
Lincs.

d. Ssp. saxonicum Ingrouille. Salt-marshes; N Essex.

e. Ssp. mutatum Ingrouille. Seaside rocks; Lannacombe, S
Devon.

f. Ssp. sarniense Ingrouille. Seaside cliffs and rocks;
frequent in CI (all main islands).

8. L. paradoxum Pugsley. Stems relatively short, with short 246
branches, smooth at least below; spikes rather lax or dense, RRR
the outer bract concealing inner bract near end of spike is
diagnostic. St David's Head, Pembs; endemic.

9. L. procerum (Salmon) Ingrouille. Stems relatively tall, 247
usually branched above only, smooth; spikes dense, with R
(3)5-9 spikelets in lowest cm; endemic.

a. Ssp. procerum. Cliffs and salt-marshes; E Ir, Wa, SW En,
and Rottingdean, E Sussex (possibly intrd in last).

b. Ssp. devoniense Ingrouille. Cliffs; near Torquay, S
Devon.

c. Ssp. cambrense Ingrouille. Limestone cliff near
Pembroke, Pembs.

10. L. britannicum Ingrouille. Stems relatively short, 246
quite well-branched, smooth at least below; spikes dense, R
with (4)6-8(10) spikelets in lowest cm; leaves shorter and
broader than in L. procerum; endemic.

a. Ssp. britannicum. Cliffs and promontories; N coast of E
& W Cornwall.

b. Ssp. coombense Ingrouille. Cliffs; S Devon and S coast

of E Cornwall.
  c. Ssp. transcanalis Ingrouille. Cliffs, pebble beach and salt-marsh; N Devon and Pembs.
  d. Ssp. celticum Ingrouille. Cliffs and salt-marshes; Anglesey, Cheshire and W Lancs.
  11. L. parvum Ingrouille. Stems very short and thin, to **246** 7(13)cm, with few branches, smooth; spikes dense, with 4–10 **RR** spikelets in lowest cm; leaves very small and acute. Limestone cliff; Saddle Point, Pembs; endemic.
  12. L. loganicum Ingrouille. Stems of medium size, well- **246** branched, rough; spikes rather lax, with 3–7 spikelets in **RR** lowest cm. On rocks and scree; S of Land's End, W Cornwall; endemic.
  13. L. transwallianum (Pugsley) Pugsley. Stems of medium **246** size, often well-branched, smooth; spikes dense, with often **RR** >10 spikelets in lowest cm; the narrow petals are diagnostic. Limestone cliffs; Giltar Point, Pembs; endemic.
  14. L. dodartiforme Ingrouille. Stems relatively tall, **247** well-branched, rough at least above; spikes rather lax, with **RR** 3–6(7) spikelets in lowest cm; the obovate-spathulate leaves to 22mm wide are diagnostic. Chalk cliffs and shingle; Dorset; endemic.
  15. L. recurvum Salmon. Stems relatively short to medium **246** size, often well-branched, rough; spikes dense, with 5–11 **R** spikelets in lowest cm; leaves mostly small, spathulate; the most westerly and most northerly sp.; endemic.
  a. Ssp. recurvum. Low limestone cliffs; tip of Portland, Dorset.
  b. Ssp. portlandicum Ingrouille. Limestone and salt-marshes; elsewhere on Portland, Dorset; Barrow Harbour and Banna Strand, N Kerry.
  c. Ssp. pseudotranswallianum Ingrouille. Limestone cliffs and rocks; Clare.
  d. Ssp. humile (Girard) Ingrouille. Cliffs and scree; E & W Donegal; St Bees Head, Cumberland; Galloway, Wigtowns.

2. ARMERIA Willd. – _Thrifts_
Leaves numerous, very narrow; aerial stems unbranched, terminating in dense hemispherical inflorescence with tubular sheath of fused bracts beneath; flowers pink (rarely white); styles pubescent below.
Both species are dimorphic and self-incompatible, with pollen- and stigma-types as in _Limonium vulgare_.

Other spp. – **A. pseudarmeria** (Murray) Mansf. (_Estoril Thrift_), from Portugal, is grown in gardens; it was formerly natd on seaside cliffs in S Hants, but is now extinct. It differs from A. _alliacea_ in being even more robust with flower-heads 3–5cm across and calyx-teeth with terminal points >1mm.

  1. A. maritima (Miller) Willd. – _Thrift_. Leaves linear,

<2mm wide, 1(-3)-veined, usually with hairs at least on margins; flower-heads 15-25mm wide; calyx teeth <1mm incl. very short mucro.

a. Ssp. maritima. Stems usually pubescent, to 30cm; bract-sheath <15mm; outermost bracts (excl. sheath) shorter than inner. Native; salt-marshes, saline turf, rocks and cliffs by sea, and inland on mountain rocks; common round coasts of BI and on mountains in N Wa, N En and Sc.

b. Ssp. elongata (Hoffm.) Bonnier. Stems glabrous, the longer 20-55cm; bract-sheath 12-25mm; outermost bracts (excl. sheath) usually as long as or slightly longer than inner. Native; lowland rough pasture; near Ancaster, S Lincs, formerly elsewhere in S Lincs and Leics.

Other variants, notably those inland on mountains or on serpentine or heavy-metal spoil-heaps, are not sufficiently distinct for taxonomic recognition.

1 x 2. A. maritima x A. arenaria occurs in W Jersey where habitats of the parents meet. It is fertile and intermediate in leaf width, pubescence, and length of outer bracts and calyx-teeth; endemic.

2. A. arenaria (Pers.) Schultes (A. alliacea auct. non **RR** (Cav.) Hoffsgg. & Link) - Jersey Thrift. Leaves narrowly oblanceolate, some >3mm wide, 3-5(7)-veined, glabrous; stems glabrous, to 60cm; flower-heads 20-30mm wide; bract-sheath 20-40mm; outermost bracts longer than inner, often ± leaf-like and longer than flower-head; calyx-teeth 1.5-2mm, incl. terminal point c.1/2 that. Native; fixed dunes; W & S Jersey.

50. PAEONIACEAE - Peony family

Perennial herbs; some roots strongly tuberous; leaves basal and spiral, (bi)pinnate to (bi)ternate, the segments entire, petiolate, exstipulate. Flowers large, solitary, terminal, bisexual, hypogynous, actinomorphic; sepals 5, free; petals 5-8, free, red; stamens very numerous, with red filaments and yellow anthers; carpels 3-5, free, each with several ovules; style 0; stigmas red, hooked or coiled; fruit a follicle. Unmistakable flowers and compound leaves.

1. PAEONIA L. - Peony
1. P. mascula (L.) Miller - Peony. Stems to 60cm, simple, erect, usually several with several basal leaves forming a clump; basal leaves <40cm, with 9-16 segments; flowers 8-14cm across; follicles 2-5cm, densely pubescent ± recurved. Intrd; natd on limestone on Steep Holm Island, N Somerset, since at least 1803, and on grassy bank, Lanarks, since 1985; S Europe.

51. ELATINACEAE - Waterwort family

Small ± aquatic annuals; leaves opposite, simple, entire, ±
petiolate, stipulate. Flowers minute, solitary in leaf-
axils, bisexual, hypogynous, actinomorphic; sepals 3-4,
membranous, free or united at base; petals 3-4, free,
pinkish-white; stamens 2x as many as petals; ovary 3-4-
celled, each with numerous ovules on axile placentation;
styles 3-4, very short; stigmas capitate; fruit a 3-4-celled
capsule.
Superficially like Portulacaceae, but differs in many floral
characters; distinguished from Caryophyllaceae by the
completely compartmented ovary and herbaceous stipules
(absent or scarious in Caryophyllaceae).

1. ELATINE L. - Waterworts
  1. E. hexandra (Lapierre) DC. - Six-stamened Waterwort. **R**
Stems procumbent, rooting at nodes, to 10(20)cm; leaves
c.6mm, elliptic to spathulate; pedicels at least as long as
flowers; sepals and petals 3(-4); stamens 6(8); capsule with
3(-4) valves. Native; in ponds and on wet mud; very local
and widely scattered in BI, mostly in W & SE. Can live for
several years submerged without flowering.
  2. E. hydropiper L. - Eight-stamened Waterwort. Like E. **RR**
hexandra but pedicels 0 or very short; sepals and petals 4;
stamens 8; capsule with 4 valves. Native; in ponds and small
lakes; rare and very local, scattered in Br and NE Ir.

52. CLUSIACEAE - St John's-wort family
(Guttiferae, Hypericaceae)

Perennial or rarely annual herbs or small shrubs; leaves
opposite, simple, ± entire, with translucent and/or coloured
glands, exstipulate; petioles absent or very short. Flowers
solitary or in terminal cymes, actinomorphic, bisexual,
hypogynous; sepals 5, free, often glandular; petals 5, free,
yellow, often glandular; stamens numerous, usually partially
fused into 3 or 5 bundles; ovary 1-, 3- or 5-celled, each
cell with many ovules; styles 3 or 5; stigmas capitate; fruit
a capsule or becoming succulent and berry-like.
Easily recognized by entire, opposite, exstipulate leaves, 5
yellow petals, and numerous stamens grouped in bundles.

1. HYPERICUM L. - St John's-worts

1 Stems and leaves conspicuously pubescent   2
1 Stems glabrous; leaves glabrous to slightly pubescent   3
  2 Stems procumbent and rooting at nodes below, soft;
    glands on sepals reddish   **18. H. elodes**
  2 Stems erect, rooting only near base, stiff;
    glands on sepals black   **16. H. hirsutum**

3 Glands on leaves, sepals and petals all translucent
  and inconspicuous, or 0   4
3 Some glands on leaves, sepals and/or petals black   10
  4 Delicate herb; leaves rarely >1.5 x 0.5cm
                              19. H. canadense
  4 Stems woody; leaves rarely <2 x 1cm   5
5 Styles 5; petals >(16)20mm; stamens c.3/4 as long as
  petals   6
5 Styles 3; petals <20mm; stamens c. as long as or
  longer than petals   7
  6 Rhizomatous; aerial stems to 60cm, not or little
    branched   **1. H. calycinum**
  6 Not rhizomatous; aerial stems to 1.7m, much
    branched   **2. H. pseudohenryi**
7 Petals >15mm; styles >3x as long as ovary at
  flowering; leaves strongly smelling of goats when
  bruised   **5. H. hircinum**
7 Petals <15mm; styles <3x as long as ovary at
  flowering; leaves often scented but without strong
  goat-like smell when bruised   8
  8 Petals shorter than to c. as long as sepals; styles
    <5mm, c.1/2 as long as ovary; fruit succulent when
    ripe   **3. H. androsaemum**
  8 Petals longer than sepals; styles >5mm, at least as
    long as ovary; fruit dry when ripe   9
9 Petals <4.5mm wide; sepals <2.5mm wide, not overlapping laterally   **6. H. xylosteifolium**
9 Petals >4.5mm wide; sepals >2mm wide, overlapping
  laterally   **4. H. x inodorum**
  10 Stems with 4 ridges, sometimes 2 strong and 2
    weak; ridges often winged   11
  10 Stems with 0-2 ridges; ridges not winged   14
11 Stems with 2 weak and 2 strong ridges
                      8. H. x desetangsii
11 Stems tetraquetrous, with 4 strong ridges   12
  12 Petals pale yellow, not red-tinged, <7.5mm, c. as
    long as sepals; stems broadly winged (0.25-0.5mm)
                      11. H. tetrapterum
  12 Petals bright yellow, sometimes red-tinged,
    >7.5mm, >2x as long as sepals; stems not or
    narrowly winged (<0.25mm)   13
13 Stems not winged; sepals obtuse   **9. H. maculatum**
13 Stems narrowly winged; sepals acute   **10. H. undulatum**
  14 Leaves at least sparsely pubescent on lowerside
                      17. H. montanum
  14 Leaves glabrous   15
15 Sepals unequal, 3 longer and wider than the other 2;
  petals <2x as long as sepals   **12. H. humifusum**
15 Sepals + equal; petals >2x as long as sepals   16
  16 Stems not ridged; leaves without black glands 17
  16 Stems with 2 ridges; leaves with black glands near
    margin   18

17 Leaves on main stem triangular-ovate, widest near
   base, + sessile, + clasping stem   **14. H. pulchrum**
17 Leaves broadly elliptic to orbicular, widest near
   middle, distinctly petiolate, not clasping stem
                                             15. H. nummularium
   18 Margins of sepals fringed with stalked black
      glands; leaves + without translucent glands
                                          13. H. linariifolium
   18 Margins of sepals without stalked black glands;
      leaves with several to numerous translucent
      glands   19
19 Margins of sepals entire, acute at apex **7. H. perforatum**
19 Margins of sepals denticulate towards apex, with an
   apical apiculus   **8. H. x desetangsii**

Other spp. - A few other spp. grown in gardens as shrubs or
rock-garden plants may be found as relics or throwouts. **H.**
'Hidcote', a sterile cultivar of unknown, probably hybrid
origin, is commonly mass-planted in public places. It
differs from H. pseudohenryi in its stamens being <1/2 (not
c.3/4) as long as petals and in its rounded (not acuminate to
obtuse) sepals. **H. forrestii** (Chitt.) N. Robson, from China,
differs from H. 'Hidcote' in its styles longer (not shorter)
than ovary and yellow (not orange) anthers; a single plant
was found in Westmorland in 1988.

1. H. calycinum L. - Rose-of-Sharon. Strongly rhizomatous
glabrous evergreen shrub with erect little-branched 4-lined
stems to 60cm; flowers 5-8cm across; black glands 0. Spreads
vegetatively but rarely sets seed. Intrd; cultd in gardens
and mass planted in parks and on roadsides; natd on hedgebanks and in shrubberies usually near gardens; scattered
through most of Br and Ir, especially S; Turkey and Bulgaria.
2. H. pseudohenryi N. Robson - Irish Tutsan. Erect semi-
evergreen glabrous shrub with much-branched 4-ridged stems to
1.7m; flowers 3-5.5cm across; black glands 0. Intrd; natd in
riverside woodland in W Cork; China.
3. H. androsaemum L. - Tutsan. Erect + deciduous glabrous
shrub with branched 2-ridged stems to 0.8m; flowers 1.5-2.5cm
across; stamens c. as long as petals; black glands 0.
Native; damp woods and shady hedgebanks; locally frequent
throughout most of BI, especially in W, but almost absent
from NE Sc; cultivated and sometimes natd.
4. H. x inodorum Miller (H. elatum Aiton; H. androsaemum x
H. hircinum) - Tall Tutsan. Erect + deciduous glabrous
shrub with much-branched 2-ridged stems to 2m; flowers
1.5-3cm across; stamens slightly longer than petals; black
glands 0; partially fertile. Intrd; grown in gardens and
mass-planted in parks and on roadsides (especially cv.
'Elstead' for its red fruits); rarely natd in shady places;
very sparsely scattered throughout much of BI, especially in
SW; probably spontaneous in SW Europe but not in BI.

5. H. hircinum L. - <u>Stinking</u> Tutsan. Erect ± deciduous glabrous shrub with much-branched 4-ridged stems to 1.5m; flowers 2.5-4cm across; stamens slightly longer than petals; black glands 0. Intrd; grown in gardens and rarely natd in shady places; scattered through much of BI, especially SW; Mediterranean. Our plant is ssp. **majus** (Aiton) N. Robson.

6. H. xylosteifolium (Spach) N. Robson - <u>Turkish</u> Tutsan. Rhizomatous ever-green glabrous shrub; stems to 1.5m, erect, much-branched, 4-lined; flowers 1.5-3cm across; stamens c. as long as petals; black glands 0. Intrd; sometimes grown in gardens and natd in 3 shady places in W Lancs and MW Yorks; Caucasus.

7. H. perforatum L. - <u>Perforate</u> St John's-wort. Erect glabrous rhizomatous perennial with 2-ridged stems to 80cm; leaves with abundant translucent glands; flowers 15-25mm across; sepals entire, acute at apex; petals bright yellow; black glands few on sepals and petals, sessile. Native; dryish grassland, banks and open woodland; common throughout most of BI and the commonest <u>Hypericum</u> in En, but rare in N Sc. Very variable, especially in leaf-shape.

8. H. x desetangsii Lamotte (<u>H. perforatum</u> x <u>H. maculatum</u>) - <u>Des</u> <u>Etangs'</u> <u>St</u> John's-wort. Partially fertile and back-crossing to give a range of intermediates; variously inter-mediate between its parents (especially in its 2-4 stem-ridges, rather few translucent leaf glands, and denticulate sepal margins with an apiculate apex); often found in absence of 1 or both parents. Native; grassland of varying dampness; sparsely scattered through En, Wa and S Sc but easily over-looked. All but 2 records refer to nothossp. **desetangsii** (2n=32), with <u>H. maculatum</u> ssp. <u>obtusiusculum</u> as 1 parent; nothossp. **carinthiacum** (A. Frohl.) N. Robson (2n=24 or 40), with <u>H. maculatum</u> ssp. <u>maculatum</u> as parent, was found in Kintyre in 1899 and Lanarks in 1984. The latter nothossp. has less widely branched stems and entire sepals.

9. H. maculatum Crantz - <u>Imperforate</u> St John's-wort. Erect rhizomatous glabrous perennial with tetraquetrous wingless stems to 60cm; leaves with 0 or very few translucent glands; flowers 15-25mm across; petals bright yellow; black glands on leaves, petals and sometimes sepals, sessile. Native; grassy, usually moist places.

a. Ssp. maculatum. Inflorescence-branches arising at c.30 degrees; black glands on petals mainly as superficial dots; sepals (1.7)2-3mm wide, entire; 2n=16. Scattered in C Sc; rare and very scattered in En and perhaps intrd there.

b. Ssp. obtusiusculum (Tourlet) Hayek (<u>H. dubium</u> Leers). Inflorescence-branches arising at c.50 degrees; black glands on petals mainly as superficial lines; sepals 1.2-2mm wide, denticulate at edges; 2n=32. Scattered over most of Br and Ir and locally frequent; much commoner than ssp. <u>maculatum</u> and often treated as a separate sp.

10. H. undulatum Schousboe ex Willd. - <u>Wavy</u> St John's-wort.   R Erect rhizomatous glabrous perennial with tetraquetrous narrowly winged <0.25mm) stems to 60cm; leaves undulate, with abundant translucent glands; flowers 12-20mm across; petals bright yellow, red-tinged on lowerside; black glands on leaves, sepals and sometimes petals, sessile. Native; marshy fields and streamsides; very local in SW En and W Wa.

11. H. tetrapterum Fries (<u>H.</u> <u>quadrangulum</u> L.) - <u>Square-stalked</u> <u>St</u> <u>John's-wort</u>. Erect rhizomatous glabrous perennial with tetraquetrous distinctly winged (0.25-0.5mm) stems to 60cm; leaves with many small translucent glands; flowers 9-13mm across; petals pale yellow; black glands on leaves and sometimes petals or sepals, sessile. Native; marshes, river-banks and damp meadows; frequent throughout BI except N Sc.

12. H. humifusum L. - <u>Trailing</u> <u>St</u> <u>John's-wort</u>. Procumbent to ascending glabrous perennial with thin 2-lined stems to 20cm; leaves <15mm, obovate or ovate to oblanceolate or lanceolate; flowers 8-12mm across; petals <2x as long as sepals, bright yellow; black glands on leaves and rather sparse on sepals and petals, sessile and/or stalked. Native; open woods, hedgebanks and dry heathland mostly on acid soils; frequent throughout most of BI but rare in N Sc.

12 x 13. H. humifusum x H. linariifolium is the probable parentage of intermediates (notably in petal length) that occur in CI with both parents. Possibly endemic.

13. H. linariifolium Vahl - <u>Toadflax-leaved</u> <u>St</u> <u>John's-wort</u>. **RR** Erect to ascending glabrous perennial with thin faintly 2-lined stems to 40cm; leaves <30mm, narrowly lanceolate to linear; flowers 15-20mm across; petals >2x as long as sepals, bright yellow; black glands sessile on leaves, stalked on sepals and petals. Native; rocky acid slopes; rare in SW En, Wa and CI.

14. H. pulchrum L. - <u>Slender</u> <u>St</u> <u>John's-wort</u>. Erect glabrous perennial with thin terete stems to 60cm; main stem-leaves triangular-ovate; flowers 12-18mm across; petals bright yellow, red-tinged on lowerside; black glands on sepals and petals, stalked. Native; dry open woodland, hedgebanks and heathland, usually on acid soils; locally common through BI.

15. H. nummularium L. - <u>Round-leaved</u> <u>St</u> <u>John's-wort</u>. Decumbent to erect glabrous perennial with thin terete stems to 30cm; leaves broadly elliptic to orbicular; flowers 18-30mm across; petals bright yellow, red tinged; black glands on sepals and petals, + stalked. Intrd; natd on quarry rock-ledge in NW Yorks since 1972; mountains of SW Europe.

16. H. hirsutum L. - <u>Hairy</u> <u>St</u> <u>John's-wort</u>. Erect pubescent perennial with terete stems to 1m; flowers 15-22mm across; petals pale yellow; black glands on sepals and petals, stalked. Native; open woodland, river-banks and damp grass-land; frequent throughout most of Br but rare in W, very local in Ir.

17. H. montanum L. - <u>Pale</u> <u>St</u> <u>John's-wort</u>. Erect perennial **R** with terete glabrous stems to 1m; petals pale yellow; black glands sessile on leaves, stalked on sepals. Native; open woodland, hedgebanks and rocky slopes, usually on calcareous soils; local but widespread in En and Wa.

18. H. elodes L. - <u>Marsh St John's-wort</u>. Densely pubescent stoloniferous perennial with ascending to erect terete flowering stems to 40cm; flowers 12-20mm across; black glands 0; red glands on sepals, stalked. Native; bogs, pond- and stream-sides on acid soil; locally frequent in suitable places throughout BI in W & S, but rare or absent in C & E.

19. H. canadense L. - <u>Irish St John's-wort</u>. Glabrous annual   **RR** or perennial with basal buds and erect very slender tetraquetrous stems to 20cm; flowers <1cm across; petals bright yellow; black glands 0. Probably intrd; very rare in 2 areas in W and SW Ir; N America.

53. TILIACEAE - <u>Lime family</u>

Trees; leaves deciduous, alternate, simple, broadly ovate, cordate, serrate, petiolate, with caducuous stipules. Flowers fragrant, 2-25 in cymes whose stalk is fused to large, narrowly oblong papery persistent bracteole dispersed with fruit, actinomorphic, bisexual, hypogynous; sepals and petals 5 each, free, creamish-yellow; stamens numerous, $\pm$ coherent in 5 bundles; ovary 5-celled, each cell with 2 ovules; style 1; stigma capitate, $\pm$ lobed; fruit a nut with 1-3 seeds.

The persistent bracteole is diagnostic.

1. TILIA L. - <u>Limes</u>

1 Leaves pubescent on lowerside; flowers 2-4(6) per
  cyme; fruit strongly ribbed   **1. T. platyphyllos**
1 Leaves glabrous except for dense hair-tufts in vein
  axils on lowerside; flowers 4-15 per cyme; fruit
  not or slightly ribbed   3
  2 Cymes held obliquely erect above foliage; leaves
    mostly 3-6cm, with scarcely prominent tertiary
    veins on upperside   **3. T. cordata**
  2 Cymes pendent among foliage; leaves mostly 6-9cm,
    with prominent tertiary veins on upperside
                                                  2. T. x vulgaris

Other spp. - Several spp. may be found planted in parks, etc.: **T.** 'Petiolaris' (<u>Pendent Silver-lime</u>), easily the commonest, has densely pubescent young twigs, whitish-tomentose lowerside to leaves and petioles >1/2 as long as leaves; **T. tomentosa** Moench (<u>Silver-lime</u>), from SE Europe, is similar to <u>T.</u> 'Petiolaris' but has short petioles <1/2 as long as leaves and much less pendent branches; and **T. x euchlora** K. Koch (<u>Caucasian Lime</u>), from Caucasus, differs from <u>T.</u> x <u>vulgaris</u> in its glossy leaf upperside.

1. T. platyphyllos Scop. - <u>Large-leaved Lime</u>. Tree to 34m;   **R**

young twigs pubescent; leaves mostly 6-12cm, thinly pubescent on lowerside, especially on veins; cymes pendent among leaves, with 2-4(6) flowers. Native; in woods and copses on base-rich soils; very local and scattered in En and Wa, more widespread as intrd plants but status often doubtful. Our plant is said to be ssp. **cordifolia** (Besser) C. Schneider, but perhaps not all intrd plants are this.

2. T. x vulgaris Hayne (<u>T.</u> x <u>europaea</u> auct. non L.; <u>T. platyphyllos</u> x <u>T. cordata</u>) - <u>Lime</u>. Tree to 46m; young twigs soon glabrous; leaves mostly 6-9cm, glabrous or + so on lowerside except for vein-axils; cymes pendent among leaves, with 4-10 flowers. Native; rare, in a few woods in C En with both parents, widely planted and sometimes natd throughout BI; 1 of the commonest planted trees. Partially fertile.

3. T. cordata Miller - <u>Small-leaved</u> <u>Lime</u>. Tree to 38m; young twigs soon glabrous; leaves mostly 3-6cm, glabrous on lowerside except for vein-axils; cymes obliquely erect, held above foliage, with 4-10(15) flowers. Native; woods on rich soils; En (mostly C) and Wa, locally common as a native and also planted and + natd more widely.

54. MALVACEAE - <u>Mallow family</u>

Annual to perennial herbs or sometimes shrubs; leaves alternate, usually palmately veined, often palmately lobed and sometimes + palmate, petiolate, stipulate. Flowers in racemes or small panicles or solitary and axillary, actinomorphic, bisexual or sometimes gynomonoecious, hypogynous, usually with an epicalyx of 3-c.13 sepal-like segments below calyx; sepals 5, free or fused below; petals 5, + free but often fused at extreme base, mostly pink to purple; stamens numerous, the filaments united below into tube, divided above with each branch bearing a 1-celled anther-lobe; ovary 5-many-celled, each cell with 1-many ovules; styles free or united below, 5-many; stigmas linear or capitate; fruit a capsule or breaking into 1-several-seeded nutlets.
Easily recognized by the stamens united into a tube round the carpels.

1 Epicalyx absent   2
1 Calyx-like epicalyx present below true calyx   4
  2 Flowers pink to purple; stigmas linear   **6. SIDALCEA**
  2 Flowers yellow; stigmas capitate   3
3 Leaves palmately veined; nutlets with several
  seeds   **7. ABUTILON**
3 Leaves pinnately veined; nutlets 1-seeded   **1. SIDA**
  4 Carpels 5; fruit a capsule   **8. HIBISCUS**
  4 Carpels > 6; fruit breaking into nutlets   5
5 Epicalyx-segments 6-10   6
5 Epicalyx-segments 3   7
  6 Staminal-tube terete, pubescent   **4. ALTHAEA**

6 Staminal-tube 5-angled, glabrous   **5. ALCEA**
7 Epicalyx-segments free to base   **2. MALVA**
7 Epicalyx-segments fused below   **3. LAVATERA**

1. SIDA L. - Queensland-hemps
Annuals to perennials, woody or not; epicalyx 0; petals
yellowish; carpels 5-14; stigmas capitate; fruit breaking
into numerous 1-seeded nutlets.

1. S. spinosa L. - Prickly Mallow. Annual to perennial; **261**
stems to 70cm, often woody at base, erect; leaves widest near
base, rounded and cordate at base; petioles <3cm, those on
larger leaves with small spine; carpels 5. Intrd; very local
casual on tips, but characteristic introduction with soya-
beans and sometimes with bird-seed; C & S En; tropics.
2. S. rhombifolia L. - Queensland-hemp. Usually woody **261**
perennial to 2m, but smaller annual here; leaves widest near
middle, narrowed to base but often minutely cordate at
extreme base; petiole <1cm; carpels 8-14. Intrd; frequent on
tips with soya-bean waste, but less common than S. spinosa; S
En; tropics.

2. MALVA L. - Mallows
Annual to perennial herbs; epicalyx of 3 segments free to
base; petals pink to purple (or white); carpels numerous;
stigmas linear; fruit breaking into numerous 1-seeded
nutlets.
Fully mature fruit is essential for correct determination,
especially of spp. 5-8.

1 Nutlets smooth or faintly reticulate with low rounded
   ridges   2
1 Nutlets strongly reticulate with sharp ridges   5
   2 Petals bright pink (or white), >16mm, >3x as long
     as sepals; upper leaves usually very deeply
     divided   3
   2 Petals pinkish- or whitish-mauve, <16mm, <3x as
     long as sepals; leaves shallowly lobed   4
3 Epicalyx-segments >3x as long as wide; calyx, epicalyx
   and pedicels with only simple hairs   **1. M. moschata**
3 Epicalyx-segments <3x as long as wide; calyx, epicalyx
   and pedicels with many stellate hairs   **2. M. alcea**
   4 Nutlets smooth, shortly pubescent; petals
     usually 2-3x as long as sepals   **7. M. neglecta**
   4 Nutlets obscurely ridged, glabrous or nearly so;
     petals usually <2x as long as sepals
                                            8. M.verticillata
5 Perennial; petals (12)20-30mm, >2x as long as sepals,
   usually bright pinkish-purple   **3. M. sylvestris**
5 Annual or biennial; petals 4-12mm, <2x as long as
   sepals, usually pale pinkish- or whitish-mauve   6
   6 Epicalyx-segments ovate-lanceolate, <3x as long

                    as wide; staminal-tube pubescent   **4. M. nicaeensis**
       6 Epicalyx-segments linear-lanceolate, >3x as long
                    as wide; staminal-tube glabrous or nearly so   7
       7 Calyx long-ciliate (longer hairs c.1mm); some pedicels
                    >1cm at fruiting; angle of nutlets between dorsal and
                    lateral surfaces sharp but not winged   **6. M. pusilla**
       7 Calyx glabrous or short-ciliate (hairs <0.5mm);
                    pedicels all <1cm at fruiting; angle of nutlets
                    between dorsal and lateral surfaces forming narrow
                    wavy wing   **5. M. parviflora**

1. M. moschata L. - Musk-mallow. Perennial with erect stems **261**
to 80cm; upper leaves usually very deeply divided, but sometimes not so; petals >16mm, bright pink to white; nutlets
smooth, with long hairs, rounded between lateral and dorsal
surfaces. Native; grassy banks and fields, especially on
rich soils; throughout most of BI but rare in N & W Sc and
parts of Ir, probably not native in most of Ir and N Br.
2. M. alcea L. - Greater Musk-mallow. Perennial with erect **261**
or spreading stems to 1.2m; similar to M. moschata but with
stellate, not simple, hairs on all vegetative parts; nutlets
glabrous or pubescent; and see key. Intrd; grassy and waste
places, sometimes natd; SE En; Europe.
3. M. sylvestris L. - Common Mallow. Spreading perennial **261**
with erect to decumbent stems to 1m; leaves with shallow
rounded lobes; petals >(12)20mm, bright pinkish-purple with
dark stripes; nutlets strongly reticulate, usually glabrous,
with sharp angle between dorsal and lateral surfaces.
Native; waste and rough ground, by roads and railways; common
throughout lowland En, Wa and CI, scattered elsewhere.
4. M. nicaeensis All. - French Mallow. Annual or biennial **261**
with ascending to decumbent stems to 50cm; similar to M.
sylvestris but petals 10-12mm, pale mauve; and see key.
Intrd; waste places, rather rare casual from wool and other
sources; mainly S En, very rare elsewhere; S Europe.
5. M. parviflora L. - Least Mallow. Annual with erect to **261**
decumbent stems to 50cm; leaves with shallow rounded lobes;
petals <5mm, pale mauve; nutlets strongly reticulate,
glabrous or pubescent, the angle between dorsal and lateral
surfaces raised to form narrow wavy wing. Intrd; waste
places, casual from several sources, especially wool; very
scattered over Br; S Europe.
6. M. pusilla Smith (M. rotundifolia L.) - Small Mallow. **261**
Similar to M. parviflora but see key; nutlets strongly
reticulate, glabrous or pubescent, the angle between dorsal
and lateral surface sharp and jagged but not winged. Intrd;
waste places, casual from several sources, especially wool,
sometimes persistent; very scattered over Br; Europe.
7. M. neglecta Wallr. - Dwarf Mallow. Similar to M. parvi- **261**
flora but petals 9-13mm; nutlets smooth, pubescent, with
sharp smooth angle between dorsal and lateral surfaces.
Native; rough and waste ground, waysides; frequent in C & S

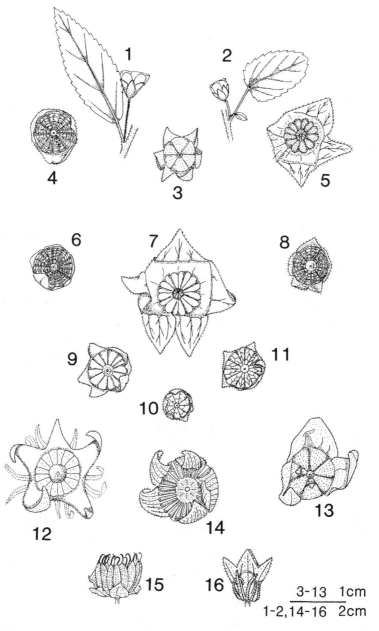

FIG 261 – Malvaceae. 1-2, flowering nodes of **Sida**. 1, **S. rhombifolia**. 2, **S. spinosa**. 3-16, fruits. 3, **Sidalcea malviflora**. 4, **Malva nicaeensis**. 5, **M. moschata**. 6, **M. parviflora**. 7, **M. alcea**. 8, **M. sylvestris**. 9, **M. neglecta**. 10, **M. verticillata**. 11, **M. pusilla**. 12, **Althaea officinalis**. 13, **Lavatera arborea**. 14, **Alcea rosea**. 15, **Abutilon theophrasti**. 16, **Hibiscus trionum**.

Br and CI, very scattered in N En, Wa and Sc and probably
intrd there, often only casual.
8. M. verticillata L. (M. crispa (L.) L.) - Chinese Mallow. 261
Similar to M. parviflora but petals 5-9mm; nutlets obscurely
ridged, + glabrous, with sharp smooth angle between dorsal
and lateral surfaces. Intrd; rough and waste ground, casual
from wool and other sources; infrequent in C & S En; E Asia.

3. LAVATERA L. - Tree-mallows
Annual to perennial, woody or not; epicalyx of 3 segments
fused below; petals pink to purple (or white); carpels
numerous; stigmas linear; fruit breaking into numerous
1-seeded nutlets.

1 Central axis of fruit expanded above to form umbrella-
   like disc concealing nutlets   **5. L. trimestris**
1 Central axis of fruit not expanded above; nutlets
   clearly visible from above unless obscured by calyx   2
   2 Epicalyx <3/4 as long as calyx   **3. L. plebeia**
   2 Epicalyx 3/4 as long as to longer than calyx   3
   3 Flowers 1 in each leaf-axil   **4 L. thuringiaca**
   3 Flowers 2-several in each leaf-axil   4
      4 Epicalyx expanding to longer than calyx in fruit;
         petals deep pinkish-purple with dark stripes
                                         1. L.arborea
      4 Epicalyx slightly shorter than to as long as calyx
         in fruit; petals lilac   **2. L. cretica**

1. L. arborea L. - Tree-mallow. Biennial with erect stems 261
to 3m, woody below; younger parts softly stellate-tomentose;   R
leaves shallowly 5-7-lobed; petals 14-20mm, deep pinkish-
purple with dark stripes. Native; rocks, cliff-bottoms and
waste ground near sea; W coast of Br from Dorset to Ayrs, CI,
scattered in Ir, probably intrd on coasts of S & E Br.
2. L. cretica L. - Smaller Tree-mallow. Annual or biennial RR
herb with erect to decumbent stems to 1(1.5)m; younger parts
stellate-pubescent; leaves shallowly 5-7-lobed; petals 10-
20mm, lilac. Native; rough and waste ground by sea; frequent
in Scillies and CI, formerly W Cornwall mainland but now
sporadic there and perhaps intrd, rare casual elsewhere.
3. L. plebeia Sims - Australian Hollyhock. Annual or
biennial herb with erect to decumbent stems to 1(2)m; younger
parts rather sparsely stellate-pubescent; leaves 3-7-lobed;
petals 12-25mm, lilac or white. Intrd; rather infrequent
wool-alien; very scattered in En; Australia.
4. L. thuringiaca L. (L. olbia auct. non L.) - Garden
Tree-mallow. Perennial (+ shrub) with woody branching erect
stems to 2.5m; younger parts softly stellate-pubescent;
leaves 3-5-lobed, the central lobe long; petals 15-30mm,
purple to pale pink. Intrd; grown in gardens and sometimes
persisting as escape or throwout; S En; SE Europe.
5. L. trimestris L. - Royal Mallow. Annual with erect or

ascending stems to 1(1.2)m; younger parts sparsely pubescent with simple and few-rayed hairs; leaves shallowly 3-7-lobed; petals 20-45mm, pink or white. Intrd; grown in gardens and sometimes found as casual, escape or throwout; S En; Mediterranean.

4. ALTHAEA L. - Marsh-mallows

Annual to perennial herbs; epicalyx of 6-10 segments fused below; petals pink to purple; staminal-tube terete, pubescent; carpels numerous; stigmas linear; fruit breaking into numerous 1-seeded nutlets.

1. A. officinalis L. - Marsh-mallow. Perennial with erect **261** stems to 1.5m; younger parts softly stellate-tomentose; **R** leaves shallowly 3-5-lobed; petals 15-20mm, pale pink. Native; brackish ditches, banks and grassland near sea; locally common round coasts of Br N to Lincs and S Wa, formerly CI, scattered elsewhere in Br and Ir but intrd.

2. A. hirsuta L. - Rough Marsh-mallow. Annual with erect to **RRR** decumbent stems to 60cm, coarsely pubescent with long simple and short stellate hairs; lower leaves shallowly, upper deeply 3-5-lobed or palmate; petals 12-16mm, lilac. Probably intrd; field and wood borders, N Somerset and W Kent (since 1792), casual or persistent in scattered places over most of En and Wa; Europe.

5. ALCEA L. - Hollyhocks

Biennial to perennial herbs; epicalyx of 6-7 segments fused below; petals various shades of red, yellow or white; staminal-tube 5-angled, glabrous; carpels numerous; stigmas linear; fruit breaking into numerous 1-seeded nutlets.

1. A. rosea L. (A. ficifolia L., Althaea rosea (L.) Cav., A. **261** ficifolia (L.) Cav., A. x cultorum Bergmans) - Hollyhock. Stems erect, to 3m; younger parts rather softly stellate-tomentose; leaves shallowly to rather deeply 3-9-lobed; petals 25-50mm. Intrd; much grown in gardens and frequent escape or throwout on tips and waste ground, usually casual; scattered over much of BI; garden origin from W Asian parents. A. ficifolia has rather deeply divided leaves and usually yellow flowers, but is not specifically distinct.

6. SIDALCEA A. Gray - Greek Mallow

Perennial herbs; epicalyx 0; petals pinkish-purple; carpels numerous; stigmas linear; fruit breaking into numerous 1-seeded nutlets.

1. S. malviflora (DC.) A. Gray ex Benth. - Greek Mallow. **261** Stems erect, to 1.5m, somewhat pubescent above; lower leaves shallowly, upper deeply 3-7-lobed or palmate; flowers in simple racemes; petals 10-25mm. Intrd; grown for ornament and sometimes persisting as throw-out or escape; very

scattered over Br; SW N America. Plants in Br are cultivars
of unknown origin and parentage; few or none are pure S.
malviflora, but probably include it in their parentage. A
plant natd by a stream in Shetland has been tentatively named
S. hendersonii S. Watson.

7. ABUTILON Miller - <u>Velvetleaf</u>
Annual herb; epicalyx 0; petals yellow; carpels numerous;
stigmas capitate; fruit composed of 5 almost separate,
several-seeded nutlets.

1. **A. theophrasti** Medikus - <u>Velvetleaf</u>. Erect annual to 1m; **261**
younger parts softly tomentose with stellate and simple
hairs; leaves suborbicular, acuminate, cordate, not lobed;
petals 7-13mm; nutlets each with slender beak 2-3mm. Intrd;
casual in waste places and tips, frequent bird-seed, oil-
seed and wool alien; scattered over En and Wa; SE Europe.

8. HIBISCUS L. - <u>Bladder</u> <u>Ketmia</u>
Annual herb; epicalyx of 10-13 segments free ± to base;
petals pale yellow with violet patch at base; carpels 5,
fused; stigmas capitate; fruit a dehiscent capsule with 5
many-seeded cells.

1. **H. trionum** L. - <u>Bladder</u> <u>Ketmia</u>. Erect to decumbent **261**
annual to 50m; rather sparsely stellate-pubescent; leaves
mostly palmate with 3(-5) lobes; petals 15-25mm; calyx
enlarging and inflated round fruit. Intrd; casual in waste
places and tips; frequent bird-seed, oil-seed and wool alien;
scattered over En, Wa and CI; SE Europe, Asia, Africa.

55. SARRACENIACEAE - <u>Pitcherplant family</u>

Herbaceous perennials; leaves all in basal rosette,
consisting of tubular insectivorous 'pitchers' with small
leaf-blade ('hood') at entrance. Flowers solitary, terminal,
on leafless pedicels from ground level, actinomorphic, bisexual, hypogynous, with 3 small bracteoles resembling
epicalyx; sepals 5, ± petaloid, free; petals 5, showy,
free; stamens numerous; ovary 5-celled, with many ovules in
each cell; style 1, short; stigma greatly expanded, peltate;
fruit a capsule.
Unmistakable pitcherplants.

1. SARRACENIA L. - <u>Pitcherplant</u>

Other spp. - **S. flava** L. (<u>Trumpets</u>), from SE N America, has
been found natd in a valley mire in S Hants; it differs from
S. purpurea in its pitchers being straight and ± erect, with
an arching hood, and in its yellow petals.

1. S. purpurea L. - <u>Pitcherplant</u>. Pitchers <(30)cm, green,
marbled with red, curved, decumbent; hood erect; pedicels
<60cm; flowers nodding, c.5cm across, with purplish-red
petals and sepals; stigma c.3cm across, greenish. Intrd;
planted in wet peat-bogs and well natd; C Ir (planted in
Roscommon in 1906, later from there to other places), a few
less permanent sites scattered in En; NE N America.

56. DROSERACEAE - <u>Sundew family</u>

Herbaceous perennials, leaves all in basal rosette, reddish
and covered with sticky hairs (insectivorous), stipulate,
with petioles. Flowers in simple cyme, on leafless peduncle
from ground level, actinomorphic, bisexual, hypogynous,
usually remaining closed; sepals 5-8, ± fused at base; petals
5-8, 4-6mm, white, free; stamens 5-8; ovary 1-celled, with
many ovules; styles 2-6, often deeply divided above or fused
below; stigmas linear; fruit a capsule.
Unmistakable insectivorous plants.

1. DROSERA L. - <u>Sundews</u>

1 Leaf-blade ± orbicular, abruptly narrowed into
  pubescent petiole   ·   **1. D. rotundifolia**
1 Leaf-blade obovate to linear-oblong, gradually
  narrowed into ± glabrous petiole   2
  2 Peduncles straight, rising from near centre of
    leaf-rosette, often much longer than leaves
                        2. D. longifolia
  2 Peduncles curved at base, arising from sides of
    leaf-rosette, often little longer than leaves
                        3. D. intermedia

Other spp. - 2 S Hemisphere spp. have been planted on bogs
in Surrey in 1970s and 1980s and are surviving. **D. capensis**
L., from S Africa, has linear-oblong leaves mostly 10-15cm,
often a short ± erect aerial stem, and purple flowers; **D.
binata** Labill., from Australia and New Zealand, has even
larger leaves 1-few times forked into linear lobes <15cm, and
white petals <1cm.

1. D. rotundifolia L. - <u>Round-leaved Sundew</u>. Inflorescences
to 10(25)cm, the peduncles straight and arising centrally as
in <u>D. longifolia</u> (see key); leaves (incl. petiole) <5cm, the
blades ± orbicular, <10 x 10mm. Native; wet acid peaty
places with little shade; in suitable places over most of BI,
often common in N & W but absent from much of C, E & S En.
1 x 2. D. rotundifolia x D. longifolia = D. x obovata Mert.
& Koch is similar to <u>D. intermedia</u> in leaf characters but has
a straight, centrally arising inflorescence and is sterile;
it is rare but has scattered records over much of W Br and

Ir, especially NW Sc.

1 x 3. D. rotundifolia x D. intermedia = D. x beleziana
Camus has a leaf-shape closer to that of D. rotundifolia but
curved, lateral inflorescences and is sterile; it has been
recorded in S Hants, W Norfolk and Cards.

2. D. longifolia L. (D. anglica Hudson) - Great Sundew.
Inflorescences to 18(30)cm; leaves (incl. petiole) <13cm, the
blades narrowly obovate to linear-oblong, <30 x 10mm.
Native; similar places (or wetter) and distribution to D.
rotundifolia, but much more local and absent from CI and most
of En, Wa and E Sc.

3. D. intermedia Hayne - Oblong-leaved Sundew. Inflorescences to 5(10)cm; leaves (incl. petiole) <5cm, the blades
obovate to narrowly so, <10 x 5mm. Native; similar places to
D. rotundifolia; very locally common in scattered areas of Br
and Ir, mostly in W.

57. CISTACEAE - Rock-rose family

Annuals or woody evergreen dwarf perennials; leaves opposite
(at least below), simple, entire, sessile to shortly
petiolate, stipulate or not. Flowers in terminal mostly
simple cymes, actinomorphic, bisexual, hypogynous; sepals 5,
free, 2 outer smaller than 3 inner (except in Cistus); petals
5, free, predominantly yellow or white; stamens numerous;
ovary 1-celled, with many ovules on 3 placentae (5-celled in
Cistus); style 0 or 1; stigma 1 (and then 3-lobed, or 5-lobed
in Cistus) or 3; fruit a capsule with 3 valves (5 in
Cistus).
Recognizable by the distinctive sepals and gynoecium, and
stellate hairs on vegetative parts.

1 Annuals; style very short or 0   **1. TUBERARIA**
1 Woody perennials; style at least as long as ovary
                                                    2. HELIANTHEMUM

Other genera - CISTUS L. differs from other 2 genera in
contrasts given in family description and is a shrub. **C.
laurifolius** L. and **C. incanus** L., from S Europe, are grown
for ornament and occasionally self-sown individuals persist
for a short while outside gardens. In both spp. the flowers
are 4-6cm across, white or pink respectively.

1. TUBERARIA (Dunal) Spach - Spotted Rock-rose
Annual with basal leaf-rosette; leaves with 3 obvious veins;
stipules 0; style 0 or very short.

1. T. guttata (L.) Fourr. - Spotted Rock-rose. Stems ±   **RR**
procumbent to erect, to 30cm but often <10cm; leaves
pubescent or sparsely so on lowerside; flowers 8-15mm across;
petals yellow, usually with brownish-red blotch at base.

Native; dry barish ground near sea; W & SW Ir, NW Wa, Jersey and Alderney. Plants from Wa and Ir differ from those in CI and France in having wider leaves, shorter internodes and flowers with bracts. They have been separated as ssp. **breweri** (Planchon) E. Warb., but are probably not worth ssp. status.

2. HELIANTHEMUM Miller - <u>Rock-roses</u>
Dwarf straggly or bushy woody perennials; leaves with 1 obvious vein; stipules present or 0; style at least as long as ovary.

1 Stipules 0; style strongly S-shaped, shorter than
  stamens; flowers <15(20)mm across   **3. H. canum**
1 Stipules present; style slightly kinked near base,
  longer than stamens; flowers mostly >(15)20mm across   2
  2 Petals white; leaves grey-tomentose on upperside
                                 2. H. apenninum
  2 Petals yellow; leaves green-pubescent on upperside
                              1. H. nummularium

1. H. nummularium (L.) Miller (<u>H.</u> chamaecistus Miller) - <u>Common</u> <u>Rock-rose</u>. Stems procumbent or decumbent, wiry, to 50cm; leaves elliptic-oblong to narrowly so, green and sparsely pubescent on upperside, whitish-tomentose on lowerside; flowers 1-12 per cyme, mostly 20-25mm across; petals yellow. Native; base-rich grassland; common in suitable places over most of Br, but absent from many areas, 1 place in E Donegal.
1 x 2. H. nummularium x H. apenninum = H. x sulphureum Willd. ex Schldl. occurs in the N Somerset and formerly in the S Devon site of <u>H.</u> apenninum with both parents; it has pale yellow petals and intermediate leaf characters and is usually highly fertile.
2. H. apenninum (L.) Miller - <u>White Rock-rose</u>. Similar to **RR** <u>H.</u> <u>nummularium</u> but leaves usually narrower (narrowly oblong to linear) and grey-tomentose on upperside; petals white. Native; in dry limestone grassland; 1 area in each of N Somerset and S Devon.
3. H. canum (L.) Hornem. - <u>Hoary Rock-rose</u>. Stems **R** procumbent to ascending, often rather twiggy below, to 25cm; leaves green on upperside, whitish-tomentose on lowerside; flowers 1-6(10) per cyme, mostly 10-15mm across; petals yellow. Native; rocky limestone pastures.
1 Leaves glabrous to subglabrous on upperside; flowers
  1-3(5) per cyme   **c. ssp. levigatum**
1 Leaves pubescent to sparsely so on upperside; flowers
  (2)3-6 per cyme   2
  2 Leaves ± persistent on lower parts of non-flowering
    shoots, usually sparsely pubescent on upperside,
    mostly >10mm   **b. ssp. piloselloides**
  2 Leaves not persistent on lower parts of non-

flowering shoots, usually pubescent on upperside,
often <10mm   **a. ssp. canum**
a. Ssp. canum (<u>H.</u> oelandicum (L.) DC. ssp. <u>canum</u> (L.)
Bonnier). Leaves more pubescent than in other sspp. Very
local in N Wa, S Wa and NW En.
b. Ssp. piloselloides (Lapeyr.) M. Proctor (<u>H.</u> oelandicum
ssp. <u>piloselloides</u> (Lapeyr.) Greuter & Burdet). Leaves
larger and less pubescent than in ssp. <u>canum</u>. W. Ir.
Possibly distinct from Pyrenean ssp. piloselloides.
c. ssp. levigatum M. Proctor. Stems shorter, leaves smaller
and less pubescent, and inflorescence fewer flowered than in
ssp. <u>canum</u>. Cronkley Fell, NW Yorks; endemic.

58. VIOLACEAE - <u>Violet family</u>

Annual or perennial herbs; leaves alternate or ± all basal,
simple, toothed, petiolate, stipulate. Flowers solitary,
axillary or from basal rosette, zygomorphic, bisexual,
hypogynous; sepals 5, free, with appendages below their
insertion; petals 5, free, the lower one with a backwards-
directed spur, blue or yellow to white; stamens 5, the 2
lower with spur inserted into petal spur; ovary 1-celled with
many ovules on 3 placentas; style 1, thickened above, usually
bent; stigma 1, capitate; fruit a capsule with 3 valves.
The well-known pansies and violets.

1. VIOLA L. - <u>Violets</u>

1 Stipules ovate to linear-lanceolate, finely toothed;
   lateral 2 petals spreading horizontally; style not
   or gradually thickened distally, often hooked (subg.
   Viola - <u>Violets</u>)   2
1 Stipules ± leaf-like, at least some deeply lobed;
   lateral 2 petals directed upwards; style with ±
   globose apical swelling with hollow in one side
   (subg. <u>Melanium</u> (DC.) Hegi - <u>Pansies</u>)   10
   2 Style straight above, with oblique apex; leaves
      orbicular, obtuse to rounded at apex **9. V. palustris**
   2 Style hooked or with lateral beak at apex; leaves
      acute to obtuse at apex   3
3 Leaves (petioles and blades) and usually capsules
   pubescent   4
3 Leaves and capsules glabrous   6
   4 Sepals acute   **3. V. rupestris**
   4 Sepals obtuse to ± rounded   5
5 Creeping stolons present; flowers sweet-scented;
   hairs on petioles mostly <0.3mm, reflexed or
   appressed   **1. V. odorata**
5 Stolons 0; flowers not scented; hairs on petioles
   mostly 0.3-1mm, patent   **2. V. hirta**
   6 Plant with basal leaf-rosette; leaves ≤1.3x as

        long as wide   7
    6 Basal leaf-rosette 0; leaves >1.4x as long as wide   8
 7 Sepal-appendages >1.5mm, corolla-spur paler than
    petals   **4. V. riviniana**
 7 Sepal-appendages <1.5mm; corolla-spur darker than
    petals   **5. V. reichenbachiana**
    8 Corolla-spur <2x as long as sepal-appendages; roots
       creeping underground, sending up stems at intervals
                                              8. V. persicifolia
    8 Corolla-spur >2x as long as sepal-appendages; all
       stems arising from central tuft   9
 9 Petals clear blue; leaves ovate to ovate-lanceolate,
    truncate to cordate at base   **6. V. canina**
 9 Petals cream to greyish-violet; leaves lanceolate to
    ovate-lanceolate, rounded to cuneate at base
                                                   7. V. lactea
    10 Spur 10-15mm   **10. V. cornuta**
    10 Spur <7mm   11
 11 Flowers usually >3.5cm vertically across, with
    strongly overlapping petals   **13. V. x wittrockiana**
 11 Flowers <2.5(3.5)cm vertically across, with not or
    slightly overlapping petals   12
    12 Plant perennial, with stems creeping underground   13
    12 Plant annual to perennial; creeping underground
       stems 0   14
 13 Flowers >(1.5)2cm vertically across; terminal segment
    of stipules scarcely wider than others, entire
                                                   11. V.lutea
 13 Flowers often <2cm vertically across; terminal
    segment of stipules distinctly wider than others,
    often slightly crenate   **12. V. tricolor**
    14 Corolla 4-8mm vertically across, concave; Scillies
       and CI only   **15. V. kitaibeliana**
    14 Corolla >8mm vertically across, ± flat when
       fresh; widespread   15
 15 Corolla 8-20mm vertically across, usually yellow or
    cream (rarely with blue), usually shorter than calyx;
    terminal segment of stipules usually strongly
    crenate, ± leaf-like; projection below stigma-
    hollow 0 or indistinct   **14. V. arvensis**
 15 Corolla 10-25(35)mm vertically across, usually with
    some blue or violet, usually longer than calyx;
    terminal segment of stipules entire to obscurely
    crenate, scarcely leaf-like; projection below stigma-
    hollow distinct   **12. V. tricolor**

1. V. odorata L. - <u>Sweet Violet</u>. Perennial with leaves and
flowers from central tuft; creeping stolons present; leaves
pubescent, broadly ovate to ovate-orbicular, deeply cordate;
flowers violet, or white with violet or purple spur, rarely
pinkish.   Native; woodland, scrub and hedgerows, mostly on
base-rich soils; most of BI, but local or very local in Sc,

Ir, Wa and CI and often not native.

1 x 2. V. odorata x V. hirta = V. scabra F. Braun (V. x permixta Jordan) is found in Br N to Yorks and in Roxburghs with the parents; it is intermediate in most characters, incl. stolon development and flower-scent, and partially fertile.

2. V. hirta L. - Hairy Violet. Perennial with leaves and flowers from central tuft; stolons 0; leaves pubescent, ovate, deeply cordate; flowers blue-violet. Native; calcareous pastures and open scrub; suitable places in Br N to C Sc, rare and scattered in C Ir. Ssp. **calcarea** (Bab.) E. Warb. differs in being smaller in most parts, especially the shorter petal-spur and narrower petals, but is probably only a var.

3. V. rupestris F.W. Schmidt - Teesdale Violet. Perennial **RR** usually with leafy stems (short or 0 in dwarfed plants) to 5(10)cm and basal leaves from central tuft, pubescent $\pm$ all over; leaves broadly ovate to ovate-orbicular, cordate; flowers pale blue-violet. Native; short turf or barish places on limestone; Durham, Westmorland and MW Yorks.

3 x 4. V. rupestris x V. riviniana = V. x burnatii Gremli occurs with the parents in Upper Teesdale (Durham); it is intermediate in most characters, including a very fine pubescence, and highly sterile.

4. V. riviniana Reichb. - Common Dog-violet. Perennial with leafy stems (short or 0 in dwarfed plants) to 20cm and basal leaves from central tuft, glabrous to slightly pubescent on most parts; leaves broadly ovate to ovate-orbicular, obtuse, deeply cordate; flowers pale blue-violet. Native; wide range of woods and grassland; common all over BI. Ssp. **minor** (Murb. ex Gregory) Valent. differs in being smaller in all parts; it is an ecotype of more exposed places, best treated as var. minor (Murbeck ex Gregory) Valent.

4 x 5. V. riviniana x V. reichenbachiana = V. x bavarica Schrank is intermediate in length of sepal-appendages but has a dark spur and is highly (but not fully) sterile; it is only sparsely scattered in En, Wa and Ir despite the frequent cohabitation of the parents.

4 x 6. V. riviniana x V. canina = V. x intersita G. Beck (V. x weinhartii W. Becker) is scattered in Br and Ir where the parents meet on heaths and dunes; it is intermediate in habit and leaf characters and highly sterile.

4 x 7. V. riviniana x V. lactea is frequent in En, Wa and Ir in the range of V. lactea; it is intermediate in habit and leaf and flower characters, and quite highly sterile. It often forms large clonal patches.

5. V. reichenbachiana Jordan ex Boreau - Early Dog-violet. $\pm$ glabrous perennial with leafy stems to 20cm and basal leaves from central tuft; leaves ovate to broadly ovate, subacute, deeply cordate; flowers bluish-mauve. Native; woods and hedgebanks; common in En, scattered in Wa and Ir, rare in Sc.

5 x 6. V. reichenbachiana x V. canina = V. x mixta A. Kerner

(V. x borussica (Borbas) W. Becker) has been recorded from S
En but is very rare as the parents rarely meet; it is intermediate and fairly highly sterile, but would be extremely
difficult to tell from V. riviniana x V. canina.

6. V. canina L. - Heath Dog-violet. Glabrous or sparsely
pubescent perennial with leafy stems to 30(40)cm; basal leaf-
rosette 0; leaves ovate to ovate-lanceolate; flowers blue
with whitish-yellow spur. Native.

a. Ssp. canina. Stems procumbent to ascending; leaves <2x
as long as wide, cordate; stipules of middle leaves <1/3 as
long as petiole. Dry or wet heaths and dunes, fens; local
throughout BI.

b. Ssp. montana (L.) Hartman. Stems erect; leaves c.2x as
long as wide, truncate to subcordate; stipules of middle
leaves ≤1/2 as long as petiole; flowers relatively large.
Fens; Hunts and Cambs.

6 x 7. V. canina x V. lactea = V. x militaris Savoure occurs
quite frequently within the range of V. lactea in En, Wa and
Ir; it is intermediate in leaf-shape and flower-colour, shows
c.10 per cent fertility and is vigorous.

6 x 8. V. canina x V. persicifolia = V. x ritschliana W.
Becker occurs in C En and C Ir with the parents; it is intermediate in leaf chracters but close to V. canina in flowers
and to V. persicifolia in habit. It is sterile and often
very vigorous.

7. V. lactea Smith - Pale Dog-violet. Subglabrous perennial   R
with leafy stems to 20cm; basal leaf-rosette 0; leaves
lanceolate to ovate-lanceolate, rounded to cuneate at base;
flowers cream to greyish-violet. Native; dry heaths; very
local in S En, S & W Wa, Man and C & S Ir.

8. V. persicifolia Schreber (V. stagnina Kit.) - Fen Violet. **RRR**
Subglabrous perennial with leafy stems to 25cm from central
tuft and from spreading roots; basal leaf-rosette 0; leaves
ovate-lanceolate, truncate to subcordate at base; flowers
bluish-white to white, with greenish spur. Native; fens;
Cambs and Hunts, formerly N to SW Yorks, C & W Ir.

9. V. palustris L. - Marsh Violet. Perennial with thin
rhizomes or sometimes stolons producing leaves and flowers;
aerial stems 0; leaves orbicular, deeply cordate; flowers
bluish-lilac with darker veins. Native; bogs, fens, marshes,
wet heaths and woods.

a. Ssp. palustris. Leaves rounded at apex, glabrous;
bracteoles below middle of pedicel. Throughout most of Br
and Ir, but absent from much of C & E En.

b. Ssp. juressi (Link ex Wein) P. Fourn. Leaves obtuse to
subacute, usually with pubescent petioles; bracteoles near
middle of pedicel. Local in Ir, S & W Wa and W En.

10. V. cornuta L. - Horned Pansy. Perennial with slender
rhizome; aerial stems to 30cm; stipules serrate or lobed to
c.1/2 way to midrib, with ± triangular apical lobe; flowers
2-4cm across, violet or lilac, fragrant; spur 10-15mm.
Intrd; grown in gardens and natd in grassy places; frequent

in Sc, rare and mostly ± casual elsewhere in Br; Pyrenees. Hybrids between this and V. x wittrockiana are grown (Bedding Viola, Violetta) and might also escape.

11. V. lutea Hudson - Mountain Pansy. Perennial with slender rhizome; aerial stems to 20cm; stipules lobed nearly to midrib, with linear-elliptic entire apical lobe; flowers (1.5)2-3.5cm across, usually yellow, less often blue or purple or blotched, spur 3-6mm; most pollen-grains with 4 pores. Native; upland pastures and rocky places, often on base-rich or heavy-metalliferous soils; upland areas of Wa, Sc and N En, very local in C & S Ir.

11 x 12. V. lutea x V. tricolor is known for certain only from Derbys and S Northumb but is probably overlooked; it is difficult to determine without detailed study. Hybrids are intermediate or closer to V. lutea in appearance and partially fertile, and probably backcross.

12. V. tricolor L. - Wild Pansy. Annual to perennial; stipules lobed ± to midrib, with narrowly elliptic to oblanceolate or narrowly obovate entire to crenate apical lobe; flowers 1-2.5(3.5)cm across, usually purple to violet or blotched with yellow, less often all yellow; spur 3-6.5mm; most pollen-grains with 4 pores. Native.

a. Ssp. tricolor. Annual, or perennial with rhizomes 0 or ill-developed. Common on waste, marginal and cultivated ground throughout most of BI. Perennial plants occurring in hilly areas of N have been referred to ssp. **saxatilis** (F.W. Schmidt) E. Warb. (V. lepida Jordan), but probably are variants of ssp. tricolor or its hybrids with V. arvensis or V. lutea.

b. Ssp. curtisii (E. Forster) Syme). Perennial with well-developed rhizomes. Frequent on maritime dunes in W Br and most of Ir, and inland by lakes in N Ir and on heaths of Norfolk and Suffolk.

12 x 14. V. tricolor x V. arvensis = V. x contempta Jordan occurs scattered throughout BI, but its frequency is uncertain. It is intermediate in most characters, especially stylar flap and corolla-size and -colour, and partially fertile. Variants of both parents, but mainly V. arvensis, resemble it.

13. V. x wittrockiana Gams ex Kappert - Garden Pansy. Probably derived from various V. tricolor x V. arvensis crosses and recognisable by flowers 3.5-10cm across with strongly overlapping petals. Intrd; much grown in gardens and parks and often escaping on rough or cultivated land or on tips; scattered throughout lowland Br, mainly in S. Wild populations often show segregation and/or backcrossing to V. tricolor or less often to V. arvensis.

14. V. arvensis Murray - Field Pansy. Annual with erect to decumbent stems to 40cm; stipules lobed nearly to base, with leaf-like elliptic to obovate (to narrowly so) crenate to serrate apical lobe; flowers 8-20mm across, usually yellow or cream, less often with violet blotches or suffusion; spur

2-4mm; most pollen-grains with 5 spores. Native; weed of cultivated and waste ground; common throughout most of BI.

15. V. kitaibeliana Schultes - <u>Dwarf Pansy</u>. Annual with **RR** erect to decumbent stems to 10cm; stipules similar to those of <u>V. arvensis</u> but much smaller; flowers 4-8mm across, usually cream suffused with violet; spur 1-2mm, often violet. Native; short turf on sandy soil by sea; Scillies and CI.

59. TAMARICACEAE - Tamarisk family

Deciduous shrubs or small trees; leaves alternate, simple, entire, small and ± scale-like, exstipulate, sessile. Flowers small, in long catkin-like racemes, actinomorphic, bisexual, hypogynous; sepals 5, free; petals 5, free, white to pink, ovate to trullate; stamens 5, joined at base to lobed or stellate nectariferous disc; ovary 1-celled, with many ovules; styles 3-4; stigmas capitate; fruit a capsule; seeds with apical tuft of hairs.
Easily recognizable by the distinctive habit and flower structure.

1. TAMARIX L. - <u>Tamarisks</u>
1. T. gallica L. (<u>T. anglica</u> Webb) - <u>Tamarisk</u>. Shrub or small tree to 3m; leaves 1.5-3mm, narrowly triangular-ovate, acute to acuminate, green to glaucous; racemes 2-5cm x 4-5mm; petals 1.5-2mm. Intrd; planted as windbreak and stabilizer in sandy places by sea, persistent and appearing natd but rarely self-sown; coasts of CI and Br N to N Wa and Suffolk, rarely elsewhere; SW Europe.
2. T. africana Poiret - <u>African Tamarisk</u>. Differs mainly in larger racemes (3-7cm x 5-9mm) and petals (2.5-3mm). Intrd; same habitats as <u>T. gallica</u>, but much rarer and never self-sown; S coast of En; Mediterranean.

60. FRANKENIACEAE - Sea-heath family

Evergreen perennial with stems woody at base; leaves opposite, simple, entire, small and heather-like, exstipulate, sessile. Flowers small, solitary, terminal or in branchlet-forks, sessile, actinomorphic, bisexual, hypogynous; sepals usually 5, fused to >1/2 way; petals usually 5, free, pink, with small scale at junction of claw and expanded limb; stamens usually 6 in 2 whorls of 3; ovary 1-celled, with many ovules; style 1, divided into 3 near apex; stigmas 3, elongate; fruit a capsule.
Easily recognisable by the distinctive leaves and flower structure.

1. FRANKENIA L. - <u>Sea-heath</u>
1. F. laevis L. - <u>Sea-heath</u>. Stems procumbent, mat- **R**

forming, to 35cm; leaves 3-7mm, with strongly revolute margins, appearing linear to narrowly oblong with widened ciliate basal part; flowers 5-6mm across. Native; on sandy or silty barish ground on drier parts of salt-marshes; coasts of CI and SE Br from Wight to N Lincs, probably intrd in Anglesey, Glam and N Devon.

61. CUCURBITACEAE - White Bryony family

Annuals or herbaceous perennials; stems scrambling or climbing, usually with tendrils; leaves alternate, simple, usually palmately veined, usually palmately lobed, exstipulate, petiolate. Flowers 1-several in axillary racemes or irregular groups, actinomorphic, epigynous, unisexual (dioecious or monoecious); hypanthium present, bearing sepals, petals and stamens; sepals 5, free; petals 5, united at least at base; stamens apparently 3 (rarely 5), 2 with 2 pollen-sacs and 1 with 1; ovary 1-celled but often appearing (2)3(-5)-celled due to deeply intrusive parietal placentae, with few to many ovules; style 1; stigmas 2-3, ± capitate; fruit succulent and indehiscent, berry-like or cucumber-like.
  Recognizable by the scrambling or climbing habit, unisexual epigynous flowers with unusual stamens, and distinctive fruit.

1 Dioecious perennials; fruit <1cm, red   **1. BRYONIA**
1 Monoecious annuals; fruit >2cm, green to yellow   2
    2 Tendrils absent; fruit densely hispid when
       ripe   **2. ECBALLIUM**
    2 Tendrils present; fruit glabrous to sparsely
       pubescent or hispid when ripe   3
3 Male flowers several together; central axis of anther
    extending beyond pollen-sacs   **3. CUCUMIS**
3 Male flowers solitary; central axis of anther not
    extending beyond pollen-sacs   4
    4 Leaves pinnately lobed; corolla divided much
       >1/2 way   **4. CITRULLUS**
    4 Leaves palmately lobed; corolla divided <1/2 way
                                                   5. CUCURBITA

Other genera - SICYOS L. resembles Cucumis in its clustered male flowers and deeply divided corolla, but has branched tendrils, bristly fruits and 5 ± adherent stamens. **S. angulatus** L. (Bur Cucumber), from E N America, occurs as an infrequent casual mainly with soya-bean waste.

1. BRYONIA L. - White Bryony
Dioecious perennials with tuberous roots; stems climbing, with simple tendrils; flowers several together; corolla divided >1/2 way, greenish-white; fruit a red globose, ±

glabrous berry.

1. B. dioica Jacq. (<u>B.</u> <u>cretica</u> L. ssp. <u>dioica</u> (Jacq.) Tutin)
- <u>White</u> <u>Bryony</u>. Stems to c.4m; leaves deeply palmately
lobed, + hispid; corolla 5-12mm; fruit 5-9mm. Native; in
scrub and hedgerows, mostly on well-drained base-rich soils;
En but absent from most of N and SW, Wa but rare except in E,
intrd elsewhere.

2. ECBALLIUM A. Rich. - <u>Squirting</u> <u>Cucumber</u>
Monecious annuals with non-tuberous roots; stems trailing;
tendrils 0; male flowers several together; female flowers
solitary; corolla divided <1/2 way, yellowish; fruit
ellipsoid to narrowly so, green, densely hispid, explosive
when ripe.

1. E. elaterium (L.) A. Rich. - <u>Squirting</u> <u>Cucumber</u>. Stems
to 50cm, hispid; leaves not or shallowly and irregularly
palmately lobed, densely and coarsely pubescent; corolla
10-20mm; fruit 3-5cm, usually pendent on + erect pedicel.
Intrd; rare casual in waste and cultivated land in CI and S
Br, sometimes persisting in Jersey and extreme S En; S
Europe.

3. CUCUMIS L. - <u>Cucumbers</u>
Monoecious annuals with non-tuberous roots; stems climbing
with simple tendrils; male flowers several together; female
flowers solitary; corolla divided <1/2 way, yellow; fruit
ellipsoid to cylindrical, very variable in surface texture,
green to yellow, glabrous to pubescent.

1. C. melo L. - <u>Melon</u>. Stems to 1m, hispid; leaves not or
shallowly palmately lobed, hispid to softly pubescent;
corolla (5)20-30mm; fruit 3->30cm, green to yellow, usually
ellipsoid, pubescent at first. Intrd; common casual on tips
and at sewerage works; mainly S En; Africa.
2. C. sativus L. - <u>Cucumber</u>. Stems to 1(2)m, hispid; leaves
shallowly to rather deeply palmately lobed, pubescent and
scabrid; corolla 20-30mm, yellow; fruit 10->30cm, green,
usually glabrous, smooth or warty, much longer than wide.
Intrd; rare casual on tips and at sewerage works, sometimes
grown as crop on field-scale; S En; India.

4. CITRULLUS Schrader - <u>Water</u> <u>Melon</u>
Monoecious annuals with non-tuberous roots; stems climbing
with simple or branched tendrils; male and female flowers
solitary; corolla divided >1/2 way, yellow; fruit globose to
slightly elongated, green to yellow, smooth, + glabrous.

1. C. lanatus (Thunb.) Matsum. & Nakai - <u>Water</u> <u>Melon</u>. Stems
to 1m, woolly when young; leaves with 3 main lobes, the
central one forming most of leaf, all 3 pinnately lobed,

pubescent when young, scabrid later; corolla 5-17mm; fruit
2->30cm. Intrd; rare casual on tips, frequent at sewerage
works; S Br; S Africa.

5. CUCURBITA L. - Marrow
Monoecious annuals with non-tuberous roots; stems climbing
with usually branched tendrils; male and female flowers
solitary; corolla divided <1/2 way, yellow; fruit globose to
cylindrical, green to yellow, smooth to warty, finally
glabrous.

Other spp. - **C. maxima** Duchesne ex Lam. (Pumpkin), from C
America, differs from C. pepo in its narrow terete (not
thickened fluted) female pedicels and usually huge
depressed-globose fruits; it is a rare casual but might
become commoner.

1. C. pepo L. - Marrow. Stems to 2m, hispid; leaves
shallowly to deeply palmately lobed, hispid; corolla 30-
110mm; fruit extremely variable, often very large. Intrd;
common casual on tips and at sewerage works, sometimes grown
on field scale; mostly S Br; N C America. Most casual plants
in Br result from use as vegetables (marrow, courgette), but
some perhaps from use as ornamentals (gourds).

62. SALICACEAE - Willow family

Deciduous trees or shrubs; leaves alternate, rarely ±
opposite, simple, usually serrate, stipulate, petiolate.
Flowers much reduced, in racemose catkins, normally
dioecious, each in axil of bract, each with either cup-like
perianth or 1-2 small nectaries at base; male flowers with
1-many stamens; female flowers with 1-celled ovary with
numerous ovules; styles 1(-2), short; stigmas 2, often bifid,
often large; fruit a 2-valved capsule; seeds with long plume
of hairs arising from base.
   Dioecious, catkin-bearing trees or shrubs with simple
flowers and plumed seeds borne in a capsule.

1 Flowers with 1-2 nectaries; cup-like perianth 0;
     bracts entire; winter buds with 1 outer scale   **2. SALIX**
1 Flowers with cup-like perianth; nectaries 0; bracts
     toothed; winter buds with several outer scales
                                                    1. POPULUS

1. POPULUS L. - Poplars
Trees; winter buds with several outer scales; flowers
appearing before leaves, with toothed or deeply divided
bracts, with cup-like perianth; nectaries 0; stamens 5-c.60.
   The most important characters are the general habit of the
tree and the shape of leaves on the dwarf lateral shoots

(short-shoots) or extension shoots (long-shoots); leaves on suckers or epicormic shoots are often very different. The sex of the tree can be important in identification.

1 Leaves of long-shoots densely tomentose on lowerside, sometimes glabrescent late in season 2
1 Leaves of long-shoots glabrous to slightly pubescent on lowerside 3
  2 Leaves of long-shoots distinctly palmately 3-5-lobed; bracts subentire to dentate; trees usually female **1. P. alba**
  2 Leaves of long-shoots not or scarcely palmately lobed, coarsely and bluntly dentate; bracts laciniate; trees mostly male **2. P. x canescens**
3 Leaves of long- and short-shoots coarsely and bluntly sinuate-dentate; bracts with long silky hairs **3. P. tremula**
3 Leaves shortly and regularly serrate; bracts glabrous or shortly pubescent 4
  4 Buds and young leaves strongly balsam-scented; leaves much paler on lowerside; petioles $\pm$ terete (Balsam-poplars) 5
  4 Not balsam-scented; leaves only slightly paler on lowerside; petioles strongly laterally compressed (Black Poplars) 7
5 Leaves cordate at base; petioles pubescent; female only; tree usually with broad crown and many suckers **6. P. candicans**
5 Leaves truncate or rounded to broadly cuneate at base (rarely subcordate); petioles glabrous; male or female; tree usually with very narrow crown and suckers 0 or few 6
  6 Tree with subfastigiate outline and curved-erect branches; female only **8. P. 'Balsam Spire'**
  6 Tree rather narrowly pyramidal, with spreading lower branches; mostly male **7. P. trichocarpa**
7 Leaves without sessile glands on lamina near top of petiole, broadly cuneate (to truncate) at base, not minutely ciliate (often slightly pubescent); tree either fastigiate with erect branches or with spreading branches and large burrs on trunk **4. P. nigra**
7 Many leaves with 1-2 small sessile glands on lamina near top of petiole, truncate or subcordate to broadly cuneate at base, minutely ciliate when young; tree narrow or broad but not fastigiate and not with burrs on trunk **5. P. x canadensis**

Other spp. - Several other spp. are grown in parks and some are now being used in groups or rows. The most common are **P. balsamifera** L. (P. tacamahacca Miller) (Eastern Balsam-poplar), from N America, differing from P. trichocarpa in its suckers, more upswept branches, rounded to subcordate

leaf-bases, and less angled young shoots; and **P. laurifolia**
Ledeb., from C Asia, differing in its spreading habit,
narrower leaves and very angled young branches. <u>P.</u>
<u>laurifolia</u> x <u>P. nigra</u> 'Italica' = **P. x berolinensis** Dippel
(<u>Berlin Poplar</u>) is subfastigiate and obviously a Black-
Balsam-poplar hybrid, differing from <u>P.</u> 'Balsam Spire' in its
cuneate leaf-base.

1. P. alba L. - <u>White Poplar</u>. Tree to 24m with broad **279**
spreading crown, many suckers and smooth grey bark; leaves
palmately lobed, densely white-tomentose on lowerside; males
very rare. Intrd; much planted and often well natd from
dense sucker growth, especially on coastal dunes; scattered
in BI, especially in C & S Br and BI; Europe.
2. P. x canescens (Aiton) Smith (<u>P.</u> x <u>hybrida</u> M. Bieb.; <u>P.</u> **279**
<u>alba</u> x <u>P. tremula</u>) - <u>Grey Poplar</u>. Tree to 37m, similar to <u>P.</u>
<u>alba</u> but more vigorous; leaves coarsely and bluntly sinuate-
dentate, densely tomentose on lowerside at first, often much
less so or subglabrous later; female rather rare; partially
fertile. Intrd; much planted but rarely natd, probably never
arising anew in BI but often alone or in groups among native
taxa, especially in damp woods; scattered throughout BI but
absent from much of Sc; Europe.
3. P. tremula L. - <u>Aspen</u>. Tree to 24m, rather similar in **279**
habit to <u>P. alba</u>; leaves suborbicular, coarsely and bluntly
sinuate-dentate, glabrous or rarely sparsely pubescent at
maturity; sucker leaves ovate-cordate, pubescent. Native;
woods and hedgerows on all sorts of soils, often forming
suckering thickets; throughout BI.
4. P. nigra L. - <u>Black-poplar</u>. Trees of various habit; **279**
leaves rhombic-ovate to trullate, broadly cuneate (to
truncate) at base, without glands, serrate with many small
teeth.
1 Tree with spreading crown; trunk with large burrs
                                                      a. ssp. betulifolia
1 Tree fastigiate, with ± erect branches; trunk
    without burrs   2
    2 Young leaves and stems sparsely pubescent; male
                                                      d. 'Plantierensis'
    2 Young leaves and stems glabrous   3
3 Tree very narrowly fastigiate; male   **b.** 'Italica'
3 Tree often more widely fastigiate; female **c.** 'Gigantea'
  a. Ssp. betulifolia (Pursh) W. Wettst. Spreading tree to
33m with burred trunk; young leaves and stems sparsely
pubescent. Native; in fields by streams and ponds, typically
in river flood-plains; scattered throughout most of En and
Wa, frequent in E Wa, C En and E Anglia, also in C Ir,
planted elsewhere, mostly male.
  b. 'Italica' (<u>P.</u> <u>nigra</u> var. <u>italica</u> Muenchh.) - <u>Lombardy-</u>
<u>poplar</u>. Narrowly fastigiate tree to 36m; trunk not burred;
shoots glabrous; male only. Intrd; much planted in parks and
as screening or windbreaks, not natd; throughout much of BI;

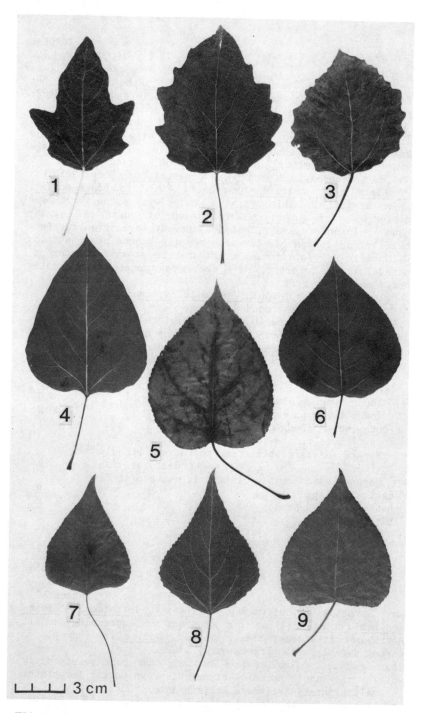

FIG 279 – Leaves of **Populus**. 1, **P. alba**. 2, **P. x canescens**. 3, **P. tremula**. 4, **P. trichocarpa**. 5, **P. candicans**. 6, **P. 'Balsam Spire'**. 7, **P. nigra**. 8, **P. x canadensis 'Marilandica'**. 9, **P. x canadensis 'Serotina'**.

garden origin.

c. 'Gigantea' - <u>Giant</u> <u>Lombardy-poplar</u>. Like 'Italica' but female and usually broader as in 'Plantierensis'. Intrd; grown in same way as 'Italica' but less common.

d. 'Plantierensis' (<u>P.</u> <u>nigra</u> ssp. <u>betulifolia</u> x <u>P.</u> 'Italica'). Similar to 'Italica' but often less narrowly fastigiate; young shoots pubescent; male only. Intrd; used similarly to 'Italica' and now as common; garden origin.

5. P. x canadensis Moench (<u>P.</u> x <u>euramericana</u> (Dode) Guinier; <u>P.</u> <u>nigra</u> x <u>P.</u> <u>deltoides</u> Marshall) - <u>Hybrid</u> <u>Black-poplar</u>. Trees of various habit, to >40m, those with spreading crowns differing from <u>P.</u> <u>nigra</u> in absence of burrs on trunks; usually largely upswept, not downcurved, lower branches (not twigs); and leaves usually with sessile glands near junction with petiole. Intrd; many cultivars are grown for ornament, screening and forestry, only the more frequent are treated here; garden origin.

<u>Multi-access</u> <u>key</u> <u>to</u> <u>cultivars</u> <u>of</u> <u>P.</u> <u>x</u> <u>canadensis</u>

Crown of tree with rounded outline	A
Crown of tree narrow (but not fastigiate)	B
Twigs of lowest branches stiff, many up-pointed	C
Twigs of lowest branches mostly pendent, often out-	
or up-turned at tip	D
	E
Tree male	F
Tree female	
Young twigs glabrous	G
Young twigs pubescent	H
Young leaves green almost from start	I
Young leaves bronze-coloured at first	J
Mature leaves mostly truncate to very broadly	
cuneate at base	K
Mature leaves mostly distinctly cuneate at base	L
Early leafing	M
Leafing in mid-season	N
Late leafing	O

AC(EF)GJKN	'Gelrica'
ACEGJKO	'Serotina'
ADFGILN	'Marilandica'
ADFGJKN	'Regenerata'
BCEGJLN	'Eugenei'
BCEHJKM	'Robusta'

a. 'Serotina' (<u>P.</u> <u>nigra</u> ssp. <u>nigra</u> x <u>P.</u> <u>deltoides</u>) - <u>Black-</u> **279** <u>Italian</u> <u>Poplar</u>. Grown for ornament and screening in all sorts of situations; the commonest <u>Populus</u> in such places throughout BI; arose France early 1700s.

b. 'Gelrica'. Similar to 'Serotina' but bark persistently whitish-grey until becoming fissured; grown mostly for timber in small plantations; arose Holland late 1800s.

c. 'Regenerata' (<u>P.</u> 'Marilandica' x <u>P.</u> 'Serotina') - <u>Railway</u> <u>Poplar</u>. Grown for ornament and screening, especially by railways in SE En; arose France 1814.

d. 'Robusta' (<u>P.</u> <u>deltoides</u> 'Cordata' x <u>P.</u> <u>nigra</u>

'Plantierensis'). Grown as timber in small plantations and as screening by roads, factories and canals; much of Br, especially S; arose France 1895.
e. 'Eugenei' (<u>P.</u> 'Regenerata' x <u>P.</u> <u>nigra</u> 'Italica') - <u>Carolina</u> <u>Poplar</u>. Grown in same situations as 'Robusta'; similar distribution but less common; arose France 1832.
f. 'Marilandica' (<u>P.</u> <u>nigra</u> ssp. <u>nigra</u> x <u>P.</u> 'Serotina'). 279 Planted as ornamental and for screening over much of Br, especially S; arose Europe c.1800.
Hybrid Black Poplars also form an extensive young natural population on damp rough ground at Hackney, Middlesex. The parentage is probably <u>P.</u> x <u>canadensis</u> 'Marilandica' x <u>P.</u> x <u>canadensis</u> 'Serotina'.
5 x 6. P. x canadensis 'Serotina' **x P. candicans** occurs in the Hackney hybrid population; the plants are obviously Black-Balsam poplar hybrids and the suggested parentage is the only likely one in the area.
6. P. candicans Aiton (<u>P.</u> <u>gileadensis</u> Roul.) - <u>Balm-of-</u> 279 <u>Gilead</u>. Spreading tree to 30m with many suckers and very viscid buds; leaves triangular-ovate, truncate to cordate at base, sparsely pubescent, crenate-serrate; female only. Intrd; planted for ornament in damp woods and by rivers and ponds, often natd by suckering; Br, mainly S. Unknown, probably hybrid, origin. 'Aurora', with large variegated leaves, is becoming much-grown in parks and by roads.
7. P. trichocarpa Torrey & A. Gray ex Hook. - <u>Western</u> 279 <u>Balsam-poplar</u>. Tree to 35m with narrow outline but spreading lower branches; leaves ovate to narrowly so, truncate or rounded (rarely subcordate) to broadly cuneate at base, subentire to very shallowly serrate; mostly male. Intrd; frequently planted for ornament and for timber in small plots, sometimes freely suckering; W N America.
8. P. 'Balsam Spire' ('Tacatricho 32'; <u>P.</u> <u>balsamifera</u> L. x 279 <u>P.</u> <u>trichocarpa</u>)- Hybrid Balsam-poplar. Tree to >30m with very narrow outline; similar to <u>P.</u> <u>trichocarpa</u> but see key; female only. Used for screening, as windbreak and in plantations for timber, now 1 of the most planted poplars in Br; arose N America 1930s.

2. SALIX L. - <u>Willows</u>
Trees or dwarf to tall shrubs; winter buds with 1 outer scale; flowers appearing before, with or after leaves, with entire bracts, with 1-2 nectaries; cup-like perianth 0; stamens 1-5(12).
Identification is often made difficult by the extensive degree of hybridization (59 combinations known in BI, of which 10 are hybrids between 3 spp.). The 13 hybrids treated fully here are not necessarily the most common, but those likely to be found without parents nearby. Most of these are crosses between the taller lowland species that have been planted for ornament and basket-work.
In many cases catkins and both young and mature leaves are

desirable for identification; keys based on catkins alone are
not satisfactory. Mature leaves should be collected in July-
September, before senescence; abnormally vigorous shoots
(e.g. suckers) should be avoided. In this account the so-
called catkin-scales (scales associated with each flower) are
termed bracts, which they truly are. 'Twigs' refers to those
that have completed 1-2 years' growth, not current growth.
Except locally, where plants have been clonally propagated,
the spp. are generally represented by both sexes roughly
equally, but many of the hybrids are much more commonly, or
only, female. Bisexual catkins are not rare, especially in
hybrids, inluding some of those noted as 'female only'.

General key
1 Shrubs <80(100)cm high (stems may be longer but
     not erect)   Key A
1 Tree or shrub >1m high   2
     2 Shrub <1.5(2)m high   Key B
     2 Tree or shrub >2m high   3
3 Leaves closely and finely serrate (>(3)5
     serrations/cm at leaf midpoint); bracts yellowish
     (except nos 14 & 15)   Key C
3 Leaves entire, obscurely crenate-serrate, or
     coarsely and irregularly serrate (<4(7) serrations/
     cm at leaf midpoint); bracts dark or dark-tipped
     (except no. 10)   4
     4 Leaves <3x as long as wide, or mostly so   Key D
     4 Leaves >3x as long as wide, or mostly so   Key E

Key A - Shrubs <80(100)cm high
1 Stems usually <10cm; leaves not or scarcely longer
     than wide, rounded to emarginate at apex; catkins
     appearing when leaves ± mature, all or most from
     terminal buds   2
1 Stems usually >10cm; leaves longer than wide, usually
     obtuse to acute at apex; catkins appearing before
     leaves mature, mostly from lateral buds   3
     2 Leaves entire, with distinctly impressed veins on
          upperside; petioles mostly >10mm   **35. S. reticulata**
     2 Leaves crenate-serrate; veins not impressed on
          leaf upperside; petioles mostly <4mm **34. S. herbacea**
3 Leaves densely pubescent to tomentose at least until
     maturity at least on lowerside   4
3 Leaves glabrous to sparsely pubescent at maturity on
     both sides   7
     4 Stipules mostly large and ± persistent at
          maturity   5
     4 Stipules small and mostly soon dropping off   6
5 Leaves ± entire; catkins densely silky with yellow
     hairs; Scottish Highlands only   **31. S. lanata**
5 Leaves crenate-serrate; catkins with greyish to
     whitish hairs; widespread   **25. S. aurita**

6 Leaves sericeous on lowerside; widespread
                                              29. S. repens
6 Leaves tomentose on lowerside; mountains of N Br
   only   **30. S. lapponum**
7 Leaves cordate at base   **26. S. eriocephala**
7 Leaves cuneate at base   8
   8 Stipules mostly large and persistent to maturity   9
   8 Stipules mostly small and soon dropping off   10
9 Leaves shining green on both sides, not blackening
   when dried; ovary usually pubescent   **33. S. myrsinites**
9 Leaves green on upperside, ± glaucous on lowerside,
   blackening when dried; ovary mostly ± glabrous
                                           27. S. myrsinifolia
   10 Leaves mostly <2.5 x 1.5cm   **32. S. arbuscula**
   10 Leaves mostly >2.5 x 1.5cm   **28. S. phylicifolia**

<u>Key B</u> - Shrubs 1-1.5(2)m high
1 Leaves glabrous to sparsely pubescent at maturity
   on both sides   2
1 Leaves densely pubescent to tomentose at least until
   maturity at least on lowerside   4
   2 Leaves cordate at base   **26. S. eriocephala**
   2 Leaves cuneate at base   3
3 Stipules mostly large and persistent at maturity;
   leaves blackening when dried; ovary mostly ±
   glabrous, with pedicel c.1mm; twigs usually
   pubescent   **27. S. myrsinifolia**
3 Stipules mostly small and soon dropping off;
   leaves not blackening when dried; ovary mostly
   pubescent, with 0 to very short pedicel; twigs
   usually ± glabrous   **28. S. phylicifolia**
   4 Leaves mostly >4x as long as wide, tomentose on
      lowerside; female only   **21. S. x fruticosa**
   4 Leaves <4x as long as wide, or if more then
      appressed-pubescent on lowerside   5
5 Stipules mostly large and ± persistent to
   maturity   **25. S. aurita**
5 Stipules small and mostly soon dropping off   6
   6 Leaves sericeous on lowerside; widespread
                                           29. S. repens
   6 Leaves tomentose on lowerside; mountains of N Br
      only   **30. S. lapponum**

<u>Key C</u> - Trees or shrubs >2m high; leaves closely and finely
      serrate; bracts yellowish (except dark-tipped in <u>S.
      daphnoides</u> and <u>S. acutifolia</u>)
1 Twigs dark purple, with conspicuous whitish bloom
   easily rubbed off; bracts dark-tipped   2
1 Twigs variously coloured, rarely dark purple, without
   bloom; bracts yellowish   3
   2 Leaves >5x as long as wide; ultimate branchlets
      ± pendent   **15. S. acutifolia**

    2 Leaves usually <5x as long as wide; branchlets not
       pendent   **14. S. daphnoides**
 3 Branchlets pendent (weeping willows)   4
 3 Branchlets erect or spreading   5
    4 Leaves pubescent to sericeous ± until maturity;
       ovary little longer than subtending bract
                                                  8. S. x sepulcralis
    4 Leaves glabrous to sparsely pubescent even when
       very young; ovary much longer than subtending
       bract   **6. S. x pendulina**
 5 Leaves elliptic to ovate, 2-4x as long as wide;
    stamens >4   **1. S. pentandra**
 5 Leaves narrowly elliptic to lanceolate, >(3)4x as long
    as wide; stamens <4   6
    6 Stipules mostly large and ± persistent to
       maturity; bark smooth, flaking off in large
       patches or peels   7
    6 Stipules mostly 0 or small and soon dropping off;
       bark furrowed, not flaking off   8
 7 Young leaves and twigs glabrous; twigs with strong
    ridges or angles   **9. S. triandra**
 7 Young leaves and twigs pubescent; twigs terete or
    obscurely ridged or angled   **10. S. x mollissima**
    8 Mature leaves sericeous, or ± glabrous and dull
       on upperside   9
    8 Mature leaves glabrous and rather glossy on
       upperside   10
 9 Leaves subglabrous on both sides at maturity
                                                   5. S. x rubens
 9 Leaves ± sericeous at least on lowerside at maturity
                                                       7. S. alba
    10 Twigs and leaves glabrous from the first   11
    10 Twigs and leaves sparsely pubescent or ciliate at
       first   12
 11 Leaves mostly with 3-6 serrations/cm; twigs very
    fragile at branches; buds shiny; male flowers with
    2-3 stamens   **4. S. fragilis**
 11 Leaves mostly with 6-10 serrations/cm; twigs scarcely
    fragile at branches; buds scarcely shiny; male
    flowers mostly with 3-4 stamens   **2. S. x meyeriana**
    12 Twigs yellowish-orange to reddish   **5. S. x rubens**
    12 Twigs brownish   13
 13 Leaves mostly with 6-10 serrations/cm; male flowers
    mostly with 3-4 stamens; male only **3. S. x ehrhartiana**
 13 Leaves mostly with 3-6 serrations/cm; male flowers
    with 2-3 stamens   **4. S. fragilis**

<u>Key D</u> - Trees or shrubs >2m high; leaves entire, obscurely
          crenate-serrate, or coarsely and irregularly serrate,
          <3x as long as wide; bracts dark or dark-tipped
 1 Catkins developing with leaves, usually with small
    leaf-like bracts on peduncle   2

1 Catkins developing before leaves on bare twigs, with
 reduced or 0 bracts on peduncle   3
 2 Stipules mostly large and persistent at maturity;
  leaves blackening when dried; ovary mostly ±
  glabrous, with pedicel c.1mm; twigs usually
  pubescent   **27. S. myrsinifolia**
 2 Stipules mostly small and soon dropping off; leaves
  not blackening when dried; ovary mostly pubescent,
  with 0 or very short pedicel; twigs usually ±
  glabrous   **28. S. phylicifolia**
3 Wood of twigs without ridges under bark   4
3 Wood of twigs with longitudinal ridges under bark 5
 4 Leaves glabrous, >2x as long as wide **11. S. purpurea**
 4 Leaves pubescent on lowerside, ≤2x as long as
  wide   **22. S. caprea**
5 Leaves very sparsely pubescent on lowerside (mostly
 on veins) at maturity; stipules small, soon falling
 off   **24. S. x laurina**
5 Leaves distinctly pubescent on lowerside at maturity;
 stipules large, persistent to maturity   6
 6 Leaves mostly >8cm; catkins mostly >3cm; female
  only   **18. S. x calodendron**
 6 Leaves mostly <8cm; catkins mostly <3cm 7
7 Leaves distinctly rugose, undulate at margin, without
 rust-coloured hairs on lowerside   **25. S. aurita**
7 Leaves scarcely rugose or undulate, usually with some
 rust-coloured hairs on lowerside   **23. S. cinerea**

Key **E** - Trees or shrubs >2m high; leaves entire, obscurely
  crenate-serrate, or coarsely and irregularly serrate,
  >3x as long as wide; bracts dark or dark-tipped
  (except yellowish in S. x mollissima)
1 Leaves glabrous to sparsely pubescent on lowerside
 at maturity   2
1 Mature leaves pubescent to tomentose on lowerside 5
 2 Leaves glabrous even when young, commonly opposite
  or subopposite; male flowers apparently with 1
  stamen   **11. S. purpurea**
 2 Leaves pubescent when young, alternate; male
  flowers with 2 free or only partially fused
  stamens   3
3 Leaves glabrous at maturity   **13. S. x forbyana**
3 Leaves sparsely pubescent on lowerside at maturity 4
 4 Catkins appearing with leaves; bracts yellowish;
  male flowers with 2-3 free stamens
         10. S. x mollissima
 4 Catkins appearing before leaves; bracts dark-
  tipped; male flowers with 2 free or partly united
  stamens   **12. S. x rubra**
5 Leaves linear or nearly so, mostly >6x as long as
 wide   **16. S. viminalis**
5 Leaves lanceolate to narrowly elliptic, mostly <6x as

long as wide   6
6 Wood of twigs without ridges under bark
                                                       17. S. x sericans
6 Wood of twigs with longitudinal ridges under bark   7
7 Leaves pubescent on lowerside at maturity, but not
   densely so or soft to touch   **20. S. x smithiana**
7 Leaves densely pubescent, velvety or silky to touch   8
8 Most leaves <4x as long as wide, not undulate at
   margin   **18. S. x calodendron**
8 Most leaves >4x as long as wide, strongly to
   slightly (or not) undulate at margin   9
9 Leaves usually slightly undulate at margin, velvety-
   tomentose on lowerside   **19. S. x stipularis**
9 Leaves usually strongly undulate at margin, pubescent
   with appressed or crisped hairs on lowerside
                                            21. S. x fruticosa

Other spp. - Several spp. are cultivated for ornament and
may rarely appear as isolated plants in the wild.   **S.
elaeagnos** Scop. (Olive Willow), from Europe, occurs as
single bushes in W Sussex and W Norfolk; it resembles S.
viminalis but has very narrow leaves tomentose (not
sericeous) on lowerside and glabrous (not pubescent) ovaries.
S. udensis Trautv. & Meyer (S. sachalinensis Schmidt), from E
Asia, has become ± natd in a woodland bog in Dunbarton where
it was thrown out in the 1960s; it resembles S. viminalis but
has leaves 6-15cm, c.6x as long as wide, shiny on upperside
and glabrous to sparsely pubescent on lowerside, and glabrous
twigs.   It differs from S. x forbyana in its longer leaves,
and it is male.   **S. babylonica** L., from China, the original
Weeping Willow, is rarely grown in BI but much mis-recorded
for its hybrids with S. alba and S. fragilis.

Subgenus 1 - SALIX (spp. 1-10). Trees or tall shrubs; leaves
finely and closely serrate, mostly lanceolate to narrowly
elliptic, acuminate; catkins long and slender, borne on leafy
peduncles arising from lateral buds; bracts yellowish;
stamens 2-8(12), free; male flowers with 2 nectaries, female
with 1-2; ovary glabrous except in some hybrids.

1. S. pentandra L. - Bay Willow.   Tree to 10(18)m; twigs **290**
reddish-brown, glossy, glabrous; leaves 5-12 x 2-5cm,
glabrous; stamens (4)5-8(12).   Native; wet ground and by
ponds and streams; N & C Br and Ir, planted and sometimes ±
wild further S.
2. S. x meyeriana Rostkov ex Willd. (S. pentandra x S. **290**
fragilis) - Shiny-leaved Willow. Tree to 15m; twigs brown,
glossy, glabrous; leaves 5-12 x 1.5-4cm, glabrous; stamens
(2)3-4(5).   Native; damp areas and by rivers; frequent in N
En, N Wa and N Ir but there and also outside range of S.
pentandra often of cultivated origin.
3. S. x ehrhartiana Smith (S. pentandra x S. alba) - **290**

Ehrhart's Willow. Tree to 15(25)m; twigs brownish, slightly glossy, glabrous; leaves 6-10 x 1.3-2.5cm, appressed-pubescent at first, becoming glabrous; stamens (2)3-4(5); male only. Probably intrd; planted and often ± wild; scattered in En, especially SE; garden origin but native in Europe.

4. S. fragilis L. (S. decipiens Hoffm.) - Crack-willow. 290 Tree to 25m; twigs very brittle at branches, often sparsely pubescent at first but glabrous later, ± glossy, pale brown; leaves 9-15 x 1.5-3(4)cm, at first sparsely pubescent, soon glabrous, more coarsely serrate than other taxa in subg. Salix; stamens 2(-3). Native; common in damp places over most of lowland BI, but often planted. Variable; the commonest variant (var. **russelliana** (Smith) Koch - Bedford Willow) is a female tree with long narrow leaves with rather uneven teeth; var. **furcata** Ser. ex Gaudin is a male tree with forked catkins and rather wide leaves; both presumably of garden origin. Native plants are var. **fragilis**, with leaves at first sparsely pubescent, and var. **decipiens** (Hoffm.) Koch, with leaves wholly glabrous; both are of both sexes.

4 x 9. S. fragilis x S. triandra = S. x alopecuroides Tausch (S. x speciosa Host) occurs very sparsely in Br and Ir; it is intermediate, with 2-3 stamens; female unknown.

5. S. x rubens Schrank (S. x basfordiana Scaling ex J. 290 Salter; S. alba x S. fragilis) - Hybrid Crack-willow. Tree to 30m; represented by whole range of variants and cultivars linking S. fragilis with S. alba var. caerulea (nothovar. **rubens**) and with S. alba var. vitellina (nothovar. **basfordiana** (Scaling ex J. Salter) Meikle); the latter have yellowish-orange twigs. Native; frequent over most of lowland BI but perhaps most often of cultivated origin.

6. S. x pendulina Wender. (S. x blanda Andersson; S. 290 fragilis x S. babylonica) - Weeping Crack-willow. Tree to 15m with distinctive weeping habit; twigs pale brown, glabrous, somewhat brittle at branches, glossy; leaves 10-12 x 1.5-2cm, glabrous almost from first; catkins female or bisexual. Intrd; frequently planted and found in wild in BI, but less common than S. x sepulcralis.

7. S. alba L. - White Willow. Tree to 33m; twigs pubescent 290 when young, becoming glabrous and ± glossy, brown, or yellow-ish to reddish (var. **vitellina** (L.) Stokes (ssp. vitellina (L.) Arcang.) - Golden Willow); leaves 5-10(12) x 0.5-1.5cm, densely whitish-pubescent at first, becoming sparsely so at maturity, or 6-12 x 1.5-2.5cm and becoming ± glabrous on upperside at maturity (var. **caerulea** (Smith) Dumort. (ssp. caerulea (Smith) Rech. f.) - Cricket-bat Willow); stamens 2. Native; marshes, wet hollows and by streams and ponds; common over most of lowland BI, rarer in W and N Sc, but often planted. S. alba 'Britzensis' is a much planted cultivar of var. vitellina with bright orange-red twigs.

8. S. x sepulcralis Simonkai (S. x chrysocoma Dode; S. alba 290 x S. babylonica) - Weeping Willow. Tree to 22m with distinctive weeping habit; twigs sparsely pubescent at first, soon glabrous and ± glossy, brownish, or yellowish (nothovar. **chrysocoma** (Dode) Meikle); leaves 7-12 x 0.7-1.8cm, sparsely pubescent at first, soon ± glabrous; stamens 2; catkins often bisexual. Intrd; much planted, often in masses and often persisting in wild places; lowland BI; garden origin.

9. S. triandra L. - Almond Willow. Small tree or shrub to 290 10m; twigs glabrous, pale brown, glossy, ridged; leaves (2)4-11(15) x 1-3(4)cm, glabrous; stamens 3. Native; damp places; frequent in C & S En, much less so in Wa, N En, Sc and Ir and probably always intrd; much planted for basketry.

9 x 11. S. triandra x S. purpurea (= S. x leiophylla auct. non Camus & A. Camus) is known only as a male from an osier-bed in Notts.

10. S. x mollissima Hoffm. ex Elwert (S. triandra x S. 290 viminalis) - Sharp-stipuled Willow. Shrub to 5m; twigs glabrous to glabrescent, brownish, dull; leaves 8-13 x 1-1.5cm; stamens 2-3. Possibly native; damp places; scattered over Br and Ir, frequent in C & S En but very local elsewhere. Nothovar. **hippophaifolia** (Thuill.) Wimmer is closer to S. viminalis, with subentire or obscurely serrate leaves sparsely pubescent on lowerside, and with catkins <3.5cm (both sexes); nothovar. **undulata** (Ehrh.) Wimmer is closer to S. triandra, with closely serrate ± glabrous leaves and longer narrower catkins (female only).

Subgenus 2. - VETRIX Dumort. (subg. Vimen Dumort., subg. Caprisalix Dumort.) (spp. 11-32). Shrubs or small trees; leaves mostly entire to coarsely and irregularly serrate, orbicular to linear; catkins short and wide, sessile or borne on leafy peduncles, arising from lateral buds; bracts dark or with dark tips; stamens 2, free or variously fused, sometimes appearing 1; nectary 1; ovary usually pubescent.

11. S. purpurea L. - Purple Willow. Shrub 1.5-5m; twigs 290 glabrous, yellowish to purplish-brown, ± glossy; leaves often opposite to subopposite, (2)3-8(10) x 0.5-3cm, glabrous, or pubescent at first, entire or obscurely serrate; stamen apparently 1. Native; damp places; scattered throughout BI, locally common, often planted.

11 x 16 x 22 x 23. S. purpurea x S. viminalis x S. caprea x S. cinerea = S. x taylorii Rech. f. has been reported, with some doubt, from Angus; endemic.

11 x 23. S. purpurea x S. cinerea = S. x pontederiana Willd. (S. x sordida A. Kerner) is very scattered in Br.

11 x 23 x 25. S. purpurea x S. cinerea x S. aurita = S. x confinis Camus & A. Camus has been recorded from Perths.

11 x 25. S. purpurea x S. aurita = S. x dichroa Doell is known from Cheviot.

11 x 25 x 28. S. purpurea x S. aurita x S. phylicifolia = S. x sesquitertia F.B. White has been found in Dumfriess; endemic.

11 x 27. S. purpurea x S. myrsinifolia = S. x beckiana Beck occurs in MW Yorks and S Northumb.

11 x 28. S. purpurea x S. phylicifolia = S. x secernata F.B. White is known from S Northumb, Dumfriess and Dunbarton; endemic.

11 x 29. S. purpurea x S. repens = S. x doniana G. Anderson ex Smith has been found in S Lancs and M Perth.

12. S. x rubra Hudson (S. purpurea x S. viminalis) - Green- 290 leaved Willow. Shrub or small tree to 7m; twigs glabrous when mature, yellowish-brown, ± glossy; leaves 4-12(15) x 0.8-1(1.5)cm, at first densely pubescent, becoming glabrous or remaining pubescent, ± entire to remotely serrate; stamens free or partly united. Native; wet places, often with parents, but also relic of cultivation; throughout most of Br and Ir.

13. S. x forbyana Smith (S. purpurea x S. viminalis x S. 290 cinerea) - Fine Osier. Erect shrub to 5m; twigs glabrous, yellowish, glossy; leaves 3-12 x 0.8-2.5cm, at first pubescent, becoming glabrous, obscurely serrate; stamens free or partly united (male very rare). Probably intrd; wet places, often as relic of osier cultivation; scattered in En, Wa, S Sc and N Ir; garden origin.

14. S. daphnoides Villars - European Violet-willow. Tall 290 shrub or slender tree to 10(18)m; twigs glabrous, violet-brown, glossy but with dense whitish bloom, not pendent; leaves (4)7-12(14) x (1)2-3(4)cm, soon glabrous, closely and finely serrate; stamens free. Intrd; widely planted (both sexes) and sometimes found in wild places; very sparsely scattered in Br; Scandinavia.

15. S. acutifolia Willd. (S. daphnoides ssp. acutifolia 290 (Willd.) Zahn) - Siberian Violet-willow. Very similar to S. daphnoides but leaves 6-16 x 1-2cm; twigs very slender, ± pendent at ends. Intrd; widely planted (male only) and found in similar places to S. daphnoides; U.S.S.R.

16. S. viminalis L. - Osier. Erect shrub or small tree to 291 6(-10)m; twigs pubescent at first, becoming glabrous, rather dull, yellowish-brown to brown; leaves 10-15(18) x 0.5-1.5(2.5)cm, densely sericeous on lowerside, dull and sparsely pubescent on upperside, with ± entire ± revolute margins; stamens free. Native; mostly damp places; common in lowland BI, but perhaps not native in N & W.

16 x 23 x 29. S. viminalis x S. cinerea x S. repens = S. x angusensis Rech. f. occurs on sand-dunes in Angus; endemic.

16 x 29. S. viminalis x S. repens = S. x friesiana Andersson occurs in S Lancs and Sutherland.

17. S. x sericans Tausch ex A. Kerner (S. x laurina auct. 291 non Smith; S. viminalis x S. caprea) - Broad-leaved Osier. Shrub or small tree to 9m; twigs at first pubescent, soon ± glabrous, ± glossy, yellowish to reddish; leaves 6-12 x 1.3-3cm, densely grey-pubescent on lowerside, glabrescent on upperside, subentire to remotely serrate; stamens free. Native; often with parents but also relic of cultivation;

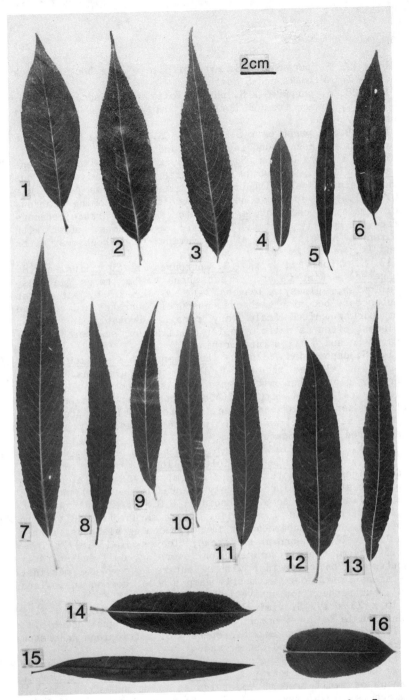

FIG 290 – Leaves of **Salix**. 1, S. pentandra. 2, S. **x** meyeriana. 3, S. **x** ehrhartiana. 4, S. purpurea. 5, S. **x** rubra. 6, S. **x** forbyana. 7, S. fragilis. 8, S. **x** rubens. 9, S. **x** pendulina. 10, S. alba. 11, S. **x** sepulcralis. 12, S. daphnoides. 13, S. acutifolia. 14, S. triandra. 15, S. **x** mollissima. 16, S. eriocephala.

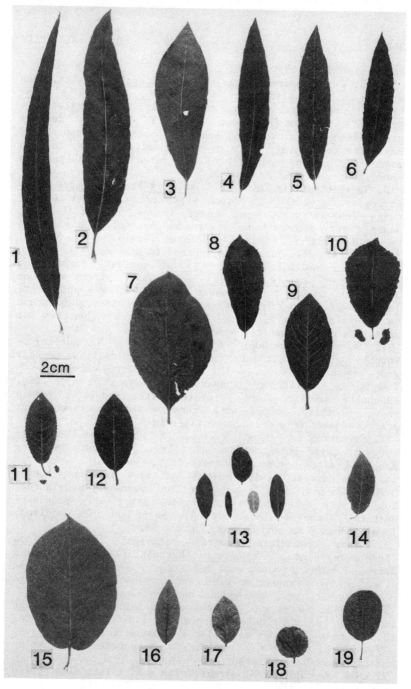

FIG 291 – Leaves of *Salix*. 1, *S. viminalis*. 2, *S. x sericans*. 3, *S. x calodendron*. 4, *S. x stipularis*. 5, *S. x smithiana*. 6, *S. x fruticosa*. 7, *S. caprea*. 8, *S. cinerea*. 9, *S. x laurina*. 10, *S. aurita*. 11, *S. myrsinifolia*. 12, *S. phylicifolia*. 13, *S. repens*. 14, *S. lapponum*. 15, *S. lanata*. 16, *S. arbuscula*. 17, *S. myrsinites*. 18, *S. herbacea*. 19, *S. reticulata*.

common in most of BI.

18. S. x calodendron Wimmer (_S._ x <u>dasyclados</u> auct. non 291
Wimmer; _S._ <u>viminalis</u> x _S._ <u>caprea</u> x _S._ <u>cinerea</u>) - <u>Holme
Willow</u>. Erect shrub or small tree to 12m; twigs persistently
densely pubescent; leaves 7-18 x 2.5-5cm, densely
grey-pubescent on lowerside, sparsely pubescent on upperside,
subentire to remotely serrate; female only. Native; in damp
places not always near any of its parents, probably often
intrd; frequent over most of BI.

19. S. x stipularis Smith (<u>S.</u> <u>viminalis</u> x _S._ <u>caprea</u> x <u>S.</u> 291
<u>aurita</u>) - <u>Eared</u> <u>Osier</u>. Erect shrub or small tree to 10m;
twigs densely greyish-pubescent; leaves 6-11 x 1-2.5cm,
densely grey-pubescent on lowerside, sparsely pubescent on
upperside, entire, often ± undulate; female only. Native;
damp places, often planted; scattered in N Br and N Ir.

20. S. x smithiana Willd. (<u>S.</u> x <u>geminata</u> Forbes; <u>S.</u> 291
<u>viminalis</u> x _S._ <u>cinerea</u>) - <u>Silky-leaved</u> <u>Osier</u>. Erect shrub or
small tree to 9m; twigs densely and persistently pubescent;
leaves 6-11 x 0.8-2.5cm, pubescent on lowerside, glabrescent
on upperside; stamens free. Native; often with parents but
sometimes not; common throughout most of BI.

21. S. x fruticosa Doell (<u>S.</u> <u>viminalis</u> x _S._ <u>aurita</u>) - 291
<u>Shrubby</u> <u>Osier</u>. Erect shrub to 5m; twigs sparsely to densely
grey-pubescent; leaves 4-10 x 0.7-2cm, grey-pubescent on
lowerside, subglabrous to sparsely pubescent on upperside,
subentire or irregularly crenate, often undulate; female
only. Native; often with parents but sometimes not;
scattered over Br and Ir, especially in N & W.

22. S. caprea L. - <u>Goat</u> <u>Willow</u>. Shrub or small tree to 291
10(19)m; twigs pubescent to sparsely so at first, becoming
glabrous, glossy or dull; stamens free. Native.

a. Ssp. caprea. Twigs quickly becoming glabrous; leaves
5-12 x 2.5-8cm, densely pubescent on lowerside, subglabrous
to sparsely pubescent on upperside, irregularly undulate-
serrate. Damp and rough ground, hedges and open woodland;
locally common to abundant throughout BI.

b. Ssp. sphacelata (Smith) Macreight (ssp. <u>sericea</u>
(Andersson) Flod., <u>S.</u> <u>coaetanea</u> (Hartman) Flod.). Twigs and
leaves remaining pubescent for longer; leaves 3-7 x 1.5-
4.5cm, densely appressed-pubescent on lowerside, entire to
obscurely serrate. Damp ground on mountains; Sc.

22 x 23. S. caprea x S. cinerea = S. x reichardtii A. Kerner
is common and variable, completely linking the parents,
through most of BI, in many places commoner than <u>S.</u> <u>caprea</u>.

22 x 25. S. caprea x S. aurita = S. x capreola J. Kerner ex
Andersson occurs scattered over Br and Ir.

22 x 27. S. caprea x S. myrsinifolia = S. x latifolia Forbes
occurs in N Br within the range of <u>S.</u> <u>myrsinifolia</u>.

22 x 28. S. caprea x S. phylicifolia occurs with the parents
in C & N Sc.

22 x 29. S. caprea x S. repens = S. x laschiana Zahn occurs
with the parents in Sc.

22 x 30. S. caprea x S. lapponum = S. x laestadiana Hartman occurs with the parents in E Perth and Angus.

22 x 33. S. caprea x S. myrsinites = S. x lintonii Camus & A. Camus occurs with the parents in Perths and Angus; endemic.

23. S. cinerea L. - <u>Grey Willow</u>. Shrub or small tree to **291** 6(15)m; twigs densely pubescent at first, often becoming glabrous or nearly so, dark reddish-brown, dull; leaves 2-9(16) x 1-3(5)cm, pubescent to densely so on lowerside, subglabrous or sparsely pubescent on upperside, subentire to ± undulate-serrate; stamens free. Native.

a. Ssp. cinerea. Twigs usually ± persistently pubescent; leaves mostly obovate or oblong, dull and ± pubescent on upperside, densely grey-pubescent on lowerside. Marshes and fens at low altitudes; the predominant ssp. in E Anglia and Lincs, but also very scattered elsewhere in Br and Ir.

b. Ssp. oleifolia Macreight (ssp. <u>atrocinerea</u> (Brot.) Silva & Sobrinho, <u>S. atrocinerea</u> Brot.). Twigs usually glabrescent; leaves mostly narrowly obovate or oblong, slightly glossy and usually nearly glabrous on upperside, grey-pubescent but with some stiff rust-coloured hairs on lowerside. Wet places, woods and marginal habitats, lowland and upland; commonest <u>Salix</u> throughout lowland BI except where replaced by ssp. <u>cinerea</u>.

23 x 25. S. cinerea x S. aurita = S. x multinervis Doell is scattered through most of BI on acid soils.

23 x 25 x 27. S. cinerea x S. aurita x S. myrsinifolia = S. x forbesiana Druce has been found in Dumfriess and W Perth.

23 x 27. S. cinerea x S. myrsinifolia (= <u>S.</u> x <u>strepida</u> Forbes non Schleicher) occurs with the parents in N BI.

23 x 29. S. cinerea x S. repens = S. x subsericea Doell occurs rather rarely with the parents in C & N Br.

24. S. x laurina Smith (<u>S.</u> x <u>wardiana</u> Leefe ex F.B. White; **291** <u>S. cinerea</u> x <u>S. phylicifolia</u>) - <u>Laurel-leaved Willow</u>. Erect shrub to 6m; twigs ± glabrous, brown, glossy; leaves 4-10 x 1.8-3.5cm, glabrous or subglabrous on upperside, sparsely pubescent on lowerside, entire to very shortly serrate; female only. Native; frequent with the parents in N BI, but also scattered elsewhere probably of garden origin.

25. S. aurita L. - <u>Eared Willow</u>. Shrub to 2(3)m; twigs **291** pubescent at first, becoming ± glabrous, dark reddish-brown, dull; leaves 2-4(6) x 1-3(4)cm, rugose, grey-pubescent on lowerside, subglabrous to sparsely pubescent on upperside, undulate and obscurely serrate, with twisted apex; stamens free. Native; heathland, scrub and rocky hills on acid soils; frequent to abundant in suitable places throughout BI.

25 x 27. S. aurita x S. myrsinifolia = S. x coriacea Forbes occurs with the parents in N BI.

25 x 27 x 28. S. aurita x S. myrsinifolia x S. phylicifolia = S. x saxetana F.B. White occurs with parents in S & C Sc.

25 x 28. S. aurita x S. phylicifolia = S. x ludificans F.B. White occurs with the parents in N Br.

25 x 29. S. aurita x S. repens = S. x ambigua Ehrh. occurs commonly throughout BI wherever the parents meet.

25 x 29 x 34. S. aurita x S. repens x S. herbacea = S. x grahamii Borrer ex Baker has been found in W Sutherland and W Donegal.

25 x 30. S. aurita x S. lapponum = S. x obtusifolia Willd. occurs with the parents in C & N Sc.

25 x 34. S. aurita x S. herbacea = S. x margarita F.B. White occurs with the parents in C & N Sc.

26. S. eriocephala Michaux (S. rigida Muhlenb., S. cordata 290 Muhlenb. non Michaux) - Heart-leaved Willow. Rhizomatous shrub to 2m; twigs very brittle at branches, glabrous, greenish-brown; leaves 5-12 x 1.5-4cm, glabrous, finely serrate, strongly cordate at base. Intrd; wet bog; well natd female patch in Warks; N America.

27. S. myrsinifolia Salisb. (S. nigricans Smith) - Dark- 291 leaved Willow. Shrub to 4m, often much less; twigs densely R pubescent at first, slowly becoming subglabrous, dull, brown or greenish; leaves 2-6.5 x 1.5-3.5cm, sparsely pubescent at first, becoming subglabrous, irregularly crenate-serrate; stamens free. Native; by ponds and streams and in damp rocky places; frequent in N En and Sc, formerly N Ir, scattered intrd plants in C & S Br, but perhaps native in CE En.

27 x 28. S. myrsinifolia x S. phylicifolia = S. x tetrapla Walker (S. x tenuifolia Smith) occurs commonly in N Br and often completely links the parents wherever they meet.

27 x 28 x 33. S. myrsinifolia x S. phylicifolia x S. myrsinites occurs with the parents in CE Sc.

27 x 29. S. myrsinifolia x S. repens = S. x felina Buser ex Camus & A. Camus occurs with the parents in E Perth.

27 x 33. S. myrsinifolia x S. myrsinites = S. x punctata Wahlenb. occurs with the parents in C Sc.

27 x 34. S. myrsinifolia x S. herbacea = S. x semireticulata F.B. White occurs with the parents in M Perth; endemic.

28. S. phylicifolia L. (S. hibernica Rech. f.) - Tea-leaved 291 Willow. Shrub to 4(5)m; twigs subglabrous to sparsely pubescent at first, soon ± glabrous, reddish-brown ± glossy; leaves 2-6 x 1-5cm, often sparsely pubescent at first, glabrous at maturity, entire to irregularly serrate; stamens free. Native; similar places to S. myrsinifolia; frequent in N En and Sc, and in Leitrim and Sligo.

28 x 29. S. phylicifolia x S. repens (= S. x schraderiana auct. non Willd.) has been found in Angus and doubtfully elsewhere in Sc; endemic.

28 x 30. S. phylicifolia x S. lapponum = S. x gillotii Camus & A. Camus has been recorded, with some doubt, from M Perth and Angus.

28 x 33. S. phylicifolia x S. myrsinites = S. x notha Andersson occurs with the parents in M Perth and Angus.

29. S. repens L. - Creeping Willow. Procumbent to erect 291 shrub to 1.5(2)m, but usually <1m; twigs glabrous to densely sericeous and variously coloured; leaves 1-3.5 x 0.4-2.5cm, glabrous to pubescent or silvery-sericeous on both sides, but usually ± pubescent on lowerside, entire to obscurely crenate-serrate; stamens free. Native; acid heaths and moors both dry and wet, fens, dunes and dune-slacks; in suitable places throughout BI. Extremely variable, especially in leaf-shape. The commonest variant (var. **repens**) is rhizomatous, with procumbent to decumbent soon subglabrous stems and small leaves soon subglabrous on upperside; it occurs mostly on heaths and moors. Var. **argentea** (Smith) Wimmer & Grab. (ssp. _argentea_ (Smith) Neumann ex Rech. f.) occurs mostly in dune-slacks and differs in its taller, strong habit with sericeous stems and larger leaves persistently densely sericeous on both sides. Var. **fusca** Wimmer & Grab. is shortly rhizomatous with ± erect stems but leaves as in var. repens, and occurs in fens in E Anglia. Numerous intermediates occur.

29 x 30. S. repens x S. lapponum = S. x pithoensis Rouy occurs with the parents in E Perth.

29 x 34. S. repens x S. herbacea = S. x cernua E.F. Linton occurs with the parents in C & N Sc.

30. S. lapponum L. - Downy Willow. Low shrub to 1(1.5)m; 291 twigs pubescent at first, glabrous and rather glossy dark   R reddish-brown later; leaves 1.5-7 x 1-2.5cm, subglabrous to pubescent on upperside, usually densely pubescent on lowerside, entire or subentire; stamens free. Native; rocky mountain slopes and cliffs; N & C mainland Sc, isolated localities in Westmorland and Dumfriess.

30 x 32. S. lapponum x S. arbuscula = S. x pseudospuria Rouy occurs with the parents in C & N Sc.

30 x 34. S. lapponum x S. herbacea = S. x sobrina F.B. White occurs wih the parents in CE Sc.

30 x 35. S. lapponum x S. reticulata = S. x boydii E.F. Linton occurs with the parents in Angus; endemic.

31. S. lanata L. - Woolly Willow. Low shrub to 1m; twigs 291 pubescent at first, soon glabrous ± glossy, brown; leaves RR 3.5-7 x 3-6.5cm, densely white-tomentose at first, becoming subglabrous with age, entire or subentire; stamens free. Native; damp mountain rock-ledges; local, highlands of C Sc.

31 x 34. S. lanata x S. herbacea = S. x sadleri Syme occurs with the parents in CE Sc.

32. S. arbuscula L. - Mountain Willow. Low shrub to 80cm; 291 twigs sparsely pubescent at first, soon glabrous, reddish-   R brown, ± glossy; leaves 1.5-3(5) x 1-1.5(3)cm, glabrous on upperside, densely appressed-pubescent on lowerside at first, becoming ± glabrous, crenate-serrate; stamens free. Native; damp rocky slopes and mountain ledges; locally abundant on mountains in S & C Sc.

32 x 34. S. arbuscula x S. herbacea = S. x simulatrix F.B. White occurs with the parents in M Perth and Argyll.

32 x 35. S. arbuscula x S. reticulata = S. x ganderi Huter ex Zahn occurs with the parents in M Perth.

Subgenus 3 - CHAMAETIA Dumort. (spp. 33-35). Dwarf shrubs; leaves entire to coarsely serrate, ovate to orbicular; catkins short and wide, on leafless peduncles, mostly arising from terminal buds; bracts pale or dark; stamens 2, free; nectaries 1-2, often lobed; ovary glabrous or pubescent.

33. S. myrsinites L. - Whortle-leaved Willow. Low spreading **291** shrub <50cm; twigs sparsely pubescent at first, soon becoming **R** glabrous, ± glossy, reddish-brown; leaves 1.5-7 x 0.5-2.5(3)cm, sparsely pubescent at first, soon glabrous, glossy on upperside, crenate-serrate; bracts dark brown; nectary 1; ovary densely pubescent with iridescent hairs. Native; rocky ledges and slopes on mountains; local in N & C Sc. Possibly better placed in subg. Vetrix.

34. S. herbacea L. - Dwarf Willow. Dwarf shrub <10cm; twigs **291** very sparsely pubescent at first, soon glabrous, glossy, reddish-brown; leaves 0.3-2 x 0.3-2cm, sparsely pubescent at first, soon glabrous, ± glossy, rounded to emarginate at apex, crenate-serrate; bracts yellowish or red-tinged; nectaries 1-2, usually lobed; ovary glabrous. Native; rock-ledges and rocky mountain-tops; locally common in Sc, N En and parts of Wa and Ir.

35. S. reticulata L. - Net-leaved Willow. Dwarf shrub **291** <20cm; twigs sparsely pubescent at first, soon glabrous, dark **R** reddish-brown; leaves 1.2-4(5) x 1-2.5(3.5)cm, densely pubescent at first, becoming glabrous at least on upperside, rounded at apex, entire; bracts purplish-brown, suborbicular; nectaries usually in form of lobed cup; ovary densely pubescent. Native; wet rock-ledges and mountain-slopes; very local in C & N mainland Sc.

62A. CAPPARACEAE

CLEOME hassleriana Chodat (Spiderflower), from S America, is an erect annual to 1m with palmate leaves often prickly on petioles and with 5-7 leaflets, showy scented flowers with 4 white or pink petals arranged in a bracteate raceme, and a narrow capsule <12cm x 5mm. It is grown as a bedding plant and sometimes occurs as a relic on tips in S Br.

63. BRASSICACEAE - Cabbage family
(Cruciferae)

Herbaceous annuals to perennials, rarely dwarf shrubs; leaves alternate, simple to pinnate, petiolate or sessile, exstipulate. Flowers usually in racemes, usually much elongating after flowering, sometimes in panicles, normally bisexual, hypogynous, usually actinomorphic; sepals 4, free; petals 4, free, rarely 0; stamens usually 6 (2 shorter outer, 4 longer inner), sometimes 4 by loss of 2 outer; ovary usually 2-celled, each with 1-many ovules; style 1; stigma 2, often bilobed, usually capitate; fruit dry, usually opening from below by 2 valves, sometimes breaking into segments transversely, sometimes indehiscent.

Easily recognized by the distinctive flowers with 4 sepals and 4 petals in decussate pairs, the (4-)6 stamens, and the characteristic fruits.

It is difficult to construct keys without both ripe fruits and a knowledge of petal colour. Caution is needed when assessing flower colour of dried specimens, as sometimes yellow can fade to white and white discolour to yellow; the keys allow for this in most but perhaps not all cases. Hairtype is also very important, and a strong lens is often needed; both leaves and stems should be examined, as hairtype can differ on these 2 organs (branched hairs are commoner on leaves). Plants that are sterile for some reason (hybridity, lack of suitable pollen, genetic mutation) are often impossible to key out, or will key out wrongly due to mis-shapen fruits or fewer than normal seeds; these possibilities must always be borne in mind. In this work 9 groups of distinctive genera are first defined; some of these rarely produce ripe fruits and would not be identifiable with the main keys which, however, do cover all the genera. Fruit lengths include beak, style and stigma unless otherwise stated.

Distinctive genera
Fresh plants smelling of garlic when crushed
                                  3. ALLIARIA, 30. THLASPI
Flowers distinctly zygomorphic, 2 petals on 1 side
  much shorter than 2 on other   **29. TEESDALIA, 31. IBERIS**
Erect plants with white scaly rhizome, pinnate basal
  leaves, pink flowers, and purple bulbils in leaf-axils
                                              14. CARDAMINE
Robust plants with strong tap-root, large dock-like leaves
  sometimes partly pinnately divided, and small white
  flowers in large panicles   **13. ARMORACIA**
Small aquatic, often submerged, plants with leaves all
  subulate and in basal rosette; inflorescence with few
  small flowers or short swollen fruit   **34. SUBULARIA**
Low straggly annuals with pinnate leaves, small
  inflorescences borne opposite the leaves, and bilobed
  2-seeded fruits   **33. CORONOPUS**
Erect plants with small yellow flowers in panicles, and
  pendent, winged, indehiscent, 1-seeded fruits **5. ISATIS**
Petals 0   **14. CARDAMINE, 26. CAPSELLA, 32. LEPIDIUM,**
                   33. CORONOPUS, 34. SUBULARIA
Petals bifid >1/3 way to base **19. BERTEROA, 22. EROPHILA**
Stems woody for most of length **16. AUBRIETA, 18. ALYSSUM,**
                                              31. IBERIS

General key
1 Fruit splitting longitudinally by 2 valves to release
  1-many seeds on each side   2
1 Fruit indehiscent, or splitting transversely, or
  splitting longitudinally but not releasing seeds
  separately   Key A
  2 Fruit >3x as long as wide (siliqua)   3
  2 Fruit <3x as long as wide (silicula)   4
3 Petals pale yellow to golden   Key B
3 Petals white, pink, purple, mauve or reddish, or
  rarely 0   Key C
  4 Fruit distinctly compressed, the septum at
    right angles to plane of compression (angusti-
    septate)   Key D
  4 Fruit distinctly compressed, the septum
    parallel with the plane of compression
    (latiseptate), or fruit scarcely compressed   Key E

Key A - Fruit indehiscent, or splitting transversely, or
      splitting longitudinally but not releasing seeds
      separately
1 Fruit pendent, flattened, winged, 1-seeded   **5. ISATIS**
1 Fruit not pendent, not winged   2
  2 Petals yellow   3
  2 Petals white to pink or purple   6
3 Fruit >12mm; fresh root smelling of radish   **46. RAPHANUS**
3 Fruit <12mm; fresh root not smelling of radish   4
  4 Fruit composed of 2 segments separated by
    transverse constriction   **44. RAPISTRUM**
  4 Fruit composed of 1 segment or with slight
    longitudinal constriction   5
5 Fruit irregularly warty, 5-10mm; at least some leaves
  pinnately lobed   **6. BUNIAS**
5 Fruit minutely tuberculate, <5mm; leaves entire to
  dentate   **25. NESLIA**
  6 Fruit clearly divided longitudinally into 2
    1-seeded halves   7
  6 Fruit clearly divided transversely or not
    obviously divided   8
7 Leaves simple; inflorescence a terminal panicle
                     32. LEPIDIUM
7 Leaves pinnate; inflorescences mostly simple leaf-
  opposed racemes   **33. CORONOPUS**
  8 Fruit 1-seeded, with sessile stigma   **45. CRAMBE**
  8 Fruit usually 2-11-seeded, if 1-seeded then
    with stigma at end of narrow beak   9
9 Plant glabrous; leaves fleshy; fresh root not
  smelling of radish   **43. CAKILE**
9 Plant with stiff hairs below; leaves not fleshy;
  fresh root smelling of radish   **46. RAPHANUS**

Key B - Fruits >3x as long as wide (siliquae), splitting

        longitudinally from base into 2 valves; flowers pale
        yellow to golden

1 Stem-leaves distinctly clasping stem at their base   2
1 Stem-leaves not clasping stem, petiolate or narrowed
  to base   7
  2 Hairs present at least on lowest leaves, branched
          15. ARABIS
  2 Hairs 0 or simple on all parts   3
3 Fruit with distinct stalk 0.5-1mm between valves and
  sepal-scars   **40. ERUCASTRUM**
3 Fruit with base of valves immediately above sepal-
  scars   4
  4 Fruit with broad-based beak $\geq$4mm   **37. BRASSICA**
  4 Fruit with narrowly cylindrical beak $\leq$4mm   5
5 Leaves all simple and entire   **35. CONRINGIA**
5 At least lower leaves pinnate or pinnately lobed   6
  6 Valves of fruit with prominent midrib; seeds in
    1 row under each valve   **11. BARBAREA**
  6 Valves of fruit with midrib not or scarcely
    discernible; seeds $\pm$ in 2 rows under each valve
          12. RORIPPA
7 Leaves linear, all in tight basal rosette   **21. DRABA**
7 Leaves rarely all linear, not in tight basal rosette   8
  8 Fruit terminated by distinct beak >3mm; beak
    either wide at base and tapered distally or wide
    for most of length, sometimes with seeds   9
  8 Fruit terminated by sessile stigmas or by short
    narrow cylindrical style $\leq$4mm   13
9 Seeds in 2 rows under each valve; petals pale yellow
  with conspicuous violet veins   **39. ERUCA**
9 Seeds in 1 row under each valve; petals without
  conspicuous violet veins   10
10 Fruit <2cm, closely appressed to stem
          42. HIRSCHFELDIA
10 Fruit >2cm, not closely appressed to stem   11
11 Each valve of fruit with 1 prominent vein; beak
  long-conical   **37. BRASSICA**
11 Each valve of fruit with 3(-7) prominent parallel
  veins; beak long-conical or flat and sword-like   12
12 Sepals erect; beak of fruit flat and sword-like
          41. COINCYA
12 Sepals patent (to erecto-patent); beak of fruit
  flat and sword-like or long-conical   **38. SINAPIS**
13 Hairs present at least below, at least some stellate
  or 2-3-armed (arms often closely appressed to plant) 14
13 Hairs 0, or present and all simple   16
  14 Leaves 2-3x finely divided almost to midrib;
    petals c. as long as sepals   **2. DESCURAINEA**
  14 Leaves entire to simply toothed or lobed; petals
    c.2x as long as sepals   15
15 Compact basal leaf-rosette present; basal leaves
  pinnately lobed   **15. ARABIS**

15 Basal leaf-rosette 0 or very ill-defined; lower
   leaves entire to toothed   **7. ERYSIMUM**
   16 Valves of fruit with midrib not or scarcely
      discernible   **12. RORIPPA**
   16 Valves of fruit with conspicuous midrib   17
17 Seeds in 2 rows under each valve   **36. DIPLOTAXIS**
17 Seeds in 1 row under each valve   18
   18 Each valve of fruit with 3 prominent parallel
      veins   **1. SISYMBRIUM**
   18 Each valve of fruit with 1 prominent vein   19
19 Seeds ± globose; flowers all without bracts; valves
   of fruit rounded   **37. BRASSICA**
19 Seeds distinctly longer than wide; lower flowers
   usually with small bracts; valves of fruit keeled at
   midrib   **40. ERUCASTRUM**

Key C - Fruits >3x as long as wide (siliquae), splitting
        longitudinally from base into 2 valves; flowers
        white, mauve, pink, purple or red
1 Fruit with flat, sword-like beak at apex; petals
  whitish with violet veins   **39. ERUCA**
1 Fruit with beak 0 or cylindrical; petals rarely
  whitish with violet veins   2
  2 Basal and lower stem-leaves pinnate or ternate   3
  2 Basal and lower stem-leaves entire to lobed, but
    lobes not reaching midrib   4
3 Valves of fruit convex (fruit little wider than
  thick), opening by separating laterally from base
                                             12. RORIPPA
3 Valves of fruit ± flat (fruit much wider than thick),
  opening by suddenly spiralling upwards from base
                                             14. CARDAMINE
  4 Petals bifid c.1/2 way to base; small annual with
    leaves confined to basal rosette   **22. EROPHILA**
  4 Petals entire or notched to <1/4 way to base;
    stems usually bearing leaves   5
5 Plant glabrous or with simple hairs only   6
5 Plant pubescent, all or some hairs branched or
  stellate, sometimes sparse   10
  6 Stem-leaves clasping stem at base   **35. CONRINGIA**
  6 Stem-leaves not clasping stem   7
7 Basal leaves cordate at base; plant smelling of
  garlic when crushed   **3. ALLIARIA**
7 Basal leaves cuneate at base; plant not smelling of
  garlic   8
  8 Petals >10mm; stigma deeply 2-lobed   **8. HESPERIS**
  8 Petals <10mm; stigma capitate   9
9 Annual lowland weed; valves of fruit convex
                                             1. SISYMBRIUM
9 Perennial of mountains of N & W; valves of fruit
  ± flat   **15. ARABIS**
10 Stigma deeply 2-lobed, the lobes visible at

```
        apex of fruit (though often closed up together)   11
   10 Stigma capitate or slightly notched                 14
11 Seeds narrowly to broadly winged                       12
11 Seeds not winged                                       13
   12 Hairs branching from basal stalk in tree-like
      form; seeds broadly winged all round     10. MATTHIOLA
   12 Hairs with 2 arms, both appressed to plant body,
      without common stalk; seeds narrowly winged, often
      not all round                            7. ERYSIMUM
13 Spreading annual; most hairs with stalk 0 and 2-4
   arms appressed close to plant surface       9. MALCOLMIA
13 Erect biennial or perennial; most hairs simple or with
   stalk and 2 arms, patent                    8. HESPERIS
   14 Fruit <8x as long as wide (excl. style), >2mm
      wide                                               15
   14 Fruit >8x as long as wide (excl. style), <2mm
      wide                                               16
15 Flowers white; fruit flat, usually twisted; stems
   erect                                        21. DRABA
15 Flowers rarely white; fruit scarcely compressed, not
   twisted; stems procumbent to decumbent      16. AUBRIETA
   16 Basal leaves pinnately lobed             15. ARABIS
   16 Basal leaves entire to dentate                     17
17 Annual; fruit cylindrical to tetraquetrous
                                            4. ARABIDOPSIS
17 Biennial to perennial; fruit strongly flattened
                                               15. ARABIS
```

Key D - Fruits <3x as long as wide (siliculae), obviously
        compressed, angustiseptate, splitting longitudinally
        from base into 2 valves to release 1 or more seeds on
        each side

```
1 Corolla distinctly zygomorphic, with 2 petals on 1
  side much shorter than 2 on other                        2
1 Corolla actinomorphic, with 4 equal petals               3
   2 Seeds 2 under each valve; style + 0; shorter
     petals c. as long as sepals            29. TEESDALIA
   2 Seed 1 under each valve; style distinct; shorter
     petals >2x as long as sepals             31. IBERIS
3 Fruit distinctly winged at edges (i.e. valve midrib)    4
3 Fruit sometimes keeled but not winged at edges (i.e.
  valve midrib)                                            6
   4 Basal leaves strongly cordate at base   30. THLASPI
   4 Basal leaves 0 or cuneate to rounded at base         5
5 Seeds 3-8 under each valve                 30. THLASPI
5 Seeds 1(2) under each valve                32. LEPIDIUM
   6 Fruit obtriangular                       26. CAPSELLA
   6 Fruit orbicular to elliptic, oblong or obovate       7
7 Flowers appearing solitary, on long pedicels
  arising from axils of basal leaves        28. IONOPSIDIUM
7 Flowers in racemes, all or mostly without bracts        8
   8 Fruits distinctly keeled at edges (i.e. valve
```

   midrib)   **32. LEPIDIUM**
   8 Fruits obtuse to rounded at edges (i.e. valve
     midrib)   9
 9 Basal leaves deeply pinnately lobed; flowers
   <2mm across   **27. HORNUNGIA**
 9 Basal leaves entire to palmately angled or shallowly
   lobed; flowers >2mm across   **23. COCHLEARIA**

Key E - Fruits <3x as long as wide (siliculae), either not
        compressed or obviously compressed and latiseptate,
        splitting longitudinally to release 1 or more seeds
        on each side
 1 Leaves all confined to basal rosette   2
 1 At least some leaves borne on stem   5
   2 Petals yellow   **21. DRABA**
   2 Petals white   3
 3 Aquatic, often submerged; leaves subulate **34. SUBULARIA**
 3 Not aquatic; leaves flat   4
   4 Annual; petals bifid c.1/2 way to base; fruit
     glabrous   **22. EROPHILA**
   4 Perennial; petals notched <1/4 way to base; fruit
     pubescent   **21. DRABA**
 5 Petals yellow   6
 5 Petals white to pink or purple   10
   6 Plant glabrous or with simple hairs   7
   6 Plant pubescent; all or many hairs branched   8
 7 Fruit not or scarcely keeled at junction of valves
                                              12. RORIPPA
 7 Fruit strongly keeled (+ winged) at junction of
   valves   **24. CAMELINA**
   8 Petals bifid nearly 1/2 way to base   **19. BERTEROA**
   8 Petals entire or notched <1/4 way to base   9
 9 Fruit not or scarcely flattened; seeds >2 under each
   valve; style >1mm in fruit   **24. CAMELINA**
 9 Fruit strongly flattened; seeds 2 under each valve;
   style <1mm   **18. ALYSSUM**
  10 Robust perennial with strong tap-root and large
     dock-like basal leaves (sometimes pinnately lobed)
                                              13. ARMORACIA
  10 Plant without dock-like basal leaves   11
 11 At least most fruits >2 x 1cm, very flat **17. LUNARIA**
 11 Fruits <2cm, <1cm wide, flat or not   12
  12 Plant glabrous or with simple hairs   13
  12 Plant with at least some branched or stellate
     hairs   14
 13 Basal leaves abruptly contracted into long petiole
                                              23. COCHLEARIA
 13 Basal leaves + sessile, gradually narrowed to base
                                              24. CAMELINA
  14 Fruit scarcely compressed   15
  14 Fruit strongly compressed   16
 15 Flowers rarely white; procumbent to decumbent much-

```
     branched perennial                      16. AUBRIETA
15 Flowers white; erect little-branched annual
                                             24. CAMELINA
   16 Petals bifid nearly 1/2 way to base    19. BERTEROA
   16 Petals entire to notched <1/4 way to base      17
17 Fruit c. as long as wide, notched at apex  18. ALYSSUM
17 Fruit longer than wide, rounded to obtuse         18
   18 Seeds >2 under each valve; basal leaf-rosette
      present at flowering                    21. DRABA
   18 Seeds 1 under each valve; basal leaf-rosette 0
      at flowering                           20. LOBULARIA
```

Other genera - Many other genera have been recorded as
casuals in BI. The following 16 genera, stated in recent
British and European Floras to occur in BI, either do not
nowadays occur or are far too rare to merit treatments here:
AETHIONEMA R. Br., **BOREAVA** Jaub. & Spach, **CALEPINA** Adans.,
CARRICHTERA DC. (cf. Vella L.), **CHORISPORA** R. Br. ex DC.,
ENARTHROCARPUS Labill., **ERUCARIA** Gaertner, **EUCLIDIUM** R. Br.,
GOLDBACHIA DC., **IONDRABA** Reichb. (cf. Biscutella L.),
LYCOCARPUS O. Schulz, **MICROSISYMBRIUM** O. Schulz, **MORICANDIA**
DC., **MYAGRUM** L., **SUCCOWIA** Medikus, **TETRACME** Bunge. Nearly 30
spp. mentioned under 'Other spp.' in various genera as too
rare to be treated here come into the same category; in many
cases the generic descriptions do not cover these, and often
even listing them has been considered too wasteful of space.

TRIBE 1 - SISYMBRIEAE (genera 1-6). Hairs branched or
unbranched, sometimes glandular; petals yellow or white;
fruit usually a beakless siliqua with small stigma, sometimes
1-2-seeded and indehiscent.

1. SISYMBRIUM L. - Rockets
Annuals or perennials; basal leaves simple, entire to very
deeply lobed; hairs unbranched; petals yellow, sometimes very
pale; fruit a beakless siliqua with convex valves; seeds in 1
row under each valve.

```
1 Leaves all entire to dentate                        2
1 Lowest leaves lobed >1/4 to midrib                  4
   2 Annual; petals <3mm, not longer than sepals or
     stamens                             7. S. erysimoides
   2 Perennial; petals >3mm, longer than sepals and
     stamens                                           3
3 Leaves pubescent on lowerside, narrowly acute to
  acuminate at apex                       1. S. strictissimum
3 Leaves ± glabrous on lowerside, subacute to apiculate
  at apex                                4. S. volgense
   4 Fruits <2cm, strongly appressed to stem
                                         8. S. officinale
   4 At least most fruits >2cm, patent to ± erect but
     not appressed to stem                            5
```

5 Upper stem-leaves pinnate with linear divisions
                                              5. S. altissimum
5 Upper stem-leaves variously divided but usually not
   to midrib and never with linear segments   6
   6 Pedicels c.1mm wide, >2/3 as wide as ripe fruit   7
   6 Pedicels c.0.3-0.6mm wide, <2/3 as wide as ripe
      fruit   8
   7 Petals <3mm, not longer than sepals or stamens; fruit
      c.2.5-5cm   **7. S. erysimoides**
   7 Petals >6mm, much longer than sepals and stamens;
      fruit c.4-11cm   **6. S. orientale**
      8 Lower part of stem with ± dense patent to reflexed
         stiff hairs >1mm   **3. S. loeselii**
      8 Lower part of stem glabrous to sparsely pubescent,
         or ± densely pubescent with hairs <0.5mm   9
   9 Petals pale yellow, <5mm; anthers <1mm   **2. S. irio**
   9 Petals bright yellow, mostly >5mm; anthers >1mm   10
      10 Rhizomatous perennial; most fruits >3.5cm
                                               4. S. volgense
      10 Annual; most fruits <3.5cm   **3. S. loeselii**

Other spp. - S. austriacum Jacq. and **S. polyceratium** L.,
from S Europe, have occurred as casuals but now either do not
occur or are extremely rare.

   1. S. strictissimum L. - <u>Perennial</u> <u>Rocket</u>. Perennial with 306
strong rootstock; stems little-branched, to 1.5m, erect;
leaves entire to dentate; fruits 3-8cm, erect to erecto-
patent. Intrd; walls, rough and waste ground; natd in a very
few places in En from Surrey to Durham, occasionally a
casual; Europe.
   2. S. irio L. - <u>London-rocket</u>. Annual; stems well-branched, 306
to 60cm, erect to ascending; lower leaves variously lobed but
some deeply; fruits 3-6cm, patent to ± erect. Intrd; walls,
roadsides and waste places; ± natd in a few places in Br and
Ir, a more frequent casual, especially wool-alien; Europe.
   3. S. loeselii L. - <u>False London-rocket</u>. Annual; stems 306
rather little-branched, to 1(1.5)m, erect; lower leaves
deeply and regularly lobed; fruits 1.5-4.5cm, erect or
erecto-patent. Intrd; waste places and tips; fairly frequent
casual in S Br and natd in London area; Europe.
   4. S. volgense M. Bieb. ex Fourn. - <u>Russian</u> <u>Mustard</u>. 306
Rhizomatous perennial; stems to 75cm, erect to ascending;
lower leaves variously lobed, often deeply, rarely not;
fruits 2.5-6cm, erect to erecto-patent. Intrd; waste ground;
natd in a few places widely scattered in En; Russia.
   5. S. altissimum L. - <u>Tall</u> <u>Rocket</u>. Annual; stems to 1m, ± 306
erect; leaves deeply divided, the upper pinnate with long
linear lobes; fruit 5-10cm, erecto-patent. Intrd; waste
places; frequently natd or persistent casual scattered over
much of BI; Europe.
   6. S. orientale L. - <u>Eastern</u> <u>Rocket</u>. Annual; stems to 1m, 306

erect to ascending; leaves deeply divided, the upper with
only 0–2 basal lobes each side; fruit 4–10cm, erecto-patent.
Intrd; waste places; frequently natd or persistent casual
scattered over much of BI; Europe.

7. S. erysimoides Desf. - <u>French</u> <u>Rocket</u>. Annual; stems to **306**
60cm, erect; leaves variously divided, often deeply, rarely
unlobed; fruits 2–5cm, patent to erecto-patent. Intrd; a
fairly regular wool-alien in En; W Mediterranean. Petals
often fade to white in dried material.

8. S. officinale (L.) Scop. - <u>Hedge</u> <u>Mustard</u>. Annual or **306**
biennial; stems to 1m, erect or spreading; leaves deeply
lobed; fruits 1–2cm, erect and closely appressed to stem.
Native; waste places, rough and cultivated ground, hedges and
roadsides; very common throughout BI except N Sc.

2. DESCURAINIA Webb & Berth. - <u>Flixweed</u>
Annuals or biennials; basal leaves 2–3-pinnate or nearly so;
hairs both unbranched and branched, ± patent; petals pale
yellow; fruit a beakless siliqua with convex valves; seeds in
1 row under each valve.

1. D. sophia (L.) Webb ex Prantl - <u>Flixweed</u>. Stems to 1m, **306**
erect, branched above; leaves 2–3x divided ± to midrib, with   **R**
linear segments; fruits 1–4cm, ± erect to erecto-patent on
long pedicels. Possibly native; roadsides, rough and waste
ground; throughout much of Br, locally frequent in E, absent
from much of N & W, very local in Ir.

3. ALLIARIA Heister ex Fabr. - <u>Garlic</u> <u>Mustard</u>
Biennials; basal leaves simple, toothed; hairs unbranched;
petals white; fruit a beakless siliqua with angled valves;
seeds in 1 row under each valve.

1. A. petiolata (M. Bieb.) Cavara & Grande - <u>Garlic</u> <u>Mustard</u>. **306**
Fresh plant smelling of garlic; stems to 1.2m, erect, little-
branched; leaves petiolate, ovate, the lower cordate,
coarsely dentate; fruits 3–7cm, erect to erecto-patent on
short thick pedicels. Native; rough ground, hedgerows and
shady places; throughout most of BI but local in Sc and Ir.

4. ARABIDOPSIS (DC.) Heynh. - <u>Thale</u> <u>Cress</u>
Annuals; basal leaves simple; hairs both unbranched and
branched; petals white; fruit a beakless siliqua with convex
to angled valves; seeds in 1 row under each valve.

1. A. thaliana (L.) Heynh. - <u>Thale</u> <u>Cress</u>. Stems erect, to **306**
30(50)cm, simple or branched, with few leaves; basal leaves
entire to dentate; stem-leaves ± entire, sessile; fruits
(5)8–15(20)mm, erecto-patent on long slender pedicels.
Native; cultivated ground and bare places on banks, walls,
rocks and waysides; throughout BI, rarer in N & W.

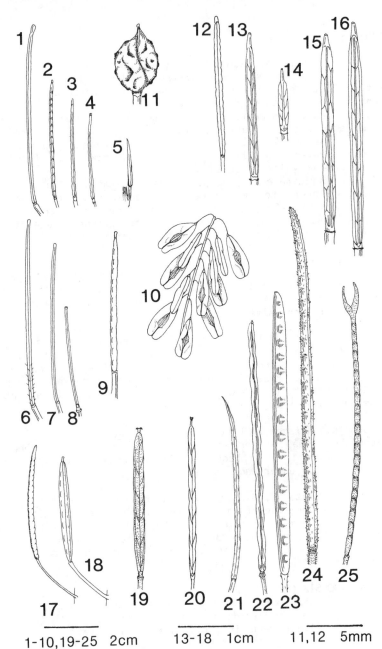

FIG 306 – Fruits of Brassicaceae. 1, Sisymbrium altissimum.
2, S. irio. 3, S. erysimoides. 4, S. loeselii. 5, S.
officinale. 6, S. orientale. 7, S. volgense. 8, S.
strictissimum. 9, Alliaria. 10, Isatis. 11, Bunias. 12,
Arabidopsis. 13, Barbarea stricta. 14, B. vulgaris. 15, B.
intermedia. 16, B. verna. 17, Descurainia. 18, Erysimum
cheiranthoides. 19, E. cheiri. 20, E. x marshallii. 21,
Malcolmia. 22, Hesperis. 23, Matthiola incana. 24, M.
sinuata. 25, M. longipetala.

1-10,19-25 2cm   13-18 1cm   11,12 5mm

5. ISATIS L. - <u>Woad</u>
Biennials to perennials; basal leaves simple; hairs 0 or
unbranched; petals yellow; fruit pendent, indehiscent, 1-
seeded, winged; style 0.

Other spp. - I. lusitanica L. (<u>I.</u> <u>aleppica</u> Scop.) formerly
occurred as a casual.

1. I. tinctoria L. - <u>Woad</u>. Stems erect, to 1.5m, well- **306**
branched; leaves lanceolate to oblanceolate, entire, clasping
the stem, ± glabrous; fruits 1-2.5cm, purple-brown when
ripe. Intrd; natd on cliffs in E Gloucs and Surrey since
before 1800, infrequent casual elsewhere, formerly commoner;
Europe.

6. BUNIAS L. - <u>Warty-cabbages</u>
Biennials to perennials; basal leaves simple; hairs mixed
unbranched, branched and thick glandular; petals yellow;
fruit irregularly warty-ovoid, indehiscent, 1-2-seeded, with
short style.

Other spp. - B. erucago L. (<u>Southern</u> <u>Warty-cabbage</u>) used to
occur as a casual but is now absent or very rare.

1. B. orientalis L. - <u>Warty-cabbage</u>. Stems erect, to 1m, **306**
well-branched; lower leaves oblanceolate, variously lobed and
toothed, at least some pinnately lobed, sparsely pubescent;
fruits 5-10mm, on long slender erecto-patent pedicels.
Intrd; natd in waste and rough grassy places scattered over
most of BI, mostly in C & S En; E Europe.

TRIBE 2 - HESPERIDEAE (genera 7-10). Hairs branched (sometimes closely appressed to plant), sometimes unbranched,
sometimes glandular; petals usually white to pink or purple,
sometimes yellow to red; fruit a beakless siliqua with
usually deeply (sometimes shallowly) 2-lobed stigma.

7. ERYSIMUM L. (<u>Cheiranthus</u> L.) - <u>Wallflowers</u>
Annuals to perennials; leaves simple; hairs 2-3 branched,
stalkless, the arms tightly appressed; petals yellow to red,
brown or purple; fruit a flattened to 4-angled siliqua with
strong midrib; stigma + capitate to 2-lobed; seeds in 1(-2)
rows under each valve, not winged or with narrow wing.

1. E. cheiranthoides L. - <u>Treacle</u> <u>Mustard</u>. Usually annual; **306**
stems erect, usually branched, to 60(100)cm; lower leaves
elliptic, entire to shallowly toothed; petals 3-6mm, yellow;
fruits 1-3cm, sparsely pubescent, erecto-patent. Intrd;
cultivated and waste ground; scattered over much of BI but
rare or local except in C & S En; Europe.
2. E. x marshallii (Henfrey) Bois (<u>E.</u> <u>allionii</u> hort. nom. **306**
illeg., <u>Cheiranthus</u> <u>allionii</u> Hort.; <u>E.</u> <u>decumbens</u> (Schleicher **309**

ex Willd.) Dennst. x E. perofskianum Fischer & C. Meyer) -
Siberian Wallflower. Usually biennial; stems often branched
below, erect to ascending, to 50cm; leaves narrowly elliptic,
remotely dentate; petals 20-30mm, orange; fruits 3-9cm,
appressed-pubescent, nearly erect. Intrd; much grown in
gardens, not natd but fairly frequent on tips etc.; S En;
probably garden origin. Distinguished from E. cheiri with
orange flowers by its bilobed stigma and tetraquetrous
fruits.
3. E. cheiri (L.) Crantz (Cheiranthus cheiri L.) - Wall- 306
flower. Usually perennial; stems well-branched and woody 309
below, erect to decumbent, to 50(80)cm; leaves narrowly
elliptic, entire; petals 20-35mm, yellow to red, brown or
purple; fruits 2.5-8cm, densely appressed-pubescent, nearly
erect. Intrd; much grown and often natd on walls, banks and
other dry places; scattered throughout BI except N Sc; E
Mediterranean.

8. HESPERIS L. - Dame's-violet
Biennial to perennial; hairs mixed simple and branched, or 0;
petals white to pink or purple; fruit a ± cylindrical
slightly constricted siliqua with ± strong midrib; stigma
2-lobed, seeds in 1 row under each valve, not winged.

Other spp. - H. laciniata All., from S Europe, formerly
occurred as a casual.

1. **H. matronalis** L. - Dame's-violet. Stems erect, to 1.2m, 306
branched above, glabrous to sparsely pubescent; stem-leaves
ovate, petiolate, sharply serrate; petals 15-25mm; fruits
2-10cm, ± glabrous, erecto-patent to patent. Intrd; much
grown and a common ± natd escape in waste places, banks,
grassland, hedges and verges; frequent throughout BI; Europe.

9. MALCOLMIA R. Br. - Virginia Stock
Annuals; basal leaves simple; hairs branched, with 2-4
appressed arms and stalk 0; petals pink to purple; fruit a
constricted siliqua with ± rounded valves with 3 veins;
stigma 2-lobed; seeds in 1 row under each valve, not winged.

1. **M. maritima** (L.) R. Br. - Virginia Stock. Stems 306
decumbent to erect, much-branched from base, to 40cm; leaves
obovate to oblanceolate, entire or toothed; petals 12-25mm;
fruits 3.5-8cm, pubescent, ± erect. Intrd; much grown in
gardens and commonly escaping on tips and in waste places; S
& C Br and CI; Balkans.

10. MATTHIOLA R. Br. - Stocks
Annuals to perennials; basal leaves simple; hairs branched,
stalked; petals white to pink or purple; fruit a not or sometimes ± constricted siliqua with 1-veined flattened to ±
rounded valves; stigma strongly 2-lobed, each lobe with

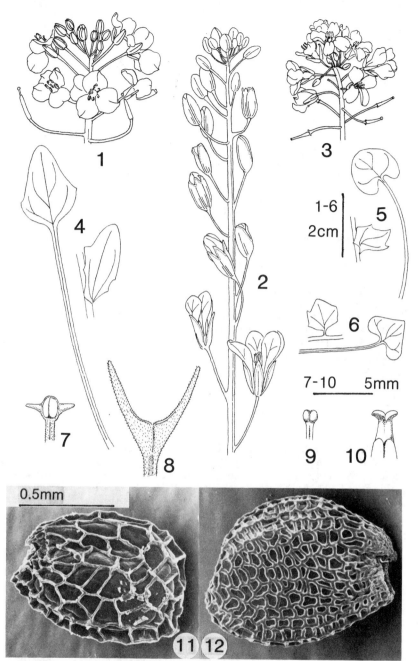

FIG 309 - **Brassicaceae.** 1-3, inflorescences of **Brassica.** 1,
B. napus. 2, **B. oleracea.** 3, **B. rapa.** 4-6, basal and stem-
leaves of **Cochlearia.** 4, **C. anglica.** 5, **C. officinalis.** 6, **C.
danica.** 7-10, stigmas. 7, **Matthiola incana.** 8, **M.
longipetala.** 9, **Erysimum x marshallii.** 10, **E. cheiri.** 11-12,
seeds of **Rorippa.** 11, **R. nasturtium-aquaticum.** 12, **R.
microphylla.**

dorsal horn-like or hump-like process; seeds in 1 row under
each valve, broadly winged

1 Fruit + cylindrical, + constricted between seeds,
   with apical horn-like processes >(2)3mm
                                        3. M. longipetala
1 Fruit + compressed, not constricted between seeds;
   with apical horn-like processes <3mm   2
   2 Leaves all entire; fruits without glands
                                        1. M. incana
   2 Lower leaves sinuate to lobed; fruits with large
      yellow to black sessile glands   **2. M. sinuata**

Other spp. - M. tricuspidata (L.) R. Br. and **M. fruticulosa**
L.) Maire (M. tristis R. Br.) have occurred as casuals, but
very rarely or not nowadays.

   1. M. incana (L.) R. Br. - Hoary Stock. Annual or 306
perennial; stems woody below, erect to ascending, branched 309
from base, to 80cm; leaves lanceolate to oblanceolate,   RR
entire; petals 20-30mm; fruits 4.5-15cm, erecto-patent.
Possibly native on sea-cliffs in S En; elsewhere a garden
escape, often natd on walls and banks etc. in En, Wa and CI.
   2. M. sinuata (L.) R. Br. - Sea Stock. Biennial; stems 306
woody below, erect, branched or not, to 60cm; leaves   RR
lanceolate to oblanceolate, the lower sinuate-lobed; petals
15-25mm; fruits 5-15cm, erecto-patent. Native; sand-dunes
and sea-cliffs; S Ir, S Wa, SW En and CI, very local and
decreasing.
   3. M. longipetala (Vent.) DC. (M. bicornis (Sibth. & Smith) 306
DC., M. oxyceras DC.) - Night-scented Stock. Annual; stems 309
much-branched, diffuse, to 50cm; leaves linear to narrowly
elliptic, entire to distantly toothed; petals 15-25mm; fruits
5-15cm, + patent, with upcurved horn-like stigmatic processes
(2)5-10mm. Intrd; grown in gardens and a frequent casual on
tips and in waste places in En; Balkans. Our plants are
mostly ssp. **bicornis** (Sibth. & Smith) P. Ball.

TRIBE 3 - ARABIDEAE (genera 11-16). Hairs branched or
unbranched, sometimes stellate, not glandular; petals yellow
or white; fruit usually a beakless silqua with small stigma,
sometimes a silicula.

11. BARBAREA R. Br. - Winter-cresses
Biennials or perennials; basal leaves pinnate; hairs 0 or
few, unbranched; petals yellow; fruit a siliqua with angled
valves; seeds in 1 row under each valve.

1 Uppermost stem-leaves simple, toothed or lobed to
  <1/2 way to midrib (ignore basal lobes); seeds 1-1.8mm 2
1 Uppermost stem-leaves + pinnate to pinnately lobed
  to >1/2 way to midrib; seeds 1.6-2.4mm   3

2 Fruit with style 2-3.5mm; flowers buds glabrous
                                          1. B. vulgaris
2 Fruit with style 0.5-1.6(2.3)mm; flower buds with
   hairs at apex of sepals   **2. B. stricta**
3 At least some fruits >4cm; fresh petals >5.6mm
                                          4. B.verna
3 Fruits <4cm; fresh petals <5.6(6.3)mm **3. B. intermedia**

1. B. vulgaris R. Br. (B. arcuata (Opiz ex J.S. & C. Presl) **306**
Reichb.) - Winter-cress. Stems erect, to 1m; basal leaves
usually at least as wide at most distal lateral leaflets as
at terminal leaflet; fruits (0.7)1.5-3.2cm with style
2-3.5(4)mm. Native; hedges, streamsides, roadsides and waste
places, often on damp soil; throughout most of BI.
2. B. stricta Andrz. - Small-flowered Winter-cress. Stems **306**
erect, to 1m; basal leaves usually much wider at terminal
leaflet than at any lateral leaflets; fruits 1.3-2.8(3.5)cm
with style 0.5-1.6(2.3)mm. Probably intrd; similar places to
B. vulgaris; very scattered and mostly casual in En, Wa and S
Sc, natd in parts of S En and S Wa; Europe.
3. B. intermedia Boreau - Medium-flowered Winter-cress. **306**
Stems erect, to 60cm; basal leaves with 3-6 pairs of lateral
leaflets; fruit 1.5-3.2(3.6)cm with style 0.6-1.7mm. Intrd;
waste, open and cultivated ground and by roads and streams;
sparsely scattered in most of BI; Europe.
4. B. verna (Miller) Asch. - American Winter-cress. Stems **306**
erect, to 75cm; basal leaves with(3)5-10 pairs of lateral
leaflets; fruit (2.8)3.5-7.1cm with style 0.6-2(2.3)mm.
Intrd; waste, cultivated and open ground and by roads;
sparsely scattered through most of BI but rare in Sc; Europe.

12. RORIPPA Scop. (Nasturtium R. Br.) - Water-cresses
Annuals to perennials; basal leaves simple and unlobed to
pinnate; hairs 0 or unbranched; petals white or yellow; fruit
a silicula or siliqua with convex valves and indistinct
midrib; seeds in (1 or)2 rows under each valve.
Hybrids are not common but are effectively spread
vegetatively and may occur in absence of both parents. R.
sylvestris, R. amphibia and R. austriaca are highly self-
sterile and therefore often do not set seed; hence sterility
is not diagnostic for hybridity in the yellow-flowered taxa.

1 Petals white   2
1 Petals yellow   4
   2 Seeds per fruit <5; pollen <40% with full
      contents   **2. R. x sterilis**
   2 Seeds per fruit >10; pollen >80% with full
      contents   3
3 Seeds in 2 rows under each valve, with c.6-12 cells
   showing across broadest width
                              1. R. nasturtium-aquaticum
3 Seeds mostly in 1 row under each valve, with c.12-20

```
cells showing across broadest width   3. R. microphylla
4 Petals c. as long as sepals; stems not rooting        5
4 Petals >1.5x as long as sepals; stems often
   rooting when contacting ground or water              6
5 Fruit 2-3x as long as pedicels; sepals <1.6mm
                                        4. R. islandica
5 Fruit <2x as long as pedicels; sepals >1.6mm
                                        5. R. palustris
   6 Stem-leaves with auricles at base, + clasping stem  7
   6 Stem-leaves with no or very small auricles, not
      clasping stem                                       9
7 Leaves toothed but not lobed; fruit (excl. style)
   <1.5x as long as wide                  11. R. austriaca
7 Leaves strongly lobed; fruit (excl. style) >1.5x as
   long as wide                                           8
   8 Petals <3.5mm when fresh; stem-leaves c.4x as long
      as wide                            6. R. x erythrocaulis
   8 Petals >3.5mm when fresh; stem-leaves 2-3x as long
      as wide                            9. R. x armoracioides
9 Upper stem-leaves toothed but not lobed; fruit (excl.
   style) 2-2.5(3)x as long as wide        10. R. amphibia
9 Upper stem-leaves lobed; fruit (excl. style) (2)2.5-
   7.5x as long as wide                                  10
   10 Stem-leaves with terminal segment <1/4 of total
      length; fruit 9-22 x <1.3mm, plus style <1.2mm
                                         7. R. sylvestris
   10 Stem-leaves with terminal segment >1/4 of total
      length; fruit 3-10 x >1.3mm, plus style >1.2mm    11
11 Fruiting pedicels mostly reflexed      8. R. x anceps
11 Fruiting pedicels patent to erecto-patent
                                         9. R. x armoracioides
```

ascending stems to 1m; leaves pinnate, entire to sinuate; petals white, 4-6mm when fresh; stamens dehiscing inwardly; fruit 11-19mm (mean per plant) x 1.9-2.7mm, with pedicels 7-12mm (mean per plant); seeds in 2 rows under each valve, with c.25-50(70) cells showing in lateral view, c.6-12 across broadest width; 2n=32. Native; in and by streams, ditches and marshes; frequent to common over most of BI.

2. **R. x sterilis** Airy Shaw - <u>Hybrid</u> <u>Water-cress</u>. Differs 318 from <u>R. nasturtium-aquaticum</u> in stamens dehiscing mostly outwardly; fruit irregularly shaped, distorted due to only 0-3 full seeds variously arranged, with pedicels 7-10mm (mean per plant); full seeds with c.50-100(120) cells showing in lateral view; 2n=48. Native; in similar places to parents but often without either near; scattered over most of BI.

3. **R. microphylla** (Boenn.) N. Hylander ex A. & D. Love 309 (<u>Nasturtium</u> <u>microphyllum</u> (Boenn.) Reichb. - <u>Narrow-fruited</u> 318 <u>Water-cress</u>. Differs from <u>R. nasturtium-aquaticum</u> in stamens dehiscing outwardly; fruit 16-23mm (mean per plant) x 1.3-

1.8mm, with pedicels 14-19mm (mean per plant); seeds in 1 row
under each valve, with c.90-120(190) cells showing in lateral
view, c.12-20 across broadest width; 2n=64. Native; similar
habitat and distribution to R. nasturtium-aquaticum but
slightly less common in S and more common in N.

4. R. islandica (Oeder ex Murray) Borbas - Northern Yellow-   R
cress. Annual or short-lived perennial with procumbent to
ascending stems to 15(30)cm; stem-leaves deeply pinnately
lobed or ± pinnate; petals yellow, 1-1.7mm when fresh; fruit
6-12 x 2-3mm, c.2-3x as long as patent to reflexed pedicels.
Native; open pond-sides and other damp places; very scattered
in W Ir, W Wa, Man and Sc, mostly near sea.

5. R. palustris (L.) Besser (R. islandica auct. non (Oeder   318
ex Murray) Borbas) - Marsh Yellow-cress. Differs from R.
islandica in often ascending to erect stems to 60cm; petals
1.5-2.8mm when fresh; fruit 4-12 x 1.7-3mm, c.0.8-2x as long
as pedicels. Native; open damp and waste ground; frequent to
common in most of BI but rare in N Sc.

5 x 7. R. palustris x R. sylvestris has been recorded from
BI on several occasions but needs confirmation.

6. R. x erythrocaulis Borbas (R. palustris x R. amphibia) -
Thames Yellow-cress. Vigorous perennial to 1m; stem-leaves
pinnately lobed; petals yellow, 2.5-3.5mm when fresh; fruit
3.5-5.5 x 1.5-2mm, much shorter than patent to reflexed
pedicels; triploid and sterile, or tetraploid and fertile.
Native; by R. Thames in Surrey and Middlesex and by R. Avon
in Warks with parents, by R. Avon in W Gloucs without
parents.

7. R. sylvestris (L.) Besser - Creeping Yellow-cress.   318
Spreading perennial with shoots arising from creeping roots;
stems to 60cm, decumbent to ± erect; stem-leaves pinnate to
deeply pinnately lobed; petals yellow, 2.2-5.5mm when fresh;
fruit 9-22 x 1-1.2mm, c.2x as long as patent to erecto-patent
pedicels. Native; in damp places and disturbed ground;
frequent in most of BI but rare in N Sc.

8. R. x anceps (Wahlenb.) Reichb. (R. sylvestris x R.
amphibia) - Hybrid Yellow-cress. Perennial with habit of R.
sylvestris; stem-leaves intermediate in lobing; petals
yellow, 3.5-5.5mm when fresh; fruit 3-10 x 1.2-2.5mm, c. as
long as mostly reflexed pedicels. Native; the commonest
hybrid yellow-cress, often fertile and back-crossing but
often not with parents; scattered in En, Wa and Ir.

9. R. x armoracioides (Tausch) Fuss (R. sylvestris x R.
austriaca) - Walthamstow Yellow-cress. Perennial with habit
of R. sylvestris; stem-leaves intermediate in lobing; petals
yellow, 3-4.5mm when fresh; fruit 3-9 x 1.5-2mm, c. as long
as patent to erecto-patent pedicels, with some full seeds.
Native; damp waste ground; S Essex and Main Argyll.

10. R. amphibia (L.) Besser - Great Yellow-cress. Perennial   318
with shoots arising from axils of creeping stems; stems to
1.2m, ascending to erect; lower stem-leaves pinnately lobed,
upper only toothed; petals yellow, 3.5-6.2mm when fresh;

fruit 2.5-6 x 2.7-3mm, much shorter than mostly reflexed
pedicels. Native; in and by rivers, ponds and ditches;
frequent in En, local in Wa and Ir, rare and probably intrd
in S Sc.
  10 x 11. R. amphibia x R. austriaca = R. x hungarica Borbas
was found in 1987 as 1 plant on a river-bank in S Essex; it
is intermediate (petals 4.8-5.7mm; sepals 3.2-3.6mm) and
sterile.
  11. R. austriaca (Crantz) Besser - Austrian Yellow-cress. **318**
Perennial with shoots arising from creeping roots; stems to
1m, ascending to erect; leaves toothed, not lobed; petals
yellow, 3-5mm when fresh; fruit 2-3 x 2-3mm, much shorter
than patent to erecto-patent pedicels. Intrd; waste ground,
and by roads and rivers; natd in scattered localities in En
and S Wa; Europe.

13. ARMORACIA P. Gaertner, Meyer & Scherb. - Horse-radish
Perennials with deep strong roots; basal leaves simple; hairs
+ 0; petals white; fruit not or very rarely ripening, the
best developed being + terete siliculae with few seeds in 2
rows under each valve.

  1. A. rusticana P. Gaertner, Meyer & Scherb. - Horse-radish. **318**
Forming extensive patches; stems to 1.2m, with narrow unlobed
leaves; basal leaves to 50cm, dock-like, deeply lobed on
young plants, later + entire; inflorescences densely branched
panicles. Intrd; grassy places and waste ground as relic of
cultivation or where dumped; frequent in En, Wa and CI,
scattered in Ir and Sc;? W Asia.

14. CARDAMINE L. (Dentaria L.) - Bitter-cresses
Annuals to perennials; basal leaves pinnate or ternate; hairs
0 or unbranched; petals white to pink or purple, sometimes 0;
fruit a compressed siliqua with valves that spring open
suddenly to release seeds; seeds in 1 row under each valve.

1 Upper stem-leaves with purple bulbils in axils
                                                1. C. bulbifera
1 Stem-leaves without axillary bulbils   2
  2 Leaves with (1-)3 leaflets   **3. C. trifolia**
  2 Lower leaves with >5 leaflets   3
3 Petals <4mm, <2x as long as sepals, white, sometimes
  0   4
3 Petals >5mm, >2x as long as sepals, white to pink or
  purple   6
  4 Stem-leaves with small basal auricles clasping
    stem   **7. C. impatiens**
  4 Stem-leaves without auricles   5
5 Most flowers with 4(5) stamens; stems with 0-4(5)
  leaves, glabrous to sparsely pubescent   **9. C. hirsuta**
5 Most flowers with 6 stamens; stems with (3)4-7(10)
  leaves, conspicuously pubescent   **8. C. flexuosa**

6 Plant with whitish succulent rhizome with fleshy
  scale-leaves   **2. C. heptaphylla**
6 Rhizome 0 or present, not whitish or with fleshy
  scale-leaves   7
7 Anthers blackish-violet; petals usually white; stigma
  tapered from style   **4. C. amara**
7 Anthers yellow; petals usually pale to deep pink;
  stigma minutely capitate   8
8 All leaves large, with apical leaflet much wider
  than lateral ones   **5. C. raphanifolia**
8 Upper leaves with much narrower leaflets than
  lower leaves, the apical leaflet not or scarcely
  wider than lateral ones   **6. C. pratensis**

1. **C. bulbifera** (L.) Crantz (<u>Dentaria</u> <u>bulbifera</u> L.) - <u>Coral-</u>   R
<u>root</u>. Perennial with white succulent rhizome; stems erect,
unbranched, to 75cm; leaves with 1-7 entire to serrate leaflets; petals 12-16mm, pinkish-purple. Native; deciduous
woodland; very local in C & SE En.
2. **C. heptaphylla** (Villars) 0. Schulz (<u>Dentaria</u> <u>pinnata</u>
(Lam.) R.Br.) - <u>Pinnate</u> <u>Coralroot</u>. Perennial with white
succulent rhizome; stems erect, unbranched, to 60cm; leaves
with 5-11 serrate leaflets; petals 14-20mm, white or pinkish-
purple. Intrd; persistent relic of cultivation in woodland;
S Essex, Durham and W Gloucs; Europe.
3. **C. trifolia** L. - <u>Trefoil</u> <u>Cress</u>. Perennial with thin
rhizome; stems erect to ascending, unbranched, to 30cm;
leaves mostly ternate, those on stem few and reduced; petals
6-11mm, white. Intrd; natd in shady place as relic of
cultivation; scattered over En; Europe.
4. **C. amara** L. - <u>Large</u> <u>Bitter-cress</u>. Perennial with short
rhizome; stems little or not branched, erect to ascending, to
60cm; leaves with 5-11 sinuate-crenate leaflets; petals
6-10mm, white or rarely pinkish-purple. Native; stream-
sides, marshes and flushes; locally common over En and Sc,
very local in Wa and N Ir.
5. **C. raphanifolia** Pourret (<u>C.</u> <u>latifolia</u> Vahl non Lej.) -
<u>Greater</u> <u>Cuckooflower</u>. Perennial with long strong rhizome;
stems little or not branched, erect, to 70cm; leaves with
3-11 sinuate-crenate leaflets; petals 8-12mm, purplish-red,
rarely white. Intrd; natd in damp often shady places; very
scattered over Br, mostly in N & W; S Europe.
6. **C. pratensis** L. (<u>C.</u> <u>nymanii</u> Gand., <u>C.</u> <u>matthioli</u> Moretti,
<u>C.</u> <u>hayneana</u> (Reichb.) Fritsch, <u>C.</u> <u>palustris</u> (Wimmer & Grab.)
Peterm., <u>C.</u> <u>rivularis</u> Schur, <u>C.</u> <u>crassifolia</u> Pourret) -
<u>Cuckooflower</u>. Perennial with 0 or short rhizome; stems
little or not branched, ascending to erect, to 60cm; leaves
very variable, with 3-c.21 entire to crenate leaflets; petals
6-18mm, pale to deep pink, rarely white. Native; wet grassy
places; common throughout BI. Extremely variable but
impossible to subdivide usefully; identity of British
variants with taxa named in Europe is very dubious. Many

clones are sterile, but reproduce by rooting from leaves.
6 x 8. C. pratensis x C. flexuosa = C. x fringsii Wirtgen f.
(C. x haussknechtiana O. Schulz) occurs in scattered
localities in Br with the parents; it is intermediate
(especially in petals) and sterile, but reproduces
vegetatively as in C. pratensis.
   7. C. impatiens L. - Narrow-leaved Bitter-cress. Annual or   **R**
biennial; stems erect, to 60cm, little or not branched, with
many leaves with 7-c.19 entire to serrate leaflets; petals
2-3mm, white, sometimes O. Native; damp woods and river-
banks; locally frequent in W & S En and Wa, very rare in Sc,
Westmeath.
   8. C. flexuosa With. - Wavy Bitter-cress. Annual to short- **318**
lived perennial; stems flexuous, ascending to erect, often
well branched, to 50cm, with (3)4-7(10) leaves with 7-c.25
entire to dentate leaflets; petals 2.5-3mm, white; 2n=32.
Native; marshes, steam-sides and sometimes cultivated ground;
common throughout BI.
   8 x 9. C. flexuosa x C. hirsuta = C. x zahlbruckneriana O.
Schulz has been recorded near the parents in Monts; it
resembles C. flexuosa but is sterile and has 2n=24.
   9. C. hirsuta L. - Hairy Bitter-cress. Annual to biennial;
stems erect to ascending, often well branched at base, to
30cm, with 0-4(5) leaves with 3-c.11 entire to dentate leaflets; petals 2-3mm, white, sometimes O; 2n=16. Native; open
and cultivated ground, rocks and walls; common throughout BI.

15. ARABIS L. (Turritis L., Cardaminopsis (C. Meyer) Hayek) -
Rock-cresses
Annuals to perennials; basal leaves simple, deeply lobed to ±
entire; hairs usually branched and unbranched, rarely O or
all simple; petals white, pale yellow or pinkish-purple;
fruit a compressed or 4-angled siliqua; seeds in 1 or 2 rows
under each valve.

1 Stem-leaves strongly clasping stem at base; auricles
   distinctly longer than stem-width   2
1 Stem-leaves not or scarcely clasping stem at base;
   auricles O or much shorter than stem-width   6
  2 Non-flowering shoots O or forming ± sessile
     rosettes in compact clump   3
  2 Non-flowering shoots elongated, some becoming
     stolons and mat-forming   5
3 Fruits >7cm, patent and curved downwards when ripe
                            4. A. turrita
3 Fruits <7cm, erect   4
  4 Petals pale yellow; seeds in 2 rows under each
     valve   **3. A. glabra**
  4 Petals white; seeds in 1 row under each valve
                            7. A. hirsuta
5 Lowest leaves with petiole c. as long as blade;
   stem-leaves with pointed to narrowly rounded

auricles longer than wide; widespread   **6. A. caucasica**
5 Lowest leaves with petiole much shorter than blade;
  stem-leaves with broadly rounded auricles c. as long
  as wide; Skye   **5. A. alpina**
  6 Ripe fruits patent to erecto-patent, arising at
    <45 degrees; basal leaves mostly deeply sinuate,
    with long petioles   7
  6 Ripe fruits erect or nearly so, arising at much
    <45 degrees; basal leaves mostly shallowly
    sinuate with short petioles   8
7 Plant with branching rhizome; mountains of N & W BI
                                                     1. A. petraea
7 Plant with only root underground; alien in C & S En
                                                     2. A. arenosa
  8 Basal leaves entire to slightly toothed; petals
    white; pedicels mostly <8mm   **7. A. hirsuta**
  8 Basal leaves sinuate-lobed; petals white, pink
    or yellow, if pure white then most pedicels >8mm   9
9 Petals pale yellow; pedicels mostly <7mm **9. A. scabra**
9 Petals white to pink; pedicels mostly >7mm
                                                     8. A. collina

1. A. petraea (L.) Lam. (<u>Cardaminopsis</u> <u>petraea</u> (L.) Hiit.) - 318
<u>Northern</u> <u>Rock-cress</u>. Loosely mat-forming perennial with **R**
branching rhizome; stems ascending to erect, to 25cm; basal
leaves long-stalked, deeply lobed; flowers few; petals white
to purplish. Native; mountain rock ledges and crevices; N
Wa, NW En, Sc, extremely local in Ir.
2. A. arenosa (L.) Scop. (<u>Cardaminopsis</u> <u>arenosa</u> (L.) Hayek)
- <u>Sand</u> <u>Rock-cress</u>. Tufted annual to perennial; stems erect,
to 40cm; basal leaves long-stalked, deeply lobed; flowers
usually numerous; petals white or pinkish-purple.   Intrd;
waste and other open ground; C & S En; Europe.
3. A. glabra (L.) Bernh. (<u>Turritis</u> <u>glabra</u> L.) - <u>Tower</u> 318
<u>Mustard</u>. Tufted biennial; stems erect, to 1m, pubescent **RR**
below, glabrous above; basal leaves entire to sinuate-lobed;
flowers numerous; petals pale yellow.   Native; dry grassy,
rocky and waste places; very local and decreasing in En,
casual in Sc.
4. A. turrita L. - <u>Tower</u> <u>Cress</u>.   Tufted biennial or
perennial; stems erect, to 70cm, pubescent; basal leaves
sinuate-toothed; flowers numerous; petals pale yellow.
Intrd; on old walls; Cambridge, formerly elsewhere; Europe.
5. A. alpina L. - <u>Alpine</u> <u>Rock-cress</u>. Mat-forming perennial **RRR**
with stolons producing rosettes; stems ± erect, to 40cm;
basal leaves short-stalked, sinuate-toothed;   flowers
numerous; petals white.   Native; rock-ledges at 820-850m;
Skye (N Ebudes).
6. A. caucasica Willd. ex Schldl. (<u>A.</u> <u>alpina</u> ssp. <u>caucasica</u>
(Willd. ex Schldl.) Briq.). - <u>Garden</u> <u>Arabis</u>. Differs from <u>A.</u>
<u>alpina</u> in its long-stalked basal leaves with usually fewer
teeth; larger petals (mostly >10mm, not <10mm) white to

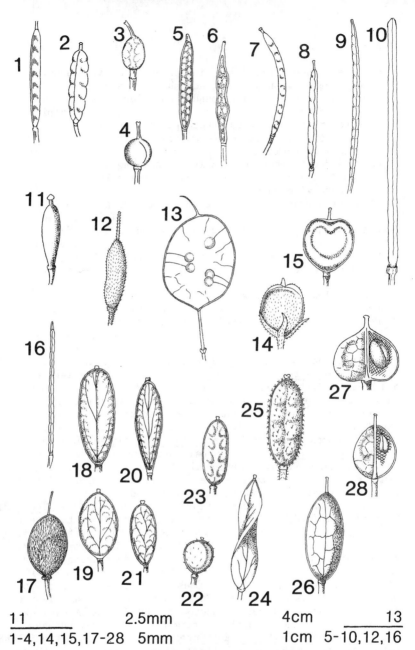

11 —————— 2.5mm   4cm —————— 13
1-4,14,15,17-28 5mm   1cm 5-10,12,16

FIG 318 – Fruits of Brassicaceae. 1, Rorippa sylvestris. 2,
R. palustris. 3, R. amphibia. 4, R. austriaca. 5, R.
nasturtium–aquaticum. 6, R. x sterilis. 7, R. microphylla. 8,
Arabis petraea. 9, A. hirsuta. 10, A. glabra. 11, Armoracia.
12, **Aubrieta**. 13, Lunaria. 14, **Alyssum** alyssoides. 15, A.
saxatile. 16, Cardamine flexuosa. 17, Berteroa. 18, Erophila
verna var. verna. 19, var. **praecox**. 20, E. majuscula. 21, E.
glabrescens. 22, Lobularia. 23, Draba muralis. 24, D. incana.
25, D. norvegica. 26, **D. aizoides**. 27, Lepidium draba ssp.
draba. 28, ssp. **chalepense**.

purplish; and leaves more densely whitish-grey (not grey-green) pubescent. Intrd; garden escape well natd on walls and limestone rocks; frequent in En and Wa, rare in Sc; S Europe.

7. A. hirsuta (L.) Scop. (A. brownii Jordan) - Hairy Rock- **318** cress. Tufted biennial to perennial; stems erect, to 60cm; basal leaves short- to rather long-stalked, ± entire to shallowly toothed; flowers numerous; petals white. Native; limestone rocks and bare places in grassland, walls; locally common throughout BI. A. brownii is endemic to dunes in W Ir; it is smaller, with hairs confined to leaf-margins, and ± unwinged (not winged) seeds, and may merit ssp. status.

8. A. collina Ten. (A. muricola Jordan, A. muralis Bertol. non Salisb., A. rosea DC.) - Rosy Cress. Tufted perennial; stems erect to ascending, to 30cm; basal leaves short-stalked, sinuate-lobed; flowers rather few; petals white or pink. Intrd; garden ecape natd on walls and banks; S & C Sc; S Europe.

9. A. scabra All. (A. stricta Hudson) - Bristol Rock-cress. **RRR** Tufted perennial; stems erect, to 25cm; basal leaves short-stalked, sinuate-lobed; flowers rather few; petals pale yellow. Native; limestone rock crevices and rubble; near Bristol (N Somerset and W Gloucs).

16. AUBRIETA Adans. - Aubretia
Perennials; leaves simple, with few deep teeth; hairs unbranched and stellate; petals mauve to purple, rarely seeds in 2 rows under each valve.

Other spp. - Some escaped cultivars might involve some other spp. in their ancestry.

1. A. deltoidea (L.) DC. - Aubretia. Mat-forming, with **318** numerous sterile shoots; flowering stems decumbent to erect; to 15cm; flowers few; petals 12-20mm. Intrd; much grown in gardens and sometimes ± natd on walls and rocky banks; scattered in Br and CI, mainly S; SE Europe.

TRIBE 4 - ALYSSEAE (genera 17-22). Hairs unbranched or branched, often stellate, not glandular; petals yellow, white or pinkish-purple; fruit usually a latiseptate silicula, sometimes a short siliqua.

17. LUNARIA L. - Honesty
Biennials; basal leaves simple, long-stalked; hairs unbranched; petals pinkish-purple, sometimes white; fruit a distinctive, large, latiseptate flat silicula; seeds in 2 rows under each valve.

Other spp. - L. rediviva L., from Europe, differing from L. annua in being a perennial with narrowly elliptic fruits and distinctly stalked upper leaves, occurs as a rare casual or garden escape.

1. L. annua L. - <u>Honesty</u>. Stems erect, to 1m; lower leaves **318**
long-stalked, ovate-cordate, sharply dentate, the upper
similar but smaller and ± sessile; petals 15-25mm; fruit
2.5-7 x 1.5-3.5cm, oblong-elliptic to suborbicular. Intrd;
very commonly grown and often escaping on tips, roadsides,
waste ground etc, occasionally persistent; scattered in Br
and CI; SE Europe.

18. ALYSSUM L. (<u>Aurinia</u> Desv.) - <u>Alisons</u>
Annuals to perennials; hairs branched and unbranched, often ±
stellate; leaves simple, ± entire; petals yellow; fruit a
latiseptate silicula, not or slightly inflated; seeds 2 under
each valve.

1. A. alyssoides (L.) L. - <u>Small Alison</u>. Densely pubescent **318**
annual or biennial with erect to ascending often well-
branched stems to 30cm; basal leaves withered by flowering
time; petals pale yellow, fading to white when dried; fruit
3-4mm, ± orbicular, pubescent, slightly inflated. Intrd;
grassy and arable fields; formerly widespread in S & E Br,
now ± confined to Suffolk; Europe.
2. A. saxatile L. (<u>Aurinia</u> <u>saxatilis</u> (L.) Desv.) - <u>Golden</u> **318**
<u>Alison</u>. Densely pubescent perennial to 45cm; stems branched
and ± woody at base, apically producing biennial flowering
stems with basal leaves persistent at flowering; petals
usually bright yellow; fruit 3.5-6mm, ± orbicular, glabrous,
± flat. Intrd; commonly grown in gardens and seeding
abundantly, sometimes escaping on to walls and dry banks;
scattered in En and Wa; C & SE Europe.

19. BERTEROA DC. - <u>Hoary Alison</u>
Annuals to perennials; hairs stellate; leaves simple, ±
entire; petals white, often discolouring yellow when dried,
bifid nearly 1/2 way to base; fruit a latiseptate silicula,
slightly inflated; seeds 2-6 under each valve.

1. B. incana (L.) DC. - <u>Hoary Alison</u>. Grey-pubescent; stems **318**
erect, to 60cm; fruit 4-8mm, with style 1.5-3.5mm. Intrd;
rough grassy waste places, waysides; scattered as casual in C
& S Br, natd in few areas in S; Europe.

20. LOBULARIA Desv. - <u>Sweet Alison</u>
Annuals to perennials; hairs 2-armed with 0 stalk; leaves
simple, entire; petals white, sometimes purplish; fruit a
latiseptate silicula, slightly inflated; seeds 1 under each
valve.

1. L. maritima (L.) Desv. - <u>Sweet Alison</u>. Plant grey- **318**
pubescent; stems much branched, decumbent to ascending, to
30cm; fruit obovate, broadly elliptic or suborbicular, 2-

3.5mm. Intrd; much grown in gardens and commonly escaping on walls and other dry places, well natd on coastal sands in S Br and CI; scattered over most of BI; S Europe.

21. DRABA L. - Whitlowgrasses

Annuals to perennials; hairs unbranched or branched, often stellate; leaves simple; petals white or yellow; fruit a latiseptate silicula or sometimes short siliqua, ± not inflated; seeds in 2 rows under each valve.

1 Petals yellow; style >1mm in fruit; hairs all
  unbranched   **1. D. aizoides**
1 Petals white; style <1mm in fruit; at least some
  hairs branched and stellate   2
  2 Annual; stem-leaves <2x as long as wide, cordate
    and clasping stem at base   **4. D. muralis**
  2 Perennial; at least some stem-leaves >2x as long
    as wide or stem-leaves 0, tapered to rounded at
    base and ± not clasping stem   3
3 Stems with 0-2(3) leaves, <8cm; fruit 3-8mm, not
  twisted   **2. D. norvegica**
3 Stems normally with >3 leaves, normally >10cm; fruit
  7-12mm, usually twisted   **3. D. incana**

1. D. aizoides L. - Yellow Whitlowgrass. Tufted perennial **318** with all leaves in basal rosettes; leaves linear, entire; **RR** stems to 15cm, erect, few-flowered; petals yellow; fruit 5-10mm, with style 1.2-3mm. Probably native; limestone rocks and walls near sea; Gower peninsula, Glam (known since 1795); European mountains.

2. D. norvegica Gunnerus (D. rupestris R. Br.) - Rock **318** Whitlowgrass. Tufted perennial with all leaves in basal **R** rosettes or 1-3 on flowering stems; leaves linear to narrowly elliptic, entire or nearly so; stems to 6cm, erect, few-flowered; petals white; fruit 3-8mm, with style <0.5mm. Native; bare places near mountain-tops, usually calcareous; very local in C & N Sc. D. rupestris, from Sc and Faroes, might be distinct.

3. D. incana L. - Hoary Whitlowgrass. Tufted perennial with **318** leafy many-flowered erect stems to 40cm and sterile basal rosettes; leaves narrowly elliptic to narrowly ovate, usually coarsely toothed; petals white; fruit 7-12mm, with style <1mm. Native; rock-ledges and soil-pockets mainly on limestone, and sand-dunes in N & W Sc; very local in N Wa, N En, N & W Ir and C & N Sc. Dwarf plants are easily confused with D. norvegica, which usually has few stellate and more simple hairs.

4. D. muralis L. - Wall Whitlowgrass. Annual with basal **318** leaf rosette and sparsely leafy many-flowered erect stem to **R** 40cm; basal leaves obovate; stem-leaves ovate-cordate, coarsely toothed; petals white; fruit 3-6mm, with style <0.5mm. Native; soil-pockets on limestone rocks and cliffs, also natd on walls and in gardens; scattered over much of BI except N Sc and CI, but perhaps native only in SW & N En.

22. EROPHILA DC. - Whitlowgrasses

Early flowering ephemerals; hairs unbranched or branched, often stellate; leaves confined to basal rosette, simple, entire or toothed; petals white, bifid ± 1/2 way to base; fruit a latiseptate silicula or sometimes short siliqua, inflated or not; seeds in 2 rows under each valve; flowering in very early spring.

A much misunderstood genus. Many traditionally used characters, such as fruit size and shape and ovule number, do not correlate well with cytological characters or breeding behaviour and separate only pure-breeding lines. This account is based upon the work of S.A. Filfilan and T.T. Elkington, which largely confirmed the earlier conclusions of O. Winge.

1 Leaves and lower parts of stems densely grey-pubescent; petioles 1/5 to 1/2 as long as laminas; seeds 0.3-0.5mm; petals bifid ≤1/2way to base
                                        1. E. majuscula

1 Leaves and lower parts of stems subglabrous to moderately pubescent, green; petioles ≥1/2 as long as laminas; seeds 0.5-0.8mm   2
  2 Petioles 0.5-1x as long as laminas; petals bifid 1/2-3/4 way to base; usually pubescent   **2. E. verna**
  2 Petioles 1.5-2.5x as long as laminas; petals bifid ≤1/2 way to base; usually subglabrous
                                    3. E. glabrescens

1. E. majuscula Jordan - Hairy Whitlowgrass. Plant densely 318 pubescent, with 2-many stems to 9cm; fruit oblong to elliptic, ± flat, 2.5- 6mm, 1.5-4x as long as wide, with 15-60(70) seeds; 2n=14. Native; all sorts of open, dry ground, especially on calcareous soils, but rarely or not on dunes; scattered sparsely through BI except N & W Sc.

2. E. verna (L.) DC. (E. spathulata Lang, E. praecox 318 (Steven) DC.) - Common Whitlowgrass. Plant sparsely to moderately pubescent, with 1-many stems to 10(25)cm; fruit oblanceolate to oblong or elliptic, or broadly elliptic to ± orbicular in var. **praecox** (Steven) Diklic (ssp. praecox (Steven) Walters), ± flat or ± inflated, 1.5-9mm, 1.5-3x as long as wide, with 15-50 seeds; 2n=30-46. Native; all sorts of open, dry ground, especially calcareous, rocks, walls, open grassland and dunes; locally common throughout BI.

3. E. glabrescens Jordan - Glabrous Whitlowgrass. Plant 318 subglabrous to sparsely pubescent with 1-many stems to 9cm; fruit oblanceolate to elliptic, ± flat, 3-6mm, 1.5-4x as long as wide, with 20-60 seeds; 2n=48-56. Native; habitat and distribution as in E. verna, but less common.

TRIBE 5 - LEPIDEAE (genera 23-34). Hairs 0, unbranched or (less often) branched; petals white, sometimes pink, yellow or 0; fruit usually angustiseptate silicula, sometimes latiseptate, sometimes indehiscent.

23. COCHLEARIA L. - Scurvygrasses.
Annuals to perennials (mostly biennials); basal leaves simple, long-stalked; hairs 0 or unbranched; petals white or mauve; fruit an inflated or ± compressed and angustiseptate silicula; seeds in 2 rows under each valve.
A very plastic difficult genus that has been interpreted in several ways; this account is based upon the work of G.M. Fearn and J.J.B. Gill.

1 Basal leaves cuneate at base; fruit compressed, angustiseptate, the septum >3x as long as wide
        1. C. anglica
1 Basal leaves cordate to very broadly cuneate at base; fruit scarcely compressed, the septum <2(3)x as long as wide   2
  2 Upper stem-leaves petiolate; flowers <5(6)mm across; fruits with <12(16) seeds   **5. C. danica**
  2 Upper stem-leaves sessile, clasping stem; flowers mostly >5mm across; fruits with <8 seeds   3
3 Perennial with ± woody base; basal leaves ± persistent, all <7(10)mm wide, coriaceous; at least some fruits acute at both ends and widest at or just below middle   **4. C. micacea**
3 Biennial to perennial; basal leaves rarely persistent, often >1cm wide, rarely coriaceous; fruits rarely acute at both ends, if so widest above middle   4
  4 Inflorescence in bud strongly concave on top; stems erect, to 30cm; petals always white   **2. C. pyrenaica**
  4 Inflorescence in bud flat to slightly concave on top; stems procumbent to erect, to 75cm; petals white or mauve   **3. C. officinalis**

1. C. anglica L. - English Scurvygrass. Biennial to 309 perennial with erect to ascending stems to 40cm; basal leaves 318 cuneate at base; stem-leaves stalked or not, the latter clasping stem; fruit 8-14mm, compressed; 2n=48. Native; muddy shores and estuaries, often in very wet places; coasts of most of BI.
1 x 3. C. anglica x C. officinalis = C. x hollandica Henrard occurs probably wherever the parents meet; it is intermediate in fruit compression and shape of basal leaves. It is vigorous and fertile, and backcrosses to both parents; 2n=36.
2. C. pyrenaica DC. (C. officinalis ssp. pyrenaica (DC.) Bonnier & Layens, C. alpina pro parte excl. typ.) - Pyrenean Scurvygrass. Biennial to perennial with usually erect stems to 40cm; basal leaves usually cordate at base; stem-leaves all or most sessile, clasping stem; fruit 3-7mm, not

compressed; 2n=12.   Native; barish damp places on upland
limestone; N En (Derbys to Cumberland), N Ebudes.   Most
records from Wa and Sc seem to be errors for the true C.
alpina.
   3. C. officinalis L. (C. alpina (Bab.) H. Watson, C. **309**
officinalis ssp. alpina (Bab.) Hook. f., C. scotica Druce, C. **318**
atlantica Pobed., C. islandica Pobed., C. groenlandica auct.
non L.) – Common Scurvygrass.   Biennial to perennial with
usually erect stems to 40(75)cm; basal leaves usually cordate
at base; stem-leaves mostly sessile and clasping stem, the
lowest often short-stalked; fruit 3-7mm, not compressed;
2n=24.   Native; salt-marshes and other habitats by sea, wet
areas on acid mountains and by salt-treated roads inland;
round the coasts of BI, on mountains in Br and Ir (but absent
from areas of En where C. pyrenaica grows), and by some roads
in En and Wa.   Very variable.   Plants on coasts of W Ir, W &
N Sc and perhaps W En and Wa are often distinct in having
very short stems, smaller flowers and smaller leaves with
truncate to broadly cuneate bases, and come under **C. scotica**;
they might be a ssp.
   3 x 5. C. officinalis x C. danica occurs probably frequently
in Br where the parents meet; it is intermediate in leaf and
flower characters and fertile.
   4. C. micacea E. Marshall – Mountain Scurvygrass.   Perennial   **RR**
with woody base producing short stolons and procumbent to
erect stems to 20cm; basal leaves cordate at base; stem-
leaves stalked below, sessile above; fruit 3-6mm, not
compressed; 2n=26.   Native; on micaceous schists above 800m;
M & E Perth, Angus and W Ross.
   5. C. danica L. – Danish Scurvygrass.   Annual to biennial **309**
with ascending to erect stems to 20cm; basal leaves cordate
at base; stem-leaves ivy-shaped, stalked or the uppermost
sessile but not clasping stem; fruit 3-6mm, not compressed;
2n=42.   Native; sandy and pebbly shores, banks and walls near
sea, and by railways and salt-treated roads inland; coasts of
most of BI and widespread inland in En, Wa and NE Ir.   Some
plants from Shetland are much more robust with very fleshy
leaves, and might deserve taxonomic recognition.

24. CAMELINA Crantz – Gold-of-pleasures
Annuals or sometimes biennials; leaves simple, lanceolate to
narrowly elliptic, entire or nearly so; hairs branched and
unbranched; petals yellow; fruit a latiseptate, + inflated, +
smooth silicula with keeled to + winged margin and long
persistent style, on patent to erecto-patent pedicels; seeds
in 2 rows under each valve.

   Other spp. – **C. alyssum** (Miller) Thell. (C. sativa ssp.
alyssum (Miller) Hegi & E. Schmid), **C. pilosa** (DC.) Vassilcz.
(C. sativa ssp. pilosa (DC.) Thell.) and **C. macrocarpa**
Wierzb. ex Reichb. have been recorded as casuals, but very
rarely if ever occur nowadays.

1. C. sativa (L.) Crantz - <u>Gold-of-pleasure</u>. Plant sub- **339**
glabrous or sometimes pubescent; stems to 70cm, erect, petals
4-5mm; fruit (5)6-9(10) x (3.5)4-5(5.5)mm (excl. style);
style <1/3x as long as rest of fruit; seeds 1.2-2(2.5)mm.
Intrd; frequent casual from many sources, especially bird-
seed, formerly common in arable fields, now mostly on tips;
scattered over Br and Ir; S & E Europe.
2. C. microcarpa Andrz. ex DC. (<u>C. sativa</u> ssp. <u>microcarpa</u> **339**
(Andrz. ex DC.) Thell.) - <u>Lesser</u> <u>Gold-of-pleasure</u>. Differs
from <u>C. sativa</u> in being always pubescent; petals 2.5-4mm;
fruit 4-7(7.5) x 2.4-4.5(5)mm (excl. style); style >1/3x as
long as rest of fruit; seeds 0.7-1.5(1.6)mm. Intrd; similar
places to <u>C. sativa</u> but much less common; sporadic in Br; S &
E Europe.

25. NESLIA Desv. - <u>Ball</u> Mustard
Annuals; leaves simple, narrowly ovate to elliptic, entire or
nearly so, clasping stem at base; hairs ± stellate; petals
yellow; fruit a latiseptate, indehiscent, reticulately
ridged, unkeeled silicula with long persistent style; seed
usually 1.
1. N. paniculata (L.) Desv. - <u>Ball</u> <u>Mustard</u>. Plant **339**
pubescent; stems to 60cm, erect; petals c.2mm; fruit 1.5-3 x
1.5-3mm on patent to erecto-patent pedicels to 12mm. Intrd;
casual from a variety of sources, once common now rare, on
tips and waste places; sporadic in Br; S & E Europe.

26. CAPSELLA Medikus - <u>Shepherd's-purses</u>
Annuals to biennials; basal leaves simple, entire to deeply
pinnately lobed; stem-leaves clasping stem at base; hairs
unbranched; petals white or sometimes red-tinged or 0; fruit
an angustiseptate, ± obtriangular, compressed silicula; seeds
in 2 rows under each valve.
Genetically similar to <u>Erophila</u> in producing many pure-
breeding lines at more than 1 ploidy level, and as with the
latter it is possible with difficulty to recognize the
different ploidy levels as spp., at least in BI.

1. C. bursa-pastoris (L.) Medikus - <u>Shepherd's-purse</u>. Plant **327**
glabrous to sparsely pubescent; stems to 40cm, usually erect;
petals white, 2-3mm, sometimes 0; fruit 5-9 x 3.5-7.5mm, ±
obtriangular, with straight to slightly convex sides,
variably emarginate at apex, with style <0.5mm, the pedicels
patent to erecto-patent; 2n=32. Native; cultivated and other
open ground; common throuhout BI. Extremely variable,
especially in leaf and fruit shape; c.25 segregates have been
recognised in BI.
1 x 2. C. bursa-pastoris x C. rubella = C. x gracilis Gren.
has been found sporadically in En with the parents; it is
intermediate in fruit characters and produces very long
inflorescences with very little seed, but note that <u>C. bursa-
pastoris</u> often forms sterile capsules in colder weather.

2. C. rubella Reuter - Pink Shepherd's-purse. Differs from 327
C. bursa-pastoris in its petals 1.5-2mm (scarcely longer than
sepals) and usually (like the sepals) red-tinged; fruit 5-7 x
4-6mm, with concave sides forming more gradually tapered
base; 2n=16. Intrd; cultivated and waste ground; sporadic in
Br and CI, mainly S & C En; S Europe.

27. HORNUNGIA Reichb. - Hutchinsia
Annuals; leaves deeply pinnately lobed or $\pm$ pinnate; hairs
stellate; petals white; fruit an angustiseptate $\pm$ compressed
silicula; seeds (1-)2 under each valve.

1. H. petraea (L.) Reichb. - Hutchinsia. Early-flowering 327
ephemeral; stems erect or ascending, to 10(15)cm; petals   R
c.1mm, scarcely longer than sepals; fruit 2-3 x 1-1.5mm, on
patent to erecto-patent pedicels 2-6mm. Native; bare places
becoming desiccated in summer, on carboniferous limestone and
calcareous dunes; very local in N & SW En, Wa and CI.

28. IONOPSIDIUM Reichb. - Violet Cress
Annuals; leaves simple, entire to 3-lobed; hairs 0; petals
pink to purple, sometimes white; fruit an angustiseptate
slightly compressed silicula; seeds 2-5 under each valve.

1. I. acaule (Desf.) Reichb. - Violet Cress. Stems $\pm$ absent 339
to short and congested; leaves appearing as if in a rosette,
ovate to orbicular, entire, c.0.5cm, on long thin petioles;
flowers solitary in leaf-axils on long thin pedicels; petals
5-8mm, $\pm$ not notched; fruit 4-6mm, with style exceeding
apical notch. Intrd; grown as garden annual and sometimes
found self-sown on tips, rough ground and roadsides; S En;
Portugal.

29. TEESDALIA R. Br. - Shepherd's Cress
Annuals; leaves deeply pinnately lobed or $\pm$ pinnate; hairs
unbranched or 0; petals white, the 2 abaxial c.2x as long as
the 2 adaxial ones; fruit an angustiseptate, compressed
silicula, $\pm$ keeled to very narrowly winged round edges; seeds
2 under each valve.

Other spp. - T. coronopifolia (Bergeret) Thell. was recorded
in error from Eigg, N Ebudes.

1. T. nudicaulis (L.) R. Br. - Shepherd's Cress. Most 327
leaves in basal rosette; stem-leves few and reduced; stems   R
erect to ascending, to 45cm, puberulent; fruit 3-5 x 2-4mm,
rounded at base, emarginate at apex, with minute style, on
patent pedicels 3-6mm. Native; open sand, gravel or shingle;
scattered very locally through BI and CI, NE Ir.

30. THLASPI L. (Pachyphragma (DC.) Reichb.) - Penny-cresses
Annuals to perennials, sometimes with rhizome; leaves simple,

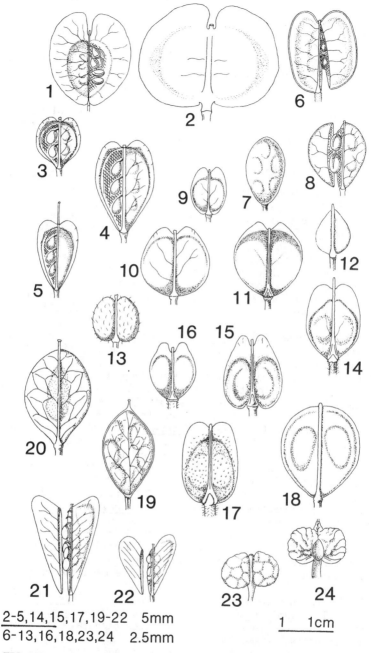

2-5,14,15,17,19-22 5mm
6-13,16,18,23,24 2.5mm

1 1cm

FIG 327 - Fruits of **Brassicaceae**. 1, Thlaspi arvense. 2, T. macrophyllum. 3, T. perfoliatum. 4, T. alliaceum. 5, T. caerulescens. 6, Teesdalia. 7, Subularia. 8, Hornungia. 9, **Lepidium** hyssopifolium. 10, L. bonariense. 11, L. virginicum. 12, L. graminifolium. 13, L. latifolium. 14, L. heterophyllum. 15, L. sativum. 16, L. ruderale. 17, L. campestre. 18, L. perfoliatum. 19, Cochlearia officinalis. 20, C. anglica. 21, Capsella bursa-pastoris. 22, C. rubella. 23, Coronopus didymus. 24, C. squamatus.

entire to dentate; hairs unbranched or 0; petals white; fruit
an angustiseptate compressed silicula with narrow to broad
wing round edges; seeds (1)3-8 under each valve.

1 Plant with strong rhizome; stem-leaves all petiolate
                                               5. T. macrophyllum
1 Plant without rhizome; stem-leaves sessile, clasping
  stem   2
  2 Fruit >10mm, with wing >1mm wide at midpoint
                                                         1. T. arvense
  2 Fruit <10mm, with wing <1mm wide at midpoint   3
3 Biennial to perennial, usually with non-flowering
  leaf-rosettes; style equalling or exceeding apical
  notch of fruit   **4. T. caerulescens**
3 Annual; style c.1/2 as long as apical notch of fruit   4
  4 Plant glabrous, not smelling of garlic; stem-
    leaves with rounded auricles   **3. T. perfoliatum**
  4 Plant smelling of garlic when crushed; stem
    sparsely pubescent at base; stem-leaves with acute
    auricles   **2. T. alliaceum**

1. T. arvense L. - Field Penny-cress. Glabrous annual, 327
stinking (but not of garlic) when crushed; stems erect, to
60cm; stem-leaves with pointed auricles; fruit 10-18mm,
broadly winged all round, with deep apical notch much
exceeding style length. Possibly native; weed of waste and
arable land; scattered over most of BI but much commoner in E
and absent from much of Sc, Wa and Ir.
2. T. alliaceum L. - Garlic Penny-cress. Annual smelling of 327
garlic when crushed; stems erect, to 60cm, pubescent at base
of stem; stem-leaves sessile, clasping stem with pointed
auricles; fruit 4-10mm, narrowly winged mainly apically, with
shallow apical notch but even shorter style. Intrd; natd
weed of arable fields and borders; E Kent, S Essex; Europe.
3. T. perfoliatum L. - Perfoliate Penny-cress. Glabrous 327
annual; stems erect, to 25cm; stem-leaves sessile, clasping   RR
stem with rounded auricles; fruit 4-6mm, rather broadly
winged apically, with wide ± deep apical notch and much
shorter style. Native; bare limestone stony ground mainly or
only in E Gloucs, N Wilts and Oxon, rare casual elsewhere in
En.
4. T. caerulescens J.S. & C. Presl (T. alpestre L. non 327
Jacq., T. calaminare (Lej.) Lej. & Courtois) - Alpine Penny-   R
Cress. Glabrous perennial or sometimes biennial with sterile
leaf-rosettes; stems erect, to 40cm; stem-leaves sessile,
clasping stem with subacute to ± rounded auricles; fruit
5-8mm, narrowly to rather broadly winged mainly apically,
with usuallly shallow apical notch equalled or exceeded by
style. Native; bare or sparsely grassed stony places mainly
on limestone naturally or artificially contaminated with lead
or zinc; extremely local and disjunct in Br from N Somerset
to C Sc. Very variable but not able to be subdivided

satisfactorily.

5. T. macrophyllum Hoffm. (Pachyphragma macrophyllum 327
(Hoffm.) N. Busch - Caucasian Penny-cress. Glabrous
perennial with strong rhizomes, smelling of garlic when
crushed; stems erect to ascending, to 40cm; stem-leaves
cordate at base, petiolate; fruit 8-12mm, broadly winged,
with deep apical notch much exceeding style length. Intrd;
natd in woodland in Herts, N Somerset and Salop; Caucasus.

31. IBERIS L. - Candytufts
Annuals to perennials; leaves simple, entire to deeply lobed;
hairs unbranched or O; inflorescence a corymb at flowering,
often elongate in fruit; petals white to purple or mauve, the
2 abaxial much longer than the 2 adaxial; fruit an angustiseptate compressed silicula with fairly narrow to broad wing
round edges; seeds 1 under each valve.

1 Perennial with stems woody at base and leaves
  evergreen   **1. I. sempervirens**
1 Annual with herbaceous stem   2
  2 Inflorescence elongating in fruit; fruits 3-6mm
                                    2. I. amara
  2 Inflorescence remaining corymbose in fruit; fruits
    7-10mm   **3. I. umbellata**

1. I. sempervirens L. - Perennial Candytuft. Glabrous 339
sprawling shrub with procumbent woody lower parts and
ascending flowering shoots to 25cm high; leaves entire,
obtuse; inflorescence elongating in fruit; fruits 4-7mm,
broadly winged especially apically. Intrd; much grown in
gardens and a persistent relic or throwout, occasionally
self-sown; S Br; S Europe.
2. I. amara L. - Wild Candytuft. Sparsely pubescent annual; 339
stems erect, with divergent branches, to 30cm; leaves entire   R
to lobed or dentate, acute to ± rounded; fruits 3-6mm,
narrowly winged. Native; bare places in grassland and arable
fields on dry calcareous soils; CS En N to Cambs, often
reported as casual elsewhere but probably mostly errors for
I. umbellata.
3. I. umbellata L. - Garden Candytuft. Glabrous annual; 339
stems erect, with divergent branches, to 70cm; leaves ±
entire, acute; fruits 7-10mm, with very broad pointed wings
apically. Intrd; much grown in gardens and a common casual
on tips and in waste places; Br and CI; S Europe.

32. LEPIDIUM L. (Cardaria Desv.) - Pepperworts
Annuals to perennials; leaves simple to 2-3-pinnate; hairs
unbranched or O; inflorescence a raceme or panicle; petals
usually white, sometimes reddish, yellowish or O; fruit an
angustiseptate compressed silicula, sometimes ± indehiscent,
strongly keeled to winged round edges; seeds 1(2) under each
valve but sometimes not developing.

1 Lower and upper stem-leaves strikingly different, the
   lower finely 2-3-pinnate, the upper ovate, entire, +
   encircling stem; petals yellow   **8. L. perfoliatum**
1 Lower and upper stem-leaves often different but with
   + gradual transition; petals white, reddish or 0   2
   2 Fruit >4mm, usually at least as long as pedicel;
      basal 1/2 of style usually fused with wings of
      fruit   3
   2 Fruit <4mm, usually shorter than pedicel; style
      free from wings of fruit or wings 0   5
3 Lower and middle stem-leaves very deeply pinnately
   lobed to 2-pinnate, not clasping stem   **1. L. sativum**
3 Lower and middle stem-leaves simple, toothed to very
   shallowly lobed, clasping stem at base   4
   4 Fruit covered with scale-like vesicles; style not
      exceeding apical notch of fruit   **2. L. campestre**
   4 Fruit with few or 0 vesicles; style exceeding
      apical notch of fruit   **3. L. heterophyllum**
5 Perennial; fruit unwinged, rounded to subacute at
   apex, with style projecting beyond   6
5 Normally annual or biennial; fruit usually winged at
   least apically, notched at apex, with style not or
   scarcely projecting beyond notch   8
   6 Fruit not or scarcely dehiscent; inflorescence a
      + corymbose panicle   **11. L. draba**
   6 Fruit readily dehicent; inflorescence a raceme or
      a racemose to pyramidal panicle   7
7 Upper stem-leaves linear; fruit 2-4mm, in lax
   branched racemes   **10. L. graminifolium**
7 Upper stem-leaves elliptic to narrowly so; fruit
   1.5-2.5mm, in congested panicles   **9. L. latifolium**
   8 Middle and upper stem-leaves deeply pinnately
      lobed to 2-pinnate   **5. L. bonariense**
   8 Middle and upper stem-leaves entire to dentate   9
9 Fruit (2)2.3-3.5mm wide; apical notch >2mm deep,
   c.1/10 of fruit length   **4. L. virginicum**
9 Fruit 1.2-2(2.3)mm wide; apical notch <2mm deep,
   <1/10 of fruit length   10
   10 Upper stem-leaves entire, rounded to subacute at
      apex; basal leaves (gone before fruiting) pinnate
      to 2-pinnate   **6. L.ruderale**
   10 Upper stem-leaves entire to dentate, acuminate to
      subacute at apex; basal leaves (gone before
      fruiting) dentate to deeply pinnately lobed
                                         7. L. hyssopifolium

Other spp. - >30 other spp. have been recorded as casuals in
BI, most as wool-aliens or seed-aliens, all rare or no longer
occurring.

1. L. sativum L. - <u>Garden</u> <u>Cress</u>.   Erect annual to 50cm; **327**
stem-leaves not clasping stem, lower ones 1-2-pinnate, upper

ones entire to lobed; fruit 5-6mm, broadly winged apically with deep notch and shorter style. Intrd; commonly grown as seedling salad-plant and a frequent contaminant of grain and bird-seed, found on tips and waysides; frequent casual in much of BI; W Asia originally.

2. L. campestre (L.) R. Br. - <u>Field</u> Pepperwort. Erect to 327 decumbent annual to biennial to 60cm; stem-leaves simple, shortly dentate, with acute auricles clasping stem; fruit 4.5-6mm, broadly winged apically with ± deep notch and shorter style. Native; open grassland, banks, walls, waysides and arable fields; scattered but locally common in much of Br and Ir, mostly in S.

3. L. heterophyllum Benth. (<u>L. smithii</u> Hook., <u>L. pratense</u> 327 (Serres ex Gren. & Godron) Rouy & Fouc., <u>L. villarsii</u> Gren. & Godron, <u>L. hirtum</u> (L.) Smith) - <u>Smith's</u> Pepperwort. Erect to decumbent perennial to 45cm; stem-leaves similar to those of <u>L. campestre</u>; fruit 4.5-6mm, broadly winged apically with rather shallow notch and longer style. Native; similar places to <u>L. campestre</u>; scattered but locally common in much of BI, mostly in W.

4. L. virginicum L. (<u>L. neglectum</u> Thell., <u>L. ramosissimum</u> 327 Nelson, <u>L. densiflorum</u> Schrader) - <u>Least</u> Pepperwort. Erect annual to biennial to 50cm; stem-leaves simple, not clasping stem, lower ones ± lobed, middle and upper ones dentate to ± entire; fruit 2.3-4mm, narrowly winged apically with shallow notch and shorter style. Intrd; rather infrequent casual probably mainly from wool and bird-seed; scattered and sporadic in Br; N America.

5. L. bonariense L. - <u>Argentine</u> Pepperwort. Erect annual or 327 biennial to 60cm; stem-leaves not clasping stem, 2-pinnate to deeply pinnately lobed; fruit 2-3.5mm, narrowly winged apically with shallow notch and shorter style. Intrd; infrequent casual on tips and in fields, mainly from wool and bird-seed; very scattered in Br, mostly S En; S America.

6. L. ruderale L. - <u>Narrow-leaved</u> Pepperwort. Erect often 327 much-branched annual or biennial to 45cm; stem-leaves not clasping stem, lower ones pinnate, upper ones simple, linear; fruit 2-2.5mm, narrowly winged apically with shallow notch and shorter style. Probably intrd; common casual of waste places, waysides and tips, natd and possibly native in open ground especially near sea; locally common in E & SE En, scattered elsewhere in Br and Guernsey, mainly En.

7. L. hyssopifolium Desv. (<u>L. africanum</u> (Burm. f.) DC., <u>L.</u> 327 <u>divaricatum</u> Sol. ssp. <u>linoides</u> (Thunb.) Thell.) - <u>African</u> Pepperwort. Erect annual or biennial to 45cm; stem-leaves not clasping stem, lower ones deeply lobed, upper ones entire to dentate, linear; fruit 2- 2.5mm, narrowly winged apically, with shallow notch and shorter style. Intrd; infrequent casual on tips and in fields, mainly from wool; very scattered in Br; S Africa.

8. L. perfoliatum L. - <u>Perfoliate</u> Pepperwort. Erect annual 327 or biennial to 45cm; basal and lower stem-leaves 2-3-pinnate,

not clasping stem, upper stem-leaves entire, ovate, ±
encircling stem; fruit 3-4.5mm, narrowly winged apically,
with shallow notch and style usually ± equalling it. Intrd;
infrequent casual on tips and by docks, mostly from grain and
and grass-seed; very scattered in Br; Europe.
9. L. latifolium L. - <u>Dittander</u>. Erect rhizomatous 327
perennial to 1.5m; basal leaves ovate, cordate to broadly   R
cuneate, serrate, with long petiole, merging into narrowly
elliptic, entire, sessile upper stem-leaves; fruit 1.5-2.5mm,
not winged or notched, with slightly protruding stigma.
Native; damp barish ground near sea; coasts of Br N to S Wa
and NE Yorks, Guernsey, casual or natd on waste land inland.
10. L. graminifolium L. - <u>Tall Pepperwort</u>. Erect perennial 327
to 60cm (but often not persistent in BI); stem-leaves
elliptic to linear, lobed to entire; fruit 2-4mm, not winged
or notched, with slightly protruding stigma. Intrd; a casual
mostly near docks, sometimes persistent but decreasing; very
sporadic in S Br; S Europe.
11. L. draba L. - <u>Hoary Cress</u>. Stems erect, to 90cm; basal
leaves cuneate, petiolate; stem-leaves sessile, elliptic,
clasping stem at base; fruit 2.5-4mm, obtuse at apex with
projecting style 0.5-1.2mm. Intrd; waste ground, by roads,
paths and railways, arable land, sandy ground near sea.
  a. Ssp. draba (<u>Cardaria</u> <u>draba</u> (L.) Desv.). Leaves greyish- 318
green; fruit at least as wide as long, truncate to cordate at
base, usually reticulately ridged. Throughout most of BI,
but absent from large areas of N Br and Ir, first record
1829; S Europe.
  b. Ssp. chalepense (L.) Thell. (<u>Cardaria</u> <u>chalepensis</u> (L.) 318
Hand.-Mazz., <u>C.</u> <u>draba</u> ssp. <u>chalepensis</u> (L.) O. Schulz).
Leaves brighter green; fruit usually longer than wide,
rounded to broadly cuneate at base, usually smooth. Rare
casual formerly natd in Staffs, E Cornwall and Cumberland,
now only at Sharpness (W Gloucs) and Middlesex; SW Asia.

33. CORONOPUS Zinn - <u>Swine-cresses</u>
Annuals or biennials with inflorescences mostly opposite
leaves; leaves all deeply pinnately lobed; hairs unbranched
or 0; petals white or 0; fruit an angustiseptate only
slightly compressed silicula, indehiscent or breaking into 2
halves; seed 1 under each valve.

1. C. squamatus (Forsskaol) Asch. - <u>Swine-cress</u>. Stems 327
glabrous or nearly so, procumbent, to 30cm; petals longer
than sepals; fertile stamens 6; pedicels shorter than fruit;
fruit 2-3mm, strongly ridged or warty, rounded to slightly
retuse at apex with style protruding, indehiscent. Probably
native; waste ground, paths and round gateways; throughout
much of BI, but rare and absent from large areas NW of Yorks
to S Wa.
2. C. didymus (L.) Smith - <u>Lesser Swine-cress</u>. Strong 327
smelling when crushed; stems usually ± pubescent, procumbent

to ascending, to 40cm; petals shorter than sepals or 0;
fertile stamens 2 or 4; pedicels longer than fruit; fruit
1.2-1.7mm, reticulate, emarginate at apex with very short
included style; breaking into 2 halves. Intrd; cultivated
and waste ground; frequent in S BI, scattered N to C Sc; S
America.

34. SUBULARIA L. - <u>Awlwort</u>
Aquatic annuals or biennials; leaves confined to basal
rosette, subulate, entire; hairs 0; petals white, rarely 0;
fruit a latiseptate ± inflated silicula; 2-7 seeds in 2 rows
under each valve.

1. S. aquatica L. - <u>Awlwort</u>. Leaves ± erect, to 4(7)cm; **327**
flowering stems erect, to 8cm, with 2-8(12) flowers; fruit **R**
(1)3-5mm. Native; in stony or gravelly base-poor lakes,
usually totally submerged, sometimes exposed in droughts;
very local in Wa, NW En, N & W Ir, Sc. Often grows with and
confused with <u>Isoetes</u>, <u>Littorella</u>, <u>Lobelia</u> dortmanna,
<u>Eleocharis</u> <u>acicularis</u> and <u>Juncus</u> <u>bulbosus</u>.

TRIBE 6 - BRASSICEAE (genera 35-46). Hairs 0 or unbranched;
petals yellow, white or pink to purple or mauve; fruit
usually a many-seeded beaked or unbeaked siliqua dehiscing
longitudinally to release seeds, sometimes dehiscing transversely but not releasing seeds separately, sometimes
indehiscent, sometimes only 1-4-seeded and transversely
dehiscent.

35. CONRINGIA Heister ex Fabr. - <u>Hare's-ear</u> Mustard
Annuals; leaves simple, entire, basal ones petiolate, upper
ones sessile, with large auricles clasping stem; petals
greenish- to yellowish-white; fruit an unbeaked, longitudinally dehiscent siliqua; seeds in 1 row under each valve.

Other spp. - **C. austriaca** (Jacq.) Sweet, from SE Europe,
differing from <u>C. orientalis</u> in its 8-angled siliqua (3
prominent veins per valve) and paler, smaller petals,
formerly occurred as a casual.

1. C. orientalis (L.) Dumort. - <u>Hare's-ear</u> Mustard. **339**
Glabrous, glaucous; stems erect, to 50cm; leaves ovate to
obovate; fruit 6-12cm, ± erect, 4-angled, each valve with 1
prominent vein. Intrd; casual in arable land and waste
places, often near sea, once frequent, now very sporadic;
scattered in Br and CI; E Mediterranean. Often overlooked as
<u>Brassica</u> <u>rapa</u>, which has beaked fruits.

36. DIPLOTAXIS DC. - <u>Wall-rockets</u>
Annuals to perennials; leaves deeply pinnately lobed,
strongly smelling when crushed; petals yellow, rarely white;
fruit an unbeaked longitudinally dehiscent siliqua; seeds in

2 rows under each valve.

Other spp. - **D. viminea** (L.) DC. and **D. catholica** (L.) DC. from S Europe, and **D. tenuisiliqua** Del. from N Africa, formerly occurred as casuals, and the white-flowered **D. erucoides** (L.) DC. (White Wall-rocket), from S Europe, was formerly natd in S En.

1. **D. tenuifolia** (L.) DC. - Perennial Wall-rocket. Glabrous **339** perennial with branching leafy stems to 80cm; leaves with lobes >3x as long as wide; petals mostly >8mm; fruit 1.5-4cm, with distinct c.0.5mm stalk between sepal-scars and base of valves, erect on erecto-patent pedicel nearly as long to longer than fruit. Possibly native; dry waste places, bare ground, banks and walls; scattered through Br and CI N to C Sc, locally common in S En and CI.

2. **D. muralis** (L.) DC. - Annual Wall-rocket. Glabrous or **339** sparsely pubescent annual or sometimes short-lived perennial with branched stems to 60cm leafy only near base; leaves with lobes <3x as long as wide; petals mostly <8mm; fruit 1.5-4cm, with base of valves immediately above sepal-scars, erecto-patent, with pedicel usually <1/2 as long as fruit. Intrd; dry waste places, rocks, walls and arable land; similar distribution to D. tenuifolia but scattered in Ir; Europe.

37. BRASSICA L. - Cabbages
Annuals to perennials; leaves crenate to deeply pinnately lobed; petals yellow; fruit a beaked or unbeaked longitud- inally dehiscent siliqua; seeds in 1 row under each valve; valves with 1 strong vein.

1 Stem-leaves distinctly clasping stem at base   2
1 Stem-leaves not clasping stem, petiolate or narrowed
    to base   4
    2 Sepals erect in flower; flowering part of
      inflorescence elongated, the buds greatly over-
      topping open flowers; plant glabrous   **1. B. oleracea**
    2 Sepals erecto-patent to patent in flower;
      flowering part of inflorescence scarcely
      elongated, the buds at c. same level as or over-
      topped by uppermost open flowers; lowest leaves
      usually with some hairs   3
3 Buds slightly overtopping open flowers, forming
  convex 'dome'; petals mostly >11mm, bright pale
  yellow   **2. B. napus**
3 Buds overtopped by open flowers, forming concave
  'bowl'; petals <12mm, bright deep yellow   **3. B. rapa**
    4 Fruit terminated by distinct ± conical beak >4mm,
      sometimes with 1-2 seeds   5
    4 Fruit terminated by slender style <4mm, not or
      scarcely wider at base   7
5 Lowest leaves with >3 pairs of lateral lobes; fruit

with beak $\geq$10mm, some with 1-2 seeds **4. B. tournefortii**
5 Lowest leaves with $\leq$3 pairs of lateral lobes;
  fruit with seedless beak $\leq$10mm   6
  6 Lowest leaves with 1-3 pairs of lateral lobes; fruit
    $\geq$3.5mm wide; sepals 4.5-7mm   **5. B. juncea**
  6 Lowest leaves with 0-1 pairs of lateral lobes; fruit
    $\leq$3.5mm wide; sepals 7-10mm   **6. B. carinata**
7 Fruit closely appressed to stem; pedicels <6mm
                                                         8. B. nigra
7 Fruit erecto-patent; pedicels >5mm   8
  8 Fruit with distinct stalk 1.5-4.5mm between sepal-
    scars and base of valves, with style $\leq$2.5mm
                                                      7. B. elongata
  8 Fruit with base of valves within 1.5mm of sepal-
    scars, with style $\geq$2.5mm   **6. B. carinata**

Other spp. - **B. fruticulosa** Cirillo, from S Europe, used to
be found as a casual but is now absent or extremely rare.

1. B. oleracea L. - <u>Cabbage</u>. Glabrous biennial to perennial **309**
to 2m; stems often decumbent and woody below, with numerous **R**
leaf-scars, erect above; roots never tuberous (unless
diseased); basal leaves crenate to deeply lobed; stem-leaves
clasping stem at base; sepals erect; petals 12-30mm; fruit
5-10cm, with a conical usually seedless beak 4-10mm; 2n=18.
Possibly native on sea-cliffs scattered round Br, mostly in
S; common casual on tips, neglected gardens and roadsides
throughout BI. Wild plants are var. **oleracea** (<u>Wild</u> Cabbage);
the commonest crop-plants are placed in var. **capitata** L.
(<u>Cabbage</u>), var. **sabauda** L. (<u>Savoy</u> Cabbage), var. **viridis** L.
(<u>Kale</u>), var. **botrytis** L. (<u>Cauliflower</u>, <u>Broccoli</u>), var.
gemmifera DC. (<u>Brussels-sprout</u>), and var. **gongylodes** L.
(<u>Kohl-rabi</u>).
2. B. napus L. - <u>Rape</u>. Annual to biennial with glaucous, $\pm$ **309**
glabrous, erect stems to 1.5m; roots sometimes tuberous; **339**
basal leaves crenate to deeply lobed, usually sparsely
pubescent; stem-leaves clasping stem at base; sepals erecto-
patent; petals 11-18mm; fruit 5-10cm, with a conical usually
seedless beak 5-15mm; 2n=38 (derived from <u>B. oleracea</u> x <u>B.
rapa</u>). Intrd; frequent relic of cultivation and from seed
importation; throughout BI and now greatly increasing.
a. Ssp. oleifera (DC.) Metzger - <u>Oil-seed</u> <u>Rape</u>. Root
slender. Grown for its oil-bearing seeds and as a seedling
salad-plant (substitute for <u>Sinapis</u> <u>alba</u>), and rarely for
fodder and green manure; now the commonest casual <u>Brassica</u> by
roads, field-margins and on tips; also a bird-seed alien and
found near oil-processing factories.
b. Ssp. rapifera Metzger (var. <u>napobrassica</u> (L.) Reichb.) -
<u>Swede</u>. Root swollen into a yellow-fleshed tuber. Grown for
its root-tubers and rarely for fodder; frequent relic of
cultivation.
Other sspp., including the weedy ssp. <u>napus</u>, may occur but

are much rarer.

2 x 3. B. napus x B. rapa = B. x harmsiana O. Schulz occurs sporadically in crops of B. napus when exposed to pollen from B. rapa; it closely resembles B. napus and can be told only by its sterility and chromosome number (2n=29).

3. B. rapa L. - Turnip. Differs from B. napus in its green **309** usually more pubescent basal leaves; petals 6.5-12mm; patent sepals; 2n=20; and see key. Probably intrd; frequent by streams and rivers, and as a relic of cultivation; throughout BI, probably less common nowadays.

a. Ssp. campestris (L.) Clapham (ssp. sylvestris (Lam.) Janchen, B. campestris L.) - Wild Turnip. Root slender; seeds <1.6mm, grey to blackish. Frequent by streams and rivers, often with B. nigra.

b. Ssp. oleifera (DC.) Metzger - Turnip-rape. Root slender; seeds mostly >1.6mm, red-brown. Grown as fodder or oil-seed crop; mostly a bird-seed or oil-processing alien.

c. Ssp. rapa - Turnip. Root swollen into a white-fleshed tuber. Grown for its root-tubers and sometimes as fodder or green manure; frequent relic of cultivation but less common than formerly.

4. B. tournefortii Gouan - Pale Cabbage. Usually annual; **339** stems erect, to 50cm; basal leaves deeply lobed with 5-10 pairs of lobes, hispid; stem-leaves much reduced, not clasping stem; sepals ± erect; fruit 3-7cm, with tapering-conical beak 10-16(20)mm with 0-2 seeds; 2n=20. Intrd; a rather frequent wool-alien, very rarely from other sources; sporadic in Br; Mediterranean region.

5. B. juncea (L.) Czernj. (B. integrifolia (West) Rupr.) - Chinese Mustard. Annual; stems erect, to 1m; basal leaves deeply lobed with 1-3 pairs of lobes, sparsely hispid; stem-leaves deeply dentate to ± entire, petiolate; sepals erecto-patent; petals 9-14mm; fruit 3-6cm x 3.5-4mm, with conical seedless beak (4)5-9(12)mm; 2n=36 (derived from B. nigra x B. rapa). Intrd; a frequent bird-seed alien, sometimes from wool and other sources; scattered in Br; S & E Asia. Often superficially closely resembling Sinapis arvensis, but see sepal and fruit characters.

6. B. carinata A. Braun (B. integrifolia auct. non (West) Rupr.) - Ethiopian Rape. Differs from B. juncea in petals 13-17mm; fruit 2.5-5.5cm x (1)2-3.5mm, with seedless beak 2.5-6(7)mm; 2n=34 (derived from B. oleracea x B. nigra); and see key. Intrd; occasional bird-seed alien; sporadic in En, formerly natd in Middlesex; Abyssinia.

7. B. elongata Ehrh. - Long-stalked Rape. Biennial to perennial; stems erect, to 1m; basal leaves pinnately lobed to shallowly toothed, hispid; stem-leaves much reduced, not clasping stem; sepals ± erect; fruit 1-2.5cm, with stalk-like base above sepal-scars 1.5-4.5mm, with style 0.5-2.5mm; 2n=22. Intrd; an occasional casual of waste places; sporadic in Br; SE Europe.

8. B. nigra (L.) Koch - Black Mustard. Annual; stems erect, **339**

to 1.2m; basal leaves pinnately lobed, hispid; stem-leaves lobed below, entire above, petiolate; sepals erecto-patent; fruit 8-25mm, with style 1.5-4mm; 2n=16. Probably native; sea-cliffs, river banks, rough ground and waste places; frequent in Br and CI N to S Sc, very scattered in Ir.

38. SINAPIS L. - Mustards
Annuals; leaves crenate to deeply pinnately lobed; sepals patent; petals yellow; fruit a longitudinally dehiscent siliqua with a distinct beak usually >1/3 as long as valves; seeds in 1 row under each valve; valves with 3(-7) strong veins.

Other spp. - **S. flexuosa** Poiret (S. hispida Schousboe), from Spain, formerly occurred as a casual; differs from S. alba in having <10 seeds per fruit and more hispid leaves and stems.

1. S. arvensis L. - Charlock. Plant to 1m, simple to well- **339** branched, ± glabrous to hispid; leaves lobed or not, if so the terminal lobe much the largest; fruit 2.5-4.5cm, with (4)8-13 seeds, with 0-1-seeded conical beak 7-16mm and 1/3-3/4 as long as valves. Probably native; arable and waste land, tips and roadsides; throughout BI.
2. S. alba L. - White Mustard. Plant to 70cm, differing **339** from S. arvensis in its deeply pinnately lobed leaves; fruit 2-4cm, with 2-8 seeds, with 0-1-seeded strongly flattened beak 10-30mm and 1-1.5x as long as valves. Intrd; S Europe.
a. Ssp. alba. Leaves deeply pinnately lobed, the terminal lobe much the largest; fruit usually hispid. Grown as fodder or green manure or for mustard-seed or seedling salad-plant, natd or casual in arable or waste land and on waysides and tips, especially on calcareous soils; scattered over most of BI but absent from much of N.
b. Ssp. dissecta (Lagasca) Bonnier (S. dissecta Lagasca). Leaves 2-pinnately lobed, the terminal lobe little larger than largest laterals; fruit glabrous to slightly pubescent. Infrequent casual on waste land; sporadic in Br.

39. ERUCA Miller - Garden Rocket
Annuals; leaves deeply pinnately lobed; sepals erect; petals white to pale yellow with conspicuous violet veins; fruit a longitudinally dehiscent siliqua with a distinct beak usually >1/3 as long as valves; seeds in 2 rows under each valve; valves with 1 strong vein.

1. E. vesicaria (L.) Cav. (E. sativa Miller) - Garden **339** Rocket. Stems usually much branched, decumbent to erect, to 1m, glabrous to hispid; fruit 1.2-2.5cm, with seedless strongly flattened beak 4-10mm. Intrd; infrequent casual, rarely persisting, on waste land; scattered over Br and CI; S Europe. Most or all plants are ssp. **sativa** (Miller) Thell., with caducous sepals of which only the outer 2 are saccate at

base; ssp. **vesicaria**, with persistent sepals all saccate at
base might sometimes occur.

40. ERUCASTRUM C. Presl - Hairy Rocket

Annuals to perennials; leaves deeply pinnately lobed; sepals
erect to ± patent; petals yellow; fruit a beakless or shortly
beaked longitudinally dehiscent siliqua slightly constricted
between seeds; seeds in 1 row under each valve; valve with 1
strong vein.

Other spp. - E. nasturtiifolium (Poiret) O. Schulz, from SW
Europe, formerly occurred as a casual; it differs from E.
gallicum in its stem-leaves having their basal lobes clasping
the stem; flowers all without bracts; sepals ± patent; and
fruit with c.0.5-1mm stalk above sepal-scars and with 3-6mm
beak with 1-2 seeds.

1. E. gallicum (Willd.) O. Schulz - Hairy Rocket. Pubescent **339**
annual or biennial; stems erect, to 60cm; stem-leaves ±
pinnate; sepals ± erect; fruit 2-4.5cm, with seedless beak
3-4mm.   Intrd;   infrequent,   formerly   frequent,   casual   of
arable and waste land, rarely persisting; scattered in BI,
mainly S; Europe.

41. COINCYA Rouy (Hutera Porta, Rhynchosinapis Hayek) -
Cabbages

Annuals to perennials; leaves pinnately lobed to pinnate;
sepals erect; petals yellow; fruit a longitudinally dehiscent
siliqua with distinct beak usually 1/5-1/3 as long as valves
and with (0)1-5 seeds; seeds in 1 row under each valve;
valves with 3 strong veins.

1. C. monensis (L.) Greuter & Burdet - see sspp. for English **339**
names.   Annuals   to   perennials,   subglabrous   to   pubescent   **RR**
below; basal leaves with 3-9 pairs of lateral lobes, the
terminal lobe not much larger than laterals; siliqua 2.5-8cm
incl. beak 5-24mm, 1.5-3mm wide.
  a. **Ssp.   monensis**   (Hutera   monensis   (L.)   Gomez-Campo,
Rhynchosinapis monensis (L.) Dandy ex Clapham) - Isle of Man
Cabbage.   Plant to 60cm; stems procumbent to ascending,
glabrous to sparsely hispid below; seeds 1.4-2mm.   Native;

FIG 339 - Fruits of **Brassicaceae**. 1, **Conringia**. 2, **Diplotaxis
tenuifolia**. 3, **D. muralis**. 4, **Eruca**. 5, **Erucastrum**. 6,
Coincya monensis. 7, **Sinapis alba**. 8, **S. arvensis**. 9,
Brassica nigra. 10, **B. tournefortii**. 11, **B. napus**. 12,
Hirschfeldia. 13, **Raphanus raphanistrum** ssp. **landra**. 14, ssp.
maritimus. 15, ssp. **raphanistrum**. 16, **R. sativus**. 17,
Rapistrum rugosum ssp. **rugosum**. 18, ssp. **linneanum**. 19, **R.
perenne**. 20, **Camelina sativa**. 21, **C. microcarpa**. 22,
Ionopsidium. 23, **Iberis sempervirens**. 24, **I. amara**. 25, **I.
umbellata**. 26, **Neslia**. 27, **Crambe maritima**. 28, **Cakile**.

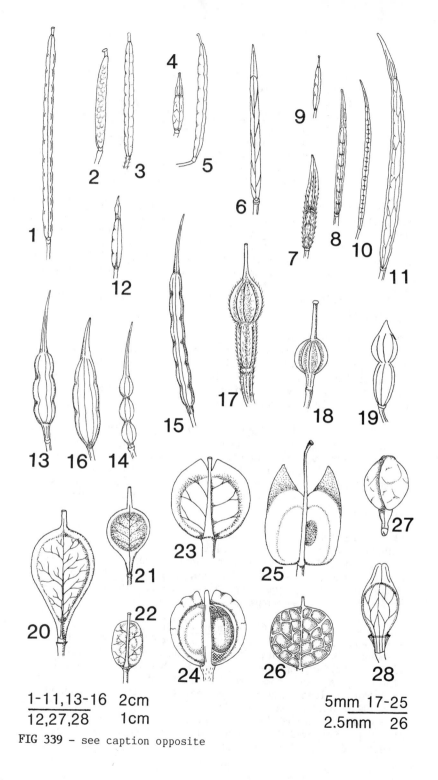

1-11,13-16 2cm
12,27,28 1cm

5mm 17-25
2.5mm 26

FIG 339 - see caption opposite

sandy ground near sea; Man and W Br from C Sc to SW En, rare
casual elsewhere; endemic.
  b. Ssp. recurvata (All.) Leadlay (C. cheiranthos (Villars)
Greuter & Burdet, Hutera cheiranthos (Villars) Gomez-Campo,
Rhynchosinapis cheiranthos (Villars) Dandy, R. erucastrum
Dandy ex Clapham excl. typ.) - Wallflower Cabbage. Plant to
1m; stems usually erect, hispid to rather sparsely so below;
seeds 0.8-1.6mm. Intrd; casual on sandy ground, waste places
and roadsides in SW Br, natd in Mons since 1975, Jersey since
1832; W Europe.
  2. C. wrightii (O. Schulz) Stace (Hutera wrightii (O. **RRR**
Schulz) Gomez-Campo, Rhynchosinapis wrightii (O. Schulz)
Dandy ex Clapham) - Lundy Cabbage. Biennials to perennials
to 1m, often woody near base, pubescent over all or most of
stem incl. inflorescence; basal leaves with 3-5(6) pairs of
lateral lobes, the terminal lobe much larger than laterals;
siliqua 2-8cm incl. beak 7-16mm, 3-4mm wide; seeds 1.2-1.9mm.
Native; cliffs and slopes on SE part of Lundy Island, N
Devon; endemic.

42. HIRSCHFELDIA Moench - Hoary Mustard
Annuals to short-lived perennials; lower leaves pinnate to
deeply pinnately lobed; sepals ± erect; petals yellow; fruit
a longitudinally dehiscent siliqua with distinct beak usually
c.1/2 as long as valves with (0)1-2 seeds; seeds in 1 row
under each valve; valves with 1-3 ± strong veins.

  1. H. incana (L.) Lagr.-Fossat - Hoary Mustard. Stems **339**
erect, to 1.2m, whitish pubescent below with short stiff
hairs; fruit 6-17mm, appressed to stem, with beak 3-5mm,
swollen round seeds and abruptly narrowed distally. Intrd;
waste places and waysides; CI and Br, especially S, often
casual but increasingly natd, also frequent wool-alien; S
Europe.

43. CAKILE Miller - Sea Rocket
Glabrous annuals; leaves entire to pinnately lobed; sepals
erect; petals mauve to pink or white; fruit breaking transversely into 2 1-seeded portions, the proximal often seedless, the distal longer and wider, keeled laterally and with
prominent veins and margin.

  1. C. maritima Scop. (C. edentula auct. non (Bigelow) Hook.) **339**
- Sea Rocket. Stems procumbent to ± erect, to 50cm; leaves ±
glaucous, ± succulent; fruit 10-25mm; proximal segment 2-
10mm; distal segment 8-17mm, ovoid. Native; near sea drift-
line on sand and sometimes shingle; around coasts of BI.
Variable in leaf- and fruit-shape; there is no consensus as
to whether our plants are subspecifically distinct from those
of Portugal and the Mediterranean. If they are, our plants
are ssp. **integrifolia** (Hornem.) N. Hylander ex Greuter &
Burdet; if not, they come under ssp. **maritima**. Records of C.

<u>edentula</u> from BI were errors.

44. RAPISTRUM Crantz - <u>Cabbages</u>

Annuals to perennials; leaves dentate to deeply pinnately lobed; sepals erecto-patent; petals yellow; fruit breaking transversely into 2 portions, the proximal 0-1(2)-seeded, the distal 1-seeded, narrowed at apex into persistent style and variously ribbed or wrinkled.

1. **R. rugosum** (L.) Bergeret (<u>R. hispanicum</u> (L.) Crantz, <u>R.</u> **339**
<u>orientale</u> (L.) Crantz, <u>R. rugosum</u> ssp. <u>orientale</u> (L.)
Arcang., spp. <u>linneanum</u> (Cosson) Rouy & Fouc.) - <u>Bastard</u>
<u>Cabbage</u>. Erect annual to 80cm, hispid at least below; fruit 4-10mm; distal segment abruptly narrowed into (0.8)1-3(5)mm style, ribbed and rugose, glabrous to densely hispid; proximal segment usually much narrower than distal, mostly 0(1)-seeded. Intrd; casual in waste and arable land, on tips and waysides; frequent and increasing in S & C Br and CI, natd in parts of S En; S Europe. Fruits very variable in relative sizes of pedicel and 2 segments, pubescence and degree of sculpturing, but the variants are probably not more than vars. All or most of our plants are referable to ssp. **linneanum**, with long thin pedicels (2.5-5 x 0.3-0.7mm) and seedless lower fruit segments.

2. **R. perenne** (L.) All. - <u>Steppe</u> <u>Cabbage</u>. Erect or **339** spreading biennial or perennial to 80cm, hispid at least below; fruit 6-10mm; distal segment gradually narrowed into 0.5-1mm style, longitudinally ribbed, glabrous; proximal segment usually similar in size to distal but less or not ribbed, mostly 1-seeded. Intrd; similar places to <u>R.</u> <u>rugosum</u>; scattered in S & C Br, much rarer than <u>R. rugosum</u> but well natd in some places; C & E Europe.

45. CRAMBE L. - <u>Sea-kale</u>

Large perennials with thick long roots; leaves large and irregularly lobed or toothed; sepals ± patent; petals white; fruit breaking transversely into 2 portions, the proximal sterile and stalk-like, the distal 1-seeded.

1. **C. maritima** L. - <u>Sea-kale</u>. Glabrous, glaucous, densely **339**
branched cabbage-like plant to 75cm; basal leaves undulate at **R**
margins; stem-leaves many, similar; distal portion of fruit 7-12mm, broadly ellipsoid to globose. Native; on sand, rocks and cliffs but mostly shingle, by sea; coasts of BI N to C Sc, but absent from many areas.

2. **C. cordifolia** Steven - <u>Greater</u> <u>Sea-Kale</u>. Sparsely pubescent erect perennial to 2m; basal leaves plane at margins; stem-leaves very few, much smaller; distal portion of fruit 4.5-5mm, ovoid. Intrd; garden throwout with persistent roots conspicuous on tips and waste ground; scattered in En; Caucasus.

46. RAPHANUS L. - <u>Radishes</u>
Annuals to perennials with distinctive radish-like smell when
crushed; leaves shallowly pinnately lobed to pinnate; sepals
erect; petals white, mauve or yellow, often with darker
veins; fruit indehiscent or transversely dehiscent into 1-8
1-seeded segments, with long persistent narrow beak.

1. R. raphanistrum L. - see sspp. for English names. Hispid
annual to perennial, with slender root; petals white, mauve
or yellow; fruit strongly constricted between seeds, at least
partly transversely dehiscent.
1 Fruit with cylindrical or oblong segments usually
   longer than wide, with beak (2.5)3-6x as long as
   apical fertile segments   **a. ssp. raphanistrum**
1 Fruit with + globose segments c. as long as wide,
   with beak 1-3(4)x as long as apical fertile segment   2
   2 Leaves with crowded lateral lobes; petals 15-22mm
      when fresh   **b. ssp. maritimus**
   2 Leaves with + distant lateral lobes; petal 8-15mm
      when fresh   **c. ssp. landra**
   a. Ssp. raphanistrum - <u>Wild Radish</u>. Annual; stems ascending 339
to erect, to 75cm, usually little branched; petals white,
mauve or yellow; fruit 3-9cm, 2-5.5mm wide, with (1)3-8(10)
segments and beak 10-30mm. Probably intrd; cultivated and
rough ground, waste places and tips; frequent throughout BI,
yellow-petalled plants commonest in N & W; Europe.
   b. Ssp. maritimus (Smith) Thell. (<u>R. maritimus</u> Smith) - <u>Sea</u> 339
<u>Radish</u>. Biennial to perennial; stems often much branched and
very leafy, to 80cm; petals usually yellow, white in CI;
fruit 1.5-4.5cm, 4.5-10mm wide, with 1-5(6) segments and beak
6-20mm. Native; sea-shores and maritime cliffs and waste
places; coasts of BI N to NW Sc, absent from most of E Br.
Intermediates (probably hybrids) with ssp. <u>raphanistrum</u> occur
in SW En and CI.
   c. Ssp. landra (Moretti ex DC.) Bonnier & Layens (<u>R. landra</u> 339
Moretti ex DC.) - <u>Mediterranean Radish</u>. Like ssp. <u>maritimus</u>
but annual to perennial, less robust; fruit 2.5-6cm, with
beak 15-40mm; and see key. Intrd; rather rare casual of
waste places; sporadic in En and Wa; Mediterranean.
   1 x 2. R. raphanistrum x R. sativus = R. x micranthus
(Uechtr.) O. Schulz occurs sporadically where <u>R. raphanistrum</u>
has occurred as a weed near <u>R. sativus</u>. It is partially
fertile, usually white-flowered, and variously intermediate,
usually with partly dehiscent fruits and thin roots.
   2. R. sativus L. - <u>Garden Radish</u>. Hispid annual or biennial 339
to 80cm, usually with swollen often reddish root; petals
white or mauve; fruit not or scarcely constricted between
seeds, indehiscent, 2-9cm x 8-15mm, with (1)5-12 seeds.
Intrd; grown as salad plant or as animal fodder (<u>Fodder</u>
<u>Radish</u>, with thin root and much-branched leafy stems) and
often escaping or persisting; fields, gardens and tips;
sporadic throughout BI; origin probably Mediterranean.

64. RESEDACEAE - Mignonette family

Herbaceous annuals to perennials; leaves alternate, simple to pinnate, sessile or petiolate, ± exstipulate (stipules often present as minute glandular teeth). Flowers in racemes, bisexual, hypogynous, zygomorphic; sepals 4-6(8), free; petals as many as sepals, free, the upper 1-2, lateral 2 and lower 1-2 different in form, the upper largest, at least some with entire proximal and deeply lobed distal regions; stamens 7-c.25, inserted on nectar-secreting disc; ovary 1-celled, open at top, composed of 3-4 carpels, with many ovules; styles 0; stigmas borne 1 on apical lobe of each carpel; fruit a capsule, open at top from start.
Easily recognized by the zygomorphic flowers with open-topped ovary and 4-6 white or yellowish distinctive petals.

Other genera - SESAMOIDES Ortega (Astrocarpus DC.) differs from Reseda in its 4-7 carpels being ± free; S. pygmaea (Scheele) Kuntze (A. sesamoides (L.) DC.) formerly occurred as a casual.

1. RESEDA L. - Mignonettes

1 At least upper and middle leaves deeply pinnately
    lobed   2
1 All leaves entire to minutely toothed, or a few with
    1-2 lateral lobes   3
    2 Carpels 3; petals yellowish; filaments falling
        after flowering; seeds smooth   **3. R. lutea**
    2 Carpels 4; petals white; filaments persistent until
        fruit ripe; seeds tuberculate   **2. R. alba**
3 Sepals and petals 4; fruits crowded, stiffly erecto-
    patent, <7mm, on pedicels <4mm; seeds smooth
                                                           1. R.luteola
3 Sepals and petals 6; fruits well spaced, pendent,
    >7mm, on pedicels >5mm; seeds rugose   4
    4 Capsules 5-8(11)mm; sepals ≤5mm at fruiting
                                                           5. R.odorata
    4 Most mature capsules 11-15mm; many sepals >5(12)mm
        at fruiting   **4. R. phyteuma**

1. R. luteola L. - Weld. Glabrous biennial; main stem stiffly erect, to 1.5m; leaves entire, linear to lanceolate or oblanceolate; sepals and (yellow) petals 4; capsule 3-6mm, of 3 carpels. Native; open grassland, disturbed, waste and arable land mostly on base-rich soils; throughout most of BI except much of N & W Sc.
2. R. alba L. - White Mignonette. Glabrous annual to perennial; stem well branched, erect to ascending, to 75cm; leaves pinnately lobed; sepals and (white) petals 5-6; capsule 6-15mm, of 4 carpels. Intrd; casual on waste ground, often near sea, occasionally persisting (natd in Mons since

1968) but less common than formerly; S Br and CI; S Europe.
3. R. lutea L. (<u>R. stricta</u> auct. non Pers.) - <u>Wild</u>
<u>Mignonette</u>. Puberulent biennial to perennial; stem well
branched, decumbent to erect, to 75cm; leaves pinnately
lobed; sepals and (yellowish) petals 6; capsule 7-20mm, of 3
carpels. Native; disturbed, waste and arable land,
especially on calcareous soils; through much of BI but much
commoner in E and absent from large areas of Sc, Wa and Ir.
4. R. phyteuma L. - <u>Corn Mignonette</u>. Puberulent annual to
biennial; stems much branched below, procumbent to ascending,
to 30cm; leaves mostly simple, sometimes some with 1-2
lateral lobes; sepals and (white) petals 6; capsule mostly
11-25mm, of 3 carpels. Intrd; rare and decreasing casual of
waste ground, ± natd in cornfields and field margins in a few
places in S En; S Europe.
5. R. odorata L. - <u>Garden Mignonette</u>. Similar to <u>R.</u>
<u>phyteuma</u> but see key. Intrd; grown in gardens for its
scented flowers and occasionally found as casual on tips or
waste places in S Br; often overlooked for <u>R. phyteuma</u> or <u>R.</u>
<u>lutea</u>; SE Mediterranean.

64A. CLETHRACEAE

CLETHRA arborea Aiton (<u>Lily-of-the-valley-tree</u>), from
Madeira, is an evergreen tree that produces self-sown
seedlings on warm walls in CI. It has terminal panicles of
white flowers similar to those of <u>Arbutus</u>, but differing in
the ± free petals and pollen grains dispersed singly.

65. EMPETRACEAE - <u>Crowberry family</u>

Dwarf, heather-like, evergreen shrubs; leaves whorled to
spiral, simple, entire, shortly petiolate, exstipulate.
Flowers in axillary clusters of 1-3 with 3 small bracts,
small and inconspicuous, bisexual to dioecious, hypogynous,
actinomorphic; sepals 3, free; petals 3, free, pinkish;
stamens 3; ovary 6-9-celled, with 1 ovule in each cell; style
1, short; stigmas in form of 6-9 broad toothed rays; fruit a
drupe with <9 seeds.
Distinctive in its heather-like habit, 3-merous flowers and
black fruits.

1. EMPETRUM L. - <u>Crowberry</u>
1. E. nigrum L. - <u>Crowberry</u>. Stems and leaf-margins
glandular when young, glabrous later; leaves strongly
revolute obscuring abaxial surface, 3-7 x 1-2mm; fruit black,
4-8mm across, subglobose. Native; peaty and rocky moors,
bogs and mountain-tops.
  a. Ssp. nigrum. Stems to 1.2m, procumbent, slender, rooting
along length; leaves ± parallel-sided, mostly 3-5x as long as

wide; flowers dioecious, rarely bisexual; 2n=26.   Frequent
and often abundant in suitable places in Br and Ir NW of line
from Devon to NE Yorks, lowland to c.800m.
   b. Ssp. hermaphroditum (Hagerup) Boecher (E. hermaphroditum
Hagerup).   Stems to 50cm, less procumbent, stiff, not
rooting; leaves with curved sides, mostly 2-4x as long as
wide; flowers bisexual (remains of stamens usually visible at
base of some fruits); 2n=52.   Usually at higher altitudes
(>650m) than E. nigrum but overlapping, often in drier
places; Caerns, Lake District, highlands of Sc.

66. ERICACEAE - Heather family

Deciduous or evergreen trees or dwarf shrubs; leaves whorled,
opposite or alternate, simple, petiolate or not, exstipulate.
Flowers   variously   arranged,   bisexual   or   functionally
dioecious, hypogynous to epigynous, actinomorphic to slightly
zygomorphic; sepals free or fused, 4-5; petals as many as
sepals, partly or completely fused or rarely free; stamens
from 1x to 2x as many as petals, usually borne on receptacle
or disc, usually with anthers opening by pores; pollen-grains
released   in   tetrads.   Ovary   4-5(10)-celled   with   axile
placentation and 1-many ovules per cell; style 1; stigma
capitate; fruit a capsule, berry or drupe.
Very variable in superficial flower characters, but usually
recognizable by the woody often evergreen habit, usually
fused petals, stamens borne on receptacle (not on corolla),
and anthers usually opening by pores. All our spp., except
Arbutus unedo, Erica terminalis and E. x darleyensis, are
calcifuges.

1 Ovary inferior; fruit a berry with persistent calyx-
   lobes at apex   **13. VACCINIUM**
1 Ovary superior; fruit various, if fleshy then calyx
   deciduous or persistent at base of fruit   2
   2 Petals free, white; leaves tomentose with rust-
      coloured hairs on lowerside   **1. LEDUM**
   2 Petals fused at least at base; leaves tomentose or
      not, if so then hairs not rust-coloured   3
3 Most leaves opposite or whorled; anthers with 0 or
   basal appendages   4
3 Leaves alternate or spiral; anthers with 0 or terminal
   appendages   7
   4 Sepals and petals 4; stamens 8; corolla persistent
      around ripe fruit   5
   4 Sepals and petals 5; stamens 5 or 10; corolla
      falling before fruiting   6
5 Leaves opposite; corolla shorter than calyx, divided
   >1/2 way to base   **11. CALLUNA**
5 Leaves mostly in whorls of 3-4(5); corolla longer than
   calyx, normally divided <1/2 way to base   **12. ERICA**

6 Leaves <15mm, with strongly revolute margins;
  stamens 5; corolla divided c.1/2 way to base,
  without pouches   **3. LOISELEURIA**
6 Leaves >15mm, with flat margins; stamens 10;
  corolla divided much <1/2 way to base, with 10
  small pouches on inside near base   **4. KALMIA**
7 Corolla campanulate, >15mm, slightly zygomorphic
                                                     2. RHODODENDRON
7 Corolla narrowed distally, <15mm, actinomorphic   8
  8 Fruit succulent or surrounded by a succulent calyx   9
  8 Fruit a dry capsule with a dry calyx   12
9 Leaves spine-tipped   **8. GAULTHERIA**
9 Leaves without spines   10
  10 Calyx becoming succulent and surrounding fruit when
     ripe   **8. GAULTHERIA**
  10 Calyx remaining dry and small at base of succulent
     fruit   11
11 Erect tree or shrub flowering in late autumn; fruit
   red, very warty   **9. ARBUTUS**
11 Procumbent shrub flowering in summer; fruit red or
   black, smooth, ± glossy   **10. ARCTOSTAPHYLOS**
   12 Petals and sepals 4; leaves white-tomentose on
      lowerside   **6. DABOECIA**
   12 Petals and sepals 5; leaves glabrous but often
      white on lowerside   13
13 Calyx and pedicels with glandular hairs; corolla
   purple; anthers without appendages   **5. PHYLLODOCE**
13 Calyx and pedicels glabrous; corolla pink; anthers
   with horn-like appendages at apex   **7. ANDROMEDA**

Other genera - **PIERIS** D. Don would key out in couplet 13 but
differs from <u>Phyllodoce</u> and <u>Andromeda</u> in its flat, crenate-
serrate leaves. **P. japonica** (Thunb.) D. Don, from Japan,
along with 2 similar spp., is grown for ornament, and
seedlings occur in a cemetery in Surrey; it is an evergreen
shrub to 2m with leaves 3-8cm and white flowers in ± pendent
panicles.

1. LEDUM L. - <u>Labrador-tea</u>
Leaves alternate, evergreen; flowers in dense terminal
racemes, petals 5, free, white; stamens (5)6-8(10); anthers
without appendages; ovary superior; fruit a capsule.

1. L. palustre L. (<u>L. groenlandicum</u> Oeder) - <u>Labrador-tea</u>. 354
Well-branched ± upright shrub to 1.2m; leaves 1.5-5cm,
entire, with strongly revolute margins, densely rusty-
tomentose on lowerside, elliptic to narrowly so; flowers
8-16mm across. Intrd; well natd in bogs and other wet peaty
ground; scattered from S En to C Sc; N America. Despite
claims that ssp. **palustre**, from N Europe, occurs in Br, all
our plants are ssp. **groenlandicum** (Oeder) Hulten or perhaps
intermediate; ssp. <u>palustre</u> has narrower, less densely

tomentose leaves and (7-)10(11) stamens.

2. RHODODENDRON L. - Rhododendrons
Leaves alternate, deciduous or evergreen; flowers in dense
terminal racemes; petals 5, fused to form campanulate lobed
corolla; stamens 5 or 10; anthers without appendages; ovary
superior; fruit a capsule.

Other spp. - A very large number of spp., hybrids and
cultivars are grown and may persist in neglected parks and
gardens; some records of the following 2 might belong to some
of these.

1. R. ponticum L. - Rhododendron. Densely branched,
spreading to upright shrub to 5m; leaves 6-20cm, evergreen, $\pm$
flat, elliptic to oblong or narrowly so, entire, glabrous;
flowers mauvish-purple, c.4-6cm across; stamens 10. Intrd;
extensively natd by seeding and suckering on sandy and peaty
soils and on rocks both in woods and in open; suitable places
throughout BI. Most or all our plants are ssp. **ponticum**,
from SE Europe and SW Asia.
2. R. luteum Sweet - Yellow Azalea. Upright shrub to 2m;
leaves 6-12cm, deciduous, flat, oblong-lanceolate, shallowly
serrate, slightly pubescent; flowers yellow, c.5cm across;
stamens 5. Intrd; natd in woods by suckering; scattered in
Br; E Europe and W Asia.

3. LOISELEURIA Desv. - Trailing Azalea
Leaves opposite, evergreen; flowers 1-several in terminal
apparent umbels; petals 5, fused to form campanulate lobed
corolla; stamens 5; anthers without appendages; ovary
superior; fruit a capsule.

1. L. procumbens (L.) Desv. - Trailing Azalea. Densely
branched domed or trailing shrub to 25cm high; leaves <1cm,
entire, elliptic to narrowly so, glabrous, with strongly
revolute margins; flowers pink, 3-6mm aross. Native; rocky
and peaty moors and mountains above 400m; locally frequent in
highlands of Sc, rare in Orkney and Shetland.

4. KALMIA L. - Sheep-laurels
Leaves mostly whorled or opposite, evergreen; flowers in
terminal or axillary small umbel-like racemes; petals 5,
fused to form saucer-shaped slightly lobed corolla with 10
small pouches on inside near base; stamens 10, enclosed in
corolla-pouches before dehiscence; anthers without appendages; ovary superior; fruit a capsule.

1. K. polifolia Wangenh. - Bog-laurel. Straggling to erect **348**
shrub to 70cm; leaves mostly opposite, 1-4cm, elliptic to
oblong or narrowly so, glabrous, entire, with revolute
margins, whitish on lowerside; flowers in subterminal leaf-

FIG 348 – Ericaceae. 1, Erica lusitanica. 2, E. x
darleyensis. 3, E. ciliaris. 4, E. vagans. 5, E. erigena. 6,
E. terminalis. 7, Kalmia angustifolia. 8, K. polifolia.

axils, pink, 1-2cm across. Intrd; natd in wet peaty bogs and moors; few places in SE & N En and C Sc; N America.

2. K. angustifolia L. - <u>Sheep-laurel</u>. Erect to ascending 348 shrub to 1m; leaves opposite or in whorls of 3, 2-7cm, elliptic to oblong, entire, glabrous, with flat or slightly revolute margins, brownish-green on lowerside; flowers in lateral axillary racemes (actually subterminal on previous year's growth), pink, 7-12mm across. Intrd; in similar places to <u>K. polifolia</u> and nearly as rare; S to N En; E N America.

5. PHYLLODOCE Salisb. - <u>Blue</u> <u>Heath</u>
Leaves alternate, evergreen; flowers few on long pedicels in subterminal clusters; petals 5, fused to form tubular corolla narrowed and shortly lobed distally; stamens 10; anthers without appendages; ovary superior; fruit a capsule.

1. P. caerulea (L.) Bab. - <u>Blue</u> <u>Heath</u>. Domed shrub to 20cm RRR high; leaves <15mm, strongly revolute at margins, minutely serrate on apparent margins, linear to linear-oblong; young leaves, pedicels and calyx glandular; flowers 7-12mm, mauvish-purple. Native; rocky moorland at 680-840m; very local in Westerness and M Perth. Irregular flowering behaviour; when sterile easily overlooked for <u>Loiseleuria</u> or <u>Empetrum</u>, with which it often grows, but details of leaves are quite different.

6. DABOECIA D. Don - <u>St Dabeoc's Heath</u>
Leaves alternate, evergreen; flowers in lax terminal racemes; petals 4, fused to form tubular corolla narrowed and shortly lobed distally; stamens 8; anthers without appendages; ovary superior; fruit a capsule.

1. D. cantabrica (Hudson) K. Koch - <u>St</u> <u>Dabeoc's Heath</u>. R Straggly or loosely domed shrub to 50(70)cm high; leaves <15mm, entire, revolute at margins, sublinear to elliptic, white-tomentose on lowerside; flowers 8-14mm, pinkish-purple. Native; peaty and rocky moorland; locally common in Connemara (E & W Mayo and W Galway).

7. ANDROMEDA L. - <u>Bog-rosemary</u>
Leaves alternate, evergreen; flowers in small terminal umbel-like clusters; petals 5, fused to form tubular corolla narrowed and shortly lobed distally; stamens 10; anthers each with 2 long terminal appendages; ovary superior; fruit a capsule.

1. A. polifolia L. - <u>Bog-rosemary</u>. Straggly glabrous shrub; R stems to 35cm; leaves 1-4cm, entire, revolute at margins, sub-linear to elliptic, white-glaucous on lowerside; flowers 5-8mm, pale pink. Native; wet peaty places; locally common, C En and C Wa to C Sc, Ir, formerly more widespread in En.

8. GAULTHERIA L. (Pernettya Gaudich.) - Aromatic Wintergreens
Leaves alternate, evergreen; flowers solitary and axillary or
in terminal and subterminal racemes; petals 5, fused to form
tubular corolla narrowed and shortly lobed distally; stamens
10; anthers with 4 short terminal appendages; ovary superior;
fruit a capsule surrounded by succulent, berry-like swollen
calyx, or a succulent berry with calyx remaining small and
dry at its base.

1 Leaves <2cm, spine-tipped; fruit a succulent berry
   with calyx remaining small and dry at its base;
   functionally dioecious   **3. G. mucronata**
1 Leaves >2cm, not spine-tipped; fruit a capsule
   surrounded by succulent berry-like swollen calyx;
   plants bisexual   2
   2 Leaves cuneate at base; flowers solitary in leaf-
      axils; fruit bright red   **2. G. procumbens**
   2 Leaves rounded to cordate at base; flowers in
      terminal and subterminal racemes; fruit purplish-
      black   **1. G. shallon**

1. G. shallon Pursh - Shallon. Thicket-forming shrub to **354**
1.5m; leaves 5-10cm, rounded to cordate at base, closely
minutely serrate; flowers in terminal and subterminal racemes
5-12cm, with dense stalked glands on pedicels and rhachis,
7-10mm, white to pink; fruit purplish-black. Intrd; planted
as cover and food for birds, natd in woodland and shrubberies
especially on sand and peat; scattered through En, Ir, Sc and
CI; W N America.
   1 x 3. G. shallon x G. mucronata (=X Gaulnettya wisleyensis
W. Marchant nom. nud.) is a well-known garden plant but has
arisen on heathy ground in S Hants (found in 1981 with both
parents). It is intermediate in leaf characters, but the
wild plant is closer to G. shallon than is the commonest
garden cultivar.
   2. G. procumbens L. - Checkerberry. Dwarf ground-covering
shrub to 15cm; leaves 2-5cm, cuneate at base, rather remotely
minutely serrate; flowers solitary in leaf-axils, with
pubescent but not glandular pedicels, 5-7mm, white to pink;
fruit bright red. Intrd; planted as ground cover and natd in
woodland; few places in Sc and perhaps En; E N America.
   3. G. mucronata (L.f.) Hook. & Arn. (Pernettya mucronata **354**
(L.f.) Gaudich. ex Sprengel) - Prickly Heath. Erect or
spreading suckering shrub to 1.5m; leaves 8-20mm, cuneate to
subcordate at base, remotely and minutely serrate; flowers
1-few in sub-terminal axillary clusters, with subglabrous to
puberulent pedicels, 2.5-5mm, white; berry white to purple.
Intrd; natd in open woodland and shrubberies on sandy soil;
scattered in Br and Ir; Chile.

9. ARBUTUS L. - Strawberry-tree
Leaves alternate, evergreen; flowers in terminal panicles;

petals 5, fused to form tubular corolla narrowed and shortly lobed distally; stamens 10; anthers with 2 long terminal appendages; ovary superior; fruit a warty, globose berry.

1. A. unedo L. - <u>Strawberry-tree</u>. Shrub or tree to 5(11)m; **RR** leaves 4-11cm, ovate to obovate or narrowly so, serrate; flowers 6-11mm, white or pink-tinged; fruit red, 1.5-2cm across; flowering in autumn together with fruits from previous year. Native; rocky ground in scrub and young woodland; S & N Kerry, W Cork and Sligo; rarely natd (bird-sown) on mostly chalk or limestone slopes in En and Wa.

10. ARCTOSTAPHYLOS Adans. (<u>Arctous</u> (A. Gray) Niedenzu) - <u>Bearberries</u>
Leaves alternate, deciduous or evergreen; flowers 1-few in terminal clusters; petals 5, fused to form tubular corolla narrowed and shortly lobed distally; stamens 10; anthers with 2 terminal appendages; ovary superior; fruit a smooth, globose, berry-like drupe.

1. A. uva-ursi (L.) Sprengel - <u>Bearberry</u>. Procumbent shrub with stems to 1.5m; leaves 1-3cm, entire, evergreen, obovate to oblanceolate; flowers 4-6mm, white, pink-tinged; stamens with reflexed appendages c. equalling anthers; fruit c.8-10mm across, bright red. Native; peaty and rocky moorland in lowlands and mountains; locally common in Sc, N En and N & W Ir.
2. A. alpinus (L.) Sprengel (<u>Arctous</u> alpinus (L.) Niedenzu) **R** - <u>Mountain</u> Bearberry. Procumbent shrub with stems to 60cm; leaves 1-2.5cm, serrate, dying in autumn but persistent until next Spring, obovate; flowers 3-5mm, white or green-tinged; stamens with erect appendages much shorter than anthers; fruit c.6-10mm across, black. Native; mountain moorland; local in N & NW Sc.

11. CALLUNA Salisb. - <u>Heather</u>
Leaves opposite, evergreen, sessile; flowers in long usually terminal racemes or panicles; petals 4, fused at basal 1/4 or less; stamens 8; anthers with 2 basal appendages; ovary superior; fruit a capsule dehiscing along line of fusion of carpels.

1. C. vulgaris (L.) Hull - <u>Heather</u>. Decumbent to erect shrub to 60(150)cm; leaves 2-3.5mm, sessile, entire, very strongly revolute at margins making the cross-section triangular, with 2 pointed auricles at base; flowers 3-4.5mm, pink (or white). Native; heaths, moors, rocky places, bogs and open woodland, mainly on sandy or peaty soil; abundant in suitable places throughout BI.

12. ERICA L. - <u>Heaths</u>
Leaves in whorls of 3-4(5), evergreen, shortly petiolate; flowers in various terminal and/or axillary clusters; petals

4, fused at basal 1/2 or more; stamens 8, anthers with 0 or 2
basal appendages; ovary superior; fruit a capsule dehiscing
between lines of fusion of carpels.

1 Anthers at least partly exserted from corolla   2
1 Anthers included in corolla   4
  2 Pedicels longer than calyx; corolla-lobes
    divergent distally; summer-flowering   **9. E. vagans**
  2 Pedicels shorter than calyx; corolla-lobes +
    parallel distally; winter- to spring-flowering   3
3 Stems to 1.2(-2)m, with well-developed main stems;
  young twigs with flanges of tissue running <1/2 way
  from leaf-bases to next lower node; flowering Mar-
  Jun   **7. E. erigena**
3 Stems to 60cm, without well-developed main stems;
  young twigs with flanges of tissue running >1/2 way
  from leaf-bases to next lower node; flowering Nov-
  Jun   **8. E. x darleyensis**
  4 Revolute leaf-margins meeting closely under leaf,
    entirely obscuring lowerside; flowers usually in
    panicles   5
  4 Revolute leaf-margins not meeting or meeting only
    distally under leaf, revealing at least proximal
    part of lowerside; flowers in racemes or apparent
    umbels   6
5 Bracteoles borne only on proximal part of pedicel,
  not overlapping calyx; shrub often >80cm, spring-
  flowering   **6. E. lusitanica**
5 Some bracteoles borne near apex of pedicel, overlapping calyx; shrub <80cm, summer-flowering
                               5. E. cinerea
  6 Flowers in terminal + elongated racemes; anthers
    with 0 basal appendages   **1. E. ciliaris**
  6 Flowers in terminal umbel-like clusters; anthers
    with basal appendages   7
7 Lowerside of leaves green; sepals glabrous or with
  only short hairs; anthers with triangular appendages
                              4. E. terminalis
7 Lowerside of leaves whitish; sepals with long hairs;
  anthers with linear appendages   8
  8 Sepals and uppersides of leaves usually glabrous
    except for long hairs; most of leaf lowerside
    exposed; ovary and fruit glabrous   **2. E. mackaiana**
  8 Sepals and uppersides of leaves usually with dense
    short hairs as well as long hairs; most of leaf
    lowerside obscured; ovary and fruit pubescent
                              3. E. tetralix

1. E. ciliaris L. - <u>Dorset Heath</u>. Straggly shrub to 60cm; **348**
leaves in whorls of 3(-4), 2-4mm, with long usually glandular **RR**
hairs; flowers in elongated racemes; corolla 8-12mm, bright
reddish-pink (or white), oblique-ended.   Native; heaths,

often damp; very locally frequent in Dorset, S Devon, W Cornwall and W Galway.

1 x 3. E. ciliaris x E. tetralix = E. x watsonii Benth. occurs throughout the range of E. ciliaris in BI; it has leaves more like those of E. tetralix and flowers nearer those of E. ciliaris and is highly sterile, though back-crossing occurs in Dorset.

2. E. mackaiana Bab. - Mackay's Heath. Straggly to compact **RR** shrub to 60cm; leaves in whorls of 4, 2-4.5mm, with long glandular hairs, with most of lowerside exposed; flowers in terminal umbel-like clusters; corolla 5-7mm, purplish-pink. Native; peaty bogs; very local in W Galway and W Donegal.

2 x 3. E. mackaiana x E. tetralix = E. x stuartii E.F. Linton (E. x praegeri Ostenf.) occurs near E. mackaiana in all its sites; it is intermediate in leaf-shape and pubescence and is completely sterile.

3. E. tetralix L. - Cross-leaved Heath. Straggly shrub to 70cm; leaves in whorls of 4, 2-5mm, with dense short and usually few long usually glandular hairs, with most (but not all) of lowerside obscured; flowers in terminal umbel-like clusters; corolla 5-9mm, pink (or white). Native; bogs and usually wet heaths and moors; suitable places throughout BI.

3 x 9. E. tetralix x E. vagans = E. x williamsii Druce has been found as small patches on several occasions on Lizard Peninsula, W Cornwall, with the parents; it resembles E. vagans in most characters but is sterile and has pubescent young leaves. It is common in cultivation.

4. E. terminalis Salisb. - Corsican Heath. Bushy or erect **348** shrub to 1m (rarely more); leaves in whorls of 4, 3-6mm, puberulent when young, without long hairs; ± glabrous later; flowers in terminal umbel-like clusters; corolla 5-7mm, bright pink. Intrd; natd on sand-dunes; Magilligan, London-derry, known since c.1900; W Mediterranean.

5. E. cinerea L. - Bell Heather. Straggly shrub to 60cm; leaves in whorls of 3, 4-7mm, glabrous; flowers in terminal racemes or panicles, or sometimes in ± umbel-like clusters; corolla 4-6mm, bright reddish-purple (or white). Native; usually dry heaths and moors; suitable places throughout BI.

6. E. lusitanica Rudolphi - Portuguese Heath. Erect shrub **348** to 2(-3)m; leaves in whorls of 3-4, 5-7mm, glabrous; flowers in terminal panicles; corolla 4-5mm, pinkish-white. Intrd; natd on heaths and railway-banks; E & W Cornwall (known since 1920) and Dorset; SW Europe.

7. E. erigena R. Ross (E. hibernica (Hook & Arn.) Syme non **348** Utinet, E. mediterranea auct. non L., E. herbacea L. ssp. **R** occidentalis (Benth.) Lainz) - Irish Heath. Erect shrub to 1.2(-2)m; leaves in whorls of 4, 5-8mm, glabrous; flowers in terminal racemes or panicles; corolla 5-7mm, purplish-pink (or white). Native; in usually well-drained parts of bogs; W Galway and W Mayo. Flowers Mar-Jun.

8. E. x darleyensis Bean (E. erigena x E. herbacea L.) - **348** Darley Dale Heath). Differs from E. erigena in key

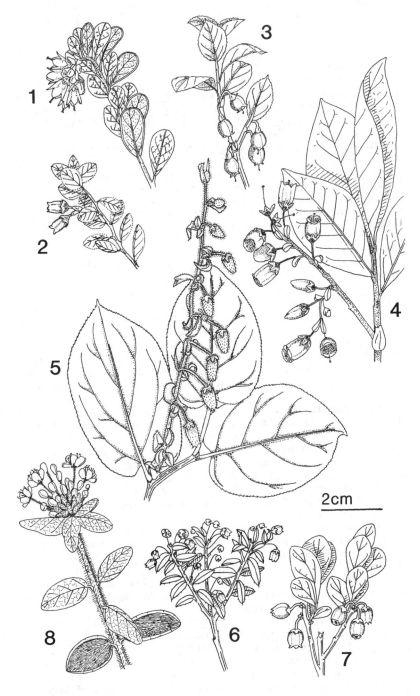

FIG. 354 - Ericaceae. 1, Vaccinium vitis-idaea. 2, V. x intermedium. 3, V. myrtillus. 4, V. corymbosum. 5, Gaultheria shallon. 6, G. mucronata. 7, Vaccinium uliginosum. 8, Ledum palustre ssp. groenlandicum.

characters. Intrd; planted in 1930s on bank in E Suffolk and
now well established; garden origin. Flowers Nov-Jun.
9. E. vagans L. - Cornish Heath. Straggly shrub to 80cm;   348
leaves in whorls of 4-5, 5-10mm, glabrous; flowers in   **RR**
subterminal racemes, on long pedicels; corolla 2.5-3.5mm,
pale pink to white. Native; dry heaths, often relatively
base-rich; W Cornwall, mostly on Lizard Peninsula; commonly
cultivated and sometimes persisting elsewhere, well natd or
perhaps native in Fermanagh.

13. VACCINIUM L. (Oxycoccus Hill) - Bilberries
Leaves alternate, deciduous or evergreen; flowers solitary or
clustered, terminal or axillary; petals 4-5, fused at base or
for most part to form variously shaped corolla; stamens 8 or
10; anthers with or without terminal appendages; ovary
inferior; fruit a berry.

1 Corolla divided >3/4 way to base; pedicels erect,
   filiform; leafy stems procumbent for most part   2
1 Corolla divided <2/3 way to base; pedicels not erect
   or filiform; leafy stems erect to decumbent   4
   2 Leaves narrowly oblong, at least some >1cm;
      bracteoles above middle of pedicel, mostly >1mm
      wide; leafy shoot continuing growth beyond flower
      cluster in same year   **3. V. macrocarpon**
   2 Leaves ovate-elliptic or narrowly so, rarely >1cm;
      bracteoles at or below middle of pedicel, <0.5mm
      wide; flowers in terminal groups of 1-c.5   3
3 Pedicels minutely puberulous   **1. V. oxycoccos**
3 Pedicels glabrous or almost so   **2. V. microcarpum**
   4 Some leaves >3cm (to 8cm)   **7. V. corymbosum**
   4 Leaves <3cm   5
5 Leaves serrate to serrulate; stems acutely angled
                                         6. V. myrtillus
5 Leaves entire to obscurely crenulate; stems terete   6
   6 Leaves evergreen; flowers in terminal racemes;
      corolla widest at mouth, divided c.1/2 way to base;
      fruit red   **4. V. vitis-idaea**
   6 Leaves deciduous; flowers in axillary clusters of
      1-4; corolla narrowed at mouth, divided <1/4 way to
      base; fruit bluish-black   **5. V. uliginosum**

1. V. oxycoccos L. (Oxycoccus palustris Pers.) - Cranberry.
Procumbent shrub with stems to 30(80)cm; leaves (2)5-10(12) x
2-6mm, entire, evergreen, often widest near or not much below
middle; flowers in terminal groups of 1-4; pedicels
puberulous; corolla bright pink, with strongly reflexed
lobes; fruit red, (6)8-10(15)mm across, globose to pear-
shaped; 2n=48. Native; bogs and very wet heaths; locally
frequent in much of Br and Ir, but absent from most of S En,
S Ir and N Sc.
2. V. microcarpum (Turcz. ex Rupr.) Schmalh. (Oxycoccus   **R**

microcarpus Turcz. ex Rupr.) - Small Cranberry. Differs from
V. oxycoccos in its leaves 2-6(8) x 1-2.5mm, often widest
near base; flowers in groups of 1-2; pedicels glabrous or
almost so; fruit 5-8(10)mm across, pear-shaped or ellipsoid;
2n=24. Native; bogs; C & N mainland Sc; S Northumb. Intermediates with V. oxycoccos occur and the 2 taxa might be
better considered as vars or sspp. of the latter.
   3. V. macrocarpon Aiton (Oxycoccus macrocarpus (Aiton)
Pursh) - American Cranberry. Differs from V. oxycoccos in
being more robust; leaves 6-18 x 2-5mm, oblong; flowers in
groups of 1-5(10); fruit 9-14(20)mm across, globose; and see
key. Intrd; grown for its fruit and natd from bird-sown
seed, mostly in peaty places; scattered in Br from S En to W
Sc; E N America.
   4. V. vitis-idaea L. - Cowberry. Erect to decumbent shrub **354**
to 30cm; young stems terete, puberulous; leaves 10-30mm,
obovate to elliptic, entire to obscurely crenulate, evergreen; flowers in short terminal racemes, pinkish white;
fruit red, 6-10mm across, globose. Native; moors and open
peaty woods; locally abundant in Br from S Wa and C En
northwards, scattered in Ir.
   4 x 6. V. vitis-idaea x V. myrtillus = V. x intermedium 354
Ruthe occurs very locally with the parents in Staffs, Derbys
and Yorks, with old or unconfirmed records elsewhere, but is
absent from most areas where the parents cohabit; it is
convincingly intermediate in all leaf, stem, flower and fruit
characters and sets some good seed.
   5. V. uliginosum L. - Bog Bilberry. Erect to ascending **354**
shrub to 50(80)cm; young stems terete, glabrous to
puberulous; leaves 8-25mm, obovate, entire, deciduous;
flowers in groups of 1-4 on short side-branches, pale pink;
fruit bluish-black with whitish bloom, 6-10mm across,
globose. Native; moors; locally common in C & N Sc, very
local in S Sc and N En. Some plants from Shetland resemble
ssp. **microphyllum** Lange, from N Europe, which differs in its
smaller parts, less revolute corolla-lobes and diploid (not
tetraploid) chromosome number; they need further study.
   6. V. myrtillus L. - Bilberry. Erect to ascending shrub to **354**
50(100)cm; stems acutely angled, glabrous; leaves 10-30mm,
ovate to elliptic, serrate to serrulate, deciduous; flowers
1(-2) in leaf-axils, pinkish-red; fruit bluish-black with
whitish bloom, 6-10mm across, + globose. Native; heaths,
moors and woods; common in suitable places throughout Br and
Ir, but absent from much of C & E En.
   7. V. corymbosum L. - Blueberry. Erect shrub to 1(-2.5)m; **354**
young stems terete, puberulous; leaves 10-80mm, elliptic,
entire to serrate, deciduous; flowers in terminal or lateral
stalked racemes, white or pink-tinged; fruit bluish-black
with whitish bloom, 5-12mm across, + globose. Intrd; natd on
heathland from bird-sown seeds; S Hants and Dorset; E N
America.

67. PYROLACEAE - <u>Wintergreen family</u>

Herbaceous rhizomatous perennials; leaves opposite, whorled,
alternate or + all basal, simple, petiolate, exstipulate.
Flowers terminal, solitary or in a raceme, bisexual,
hypogynous, actinomorphic; sepals 5, + free, petals 5, free,
+ white; stamens 10, with anthers opening by 2 apparently
apical pores; ovary incompletely 5-celled with axile
placentation and many ovules per cell; style 1; stigma 5-
lobed to + capitate; fruit a capsule dehiscing between lines
of fusion of carpels.
Herbaceous plants with + white petals and with flowers
similar in structure to those of Ericaceae.

1 Flowers solitary   3. MONESES
1 Flowers in terminal raceme   2
   2 Flowers all turned to 1 side of axis; anther pores
      borne on main body of anther; petioles <2cm
                                                      2. ORTHILIA
   2 Flowers facing all directions; anther pores borne
      at end of short tubular anther outgrowths; longest
      petioles >2cm   1. PYROLA

1. PYROLA L. - <u>Wintergreens</u>
Flowers in terminal raceme, facing all directions; anthers
with pores borne on very short tubes; pollen-grains released
in tetrads.

1 Style strongly curved; flowers saucer- to cup-shaped
                                               3. P. rotundifolia
1 Style + straight; flowers + globose   2
   2 Style 1-2mm, included in flower, not widened below
      stigma   1. P. minor
   2 Style 4-6mm, just exserted from flower, widened
      immediately below stigma-lobes   2. P. media

1. P. minor L. - <u>Common Wintergreen</u>. Flowering stem erect,
to 20(30)cm, with all leaves near base; leaves 2.5-6cm,
ovate-elliptic, obtuse to rounded at apex, crenate-serrate,
with petioles slightly shorter; flowers 4-7mm across; petals
pinkish-white; style 1-2mm, straight, not widened below
stigma. Native; mostly on leaf-mould in woods in S, also on
damp rock-ledges and peaty moors in N, rarely sand-dunes;
very scattered over most of Br and Ir.
2. P. media Sw. - <u>Intermediate Wintergreen</u>. Differs from P.   R
<u>minor</u> in leaves broadly elliptic to orbicular, with petioles
c. as long or longer; flowers 7-11mm across; style 4-6mm,
widened immediately below stigma. Native; humus-rich moors
and woods; frequent in C & N Sc (not outer islands),
extremely local in N Ir, S Sc and N En.
3. P. rotundifolia L. - <u>Round-leaved Wintergreen</u>. Differs   R
from P. <u>minor</u> in leaves ovate-elliptic to orbicular, with

petioles longer; flowers 8-12mm across; petals white; style 4-10mm, strongly curved, widened immediately below stigma. Native. The 2 sspp. are of doubtful value; all characters, except possibly those of the sepals, break down.

a. Ssp. rotundifolia. Leaves rarely orbicular; scale-leaves on stems above true leaves 1-2; pedicels 4-8mm; sepals triangular-lanceolate, acute; anthers 2.2-2.8mm; style 6-10mm. Damp rock-ledges, woods, bogs and fens; very local in En, Sc and C Ir.

b. Ssp. maritima (Kenyon) E. Warb. Leaves usually orbicular; scale-leaves 2-5; pedicels 2-5mm; sepals oblong-lanceolate, obtuse; anthers 1.9-2.4mm; style 4-6mm. Damp hollows in sand-dunes; W coast of Br from Cumberland to N Devon, Co Wexford, doubtfully reported from E Br.

2. ORTHILIA Raf. (Ramischia Opiz ex Garcke) - Serrated Wintergreen
Stem-leaves alternate; flowers in terminal raceme, all turned to 1 side; anthers with pores borne on main body of anther; pollen-grains released singly.

1. O. secunda (L.) House (Ramischia secunda (L.) Garcke,   R
Pyrola secunda L.) - Serrated Wintergreen. Flowering stem erect, often curved at top, to 10(20)cm, with all leaves near base; leaves 2-4cm, ovate to elliptic-ovate, acute to obtuse at apex, finely serrate, with shorter petiole; flowers 4.5-6mm across; petals greenish-white; style 4.5-6mm, straight. Native; woods and damp rock-ledges; rather local from N En to N Sc (not outer islands), very local in Wa and N & C Ir.

3. MONESES Salisb. - One-flowered Wintergreen
Stem-leaves opposite or in whorls of 3; flowers single, terminal, pendent or turned to 1 side; anthers with pores borne on short tubes; pollen-grains released in tetrads.

1. M. uniflora (L.) A. Gray - One-flowered Wintergreen.   RR
Flowering stem erect, to 10(15)cm, with all leaves near base; leaves 6-25mm, orbicular, crenate-serrate, rounded at apex, with shorter petiole; flowers 12-20mm across; petals white; style 5-7mm, straight. Native; on leaf-litter in pinewoods; very local in NE Sc.

68. MONOTROPACEAE - Bird's-nest family

Saprophytic ± chlorphyll-less herbaceous perennials; leaves alternate, sessile, scale-like. Flowers in terminal raceme, bisexual, hypogynous, actinomorphic; sepals 4-5, free; petals 4-5, free, pale brownish-yellow; stamens 8 or 10, opening by longitudinal slits; ovary and fruit as in Pyrolaceae.
Easily recognized by its brownish-yellow aerial parts (utterly different in flower structure from the similarly

chlorophyll-less <u>Neottia</u>, <u>Orobanche</u> and <u>Lathraea</u>), but very
close to and perhaps best in Pyrolaceae.

1. MONOTROPA L. - <u>Yellow Bird's-nest</u>
  1. M. hypopitys L. - <u>Yellow Bird's-nest</u>. Stems to 30cm,
pendent at apex in flower, erect in fruit, arising from dense
± globose underground organ; scale-leaves numerous, brownish-
yellow, to 13mm; flowers pale dull yellow, usually glabrous
on outside. Native; on leaf-litter in woods (especially
<u>Pinus</u> and <u>Fagus</u>) and on sand-dunes.
  a. Ssp. hypopitys. Flowers ≤11; petals 9-13mm; stamens,
carpels and inside of petals pubescent; style equalling or
longer than ovary; 2n=48. Very scattered in En, ? elsewhere.
  b. Ssp. hypophegea (Wallr.) Holmboe (<u>M.</u> <u>hypophegea</u> Wallr.,
<u>M.</u> <u>hypopitys</u> var. <u>glabra</u> Roth). Flowers ≤8; petals 8-10mm;
ovary glabrous; stamens, style and inside of petals glabrous
or pubescent; style equalling or shorter than ovary; 2n=16.
Local in Br and Ir and absent from much of N & W. General
pubescence of inside of flowers does not adequately separate
the 2 sspp.; so-called intermediates resemble ssp. <u>hypophegea</u>
in all characters (incl. chromosome number) except pubescence
and are referable to that ssp.

69. DIAPENSIACEAE - <u>Diapensia family</u>

Cushion-like, evergreen dwarf shrub; leaves alternate,
simple, entire, tapered to base, exstipulate. Flowers
solitary and terminal; bisexual, hypogynous, actinomorphic;
sepals 5, ± free; petals 5, fused to about 1/2 way, white;
stamens 5, alternating with petals; ovary 3-celled with axile
placentation and many ovules per cell; style 1; stigma
capitate or ± 3-lobed; fruit a capsule.
Easily recognisable by its habit, 5 stamens and 3-locular
ovary.

1. DIAPENSIA L. - <u>Diapensia</u>
  1. D. lapponica L. - <u>Diapensia</u>. Plant forming dome-shaped **RRR**
cushions to 6cm high; leaves 5-10mm, obovate, obtuse; flowers
1-2cm across, on stems to 3cm. Native; exposed mountains at
760-850m; on 2 hills 24km apart NW of Fort William, Westerness, first found in 1951.

69A. MYRSINACEAE

MYRSINE africana L., from Africa to Nepal, is an evergreen
shrub to 1.5(2)m that is used as a hedging plant in Guernsey,
CI, where it survives after neglect. It has small obovate
leaves toothed distally and gland-dotted, clusters of 3-6
subsessile very small dioecious flowers with 4-5 petals fused
into a very short tube, and a 1-seeded pale blue fleshy

fruit.

70. PRIMULACEAE - Primrose family

Herbaceous annuals or perennials; leaves variously arranged,
simple to pinnate, petiolate or not, exstipulate. Flowers
variously arranged, bisexual, hypogynous (semi-epigynous in
Samolus), actinomorphic; sepals free to fused, 5(-9); petals
fused at least at base, as many as sepals (0 in Glaux);
stamens as many as sepals, borne on corolla if present; ovary
1-celled with free-central placentation and many ovules;
style 1; stigma capitate; fruit a capsule.
Very variable in superficial flower characters, but usually
recognizable by the herbaceous habit, fused petals, and one-
celled ovary with free-central placentation.

```
1  Leaves pinnate; plant a submerged aquatic    2. HOTTONIA
1  Leaves simple; plant not submerged                        2
   2  All leaves basal                                       3
   2  Some or all leaves on stems                            4
3  Corolla-lobes patent to erecto-patent; underground
   tuber 0                                        1. PRIMULA
3  Corolla-lobes strongly reflexed; underground tuber
   present                                        3. CYCLAMEN
   4  Corolla yellow                             4. LYSIMACHIA
   4  Corolla white to red, blue or purple, or 0           5
5  All or most leaves in single apparent whorl at top
   of stem; corolla-lobes mostly 6-7            5. TRIENTALIS
5  Leaves opposite or alternate along stem; corolla-
   lobes 5                                                  6
   6  Ovary 1/2-inferior; corolla white, longer than
      calyx                                        8. SAMOLUS
   6  Ovary superior; corolla usually coloured or 0 or
      shorter than calyx                                    7
7  Corolla present; calyx-lobes ± free; capsule dehiscing
   transversely                                  6. ANAGALLIS
7  Corolla 0; calyx-lobes fused for >1/4; capsule
   dehiscing longitudinally                        7. GLAUX
```

1. PRIMULA L. - Primroses
Perennials; leaves all basal, simple; calyx-tube longer than
lobes; corolla-lobes 5, patent to erecto-patent; corolla-
tube c. as long as lobes or longer; capsule dehiscing by 5
teeth or valves.
 Most spp. are usually heterostylous, some plants having
stigmas higher than anthers (pin-eyed) and others vice versa
(thrum-eyed).

```
1  Pale mealy coating present on various parts of leaves,
   scape, pedicels and flowers                              2
1  Mealy coating 0                                          6
```

2 Corolla $\geq$(12)15mm across, yellow   3
2 Corolla $\leq$15mm across, lilac to purple, rarely
  white   5
3 Leaves with narrow whitish border with short-stalked
  glands, often mealy or puberulent on upperside
                             6. P. auricula
3 Leaves green to margin, glabrous, not mealy   4
4 Leaves cuneate at base   **8. P. sikkimensis**
4 Leaves truncate to cordate at base   **7. P. florindae**
5 Flowers heterostylous; corolla-lobes usually lilac,
  with gaps between at least near base (N En)
                               4. P. farinosa
5 Flowers homostylous (anthers and stigma $\pm$ at same
  level); corolla-lobes usually purple, overlapping
  or contiguous (N Sc)   **5. P. scotica**
6 Flowers borne in 2 or more whorled tiers up scape;
  corolla usually purplish-red   **9. P. japonica**
6 Flowers borne singly from base of plant or in a
  single umbel on scape; corolla usually yellow   7
7 Pedicels with long shaggy hairs; ripe capsules lying
  near or on ground, with viscid seeds; flowers usually
  borne singly from base of plant   **1. P. vulgaris**
7 Pedicels with short fine hairs; ripe capsules held $\pm$
  erect, with dry seeds; flowers borne in umbel on
  scape   8
8 Corolla usually <15mm across, with folds in throat;
  calyx uniformly pale green, with acute or obtuse
  and apiculate teeth; capsule enclosed in calyx
                                3. P. veris
8 Corolla usually >15mm across, without folds in
  throat; calyx pale green with dark green midribs,
  with $\pm$ acuminate teeth; capsule c. as long as
  or longer than calyx   **2. P. elatior**

Other spp. - Several spp. are grown in gardens or are
planted outside and may persist for a while without
attention. **P. pulverulenta** Duthie, from W China, resembles
P. japonica but is mealy; and **P. 'Wanda'** (= P. juliae Krecz.
x P. vulgaris), of garden origin, resembles P. vulgaris but
has wine-red flowers and is $\pm$ glabrous.

1. P. vulgaris Hudson (P. acaulis (L.) Hill) - Primrose.
Plant with long shaggy hairs; leaves gradually tapered to
base; flowers borne singly on pedicels to 12cm arising from
leaf-rosette, or rarely in an umbel on a scape, c.2-4cm
across, usually pale yellow, rarely white or reddish-pink;
calyx $\pm$ uniformly pale green. Native; woods, hedgebanks and
in damper areas in grassland, often on heavy soils; throughout BI but rare or 0 locally.
1 x 2. P. vulgaris x P. elatior = **P. x digenea** A. Kerner
occurs frequently around the area of P. elatior in E Anglia.
It is intermediate in leaf, flower and pubescence characters;

the flowers are normally borne in an umbel on a scape. It is
fertile and hybrid swarms arise.

**1 x 2 x 3. P. vulgaris x P. elatior x P. veris = P. x
murbeckii** Lindq. occurs very rarely near P. elatior in W
Suffolk, both from wild parents and by pollination of wild P.
elatior by garden (sometimes purple-flowered) P. x polyantha.

1 x 3. P. vulgaris x P. veris = P. x polyantha Miller (P. x
tommasinii Gren. & Godron, P. x variabilis Goupil non Bast.)
occurs sporadically where the parents meet over most of BI,
often in scrubby areas. It is intermediate in leaf, flower
and pubescence characters; the flowers are normally borne in
an umbel on a scape. It is partially fertile but back-
crossing and introgression are rare. The garden Polyanthus
is probably of this origin, and occasionally this escapes
from cultivation or persists where planted; such plants may
vary considerably in flower colour, from white to orange,
purple or mauve. This hybrid, P. elatior, and plants of P.
vulgaris with a scape are often confused; careful attention
to leaves, flowers and pubescence is needed to avoid this.

2. P. elatior (L.) Hill - Oxlip. Plant with rather short   **R**
crisped hairs; leaves less gradually narrowed at base than in
P. vulgaris and often + abruptly so; flowers in an umbel on a
scape to 30cm, c.15-20mm across, pale yellow; calyx with
darker green midribs. Native; very locally abundant in woods
on clay in E Anglia, in area + lacking P. vulgaris, and 2
small outlying areas in Bucks.

2 x 3. P. elatior x P. veris = P. x media Peterm. occurs
rarely in the area of P. elatior in E Anglia; it is intermediate in diagnostic characters and partially fertile, but
backcrossing has not been detected.

3. P. veris L. - Cowslip. Plant with short straight hairs;
leaves abruptly contracted to petiole at base; flowers in an
umbel on a scape to 30cm, c.8-15mm across, deep or brownish-
yellow, rarely reddish; calyx uniformly pale green. Native;
locally common in grassy places usually on light base-rich
soils; throughout most of BI, but rare in much of N.

4. P. farinosa L. - Bird's-eye Primrose. Much of plant with **R**
mealy covering, but hairs 0; leaves gradually tapered at
base, obovate; flowers in an umbel on a scape to 15cm, c.7-
15mm across, usually pinkish-lilac; heterostylous; 2n=18.
Native; damp grassy, stony or peaty ground on limestone;
locally frequent in N En from Yorks and Lancs to Cumberland,
formerly S Sc.

5. P. scotica Hook. - Scottish Primrose. Sometimes   **R**
biennial; differs from P. farinosa in its usually wider
leaves, longer narrower corolla-tube, and more overlapping
and purple corolla-lobes; homostylous; 2n=54. Native; damp
grassy places near sea on cliffs, dunes and pastures; W
Sutherland, Caithness and Orkney; endemic.

6. P. auricula L. - Auricula. Inflorescence with mealy
covering; leaves gradually or abruptly contracted to short
petiole, mealy or not, with short-stalked glandular hairs;

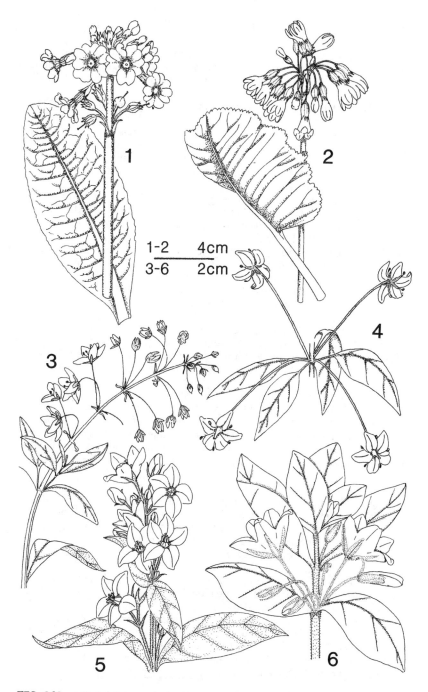

FIG 363 - Primulaceae. 1, Primula japonica. 2, P. florindae.
3, Lysimachia terrestris. 4, L. ciliata. 5, L. vulgaris. 6,
L. punctata.

flowers in an umbel on a scape to 15cm, c.(12)15-25mm across, dull yellow (other colours in garden plants). Intrd; planted on rock-ledge in Caenlochan Glen, Angus, known since 1880; mountains of CS Europe.

7. **P. florindae** Kingdon-Ward - <u>Tibetan</u> <u>Cowslip</u>. **363** Inflorescence with mealy covering; leaves abruptly contracted to long petiole, with cordate or truncate base, not mealy, with 0 hairs; flowers in an umbel on a scape to 60cm, c.(12)15-30mm across, yellow. Intrd; planted by ponds and streams and in marshes; natd in several places in N En and Sc; Tibet.

8. **P. sikkimensis** Hook. f. - <u>Sikkim</u> <u>Cowslip</u>. Differs from <u>P. florindae</u> in its leaves gradually tapered to cuneate base. Intrd; planted on mountains in Caerns in 1921 and still there, often not flowering; Himalayas to W China.

9. **P. japonica** A. Gray - <u>Japanese</u> <u>Cowslip</u>. Plant not mealy, **363** glabrous; leaves gradually tapered to base; flowers borne in 2-several tiered whorls on a scape to 50cm, c.15-25mm across, purplish-red. Intrd; planted in shady moist places; natd in a few places in En and W Ir; Japan.

2. **HOTTONIA** L. - <u>Water-violet</u>
Perennials with submerged vegetative parts and emergent inflorescences; leaves ± whorled, pinnate with linear lobes; calyx divided nearly to base; corolla lobes 5, patent, longer than tube; capsule dehiscing by 5 valves.

1. **H. palustris** L. - <u>Water-violet</u>. Stems to 1m or more, floating under water, rooting at nodes; flowers on tiered whorls on erect peduncle, 15-25mm across, lilac with yellow throat. Native; shallow ponds and ditches; scattered in En and Wa, locally common E En, very locally natd in Ir.

3. **CYCLAMEN** L. - <u>Cyclamen</u>
Perennials with large underground tuber; leaves all basal, simple; calyx-tube shorter than lobes; corolla-lobes 5, strongly reflexed, longer than tube; capsule dehiscing by 5 valves which become reflexed and on pedicel which becomes tightly spiralled at maturity.

Other spp. - Several spp. are grown in 'wild' gardens and may locally persist after neglect. **C. graecum** Link, from Greece, is an Autumn-flowering sp. differing from <u>C. hederifolium</u> in the tuber rooting only from lowerside (not mainly from upperside) and the leaves not or scarcely angled. **C. repandum** Sibth. & Smith is a Spring-flowering sp. from C & E Mediterranean; its leaves and flowers resemble those of <u>C. hederifolium</u> and its corm that of <u>C. graecum</u>.

1. **C. hederifolium** Aiton (<u>C. neapolitanum</u> Ten.) - <u>Cyclamen</u>. Leaves 2.5-8cm, broadly ovate, cordate, dentate, angled at margin, with long petiole; flowers borne singly on pedicels

to 30cm arising from leaf-rosette, appearing in late Summer
and Autumn before leaves; corolla-lobes pale pink or rarely
white with dark basal blotches, 12-25mm. Intrd; natd in
woods and hedgerows; very scattered in CI and Br N to S Sc,
known in E Kent since 1778; S Europe.
2. C. coum Miller - <u>Eastern</u> <u>Cyclamen</u>. Differs from <u>C</u>.
<u>hederifolium</u> in corm rooting from lowerside; leaves not
angled, usually ± reniform; flowers appearing in Spring with
the leaves; corolla-lobes bright pink, 7-15mm. Intrd; natd
on roadside verge; Surrey; E Mediterranean.

4. LYSIMACHIA L. (<u>Naumburgia</u> Moench) - <u>Loosestrifes</u>
Perennials; leaves opposite or whorled, simple; calyx divided
nearly to base; corolla-lobes 5-7, patent to ± erect, longer
than tube, yellow; capsule opening by 5 valves.

1 Stems procumbent or decumbent; flowers borne singly
  in axils of normal leaves   2
1 Stems erect; at least some main leaf-axils with >1
  flower, or flowers borne in terminal or axillary
  racemes   3
  2 Leaves obtuse to rounded at apex, dotted with
    usually black glands; calyx-lobes ovate
                                             2. L. nummularia
  2 Leaves acute to subacute at apex, not glandular;
    calyx-lobes subulate to linear   **1. L. nemorum**
3 Flowers all borne in axils of much reduced bracts
  in terminal raceme; bulbils up to 2cm x 2mm, usually
  produced in leaf-axils late in season   **6. L. terrestris**
3 At least lower flower clusters or racemes borne in
  axils of ± normal leaves; bulbils 0   4
  4 Flowers <10mm across, borne in dense axillary
    racemes; petals linear   **7. L. thyrsiflora**
  4 Flowers >15mm across, borne in few-flowered
    axillary clusters or in terminal panicles   5
5 Pedicels >2cm; calyx and leaves glabrous   **4. L. ciliata**
5 Pedicels ≤2cm; calyx glandular-pubescent; leaves
  pubescent   6
  6 Corolla-lobes glandular-pubescent at margins;
    calyx-teeth uniformly green   **5. L. punctata**
  6 Corolla-lobes glabrous at margins; calyx-teeth with
    conspicuous orange margin   **3. L. vulgaris**

1. L. nemorum L. - <u>Yellow</u> <u>Pimpernel</u>. Stems to 40cm,
procumbent to decumbent; leaves 1-3cm, opposite, ovate to
broadly so, glabrous; flowers solitary in leaf-axils; corolla
5-8mm, glabrous, with 5 lobes. Native; woods and copses;
throughout most of BI.
2. L. nummularia L. - <u>Creeping-Jenny</u>. Stems to 60cm,
procumbent; leaves 1.5-3cm, opposite, broadly ovate-elliptic
to orbicular, dotted with black glands, otherwise glabrous;
flowers solitary in leaf-axils; corolla 8-18mm, dotted with

black glands, with 5 lobes. Native; damp places, often in
shade; throughout most of BI N to C Sc, but a natd garden
escape in many localities, especially in N.
3. L. vulgaris L. - <u>Yellow</u> Loosestrife. Stems to 1.5m, 363
erect; leaves 3-10cm, lanceolate to narrowly elliptic,
opposite or in whorls of 3-4, pubescent and with minute
glands; flowers in terminal panicles composed of axillary
flower-clusters; corolla 8-15mm, dotted with pale glands,
with 5 lobes. Native; ditches, marshes and by lakes and
rivers; scattered through most of BI except N Sc.
4. L. ciliata L. - <u>Fringed</u> Loosestrife. Stems to 1.2m, 363
erect; leaves 4-12cm, lanceolate to ovate, opposite or in
whorls of 3-4, ± glabrous; flowers in loose axillary groups;
corolla 9-15mm, densely covered with sessile glands towards
base, with 5 lobes. Intrd; natd in rough ground and damp or
shady places; rare garden escape scattered in Br mainly in N
En and Sc; N America.
5. L. punctata L. - <u>Dotted</u> Loosestrife. Stems to 1.2m, 363
erect; leaves 4-12cm, lanceolate to ovate, opposite or in
whorls of 3-4, pubescent and gland-dotted; flowers in dense
axillary groups; corolla 10-16mm, with dense short-stalked
glands especially at margin, with 5 lobes. Intrd; natd in
rough ground and damp places; common and increasing garden
escape scattered over most of BI; SE Europe.
6. L. terrestris (L.) Britton, Sterns & Pogg. - <u>Lake Loose-</u> 363
<u>strife</u>. Stems to 80cm, erect; leaves 2-8cm, narrowly
elliptic to lanceolate, opposite, dotted with black glands,
otherwise glabrous; flowers (often not produced) in axils of
much reduced bracts in terminal raceme; corolla 4-7mm, with
black glandular streaks, with 5 lobes. Intrd; damp places on
shore of Lake Windermere, Westmorland; N America.
7. L. thyrsiflora L. (<u>Naumburgia thyrsiflora</u> (L.) Reichb.) -   R
<u>Tufted</u> Loosestrife. Stems to 70cm, erect; leaves 7-10cm,
linear-lanceolate to narrowly elliptic, opposite, dotted with
black glands but otherwise glabrous; flowers in dense,
stalked, axillary racemes with much reduced bracts; corolla
4-6mm, with black glandular dots, with 5-7 lobes.
Native; wet places in marshes and by ditches and canals;
scattered in N En and C & S Sc, E Donegal.

5. TRIENTALIS L. - <u>Chickweed Wintergreens</u>
Glabrous perennials; leaves mostly in an apparent whorl at
apex of stem, simple; calyx divided almost to base; corolla-
lobes (5)6-7(9), erecto-patent, longer than tube, white;
capsule opening by 5 valves.

Other spp. - **T. borealis** Raf., from N America, is sold in
nurseries and has been found to persist in and near them in W
Kent and S Hants; easily confused with <u>T. europaea</u> but has
the larger leaves up to 8(10)cm, ovate to elliptic and
tapering acute to acuminate, with only scale-leaves up stem.

1. T. europaea L. - <u>Chickweed</u> <u>Wintergreen</u>. Stems to
20(25)cm, simple, erect; leaves (3)5-8(10), 1-5(8)cm, the
larger obovate, acute to subobtuse, a few much reduced but
not scale-like up the stem; flowers 1-2, each on erect
pedicel 2-7cm; corolla 6-10mm. Native; on humus in open
pine-woods and heather-moors; Lancs and Yorks to N Sc, E
Suffolk, locally common in Sc.

6. ANAGALLIS L. (<u>Centunculus</u> L.) - <u>Pimpernels</u>
Glabrous annuals or perennials; leaves opposite or alternate,
simple; calyx divided ± to base; corolla-lobes 5, erect to
patent scarcely longer to much longer than tube; capsule
opening by transverse line of dehiscence.

1 Upper leaves alternate; corolla divided <3/4 way to
  base, <2mm, much shorter than calyx-lobes
                                        3. A. minima
1 Leaves all opposite; corolla divided almost to base,
  the lobes >3mm, c. as long as or longer than calyx-
  lobes   2
  2 Stems procumbent, rooting at nodes; corolla-lobes
    >2x as long as calyx-lobes; leaves suborbicular
                                        1. A. tenella
  2 Stems decumbent to ascending, not rooting at nodes;
    corolla-lobes <2x as long as calyx-lobes; leaves
    ovate   **2. A. arvensis**

1. A. tenella (L.) L. - <u>Bog</u> <u>Pimpernel</u>. Perennial; stems to
20cm, procumbent; leaves <1cm, opposite; flowers solitary in
leaf-axils on erect pedicels to 3.5cm; corolla 6-10mm, pale
pink, the lobes erecto-patent, entire, glabrous. Native;
bogs and damp peaty ground; scattered over BI, common in
parts of W, absent from much of E.
2. A. arvensis L. - see sspp. for English names. Usually
annual; stems to 40cm, not rooting at nodes; leaves often
>1cm, opposite,; flowers solitary in leaf-axils on pedicels
to 3.5cm; corolla 4-7(10)mm, the lobes patent, denticulate or
with glands at margin. Native; arable and waste land and
open ground.
  a. Ssp. arvensis - <u>Scarlet</u> <u>Pimpernel</u>. Corolla-lobes usually 368
red, sometimes variously pink or white or blue, entire to
crenulate, with numerous minute hairs with 3 cells (incl.
basal one), the most distal globose and glandular. Most of
BI, common in S, rare in N Sc.
  b. Ssp. caerulea Hartman (<u>A.</u> <u>foemina</u> Miller, <u>A.</u> <u>arvensis</u> 368
ssp. <u>foemina</u> (Miller) Schinz & Thell.) - <u>Blue</u> <u>Pimpernel</u>. R
Corolla-lobes blue, crenulate to denticulate, with sparse
minute hairs with 4 cells (incl. basal one), the most distal
ellipsoid and glandular. Much rarer than ssp. <u>arvensis</u>;
usually in arable land and mostly in C & S En.
  Sterile red-flowered hybrids (<u>A.</u> x <u>doerfleri</u> Ronn.) occur
between the 2 sspp. but are very rare.

3. A. minima (L.) E.H. Krause (<u>Centunculus minimus</u> L.) - <u>Chaffweed</u>. Annual; stems erect to decumbent, to 5(8)cm; leaves 3-5mm, the upper ones alternate; flowers solitary in leaf-axils, subsessile; corolla <2mm, white to pink, the lobes ± erect, entire, glabrous. Native; bare damp sandy ground on heaths and in woodland rides; scattered over most of BI, mainly coastal in W.

7. GLAUX L. - <u>Sea-milkwort</u>
Glabrous slightly succulent rhizomatous perennials; leaves opposite, simple; calyx divided c.1/2 way to base into 5 lobes; corolla 0; capsule opening by 5 valves.

1. G. maritima L. - <u>Sea-milkwort</u>. Stems procumbent to suberect, to 30cm; leaves 4-12mm, often reduced below; flowers solitary in leaf-axils, sessile; calyx 3-5mm, white to pink. Native; saline sandy, muddy, rocky or grassy places; round coasts of BI and in a few inland salt-marshes.

8. SAMOLUS L. - <u>Brookweed</u>
Glabrous perennials; leaves in basal rosette and alternate up stem, simple; calyx-lobes ± free but fused to ovary; corolla-lobes 5, erect, longer than tube, white; capsule opening by 5 teeth.

1. S. valerandi L. - <u>Brookweed</u>. Stems to 45cm, erect; leaves 1-8cm, obovate; flowers 2-4mm, in terminal and axillary racemes, each with 1 small bracteole on pedicel. Native; wet places, especially by streams and flushes near the sea; coasts of BI, except NE Sc, frequent in E En but otherwise rare inland.

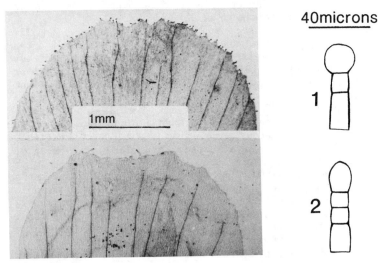

FIG 368 - Apex of corolla-lobe and single glandular hair of **Anagallis arvensis**. 1, ssp. **arvensis**. 2, ssp. **caerulea**. Drawings by C.A. Stace.

71. PITTOSPORACEAE - Pittosporum family

Shrubs; leaves evergreen, alternate, simple, entire, exstipulate, petiolate. Flowers solitary in leaf-axils or few
in terminal clusters, bisexual or monoecious, hypogynous,
actinomorphic; sepals 5, fused at base; petals 5, free,
purplish; stamens 5; ovary 1-celled with numerous ovules on
2-4 parietal placentas; style 1; stigma capitate; fruit a
2-4-celled capsule.
Easily recognized by the evergreen entire leaves, purple
petals and 1-celled, 2-4-carpellary ovary.

1. PITTOSPORUM Banks ex Sol.- Pittosporums
1. P. **crassifolium** Banks & Sol. ex A. Cunn. - Karo. Dense 370
shrub or tree to 5(8)m; leaves 5-8cm, obovate-oblong, upperside dark green, lowerside white-tomentose, with revolute
margin; flowers male and female mixed in terminal clusters,
10-15mm; capsule with 3(-4) valves. Intrd; planted as screen
or windbreak by sea; persisting and sometimes self-sown in
Scillies, rarely in W Cornwall and Jersey; New Zealand.
Resembles Olearia traversii (Asteraceae) when sterile but
leaves are alternate and 1st-year twigs + terete.
2. P. **tenuifolium** Gaertner - Kohuhu. Shrub or tree to 370
5(10)m; leaves 1-7cm, elliptic-oblong, mid-green on both
surfaces, + glabrous when mature, with undulate margin;
flowers bisexual, solitary, axillary, 10-15mm; capsule with 2
valves. Intrd; planted for ornament or screen by sea;
self-sown in W Cornwall; New Zealand. Resembles Olearia
paniculata (Asteraceae) when sterile but latter has thicker
leaves with white lowerside.

72. HYDRANGEACEAE - Mock-orange family

Shrubs; leaves deciduous, opposite, simple, exstipulate,
petiolate. Flowers in raceme-like cymes terminal on lateral
shoots (in terminal corymbs in Hydrangea), bisexual,
1/2-epigynous (epigynous in Hydrangea), actinomorphic; sepals
4(-6), free; petals 4(-6), free; stamens numerous (9-20in
Hydrangea); ovary 4(-6)-celled (2-3-celled in Hydrangea),
with numerous ovules on axile placenta; styles 4(-6), united
below (2-3, free in Hydrangea); stigmas clavate; fruit a
4(-6)-valved capsule (2-3-celled, dehiscent at apex in
Hydrangea).
The only opposite-leaved shrubs with large white flowers
with free petals, 9-numerous stamens and 1/2-inferior or
inferior ovary.

Other genera - HYDRANGEA L. (Hydrangeas) differs from
Philadelphus as above and in the small fertile flowers
surrounded by fewer large sterile ones. Several spp. are
grown in gardens and some persist or spread on to walls. **H.**

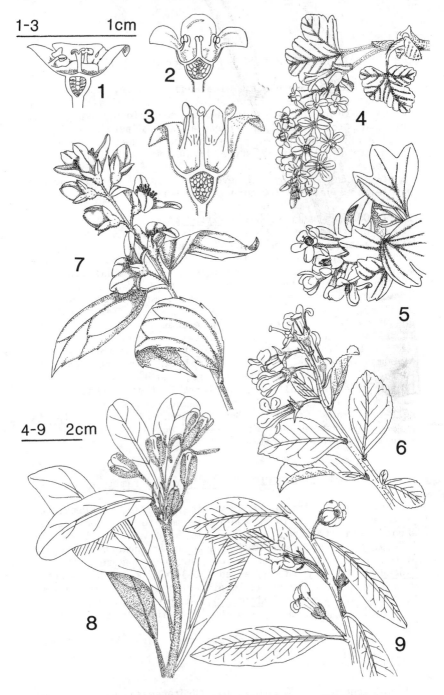

FIG 370 – Grossulariaceae, Hydrangeaceae, Pittosporaceae.
1-3, flowers of **Ribes**. 1, **R. rubrum**. 2, **R. spicatum**. 3, **R. nigrum**. 4, **Ribes sanguineum**. 5, **R. odoratum**. 6, **Escallonia macrantha**. 7, **Philadelphus x virginalis**. 8, **Pittosporum crassifolium**. 9, **P. tenuifolium**.

petiolaris Siebold & Zucc. and **H. anomala** D. Don both climb by aerial rootlets like Hedera; H. petiolaris, from Japan, has 15-20 stamens, and H. anomala, from China, has 9-15 stamens and more deeply toothed leaves. **H. sargentiana** Rehder, from China, is an erect shrub to 3m with stems densely covered in hairs and bristles and pubescent leaves.

1. PHILADELPHUS L. - Mock-oranges

Other spp. - Several other spp. from N America and China and hybrids often of complex parentage are grown in gardens, and some wild plants might be referable to them. Small-flowered, very fragrant, glabrous plants with almost entire leaves are probably hybrids involving **P. microphyllus** A. Gray.

1. P. coronarius L. - Mock-orange. Shrub to 3m; leaves 5-10cm, ovate to elliptic-oblong, serrulate, glabrous to sparsely pubescent on lowerside; flowers 2.5-5cm across, fragrant, glabrous, with ± patent petals. Intrd; commonly grown and frequently found as a relic in hedges and copses but rarely self-sown; scattered in En; Europe.

2. P. x virginalis Rehder (?P. coronarius x P. microphyllus **370** x P. pubescens Lois.). Differs from P. coronarius in leaves pubescent on lowerside; calyx pubescent; flowers cup-shaped, often flore pleno. Intrd; now commoner than P. coronarius in gardens and most recent relics in the wild are probably this hybrid, but distribution unknown; garden origin.

73. GROSSULARIACEAE - Gooseberry family
(Escalloniaceae)

Shrubs; leaves alternate, simple, often lobed, exstipulate, petiolate. Flowers solitary or in racemes or panicles, bisexual to dioecious, epigynous, with hypanthium, actinomorphic; sepals 5, arising from hypanthium; petals 5, free; stamens 5; ovary 1-locular, with numerous ovules on 2 parietal placentas; styles 2 with capitate stigmas or style 1 with bilobed or capitate stigma; fruit a berry or capsule. Distinguishable by shrubby habit, 5 sepals, petals and stamens arising from hypanthium, and inferior 1-celled ovary with 2 parietal placentas.

1 Leaves evergreen, not lobed; fruit a capsule; petals longer than sepals **1. ESCALLONIA**
1 Leaves deciduous, palmately lobed; fruit a berry; petals shorter than sepals **2. RIBES**

1. ESCALLONIA Mutis ex L.f. - Escallonia
Leaves evergreen, simple, serrate; flowers in terminal racemes or panicles; petals much longer than calyx, with distinct claw; fruit a capsule.

Other spp. - Many spp., hybrids and cultivars are grown;
some wild plants might not be E. macrantha, but this needs
checking. The most likely extra taxon is **E. x langleyensis**
Veitch (E. macrantha x E. virgata (Ruiz Lopez & Pavon)
Pers.), of garden origin, with petals with claws shorter than
limbs and not forming tube and usually pale pink.

1. E. macrantha Hook. & Arn. (E. rubra (Ruiz Lopez & Pavon) **370**
Pers. var. macrantha (Hook. & Arn.) Reiche) - Escallonia.
Shrub to 3m (rarely more); young growth glandular-viscid;
leaves 1-8cm, elliptic to obovate, gland-dotted, scented when
crushed; flowers 12-20mm; petals pink to deep red, with claws
longer than limbs and arranged into false corolla-tube.
Intrd; planted for hedging and ornament near sea; common
persistent relic in SW En, S Wa, W Ir and CI, very rarely
self-sown; Chile.

2. RIBES L. - Gooseberries
Leaves deciduous, palmately lobed, variously toothed; flowers
solitary or in racemes, on short lateral branches; petals
shorter than sepals, not forming a tube; fruit a berry.

1 Spines present on branches; flowers solitary or in
   short racemes of 2(-3)   **7. R. uva-crispa**
1 Spines 0; flowers in racemes of >4   2
   2 Flowers bright pink to red, bright yellow, or
      (rarely) white; hypanthium tubular, longer than
      wide   3
   2 Flowers green to yellowish-green, sometimes tinged
      purplish; hypanthium disk- to cup-shaped, wider
      than long   4
3 Flowers bright pink to red, rarely white; leaves
   pubescent, scented when crushed   **4. R. sanguineum**
3 Flowers bright yellow; leaves glabrous, not scented
                                            5. R. odoratum
   4 Leaves with sessile orange glands on lowerside,
      scented when crushed; fruit black   **3. R. nigrum**
   4 Leaves with mostly stalked reddish glands, not
      scented; fruit red or rarely whitish   5
5 Dioecious; bracts >4mm   **6. R. alpinum**
5 Flowers bisexual; bracts <2mm   6
   6 Hypanthium cup-shaped; anther-lobes contiguous
                                            2. R. spicatum
   6 Hypanthium saucer-shaped; anther-lobes distinctly
      separated by connective   **1. R. rubrum**

1. R. rubrum L. (R. sylvestre (Lam.) Mert. & Koch) - Red **370**
Currant. Shrub to 2m; leaves 3-10cm, glabrous to sparsely
pubescent, not scented; flowers in pendent racemes, 4-6mm
across, greenish-yellow, ± glabrous; berry 6-10mm across, red
or rarely whitish. Probably intrd; woods, hedges and scrub,
much grown and often obviously relict or escaped; throughout

most of BI but rare in many places; Europe.

2. R. spicatum Robson - <u>Downy</u> <u>Currant</u>. Differs from <u>R. **370**
rubrum</u> in key characters; usually more pubescent but other **R**
differences are not constant. Native; woods on limestone,
mostly in uplands; very local from Lancs and Yorks to
Caithness.

3. R. nigrum L. - <u>Black</u> <u>Currant</u>. Shrub to 2m; leaves as in **370**
<u>R. rubrum</u> but with more pointed main lobes, scented when
crushed; flowers in pendent racemes, 6-10mm across, greenish-
yellow or tinged with purple; hypanthium deeply cup-shaped,
pubescent; berry 10-15mm across, black, without bloom.
Probably intrd; woods, hedges and shady streamsides, much
grown and usually obviously relict or escaped; throughout
most of BI; Europe.

4. R. sanguineum Pursh - <u>Flowering</u> <u>Currant</u>. Shrub to 2.5m; **370**
leaves 3-10cm, pubescent, scented when crushed; flowers in
pendent racemes, 6-10mm across, bright pink to red, rarely
white; berry 6-10mm across, purplish-black with whitish
bloom. Intrd; much grown in gardens and a frequent relic,
sometimes self-sown, especially in W; scattered throughout
BI; W N America.

5. R. odoratum Wendl. f. (<u>R. aureum</u> auct. non Pursh) - **370**
<u>Buffalo</u> <u>Currant</u>. Shrub to 2.5m; leaves 2-5cm, glabrous, not
scented; flowers in pendent racemes, 6-10mm across, bright
yellow, fragrant, glabrous; berry 6-10mm across, dark red to
purplish-black, without bloom. Intrd; commonly grown and
sometimes found relict or self-sown in hedgerows, roadsides
and scrub; scattered in S En; C USA.

6. R. alpinum L. - <u>Mountain</u> <u>Currant</u>. Shrub to 2m, some- **R**
times pendent on rock-faces; leaves 2-5cm, sparsely
pubescent, not scented; flowers in ± erect racemes,
dioecious, the female fewer per raceme, 1.5-3mm across,
greenish-yellow, glabrous to sparsely pubescent; berry 6-10mm
across, red. Native; limestone woods, often on rocks or
cliffs, also an escape in other shady places; native for
certain only N Wa and N En, common only in Peak District,
widely spread as escape in Br except extreme N & S.

7. R. uva-crispa L. - <u>Gooseberry</u>. Spiny shrub to 1(-1.5)m;
leaves 2-5cm, usually pubescent, not scented; flowers 1-3 in
stiff groups, 6-12mm across, greenish-yellow or red-tinged,
pubescent; berry 10-20mm across, greenish-yellow, sometimes
reddish, much larger in cultivars. Probably native; hedges,
scrub and open woods, often obviously relict or escaped; over
most of BI but not native in Ir.

74. CRASSULACEAE - Stonecrop family

Annual to perennial herbs or rarely woody; leaves often
succulent, spiral or alternate, less often opposite, simple,
sessile or petiolate, exstipulate. Flowers usually in
terminal cymes, less often in terminal racemes or solitary

and axillary, bisexual or rarely dioecious, hypogynous, actinomorphic; sepals free to fused, 3–c.18 (mostly 5); petals free to fused, as many as sepals; stamens as many or 2x as many as petals; carpels as many as petals, free or slightly fused at base, with 2–many ovules, tapering to small stigma; fruit a group of follicles.

Easily recognized by the free (or ± free) carpels as many as sepals and petals, stamens as many or 2x as many as petals, and usually succulent leaves.

1 Petals fused to form tube for >1/2 their length;
   basal leaves peltate   **2. UMBILICUS**
1 Petals free or fused only at base; leaves not peltate 2
   2 Stamens as many as petals; leaves opposite
                                                        1. CRASSULA

   2 Stamens 2x as many as petals; leaves usually
      alternate or spiral   3
3 Flowers with 4–5 petals and sepals   **5. SEDUM**
3 Flowers with 6 or more petals and sepals   4
   4 Leaves about as thick as wide   **5. SEDUM**
   4 Leaves distinctly wider than thick, distinctly
      flat on upperside   5
5 Petals yellow   **4. AEONIUM**
5 Petals dull pink to purplish   **3. SEMPERVIVUM**

1. CRASSULA L. (<u>Tillaea</u> L.) – <u>Pigmyweeds</u>
Aquatic or terrestrial annuals to perennials, glabrous or nearly so; leaves opposite, often fused in pairs at base, succulent or ± so, entire; flowers <5mm, 3–5-merous; petals free or ± so, white to pink; stamens as many as petals.

1 Leaves strongly succulent, obovate; flowers in heads
   on common peduncle; petals 5, with apical dorsal
   appendage   **5. C. pubescens**
1 Leaves weakly succulent, linear to ovate or elliptic;
   flowers borne singly in leaf-axils; petals 3–4,
   without apical appendage   2
   2 Flowers sessile or ± so (pedicels <1mm)   3
   2 Flowers on >2mm pedicels   4
3 Leaves 1–2mm; petals mostly 3, shorter than sepals
                                                      1. C. tillaea
3 Leaves 3–5mm; petals 4, longer than sepals
                                                      2. C. aquatica
   4 Terrestrial; petals shorter than to c. as long as
      sepals; stems to 12cm   **4. C. decumbens**
   4 Aquatic or on mud; petals longer than sepals;
      stems often >12cm   **3. C. helmsii**

1. C. tillaea Lester-Garl. (<u>Tillaea</u> <u>muscosa</u> L.) – <u>Mossy</u>   R
<u>Stonecrop</u>. Annual; stems procumbent to ascending, to 5cm; leaves closely set on stem, usually red, 1–2mm, ovate or elliptic; flowers 1–2mm, solitary in leaf-axils, sessile or ±

so, 3(–4)-merous. Native; sandy or gravelly ground in open places; S En, E Anglia, Notts, CI.

2. C. aquatica (L.) Schoenl. (<u>Tillaea</u> <u>aquatica</u> L.) – **379** <u>Pigmyweed</u>. Annual; stems procumbent to decumbent, lax, to **RRR** 5cm; leaves ± distant on stem, 3–5mm, linear; flowers 1–2mm, solitary in leaf-axils, sessile or ± so, 4-merous. Probably native; muddy pool-margin, MW Yorks, found 1921, gone by 1945; on mud by water, Westerness, found 1969.

3. C. helmsii (Kirk) Cockayne (<u>C.</u> <u>recurva</u> (Hook. f.) Ostenf. **379** non N.E. Br., <u>Tillaea</u> <u>recurva</u> (Hook. f.) Hook. f.) – <u>New</u> <u>Zealand</u> <u>Pigmyweed</u>. Perennial; stems trailing in water or ascending from it or decumbent in mud, lax, to 30cm; leaves ± distant on stem, 4–15(20)mm, linear to lanceolate; flowers 1–2mm, solitary in leaf-axils, on pedicels 2–8mm, 4-merous. Intrd; grown by aquarists and discarded or planted in ponds; well natd in many places in S En and CI, scattered N to C Sc, Co Down, rapidly spreading; Australia and New Zealand.

4. C. decumbens Thunb. (<u>C.</u> <u>macrantha</u> (Hook. f.) Diels & **379** Pritzel) – <u>Scilly</u> <u>Pigmyweed</u>. Annual; stems decumbent to ascending, to 12cm; leaves ± distant on stem, 4–7mm, linear-lanceolate; flowers 2–4mm, solitary in leaf-axils, on pedicels 4–10(15)mm, 4-merous. Intrd; natd weed in damp sandy bulbfields and tracksides in Scilly (first found 1959), occasional wool-alien in S En; S Africa, Australasia.

5. C. pubescens Thunb. (<u>C.</u> <u>radicans</u> (Haw.) D. Dietr.) – **379** <u>Jersey</u> <u>Pigmyweed</u>. Perennial; stems erect to procumbent and stoloniferous, succulent, to 20cm; leaves crowded in 1.5–2.5cm across rosettes near stem tips, 3–10mm, obovate, very succulent; flowers 3–5mm, 6–10 subsessile in crowded heads on 3–4cm axillary peduncles, 5-merous. Intrd; natd on sandy ground in 1 place in Jersey, found 1970; S Africa. Our plant is probably ssp. **radicans** (Haw.) Toelken, but its identity needs checking.

2. UMBILICUS DC. – <u>Navelwort</u>

Glabrous perennials; leaves alternate on stem and in basal rosette, succulent, the lower and basal ones peltate, crenate; flowers >5mm, 5-merous; petals fused >1/2 way from base; stamens 2x as many as petals, fused to corolla-tube.

1. U. rupestris (Salisb.) Dandy – <u>Navelwort</u>. Stem usually erect, to 30(50)cm; basal leaves orbicular, 1–7cm across, with long petiole; flowers in simple raceme occupying most of stem, 7–10mm, greenish-white to pinkish-brown. Native; rocks, walls and stony hedgebanks; frequent in Ir, CI and W Br N to C Sc, rare or absent in E & C Br.

3. SEMPERVIVUM L. – <u>House-leek</u>

Glandular-pubescent perennials; leaves in dense basal rosette and alternate up stem, succulent, entire; flowers >5mm, 8–18(mostly 13)-merous; petals ± free, dull pink to purplish; stamens 2x as many as petals.

1. S. tectorum L. - <u>House-leek</u>. Stem erect, to 40(60)cm, arising from centre of mature rosette which then dies; leaves 2-4(6)cm, oblong-lanceolate, often reddish; flowers in group of cymes at stem apex, 1.5-3cm across; petals 8-12mm, narrow. Intrd; grown on wall-tops and roofs, rarely sand-dunes, very persistent but hardly natd; scattered over Br; Europe.

4. AEONIUM Webb & Berth. - <u>Aeonium</u>
Almost glabrous perennials, sometimes woody below; leaves in large dense rosette and alternate up stem, succulent, ± entire; flowers >5mm, 8-11-merous; petals ± free, yellow; stamens 2x as many as petals.

Other spp. - A few other spp. are grown in SW En and are very noticeable, but not truly natd. **A. arboreum** (L.) Webb & Berth., from Morocco, differs from <u>A. cuneatum</u> in having the smaller leaf-rosettes (10-18cm across) borne at the apex of thick branching woody stems.

1. A. cuneatum Webb & Berth. - <u>Aeonium</u>. Plant mostly herbaceous, woody very near base, leaf-rosettes near ground, to 50cm across; flowering stems to 80(120)cm, ± erect; leaves up to 30 x 80cm, oblong-oblanceolate, bright green; flowers in dense mass of cymes at stem-apex, 1-2cm across. Intrd; grown on walls in SW En and conspicuous and ± natd in Scilly; Canary Isles.

5. SEDUM L. (<u>Rhodiola</u> L.) - <u>Stonecrops</u>
Annuals or (usually) perennials; leaves alternate,,sometimes crowded and ± in a rosette, succulent, entire or toothed; flowers >3mm, 4-9-merous; petals free, various in colour; stamens 2x (or c.2x) as many as petals.

```
1  Leaves distinctly dorsiventral, >3x as wide as thick    2
1  Leaves <3x as wide as thick, ± terete or flattened
     on upperside                                          9
   2  Rhizome thick, succulent, scaly; flowers
      dioecious, usually 4-merous              1. S. rosea
   2  Rhizome 0, or not scaly and non-succulent; flowers
      bisexual, 5(-6)-merous                               3
3  Petals yellow; stems woody, perennial                   4
3  Petals pink to purplish-red, rarely white; stems not
   woody, herbaceous                                       5
   4  Leaves mostly >4.5cm, 3-4x as long as wide; petals
      c.4x as long as wide                    2. S. praealtum
   4  Leaves mostly <4.5cm, c.2x as long as wide; petals
      c.3x as long as wide                    3. S. confusum
5  Stamens distinctly longer than petals and sepals
                                              4. S. spectabile
5  Stamens shorter than to c. as long as petals           6
   6  Petals 3-5mm; stems rooting only near base           7
   6  Petals 5-12mm; stems rooting along length            8
```

7 Leaves entire; non-flowering shoots procumbent; roots
   not tuberous   **6. S. anacampseros**
7 Leaves toothed, non-flowering shoots ± erect; roots
   tuberous   **5. S. telephium**
   8 Petals mostly >8mm; leaves narrowed to base but
      scarcely petiolate; flowers mostly pedicellate
                                         7. S. spurium
   8 Petals mostly <8mm; leaves with distinct petiole;
      flowers ± sessile   **8. S. stoloniferum**
9 Petals yellow   10
9 Petals white to pink or red   14
   10 Most leaves >7mm, acute or apiculate; ripe
      follicles erect; flowers 5-9-merous   11
   10 Leaves <7mm, obtuse; ripe follicles ± patent;
      flowers 5-merous   13
11 Sterile shoots with terminal tassel-like cluster of
   living leaves and persistent dead ones below; leaves
   flattened, abruptly apiculate; filaments and
   follicles smooth; sepals subacute to obtuse
                                         11. S. forsterianum
11 Sterile shoots with long terminal region of living
   leaves; dead leaves not persistent; leaves subterete,
   acute to acuminate; base of filaments and inner side
   of follicles minutely papillose; sepals acute   12
   12 Inflorescence erect in bud; leaves mostly >3mm
      wide   **9. S. nicaeense**
   12 Inflorescence pendent in bud; leaves mostly <3mm
      wide   **10 S. rupestre**
13 Leaves ovoid, broadest near base, with acrid taste
   when fresh   **12. S. acre**
13 Leaves ± cylindrical, ± parallel-sided, not acrid
                                         13. S. sexangulare
   14 Leaves, pedicels and sepals with small glandular
      hairs   15
   14 Plant glabrous   17
15 Leaves mostly opposite   **17. S. dasyphyllum**
15 Leaves mostly alternate   16
   16 Petals 5, pink; ripe follicles erect **18. S. villosum**
   16 Petals mostly 6-7, white, with pink midrib; ripe
      follicles patent   **19. S. hispanicum**
17 Inflorescence with 2(-3) main branches each with 3-6
   flowers   **16. S. anglicum**
17 Inflorescence dense, corymbose, with several branches 18
   18 Inflorescence <20-flowered   **15. S. lydium**
   18 Inflorescence >20-flowered   **14. S. album**

Other spp. - Several other spp. are grown in gardens and may
be persistent relics or throwouts. **S. spathulifolium** Hook.
(Colorado Stonecrop), from N America, has procumbent rooting
stems, flat, entire, white-bloomed leaves and yellow flowers.
S. hybridum L., from E Asia, resembles S. spurium with yellow
flowers but has alternate coarsely toothed leaves and a woody

rhizome. **S. kamtschaticum** Fischer & C. Meyer ssp.
middendorfianum (Maxim.) Frod. (S. middendorfianum Maxim.),
from E Asia, is related to S. hybridum but has erect non-
rooting stems to 30cm. **S. cepaea** L., from S Europe, is a
white-flowered annual with erect stems to 25cm ± entirely
occupied by the pyramidal diffuse inflorescence; it was
formerly natd in Bucks.

1. S. rosea (L.) Scop. (Rhodiola rosea L.) - Roseroot.
Glabrous rhizomatous perennial; stems erect, to 35cm; leaves
alternate, 1-4cm, dorsiventral, serrate; flowers in dense
subcorymbose panicle; petals 4(-5), 2-4mm, greenish-yellow.
Native; mountain rocks and sea cliffs; Wa, Ir, Sc, N En, rare
garden escape elsewhere.

2. S. praealtum A. DC. (S. dendroideum auct. non Sesse &
Mocino ex DC.) - Greater Mexican-stonecrop. Bushy, glabrous,
evergreen shrub to 75cm; leaves near ends of branches,
alternate, 5-7cm, dorsiventral but very thick, entire;
flowers in large panicle; petals 5, 6-9mm, bright yellow.
Intrd; natd on cliffs in Jersey since 1920, occasional relic
on banks in Guernsey; Mexico.

3. S. confusum Hemsley - Lesser Mexican-stonecrop. Differs
from S. praealtum in smaller size (to 40cm); shorter leaves
(see key) with rounded (not obtuse) apex; smaller denser
inflorescence (c.5cm, not c. 10cm) and shorter petals (see
key). Intrd; persistent spreading relic of cultivation on
banks in Guernsey and W Cornwall; Mexico.

4. S. spectabile Boreau - Butterfly Stonecrop. Glabrous **379**
perennial; stems erect, to 50cm; leaves opposite or in whorls
of 3, 4-10cm, dorsiventral, crenate-serrate; flowers in large
corymbose panicle; petals 5, 5-8.5mm, pale pink to purplish-
red. Intrd; common in gardens and a persistent relic or
throw-out; S & C En, natd in woodland in N Wilts since 1930;
China and Japan.

5. S. telephium L. - Orpine. Glabrous perennial; stems
erect, to 60cm; leaves alternate, 2-8cm, dorsiventral,
serrate; flowers in rounded panicle; petals 5, 3-5mm,
reddish-purple, rarely white. Native; woods, hedgebanks and
rocky places; local throughout most Br, only as escape in Ir
and parts of Br. Distribution of sspp. unknown; the leaf
characters are of doubtful value and the sspp. need
reappraisal. If the follicle character is reliable ssp.
fabaria is much the commoner (or perhaps the only one) in BI.
a. Ssp. telephium (ssp. purpurascens Syme). Follicles with
groove on back; leaves usually sessile, tapering to ±
truncate base.
b. Ssp. fabaria (Koch) Kirschl. Follicles not grooved;
leaves tapering to cuneate base, the lower often petiolate.
6. S. anacampseros L. - Love-restoring Stonecrop. Glabrous **379**
perennial; stems ascending, to 25cm; sterile shoots
procumbent; leaves alternate, 1.5-3cm, dorsiventral, entire;
flowers in rounded panicle; petals (4-)5, 4-5mm, pink to

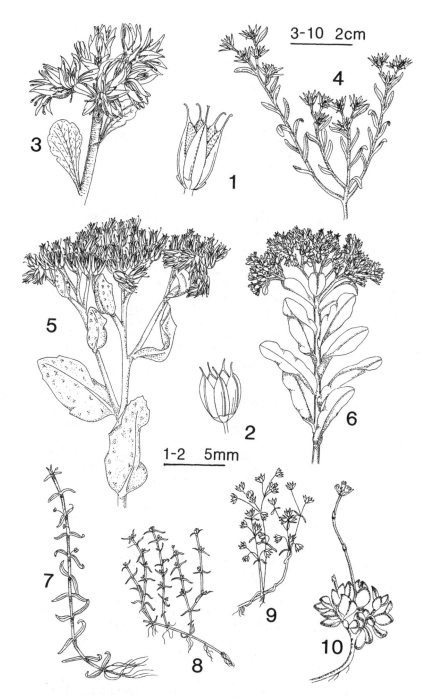

FIG 379 – Crassulaceae. 1-2, fruits of Sedum. 1, S. **rupestre**. 2, S. **forsterianum**. 3, **Sedum spurium**. 4, S. **hispanicum**. 5, S. **spectabile**. 6, S. **anacampseros**. 7. **Crassula helmsii**. 8, C. **aquatica**. 9, C. **decumbens**. 10, C. **pubescens**.

mauve. Intrd; a garden relic or throw-out; sporadic in S & C
En; S Europe.

7. S. spurium M. Bieb. - <u>Caucasian-stonecrop</u>. Glabrous **379**
perennial; stems decumbent or procumbent, rooting along
length, to 20cm; leaves usually opposite, 1.5-3cm,
dorsiventral, crenate-serrate, strongly papillose at margin;
flowers in corymbose panicle; petals 5, 8-12mm, pink to
reddish-purple, rarely white. Intrd; commonly grown in
gardens and very persistent as escape, relic or throw-out;
very scattered throughout BI except N Sc; Caucasus.

8. S. stoloniferum S. Gmelin - <u>Lesser Caucasian-stonecrop</u>.
Differs from <u>S. spurium</u> in being smaller; leaves obscurely
papillose at margin; and see key. Intrd; similar places to
<u>S. spurium</u> but much rarer; S En and CI; Caucasus.

9. S. nicaeense All. (<u>S. sediforme</u> (Jacq.) Pau) - <u>Pale</u>
<u>Stonecrop</u>. Glabrous perennial like a large <u>S. rupestre</u> but
stems to 50cm; sepals obtuse and mucronate; petals 5-8, very
pale to greenish-yellow; and see key. Intrd; dry sunny
banks; well natd by road in W Kent; Mediterranean.

10. S. rupestre L. (<u>S. reflexum</u> L.) - <u>Reflexed Stonecrop</u>. **379**
Glabrous perennial; stems erect to ascending, to 35cm;
sterile shoots decumbent, rooting; leaves spiral, 12-20mm, ±
terete, linear; flowers in corymbose to rounded panicle;
petals (5)6-7(9), 6-7mm, bright yellow. Intrd; natd on
walls, rocks and stony banks; locally common over BI except N
Sc; Europe.

11. S. forsterianum Smith (<u>S. elegans</u> Lej., <u>S. rupestre</u> **379**
auct. non L.) - <u>Rock Stonecrop</u>. Usually less robust than <u>S.</u>   **R**
<u>rupestre</u> and differs constantly in all key characters.
Native; rocks and screes, either dry in open or wet in woods;
local in Wa and SW En, also grown in gardens and natd as for
<u>S. rupestre</u>. Populations differ in various characters but do
not fit into 2 sspp. (ssp. **forsterianum** and ssp. **elegans**
(Lej.) E. Warb.) as sometimes claimed.

12. S. acre L. - <u>Biting Stonecrop</u>. Glabrous, acrid-tasting
perennial; stems procumbent, rooting, sending up ascending to
erect flowering stems to 29cm and shorter sterile shoots;
leaves alternate, 3-5mm, ± terete, ovoid; flowers in small
cymes; petals 5, 6-8mm, bright yellow. Native; walls, rocks,
open grassland and maritime sand and shingle; throughout most
of BI.

13. S. sexangulare L. - <u>Tasteless Stonecrop</u>. Differs from
<u>S. acre</u> in being often taller (to 25cm); leaves 3-6mm; petals
4-6mm; and see key. Intrd; natd on walls and rocks;
scattered in En and Wa; Europe.

14. S. album L. - <u>White Stonecrop</u>. Glabrous perennial;
stems procumbent, rooting, sending up ascending to erect
flowering stems to 20cm and shorter sterile shoots; leaves
alternate, 4-12mm, ± terete, ovoid to cylindrical; flowers in
dense subcorymbose panicle; petals 5, 2-4mm, white, sometimes
tinged pink. Probably intrd; walls, rocks and stony ground;
scattered through most of BI, possibly native in SW and WC

En; Europe.   Ssp. **micranthum** (Bast.) Syme, with smaller
flowers and shorter ovoid leaves than ssp. **album,** is only an
extreme variant probably not worth ssp. rank.
15. S. lydium Boiss. - <u>Least</u> Stonecrop. Glabrous perennial;
similar vegetatively to <u>S. sexangulare</u> but leaves not spurred
at base as in latter; flowers in small dense corymbose
panicle 1-2cm across; petals 5, 2-4mm, white with red midrib.
Intrd; garden outcast or escape natd in a few places in En
and Ir; Turkey.
16. S. anglicum Hudson - <u>English</u> Stonecrop.   Glabrous
perennial; similar to <u>S. acre</u> but leaves grey-green often
tinged red (not bright green) and not acrid; petals 2.5-4.5m,
white tinged pink; follicles ± erect (not ± patent). Native;
rocks, sand and shingle; common in much of CI, Ir and W Br,
very local and mainly coastal in C & E Br.
17. S. dasyphyllum L. - <u>Thick-leaved</u> Stonecrop.   Glandular-
pubescent perennial; stems ascending, rooting at base, to
10cm; leaves mostly opposite, 3-6mm, ± terete, ovoid to
ellipsoid; flowers in rather few-flowered cymes; petals 5-6,
2.5-4mm, white tinged pink. Intrd; natd on walls and rocks;
very scattered in En, Wa and Ir; Europe.
18. S. villosum L. - <u>Hairy</u> Stonecrop.   Glandular-pubescent   R
biennial to perennial; stems erect to ascending, rooting near
base, to 10(15)cm; leaves alternate, 3-8mm, semiterete (flat
on upperside), often reddish, linear-oblong; flowers in
rather few-flowered cymes; petals 5, 3.5-5mm, pink. Native;
steamsides and stony flushes in hilly areas; N En, C & S Sc.
19. S. hispanicum L. - <u>Spanish</u> Stonecrop. Glandular- 379
pubescent, usually perennial; stems decumbent to ± erect, to
10cm; leaves alternate, 4-15mm, semi-terete; linear to
narrowly ellipsoid; flowers in diffuse groups of cymes;
petals (5)6-7(9), 3.5-6mm, white with red midrib.   Intrd;
natd on walls and stony ground; rare in S & C En; SE Europe.

75. SAXIFRAGACEAE - <u>Saxifrage family</u>
(Parnassiaceae)

Annual to perennial herbs, rarely woody at base; leaves
alternate or all basal, rarely opposite, simple to compound,
sessile or petiolate, stipulate or not. Flowers solitary or
in various cymes or racemes, bisexual or sometimes unisexual
in various arrangements, variously hypogynous, perigynous or
epigynous, actinomorphic or less often zygomorphic; sepals
4-5, free to fused at base but often borne on hypanthium;
petals 4-5, free, less often 0; stamens 3, 5, 8 or 10;
carpels normally 2, sometimes 4, fused only at base or for
varying distances ± to top, with many ovules on axile or
parietal placentas, tapering to small stigma; fruit 2
follicles variously fused to form a capsule.
Very variable in vegetative and floral characters, but
distinguishable by the 2 carpels fused only at base or for

varying distances to apex; <u>Parnassia</u> is distinct in its
staminodes.

```
1  Leaves compound, or simple and divided >1/2 way to
   base                                                         2
1  Leaves simple, divided <1/2 way to base                      4
   2  Leaves (incl. petioles) <5cm; inflorescence few-
      flowered, <5cm                                   5. SAXIFRAGA
   2  Leaves >10cm; inflorescence many-flowered, >5cm           3
3  Leaves palmate, with 5-9 leaflets                   2. RODGERSIA
3  Leaves ternate, the main divisions ternate to pinnate
                                                        1. ASTILBE
   4  Leaves peltate                                   4. DARMERA
   4  Leaves not peltate (petiole joining lamina at edge) 5
5  Petals 0; sepals 4                             9. CHRYSOSPLENIUM
5  Petals present; sepals 5                                     6
   6  Five large divided staminodes present, alternating
      with 5 stamens; flowering stems with 1 flower and
      1 leaf                                        10. PARNASSIA
   6  Staminodes 0; flowering stems with >1 flower and/
      or >1 leaf                                                7
7  Stamens as many as sepals or fewer                           8
7  Stamens 2x as many as sepals                                 9
   8  Petals 4, brown; stamens 3                        7. TOLMIEA
   8  Petals 5, pink to red; stamens 5                 6. HEUCHERA
9  Petals fringed with long narrow lobes               8. TELLIMA
9  Petals entire to minutely toothed                            10
   10 Thick surface rhizome present; at least some
      leaves >10cm; petals pink to red                 3. BERGENIA
   10 Rhizome 0 or thin; all leaves <10cm (if petals
      pink or red then leaves <1cm)                   5. SAXIFRAGA
```

1. ASTILBE Buch.-Ham. ex D. Don - <u>False-buck's-beards</u>
Perennials; leaves ternate, the primary divisions ternate to
pinnate; inflorescence a many-flowered terminal panicle;
flowers bisexual or unisexual (dioecious to polygamous), ±
hypogynous; sepals 5; petals 0 or 5; stamens (5-)10; carpels
2(-3), ± free to united at base to form 2-celled ovary with
axile placentation.
   Often confused with <u>Spiraea</u> (always shrubs) or <u>Aruncus</u>
(carpels 3, stamens >10) (both Rosaceae), but differs in
floral details. Female plants have short sterile stamens.

```
1  Petals 0; plant usually >1m                     3. A. rivularis
1  Petals 5, longer than sepals; plant <1m                      2
   2  Stems with short whitish to brown hairs; petals
      white to pale pink                           1. A. japonica
   2  Stems with dense long shaggy brown hairs; petals
      pink to red                               2. A. x arendsii
```

Other spp. - Several spp. and hybrids are grown in gardens
and some wild plants might be other than the following.

1. A. japonica (Morren & Decne.) A. Gray - <u>False-buck's-</u><u>beard</u>. Stems to 80cm, erect, rather sparsely pubescent with very short white to brown hairs; leaves 2-3-ternate; inflorescence up to 30cm; petals 1.5-2x as long as sepals, linear, white or pale pink. Intrd; grown in gardens and natd in wild usually in damp places; very scattered in Br, mostly in N En and Sc; Japan.

2. A. x arendsii Arends (<u>A.</u> x <u>rosea</u> Hort.; ?<u>A.</u> <u>chinensis</u> 384 (Maxim.) Franchet x <u>A.</u> <u>japonica</u>) - <u>Red</u> <u>False-buck's-beard</u>. Differs from <u>A.</u> <u>japonica</u> in stems with dense long shaggy brown hairs; primary leaf divisions sometimes pinnate; inflorescence denser; petals 2-3x as long as sepals, pink to red. Intrd; similar situations to <u>A.</u> <u>japonica</u>; very scattered in En and Sc; garden origin.

3. A. rivularis Buch.-Ham. ex D. Don - <u>Tall</u> <u>False-buck's-beard</u>. Stems to 1.6m, erect, glabrous to sparsely pubescent with very short brownish hairs; leaves 2-3-ternate or -pinnate, with sparse but long, wispy brown hairs at branch-points; inflorescence up to 60cm. Intrd; grown rather rarely in gardens, natd by steams and tracks in forest plantations in Kintyre; C Asia.

2. RODGERSIA A. Gray - <u>Rodgersia</u>
Perennials with short stout rhizome; leaves palmate with 5-9 simple leaflets; inflorescence a many-flowered terminal panicle; flowers bisexual, shallowly perigynous; sepals 5; petals 5; stamens 10; carpels 2, united for most part to form 2-celled ovary with axile placentation.

1. R. podophylla A. Gray - <u>Rodgersia</u>. Stems to 1.3m, erect, pubescent; leaves with long petiole and 5-9 distally acutely lobed leaflets each up to 30cm; inflorescence up to 25cm; flowers c.5mm across, yellowish-white. Intrd; planted in damp places by ponds and rivers, sometimes persistent and spreading vegetatively (rarely sets seed); natd in Surrey, Monts and Dunbarton, perhaps elsewhere; Korea and Japan.

3. BERGENIA Moench - <u>Elephant-ears</u>
Glabrous perennials with stout scaly rhizome usually on soil surface; leaves large, simple, thick, serrate-crenate; inflorescence a many-flowered panicle on erect leafless stem; flowers perigynous with cup-shaped hypanthium; sepals 5; petals 5, pink; stamens 10; carpels 2, united only at base to form 2-celled ovary with axile placentation.

Other spp. - The following is the most commonly cultivated sp., but some escapes or relics might be the related **B. cordifolia** (Haw.) Sternb., from Siberia, which differs in its ± orbicular cordate leaves, less pendent flowers and petals with ± orbicular limb, or the hybrid between the 2.

1. B. crassifolia (L.) Fritsch - <u>Elephant-ears</u>. Leaves

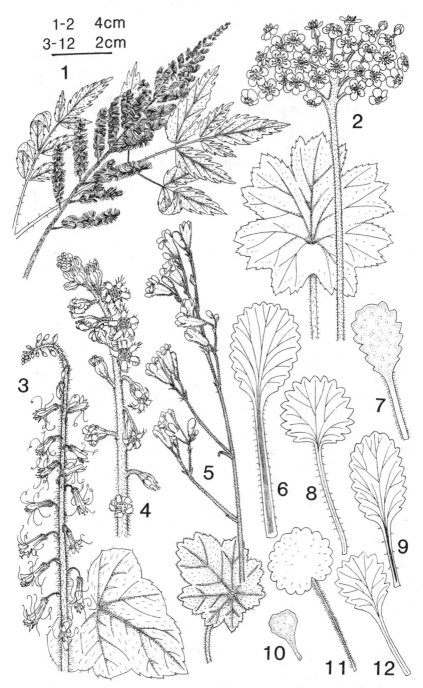

FIG 384 – Saxifragaceae. 1, Astilbe x arendsii. 2, Darmera. 3, Tolmiea. 4, Tellima. 5, Heuchera. 6–12, leaves of Saxifraga. 6, S. x geum. 7, S. umbrosa. 8, S. x polita. 9, S. x urbium. 10, S. cuneifolia. 11, S. hirsuta. 12, S. spathularis.

ovate to obovate, 6-20cm with petiole c.1/2 as long, cuneate to subcordate at base; flowering stem 10-40cm; flowers 15-25mm across, ± pendent; petals with ovate to obovate limb. Intrd; very persistent garden relic or throwout; scattered in Br and CI; Siberia. Rarely sets seed.

4. DARMERA Voss ex Post & Kuntze (<u>Peltiphyllum</u> (Engl.) Engl. non <u>Peltophyllum</u> Gardner) - <u>Indian-rhubarb</u>
Pubescent perennials with stout rhizome; leaves simple, peltate, palmately lobed and sharply serrate; inflorescence a ± corymbose panicle on erect leafless stem; flowers almost hypogynous; sepals 5; petals 5, pink to whitish; stamens 10; carpels 2, ± free.

1. D. peltata (Torrey ex Benth.) Voss ex Post & Kuntze 384 (<u>Peltiphyllum</u> <u>peltatum</u> (Torrey ex Benth.) Engl.) - <u>Indian-rhubarb</u>. Leaves 5-40cm across, on petioles up to 1m; flowering stem to 1(1.5)m, branched only near top; flowers 10-15mm across. Intrd; grown in damp places and natd where planted or outcast; scattered over Br and Ir; W USA. Rarely sets seed.

5. SAXIFRAGA L. - <u>Saxifrages</u>
Pubescent annuals or perennials; leaves simple to almost compound; flowers in simple or compound cymes, sometimes solitary; ovary superior to ± completely inferior; hypanthium ± 0; sepals 5; petals 5; stamens 10; carpels 2, fused at least at base to form 2-celled ovary with axile placentation.

1 Leaves opposite; petals purple   **14. S. oppositifolia**
1 Leaves alternate (spiral); petals yellow or white   2
  2 Some or all flowers replaced by reddish bulbils
                       17. S. cernua
  2 Bulbils in inflorescence 0   3
3 Petals bright yellow   4
3 Petals white to cream   6
  4 Annuals; leaves ± orbicular, abruptly contracted
    to petiole   **2. S. cymbalaria**
  4 Perennials; leaves linear to oblanceolate,
    gradually contracted to petiole or ± sessile   5
5 Ovary superior; flowers 1(-3) per stem   **1. S. hirculus**
5 Ovary semi-inferior; flowers usually >3 per stem
                       15. S. aizoides
  6 Flowers strongly zygomorphic, the lower petals
    >2x as long as 3 upper ones   **5. S. stolonifera**
  6 Flowers actinomorphic or ± so   7
7 Ovary superior, the sepals arising from underneath it 8
7 Ovary partly inferior, the sepals arising from its
    side or top   16
  8 Leaves ± sessile or with petiole <1/2 as long
    as blade   **4. S. stellaris**

  8 Basal leaves with petiole almost as long as to
    longer than blade   9
  9 Stems leafy; sepals erecto-patent   **6. S. rotundifolia**
  9 Stems leafless; sepals reflexed   10
 10 Petals without red spots; leaf-laminas entire for
    ≥ basal 1/3 of margin   **7. S. cuneifolia**
 10 Petals usually with small red spots; leaf-laminas
    toothed for ≥ apical 3/4 of margin   11
 11 Petioles subterete; laminas sparsely pubescent over
    both surfaces, cordate at base   **13. S. hirsuta**
 11 Petioles distinctly flattened; laminas glabrous or
    nearly so at least on lowerside   12
 12 Petioles densely pubescent on lateral margins;
    glabrous on upperside   **8. S. umbrosa**
 12 Petioles ± glabrous to rather sparsely pubescent   13
 13 Laminas with acute teeth; petioles subglabrous
                                             11. S. spathularis
 13 Laminas with subacute to rounded teeth, and/or
    petioles distinctly ciliate   14
 14 Petioles usually much longer than lamina; laminas
    sparsely pubescent to glabrous   15
 14 Petioles usually c. as long as to slightly longer
    than laminas; laminas glabrous   **9. S. x urbium**
 15 Leaves mostly c. as long as wide, with scarcely
    visible translucent border (<0.2mm) and rather sharp
    teeth   **12. S. x polita**
 15 Many leaves distinctly longer than wide, with
    conspicuous translucent border (≥0.2mm) and low,
    blunt teeth   **10. S. x geum**
    16 Leaves all in basal rosette   **3. S. nivalis**
    16 Stem-leaves present   17
 17 Annual, without perennating organs   **22. S. tridactylites**
 17 Perennial, with sterile rosettes, stolons, rhizomes
    or basal bulbils   18
    18 Basal bulbils usually present; stolons terminating
       in leafy rosettes 0   19
    18 Bulbils 0; procumbent stolons terminating in dense
       leafy rosettes present   20
 19 Flowers mostly >3 per stem; petals >(6)9mm; basal
    leaves mostly ≥7-lobed   **18. S. granulata**
 19 Flowers 1-3 per stem; petals <6mm; basal leaves
    3-7-lobed   **16. S. rivularis**
    20 Leaf-lobes acuminate to narrowly acute, apiculate
       to aristate; flower-buds ± pendent   **19. S. hypnoides**
    20 Leaf-lobes rounded, obtuse or acute, shortly
       mucronate or not; flower-buds erect   21
 21 Leaf-lobes rounded, obtuse or subacute; petals dull
    creamy- or greenish-white; Sc and Wa   **21. S. cespitosa**
 21 Leaf-lobes subacute to acute; petals pure white;
    Ir and Wa   **20. S. rosacea**

  1. **S. hirculus** L. - <u>Marsh Saxifrage</u>. Stoloniferous **RR**

perennial; stems ascending to erect, leafy, to 20cm; leaves
lanceolate to oblanceolate, the lowest tapered to long
petiole, entire; flowers 1(3), terminal; petals yellow,
10-15mm; ovary superior. Native; wet places on moors; very
local in N En, N Ir and S & C Sc, decreasing and formerly
more widespread.

2. S. cymbalaria L. - <u>Celandine</u> Saxifrage. Annual; stems
decumbent to suberect, to 20cm; leaves orbicular to
transversely elliptic, cordate to rounded at base, with 3-7
broad shallow lobes, with long petiole; flowers in sparse
diffuse cymes; petals yellow, 3-5mm; ovary superior. Intrd;
weed escaping locally from nurseries and gardens in shady
places; scattered in Br, Co Antrim; E Mediterranean. Our
plant is var. **huetiana** (Boiss.) Engl. & Irmscher.

3. S. nivalis L. - <u>Alpine</u> Saxifrage. Perennial with basal   R
leaf-rosette and leafless erect stem to 15cm; leaves obovate
to suborbicular, on long petioles, closely crenate-serrate;
flowers in dense panicle; petals white, 2-3mm; ovary semi-
inferior. Native; mountain rocks and cliffs; very local in N
Wa, NW En, NW Ir and Sc.

4. S. stellaris L. - <u>Starry</u> Saxifrage. Stoloniferous
perennial with basal leaf-rosette and ± leafless erect stem
to 20cm; leaves obovate or obtrullate to narrowly so,
scarcely petiolate, remotely serrate or dentate; flowers in
loose panicle or cyme; petals white, 4-6mm; ovary superior.
Native; wet rocks and stony places, in flushes and by streams
in mountains; frequent in N Wa, N En and Sc, local in Ir.

5. S. stolonifera Curtis (<u>S.</u> <u>sarmentosa</u> Schreber) -
<u>Strawberry</u> Saxifrage. Perennial with long thin stolons
producing new plants at apex; leaves orbicular, cordate,
crenate-dentate, with long petioles; flowers in large diffuse
panicles on leafless stems to 60cm; petals white, the 2 lower
petals much longer than 3 upper; ovary <1/2 superior. Intrd;
natd on shady walls in Cornwall; Japan and China.

6. S. rotundifolia L. - <u>Round-leaved</u> Saxifrage. Perennial
with basal leaf-rosette and erect leafy stem to 40cm; leaves
orbicular, thin, cordate, coarsely dentate, basal ones with
long petiole; flowers in loose panicle; petals white with red
spots, 6-11mm; ovary superior. Intrd; grown in gardens and
natd by shady streams in N En and Sc; Europe.

7. S. cuneifolia L. - <u>Lesser</u> Londonpride. Stoloniferous **384**
perennial with basal leaf-rosette and leafless erect stem to
30cm; leaves suborbicular-spathulate, thick, on long
petioles, dentate in distal 1/2; flowers in loose panicle;
petals white, 2.5-4mm; ovary superior. Intrd; natd on old
walls; NW Yorks, rare short-lived escape elsewhere; S Europe.

8. S. umbrosa L. - <u>Pyrenean</u> Saxifrage. Differs from <u>S.</u> **384**
<u>spathularis</u> in larger obovate-oblong crenate-dentate leaves
with 9-21 obtuse to rounded teeth and translucent border
0.25-0.4mm; petioles densely pubescent on margins; petals
3-4mm. Intrd; formerly much grown in gardens, natd on shady
limestone rocks; Yorks, perhaps Derbys and Dunbarton;

388   SAXIFRAGACEAE

Pyrenees. Many records refer to S. x urbium.
9. S. x urbium D. Webb (S. umbrosa x S. spathularis) – **384**
Londonpride. Differs from S. spathularis in larger less
orbicular leaves with usually 19-25 subacute to obtuse teeth
and translucent border 0.2-0.35mm; petioles pubescent on
margins; petals 4-5mm. Intrd; much grown in gardens, natd in
waste places, in woods, by streams and on walls and rocks;
throughout BI; garden origin. Usually sterile but fertile
plants exist.
10. S. x geum L. (S. umbrosa x S. hirsuta) – Scarce **384**
Londonpride. Fertile and forms spectrum connecting parents
in all leaf characters. Intrd; rather rare in gardens and
rarely natd in shady and damp, often rocky places; W Cornwall
and SE & MW Yorks; Pyrenees. Many records refer to S.
hirsuta or S. x urbium.
11. S. spathularis Brot. – St Patrick's-cabbage. Stolon- **384**
iferous perennial with basal leaf-rosette and leafless erect **R**
stem to 40cm; leaves ± orbicular to obovate, thick, with
translucent border <0.2mm, glabrous, acutely dentate with
usually 9-15 teeth; petiole long, flattened, pubescent only
at base; flowers in loose panicle; petals white with red
spots, 3-5mm; ovary superior. Native; damp rocks in
mountains; locally common in W & SW Ir, rare elsewhere in Ir.
12. S. x polita (Haw.) Link (S. spathularis x S. hirsuta) – **384**
False Londonpride. Fertile and forms spectrum connecting the
parents; intermediate in leaf shape, toothing and pubescence
and in petiole shape and pubescence. Native; mostly with
parents and commoner than S. hirsuta in SW Ir, frequent in W
Mayo and W Galway without S. hirsuta and sometimes without S.
spathularis, also natd garden escape in S Wa, N En and C Sc.
13. S. hirsuta L. (S. geum L. 1762 non L. 1753) – Kidney **384**
Saxifrage. Stoloniferous perennial with basal leaf-rosette **R**
and leafless erect stem to 40cm; leaves orbicular, thin,
cordate, pubescent on both sides, crenate-dentate, with (11)
15-25 ± rounded teeth; petiole long, scarcely flattened,
pubescent all round; flowers in loose panicle; petals white
with red spots, 3-5mm; ovary superior. Native; damp rocks in
mountains; locally common in N & S Kerry and W Cork, grown in
gardens and natd in S & C Sc, N En and Man.
14. S. oppositifolia L. – Purple Saxifrage. Procumbent mat-
forming perennial; leaves opposite, ovate to obovate, entire;
flowers solitary, terminal on shortly upturned stems; petals
purple, 5-10mm; ovary semi-inferior. Native; damp mountain
rocks and scree; locally common in N & S Wa, N & W Ir, NW En
and Sc.
15. S. aizoides L. – Yellow Saxifrage. Perennial, with
sterile and fertile decumbent to ascending stems to 25cm;
leaves linear to narrowly oblong-elliptic, scarcely
petiolate, with small distant teeth; flowers in diffuse cyme;
petals yellow, 3-6mm; ovary semi-inferior. Native; wet rocks
and streamsides in mountains, down to sea-level on dunes in N
Sc; locally common in N En and N & C Sc, rare in N Ir,

extinct in Wa.

16. S. rivularis L. - <u>Highland Saxifrage</u>. Stoloniferous **RR**
perennial usually with basal bulbils; stems slender,
ascending, to 12cm, with few leaves; leaves orbicular to
transversely elliptic, cordate, with 3-7 deep lobes; flowers
1-3 per stem; petals white, 3-5mm; ovary semi-inferior.
Native; wet mountain rocks above 900m; rare in C & N Sc.

17. S. cernua L. - <u>Drooping Saxifrage</u>. Perennial with basal **RRR**
leaf-rosette bearing axillary bulbils; stem erect, to 15cm,
with axillary bulbils replacing all flowers or sometimes all
but the terminal one; leaves orbicular to transversely
elliptic, cordate, with 3-5 rather deep lobes; petals white,
up to 12mm; ovary nearly superior. Native; basic mountain
rocks above 900m; very rare in C Sc.

18. S. granulata L. - <u>Meadow Saxifrage</u>. Perennial with
basal leaf-rosette bearing axillary bulbils; stems erect, to
50cm, leafy; leaves ± orbicular, cordate, mostly >7-lobed,
the basal with long petioles; flowers in loose cyme, gynodioecious; petals white, (6)9-16(20)mm; ovary semi-inferior.
Native; moist but well-drained base-rich grassland; locally
common throughout most of Br, very rare in E Ir, also natd
garden escape (often as <u>flore pleno</u>) elsewhere in BI.

19. S. hypnoides L. (<u>S. platypetala</u> Smith) - <u>Mossy</u>
<u>Saxifrage</u>. Stoloniferous laxly mat-forming perennial;
differs from <u>S. rosacea</u> ssp. <u>rosacea</u> in linear leaf-lobes
acuminate and ± aristate at apex; flower-buds pendent.
Native; damp rock-ledges, boulders and dunes and by mountain
streams; locally common in Br from Derbys, N Somerset and Wa
northwards, very local in Ir, rare garden escape elsewhere.

19 x 20. S. hypnoides x S. rosacea is intermediate in habit
and leaf characters and occurs with the parents in S
Tipperary and Clare; some persistent garden escapes in Br
might be of the same parentage. Certain examples of <u>S.</u>
<u>hypnoides</u> in N Wa also suggest past hybridization with <u>S.</u>
<u>rosacea</u>. Endemic.

19 x 22. S. hypnoides x S. tridactylites was found with the
parents in 1906 on limestone in MW Yorks; endemic.

20. S. rosacea Moench (<u>S. decipiens</u> Ehrh.) - <u>Irish</u>
<u>Saxifrage</u>. Stoloniferous cushion- or mat-forming perennial
with leaf-rosettes and erect nearly leafless stems to 20cm;
leaves with 3-5(7)lobes acute to subacute and often mucronate
at apex; flowers 2-5(8) in lax cyme; petals white, 6-8mm;
ovary semi-inferior. Native; damp cliffs, rocks and stream-
sides on mountains. Very variable; 1 population deserves
spp. rank.

   a. Ssp. rosacea. Leaves with glandular and non-glandular **R**
hairs, many of which exceed 0.5mm. Locally common in S & W
Ir, 1 place in Caerns, possibly formerly in Angus; natd in N
Somerset.

   b. Ssp. hartii (D. Webb) D. Webb (<u>S. hartii</u> D. Webb). **RRR**
Leaves with hairs all glandular and <0.5mm; plant more
robust. Arranmore Island, W Donegal; endemic.

21. S. cespitosa L. - <u>Tufted</u> <u>Saxifrage</u>. Stoloniferous **RRR**
cushion-forming perennial with basal leaf-rosettes and erect
nearly leafless stems to 10cm; leaves (entire to) 3(-5)-
lobed, with many short glandular hairs, the lobes rounded to
subacute at apex; flowers 1-3(5) in lax cyme; petals off-
white, 4-5mm; ovary semi-inferior. Native; mountain rocks
above 600m; rare and very local in C Sc and N Wa.

22. S. tridactylites L. - <u>Rue-leaved</u> <u>Saxifrage</u>. Annual;
stems erect, to 10(16)cm; leaves oblanceolate, the lower ones
mostly with 3-5 deep lobes, otherwise entire; flowers in
diffuse cymes; petals white, 2-3mm; ovary semi-inferior.
Native; bare dry ground on walls, rocks and sand, mostly
calcareous; locally common throughout most of BI.

6. HEUCHERA L. - <u>Coral-bells</u>
Pubescent perennials; leaves all basal, palmately lobed,
serrate; inflorescence a panicle on erect leafless stem;
flowers 1/2 or more inferior with campanulate hypanthium
above; sepals 5; petals 5; stamens 5; carpels 2, fused to
form 1-celled ovary with parietal placentation.

Other spp. - Some wild plants might be hybrids of the
following with other spp. of this or related genera.

1. H. sanguinea Engelm. - <u>Coral-bells</u>. Leaves 2-6cm, **384**
broadly ovate to orbicular, cordate, on petioles 7-15cm;
flowering stem to 50cm; flowers bright pinkish-red, 6-13mm;
petals smaller than sepals. Intrd; grown in gardens and
persistent on waste ground and tips; scattered in SE En; S N
America. Rarely sets seed.

7. TOLMIEA Torrey & A. Gray - <u>Pick-a-back-plant</u>
Pubescent perennials; leaves mostly basal, palmately lobed,
serrate, producing plantlets at junction with petiole in
moist conditions; inflorescence a simple raceme on ± leafy
stem; flowers zygomorphic, perigynous, with tubular
hypanthium; sepals 5; petals 4(-5), filiform, the lowest
usually missing; stamens 3, opposite the 3 upper sepals;
carpels 2, fused for most part to form 1-celled ovary with
parietal placentation.

1. T. menziesii (Pursh) Torrey & A. Gray - <u>Pick-a-back-</u> **384**
<u>plant</u>. Leaves 4-10cm, broadly ovate to ± orbicular, cordate,
on petioles 10-30cm; flowering stems to 70cm; flowers brown,
6-15mm; petals much narrower than sepals. Intrd; grown as
pot-plant and in gardens, natd in damp shady places and
persistent on tips and waste ground; scattered through most
of Br; W N America. Rarely sets seed.

8. TELLIMA R. Br. - <u>Fringe-cups</u>
Pubescent perennials; leaves mostly basal, palmately lobed,
serrate; inflorescence a simple raceme on ± leafy stem;

flowers 1/4-1/2-inferior, with campanulate hypanthium above;
sepals 5; petals 5, broad and fringed with long narrow lobes;
stamens 10; carpels 2, fused for most part to form 1-celled
ovary with parietal placentation.

1. T. grandiflora (Pursh) Douglas ex Lindley - <u>Fringe-cups</u>.   **384**
Leaves 4-10cm, broadly ovate to orbicular, cordate, on
petioles 5-20cm; flowering stems to 70cm; flowers green,
usually pink- or red-tinged, 6-15mm. Intrd; grown in gardens
and well natd in woods and damp hedgerows; scattered through
most of Br; W N America. Sets abundant seed.

9. CHRYSOSPLENIUM - L. - <u>Golden-saxifrages</u>
Sparsely pubescent perennials with procumbent sterile and
erect leafy flowering stems; leaves orbicular, crenate,
petiolate; inflorescence of dichotomous subcorymbose cymes;
flowers golden-yellow, epigynous; hypanthium 0; sepals 4;
petals 0; stamens 8; carpels 2, fused to form 1-celled ovary
with parietal placentation.

1. C. oppositifolium L. - <u>Opposite-leaved</u> <u>Golden-saxifrage</u>.
Sterile shoots leafy; flowering stems to 15cm; leaves
opposite, up to 2cm, cuneate to rounded at base, with petiole
up to as long as blade. Native; wet places by streams, in
flushes and boggy woods, on mountain ledges; throughout BI
but rare in parts of E En.
2. C. alternifolium L. - <u>Alternate-leaved</u> <u>Golden-saxifrage</u>.
Sterile shoots with only scale-leaves; flowering stems to
20cm; leaves alternate, up to 2.5cm, cordate at base, the
lowest with petiole much longer than blade. Native; similar
places to <u>C. oppositifolium</u>; local over most of Br.

10. PARNASSIA L. - <u>Grass-of-Parnassus</u>
Glabrous perennials; leaves mostly basal, simple, entire,
petiolate; flower solitary, terminal, hypogynous; sepals 5;
petals 5, white; stamens 5, alternating 5 large much divided
staminodes; carpels 4, fused to form ovary 1-celled with
parietal placentation at least below.

1. P. palustris L. - <u>Grass-of-Parnassus</u>. Flowering stems to
30cm, often much less, with 1 sessile leaf; basal leaves
1-5cm, ovate, cordate, petiolate; flowers 15-30mm across.
Native; marshes, damp grassland and dune-slacks; local in
much of Br and Ir, but absent or very rare in most of S Wa
and S En.

76. ROSACEAE - <u>Rose family</u>

Trees, shrubs or annual to perennial herbs; leaves alternate,
simple to compound, usually petiolate, stipulate (stipules
often caducous) or rarely exstipulate. Flowers sometimes

solitary, usually in various, often dense and compound, cymes
or racemes, bisexual or sometimes unisexual in various
arrangements, usually perigynous or epigynous, the hypanthium
variously developed but sometimes $\pm$ 0 and flowers $\pm$
hypogynous, actinomorphic; sepals (4)5(-10), free but usually
borne on hypanthium; epicalyx often present outside calyx;
petals (4)5(-16), free, sometimes 0; stamens usually 2-4x as
many as sepals, sometimes more or fewer; carpels 1 to many,
free or fused, each usually with 2, sometimes 1 or >2,
ovules; style 1 per carpel, usually free, sometimes $\pm$ fused
into column; stigmas capitate; fruit of 1 or more achenes,
drupes, follicles or rarely a capsule, in epigynous flowers
surrounded by succulent or pithy hypanthium.
   Extremely variable in most characters, but usually
recognizable by the alternate stipulate leaves (stipules
often present only in young state or on leader shoots in
woody spp.); the perigynous or epigynous flowers with usually
5 free petals, numerous stamens, and 1-many free or fused
2-ovuled carpels each with a separate style; and the
1-2-seeded fruits often aggregated into false-fruits.
Exceptions to virtually all the above occur.

General key
1 Herbs; stems annual, sometimes woody at base   Key C
1 Trees or shrubs; stems woody, biennial to perennial   2
   2 Flowers hypogynous to perigynous; hypanthium
      variously developed and often enclosing carpels,
      but not fused with them; fruit dry or 1-several
      drupes   Key A
   2 Flowers epigynous; hypanthium completely enclosing
      carpels, becoming fleshy and fused with them at
      fruiting   Key B

Key A - Woody plants with hypogynous to perigynous flowers
1 Fruits a cluster of small drupes on strongly convex
   receptacle with $\pm$ flat hypanthium around   **8. RUBUS**
1 Fruits dry, or if drupes then hypanthium cup-shaped
   (and drupe usually 1)   2
   2 Leaves simple   3
   2 Leaves pinnate   9
3 Petals yellow   **7. KERRIA**
3 Petals white to bright pink   4
   4 Petals >6; fruit terminated by long feathery
      appendage (style)   **14. DRYAS**
   4 Petals normally 5; fruit without feathery
      appendage   5
5 Fruit 1-5 fleshy drupes   6
5 Fruit dry achenes or follicles   7
   6 Carpel, style and drupe 1; leaves serrate to
      crenate   **22. PRUNUS**
   6 Carpels and styles 5; drupes 1-5; leaves entire
                                                 23. OEMLERIA

7 Carpels with 1-2 ovules; fruits indehiscent, 1-seeded
  (achenes) **5. HOLODISCUS**
7 Carpels with >2 ovules; fruits dehiscent, >1-seeded
  (follicles) 8
  8 Exstipulate; carpels free; follicles not inflated,
    dehiscing along 1 side **3. SPIRAEA**
  8 Stipulate; carpels fused at base; follicles
    inflated, dehiscing along 2 sides **2. PHYSOCARPUS**
9 Stems spiny; fruits enclosed in succulent hypanthium
  21. ROSA
9 Stems not spiny; fruits dry, not enclosed in
  hypanthium 10
  10 Flowers in dense globose heads; petals 0;
    hypanthium usually with spines at maturity
    18. ACAENA
  10 Flowers not in dense globose heads; petals
    present; hypanthium never spiny 11
11 Flowers <15mm across, many in each group; petals
  white; leaflets >9; fruit follicles **1. SORBARIA**
11 Flowers >15mm across, 1-few in each group; petals
  yellow; leaflets <9; fruit achenes **9. POTENTILLA**

Key B - Woody plants with epigynous flowers
1 Walls of carpels (within hypanthium) becoming stony
  at fruiting 2
1 Walls of carpels (within hypanthium) becoming
  cartilaginous at fruiting 5
  2 Flowers > 2cm across; sepals >1cm; fruit
    >2cm, brown **34. MESPILUS**
  2 Flowers <2cm across; sepals <1cm; fruit <2cm,
    orange to red or purple to black 3
3 Stems not spiny; leaves entire **32. COTONEASTER**
3 Stems usually spiny; leaves toothed or lobed 4
  4 Leaves evergreen; stipules minute, caducous
    33. PYRACANTHA
  4 Leaves deciduous; stipules persistent at least on
    leading shoots **35. CRATAEGUS**
5 Carpels with >4 ovules, later usually with >2 seeds 6
5 Carpels with 1-2 ovules and seeds 7
  6 Leaves entire, dull green on upperside, pubescent
    on lowerside; styles free **24. CYDONIA**
  6 Leaves serrate, shiny green on upperside, glabrous;
    styles fused below **25. CHAENOMELES**
7 Flowers in racemes **30. AMELANCHIER**
7 Flowers in umbels or corymbs 8
  8 Inflorescence a simple umbel or corymb, rarely
    producing >3 fruits 9
  8 Inflorescence a branched corymb, usually producing
    >3 fruits 10
9 Flesh (hypanthium) of fruit with gritty groups of
  stone-cells; styles free; fruit usually pear-shaped
  26. PYRUS

9 Flesh of fruit without stone-cells; styles fused at
   base; fruit usually apple-shaped   **27. MALUS**
   10 Leaves entire, evergreen; hypanthium not enclosing
      apex of carpels and easily separated from carpels
      at maturity   **31. PHOTINIA**
   10 Leaves serrate to pinnate, deciduous; hypanthium
      wholly enclosing and strongly adherent to carpels
      at maturity   11
11 Styles 2-4(5); leaves simple or pinnate, the main
   veins running straight to leaf margin   **30. SORBUS**
11 Styles 5; leaves simple, the main veins curved near
   apex and running parallel to (not reaching) leaf
   margin   **29. ARONIA**

Key C - Herbaceous plants
1 Petals 0   2
1 Petals present   6
   2 Leaves simple or palmate   3
   2 Leaves pinnate   5
3 Annuals; flowers in leaf-opposed clusters; stamens
   1(-2)   **20. APHANES**
3 Perennials; flowers in terminal clusters; stamens
   4-5(10)   4
   4 Carpel 1; leaves palmate or palmately lobed
                                               19. ALCHEMILLA
   4 Carpels 5-12; leaves ternate   **10. SIBBALDIA**
5 Hypanthium usually with 4 spines at apex; stamens 2;
   plant with at least some procumbent stems **18. ACAENA**
5 Hypanthium not spiny; stamens 4-many; stems erect
                                               17. SANGUISORBA
   6 Petals >6   **14. DRYAS**
   6 Petals 4-6   7
7 Epicalyx 0   8
7 Calyx-like epicalyx present behind true sepals   11
   8 Petals yellow; fruit with hooked bristles
                                               15. AGRIMONIA
   8 Petals white to red or purple; fruit without
      bristles   9
9 Stipules 0; carpels 3, several-seeded   **4. ARUNCUS**
9 Stipules present; carpels >5, 1-seeded   10
   10 Flowers numerous in inflorescence; fruit a head
      of achenes   **6. FILIPENDULA**
   10 Flowers 1-6(10) in inflorescence; fruit a head of
      drupes   **8. RUBUS**
11 Leaves pinnate   12
11 Leaves palmate or ternate   14
   12 Carpels and achenes enclosed in hypanthium;
      stamens 5-10   **16. AREMONIA**
   12 Hypanthium not covering carpels and achenes;
      stamens 10-numerous   13
13 Styles strongly hooked, persistent in fruit **13. GEUM**

13 Styles not hooked, deciduous before fruiting
>   **9. POTENTILLA**
> 14 Receptacle becoming red and succulent in fruit   15
> 14 Receptacle remaining dry in fruit   16
> 15 Petals yellow; epicalyx segments serrate at apex
>> **12. DUCHESNEA**
> 15 Petals white to pinkish; epicalyx segments entire
>> **12. FRAGARIA**
>> 16 Petals <2mm; leaflets with (usually) 3 teeth at
>> extreme apex   **10. SIBBALDIA**
>> 16 Petals >(1.5)2mm; if <2.5mm then leaflets serrated
>> round apical >1/2 of margin   **9. POTENTILLA**

Other genera - 2 woody genera of Spiraeoideae are often
grown in gardens and mass-planted in public places, and may
occasionally persist. **STEPHANANDRA** Siebold & Zucc.
(represented mainly by **S. incisa** (Thunb.) Zabel) and **NEILLIA**
D. Don (mainly **N. sinensis** Oliver) are both Asian and have
simple, lobed, stipulate leaves, and differ from <u>Physocarpus</u>
in having non-inflated follicles dehiscent down 1 side only.
<u>Stephanandra</u> has flowers in corymbs and 1 1-2-seeded
follicle; <u>Neillia</u> has racemes and 1-2 usually 5-seeded
follicles.

SUBFAMILY 1 - SPIRAEOIDEAE (genera 1-5). Stipulate or exstipulate shrubs, sometimes herbs; hypanthium cup- or saucer-
shaped, not enclosing carpels and fused to them only at base;
epicalyx 0; stamens >10; flowers 5-merous; carpels 3-5, often
>2-seeded; fruit dry, a group of follicles or sometimes of
achenes or a capsule; chromosome base-number 8 or 9.

1. SORBARIA (Ser. ex DC.) A. Braun - <u>Sorbarias</u>
Deciduous shrubs; leaves pinnate with > 11 leaflets,
stipulate; flowers crowded in large terminal panicles; petals
white; carpels 5; fruit several-seeded follicles dehiscing
along 1 margin.
The spp. are often mis-identified.

1 Panicle rather dense; follicles pubescent, on erect
  stalks   **1. S. sorbifolia**
1 Panicle rather diffuse; follicles glabrous, on
  recurved stalks   2
> 2 Styles arising from apex of carpel; longest
>   stamens c. as long as petals   **2. S. tomentosa**
> 2 Style arising from well below apex of carpel;
>   longest stamens nearly 2x as long as petals
>>   **3. S. kirilowii**

Other spp. - **S. grandiflora** (Sweet) Maxim., from E Siberia,
has recently been found ± wild in W Kent; it differs from <u>S.</u>
<u>sorbifolia</u> in its stem to only 1m, smaller leaves, and
longest stamens c. as long as (not 2x as long as) petals.

1. S. sorbifolia (L.) A. Braun - <u>Sorbaria</u>. Stems to 2.5m, ± **398**
erect; leaflets 2-9cm, lanceolate, acuminate, 2-serrate;
panicle 7-34cm, with dense flowers and branches arising at
narrow angle.   Intrd; commonly grown and extensively
suckering, natd on walls and waste ground; SE En; N Asia.
2. S. tomentosa (Lindley) Rehder (<u>S.</u> <u>aitchisonii</u> (Hemsley) **398**
Hemsley ex Rehder) - <u>Himalayan Sorbaria</u>. Differs from <u>S.</u>
<u>kirilowii</u> in leaves glabrous to pubescent along veins on
lowerside, without stellate hairs; and see key. Intrd; grown
in gardens and sometimes natd on walls and in scrub; C & S
Sc, C & S En; W. Himalayas.
3. S. kirilowii (Regel) Maxim. (<u>S.</u> <u>arborea</u> C. Schneider, <u>S.</u>
<u>assurgens</u> A. Vilm. & Bois) - <u>Chinese Sorbaria</u>. Stems to 6m,
spreading; leaflets 3-13cm, lanceolate, acuminate, 2-serrate,
± glabrous to sparsely pubescent on lowerside with simple
hairs in vein axils and sometimes more extensive stellate
hairs; panicle 11-42cm, rather diffuse with widely spreading
branches. Intrd; grown in gardens and sometimes natd; SE En,
Jersey; China.

2. PHYSOCARPUS (Cambess.) Maxim. - <u>Ninebark</u>
Deciduous shrubs; leaves simple, stipulate when young;
flowers crowded in ± corymbose racemes; petals white to
pinkish; carpels (3)4-5; fruit 2- to several-seeded inflated
follicles dehiscing along both margins.

1. P. opulifolius (L.) Maxim. - <u>Ninebark</u>. Stems to 2(3)m, **398**
erect to arching; leaves broadly ovate, with 3-5 serrate
lobes, ± glabrous; flowers c.1m across; fruit on long
pedicels, erect, glabrous.   Intrd; natd as relic of
cultivation in shrubberies, rough ground and by streams; C &
N En and Sc, Tyrone; E N America.

3. SPIRAEA L. - <u>Brideworts</u>
Deciduous shrubs; leaves simple, exstipulate; flowers in
dense or fairly dense panicles or corymbs; petals white to
pinkish-purple; carpels 5; fruit several-seeded follicles
dehiscing along 1 margin.

```
1  Inflorescence hemispherical to flat-topped, at least
     as wide as long                                        2
1  Inflorescence cylindrical to conical, longer than wide   5
     2  Inflorescence a simple corymb                       3
     2  Inflorescence a compound corymb                     4
3  Leaves cuneate at base, obovate, at least some 3(-5)-
     lobed distally; petals longer than stamens
                                       12. S. x vanhouttei
3  Leaves rounded at base, ovate, not distinctly lobed;
     petals shorter than stamens       11. S. chamaedryfolia
     4  Petals white; branchlets strongly angled; leaves
          mostly <2cm; stamens <1.2x as long as petals;
          follicles pubescent              10. S. canescens
```

4 Petals pink; branchlets ± terete; leaves mostly
>2cm; stamens >1.2x as long as petals; follicles
glabrous   **9. S. japonica**
5 Leaves glabrous or nearly so; sepals erect in fruit   6
5 Leaves tomentose to slightly pubescent on lowerside;
sepals reflexed in fruit   8
6 Panicles broadly conical, with long branches near
base; leaves mostly widest above mid-way; petals
usually white; stamens c. as long as petals
                                            4. S. alba
6 Panicles ± cylindrical, with short branches;
leaves mostly widest below mid-way; petals
usually pink; stamens usually distinctly longer
than petals   7
7 Panicle-branches usually pubescent; leaves usually
lanceolate; petals usually bright pink; pollen >90
per cent fertile   **1. S. salicifolia**
7 Panicle-branches usually sparsely pubescent; leaves
usually narrowly ovate; petals usually pale pink;
pollen <20% fertile   **2. S. x rosalba**
8 Panicles usually both terminal and lateral, the
terminal ones broadly conical, c. as long as wide;
leaves of flowering shoots mostly <4cm
                                        8. S. x brachybotrys
8 Panicles terminal only, much longer than wide;
leaves of flowering shoots mostly >4cm   9
9 Leaves tomentose on lowerside   10
9 Leaves pubescent on lowerside, the leaf showing
through   11
10 Leaves oblong, white- to pale grey-tomentose on
lowerside, ± entire in proximal c.1/2 of margin;
follicles glabrous   **6. S. douglasii**
10 Leaves ovate, grey- to buff-tomentose on lowerside,
serrate for distal >2/3 of margin; follicles
pubescent   **7. S. tomentosa**
11 Leaves oblong, ± entire in proximal c.1/2 of
margin; pollen >90 per cent fertile   **6. S. douglasii**
11 Leaves ovate, serrate for distal >2/3 of margin;
pollen <20 per cent fertile   12
12 Panicles narrowly conical; petals often very pale
pink   **5. S. x billardii**
12 Panicles subcylindrical; petals pink
                                      3. S. x pseudosalicifolia

Other spp. - The following relatives of S. chamaedryfolia
may appear as relics in neglected gardens, hedges and woodland. **S. x arguta** Zabel (S. thunbergii Sieb. x S. x
multiflora Zabel) (Bridal-spray) is a garden hybrid with
small umbels distributed all along branches and oblanceolate
leaves 1-4cm. **S. media** Schmidt, from E Europe, has much
larger subglobose umbels and elliptic leaves 2-5cm.

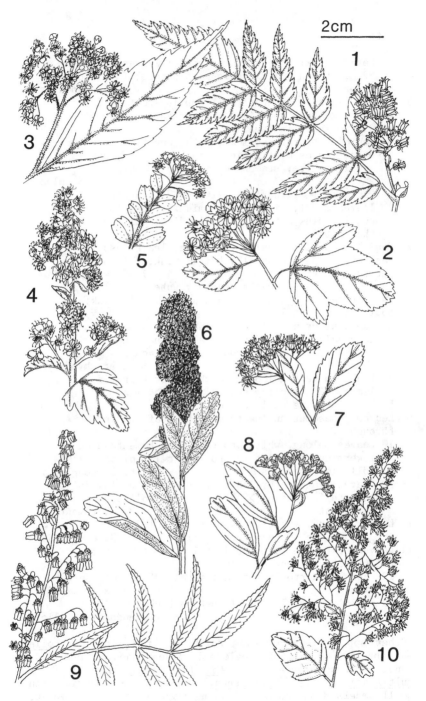

FIG 398 – Rosaceae: Spiraeoideae. 1, *Sorbaria sorbifolia*. 2, **Physocarpus**. 3, *Spiraea japonica*. 4, *S. alba*. 5, *S. canescens*. 6, *S. douglasii*. 7, *S. chamaedryfolia*. 8, *S. x vanhouttei*. 9, *Sorbaria tomentosa*. 10, **Holodiscus**.

1. 1. **S. salicifolia** L. - <u>Bridewort</u>. Strongly suckering;
stems ± erect, to 2m; leaves 4-8cm, elliptic-oblong to
lanceolate, sharply serrate except at extreme proximal end,
glabrous; panicle ± cylindrical; petals pink. Intrd;
formerly much grown, natd in hedges and rough ground;
throughout BI but over-recorded and now very rare; C Europe.
2. **S. x rosalba** Dippel (<u>S. x rubella</u> Dippel; <u>S. salicifolia</u>
x <u>S. alba</u>) - <u>Intermediate Bridewort</u>. Differs from <u>S.</u>
<u>salicifolia</u> in narrowly conical panicle; usually very pale
pink petals; and see key. Intrd; garden hybrid natd in
hedges, etc.; probably throughout Br and commoner than <u>S.</u>
<u>salicifolia</u> in Sc.
3. **S. x pseudosalicifolia** Silverside (<u>S.</u> x <u>billardii</u> auct.
non Herincq; <u>S. salicifolia</u> x <u>S. douglasii</u>) - <u>Confused</u>
<u>Bridewort</u>. Intermediate in leaf shape, serration and
pubescence; inflorescence ± cylindrical; petals pink. Intrd;
garden hybrid natd in hedges, etc.; throughout Br.
4. **S. alba** Duroi (<u>S. latifolia</u> (Aiton) Borkh.) - <u>Pale</u> 398
<u>Bridewort</u>. Differs from <u>S. salicifolia</u> in leaf- and
panicle-shape (see key); petals usually white, sometimes
pink. Intrd; natd from gardens in hedges, etc.; frequent in
Sc, N Ir and probably N En; E N America.
5. **S. x billardii** Herincq (<u>S. alba</u> x <u>S. douglasii</u>) -
<u>Billard's Bridewort</u>. Intermediate in leaf shape, pubescence
and serration, panicle shape and flower colour (see key).
Intrd; garden hybrid natd in hedges, etc.; rather rare in Br.
6. **S. douglasii** Hook. - <u>Steeplebush</u>. Strongly suckering; 398
stems ± erect, to 2m; leaves 4-8cm, oblong, sharply serrate
in distal 1/2, ± entire in proximal 1/2; panicle ±
cylindrical; petals pink. Intrd; commonly grown in gardens,
natd in hedges, etc.; commoner than <u>S. salicifolia</u> in Br and
under-recorded, ?Ir; W N America.
   a. **Ssp. douglasii**. Leaves whitish- or pale greyish-
tomentose on lowerside. Common.
   b. **Ssp. menziesii** (Hook.) Calder & Roy L. Taylor. Leaves
subglabrous to pubescent on veins on lowerside. Very rarely
natd.
7. **S. tomentosa** L. - <u>Hardhack</u>. Differs from <u>S. douglasii</u> in
leaf shape, pubescence and serration, and follicle pubescence
(see key). Intrd; grown in gardens, rather rarely natd in
En; E N America.
8. **S. x brachybotrys** Lange (<u>S. douglasii</u> x <u>S. canescens</u>) -
<u>Lange's Spiraea</u>. Stems ± erect, to 2m; leaves 1.5-4cm,
elliptic-oblong, toothed only near apex, greyish-tomentose to
-pubescent on lowerside; panicle broadly conical; petals pale
pink. Intrd; garden hybrid rarely natd in Sc.
9. **S. japonica** L. f. - <u>Japanese Spiraea</u>. Stems ± erect, to 398
1.5m; leaves 5-12cm, elliptic to narrowly so, coarsely
serrate, glabrous to slightly pubescent on veins; panicle ±
flat-topped; petals pink. Intrd; grown in gardens, natd in
hedges, etc.; occasional in Br and CI; Japan. **S.** 'Anthony
Waterer' is the commonest clone in gardens and might be so in

the wild too.

10. S. canescens D. Don - Himalayan Spiraea. Stems arching, **398**
to 2m; leaves 1-2.5cm, obovate, crenate-dentate in distal
1/2, pubescent on lowerside; panicle round-topped; petals
white. Intrd; grown in gardens, natd in scrub; occasional in
S Br; Himalayas.

11. S. chamaedryfolia L. - Elm-leaved Spiraea. Stems **398**
arching, to 2m; leaves 2-7cm, ovate, coarsely serrate except
near base, glabrous or nearly so; corymb simple, round-
topped; petals white. Intrd; grown in gardens, natd in river
gorge, Angus since before 1966; SE Europe. Our plant is ssp.
ulmifolia (Scop.) J. Duvign.

12. S. x vanhouttei (Briot) Zabel (S. cantoniensis Lour. x **398**
S. trilobata L.) - Van Houtte's Spiraea. Stems arching, to
2m; leaves 1.5-4cm, obovate, coarsely serrate and at least
some 3(-5)-lobed in distal part, glabrous; corymb simple,
nearly flat-topped; petals white. Intrd; grown in gardens;
natd in hedges, etc.; scattered in En; garden origin.

4. ARUNCUS L. - Buck's-beard

Herbaceous perennials; leaves 2-3-pinnate, exstipulate;
flowers + dioecious, in very dense terminal panicles; petals
white; carpels 3; fruit 2- to several-seeded follicles
dehiscing along 1 margin.

1. A. dioicus (Walter) Fern. (A. sylvestris Kostel., A.
vulgaris Raf.) - Buck's-beard. Stems to 2m, erect; basal
leaves to 1m, the leaflets ovate, sharply 2-serrate, glabrous
or nearly so; inflorescence up to 50cm, much-branched;
pedicels strongly recurved in fruit; follicles glabrous.
Intrd; much planted in gardens, very persistent in woodland
and by water, but + never seeding; scattered throughout En
and Sc, especially N; Europe.

5. HOLODISCUS (K. Koch) Maxim. - Ocean-spray

Deciduous shrubs; leaves simple, exstipulate; flowers in
dense terminal panicles; petals creamy-white; carpels 5;
fruit 1-seeded indehiscent achenes.

1. H. discolor (Pursh) Maxim. - Ocean-spray. Stems to 4m, **398**
erect to arching; leaves ovate to broadly so, coarsely
toothed and + lobed, greyish-pubescent to -tomentose on
lowerside; panicles arching to pendent, up to 30cm; achenes
pubescent. Intrd; grown in gardens, natd in hedges and scrub
and on walls; scattered in En and Sc; W N America.

SUBFAMILY 2 - ROSOIDEAE (genera 6-21). Stipulate herbs,

sometimes shrubs; hypanthium variably concave, sometimes
flat, not or sometimes enclosing carpels but fused to them
only at base; epicalyx present or 0; stamens 1-numerous;
flowers mostly 4-6-merous; carpels 1-numerous, 1-2-seeded;
fruit a head of achenes or drupes, sometimes borne on

succulent receptacle or surrounded by succulent or dry
hypanthium; chromosome base-number 7, 8 or 9.

6. FILIPENDULA Miller - <u>Meadowsweets</u>
Herbaceous perennials; leaves pinnate or reduced to terminal
lobe only; flowers in terminal ± flat-topped panicles, 5-6-
merous; epicalyx 0; hypanthium ± flat to saucer-shaped;
stamens numerous; carpels 4-12; fruit a head of achenes each
with 1-2 seeds.

1 Basal leaves with terminal and 8-30 pairs of main
  leaflets all 0.5-2cm; petals usually 6   **1. F. vulgaris**
1 Basal leaves with terminal and 0-5 pairs of main
  leaflets all >2cm; petals usually 5   2
  2 Carpels 6-10, spirally twisted, glabrous; leaves
    with 2-5 pairs of large lateral leaflets
                                2. F. ulmaria
  2 Carpels 4-6, straight, ciliate on edges; leaves
    without large lateral leaflets   **3. F. kamtschatica**

1. F. vulgaris Moench - <u>Dropwort</u>. Stems erect, to
50(100)cm; leaves with 8-30 pairs of main leaflets and
smaller ones between, glabrous; petals 5-9mm, creamy-white;
achenes 6-12, erect, pubescent. Native; calcareous
grassland; very locally frequent in Br N to C Sc, grown in
gardens and sometimes natd elsewhere.
2. F. ulmaria (L.) Maxim. - <u>Meadowsweet</u>. Stems erect, to
1.2m; leaves with 2-5 pairs of main leaflets and smaller ones
between, tomentose to glabrous on lowerside; flowers >5mm
across, white. Native; all sorts of wet and damp places;
common throughout BI.
3. F. kamtschatica (Pallas) Maxim. - <u>Giant</u> <u>Meadowsweet</u>.
Stems erect, to 3m; leaves densely pubescent on lowerside,
with 1 large, lobed terminal leaflet and 1-2 pairs of very
small laterals; flowers <5mm across, pink or white. Intrd;
planted in damp places and sometimes natd; very scattered in
Br, especially Sc and N En, W Ir; E Asia. Plants with pink
to red flowers are often referred to **F. purpurea** Maxim., from
Japan, but this is totally glabrous and our plants are
probably cultivars of <u>F. kamtschatica</u> or hybrids of it. <u>F.</u>
<u>purpurea</u> is grown in gardens.

7. KERRIA DC. - <u>Kerria</u>
Deciduous shrubs; leaves simple; flowers solitary, terminal
on lateral branches, 5-merous; epicalyx 0; hypanthium ± flat;
stamens numerous; carpels 5-8; fruit a head of achenes.

1. K. japonica (L.) DC. - <u>Kerria</u>. Stems to 2.5m, erect,
bright green above; leaves ovate to lanceolate, coarsely
serrate, sparsely pubescent; flowers 2.5-5cm across, yellow.
Intrd; much grown in gardens, persistent in neglected
shrubberies and old garden sites, usually as <u>flore pleno</u>, but

self-sown in Middlesex; very scattered in Br; China.

8. RUBUS L. - <u>Brambles</u>
Deciduous or semi-evergreen shrubs, often spiny, or herbaceous perennials; leaves simple, pinnate or palmate; flowers solitary or in racemes or panicles, usually 5-merous; epicalyx 0; hypanthium flat, with receptacle usually extended upwards from centre; stamens numerous; carpels usually numerous; fruit a head of (1)2-many 1-seeded drupes.
Most of the taxa of subg. <u>Rubus</u> (taxa 12-25) form an extremely complex, largely apomictic group (sect. <u>Glandulosus</u>), often known collectively as **R. fruticosus** L. agg. 2 other sections (<u>Rubus</u> and <u>Corylifolii</u>) are often included within this complex, but they are probably derived from ancient hybrids between it and <u>R. idaeus</u> and <u>R. caesius</u> respectively, and are here treated separately. Over 400 microspp. have been recognized in BI in these 3 sections together; <u>R. caesius</u> is the only other member of subg. <u>Rubus</u> in BI. In this work the microspp. of sects. <u>Rubus</u>, <u>Corylifolii</u> and <u>Glandulosus</u> are not treated in full but, following the view of A. Newton and E.S. Edees, 11 rather ill-defined series, representing the main nodes in the spectrum of variation, are recognized in sect. <u>Glandulosus</u>. These are keyed out in couplets 15-24, but most of the characters used are relative and a high level of success will be achieved only after much experience.

1	Leaves simple	2
1	Leaves pinnate, palmate or ternate	5
	2 Stipules arising direct from stem; receptacle strongly convex	3
	2 Stipules fused to petiole proximally; receptacle flat	4
3	Stems herbaceous, erect; flowers solitary, dioecious	**1. R. chamaemorus**
3	Stems arching or procumbent, rooting at tips; flowers in clusters, bisexual	**2. R. tricolor**
4	Petals white	**6. R. parviflorus**
4	Petals reddish-purple	**5. R. odoratus**
5	Stipules arising direct from stem; stems all annual, producing flowers in 1st year; receptacle flat	6
5	Stipules fused to petiole proximally; stems at least biennial, producing flowers in 2nd year; receptacle conical	7
	6 Stoloniferous, without rhizomes; petals <6mm, white	**3. R. saxatilis**
	6 Rhizomatous, without stolons; petals >6mm, pink	**4. R. arcticus**
7	Stems densely covered with white bloom	**11. R. cockburnianus**
7	Stems green to red, sometimes glaucous but not with dense white bloom	8

8 Stems usually not rooting at tips; leaves ternate
   or pinnate; fruit usually red to orange, usually
   separating from receptacle at maturity   9
8 Stems usually rooting at tips; leaves ternate or
   palmate; fruit usually black, coming away with
   extension of receptacle at maturity (subg. Rubus)   12
9 At least lower leaves pinnate with 5(-7) leaflets;
   petals white   10
9 Leaves normally ternate; petals pink to purple   11
   10 Stems ± erect; fruit separating from receptacle
      at maturity; stem with weak prickles   7. R. idaeus
   10 Stems arching and eventually rooting at tips;
      fruit coming away with extension of receptacle at
      maturity; stem with moderate prickles
                                      10. R. loganobaccus
11 Leaves ± glabrous; flowers >2cm across, 1-few per
   group, appearing in Spring; glandular hairs 0
                                      9. R. spectabilis
11 Leaves white-tomentose on lowerside; flowers <1.5cm
   across, many per group, appearing in Summer;
   glandular bristles dense on stems and flower-stalks
                                      8. R. phoenicolasius
   12 Fruit with glaucous bloom, composed of rather few,
      large, only loosely coherent drupes; leaflets 3,
      the lateral ones ± sessile   25. R. caesius
   12 Fruit without glaucous bloom, usually composed of
      usually many tightly coherent drupes; leaflets
      often >3   13
13 Leaflets ± overlapping, the basal pair ± sessile;
   stipules lanceolate; inflorescence usually a ±
   simple corymb   24. R. sect. Corylifolii
13 Leaflets mostly ± not overlapping, the basal pair
   usually stalked; stipules linear; inflorescence
   compound or ± racemose   14
   14 Stems suberect, usually not rooting at tips;
      suckers often produced from roots; fruits often
      red to purple, sometimes black   12. R. sect. Rubus
   14 Stems procumbent to arching, usually rooting at
      tips; suckers not produced; fruits black (sect.
      Glandulosus)   15
15 Stalked glands 0 or rare and inconspicuous on 1st-
   year stems   16
15 Stalked glands on 1st-year stems obvious   18
   16 Leaflets variously pubescent on lowerside but
      not, or only the upper ones, tomentose
                                  13. R. series Sylvatici
   16 Leaflets all tomentose on lowerside   17
17 Leaflets greyish-white-tomentose on lowerside; few
   stalked glands sometimes present in inflorescence
                                  14. R. series Rhamnifolii
17 Leaflets chalky-white-tomentose on lowerside; stalked
   glands 0   16. R. series Discolores

18 Stamens shorter than styles; stalked glands usually
   few   **15. R. series Sprengeliani**
18 Stamens as long as or longer than styles; stalked
   glands usually many   19
19 Main prickles ± confined to angles of stem, distinct
   from smaller pricklets and acicles   20
19 Main prickles occurring all round stem, grading into
   pricklets and acicles   23
20 Stems conspicuously pubescent, the stalked glands
   less conspicuous   **17. R. series Vestiti**
20 Stems glabrous to pubescent, the stalked glands
   more conspicuous than hairs   21
21 Terminal leaflet obovate, often broadly so, with
   short cuspidate apex and serrulate margin
                                       18. R. series Mucronati
21 Terminal leaflet usually ovate to elliptic, usually
   less abruptly narrowed at apex, usually more serrate
   on margin   22
22 Broad-based pricklets, acicles and stalked glands
   on stems subequal   **21. R. series Radulae**
22 Pricklets, acicles and stalked glands unequal
                                       19. R. series Micantes
23 Prickles and stalked glands in very variable
   quantities on same plant   **20. R. series Anisacanthi**
23 Prickles and stalked glands not markedly variable in
   quantity on same plant   24
24 Prickles strong; pricklets often more numerous
   than stalked glands   **22. R. series Hystrices**
24 Prickles weak; stalked glands and acicles often
   very numerous   **23. R. series Glandulosi**

1. R. chamaemorus L. - <u>Cloudberry</u>. Stems to 20cm, erect,
annual, clothed only with hairs; leaves simple, ± orbicular,
palmately 5-7-lobed; flowers dioecious, solitary, terminal,
white, 20-30mm across; fruit orange when ripe, of 4-20 large
drupes. Native; peaty moors and bogs on mountains; Br N from
N Wa and Derbys, Tyrone.
2. R. tricolor Focke - <u>Chinese</u> <u>Bramble</u>. Stems to several m, **405**
trailing, rooting at tips, with dense brownish bristles;
leaves simple, ovate, pinnately 3-5-lobed; flowers in
racemes, 20-25mm across, white; fruit red, but plant often
not flowering. Intrd; grown as ground-cover but spreading
and natd in hedges and shrubberies; very scattered in En;
China.
3. R. saxatilis L. - <u>Stone</u> <u>Bramble</u>. Flowering stems to
40cm, erect, with hairs and very weak prickles; stolons
longer, often rooting at apex but the rest dying in winter;
leaves ternate, with ovate to elliptic leaflets; flowers few,
in terminal corymbs, white, 8-15mm across; fruit red, of 1-6
drupes. Native; woods, screes and mountain slopes on basic
soils; rather scattered in Ir, Sc, Wa and C, N & W En.
4. R. arcticus L. - <u>Arctic</u> <u>Bramble</u>. Stems to 30cm, erect, **RR**

FIG 405 – Leaves of **Rubus**. 1, **R. parviflorus**. 2, **R. tricolor**. 3, **R. phoenicolasius**. 4, **R. spectabilis**. 5, **R. caesius**. 6, **R. idaeus**. 7, **R. loganobaccus**. 8, **R. cockburnianus**.

annual, ± glabrous; leaves ternate, with ovate to obovate leaflets; flowers 1-3, terminal, pink, 15-25mm across; fruit dark red, of numerous drupes. Probably once native; several records from highlands of Sc, the last in 1841.

5. R. odoratus L. - Purple-flowered Raspberry. Stems to 3m, erect, with glandular and non-glandular hairs; leaves simple, ± orbicular, 5-lobed; flowers in panicles of usually >6, purple, 3-5cm across; fruit red. Intrd; grown for ornament, occasionally natd in rough places; very scattered in S En; E N America.

6. R. parviflorus Nutt. - Thimbleberry. Stems to 2m, erect, 405 with glandular and non-glandular hairs; leaves simple, ± orbicular, 3-5-lobed; flowers in corymbs of 3-6(10), white, 3-6cm across; fruit red. Intrd; grown for ornament, occasionally natd in rough ground; very scattered in En, Sc and N Kerry; W N America.

7. R. idaeus L. - Raspberry. Stems to 1.5(2.5)m, erect, 405 with few to numerous weak prickles, otherwise ± glabrous to pubescent; leaves pinnate, with 3-7 ovate leaflets white-tomentose on lowerside; flowers few in racemes, white, c.1cm across, in some plants male only; fruit red, rarely yellow or white. Native; woods, heaths and marginal ground; frequent throughout BI, but only escape from cultivation in some places.

7 x 8. R. idaeus x R. phoenicolasius = R. x paxii Focke was found in 1930 in S Lancs.

7 x 25. R. idaeus x R. caesius = R. x pseudoidaeus (Weihe) Lej. (R. x idaeoides Ruthe) is very sparsely scattered in En, N Tipperary; it resembles R. caesius in habit and stem characters and R. idaeus in leaf characters. It is largely sterile.

8. R. phoenicolasius Maxim. - Japanese Wineberry. Stems to 405 2(3)m, erect and spreading, with dense reddish glandular bristles and sparse weak spines; leaves with 3(-5) ovate leaflets white-tomentose on lowerside; flowers several in racemes, pink, c.1cm across; fruit red. Intrd; grown for ornament and fruit and natd in rough places and scrub; scattered in S Br; E Asia.

9. R. spectabilis Pursh - Salmonberry. Stems to 2m, ± 405 erect, with weak spines mostly below, otherwise ± glabrous; leaves ternate, with ovate ± glabrous leaflets; flowers usually solitary on lateral branches, pink, 2-3cm across; fruit orange. Intrd; grown for ornament and natd in woods and hedgerows; scattered throughout BI; W N America.

10. R. loganobaccus L. Bailey - Loganberry. Stems to 405 several m, arching and finally rooting at tips, with numerous moderate prickles; leaves pinnate, with 5 ovate leaflets white-tomentose on lowerside; flowers in panicles, white, c.2-3cm across; fruit dark purplish-red. Intrd; grown in gardens for fruit and frequently bird-sown or escaped in hedges and waste places; scattered throughout Br and CI; garden origin (R. idaeus x R. vitifolius Cham. & Schldl.

(Rubus sect. Ursini Focke)) in 1881.
11. R. cockburnianus Hemsley – White-stemmed Bramble. Stems **405** to 5m, erect and arching, with few strong prickles, densely covered in white bloom; leaves pinnate, with 7–11 narrowly elliptic-ovate leaflets grey-pubescent on lowerside; flowers in panicles, purplish, 1–1.8cm across; fruit black. Intrd; grown in gardens for ornamental stems, natd rarely En and Sc; China.

12–25. R. subg. Rubus. Here treated as composed of R. caesius and R. sects Glandulosus (with 11 series), Rubus and Corylifolii. R. ulmifolius is a sexual diploid, but all other microspp. are facultative apomicts (triploids, tetraploids, pentaploids and hexaploids) that require pollination in order to produce even apomictic seed (pseudogamy). Sexual seed can also be produced, and hybridization is common.

12. R. sect. Rubus (sect. Suberecti Lindley). Variously intermediate between R. idaeus and R. fruticosus agg.: suckers often produced from roots; stems usually suberect, often not rooting at tips; fruit red to black. Native; sunny and partly shaded places; throughout BI. 20 microspp. currently placed here, incl. the natd N American spp. **R. canadensis** L., **R. pergratus** Blanchard, and **R. allegheniensis** Porter. A natural recent hybrid between R. idaeus and R. fruiticosus agg. was recorded from Berks in 1922; it had stems rooting at the tips and red fruit tasting of raspberries and separating from the receptacle (not so in sect. Rubus). R. loganobaccus (q.v.) and more recent crop species (e.g. Tayberry, Boysenberry) are also of similar parentage.

13–23. R. sect. Glandulosus Wimmer & Grab. (subsect. Hiemales E.H. Krause; R. fruticosus L. agg.) – Bramble. Stems procumbent to arching, (potentially) rooting at tips, with well-developed prickles and varying quantities of pricklets, acicles, stalked glands and hairs; roots not producing suckers; leaves ternate to 5–7-palmate, with basal pair of leaflets usually stalked; stipules linear; flowers in compound or racemose inflorescences, white to pink; fruit black, not bloomed. Native; all sorts of habitats both natural and man-made, but much less common on calcareous soils; throughout BI, but very local in N Sc. Very variable; almost all characters of inflorescence and vegetative parts are of value in distinguishing the microspp. and series, but probably the indumentum and armature are the most important.

13. R. series Sylvatici (P.J. Mueller) Focke (sect. Sylvatici P.J. Mueller, subsect. Virescentes Genev.). 56 microspp. currently placed here, incl. the distinctive alien **R. laciniatus** Willd. (origin unknown).

14. R. series Rhamnifolii (Bab.) Focke (sect. Sylvatici subsect. Discoloroides Genev. ex Sudre). 41 microspp. currently placed here, incl. the probably alien **R. elegantispinosus** (Schum.) H.E. Weber, from Europe.

15. R. series Sprengeliani Focke (sect. <u>Sprengeliani</u> (Focke) W.C.R. Watson nom. inval.). 4 microspp. currently placed here.

16. R. series Discolores (P.J. Mueller) Focke (Sect. <u>Discolores</u> P.J. Mueller). 11 microspp. currently placed here, incl. the probably alien **R. armeniacus** Focke (<u>R. procerus</u> auct.), from Europe, much grown for its fruit as 'Himalayan Giant'. 1 sp., **R. ulmifolius** Schott (<u>R. inermis</u> auct. non Pourret), is the only diploid and only sexual sp. of subg. <u>Rubus</u> in BI. It is widespread in BI, especially on chalk and clay where few other spp. occur, and hybridizes with many other spp. as pollen parents; hybrids may be fertile or sterile.

17. R. series Vestiti (Focke) Focke (group. <u>Vestiti</u> Focke). 20 microspp. currently placed here.

18. R. series Mucronati (Focke) H.E. Weber (sect. <u>Rotundifolii</u> W.C.R. Watson nom. inval., sect. <u>Appendiculati</u> (Genev.) Sudre). 11 microspp. currently placed here.

19. R. series Micantes Sudre ex Bouvet (Series <u>Apiculati</u> Focke pro parte, series <u>Grandifolii</u> Focke). 28 microspp. currently placed here.

20. R. series Anisacanthi H.E. Weber (series <u>Apiculati</u> Focke pro parte, series <u>Dispares</u> W.C.R. Watson pro parte nom. inval.). 18 microspp. currently placed here.

21. R. series Radulae (Focke) Focke (subsect. <u>Rudes</u> Sudre pro parte). 40 microspp. currently placed here.

22. R. series Hystrices Focke. 33 microspp. currently placed here.

23. R. series Glandulosi (Wimmer & Grab.) Focke (sect. <u>Glandulosus</u> Wimmer & Grab., series <u>Euglandulosi</u> W.C.R. Watson nom. inval.). 7 microspp. currently placed here.

24. R. sect. Corylifolii Lindley (sect. <u>Triviales</u> P.J. Mueller). Variously intermediate between <u>R. caesius</u> and <u>R. fruticosus</u> agg.: stems low-arching to procumbent, (potentially) rooting at tips, usually glaucous, with variable indumentum and armature but usually only moderate prickles; leaflets 3-5, the basal pair sessile or nearly so; flowers usually large; fruits black, not glaucous, often with large, few drupelets. Native; open places; throughout range of <u>R. caesius</u>. 19 microspp. are currently placed here. Some of these, and other very local, un-named plants, have highly imperfect fruits and are probably recent hybrids.

25. R. caesius L. - <u>Dewberry</u>. Stems low-arching to **405** procumbent, rooting at tips, with glaucous bloom, with moderate prickles but 0 or few hairs or glands; leaves ternate, the leaflets ovate, ± glabrous to pubescent on lowerside, with lanceolate stipules; flowers few in corymbs, white, 2-2.5cm across; fruits with large, few drupelets, black, with glaucous bloom. Native; disturbed ground, grassland, scrub and sand-dunes, often on clayey or basic soils; throughout C & S BI but rare and very local in N and much of W. Apomictic.

9. POTENTILLA L. (<u>Comarum</u> L.) - <u>Cinquefoils</u>

Herbaceous (annuals to) perennials or rarely deciduous
shrubs; leaves pinnate, ternate or palmate, or the upper ones
simple; flowers solitary or few in cymes, (4-)5-merous;
epicalyx present; hypanthium flat to saucer-shaped, with
receptacle slightly to strongly convex; stamens $\geq$(5)10;
carpels 4-numerous; fruit a head of achenes.

1	Lower leaves pinnate	2
1	Lower leaves ternate or palmate	5
	2 Shrub; leaflets entire; achenes pubescent	
		1. P. fruticosa
	2 Herb; leaflets toothed; achenes glabrous	3
3	Petals purple; plant with long woody rhizome	
		2. P. palustris
3	Petals yellow or white; plant without long rhizome	4
	4 Petals white; flowers in terminal cyme	
		4. P. rupestris
	4 Petals yellow; flowers solitary, axillary	
		3. P. anserina
5	Petals white; achenes pubescent on 1 side	
		17. P. sterilis
5	Petals yellow; achenes glabrous (but receptacle often pubescent)	6
	6 Leaves grey- to white-tomentose on lowerside	7
	6 Leaf-surface visible through pubescence on lowerside	8
7	Leaves grey-tomentose on lowerside, with flat margin; upper part of stem with mixed straight and woolly hairs; petals 5-7mm	**6. P. inclinata**
7	Leaves white-tomentose on lowerside, with revolute margin visible from lowerside as narrow green edge; upper part of stem with dense woolly hairs; petals 4-5mm	**5. P. argentea**
	8 At least some flowers with 4 petals and sepals	9
	8 All flowers with 5 petals and sepals	11
9	Plant highly sterile (with 0-few achenes per flower); petioles of stem-leaves >1cm, all $\pm$ same length	**15. P. x mixta**
9	Plant fertile (many achenes per flower); stem-leaves $\pm$ sessile to petiolate, if latter petioles decreasing markedly in size towards stem apex	10
	10 Carpels <20; stem-leaves sessile or with stalks <5mm; all or nearly all leaves ternate (ignore stipules); stems not rooting at nodes; flowers $\pm$ all 4-merous	**13. P. erecta**
	10 Carpels >20, lower stem-leaves with stalks >10mm; some leaves with 4-5 leaflets; stems rooting at nodes late in season; some flowers 5-merous	**14. P. anglica**
11	Flowers solitary in leaf-axils; main stems procumbent and rooting at nodes	**16. P. reptans**

11 Flowers (often few) in terminal cymes; main stems not
    rooting at nodes   12
    12 Flowering stem arising laterally from side of
       terminal leaf-rosette, usually <2mm wide, with
       only reduced (usually simple) leaves   13
    12 Flowering stem arising from centre of leaf-
       rosette (the latter often withered by flowering-
       time), usually >2mm wide, with several well-
       developed leaves   14
13 Vegetative stems long, procumbent, often rooting,
   mat-forming; free part of stipules of basal leaves
   linear-triangular; flowers mostly <15mm across
                                              12. P. neumanniana
13 Vegetative stems short, not rooting or mat-forming;
   free part of stipules of basal leaves ovate-
   lanceolate; flowers mostly >15mm across   **11. P. crantzii**
    14 Petals >6mm, longer than sepals   **7. P. recta**
    14 Petals <5mm, shorter than to as long as sepals   15
15 Petals c.1/2 as long as sepals; stamens 5-10; achenes
   <0.8mm, smooth   **10. P. rivalis**
15 Petals c.3/4-1x as long as sepals; stamens c.20;
   achenes >0.8mm, minutely rugose   16
    16 Leaves ± all ternate; epicalyx-segments longer
       than sepals in fruit   **9. P. norvegica**
    16 Most lower leaves with 5 leaflets; epicalyx-
       segments shorter than sepals in fruit
                                               8. P. intermedia

Other spp. - **P. thuringiaca** Bernh. ex Link, from C & S
Europe, was formerly ± natd in a few places; it is similar to
<u>P. crantzii</u> but more robust with narrower leaflets with the
terminal tooth shorter than the 2 on either side.

1. P. fruticosa L. - <u>Shrubby Cinquefoil</u>. Deciduous erect or **RR**
spreading shrub to 1m; leaves pinnate, with (3)5-7(9)
leaflets; flowers 1-many in cymes, functionally dioecious;
petals yellow, 6-16mm, much longer than sepals. Native;
rock-ledges, river- and lake-margins in full sun; extremely
local in 2 areas of N En and the Burren area of W Ir.
2. P. palustris (L.) Scop. (<u>Comarum palustre</u> L.) - <u>Marsh
Cinquefoil</u>. Herbaceous perennial with long woody rhizome and
ascending stems to 50cm; leaves pinnate, with (3)5-7
leaflets; flowers 1-few in terminal cymes; petals purple,
much shorter than sepals. Native; fens, marshes and bogs;
common over most of BI but very local in S & C Br and CI.
3. P. anserina L. - <u>Silverweed</u>. Perennial with long
procumbent stolons and terminal leaf-rosettes; leaves
pinnate, with (3)7-12 pairs of narrowly elliptic-oblong main
leaflets alternating with small ones; flowers solitary in
leaf-axils on stolons, on erect pedicels to 25cm; petals
yellow, 7-10mm, c.2x as long as sepals. Native; waste
places, waysides, pastures and sand-dunes; common throughout

BI.

4. P. rupestris L. - <u>Rock Cinquefoil</u>. Perennial with erect **RRR**
flowering stems to 60m arising from leaf-rosette; leaves
pinnate, with 2-4 pairs of broadly elliptic leaflets; flowers
in terminal cymes; petals white, longer than sepals. Native;
on basic rocks; 1 site each in Monts and Rads, 2 in E
Sutherland, occasional escape from cultivation elsewhere.

5. P. argentea L. - <u>Hoary Cinquefoil</u>. Perennial with **412**
decumbent to ascending stems to 30cm arising from leaf- **R**
rosette; leaves palmate with 5 leaflets white-tomentose on
lowerside; flowers in terminal cymes; petals yellow, 4-5mm,
c. as long as sepals. Native; sandy grassland and waste
ground; local and decreasing in CI and Br N to C Sc, common
only in E En.

6. P. inclinata Villars (<u>P.</u> <u>canescens</u> Besser) - <u>Grey</u> **412**
<u>Cinquefoil</u>. Perennial with erect to ascending stems to 50cm
arising from leaf-rosette; differs from <u>P.</u> argentea in larger
size and see key. Intrd; waste places; occasional casual in
Br, mainly SE En; C & S Europe.

7. P. recta L. - <u>Sulphur Cinquefoil</u>. Perennial with erect **412**
stems to 10cm arising from leaf-rosette; leaves palmate with
5-7 leaflets; flowers in terminal cymes; petals yellow, 6-
12mm, longer than sepals. Intrd; grown in gardens and natd
in waste ground, roadside banks and grassy places; scattered
over C & S En, rare casual elsewhere; S Europe.

8. P. intermedia L. - <u>Russian Cinquefoil</u>. Biennial or **412**
perennial with erect to ascending stems to 50cm arising from
leaf-rosette; leaves palmate, the lower with 5 leaflets;
flowers in terminal cymes; petals yellow, 4-5mm, c. as long
as sepals. Intrd; casual in waste and grassy places, sometimes ± natd; scattered in En; Russia.

9. P. norvegica L. - <u>Ternate-leaved Cinquefoil</u>. Annual to **412**
short-lived perennial with erect to ascending stems to 50cm;
leaves ternate; flowers in terminal cymes; petals yellow,
4-5mm, c. as long as sepals or shorter. Intrd; casual or
sometimes natd in waste places; very scattered in En and Wa,
rare casual in Sc and N Ir; N & C Europe.

10. P. rivalis Nutt. ex Torrey & A. Gray - <u>Brook Cinquefoil</u>.
Annual to biennial with erect to ascending stems to 50cm;
leaves with 3-5 leaflets; flowers in terminal cymes; petals
yellow, 1.5-3mm, shorter than sepals. Intrd; natd by pool at
1 site in Salop since at least 1976; N America.

11. P. crantzii (Crantz) G. Beck ex Fritsch - <u>Alpine</u> **R**
<u>Cinquefoil</u>. Perennial with terminal leaf-rosettes and
ascending flowering stems to 20cm arising from side; basal
leaves palmate, with (3-)5 leaflets; flowers 1-few in cymes,
c.15-20mm across; petals yellow, longer than sepals. Native;
sparse basic grassland, rocky places and crevices on
mountains; very local in N Wa, N En and Sc.

11 x 12. P. crantzii x P. neumanniana (= ?<u>P.</u> x <u>beckii</u> Murr)
might be the identity of intermediate plants in MW Yorks, S
Aberdeen and Cheviot, the last far from either species.

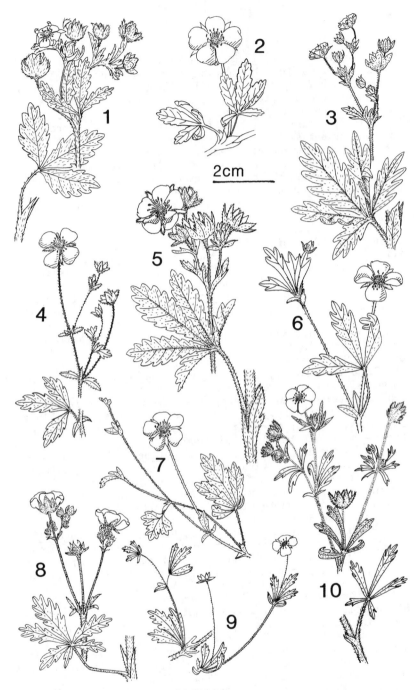

FIG 412 - Potentilla. 1, P. norvegica. 2, P. reptans. 3, P. intermedia. 4, P. x suberecta. 5, P. recta. 6, P. anglica. 7, P. x mixta. 8, P. inclinata. 9, P. erecta. 10, P. argentea.

Intermediacy is shown in stipule-shape, flower-size and growth-habit.

12. P. neumanniana Reichb. (<u>P.</u> <u>tabernaemontani</u> Asch., <u>P.</u> <u>verna</u> auct. non L.) - <u>Spring Cinquefoil</u>. Perennial with terminal leaf-rosettes, ± woody procumbent stolons, and ascending flowering stems to 10cm arising from side; basal leaves palmate, with 5-7 leaflets; flowers 1-few in cymes, c.10-15mm across, otherwise as in <u>P. crantzii</u>. Native; dry basic grassland and rocky slopes; very local in Br. Differs from dwarfed <u>P. reptans</u> in sterile state by basal leaflets arising direct from top of petiole, not from near base of sub-basal pair.   **R**

13. P. erecta (L.) Raeusch. - <u>Tormentil</u>. Perennial with **412** basal leaf-rosette (often withered by flowering) and erect to procumbent non-rooting flowering stems to 45cm; leaves ternate, sessile or nearly so, with 2 stipules resembling small leaflets; flowers all or nearly all with 4 petals, few to many in loose terminal cymes, 7-15mm across; carpels 4-20. Native; grassland and dwarf-shrubland on heaths, moors, bogs, mountains, roadsides and pastures, mostly on acid soils but sometimes on limestone.

a. Ssp. erecta. Stems to 25cm; stem-leaves serrate in distal 1/2 only, with teeth <1.5mm, the uppermost leaf c.6-16mm; petals 2.5-4.5mm; fruiting pedicels 6-30mm. Common throughout BI, mostly in lowlands.

b. Ssp. strictissima (Zimm.) A. Richards. More robust, with more coarsely dentate leaves; stems 15-45 cm; stem-leaves serrate for most of length, with teeth >1.5mm, the uppermost leaf 12-30mm; petals 4-6mm; fruiting pedicels (12)20-50mm. Upland areas of BI N from S Wa and S Ir.

13 x 14. P. erecta x P. anglica = P. x suberecta Zimm. **412** occurs frequently with the parents in BI. It resembles <u>P. erecta</u> in habit but the stems may rarely root at nodes late in the season and it is intermediate in leaflet-, petal- and carpel-number, petiole-length and flower-size. It is partially fertile, with <10 achenes per flower.

14. P. anglica Laich. (<u>P. procumbens</u> Sibth. nom. illeg.) - **412** <u>Trailing Tormentil</u>. Perennial, with persistent basal leaf-rosette and decumbent to procumbent stems to 80cm rooting at nodes late in season; lower stem-leaves with 3-5 leaflets, with petioles 1-2cm (upper leaves with markedly shorter petioles); flowers solitary in stem-leaf axils, some with 4 others with 5 petals, 12-18mm across; carpels 20-50. Native; wood-borders, heaths and dry banks; scattered throughout BI N to C Sc, but over-recorded for <u>P. x mixta</u>.

15. P. x mixta Nolte ex Reichb. (<u>P. x italica</u> Lehm.) - **412** <u>Hybrid Cinquefoil</u>. Differs from <u>P. reptans</u> in being less robust with some leaves with 3 or 4 leaflets and some flowers with 4 petals and sepals; from <u>P. anglica</u> in having leaves all with ± same length petioles; and from both by its high sterility. Native; frequent throughout most of BI (commoner than <u>P. anglica</u>), often in absence of parents. This taxon is derived from both P. erecta x P. reptans and P. anglica x P. reptans.

16. P. reptans L. - Creeping Cinquefoil. Perennial with 412 persistent basal leaf-rosette and procumbent rooting flowering stems to 1m; stem-leaves all palmate with 5 leaflets and petioles all >1cm and ± equal-lengthed; flowers solitary in stem-leaf axils, all with 5 petals and sepals, 15-25mm across; carpels 60-120. Native; rough ground, hedgebanks, sand-dunes and open grassland; common in BI N to S Sc, very local further N.

17. P. sterilis (L.) Garcke - Barren Strawberry. Perennial with procumbent stolons and terminal leaf-rosettes; leaves ternate, with broadly obovate leaflets; flowers 1-3 in cymes on decumbent axillary flowering stems to 15cm; petals white, c. as long as sepals. Native; wood-margins and clearings, scrub and hedgebanks; common throughout BI except extreme N. Distinguished from Fragaria vesca by the grey-green leaflets with terminal tooth shorter than adjacent 2, and petals with wide gaps between.

10. SIBBALDIA L. - Sibbaldia
Herbaceous perennials; differ from Potentilla in leaves ternate; flowers 5-merous, in compact heads; petals ≤2mm or 0; stamens (4)5(-10); carpels 5-12.

1. S. procumbens L. (Potentilla sibbaldi Haller f.) -   R Sibbaldia. Leaves in basal rosette with obovate-obtriangular leaflets mostly with 3 apical teeth; flowering stems 1-5cm, procumbent to ascending; flowers densely clustered, c.5mm across; petals 0 or 5, 1.5-2mm, pale yellow. Native; grassy and rocky slopes and rock-crevices above 470m; frequent in C & N Sc, Westmorland record probably an error.

11. FRAGARIA L. - Strawberries
Herbaceous perennials, usually stoloniferous; leaves ternate; flowers in cymes on ± leafless stems arising from axils of leaf-rosette, 5-merous; epicalyx present, with entire segments; hypanthium ± flat, with strongly convex receptacle; petals white or flushed pink; stamens and carpels numerous; fruit a head of achenes borne on outside of enlarged, red, succulent receptacle.

1 leaflets glabrous or nearly so on upperside; most fruiting heads >15mm wide, usually with sepals appressed; achenes sunk into surface of ripe receptacle   **3. F. x ananassa**
1 Leaflets pubescent to sparsely so on upperside; fruiting heads <15mm wide, with sepals not appressed; achenes prominent from surface of ripe receptacle   2
  2 Flowers bisexual; uppermost pedicel in each cyme with apically directed hairs at fruiting; leaves rather glossy on upperside when fresh   **1. F. vesca**

2 Flowers functionally dioecious; uppermost pedicel
  in each cyme with many patent hairs at fruiting;
  leaves dull on upperside when fresh   **2. F. muricata**

1. F. vesca L. - <u>Wild</u> <u>Strawberry</u>. Stolons 0 to abundant;
leaflets 1-6cm, elliptic-obovate, acute at base; flowering
stems about as long as rosette-leaves, to 30cm; flowers
10-20mm across; fruit c.1cm across, with achenes raised above
surface of ripe receptacle. Native; woods, scrub and hedgerows; common throughout BI. A robust variant without stolons
and with flower and fruit produced continuously until the
frosts is known as <u>Alpine</u> <u>Strawberry</u>, and may escape. See
<u>Potentilla</u> <u>sterilis</u> for differences.

2. F. muricata Miller (<u>F.</u> <u>vesca</u> race <u>moschata</u> Duchesne nom.
inval.) - <u>Hautbois</u> <u>Strawberry</u>. Differs from <u>F.</u> <u>vesca</u> in
larger size (stems to 40cm, longer than leaves); 0 or few
stolons; terminal leaflet usually subacute to obtuse at base;
flowers 15-30mm across; and see key. Intrd; formerly
cultivated, natd in scrub and hedgerows; scattered throughout
Br but now rare and over-recorded for large <u>F.</u> <u>vesca</u> or small
<u>F.</u> <u>x</u> <u>ananassa</u>; C Europe.

3. F. x ananassa (Weston) Lois., Vilm., Nois. & J. Deville -
<u>Garden</u> <u>Strawberry</u>. Differs from <u>F.</u> <u>vesca</u> in larger size of
parts; many stolons; terminal leaflet usually obtuse to
rounded at base; flowers 20-35mm across; and see key. Intrd;
much cultivated, frequently natd in waste places and
field-borders; scattered throughout BI; garden origin.

12. DUCHESNEA Smith - <u>Yellow-flowered</u> <u>Strawberry</u>
Stoloniferous perennials; differ from <u>Fragaria</u> in solitary
flowers; epicalyx-segments 3-toothed at apex, much larger
than sepals; petals yellow.

1. D. indica (Andrews) Focke - <u>Yellow-flowered</u> <u>Strawberry</u>.
Stolons to 50cm, slender, bearing ternate leaves and solitary
axillary flowers on erect pedicels 3-10cm; flowers c.10-18mm
across; fruit ± globose, 8-16mm across, dry, tasteless.
Intrd; cultivated as curiosity and occasionally escapes; very
scattered in Br; S & E Asia.

13. GEUM L. - <u>Avens</u>
Herbaceous perennials; leaves pinnate; flowers in terminal
cymes on stems arising from leaf-rosette, rarely solitary,
5-(rarely more)merous; epicalyx present; hypanthium ± flat to
saucer-shaped, with strongly convex rceptacle; petals yellow
to purple; stamens and carpels numerous; fruit a head of
achenes with long styles terminating in hook (after apical
segment has fallen off).

1 Petals creamy-pink to purplish, with long claw at
  base; ripe achenes carried up from flower centre on
  stalk >5mm   **1. G. rivale**

1 Petals yellow, scarcely clawed at base; ripe achenes
   ± sessile in flower centre   2
   2 Achenes <150, in globose head, with basal part of
      style glabrous   **2. G. urbanum**
   2 Achenes >150, in ovoid head, with basal part of
      style glandular   **3. G. macrophyllum**

Other spp. - Garden plants resembling G. rivale but with
bright red petals are cultivars of **G. chiloense** Balbis (G.
coccineum auct. non Sibth. & Smith), from Chile; they
occasionally persist on tips for a while.

1. G. rivale L. - Water Avens. Stems to 50cm, erect;
stipules mostly <1cm; flowers pendent, with erect, creamy-
pink to pinkish-purple petals 8-15mm; achenes in globose
long-stalked head above sepals with style with long hairs and
short glands near base. Native; marshes, stream-sides,
mountain rock-ledges and open woodland; throughout Br and Ir
but very local in S and absent from large areas, also garden
escape in non-native areas. Dwarf plants with large flowers
and more deeply and roughly incised leaves from N Sc have
been called ssp. **islandicum** A. & D. Love, but all intermediates occur.
1 x 2. G. rivale x G. urbanum = G. x intermedium Ehrh. is
common wherever the parents meet. It is intermediate in all
respects and highly fertile, forming a complete spectrum
between the parents.
2. G. urbanum L. - Wood Avens. Stems to 70cm, erect or
spreading; stipules 1-3cm; flowers erect, with patent, yellow
petals 4-7mm; achenes in sessile globose head, with glabrous
style. Native; woods and hedgerows; common throughout BI
except far N.
3. G. macrophyllum Willd. - Large-leaved Avens. Stems to
1m, erect; stipules 6-15mm; flowers erect, with patent,
yellow petals 3.5-7mm; achenes in sessile oblong head, with
style with short glands near base. Intrd; natd in woods and
by paths; several parts of Sc; N America and NE Asia.

14. DRYAS L. - Mountain Avens
Perennials, woody at base, herbaceous above; leaves ± evergreen, simple, bluntly serrate; flowers solitary, axillary,
7-10-merous; epicalyx 0; hypanthium saucer-shaped, with
convex receptacle; petals white, mostly 8; stamens and
carpels numerous; fruit a head of achenes with long feathery
styles.

1. D. octopetala L. - Mountain Avens. Stems to 50cm,   R
procumbent, rooting; leaves 6-25mm, dark glossy green on
upperside, white-tomentose on lowerside; petals 7-17mm;
styles in fruit 2-3cm. Native; base-rich rock-crevices and
-ledges on mountains; very local in NW Wa, N Ir and N En,
local in MW Ir and N & W Sc.

15. AGRIMONIA L. - Agrimonies

Herbaceous perennials; leaves pinnate; flowers in long terminal racemes, 5-merous; epicalyx 0; hypanthium deeply concave, narrow at mouth, surrounding carpels; petals yellow; stamens 5-20; carpels 2; fruit of 1-2 achenes enclosed in woody hypanthium which has ring of hooked bristles distally. Vegetative differences between the 2 spp. are often exaggerated; A. eupatoria is fragrant when crushed and is often glandular on leaf lowersides, though glands may be concealed by denser pubescence.

1. A. eupatoria L. - Agrimony. Stems ± erect, to 1m, **418** diffusely branched, pubescent with long and short non-glandular hairs; main leaflets 3-6 pairs, rather bluntly serrate, with 0 to rather few sessile shining glands on lowerside; fruiting hypanthium obconical, deeply grooved ± to apex, with outermost bristles patent to erecto-patent; 2n=28. Native; grassy places in fields and hedgerows; throughout BI except most of N Sc.

1 x 2. A. eupatoria x A. procera = A. x wirtgenii Asch. & Graebner is intermediate in indumentum (glands and hairs) and leaflet toothing; it does not form fruit; 2n=42. It was found in the late 1940s in S Northumb, but other records have not been confirmed.

2. A. procera Wallr. (A. odorata auct. non (L.) Miller, A. **418** repens auct. non L.) - Fragrant Agrimony. Differs from A. eupatoria in more leafy stems with ± all hairs >2mm; more deeply and acutely serrate, less pubescent leaflets with abundant sessile glands on lowerside; fruiting hypanthium campanulate, with grooves extending <3/4 way to apex and outermost bristles reflexed; 2n=56. Native; same habitats and distribution as A. eupatoria but much more scattered.

16. AREMONIA Necker ex Nestler - Bastard Agrimony

Herbaceous perennials; leaves pinnate; flowers in small terminal cymes, 5-merous; epicalyx of 5 lobes, surrounded by 8-12 fused bracts; hypanthium deeply concave, surrounding carpels; petals yellow; stamens 5-10; carpels 2; fruit of 1-2 achenes enclosed in woody pubescent hypanthium without hooked bristles but concealed by bracts.

1. A. agrimonioides (L.) DC. (Agrimonia agrimonioides L.) - Bastard Agrimony. Stems to 30cm, decumbent, scarcely or not longer than basal leaves; leaves 2-3 pairs of main leaflets, the apical 3 much the largest; fruiting hypanthium 5-6mm, ± globose. Intrd; natd in woods and shady roadsides; C Sc; S & C Europe.

17. SANGUISORBA L. (Poterium L.) - Burnets

Herbaceous perennials; leaves pinnate; flowers in very dense terminal spikes, all bisexual or bisexual and unisexual mixed; sepals 4; epicalyx 0; hypanthium deeply concave,

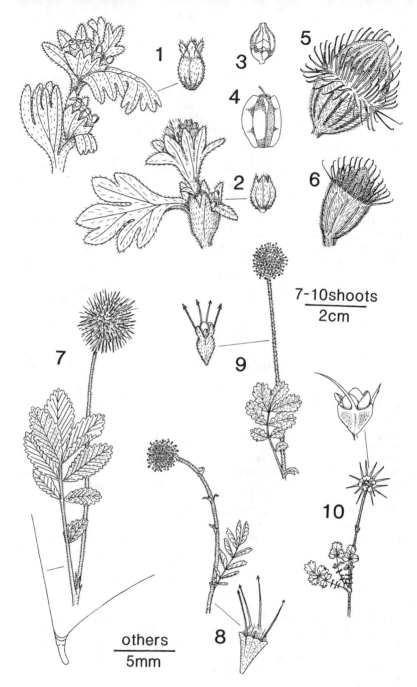

FIG 418 – Rosaceae. 1–2, shoots and fruiting hypanthia of **Aphanes**. 1, **A. arvensis**. 2, **A. inexspectata**. 3–4, fruiting hypanthia of **Sanguisorba minor**. 3, ssp. **minor**. 4, ssp. **muricata**. 5–6, fruiting hypanthia of **Agrimonia**. 5, **A. procera**. 6, **A. eupatoria**. 7–10, shoots and fruiting hypanthia of **Acaena**. 7, **A. ovalifolia**. 8, **A. novae–zelandiae**. 9, **A. anserinifolia**. 10, **A. inermis**.

surrounding carpels; petals 0; stamens 4 or numerous; carpels 1-2; stigmas papillate or tasselled; fruit 1-2 achenes enclosed in hard hypanthium.

1 Upper flowers female, others bisexual or male; stamens
  numerous; stigmas 2, tasselled   **3. S. minor**
1 All flowers bisexual; stamens 4; stigma 1, papillate 2
2 Flower-heads 1-3cm; sepals and stamens dull
  purplish; stamens c.4mm   **1. S. officinalis**
2 Flower-heads 3-16cm; sepals green and white;
  stamens white, c.10mm   **2. S. canadensis**

1. S. officinalis L. (Poterium officinale (L.) A. Gray) - Great Burnet. Stems to 1.2(1.7)m, erect; basal leaves with 3-7 pairs of leaflets; flower-heads subglobose to oblong-ellipsoid, purplish. Native; damp unimproved grassland; locally frequent in Br N to C Sc, N Ir.

2. S. canadensis L. - White Burnet. Stems to 2m, erect; basal leaves with 5-10 pairs of leaflets; flower-heads cylindrical, white. Intrd; grown in gardens and sometimes natd; frequent in C Sc, very scattered elsewhere in Sc and En; N America.

3. S. minor Scop. (Poterium sanguisorba L.) - see sspp. for English names. Stems erect, to 50(80)cm; leaves with 3-12 pairs of leaflets; flower-heads 6-15mm, globose or slightly elongated, sepals green and purple; stamens yellow, c.6mm.

a. Ssp. minor - Salad Burnet. Fruiting hypanthium 3-4.5mm, **418** with thickened but scarcely winged ridges on angles, the faces distinctly reticulate with finer ridges. Native; calcareous or sometimes neutral grassland and rocky places; locally common throughout BI N to C Sc.

b. Ssp. muricata (Gremli) Briq. (Poterium sanguisorba ssp. **418** muricatum (Gremli) Rouy & Camus, P. polygamum Waldst. & Kit.) - Fodder Burnet. Usually more robust and leafy with more deeply and sharply toothed leaflets; fruiting hypanthium 3-5mm, with often undulate wings on angles, the faces smooth to irregularly rugose. Intrd; formerly grown for fodder and still natd in grassy places; very scattered in BI, frequent only in C & S En; S Europe.

18. ACAENA Mutis ex L. - Pirri-pirri-burs
Perennials, with stems woody at base but herbaceous distally; leaves pinnate; flowers in globose heads, bisexual; sepals 4; epicalyx 0; hypanthium deeply concave, surrounding carpels; petals 0; stamens 2; carpels 1-2; stigmas long-papillate; fruit 1-2 achenes enclosed in dry hypanthium which usually develops barbed spines at apex.

1 Apical pair of leaflets c. as long as wide; spines on
  hypanthium 0, or imperfect, or not barbed **4. A. inermis**
1 Apical pair of leaflets 1.2-2.5x as long as wide;
  spines on hypanthium 2-4, barbed at apex   2

2 Apical pair of leaflets each with (11)17-23 teeth;
   hypanthium with 2 spines   **3. A. ovalifolia**
2 Apical pair of leaflets each with 5-12(15) teeth;
   hypanthium with up to 4 spines   3
3 Leaflets glossy green on upperside; apical pair of
   leaflets 1.8-2.5x as long as wide **1. A. novae-zelandiae**
3 Leaflets matt green on upperside, often edged and
   veined with brown; apical pair of leaflets 1.2-2x
   as long as wide   **2. A. anserinifolia**

Other spp. - Of several spp. reported as casuals or escapes,
A. caesiiglauca (Bitter) Bergmans and **A. magellanica** (Lam.)
Vahl might become established. Both resemble A. inermis in
their glaucous leaves with broad apical leaflets, but have
2-4 well-developed barbed spines. A. caesiiglauca, from New
Zealand, has only 7-9 leaflets; A. magellanica, from extreme
S America, has mostly >11 leaflets.

1. A. novae-zelandiae Kirk (A. anserinifolia auct. non 418
(Forster & G. Forster) Druce) - Pirri-pirri-bur. Woody stems
procumbent, mat-forming; herbaceous stems ascending, to 15cm;
leaflets 9-13((15), the apical pair 5-20mm; carpels and
stigma 1; spines mostly 6-10mm, often 1 or 2 much shorter.
Intrd; grown in gardens and intrd with shoddy, now well natd
on barish ground; S & E Br N to SE Sc, very scattered in Ir;
New Zealand and Australia.
2. A. anserinifolia (Forster & G. Forster) Druce (A. pusilla 418
(Bitter) Allan) - Bronze Pirri-pirri-bur. Differs from A.
novae-zelandiae in key characters; the bronzy-tinged foliage
is distinctive; apical pair of leaflets 3-10mm; spines mostly
3.5-6mm, often 1 or 2 much shorter.   Intrd; probably
originally a wool-alien, now natd on barish ground; very few
places from S En to C Sc, N Ir; New Zealand. Dwarf plants
with small parts and relatively short apical leaflets occur
in W Galway and W Mayo; they are referable to A. pusilla,
which is probably a montane ecotype of A. anserinifolia.
2 x 4. A. anserinifolia x A. inermis was found in 1967 at
Batworthy Brook, Dartmoor, S Devon, and is grown in gardens;
most flowers have 2 stigmas but develop spines.
3. A. ovalifolia Ruiz Lopez & Pavon - Two-spined Acaena. 418
Differs from A. novae-zelandiae in key characters; leaflets
(7-)9(11), the apical pair 10-30mm, 1.7-2x as long as wide.
Intrd; natd on barish ground; very few places from S En to N
Sc, SE Ir; S America.
4. A. inermis Hook. f. (A. microphylla auct. non Hook. f.) - 418
Spineless Acaena.   Herbaceous stems to 6cm; leaflets
(7)11-13, bluish grey-green tinged brown or orange, the
apical pair 2-8mm, c. as long as wide, with 5-10 teeth;
carpels and stigmas 2; spines usually 0, sometimes 1-4, not
or scarcely barbed.   Intrd; grown in gardens and natd on
barish ground; very scattered in Sc; New Zealand.

19. ALCHEMILLA L. - Lady's-mantles

Herbaceous perennials; leaves palmate or simple and palmately lobed; flowers in terminal compound cymes, bisexual; sepals 4; epicalyx present; hypanthium deeply concave, surrounding carpel; petals 0; stamens 4; carpel 1; fruit an achene enclosed in dry hypanthium.

All Br spp. are obligate apomicts and many differ only by the small characters often diagnosing agamospecies. The most important characters are degree and distribution of pubescence; shape of leaves, especially number, shape and toothing of leaf-lobes; and shape of sinuses between leaf-lobes and at base of lamina. The terms small, medium and large are relative and often help to distinguish taxa in 1 locality; in general 'small' means stems usually <20cm, leaves usually <3cm; 'medium' means stems usually <50cm, leaves usually <5cm; 'large' means stems up to 60(80)cm, leaves up to 7(10)cm. The terms leaves and petioles refer to those of the basal rosette, excluding the first-formed ones.

1 Leaves palmately divided >1/2 way to base,
  densely silver-sericeous on lowerside   2
1 Leaves palmately divided <1/2 way to base,
  variously pubescent but never silver-sericeous   3
  2 Leaves divided ± to base, the leaflets mostly
    <6mm wide   **1. A. alpina**
  2 Leaves divided 3/5 to 4/5 to base, the lobes
    mostly >6mm wide   **2. A. conjuncta**
3 Petioles and stem glabrous, subglabrous or with
  appressed or subappressed hairs   4
3 Petioles and lower part of stem with erecto-patent,
  patent or reflexed hairs   7
  4 Stems up to and including 1st inflorescence-
    branches pubescent; leaves pubescent on upperside   5
  4 Stems pubescent only on lowest 2 or 3 internodes;
    leaves glabrous on upperside   6
5 Hypanthium tapered to cuneate base   **8. A. gracilis**
5 Hypanthium rounded at base   **12. A. glomerulans**
  6 Sinuses between leaf-lobes toothed to base; teeth
    in middle of each side of leaf-lobes larger than
    those above or below   **14. A. glabra**
  6 Sinuses between leaf-lobes with toothless region
    at base c.2x as long as a tooth; teeth on leaf-
    lobes subequal   **13. A. wichurae**
7 Epicalyx-segments c. as long as sepals; hypanthium
  much shorter than mature achene   **15. A. mollis**
7 Epicalyx-segments distinctly shorter than sepals;
  hypanthium as long as mature achene   8
  8 Pedicels and hypanthia both pubescent, sometimes
    sparsely   9
  8 Pedicels glabrous; hypanthia usually glabrous   11
9 Leaf-lobes usually 5, with sinuses between them with
  toothless region at base ≥2x as long as a tooth;

stems rarely >5cm   **11. A. minima**
9 Leaf-lobes (5)7-9, with sinuses between them toothed
   to base; stems usually >5cm   10
   10 Base of stems and petioles tinged wine-red; leaf
      basal sinus open (>45 degrees)
                     10b. A. filicaulis ssp. vestita
   10 Base of stems and petioles brownish; leaf basal
      sinus very narrow (<30 degrees) to closed
                       3. A. glaucescens
11 Leaves glabrous on upperside (or with very sparse
   hairs in folds)   12
11 Leaves pubescent on upperside, sometimes only in
   folds   13
   12 Leaf-lobes rounded, with subequal teeth
                     9. A. xanthochlora
   12 Leaf-lobes ± straight-sided then rounded to
      subtruncate at apex, with teeth in middle of each
      side larger than those above or below
                     7. A. acutiloba
13 Leaves ± densely pubescent on both surfaces   14
13 Leaves rather sparsely pubescent on upperside, with
   hairs usually ± confined to folds (or frequent only
   there)   16
   14 Hypanthium tapered to cuneate base; stem-hairs
      mostly erecto-patent   **8. A. gracilis**
   14 Hypanthium rounded at base; stem-hairs patent to
      reflexed   15
15 Flowers mostly >2.5mm across; petioles and stems with
   patent hairs   **4. A. monticola**
15 Flowers mostly <2.5mm across; petioles and stems with
   many reflexed hairs   **5. A. tytthantha**
   16 Leaf-lobes ± straight-sided then rounded to
      subtruncate at apex, with teeth in middle of each
      side larger than those above or below
                     7. A. acutiloba
   16 Leaf-lobes ± rounded, with teeth ± equal to
      unequal but not with large ones in middle of each
      side   17
17 Leaf-lobes usually 5, with sinuses between them with
   toothless region at base >2x as long as a tooth;
   stems rarely >5cm   **11. A. minima**
17 Leaf-lobes (5)7-9, with sinuses between them toothed
   to base; stems usually >5cm   18
   18 Hypanthium often with some hairs; stems and
      petioles with patent hairs, tinged wine-red at
      base   **10a. A. filicaulis ssp. filicaulis**
   18 Hypanthium glabrous; stems and petioles usually
      with some reflexed hairs, brownish at base
                     6. A. subcrenata

1. A. alpina L. - <u>Alpine</u> Lady's-mantle.   Stems to 20cm, **424**
ascending; leaves palmate, with 5-7 narrowly oblong-elliptic

leaflets, densely silver-sericeous on lowerside. Native;
grassland, scree and rock-crevices on mountains; Lake
District, C & N Sc, Kerry and Wicklow, intrd in Derbys.

2. A. conjuncta Bab. - <u>Silver Lady's-mantle</u>. Stems to 30cm, 424
ascending; leaves simple, palmately lobed 3/5 to 4/5 to base,
with 7-9 elliptic lobes, densely silver-sericeous on lowerside. Intrd; much grown in gardens and sometimes escaping;
scattered in N & C Br, with some very old records in N Wa and
Sc once considered native but probably planted; Alps.

3-15. A. vulgaris L. agg. - <u>Lady's-mantle</u>. Stems to
60(80)cm, often much less, decumbent to ascending; leaves up
to 7(10)cm, often much less, simple, palmately lobed <1/2 way
to base, glabrous to densely pubescent. Damp rich grassland,
woodland margins and rides, rock-ledges; throughout most of
Br and Ir but rare or absent in most of SE En.

3. A. glaucescens Wallr. (<u>A. minor</u> auct. non Huds.). Plant 424
small; whole plant with + dense patent to erecto-patent   RR
hairs. Native; very local in N En, NW Sc, NW Ir, rare escape
elsewhere in En.

4. A. monticola Opiz. Plant medium; petioles, leaves, stems 424
and lowest inflorescence-branches with + dense patent hairs; RR
ultimate inflorescence-branches and pedicels glabrous;
hypanthium glabrous to sparsely pubescent. Native; very
local in NW Yorks and Durham, rare escape elsewhere in En.

5. A. tytthantha Juz. Plant medium; stems and petioles with 424
patent and reflexed hairs; leaves with dense patent hairs;
upper part of inflorescence, pedicels and hypanthia glabrous.
Intrd; natd in C & S Sc, probably from botanic gardens;
Crimea.

6. A. subcrenata Buser. Plant medium; petioles and lower 424
parts of stems with patent and usually reflexed hairs; upper RR
part of stem and whole inflorescence glabrous; leaves
pubescent, rather sparsely so on upperside. Native; very
local in NW Yorks and Co Durham, 1st found 1951.

7. A. acutiloba Opiz. Plant large; petioles and lower parts 424
of stems with dense patent hairs; upper part of stem and   RR
whole inflorescence glabrous; leaves densely pubescent on
lowerside, glabrous to sparsely pubescent on upperside.
Native; quite widespread in Co Durham only; discovered 1946.

8. A. gracilis Opiz. Plant medium; leaves, petioles, stems 424
and lower inflorescence-branches with erecto-patent to   RR
subappressed hairs; ultimate inflorescence-branches, pedicels
and hypanthia glabrous. Native; very local in S Northumb,
discovered 1976, also ?Co Durham.

9. A. xanthochlora Rothm. Plant medium; stems, petioles and 424
leaf lowerside with dense patent or erecto-patent hairs;
inflorescence glabrous or nearly so; leaf upperside glabrous
or with very sparse hairs in folds. Native; + throughout
range of agg., common in N, very local in SE En and S Ir.

10. A. filicaulis Buser. Plant small to medium; lower part 424
of stem, petioles and leaves with + dense patent hairs;
hypanthia glabrous or pubescent. Intermediates between the 2

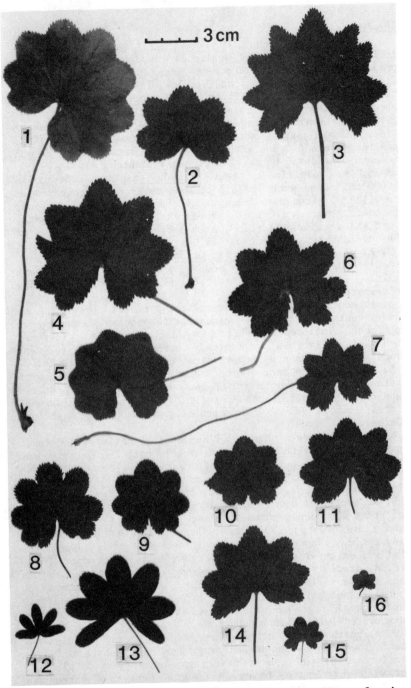

FIG 424 — Leaves of **Alchemilla**. 1, **A. mollis**. 2, **A. xanthochlora**. 3, **A. acutiloba**. 4, **A. glomerulans**. 5, **A. tytthantha**. 6, **A. subcrenata**. 7, **A. gracilis**. 8, **A. wichurae**. 9, **A. monticola**. 10, **A. glaucescens**. 11, **A. glabra**. 12, **A. alpina**. 13, **A. conjuncta**. 14–15, **A. filicaulis**. 16, **A. minima**.

sspp. occur in Sc.

a. Ssp. filicaulis. Upper part of stem and whole inflorescence except hypanthia glabrous; leaf upperside rather sparsely pubescent. Native; rather scattered from Wa and C En to N Sc, NW Ir, probably under-recorded.

b. Ssp. vestita (Buser) Bradshaw (<u>A. vestita</u> (Buser) Raunk.). Upper part of stem, whole inflorescence and leaf upperside rather densely pubescent. Native; distribution of agg., the commonest taxon in genus.

11. A. minima Walters. Plant small; whole plant except **424** sometimes pedicels pubescent, but more sparsely so than in <u>A.</u> **RR** <u>filicaulis</u> ssp. <u>vestita</u>. Native; very local on 2 hills in MW Yorks, described 1949; endemic.

12. A. glomerulans Buser. Plant medium; stems, petioles and **424** leaves with subappressed hairs; whole inflorescence glabrous **R** to subglabrous. Native; local in N En and Sc.

13. A. wichurae (Buser) Stefansson. Plant rather small; **424** lowest 2(-3) stem-internodes and petioles appressed- **R** pubescent; leaves pubescent on lowerside veins, glabrous otherwise; middle and upper parts of stem and whole inflorescence glabrous. Native; local in N En and Sc, intrd in NW Wa.

14. A. glabra Neyg. (<u>A. obtusa</u> auct. non Buser). Plant **424** large; pubescence distributed as in <u>A. wichurae</u> but sparser, sometimes only on lowest 1-2 stem-internodes. Native; almost throughout range of agg., but almost absent from En S of Peak District.

15. A. mollis (Buser) Rothm. Plant large to very large; **424** whole plant with dense patent hairs except rather sparse on hypanthia and 0 on pedicels. Intrd; much grown in gardens and prolifically seeding; ± natd in very scattered places nearly throughout Br; Carpathians.

20. APHANES L. - <u>Parsley-pierts</u>
Annuals; leaves deeply palmately lobed; flowers in small dense, leaf-opposed clusters, bisexual; sepals 4; epicalyx present; hypanthium deeply concave, surrounding carpel; petals 0; stamen 1(-2); carpel 1; fruit an achene enclosed in dry hypanthium.

1. A. arvensis L. (<u>Alchemilla arvensis</u> (L.) Scop.) - **418** <u>Parsley-piert</u>. Stems decumbent to nearly erect, to 10(20)cm; leaves up to 1cm; stipules at fruiting nodes fused into a leaf-like cup with ovate-triangular teeth at apex c.1/2 as long as entire portion; fruiting hypanthium 2-2.6mm incl. erect sepals c.0.6-0.8mm, with a slight constriction where hypanthium and sepals meet, reaching ± to apex of stipules at maturity. Native; cultivated and other bare ground on well-drained soils; frequent ± throughout BI.

2. A. inexspectata Lippert (<u>A. microcarpa</u> auct. non (Boiss. **418** & Reuter) Rothm., <u>Alchemilla microcarpa</u> auct. non Boiss. & Reuter) - <u>Slender Parsley-piert</u>. Differs from <u>A. arvensis</u> in

stipule-teeth at fruiting nodes ovate-oblong, c. as long as
entire portion; fruiting hypanthium 1.4-1.9mm incl.
convergent sepals c.0.3-0.5mm, the sepals continuing curved
outline of hypanthium, falling well short of apex of stipules
at maturity. Native; similar places to A. arvensis but
rarely on base-rich soils and less often in arable ground;
frequent ± throughout BI.

21. ROSA L. - Roses
Deciduous or rarely evergreen shrubs with spiny stems and
petioles; leaves pinnate; flowers solitary or in few-(many-)
flowered corymbs, usually 5-merous; epicalyx 0; hypanthium
deeply concave, surrounding carpels, narrowed at apex;
stamens and carpels numerous; fruit a head of achenes
enclosed by succulent hypanthium.
An extremely complex genus, much hybridized and selected in
cultivation. Half of our spp. (sect. Caninae DC., nos. 11-
20) contribute unbalanced gametes, the male one having 7
chromosomes and the female ones 21, 28 or 35 (from plants
with 2n=28, 35, 42 respectively). Hybrids are very common
and reciprocal crosses can produce very different offspring.
The ripe fruits and leaves provide the most important
characters, but a collection from the same plant at flowering
is often also desirable. 'Disc' refers to the thickened rim
at top of the hypanthium at fruiting, in centre of which is
the 'orifice' through which the styles project. Only the
spp. are keyed here, but even for these access to accurately
named material is often a prerequisite for successful
determination; in some areas hybrids are commoner than either
parent, but can often be determined if the parents present in
the area are known. The classification of sect. Caninae
adopted here reflects the views of G.G. Graham and A.L.
Primavesi.
The application of many binomials to hybrid combinations is
uncertain; further literature research and typification will
undoubtedly uncover names earlier than several of those used
here.

1 Styles exserted and fused into a column, sometimes
   becoming free at fruiting   2
1 Styles exserted or not, free (may appear fused in
   dried material)   6
   2 Leaflets 3(-5)   **2. R. setigera**
   2 Leaflets 5-9   3
3 Styles pubescent; semi-evergreen; stems ±
   procumbent   **3. R. luciae**
3 Styles glabrous; deciduous; stems trailing to
   strongly arching   4
   4 Flowers 2-3cm across, in groups of >(6)10;
      stipules lobed >1/2 way to petiole   **1. R. multiflora**
   4 Flowers mostly 3-5cm across, in groups of 1-6(10);
      stipules not lobed or lobed <1/2 way to petiole   5

5 Styles as long as stamens; top of hypanthium flat;
inner sepals entire, outer with very few lobes
4. R. arvensis
5 Styles shorter than stamens; top of hypanthium
conical; sepals pinnately lobed **11. R. stylosa**
6 Sepals entire, ± tapering to apex, erect or
suberect and persistent until after fruit ripe 7
6 At least some sepals lobed or with strongly
expanded tips, if ± entire then patent to
reflexed and/or falling before fruit ripe 9
7 Fruit blackish when ripe; flowers all solitary,
without bract **5. R. pimpinellifolia**
7 Fruit red when ripe; flowers 1-several, with 1 or
more much reduced leaves (bracts) at base of pedicels 8
8 Stems and prickles tomentose; flowers 6-9cm
across; pedicels tomentose
(see text) **6. R. rugosa** and **7. R. 'Hollandica'**
8 Stems and prickles ± glabrous; flowers 3-5cm
across; pedicels glandular-pubescent **17. R. mollis**
9 Leaflets 3-5; flowers mostly solitary, without bract,
5-7(9)cm across **10. R. gallica**
9 Leaflets (3)5-9; flowers 1-several, with 1 or more
much reduced leaves (bracts) at base of pedicels,
(2)3-5(6)cm across 10
10 Sepals entire or some with few very narrow
lateral lobes, some or all falling before fruit
ripe 11
10 Outer sepals on ± all flowers with lateral
lobes; sepals falling early or persistent 12
11 Leaves strongly red-tinged; petals usually shorter
than sepals; pedicels, fruits and sepals glabrous to
very sparsely glandular-pubescent **8. R. glauca**
11 Leaves green; petals longer than sepals; pedicels,
fruits and sepals densely glandular-pubescent
9. R. virginiana
12 Leaflets glabrous, sometimes with few stalked
glands on midrib but without eglandular hairs 13
12 Leaflets tomentose to pubescent with eglandular
hairs on lowerside, at least on midrib 14
13 Orifice of disc c.1/5 its total width; styles
glabrous, pubescent or woolly, forming ± loose
group; sepals mostly patent to reflexed after
flowering, falling before fruit ripe **12. R. canina**
13 Orifice of disc c.1/3 its total width; styles
woolly, forming dense mass ± obscuring disc; sepals
mostly erect to erecto-patent after flowering,
usually persistent until fruit ripe
13b. R. caesia ssp. glauca
14 Leaflets with prominent, ± sticky, sessile and
short-stalked glands on lowerside, giving fresh
leaf fruity smell when rubbed, pubescent to
sparsely so with eglandular hairs on lowerside 15

14 Leaflets with 0 or ± inconspicuous glands on
   lowerside, the glands ± confined to veins or if
   over whole surface then with no or resinous smell
   and leaves usually ± tomentose on lowerside   17
15 Pedicels glabrous; leaflets cuneate at base
                                             20. R. agrestis
15 Pedicels glandular-pubescent; leaflets rounded at
   base   16
   16 Stems erect; prickles unequal; styles pubescent;
      sepals mostly erect to patent, persistent until
      fruit reddens   **18. R. rubiginosa**
   16 Stems arching; prickles ± equal; styles glabrous
      or nearly so; sepals mostly reflexed, falling
      before fruit reddens   **19. R. micrantha**
17 Leaflets without glands or with few on midrib on
   lowerside, 1-2-serrate with teeth not or variably
   gland-tipped   18
17 Leaflets glandular on lowerside, at least on midrib
   and lateral veins, 2-serrate with gland-tipped teeth 20
   18 Orifice of disc c.1/3 its total width; styles
      woolly, forming dense mass ± obscuring disc;
      sepals mostly erect to erecto-patent after
      flowering, usually persistent until fruit ripe
                                      13a. R. caesia ssp. caesia
   18 Orifice of disc c.1/5 its total width; styles
      glabrous, pubescent or woolly, forming ± loose
      group; sepals mostly patent to reflexed after
      flowering, falling before fruit ripe   19
19 Lobes on outer sepals narrow, usually entire;
   prickles moderately hooked, longer than width of
   base; leaves usually eglandular on lowerside
                                             12. R. canina
19 Lobes on outer sepals broad, usually lobed or
   toothed; prickles strongly hooked, c. as long as
   width of base; leaves usually glandular on lowerside of midrib   **14. R. obtusifolia**
   20 Orifice of disc c.1/5 its total width; styles
      glabrous to pubescent   21
   20 Orifice of disc c.1/3-1/2 its total width; styles
      woolly   22
21 Prickles strongly hooked, c. as long as width of
   base; pedicels 5-15mm; pedicels, fruits and sepals
   glabrous or sparsely glandular-pubescent
                                       14. R. obtusifolia
21 Prickles ± straight to arched, longer than width
   of base; pedicels (10)15-25mm; pedicels, fruits and
   sepals glandular-pubescent to densely so
                                        15. R. tomentosa
   22 Sepals erect or suberect, entire or with few
      lateral lobes, persistent until fruit decays;
      orifice of disc c.2/5-1/2 its total width;
      prickles ± all straight   **17. R. mollis**

22 Sepals erect to erecto-patent, with lateral lobes
falling from ripe fruit; orifice of disc c.1/3
its total width; at least some prickles
curved
16. R. sherardii

Other spp. – Innumerable spp., cultivars and (often complex)
hybrids are grown in gardens and may be very persistent in
hedges, waste ground, parks and estates. The cultivars and
hybrids are best named as cultivars without reference to
their parentage, e.g. Rosa 'Queen Elizabeth'; in many cases
the parentage is in fact unknown. These cultivars are not
included in this work; they include nearly all flore pleno
plants (but spp. 1, 3, 6 and 9 also occur in this form). **R.
sempervirens** L. (sect. Synstylae) (Evergreen Rose) was
formerly natd in Worcs; it is evergreen with glabrous styles,
red fruits, leaves with 5-7 leaflets and unlobed stipules,
and flowers 2.5-6cm across. **R. majalis** Herrm. (R. cinnamomea
L. 1759 non 1753) and **R. pendulina** L. (R. cinnamomea L. 1753
non 1759) (sect. Cinnamomeae) are sometimes found in
semi-wild places; the former has depressed-globose fruits
with glabrous pedicels; the latter ovoid to obovoid fruits
with glandular-pubescent pedicels recurved in fruit. **R.
sericea** Lindley (R. omeiensis Rolfe) (sect. Pimpinellifoliae) and **R. wilmottiae** Hemsley (sect. Gymnocarpae Thory)
have recently been recorded from hedges in MW Yorks; the
former has very broad-based spines, 9-13 leaflets and
solitary white flowers with 4 petals; the latter has
straight, paired prickles, 7-9 glabrous leaflets <15mm, and
solitary pinkish-purple flowers. **R. nitida** Willd. (sect.
Carolinae) differs from R. virginiana in its narrower
leaflets and abundant hairs and stalked glands; it has been
doubtfully recorded in similar situations.

Section 1 – SYNSTYLAE DC. (spp. 1-4). Leaflets 3-9; flowers
1-many, with bracts; sepals entire or nearly so, falling
before fruit ripe; styles fused into column ± as long as
stamens.

1. R. multiflora Thunb. ex Murray – Many-flowered Rose. **433**
Scrambler, to 5m; leaves with 7-9 leaflets, pubescent to
sparsely so but eglandular on lowerside, 1-serrate with
glandular teeth; flowers usually >10, usually white, 2-3cm
across; fruits ovoid to globose, <1cm. Intrd; grown in
gardens for ornament and as stock for rambler-roses; natd in
hedges and copses; scattered in CI and Br; E Asia.
1 x 18. R. multiflora x R. rubiginosa occurs on the Common,
Herm, CI; ?endemic (1 x 19 only).
2. R. setigera Michaux – Prairie Rose. Scrambler, to 5m; **433**
leaves with 3(-5) leaflets, glabrous to ± pubescent but
eglandular or ± so on lowerside, 1-serrate with eglandular
teeth; flowers few-several, white to pink, 4-6cm across;
fruits ± globose, c.1cm. Intrd; well natd in scrub; Jersey

and Guernsey; E & C N America.

3. R. luciae Franchet & Rochebr. (R. wichuraiana Crepin) - **433**
Memorial Rose. Semi-evergreen, ± procumbent, to 4m; leaves
with 5-9 leaflets, ± glabrous, 1-serrate, eglandular; flowers
few-many, white, 3-5cm across; fruits ovoid to globose,
c.1cm. Intrd; grown as several cultivars (and hybrids) and
natd in open places, low scrub and beaches mostly near sea;
Dunbarton, E Kent, Guernsey and Jersey; E Asia.

4. R. arvensis Hudson - Field-rose. Weakly trailing, to **433**
1(2)m; leaves with 5-7 leaflets, glabrous to very sparsely
pubescent on lowerside, usually 1-serrate, usually glandless;
flowers 1-few, white, 3-5cm across; fruit globose to
ellipsoid, 1-1.5m. Native; low scrub, hedgerows, woods and
open places; frequent in C & S Br and Ir, rare in N Ir and N
En and ± absent from Sc.

The following hybrids occur; those with R. arvensis as
female parent usually resemble that sp. in habit but have
pink petals and other characters indicating the male parent;
those with a sp. of Caninae as female parent usually resemble
the latter in habit but reveal various features of R.
arvensis on inspection.

4 x 11. R. arvensis x R. stylosa = R. x pseudorusticana
Crepin ex Rogers (R. x bibracteoides Wolley-Dod) (4 x 11 and
11 x 4).

4 x 12. R. arvensis x R. canina = R. x verticillacantha
Merat (R. x kosinsciana Besser, R. x wheldonii Wolley-Dod, R.
x deseglisei Boreau) (4 x 12 and 12 x 4).

4 x 13. R. arvensis x R. caesia ssp. **glauca** (13 x 4 only).

4 x 14. R. arvensis x R. obtusifolia = R. x rouyana Duffort
ex Rouy (14 x 4 only).

4 x 16. R. arvensis x R. sherardii (4 x 16 and 16 x 4).

4 x 18. R. arvensis x R. rubiginosa = R. x consanguinea
Gren. (4 x 18 and 18 x 4).

4 x 19. R. arvensis x R. micrantha = R. x inelegans Wolley-
Dod (4 x 19 and 19 x 4).

Section 2 - PIMPINELLIFOLIAE DC. (sp. 5). Leaflets 5-11(13);
flower 1, without bracts; sepals entire, persistent until
fruit decay; styles free.

5. R. pimpinellifolia L. - Burnet Rose. Strongly suckering; **433**
stems erect, to 50(100)cm, with numerous slender spines and
bristles; leaves glabrous or sparsely pubescent on lowerside,
1-2-serrate, with 0 or sparse glands; flowers white, 2-4cm
across; fruit ± globose, blackish-purple, rather dry, with
erect sepals, 10-15mm. Native; dry sandy places near sea, on
inland heaths and limestone; round most coasts of BI, very
local inland.

The following hybrids occur; they can be recognized in the
same way as R. arvensis hybrids with spp. of Caninae.

5 x 12. R. pimpinellifolia x R. canina = R. x hibernica
Templeton (5 x 12 and 12 x 5).

5 x 13. R. pimpinellifolia x R. caesia = R. x margerisonii
(Wolley-Dod) Wolley-Dod (R. x setonensis Wolley-Dod) (5 x 13
and 13 x 5; both sspp. of R. caesia).
5 x 15. R. pimpinellifolia x R. tomentosa = R. x coronata
Crepin ex Reuter (5 x 15 and 15 x 5).
5 x 16. R. pimpinellifolia x R. sherardii = R. x involuta
Smith (5 x 16 and 16 x 5).
5 x 17. R. pimpinellifolia x R. mollis = R. x sabinii J.
Woods (5 x 17 and 17 x 5).
5 x 18. R. pimpinellifolia x R. rubiginosa = R. x cantiana
(Wolley-Dod) Wolley-Dod (R. x moorei (Baker) Wolley-Dod) (5 x
18 only).

Section 3 - CINNAMOMEAE Ser. (subg. Cassiorhodon Dumort.)
(spp. 6-8). Leaflets 5-9; flowers 1-several, with bracts;
sepals entire or nearly so, falling before or persistent
until after fruit ripe; styles free.

6. R. rugosa Thunb. ex Murray - **Japanese** Rose. Strongly 433
suckering; stems erect, to 1.5(2)m, tomentose, with numerous
tomentose spines and bristles; leaflets bullate, green,
rather shiny, tomentose to pubescent and glandular on
lowerside, 1-serrate; flowers 1-3, white to red, 6-9cm
across; fruit globose to depressed-globose, 1.5-2.5cm.
Intrd; much grown for ornament and as stock for cultivars,
natd on dunes, rough ground and banks, often mass-planted;
scattered through much of BI; E Asia.
R. rugosa occasionally forms spontaneous hybrids with native
spp., but the other sp. is often very difficult to determine;
such hybrids sometimes resemble R. 'Hollandica'. R. rugosa x
R. canina = R. x praegeri Wolley-Dod was found in Co Antrim
in 1927, and hybrids possibly involving R. mollis and R.
caesia have been recorded in En and Sc.

7. R. 'Hollandica' - **Dutch** Rose. Differs from R. rugosa in 433
lighter green, matt, scarcely bullate, pubescent to sparsely
pubescent leaves; flowers 2-5 together, with dark or mauvish-
red petals; and globose to ovoid fruits 1-1.5cm and often ±
pendent. Intrd; much used as rootstock, ± natd outcast or
relic in hedges and waste ground; scattered in Br from C En
to Shetland. Garden origin; hybrid between R. rugosa and
some other sp.
8. R. glauca Pourret non Villars ex Lois. (R. rubrifolia 433
Villars nom. illegit.) - **Red-leaved** Rose. Stems erect, to
3m, glabrous, with few prickles, with numerous acicles on
suckers; leaflets ± flat, red-tinged, glabrous, 1-serrate,
glandless; flowers (1)few-several, deep pink, 3-4.5cm across;
fruit ellipsoid to subglobose, 12-18mm. Intrd; much grown
and natd via birds; very scattered in Sc and En, but probably
under-recorded; C Europe.

Section 4 - CAROLINAE Crepin (sp. 9). Leaflets (5)7-9:

flowers 1-few, with bracts; sepals entire or nearly so,
falling before fruit ripe; styles free.

9. R. virginiana Herrm. (R. lucida Ehrh.) - Virginian Rose. **433**
Variably suckering; stems ± erect, to 2m, with few curved
spines and 0 or few hairs and stalked glands; leaflets
glabrous or sparsely pubescent on lowerside, 1-serrate,
eglandular; flowers pink or white, 4-6cm across; fruit subglobose, 1-1.5cm. Intrd; formerly much grown and still natd
in scrub and hedgerows; scattered in En and Wa; E N America.

Section 5 - ROSA (Gallicanae DC.) (sp. 10). Leaflets 3-5;
flower mostly 1, without bracts; sepals pinnately lobed,
falling before fruit ripe; styles free.

10. R. gallica L. - Red Rose (of Lancaster). Strongly **433**
suckering; stems erect, to 1m, with slender prickles and
glandular hairs; leaflets pubescent and glandular on
lowerside, mostly 1-serrate; flowers pink to red, 5-7(9)cm
across; fruit ellipsoid to globose, 1-1.5cm. Intrd; formerly
much grown and still natd in scrub and hedges; very scattered
in En and Guernsey; Europe. Its hybrid derivatives **R.
centifolia** L. and **R. damascena** Miller are also sometimes
found in semi-wild places.
R. x alba L. (R. x collina Jacq. non J. Woods) - (White Rose
(of York)) is a hybrid of R. gallica and R. arvensis and/or
R. canina; it has ± double, white flowers and there are a few
records of it in the wild.

Section 6 - CANINAE DC. (spp. 11-20). Leaflets (3)5-7(9);
flowers 1-several, with bracts; sepals usually pinnately
lobed, falling before or persistent until after fruit ripe;
styles free or (R. stylosa only) fused into column shorter
than stamens.

11. R. stylosa Desv. - Short-styled Field-rose. Stems **433**
arching, to 3(4)m, with hooked prickles; leaflets pubescent
on lowerside, 1-serrate, eglandular; flowers white to pale
pink, 3-5cm across; fruit ovoid, 1-1.5cm, ± glabrous,
pedicels 2-4cm, usually glandular-pubescent; sepals reflexed
after flowering, falling before fruit ripe. Native; hedges,
scrub, wood-borders; Br and Ir S of a line from Offaly to E
Suffolk, but now rare.
    Hybrids with R. arvensis (q.v.) and spp. of Caninae occur:
11 x 12. R. stylosa x R. canina = **R. x andegavensis** Bast.
(R. x rufescens Wolley-Dod) (11 x 12 and 12 x 11).
11 x 14. R. stylosa x R. obtusifolia (14 x 11 only).
11 x 18. R. stylosa x R. rubiginosa (18 x 11 only).
11 x 20. R. stylosa x R. agrestis (=R. x belnensis auct. non
Ozan) (11 x 20 and 20 x 11).
12. R. canina L. (R. dumetorum auct. non Thuill., R. **433**
squarrosa auct., ?(Rau) Boreau, R. corymbifera Borkh.) -

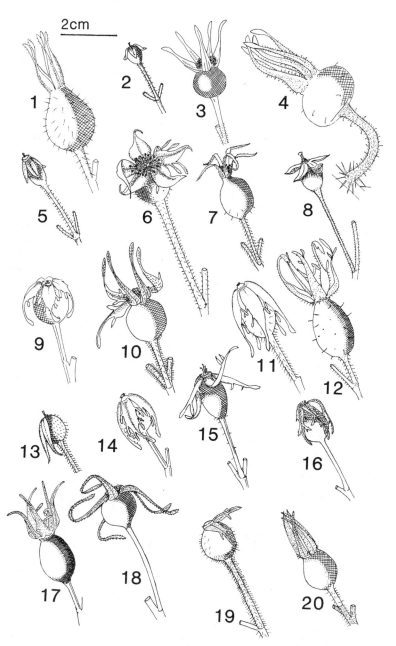

FIG 433 - Fruits of Rosa. 1, **R. mollis.** 2, **R. multiflora.** 3, **R. pimpinellifolia.** 4, **R. rugosa.** 5, **R. luciae.** 6, **R. gallica.** 7, **R. glauca.** 8, **R. arvensis.** 9, **R. canina.** 10, **R. rubiginosa.** 11, **R. micrantha.** 12, **R. sherardii.** 13, **R. setigera.** 14, **R. obtusifolia.** 15, **R. stylosa.** 16, **R. agrestis.** 17, **R. caesia.** 18, **R. virginiana.** 19, **R. tomentosa.** 20, **R. 'Hollandica'.**

Dog-rose. Stems arching, to 3(4)m, with usually strongly curved to hooked prickles; leaflets glabrous to pubescent on lowerside, 1-2-serrate, eglandular or with glandular teeth or sometimes some glands on lowerside veins; flowers white to pink, 3-5cm across; fruit (globose to) ovoid, 1.5-2cm, usually glabrous; pedicels 0.5-2cm, glabrous to sparsely glandular-pubescent; sepals reflexed after flowering, falling before fruit ripe. Native; hedges, scrub, wood-borders; common throughout most of BI except sparse in Sc and absent from much of N. Extremely variable. Pubescent plants (<u>R. dumetorum</u> auct., <u>R. corymbifera</u> Borkh.) are often separated as a sp., but every intermediate exists; they are best treated as an informal group 'Pubescentes'. 3 other groups have been recognized: '<u>Lutetianae</u>' (<u>R. canina</u> sensu stricto), with 1-serrate leaves and with glands confined to bracts; '<u>Dumales</u>' (<u>R. dumalis</u> auct. non Bechst.), with 2-serrate leaves and with glands on leaf-teeth, bracts, stipules and leaf-rhachis; and '<u>Transitoriae</u>', intermediate. Pubescent plants are usually 1-serrate, but often 2-serrate though rarely with glandular teeth. The numerous intermediates and poor character-correlation suggests only 1 sp. should be recognized.

<u>R. canina</u> hybridises with all other native spp. of <u>Rosa</u>:

12 x 13. R. canina x R. caesia = R. x dumalis Bechst. (<u>R.</u> x <u>subcanina</u> (Christ) Dalla Torre & Sarnth., <u>R.</u> x <u>subcollina</u> (Christ) Dalla Torre & Sarnth.) (12 x 13 and 13 x 12; both sspp. of <u>R. caesia</u>).

12 x 14. R. canina x R. obtusifolia = R. x dumetorum Thuill. (<u>R.</u> x <u>subobtusifolia</u> Wolley-Dod, <u>R.</u> x <u>concinnoides</u> Wolley-Dod) (12 x 14 and 14 x 12).

12 x 15. R. canina x R. tomentosa = R. x scabriuscula Smith (<u>R.</u> x <u>curvispina</u> Wolley-Dod, <u>R.</u> x <u>aberrans</u> Wolley-Dod) (12 x 15 and 15 x 12).

12 x 16. R. canina x R. sherardii (12 x 16 and 16 x 12).

12 x 17. R. canina x R. mollis = R. x molletorum Heslop-Harrison (12 x 17 and 17 x 12).

12 x 18. R. canina x R. rubiginosa = R. x nitidula Besser (<u>R.</u> x <u>latebrosa</u> Desegl., <u>R.</u> x <u>latens</u> Wolley-Dod) (12 x 18 and 18 x 12).

12 x 19. R. canina x R. micrantha = R. x toddiae Wolley-Dod (12 x 19 and 19 x 12).

12 x 20. R. canina x R. agrestis = R. x belnensis Ozan (20 x 12 only).

13. R. caesia Smith (<u>R.</u> <u>dumalis</u> auct. non Bechst.) - see **433** sspp. for English names. Stems arching, to 2(3)m, with strongly curved to hooked prickles; leaflets flat, mostly 1-serrate, with few or 0 glands; flowers and fruit as in <u>R. canina</u> except as in key (couplets 13 or 18); pedicels 0.5-1.5cm. Native; hedges, scrub and wood-borders; throughout most of N 1/2 of BI, rare and very scattered in S. Relative distribution of 2 sspp. uncertain.

a. Ssp. caesia (<u>R. coriifolia</u> Fries) - <u>Hairy Dog-rose</u>.

Stems green or somewhat red; leaflets rugose, scarcely glaucous, pubescent on lowerside.

b. Ssp. glauca (Nyman) G.G. Graham & Primavesi (R. afzeliana Fries, R. vosagiaca N. Desp., R. glauca Villars ex Lois. non Pourret) - Glaucous Dog-rose. Stems often strongly red-coloured and glaucous; leaflets scarcely rugose, glabrous, glaucous. Isolated bushes have a characteristic open, not dense, appearance.

The following additional hybrids have been recorded:

13 x 14. R. caesia x R. obtusifolia (14 x 13 only; ssp. glauca only).

13 x 15. R. caesia x R. tomentosa = R. x rogersii Wolley-Dod (13 x 15 and 15 x 13; ssp. glauca only).

13 x 16. R. caesia x R. sherardii (13 x 16 and 16 x 13; both sspp. of R. caesia).

13 x 17. R. caesia x R. mollis = R. x glaucoides Wolley-Dod (13 x 17 and 17 x 13; both sspp. of R. caesia).

13 x 18. R. caesia x R. rubiginosa (=R. x obovata (Baker) Ley non Raf.) (13 x 18 and 18 x 13; both sspp. of R. caesia).

13 x 19 R. caesia x R. micrantha = R. x longicolla Ravaud ex Rouy (19 x 13 only; ssp. glauca only).

14. R. obtusifolia Desv. - Round-leaved Dog-rose. Stems **433** arching, to 2(3)m, with strongly and abruptly hooked prickles; leaflets pubescent on lowerside (often also on upperside), (1-)2-serrate with glandular teeth and often glandular on lowerside, more rounded overall and at base than in R. canina; flowers white (to pale pink), 3-5cm across; fruit globose to ovoid 1-1.6(2)cm, usually glabrous; pedicels (0.6)1-1.6cm, usually glabrous; sepals strongly lobed, reflexed after flowering, falling before fruit ripe. Native; hedges and scrub; En, Wa and Ir, but rare in many areas.

The following additional hybrids have been recorded:

14 x 15. R. obtusifolia x R. tomentosa (14 x 15 and 15 x 14).

14 x 18. R. obtusifolia x R. rubiginosa = R. x tomentelliformis Wolley-Dod (14 x 18 only).

15. R. tomentosa Smith (R. scabriuscula auct. non Smith) - **433** Harsh Downy-rose. Stems arching, to 2m, with ± straight to arching prickles; leaflets pubescent to tomentose on lower-side (often also on upperside), 2-serrate with glandular teeth, with glands on lowerside; flowers white to pink, 3-5cm across; fruit ovoid, 1-2cm, (glabrous or) glandular-pubescent; pedicels (1)1.5-2.5cm, glandular-pubescent; sepals pinnately lobed, suberect to patent, falling before fruit ripe. Native; hedges, scrub and open woods; frequent in BI N to S Sc, very rare in C & N Sc.

The following additional hybrids have been recorded:

15 x 16. R. tomentosa x R. sherardii = R. x suberectiformis Wolley-Dod (16 x 15 only).

15 x 17. R. tomentosa x R. mollis (17 x 15 only).

15 x 18. R. tomentosa x R. rubiginosa = R. x avrayensis Rouy

(15 x 18 and 18 x 15).
15 x 20. R. tomentosa x R. agrestis (15 x 20 only).
16. R. sherardii Davies - <u>Sherard's</u> <u>Downy-rose</u>.   Differs **433** from R. tomentosa in more compact habit, to 1.5(2)m; prickles more slender; flowers usually deep pink; fruit usually globose to obovoid; pedicels 0.5-1.5cm; sepals erect to suberect, persistent until fruit ripe but falling before its decay; and see key (couplet 20). Native; scrub, hedges and open woods, throughout most of Br and Ir but common in Sc and very rare in S En.
The following additional hybrids have been recorded:
16 x 17. R. sherardii x R. mollis = ?R. x shoolbredii Wolley-Dod (16 x 17 and 17 x 16).
16 x 18. R. sherardii x R. rubiginosa = R. x suberecta (J. Woods) Ley (R. x <u>burdonii</u> Wolley-Dod) (16 x 18 and 18 x 16).
16 x 19. R. sherardii x R. micrantha (16 x 19 only).
16 x 20. R. sherardii x R. agrestis (16 x 20 and 20 x 16).
17. R. mollis Smith (R. <u>villosa</u> auct. non L.) - <u>Soft Downy-</u> **433** <u>rose</u>. Differs from R. tomentosa and R. sherardii in more compact habit with erect stems; prickles more slender, straight; flowers usually very deep pink; fruit usually globose; pedicels 0.5-1.5cm; sepals erect, simple to slightly lobed, persisting until fruit decays; and see key (couplets 20 & 22). Native; similar habitats and range to R. sherardii, but S to only Glam and Derbys.
The following additional hybrid has been recorded:
17 x 18. R. mollis x R. rubiginosa = R. x molliformis Wolley-Dod (17 x 18 and 18 x 17).
18. R. rubiginosa L. - <u>Sweet-briar</u>. Stems erect, to 2m, **433** with hooked, unequal prickles; leaflets pubescent at least on veins and glandular on lowerside, ± 2-serrate with glandular teeth; flowers pink, 2.5-4cm across; fruit subglobose to ovoid, 1-1.5cm, glabrous or sparsely glandular-pubescent; pedicels 0.8-1.5cm, glandular-pubescent; sepals erect to patent, persistent until fruit reddens. Native; mostly in scrub on calcareous soils; scattered throughout BI, ± common on chalk in SE En.
The following additional hybrid has been recorded:
18 x 19. R. rubiginosa x R. micrantha = R. x bigeneris Duffort ex Rouy (19 x 18 only).
19. R. micrantha Borrer ex Smith - <u>Small-flowered Sweet-</u> **433** <u>briar</u>.   Differs from R. rubiginosa in flowers 2-3.5cm; pedicels 1-2cm, rarely glabrous; and see key (couplet 16). Native; similar places to R. rubiginosa but often not on calcareous soils; scattered throughout En and Wa, local in CI, SW Ir, rare in Sc.
The following additional hybrid has been recorded:
19 x 20. R. micrantha x R. agrestis = R. x bishopii Wolley-Dod (20 x 19 only).
20. R. agrestis Savi (R. <u>elliptica</u> auct. non Tausch) - **433** <u>Small-leaved</u> <u>Sweet-briar</u>. Differs from R. rubiginosa in stems arching; prickles ± equal; flowers white to pale pink;

fruit usually glabrous; pedicels 1-2cm, glabrous; sepals usually reflexed, falling before fruit reddens. Native; scrub, mostly on calcareous soils; very scattered and mostly rare in Br and Ir N to S Sc, frequent mainly in parts of Ir. Plants identified as **R. elliptica** Tausch were probably R. agrestis and/or hybrids of the latter or perhaps of R. micrantha.

SUBFAMILY 3 - PRUNOIDEAE (genera 22-23).

Stipulate trees or shrubs; hypanthium concave, not enclosing carpels and fixed to them only at base; epicalyx 0; stamens >10; flowers 5-merous; carpels 1(or 5), 1-2-seeded; fruit 1(-5) drupes; chromosome base-number 8.

22. PRUNUS L. - Cherries

Trees or shrubs; leaves simple, serrate to crenate; flowers solitary or in racemes, corymbs or umbels; carpel 1; fruit a drupe.

```
1  Flowers usually >10 in elongated racemes                  2
1  Flowers solitary or 2-c.10 in umbels or corymbs           5
   2  Leaves coriaceous, evergreen; racemes without
      leaves at base; fruit conical-acute at apex            3
   2  Leaves herbaceous, deciduous; racemes usually with
      1-few leaves (often reduced) near base; fruit
      rounded to apiculate at apex                           4
3  Petioles and 1st-year stems deep red; leaves serrate;
   racemes mostly longer than leaves         14. P. lusitanica
3  Petioles and 1st-year stems green; leaves crenate to
   obscurely serrate; racemes mostly shorter than
   leaves                                    15. P. laurocerasus
   4  Petals >5mm; sepals falling before fruit ripe
                                             12. P. padus
   4  Petals <5mm; sepals persistent until fruit ripe
                                             13. P. serotina
5  Ovary and fruit pubescent; leaves mostly lanceolate
   to oblanceolate                                           6
5  Ovary and fruit glabrous; leaves mostly ovate to
   obovate                                                   7
   6  Drupe becoming dry and splitting at maturity, much
      wider than thick, with pitted stone      2. P. dulcis
   6  Drupe becoming very succulent and not splitting
      at maturity, ± globose, with deeply grooved
      stone                                     1. P. persica
7  Flowers in short subcorymbose racemes with green
   bracts on proximal part of axis             8. P. mahaleb
7  Flowers solitary or in umbels, or rarely very short
   corymbose racemes without green bracts on axis           8
   8  Flowers (1)2-6(10) together, with group of large
      (>5mm) often green or reddish bud-scales at base
      of cluster (not in P. pensylvanica); ripe fruit
      not pruinose, with longer pedicel (cherries)          9
```

  8 Flowers 1-3 together, with 0 or small (<3mm) brown
     bud-scales at base of cluster; ripe fruit
     pruinose, with usually shorter pedicel (plums)   13
9 Leaves with acuminate to aristate teeth   10
9 Leaves with acute to obtuse teeth   11
  10 Leaves <8cm, with acuminate teeth; flowers not
     pendent   **11. P. incisa**
  10 Some leaves >8cm, with aristate teeth; flowers
     usually pendent   **10. P. serrulata**
11 Bud-scales at base of inflorescence falling before
    flowers fully open; flowers 1-2cm across
                         9. P. pensylvanica
11 Bud-scales at base of inflorescence persistent;
    flowers 2-3.5cm across   12
  12 Hypanthium cup-shaped, not constricted at opening;
     some bud-scales at base of flowers usually green
     and leaf-like; never a large tree   **7. P. cerasus**
  12 Hypanthium bowl-shaped, constricted at opening;
     bud-scales not green and leaf-like; often
     a large tree   **6. P. avium**
13 1st-year twigs green, shiny, glabrous   **3. P. cerasifera**
13 1st-year twigs brown to grey, dull, often pubescent   14
  14 Fruit <2cm, blue-black; flowers appearing before
     leaves; petals 5-8mm; twigs very spiny **4. P. spinosa**
  14 Fruit usually >2cm, blue-black, red or yellow-
     green; flowers appearing with leaves; petals
     7-12mm; twigs not or sparsely spiny **5. P. domestica**

Other spp. - Many spp. and cultivars are grown for ornament
and occasionally persist in semi-wild conditions; most
cultivars are grafted on to stock of wild British spp. and
often do not produce fruit, so do not regenerate sexually or
vegetatively.

  1. P. persica (L.) Batsch - <u>Peach</u>. Deciduous tree to 6m;
leaves lanceolate to oblanceolate, 5-15cm; petioles mostly
1-1.5cm; flowers 1(-2); petals 10-20mm, deep pink; fruit the
familiar peach. Intrd; frequent on tips and waste ground in
towns, from discarded stones, but rarely reaching maturity; S
& C En; China.
  2. P. dulcis (Miller) D. Webb (<u>P. amygdalus</u> Batsch) -
<u>Almond</u>. Vegetatively similar to <u>P. persica</u> but tree to 8m;
leaves 4-12cm; petioles mostly 1.5-2.5cm; flowers (1-)2;
petals white to deep pink; fruit the familiar almond. Intrd;
extremely common street- and park-tree, also frequent on tips
and waste places in S Br; SW Asia.
  3. P. cerasifera Ehrh. - <u>Cherry Plum</u>. Deciduous, sometimes
spiny shrub or tree to 8(12)m; leaves 3-7cm, ovate to
obovate, green or purplish; flowers appearing with or before
leaves, the earliest in the genus, 1-2(3); petals white or
pink; fruit ± globose, 2-3cm, dark red or yellow, scarcely
bloomed. Intrd; common in hedges and as street-tree, planted

for hedging and ornament (mostly purplish-leaved var.
pissardii (Carriere) L. Bailey (P. 'Atropurpurea')),
spreading by suckers and often used as stock for other Prunus
cultivars, not commonly fruiting; CI and Br N to C Sc; SE
Europe and SW Asia.

4. **P. spinosa** L. - Blackthorn. Deciduous, dense, spiny
shrub to 4m; leaves obovate to oblanceolate, 1-3(4)cm;
flowers appearing before leaves, 1(-2); petals white; fruit
nearly globose, 8-15mm, bluish-black with dense bloom.
Native; hedges, scrub and woods; common almost throughout BI.

4 x 5. P. spinosa x P. domestica = P. x fruticans Weihe
occurs in hedges sporadically throughout En; it is intermediate, fertile and variable. P. spinosa var. macrocarpa
Wallr. probably belongs here.

5. **P. domestica** L. (P. insititia L.) - Wild Plum. Deciduous
large shrub or tree to 8(12)m; leaves 3-8cm, obovate to
elliptic; flowers appearing with leaves, mostly 2-3; petals
white; fruit varied in size, shape and colour, with weak to
strong bloom. Intrd; hedges, copses, scrub and waste ground;
throughout most of BI; SW Asia. 2 or 3 sspp. are often
recognized: ssp. **domestica** (Plum) with sparsely pubescent
spineless twigs and usually large fruits with very flattened
stone; ssp. **insititia** (L.) Bonnier & Layens (Bullace,
Damson), with densely pubescent often spiny twigs and small
fruits with less flattened stone; and ssp. **italica** (Borkh.)
Gams ex Hegi (P. x italica Borkh.) (Greengage), +
intermediate. However, these have been so much hybridized
that character-correlation has partly broken down and the
sspp. are often scarcely discernible.

6. **P. avium** (L.) L. - Wild Cherry. Deciduous tree to 31m;
leaves 6-15cm, obovate to elliptic; flowers 2-6 in umbels, on
pedicels 15-45mm, usually shallowly cup-shaped; fruit +
globose, 9-12mm, black, red or yellow. Native; hedgerows,
wood-borders and copses; throughout BI.

7. **P. cerasus** L. - Dwarf Cherry. Deciduous shrub or small
tree to 8m; differs from P. avium in saucer-shaped flowers
fewer per umbel (mostly 2-4) with shorter pedicels (1-4cm)
hence mixed more among the leaves than projecting beyond;
fruit bright red, never sweet; and see key. Intrd; hedges
and copses; throughout most of BI N to C Sc, but much
confused with P. avium and distribution uncertain; SW Asia.

8. **P. mahaleb** L. - St Lucie Cherry. Deciduous shrub or **440**
small tree to 6(10)m; leaves 3-7cm, ovate to broadly so,
rounded to subcordate at base; flowers up to 10 in subcorymbose racemes on pedicels up to 15mm; petals white; fruit
<1cm, broadly ovoid, black. Intrd; well natd by railways, in
grassland and in woods; several places in S En; S Europe.

9. **P. pensylvanica** L.f. - Pin Cherry. Deciduous tree to
12m; leaves 6-12cm, ovate to obovate or narrowly so, cuneate
to rounded at base; flowers up to 6(10) in umbels or very
short corymbose racemes on pedicels up to 15mm; petals white;
fruit c.6mm, subglobose, red. Intrd; natd in and by woodland

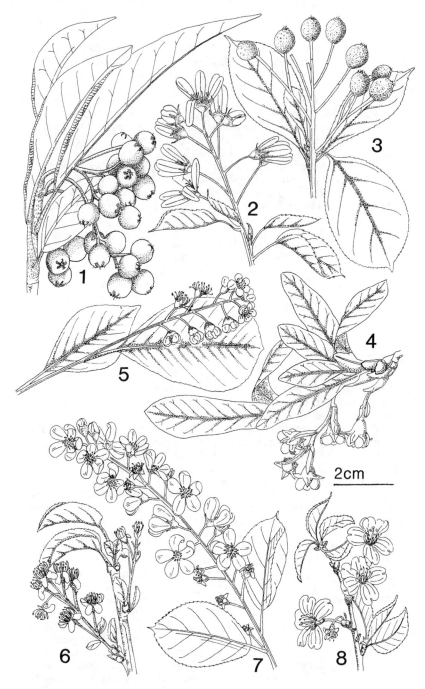

FIG 440 – Rosaceae. 1, Photinia. 2, Amelanchier. 3, Aronia melanocarpa. 4, Oemleria. 5, Prunus serotina. 6, P. mahaleb. 7, P. padus. 8, P. incisa.

(self-sown trees to 9m) in Surrey; N America.

10. P. serrulata Lindley - <u>Japanese</u> Cherry. Deciduous tree to 12m; leaves 5-12cm, ovate to obovate or narrowly so, long-acuminate, with aristate teeth; flowers 2-6 in umbels; petals white or pink; fruit ± globose, black, but rarely formed. Intrd; much planted by roads and in parks, and often found as relic in wild places in much of BI; Japan and China. Many cultivars, often <u>flore</u> <u>pleno</u> and usually grafted on to P. avium as stock; not natd.

11. P. incisa Thunb. ex Murray - <u>Fuji</u> <u>Cherry</u>. Deciduous **440** shrub or small tree to 6(10)m; leaves 4-8cm, ovate to narrowly so, long-acuminate, 2-serrate, with acuminate teeth; flowers 1-4, mostly 3; petals pink; fruit 6-8mm, ovoid, purplish-black. Intrd; natd in oakwoods on clay-with-flints, Chinnor Hill, Oxon; Japan.

12. P. padus L. - <u>Bird</u> Cherry. Deciduous shrub or tree to **440** 19m; leaves 5-10cm, obovate to elliptic, glabrous or with white hairs in tufts along lowerside midrib; flowers in elongate ± erect to pendent racemes; pedicels 8-15mm in flower; petals 6-9mm, white; fruit 6-8mm, ± globose, shiny-black. Native; woods and scrub; Br from C En and S Wa to N Sc, very scattered in Ir, also much planted and natd in S & C En.

13. P. serotina Ehrh. - <u>Rum</u> Cherry. Differs from P. padus **440** in leaves glabrous or with rows of brown hairs along lowerside midrib; shorter stiffer racemes with closer-packed flowers; pedicels 3-7mm in flower (-10mm in fruit); petals 3-4.5mm; fruit 8-10mm, purplish-black. Intrd; natd in woods and commons; S En; E N America.

14. P. lusitanica L. - <u>Portugal</u> Laurel. Evergreen shrub or tree to 12m; leaves narrowly ovate to oblong-ovate, 6-13cm; flowers in 10-25cm racemes; petals white; fruit conical-ovoid, mostly 8-10mm, purplish-black. Intrd; commonly planted, sometimes natd in woods, shrubberies and waste land; scattered in Br N to C Sc; SW Europe.

15. P. laurocerasus L. - <u>Cherry</u> Laurel. Evergreen shrub or tree to 10m; leaves usually oblong-ovate, 5-15cm; flowers in 7-13cm racemes; petals white; fruit conical-ovoid, mostly 10-12mm, purplish-black to ± black. Intrd; abundantly planted, frequently natd in woods and shrubberies, throughout most of BI; SE Europe.

23. OEMLERIA Reichb. (<u>Osmaronia</u> E. Greene) - <u>Oso-berry</u>
Deciduous shrubs; leaves simple, entire; flowers dioecious or partially so, in racemes; carpels 5; fruit a cluster of 1-5 drupes.

1. O. cerasiformis (Torrey & A. Gray ex Hook. & Arn.) Landon **440**
(<u>Osmaronia</u> <u>cerasiformis</u> (Torrey & A. Gray ex Hook. & Arn.) E. Greene) - <u>Oso-berry</u>. Suckering shrub with erect stems to 2(3)m; leaves lanceolate to narrowly elliptic, 7-11cm; flowers c.5-10 in short pendent racemes; petals greenish-

white; fruit 10-15mm, bluish-black, bloomed, rarely produced.
Intrd; natd on rough ground in Middlesex and W Kent; W N
America.

SUBFAMILY 4 - MALOIDEAE (Pomoideae) (genera 24-35). Trees or
shrubs with stipules; hypanthium concave, enclosing carpels
and fused to them all round; epicalyx 0; stamens >10; flowers
5-merous; carpels 1-5, 1-2(several)-seeded; fruit consisting
of fused carpels surrounded by succulent hypanthium;
chromosome base-number 17.

24. CYDONIA Miller - Quince
Leaves simple, entire, deciduous; flowers solitary; stamens
15-25; carpels 5, with free styles and numerous ovules, the
walls cartilaginous in fruit.

1. C. oblonga Miller - Quince. Spineless shrub or tree to
3(6)m; leaves 5-10cm, ovate, grey-tomentose on lowerside;
flowers 4-5cm across, white or pink; fruit up to 12cm,
yellow. Intrd; not or rarely regenerating but very
persistent in hedges and woods, sometimes used as stock for
Pyrus; scattered in S Br; Asia.

25. CHAENOMELES Lindley - Japanese Quince
Leaves simple, serrate, deciduous, flowers in clusters of
1-4; stamens 40-60; carpels 5, with styles fused at base and
numerous ovules, the walls cartilaginous in fruit.

1. C. speciosa (Sweet) Nakai - Japanese Quince. Open spiny
shrub to 3m; leaves 4-10cm, ovate, glabrous; flowers 3.5-
4.5cm across, usually red; fruit up to 8cm, yellowish-green.
Intrd; not or rarely regenerating but very persistent in
hedges and woods; scattered in CI and S En; China.

26. PYRUS L. - Pears
Leaves simple, serrate to crenate, deciduous; flowers in
simple corymbs; petals white; stamens 20-30; carpels 2-5,
with free styles and 2 ovules, the walls cartilaginous in
fruit; fruit with groups of gritty stone-cells in hypanthium.

1 Fruit >(5)6cm, soft and sweet when mature, usually
   pear-shaped   **3. P. communis**
1 Fruit <5cm, hard and sour even when mature, not
   pear-shaped   2
    2 Fruit with caducous calyx, <1.5(2)cm; petals 6-
       10mm; inflorescence rhachis >1cm   **1. P. cordata**
    2 Fruit with persistent calyx, 1.5-4cm; petals 10-
       17mm; inflorescence rhachis <1cm   **2. P. pyraster**

1. P. cordata Desv. - Plymouth Pear. Spiny shrub to 4m; **RRR**
leaves 1-4.5cm, ovate or broadly so; fruits 8-18mm, globose
or obovoid, brownish-red. Possibly native; 2 hedges near

Plymouth, S Devon (known since 1870), 3 sites near Truro, W Cornwall (found 1989).

2. P. pyraster (L.) Burgsd. - Wild Pear. Usually ± spiny shrub or tree to 15m; leaves 2.5-7cm, ovate to broadly so; fruits 1.5-4cm, globose to obovoid or obconical, yellow to reddish- or dark-brown. Intrd; hedges and wood-margins; scattered through S & C Br, CI and Ir; Europe. 2 sspp. recently recognized (ssp. **pyraster**, with ± glabrous leaves and globose to obconical fruits; ssp. **achras** (Wallr.) Terpo, with pubescent leaves and obconical to obovoid fruits) seem scarcely distinct in practice. Many wild pears are probably stocks of formerly cultivated trees.

3. P. communis L. - Pear. Differs from P. pyraster in usually non-spiny tree to 20m; fruit >(5)6cm, edible, usually pear-shaped. Intrd; very commonly grown and found in the wild in hedges and waste ground from old trees or discarded seeds; garden origin. Often scarcely distinguishable from P. pyraster except in fruit.

27. MALUS Miller - Apples

Leaves simple, serrate, deciduous; flowers in simple corymbs; stamens 15-50; carpels 3-5, with styles fused below and with 2 ovules, the walls cartilaginous in fruit; fruit without groups of gritty stone-cells.

1 Leaves purplish-green to purple; fruit dark reddish-
  purple, c. as long as pedicel   **3. M. x purpurea**
1 Leaves green; fruits green to yellow or red, longer
  than pedicel   2
  2 Leaves glabrous when mature; pedicels and outside
    of calyx glabrous   **1. M. sylvestris**
  2 Leaves pubescent on lowerside; pedicels and outside
    of calyx pubescent   **2. M. domestica**

Other spp. - Many spp. and cultivars are grown for their ornamental flowers or fruit, and sometimes persist in wild places or produce odd seedlings. The following are street trees with pedicels longer than fruit: **M. baccata** (L.) Borkh. (Siberian Crab) with white petals, mostly 5 styles and red fruits <1cm; and **M. floribunda** Sieb. ex Van Houtte (Japanese Crab), with pink and white petals, mostly 4 styles, and red or yellow fruits >1cm. Many taxa are best known by their cultivar names only, e.g. M. 'John Downie'.

1. M. sylvestris (L.) Miller - Crab Apple. Tree to 10m, often spiny; twigs glabrous; leaves 3-5cm, ovate to elliptic, glabrous when mature; petiole 1.5-3cm; pedicels and outside of calyx glabrous; petals pinkish-white; fruit apple-shaped, yellowish-green, c.2-3cm. Native; woods, hedges and scrub; probably throughout BI except N Sc, but much over-recorded for M. domestica.

2. M. domestica Borkh. (M. sylvestris ssp. mitis (Wallr.)

Mansf.) - Apple. Tree to 10(20)m, not spiny; similar to M. sylvestris but larger in most parts; leaves up to 15cm, with relatively shorter petiole; fruit up to 12cm, variously coloured. Intrd; much grown and often natd in hedges, scrub and waste ground; throughout BI and much commoner than M. sylvestris. Self-sown plants usually have small, yellowish, sour fruits.

3. M. x purpurea (Barbier) Rehder (M. niedzwetzkyana Dieck x M. atrosanguinea (Spaeth) Schneider) - Purple Crab. Tree to 11m, not spiny; leaves up to 10cm, ovate to elliptic, pubescent; petiole 2-3.5cm; petals deep pink; fruit ± globose, dull reddish-purple, 2.5-3.5cm. Intrd; grown in parks and by roads, rarely self-sown; natd in W Kent; garden origin c.1900.

28. SORBUS L. - Whitebeams

Leaves pinnate, or simple and serrate to pinnately lobed, deciduous; flowers in compound corymbs; petals white; stamens 15-25; carpels 2-4(5), with styles free or fused below and 2 ovules, the walls cartilaginous in fruit.

A difficult genus, consisting of several well-defined sexual species and a number of apomictic ones, many of which are of hybrid origin. The apomicts mostly fall into 3 groups: those similar to S. aria (S. aria agg.); those intermediate between either S. torminalis or S. aria agg. and S. aucuparia (S. intermedia agg.); and those intermediate between S. aria agg. and S. torminalis (S. latifolia agg.). Most diagnostic characters concern leaves and fruits, both of which are required for identification by beginners. 'Leaves' refers to the broader leaves of the short-shoots (not those on leading shoots). It is important that plants are surveyed as fully as possible and that the means of each character are used for identification.

1 Leaves pinnate, at least proximally, with >1 pair of
   completely free leaflets   2
1 Leaves not to deeply lobed, but without free leaflets   4
   2 Leaves completely pinnate, with >4 pairs of
     leaflets and a single terminal leaflet
                                                      1. S. aucuparia
   2 Leaves pinnate only proximally, with 1-2(3) pairs
     of leaflets and a distal lobed part of leaf (if
     <5 pairs of leaflets, see 1x7 and 1x9)   3
3 Leaves 5.5-8.5cm, rather thinly grey-tomentose on
   lowerside, with mostly 1 pair of free leaflets; fruit
   longer than wide; petals c.4mm   **2. S. pseudofennica**
3 Leaves 7.5-10.5cm, densely whitish-grey-tomentose on
   lowerside, with mostly 2 pairs of free leaflets;
   fruit globose; petals c.6mm   **3. S. hybrida**
   4 Mature leaves sparsely pubescent on lowerside, at
     least some lobed >1/3 way to midrib proximally;
     fruit brown   **24. S. torminalis**

4 Mature leaves tomentose on lowerside (wearing off
  with age), if lobed >1/3 way to midrib then fruit
  red   5
5 Leaves white-tomentose on lowerside; fruit red
                         9–17. S. aria agg.
5 Leaves grey- to yellowish-tomentose on lowerside;
  fruit red, orange or brown   6
  6 Fruit red   4–8. S. intermedia agg.
  6 Fruit orange to brown   18–23. S. latifolia agg.

Other spp. – Many spp. are grown for ornament and may
persist in wild places, especially **S. hupehensis** C. Schneider
(Hupeh Rowan), from China, with pinnate leaves and ± white
fruit; **S. domestica** L. (Service-tree), from S Europe,
resembling S. aucuparia but with fissured (not smooth) trunk,
5 styles, and greenish or brownish fruit rarely <2cm; and
several taxa in S. aria agg.

1. S. aucuparia L. – Rowan. Tree to 18m; leaves pinnate,
with (4)5–7(9) pairs of leaflets; styles 3–4; fruit 6–
9(14)mm, scarlet (to yellow in cultivars). Native; woods,
moors, rocky places except on heavy soils; throughout Br and
Ir, intrd in CI.

1 x 7. S. aucuparia x S. intermedia occurs as occasional
spontaneous trees near the parents in Br N to Staffs and
Merioneth. It has partially pinnate ovate-oblong leaves with
usually (1)2–3(5) pairs of free, well-separated leaflets and
a total of 10–12 pairs of lateral veins, sterile pollen, and
scarlet fruits with no seed; 2n=51.

1 x 9. S. aucuparia x S. aria = S. x thuringiaca (Ilse) **446**
Fritsch (S. x semipinnata (Roth) Hedlund non Borbas) occurs
as occasional spontaneous trees near the parents scattered
over Br mostly in N, and as planted trees. It closely
resembles S. aucuparia x S. intermedia but has brownish-red
fruits with some viable seed, some fertile pollen, and
usually wider, more oblong leaves; 2n=34.

2. S. pseudofennica E. Warb. – Arran Service-tree. Tree to **446**
7m; leaves partially pinnate, differing from those of above 2 **RR**
hybrids in having total of 7–9(10) pairs of lateral veins and
1(-2) pairs of broader ± overlapping free leaflets; styles
2–3; fruits scarlet; fully fertile. Native; steep granite
stream-bank; Glen Catacol, Arran (Clyde Is); endemic.

3. S. hybrida L. – Swedish Service-tree. Tree to 10m; **446**
leaves partially pinnate, differing from those of above 2
hybrids in having total of 8–10 pairs of lateral veins and
(1-)2(-3) pairs of free well-separated leaflets; styles 2–3;
fruits scarlet; fully fertile. Intrd; frequently grown in
gardens and parks and sometimes self-sown; natd in W Kent and
N Aberdeen; Scandinavia.

4–8. S. intermedia agg. – Swedish Whitebeam. Leaves
variable but often with lowest 3-6 pairs of lateral veins
ending in distinct leaf-lobe divided >1/6 way to midrib,

FIG 446 – Leaves of **Sorbus**. 1, **S. aria**. 2, **S. leptophylla**. 3, **S. eminens**. 4, **S. hibernica**. 5, **S. wilmottiana**. 6, **S. lancastriensis**. 7, **S. rupicola**. 8, **S. vexans**. 9, **S. porrigentiformis**. 10, **S. pseudofennica**. 11, **S. hybrida**. 12, **S. x thuringiaca**. 13, **S. x vagensis**.

FIG 447 – Leaves of **Sorbus**. 1, S. **latifolia**. 2, S. **decipiens**. 3, S. **bristoliensis**. 4, S. **devoniensis**. 5, S. **croceocarpa**. 6, S. **leyana**. 7, S. **intermedia**. 8, S. **minima**. 9, S. **subcuneata**. 10, S. **anglica** (Llangollen). 11, S. **arranensis**. 12, S. **anglica** (typical).

grey- to yellowish-tomentose on lowerside; fruit scarlet to crimson; styles 2.

Multi-access key to spp. of S. intermedia agg.

Leaves mostly with 7-8 pairs of lateral veins	A
Leaves mostly with 8-10 pairs of lateral veins	B
Leaves mostly lobed >1/3 way to midrib	C
Leaves mostly lobed <1/3 way to midrib	D
Petals mostly c.4mm	E
Petals mostly c.5mm	F
Petals mostly c.6mm	G
Fruit mostly 6-8mm	H
Fruit mostly 8-11mm	I
Fruit mostly 11-15mm	J
Fruit ± globose or wider than long	K
Fruit longer than wide	L

ACEIL	4. S. arranensis
BCFIK	5. S. leyana
BDEHK	6. S. minima
BDG(IJ)K	8. S. anglica
BDGJL	7. S. intermedia

4. S. arranensis Hedlund. Tree to 7.5m. Native; steep **447** granite stream-banks; Arran (Clyde Is); endemic. **RR**

5. S. leyana Wilm. Shrub to 3m. Native; carboniferous **447** limestone crags; near Merthyr Tydfil (Brecs); endemic. **RR**

6. S. minima (Ley) Hedlund. Shrub to 3m. Native; **447** carboniferous limestone crags; near Crickhowell (Brecs); **RR** endemic.

7. S. intermedia (Ehrh.) Pers. Tree to 10(18)m. Intrd; **447** much planted and frequently self-sown in copses and on rough ground; scattered over most of Br and Ir; Baltic region.

8. S. anglica Hedlund. Shrub to 3m. Native; woods and **447** rocky places mostly on carboniferous limestone; very local in **RR** Wa, SW En and N Kerry; endemic. A variant from near Llangollen (Denbs) with more narrowly cuneate leaf-bases **447** might be a separate sp.

9-17. S. aria agg. - Common Whitebeam. Leaves toothed, rarely with lobes >1/5 way to midrib, white-tomentose on lowerside; fruit scarlet to crimson; styles 2.

Multi-access key to spp. of S. aria agg.

Fruit longer than wide	A
Fruit ± globose or wider than long	B
Fruit mostly 8-12mm	C
Fruit mostly 12-15mm	D
Fruit mostly 15-22mm	E
Fruit scarlet	F
Fruit crimson or pinkish-scarlet	G
Leaves mostly <1.5x as long as wide	H
Leaves mostly 1.5-2x as long as wide	I
Leaves mostly >2x as long as wide	J
Leaves mostly with >10 pairs of lateral veins	K
Leaves mostly with <10 pairs of lateral veins	L

(AB)(CD)F(HIJ)K	9. S. aria

```
A(CD)GIL                                11. S. wilmottiana
ADFIL                                      17. S. vexans
AEFIK                                   10. S. leptophylla
AEGHK                                      12. S. eminens
BCG(HI)L                             14. S. porrigentiformis
B(DE)GH(KL)    Ireland only             13. S. hibernica
BDG(HI)L       NW England only
               Leaves <1.75(1.9)x as long as wide
                                        15. S.lancastriensis
BDG(IJ)L       Leaves >(1.7)1.8x as long as wide
                                        16. S. rupicola
```

9. S. aria (L.) Crantz. Tree to 15(23)m. Native; woods, **446** scrub, rocky places, mostly on calcareous soils; probably only in En N to Beds and E Gloucs, and in Galway, but commonly planted and ± natd throughout most of Br and Ir. Sexual and very variable, overlapping to some extent with most other spp. in the agg.

9 x 16. S. aria x S. rupicola has been reported from N Somerset and W Gloucs.

9 x 24. S. aria x S. torminalis = S. x vagensis Wilm. (?S. x **446** rotundifolia (Bechst.) Hedlund) occurs as occasional trees near the parents in W Gloucs, Mons and Herefs. It is rather variable and both fertile and sterile trees are known. It has brownish-orange fruits and differs from S. devoniensis and S. subcuneata in the longer than wide (not subglobose) fruit with few small (not many large) lenticels.

10. S. leptophylla E. Warb. Shrub to 3m; leaves more deeply **446** toothed than in S. aria. Native; carboniferous limestone **RR** crags; 2 areas in Brecs; endemic.

11. S. wilmottiana E. Warb. Shrub or small tree to 6m. **446** Native; rocky carboniferous limestone woodland and scrub in **RR** Avon Gorge (N Somerset and W Gloucs); endemic.

12. S. eminens E. Warb. Shrub or small tree to 6m. Native; **446** rocky carboniferous limestone woodland; Avon Gorge and Wye **RR** Valley (N Somerset, W Gloucs, Mons and Herefs); endemic.

13. S. hibernica E. Warb. Tree to 6m. Native; rocky **446** carboniferous limestone scrub; Ir, mostly C; endemic. **R**

14. S. porrigentiformis E. Warb. Shrub or small tree to 5m. **446** Native; rocky limestone woods; SW Br from S Devon to Brecs; **R** endemic.

14 x 24. S. porrigentiformis x S. torminalis has been reported from W Gloucs.

15. S. lancastriensis E. Warb. Shrub or small tree to 5m. **446** Native; rocky scrub and woodland, usually on carboniferous **RR** limestone; W Lancs and Westmorland; endemic.

16. S. rupicola (Syme) Hedlund. Shrub or small tree to 6m. **446** Native, rocky woodland, scrub and cliffs, usually on **R** limestone; scattered over Br nd Ir but not in E or SC En.

16 x 24. S. rupicola x S. torminalis has been recorded from Glam and Brecs.

17. S. vexans E. Warb. Small tree to 6m. Native; rocky **446** woods near coast of Bristol Channel (N Devon and S Somerset), **RR**

450 ROSACEAE

not on limestone; endemic.
18–23. S. latifolia agg. – <u>Broad-leaved</u> <u>Whitebeam</u>. Leaves
with ± triangular, acute or acuminate lobes usually divided
>1/6 way to midrib, grey-tomentose on lowerside; fruit orange
to orange-brown, often ± brown when fully ripe; styles 2,
often fused at base.
<u>Multi-access</u> <u>key</u> <u>to</u> spp. <u>of</u> <u>S</u>. <u>latifolia</u> <u>agg.</u>

Fruit orange	A
Fruit orange-brown, sometimes turning brown	B
Fruits ≤13mm	C
Some fruits >13mm	D
Fruits mostly longer than wide	E
Fruits mostly subglobose	F
Anthers cream	G
Anthers pink	H
Leaves mostly with ≥10 pairs of lateral veins	I
Leaves mostly with ≤10 pairs of lateral veins	J
Leaves mostly ≥1.3x as long as wide	K
Leaves mostly ≤1.3x as long as wide	L
Largest leaf-lobes divided ≤1/6 way to midrib	M
Largest leaf-lobes divided ≥1/6 way to midrib	N

A(CD)EGIKN **18. S. decipiens**
ACEHJKM **22. S. bristoliensis**
AD(EF)GJLN   Leaf l/w ratio 1.1–1.3   **23. S. latifolia**
ADFG(IJ)(KL)M Leaf l/w ratio 1.2–1.6 **21. S. croceocarpa**
B(CD)FGJKM   Leaf l/w ratio 1.3–1.6
            Fruit with many large lenticels
                        20. S. devoniensis
BCEGJ(KL)N   Leaf l/w ratio 1.2–1.7
            Fruit with few small lenticels
                    9 x 24. S. aria x S. torminalis
BCFGJKN   Leaf l/w ratio 1.6–1.9
            Fruit with many large lenticels
                        19. S. subcuneata

18. S. decipiens (Bechst.) Irmisch. Tree to 10m. Intrd; 447
natd in Avon Gorge (N Somerset) and at Achnashellach (W
Ross), and planted elsewhere; C Europe.
19. S. subcuneata Wilm. Tree to 10m. Native; open rocky 447
Quercus woods near coast; N Devon and S Somerset; endemic.   **RR**
20. S. devoniensis E. Warb. Tree to 15m. Native; woods and 447
hedges on well-drained soils; widespread in Devon, very local   **R**
in E Cornwall, Man and SE & NE Ir; endemic.
21. S. croceocarpa Sell. Tree to 21m. Intrd; frequently 447
planted, natd in Br from S En to C Sc; origin unknown.
22. S. bristoliensis Wilm. Tree to 10m. Native; rocky 447
woods and scrub on carboniferous limestone; Avon Gorge (N **RR**
Somerset and W Gloucs); endemic.
23. S. latifolia (Lam.) Pers. Tree to 20m; frequently 447
planted, natd in Br from S En to C Sc; SW Europe.
24. S. torminalis (L.) Crantz – <u>Wild</u> <u>Service-tree</u>. Tree to
27m; leaves broadly ovate, 0.9–1.3x as long as wide, with
triangular acute lobes divided up to 1/2 way to midrib,

slightly pubescent on lowerside; fruit brown, longer than
wide; styles 2, fused to c.1/2 way. Native; woods, scrub and
hedgerows mostly on clay or limestone; thoughout most of En
and Wa, but local.

29. ARONIA Medikus - Chokeberries
Leaves simple, serrate, deciduous; flowers in compound
corymbs; petals white; stamens c.20; carpels 5, with 5 styles
fused near base, the walls cartilaginous at fruiting.
Very few records, but likely to increase in future.

1. **A. arbutifolia** (L.) Pers. - Red Chokeberry. Suckering
shrub to 3m; leaves narrowly obovate to elliptic, densely
pubescent on lowerside; fruit red. Intrd; natd in woodland
on sandy soil; Surrey; E N America.
2. **A. melanocarpa** (Michaux) Elliott - Black Chokeberry. **440**
Suckering shrub to 1.5m; leaves obovate, subglabrous; fruit
black. Intrd, natd in boggy area; Caerns; E N America.
The hybrid between the above 2 (**A. x prunifolia** (Marshall)
Rehder), with purple fruits, is grown and should be sought in
the wild.

30. AMELANCHIER Medikus - Juneberry
Leaves simple, serrate, deciduous; flowers in racemes; petals
white; stamens 10-20; carpels 5, with 5 styles fused near
base, the walls cartilaginous at fruiting.

1. **A. lamarckii** F.-G. Schroeder (A. laevis auct. non Wieg., **440**
A. confusa auct. non N. Hylander, A. grandiflora auct. non
Rehder, A. intermedia auct. non Spach, A. canadensis auct.
non (L.) Medikus) - Juneberry. Tree to 10m; leaves mostly
oblong, subglabrous; fruit purplish-black, often not formed.
Intrd; natd on mainly sandy soils in woodland and scrub;
frequent in SC & SE En, very sparsely scattered elsewhere in
En and in Jersey; N America.

31. PHOTINIA Lindley (Stranvaesia Lindley) - Stranvaesia
Leaves simple, entire, evergreen; flowers in compound
corymbs; petals white; stamens c.20; carpels 5, with 5 styles
fused to c.1/2 way, the walls cartilaginous at fruiting;
hypanthium not quite reaching apex of carpels at fruiting.

1. **P. davidiana** (Decne.) Cardot (S. davidiana Decne.) - **440**
Stranvaesia. Shrub or tree to 3(8)m; leaves lanceolate to
oblanceolate, subglabrous; fruit scarlet. Intrd; grown in
gardens, bird-sown in rough ground in S, planted in forestry
plantations in N; extremely scattered in En and Sc; China.

32. COTONEASTER Medikus - Cotoneasters
Leaves simple, entire; flowers in compound corymbs, small
clusters, or solitary; stamens 10-20; carpels 1-5, with 1-5 ±
free styles, the walls stony at fruiting.

The genus parallels Sorbus in that it contains both very
variable sexual spp. and much less variable apomictic spp.;
the precise extent of apomixis is unknown. Much misidentification has occurred.
A large genus becoming increasingly natd via bird-sown seed
from garden or roadside ornamentals. Until familiar with the
genus, flowers, ripe fruit and Summer leaves (after flowering
but before fruit ripe) are necessary for determination, and a
knowledge of the degree of leaf retention in winter is also
desirable. Fruit colours are often diagnostic, but are
difficult to describe, so only 5 colours are defined here:
dark purple to black; yellow (to yellowish-orange); orange-
red; bright red; and crimson (a deep red mildly tinted with
blue). The number of carpels ('stones') per fruit should be
counted in at least 5 fruits; closely adherent stones are
counted as separate. Leaf-sizes and pubescence refer to
those of fully grown Summer leaves; at flowering they may be
much smaller and more densely pubescent in deciduous spp.
The following keys first use 7 characters to separate 10
groups in a multi-access key. 9 of these 10 groups are then
dealt with by dichotomous keys; the 10th group contains only
1 sp.

Multi-access general key
Petals patent, usually white   A
Petals erect to erecto-patent, usually pink   B
   Leaves deciduous or mostly so, most dropped by Jan   C
   Leaves evergreen   D
Fruit dark purple to black   E
Fruit yellow to bright red or crimson   F
   Veins deeply impressed on leaf upperside   G
   Veins not or slightly impressed on leaf upperside   H
Summer leaves densely pubescent to tomentose on
         lowerside, largely or wholly obscuring surface   I
Summer leaves glabrous to pubescent on lowerside,
                       leaving most of surface exposed   J
   Hypanthium and calyx glabrous to sparsely pubescent   K
   Hypanthium and calyx pubescent   L
   Hypanthium and calyx densely pubescent to tomentose   M
Flowers mostly 1-2(3) together   N
Flowers mostly 3-10(12) together   O
Flowers mostly (10)12-many together   P

ACEH(IJ)(KLM)(OP)   Key A
ACFHJKP   1. C. multiflorus
ACF(GH)(IJ)MP   Key B
ADF(GH)(IJ)M(OP)   Key C
ADFH(IJ)(KL)(NO)   Key D
BCE(GH)J(KLM)(NOP)   Key E
BCF(GH)I(KLM)O   Key F
BCFGJ(KL)P   Key G
BCFHJ(KL)(NO)   Key H

Key A - Petals patent; leaves deciduous; fruit dark purple to black
1 Leaves <5(6)cm, <1.25x as long as wide; flowers mostly <20 per inflorescence   2
1 Some leaves >6cm, >1.25x as long as wide; flowers mostly >(15)20 per inflorescence   3
　　2 Shrub to 2m; leaves 1.5-4cm; flowers 4-10 per inflorescence; hypanthium and calyx pubescent
　　　　　　　　　　　　　　　　　　　　　　　　3. C. hissaricus
　　2 Shrub or tree to 6m; leaves 2.5-5(6)cm; flowers 5-20 per inflorescence; hypanthium and calyx tomentose   **2. C. elliptica**
3 Hypanthium and calyx densely pubescent to tomentose; leaves sparsely to densely pubescent on lowerside in Summer; midrib lowersides, peduncles and pedicels pubescent to densely so at fruiting   4
3 Hypanthium and calyx sparsely pubescent to pubescent; leaves subglabrous on lowerside in Summer; midrib lowersides, peduncles and pedicels very sparsely pubescent at fruiting   5
　　4 Flowers mostly <15 per inflorescence; fruiting inflorescences mostly <3 x 3cm   **5. C. affinis**
　　4 Flowers mostly >15 per inflorescence; fruiting inflorescences mostly >3 x 3cm   **4. C. ignotus**
5 Larger leaves mostly <2x as long as wide, very obtuse to rounded and apiculate at apex   **6. C. obtusus**
5 Larger leaves mostly >2x as long as wide, subacute to obtuse and apiculate at apex   6
　　6 Fruits with dense whitish bloom, with very open apex exposing stones   **7. C. bacillaris**
　　6 Fruits without or with sparse whitish bloom, with scarcely open apex ± not exposing stones
　　　　　　　　　　　　　　　　　　　　　　　　　8. C. transens

Key B - Petals patent; leaves deciduous to ± so; fruit yellow to red (see also C. multiflorus)
1 Leaves flat on upperside, very dull, the veins not impressed; fruits 4-6mm   **9. C. frigidus**
1 Leaves with slightly to strongly impressed veins on upperside, usually somewhat shiny; most fruits >6mm   2
　　2 Leaves mostly 2.5-3x as long as wide; flowers >20 per inflorescence; fruits with 2(-5) stones, 5-8mm   **10. C. x watereri**
　　2 Leaves mostly 2-2.5x as long as wide; flowers <15(20) per inflorescence; fruits with 1-2 stones, 6-10mm   3
3 Leaves pubescent to densely so on lowerside; calyx and hypanthium densely pubescent to tomentose **5. C. affinis**
3 Leaves sparsely pubescent to subglabrous on lowerside; calyx and hypanthium sparsely pubescent **8. C. transens**

Key C - Petals patent; leaves evergreen or ± so; fruit yellow
      to red; hypanthium and calyx densely pubescent to
      tomentose
1 Leaves most or all <3cm; flowers <15 per
   inflorescence   14. C. pannosus
1 Leaves most or all >3cm; flowers >15 per
   inflorescence   2
    2 Leaves flat on upperside, the veins not impressed,
       usually semi-deciduous   9. C. frigidus
    2 Leaves with somewhat to strongly impressed veins
       on upperside, usually ± evergreen   3
3 Leaves obtuse to rounded at apex; fruits usually
   with 2 stones   15. C. lacteus
3 Leaves acute to subacute at apex; fruits with 2-5
   stones   4
    4 Leaves oblanceolate to narrowly elliptic; fruits
       mostly 4-5mm, often slightly wider than long,
       with 3-5 stones   11. C. salicifolius
    4 Leaves narrowly obovate to elliptic; fruits mostly
       5-8mm, often slightly longer than wide, with
       2-3(5) stones   10. C. x watereri

Key D - Petals patent; leaves evergreen; fruit red;
      hypanthium and calyx glabrous to sparsely pubescent
      or pubescent
1 Some leaves >1.5cm   2
1 Leaves all <1(1.5)cm   4
    2 Some leaves >2cm; stems all procumbent; fruits
       often with 5 stones   12. C. dammeri
    2 Leaves all <2cm; at least some stems erect or
       arching; fruits with 2-4 stones   3
3 Stems arching and rooting, with procumbent ends;
   leaves obtuse to rounded at apex; fruits bright red,
   with 2-4 stones   13. C. x suecicus
3 Stems stiffly erect and wide-branching; leaves obtuse
   to acute at apex; fruits crimson, with 2 stones
                                               16. C. buxifolius
    4 Leaves obtuse to acute at apex, some usually
       >12mm; flowers 1-several per inflorescence
                                               16. C. buxifolius
    4 Leaves obtuse to rounded at apex, usually <12mm;
       flowers usually all solitary   5
5 Fruits crimson   6
5 Fruits red to orange-red   8
    6 Leaves mid-green and dull on upperside, glabrous
       to sparsely pubescent on lowerside   17. C. congestus
    6 Leaves dark green and very shiny on upperside,
       pubescent to rather sparsely so on lowerside   7
7 Leaves 1-3mm wide, with strongly revolute margins;
   petioles <1.5mm; fruits mostly 4-5 x 5-6mm
                                               19. C. linearifolius
7 Leaves 2-7mm wide, with weakly or non-revolute

margins; petioles >1.5mm; fruits mostly 5-8 x 6-9mm
                                        18. C. integrifolius
8 Leaves mostly >2x as long as wide; stiffly erect
   to spreading, to 1m   **20. C. conspicuus**
8 Leaves mostly <2x as long as wide; procumbent, to
   50cm, rooting along length   **21. C. cashmiriensis**

<u>Key</u> <u>E</u> - Petals erect to erecto-patent; leaves deciduous;
         fruit dark purple to black
1 Leaves <2cm, obtuse at apex; flowers 1-4 together
                                        29. C. nitens
1 Leaves all or nearly all >2cm, acute to acuminate
   at apex; flowers 3-30 together   2
   2 Leaves strongly bullate with deeply impressed
     veins on upperside, usually some (often most)
     >8cm   **38. C. moupinensis**
   2 Leaves not bullate, with not or slightly impressed
     veins on upperside, rarely >8cm   3
3 Calyx and hypanthium glabrous or nearly so on outside
   at flowering, glabrous at fruiting **30. C. lucidus**
3 Calyx and hypanthium densely pubescent to tomentose
   at flowering, pubescent to sparsely so at fruiting   4
   4 Leaves tapering-acuminate, with not or scarcely
     impressed lateral veins on upperside
                                        32. ·C. laetevirens
   4 Leaves acute to shortly and ± abruptly acuminate,
     with distinctly though slightly impressed lateral
     veins on upperside   **31. C. villosulus**

<u>Key</u> <u>F</u> - Petals erect to erecto-patent; leaves deciduous;
         fruit red; leaves densely pubescent to tomentose on
         lowerside
1 Hypanthium and calyx glabrous or nearly so
                                        33. C. integerrimus
1 Hypanthium and calyx pubescent to tomentose, the
   pubescence persisting on calyx until fruit ripe   2
   2 Fruits strongly obovoid to almost pear-shaped, ±
     pendent, with 2 stones; some leaves often >3cm
                                        45. C. zabelii
   2 Fruits subglobose to ellipsoid or slightly obovoid,
     held stiffly, with 3-4 stones; leaves ± all
     <2.5(3)cm   3
3 Fruits bright red, 6-8mm, ± globose; branches
   arching-erect, long and whip-like; plant usually
   >1.5m   **39. C. dielsianus**
3 Fruits orange-red, 8-10mm, usually slightly longer
   than wide; branches stiffly spreading; plant usually
   <1.5m   **40. C. splendens**

<u>Key</u> <u>G</u> - Petals erect; leaves deciduous; fruit red; leaves
         pubescent to sparsely so on lowerside, bullate or at
         least with veins deeply impressed on upperside

1 Leaves 3.5–7cm, weakly bullate, pubescent on lowerside
   at flowering; fruits mostly <8mm   **36. C. bullatus**
1 Leaves 5–12cm, strongly bullate, sparsely pubescent
   on lowerside at flowering; fruits mostly >8mm
                                           37. C. rehderi

Key **H** – Petals erect to erecto-patent; leaves deciduous;
    fruit red to orange-red; leaves pubescent to sparsely
    so on lowerside, not bullate and with veins not or
    scarcely impressed on upperside
1 Plant usually >1m; leaves most or all >1.3x as long
   as wide, acute or acuminate to obtuse at apex   2
1 Plant rarely >1m (except in C. nitidus) unless
   supported; leaves most or all <1.3x as long as wide,
   mostly rounded to broadly obtuse (sometimes subacute
   or apiculate) at apex   4
   2 Fruits parallel-sided, oblong in side view
      (sausage-shaped)   **28. C. divaricatus**
   2 Fruits subglobose to broadly ellipsoid or obovoid,
      with curved sides   3
3 Fruits orange-red; leaves <3cm   **35. C. simonsii**
3 Fruits bright red; most leaves >3cm   **34. C. mucronatus**
   4 Leaves matt or ± so on upperside, ± undulate;
      fruits bright red   5
   4 Leaves shiny on upperside, undulate or flat;
      fruits orange-red to red   6
5 Leaves mostly 1–2.5cm; fruits mostly 10–12mm;
   plant to 1m   **27. C. nanshan**
5 Leaves mostly 0.5–1.5cm; fruits mostly 6–7mm; plant
   to 25cm   **26. C. adpressus**
   6 Branches forming ± regular herring-bone pattern;
      leaves flat   7
   6 Branches irregular, not forming herring-bone
      pattern; leaves flat or undulate   8
7 Leaves mostly 1–2cm; fruits mostly 6–8mm; calyx
   glabrous on surface   **24. C. hjelmqvistii**
7 Leaves mostly 0.6–1.2cm; fruits mostly 4–6mm; calyx
   pubescent on surface even in fruit   **23. C. horizontalis**
   8 Leaves flat; flowers solitary, pendent; petals
      red at base   **22. C. nitidus**
   8 Leaves slightly undulate; flowers (1–)3 together,
      erect; petals purplish-black at base
                                           25. C. atropurpureus

Key **I** – Petals erect to erecto-patent; leaves evergreen;
      fruit orange-red to red
1 Most leaves >2cm   2
1 Most leaves <2cm   3
   2 Leaves 2–3.5cm; calyx-lobes acuminate, with hair-
      less points c.0.5mm; fruit orange-red, 6–9mm,
      broadly obovoid   **41. C. franchetii**
   2 Leaves 2.5–5cm; calyx-lobes apiculate, with hair-

less points c.0.2mm; fruit red, 8-10mm, subglobose
                                        42. C. sternianus
3 Veins deeply impressed on leaf upperside; flowers
  1-4 together; fruit with 3-4 stones   **43. C. insculptus**
3 Veins not or scarcely impressed on leaf upperside;
  flowers 6-10 together; fruit with 2-3 stones
                                        44. C. amoenus

Other spp. - Probably 60-80 spp. are widely cultivated and
most of them can be expected to become natd as the fruits are
so attractive to birds. Natd plants occur wherever birds
defecate, notably on waste and rough ground, on banks and
walls, in open woodland and scrub, and on grassland becoming
colonized by shrubs, especially on chalk and limestone.

Section 1 - CHAENOPETALUM Koehne (spp. 1-21). Petals patent,
white; flowers often in groups of 15-many (often not); leaves
often evergreen (often not); fruits with 1-2(3) stones
(except 3-5 in spp. 10-12).

1. C. multiflorus Bunge - Many-flowered Cotoneaster. Shrub
or tree to 3(5)m with arching branches; leaves 2-5cm, flat,
very sparsely pubescent on lowerside; inflorescences >10-
flowered; fruits red, 8-11mm, subglobose, usually with 2 ±
fused stones. Intrd; natd on chalk in Beds; W Siberia.
2. C. ellipticus (Lindley) Loudon (C. lindleyi Steudel nom.
illeg., C. insignis Pojark.) - Lindley's Cotoneaster. Shrub
or tree to 6m with erect to arching branches; leaves 2.5-6cm,
flat, pubescent to sparsely so on lowerside; inflorescences
5-15(20)-flowered; fruits bluish-black, 7-9mm, subglobose,
with 1(-2) stones. Intrd; natd in SE En; Himalayas.
3. C. hissaricus Pojark. - Round-leaved Cotoneaster. Shrub
1.5-2m with rigid close branches; leaves 1.5-4cm, flat,
pubescent on lowerside; inflorescences 4-10-flowered; fruits
dark purple to bluish-black, 6-9mm, subglobose, with 1(-2)
stones. Intrd; natd in SE En; Himalayas.
4. C. ignotus Klotz (C. hissaricus auct. non Pojark.) -
Black-grape Cotoneaster. Shrub or tree to 6m with erect to
arching branches; leaves 3.5-8cm, flat, pubescent on lowerside; inflorescences mostly >15-flowered; fruits bluish-
black, 7-8mm, subglobose. Intrd; natd in SE En; Himalayas.
5. C. affinis Lindley - Purpleberry Cotoneaster. Shrub or
tree to 5m, with erect to arching branches; leaves 4-9cm,
flat, pubescent on lowerside; inflorescences <15-flowered;
fruits red, becoming purplish-black when fully ripe, 7-9mm,
subglobose. Intrd; natd in S En; Himalayas.
6. C. obtusus Wallich ex Lindley (C. cooperi auct. non
Marquand) - Dartford Cotoneaster. Shrub or tree to 5m, with
erect to arching branches; leaves 3.5-9cm, flat, subglabrous
on lowerside; inflorescences 5-15-flowered; fruits black,
6-8mm, subglobose. Intrd; natd in W Kent; Himalayas.
7. C. bacillaris Wallich ex Lindley (C. affinis var.

bacillaris (Wallich ex Lindley) C. Schneider) – Open-fruited
Cotoneaster. Shrub or tree to 5m, with widely arching
branches; leaves 3–8.5cm, flat, very sparsely pubescent on
lowerside; inflorescences 5–20-flowered; fruits purplish-
black, 6–10mm, broadly obovoid, the flesh not concealing
stones at apex. Intrd; natd in SE En; Himalayas.
    8. C. transens Klotz – Godalming Cotoneaster. Differs from
C. bacillaris in leaves slightly thicker; fruits red,
becoming brownish-black when fully ripe, with 1–2 (not 2)
stones; and see Key A (couplet 6). Intrd; natd in Beds and
Surrey; China.
    9. C. frigidus Wallich ex Lindley – Tree Cotoneaster. Erect
shrub or strong tree to 8(18)m, deciduous to semi-evergreen;
leaves 6–13cm, flat, pubescent on lowerside; inflorescences
usually >20-flowered; fruits usually red, sometimes orange,
yellow or crimson, 4–6mm, depressed-globose. Intrd; natd in
many parts of Br and Ir; Himalayas.
    10. C. x watereri Exell (C. frigidus x C. salicifolius) – **459**
Waterer's Cotoneaster. Usually semi-evergreen erect shrubs
to 8m; many cultivars and their seedlings variously intermediate between the parents, with range in fruit colour as
for C. frigidus, but fruits often larger than in either
parent (5–8mm). Intrd; natd in many parts of Br and Ir;
garden origin.
    11. C. salicifolius Franchet – Willow-leaved Cotoneaster.
Erect, arching or ± procumbent evergreen shrub to 5m; leaves
3–8cm, shiny and ± bullate on upperside, tomentose on lowerside; inflorescences usually >20-flowered; fruits red, 4–5mm,
subglobose to depressed-globose. Intrd; natd in S Br; China.
    12. C. dammeri C. Schneider – Bearberry Cotoneaster.
Procumbent shrub with branches to 3m; leaves 1.5–3cm, shiny
and flat on upperside, glabrous to sparsely pubescent on
lowerside; inflorescences 1–2(4)-flowered; fruits bright red,
6–8mm, globose to very broadly obovoid. Intrd; natd rarely
in Br, mostly S En; China.
    13. C. x suecicus Klotz (?C. dammeri x C. conspicuus) –
Swedish Cotoneaster. Stems arching to 50cm high, trailing to
2m; leaves 1–2cm, otherwise similar to those of C. dammeri;
inflorescences 1–4(6)-flowered; fruits as in C. dammeri but
less bright red. Intrd; natd in S En; garden origin.
Cultivars 'Skogholm' and 'Coral Beauty', often mass-planted,
belong here; seedlings are often very like their parent.
    14. C. pannosus Franchet – Silverleaf Cotoneaster. Stems
long, slender, slightly arching, to 3m; leaves 1–2.5cm, dull
and nearly flat on upperside, tomentose on lowerside;
inflorescences 6–12-flowered; fruits dull red, 5–8mm,
subglobose to ellipsoid, often with conspicuous erect sepals.
Intrd; natd in W Kent; China.
    15. C. lacteus W. Smith – Late Cotoneaster. Spreading shrub **459**
to 4m; leaves 3.5–9cm, slightly shiny and with deeply
impressed veins on upperside, tomentose on lowerside;
inflorescences >20-flowered; fruits red to crimson-red,

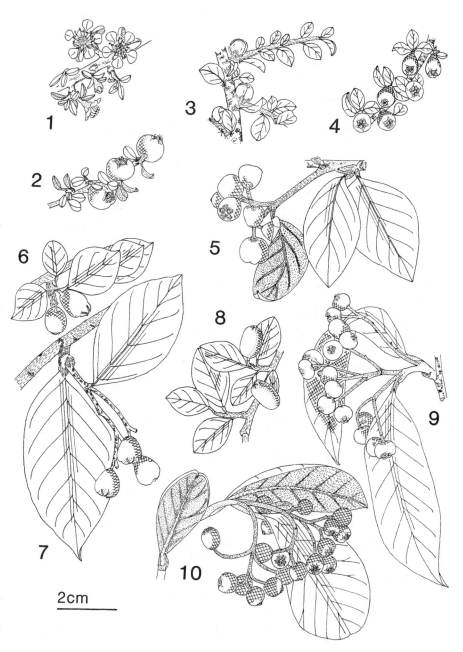

FIG 459 — Cotoneaster. 1-2, C. integrifolius. 3-4, C. horizontalis. 5, C. sternianus. 6, C. simonsii. 7, C. rehderi. 8, C. divaricatus. 9, C. x watereri. 10, C. lacteus.

2cm

5-6mm, $\pm$ globose. Intrd; natd in S En, planted as field-hedges in E Anglia; China. The latest flowering and fruiting sp.; fruits rarely ripen before Nov.

16. C. buxifolius Wallich ex Lindley - Box-leaved Cotoneaster. Stiffly branched shrub to 1(2)m; leaves 0.5-2cm, shiny and $\pm$ flat on upperside, sparsely to densely pubescent on lowerside; inflorescences 1-8-flowered; fruits crimson, 7-10mm, subglobose. Intrd; natd in S En; S India.

17. C. congestus Baker - Congested Cotoneaster. Tightly branched shrub to 70cm; leaves 0.5-1.2cm, dull and slightly rugose on upperside, very sparsely pubescent on lowerside; flowers solitary; fruits crimson, 6-7mm, depressed-globose. Intrd; natd in S Hants and W Kent; Himalayas.

18. C. integrifolius (Roxb.) Klotz (C. thymifolius Lindley, 459 C. microphyllus auct. non Wallich ex Lindley) - Small-leaved Cotoneaster. Procumbent to arching shrub to 1m; leaves 0.4-0.8(1.5)cm, very shiny and flat on upperside, pubescent to sparsely so on lowerside; flowers all or $\pm$ all solitary; fruits crimson, 5-8mm, depressed-globose to globose. Intrd; frequently natd over much of BI; Himalayas to W China. Very commonly grown, $\pm$ always as C. microphyllus, which has not been found natd.

19. C. linearifolius (Klotz) Klotz (C. thymifolius auct. non Lindley) - Thyme-leaved Cotoneaster. Differs from C. integrifolius in procumbent shrub to 60cm; and see Key D (couplet 7). Intrd; natd in SE En; Himalayas.

20. C. conspicuus Marquand - Tibetan Cotoneaster. Stiffly erect to spreading shrub to 1m; differs from C. integrifolius in leaves much less shiny on upperside; and fruits bright red, 6-9mm. Intrd; natd in SE En; Tibet.

21. C. cashmiriensis Klotz (C. cochleatus auct. non (Franchet) Klotz)- Kashmir Cotoneaster. Differs from C. conspicuus as in Key D (couplet 8). Intrd; natd in E Kent; Kashmir. Commonly grown, usually as C. cochleatus.

Section 2 - COTONEASTER (Orthopetalum Koehne) (spp. 22-45). Petals erect to suberect, pink or pink-tinged; flowers in groups of 1-10(12) (except more in spp. 34-36); leaves deciduous (except in spp. 39-42); fruits with 2-5 stones.

22. C. nitidus Jacques (C. distichus Lange) - Distichous Cotoneaster. Erect shrub to 2m, with wide-spreading branches; leaves 5-13mm, shiny and flat on upperside, sub-glabrous on lowerside; flowers solitary; fruits red to orange-red, 7-11mm, broadly obovate to subglobose. Intrd; natd in S Hants; Himalayas.

23. C. horizontalis Decne. - Wall Cotoneaster. Arching to 459 horizontal shrub to 1(3)m, often vertical on walls, with very regular herring-bone branching; leaves 0.6-1.2cm, shiny and flat on upperside, subglabrous on lowerside; flowers 1-2 together; fruits red to orange-red, 4-6mm, $\pm$ globose. Intrd; commonly natd in much of BI; W China.

24. C. hjelmqvistii Flinck & Hylmoe (<u>C. horizontalis</u> 'Robustus') – <u>Hjelmqvist's</u> Cotoneaster. Differs from <u>C. horizontalis</u> in stronger growth to 1.5(4.5)m with less regular herring-bone branches; larger, ± orbicular leaves to 1.5(2)cm; and fruits 6–8mm. Intrd; natd in W Kent and W Cornwall; ?W China.

25. C. atropurpureus Flinck & Hylmoe (<u>C. horizontalis</u> 'Prostratus') – <u>Purple-flowered</u> Cotoneaster. Differs from <u>C. horizontalis</u> in branching irregular; leaves 0.9–1.4cm, thin, slightly undulate at margin, obtuse to truncate at apex; flowers (1–)3 together; and fruits mostly 6–9mm. Intrd; natd in Lanarks; China.

26. C. adpressus Bois – <u>Creeping</u> Cotoneaster. Usually procumbent shrub to 25cm, with irregular branching; leaves 0.5–1.5cm, matt on upperside, ± undulate, very sparsely pubescent on lowerside; flowers 1–2 together; fruits bright red, 6–7mm, subglobose. Intrd; natd in W Kent and Lanarks; W China.

27. C. nanshan A. Vilm. ex Mottet (<u>C. adpressus</u> var. <u>praecox</u> Bois & Berthault) – <u>Dwarf</u> Cotoneaster. Differs from <u>C. adpressus</u> in spreading shrub to 50(100)cm; leaves mostly 1–2.5cm; fruits 9–12mm, very broadly ellipsoid. Intrd; natd in W Kent; W China.

28. C. divaricatus Rehder & E. Wilson – <u>Spreading</u> 459 Cotoneaster. Shrub to 2m with wide spreading or arching branches; leaves 0.8–2.5cm, flat and shiny on upperside, very sparsely pubescent on lowerside; flowers 2–3(4) together; fruits (7)9–12mm, bright red, oblong-ellipsoid. Intrd; natd in S & SE En; China.

29. C. nitens Rehder & E. Wilson – <u>Few-flowered</u> Cotoneaster. Erect densely branched shrub to 2.5m; leaves 0.8–2.2cm, flat and shiny on upperside, subglabrous on lowerside; flowers 1–4 together; fruits 7–9mm, black, subglobose. Intrd; natd in SE En; China.

30. C. lucidus Schldl. – <u>Shiny</u> Cotoneaster. Spreading to erect shrub to 2m; leaves 2–5(7)cm, shiny and with slightly impressed veins on upperside, sparsely pubescent on lowerside, with brilliant Autumn colours; flowers 3–15 together; fruits 8–10mm, shiny-black, subglobose. Intrd; natd in scattered places in En; Siberia and Mongolia.

31. C. villosulus (Rehder & E. Wilson) Flinck & Hylmoe (<u>C. acutifolius</u> auct. non Turcz.) – <u>Lleyn</u> Cotoneaster. Erect shrub to 4m; leaves 3–8(10)cm, slightly shiny on upperside with impressed veins, sparsely pubescent on lowerside; flowers 3–15 together; fruits 8–10mm, shiny-black, broadly obovoid. Intrd; natd in scattered places in En and Wa; China.

32. C. laetevirens (Rehder & E. Wilson) Klotz (<u>C. ambiguus</u> auct. non Rehder & E. Wilson) – <u>Ampfield</u> Cotoneaster. Differs from <u>C. villosulus</u> in shrub to 3m; leaves ± matt on upperside with scarcely impressed veins; and see Key E (couplet 4). Intrd; natd in S Hants; China.

33. C. integerrimus Medikus - <u>Wild</u> Cotoneaster. Irregularly **RRR** branched spreading shrub to 1.5m; leaves 1-4cm, flat and matt on upperside, tomentose on lowerside; flowers 1-4(7) together; fruits 7-11mm, red, globose. Native; very few plants on limestone of Great Orme's Head, Caerns, known since 1783.

34. C. mucronatus Franchet - <u>Mucronate</u> Cotoneaster. Erect shrub to 4m; leaves 2-6cm, flat and matt on upperside but often with undulate edges, pubescent on lowerside; flowers 2-4 together; fruits 8-12mm, red, broadly ellipsoid. Intrd; natd in Middlesex; China.

35. C. simonsii Baker - <u>Himalayan</u> Cotoneaster. Erect shrub **459** to 3(4)m; leaves 1.5-2.5(3)cm, shiny and flat on upperside, rather sparsely pubescent on lowerside; flowers 1-4 together; fruits (6)8-11mm, orange-red, globose to broadly obovoid. Intrd; natd rather commonly thoughout most of BI; Himalayas.

36. C. bullatus Bois - <u>Hollyberry</u> Cotoneaster. Arching shrub to 4m; leaves 3.5-7cm, bullate on upperside, pubescent on lowerside; flowers 12-30 together; fruits mostly 6-8mm, bright shiny red, subglobose to obovoid. Intrd; natd frequently in Br and Ir; China. Fruits ripen very early (Aug).

37. C. rehderi Pojark. (<u>C.</u>, bullatus var. <u>macrophyllus</u> **459** Rehder & E. Wilson) - <u>Bullate</u> Cotoneaster. Arching shrub to 4.5m; leaves 5-12cm, extremely bullate on upperside, rather sparsely pubescent on lowerside; flowers and fruits as in <u>C.</u> <u>bullatus</u> except calyx less pubescent and fruits mostly 8-11mm. Intrd; natd frequently in Br and Ir; China. Fruits also ripen in Aug.

38. C. moupinensis Franchet (<u>C. foveolatus</u> auct. non Rehder & E. Wilson) - <u>Moupin</u> Cotoneaster. Differs from <u>C. rehderi</u> in shrub to 3m; leaves 4-10cm; and fruits purplish-black, 6-10mm. Intrd; natd in NW En; China.

39. C. dielsianus E. Pritzel ex Diels - <u>Diels'</u> Cotoneaster. Arching shrub to 2(3)m; leaves 1.2-2.5(3)cm, with veins slightly impressed on upperside, greyish- or greenish-tomentose on lowerside; flowers 3-7 together; fruits 6-8mm, bright red, subglobose. Intrd; natd rather frequently in Br; China. Often confused with <u>C. franchetii</u> but deciduous and fruits bright red.

40. C. splendens Flinck & Hylmoe - <u>Showy</u> Cotoneaster. Widely spreading shrub to 1(2)m; leaves 1-2(2.5)cm, ± flat on upperside, greyish- or yellowish-tomentose on lowerside; flowers 2-8 together; fruits 8-11mm, orange to orange-red, broadly obovoid to subglobose. Intrd; natd in SE En; W China.

41. C. franchetii Bois - <u>Franchet's</u> Cotoneaster. Arching shrub to 3(4)m; leaves 2-3.5cm, with veins deeply impressed on upperside, yellowish-tomentose on lowerside; flowers 5-15 together; fruits 6-8mm, orange-red, broadly obovoid to subglobose. Intrd; natd in scattered places in Br; W China.

42. C. sternianus (Turrill) Boom - <u>Stern's</u> Cotoneaster. **459**

Differs from C. franchetii as in Key I (couplet 2). Intrd;
natd in scattered places in Br; Burma. Most plants
determined as C. franchetii are this.
43. C. insculptus Diels - Engraved Cotoneaster. Shrub to
3m, with spreading branches; leaves 1-2(2.5)cm, with veins
deeply impressed on upperside, white- to grey-tomentose on
lowerside; flowers 1-4 together; fruits 6-10mm, orange-red,
broadly obovoid. Intrd; natd in Caerns; China.
44. C. amoenus E. Wilson - Beautiful Cotoneaster. Densely
branched shrub to 1.5m; leaves 1-2cm, ± flat on upperside,
white- to grey-tomentose on lowerside; flowers 6-10 together;
fruits 5-6mm, red, subglobose, often with conspicuous erect
sepals as in C. pannosus. Intrd; natd in S & SE En; China.
45. C. zabelii C. Schneider - Cherryred Cotoneaster. Erect
shrub to 3m; leaves 1.5-4cm, with veins slightly impressed on
upperside, densely pubescent to ± tomentose on lowerside;
flowers 4-10 together; fruits bright red, strongly obovoid to
± pear-shaped. Intrd; natd in W Kent; China.

33. PYRACANTHA M. Roemer - Firethorns
Leaves simple, serrate, evergreen; flowers in compound
corymbs; petals white; stamens 20; carpels 5, with 5 free
styles, the walls stony at fruiting.

1. P. coccinea M. Roemer - Firethorn. Spiny shrub to 2(6)m;
leaves 2-7cm, narrowly obovate to oblanceolate, with
pubescent petioles; inflorescence-stalks pubescent; fruits
globose to depressed-globose, yellow, orange or scarlet.
Intrd; very commonly grown and natd as bird-sown plants on
banks and walls and in rough ground; frequent in S & C Br; S
Europe.
2. P. rogersiana (A.B. Jackson) Coltman-Rogers - Asian
Firethorn. Differs from P. coccinea in ± glabrous petioles
and inflorescence branches and smaller leaves up to 3(5)cm.
Intrd; commonly grown but known natd only in E Kent, probably
overlooked; China. Some cultivars are P. coccinea x P.
rogersiana and these might become natd too.

34. MESPILUS L. - Medlar
Leaves simple, ± entire, deciduous; flowers solitary; petals
white; stamens 30-40; carpels 5, with 5 free styles, the
walls stony at fruiting.

1. M. germanica L. - Medlar. Shrub or tree to 9m, sometimes
spiny; leaves 5-12cm, elliptic-oblong, pubescent; flowers
3-5cm across excl. sepals which project beyond petals; fruit
subglobose, 2-3cm (to 6cm in cultivars), with persistent
erect to erecto-patent sepals. Intrd; natd in hedges for at
least 4 centuries; local in CI and S Br, sporadic in C & N
En; SE Europe.

34 x 35. MESPILUS X CRATAEGUS = X CRATAEMESPILUS Camus - Haw-medlar

34/1 x 35/1. x C. grandiflora (Smith) Camus (M. germanica x C. laevigata) occurs sporadically as isolated trees of uncertain origin very scattered through Br; leaves elliptic to obovate, serrate, usually slightly lobed; fruits brown, obovoid, <2cm, with 2-3 styles and carpels, some usually 2-3 together; sepals persistent but <1cm. This binomial might apply to M. germanica x C. monogyna (x C. gillotii G. Beck) which possibly also occurs. Not to be confused with the graft-hybrid **+ Crataegomespilus dardarii** Simon-Louis ex Ballair, which has very variable leaves on one plant and usually branches of the pure spp. here and there, and clustered fruits with sepals as in Mespilus.

35. CRATAEGUS L. - Hawthorns
Leaves simple, serrate, lobed or not, deciduous; flowers in corymbs; petals white to pink; stamens (5)10-20; carpels 1-5, with as many free styles, the walls stony at fruiting.

1 At least some leaves usually lobed >1/3 way to midrib; both apices of lobes and sinuses between lobes on at least some leaves reached by major vein from midrib 2
1 Leaves not lobed or lobed usually <1/3 way to midrib; only apices of lobes or main teeth (not sinuses) reached by major vein 5
  2 Leaves densely pubescent; styles and nutlets (3)4-5 **10. C. laciniata**
  2 Leaves glabrous to sparsely pubescent; styles and nutlets 1-3 3
3 Leaves varying from unlobed to deeply lobed on 1 tree, some narrowly oblong to oblanceolate and entire in basal 1/2 **9. C. heterophylla**
3 Leaves not strongly heterophyllous, none narrowly oblong, unlobed and entire in basal 1/2 4
  4 Styles and nutlets 1(-2); deepest sinus between leaf-lobes reaching >2/3 way to midrib; leaf-lobes (3)5-7, the lowest pair acute **7. C. monogyna**
  4 Styles and nutlets (1)2-3; deepest sinus between leaf-lobes reaching <2/3 way to midrib; leaf-lobes usually 3, the lateral pair obtuse **8. C. laevigata**
5 Styles and nutlets mostly 2-3 6
5 Styles and nutlets mostly 4-5 8
  6 Leaves and inflorescences glabrous; nutlets without hollows on inner surfaces **4. C. crus-galli**
  6 Leaves pubescent on lowerside veins; inflorescence-stalks pubescent; nutlets with hollows on inner surfaces 7
7 Stamens 10-15; leaves cuneate (<90 degrees) at base **5. C. x persimilis**
7 Stamens 15-20; leaves broadly cuneate (>90 degrees) at base **6. C. succulenta**

8 Leaves pubescent on lowerside at first, only on
  veins later; inflorescence-stalks tomentose;
  stamens 10   **1. C. submollis**
8 Leaves glabrous to pubescent only on veins at
  first, glabrous later; inflorescence-stalks
  glabrous or with sparse shaggy hairs   9
9 Stamens 20; fruits subglobose; flowers >2cm across
                                            2. C. coccinioides
9 Stamens 10; fruits ellipsoid to pear-shaped; flowers
  <2cm across   **3. C. pedicellata**

Other spp. - Increasing numbers of spp. are grown for
ornament and more might become natd, especially N American
spp. with scarcely lobed leaves and large fruits; some are
hybrids and do not breed true. Odd individuals in woods and
hedges may be bird-sown but are often planted, frequently as
grafts on common spp. There are odd records for at least 10
spp. **C. punctata** Jacq., from N America, related to C. crus-
galli, was formerly natd.

1. C. submollis Sarg. (C. coccinea auct. non L.) - Hairy
Cockspur-thorn. Tree to 8m; spines 5-7cm, thin; leaves
ovate, doubly serrate and shallowly lobed; fruits pear-
shaped, 15-20mm. Intrd; frequently planted, self-sown in
hedges and rough ground in SE En; E N America.
2. C. coccinioides Ashe - Large-flowered Cockspur-thorn.
Tree to 7m; spines 3-5cm, strong; leaves triangular-ovate,
doubly serrate and shallowly lobed, with acuminate teeth;
fruits subglobose, 15-20mm. Intrd; planted and natd as for
C. submollis; C USA.
3. C. pedicellata Sarg. (C. coccinea auct. non L.) - Pear-
fruited Cockspur-thorn. Tree to 7m; spines 3-5cm, medium;
leaves broadly ovate, doubly serrate and shallowly lobed;
fruits pear-shaped to ellipsoid, 15-20mm. Intrd; planted and
natd as for C. submollis; E N America.
4. C. crus-galli L. - Cockspur-thorn. Tree to 6m; spines
3-8cm, medium; leaves elliptic to obovate, doubly serrate,
not lobed; fruits subglobose, c.10mm. Intrd; formerly much
planted and still natd in hedges in S & C Br, over-recorded
for C. persimilis. E N America.
5. C. persimilis Sarg. (C. prunifolia Pers. non (Marshall)
Baumg.) - Broad-leaved Cockspur-thorn. Differs from C.
crus-galli in pubescence on leaves and inflorescence, wider
leaves, and fruit falling in Autumn (not Spring). Intrd;
abundantly planted, sometimes self-sown in S En, often
misnamed C. crus-galli; our plant is cultivar 'Prunifolia'.
6. C. succulenta Schrader - Round-fruited Cockspur-thorn.
Tree to 6m; spines 3-5cm, strong; leaves broadly obovate to
elliptic, doubly serrate, slightly lobed near apex; fruits
globose, 12-15mm. Intrd; frequently planted, natd in woods
and hedges; S En; E N America.
7. C. monogyna Jacq. - Hawthorn. Shrub or tree to 10(15)m;

spines 1-2.5cm, strong to medium; leaves ovate to broadly so, (3)5-7-lobed >2/3 way to midrib, the lobes acutely serrate near apex; fruits subglobose to broadly ellipsoid, (6)8-10(13)mm. Native; wood-borders, scrub and hedges; abundant throughout BI. Our plants have been referred to ssp. **nordica** Franco. Very varied in growth-form, from procumbent shrub to erect tree.

7 x 8. C. monogyna x C. laevigata = C. x macrocarpa Hegetschw. (C. x ovalis Kittel, C. x media auct. non Bechst.) is common throughout the range of C. laevigata and even beyond (where the latter presumably once occurred); natd in Co Antrim. It is fertile and covers the whole spectrum of intermediacy.

8. C. laevigata (Poiret) DC. (C. oxyacanthoides Thuill.) - **Midland Hawthorn.** Tree to 10m; differs from C. monogyna in less stiff, less spiny twigs; and see key. Native; woods, often well shaded, and hedges; common in C & SE En, scattered W to Wa and N to N En, but extent of native area uncertain, natd in SW & NE Ir.

9. C. heterophylla Fluegge - Various-leaved Hawthorn. Tree to 6m; spines very sparse; leaves strongly heterophyllous, narrowly oblong to elliptic, some with 3-5 forward-directed lobes, some unlobed; fruits oblong, 10-18mm, bright red, with 1 nutlet and style. Intrd; planted in parks, rarely self-sown; natd in cemetery in Middlesex; Caucasus.

10. C. laciniata Ucria (C. orientalis Pallas ex M. Bieb.) - Oriental Hawthorn. Tree to 6m; spines very sparse; leaves obovate to oblong-obovate, 3-7-lobed >3/4 (often nearly whole) way to midrib, the lobes narrow with few apical teeth; fruits globose to pear-shaped, 15-20mm, brick-red to orange. Intrd; frequently grown, natd in hedges and on banks in S En; SE Europe and SW Asia.

77. MIMOSACEAE - Australian Blackwood family
(Leguminosae subfam. Mimosoideae)

Suckering trees; leaves alternate, of 2 sorts, the juvenile bipinnate with numerous leaflets, the adult simple; stipules ± 0. Flowers in dense racemes of spherical heads, bisexual, hypogynous, actinomorphic; sepals mostly 5, fused into tube; petals mostly 5, fused into tube proximally; stamens numerous, longer than petals; carpel 1, with several ovules in row; style 1; stigma capitate; fruit a legume.

At once recognizable by the small spherical pom-poms of flowers of which the stamens are the most conspicuous part.

Other genera - ALBIZIA Durazz. differs from Acacia in the stamens fused below and in the tuft-like, not spherical, flower clusters. **A. lophantha** Benth., from Australia, with large bipinnate leaves, is planted for ornament in Scillies and gives rise to seedlings.

1. ACACIA Miller - Australian Blackwood

Other spp. - A few spp. are grown for ornament in SW En and may sometimes spread very locally by suckers or produce young seedlings, but are not truly natd.

1. A. melanoxylon R. Br. - <u>Australian</u> <u>Blackwood</u>. Tree to 15m; adult leaves 6-13(20)cm, lanceolate to oblanceolate, slightly curved, with 3-5 longitudinal veins; flower-heads c.10mm across, few per raceme, cream to yellow. Intrd; grown for ornament; locally ± natd in S Devon and Scillies; Australia.

77A. CAESALPINIACEAE
(Leguminosae subfam. Caesalpinioideae)

CASSIA occidentalis L. (<u>Coffee</u> <u>Senna</u>), from the tropics, is a fairly frequent casual from soya-bean waste, but rarely reaches flowering. It is an erect herbaceous perennial with paripinnate leaves up to 20cm and yellow flowers up to 3cm across in racemes; sepals 5; petals 5, slightly zygomorphic; stamens 10, varying from 2 large to 2 very reduced; carpel 1, with many ovules; fruit a ± indehiscent legume 5-13cm.

78. FABACEAE - Pea family
(Papilionaceae; Leguminosae subfam. Papilionoideae, subfam. Lotoideae)

Annual to perennial herbs, shrubs or trees, sometimes spiny; leaves alternate, simple to palmate or pinnate, often with tendrils, usually stipulate. Flowers solitary or variously grouped, bisexual, hypogynous, zygomorphic, always like that of the pea in organization; sepals usually 5, usually fused into tube, often differentiated into upper lip of 2 sepals and lower lip of 3 sepals; petals 5, the upper (<u>standard</u>), 2 free laterals (<u>wings</u>) and 2 fused lower (<u>keel</u>), the last ± concealing the stamens and carpel; stamens 10, usually all fused into tube below or the uppermost free and the 9 lower fused, rarely all 10 free; carpel 1, with 1-many ovules in row; style 1, stigma capitate; fruit basically a legume, but very variably modified, usually dehiscent along 2 sides but often a schizocarp (breaking transversely into 1-seeded units).

The flowers, like those of a pea, are diagnostic; fruits and leaves are very variable.

<u>General key</u>
1 Leaves simple, sometimes reduced to a tendril, spine or scale, sometimes 0   <u>Key A</u>
1 At least some leaves with at least 2 leaflets   2

2 Leaves with 1-many pairs of leaflets, with or
  without an odd terminal leaflet, if with then
  pairs of leaflets >1, if without then often
  with tendrils   Key C
2 Leaves ternate or palmate, without tendrils   Key B

Key A - Leaves simple, sometimes reduced to tendril, spine or
        scale, or 0
1 Herbaceous annuals or perennials   2
1 Woody shrubs   5
  2 Fruit opening along 2 sides like a pea-pod
                                               21. LATHYRUS
  2 Fruit indehiscent, or breaking transversely between
    seeds   3
3 Fruit 1(-2)-seeded, enclosed in calyx   11. ANTHYLLIS
3 Fruit >2-seeded, exserted from calyx   4
  4 Plant glabrous; fruit curved, smooth   15. CORONILLA
  4 Plant slightly pubescent; fruit spiralled, with
    tubercles or weak spines   18. SCORPIURUS
5 Plant spiny, at least some spines branched   6
5 Spines 0 or simple   7
  6 Upper calyx-lip with 2 short teeth; small
    bracteole present on either side of flower   35. ULEX
  6 Upper calyx-lip divided >1/3 way to base;
    bracteoles 0   34. GENISTA
7 Flowers white   32. CYTISUS
7 Flowers yellow to reddish   8
  8 Twigs strongly angled or grooved; upper calyx-lip
    with 2 short teeth (divided <1/5 way to base)
                                                32. CYTISUS
  8 Twigs finely grooved; upper calyx-lip divided
    >1/4 way to base   9
9 Spines 0; upper calyx-lip divided nearly to base, the
  2 halves inclined downwards near lower lip 33. SPARTIUM
9 Spines 0 or present; calyx with distinct upper and
  lower lips, the upper divided 1/4 to 3/4 way to base
                                               34. GENISTA

Key B - Leaves ternate or palmate, without tendrils
1 Leaves palmate, with >4 leaflets   30. LUPINUS
1 Leaves ternate   2
  2 Woody trees or shrubs   3
  2 Herbaceous annuals or perennials   8
3 Stems spiny; corolla yellow   35. ULEX
3 Stems not spiny, or if spiny corolla not yellow   4
  4 Fruit spiral   27. MEDICAGO
  4 Fruit ± straight   5
5 Leaflets toothed; flowers usually pinkish-purple,
  rarely white   24. ONONIS
5 Leaflets entire; flowers usually yellow, rarely white 6
  6 Leaflets mostly >3cm; racemes pendent   31. LABURNUM
  6 Leaflets mostly <3cm; flowers 1-few or in ± erect

```
            racemes                                            7
7  Upper lip of calyx deeply bifid              34. GENISTA
7  Upper lip of calyx with 2 short teeth        32. CYTISUS
   8  Main lateral veins of leaflets running whole way
      to margin; leaflets often toothed                        9
   8  Main lateral veins of leaflets not reaching
      margin; leaflets rarely toothed                         17
9  Calyx with glandular (and often non-glandular)
   hairs; all 10 stamens fused into tube        24. ONONIS
9  Calyx without glandular hairs; 9 stamens fused into
   tube, the 10th free                                        10
   10  Flowers in elongated racemes; fruit <7mm, all
       with 1-2 seeds                                         11
   10  Flowers few, or in short dense heads, or if in
       elongated racemes then fruits >7mm and at least
       some with >2 seeds                                     12
11 Flowers yellow or white; fruits exserted from calyx-
   tube                                         25. MELILOTUS
11 Flowers cream or pink to purple; fruits included in
   calyx-tube                                   28. TRIFOLIUM
   12  Fruits spiralled into >1/2 complete coil, often
       spiny                                    27. MEDICAGO
   12  Fruits straight to curved (<1/2 complete coil),
       never spiny                                            13
13 Fruits >3cm, plus beak >1cm                  26. TRIGONELLA
13 Fruits <3cm, incl. beak <1cm                              14
   14  Flowers yellow; fruits + curved, >7mm, at least
       some >2-seeded                                         15
   14  Flowers not yellow, or if yellow then fruits
       straight, <7mm, 1-2-seeded                            16
15 Ripe fruits pendent; plant annual            26. TRIGONELLA
15 Ripe fruits + erect; plant perennial         27. MEDICAGO
   16  Fruits inflated, exserted from calyx and forming a
       compact naked head                       26. TRIGONELLA
   16  Fruits not or scarcely inflated, usually at least
       partly covered by calyx or persistent corolla,
       not forming a compact naked head         28. TRIFOLIUM
17 Leaflets with small stipule-like outgrowths at base    18
17 Leaflets without stipule-like outgrowths at base       20
   18  Fruit erect to patent; stipules broadly ovate

                                                3. VIGNA
   18  Fruit pendent; stipules triangular to narrowly so  19
19 Corolla c. as long as calyx; common peduncle 0 to
   very short; plant with brown patent hairs    4. GLYCINE
19 Corolla much longer than calyx; common peduncle long;
   plant glabrous to rather sparsely pubescent with
   whitish hairs                                2. PHASEOLUS
   20  Petals blue to white; leaflets dentate   5. PSORALEA
   20  Petals yellow; leaflets entire                        21
21 All 10 stamens free                          29. THERMOPSIS
21 9(-10) stamens fused into a tube                          22
   22  Fruit 1(-2)-seeded, enclosed in calyx    11. ANTHYLLIS
```

22 Fruit >2-seeded, exserted   23
23 Fruit curved, breaking transversely between seeds
   at maturity   **15. CORONILLA**
23 Fruit ± straight, dehiscing longitudinally along 2
   sides, sometimes tardily   24
   24 Fruit with 4 longitudinal wings   **13. TETRAGONOLOBUS**
   24 Fruit not winged   **12. LOTUS**

<u>Key C</u> – Leaves with 1-many pairs of leaflets, if with 1 then
            without odd terminal leaflet, often with tendrils
1 Leaves with even no. of leaflets, terminated by point
  or tendril   2
1 Leaves with odd no. of leaflets, terminated by single
  leaflet   6
  2 Stem winged, and/or leaflets parallel-veined
                                        21. LATHYRUS
  2 Stem not or scarcely winged; leaflets pinnately
    veined   3
3 At least some stipules >2cm, larger than leaflets
                                           22. PISUM
3 Stipules <2cm, smaller than leaflets   4
  4 Calyx-teeth equal, >2x as long as tube   **20. LENS**
  4 Calyx-teeth usually unequal, 2-5 of them <2x as
    long as tube   5
5 Style glabrous, or pubescent all round, or pubescent
  only on lowerside   **19. VICIA**
5 Style pubescent only on upperside   **21. LATHYRUS**
  6 Woody shrubs or trees   7
  6 Herbaceous, sometimes ± woody at base   10
7 Tree; corolla white   **1. ROBINIA**
7 Shrub; corolla pale yellow to orange   8
  8 Fruit strongly inflated, indehiscent or dehiscing
    longitudinally; flowers in racemes   **7. COLUTEA**
  8 Fruit ± not inflated, breaking transversely
    between seeds; flowers in umbels   9
9 Claws of wings and standard 2-3x as long as calyx;
  fruits 5-11cm; stems ridged or furrowed **16. HIPPOCREPIS**
9 Claws of wings and standard 1-1.3x as long as calyx;
  fruits 1-5cm; stems terete   **15. CORONILLA**
  10 Flowers in racemes   11
  10 Flowers solitary or in umbels   14
11 Corolla pink to purple; fruits with 1 seed
                                        10. ONOBRYCHIS
11 Corolla white, yellow or blue; fruits with >2 seeds   12
  12 Keel beaked at apex   **9. OXYTROPIS**
  12 Keel subacute to rounded at apex   13
13 Fruit not partitioned internally, terete, glabrous;
   uppermost stamen fused to stamen-tube for part of
   its length   **6. GALEGA**
13 Fruit longitudinally partitioned by internal membrane
   variously shaped but not terete and glabrous; uppermost stamen free   **8. ASTRAGALUS**

14 Fruit with 2 longitudinal sutures, usually
   dehiscing along them, without transverse sutures   15
14 Fruit breaking along transverse sutures between
   seeds   17
15 Fruits >3x as long as wide, with >3 seeds   **12. LOTUS**
15 Fruits <3x as long as wide, with 1-2 seeds   16
16 Fruits longer than calyx; leaflets toothed   **23. CICER**
16 Fruits enclosed within calyx; leaflets entire
      11. ANTHYLLIS
17 Fruit-segments and seeds horseshoe-shaped
      16. HIPPOCREPIS
17 Fruit-segments and seeds oblong-ellipsoid   18
18 Strong perennials; flowers usually >10 per
   inflorescence, 8-15mm   **17. SECURIGERA**
18 Annuals; flowers <8(12) per inflorescence,
   3-9mm   **14. ORNITHOPUS**

Other genera - DORYCNIUM Miller (Loteae), from S Europe,
differs from <u>Lotus</u> in the tall semi-shrubby habit and densely
crowded white to pink small flowers with a dark red to black
keel; **D. rectum** (L.) Ser. was formerly natd in Surrey and **D.
pentaphyllum** Scop. (<u>D. gracile</u> Jordan) in E Kent.   **SESBANIA**
Adans. (Sesbanieae) is a tall erect glabrous annual with
pinnate leaves with very numerous leaflets and few yellow
flowers in axillary racemes; **S. exaltata** (Raf.) Cory
(<u>Colorado</u> River-hemp), from S N America, occurs occasionally
as a casual from various food-plant sources.   **ARACHIS** L.,
(Aeschynomeneae), originally from Brazil, is a small annual
with pinnate leaves with 2 pairs of large leaflets and
solitary yellow flowers whose stalks elongate and bury under
the ground where the fruit develops; **A. hypogaea** L.
(Ground-nut) is occasional on rubbish-tips but usually does
not reach flowering.

TRIBE 1 - ROBINEAE (genus 1). Trees; leaves imparipinnate,
with entire leaflets; flowers in pendent racemes; 9 stamens
forming tube, 10th free; fruit longitudinally dehiscent,
>3-seeded.

1. ROBINIA L. - <u>False-acacia</u>
1. R. pseudoacacia L. - <u>False-acacia</u>. Deciduous tree to
29m; twigs with stipular spines; leaves ± glabrous, with
elliptic leaflets; flowers numerous, white; fruit 5-10cm,
with 3-10 seeds. Intrd; banks, scrub and woodland, natd by
suckering and (less often) seeding; scattered over most of BI
but natd ± only in S; N America.

TRIBE 2 - PHASEOLEAE (genera 2-4). Herbs; leaves ternate,
with entire leaflets, with stipule-like outgrowths below each
leaflet; flowers in axillary racemes or clusters; 9 stamens
forming tube, 10th free or also part of tube; fruit
longitudinally dehiscent, often tardily so, >2-seeded.

2. PHASEOLUS L. - Beans
Herbs, often climbing, subglabrous to shortly or appressed-pubescent; flowers in racemes with long common peduncle; corolla much longer than calyx; fruit pendent, many-seeded.

1. P. **vulgaris** L. - French Bean. Annual, with spirally climbing stem to 3m or (usually) not climbing; flowers up to 6, white, sometimes purplish; cotyledons borne above ground. Intrd; much grown as vegetable and frequent on tips and in waste places; scattered throughout lowland Br; S America.
2. P. **coccineus** L. - Runner Bean. Perennial with tuberous roots but rarely surviving winter, with climbing stem to 5m; flowers up to 20, bright red, rarely white; cotyledons subterranean. Intrd; much grown as vegetable and casual as for P. vulgaris; tropical America.

3. VIGNA Savi - Mung-bean
Herbs, not climbing, with patent hairs; flowers with long common peduncle; corolla much longer than calyx; fruit erect to patent, (3)8-14-seeded.

1. V. **radiata** (L.) Wilczek (V. mungo auct. non (L.) Hepper, Phaseolus aureus Roxb.) - Mung-bean. Differs from Glycine max in generic characters. Intrd; occurrence as for G. max; Asia.

4. GLYCINE Willd. - Soya-bean
Herbs, not climbing, with long patent brown hairs; flowers with 0 or very short common peduncle; corolla c. as long as calyx; fruit pendent, 2-4-seeded.

1. G. **max** (L.) Merr. (G. soja auct. non Siebold & Zucc.) - Soya-bean. Erect annual to 60(100)cm; flowers up to 8, often many fewer, white to purple, inconspicuous. Intrd; where seed is spilled on tips, waste land and near docks and factories; very scattered in S Br; origin uncertain.

TRIBE 3 - PSORALEEAE (genus 5). Herbs; leaves ternate, with dentate leaflets; flowers in erect axillary racemes; 10 stamens forming tube; fruit 1-seeded, indehiscent.

5. PSORALEA L. - Scurfy Pea
1. P. **americana** L. - Scurfy Pea. Erect perennial (but not **473** surviving winter) to 50cm; leaflets ovate-orbicular, 8-50mm; **481** flowers fairly dense, c.8mm, white tinged violet. Intrd; bird-seed alien on tips; sporadic in S Br; W Mediterranean.

TRIBE 4 - GALEGEAE (genera 6-9). Herbs or shrubs; leaves imparipinnate, with entire leaflets; flowers in axillary racemes; 9 stamens forming tube, 10th free or fused to others for c.1/2 way; fruit longitudinally dehiscent or ± indehiscent, few- to many-seeded.

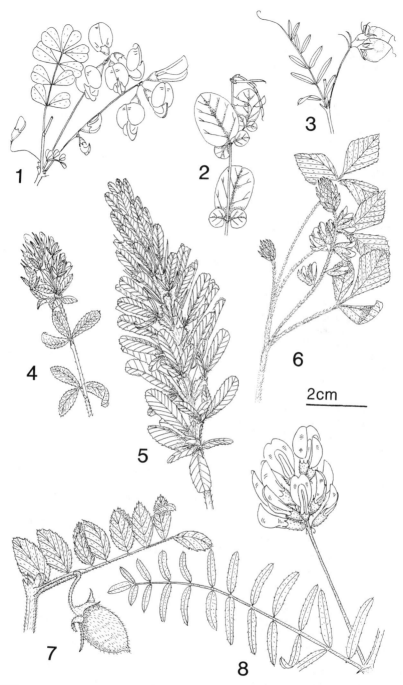

FIG 473 - Fabaceae. 1, **Hippocrepis emerus**. 2, **Coronilla scorpioides**. 3, **Lens culinaris**. 4, **Ononis mitissima**. 5, O. baetica. 6, **Psoralea americana**. 7, **Cicer arietinum**. 8, **Astragalus cicer**.

6. GALEGA L. - <u>Goat's-rue</u>
Flowers numerous in erect racemes; 10th stamen partly fused
to other 9; fruit ± terete, erect, not inflated, dehiscent.

1. G. officinalis L. - <u>Goat's-rue</u>. Erect, glabrous or **481**
sparsely pubescent perennial to 1.5m; leaves with 9-17
oblong-elliptic to ovate or narrowly so leaflets; flowers
white to bluish-mauve. Intrd; much grown for ornament and
frequent on tips and in waste and grassy places; frequent in
S & C Br; Europe.

7. COLUTEA L. - <u>Bladder-sennas</u>
Flowers c.2-8 in ± erect racemes; 10th stamen free; fruit
greatly inflated, ± pendent.

1. C. arborescens L. - <u>Bladder-senna</u>. Deciduous shrub to **481**
4m; leaves with 7-13, elliptic to broadly elliptic leaflets,
sparsely pubescent; flowers pale to deep yellow, with beakless keel; fruit 5-7cm, indehiscent. Intrd; much grown in
gardens and natd in waste and grassy places, on roadsides and
railway banks; frequent in S Br, especially SE En; Europe.
2. C. x media Willd. (<u>C.</u> <u>arborescens</u> x <u>C.</u> <u>orientalis</u> Miller)
- <u>Orange</u> <u>Bladder-senna</u>. Differs from <u>C.</u> <u>arborescens</u> in
orange-bronze flowers with beaked keel; fruits splitting open
at apex. Intrd; commonly grown and natd as for <u>C.</u>
<u>arborescens</u>; W Kent, probably overlooked; garden origin.

8. ASTRAGALUS L. - <u>Milk-vetches</u>
Herbaceous perennials (rarely annuals); flowers in ± erect
racemes; keel not beaked; 10th stamen free; fruit inflated to
not so, variable.

```
1  Corolla blue to purple (rarely white)                  2
1  Corolla whitish-cream to yellow                        3
   2  Stipules free from each other; flowers patent to
      reflexed; fruit with dark, appressed hairs
                                            3. A. alpinus
   2  Stipules fused below; flowers erect to erecto-
      patent; fruit with whitish, spreading hairs
                                            2. A. danicus
3  Leaflets mostly <17; fruits >2cm    4. A. glycyphyllos
3  Leaflets mostly >17; fruits <2cm                       4
   4  Fruits ovoid-globose, 10-15mm; calyx 7-10mm;
      standard 14-16mm                        1. A. cicer
   4  Fruits oblong, compressed, 8-10mm; calyx 4-5mm;
      standard 9-12mm                       5. A. odoratus
```

Other spp. - **A. boeticus** L., from Mediterranean, was
formerly natd in Gloucs; it is an erect annual with yellow
flowers and trigonous fruits 2-4cm.

1. A. cicer L. - <u>Chick-pea Milk-vetch</u>. Spreading to sub- **473**

erect perennial to 60(100)cm; leaves with 17-31 leaflets; **481**
flowers 12-16mm, yellow; fruit 10-15mm, strongly inflated,
pubescent with mostly blackish ± patent hairs. Intrd; natd
on hedgebank near granary in Midlothian since 1920s; Europe.

2. A. danicus Retz. - <u>Purple Milk-vetch</u>. Ascending **481**
perennial to 30cm; leaves with 13-27 leaflets; flowers 15-
18mm, bluish-purple, rarely white; fruit 7-9mm, ± inflated,
pubescent with white patent hairs. Native; short grass on
calcareous well-drained soils; local in E Br from Beds to E
Sutherland, extremely local and scattered elsewhere in En and
Sc, and Aran Isles (Clare).

3. A. alpinus L. - <u>Alpine Milk-vetch</u>. Differs from **A.** **RR**
danicus in flowers 10-14mm, pale blue tipped bluish-purple;
fruit 8-12mm; and see key. Native; grassy rocky places on
mountains at 700-800m; very rare, 4 places in C Sc.

4. A. glycyphyllos L. - <u>Wild Liquorice</u>. Sprawling perennial **481**
to 1(1.5)m; leaves with 7-15 leaflets; flowers 11-15mm,
cream; fruit 25-40mm, ± laterally compressed, glabrous.
Native; grassy places and scrub, mostly on calcareous soils;
very scattered in Br N to Moray.

5. A. odoratus Lam. - <u>Lesser Milk-vetch</u>. Erect to ascending **481**
perennial to 30cm; leaves with 19-29 leaflets; flowers 9-
12mm, whitish-cream to yellow; fruit 8-10mm, ± laterally
compressed, sparsely appressed-pubescent. Intrd; natd in
grassy places; very scattered in C & S En; E Mediterranean.

9. OXYTROPIS DC. - <u>Oxytropises</u>
Herbaceous perennials; flowers in erect racemes; keel beaked
at apex; 10th stamen free; fruit elongated, grooved
abaxially, slightly inflated, erect.

1. O. halleri Bunge ex Koch - <u>Purple Oxytropis</u>. Pubescent **RR**
perennial with leaves and leafless peduncle to 30cm arising
from dense tuft; leaves with 21-31 leaflets; flowers 15-20mm,
usually pale purple, rarely white; fruit 15-20(25)mm,
pubescent, divided internally by septa from both adaxial and
abaxial sutures and ± bilocular. Native; grassy rocky
places; very local in SW, C & N mainland Sc.

2. O. campestris (L.) DC. - <u>Yellow Oxytropis</u>. Differs from **481**
O. halleri in pale yellow flowers often strongly tinged with **RR**
purple; fruit 14-18mm, divided internally by septum from
abaxial suture only and semi-bilocular. Native; cliffs and
rock-ledges; rare and extremely local in C and SW Sc. Other
characters used to separate the spp. are unreliable; even
flower colour is fallible.

TRIBE 5 - HEDYSAREAE (genus 10). Herbaceous perennials;
leaves imparipinnate, with entire leaflets; flowers in
axillary racemes; 9 stamens forming tube, 10th free; fruit
indehiscent, 1-seeded.

10. ONOBRYCHIS Miller - <u>Sainfoin</u>
   1. O. viciifolia Scop. - <u>Sainfoin</u>. Stems suberect, to **481**
60(80)cm; leaves with 13-29 leaflets; flowers numerous in
erect racemes, pinkish-red, with wings <1/2 as long as other
petals; fruit 5-8mm, pubescent, compressed, reticulately
ridged, toothed. Possibly native; grassland and bare places
mostly on chalk and limestone; locally frequent in Br N to
Yorks, scattered casual or natd alien elsewhere in Br and Ir.

TRIBE 6 - LOTEAE (genera 11-13). Annuals or herbaceous
perennials; leaves ternate to imparipinnate, with entire
leaflets; flowers solitary or in umbels; 9 stamens forming
tube, 10th free or variably and loosely fused with others;
fruit dehiscent or not, 1-many-seeded.

11. ANTHYLLIS L. - <u>Kidney</u> <u>Vetch</u>
Lower leaves with large terminal and 0-3 pairs of small
lateral leaflets, grading variably to upper leaves with <15
+ equal leaflets; stipules small, caducous; 10th stamen
variably and loosely fused with other 9; calyx inflated,
enclosing indehiscent 1(-2)-seeded fruit.

   1. A. vulneraria L. - <u>Kidney</u> <u>Vetch</u>. Erect to procumbent
perennial to 60cm; inflorescences usually paired, with leaf-
like bracts at base; calyx densely white-pubescent, dispersed
with fruit; petals yellow to red. The following sspp. are
not well differentiated and a population should be sampled to
obtain a set of characters. Perhaps only 2 sspp. (a-c and
d-e) should be recognised.
1 Calyx (4.5)5-7mm wide, the lateral teeth not
   appressed to upper ones; upper leaves with large
   terminal and (0)1-4 pairs of smaller lateral leaflets   2
1 Calyx 2-4(5)mm wide, the lateral teeth appressed to
   upper ones and + obscured; upper leaves with 4-7
   pairs of lateral leaflets scarcely smaller than
   terminal one   3
   2 Calyx-hairs appressed, sparse   **d. ssp. carpatica**
   2 Calyx-hairs + patent   **e. ssp. lapponica**
   3 Stem-hairs all patent   **c. ssp. corbierei**
   3 At least upper part of stem with appressed hairs   4
      4 Calyx usually with red tip; stems usually with
         appressed hairs throughout or semi-appressed ones
         at base   **a. ssp. vulneraria**
      4 Calyx usually without red tip; lower part of stems
         with patent hairs   **b. ssp. polyphylla**
   a. Ssp. vulneraria. Stems appressed-pubescent throughout or
with semi-appressed or rarely patent hairs at base; upper
leaves with 4-7 pairs of leaflets scarcely or not smaller
than terminal one; calyx usually red-tipped; corolla yellow
to orange, rarely red. Native; grassland, dunes, cliff-tops,
waste ground, usually calcareous; locally common throughout
BI. Red-flowered plants on sea-cliffs in W Cornwall and

Pembs are var. **coccinea** L. Pink-flowered plants occur in Caerns. Plants with stem-branches exceeding their subtending leaves (not so in var. **vulneraria**) on coasts of En, Wa, Ir and CI are var. **langei** Jalas (A. maritima auct. non Schweigger ex Hagen); they are intermediate between sspp. vulneraria and **iberica** (W. Becker) Jalas, from W and SW Europe.

b. Ssp. polyphylla (DC.) Nyman. Stems with some patent hairs near base, with only appressed hairs above; leaves as in ssp. vulneraria but generally larger, hence plant more leafy; calyx usually not red-tipped; corolla pale yellow. Intrd; natd on grassy banks in E Sc; C & E Europe.

c. Ssp. corbierei (Salmon & Travis) Cullen (var. sericea Breb.). Stems with all hairs patent; leaves ± succulent, the upper with 4-5 pairs of leaflets scarcely or not smaller than terminal one; calyx not red-tipped; corolla yellow. Native; sea-cliffs in Anglesey, W Cornwall and CI (Sark and Guernsey); endemic.

d. Ssp. carpatica (Pant.) Nyman (ssp. vulgaris (Koch) Corbiere, ssp. pseudovulneraria (Sagorski) J. Duvign.). Stems sparsely appressed-pubescent; upper leaves with large terminal leaflet and (0)1-4 pairs of smaller lateral leaflets; calyx red-tipped or not; corolla pale yellow. Intrd; natd in marginal and disturbed ground; scattered over Br, Ir and Man; C Europe. Our plant is var. **pseudovulneraria** (Sagorski) Cullen.

e. Ssp. lapponica (N. Hylander) Jalas. Stems appressed-pubescent or with patent hairs below; upper leaves as in ssp. carpatica; calyx usually extensively red-tipped; corolla yellow. Native; banks, cliffs and rock-ledges; mountainous areas of N En, W & NW Ir and Sc.

12. LOTUS L. - Bird's-foot-trefoils
Leaves with 5 leaflets, the lowest 2 at base of rhachis and resembling stipules; stipules minute, soon caducous or withering; 10th stamen free; calyx not enclosing fruit; fruit several-seeded, longitudinally dehiscent, not ridged or angled.

1 Annuals, with dense patent hairs; fruit mostly <2mm
  wide   2
1 Perennials, glabrous to variously pubescent; fruit
  mostly >2mm wide   3
  2 Fruit 12-30mm, >3x as long as calyx, with >12
    seeds; keel with ± right-angled bend c.1/2 way
    along lower edge of limb   **5. L. angustissimus**
  2 Fruit 6-15mm, <3x as long as calyx, with <12
    seeds; keel with obtuse-angled bend near base of
    lower edge of limb   **4. L. subbiflorus**
3 Stem hollow; calyx-teeth recurved in bud, the upper
  2 with acute sinus between them   **3. L. pedunculatus**
3 Stem solid or rarely hollow; calyx-teeth erect in

bud, the upper 2 with obtuse sinus between them   4
4 Leaflets of upper leaves mostly >4x as long as
  wide, acute to acuminate   **1. L. glaber**
4 Leaflets of upper leaves mostly <3(4)x as long as
  wide, subacute to obtuse   **2. L. corniculatus**

1. L. glaber Miller (<u>L. tenuis</u> Waldst. & Kit. ex Willd., <u>L.</u> <u>corniculatus</u> ssp. <u>tenuis</u> (Waldst. & Kit. ex Willd.) Syme, ssp. <u>tenuifolius</u> (L.) Hartman) - <u>Narrow-leaved</u> <u>Bird's-foot-</u> <u>trefoil.</u> Glabrous to sparsely pubescent perennial with sprawling to suberect stems to 90cm; leaflets linear to linear-lanceolate; flowers (1)2- 4(6), 6-12mm, yellow; fruit 15-30mm. Native; dry grassy places; scattered in Br N to C Sc, N & E Ir, CI, frequent only in S & E En and probably intrd in Ir, Sc and much of N & W En.

2. L. corniculatus L. - <u>Common</u> <u>Bird's-foot-trefoil.</u> Glabrous to sparsely pubescent perennial with procumbent to ascending stems to 50cm; leaflets lanceolate or oblanceolate to suborbicular; flowers (1)2-7, 10-16mm, yellow to orange, often streaked with red; fruit 15-30mm. Native; grassy and barish places, mostly on well-drained soils; common throughout BI. Native plants have solid stems; robust erect plants with large leaflets, light-coloured keels and sometimes hollow stems, mostly on roadsides, are probably intrd and referable to var. **sativus** Chrtkova.

3. L. pedunculatus Cav. (<u>L. uliginosus</u> Schk.) - <u>Greater</u> <u>Bird's-foot-trefoil.</u> Glabrous to pubescent perennial with erect to ascending stems to 1m; leaflets ovate to obovate; flowers 5-12, 10-18mm, yellow; fruits 15-35cm. Native; damp grassy places, marshes and pond-sides; frequent throughout most of BI.

4. L. subbiflorus Lag. (<u>L.</u> <u>suaveolens</u> Pers., <u>L. parviflorus</u>   **R** auct. non Desf., <u>L. hispidus</u> auct. non Desf. ex DC.) - <u>Hairy</u> <u>Bird's-foot-trefoil.</u> Pubescent annual with procumbent to decumbent stems to 30(80)cm; leaflets narrowly ovate to obovate; flowers (1)2-4, 5-10mm, yellow. Native; dry grassy places near sea; very local in SW En, SW Wa, SW Ir, CI.

5. L. angustissimus L. - <u>Slender</u> <u>Bird's-foot-trefoil.</u>   **RR** Pubescent annual; differs from <u>L. subbiflorus</u> in flowers 1-3, 5-12mm; and see key. Native; dry grassy places near sea; very local in S En and CI.

13. TETRAGONOLOBUS Scop. - <u>Dragon's-teeth</u>
Leaves with 3 leaflets; stipules herbaceous, persistent; 10th stamen free; calyx not enclosing fruit; fruits several-seeded, tardily longitudinally dehiscent, tetraquetrous with wing c.1mm wide on each angle.

1. T. maritimus (L.) Roth - <u>Dragon's-teeth.</u> Sparsely **481** pubescent perennial with decumbent stems to 30cm; leaflets obovate to oblanceolate; flowers solitary, with leaf-like bract at base, on peduncle longer than subtending leaf,

TETRAGONOLOBUS

25-30mm, yellow with brownish streaks; fruits 25-60mm. Intrd; very locally but well natd in rough calcareous grassland; S En, rare casual elsewhere in S & C Br, known since 1875; C & S Europe.

TRIBE 7 - CORONILLEAE (genera 14-18). Herbaceous annuals or perennials or shrubs; leaves simple, ternate or impari-pinnate, with entire leaflets; flowers solitary or in umbels; 9 stamens forming tube, 10th free; fruit dehiscing trans-versely between seeds or ± indehiscent, few- to many-seeded.

14. ORNITHOPUS L. - Bird's-foots
Annuals; leaves imparipinnate with >3 pairs of lateral leaflets; fruits curved to ± straight, beaked, not or slightly constricted between the 3-12 cylindrical to oblong-ellipsoid segments.

1 Flower-heads without bract at base, or with minute
   scarious bracts   **4. O. pinnatus**
1 Flower-heads with leaf-like bract at base   2
   2 Corolla yellow; fruits not or scarcely constricted
      between segments   **1. O. compressus**
   2 Corolla white to pink; fruits distinctly
      constricted between segments   3
3 Corolla >5.5mm; bract c.1/2 as long as flowers
                                            2. O. sativus
3 Corolla <5.5mm; bract at least as long as flowers
                                         3. O. perpusillus

1. O. compressus L. - Yellow Serradella. Pubescent; stems **481** to 50cm, procumbent to decumbent; leaves with 13-37 leaflets; flowers 3-5, with leafy bracts at least as long, yellow, 5-6mm; fruits 2-5cm, curved, ± compressed, not or scarcely constricted between the 5-8 segments. Intrd; barish sandy banks; 1 locality in W Kent sporadically since 1957, rare casual in S Br and CI; S Europe.
2. O. sativus Brot. (O. roseus Dufour) - Serradella. Pubescent; stems to 70cm, procumbent to ascending; leaves with 13-37 leaflets; flowers 2-5, with leafy bracts c.1/2 as long, white to pink, 6-9mm; fruits 1.2-2.5cm, ± straight, compressed, slightly constricted between the 3-7 segments. Intrd; planted and natd on china-clay waste in E Cornwall since 1978, rare casual in S Br; SW Europe.
3. O. perpusillus L. - Bird's-foot. Pubescent; stems to **481** 30cm, procumbent to decumbent; leaves with 9-27 leaflets; flowers 3-8, with leafy bracts at least as long, white to pink, 3-5mm; fruits 1-2cm, curved, compressed, slightly constricted between the 4-9 segments. Native; dry barish sandy and gravelly ground; locally common in much of BI, especially S & E, but absent from much of Ir and Sc.
4. O. pinnatus (Miller) Druce - Orange Bird's-foot. **RR** Glabrous to sparsely pubescent; stems to 50cm, decumbent to

ascending; leaves with 5-15 leaflets; flowers 1-2(5), with 0 or minute bracts, yellow veined with red, 6-8mm; fruits 2-3.5cm, curved, terete, very slightly constricted between the 5-8(12) segments. Native; short turf or open ground on sandy soil; CI (all islands) and Scillies.

15. CORONILLA L. - Scorpion-vetches
Annuals or shrubs, with terete branches; leaves simple, ternate or imparipinnate; fruits curved to ± straight, beaked, not or scarcely constricted between the (1)2-11 ± cylindrical segments.

1. C. valentina L. (C. glauca L.) - Shrubby Scorpion-vetch. 481
Shrub to 1m; leaves with 5-13 leaflets, each 10-20mm, obovate; flowers 4-8(12), 7-12(14)mm, yellow; fruits 1-5cm, with (1)2-4(10) segments. Intrd; natd on cliffs near Torquay, S Devon, and on promenade, Eastbourne, E Sussex; Mediterranean. Our plant is ssp. **glauca** (L.) Battand.
2. C. scorpioides (L.) Koch - Annual Scorpion-vetch. 473
Glaucous, glabrous, erect to ascending annual to 40cm; leaves 481
with 1-3 leaflets, rarely all with 1, the terminal leaflet much the largest (up to 4cm), broadly elliptic; flowers 2-5, 3-8mm, yellow; fruits 2-6cm, with 2-11 segments. Intrd; fairly frequent casual, mostly from bird-seed; S Br; S Europe.

16. HIPPOCREPIS L. - Horseshoe Vetches
Herbaceous perennials or shrubs, with furrowed or ridged stems; leaves imparipinnate; fruits ± straight to curved, beaked, with horseshoe-shaped segments, strongly compressed.

1. H. emerus (L.) Lassen (Coronilla emerus L.) - Scorpion 473
Senna. Shrub to 1.5(2)m; leaves with 5-9 leaflets, each 481
10-20mm, obovate; flowers 2-5(7), (12)14-20mm, yellow; fruits 5-11cm, with 3-12 segments. Intrd; natd on roadsides and banks in several places in En N to Lincs; Europe.
2. H. comosa L. - Horseshoe Vetch. Herbaceous perennial; 481
stems procumbent to suberect, to 30(50)cm; leaves with 7-25 leaflets, each (2)4-8(16)mm, oblong to obovate; flowers (2)4-8(12), 5-10(14)mm, yellow; fruits 10-30mm, with 3-6 segments. Native; dry calcareous grassland and cliff-tops; local in Br N to Westmorland, frequent in S & E En, NW Jersey.

17. SECURIGERA DC. - Crown Vetch
Herbaceous perennials with furrowed or ridged stems; leaves imparipinnate; fruits ± straight to slightly curved, beaked, not or scarcely constricted between the 3-8(12) cylindrical segments.

1. S. varia (L.) Lassen (Coronilla varia L.) - Crown Vetch. 481
Stems sprawling, to 1.2m; leaves with 11-25 leaflets, each

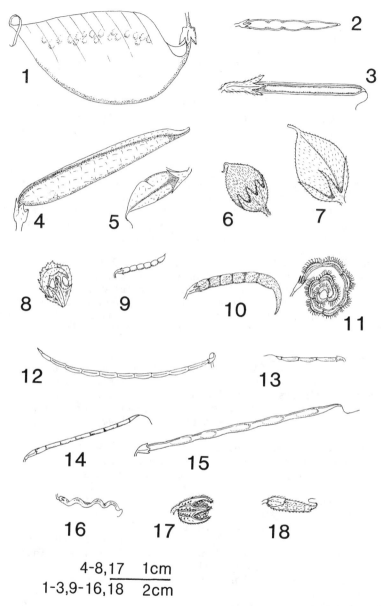

FIG 481 – Fruits of **Fabaceae**. 1, **Colutea arborescens**. 2, **Galega**. 3, **Tetragonolobus**. 4, **Astragalus glycyphyllos**. 5, **A. odoratus**. 6, **A. danicus**. 7, **A. cicer**. 8, **Onobrychis**. 9, **Ornithopus perpusillus**. 10, **O. compressus**. 11, **Scorpiurus**. 12, **Coronilla scorpioides**. 13, **C. valentina**. 14, **Securigera**. 15, **Hippocrepis emerus**. 16, **H. comosa**. 17, **Psoralea**. 18, **Oxytropis campestris**.

6-20mm, elliptic to oblong; flowers (5)10-20, 8-15mm, white
to pink or purple; fruits 2-6(8)cm, with 3-8(12) segments.
Intrd; natd in grassy places and rough ground; scattered
through Br N to C Sc, E Ir, Guernsey; Europe.

18. SCORPIURUS L. - Caterpillar-plant
Annuals with furrowed or ridged stems; leaves simple; fruits
much curved, spiralled or variously contorted, longitudinally
ridged, variously ornamented with spines and tubercles,
beaked, indehiscent or tardily dehiscent.

1. S. muricatus L. (S. sulcatus L., S. subvillosus L.) - **481**
Caterpillar-plant. Stems procumbent to suberect, to 80cm;
leaves obovate to oblanceolate; flowers (1)2-5, 5-12mm,
yellow, often red-tinged; fruits with 3-10 segments. Intrd;
bird-seed or wool-alien on tips, rough ground and in gardens
and parks; C & S Br, mainly S En; S Europe.

TRIBE 8 - FABEAE (Vicieae) (genera 19-22). Herbaceous
annuals or perennials; leaves usually paripinnate, often with
tendril(s) at apex, rarely simple or reduced to a tendril,
with usually entire (rarely dentate) leaflets; flowers
solitary or in axillary racemes; 9 stamens forming tube, 10th
free; fruit dehiscing longitudinally, (1)2-many seeded.

19. VICIA L. - Vetches
Stem ± not winged (often ridged); leaves paripinnate with
(1)2-many pairs of pinnately-veined leaflets, usually with
terminal tendril(s); stipules smaller than leaflets; at least
2 calyx-teeth <2x as long as tube; style glabrous, or
pubescent all round, or pubescent only on lowerside.

1 All leaves without tendrils, terminated by small point 2
1 At least upper leaves terminated by tendril(s)   4
  2 Perennials; flowers ≥6; peduncles >3cm; leaflets
    >5 pairs   **1. V. orobus**
  2 Annuals; flowers ≤6; peduncles 0 or <2cm;
    leaflets <5 pairs   3
3 Flowers <1cm; leaflets <2cm; fruits <3cm
                         13. V. lathyroides
3 Flowers >1cm; leaflets >2cm; fruit >5cm   **17. V. faba**
  4 Peduncle 0, or shorter than each flower   5
  4 Peduncle longer than each flower   11
5 Standard pubescent on back   **11. V. pannonica**
5 Standard glabrous on back   6
  6 Flowers 6-9mm, solitary; seeds tuberculate
                         13. V. lathyroides
  6 Flowers 9-25(30)mm, 1-several; flowers rarely
    cleistogamous and <9mm but then seeds smooth   7
7 Leaflets 1-3 pairs; stipules c.1cm   8
7 At least some leaves with >3 pairs leaflets; stipules
  <8mm   9

8 Fruit pubescent all over, <u><</u>1cm wide; leaflets
   >2x as long as wide   **15. V. bithynica**
8 Fruit pubescent only along 2 sutures, >1cm wide;
   leaflets <2x as long as wide   **16. V. narbonensis**
9 Perennial; seeds with hilum >1/2 total circumference;
   lower calyx-teeth longer than upper but shorter than
   tube   **10. V. sepium**
9 Annual; seeds with hilum <1/2 total circumference;
   all calyx-teeth equal, or unequal and lower longer
   than tube   10
10 Calyx-teeth unequal (the lower longer); seeds
   with hilum 1/3-1/2 total circumference; corolla
   usually yellow   **14. V. lutea**
10 Calyx-teeth <u>+</u> equal; seeds with hilum 1/4-1/3
   total circumference; corolla usually pink to
   purple   **12. V. sativa**
11 Flowers 2-8(9)mm, white to purple (not blue), 1-8 per
   raceme   12
11 Flowers 8-20mm, if <10mm then blue and >8 per raceme 14
12 Calyx-teeth equal, all at least as long as tube;
   fruits usually 2-seeded   **7. V. hirsuta**
12 Calyx-teeth unequal, at least the upper shorter
   than tube; fruits 3-6-seeded   13
13 Flowers 1-2; seeds (3)4(-5), with hilum >2x as long
   as wide and c.1/5 seed circumference   **9. V. tetrasperma**
13 Flowers 1-4(5); seeds 4-6(8), with hilum little longer
   than wide and <1/3 seed circumference **8. V. parviflora**
14 Leaves with 2-3 pairs leaflets; flowers 1-3
                                          15. V. bithynica
14 Leaves with >4 pairs leaflets; flowers usually >4 15
15 Standard with limb c.1/2 as long as claw; calyx very
   asymmetrical at base, with large bulge on upperside 16
15 Standard with limb c. as long as or longer than claw;
   calyx only slightly asymmetrical at base   17
16 Corolla reddish-purple with blackish tip
                                          6. V. benghalensis
16 Corolla blue to violet or purple, sometimes with
   white or yellow wings   **5. V. villosa**
17 Lower lobe of stipules strongly toothed; corolla
   white with blue or purple veins; seeds with hilum
   >1/2 total circumference   **4. V. sylvatica**
17 Lower lobe of stipules entire; corolla blue to purple
   or violet; seeds with hilum <1/2 total circumference 18
18 Corolla 8-12(13)mm; limb of standard c. as long
   as claw; seeds with hilum 1/4 to 1/3 total
   circumference   **2. V. cracca**
18 Corolla (10)12-18mm; limb of standard longer than
   claw; seeds with hilum 1/5 to 1/4 total
   circumference   **3. V. tenuifolia**

Other spp. - **V. cassubica** L. (<u>Danzig</u> <u>Vetch</u>), from Europe,
resembles <u>V. cracca</u> but has fruits 6-8mm wide (not 4-6mm) and

with 1-3 seeds; it was natd in W Kent from 1931 to 1964. **V. hybrida** L. (Hairy Yellow-vetch), from S Europe, resembles V. pannonica but has unequal (not subequal) calyx-teeth, hilum c.1mm (not c.2mm), and yellow corolla 18-30mm; it was formerly natd in N Somerset. Both are still rare casuals.

1. **V. orobus** DC. - Wood Bitter-vetch. Perennial with erect   **R**
stems to 60cm; leaflets 6-15 pairs; tendrils 0; flowers 6-20,
12-15(20)mm, white with purple veins; fruits 20-30mm, with
4-5 seeds. Native; grassy and rocky places and scrub;
scattered through W En, Sc, Wa and C & N Ir.

2. **V. cracca** L. - Tufted Vetch. Scrambling or climbing
perennial to 2m; leaflets 5-15 pairs; tendrils branched;
flowers 10-30(40), 8-12(13)mm, bluish-violet; fruits 10-25mm,
2-6(8)-seeded. Native; grassy and bushy places and hedgerows; common throughout BI.

3. **V. tenuifolia** Roth - Fine-leaved Vetch. Differs from V.
cracca in flowers (10)12-18mm, bluish-lilac to purple; and
see key. Intrd; natd in grassy places and rough ground;
scattered throughout most of En and Sc, often only casual;
Europe. Perhaps only a ssp. of V. cracca.

4. **V. sylvatica** L. - Wood Vetch. Scrambling or climbing
perennial to 2m; leaflets 5-12 pairs; tendrils branched;
flowers 5-15(20), 12-20mm, white with purple veins; fruits
25-30mm, 4-5-seeded. Native; open woods and wood-borders,
scree, scrub, maritime cliffs and shingle; scattered
throughout much of Br and Ir but local.

5. **V. villosa** Roth (**V.** varia Host, **V.** dasycarpa auct.,
?Ten.) - Fodder Vetch. Scrambling or climbing annual to 2m;
leaflets 4-12 pairs; tendrils branched; flowers 10-30,
10-20mm, blue to purple or violet, wings sometimes white or
yellow; fruits 20-40mm, 2-8-seeded. Intrd; natd in grassy
places, tips, rough and waste ground, more often casual;
throughout much of Br; Europe. Several sspp., based on
pubescence, flower number, size and colour, and calyx-teeth
size, are often recognized; 5 have been recorded from Br but
are of doubtful taxonomic value.

6. **V. benghalensis** L. - Purple Vetch. Differs from V.
villosa in stems to 80cm; leaflets 5-9 pairs; flowers 2-20,
reddish-purple with blackish tip; fruits 3-5-seeded. Intrd;
occasional casual and sometimes briefly persisting on tips
and rough and waste ground; sporadic since c.1910 in C & S
Br; Mediterranean.

7. **V. hirsuta** (L.) Gray - Hairy Tare. Scrambling annual to
80cm; leaflets 4-10 pairs; tendrils branched; flowers (1)2-
7(9), (2)3-5mm, whitish tinged purple; fruits 6-11mm, almost
all 2-seeded. Native; rough ground and grassy places;
throughout lowland BI but rare in N Sc.

8. **V. parviflora** Cav. (**V.** laxiflora Brot. nom. illeg., **V.**   **R**
tenuissima auct. non (M. Bieb.) Schinz & Thell., **V.**
tetrasperma ssp. gracilis Hook. f.) - Slender Tare.
Scrambling annual to 60cm; leaflets 2-4(5) pairs; tendrils

usually simple; flowers 1-4(5), (5)6-9mm, pale bluish-purple;
fruits 12-17mm, 4-6(8)-seeded. Native; grassy places; local
in S En N to Hunts, 1 record (?intrd) in N Lincs.

9. V. tetrasperma (L.) Schreber - Smooth Tare. Differs from
V. parviflora in leaflets 3-6(8) pairs; flowers 1-2(4),
4-8mm; fruits 9-16mm, (3)4(-5)-seeded; peduncles shorter than
to as long as (not longer than) leaves. Native; grassy
places, CI, En and Wa, very scattered and mostly intrd in Sc
and Ir.

10. V. sepium L. - Bush Vetch. Climbing or sprawling
perennial to 60(100)cm; leaflets (3)5-9 pairs; tendrils
branched; flowers 2-6, 12-15mm, dull purple; fruits 20-35mm,
3-10-seeded. Native; grassy places, hedges, scrub and wood-
borders; throughout BI.

11. V. pannonica Crantz - Hungarian Vetch. Climbing annual
to 60cm; leaflets 4-19 pairs; tendrils branched; flowers 1-4,
14-22mm, dirty brownish-yellow; fruits 20-35mm, 2-8-seeded,
appressed-pubescent. Intrd; frequent casual in waste places
and roadside banks in En and Wa, natd in W Kent since 1971;
Europe. Natd and most casual plants are ssp. **pannonica**.
Ssp. **striata** (M. Bieb.) Nyman, with purplish flowers, is a
rare casual.

12. V. sativa L. - Common Vetch. Climbing, sprawling or
procumbent annual to 1.5m; leaflets 3-8 pairs; tendrils
usually branched; flowers 1-2(4), variously pink to purple,
rarely white or yellow; fruits 4-12-seeded, glabrous to
sparsely pubescent.

1 Plant heterophyllous, the leaflets of upper leaves
  much (and abruptly) narrower than those of lower
  leaves; flowers + concolorous, usually bright
  pinkish-purple   **a. ssp. nigra**
1 Plant + isophyllous, the leaflets of upper leaves
  little (and gradually) narrower than those of lower
  leaves; flowers bicolorous, the standard much paler
  than wings   2
  2 Fruits smooth, usually glabrous, brown to black
                             b. ssp. segetalis
  2 Fruits slightly constricted between seeds, often
    pubescent, yellowish to brown   **c. ssp. sativa**

a. Ssp. nigra (L.) Ehrh. (ssp. angustifolia (L.) Gaudin, V.
angustifolia L. ssp. angustifolia). Plant slender, often
procumbent, to 75cm, strongly heterophyllous; flowers
14-19mm, + concolorous, usually bright pink; fruits 23-38mm,
brown to black, smooth, glabrous. Native; sandy banks,
heathland, maritime sand and shingle; throughout most of BI,
but often intrd.

b. Ssp. segetalis (Thuill.) Gaudin (V. angustifolia ssp.
segetalis (Thuill.) Arcang.). Plant more robust, to 1m, +
isophyllous; flowers 9-26mm, bicolorous; fruits 28-70mm, as
in ssp. nigra. Probably intrd; in grassy and rough places
and field-borders (sometimes cultivated for fodder);
commonest ssp. throughout most of BI; Europe.

c. Ssp. sativa. Very robust, to 1.5m, ± isophyllous; flowers 11-26(30)mm, bicolorous; fruits 36-70(80)mm, yellowish to brown, constricted between seeds, often pubescent. Intrd; uncommon casual, rarely persisting, in waste places, field-borders (formerly cultivated for fodder); very scattered in Br, formerly CI and Ir; Europe.

13. V. lathyroides L. - Spring Vetch. Procumbent to weakly climbing annual to 20cm; leaflets 2-4 pairs; tendrils simple or scarcely developed; flowers solitary, (5)6-9mm, dull purple; fruits 15-30mm, 6-12-seeded. Native; maritime sand and inland sandy heaths; scattered over most of BI except W & S Ir and NW Sc. Possibly over-recorded.

14. V. lutea L. (V. laevigata Smith) - Yellow-vetch.   **R** Procumbent to sprawling annual to 60cm; leaflets 3-8 pairs; tendrils branched; flowers 1-2, 15-25(30)mm, pale dull yellow; fruits 20-40mm, 4-8(10)-seeded. Native; maritime shingle and cliffs; very scattered round coasts of CI and Br N to C Sc, rather frequent casual inland in Br.

15. V. bithynica (L.) L. - Bithynian Vetch. Climbing or   **R** scrambling annual to 60cm; leaflets (1)2-3 pairs; tendrils branched; flowers 1-2-(3), on very short to long peduncles, 16-20mm, standard purple, wings and keel white or very pale; fruits 25-50mm, 4-8-seeded. Probably native; scrub, rough grassland and hedges; rare and decreasing by coast in CI and S Br, very scattered N to S Sc but mainly casual or intrd.

16. V. narbonensis L. - Narbonne Vetch. Erect climbing annual to 60cm; leaflets 1-3 pairs, often dentate; tendrils branched, usually only on upper leaves; flowers 1-3(6), 10-30mm, purple; fruits 30-50(70)mm, 4-7-seeded. Intrd; rather frequent casual in waste land from grain; scattered in S Br; S Europe.

17. V. faba L. - Broad Bean. Erect annual to 1m; leaflets (1)2-3 pairs; tendrils 0, replaced by weak point; flowers 1-6, usually white with blackish wings; fruits up to 30cm, 4-8-seeded. Intrd; widely cultivated as crop and common casual mainly in S Br; origin uncertain.

20. LENS Miller - Lentil
Stems scarcely winged, markedly ridged; leaves paripinnate, with 3-8 pairs of pinnately-veined leaflets; simple tendril present on upper leaves; stipules smaller than leaflets; calyx-teeth all >2x as long as tube; style pubescent on upperside.

1. L. culinaris Medikus - Lentil. Weakly climbing annual to **473** 40cm; leaflets (3)5-8 pairs; flowers 1-3, white to pale mauve; fruits (10)12-14(16)mm, nearly as deep, strongly compressed, 1-2(3)-seeded. Intrd; fairly frequent grain-alien on tips and rough ground; scattered in S Br; SW Asia.

21. LATHYRUS L. - Peas
Stems angled or winged; leaves usually paripinnate with 1-

many pairs of pinnately- or parallel-veined leaflets,
sometimes reduced to a simple blade or a simple tendril;
terminal tendril present or 0; stipules variable; calyx-teeth
variable; style pubescent only on upperside.

1 Leaves reduced to single blade or tendril   2
1 Leaves with 1-many pairs of leaflets; terminal
  tendril present or 0   3
  2 Leaf reduced to single blade; stipules <3mm;
    flowers reddish   **15. L. nissolia**
  2 Leaf reduced to simple tendril; stipules >3mm,
    leaf-like; flowers yellow   **16. L. aphaca**
3 Leaves (+) all with >1 pair of leaflets   4
3 Leaves (+) all with 1 pair of leaflets   7
  4 Stem angled, not winged   5
  4 Stem, at least above, with wings >1/2 as wide as
    stem   6
5 Stipules ovate-triangular; procumbent plant of
  maritime shingle; leaves usually with tendrils
         2. L. japonicus
5 Stipules lanceolate; erect garden escape; leaves
  without tendrils   **1. L. niger**
  6 Tendrils 0; fruit scarcely compressed **3. L.linifolius**
  6 Tendrils well developed; fruit strongly compressed
         5. L. palustris
7 Stem angled, not winged   8
7 Stem, at least above, with wings >1/2 as wide as stem 10
  8 Flowers yellow; leaflets acute; fruits strongly
    compressed   **4. L. pratensis**
  8 Flowers pink to purple; leaflets obtuse to rounded
    or retuse at apex; fruits scarcely compressed   9
9 Plant minutely pubescent; flowers >25mm
         7. L. grandiflorus
9 Plant glabrous to very sparsely pubescent; flowers
  <25mm   **6. L. tuberosus**
  10 Perennials; flowers 3-many   11
  10 Easily uprooted annuals; flowers 1-3(4)   13
11 Stipules <1/2 as wide as stem; all calyx-teeth
  shorter than tube   **8. L. sylvestris**
11 Stipules >1/2 as wide as stem (often as wide); lowest
  calyx-tooth as long as or longer than tube   12
  12 Flowers 12-22mm; leaflets >4x as long as wide
         10. L. heterophyllus
  12 Flowers 15-30mm; leaflets <4x as long as wide
         9. L. latifolius
13 Plant glabrous (or with glands on young fruits)   14
13 Plant pubescent at least on pedicels, calyx and
  fruits   15
  14 Corolla white, pink or blue; fruit with 2 narrow
    wings along dorsal suture   **12. L. sativus**
  14 Corolla yellow; fruit not winged   **13. L. annuus**
15 Flowers >20mm; leaflets ovate-oblong   **11. L. odoratus**

15 Flowers <20mm; leaflets linear-oblong   **14. L. hirsutus**

Other spp. - **L. vernus** (L.) Bernh. (Spring Pea), from Europe, is grown in gardens and sometimes occurs as a relic; it would key out as L. niger but has stems <40cm, leaflets with acuminate (not obtuse) apex, and brown (not black) fruits.

1. L. niger (L.) Bernh. - Black Pea. Erect perennial to 80cm; stems not winged; leaflets 3-6(10) pairs, lanceolate to elliptic; tendrils 0; flowers 2-10, 10-15mm, purple becoming bluish. Intrd; garden escape natd in grassy, rocky and scrubby places; rare in Sc and SE En; Europe.

2. L. japonicus Willd. (L. maritimus (L.) Bigelow) - Sea   **R**
Pea. Procumbent perennial to 90cm; stems not winged; leaflets 2-5(6) pairs, elliptic; tendrils simple or branched, sometimes 0; flowers 2-10(15), (12)15-25mm, purple becoming blue. Native; maritime shingle or rarely sand, very local and decreasing on coasts of Br from Scillies and Kent to Shetland, frequent in SE En, Guernsey. Our plant is ssp. **maritimus** (L.) P. Ball.

3. L. linifolius (Reichard) Baessler (L. montanus Bernh.) - Bitter-vetch. Erect perennial to 40cm; stems winged; leaflets 2-4 pairs, linear to elliptic; tendrils 0; flowers 2-6, 10-16mm, reddish-purple turning blue. Native; wood-borders, hedgerows, scrub; throughout Br and Ir but absent from much of CE En and C Ir. Our plant is var. **montanus** (Bernh.) Baessler.

4. L. pratensis L. - Meadow Vetchling. Climbing perennial to 1.2m; stems not winged; leaflets 1 pair, linear-lanceolate to elliptic; tendrils simple or branched; flowers (2)5-12, 10-18mm, yellow. Native; grassy places and rough ground; common throughout BI.

5. L. palustris L. - Marsh Pea. Climbing perennial to 1.2m;   **R**
stems winged; leaflets 2-3(5) pairs, narrowly elliptic-oblong; tendrils branched; flowers 2-6(8), 12-20mm, purplish-blue. Native; fens and tall damp grassland; very locally scattered in En, Wa, SW Sc and Ir, mainly E Anglia, decreasing.

6. L. tuberosus L. - Tuberous Pea. Climbing perennial to 1.2m; stems not winged; leaflets 1 pair, elliptic to narrowly obovate; tendrils branched; flowers 2-7, 12-20mm, reddish-purple. Intrd; cornfields, hedgerows and roadsides; natd in N Essex since 1859, scattered through Br N to C Sc but casual or shortly persisting only in most places; Europe.

7. L. grandiflorus Smith (L. tingitanus auct. non L.) - Two-flowered Everlasting-pea. Climbing perennial to 2m; stems not winged; leaflets 1 pair, obovate; tendrils branched; flowers 1-3(4), 25-35mm, pinkish-purple. Intrd; persistent garden escape in hedges, waste ground near old gardens; scattered in Br N to C Sc, Guernsey; C Mediterranean.

8. L. sylvestris L. - Narrow-leaved Everlasting-pea.

Climbing or scrambling perennial to 2m; stems winged; leaflets 1 pair, linear to narrowly elliptic (>4x as long as wide); tendrils branched; flowers 3-8(12), 12-20mm, dull pinkish-purple. Native; scrub, wood-borders, hedgerows, rough ground; scattered through Br N to C Sc, but intrd in many places, especially in N.

9. L. latifolius L. - <u>Broad-leaved</u> Everlasting-pea. Differs from <u>L. sylvestris</u> in stems to 3m; leaflets narrowly to broadly elliptic (<4x as long as wide); flowers 3-12, 15-30mm, brighter coloured; and see key. Intrd; very persistent garden escape in hedges, roadsides and railway banks, and rough ground; scattered through Br N to C Sc; Europe.

10. L. heterophyllus L. - <u>Norfolk</u> Everlasting-pea. Differs from <u>L. sylvestris</u> in flowers 12-22mm, brighter coloured; and see key. Intrd; natd since 1949 in damp hollows of sand-dunes in W Norfolk, perhaps nowhere else; Europe. On the Continent plants mostly have upper leaves with 2-3 pairs leaflets, but our plant (var. **unijugus** Koch) has only 1 pair.

11. L. odoratus L. - <u>Sweet</u> Pea. Climbing annual to 2.5m; stems winged; leaflets 1 pair, ovate-oblong; tendrils branched; flowers 1-3(4), 20-35mm, white, pink or mauve to purple. Intrd; frequent casual on tips and in waste places; En and Wa, mostly S; S Italy.

12. L. sativus L. - <u>Indian</u> Pea. Scrambling or climbing annual to 1m; stems winged; leaflets 1 pair, linear; tendrils branched; flowers solitary, 12-24mm, white, pink or bluish. Intrd; fairly frequent on tips, mostly from bird-seed; sporadic in S En; Mediterranean.

13. L. annuus L. - <u>Fodder</u> Pea. Scrambling or climbing annual to 1m; stems winged; leaflets 1 pair, linear; tendrils branched; flowers 1-3, 12-18mm, yellow to orange-yellow. Intrd; tips and waste places from bird-seed and other sources; sporadic in S En; Mediterranean.

14. L. hirsutus L. - <u>Hairy</u> Vetchling. Scrambling or climbing annual to 1m; stems winged; leaflets 1 pair, linear to narrowly elliptic-oblong; tendrils branched; flowers 1-3, 8-15mm, purple with bluish wings. Intrd; grassy and rough ground and waste places, formerly natd but now perhaps only casual; very scattered in En and S Sc; Europe.

15. L. nissolia L. - <u>Grass</u> Vetchling. Erect or ascending annual to 90cm; stems not winged; leaves reduced to a simple grass-like blade; stipules to 2mm; tendrils 0; flowers 1-2, 8-18mm, crimson. Native; grassy places; local in En and S Wa N to N Lincs, rare casual elsewhere, frequent only in SE En.

16. L. aphaca L. - <u>Yellow</u> Vetchling. Scrambling annual to 40(100)cm; stems not winged; leaves on mature plants reduced to simple tendril with leaf-like ovate-hastate stipules 6-50mm; flowers solitary, 10-13mm, yellow. Doubtfully native; dry banks, grassy places and rough ground; locally natd in S & SE En, casual in Br N to C Sc, Ir; Europe.

R

22. PISUM L. - <u>Garden</u> <u>Pea</u>
Stems not winged, + terete; leaves paripinnate with 1-3 pairs
of pinnately-veined leaflets, with terminal branched tendril;
stipules larger than leaflets; calyx-teeth broad, + leafy;
style pubescent only on upperside, proximal part with
reflexed margins.

1. P. sativum L. (<u>P.</u> <u>arvense</u> L.) - <u>Garden</u> <u>Pea</u>. Climbing or
sprawling annual to 2m (usually much less); flowers 1-3,
15-35mm, white to purple; fruits 3-12cm, not compressed when
ripe. Intrd; grown on field-scale and a common casual by
roads and fields, in waste places and on tips; scattered
throughout much of BI; S Europe. Purple-flowered plants
(var. **arvense** (L.) Poiret) usually have smaller parts and are
grown for fodder (<u>Field</u> <u>Pea</u>).

TRIBE 9 - CICEREAE (genus 23). Annuals; leaves imparipinnate, with sharply serrate leaflets; tendrils 0; flowers
solitary, axillary; 9 stamens forming tube, 10th free; fruit
dehiscing longitudinally, 1-2-seeded.

23. CICER L. - <u>Chick</u> <u>Pea</u>
1. C. arietinum L. - <u>Chick</u> <u>Pea</u>. Erect, glandular-pubescent **473**
annual to 35(60)cm; leaflets 3-8 pairs, ovate to obovate;
calyx-teeth subequal, c.2x as long as tube; corolla 10-12mm,
white to pale purple; fruits 17-30mm, inflated. Intrd; tips
and waste places from seed imported as food; scattered in Br,
mainly En; Mediterranean.

TRIBE 10 - TRIFOLIEAE (genera 24-28). Annual or perennial
herbs or rarely shrubs; leaves ternate, main lateral veins of
leaflets running whole way to often toothed margin; flowers
solitary or in racemes (often greatly condensed); 9 stamens
forming tube, 10th free, or all 10 forming tube; fruits
indehiscent to longitudinally dehiscent, with 1-many seeds.

24. ONONIS L. - <u>Restharrows</u>
Flowers solitary or in terminal racemes; calyx with glandular
hairs; all 10 stamens forming tube; fruits straight, 1-many
seeded, dehiscing longitudinally.

1 Perennials with stems woody at least below, sometimes
  spiny   2
1 Herbaceous annuals, not spiny   4
  2 Corolla yellow, often streaked red; fruits 12-25mm,
    with 4-10 seeds   **1. O. natrix**
  2 Corolla pink, rarely white; fruits 5-10mm, with
    1-2(4) seeds   3
3 Stems equally hairy all round, procumbent to
  ascending; leaflets <3x as long as wide, obtuse to
  emarginate   **4. O. repens**
3 Stems mainly hairy along 1 side or 2 opposite

sides, ascending to erect; leaflets >3x as long as
wide, acute to subacute   **3. O. spinosa**
4 Fruits >6-seeded; flowers 5-10mm; pedicels >5mm
                                           2. O. reclinata
  4 Fruits 2-3-seeded; flowers >10mm; pedicels <2mm   5
5 Corolla 10-12mm; calyx-tube glabrous; calyx-teeth
  triangular to narrowly so, then finely acuminate,
  shortly glandular-pubescent   **5. O. mitissima**
5 Corolla 13-16mm; calyx-tube pubescent; calyx-teeth
  setaceous + whole length, with short glandular
  and long non-glandular hairs   **6. O. baetica**

1. O. natrix L. - <u>Yellow</u> Restharrow. Dwarf shrub to 60cm,
wihout spines; leaves mostly with 3 leaflets; flowers in lax
leafy panicles, 6-15(20)mm, yellow, often with red streaks;
fruits 12-25mm, 4-10-seeded. Intrd; natd since 1947 in rough
ground in Berks; S & W Europe.
2. O. reclinata L. - <u>Small</u> Restharrow. Erect to procumbent **RRR**
annual to 15cm; leaves all ternate; flowers in loose leafy
terminal racemes, 5-10mm, pink; fruits 8-14mm, pendent,
10-20-seeded. Native; barish sand or limestone; rare and
very local in S Devon, Glam, Pembs, Wigtowns, Guernsey and
Alderney.
3. O. spinosa L. (<u>O. campestris</u> Koch, <u>O. repens</u> ssp. <u>spinosa</u>
Greuter) - <u>Spiny</u> Restharrow. Erect or ascending usually
spiny shrub to 70cm; differs from <u>O. repens</u> in flowers
10-20mm; fruits 6-10mm; seeds 2(-4); and see key. Native;
grassy places and rough ground on mostly well-drained soils;
locally frequent in Br N to S Sc, mostly S & C En.
3 x 4. O. spinosa x O. repens = O. x pseudohircina Schur
occurs with both parents in a few places from Cambs to
Durham; it is intermediate in leaflet-shape, stem-pubescence
and fruit-size, and fertile.
4. O. repens L. (<u>O. spinosa</u> ssp. <u>maritima</u> (Dumort.) P.
Fourn.) - <u>Common</u> Restharrow. Rhizomatous perennial; stems
procumbent to ascending, woody at base, to 60cm, with or
without spines; leaves usually with 1 and 3 leaflets on same
plant; flowers in loose leafy terminal racemes, (7)12-20mm,
pink; fruits 5-8mm, 1-2-seeded. Native; rough grassy places
on well-drained soils, especially coastal; locally common in
BI except N & NW Sc. Small, densely pubescent to tomentose
plants with flowers 7-12mm (not 12-20mm) growing on maritime
sands in N & S Devon are referable to ssp. **maritima** (Gren. &
Godron) Asch. & Graebner; their taxonomic status needs
investigating.
5. O. mitissima L. - <u>Mediterranean</u> Restharrow. Erect to **473**
procumbent annual to 60cm; leaves all ternate; flowers in +
dense terminal racemes, 10-12mm, pink; lower bracts often
ternate, the upper ones or all without leaflets; fruit 5-6mm,
2-3-seeded. Intrd; occasional bird-seed alien on tips;
sporadic in S En; Mediterranean.
6. O. baetica Clemente (<u>O. salzmanniana</u> Boiss. & Reuter) - **473**

Salzmann's Restharrow. Erect to procumbent annual to 60cm; lower leaves with 1, upper with 3 leaflets; flowers in ± dense ± leafy terminal racemes, 13-16mm, pink; lower bracts with 3, upper with 1 leaflet; fruits 6-10mm, 2-3-seeded. Intrd; frequent alien on tips and waste places, mostly from bird-seed; sporadic in S En; W Mediterranean.

25. MELILOTUS Miller - Melilots
Flowers in elongated racemes; calyx without glandular hairs; 9 stamens forming tube, 10th free; fruits straight, 1-2-seeded, indehiscent or very tardily dehiscent longitudinally.

1 Corolla white   **2. M. albus**
1 Corolla yellow   2
  2 Fruits with strong concentric ridges; wings
    shorter than keel   **5. M. sulcatus**
  2 Fruits with weak to strong transverse or reticulate
    ridges; wings as long as or longer than keel   3
3 Flowers 2-3.5mm; fruits <3mm   **4. M. indicus**
3 Flowers >4mm; fruits >3mm   4
  4 Fruits >5mm, mostly 2-seeded, black when ripe,
    pubescent; keel ± equalling wings   **1. M. altissimus**
  4 Fruits <5mm, mostly 1-seeded, brown when ripe,
    glabrous; keel shorter than wings   **3. M. officinalis**

1. M. altissimus Thuill. - Tall Melilot. Erect biennial or **493** perennial to 1.5m; flowers 5-7mm, yellow; standard, wings and keel ± equal; fruits 5-7mm, reticulately or transversely ridged, black when ripe, pubescent. Intrd; natd in open grassland and rough ground, casual in waste places; frequent in S & C En, scattered W to E & S Ir and N to C Sc; Europe.
2. M. albus Medikus - White Melilot. Erect annual or **493** biennial to 1.5m; flowers 4-5mm, white; wings and keel ± equal, standard longer; fruits 3-5mm, reticulately or transversely ridged, brown when ripe, glabrous. Intrd; similar distribution to M. altissima, CI; Europe.
3. M. officinalis (L.) Lam. - Ribbed Melilot. Erect to **493** decumbent biennial to 1.5m; flowers 4-7mm, yellow; standard and wings ± equal, keel shorter; fruits 3-5mm, transversely ridged, brown when ripe, glabrous. Intrd; similar distribution to M. altissima, CI; Europe.
4. M. indicus (L.) All. - Small Melilot. Erect to ascending **493** annual to 40cm; flowers 2-3.5mm, yellow; wings and keel ± equal, standard longer; fruits 1.5-3mm, reticulately or transversely ridged, olive-green when ripe, glabrous. Intrd; rough ground and waste places, usually casual (often from bird-seed or wool) but sometimes natd; scattered through BI N to C Sc, locally common in CI; S Europe.
5. M. sulcatus Desf. - Furrowed Melilot. Usually erect **493** annual to 40cm; flowers 3-4.5mm, yellow; standard and keel ± equal, wings much shorter; fruits 2.5-4mm, concentrically ridged, yellowish- or orangy-brown when ripe, glabrous.

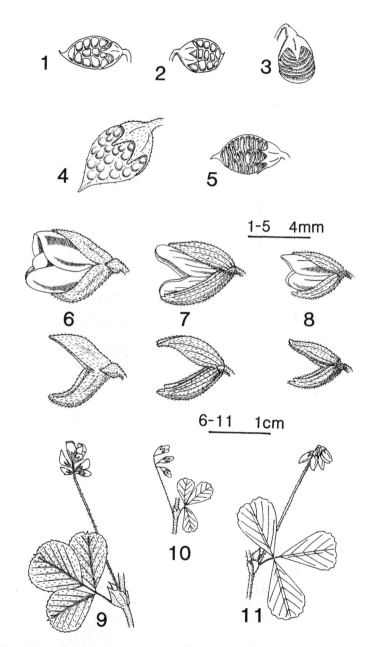

FIG 493 - Fabaceae. 1-5, fruits of **Melilotus**. 1, **M. albus**.
2, **M. indicus**. 3, **M. sulcatus**. 4, **M. altissimus**.
5, **M. officinalis**. 6-8, flowers and calyces of **Ulex**. 6, **U. europaeus**. 7, **U. gallii**. 8, **U. minor**. 9-11, flowering nodes of **Medicago** and **Trifolium**. 9, **M. lupulina**. 10, **T. micranthum**.
11, **T. dubium**.

Intrd; fairly frequent bird-seed alien on tips and waste
land; sporadic in S En; Mediterranean.

26. TRIGONELLA L. - Fenugreeks
Annuals; flowers solitary in leaf-axils or in axillary
racemes; calyx without glandular hairs; 9 stamens forming
tube, 10th free; fruits straight to curved, (1)2-many-seeded,
dehiscent longitudinally, often tardily.

1 Flowers >10mm, 1(-2) in leaf-axils; fruits >6cm,
   >9-seeded   **3. T. foenum-graecum**
1 Flowers <8mm, in axillary racemes; fruits <2cm,
   <9-seeded   2
   2 Racemes elongated; corolla yellow; fruits >8mm,
      4-8-seeded   **1. T. corniculata**
   2 Racemes subcapitate; corolla white to blue; fruits
      <8mm, 1-3-seeded   **2. T. caerulea**

1. T. corniculata (L.) L. - Sickle-fruited Fenugreek. Stems 497
erect to procumbent, to 50cm; flowers yellow, in elongated
Melilotus-like racemes, 5-7mm; fruits 10-18mm (excl. beak),
curved, reflexed, linear, transversely ridged.   Intrd;
frequent bird-seed alien on tips and waste ground; S Br;
Mediterranean.
2. T. caerulea (L.) Ser. - Blue Fenugreek. Stems erect to 497
decumbent, to 50cm; flowers white to blue, in short
subcapitate racemes, 5-7mm; fruits 3.5-5mm (excl. beak),
straight to slightly curved, patent, obovoid, mainly
longitudinally ridged. Intrd; rather sparse bird-seed alien
on tips and waste land, formerly commoner; S & C Br; E,
Mediterranean.
3. T. foenum-graecum L. - Fenugreek. Stems erect to 497
spreading, to 50cm; flowers yellowish-white, 1(-2) in leaf-
axils, 12-18mm; fruits 50-110mm plus beak 10-35mm, straight
or curved, patent, linear, mainly longitudinally ridged.
Intrd; frequent bird-seed and spice alien on tips and waste
land, now grown on small scale for seedlings and seed;
sporadic in C & S En; E Mediterranean.

27. MEDICAGO L. - Medicks
Flowers 1-many in axillary racemes; calyx without glandular
hairs; 9 stamens forming tube, 10th free; fruits slightly
curved to spiral with several complete turns, 1-many-seeded,
indehiscent, often spiny.
Spp. 3-8 have quite distinct fruits but the differences are
difficult to describe; the illustrations should be consulted.
Spine length is of virtually no value; in most spp. they can
be very short to longer than the coil diameter, and rare
spineless or tuberculate variants exist. The margin (outer
edge) of each coil is occupied by a variously thickened vein
or 'border'. Just inside this, on each face of the coil, is
a variously thickened 'submarginal vein', and between these 2

thickened veins, in each face, is the (usually channelled) 'submarginal border'. The base of each spine straddles this border and originates from both the thickened veins. The 'face' of each coil between the coil centre and the submarginal border is often characteristically veined.

1 Fruits curved to spiralled, without spines or tubercles; if flowers yellow then fruits curved to spiralled up to 1.5 turns   2
1 Fruits spiralled >(1.5)2.5 turns, usually spiny, rarely tuberculate or smooth; flowers yellow;   3
  2 Fruits <3mm in longest plane; flowers <4mm, yellow; seed 1   **1. M. lupulina**
  2 Fruits >4mm in longest plane; flowers >6mm, yellow, white, green or purplish; seeds >1   **2. M. sativa**
3 Leaflets usually each with dark blotch; fruit border grooved along centre, hence (with the 2 submarginal borders) forming 3 grooves at edge of each coil   **8. M. arabica**
3 Leaflets not dark-blotched; fruit border not grooved, hence edge of each coil with 0 or 2 grooves   4
  4 Fruit border thinner than submarginal veins, scarcely contributing to origin of spines; spines ± not grooved at base; fruits with sparse long hairs   **3. M. truncatula**
  4 Fruit border as thick or thicker than submarginal veins, contributing strongly to origin of spines; spines hence deeply grooved at least at base; fruits usually ± glabrous   5
5 Fruit border 0.5-0.8mm thick in lateral view, at least as thick as rest of each coil and ± obscuring it in lateral view   **6. M. praecox**
5 Fruit border <0.4mm thick in lateral view, thinner than rest of each coil and not completely obscuring it in lateral view   6
  6 Stipules entire to denticulate, with teeth much shorter than entire part; leaves and stems ± densely pubescent   **5. M. minima**
  6 Stipules deeply incised, with teeth much longer than entire part; leaves and stems glabrous to sparsely pubescent   7
7 Submarginal border broadly grooved, wider than border; fruit face with curved veins anastomosing to form reticulum adjacent to submarginal vein; wings longer than keel   **7. M. polymorpha**
7 Submarginal border narrowly grooved, narrower than border; fruit face with sigmoid veins not or scarcely anastomosing before joining submarginal vein; keel longer than wings   **4. M. laciniata**

Other spp. - 25-30 other spp., mainly spiny-fruited, have been found as aliens in wool and other sources, but all less

commonly than the 6 treated here. 1 plant of **M. arborea** L.
(Tree Medick), from Mediterranean, has survived on a cliff in
N Somerset since 1973; it is a silvery-appressed-pubescent
shrub to 1.5m with spineless fruits coiled in c.1 complete
turn.

1. M. lupulina L. - Black Medick. Procumbent to scrambling **493**
annual or short-lived perennial to 80cm; leaves pubescent; **497**
stipules shallowly serrate; flowers numerous in compact short
racemes, yellow, 2-3mm; fruits in c.1 coil, spineless,
pubescent or glabrous, 1.5-3mm across, black when ripe,
1-seeded. Native; grassy places and rough ground; common
throughout most of BI except parts of Sc and N Ir. See
Trifolium dubium for differences.

2. M. sativa L. - see sspp. for English names. Erect to
decumbent perennial to 90cm, leaves pubescent; stipules
entire to shallowly serrate; flowers numerous in $\pm$ compact
racemes, 5-12mm; fruits slightly curved (8-11mm long) to
spiralled (5-9mm across) in 2-3(4) turns, spineless,
pubescent or glabrous, 2-20-seeded.
1 Fruits spiralled for 2-4 complete turns; coils $\pm$
  closed in centre; flowers mauve to violet **c. ssp. sativa**
1 Fruits curved or spiralled $\leq$1.5 complete turns;
  coils $\pm$ open in centre; flowers yellow or white to
  purple   2
  2 Fruits nearly straight to curved in arc $\leq$1/2
    circle; flowers yellow   **a. ssp. falcata**
  2 Fruits curved or spiralled 0.5-1.5 complete turns;
    flowers yellow or other colours   **b. ssp. varia**
  a. Ssp. falcata (L.) Arcang. (M. falcata L.) - Sickle **497**
Medick. Flowers yellow, 6-9mm; fruits nearly straight to   **R**
curved in arc up to 1/2 circle; seeds 2-5. Native; grassy
places and rough or waste ground; native locally in E Anglia,
casual or sometimes natd elsewhere in Br scattered N to C Sc,
mostly in SE En.
  b. Ssp. varia (Martyn) Arcang. (M. x varia Martyn) - Sand
Lucerne. Flowers variously yellow, pale mauve to purple,
green or blackish, 7-10mm; fruits curved or spiralled for
0.5-1.5 complete turns; seeds 3-8, or abortive. Native;
established or casual on sandy or rough ground, arising in
situ or intrd as hybrid seed; scattered in Br N to C Sc,
especially E Anglia, Dublin. Of hybrid origin from other 2
sspp., but deserves ssp. status due to its independent
existence in many localities and use as a crop in Europe;
partly fertile and back-crosses.
  c. Ssp. sativa - Lucerne. Flowers pale mauve to violet, **497**
8-11mm; fruits spiralled for 2-3(4) complete turns; seeds
usually 10-20. Intrd; formerly much planted as crop, now
less so, common relic of cultivation, on field-margins,
roadsides, rough grassland and waste places; common in CI and
S & C En, scattered throughout rest of BI except W and N Sc;
Mediterranean.

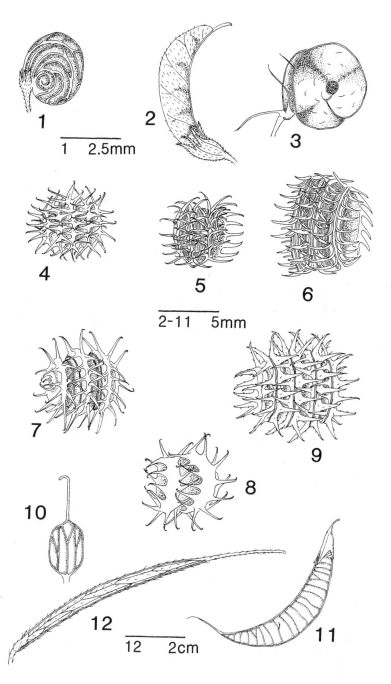

FIG 497 – Fruits of Fabaceae. 1, **Medicago lupulina**. 2–3, **M. sativa**. 2, ssp. **falcata**. 3, ssp. **sativa**. 4, **M. laciniata**. 5, **M. minima**. 6, **M. arabica**. 7, **M. polymorpha**. 8, **M. praecox**. 9, **M. truncatula**. 10, **Trigonella caerulea**. 11, **T. corniculata**. 12, **T. foenum-graecum**.

3. M. truncatula Gaertner (<u>M.</u> <u>tribuloides</u> Desr.) - <u>Strong-</u> 497
<u>spined</u> <u>Medick</u>. Procumbent to scrambling annual to 50cm, with
long hairs; stipules deeply dentate to laciniate; flowers
1-3, 5-6mm, yellow; fruits of 3-6 coils, with sparse long
hairs, with very stout ungrooved spines. Intrd; casual in
waste places near origin of seed; rather scarce wool-alien,
common tan-bark alien; scattered and sporadic in Br;
Mediterranean.

4. M. laciniata (L.) Miller (<u>M.</u> <u>aschersoniana</u> Urban) - 497
<u>Tattered</u> <u>Medick</u>. Procumbent annual to 40cm, sparsely
pubescent; stipules laciniate; flowers 1-2, 3.5-5mm, yellow;
fruits of 3-7 coils, usually ± glabrous, usually with
moderate spines grooved near base. Intrd; tips and rough
ground, common wool-alien and from other sources, rarely
persisting; sporadic throughout most of Br; N Africa.

5. M. minima (L.) L. - <u>Bur</u> <u>Medick</u>. Procumbent annual to 497
20(40)cm, densely pubescent; stipules entire to denticulate;   R
flowers 1-6, 2.5-4.5mm, yellow; fruit of 3-5 coils, often
sparsely pubescent, usually with slender spines grooved near
base. Native; sandy heaths and dunes and shingle by sea;
very local in E En from Kent to Norfolk, Jersey; common
casual throughout most of Br, often as wool-alien.

6. M. praecox DC. - <u>Early</u> <u>Medick</u>. Procumbent annual to 497
20cm, sparsely pubescent; stipules laciniate; flowers 1-2,
2-5mm, yellow; fruits of 2.5-5 coils, usually ± glabrous,
with moderate spines grooved near base. Intrd; casual on
tips and rough ground, mostly from wool; sporadic throughout
most of Br; Mediterranean.

7. M. polymorpha L. (<u>M.</u> <u>hispida</u> Gaertner nom. illeg., <u>M.</u> 497
<u>nigra</u> (L.) Krocker) - <u>Toothed</u> <u>Medick</u>. Procumbent to   R
scrambling annual to 60cm, ± glabrous when mature; stipules
laciniate; flowers 1-8, 3-4.5mm, yellow; fruits of 1.5-5(6)
coils, glabrous, usually with rather slender spines grooved
near base. Native; very local in sandy ground near sea; CI,
S En N to N Devon and N Norfolk; common casual throughout BI
in waste places. The sp. most often lacking fruit spines.

8. M. arabica (L.) Hudson - <u>Spotted</u> <u>Medick</u>. Procumbent to 497
scrambling annual to 60cm, ± glabrous when mature; stipules
dentate usually <1/2 way to midrib; flowers 1-5, 4-6mm,
yellow; fruits of 3-5(6) coils, glabrous, usually with rather
slender spines deeply grooved >1/2 way. Native; grassy and
barish places, especially near sea; CI, S & C Br; throughout
BI in waste places as common casual.

28. TRIFOLIUM L. (see sections for synonyms) - <u>Clovers</u>
Flowers 1-many in axillary congested racemes; calyx without
glandular hairs; 9 stamens forming tube, 10th free; fruits
indehiscent, longitudinally dehiscent, or dehiscent by apex
falling off, ± straight, 1-9-seeded, often partly or wholly
enclosed in calyx or persistent corolla.
The keys require flowering and fruiting racemes, usually
both present for a long period in the season. Calyx-tube

vein-number is important; in spp. with densely pubescent
calyx-tubes it is best observed by splitting the tube up one
side and observing the veins from the inside.

General key
1 Racemes with sterile corolla-less flowers mixed with
  normal ones, becoming turned down and thrust into
  ground as fruit ripens   **32. T. subterraneum**
1 Racemes wholly of sexual flowers with corollas, not
  becoming subterranean   2
  2 Calyx becoming greatly inflated in fruit, the
    inflation confined to upper lip with 2 teeth   Key A
  2 Calyx not becoming inflated in fruit, or only
    moderately and symmetrically so   3
3 Calyx-tube with 5 veins; corolla yellow, ≤8mm; fruit
  stalked, 1-seeded   Key B
3 Calyx-tube with (5)10-20 veins; corolla mostly white
  or pink to purple, if yellow then >8mm; fruit sessile,
  1-9-seeded   4
  4 Throat of calyx (remove corolla) at least partly
    closed by a thickening or a ring of hairs; flowers
    without bracts, + sessile   Key C
  4 Throat of calyx glabrous and open; each flower
    with small bract at base, or the bracts variably
    fused; flowers sessile or distinctly pedicellate
                                                        Key D

Key A - Calyx asymmetrically strongly inflated in fruit
        (sect. Vesicaria)
1 Perennials; stems rooting at nodes; flowers with
  standard uppermost   **9. T. fragiferum**
1 Annuals; stems not rooting at nodes; corolla upside-
  down so that standard is lowermost   2
  2 Fruiting calyx densely woolly, with calyx-teeth +
    completely obscured; leaflets 4-12mm
                                              11. T. tomentosum
  2 Fruiting calyx sparsely to densely pubescent, with
    2 prominent divergent calyx-teeth at apex; leaflets
    10-20(25)mm   **10. T. resupinatum**

Key B - Calyx not inflated or moderately and symmetrically
        so, with 5 veins; corolla yellow, <1cm (sect.
        Chronosemium)
1 Apical leaflet with stalk >0.5mm, distinctly longer
  than that of lateral leaflets   2
1 All leaflets with stalks <0.5mm   3
  2 Corolla 3-4mm, with standard folded longitudinally
    over fruit; racemes mostly <25-flowered
                                              14. T. dubium
  2 Corolla 4-7mm, with standard + flat over fruit;
    racemes mostly >25-flowered   **13. T. campestre**
3 Corolla 5-8mm; pedicels c.1mm; racemes mostly >20-

flowered   **12. T. aureum**
3 Corolla 1.5–3mm; pedicels c.1.5mm; racemes <10-
 flowered   **15. T. micranthum**

Key C – Calyx not inflated or moderately and symmetrically
    so, with 10–20 veins, its throat ± closed by
    thickening or hairs; bracts 0 (sect. Trifolium)
1 Firmly rooted perennials with habit of T. pratense or
 rhizomatous; flowers >12mm   2
1 Easily uprooted annuals; flowers often <12mm   5
  2 Free part of stipules of stem leaves triangular-
    ovate, abruptly narrowed to brown bristle-like
    point   **16. T. pratense**
  2 Free part of stipules of stem-leaves linear to
    lanceolate, green ± to apex   3
3 Corolla reddish-purple (rare albinos); lowest calyx-
 lobe c.1.5x as long as upper lobes and as calyx-tube
                                              17. T. medium
3 Corolla whitish-yellow, (?sometimes pale pink); lowest
 calyx-lobe >2x as long as upper lobes and as calyx-
 tube   4
  4 Flowers 15–18mm; leaflets elliptic to oblong or
    obovate; racemes ± subsessile   **18. T. ochroleucon**
  4 Flowers 20–25mm; leaflets elliptic to linear-
    oblong; some racemes pedunculate   **19. T. pannonicum**
5 Calyx-tube with c.20 veins   6
5 Calyx-tube with 10 veins   7
  6 Calyx-tube with many long hairs; corolla >10mm
                                              25. T. hirtum
  6 Calyx-tube without long hairs, usually glabrous;
    corolla <10mm   **26. T. lappaceum**
7 Free part of stipules of stem-leaves ovate-triangular,
 abruptly narrowed to obtuse, acute or shortly
 apiculate apex   8
7 Free part of stipules of stem-leaves gradually tapered
 to long narrow apex   9
  8 Racemes ± globose; stipules sharply serrate
                                              20. T. stellatum
  8 Racemes cylindrical; stipules obscurely denticulate
                                              21. T. incarnatum
9 Racemes ± cylindrical in fruit, >2x as long as wide   10
9 Racemes globose to ovoid in fruit <1.5(2)x as long as
 wide   11
  10 Flowers 10–13mm; calyx-lobes unequal, the lowest
     much the longest   **28. T. angustifolium**
  10 Flowers 3–6mm; calyx-lobes sub-equal   **27. T. arvense**
11 At least the 2 uppermost leaves on each stem-branch
 exactly opposite; flowers (5)7–12mm   12
11 All leaves alternate (sometimes 2 uppermost close
 together, but not exactly opposite); flowers 4–7mm   14
  12 Calyx with ring of hairs in throat; fruit
     exserted from calyx-tube   **30. T. alexandrinum**

12 Calyx with bilateral swellings in throat,
   completely closed; fruit enclosed in calyx-tube   13
13 Calyx-lobes with 1 vein, or with 3 veins just at
   base   **31. T. echinatum**
13 Calyx-lobes with 3 veins for >1/2 length
                            29. T. squamosum
  14 Leaflets with lateral veins thickened and arched-
     recurved at leaf-margin   **24. T. scabrum**
  14 Leaflets with lateral veins thin and straight or
     slightly forward-curved at leaf-margin   15
15 Leaflets ± glabrous on upperside; fruiting calyx-
   tube not inflated   **23. T. bocconei**
15 Leaflets pubescent on upperside; fruiting calyx-tube
   inflated   **22. T. striatum**

Key D - Calyx not inflated or slightly and symmetrically so,
    with (5)10(12) veins, its throat open; bracts present
    (sects. Lotoidea & Paramesus)
1 Flowers 1-4(5), with pedicels ≥1mm; seeds 5-9
                    1. T. ornithopodioides
1 Flowers (3)10-many, if <10 then ± sessile; seeds 1-4   2
  2 Flowers ± sessile, not reflexed after pollination;
    fruit shorter than calyx (incl. lobes)   3
  2 Flowers with pedicels ≥1mm, strongly reflexed
    after pollination; fruit longer than calyx (incl.
    lobes)   5
3 Stipules serrate; teeth of leaflets and stipules
   gland-tipped   **8. T. strictum**
3 Stipules usually entire, sometimes serrate; teeth of
   stipules and leaflets not gland-tipped   4
  4 Corolla 3-4mm, shorter than calyx (incl. lobes),
    whitish; many (often ± all) racemes congested at
    base of plant   **7. T. suffocatum**
  4 Corolla 4-7mm, longer than calyx (incl. lobes),
    purplish; racemes dispersed along stems
                    6. T. glomeratum
5 Flowers 4-5mm; upper racemes with peduncles <1cm,
   often subsessile; annual   **5. T. cernuum**
5 Flowers (5)7-13mm; all racemes with peduncles >1cm;
   perennial   6
  6 Stems not rooting at nodes, usually erect to
    ascending   **4. T. hybridum**
  6 Stems procumbent, rooting at nodes   7
7 Leaflets usually >10mm, obovate, often with light or
  dark markings, with veins translucent when fresh;
  petioles glabrous; calyx-lobes triangular-lanceolate;
  standard rounded at apex   **2. T. repens**
7 Leaflets usually <10mm, suborbicular, without light
  or dark markings, with veins not translucent when
  fresh; petioles sparsely pubescent; calyx-lobes
  triangular-ovate; standard emarginate at apex
                   3. T. occidentale

Section 1 - LOTOIDEA Crantz (sect. Ornithopoda (Mall.) Tutin
nom. inval., sect. Trifoliastrum Gray, Amoria C. Presl,
Falcatula Brot.) (spp. 1-7). Flowers all fertile, not
becoming subterranean; calyx not becoming inflated in fruit,
with (5)10(-12) veins, glabrous and open at throat; fruits
sessile, 1-9-seeded; bracts present; leaves without glands.

1. T. ornithopodioides L. (Falcatula ornithopodioides (L.) 504
Brot. ex Bab.) - Bird's-foot Clover. Procumbent ± glabrous R
annual to 20cm; racemes 1-4(5)-flowered, stalked, axillary;
flowers 6-8mm, pink to white or both; fruits exserted,
5-9-seeded. Native; sandy semi-open ground, mainly near sea;
scattered round coasts of BI N to Norfolk, Flints and Dublin.
2. T. repens L. (Amoria repens (L.) C. Presl) - White 504
Clover. Subglabrous perennial with procumbent stems to 50cm
rooting at nodes; racemes ± globose, on erect axillary
peduncles up to 20cm; flowers 7-12mm, usually white,
sometimes pale pink, rarely red or mauve; fruits exserted,
3-4-seeded; flowers scented when fresh, 2n=32. Native;
grassy and rough ground; common throughout BI.
3. T. occidentale Coombe (?T. prostratum Biasol.) - Western 504
Clover. Differs from T. repens in less robust; leaflets R
thicker; flowers not scented; 2n=16; and see key. Native;
short turf by sea; CI, SW En, S Wa, E Ir.
4. T. hybridum L. (Amoria hybrida (L.) C. Presl) - Alsike 504
Clover. Erect to decumbent subglabrous perennial to
40(70)cm; racemes ± globose, mostly axillary, stalked;
flowers 7-10mm, usually white and pink, sometimes either one;
fruits slightly exserted, 2-4-seeded. Intrd; commonly natd
in grassy and rough ground; throughout most of BI; Europe.
Ssp. **hybridum** is the cultivated plant. Ssp. **elegans** (Savi)
Asch. & Graebner, with solid (not hollow), less erect stems
and smaller racemes (<2cm across) is probably the wild
progenitor, but doubtfully a ssp.
5. T. cernuum Brot. - Nodding Clover. Procumbent to 504
ascending subglabrous annual to 25cm; racemes ± globose,
mostly axillary, the lower stalked, the upper subsessile;
flowers 3-5mm, pink; fruit exserted, 1-4 seeded. Intrd;
casual from wool and other sources, sometimes persisting for
few years; sporadic in C & S En; SW Europe.
6. T. glomeratum L. (Amoria glomerata (L.) Sojak) - 504
Clustered Clover. Procumbent to ascending subglabrous R
annual to 25cm; racemes ± globose, terminal and axillary,
sessile; flowers 4-5mm, purple; fruit enclosed in calyx,
(1-)2-seeded. Native; grassy places on sandy soil mostly
near sea; CI, S & E coast of En N to Norfolk, formerly
commoner, rare casual elsewhere in En.
7. T. suffocatum L. (Amoria suffocata (L.) Sojak) - 504
Suffocated Clover. Procumbent subglabrous annual to 3(8)cm; R
racemes ± globose, terminal and axillary, sessile, most or
all densely crowded near root; flowers 3-4mm, whitish; fruit
± enclosed in calyx, 2-seeded. Native; similar places and

distribution to T. glomeratum, but N to SE Yorks and formerly Flints.

Section 2 - PARAMESUS (C. Presl) Endl. (sect. Involucrarium auct. non Hook., Paramesus C. Presl) (sp. 8). Flowers all fertile, not becoming subterranean; calyx slightly and symmetrically swollen in fruit, with 10 veins, glabrous and open at throat; fruits sessile, 2-seeded; bracts present; leaf-teeth gland-tipped.

8. T. strictum L. (Paramesus strictus (L.) C. Presl) - **504** Upright Clover. Erect to ascending glabrous annual to **RR** 15(25)cm; racemes ± globose, terminal and axillary, stalked; flowers 5-7mm, pinkish-purple; fruit ± enclosed in calyx, 2-seeded. Native; rocky grassy places, extremely local; Jersey, W Cornwall, Rads, formerly Guernsey.

Section 3 - VESICARIA Crantz (sect. Vesicastrum Ser., sect. Fragifera Koch, Galearia C. Presl) (spp. 9-11). Flowers all fertile, not becoming subterranean; calyx becoming asymmetrically much inflated in fruit, with reticulate network of veins at fruiting, glabrous and open at throat at flowering but becoming narrowed in fruit; fruits sessile, 1-2-seeded; bracts present; leaves without glands.

9. T. fragiferum L. (Galearia fragifera (L.) C. Presl) - **504** Strawberry Clover. Subglabrous perennial with procumbent stems to 30cm rooting at nodes; racemes ± globose, on erect axillary peduncles up to 10(20)cm; flowers 5-7mm, pink; fruits enclosed in pubescent calyx, (1-)2-seeded. Native; grassy places, often on heavy or brackish soils; rather scattered but locally common in BI N to S Sc. Vegetatively resembles T. repens but lateral veins of leaflets are thickened and recurved distally almost as in T. scabrum (not so in T. repens), and stipules are gradually narrowed, not abruptly narrowed, to apex. Ssp. **fragiferum** is the common plant. Ssp. **bonannii** (C. Presl) Sojak occurs in S En; it is said to have smaller calyces (4-6mm, not 8-10mm, in fruit) with exserted corollas and slightly elongated larger racemes, but its status is very doubtful.

10. T. resupinatum L. (Galearia resupinata (L.) C. Presl) - **504** Reversed Clover. Glabrous, procumbent to suberect annual to 30cm; racemes ± globose, terminal and axillary, stalked; flowers 5-8mm, pink to purple; fruits enclosed in pubescent calyx, 1-seeded. Intrd; rather frequent casual from wool and other sources, sometimes persisting; mainly S Br; S Europe.

11. T. tomentosum L. - Woolly Clover. Differs from T. **504** resupinatum in stems procumbent, to 15cm; racemes subsessile; flowers 3-6mm; fruits 1-2-seeded; and see key. Intrd; casual from wool and other sources; sporadic mainly in S Br; Mediterranean.

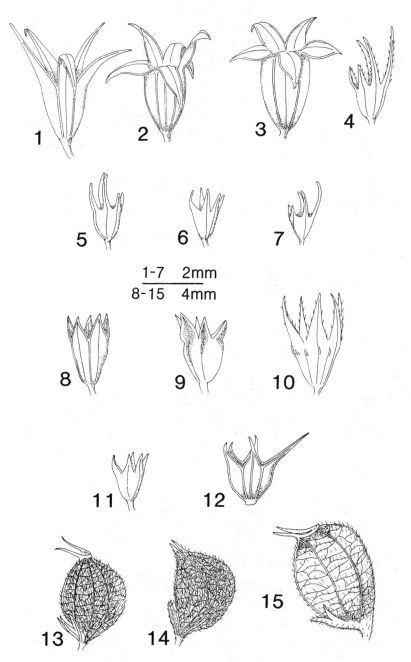

FIG 504 – Fruiting calyces of **Trifolium**. 1, **T. cernuum**. 2, **T. suffocatum**. 3, **T. glomeratum**. 4, **T. aureum**. 5, **T. dubium**. 6, **T. micranthum**. 7, **T. campestre**. 8, **T. repens**. 9, **T. occidentale**. 10, **T. ornithopodioides**. 11, **T. hybridum**. 12, **T. strictum**. 13, **T. resupinatum**. 14, **T. tomentosum**. 15. **T. fragiferum**.

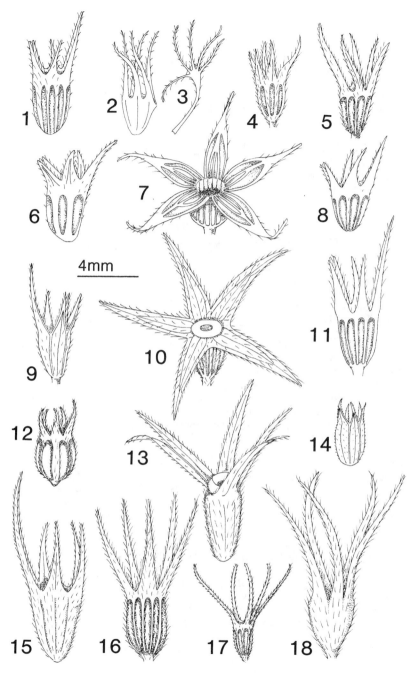

FIG 505 – Fruiting calyces of **Trifolium**. 1, **T. ochroleucon**.
2–3, **T. subterraneum** (fertile, sterile). 4, **T. scabrum**. 5, **T.
alexandrinum**. 6, **T. squamosum**. 7, **T. lappaceum**. 8, **T.
echinatum**. 9, **T. pratense**. 10, **T. stellatum**. 11, **T. medium**.
12, **T. striatum**. 13, **T. angustifolium**. 14, **T. bocconei**. 15,
T. pannonicum. 16, **T. incarnatum**. 17, **T. arvense**. 18, **T.
hirtum**.

Section 4 - CHRONOSEMIUM Ser. (Chrysaspis Desv.) (spp.
12-15). Flowers all fertile, not becoming subterranean;
calyx not becoming inflated in fruit, with 5 veins, with
glabrous and open throat; fruits stalked, exserted from
calyx, protected by standard, 1-2-seeded; bracts 0 or minute;
leaves without glands.

12. **T. aureum** Pollich (Chrysaspis aurea (Pollich) Greene) - 504
Large Trefoil. Erect to ascending, sparsely pubescent annual
to 30cm; racemes 12-20mm, globose to obovoid, terminal and
axillary, stalked; flowers >20, 5-8mm, golden-yellow; fruits
1-seeded. Intrd; natd in grassy and rough ground, formerly
frequent, now very local; scattered throughout much of Br,
but now well natd + only in C & S Sc; Europe.

13. **T. campestre** Schreber (Chrysaspis campestris (Schreber) 504
Desv.) - Hop Trefoil. Procumbent to suberect sparsely
pubescent annual to 30cm; racemes 8-15mm, + globose, terminal
and axillary, stalked; flowers >20, 4-7mm, pale yellow;
fruits 1-seeded. Native; grassy and barish places; frequent
throughout most of BI.

14. **T. dubium** Sibth. (Chrysaspis dubia (Sibth.) Desv.) - 493
Lesser Trefoil. Procumbent to suberect very sparsely 504
pubescent annual to 25cm; racemes 5-9mm, + globose, terminal
and axillary, stalked; flowers (3)5-20, 3-4mm, yellow; fruits
1-seeded; pedicels c.1mm. Native; grassy and open ground;
common + throughout BI. Often confused with Medicago
lupulina before fruiting, but has glabrous (not pubescent)
calyx, subglabrous (not pubescent) leaflets without (not
with) an apical apiculus, and usually deeper yellow flowers
with persistent corolla.

15. **T. micranthum** Viv. (Chrysaspis micrantha (Viv.) 493
Hendrych) - Slender Trefoil. Procumbent to ascending 504
subglabrous annual to 15cm; differs from T. dubium in racemes
c.4mm; flowers (1)2-6(10), 1.5-3mm; fruits 1-2-seeded;
pedicels c.1.5mm. Native; short turf, especially close-cut
lawns; common in CI and SE En, scattered to W Ir and S Sc,
casual in C & N Sc.

Section 5 - TRIFOLIUM (spp. 16-31). Flowers all fertile, not
becoming subterranean; calyx not becoming inflated in fruit
or becoming only moderately and symmetrically so, with 10-20
veins, with throat wholly or partly closed by ring of hairs
or by thickenings; fruit sessile, enclosed in calyx, 1-
seeded; bracts 0; leaves without glands.

16. **T. pratense** L. - Red Clover. Decumbent to erect, 504
pubescent, tufted perennial to 60cm; leaflets elliptic to
obovate; racemes globose to ellipsoid, terminal, + sessile;
flowers 12-18mm, pinkish-purple, sometimes pale pink, white
or cream. Native; grassy places, waste and rough ground;
common throughout BI. Agricultural variants are more robust
and have hollow stems and less denticulate leaflets; they

have been separated as var. **sativum** Schreber.

17. T. medium L. - <u>Zigzag</u> <u>Clover</u>. Ascending, sparsely **505** pubescent, rhizomatous perennial to 50cm; leaflets narrowly elliptic; racemes globose to obovoid, terminal, stalked at maturity; flowers 12-20mm, reddish-purple. Native; grassy places, hedgerows and wood-borders; frequent + throughout BI. Easily told from <u>T. pratense</u> by the stipules (see key), rhizomes and narrower, less pubescent leaflets.

18. T. ochroleucon Hudson - <u>Sulphur</u> <u>Clover</u>. Ascending to **505** erect, pubescent, tufted but shortly rhizomatous perennial to **R** 50cm; leaflets elliptic to obovate; racemes globose to ellipsoid, terminal, shortly stalked or subsessile; flowers 15-20mm, whitish-yellow. Native; grassy places on heavy soils; very local and decreasing in E Anglia W to Northants, casual elsewhere in Br.

19. T. pannonicum Jacq. - <u>Hungarian</u> <u>Clover</u>. Differs from <u>T.</u> **505** <u>ochroleucon</u> in stems + erect, to 60(80)cm; and see key. Intrd; rather scarce casual mainly as grass-seed contaminant, rarely + natd; sporadic in S En; SE Europe.

20. T. stellatum L. - <u>Starry</u> <u>Clover</u>. Erect pubescent annual **505** to 20(30)cm; leaflets obovate-cordate; racemes + globose, terminal, stalked; flowers 12-18mm, pink. Intrd; natd since at least 1804 on shingle in W Sussex, infrequent casual elsewhere in S Br; Mediterranean. The distinctive strongly patent calyx-lobes appear only in fruit.

21. T. incarnatum L. - see sspp. for English names. Erect **505** decumbent pubescent annual to (30)50cm; leaflets obovate-cordate; racemes cylindrical, up to 6cm, terminal, stalked; flowers 9-15mm.

a. Ssp. incarnatum - <u>Crimson</u> <u>Clover</u>. Usually erect, with + patent hairs; flowers crimson. Intrd; formerly much grown for fodder and a common casual often natd, now very rarely grown and uncommon in the wild; sporadic over CI and Br; S Europe.

b. Ssp. molinerii (Balbis ex Hornem.) Syme (<u>T. molinerii</u> **RR** Balbis ex Hornem.) - <u>Long-headed</u> <u>Clover</u>. Usually decumbent to ascending to 20cm, with + appressed hairs; flowers yellowish-white to pale pink. Native; short grassland near sea; Jersey, Lizard Peninsula (W Cornwall).

22. T. striatum L. - <u>Knotted</u> <u>Clover</u>. Procumbent to erect **505** pubescent annual to 30cm; leaflets obovate; racemes ovoid, terminal and axillary, sessile; flowers 4-7mm, pink. Native; short grassland and open places on sandy ground, especially near sea; locally frequent through Br N to CE Sc, E Ir, CI.

23. T. bocconei Savi - <u>Twin-headed</u> <u>Clover</u>. Erect to **505** ascending pubescent annual to 20(30)cm; leaflets narrowly **RR** obovate; racemes ovoid, terminal and axillary, sessile; flowers 4-6mm, white to pink. Native; short turf near sea; Jersey, Lizard Peninsula (W Cornwall).

24. T. scabrum L. - <u>Rough</u> <u>Clover</u>. Procumbent to erect **505** pubescent annual to 20cm; leaflets obovate; racemes ovoid, terminal and axillary, sessile; flowers 4-7mm, white to pale pinkish. Native; similar habitats and distribution to T. striatum, and often with it.

25. T. hirtum All. - Rose Clover. Ascending pubescent **505** annual to 35cm; leaflets obovate; racemes globose, ± all terminal, sessile; flowers 12-17mm, purple. Intrd; tips and waste ground, casual mainly from wool; sporadic in C & S En; S Europe.

26. T. lappaceum L. - Bur Clover. Erect or ascending **505** sparsely pubescent annual to 35cm; leaflets obovate to narrowly so; racemes globose to ovoid, mostly terminal, subsessile at first, becoming stalked; flowers 4-8mm, pink. Intrd; tips and waste places, casual mainly from bird-seed; sporadic in S Br; S Europe.

27. T. arvense L. - Hare's-foot Clover. Erect or ascending **505** pubescent annual to 20(40)cm; leaflets linear-oblong; racemes ovoid to oblong at first, soon elongating to cylindrical, mostly terminal, stalked; flowers 3-6mm, white to pink. Native; barish ground on sandy soils; locally frequent in BI N to C Sc, absent from most of Sc and Ir.

28. T. angustifolium L. - Narrow Clover. Erect, appressed- **505** pubescent annual to 50cm; leaflets linear to linear-lanceolate; racemes ovoid at first, soon elongating to cylindrical, mostly terminal, stalked; flowers 10-13mm, pink. Intrd; tips and waste ground, casual from wool and other sources; sporadic in Br; S Europe.

29. T. squamosum L. - Sea Clover. Erect to ascending rather **505** sparsely pubescent annual to 40cm; leaflets linear-oblong to   **R** oblanceolate; racemes ovoid, mostly terminal, shortly stalked; flowers (5)7-9mm, pink. Native; short, often brackish, turf by sea; very local in Br N to S Wa and N Lincs, Guernsey.

30. T. alexandrinum L. - Egyptian Clover. Erect to **505** ascending pubescent (to sparsely so) annual to 60cm; leaflets elliptic or oblong to narrowly so; racemes ovoid, terminal and axillary, stalked; flowers 8-13mm, cream. Intrd; casual mainly from grass-seed mixtures, in newly sown grass by roads and in parks; Guernsey, rare in S En; SE Mediterranean. The much rarer **T. constantinopolitanum** Ser., from SW Asia, is often confused with T. alexandrinum, but would key out as T. echinatum, from which it differs in the calyx-tube at fruiting being constricted, not gradually widened, distally.

31. T. echinatum M. Bieb. - Hedgehog Clover. Procumbent to **505** erect, sparsely pubescent annual to 50cm; leaflets obovate or elliptic to narrowly so; racemes ovoid, terminal and axillary, stalked; flowers 8-15mm, cream, often purple-tinged. Intrd; casual of tips and rough ground, mostly from bird-seed; sporadic in En; SE Europe and SW Asia.

Section 6 - TRICHOCEPHALUM Koch (Calycomorphum C. Presl) (sp. 32). Racemes with sterile corolla-less flowers mixed with normal ones, becoming turned down and thrust into ground at fruiting; calyx becoming symmetrically swollen in fruit, with

>20 very fine veins, with glabrous and open throat; fruit sessile, 1-seeded; bracts 0; leaves without glands.

32. T. subterraneum L. (Calycomorphum subterraneum (L.) C. 505 Presl). - Subterranean Clover. Decumbent to procumbent pubescent annual to 20(80)cm; racemes axillary, with 2-5(7) fertile flowers and numerous sterile ones consisting only of palmately-divided calyx, long-stalked; fertile flowers 8-14mm, whitish. Native; short turf and barish places on sandy soils, especially by sea; scattered in Br N to N Lincs and Anglesey, CI, Wicklow; also frequent wool-alien in Br, and then often very robust (var. **oxaloides** (Bunge) Rouy).

TRIBE 11 - THERMOPSIDEAE (genus 29). Rhizomatous perennial herbs; leaves ternate, with entire leaflets; flowers in erect raceme with small ovate bracts; all 10 stamens free; fruits longitudinally dehiscent, with 2-7 seeds.

29. THERMOPSIS R. Br. - False Lupin
1. T. montana Nutt. - False Lupin. Stems erect, to 70(100)cm, pubescent; leaflets up to 10cm, elliptic to narrowly so; stipules ovate to lanceolate; flowers 5-60 per raceme, yellow, 20-25mm; fruits erect, pubescent, 4-7cm. Intrd; grown in gardens and natd on old sites or rough grassy places; Shetland from c.1978, formerly Northants; W N America.

TRIBE 12 - GENISTEAE (genera 30-35). Perennial or annual herbs or woody shrubs or trees; leaves simple, entire, ternate or palmate; flowers in axillary or terminal racemes, sometimes reduced to 1 or 2; all 10 stamens forming tube; fruits longitudinally dehiscent, with 2-many seeds.

30. LUPINUS L. - Lupins
Herbaceous annuals or perennials, sometimes woody at base; leaves palmate, with long petiole; flowers in terminal erect racemes, variously coloured; fruits several-seeded, erecto-patent.

1 Stems woody towards base, not dying down to ground in
    winter   **1. L. arboreus**
1 Stems herbaceous, dying down to ground in winter   2
    2 Annuals, easily uprooted   3
    2 Tuft-forming perennials   4
3 Leaflets oblanceolate; upper lip of calyx very
    shallowly 2-lobed or ± entire; seeds mostly >7mm
                                                    5. L. albus
3 Leaflets linear; upper lip of calyx deeply bifid;
    seeds <7mm   **6. L. angustifolius**
    4 Basal leaves absent at flowering time; upper part
      of stem and petioles usually shaggy-hairy; lower
      lip of calyx 7-13mm   **4. L. nootkatensis**

4 Basal leaves present at flowering time; stem and
  petioles with rather sparse short hairs; lower lip
  of calyx 3-8mm   5
5 Stems unbranched, with 1 inflorescence; flowers blue;
  leaflets obtuse to acute; lower lip of calyx 3-6mm
                                              3. L. polyphyllus
5 Stems mostly branched, with >1 inflorescence; flowers
  various shades of blue, pink, purple, yellow or
  white; leaflets acute to acuminate; lower lip of
  calyx 5-8mm   **2. L. x regalis**

1. L. arboreus Sims - Tree Lupin. Stems erect, much
branched, to 2m, with rather sparse short hairs; leaflets
5-10(12), mostly <6cm, oblanceolate to narrowly so; lower lip
of calyx 7-11mm; corolla usually yellow, often blue-tinged,
sometimes blue or whitish. Intrd; garden escape natd in
waste places and on maritime shingle sand; very scattered
mostly on coasts of Br N to C Sc, S Ir, CI; California.
Killed or cut back in heavy frosts, so populations fluctuate.
2. L. x regalis Bergmans (L. arboreus x L. polyphyllus) -
Russell Lupin. Much closer to L. polyphyllus; stems erect,
usually at least sparingly branched, to 1.5m, with rather
sparse short hairs; leaflets (7)9-15, mostly >6cm,
oblanceolate to narrowly elliptic; lower lip of calyx 5-8mm;
corolla various colours. Intrd; commonly grown and
frequently well natd on rough ground, banks of roads and
railways; throughout lowland Br; garden origin, but
spontaneous hybrids occur in Moray, Wigtowns and Midlothian
with the parents, and perhaps also backcrosses from natd L. x
regalis to L. arboreus. Garden escapes are usually recorded
as L. polyphyllus, but this is now very rarely grown and
probably all records outside Sc are L. x regalis.
2 x 4. L. x regalis x L. nootkatensis probably occurs on
river-shingle with the parents in Moray and M Perth, but is
difficult to identify; it is intermediate and fertile.
3. L. polyphyllus Lindley - Garden Lupin. Differs from L. x
regalis in key characters. Intrd; formerly grown in gardens,
natd by rivers and railways and sometimes in waste places;
scattered in Sc; W N America. Flowers ?sometimes pink.
**3 x 4. L. polyphyllus x L. nootkatensis = L. x
pseudopolyphyllus** C.P. Smith occurs with the parents on
river-shingle in Moray and M Perth; it is intermediate and
fertile.
4. L. nootkatensis Donn ex Sims - Nootka Lupin. Stems
erect, to 1m, usually with long shaggy hairs; leaflets
6-9(12), mostly <6cm, oblanceolate; lower lip of calyx
7-13mm; corolla bluish-purple, sometimes whitish tinged
purplish. Intrd; natd on riverside shingle and moorland
since at least 1862; C & N Sc from Perths to Orkney,
especially by R. Tay and R. Dee, NW Ir; NW N America and NE
Asia.
5. L. albus L. - White Lupin. Stems erect to ascending,

sparingly branched, to 60(100)cm, moderately pubescent;
leaflets 5-9, <6cm, oblanceolate to narrowly obovate; lower
lip of calyx 8-9mm; corolla white, usually variably tinged
bluish-violet. Intrd; becoming grown as a seed-crop and
appearing on tips, at docks and in waste places; sporadic in
S En; Mediterranean. Will become much commoner.
6. L. angustifolius L. - Narrow-leaved Lupin. Stems erect,
well branched, to 60(100)cm, with rather sparse short hairs;
leaflets 5-9, <4cm, linear; lower lip of calyx 6-7mm; corolla
blue. Intrd; imported for grain and being used in trials as
seed-crop, scarce casual at docks and waste places; sporadic
in S Br; Mediterranean. Formerly grown for fodder.

31. LABURNUM Fabr. - Laburnums
Deciduous non-spiny trees; leaves ternate, with long petiole;
flowers yellow, in pendent racemes on short-shoots; fruits
several-seeded, pendent.

1. L. anagyroides Medikus - Laburnum. Tree to 8m; young
twigs, petioles, leaf lowerside, peduncle, pedicels and young
fruits appressed-pubescent, densely silvery so when very
young; racemes mostly 20-30cm; flowers c.16-20mm; fruits not
winged dorsally. Intrd; much planted and frequently
self-sown in rough ground and on banks by roads and railways;
scattered through BI; mountains of SC Europe.
1 x 2. L. anagyroides x L. alpinum = L. x watereri (Wettst.)
Dippel (L. x vossii Hort.) is now more commonly planted than
either sp. but much less or never self-sown due to sterility;
it is intermediate in pubescence, has the longer racemes with
more flowers of L. alpinum, the larger flowers of L.
anagyroides, and rarely forms fruits.
2. L. alpinum (Miller) Bercht. & J.S. Presl - Scottish
Laburnum. Tree to 13m; differs from L. anagyroides in
glabrous to sparsely pubescent parts; racemes mostly 25-35cm
with more flowers mostly 13-17mm; fruits with dorsal ridge or
wing c.1mm wide. Intrd; similar occurrence to L. anagyroides
but much confused and exact relative abundance unknown,
perhaps commoner in N Br; mountains of SC Europe.

32. CYTISUS Desf. (Sarothamnus Wimm., Lembotropis Griseb.) -
Brooms
Non-spiny shrubs; leaves simple or ternate, sessile or
petiolate; flowers yellow, in terminal or axillary racemes or
1-few in leaf-axils; calyx with upper lip 2-toothed or deeply
bifid; fruits several- to many-seeded, erect to patent.

1 Flowers in terminal racemes; all leaves ternate, most
  with petioles >1cm   **1. C. nigricans**
1 Flowers 1-few in lateral groups; upper leaves usually
  simple, lower with petioles <1cm   2
  2 Flowers white, 8-12mm; standard and keel pubescent;

upper calyx-teeth >1/2 as long as tube
                                        2. C. multiflorus
2 Flowers yellow to red, 10-25mm; corolla glabrous,
   upper calyx-teeth much <1/2 as long as tube   3
3 Fruits covered with dense white shaggy hairs; twigs
   c.10-angled, fragile   **3. C. striatus**
3 Fruits with long hairs on sutures, + glabrous on
   faces; twigs c.5-angled, pliable   **4. C. scoparius**

1. C. nigricans L. (Lembotropis nigricans (L.) Griseb.) - 513
Black Broom. Stems erect, to 1.5m, terete to obscurely
ridged, pubescent; leaves all ternate, most with petioles
>1cm, flowers in terminal leafless racemes, 7-12mm, yellow;
fruits 1.5-3.5cm, appressed-pubescent. Intrd; natd on waste
ground by railway and in gravel-pits since 1970; Middlesex
and E Kent; C & SE Europe.
2. C. multiflorus (L'Her. ex Aiton) Sweet - White Broom. 513
Stems erect or arching, to 2m, deeply c.8-ridged, glabrous
after 1st year; leaves ternate below, simple above, sessile
or nearly so; flowers 1-3 in lateral groups, 8-12mm, white;
fruits 1-3cm, appressed-pubescent. Intrd; natd on banks of
roads and railways; very scattered in N Wa and S En; Iberian
Peninsula.
3. C. striatus (Hill) Rothm. - Hairy-fruited Broom. Stems 513
erect or arching, to 3m, deeply c.10-ridged, glabrous after
1st year; leaves ternate and petiolate below, simple above;
flowers 1-2 in lateral groups, 10-20mm, yellow (usually paler
than in C. scoparius); calyx usually pubescent; fruits 1.5-
3.5cm, slightly compressed when mature, long-patent-pubescent
all over. Intrd; natd on roadside banks; very scattered in
En, Sc and Wa; Iberian Peninsula.
4. C. scoparius (L.) Link (Sarothamnus scoparius (L.) Wimmer
ex Koch) - Broom. Stems to 2.5m, deeply c.5-ridged, glabrous
after 1st year; leaves as in C. striatus; flowers 1-2 in
lateral groups, 15-20mm, deep yellow or with dark red to
mauve areas; calyx usually glabrous; fruits 2.5-5cm, strongly
compressed, long-patent-pubescent on sutures, glabrous on
faces. Native.
   a. Ssp. scoparius. Plant erect or arching, to 2.5m high.
Calcifuge of heathland, sandy banks, open woodland, rough
ground; throughout most of BI.
   b. Ssp. maritimus (Rouy) Heyw. (Sarothamnus scoparius ssp.
maritimus (Rouy) Ulbr.). Plant + procumbent, to 50cm high;
young branches more densely silky-pubescent than in ssp.
scoparius. Maritime cliffs and perhaps shingle; W Wa, SW Ir,
SW En, CI; plants on shingle at Dungeness (E Kent) do not
come quite true from seed.

33. SPARTIUM L. - Spanish Broom
Non-spiny shrubs; leaves simple, shortly petiolate to
sessile; flowers yellow, in leafless terminal racemes; calyx
with upper lip divided nearly to base; fruits many-seeded,

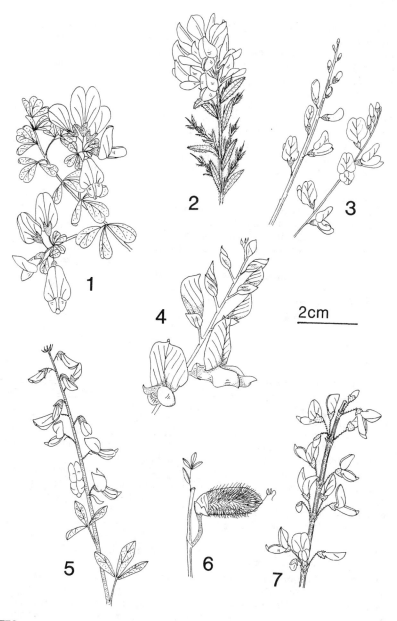

FIG 513 - Fabaceae. 1, Genista monspessulana. 2, G. hispanica. 3, G. aetnensis. 4, Spartium junceum. 5, Cytisus nigricans. 6, C. striatus. 7, C. multiflorus.

514

FABACEAE

erecto-patent.

1. S. junceum L. - <u>Spanish Broom</u>. Stems erect, to 3m, **513** terete with wide soft pith, glabrous; leaves linear to oblanceolate; flowers 20-28mm, yellow; fruits 4-10cm, strongly compressed, pubescent. Intrd; natd on sandy roadside banks and rough ground; scattered in C & S Br; Mediterranean.

34. GENISTA L. (<u>Teline</u> Medikus) - <u>Greenweeds</u>
Spiny or non-spiny shrubs; leaves simple or ternate, sessile or shortly petiolate; flowers yellow, in terminal racemes or lateral clusters; calyx with upper lip divided 1/4 to 3/4 way to base; fruits 1-several seeded, erect to patent.

```
1 Leaves all ternate                          1. G. monspessulana
1 Leaves all simple, or 0 due to early fall              2
  2 Plant with branched spines                5. G. hispanica
  2 Plant with 0 or simple spines                        3
3 Seeds 1-2; leaves most or all fallen by flowering
  time                                        6. G. aetnensis
3 Seeds 3-12; leaves present at flowering time          4
  4 Calyx, corolla and leaves pubescent       3. G. pilosa
  4 Calyx, corolla and leaves + glabrous                5
5 Flowers 7-10mm; fruits 12-20mm; plant usually spiny
                                              4. G. anglica
5 Flowers 10-15mm; fruits (15)20-30mm; plant never
  spiny                                       2. G. tinctoria
```

1. G. monspessulana (L.) L. Johnson (<u>Cytisus monspessulanus</u> **513** L., <u>Teline monspessulana</u> (L.) K. Koch) - <u>Montpellier Broom</u>. Stems not spiny, erect, to 2(3)m; leaves ternate, pubescent, obovate, petiolate; flowers 9-12mm, in lateral clusters, sparsely sericeous on keel; fruits 15-25mm, densely pubescent, 3-6-seeded. Intrd; natd on banks, roadsides and rough ground; scattered in CI and S En; Mediterranean.
2. G. tinctoria L. - <u>Dyer's Greenweed</u>. Stems not spiny, erect to procumbent, to 60(100)cm; leaves simple, glabrous, sessile; flowers 10-15mm, in terminal and lateral clusters, + glabrous; fruits 15-30mm, (3)4-10-seeded. Native.
a. Ssp. tinctoria. Plant usually erect to ascending; leaves narrowly elliptic, mostly >4x as long as wide; fruits glabrous. Grassy places, banks and rough ground; locally common in Br N to S Sc, Jersey (not Cornwall).
b. Ssp. littoralis (Corbiere) Rothm. Plant procumbent; leaves elliptic-oblong, mostly <4x as long as wide; fruits pubescent or glabrous. Grassy cliff-tops; N Devon, Cornwall, SW Pembs. On cliffs in Cornwall and N Devon procumbent plants with broad leaves but glabrous fruits occur; these are often referred to ssp. <u>tinctoria</u> but are probably better placed in ssp. <u>littoralis</u>.
3. G. pilosa L. - <u>Hairy Greenweed</u>. Stems not spiny, **RR**

procumbent or scrambling, to 50cm; leaves simple, pubescent, ovate to obovate, sessile; flowers 7-11mm, in terminal and lateral clusters, pubescent on standard and keel; fruits 14-22mm, pubescent, 3-8-seeded.   Native; cliff-tops and heathland; extremely local and decreasing in W Cornwall, Pembs, Merioneth, Brecs and E Sussex, formerly W Suffolk and W Kent.

4. G. anglica L. - Petty Whin. Stems usually with simple spines, sometimes without (var. **subinermis** (Le Grand) Rouy), erect to spreading, to 1m; leaves simple, glabrous, elliptic (those on spines ± linear), sessile; flowers 7-10mm, in terminal and lateral clusters, glabrous; fruits 12-20mm, glabrous, (3)4-12-seeded. Native; sandy and peaty heaths and moors; in suitable places throughout most of Br.

5. G. hispanica L. - Spanish Gorse. Stems with branched **513** spines, erect, to 70cm; leaves simple, pubescent, lanceolate to oblanceolate, sessile; flowers 8-14mm, in short dense terminal racemes, pubescent on keel; fruits 8-11mm, glabrous to pubescent, 1-2-seeded.   Intrd; natd on sandy and rocky hills and roadsides; Cards (since 1927) and S En; SW Europe. Our plant is ssp. **occidentalis** Rouy.

6. G. aetnensis (Raf. ex Biv.) DC. - Mount Etna Broom. **513** Stems not spiny, erect, to 3(5)m; leaves simple, pubescent, <5mm, on young growth only; flowers 8-13mm, in lax terminal racemes, pubescent on keel; fruits 8-14mm, pubescent, 1-2-seeded.   Intrd; natd on waste ground; W Kent; Sicily and Sardinia.

35. ULEX L. - Gorses
Shrubs with branched spines; leaves ternate on young plants, simple and reduced to scales or weak spines on mature plants; flowers yellow, 1-few in lateral clusters, with small bracteole on each side between calyx-lips; calyx with upper lip shortly 2-toothed; fruits (1)2-6(8)-seeded, >1/2 enclosed in calyx.

1 Bracteoles 1.8-4.5 x 1.5-4mm; >2x as wide as
     pedicels   **1. U. europaeus**
1 Bracteoles <1.5 x 1mm, <2x as wide as pedicels   2
   2 Teeth of lower calyx-lip parallel to convergent;
      calyx 9-13(15)mm; standard (12)13-18(22)mm
                                              2. U. gallii
   2 Teeth of lower calyx-lip divergent; calyx
      5-9.5(10)mm; standard 7-12(13)mm   **3. U. minor**

1. U. europaeus L. - Gorse. Densely spiny spreading shrub **493** to 2(2.5)m; spines very strong, deeply grooved; flowering mainly winter-spring (autumn flowers accompanied by many buds); calyx 10-16(20)mm, with convergent teeth; standard 12-18mm; ovules (8)9-14(16); fruits (12)14-17(19)mm, dehiscing in summer.   Native; grassy places, heathland, open woods, mostly on sandy or peaty soil; throughout BI.

1 x 2. U. europaeus x U. gallii occurs in W Br and Ir where the 2 spp. overlap, but precise range unknown; it is highly fertile and variously intermediate, especially in calyx, corolla and bracteole size, and ovule number (7-10), and flowers in autumn and winter.

2. U. gallii Planchon - Western Gorse. Densely spiny **493** spreading shrub to 1.5(2)m; spines moderately grooved; flowering mainly summer (autumn flowers accompanied by many withered ones); bracteoles 0.5-0.8 x 0.6-0.8mm; ovules (3)4-6(7); fruits (8)9-13(14)mm, dehiscing in spring. Native; habitat as for U. europaeus; mainly in CI, W 1/2 of En, Wa, Man, SW Sc, C & S Ir, but scattered in E En and E Sc.

3. U. minor Roth - Dwarf Gorse. Densely spiny spreading **493** shrub to 1(1.5)m; spines weakly grooved; flowering as in U. **R** gallii; bracteoles 0.6-0.8 x 0.4-0.6mm; ovules (3)4-6(7); fruits 6-8.5mm, dehiscing in spring. Native; heaths; mainly in S En from E Kent to Dorset and Wilts, very scattered N to Norfolk and Notts, Cumberland.

79. ELAEAGNACEAE - Sea-buckthorn family

Deciduous (or ± so), usually spiny shrubs; leaves alternate, simple, entire, ± sessile to petiolate, exstipulate, with dense silvery to reddish-brown scales. Flowers in small axillary clusters, dioecious or bisexual, deeply perigynous, actinomorphic; sepals 2 or 4, fused below; petals 0; stamens as many as sepals, inserted on base of calyx-tube; ovary 1-celled, with 1 basal ovule; style 1, long; stigma linear to capitate; fruit a berry-like achene surrounded by fleshy hypanthium.
Easily recognized by the entire, silver-scaly leaves, petal-less, 2- or 4-sepalled flowers, and fleshy 1-seeded fruits.

1 Leaves linear-lanceolate; petiole <2mm; flowers
   dioecious, appearing before leaves   **1. HIPPOPHAE**
1 Leaves elliptic; petiole >5mm; flowers bisexual,
   appearing with leaves   **2. ELAEAGNUS**

1. HIPPOPHAE L. - Sea-buckthorn
Flowers dioecious; sepals and stamens 2.

1. H. rhamnoides L. - Sea-buckthorn. Spiny, spreading and **R** suckering shrub to 3(9)m; leaves linear-lanceolate, 2-8 x 0.2-1.3cm; fruits translucent-orange with sparse scales when ripe, broadly ellipsoid to globose, 6-10mm. Native; dunes and other sandy places by sea; round coasts of Br and Ir but perhaps native only in E En; widely planted by sea and along roads inland and often self-sown.

2. ELAEAGNUS L. - <u>Elaeagnus</u>
Flowers bisexual; sepals and stamens 4.
Several of the cultivated spp. are evergreen; E. umbellata
may be semi-evergreen.

1. E. umbellata Thunb. - <u>Elaeagnus</u>. Spreading, often spiny
shrub to 3(6)m; leaves elliptic to ovate, 3-10 x 2-4cm;
fruits very scaly at first, then red, + globose, 6-8mm.
Intrd; frequently grown in gardens and by roads; bird-sown
plants natd in scattered places in S Br and Guernsey; E Asia.

80. HALORAGACEAE - Water-milfoil family

Perennial, mainly subaquatic herbs with weak trailing stems;
leaves simple and opposite, or in whorls of 3-6 and finely
pinnate, + sessile, exstipulate.   Flowers small and
inconspicuous, whorled, opposite or alternate in terminal
spikes, variously dioecious to bisexual, epigynous, with 2
minute bracteoles and 1 bract at base; sepals 4, minute, +
free; petals 4, free, small in male or bisexual flowers,
minute or 0 in female flowers; stamens (4-)8 (0 in female
flowers); cvary 1- or 4-celled, with 1 apical ovule in each
cell (0 in male flowers); styles 4 or 0; stigmas 4, clavate
to feathery; fruits a nut or a group of <4 small nuts.
The inconspicuous flowers with 4 sepals, 4 petals and 8
stamens, and fruits of 1 or 4 nutlets are diagnostic.

1 Leaves opposite, simple, not or slightly toothed
                                                   1. HALORAGIS
1 Leaves in whorls of 3-6, finely pinnate **2. MYRIOPHYLLUM**

1. HALORAGIS Forster & G. Forster - <u>Creeping Raspwort</u>
Plant growing on peat surface; leaves opposite, simple,
entire or weakly serrate; ovary 1-celled; fruit a single
nutlet.

1. H. micrantha (Thunb.) R. Br. ex Siebold & Zucc. -
<u>Creeping Raspwort</u>. Stems decumbent, to 20cm; leaves broadly
ovate to suborbicular, 3-10mm, subsessile; fruits c.0.5-1mm
incl. persistent sepals.   Intrd; natd on bare peat in bog in
W Galway, discovered 1988; SE Asia to Australasia.

2. MYRIOPHYLLUM L. - <u>Water-milfoils</u>
Plants normally subaquatic; leaves in whorls of 3-6, finely
pinnate; ovary 4-celled; fruit a group of <4 nutlets.

1 Leaves many or mostly 5 in a whorl; uppermost bracts
  deeply serrate to pinnately dissected   2
1 Leaves (3)4(-5) in a whorl; uppermost bracts simple,
  entire or minutely serrate   3
  2 Emergent leaves with dense sessile glands; only

female flowers present, whitish   **2. M. aquaticum**
2 Emergent leaves with sparse sessile glands; each
   plant with upper flowers male, lower female, and
   usually some bisexual between; flowers reddish
                                   1. M. verticillatum
3 All flowers in whorls, reddish; leaves with 13-38
   segments; usually in base-rich water   **3. M. spicatum**
3 Upper flowers opposite or alternate, all yellowish;
   leaves with 6-18 segments; usually in base-poor
   water   **4. M. alterniflorum**

Other spp. - 2 other spp. formerly occurred as natd aliens.
M. heterophyllum Michaux, from E N America, occurred until
recently in a canal in SW Yorks; it differs from all other
spp. in its 4(not 8) stamens and simple, entire to serrate
emergent leaves. **M. verrucosum** Lindley, from Australia,
occurred in gravel-pits in Beds as a wool-alien; it most
closely resembles M. spicatum but has leaves mostly in threes
and rarely >1cm and flowers all bisexual.

1. **M. verticillatum** L. - Whorled Water-milfoil. Stems to   R
3m; leaves (4)5(-6) in a whorl, with 24-35 segments; male,
female and usually bisexual flowers present, mostly 5 in a
whorl. Native mostly base-rich ponds, lakes and slow rivers
in lowlands; very scattered in En, Wa and Ir, mostly E En.
2. **M. aquaticum** (Vell. Conc.) Verdc. (M. brasiliense Camb.)
- Parrot's-feather. Stems to 2m; leaves 4-6 in a whorl, with
8-30 segments; dioecious, female only in Br; flowers mostly
4-6 in a whorl. Intrd; commonly grown by aquarists and
becoming natd where thrown out; scattered in S En, S Wa and
Jersey; S America.
3. **M. spicatum** L. - Spiked Water-milfoil. Stems to 2.5m;
leaves (3)4(-5) in a whorl, with 13-38 segments; spikes erect
from first; male, female and bisexual flowers present, 4 in a
whorl. Native; mostly base-rich ponds, lakes, slow rivers
and ditches, mostly lowland; locally common over BI.
4. **M. alterniflorum** DC. - Alternate Water-milfoil. Stems to
1.2m; leaves (3-)4 in a whorl, with 6-18 segment; spikes
pendent at first, then erect; male, female and bisexual
flowers present, mostly 2-4-whorled below, the upper ones
(male) opposite to alternate. Native; mostly base-poor
lakes, ponds, slow streams and ditches, often upland; locally
frequent over BI, mostly in N & W.

81. GUNNERACEAE - Giant-rhubarb family

Huge herbaceous perennials; stems wholly rhizomatous; leaves
alternate but clustered, rhubarb-like, simple, palmately
5-9-lobed with jagged-serrate lobes, with long stout
petioles, exstipulate. Flowers small, in huge compound erect
catkin-like panicles, usually male, female and bisexual

mixed, epigynous; sepals 2, minute, free; petals 2, free,
small; stamens (1-)2; ovary 1-celled, with 1 apical ovule;
style +0; stigmas 2, linear; fruit a small drupe.
Unique in its huge rhubarb-like leaves and erect, compact,
many-flowered inflorescences.

1. GUNNERA L. - <u>Giant-rhubarbs</u>
  1. G. tinctoria (Molina) Mirbel - <u>Giant-rhubarb</u>. Leaves
<u><</u>2m across, cordate at base; petioles <u><</u>1.5m, with pale
bristles and weak spines; inflorescences <u><</u>1m, with stout
branches <8cm. Intrd; planted by lakes etc. and often self-
sown where long-established; natd in scattered places
throughout much of lowland BI; W S America.
  2. G. manicata Linden ex Andre - <u>Brazilian Giant-rhubarb</u>.
Leaves often >2m across, peltate; petioles <u><</u>2m, with
reddish bristles and spines; inflorescences <u><</u>1.2m, with
slender branches >10cm. Intrd; planted as for G. tinctoria
and fertile but not recorded as self-sown; scarcely natd but
persistent throughout much of lowland BI; S Brazil.

82. LYTHRACEAE - <u>Purple-loosestrife family</u>

Annuals or herbaceous perennials; leaves opposite or in
whorls of 3, or upper ones alternate, simple, entire, sessile
or petiolate, exstipulate. Flowers solitary or clustered, in
leaf axils towards stem-apex, bisexual, perigynous, actinomorphic, monomorphic or trimorphic; hypanthium tubular to
funnel-or cup-shaped, with (4-)6 sepals and (4-)6 epicalyx-
segments at apex; petals usually 6, sometimes 0-5, free,
borne near apex of hypanthium; stamens 6 or 12 (sometimes
fewer); ovary 2-celled, each cell with many ovules on axile
placentas; style 1; stigma 1, capitate; fruit a capsule,
opening by 2 valves.
Distinguished by the perigynous flowers bearing 6 petals,
sepals and epicalyx-segments near the hypanthium mouth.

1. LYTHRUM L. (<u>Peplis</u> L.) - <u>Purple-loosestrifes</u>

1 Petals >4mm; stigma or some stamens exceeding sepals;
  flowers trimorphic   2
1 Petals <4mm; stigma and stamens not reaching apex of
  sepals; flowers monomorphic   3
  2 Petals >7mm; flowers clustered in whorls
                                                1. L. salicaria
  2 Petals <7mm; flowers 1-2 in each axil   **2. L. junceum**
3 Leaves obovate-spathulate; hypanthium funnel- to
  cup-shaped; capsule subglobose   **4. L. portula**
3 Leaves linear-oblong; hypanthium tubular; capsule
  cylindrical   **3. L. hyssopifolia**

1. L. salicaria L. - <u>Purple-loosestrife</u>. Erect perennial to 1.5m; leaves sessile, lanceolate, all opposite or whorled; flowers clustered in whorls, trimorphic (styles long, medium or short; six stamens at each of other 2 levels); petals 8-10mm, purple. Native; by water and in marshes and fens; common throughout most of BI except N Sc.

2. L. junceum Banks & Sol. - <u>False</u> <u>Grass-poly</u>. Erect to decumbent perennial (but annual in Br) to 70cm; leaves sessile, linear- to elliptic-oblong, opposite below, alternate above; flowers 1(-2) in leaf-axils, trimorphic; petals 5-6mm, purple. Intrd; bird-seed alien in parks and waste places, sometimes from other sources; sporadic but frequent in S & C Br; Mediterranean. Often misdetermined as next sp.

3. L. hyssopifolia L. - <u>Grass-poly</u>. Erect to decumbent **RRR** usually annual to 25cm; leaves sessile, linear to linear-oblong, opposite below, alternate above; flowers 1(-2) in leaf-axils, monomorphic; petals 2-3mm, pink; stamens usually 4-6. Native; seasonally wet bare ground; extremely local in S En and Jersey, rare casual elsewhere.

4. L. portula (L.) D. Webb (<u>Peplis</u> <u>portula</u> L.) - <u>Water-purslane</u>. Procumbent annual to 25cm, rooting at nodes; leaves obovate, petiolate, all opposite; flowers 1 per leaf-axil, monomorphic; petals 0-6, c.1mm, purplish; stamens usually 6. Native; open or bare ground by or in water, or in damp trackways; scattered throughout most of BI. Epicalyx-segments vary from c.0.3 to 2mm; plants at the latter end of the range have been called ssp. **longidentata** (Gay) Sell.

83. THYMELAEACEAE - <u>Mezereon family</u>

Early-flowering, glabrous, poisonous shrubs; leaves alternate, simple, entire, sessile or with short petiole, exstipulate. Flowers clustered in leaf-axils, bisexual, perigynous, actinomorphic or ± so; hypanthium tubular, with 4 concolorous sepals at apex; petals 0; stamens 8; ovary 1-celled, with 1 apical ovule; style 0 or short; stigma 1, capitate; fruit a 1-seeded drupe.

Distinctive in the perigynous flowers with 4 sepals, 0 petals and ovary with 1 ovule.

1. DAPHNE L. - <u>Mezereons</u>

1. D. mezereum L. - <u>Mezereon</u>. Erect, deciduous shrub to 2m; **R** leaves light green; flowers bright pink, in groups of 2-4 in axils of last year's fallen leaves, pubescent on outside, very fragrant; fruit ellipsoid, 8-12mm, bright red. Probably native in some places but often bird-sown; calcareous woods; very local in Br N to Yorks and Lancs, sporadic alien elsewhere.

1 x 2. D. mezereum x D. laureola = D. x houtteana Lindley & Paxton has been found with the parents in N Somerset, W Sussex and MW Yorks, but not since 1954; it is sterile and

intermediate in leaf-retention and flower-colour; endemic, but a well-known garden plant.
2. **D. laureola** L. - <u>Spurge-laurel</u>. Erect to decumbent evergreen shrub to 1.5m; leaves dark green; flowers yellowish-green, in racemes of 2-10 in axils of leaves, glabrous, not or slightly scented; fruit ellipsoid, 10-13mm, black. Native; woods mostly on calcareous or clayey soils; locally frequent in En, Wa and CI, sporadic alien elsewhere.

84. MYRTACEAE - <u>Myrtle</u> family

Evergreen trees or shrubs; leaves opposite or alternate, simple, entire or nearly so, sessile or petiolate, exstipulate, with aromatic glands. Flowers solitary or clustered, axillary or terminal, bisexual, epigynous to semi-epigynous, actinomorphic; hypanthium not or slightly extended above ovary, bearing 4-5 sepals and petals; sepals minute or ± free; petals free or united into hood covering unopened flower; stamens numerous; ovary 2-5-celled with numerous ovules on axile placentas; style 1; stigma 1, capitate; fruit a many-seeded capsule or berry.
Recognized by the combination of evergreen, aromatic, entire leaves, inferior ovary, numerous stamens and 4-5 petals.

1 All leaves opposite; ovary (2-)3-celled; fruit a
  berry   **3. AMOMYRTUS**
1 Adult leaves alternate; ovary 4-5-celled; fruit a
  capsule   2
  2 Leaves <2cm; petals free; shrubs   **1. LEPTOSPERMUM**
  2 Leaves >3cm; petals united into a hood covering
    flower in bud; trees   **2. EUCALYPTUS**

1. LEPTOSPERMUM Forster & G. Forster - <u>Tea-trees</u>
All leaves alternate, <2cm; flowers solitary, 0.8-1.8cm across; petals 5, white or sometimes pink, free; ovary 5-celled; fruit a woody capsule, 6-10mm wide.

1. **L. scoparium** Forster & G. Forster - <u>Broom Tea-tree</u>. Shrub to 5m, glabrous or ± so; leaves 4-20 x 2-6mm, tapered to sharp point, borne rather densely on twigs; sepals falling as soon as flowers fade. Intrd; confined by frost to extreme SW BI but there freely self-sowing; natd on Tresco, Scillies; Australia and New Zealand.
2. **L. lanigerum** (Aiton) Smith - <u>Woolly Tea-tree</u>. Differs from L. scoparium in leaves 4-15 x 2-4mm, abruptly pointed and scarcely sharply so; sepals, hypanthium and leaf-lower-side white-pubescent; sepals persistent on fruit. Intrd; natd on Tresco but less so than L. scoparium; Australia.

2. EUCALYPTUS L'Her. - <u>Gums</u>
Leaves on adult shoots alternate, those on juvenile shoots

opposite and of very different shape, >3cm; flowers solitary
or in small umbel-like clusters; petals united into a hood
covering flower in bud and breaking off transversely; ovary
mostly 4-celled; fruit a woody capsule.

```
1  Flowers and fruits solitary              3. E. globulus
1  Flowers and fruits in groups of 3 or more            2
   2  Flowers in groups of 5-12            6. E. pulchella
   2  Flowers in groups of 3                             3
3  Juvenile leaves lanceolate to narrowly elliptic; fruit
   5-6mm                                    5. E. viminalis
3  Juvenile leaves orbicular to elliptic or ovate; fruit
   >7mm                                                  4
   4  Flower buds and fruit with 2-4 longitudinal ribs,
      sessile                              4. E. johnstonii
   4  Flower buds and fruit + terete, stalked            5
5  Mature leaves 4-7cm; flower buds 6-8mm; fruit 7-10mm
                                              1. E. gunnii
5  Mature leaves 8-18cm; flower buds c.12mm; fruit
   c.17mm                                   2. E. urnigera
```

Other spp. - c.35 spp. have been tried for forestry
purposes, mainly in Ir, of which experimental plots of c.10
still exist. The commonest of these are 1, 2, 4 and 5 below,
but they rarely produce self-sown offspring. Of the others,
E. dalrympleana Maiden (Broad-leaved Kindlingbark), from
Tasmania and SE Australia, has been planted more recently and
might become as frequent; it differs from E. gunnii in its
sessile flowers and fruits and mature leaves 10-22cm.

1. E. gunnii Hook. f. - Cider Gum. Tree to 30m; juvenile
leaves orbicular to elliptic; mature leaves 4-7cm, lanceolate; flower buds 6-8mm, shortly pedicellate; fruit 7-10 x
8-9mm, hemispherial to campanulate, + terete. Intrd; planted
for small-scale forestry and for ornament widely in Ir and S
& W Br, persistent also in E N Essex since 1887; the most
hardy sp., sometimes self-sowing; Tasmania.
2. E. urnigera Hook. f. - Urn-fruited Gum. Tree to 30m;
juvenile leaves orbicular to ovate; mature leaves 8-18cm,
narrowly lanceolate, dark green; flower buds c.12mm,
pedicellate; fruit c.17 x 10mm, bowl-shaped and narrowed
distally, + terete. Intrd; planted for small-scale forestry
in Ir; Tasmania.
3. E. globulus Labill. - Southern Blue-gum. Tree to 45m;
juvenile leaves ovate to broadly lanceolate; mature leaves
10-30cm, lanceolate to falcate, dark green; flower buds up to
30mm, sessile; fruit 10-15 x 15-30mm, obconical, 4-ribbed.
Intrd; planted for ornament and rarely forestry, self-sown in
W Ir; Tasmania.
4. E. johnstonii Maiden (E. muelleri T. Moore non Miq. nec
Naudin) - Johnston's Gum. Tree to 50m; juvenile leaves
orbicular; mature leaves 5-10cm, oblong-ovate to lanceolate;

flower buds c.14 x 10mm, sessile; fruit c.10 x 12mm, hemispherical to obconical, 2-4-ribbed. Intrd; planted for small-
scale forestry in Ir; Tasmania.
5. E. viminalis Labill. - Ribbon Gum. Tree to 50m; juvenile
leaves ovate to narrowly so; mature leaves 11-18cm, linear-
lanceolate, pale green; flower buds 5-8mm, sessile to shortly
pedicellate; fruit 5-8 x 5-8mm, spherical to obconical, +
terete. Intrd; planted for small-scale forestry in Ir;
Tasmania and SE Australia.
6. E. pulchella Desf. (E. linearis Dehnh.) - White
Peppermint-gum. Tree to 15m; juvenile leaves linear to
linear-lanceolate; mature leaves 5-12cm, linear to narrowly
lanceolate; flower-buds 4-5mm, shortly pedicellate; fruit
4-5.5 x 5-6.5mm, obovoid, + terete. Intrd; planted for
ornament in extreme SW En, self-sown in Scillies; Tasmania.

3. AMOMYRTUS (Burret) Legrand & Kausel - Myrtles
Leaves all opposite; flowers 1(-3) in leaf-axils; petals
4(-5), white, free; ovary (2-)3-celled; fruit a berry.

1. A. luma (Molina) Legrand & Kausel (Myrtus luma Molina,
Eugenia apiculata DC.) - Chilean Myrtle. Shrub or tree
rarely to 18m; leaves 1.5-3cm; flowers 2-3cm across, with
strongly concave petals; fruit dark purple, globose, 6-10mm.
Intrd; thriving only in very mild areas but self-sown in
semi-natural woodland; SW En, Guernsey, SW Ir; S America.

85. ONAGRACEAE - Willowherb family

Herbaceous annuals, biennials or perennials or rarely shrubs;
leaves opposite or alternate, simple, sessile or petiolate,
exstipulate or with small caducous stipules. Flowers
solitary and axillary or in terminal racemes, bisexual,
epigynous, actinomorphic to weakly zygomorphic; hypanthium 0
or a short to very long tube, with 2 or 4 free sepals at
apex; petals 2 or 4 or rarely 0, free; stamens 2, 4 or 8;
ovary 1-, 2- or 4-celled, each cell with 1-many ovules on
axile placentas; style 1; stigma capitate to clavate or 4-
lobed; fruit a 4-celled capsule or berry, or 1-2-seeded nut.
  Distinguished by the combination of epigynous flowers with 2
or 4 sepals, 2, 4 or 8 stamens, and (1)2-4-celled ovary.

1 Sepals 2; petals 2; stamens 2; ovary 1-2-celled;
    fruit with hooked bristles   **7. CIRCAEA**
1 Sepals 4; petals 4 (rarely 0); stamens 4 or 8;
    ovary 4-celled; fruit without hooked bristles   2
    2 Shrubs; fruit a berry   **6. FUCHSIA**
    2 Herbs; fruit a capsule   3
3 Petals 0; stamens 4   **3. LUDWIGIA**
3 Petals 4; stamens 8   4
    4 Petals yellow (sometimes streaked or tinged

         reddish); hypanthium >20mm   **4. OENOTHERA**
    4 Petals pink, red or purple, sometimes white, never
       yellow; hypanthium <11mm   5
 5 Seeds without hairy plume; hypanthium 2-11mm; all
    leaves alternate   **5. CLARKIA**
 5 Seeds with hairy plume; hypanthium 0-c.3mm; lower
    leaves often opposite   6
    6 At least lowest leaves opposite; flowers + erect
       when open, actinomorphic   **1. EPILOBIUM**
    6 All leaves alternate; flowers held horizontally,
       slightly zygomorphic   **2. CHAMERION**

1. EPILOBIUM L. - Willowherbs

Perennial herbs; at least lower leaves opposite or rarely
whorled; flowers in loose terminal racemes, actinomorphic;
hypanthium very short; sepals 4; petals 4, pink to purple,
rarely white; stamens 8; ovary 4-celled; fruit a linear
capsule; seeds with a hairy plume.
   Plants vary greatly in stature, leaf-size, and degree of
branching and of pubescence, but the type of hairs and
certain aspects of leaf-shape are relatively constant.
Seed-coat ornamentation is highly diagnostic, as is the
presence of a terminal appendage, but a high magnification
($x \geq 20$) is required.
   Hybrids occur commonly where 2 or more spp. occur together,
especially in quantity for several years in disturbed ground.
Hybrids are often recognizable by their larger and more-
branched stature, longer flowering season, unusually large or
small flowers markedly more darkly coloured at petal-tips,
and partially or entirely abortive fruits. Most seeds are
abortive but some are fertile; back-crossing and even triple-
hybrids rarely occur. Most hybrids are variously intermediate in diagnostic characters, notably stigma-form and
type of pubescence.

Multi-access Key to spp. of Epilobium

Stigma 4-lobed	A
Stigma clavate	B
Stem-hairs all + appressed	C
Some stem-hairs patent or otherwise spreading, at	
least near stem apex	D
Spreading hairs 0 or all glandular	E
Some spreading hairs eglandular	F
Seeds truncate to gradually rounded at hairy end	G
Seeds with extra appendage 0.05-0.2mm long at	
hairy end	H
Seeds minutely uniformly papillose	I
Seeds with longitudinal papillose ridges	J
Seeds obscurely reticulate, not papillose	K
Stems decumbent to erect, at least stem apex and	
inflorescence axis upturned	L
Stems procumbent, only individual flowers erect	M

ADFGIL Petals 10-16mm, purplish-pink; leaves
        slightly clasping stem at base, white
        subterranean rhizomes produced   **1. E. hirsutum**
        Petals 5-9mm, paler; leaves not clasping
        stem at base; green surface short stolons
        produced   **2. E.parviflorum**
ADEGIL Lower leaves ovate, rounded at base, abruptly
        delimited from petiole 2-6mm   **3. E. montanum**
        Lower leaves narrowly elliptic, cuneate at
        base, gradually narrowed to petiole
        3-10mm   **4. E. lanceolatum**
BCEGIL Patent glandular hairs 0; perennating by
        + sessile lax leaf-rosettes; capsules
        (5.5)6.5-8(10)cm   **5. E. tetragonum**
        Patent glandular hairs present on hypanthium
        and sometimes capsule; perennating by
        elongated leafy stolons; capsules
        (3)4-6(6.5)cm   **6. E. obscurum**
BDEGIL Petioles 4-15mm; plant perennating by
        sessile leafy rosettes   **7. E. roseum**
BDEHIL Petioles <4mm; plant perennating by
        long slender stolons ending in tight bud
                                          9. E. palustre
BDEHJL Petioles <4mm; plant perennating by
        sessile leafy rosettes   **8. E. ciliatum**
B(CD)EHKL Leaves narrowly elliptic-oblong, entire
        to denticulate; stolons on soil surface,
        with green leaves   **10. E. anagallidifolium**
        Leaves ovate to lanceolate-ovate,
        distinctly dentate; stolons below soil
        surface, with yellowish scale-leaves
                                       11. E. alsinifolium
B(CD)EGIM Leaves entire or nearly so, usually
        green and with obscure veins on upper-
        side   **12. E. brunnescens**
        Leaves distinctly dentate, green, with
        obscure veins on upperside
                                      13. E. pedunculare
B(CD)EGKM Leaves entire, bronzy, with + prominent
        veins on upperside   **14. E. komarovianum**

1. E. hirsutum L. - <u>Great</u> <u>Willowherb</u>. Rhizomatous; stems
erect, to 1.8m, densely shaggy-pubescent with spreading non-
glandular and glandular hairs; leaves lanceolate to narrowly
oblong, + clasping stem and slightly decurrent at base;
petals usually bright pinkish-purple, (6)10-16(18)mm.
Native; in all sorts of wet or damp places; common throughout
lowland BI except N & C Sc.
1 x 2. E. hirsutum x E. parviflorum = E. x subhirsutum
Gennari has been found scattered in En, Wa, N Ir and
Guernsey.
1 x 3. E. hirsutum x E. montanum = E. x erroneum Hausskn.

has been found scattered in En, Wa and C Sc.
1 x 5. E. hirsutum x E. tetragonum = E. x brevipilum
Hausskn. has been found scattered in En.
1 x 7. E. hirsutum x E. roseum = E. x goerzii Rubner has
been found scattered in En.
1 x 8. E. hirsutum x E. ciliatum = E. x novae–civitatis
Smejkal has been found scattered in En, C Sc and Co Dublin.
1 x 9. E. hirsutum x E. palustre = E. x waterfallii E.
Marshall has been found scattered in En.
2. E. parviflorum Schreber - Hoary Willowherb. Perennating
by leaf rosettes or short leafy stolons; stems erect, to
75cm, densely rather matted-pubescent with spreading non-
glandular and glandular hairs; leaves oblong-lanceolate to
narrowly so, sessile, rounded at base and not clasping stem;
petals pale pinkish-purple, 5-9mm. Native; in all sorts of
wet or damp places; frequent throughout lowland BI.
2 x 3. E. parviflorum x E. montanum = E. x limosum Schur has
been found scattered in En, Wa and N Ir, and is one of the
commonest hybrids. E. parviflorum x E. montanum x E.
obscurum and E. parviflorum x E. montanum x E. roseum
probably also occur.
2 x 4. E. parviflorum x E. lanceolatum = E. x aschersonianum
Hausskn. was found last century in S Devon.
2 x 5. E. parviflorum x E. tetragonum = E. x palatinum F.
Schultz (E. x weissenbergense F. Schultz) has been found
scattered in S & C En. E. parviflorum x E. tetragonum x E.
obscurum possibly also occurs.
2 x 6. E. parviflorum x E. obscurum = E. x dacicum Borbas
has been found scattered in much of BI.
2 x 7. E. parviflorum x E. roseum = E. x persicinum Reichb.
has been found scattered in En and Wa.
2 x 8. E. parviflorum x E. ciliatum has been found scattered
in En and Guernsey.
2 x 9. E. parviflorum x E. palustre = E. x rivulare Wahlenb.
has been found scattered in Br and N Ir.
3. E. montanum L. - Broad-leaved Willowherb. Perennating by 527
subsessile leafy buds; stems erect, to 75cm, rather sparsely
pubescent with $\pm$ appressed non-glandular and $\pm$ patent glandular hairs; leaves ovate or narrowly so, rounded at base with
petiole 2-6mm; petals 8-10mm, pink. Native; shady places,
walls, rocks and cultivated ground; common throughout BI.
3 x 4. E. montanum x E. lanceolatum = E. x neogradense
Borbas has been found scattered in S En and Mons.
3 x 5. E. montanum x E. tetragonum = E. x haussknechtianum
Borbas (E. x beckhausii Hausskn.) has been found scattered in
S En.
3 x 6. E. montanum x E. obscurum = E. x aggregatum Celak.
has been found scattered in BI, and is one of the commonest
hybrids.
3 x 7. E. montanum x E. roseum = E. x mutabile Boiss. &
Reuter has been found scattered in Br and Co Dublin.
3 x 8. E. montanum x E. ciliatum has been found scattered in

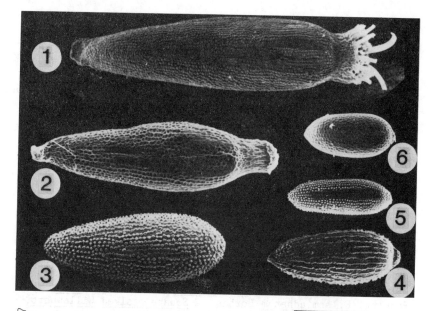

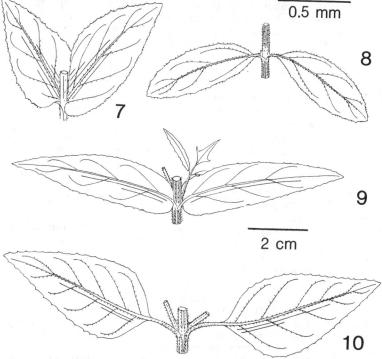

FIG 527 – 1-6, Seeds of **Epilobium**. 1, **E. palustre**. 2, **E. alsinifolium**. 3, **E. montanum**. 4, **E. ciliatum**. 5, **E. brunnescens**. 6, **E. komarovianum**. 7-10, Leaves of **Epilobium**. 7, **E. montanum**. 8, **E. lanceolatum**. 9, **E. ciliatum**. 10, **E. roseum**.

Br, where it is now probably the commonest hybrid, and E Ir.

3 x 11. E. montanum x E. alsinifolium = E. x grenieri Rouy &
Camus has been found in N En & Sc.

4. E. lanceolatum Sebast. & Mauri - Spear-leaved Willowherb. 527
Perennating by subsessile leafy buds; stems erect, to 60cm,   R
with pubescence as in E. montanum; leaves narrowly elliptic,
cuneate at base with petiole 3-10mm; petals 6-8mm, pink.
Native; waysides, walls and waste places; locally frequent in
S En, S Wa and CI.

4 x 5. E. lanceolatum x E. tetragonum = E. x fallacinum
Hausskn. (E. x ambigens Hausskn.) has been found very rarely
in S En and Guernsey.

4 x 6. E. lanceolatum x E. obscurum = E. lamotteanum
Hausskn. has been found scattered in S En.

4 x 7. E. lanceolatum x E. roseum = E. x abortivum Hausskn.
has been found rarely in S & C En.

4 x 8. E. lanceolatum x E. ciliatum has been found rarely in
S En; endemic?

5. E. tetragonum L. (E. adnatum Griseb., E. lamyi F.
Schultz, E. semiadnatum Borbas) - Square-stalked Willowherb.
Perennating by subsessile loose leaf-rosettes; stems erect,
to 75cm; stems and inflorescences with usually dense, white,
appressed hairs alone; leaves narrowly oblong to oblong-
lanceolate; petals 5-7mm, pale purplish-pink. Native; hedgerows, open woods, by water, cultivated and waste ground;
locally common in CI, S & C Br, very scattered in Ir and N
Br. Plants sometimes separated as ssp. **lamyi** (F. Schultz)
Nyman have shortly petiolate, non-decurrent leaves, rather
than ± sessile leaves ± decurrent to stem-ridges, but all
intermediates occur.

5 x 6. E. tetragonum x E. obscurum = E. x semiobscurum
Borbas (E. x thuringiacum Hausskn.) has been found scattered
in En, Wa and Jersey.

5 x 7. E. tetragonum x E. roseum = E. x borbasianum Hausskn.
(E. x dufftii Hausskn.) has been found rarely in S En.

5 x 8. E. tetragonum x E. ciliatum has been recorded from
Surrey and Bucks.

5 x 9. E. tetragonum x E. palustre = E. x laschianum
Hausskn. (E. x probstii A. Leveille) has been recorded from E
Kent and Surrey.

6. E. obscurum Schreber - Short-fruited Willowherb.
Perennating by ± elongated leafy stolons; stems erect, to
75cm; pubescence as in E. tetragonum but patent glandular
hairs present on hypanthium and sometimes a few on fruit;
leaves narrowly elliptic-ovate to lanceolate, ± sessile and ±
decurrent on to stem; petals 4-7mm, pinkish-purple. Native;
same habitats as E. tetragonum; frequent ± throughout BI.

6 x 7. E. obscurum x E. roseum = E. x brachiatum Celak. has
been found scattered in En.

6 x 8. E. obscurum x E. ciliatum has been found scattered in
Br, where it is now one of the commonest hybrids.

6 x 9. E. obscurum x E. palustre = E. x schmidtianum Rostkov has been found scattered in Br and Ir.

6 x 10. E. obscurum x E. anagallidifolium = E. x marshallianum Hausskn. has been found in Stirlings and W Sutherland.

6 x 11. E. obscurum x E. alsinifolium = C. x rivulicola Hausskn. has been found in W Perth and Banffs.

7. E. roseum Schreber - Pale Willowherb. Perennating by 527 subsessile leaf-rosettes; stems erect, to 75cm, with often abundant but sometimes very sparse patent glandular and appressed non-glandular hairs; leaves ovate-elliptic to narrowly so, gradually narrowed to petiole 4-15mm; petals 4-7mm, usually pale pink. Native; shady places, damp ground, cultivated and waste land; scattered throughout most of Br and Ir and locally frequent, apparently decreasing.

7 x 8. E. roseum x E. ciliatum has been found scattered in En and Sc; endemic?

7 x 9. E. roseum x E. palustre = E. x purpureum Fries was collected once in Surrey.

8. E. ciliatum Raf. (E. adenocaulon Hausskn.) - American 527 Willowherb. Perennating by ± sessile leaf-rosettes; stems erect, to 75(100)cm, with pubescence as in E. roseum; leaves oblong-lanceolate, rounded to subcordate at base with petiole 1.5-4mm; petals 3-6mm, pinkish-purple. Intrd; waste and cultivated ground, by roads, rivers and railways, on walls; 1st found in 1891, still spreading, now over most of BI and the commonest sp. in much of S Br; N America. Some claim **E. adenocaulon** (incl. all our plants) is distinct, but this name seems predated by **E. watsonii** Barbey.

8 x 9. E. ciliatum x E. palustre has been recorded from N Hants and Co Down.

9. E. palustre L. - Marsh Willowherb. Perennating by small 527 tight buds on ends of long filiform rhizomes; stems erect, to 60cm, with often sparse appressed (or ± so) non-glandular hairs and fewer patent glandular hairs near top, with scarcely any raised lines; leaves lanceolate to linear-lanceolate, narrowed at base, sessile or with petiole up to 4mm; petals pale pink, 4-7mm. Native; marshes, fens and ditches, often with E. parviflorum; frequent throughout BI.

9 x 10. E. palustre x E. anagallidifolium has been found in Easterness and W Sutherland.

9 x 11. E. palustre x E. alsinifolium = E. x haynaldianum Hausskn. has been found scattered in N En and Sc.

10. E. anagallidifolium Lam. - Alpine Willowherb. Perennating by leafy stolons; stems ascending to decumbent, to 20cm, with sparse appressed non-glandular hairs; patent glandular hairs very sparse, mostly on hypanthium and capsules; leaves narrowly elliptic-oblong, gradually narrowed to short petiole-like base; petals 3-4.5mm, pinkish-purple. Native; mountain flushes and stream-sides; locally frequent in N En and Sc.

10 x 11. E. anagallidifolium x E. alsinifolium = E. x boissieri Hausskn. has been found on mountains in C & N Sc.

11. E. alsinifolium Villars - <u>Chickweed</u> <u>Willowherb</u>. Perenn- 527
ating by slender ± subterranean stolons; stems (erect or) **R**
ascending to decumbent, to 25(30)cm; pubescence as in <u>E.</u>
<u>anagallidifolium</u> or with slightly more glandular hairs;
leaves lanceolate-ovate to lanceolate, rounded to short
petiole at base; petals 7-10mm, reddish-purple. Native;
similar habitats to <u>E.</u> <u>anagallidifolium</u> and similar
distribution, but also Caerns.
   12. E. brunnescens (Cockayne) Raven & Engelhorn (<u>E.</u> 527
<u>pedunculare</u> auct. non Cunn., <u>E.</u> <u>nerteroides</u> auct. non Cunn.)
- <u>New Zealand Willowherb</u>. Stems procumbent, to 20cm, with
sparse minute glandular and non-glandular hairs; leaves ±
orbicular, often purplish on lowerside, 3-7(10)mm, with
petioles 0.5-3mm; flowers solitary and erect in leaf-axils;
petals 2.5-4mm, white to pale pink. Intrd; all sorts of damp
barish ground, especially gravelly hillsides, railway
sidings, waste-tips; first collected in 1908, still
spreading, now over most of BI; New Zealand.
   12 x ?3. E. brunnescens x ?E. montanum was found in 1981 on
a spoil-heap in Co Antrim; possibly some other erect sp. than
<u>E. montanum</u> is involved; endemic.
   13. E. pedunculare Cunn. (<u>E.</u> <u>linnaeoides</u> Hook. f.) - <u>Rockery</u>
<u>Willowherb</u>. Similar to a robust <u>E.</u> <u>brunnescens</u>, but with
leaves 3-10(14)mm, acutely dentate, not purplish on lower-
side; petals 3-5mm. Intrd; weed of barish damp ground natd
by roads in W Galway and W Mayo, 1st found 1953, rare garden
weed in En nd Sc; New Zealand.
   14. E. komarovianum A. Leveille (<u>E.</u> <u>inornatum</u> Melville) - 527
<u>Bronzy</u> <u>Willowherb</u>. Similar to a small <u>E.</u> <u>brunnescens</u>, but
with entire, broadly ovate-elliptic leaves 2-6(10)mm, usually
green on lowerside, bronzy on upperside. Intrd; barely natd
garden weed in few places in En, Sc and N Ir; New Zealand.

   2. CHAMERION (Raf.) Raf. (<u>Chamaenerion</u> Seguier nom. illeg.) -
<u>Rosebay</u> <u>Willowherb</u>
Perennial rhizomatous herbs; all leaves alternate; flowers in
dense terminal racemes, slightly zygomorphic; hypanthium ±0;
sepals 4; petals 4, pinkish-purple, rarely white; stamens 8;
ovary 4-celled; fruit a linear capsule; seeds with a hairy
plume.

   Other spp. - C. dodonaei (Villars) Holub (<u>Chamaenerion</u>
<u>dodonaei</u> (Villars) Schur, <u>Epilobium</u> <u>dodonaei</u> Villars), from C
& S Europe, differs in its linear leaves but is only a very
rare casual or escape.

   1. C. angustifolium (L.) Holub (<u>Epilobium</u> <u>angustifolium</u> L.,
<u>Chamaenerion</u> <u>angustifolium</u> (L.) Scop.) - <u>Rosebay</u> <u>Willowherb</u>.
Stems erect, to 1.5m; leaves narrowly elliptic to oblong-
lanceolate, ± sessile; flowers held horizontally; 2 upper
petals wider than 2 lower ones; style deflexed; stigma
4-lobed.   Native;   waste   ground,   woodland-clearings,

embankments, rocky places and screes on mountains; throughout BI and often abundant.

3. LUDWIGIA L. - <u>Hampshire-purslane</u>

Annual to perennial ± aquatic herbs; all leaves opposite; flowers solitary in leaf-axils, actinomorphic; hypanthium 0; sepals 4; petals 0; stamens 4; ovary 4-celled; fruit a short, cylindrical, scarcely dehiscent capsule retaining sepals; seeds not plumed.

1. L. palustris (L.) Elliott - <u>Hampshire-purslane</u>. Stems **RR** procumbent to decumbent or ascending, sometimes floating at ends, to 30(60)cm; leaves ovate-elliptic, petiolate; flowers 2-5mm, inconspicuous; stigma capitate. Native; acid pools; extremely local in New Forest (S Hants) and Epping Forest (S Essex), formerly E & W Sussex and Jersey.

4. OENOTHERA L. - <u>Evening-primroses</u>

Annual to biennial (rarely perennial) herbs; all leaves alternate; flowers in terminal racemes, ± actinomorphic; hypanthium 20-45mm, a narrow tube; sepals 4; petals 4, yellow, sometimes streaked or tinged red; stamens 8; ovary 4-celled; fruit a long ± cylindrical capsule; seeds not plumed.

A critical genus where species limits are a matter of opinion; that of K. Rostanski is followed here. Flower measurements refer to the lower (first opened) ones of the inflorescence (later ones may be much smaller); fruit pubescence should also be noted on the lower part of the inflorescence. Hybrids often occur wherever 2 or more spp. occur together; reciprocal crosses often have different characters. The hybrids are fertile and may backcross and form triple-hybrids (e.g. <u>O. biennis</u> x <u>O. glazioviana</u> x <u>O. cambrica</u>), but these are very difficult to identify.

1 Capsule widest near apex, c.2-4mm wide near base;
  petals tinged reddish when withering; seeds not
  angled   **5. O. stricta**
1 Capsule widest near base, there c.6-8mm wide;
  petals always yellow; seeds sharply angled   2
  2 Petals 3-5cm; style longer than stamens, the
    stigmas held above anthers   **1. O. glazioviana**
  2 Petals 1.5-3cm; style shorter than stamens, the
    stigmas held ± at same level as anthers   3
3 Green parts of stems and fruits without red bulbous-
  based hairs   4
3 Green parts of stems and fruits with red bulbous-
  based hairs   5
  4 Petals 1.5-3cm, wider than long; capsules
    2-3(3.5)cm, all with glandular hairs   **3. O. biennis**
  4 Petals 2-3cm, c. as wide as long; capsules

3-4(4.5)cm, only upper with glandular hairs
4. O. cambrica
5   Sepals red-striped; rhachis reddish towards apex;
    capsules 2-3cm, all with glandular hairs   **2. O. fallax**
5   Sepals and rhachis green; capsules 3-4(4.5)cm, only
    upper with glandular hairs   **4. O. cambrica**

Other spp. - 10 other spp. have been recorded as casuals;
the 3 following are the least rare. **O. perangusta** Gates and
O. rubricaulis Kleb. both have petals 10-20mm but differ from
O. biennis in having red bulbous-based hairs on green parts
of stems and capsules. O. perangusta, from N America, has
the hypanthium c.30-32mm and glabrous patches on capsules; O.
rubricaulis, probably originating in Europe, has the
hypanthium 15-25mm and capsules pubescent all over. **O. rosea**
L'Her. ex Aiton, from C America, has a capsule with ±1mm wide
wings, pink to violet petals only 5-10mm and a hypanthium
<10mm. Several spp. have been recorded erroneously (e.g. **O.
grandiflora** Aiton and **O. ammophila** Focke), and several much
over-recorded as errors for the 5 below (e.g. **O. parviflora**
L. for O. cambrica). The genus is now being grown as an oil-
seed crop, and other alien spp. may occur in future; O.
glazioviana, O. biennis, O. parviflora and **O. renneri** H.
Scholz seem to be those most utilized.

1. O. glazioviana Micheli ex C. Martius (O. erythrosepala 533
Borbas) - Large-flowered Evening-primrose. Stems erect, to
1.8m; green parts of stems and fruits with many red
bulbous-based hairs; rhachis red towards apex; sepals
red-striped; petals 3-5cm, wider than long. Intrd; natd on
sand-dunes, waste ground, waysides; common in suitable places
in BI except N & C Sc; N America.
  1 x 2. O. glazioviana x O. fallax, effectively a backcross,
occurs as both 1 x 2 and 2 x 1 with the parents in S Lancs.
  1 x 3. O. glazioviana x O. biennis occurs as 3 x 1 with the
parents in scattered places in En, S Wa and Jersey. The
reciprocal, 1 x 3, is here treated as 2. O. fallax.
  1 x 4. O. glazioviana x O. cambrica = O. x britannica 533
Rostanski occurs as 1 x 4 and 4 x 1 with the parents in C & S
En and S Wa; ?endemic.
  2. O. fallax Renner - Intermediate Evening-primrose. Stems 533
erect, to 1.5m; differs from O. glazioviana in shorter style
with stigmas at level of anthers; petals 2-3cm. Intrd or
originating in situ; same habitats as O. glazioviana;
scattered in Br N to C Sc. A stable derivative of female O.
glazioviana x male O. biennis, now sometimes found in absence
of both parents; ± constant in appearance, unlike the
reciprocal cross, which has petals often >3cm and style often
exceeding stamens.
  2 x 3. O. fallax x O. biennis, effectively a backcross,
occurs as both 2 x 3 and 3 x 2 with the parents in S Lancs
and Warks.

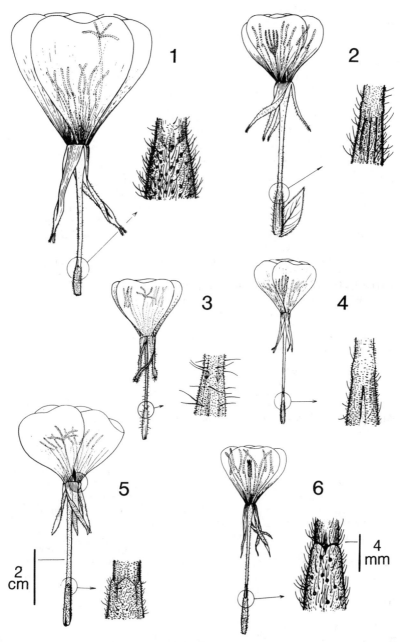

FIG 533 – Flowers of **Oenothera**. 1, **O. glazioviana**. 2, **O. fallax**. 3, **O. stricta**. 4, **O. biennis**. 5, **O. x britannica**. 6, **O. cambrica**. Drawn by J. Zygmunt.

2 x 4. O. fallax x O. cambrica, effectively a triple-hybrid, occurs as both 2 x 4 and 4 x 2 in N Hants and Warks.

3. O. biennis L. - Common Evening-primrose. Stems erect, to 533 1.5m; green parts of stems and fruits without red bulbous-based hairs (these present on red blotches on stems); rhachis green at tip, sepals green; petals 1.5-3cm, wider than long. Intrd; same habitats as O. glazioviana; frequent in suitable places in Br N to C Sc, perhaps CI; originated in Europe.

3 x 4. O. biennis x O. cambrica occurs as both 3 x 4 and 4 x 3 with the parents in C & S En and S Wa.

4. O. cambrica Rostanski (O. novae-scotiae auct. non Gates) 533 - Small-flowered Evening-primrose. Stems erect, to 1.2m; green parts of stems and fruits usually with, sometimes without (but probably because of hybridization with O. biennis), red bulbous-based hairs; rhachis green at tip; sepals green; petals 2-3cm, c. as wide as long. Intrd; natd on maritime sandy places and inland where intrd from coast; frequent by coast in Br N to Yorks and Cumberland, CI; probably N America.

5. O. stricta Ledeb. ex Link - Fragrant Evening-primrose. 533 Stems erect to ascending, to 1m; green parts of stems and fruits without red bulbous-based hairs; rhachis usually reddened towards tip; sepals red-striped and -suffused; petals 1.5-3.5cm, c. as wide as long; style c. as long as stamens. Intrd; natd in sandy places mostly on coasts; locally frequent in CI and Br N to Norfolk and S Wa, rare casual further N; Chile.

5. CLARKIA Pursh (Godetia Spach) - Clarkias
Annuals; all leaves alternate; flowers in loose terminal racemes, actinomorphic; hypanthium 2-11mm, narrowly tubular; sepals 4; petals 4, pink to purple, rarely white; stamens 8; ovary 4-celled; fruit a linear capsule; seeds not plumed.

1. C. unguiculata Lindley - Clarkia. Stems erect to ascending, to 50(80)cm; leaves lanceolate to ovate, glabrous; flower-buds pendent; hypanthium 2-5mm; sepals 10-16mm; petals 1-2cm, with claw c. as long as limb; often flore pleno. Intrd; commonly grown in gardens and frequent casual on tips, in parks and waste places; scattered in Br; California.

2. C. amoena (Lehm.) Nelson & J.F. Macbr. - Godetia. Stems erect to ascending, to 50(80)cm; leaves lanceolate, ± puberulent; flower-buds erect; hypanthium 5-30mm; sepals 12-30mm; petals 1-4(6)cm, with claw much shorter than limb; often flore pleno. Intrd; found as for C. unguiculata; W N America.

6. FUCHSIA L. - Fuchsias
Deciduous shrubs; all leaves opposite or sometimes in whorls of 3-4; flower solitary in leaf-axils, actinomorphic, pendent on long pedicels; hypanthium 5-16mm, broadly tubular; sepals 4; petals 4, pink to purple or violet, rarly white; stamens

8; ovary 4-celled; fruit a ± cylindrical black berry.

1. F. magellanica Lam. - Fuchsia. Spreading shrub to
1.5(3)m; leaves ovate to elliptic, 2.5-5.5cm; sepals 12-24 x
4-10mm, bright red; petals 6-12mm, violet; berry black, 15-
22mm. Intrd; planted as hedging in Ir and W Br; natd in Ir
(mainly S & W) and W Br N to Orkney, planted or non-
persistent outcast elsewhere, scarcely natd in CI; S Chile
and Argentina. More than 1 taxon occurs, but they have not
been worked out. Possibly the commonest taxon natd in W BI
is **F.** 'Riccartonii', also known as var. **macrostema** (Ruiz
Lopez & Pavon) Munz, with fatter buds and wider sepals than
F. magellanica, and not seeding; it is of garden origin. The
narrow-sepalled plant is sometimes self-sown in W Ir.
2. F. 'Corallina' (?F. 'Exoniensis') - Large-flowered
Fuchsia. Differs from F. magellanica in larger parts: leaves
3-8cm; sepals 25-40mm; petals 15-25mm. Intrd; natd as for F.
magellanica in Lleyn (Caerns) and Lundy Island (N Devon);
garden origin.

7. CIRCAEA L. - Enchanter's-nightshades
Perennial rhizomatous and/or stoloniferous herbs; all leaves
opposite; flowers in loose terminal racemes, ± zygomorphic, ±
horizontally held; hypanthium very short; sepals 2; petals 2,
deeply 2-lobed, white or pinkish; stamens 2; ovary 1-2-
celled, each cell with 1 seed; fruit a 1-2-seeded achene.

1 Open flowers crowded at inflorescence apex; pedicels,
  hypanthia and sepals glabrous; ovary 1-celled
                                              3. C. alpina
1 Open flowers on elongated raceme; pedicels, hypanthia
  and sepals with glandular hairs; ovary 2-celled   2
    2 Stolons produced from lower leaf-axils; petioles
      pubescent on upperside, subglabrous on lowerside;
      ovary with 1 large and 1 small cell; fruit not
      ripening   **2. C. x intermedia**
    2 Stolons 0; petioles pubescent all round; ovary
      with 2 equal cells; fruit ripening   **1. C. lutetiana**

1. C. lutetiana L. - Enchanter's-nightshade. Stems erect,
to 60cm; leaves 4-10cm, ovate, truncate to weakly cordate at
base, acuminate at apex, remotely denticulate; hypanthium 1-
1.2mm; nectariferous disc conspicuous round style-base, 0.2-
0.4mm high; petals 2-4mm; filaments 2.5-5.5mm; fruits 3-4 x
2-2.5mm. Native; woods and hedgerows etc; common throughout
BI but rare in N Sc.
2. C. x intermedia Ehrh. (C. lutetiana x C. alpina) - Upland
Enchanter's-nightshade. Stems erect, to 45cm; differs from
C. lutetiana in leaves cordate at base, abruptly acuminate,
dentate; hypanthium 0.5-1.2mm; nectariferous disc
inconspicuous around style-base, ≤0.2mm high; petals 1.8-4mm;
filaments 2-5mm; fruits ≤2 x 1.2mm; and see key. Native;

woods and shady rocky places, often on mountains; locally
frequent in N & W Br S to S Wa and Derbys, very scattered in
N & C Ir, often in absence of C. alpina or both parents.
   3. C. alpina L. - Alpine Enchanter's-nightshade. Stems   R
erect, to 30cm; stolons produced from lower leaf-axils;
leaves cordate at base, acute to abruptly acuminate, strongly
dentate; hypanthium 0.1-0.2mm; disc 0; petals 0.6-1.4mm;
filaments 1-1.5mm; fruits c.2 x 1mm. Native; same habitats
as C. x intermedia, much over-recorded for it, but more
upland; very scattered in Wa, Lake District and W Sc.

86. CORNACEAE - Dogwood family

Deciduous or evergreen shrubs or perennial herbs; leaves
opposite or alternate, simple, sessile or petiolate,
exstipulate. Flowers in terminal panicles, often ± umbel-
like, bisexual or dioecious, epigynous, actinomorphic;
hypanthium 0; sepals and petals 4 or 5, free; stamens 4 or 5,
alternate with petals; ovary 1-2-celled, each cell with 1
apical ovule, or ovary 3-celled with 2 cells empty and 1 with
apical ovule; styles 1 or 3; stigma capitate; fruit a drupe
with 1 1-2-celled stone.
   Distinguished by the usually opposite leaves, inferior ovary
with 1 ovule per cell, 4 or 5 free petals, and drupe.

1 Leaves deciduous, not thick and glossy; flowers
   bisexual   **1. CORNUS**
1 Leaves evergreen, thick and glossy; dioecious   2
   2 Leaves opposite; petals dark purple, 4   **2. AUCUBA**
   2 Leaves alternate; petals yellowish-green, 5
                                               3. GRISELINIA

1. CORNUS L. (Swida Opiz, Thelycrania (Dumort.) Fourr.,
Chamaepericlymenum Hill) - Dogwoods
Perennial herbs or deciduous shrubs; leaves opposite, ±
entire; flowers in corymbs or umbels, bisexual; sepals and
petals 4; style 1; ovary 2-celled; fruit a drupe with 1
2-celled stone.

1 Rhizomatous herbs; petals purple   **5. C. suecica**
1 Shrubs; petals white to yellow   2
   2 Inflorescences appearing before leaves, with 4
      yellow petal-like bracts at base   **4. C. mas**
   2 Inflorescences appearing after leaves, without
      petal-like bracts at base; petals whitish   3
3 Fruit purplish-black; leaves rarely with >5 pairs
   of lateral veins; petals 4-7mm   **1. C. sanguinea**
3 Fruit white to cream; many larger leaves with 6(7)
   pairs of lateral veins; petals 2-4mm   4
   4 Stone ellipsoid, tapered to flat base; leaves
      shortly and abruptly acuminate to acute   **3. C. alba**

4 Stone subglobose, rounded at base; leaves tapering
acuminate   **3. C. sericea**

1. C. sanguinea L. (<u>Thelycrania</u> <u>sanguinea</u> (L.) Fourr., <u>Swida</u>
<u>sanguinea</u> (L.) Opiz) - <u>Dogwood</u>. Shrub to 4m; bark of 1st
year twigs dark red (at least on 1 side) in winter; leaves
ovate to elliptic, 4-8cm, abruptly acuminate; fruit purplish-
black, 5-8mm. Native; woods and scrub on limestone or
base-rich clays; common in most of S & C lowland Br, very
local in S Ir, escape elsewhere.
2. C. sericea L. (<u>C.</u> <u>stolonifera</u> Michaux, <u>Thelycrania</u>
<u>sericea</u> (L.) Dandy, <u>Swida</u> <u>sericea</u> (L.) Holub) - <u>Red-osier</u>
<u>Dogwood</u>. Shrub to 3m; bark of 1st year twigs dark red or
greenish-yellow in winter; leaves ovate to elliptic, 4-10cm,
acuminate; fruit white to cream, 4-7mm, often not ripening.
Intrd; much grown in parks and on roadsides, frequently natd
by suckers; scattered in most of lowland BI; N America.
3. C. alba L. (<u>Thelycrania</u> <u>alba</u> (L.) Pojark.) - <u>White</u>
<u>Dogwood</u>. Shrub to 3m; bark of 1st year twigs usually bright
red in winter; differs from <u>C.</u> <u>sericea</u> in key characters, and
less extensively suckering. Intrd; grown as for <u>C.</u> <u>sericea</u>,
but less well natd; very scattered in lowland Br; E Asia.
Possibly not a separate sp. from <u>C.</u> <u>sericea</u>.
4. C. mas L. - <u>Cornelian-cherry</u>. Shrub or small tree to
4(8)m; bark of 1st year twigs dull greenish-grey in winter;
leaves 4-10cm, ovate to elliptic, acute to abruptly
acuminate; fruit 12-15mm, red, usually not ripening. Intrd;
grown in hedges and roadside verges and often long peristent,
but scarcely natd; scattered in S Br, but rarely N to C Sc;
Europe.
5. C. suecica L. (<u>Chamaepericlymenum</u> <u>suecicum</u> (L.) Asch. &
Graebner) - <u>Dwarf</u> <u>Cornel</u>. Stems erect, to 20cm, with
terminal inflorescence subtended by 4 white bracts longer
than flower-cluster; leaves ± sessile, ovate to elliptic,
1-3cm, with 3-5 veins all from base; fruits 5-10mm, bright
red. Native; upland moors among low shrubs; extremely local
in N En, locally frequent in W & C mainland Sc.

2. AUCUBA Thunb. - <u>Spotted-laurel</u>
Evergreen shrubs; leaves opposite, entire to remotely
serrate; flowers dioecious, the male in erect terminal
panicles, the female in small terminal clusters; sepals and
petals 4; ovary 1-celled; style 1; fruit a 1-seeded drupe.

1. A. japonica Thunb. - <u>Spotted-laurel</u>. Shrub to 5m; leaves
8-20cm, lanceolate to narrowly ovate, tapering-acute, dark
green but often with yellow blotches; drupes ellipsoid,
10-15mm, bright scarlet. Intrd; very commonly planted in
shrubberies but rarely self-sown; very scattered in Br,
mainly W, N to MW Sc; Japan.

3. GRISELINIA G. Forster - <u>New</u> <u>Zealand</u> <u>Broadleaf</u>
Evergreen shrubs; leaves alternate, entire; flowers
dioecious, in axillary racemes or panicles; sepals and petals
5; ovary 1-2-celled; styles 3; fruit a 1-seeded drupe.

1. G. littoralis (Raoul) Raoul - <u>New</u> <u>Zealand</u> <u>Broadleaf</u>.
Shrub to 3m or rarely tree to 15m; leaves 3-10cm, broadly
ovate to broadly elliptic, rather yellowish-green; drupes
6-7mm, dark purple. Intrd; commonly planted in S & W,
especially near sea, persistent and sometimes self-sown; very
scattered in S & W Br N to C Sc, Co Down; New Zealand.

87. SANTALACEAE - <u>Bastard-toadflax</u> <u>family</u>

Semi-parasitic herbaceous perennials; leaves alternate,
simple, entire, sessile, exstipulate. Flowers each with 3
bracteoles, in simple or branched terminal raceme-like cymes,
bisexual, epigynous, actinomorphic; hypanthium short, funnel-
shaped, with 5 free tepals at apex; stamens 5; ovary
1-celled, with 3 ovules; style 1, stigma capitate; fruit a
1-seeded nut.
Easily recognized by the habit, and the epigynous flowers
with 5 tepals which are retained on the nut.

1. THESIUM L. - <u>Bastard-toadflax</u>
1. T. humifusum DC. - <u>Bastard-toadflax</u>. Stems procumbent to   **R**
weakly ascending, to 20cm; leaves yellowish-green, linear,
1-veined, 5-25mm; flowers yellowish-green, 2-3mm, elongating
to c.4mm in fruit. Native; chalk and limestone grassland, a
root parasite on various herbs; very local in En N to S Lincs
and E Gloucs, Jersey and Alderney.

88. VISCACEAE - <u>Mistletoe</u> <u>family</u>
(Loranthaceae <u>pro</u> <u>parte</u>)

Semi-parasitic evergreen shrubs growing on tree-branches;
leaves opposite, simple, + sessile, exstipulate. Flowers 3-5
in tiny apical cymes, + sessile, dioecious, epigynous,
actinomorphic; hypanthium + 0; tepals (sepaloid petals) 4,
nearly free (female flowers also with 4 rudimentary sepals);
male flowers with 4 anthers sessile on tepals; female flowers
with 1-celled ovary with 2 undifferentiated ovules fused to
massive placenta and 1 capitate sessile stigma; fruit a 1-
seeded berry.
Unmistakable in habit.

1. VISCUM L. - <u>Mistletoe</u>
1. V. album L. - <u>Mistletoe</u>. Stems green, divergently
branching to form + spherical loose mass to 2m across; leaves
yellowish-green, 2-8cm, oblanceolate to narrowly obovate,

rounded at apex; flowers very inconspicuous, Feb-Apr; fruit white, globose, 6-10mm, Nov-Dec.   Native on many spp. of tree, especially <u>Malus</u>, <u>Tilia</u>, <u>Crataegus</u> and <u>Populus</u>; En and Wa N to Yorks, mostly local but common in S part of En/Wa borders, rare introduction elsewhere.

89. CELASTRACEAE - <u>Spindle family</u>

Evergreen or deciduous shrubs; leaves opposite, simple, petiolate, exstipulate. Flowers greenish-yellow, small, in axillary cymes, variously dioecious to bisexual, hypogynous to slightly epigynous, actinomorphic; sepals 4-5, $\pm$ free; petals 4-5, free; stamens 4-5; ovary 4-5-celled, each cell with (1-)2 ovules on axile placenta; style 1; stigma capitate; fruit a pink to red $\pm$ succulent 4-5-angled dehiscent capsule with 1 seed in each cell; seeds covered in orange aril.

The distinctive fruit is diagnostic; otherwise distinguished by the shrubby habit, opposite leaves and $\pm$ hypogynous flowers with 4-5 sepals, petals, stamens and fused carpels, and with a large nectar-secreting disc at base.

1. EUONYMUS L. - <u>Spindles</u>

1 Leaves leathery, evergreen; fruit with rounded lobes
                                                        3. E. japonicus
1 Leaves thin, deciduous; fruit with obtuse to winged
   lobes   2
    2 Terminal buds (Jul-Mar) <5mm; flowers mostly with
       4 sepals and stamens; fruits with mostly 4 obtuse
       angles   **1. E. europaeus**
    2 Terminal buds (Jul-Mar) >5mm; flowers mostly with
       5 sepals and stamens; fruits with mostly 5 winged
       angles   **2. E. latifolius**

Other spp. - A single bush of **E. hamiltonianus** Wall., from Himalayas, has persisted on a common in Surrey; it differs from <u>E. europaeus</u> in being semi-evergreen and having purple (not yellow) anthers.

1. E. europaeus L. - <u>Spindle</u>. Much-branched shrub or small tree to 5(8)m; leaves deciduous, 3-8(12)cm, elliptic to narrowly so or narrowly obovate, acuminate, entire to serrulate; fruits with mostly 4 obtuse lobes, 8-15mm across. Native; hedges, scrub and open woods on calcareous or base-rich soils; frequent in Br and Ir N to C Sc, planted elsewhere.
2. E. latifolius (L.) Miller - <u>Large-leaved</u> <u>Spindle</u>. Differs from <u>E. europaeus</u> in leaves 7-16cm, elliptic to obovate, rather abruptly acuminate, serrulate; fruits 15-25mm across; and see key.   Intrd; planted in hedges and gardens

and sometimes natd by bird-sown seeds; scattered in C & S En;
Europe.
3. E. japonicus L. fil. - <u>Evergreen</u> Spindle. Bushy shrub or
small tree to 5(8)m; leaves evergreen, 2-7cm, usually
obovate, obtuse to very abruptly acuminate, crenate-serrate,
often variegated; fruits with 4 rounded lobes, 6-10mm across.
Intrd; much planted for hedging, especially near sea in S,
very persistent but rarely self-sown; frequent relic by sea
in S & W Br N to S Wa, CI, rare elsewhere; Japan.

90. AQUIFOLIACEAE - Holly family

Evergreen trees or shrubs; leaves alternate, simple, usually
at least some with very spiny margins, petiolate,
exstipulate. Flowers in small axillary cymes, usually
dioecious, hypogynous, actinomorphic; sepals 4, free; petals
4, fused at base to ± free, white; male flowers with 4
stamens; female flowers with 4 usually abortive stamens and a
4-celled ovary with 1(-2) apical ovules per cell; stigma
4-lobed, sessile; fruit a (2-)4-seeded drupe.
Unmistakable in habit, even in entire-leaved variants.

1. ILEX L. - <u>Hollies</u>
1. I. aquifolium L. - <u>Holly</u>. Shrub or tree to 23m; leaves
glossy, 5-12cm, ovate to elliptic, at least the lower
undulate and strongly spinose at margins; fruits 6-10mm, ±
globose, scarlet, sometimes yellow or orange. Native; woods,
hedges and scrub; common almost throughout BI. Many
cultivars exist, some with variegated and/or ± spineless
leaves.
2. I. x altaclerensis (Loudon) Dallimore (<u>I.</u> aquifolium x <u>I.</u>
perado Aiton) - <u>Highclere</u> Holly. Differs from <u>I.</u> aquifolium
in its usually slightly larger flowers, fruit and leaves;
leaves mostly <2x as long as wide, ± flat, without lateral
spines or with few ± forwardly-pointed ones. Intrd; planted
as many cultivars, often variegated, and occurring in hedges
and woodland as relics or bird-sown plants, the latter often
not coming true from seed; garden origin. Merging into <u>I.</u>
aquifolium.

91. BUXACEAE - Box family

Evergreen shrubs or small trees; leaves opposite, simple,
entire, petiolate, exstipulate. Flowers in small axillary
yellowish clusters with 1 terminal female and several male
lateral, hypogynous, actinomorphic; male flowers with 4
sepals and 4 stamens; female flowers with several sepal-like
bracteoles and 3-celled ovary with 2 apical ovules per cell;
styles 3, short; stigmas 3, bilobed; fruit a 3-celled capsule
with 2 seeds per cell.

Recognized by the small leathery opposite leaves and
distinctively arranged unisexual flowers with 4 stamens and
3-celled ovary.

1. BUXUS L. - <u>Box</u>
  1. B. sempervirens L. - <u>Box</u>. Shrub or small tree to 5(11)m;   **RR**
leaves 1-2.5cm, elliptic to oblong, rounded to retuse at
apex; fruits 7-11mm, with persistent styles as 3 horns.
Native; woods and scrub on chalk and limestone; extremely
local in W Kent, Surrey, Berks, Bucks and W Gloucs, rarely
natd in hedges and woods elsewhere in S En.

92. EUPHORBIACEAE - <u>Spurge family</u>

Annual to perennial herbs or rarely woody annuals, often with
white latex; leaves opposite or alternate, sometimes whorled,
simple, entire or serrate, rarely palmately lobed, petiolate
or sessile, stipulate or exstipulate. Flowers variously
arranged, monoecious or dioecious, hypogynous, actinomorphic;
perianth 0 or of 3(-5) sepal-like free lobes; male flowers
with 1-many stamens, the filaments simple, jointed or
branched; female flowers with 2-3-celled ovary with 1 ovule
per cell; styles 2-3; stigmas strongly papillose or branched;
fruit a 2-3-celled capsule.
3 extremely distinct genera with superficially little in
common, but all with monoecious or dioecious flowers with
2-3-celled ovary with 1 ovule per cell and 2-3 strongly
papillose or branched stigmas.

1 Leaves palmately lobed   **2. RICINUS**
1 Leaves not lobed, entire to serrate   2
  2 Plant with copious white latex, monoecious; ovary
    and fruit 3-celled; stamen 1   **3. EUPHORBIA**
  2 Plant with watery sap, usually dioecious; ovary
    and fruit 2-celled; stamens numerous   **1. MERCURIALIS**

1. MERCURIALIS L. - <u>Mercuries</u>
Herbs with watery sap; leaves opposite, unlobed, serrate;
flowers usually dioecious, the male in catkin-like ± erect
axillary spikes, the female in smaller axillary clusters;
tepals 3, green; stamens numerous, with free, simple
filaments; ovaries 2-celled.

  1. M. perennis L. - <u>Dog's</u> <u>Mercury</u>. Rhizomatous perennial;
stems erect, simple, to 40cm, pubescent; leaves pubescent,
ovate to elliptic or narrowly so, 3-8cm; male spikes up to
c.12cm; female flower-clusters on stalks usually >1cm.
Native; woods, hedgerows and shady places among rocks; common
over much of Br but absent from Man, Hebrides, Orkney and
Shetland, very local in Jersey and Ir (mainly intrd).
  2. M. annua L. - <u>Annual</u> <u>Mercury</u>. Differs from <u>M.</u> <u>perennis</u>

in annual with fibrous root system; stems often branched; stems and leaves glabrous or nearly so, usually paler green; female flowers fewer and subsessile. Possibly native; cultivated ground and waste places; frequent in S En and CI, scattered in Br N to C Sc and in Ir.

2. RICINUS L. - <u>Castor-oil-plant</u>
Annual herb or shrub with watery sap; leaves alternate, lobed, serrate; flowers monoecious, in branched axillary groups, the male above the female; tepals 3-5, membranous; stamens numerous, on branched filaments; ovary 3-celled.

1. R. communis L. - <u>Castor-oil-plant</u>. Stems simple or branched, to 2(4)m; leaves long-petiolate, peltate, palmately lobed, up to 60cm.   Intrd; casual on tips and in waste places, often not reaching fruiting or even flowering, as garden throwout or oil-seed alien; scattered in S Br and CI; tropics.

3. EUPHORBIA L. - <u>Spurges</u>
Annual, biennial or perennial herbs with white latex; leaves opposite, alternate or whorled, unlobed, entire or serrate; flowers monoecious, in distinctive small units composed of 1 female and few male flowers together in a cup-shaped cyathium, which has 4-5 conspicuous glands at top; the cyathia solitary in leaf-axils, or (usually) in terminal compound cymes with paired or whorled branches each subtended by a bract which is leafy but often different in shape from the leaves; perianth 0; stamen 1, with jointed filament; ovary 3-celled.

```
1  Plants usually procumbent; stipules present; bracts
   and leaves similar, markedly unequal at base              2
1  Plants usually erect; stipules 0; bracts and leaves
   often different, + equal at base                          3
   2  Stems and capsules glabrous            1. E. peplis
   2  Stems and capsules pubescent           2. E. maculata
3  Leaves on main stems opposite             9. E. lathyris
3  Leaves on main stems alternate                           4
   4  Glands on cyathia rounded on outer edge               5
   4  Glands on cyathia concave on outer edge, prolonged
      into 2 points                                         10
5  Ovary and capsule smooth to granulose, but sometimes
   pubescent                                                6
5  Ovary and capsule conspicuously warty or papillose       7
   6  Ovary and capsule pubescent; leaves pubescent at
      least on lowerside, oblong to oblong-lanceolate
                                         3. E. corallioides
   6  Ovary and capsule glabrous; leaves glabrous,
      obovate                             8. E. helioscopia
7  Annuals with simple root system                          8
7  Rhizomatous perennials                                   9
```

8 Capsules with hemispherical papillae; umbel with
   5 main branches, the bracts at that node similar
   to leaves below but markedly different from bracts
   at next higher node   **6. E. platyphyllos**
8 Capsules with cylindrical papillae; umbel with 2-5
   main branches, the bracts at that node
   intermediate between leaves below and bracts at
   next higher node   **7. E. serrulata**
9 Stems without scales near base; capsule 5-6mm; bracts
   yellowish   **4. E. hyberna**
9 Stems with scales near base; capsule (2)3-4mm;
   bracts green   **5. E. dulcis**
   10 Opposite pairs of bracts fused at base; stems
      pubescent   11
   10 Opposite pairs of bracts not fused at base;
      stems glabrous   12
11 Capsule glabrous; primary branches of topmost whorl
   of inflorescence 4-12   **17. E. amygdaloides**
11 Capsule densely pubescent; primary branches of topmost whorl of inflorescence 10-20   **18. E. characias**
   12 Rhizomatous perennials   13
   12 Annuals to perennials; rhizomes 0   15
13 Leaves <2(3)mm wide   **16. E. cyparissias**
13 Some or all leaves >(3)4mm wide   14
   14 Leaves oblanceolate, very gradually tapered to
      base, rounded to obtuse at apex   **15. E. esula**
   14 Leaves linear to lanceolate or sometimes ±
      oblanceolate, not or abruptly narrowed to base,
      acute to subacute at apex   **14. E. x pseudovirgata**
15 Annuals with thin leaves, rarely on maritime sands;
   bracts and leaves similar   16
15 Biennials to perennials with ± succulent leaves, on
   maritime sands; bracts and leaves markedly different 17
   16 Leaves linear to narrowly oblong, sessile
                                         10. E. exigua
   16 Leaves ovate to obovate, petiolate   **11. E. peplus**
17 Midrib prominent on leaf lowerside; seeds pitted
                                         12. E. portlandica
17 Midrib obscure on leaf lowerside; seeds smooth
                                         13. E. paralias

Other spp. - **E. villosa** Waldst. & Kit. ex Willd. (<u>E. pilosa</u>
auct. non L.) (<u>Hairy</u> <u>Spurge</u>), from Europe, was natd near
Bath, N Somerset, from 1576 to c.1924; it would key to
couplet 6 but is a rhizomatous perennial with stems scaly
near ground. **E. ceratocarpa** Ten., from Italy, was once natd
in Glam; it would key to couplet 9 but is glabrous and has
flat-conical papillae on capsule.

1. E. peplis L. - <u>Purple</u> <u>Spurge</u>. Glabrous, often purplish, **RRR**
annual with procumbent stems to 10cm; cyathia solitary in
stem-forks and leaf-axils; capsules smooth, glabrous; seeds

fruits 5mm
plants 2cm

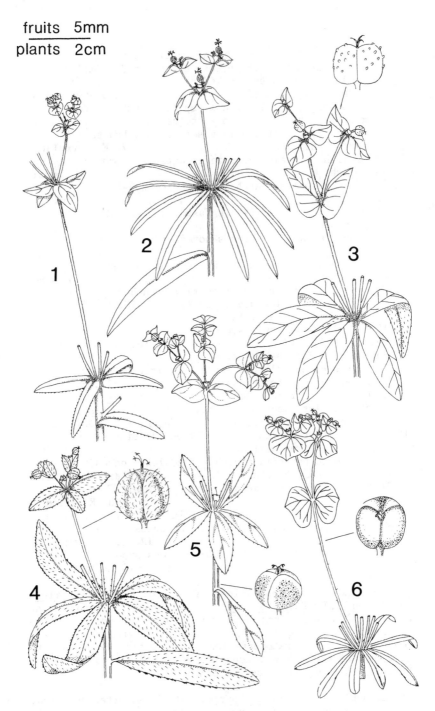

FIG 544 – Shoots and fruits of **Euphorbia**. 1, **E. serrulata**. 2, **E. x pseudovirgata**. 3, **E. dulcis**. 4, **E. corallioides**. 5, **E. platyphyllos**. 6, **E. esula**.

smooth. Native; on sandy or shingly beaches, probably
extinct; last record 1976 in Alderney; formerly in S Br from
E Kent to Cards, Waterford, CI.

2. E. maculata L. - <u>Spotted Spurge</u>. Pubescent annual, often
with dark blotches on leaves, with procumbent stems to 50cm;
cyathia 1-few in leaf-axils and stem-forks; capsules smooth,
pubescent; seeds with shallow transverse furrows. Intrd; +
natd weed of nurseries and quarry; very scattered in En, S
Wa, Jersey; N. America, natd in Europe.

3. E. corallioides L. - <u>Coral Spurge</u>. Pubescent perennial 544
with 0 or weak rhizomes; stems erect, to 60cm; leaves oblong
to oblong-oblanceolate, + sessile, serrulate; capsules finely
granulose, pubescent; seeds smooth. Intrd; natd in woods and
hedgerows at 1 site in W Sussex since c.1808, formerly in E
Sussex and Oxon; Italy.

4. E. hyberna L. - <u>Irish Spurge</u>. Sparsely pubescent or   R
rarely glabrous perennial with strong rhizomes; stems erect,
to 60cm; leaves narrowly elliptic-oblong, sessile, entire;
capsules with prominent cylindrical papillae, glabrous; seeds
smooth. Native; woods, hedgerows, grassy places and stream-
banks; extremely local in W Cornwall, N Devon and S Somerset,
frequent in most of SW Ir.

5. E. dulcis L. - <u>Sweet Spurge</u>. Sparsely pubescent to sub- 544
glabrous perennial with strong rhizomes; stems erect, to
50cm; leaves oblong- or elliptic-oblanceolate, tapered to +
sessile base, entire to serrulate; capsules with prominent
cylindrical papillae, glabrous; seeds smooth. Intrd; natd in
shady places; very scattered in Br; Europe.

6. E. platyphyllos L. - <u>Broad-leaved Spurge</u>. Glabrous or 544
pubescent erect annual to 80cm; leaves obovate to elliptic,   R
sessile, cordate at extreme base, serrulate; capsules with
small hemispherical papillae, glabrous; seeds smooth.
Native; cultivated and rough ground; formerly locally
frequent in S & E Br, now very local in S En and S Wa.

7. E. serrulata Thuill. (<u>E. stricta</u> L. nom. illeg.) - 544
<u>Upright Spurge</u>. Glabrous erect annual to 80cm; leaves as in RR
<u>E. platyphyllos</u> but narrower though often cordate at base;
capsules with prominent cylindrical papillae, glabrous; seeds
smooth. Native; limestone woods in c.10 places in W Gloucs
and Mons, natd in S Somerset.

8. E. helioscopia L. - <u>Sun Spurge</u>. Glabrous (or + so) erect
annual to 50cm; leaves obovate, tapered to base, serrulate;
capsules smooth, glabrous; seeds with reticulate ridges.
Native; cultivated ground and waste places; common over
lowland BI.

9. E. lathyris L. - <u>Caper Spurge</u>. Glabrous biennial; stems
to 1m in 1st year, producing inflorescence from top and to 2m
in 2nd year; leaves linear to narrowly oblong-lanceolate,
sessile, entire, + glaucous; capsules smooth, glabrous; seeds
rugose. Possibly native in shady places in S En, frequent
casual or natd alien in waste places and gardens over much of
Br and CI.

10. E. exigua L. - <u>Dwarf</u> <u>Spurge</u>. Glabrous erect annual to 20(30)cm; leaves linear to narrowly oblong, sessile, entire, ± glaucous; capsules smooth, with ridge on midline of each valve, glabrous; seeds closely rugose. Probably native; arable land, rarely elsewhere; common in S & E En, scattered elsewhere in most of BI except N Sc.

11. E. peplus L. - <u>Petty</u> <u>Spurge</u>. Glabrous erect annual to 30(40)cm; leaves ovate to obovate, ± petiolate, entire; capsules smooth, with 2 narrow wings near midline of each valve, glabrous; seeds pitted. Native; cultivated and waste ground; common throughout most of BI.

12. E. portlandica L. - <u>Portland</u> <u>Spurge</u>. Glabrous, erect   R biennial to perennial to 40(50)cm; rhizomes 0; leaves slightly succulent, obovate to oblanceolate, sessile, entire; inflorescence arising from top of last year's stems; capsules granulose near midline of each valve, glabrous; seeds pitted. Native; maritime sand-dunes; rather local on coasts of Ir, CI and S & W Br from S Hants to S Ebudes.

12 x 13. E. portlandica x E. paralias has been found with both parents in Wa and Wexford,; it is partially sterile and intermediate in leaf characters; endemic.

13. E. paralias L. - <u>Sea</u> <u>Spurge</u>. Differs from <u>E.</u>   R <u>portlandica</u> in more robust stems to 60cm; leaves more succulent, ovate-oblong; capsules rugose-granulose; seeds smooth; and see key. Native; maritime sand-dunes; similar distribution to <u>E. portlandica</u> and often with it, but less far N in W Br (to Wigtowns) and on S & E coast of Br N to W Norfolk.

14. E. x pseudovirgata (Schur) Soo (<u>E.</u> <u>uralensis</u> auct. non 544 Fischer ex Link, <u>E.</u> <u>esula</u> ssp. <u>tommasiniana</u> auct. non (Bertol.) Kuzm., <u>E.</u> <u>virgata</u> auct. non Waldst. & Kit. nec Desf.; <u>E.</u> <u>esula</u> x <u>E.</u> <u>waldsteinii</u> (Sojak) R.-Smith) - <u>Twiggy</u> <u>Spurge</u>. Glabrous, rhizomatous perennial; stems erect, to 1m; leaves linear to lanceolate or sometimes ± oblanceolate, not or abruptly narrowed to base, acute to subacute, entire; capsules granulose near midline of each valve, glabrous; seeds smooth, fertile. Intrd; natd in hedgerows, grassy and waste places; frequent in Br N to C Sc, Jersey; Europe. Records of **E. waldsteinii** (<u>E.</u> <u>virgata</u> Waldst. & Kit. non Desf.) are errors for this hybrid.

15. E. esula L. - <u>Leafy</u> <u>Spurge</u>. Differs from <u>E.</u> x 544 <u>pseudovirgata</u> in more delicate habit; and leaves (see key). Intrd; similar places to <u>E.</u> x <u>pseudovirgata</u> but rarer; scattered in Br but more frequent in parts of C Sc; Europe.

15 x 16. E. esula x E. cyparissias = E. x pseudoesula Schur has been recorded from W Suffolk, Surrey and S Wa; it is intermediate in leaf characters.

16. E. cyparissias L. - <u>Cypress</u> <u>Spurge</u>. Glabrous, rhizomatous perennial; stems erect, to 50cm; leaves linear, crowded (especially on lateral branches), scarcely narrowed to base, sessile, entire; capsule and seeds as in <u>E. x pseudovirgata</u>. Possibly native in chalk grassland in E Kent and perhaps elsewhere in SE En, natd in rough grassland and waste places scattered throughout Br and CI.

17. E. amygdaloides L. - <u>Wood</u> <u>Spurge</u>. Pubescent, tufted or rhizomatous perennial; stems biennial, with inflorescences arising from tops in 2nd year, to 90cm; leaves of 1st-year stems obovate to oblanceolate or narrowly elliptic, tapered to base, entire; capsule smooth or minutely punctate, glabrous; seeds smooth.

a. Ssp. amygdaloides. Rhizomes short or 0; leaves of 1st-year stems herbaceous, dull, pale- to mid-green, pubescent on lowerside and margins. Native; woods and shady hedgerows; common in CI and much of S Br N to Flints and E Norfolk, rare alien further N and in Ir.

b. Ssp. robbiae (Turrill) Stace (<u>E.</u> <u>robbiae</u> Turrill, <u>E.</u> <u>amygdaloides</u> var. <u>robbiae</u> (Turrill) R.-Smith). Rhizomes long; leaves of 1st-year stems ± coriaceous, ± shiny, dark green, ± glabrous. Intrd; natd in woods and other shady places; scattered in SE En; NW Turkey.

18. E. characias L. - <u>Mediterranean</u> <u>Spurge</u>. Densely pubescent, tufted perennial; stems biennial, with inflorescences arising from tops in 2nd year, to 1.5m; leaves of 1st-year stems oblanceolate, entire, pubescent; capsule smooth, pubescent; seeds smooth. Intrd; grown in gardens, natd on old garden sites and waste ground.

a. Ssp. characias. Glands on cyathia dark reddish-brown, with short points. Surrey; W Mediterranean.

b. Ssp. wulfenii (Hoppe ex Koch) R.-Smith. Glands on cyathia yellowish, with long points. N Somerset; E Mediterranean.

93. RHAMNACEAE - <u>Buckthorn family</u>

Evergreen or deciduous shrubs or small trees; leaves alternate or subopposite, simple, petiolate, stipulate. Flowers small, yellowish-green, in axillary cymes or solitary in leaf-axils, variously dioecious to bisexual, perigynous, actinomorphic; hypanthium short, ± campanulate; sepals 4-5, free; petals 0 or 4-5, free; stamens 4-5, abortive in female flowers; ovary 2-4-celled, each cell with 1 basal ovule; style 1 with 2-3-lobed capitate stigma or divided into 2-4 distally, each arm with capitate stigma; fruit a berry with 2-4 seeds, eventually black.

Recognizable by the shrubby habit, simple stipulate leaves, small 4-5-merous perigynous flowers, 2-4-celled ovary with 1 basal ovule per cell, and black berry.

1 Leaves serrate; winter buds with scales   **1. RHAMNUS**
1 Leaves entire; winter buds without scales   **2. FRANGULA**

1. RHAMNUS L. - <u>Buckthorns</u>
Evergreen or deciduous shrubs with alternate or subopposite

serrate leaves; flowers 4-5-merous; style divided into 3 or 4
distally.

1. R. cathartica L. - Buckthorn. Deciduous, usually spiny **555**
shrub to 8m; leaves 4-9cm, mostly with 2-4(5) pairs of major
lateral veins; petiole 6-25mm; sepals and petals mostly 4;
fruit c.6-10mm, $\pm$ globose, with 3-4 seeds. Native; hedgerows, scrub and open woods on peat and base-rich soils;
locally common in En, scattered in Wa and Ir, rare escape
elsewhere.
2. R. alaternus L. - Mediterranean Buckthorn. Evergreen, **555**
non-spiny shrub to 5m; leaves 1-6cm, mostly with 3-6 pairs of
major lateral veins; petiole 3-10mm; sepals 5; petals 0;
fruit 4-6mm, obovoid, with 2-3 seeds. Intrd; well natd in
scrub near sea in Caerns and Denbs; Mediterranean.

2. FRANGULA Miller - Alder Buckthorn
Deciduous shrubs with alternate entire leaves; flowers
5-merous; style not divided, with 2-3-lobed stigma.

1. F. alnus Miller - Alder Buckthorn. Non-spiny shrub to **555**
5m; leaves 2-7cm, mostly with 6-10 pairs of major lateral
veins; petiole 8-14mm; fruit 6-10mm, obovoid, with 2-3 seeds.
Native; scrub, bogs and open woods usually on damp peaty
soils, often base-poor but not always; locally common in En
and Wa, commoner in W than Rhamnus cathartica, very scattered
in Ir and Sc.

94. VITACEAE - Grape-vine family

Deciduous woody climbers with leaf-opposed tendrils; leaves
alternate, simple and palmately lobed or palmate, petiolate,
stipulate. Flowers small, reddish to greenish, in leaf-
opposed cymes, bisexual or mostly so, hypogynous, actinomorphic; sepals 5, very short, fused into $\pm$ lobed rim; petals
5, free or fused distally; stamens 5; ovary 2-celled, each
cell with 2 nearly basal ovules; style 1; stigma capitate;
fruit a berry with up to 4 seeds.
The woody climbing habit with leaf-opposed tendrils and
palmate or palmately-lobed leaves is diagnostic; differs from
Rhamnaceae also in hypogynous flowers with fused sepals in
leaf-opposed cymes.

1 Leaves simple; tendrils not ending in discs; petals
   fused distally, falling as flowers open   **1. VITIS**
1 Leaves palmate or simple, if simple then tendrils
   ending in discs; petals free   **2. PARTHENOCISSUS**

1. VITIS L. - Grape-vine
Leaves simple, palmately lobed; petals fused distally,
forming cap in bud which drops as flowers open.

Other spp. - **V. coignetiae** Pull., from Japan, is grown in gardens and has been found as a relic in SE En; it has scarcely lobed leaves and reddish-brown hairs on leaves and stems.

1. V. vinifera L. - Grape-vine. Woody vine potentially >10m; tendrils branched, lacking discs; leaves orbicular, cordate, with 5-7 palmate lobes; fruit green to red or black, up to 2cm, broadly ellipsoid. Intrd; increasingly grown on field-scale in S En, natd in hedges and scrub and by tips; scattered in CI, S En and S Wa; Europe.

2. PARTHENOCISSUS Planchon - Virginia-creepers
Leaves simple and palmately lobed or palmate; petals free, remaining for while after flowers open.

```
1  At least some leaves simple, 3-lobed  3. P. tricuspidata
1  All leaves palmate, most or all with 5 leaflets        2
   2  Tendrils with 5-8(12) branches each ending in
      adhesive disc                        1. P. quinquefolia
   2  Tendrils with 3-5 branches not ending in adhesive
      disc                                      2. P. inserta
```

1. P. quinquefolia (L.) Planchon - Virginia-creeper. Woody vine potentially >20m; leaves palmate, the (3-)5(7) stalked leaflets dull green on lowerside; fruit bluish-black, <1cm, globose. Intrd; much planted and natd on old walls and tips and in hedges and scrub; scattered in C & S Br; N America.
2. P. inserta (A. Kerner) Fritsch - False Virginia-creeper. Differs from P. quinquefolia in leaves more acutely serrate, shiny green on lowerside; and see key. Intrd; similar places to P. quinquefolia but rarer; scattered in S Br; N America.
3. P. tricuspidata (Siebold & Zucc.) Planchon - Boston-ivy. Differs from P. quinquefolia in most leaves simple and 3-lobed (some simple and unlobed, often some palmate with 3 leaflets). Intd; similar places to P. quinquefolia but rarer; scattered in SE En, Jersey; E Asia.

95. LINACEAE - Flax family

Herbaceous annuals or perennials; leaves opposite or alternate, simple, entire, sessile, exstipulate. Flowers in terminal cymes, bisexual, hypogynous, actinomorphic; sepals 4-5, free; petals 4-5, free; stamens 4 without staminodes or 5 usually alternating with filiform staminodes; ovary 4-5-celled with 2 ovules per cell on axile placenta, or ± 8- or 10-celled with 1 ovule per cell; styles 4 or 5; stigmas capitate; fruit a capsule opening by 8 or 10 valves.
Distinguished by the 4-5 free sepals, petals and stamens, entire exstipulate leaves, and 8- or 10-valved capsule with 8 or 10 seeds.

1 Sepals, petals and stamens 5; sepals entire to
   minutely serrate at apex; capsule with 10 valves
                                              1. LINUM
1 Sepals, petals and stamens 4; sepals deeply 2-4-
   toothed at apex; capsule with 8 valves   **2. RADIOLA**

1. LINUM L. - <u>Flaxes</u>
Glabrous annuals to perennials; leaves opposite or alternate;
sepals, petals and stamens 5; sepals entire to minutely
serrate at apex; petals white or blue, much longer than
sepals; capsule with 10 valves.

1 Leaves opposite; petals white, <7mm   **4. L. catharticum**
1 Leaves alternate; petals usually blue, >7mm   2
   2 Sepals c.1/2 as long as ripe capsule, at least
      the 2 inner rounded and apiculate at apex; stigmas
      capitate, either higher or lower than anthers
                                              3. L. perenne
   2 Sepals c. as long as ripe capsule, all abruptly
      acuminate at apex; stigmas elongate-clavate, c.
      as high as anthers   3
3 Stems usually >1; sepals and capsules 4-6mm
                                              1. L. bienne
3 Stem usually 1; sepals and capsules 6-9mm
                                           2. L. usitatissimum

1. L. bienne Miller - <u>Pale Flax</u>. (Annual,) biennial or
perennial; stems several, ascending to erect, to 60cm; leaves
linear to narrowly elliptic-oblong, 0.5-1.5mm wide, 1-3-
veined; sepals 4-6mm; petals usually blue, 8-12mm; capsule
4-6mm. Native; dry grassy places; local in BI and mostly
coastal in W, S from Notts, Man and Meath.
   2. L. usitatissimum L. - <u>Flax</u>. Annual; stem usually 1,
erect, to 85cm; differs from <u>L. bienne</u> in leaves 1.5-3(4)mm
wide, 3-veined; sepals 6-9mm; petals 12-20mm; capsule 6-9mm.
Intrd; formerly much grown for linen (tall unbranched
cultivars) or linseed-oil (shorter branched cultivars) and a
frequent casual by fields, now rarely grown but a frequent
casual from bird-seed on tips and in fields left for game;
cultivated origin.
   3. L. perenne L. (<u>L. anglicum</u> Miller) - <u>Perennial Flax</u>.   R
Perennial; stems >1, decumbent to suberect, to 60cm; differs
from <u>L. bienne</u> in leaves 1-3.5mm wide; sepals 3.5-6.5mm;
petals 13-20mm; capsule 5.5-7.5mm. Native; calcareous grassland; very local in mainly E En from N Essex to Durham and
Kirkcudbrights. Our plant is the endemic ssp. **anglicum**
(Miller) Ock.
   4. L. catharticum L. - <u>Fairy Flax</u>. Annual; stems erect, to
25cm; leaves elliptic-oblong, 1-veined; sepals 2-3mm, acute
to acuminate; petals white, 4-6mm; capsule 2-3mm. Native;
dry calcareous or sandy soils, also moorland and mountains;
frequent throughout BI.

2. RADIOLA Hill - <u>Allseed</u>
Annuals; leaves opposite; sepals, petals and stamens 8;
sepals deeply 2-4-toothed at apex; petals white, c. as long
as sepals; capsule with 8 valves.

1. R. linoides Roth - <u>Allseed</u>. Stems much branched,
extremely slender, ± erect, to 6(10)cm; leaves elliptic, 1-
veined; sepals and petals c.1mm; capsule 0.7-1mm. Native;
seasonally damp, bare, peaty or sandy, acid ground in open
places or in woodland rides; scattered over most BI but
mostly near coast.

96. POLYGALACEAE - <u>Milkwort family</u>

Small herbaceous perennials often woody at base; leaves
opposite or alternate, simple, entire, sessile or shortly
petiolate. Flowers in usually terminal racemes,
bisexual, hypogynous, zygomorphic; sepals 5, free, the 2
inner much larger than the 3 outer; petals 3, fused, 2 upper
entire, 1 lower dissected distally; stamens 8, their
filaments fused into a cleft tube that is also fused to the
petals; ovary 2-celled with 1 apical ovule per cell; style
1, with stigma and a sterile lobe at apex; fruit a 2-seeded
capsule.
The strange flowers, with 3 petals and 8 stamens fused
together, are unique.

1. POLYGALA L. - <u>Milkworts</u>

1 Leaves near base of stems smaller than those above,
   ± acute, not congested into a rosette; inner sepals
   with veins anastomosing around edges   2
1 Leaves near base of stems larger than those above, ±
   obtuse, congested into a rosette; inner sepals with
   veins not anastomosing or sparingly so and not around
   edges   3
   2 Lower stem-leaves (sometimes lost by fruiting -
      see scars left) opposite   **2. P. serpyllifolia**
   2 All leaves alternate   **1. P. vulgaris**
3 Flowers 6-7mm; stems with ± leafless portion below
   leaf-rosette   **3. P. calcarea**
3 Flowers 2-5mm; stems with leaf-rosette at or very
   near base   **4. P. amarella**

1. P. vulgaris L. - <u>Common Milkwort</u>. Stems woody at base,
procumbent or scrambling to erect, to 30cm; flowers various
shades of blue, pink or white, 4-7mm, mostly >10 per main
raceme; inner sepals slightly shorter than corolla, with
anastomosing well-branched veins, c.3/4 as wide to slightly
wider than capsule, acute to rounded-apiculate. Native;
calcareous or acid grassland, heathland and dunes.

a. Ssp. vulgaris. Inner sepals 6-8.5 x 3.5-5mm, c. as wide as capsule, with 6-20 inter-veinlet areolae; style c. as long as fruit apical notch. Frequent throughout BI.

b. Ssp. collina (Reichb.) Borbas (P. oxyptera auct. non Reichb.). Inner sepals 4-6 x 2-3.5mm, distinctly narrower than capsule, with 8-16(22) inter-veinlet areolae; style longer than fruit apical notch. Scattered throughout much of Br but distribution very uncertain.

1 x 3. P. vulgaris x P. calcarea has been recorded from scattered localities in S En; it is intermediate and sterile; endemic.

1 x 4. P. vulgaris x P. amarella = P. x skrivanekii Podp. has been found in E Kent with both parents; it has large lower leaves with a bitter taste as in P. amarella, but is much more vigorous, has intermediate corolla-size, and is partially fertile.

2. P. serpyllifolia Hose - Heath Milkwort. Stems not or scarcely woody at base, procumbent to scrambling, to 25cm; flowers various shades of blue, pink or white, 4.5-6mm, mostly <10 per main raceme; inner sepals like those of P. vulgaris but usually ± acute. Native; acid grassland and heathland; frequent throughout BI.

3. P. calcarea F. Schultz - Chalk Milkwort. Stems woody at   **R** base, procumbent below leaf-rosette then erect to ascending, to 20cm; flowers usually blue, rarely pink or white, 3-6mm, 6-20 per main raceme; inner sepals slightly shorter than corolla, with rather sparsely branched, not or sparingly anastomosing veins, c.3/4 as wide as capsule, obtuse. Native; chalk and limestone grassland; local in S En N to S Lincs.

4. P. amarella Crantz (P. amara auct. non L., P. austriaca   **RR** Crantz) - Dwarf Milkwort. Plants bitter-tasting; stems ± woody at base, erect to ascending, to 10(16)cm; flowers blue or pink in N, blue or greyish-white in S, 2-5.5mm, 7-30 per main raceme; inner sepals longer than corolla, with sparsely-branched, non-anastomosing veins, c.1/2 as wide as capsule, acute to subacute. Native; chalk and limestone grassland; very local in E & W Kent, MW & NW Yorks, Durham and Westmorland.

97. STAPHYLEACEAE - Bladdernut family

Deciduous shrubs; leaves opposite, pinnate, petiolate, stipulate when young. Flowers in small terminal panicles, bisexual, hypogynous, actinomorphic; sepals 5, free, petal-like; petals 5, free; stamens 5; ovary 2-3-celled with carpels free distally, ovules numerous on axile placentas; styles 2-3; stigmas capitate; fruit a much-inflated 2-3-celled capsule with many seeds.

Vegetatively very like Sambucus nigra, but without the characteristic smell to the crushed leaves and with diagnostic bladder-like 2-3-lobed capsules.

1. STAPHYLEA L. - Bladdernut
1. S. pinnata L. - Bladdernut. Shrub to 5m; leaflets
(3-)5(7), 5-10cm, ovate, acuminate, glabrous; flowers in
pendent panicles 5-10cm, 6-12mm, whitish; fruit 2.5-4cm, subglobose. Intrd; natd in hedges and in banks; rare in S En,
formerly commoner, now supplanted in gardens by other spp.; C
Europe.

98. SAPINDACEAE - Pride-of-India family

Deciduous trees or shrubs; leaves alternate, pinnate to 2-
pinnate, petiolate, exstipulate. Flowers in large terminal
panicles, functionally monoecious, hypogynous, zygomorphic;
sepals 5, unequal, fused proximally; petals 4, free, all
upturned, with basal appendages; stamens 8, with hairy
filaments; ovary 3-celled, each cell with 1 ovule on axile
placenta; style simple; stigmas 3, minute; fruit a much-
inflated 3-celled capsule with 3 seeds.
  Resembles Staphyleaceae in its pinnate leaves and inflated
capsule, but leaves are alternate and flowers differ in many
features (see also Colutea, Fabaceae).

1. KOELREUTERIA Laxm. - Pride-of-India
1. K. paniculata Laxm. - Pride-of-India. Tree to 16m;
leaves mostly pinnate, 15-50cm, with 9-15 ovate, serrate
leaflets; flowers numerous, bright yellow, 10-15mm across;
fruits 3-5cm, ovoid-conical. Intrd; frequently grown in S &
SE En, natd saplings on waste land in Surrey and Middlesex;
China.

99. HIPPOCASTANACEAE - Horse-chestnut family

Deciduous trees; leaves opposite, palmate, petiolate,
exstipulate. Flowers in large terminal panicles, bisexual
and male in each panicle, hypogynous, zygomorphic; sepals 5,
fused for most part; petals (4-)5, unequal, free; stamens
5-9; ovary 3-celled, each cell with 2 ovules on axile
placenta; style simple; stigma minute; fruit a large capsule
with 3 valves and 1(-3) large seeds.
  The only trees with opposite, palmate leaves; the fruits and
flowers are also unique. Resemblance of the fruits to those
of Castanea (Fagaceae) is purely superficial; in the latter
the prickly husk is a cupule containing fruits (nuts).

1. AESCULUS L. - Horse-chestnuts
1. A. hippocastanum L. - Horse-chestnut. Wide-spreading
tree to 39m; winter-buds large, very sticky; leaflets 5-7,
obovate, 10-25cm: petiole 5-20cm; flowers white with yellow

to pink blotch at base of petals, in stiffly erect conical to
cylindrical panicle 15-30cm; fruits ± globose, 5-8cm, with
numerous conical, pointed, protuberances. Intrd; abundantly
planted for ornament and often self-sown in grassy places,
copses and rough ground; throughout lowland BI; Balkans.
2. A. carnea Zeyher - Red Horse-chestnut. Tree to 28m;
differs from A. hippocastanum in smaller parts; winter buds
scarcely or not sticky; flowers bright pink to red; fruits
smooth or with very few blunt protuberances. Intrd; much
planted for ornament, often grafted on to A. hippocastanum,
in parks and by roads, self-sown in N Hants; garden origin
from A. hippocastanum x A. pavia L.

100. ACERACEAE - Maple family

Deciduous trees; leaves opposite, palmately lobed or ternate
to pinnate (rarely simple and unlobed), petiolate,
exstipulate. Flowers in terminal corymbs or raceme-like
panicles, mostly functionally monoecious or dioecious,
hypogynous or the male slightly perigynous, actinomorphic;
sepals (4-)5, free; petals 0 or (4-)5, free, ± sepaloid;
stamens usually 8, ovary 2-celled (rarely more-celled), ±
flattened, with 2 basal ovules per cell; styles 2; stigmas
long, linear, 1-sided; fruits of 2 parts, each 1-seeded with
a long wing developed from the style.
  The fruit is diagnostic; the opposite, palmately lobed
leaves of all spp. except A. negundo (and A. tataricum) occur
elsewhere in only Viburnum opulus (Caprifoliaceae).

1. ACER L. - Maples

1 Leaves ternate or pinnate; trees dioecious
                                          6. A. negundo
1 Leaves simple, palmately lobed; trees usually
  bisexual   2
  2 Leaves white on lowerside; flowers in small stiff
    compact clusters   **5. A. saccharinum**
  2 Leaves green on lowerside; flowers in erect or
    pendent panicles   3
3 Leaf-lobes serrate; panicles pendent
                                       4. A. pseudoplatanus
3 Leaf-lobes entire to irregularly dentate; flowers in
  stiff ± corymbose panicles   4
  4 Leaf-lobes obtuse; body of fruit convex
                                          3. A. campestre
  4 Leaf-lobes acuminate; body of fruit flat   5
5 Leaf-lobes entire   **2. A. cappadocicum**
5 Leaf-lobes with few acuminate teeth or sub-lobes
                                          1. A. platanoides

Other spp. - Many other spp. are grown in parks and by

FIG 555 – Leaves of **Acer** and Rhamnaceae. 1, **Acer saccharinum**. 2, **A. cappadocicum**. 3, **A. platanoides**. 4, **A. pseudoplatanus**. 5, **A. negundo**. 6, **A. campestre**. 7, **Rhamnus cathartica**. 8, **R. alaternus**. 9, **Frangula alnus**.

roads, and some occasionally produce seedlings in shrubberies
etc. In the latter category are **A. mono** Maxim., from E Asia,
like A. cappadocicum but with twigs rough as in A.
platanoides (not remaining smooth for some years); and **A.
tataricum** L. (Tartar Maple), from SE Europe and SW Asia, with
usually unlobed, ovate-oblong, serrate leaves.

1. **A. platanoides** L. - Norway Maple. Tree to 30m; leaves 555
simple, with 5-7 acuminate lobes each with few large
acuminate teeth or sub-lobes; flowers in $\pm$ erect,
yellowish-green corymbs appearing $\pm$ before leaves; fruits
with widely divergent to $\pm$ horizontal wings. Intrd;
abundantly planted and often self-sown in rough grassland,
scrub, hedges and woodland; throughout lowland BI; Europe.

2. **A. cappadocicum** Gled. (A. pictum auct. non Thunb.) - 555
Cappadocian Maple. Tree to 26m; leaves simple, with 5-7
acuminate, entire lobes; flowers in $\pm$ erect, yellowish-green,
subcorymbose panicles appearing $\pm$ before leaves; fruits as in
A. platanoides. Intrd; frequently planted in parks and by
roads; self-sown and extensively suckering in Surrey and W
Kent; SW Asia.

3. **A. campestre** L. - Field Maple. Tree to 25m; leaves 555
simple, with 3-5 obtuse to rounded lobes each entire or with
few obtuse to rounded teeth or sub-lobes; flowers in $\pm$ erect,
yellowish-green, subcorymbose panicles appearing with leaves;
fruits with $\pm$ horizontal wings. Native; woods, scrub and
hedgerows on calcareous or clay soils; common in En and Wa N
to Co Durham, planted elsewhere.

4. **A. pseudoplatanus** L. - Sycamore. Tree to 35m; leaves 555
simple, with usually 5 $\pm$ acute coarsely serrate lobes widest
at base; flowers in $\pm$ cylindrical, pendent, yellowish-green
panicles appearing with leaves; fruits with wings diverging
at c.90 degrees. Intrd; fully natd and 1 of the most
abundant trees in wide range of habitats throughout BI;
Europe.

5. **A. saccharinum** L. - Silver Maple. Tree to 31m; leaves 555
with 5 acute irregularly toothed lobes narrowed at base;
flowers in small, stiff, compact yellowish-green clusters
appearing well before leaves, males and females in separate
clusters on same or different trees; petals 0; fruits with
widely divergent to $\pm$ horizontal wings. Intrd; much (and
increasingly) planted for ornament in parks and by roads;
rarely setting seed but self-sown in Surrey; N America.

6. **A. negundo** L. - Ashleaf Maple. Tree to 17m; leaves 555
ternate to pinnate with 3-5(7) ovate, acute, slightly toothed
leaflets; male flowers in corymbs with pendent stamens, the
female in small pendent racemes, both appearing well before
leaves; petals 0; fruits with wings diverging at <90 degrees.
Intrd; commonly planted for ornament in parks and by roads
and railways; sometimes self-sown where both sexes occur in
SE En; N America.

101. ANACARDIACEAE - Sumach family

Deciduous shrubs; leaves alternate, pinnate or simple, petiolate, exstipulate. Flowers in large terminal panicles, dioecious or variously mixed, hypogynous, actinomorphic, small; sepals 5, fused at base; petals 5, ± free; stamens 5; ovary 1-celled, with 1 basal ovule; styles 3; stigmas capitate; fruit a small 1-seeded drupe.

Rhus is distinct in its thick, pithy, pubescent, little-branched stems with large pinnate leaves and large reddish inflorescences.

Other genera - COTINUS Miller differs from Rhus in its simple, entire, rounded leaves (often purple); thin, glabrous twigs; and long hairy pedicels. **C. coggygria** Scop. (Rhus cotinus L.) (Smoke-tree), from S Europe, is commonly grown in gardens and in parks and on road and railway banks, and very occasionally seedlings are found.

1. RHUS L. - Stag's-horn Sumach
1. R. hirta (L.) Sudw. (R. typhina L.) - Stag's-horn Sumach. Shrub to 5(10)m; twigs densely pubescent; leaflets (7)11-15(21), oblong-lanceolate, acute to acuminate, serrate, 5-12cm; inflorescence stiffly erect, 10-20cm, greenish in flower, then deep red. Intrd; much planted on verges and banks by roads and railways:; extensively suckering but very rarely or never self-sown in S Br; N America.

102. SIMAROUBACEAE - Tree-of-heaven family

Deciduous trees; leaves alternate, pinnate, petiolate, ex-stipulate. Flowers in terminal panicles, small, functionally monoecious and bisexual mixed, hypogynous, actinomorphic; sepals 5, fused proximally; petals 5, free; stamens 10; ovary of 5(-6) carpels loosely fused, each carpel with 1 axile ovule; styles 5(-6); stigmas peltate; fruit a group of 1-5(6) long, winged achenes.

Leaves and fruits resemble those of Fraxinus, but leaves are alternate and fruits are usually >1 per flower and with seeds in middle (not at base) of wing.

1. AILANTHUS Desf. - Tree-of-heaven
1. A. altissima (Miller) Swingle - Tree-of-heaven. Tree to 26m; leaves up to 90cm with up to 41 ovate-lanceolate, acuminate, serrate leaflets; panicles 10-20cm, greenish-white; achenes pendent, 3-4cm, reddish then whitish. Intrd; much planted in SE En, especially in Greater London and there frequently extensively suckering and self-sown; China.

102A. RUTACEAE

CITRUS spp. (Orange, Lemon, etc.) are often found as unidentifiable seedlings on rubbish tips; they can be told by their
glossy simple leaves articulated upon the winged petioles.
CHOISYA ternata Kunth (Mexican Orange), from Mexico, is a
much-grown small evergreen shrub with glossy, glabrous,
ternate leaves and white 5-petalled sweetly scented flowers
2-3cm across; it sometimes produces seedlings in shrubberies
and waste places in W Kent. **RUTA graveolens** L. (Rue), from E
Mediterranean, is a much-grown small deciduous shrub with
bluish-grey-green deeply 2-3-pinnately lobed leaves with
strong distinctive smell and numerous yellow 4-5-petalled
flowers 15-20mm across; it sometimes produces seedlings in
shrubberies, waste places and wall-cracks in SE En.

103. OXALIDACEAE - Wood-sorrel family

Perennial, rarely annual, often slightly succulent herbs,
often with bulbs and/or rhizomes; leaves all basal or
alternate, usually ternate, sometimes palmate, petiolate,
exstipulate or stipulate. Flowers 1-several in axillary,
often umbellate cymes, bisexual, hypogynous, actinomorphic,
often trimorphic; sepals 5, free; petals 5, free or ± so;
stamens 10, sometimes not all with anthers; ovary 5-celled,
each cell with many ovules on axile placentas; styles 5;
stigmas minute; fruit a 5-celled capsule.
The ternate or less often palmate leaves and conspicuous
actinomorphic flowers are diagnostic.

1. OXALIS L. - Wood-sorrels
Most bulbous spp. are trimorphic and self-incompatible, and
reproduce mainly or wholly vegetatively. The different
clones may show morphological differences, but those present
in BI represent only a small part of their whole range and a
relatively broad view of sp. limits is taken here. Many spp.
are only marginally natd, occurring ± wholly in cultivated
ground, but can be very persistent. Corolla colours given
are those in the fresh state; after drying the red/pink
colours often fade or become more bluish.

```
1 Petals yellow                                              2
1 Petals red, pink, mauve or white                           8
    2 3 sepals cordate; leaves succulent
                                          7. O. megalorrhiza
    2 No sepals cordate; leaves thin, not succulent          3
3 Aerial stem 0; bulbils present at or below soil level
                                          13. O. pes-caprae
3 Aerial stems present; bulbils 0                            4
    4 Stems procumbent, rooting freely at nodes              5
    4 Stems decumbent to erect, not or very sparsely
```

      rooting   6
5 Inflorescences always 1-flowered; capsules 3-4.5mm,
  with 3-4 seeds per cell; usually 5 stamens with
  and 5 without anthers   **4. O. exilis**
5 At least most inflorescences 2-8(12)-flowered;
  capsules (4)8-20mm, with >4 seeds per cell; usually
  all 10 stamens with anthers   **3. O. corniculata**
  6 Capsules <2x as long as wide; petals 10-15mm,
    with purple veins   **1. O. valdiviensis**
  6 Capsules >3x as long as wide; petals 5-11mm, not
    purple-veined   7
7 Pedicels (but not capsules) patent or reflexed in
  fruit; inflorescence an umbel; vegetative parts
  with only white simple hairs   **5. O. dillenii**
7 Pedicels erect in fruit; inflorescence cymose;
  vegetative parts with translucent septate hairs as
  well as white simples ones   **6. O. stricta**
  8 Stem aerial, + erect   9
  8 Stem 0 or a rhizome at or below soil level   10
9 Bulbs 0; inflorescences >1-flowered   **2. O. rosea**
9 Stem arising from bulb and producing axillary aerial
  bulbs; inflorescences 1-flowered   **14. O. incarnata**
  10 Bulbs 0; stem a rhizome   11
  10 Leaves arising from bulb at or below soil level;
     bulb often producing thin rhizomes   12
11 Flowers solitary; rhizome slender, with distant
   succulent scales   **9. O. acetosella**
11 Flowers in umbels; rhizome thick, with dense papery
   scales   **8. O. articulata**
   12 Leaves with 4 leaflets   **12. O. tetraphylla**
   12 Leaves with 3 leaflets   13
13 Leaflets widest about the middle, with submarginal
   orange or dark dots on lowerside   **10. O. debilis**
13 Leaflets widest at or near apex, without submarginal
   dots   **11. O. latifolia**

Other spp. - A pink-flowered sp. with a basal bulb and 5-10
narrow leaflets occurs in flower-beds at Douglas, Man; it has
been variously determined as **O. lasiandra** Zucc. or **O.
decaphylla** Kunth, from Mexico. Another pink-flowered bulbous
plant found in gardens in Wigtowns and Co Antrim has 3
leaflets and has been variously determined as **O. bulbifera**
Knuth, **O. drummondii** A. Gray or a variant of O. tetraphylla;
it might even be an unusual clone of O. latifolia.

1. O. valdiviensis Barneoud - Chilean Yellow-sorrel. Bulbs
0; stems erect to decumbent, to 30cm, not or little branched,
rather succulent; leaflets 3, obcordate, with rounded margins
and acute sinus, glabrous; flowers yellow with purple veins,
10-15mm, in forked cymes. Intrd; garden weed, reproducing by
seed; very scattered in C & S En; Chile.
2. O. rosea Jacq. - Annual Pink-sorrel. Bulbs 0; stems **561**

erect to ascending, to 25cm, branched or not; leaflets 3, obcordate, with rounded margins and obtuse sinus, subglabrous; flowers pale mauve-pink, 11-17mm, in sparsely branched cymes. Intrd; garden weed, reproducing by seed; CI and W Cornwall; Chile.

3. O. corniculata L. - Procumbent Yellow-sorrel. Bulbs 0; stems mostly procumbent, to 50cm, rooting at nodes, much-branched; leaflets 3, obcordate, with rounded margins and acute sinus, sparsely pubescent especially on margins, often purple; flowers yellow, 4-7.5mm, (1)2-8(12) in umbels; pedicels patent or reflexed in fruit; capsule with dense appressed hairs; seeds >4 per cell. Intrd; pernicious weed of gardens, paths, walls and waste ground, reproducing mainly by seed; common in most of En, Wa, Man and CI, very scattered in Sc and Ir; warmer parts of world.

4. O. exilis Cunn. (O. corniculata var. microphylla Hook. f.) - Least Yellow-sorrel. Differs from O. corniculata in filiform stems and all smaller parts; leaves always green; seeds 2-4 per cell; and see key. Intrd; similar places and distribution to O. corniculata, usually but not everywhere less common; New Zealand and Tasmania.

5. O. dillenii Jacq. (O. stricta auct. non L.) - Sussex Yellow-sorrel. Stems erect to decumbent, to 20cm, little-branched; leaves as in O. corniculata but always green; flowers yellow, 6-11mm, (1)2-3(4) in umbels; pedicels and capsule as in O. corniculata. Intrd; weed in sandy arable fields, reproducing by seed; known since 1950 in W Sussex; N America.

6. O. stricta L. (O. fontana Bunge, O. europaea Jordan) - Upright Yellow-sorrel. Differs from O. dillenii in stems to 40cm; thin subterranean rhizomes sometimes produced; leaves sometimes purple; flowers 5-9(15)mm, 2-5 in umbels; seeds without (not with) white patches; and see key. Intrd; weed of gardens and arable fields, reproducing mainly by seed; scattered over most of BI, common in parts of S En; N America.

7. O. megalorrhiza Jacq. (O. carnosa auct. non Molina) - Fleshy Yellow-sorrel. Stems erect to ascending, subterranean and aerial, to 20cm, thick and succulent, ± unbranched; leaflets 3, succulent, obovate, rounded to very shallowly notched at apex, glabrous, covered with translucent cells on lowerside; flowers bright yellow, 12-15mm, 1-3(5) in umbels. Intrd; natd on walls and banks only in Scillies, known since c.1936; very frost sensitive; Chile.

8. O. articulata Savigny (O. floribunda auct. non Lehm.) - 561 Pink-sorrel. Stem a thick brown-scaly horizontal to oblique rhizome; leaflets 3, obcordate, with rounded margins and acute sinus, appressed-pubescent, with orange warts especially near margin; petioles up to 25cm; peduncles up to 35cm; flowers deep pink, rarely white or pale pink, 10-15mm, c.3-25 in (sometimes partly compound) umbels. Intrd; much grown in gardens and established escape in waste and stony

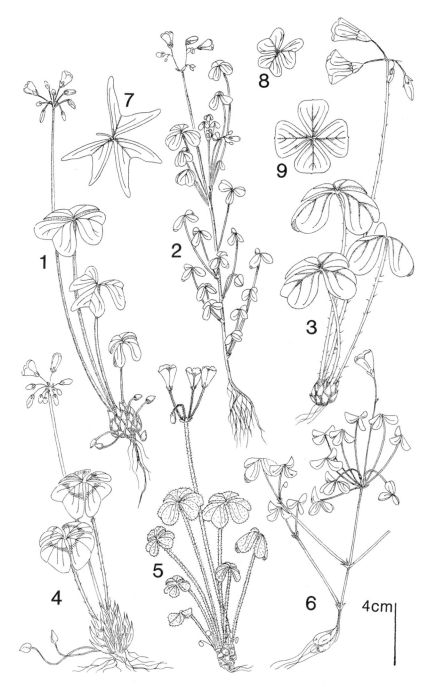

FIG 561 - 1-6, Plants of **Oxalis**. 1, **O. latifolia** (ex Jersey).
2, **O. rosea**. 3, **O. debilis**. 4, **O. tetraphylla**. 5, **O.
articulata**. 6, **O. incarnata**. 7-9, Leaves of **Oxalis**. 7, **O.
vespertilionis**. 8, **O. latifolia** (ex Cornwall). 9, **O. deppei**.

and sandy ground on roadsides, banks and seashores, often in closed vegetation, reproducing by seed and rhizomes; frequent in SW En and CI, scattered in C & S Br and Ir; E S America. Ssp. **rubra** (St Hil.) Lourt. differs in its shorter hairs, elliptic (not linear-lanceolate) sepals, and more pubescent petals; some of our plants might belong to it.

9. O. acetosella L. - <u>Wood-sorrel</u>. Stems horizontal thin rhizomes with distant succulent scale leaves; leaflets 3, obcordate, with rounded margins and acute to obtuse sinus, sparsely appressed-pubescent; petioles and peduncles up to 10cm; flowers white with mauve veins, rarely pink or mauve, 8-15mm, solitary. Native; woods, hedgebanks, shady rocks, often on humus; common + throughout BI.

10. O. debilis Kunth (<u>O.</u> <u>corymbosa</u> DC.) - <u>Large-flowered</u> 561 <u>Pink-sorrel</u>. Leaves and flowers arising from scaly bulb which produces many sessile bulblets and 0-2 swollen succulent roots; leaflets 3, obcordate, with acute to closed sinus and rounded margins, very sparsely appressed-pubescent on lowerside, with orange or dark warts only or especially near margin; petioles up to 15(20)cm; peduncles up to 20(30)cm; flowers pinkish-mauve, 15-20mm, c.3-25 in contracted cymes. Intrd; weed of gardens and other open ground, reproducing by bulblets only; frequent in S En and CI, scattered elsewhere in C & S Br; S America. The above description applies to var. **corymbosa** (DC.) Lourt.; var. **debilis** differs in its salmon-pink to brick-red flowers, smaller bulblets and glandular hairs present on styles; it occurs in a few gardens in S En.

11. O. latifolia Kunth (<u>O.</u> <u>vespertilionis</u> Zucc.) - <u>Garden</u> 561 <u>Pink-sorrel</u>. Differs from <u>O.</u> <u>debilis</u> in bulblets often formed at end of rhizomes to 3cm; leaflets with variously rounded to + pointed lobes, with obtuse to rounded sinus, without warts; flowers pink, sometimes white, 8-13mm. Intrd; habitat and distribution of <u>O.</u> <u>debilis</u>; C & S America. Variable in leaf-shape: the commonest plants have 'fishtail-shaped' (obsagittate) leaflets with elongated lobes rounded at ends and rounded to obtuse sinus; some plants from Devon, Cornwall and Guernsey have much more rounded leaflet-lobes with obtuse sinus; plants known as **O. vespertilionis** (very 561 rare in Br) have 'V-shaped leaflets' with very elongated lobes often subacute at extreme ends and rounded sinus.

12. O. tetraphylla Cav. (<u>O.</u> <u>deppei</u> Lodd. ex Sweet) - <u>Four-</u> 561 <u>leaved</u> <u>Pink-sorrel</u>. Differs from <u>O.</u> <u>debilis</u> in bulblets often formed at end of rhizomes to 6cm; leaflets 4, obsagittate, with ends of lobes rounded and sinus very obtuse, without warts; flowers pinkish-purple. Intrd; weed of gardens and arable land, reproducing by bulblets only; Jersey and Scillies; Mexico. **O. deppei** possibly occurs also; 561 it differs in its scarcely or not lobed leaflets with + truncate apex, brick-red flowers, and sessile bulblets, and might be a separate sp.

13. O. pes-caprae L. - <u>Bermuda-buttercup</u>. Leaves and

flowers arising at soil level among group of bulblets at top
of short underground stem arising from deep-seated main bulb;
leaflets 3, obcordate, with rounded margins and acute to
subacute sinus, appressed-pubescent on lowerside; petioles
and peduncles up to 30cm; flowers yellow, 20-25mm, c.3-25 in
umbels. Intrd; common weed of arable land (often bulb-
fields) in Scillies, rare in S Devon and CI, reproducing by
bulblets and swollen roots; S Africa.

14. 0. incarnata L. - Pale Pink-sorrel. Bulb producing **561**
annual, erect, branched stem to 20cm with axillary sessile
bulblets; leaflets 3, obcordate, with rounded margins and
obtuse to subacute sinus, + glabrous; flowers pale mauvish-
pink with darker veins, 12-20mm, solitary. Intrd; weed of
cultivated ground, walls and banks; frequent in SW En and CI,
scattered elsewhere in BI; S Africa.

104. GERANIACEAE - Crane's-bill family

Herbaceous annuals to perennials, sometimes woody below;
leaves alternate, simple and variously (often very deeply)
palmately or pinnately lobed, or palmate to pinnate, at least
the lower petiolate, stipulate. Flowers 1-several in
terminal or axillary, often umbellate cymes, bisexual,
hypogynous, actinomorphic to (Pelargonium) zygomorphic;
sepals 5, free but in Pelargonium the upper one with a spur
fused to pedicel; petals 5, free, rarely 0; stamens 5, or 10
but sometimes 3 or 5 without anthers, or rarely 15; ovary 5-
celled, each cell with 2 ovules, elongated distally into a
sterile column; style 1; stigmas 5, linear; fruit a dry
5-celled schizocarp, with 1 seed per cell, each mericarp and
its apical sterile beak separating from the column and
splitting open at maturity in a variety of ways (see below).
The distinctive 5-seeded fruits and usually conspicuous
actinomorphic (or nearly so) flowers are diagnostic.

1 Corolla zygomorphic, the 2 upper petals much wider
   than others; uppermost sepal with spur tightly fused
   to pedicel (but spur inconspicuous)   **4. PELARGONIUM**
1 Corolla actinomorphic to weakly zygomorphic, the
   petals scarcely different in width; sepals not
   spurred   2
   2 Stamens 15, all with anthers, fused to c.1/2 way
      into 5 groups of 3 each   **2. MONSONIA**
   2 Stamens 10, or 5 alternating with anther-less
      staminodes, not fused   3
3 Beaks of mericarps curved or loosely spirally twisted
   for 1-2 turns at maturity; leaves palmate or palmately
   lobed   **1. GERANIUM**
3 Beaks of mericarps becoming tightly spiralled along
   their own axis at maturity; leaves pinnate or pinnately
   lobed   **2. ERODIUM**

1. GERANIUM L. – Crane's-bills

Annuals to perennials; leaves simple and palmately lobed, or
palmate; flowers actinomorphic or ± so; stamens 10, free,
sometimes the outer 5 anther-less; fruits dispersing
variously, but only in G. phaeum with the mericarp separating
whole from the column and its beak twisting spirally, and
then the spirals only 1-2 and in large loops.
'Leaves' refers to lower leaves with long petioles. Most
taxa occur as rare white-petalled mutants; this aspect of
variation is not mentioned for each sp.

1 Petals narrowed at base to distinct claw 1/2 as long as limb to longer than limb	2
1 Petals without claw or with claw <1/2 as long as limb	8
2 Most petals >14mm	3
2 Most petals <14mm	4
3 Perennial with thick rhizome	**19. G. macrorrhizum**
3 Biennial with thin fibrous roots	**23. G. rubescens**
4 Leaves divided <3/4 to base	5
4 Leaves divided ± to base (ternate to palmate)	7
5 Leaves glossy green to purple, very sparsely pubescent; sepals strongly keeled on back	**20. G. lucidum**
5 Leaves grey-green, pubescent; sepals rounded on back	6
6 Mericarps appressed-pubescent, smooth; outer 5 stamens lacking anthers	**17. G. pusillum**
6 Mericarps (excluding beaks) glabrous, usually ridged; all stamens with anthers	**18. G. molle**
7 Anthers orange or purple (pale yellow in albinos); petals 8-14mm; mericarps with sparse fine ridges and 0-1(2) deep collar-like ridges at apex	**21. G. robertianum**
7 Anthers yellow; petals 5-9mm; mericarps with dense wrinkle-like ridges and 2-3(4) deep collar-like ridges at apex	**22. G. purpureum**
8 Annuals to biennials, or sometimes perennials ± without rhizome; petals mostly <10mm, rarely to 22mm	9
8 Perennials with distinct thick and/or elongated rhizome; petals >10mm, rarely less	17
9 Petals >15mm	10
9 Petals <10mm	11
10 Plant usually <60cm; flowers <3cm across; petal-claw 6-7mm	**23. G. rubescens**
10 Plant usually >60cm; flowers >3cm across; petal-claw 2-2.5mm	**24. G. maderense**
11 Seeds smooth	12
11 Seeds pitted or reticulately ridged	14
12 Mericarps (excl. beaks) glabrous, usually ridged	**18. G. molle**
12 Mericarps pubescent, smooth	13

13 Petals <5mm; sepals <3mm; outer 5 stamens lacking
   anthers   **17. G. pusillum**
13 Petals >6mm; sepals >3mm; all 10 stamens with
   anthers   **16. G. pyrenaicum**
   14 Petals ± rounded at apex; leaves divided <3/4
      to base; sepals with apiculus <0.5mm
                                                5. G. rotundifolium
   14 Petals distinctly notched at apex; leaves divided
      ≥3/4 to base; sepals with apiculus >0.5mm   15
15 Mericarps glabrous to sparsely pubescent; most
   pedicels >2.5cm   **11. G. columbinum**
15 Mericarps pubescent; pedicels ≤2.5cm   16
   16 Leaves divided almost to base; stalked glands
      frequent on upper parts of plant   **12. G. dissectum**
   16 Leaves divided c.3/4 to 7/8 to base; stalked
      glands 0   **13. G. submolle**
17 Flowers all solitary on pedicel + peduncle
                                                10. G. sanguineum
17 At least most flowers in pairs, with 2 pedicels on
   a common peduncle   18
   18 Mericarps pointed at base, with 2-4 collar-like
      ridges at apex; petals usually apiculate or with
      small triangular point at apex, sometimes ruffled
      or subentire   19
   18 Mericarps rounded at base, with 0-1 collar-like
      ridge at apex; petals rounded to notched at apex 20
19 Petals c. as long as wide, patent to slightly
   reflexed, usually purplish-black   **25. G. phaeum**
19 Petals c.1.5x as long as wide, strongly reflexed,
   never purplish-black   **26. G. x monacense**
   20 Stalked glands 0 or <0.3mm long   21
   20 Sepals and pedicels, and usually peduncles and
      upperparts of stems, with stalked glands
      >0.3mm long   26
21 Base of mericarps without tuft of bristles on
   inside; petals violet-blue   **14. G. ibericum**
21 Base of mericarps with tuft of apically pointed
   bristles on inside directed on to seed or into
   cavity; petals whitish to bright- or purplish-pink 22
   22 Hairs on pedicels, peduncles and upper parts of
      stems <0.2mm, appressed; main leaf-lobes toothed,
      with teeth up to c.5mm long   **4. G. nodosum**
   22 Pedicels, peduncles and upper parts of stem with
      patent hairs >0.5mm; main leaf-lobes with sub-
      lobes or deep teeth >(5)10mm long   23
23 Petals with veins darker than ground-colour; beaks
   of mericarps with hairs c.0.1mm   24
23 Petals with veins paler than or same colour as
   ground-colour; beaks of mericarps with some hairs
   >(0.2)0.5mm   25
   24 Petals curved outwards at apex (flowers trumpet-
      shaped), with white to very pale pink ground-

colour

3. G. versicolor

24 Petals not curved outwards at apex (flowers
funnel-shaped), ground-colour pink **2. G. x oxonianum**
25 Fruit with style (between tip of column and base of
stigmas) 2.5-3(4)mm; petals deep bright pink

1. G. endressii

25 Fruit with style (3)4-6mm; petals usually mid-pink,
often variable on 1 plant **2. G. x oxonianum**
26 Stalked glands sparse, confined to floral parts and
pedicels **Return to 25**
26 Stalked glands abundant on pedicels, peduncles
and upper parts of stems 27
27 Petals conspicuously notched; fruit never ripening

15 G. x magnificum

27 Petals rounded at apex; fruit normally ripening 28
28 Petals blue to violet-blue, with white base;
flowers and immature fruits pointing sideways or
+ downwards; fruits with styles >4mm 29
28 Petals pinkish-purple to magenta, with black or
white base; flowers and immature fruits pointing
obliquely or vertically upwards; fruits with
styles <4mm 30
29 Sepals with point >1/5 as long as main part; leaves
divided >5/6 to base **8. G. pratense**
29 Sepals with point <1/5 as long as main part;
leaves divided <5/6 to base **9. G. himalayense**
30 Petals magenta with black base, >16mm, with few
hairs on either side at base **7. G. psilostemon**
30 Petals pinkish-purple with white base, <16mm,
with abundant hairs at base on upperside

6. G. sylvaticum

Subgenus 1 - GERANIUM (spp. 1-15). Mature fruit dehiscing by
mericarps springing from column and ejecting seeds
explosively; beak becoming curved, remaining attached to
pericarp, and remaining attached or not to column; mericarps
rounded at base; petals + without claw.

1. G. endressii Gay - French Crane's-bill. Extensively
rhizomatous + erect perennial to 70cm; leaves 5-lobed c.4/5
to 5/6 to base; stalked glands usually present on sepals and
tops of pedicels; petals 16-22mm, deep bright pink, slightly
retuse at apex. Intrd; much grown in gardens, frequently
natd in grassy places and waste ground; scattered over most
of Br, rare in Ir; Pyrenees.
2. G. x oxonianum Yeo (G. endressii x G. versicolor) -
Druce's Crane's-bill. Fertile hybrid segregating and forming
spectrum between 2 parents; petals 20-26mm, pale to deep
pink, with or without dark veins, retuse at apex; and see key
(couplets 23 & 24). Intrd; often grown in gardens, natd
independently of parents in grassy places; distributed as for
G. endressii but rarer, also Guernsey; garden origin.

3. G. versicolor L. - <u>Pencilled</u> <u>Crane's-bill</u>. Shortly rhizomatous ± erect perennial to 60cm; leaves 5-lobed c.2/3 to 4/5 to base, the lobes less dissected than in <u>G. endressii</u>; stalked glands 0; petals 13-18mm, white or nearly so with magenta veins, retuse at apex. Intrd; grown and natd as for <u>G. endressii</u>, but rarer in both places; scattered in C & S Br, Ir and CI; C Mediterranean.

4. G. nodosum L. - <u>Knotted</u> <u>Crane's-bill</u>. Shortly rhizomatous ± erect perennial to 50cm; leaves 5-lobed c.2/3 to base, the lobes less dissected than in <u>G. versicolor</u>; stalked glands 0; petals 13-18mm, purplish-pink with darker veins, retuse at apex. Intrd; grown in gardens, natd in hedgerows and woodland; rare and very scattered in C & S Br; S Europe.

5. G. rotundifolium L. - <u>Round-leaved</u> <u>Crane's-bill</u>. Erect **569** to decumbent annual to 40cm; leaves 5-9-lobed ≤1/2 to base; glandular hairs abundant on most of plant; petals 5-7mm, pink, rounded or slightly retuse at apex. Native; banks, walls and stony ground; local in C & S En, S Wa, S Ir and CI.

6. G. sylvaticum L. - <u>Wood</u> <u>Crane's-bill</u>. Compact ± erect perennial to 70cm; leaves (5-)7(9)-lobed 3/4 to 4/5 to base; glandular hairs abundant on most of plant; petals 12-16mm, purplish-pink to mauvish with white base, rounded or slightly retuse at apex. Native; woods and hedges in lowlands, rock-ledges and meadows in uplands; locally common in N Br S to Yorks, very local in C En, Wa and N Ir, rarely natd elsewhere.

7. G. psilostemon Ledeb. - <u>Armenian</u> <u>Crane's-bill</u>. Compact ± erect perennial to 120cm; leaves 7-lobed c.4/5 to base; glandular hairs abundant on most of plant; petals 16-20mm, magenta with black base, rounded or slightly retuse at apex. Intrd; commonly grown in gardens, natd in grassy places; extremely scattered in En and Sc; SW Caucasus and NE Turkey.

8. G. pratense L. - <u>Meadow</u> <u>Crane's-bill</u>. Compact ± erect perennial to 1m; leaves 7-9-lobed ≥6/7 to base; glandular hairs abundant on most of plant; petals 16-24mm, blue to violet-blue, rounded at apex. Native; meadows, roadsides, open woodland, often in damp places; abundant in much of Br but absent from N Sc and parts of Wa and S En, very local in N Ir, occasionally natd elsewhere.

9. G. himalayense Klotzsch - <u>Himalayan</u> <u>Crane's-bill</u>. Extensively rhizomatous ± erect perennial to 60cm; leaves 7-lobed c.4/5 to base, with much more obtuse teeth than in <u>G. pratense</u>; stalked glands abundant on most of plant; petals 20-30mm, blue to violet-blue, rounded at apex. Intrd; grown in gardens, natd in grassy places; extremely scattered in En and Sc, perhaps overlooked for <u>G. pratense</u> or <u>G.</u> x <u>magnificum</u>; Himalayas. **G.** 'Johnson's Blue', also grown in gardens, is the sterile hybrid <u>G. pratense</u> x <u>G. himalayense</u>, and some of our plants might be this.

10. G. sanguineum L. - <u>Bloody</u> <u>Crane's-bill</u>. Shortly rhizomatous, erect to procumbent perennial to 40cm; leaves 5-7-lobed ≥3/4 to base, the lobes deeply sublobed; stalked glands 0; petals 14–22mm, bright purplish-red, sometimes (var. **striatum** Weston (var. <u>lancastriense</u> (Miller) Gray)) pink, retuse at apex. Native; grassland, rocky places, sand-dunes, open woods on calcareous soils; local in N & W Br and C Ir, mainly coastal, natd from gardens elsewhere.

11. G. columbinum L. – <u>Long-stalked Crane's-bill</u>. Erect to **569** ascending or scrambling annual to 60cm; leaves 5–7-lobed almost to base, the lobes deeply sublobed; stalked glands 0; petals 7–10mm, reddish-pink, rounded, truncate or shallowly retuse at apex. Native; grassy places, banks and scrub, mostly on calcareous soils; locally frequent in much of BI, but rare in N.

12. G. dissectum L. – <u>Cut-leaved Cranesbill</u>. Erect to **569** procumbent annual to 60cm; leaves (5–)7-lobed almost to base, the lobes deeply sublobed; stalked glands frequent on upper parts; petals 4.5–6mm, pink, deeply retuse at apex. Native; grassy and stony ground, waste places and cultivated ground; common throughout most of BI.

13. G. submolle Steudel (<u>G. core-core</u> auct. non Steudel) – **569** <u>Alderney Crane's-bill</u>. Perennial without rhizome, decumbent to ascending, to 60cm; leaves 5–7-lobed c.3/4 to 7/8 to base, the lobes deeply sublobed; stalked glands 0; petals 4.5–6.5mm, pink, retuse at apex. Intrd; hedgerows and grassy and waste places; natd in Guernsey since 1926, Alderney since 1938, rare casual in Jersey; S America. The identity of our plant is far from certain; other names than the 2 above have been suggested.

14. G. ibericum Cav. – <u>Caucasian Crane's-bill</u>. Shortly rhizomatous to compact ± erect perennial to 50cm; leaves 9–11-lobed 2/3 to 7/8 to base; stalked glands 0; petals 20–24mm, violet-blue, retuse at apex. Intrd; rarely grown in gardens, natd for many years in old churchyard, Cards; Caucasus.

15. G. x magnificum N. Hylander – (<u>G. ibericum</u> x <u>G. platypetalum</u> Fischer & Meyer) – <u>Purple Crane's-bill</u>. Differs from <u>G. ibericum</u> in stems to 75cm; leaves lobed rarely >4/5 to base; stalked glands abundant on most parts; petals 20–25mm, purplish-violet. Intrd; much grown in gardens, frequently natd in grassy places, by roads and on waste land, perhaps under-recorded for <u>G. pratense</u>; scattered throughout much of Br, Guernsey; garden origin.

<u>Subgenus</u> 2 – <u>ROBERTIUM</u> Picard (spp. 16–23). Mature fruit dehiscing by mericarps containing seeds being explosively ejected from column and breaking away from beak; beak becoming curved, usually dropping away from column; mericarps rounded at base; petals usually with distinct claw.

16. G. pyrenaicum Burm. f. – <u>Hedgerow Crane's-bill</u>. Perenn- **569** ial without rhizome, erect to ascending, to 60cm; leaves (5–)7(9)-lobed c.2/3 to base; petals 7–10mm, pinkish-purple, retuse at apex, with claw <1/2 as long as limb; mericarps

FIG 569 – **Geranium.** 1, labelled fruit of **G. rotundifolium.**
2–6, mericarps. 2, **G. lucidum.** 3, **G. pyrenaicum.** 4, **G.
rubescens.** 5, **G. robertianum.** 6, **G. purpureum.** 7–18, seeds
and mericarps. 7–8, **G. columbinum.** 9–10, **G. rotundifolium.**
11–12, **G. dissectum.** 13–14, **G. submolle.** 15–16, **G. pusillum.**
17–18, **G. molle.**

appressed-pubescent, nearly smooth; seeds smooth. Possibly
native; hedgerows, grassy places and rough ground; locally
frequent in much of BI but absent from much of N & W.

17. G. pusillum L. - <u>Small-flowered</u> Crane's-bill. Decumbent 569
to ascending annual to 40cm; leaves 7-9-lobed 1/2 to 2/3 to
base; pedicels with hairs all short; petals 2.5-4mm, mauvish-
pink, retuse at apex, with claw <1/2 as long as limb;
mericarps appressed-pubescent, + smooth; seeds smooth.
Native; cultivated and waste land and barish places among
grass; frequent to scattered throughout most of BI.

18. G. molle L. - <u>Dove's-foot</u> Crane's-bill. Differs from <u>G.</u> 569
<u>pusillum</u> in pedicels with some long as well as many short
hairs; petals 4-6mm, more deeply retuse at apex; and see key
(couplet 6), but mericarps rarely smooth (var. **aequale** Bab.).
Native; similar places to <u>G. pusillum</u>; common all over BI.

19. G. macrorrhizum L. - <u>Rock</u> Crane's-bill. Shortly
rhizomatous perennial, + erect, to 50cm; leaves 7-lobed 2/3
to 5/6 to base; petals 12-18mm, pinkish-purple, rounded at
apex, with claw c. as long as limb; mericarps glabrous, with
sharp horizontal ridges near apex; seeds smooth. Intrd;
commonly grown in gardens, natd on walls and in grassy
places; extremely few places in En and Sc, natd since 1890 in
S Devon; S Europe.

20. G. lucidum L. - <u>Shining</u> Crane's-bill. Very sparsely 569
pubescent, shining, often red, erect to ascending annual to
40cm; leaves 5-lobed c.2/3 to base; petals 8-10mm, deep pink,
rounded at apex, with claw longer than limb; mericarps
glabrous except puberulent at apex, ridged; seeds smooth.
Native; bare ground, rocks, walls and stony banks, mostly on
calcareous ground; locally common in most of BI except N Sc,
also garden escape in S En.

21. G. robertianum L. - <u>Herb-Robert</u>. Procumbent to erect, 569
strong-smelling, often red annual to biennial to 50cm; leaves
ternate to palmate with 3-5 leaflets, the leaflets much
divided + to midribs; petals 8-14mm, usually deep pink,
rounded at apex, with claw c. as long as limb; mericarps
glabrous to sparsely pubescent, with rather sparse branching
ridges and 0-1(2) collar-like ridges at apex; seeds smooth.
Native; woods, hedgerows, banks, scree and maritime shingle;
common throughout BI. Very variable and often divided into
sspp., but characters used for these cut across others and
probably do not define meaningful taxa. Most often
segregated is ssp. **maritimum** (Bab.) H.G. Baker, around coasts
of BI (especially S & W) on shingle, with procumbent habit,
often glabrous stems and leaves, usually glabrous fruits, and
slightly smaller flowers, but only the prostrateness is
constant. Ssp. **celticum** Ostenf. occurs in crevices of
limestone rocks by or near coast in S Wa and CW Ir; it is
distinct in the restriction of anthocyanin to the stem-nodes
and petiole bases, and usually has large pale flowers, red or
purple anthers, and large pubescent fruits.

21 x 22. G. robertianum x G. purpureum might occur with the parents on maritime shingle and it is possible that G. purpureum ssp. forsteri arose from this cross.

22. G. purpureum Villars (G. robertianum ssp. purpureum **569** (Villars) Nyman) - Little-Robin. Differs from G. robertianum **RR** in key characters (couplet 7). Native; rocky and stony places, and on shingle and cliffs, usually near sea; very local in SW Br from W Sussex to Carms, M Cork, Co Waterford, CI; over-recorded for G. robertianum ssp. maritimum. Procumbent to decumbent plants of maritime shingle have been separated as ssp. **forsteri** (Wilm.) H.G. Baker; they approach G. robertianum in fruit characters.

23. G. rubescens Yeo - Greater Herb-Robert. Erect, reddish, **569** faintly scented biennial to 60cm due to long petioles and large leaves, but with stem usually much shorter; differs from G. robertianum in larger leaves and flowers and fruits; petals 18-22mm, with claw c.1/2 as long as limb; fruits glabrous, with dense reticulate ridges. Intrd; natd on rough ground in Guernsey since 1968; Madeira.

24. G. maderense Yeo - Giant Herb-Robert. Erect perennial to 1(2)m, like a giant S. rubescens with stem up to 5cm wide; petioles >30cm; leaves >25cm; inflorescence >50cm across; flowers >3cm across; and petals 15-22 x 10-18mm with claw only 2-2.5mm. Intrd; well natd on cliffs in dense low vegetation in Scillies, Guernsey; Madeira.

Subgenus 3 - ERODIOIDEA Yeo (spp. 24-25). Mature fruit dehiscing by mericarps containing seeds and attached to beaks being explosively ejected from column; beak attaining 1-2 loose spirals; mericarps pointed at base and with transverse ridges near apex; petals without claw.

25. G. phaeum L. - Dusky Crane's-bill. Shortly rhizomatous to compact, ± erect perennial to 80cm; leaves 7(-9)-lobed c.2/3 to base; stalked glands abundant on most parts of plant but only c.0.1mm long, with much longer simple hairs; petals 8-12(14)mm, often with small point at apex, usually dark purplish-black, rarely pinkish-mauve (var. **lividum** (L.'Her.) Pers.) or white. Intrd; commonly grown, natd in shady places in hedges and wood-borders; scattered throughout most of Br, very local in Ir; C Europe

26. G. x monacense Harz (G. reflexum auct. non L.; G. phaeum x G. reflexum L.) - Munich Crane's-bill. Differs from G. phaeum in petals much narrower, more reflexed and with more hairs at base and resembling those of G. phaeum var. lividum (presumably 1 parent) in colour. Intrd; natd on roadside verge in E Sussex since 1975, Midlothian, perhaps overlooked for G. phaeum; garden origin.

2. MONSONIA L. - Dysentery-herbs
Annuals; leaves simple, pinnately toothed; flowers actino-morphic, stamens 15, all with anthers, with filaments fused for c.1/2 their length into 5 groups of 3; seeds dispersed

inside mericarps with beaks attached, the beaks tightly
spiralled (twisted) on their own axes.

1. M. brevirostrata Knuth – <u>Short-fruited Dysentery-herb</u>.
Sparsely pubescent, branched, ascending annual to 40cm;
leaves 16–30mm, lanceolate to narrowly ovate; flowers 1–3 on
common peduncle; petals 5–8mm, slightly longer than sepals,
rounded at apex, bluish; fruits pubescent, 22–30mm. Intrd;
fairly frequent wool-alien in fields and waste places;
scattered in En; S Africa.

3. ERODIUM L'Her. – <u>Stork's-bills</u>
Annuals to perennials; leaves simple and pinnately lobed or
pinnate; flowers actinomorphic or slightly zygomorphic;
stamens 5, free, alternating with staminodes; fruits
dispersing as in <u>Monsonia</u>, with 2 pits at apex of mericarp, 1
either side of base of beak.

1 Leaves pinnate   2
1 Leaves simple, shallowly to deeply lobed, or if ±
   compound then ternate or palmate   4
   2 Primary leaflets divided <3/4 to midrib; apical
      pits of mericarp with sessile glands **8. E. moschatum**
   2 Primary leaflets divided nearly to base; apical
      pits of mericarp glandless   3
3 Apical pits of mericarp separated from main part of
   mericarp by sharp ridge and groove, not overarched by
   hairs; flowers 3–7 per peduncle, mostly >10mm across
                                              9. E. cicutarium
3 Apical pits of mericarp not delimited by sharp ridge
   and groove, overarched by hairs from main part of
   mericarp; flowers 2–4(5) per peduncle, mostly <10mm
   across   **10. E. lebelii**
   4 Beak of fruit 0.8–1cm; petals not exceeding
      sepals or 0   **3. E. maritimum**
   4 Beak of fruit >1.5cm; petals exceeding sepals   5
5 Petals pinkish-purple; fruits with column <4cm;
   apical pits of mericarps delimited by very distinct
   groove and with conspicuous sessile glands, or with
   no groove and no glands   6
5 Petals blue to violet-purple; fruits with column
   >4cm; apical pits of mericarps delimited by distinct
   groove, without glands   7
   6 Hairs on sepals and pedicels eglandular,
      appressed; apical pits of mericarp without glands,
      not delimited by groove   **1. E. chium**
   6 Most hairs on sepals and pedicels glandular,
      patent; apical pits of mericarp with sessile
      glands, delimited by very distinct furrow
                                              2. E. malachoides
7 Lower leaves pinnately lobed, with >2 basal pairs
   of lobes not very different in size   8

7 Lower leaves ternately lobed, with 1 pair of basal
   lobes greatly exceeding all others   9
  8 Apical pits of mericarp with sparse bristles,
    bounded below by 1 blunt-rimmed groove
                          5. E. brachycarpum
  8 Apical pits of mericarps completely glabrous,
    bounded below by (1)2-3 sharp-rimmed grooves
                            4. E. botrys
9 Sepals and pedicels with many patent glandular
  hairs   **7b. E. cygnorum ssp. glandulosum**
9 Sepals and pedicels without glandular hairs   10
  10 Pedicels glabrous or with hairs near apex only;
    sepals with only ± appressed hairs <0.3mm
                   7a. E. cygnorum ssp. cygnorum
  10 Pedicels sparsely pubescent along length; sepals
    with short appressed and some longer ± patent
    hairs >0.5mm
                          6. E. crinitum

Other spp. – Over 29 alien spp. have been recorded, many
as wool-aliens, of which the genus _Erodium_ is one of the most
characteristic components. The 6 spp. treated here are by
far the most common.

1. E. chium (L.) Willd. – <u>Three-lobed</u> Stork's-bill. **574**
Suberect to ascending annual to 40cm; leaves simple, the
lowest deeply 3-lobed; sepals and pedicels with ± appressed
eglandular hairs; petals pinkish-purple; mericarps with beak
2-4cm, with eglandular pits not delimited by groove. Intrd;
infrequent wool-alien; scattered in En and Wa; Mediterranean.
2. E. malachoides (L.) L'Her. – <u>Soft</u> Stork's-bill. Suberect **574**
to ascending annual to 40cm; leaves simple, the lowest
toothed to shallowly 3-lobed; sepals and pedicels with patent
glandular and some eglandular hairs; petals pinkish-purple;
mericarps with beak 2-3.5cm, with conspicuously glandular
pits delimited by deep groove. Intrd; infrequent wool-alien;
scattered in En and Wa; S Europe.
3. E. maritimum (L.) L'Her. – <u>Sea</u> Stork's-bill. Procumbent **574**
to decumbent annual to 10(20)cm; leaves simple, the lowest **R**
toothed to shallowly pinnately lobed; sepals and pedicels
with erecto-patent eglandular hairs; petals usually 0, or
pink and not exceeding calyx; mericarps with beak 0.8-1cm,
with eglandular pits overarched by hairs from below and
delimited by distinct groove. Native; fixed maritime dunes
and barish places in short grassland, rarely inland; coasts
of CI, E & S Ir, W Br from Dorset to Wigtowns, inland in
Worcs and Co Durham, formerly more widespread.
4. E. botrys (Cav.) Bertol. – <u>Mediterranean</u> Stork's-bill. **574**
Suberect to ascending annual to 50cm; leaves simple, deeply
pinnately lobed; sepals and pedicels with numerous patent
glandular hairs and some ± patent eglandular hairs; petals
bluish; mericarps with beak 6-9(11)cm, with glabrous
eglandular pits delimited by usually 2-3 sharp-rimmed

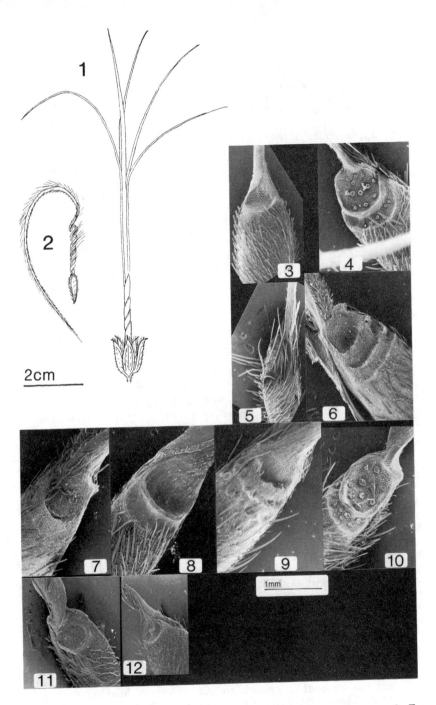

FIG 574 — *Erodium*. 1–2, fruit and single mericarp of *E. botrys*. 3–12, apical pit region of mericarps. 3, *E. chium*. 4, *E. malachoides*. 5, *E. maritimum*. 6, *E. botrys*. 7, *E. brachycarpum*. 8, *E. crinitum*. 9, *E. cygnorum*. 10, *E. moschatum*. 11, *E. cicutarium*. 12, *E. lebelii*.

grooves. Intrd; common wool-alien; scattered in Br; Mediterranean.

5. E. brachycarpum (Godron) Thell. (E. obtusiplicatum **574** (Maire, Weiller & Wilczek) Howell) - Hairy-pitted Stork's-bill. Differs from E. botrys in usually slightly shorter mericarps and beaks (but with much overlap); apical pits with some bristles and usually 1 blunt-rimmed groove. Intrd; frequent wool-alien; scattered in En, perhaps overlooked for E. botrys; W Mediterranean. A doubtfully distinct sp.; only the key character is reliable.

6. E. crinitum Carolin - Eastern Stork's-bill. Decumbent to **574** ascending annual with thick ± succulent root (perennial in native area) to 50cm; leaves simple with 3 very deep lobes or ± ternate; sepals and pedicels with erecto-patent eglandular hairs; petals bluish, with yellow or white veins; mericarps with beak 4-7cm, with eglandular pits delimited by groove. Intrd; frequent wool-alien; scattered in En; C & E Australia.

7. E. cygnorum Nees - Western Stork's-bill. Habit as in **E.** **574** crinitum; leaves simple with 3 deep lobes; mericarps with beak 5-10cm, with eglandular pits delimited by groove. Intrd; frequent wool-alien; very scattered in En; W Australia. Both sspp. (possibly best as spp.) are frequent.

a. Ssp. cygnorum. Petals bluish, with yellow or white veins; mericarps rather sparsely pubescent; and see key.

b. Ssp. glandulosum Carolin. Petals bluish, with red veins; mericarps densely pubescent; and see key.

8. E. moschatum (L.) L'Her. - Musk Stork's-bill. Suberect **574** to procumbent annual to 60cm; leaves pinnate, with toothed or **R** shallowly lobed leaflets, smelling musky when bruised; sepals and pedicels with patent glandular hairs; petals pinkish-purple; mericarps with beak 2-4.5cm, with glandular pits delimited by deep groove. Native; rough ground and barish places in short grassland, mainly near sea, also frequent wool-alien; coasts of CI, W Br from S Devon to Man, Ir, widespread in Br as casual.

9. E. cicutarium (L.) L'Her. - Common Stork's-bill. Sub- **574** erect to procumbent annual to 60cm; leaves pinnate, with deeply pinnately lobed to ± pinnate leaflets; sepals and pedicels variably with glandular and/or eglandular hairs; petals pinkish-purple, rarely white, the upper 2 often with black basal spot; mericarps with beak 1.5-4cm, with egland-ular pits not overarched by hairs from below and delimited by furrow; 2n=40. Native; barish places in grassland, waste and rough ground, on sandy or chalky soils, also common wool-alien; locally scattered over most of BI. Very variable, especially as wool-alien and on coastal dunes; ssp. **dunense** Andreas is a dwarf coastal ecotype (sometimes wrongly determined as E. lebelii or as a hybrid), but intermediates are too common for its recognition.

9 x 10. E. cicutarium x E. lebelii = E. x anaristatum Andreas is known from coastal dunes in Wa and S Lancs; it is intermediate and sterile, with 2n=30.

10. E. lebelii Jordan (E. glutinosum Dumort., E. cicutarium 574
ssp. bipinnatum auct. non (Cav.) Tourlet) - Sticky Stork's-   R
bill. Differs from E. cicutarium in stems to 15(25)cm;
sepals and pedicels always with dense patent glandular hairs;
petals pale pink to white; mericarps with beak up to 2.2cm;
and see key. Native; barish places on fixed dunes; coasts of
Br from E Kent to Wigtowns, Ir from W Cork to Co Antrim,
Jersey and Sark.

4. PELARGONIUM L'Her. - Geraniums
Perennials, woody or + so at base; leaves simple, tomentose,
very shallowly to deeply palmately lobed; flowers zygomorphic, with the upper 2 petals wider than the others and
the upper sepal with a backward-directed spur tightly fused
to pedicel; stamens 10, free, but only 7 with anthers; fruits
dispersing as in Monsonia.

Other spp. - The familar Scarlet Geranium, with characteristically scented leaves with zonal dark markings, is **P. x
hybridum** Aiton (P. inquinans Knuth x P. zonale (L.) Aiton),
of garden origin; it sometimes persists until the frosts on
rubbish-tips and waste ground where discarded.

1. P. tomentosum Jacq. - Peppermint-scented Geranium.
Scrambling to 1m; leaves up to 12cm, peppermint-scented,
deeply palmately lobed; flowers up to 20 in umbel; petals
whitish-pink, red at base, 6-9mm, 1.25-2x as long as sepals;
fruit column 15-25mm. Intrd; natd at woodland edge and among
Pteridium, Scillies; S Africa.

105. LIMNANTHACEAE - Meadow-foam family

Annual, slightly succulent, + glabrous herbs; leaves
alternate, pinnate or deeply pinnately lobed, petiolate,
exstipulate. Flowers solitary in leaf-axils, bisexual,
hypogynous, actinomorphic; sepals 5, free; petals 5, free;
stamens 10; gynoecium similar to that of Geraniaceae but
ovary with 1 ovule per cell, cells of fruit remaining closed
and dispersed as separate whole mericarps after column has
withered.
The buttercup-like flowers with quite different stamens and
carpels are distinctive; differs from Geraniaceae in lacking
stipules and in above gynoecium characters.

1. LIMNANTHES R.Br. - Meadow-foam
1. L. douglasii R. Br. - Meadow-foam. Stems erect to
ascending, to 35cm; flowers 16-25mm across, fragrant; petals
white with conspicuous yellow basal part; beak of fruit
5-9mm. Intrd; commonly grown for ornament and occasional on
tips and by roads and lakes; scattered in Br; California.

106. TROPAEOLACEAE - Nasturtium family

Annual or perennial, sometimes climbing, somewhat succulent herbs; leaves alternate, peltate, simple to very deeply palmately lobed (± palmate), petiolate, exstipulate. Flowers solitary in leaf-axils, bisexual, slightly perigynous, zygomorphic, yellow to red, showy; sepals 5, ± free, the upper 1 or 3 with single backward-directed free spur; petals 5, free, the upper 2 different from lower 3; stamens 8; ovary 3-celled, each cell with 1 apical ovule; style 1; stigmas 3, linear; fruit a 3-celled schizocarp breaking into 3 indehiscent ± succulent segments.

The zygomorphic flowers with 5 free petals and sepals, 8 stamens and 3-celled ovary are unmistakable.

1. TROPAEOLUM L. - Nasturtiums

Other spp. - T. peregrinum L. (T. canariense Hort.) (Canary-creeper), from S America, differs from T. speciosum in its yellow flowers and 5-lobed leaves; it has occasionally been reported to stray outside gardens.

1. T. speciosum Poeppig & Endl. - Flame Nasturtium. Rhizomatous perennial to several m, climbing by long petioles; leaves deeply 5-6-lobed, almost palmate; flowers scarlet, c.18-25mm across, with spur c.20-30mm. Intrd; natd clambering through bushes and in hedges; E Sussex, Co Antrim and Co Londonderry; Chile.

2. T. majus L. - Nasturtium. Annual of many cultivars, remaining dwarf or ± climbing to 2m; leaves slightly angled, otherwise entire; flowers various shades of yellow or red, usually orange, c.25-60mm across, with spur c.25-40mm. Intrd, commonly cultivated, frequent casual on tips and waste ground; scattered in Br and CI; Peru.

107. BALSAMINACEAE - Balsam family

Glabrous, somewhat succulent, annual herbs; leaves alternate, opposite or in whorls of 3, simple, petiolate, exstipulate or with stipules represented by basal glands. Flowers 1-several in axillary or terminal cymes, bisexual, sometimes some cleistogamous, hypogynous, zygomorphic; sepals 3, ± petaloid, free, the lowest sac-like with a conspicuous backward-directed or variously bent spur; petals 5, but apparently 3, the uppermost free, each of the 2 lower ones fused to each of the 2 lateral ones to form 2 compound flanges; stamens 5, with filaments fused distally and anthers fused round ovary; ovary 5-celled, each cell with numerous ovules on axile placentas; style 1; stigmas 1 or 5, minute to linear; fruit an explosive 5-celled capsule with many seeds.

The flowers, with spurred lower sepal and 2 others, fused

pairs of lateral petals, and distally fused stamens, are unique.

1. IMPATIENS L. - Balsams

1 Flowers deep pinkish-purple to white; leaves opposite
  or in whorls   **4. I. glandulifera**
1 Flowers pale yellow to orange   2
  2 Flowers (incl. spur) <1.5(2)cm; larger leaves with
    >20 teeth on each side   **3. I. parviflora**
  2 Flowers (incl. spur) all or many >2cm; larger leaves
    with <20 teeth on each side   3
3 Flowers yellow; sepal-spur held at c.90 degrees
  to rest of sepal in fresh state   **1. I. noli-tangere**
3 Flowers orange; sepal-spur held ± parallel to rest
  of sepal in fresh state   **2. I. capensis**

Other spp. - **I. balfourii** Hook. f., from Himalayas, has large pink flowers but differs from I. glandulifera in alternate leaves and sepal-spur >10mm and much less bent; it is grown in gardens and has been reported as an escape.

1. I. noli-tangere L. - Touch-me-not Balsam. Stems erect, **R** to 1m; flowers mostly 2-3.5cm, yellow with small brownish spots; lowest sepal 10-20mm plus spur 6-12mm held at c.90 degrees. Native; damp places in woods; very local in Br N to C Sc, frequent in Lake District, probably only there native.
2. I. capensis Meerb. - Orange Balsam. Stems erect, to 1.5m; flowers mostly 2-3.5cm, orange with large brownish blotches; lowest sepal 10-20mm plus spur 5-10mm held parallel to rest of sepal. Intrd; banks of rivers and canals; locally frequent in Br N to SW Yorks; N America.
3. I. parviflora DC. - Small Balsam. Stems erect, to 60(100)cm; flowers (incl. spur) mostly 0.6-1.8cm, pale yellow; lowest sepal gradually tapered into ± straight spur, in total 4-12mm. Intrd; damp shady places in woods and hedges and disturbed or cultivated ground; scattered through most of Br, locally common; C Asia.
4. I. glandulifera Royle - Indian Balsam. Stems erect, to 2m; flowers 2.5-4cm, deep to pale pinkish-purple or white; lowest sepal 12-20mm plus spur 2-7mm bent at ≥90 degrees. Intrd; banks of rivers and canals, damp places and waste ground; locally common throughout most of BI; Himalayas.

108. ARALIACEAE - Ivy family

Evergreen, woody climbers, herbaceous perennials or deciduous shrubs; leaves alternate, simple and usually palmately lobed or 1-2-pinnate, petiolate, exstipulate. Flowers in terminal umbels often grouped into panicles, bisexual, epigynous, actinomorphic; hypanthium 0; sepals represented by 5 small

teeth near top of ovary; petals 5, free; stamens 5; ovary
5-celled, with 1 apical ovule per cell; styles 1 or 5;
stigma(s) minute; fruit a usually black berry with 5 seeds.
The umbellate inflorescence and 5-celled berries are
diagnostic; the growth-forms of Hedera and Aralia are
unmistakable.

1 Evergreen; leaves simple, usually palmately lobed
                                              1. HEDERA
1 Deciduous or herbaceous; leaves 1-2-pinnate **2. ARALIA**

1. HEDERA L. - Ivies
Evergreen non-spiny, woody climbers with numerous short roots
borne along climbing stems; leaves simple, usually palmately
lobed; style 1; fruits black unless otherwise stated.
In all spp. leaves on creeping or climbing stems differ from
those on non-rooting, flowering stems; the former are more
lobed, and are those referred to in key and descriptions as
they and their hairs are the diagnostic ones.

1 Hairs on leaves and young stems semi-peltate, with
  rays fused for 1/4 to 1/2 their length; leaves
  scarcely or not lobed, the larger >15cm wide
                                          1. H. colchica
1 Hairs on leaves and young stems stellate, with rays
  fused only at extreme base; at least some leaves well
  lobed, rarely >15cm wide   2
  2 Hairs whitish, with rays lying parallel to leaf
    surface and also projecting away from it; leaves
    usually <8cm across, often lobed >1/2 way to base
                                  2a. H. helix ssp. helix
  2 Hairs often pale yellowish-brown, with rays ± all
    lying parallel to leaf surface; leaves often >8cm
    across, usually lobed <1/2 way to base
                              2b. H. helix ssp. hibernica

Other spp. - Many cultivars of all 3 above taxa are grown,
often obscuring their morphological and distributional
limits. 2 other taxa are also grown and may persist in wild
places. **H. canariensis** Willd. (Canary Ivy), from Macaronesia
and Portugal, has reddish-brown hairs with all rays lying
parallel to leaf and fused for c.1/4 their length, and
scarcely lobed ± reniform leaves; it is mostly grown as a
variegated cultivar. **H. helix** spp. **poetarum** Nyman, from C &
E Mediterranean, has lighter green, less deeply lobed leaves
than ssp. helix and slightly larger yellowish berries.

1. H. colchica (K. Koch) K. Koch - Persian Ivy. Stems
creeping, scrambling and weakly climbing to c.10m; leaves
mostly 15-25cm across, unlobed or slightly 3-5-lobed, entire
or denticulate, with reddish-brown semi-peltate hairs with
mostly 15-25 rays held parallel to surface; 2n=192. Intrd;

much grown in shrubberies and on walls and often
long-persistent and spreading; scattered over Br; Caucasus.
2. H. helix L. - Ivy. Stems variously creeping and
climbing; leaves rarely >15cm wide, distinctly 3-5-lobed,
with white to pale yellowish-brown stellate hairs (avoid old
or rubbed leaves). Native; on trees, banks, rocks and
sprawling over the ground.
a. Ssp. helix - Common Ivy. Often strongly climbing up
rocks and trees; leaves usually dark green, often with light
marbling, often lobed >1/2 way to base; 2n=48; and see key.
common over E, C and N Br, uncertain further W.
b. Ssp. hibernica (Kirchner) D. McClint. (H. hibernica
Kirchner) - Atlantic Ivy. Not or rather weakly climbing;
leaves usually mid-green, rarely marbled, usually lobed <1/2
way to base, 2n=96; and see key. The commoner taxon in CI,
Ir, and W & SW Br N to SW Sc, but with much overlap. H.
'Hibernica' (Irish Ivy) is a well-known cultivar.

2. ARALIA L. - Angelica-trees
Herbaceous perennials or deciduous, spiny shrubs, without
roots along stems; leaves 1-2-pinnate, mostly >50cm; styles
5; fruits dark purplish to black.

1 Spineless herbaceous perennial to 2m   **3. A. racemosa**
1 Spiny deciduous shrubs often >2m   2
   2 Inflorescence a + conical panicle; leaves glabrous
      on lowerside except on veins   **1. A. chinensis**
   2 Inflorescence a + umbrella-shaped panicle;
      leaves pubescent on lowerside   **2. A. elata**

1. A. chinensis L. - Chinese Angelica-tree. Suckering shrub
to 3(6)m, with thick spiny branches; leaves 2-pinnate, with
ovate, serrate leaflets; petioles spiny; inflorescence a
panicle of umbels. Intrd; planted on banks and in
shrubberies and spreading by suckers; natd in W Kent and
Midlothian; China. Our plant is var. nuda Nakai.
2. A. elata (Miq.) Seemann - Japanese Angelica-tree. Shrub
or tree to 5(10)m; differs from A. chinensis in leaflets
narrower and less closely serrate; stems more spiny; and see
key. Intrd; natd as for A. chinensis; Jersey; far E Asia.
3. A. racemosa L. - American-spikenard. Herbaceous, spineless perennial to 2m; leaves 1-2-pinnate, with ovate, serrate
leaflets; inflorescence an elongated panicle of umbels.
Intrd; planted in shrubberies and open woodland; natd in
Salop; E N America.

109. APIACEAE - Carrot family
(Umbelliferae, Hydrocotylaceae)

Herbaceous annuals to perennials, rarely shrubs; leaves
alternate, simple to palmate, pinnate or ternate (often

several times so), sessile or petiolate; stipules present in Hydrocotyle, 0 in others but petiole often much widened and sheathing stem at base.   Flowers in terminal and lateral umbels, the umbels sometimes simple but usually compound, in Hydrocotyle often with whorls of flowers below main umbel, bisexual or frequently functionally andromonoecious, rarely dioecious, epigynous, actinomorphic or with zygomorphic petals (those away from centre of umbel larger); hypanthium 0; sepals 5, usually represented by small teeth near top of ovary, or 0; petals 5, usually white, pink or yellow; stamens 5; ovary 2-celled, with 1 apical ovule per cell; styles 2, often arising from swelling (stylopodium) on top of ovary; stigma minute to capitate; fruit a dry 2-celled schizocarp, the 2 mericarps usually separating from the sterile carpophore (or carpophore lacking) but each remaining indehiscent at maturity.

Most spp. are unmistakable 'umbellifers', with compound (less often simple) umbels and distinctive fruits; genera not so conforming (Eryngium, Hydrocotyle) are individually highly distinctive.   Too many genera are recognized.   Good candidates for amalgamation are Chaerophyllum, Anthriscus and Myrrhis; and Peucedanum, Pastinaca, Ligusticum, Levisticum, Selinum and Angelica, among others.

Fully ripe fruits are extremely important for identification.   Important points are the degree and pattern of longitudinal ridging and the shape and position of sub-surface oil-bodies on the mericarps.   The face of the mericarps where they join is the commissure, and the outer face is dorsal. Fruits may be compressed to varying degrees: either dorsally (with commissure as wide as fruit) or laterally (with commissure through short axis).   Bracteoles (when present) subtend the flowers at the base of the pedicels; bracts (when present) subtend the main branches (rays) of the compound umbel.   Fruit length excludes the stylopodium; width is that in widest view.

In this work 7 distinctive groups are first defined, followed by the keys which cover all taxa except for male plants of Trinia. The list of distinctive taxa and then the keys should be tackled successively.

Distinctive genera

Fresh plants smelling of aniseed when crushed
                    8. MYRRHIS, 21. FOENICULUM, 22. ANETHUM
Plant with subterranean swollen tubers (not swollen tap-
    roots)   **11. BUNIUM, 12. CONOPODIUM, 19. OENANTHE**
Base of stem sheathed by mass of fibres (remains of old
    petioles)   **3. ASTRANTIA, 13. PIMPINELLA, 18. SESELI,
          23. SILAUM, 24. MEUM, 28. TRINIA, 38. CARUM,
                                           43. PEUCEDANUM**
Plant entirely male or female   **28. TRINIA**
Stem procumbent; leaves ± orbicular, shallowly lobed,
    with stipules   **1. HYDROCOTYLE**

Plant with subaquatic leaves much more finely divided than
aerial ones   **15. SIUM, 19. OENANTHE, 30. APIUM**
Leaves all simple, not divided or divided <1/2 way to
base   **1. HYDROCOTYLE, 4. ERYNGIUM, 10. SMYRNIUM,
27. BUPLEURUM**

Leaves divided >1/2 way to base but at most ternate to
palmate or ternately to palmately lobed, the leaflets/
lobes toothed to shallowly lobed   **2. SANICULA,
3, ASTRANTIA, 4. ERYNGIUM, 14. AEGOPODIUM, 37. FALCARIA,
40. LIGUSTICUM, 43. PEUCEDANUM**
Inflorescence with lobed or divided bracts   **2. SANICULA,
4. ERYNGIUM, 16. BERULA, 28. TRINIA, 29. CUMINUM,
31. TRACHYSPERMUM, 32. PETROSELINUM, 36. AMMI, 38. CARUM,
43. PEUCEDANUM, 48. DAUCUS**
Fruits with hooked or barbed bristles   **2. SANICULA,
4. ERYNGIUM, 6. ANTHRISCUS, 47. TORILIS, 48. DAUCUS**
Fruits tuberculate (not smooth, scaly, spiny or pubescent)
27. BUPLEURUM, 31. TRACHYSPERMUM, 47. TORILIS

General key
1 Leaves all simple and entire   **27. BUPLEURUM**
1 Leaves simple to compound; if simple at least toothed   2
  2 Stem-leaves spiny   **4. ERYNGIUM**
  2 Leaves not spiny   3
3 Leaves stipulate, all simple, ± orbicular,
shallowly lobed, with long petiole   **1. HYDROCOTYLE**
3 Leaves exstipulate, if all simple then either lobed
>1/2 way to base or sessile   4
  4 Leaves all simple, with small teeth only
10. SMYRNIUM
  4 Leaves simple to compound, if all simple then
divided >1/2 way to base   5
5 Basal leaves all ternately or palmately lobed almost
to base   6
5 Basal leaves ternate, palmate or pinnate, often
compoundly so   7
  6 Flowers subsessile, with inconspicuous bracteoles;
fruits with hooked bristles   **2. SANICULA**
  6 Flowers distinctly pedicellate, with bracteoles
at least as long; fruits covered with bifid scales
3. ASTRANTIA
7 Fruit with apical beak >2x as long as seed-bearing
part   **7. SCANDIX**
7 Fruit beakless or with beak shorter than seed-bearing
part   8
  8 Stem and often basal leaves with white, flexuous
subterranean part arising from brown tuber   9
  8 Stem and basal leaves (if present) arising from
roots at ground level, or from rhizome, or plant
aquatic   10
9 Stem hollow at fruiting; fruit with ± erect styles
gradually narrowed from stylopodium   **12. CONOPODIUM**

9 Stem solid at fruiting; fruit with reflexed styles
  suddenly contracted from stylopodium   **11. BUNIUM**
  10 Petals yellow   Key A
  10 Petals white to pink or purplish, or greenish-
    white   11
11 Fruits with spikes, bristles, hairs or conspicuous
  tubercles   Key B
11 Fruits glabrous, ± smooth   12
  12 Fruits strongly compressed dorsally, distinctly
    wider in dorsal view than in lateral view Key C
  12 Fruits not compressed dorsally or scarcely so 13
13 Base of stem sheathed by mass of fibres (remains of
  old petioles)   14
13 Basal mass of petiole-fibres 0   17
  14 Basal leaves 1-pinnate, each lobe palmately
    divided ± to base into filiform segments
    appearing as if whorled   **38. CARUM**
  14 Basal leaves 1-4-pinnate, if 1-pinnate then lobes
    not divided ± to base into filiform segments 15
15 Fruit 4-10mm; basal leaves 3-4-pinnate with filiform
  ultimate lobes   **24. MEUM**
15 Fruit 2-4mm; basal leaves 1-3-pinnate with ovate to
  linear ultimate lobes   16
  16 Plant usually dioecious; all leaf-segments linear;
    umbels with <10 rays   **28. TRINIA**
  16 Plant bisexual; leaf-segments variable, some often
    ovate, very rarely all linear; umbels usually with
    ≥10 rays   **13. PIMPINELLA**
17 Fruits >(2)2.5x as long as widest width Key D
17 Fruits ≤2x as long as widest width   18
  18 Sepals 0, or minute teeth, or vestigial rim, not or
    scarcely visible at top of fruit Key F
  18 Sepals ≥0.2mm, distinctly visible at top of
    fruit   Key E

Key A - Petals yellow
1 Fruits strongly compressed dorsally, i.e. distinctly
  wider in dorsal view than in lateral view 2
1 Fruits not compressed dorsally, or scarcely so 4
  2 Bracts >3; mericarps each with 2 lateral and 3
    dorsal prominent ridges   **42. LEVISTICUM**
  2 Bracts 0-3; mericarps with obscure ridges 3
3 Leaves 1-pinnate, with ovate lobes **44. PASTINACA**
3 Leaves 2-several times ternate, with linear lobes
    43. PEUCEDANUM
  4 Ultimate leaf-lobes filiform, <0.5mm wide 5
  4 Ultimate leaf-lobes flat, linear to ovate, >0.5mm
    wide   7
5 Firmly rooted perennials   **21. FOENICULUM**
5 Easily uprooted annuals   6
  6 Fruits dorsally compressed, with lateral wings;
    fresh plant smelling of aniseed **22. ANETHUM**

```
    6  Fruits laterally compressed, not winged; fresh
       plant not smelling of aniseed          33. RIDOLFIA
 7  Fruit laterally compressed, c.2x as wide in lateral
    view as in dorsal view                                    8
 7  Fruit not or scarcely compressed, c. as wide in
    lateral view as in dorsal view                            9
       8  Bracts and bracteoles 0-2, very short; mericarps
          with 3 acute dorsal ridges          19. SMYRNIUM
       8  Bracts 1-3, often lobed; bracteoles >3; mericarps
          with 3 rounded dorsal ridges        32. PETROSELINUM
 9  Leaves succulent; bracts >4; mericarps not winged
                                              17. CRITHMUM
 9  Leaves not succulent; bracts 0-3; mericarps with
       narrow lateral wings                   23. SILAUM
```

Key B - Petals not yellow; fruits with spines, bristles,
        hairs or conspicuous tubercles

```
 1  Fruits strongly compressed dorsally, i.e. distinctly
    wider in dorsal view than in lateral view (incl.
    projections)                                              2
 1  Fruits not compressed dorsally, or scarcely so           3
       2  Stems 1.5-5.5m; umbels with >30 rays; fruits >8mm
                                              45. HERACLEUM
       2  Stems <1.5m; umbels with <20 rays; fruits <7mm
                                              46. TORDYLIUM
 3  Fruits with conspicuous tubercles     31. TRACHYSPERMUM
 3  Fruit with spines, bristles or hairs                     4
       4  Fruits with hairs or weak or minute bristles       5
       4  Fruits with usually stout, terminally hooked or
          barbed, spines                                     7
 5  Fruits >3x as long as wide; fresh plant smelling of
    aniseed                                   8. MYRRHIS
 5  Fruits <3x as long as wide; fresh plant not smelling
    of aniseed                                               6
       6  Slender annual without basal fibres; bracts and
          bracteoles each <5; fruits >3.5mm   29. CUMINUM
       6  Perennial with base of stem sheathed by mass of
          fibres; bracts and bracteoles each >5; fruits
          <3.5mm                              18. SESELI
 7  Bracts deeply pinnately or ternately divided 48. DAUCUS
 7  Bracts 0 or simple                                       8
       8  Fruits without beak, with spines up to base of
          stylopodium, with persistent sepals  47. TORILIS
       8  Fruits with spine-less but ridged beak below
          stylopodium; sepals 0              6. ANTHRISCUS
```

Key C - Petals not yellow; fruits without spines, bristles,
        hairs or conspicuous warts, strongly dorsally
        compressed (see also Levisticum and Pastinaca -
        petals yellow, Key A)

```
 1  Leaves and/or stems pubescent to hispid   45. HERACLEUM
 1  Leaves and main parts of stems glabrous or nearly so,
```

    sometimes coarsely papillose   2
2 Easily uprooted annuals   3
2 Firmly rooted perennials   4
3 Fruits ovoid, with strong ridges; outer petals <1.5mm;
  sepals 0   **20. AETHUSA**
3 Fruits ± globose, scarcely ridged; outer petals >2mm;
  sepals persistent   **9. CORIANDRUM**
    4 Leaf-lobes <2 x 1cm   5
    4 Most leaf-lobes >2 x 1cm   6
5 Stems solid; bracts 0 or few and soon falling;
  bracteoles not or weakly reflexed   **39. SELINUM**
5 Stems hollow; bracts >3, reflexed; bracteoles reflexed
                        43. PEUCEDANUM
    6 Larger leaves 2-3-pinnate (smaller ones sometimes
        2-3-ternate or pinnate-ternate)   **41. ANGELICA**
    6 All leaves 1-2-ternate   7
7 Umbels with <20 rays; bracts 1-5; fruits with 3
  prominent acute dorsal ridges   **40. LIGUSTICUM**
7 Umbels with >20 rays; bracts 0(-2); fruits with low,
  obtuse dorsal ridges   **43. PEUCEDANUM**

<u>Key D</u> - Petals not yellow; fruits without spines, bristles,
    hairs or conspicuous warts, not or scarcely dorsally
    compressed, >(2)2.5x as long as wide
1 Sepals 0 or minute or a vestigial rim, not or scarcely
  visible at top of fruit   2
1 Sepals >0.2mm, distinctly visible at top of fruit   4
    2 Fruit not ridged in mid or basal regions; stems
      hollow   **6. ANTHRISCUS**
    2 Fruit ridged along length; stems solid or hollow   3
3 Stems solid; fruits with low, rounded ridges; fresh
  plant not aniseed-scented   **5. CHAEROPHYLLUM**
3 Stems hollow; fruits with sharp, prominent ridges;
  fresh plant aniseed-scented   **8. MYRRHIS**
    4 Umbels with 1-5 rays; at least some bracts >1/2
      as long as rays   **29. CUMINUM**
    4 Umbels with >6 rays, if fewer then bracts 0 or much
      <1/2 as long as rays   5
5 Fruits laterally compressed; lobes of lower leaves
  linear-lanceolate, >5cm, with regularly and sharply
  serrate margins; not aquatic   **37. FALCARIA**
5 Fruits not compressed or slightly dorsally so; leaf-
  lobes various, but if linear or lanceolate and >5cm
  then entire to distantly and irregularly toothed;
  often aquatic   **19. OENANTHE**

<u>Key E</u> - Petals not yellow; fruits without spines, bristles,
    hairs or conspicuous warts, not or scarcely dorsally
    compressed, <2x as long as wide; sepals >0.2mm,
    distinctly visible at top of fruit
1 Fruits subglobose, in lateral view c. as wide as long   2
1 Fruits in lateral view distinctly longer than wide   4

 2 Easily uprooted annuals; mericarps remaining fused
    even at maturity   **9. CORIANDRUM**
 2 Firmly rooted perennials; mericarps splitting apart
    at maturity   3
 3 Stems solid; fruits >2.5mm; petioles not widened at
    base, nor sheathing lateral stems   **25. PHYSOSPERMUM**
 3 Stems hollow; fruits <2.5mm; petioles widened at base
    and sheathing lateral stems   **35. CICUTA**
     4 Not aquatic; lobes of lower leaves linear-
        lanceolate, >5cm, with regularly and sharply
        serrate margins   **37. FALCARIA**
     4 Often aquatic; leaf-lobes various, but if linear
        or lanceolate and >5cm then entire to distantly or
        irregularly toothed   5
 5 Fruits laterally compressed; lobes of lower leaves
    mostly >4 x 2cm   **15. SIUM**
 5 Fruits not compressed or slightly dorsally so; lobes
    of lower leaves <4 x 2cm   **19. OENANTHE**

Key F - Petals not yellow; fruits without spines, bristles,
        hairs or conspicuous warts, not or scarcely dorsally
        compressed, <2x as long as wide, with sepals 0 or
        scarcely visible
 1 Lower leaves simply pinnate, the lobes not divided as
    far as midrib   2
 1 Lower leaves 2-4-pinnate or 1-2-ternate   8
     2 Bracts 0(-2), if constantly present then stem
        mostly procumbent with ± only leaves and peduncles
        erect   3
     2 Bracts 2-c.8; at least apical part of stem erect to
        ascending   5
 3 Plant often aquatic, at least lower part of stem
    procumbent and rooting   **30. APIUM**
 3 Plant ± never aquatic; stem erect, not rooting   4
     4 Fruit <2mm   **30. APIUM**
     4 Fruit >2mm   **13. PIMPINELLA**
 5 Styles at fruiting at least as long as stylopodium   6
 5 Styles at fruiting much shorter than stylopodium   7
     6 Lowest leaves with 2-5 pairs of leaflets each
        3-6cm; all bracts <1/2 as long as all rays **34. SISON**
     6 Lowest leaves with 4-12 pairs of leaflets each
        0.5-3.5cm; longest bracts >1/2 as long as shortest
        rays   **32. PETROSELINUM**
 7 Stems solid; bracts divided to base into linear to
    filiform lobes, the longest >1/2 as long as rays;
    upper leaves >1-pinnate   **36. AMMI**
 7 Stems hollow; bracts lobed but usually not to base,
    <1/2 as long as rays; all leaves 1-pinnate **16. BERULA**
     8 Bracts >4   9
     8 Bracts 0-2(3)   10
 9 Bracts <1/4 as long as rays, undivided; stems hollow;
    fruits with prominent ± wavy-edged ridges **26. CONIUM**

9 Bracts >1/2 as long as rays, deeply divided; stems
    solid; fruits with low smooth ridges   **36. AMMI**
    10 Plant in water or on mud; stems procumbent and
      rooting at least near base; styles at fruiting much
      shorter than stylopodium   **30. APIUM**
    10 Plant not aquatic; stems erect but sometimes
      rhizomes produced; styles at fruiting at least as
      long as stylopodium   11
11 Plant rhizomatous; leaves 1-2-ternate   **14. AEGOPODIUM**
11 Plant not rhizomatous; leaves 2-3-pinnate   12
    12 Easily uprooted annuals; bracteoles long, strongly
      reflexed   **20. AETHUSA**
    12 Firmly rooted biennials or perennials; bracteoles
      0 or scarcely reflexed   13
13 Usually some umbels with 1-2(3) bracts; all ultimate
    leaf-lobes linear to linear-lanceolate; styles $\pm$
    appressed to stylopodium   **38. CARUM**
13 Bracts 0; usually at least some leaf-lobes ovate, if
    all linear to linear-lanceolate then styles not
    appressed to stylopodium   **13. PIMPINELLA**

Other genera - **CAUCALIS** L. and **TURGENIA** Hoffm. are both
annuals from S Europe with beakless, spiny fruits, allied to
<u>Torilis</u>. They differ in having fruits with non-persistent
sepals and with broad-based spines arranged in rows on the
ridges. **C. platycarpos** L. (<u>Small Bur-parsley</u>) has 0(-2)
bracts and $\pm$ actinomorphic corollas; **T. latifolia** (L.) Hoffm.
(<u>Caucalis latifolia</u> L.) (<u>Greater Bur-parsley</u>) has (2)3-5
bracts and strongly zygomorphic corollas. Both are now
extremely rare casuals, formerly more common. **LASER** Borkh.
resembles <u>Tordylium</u> in its distinctive thick-winged, strongly
dorsally-compressed fruits, but differs in its 2-3-ternate
leaves and glabrous fruits. **L. trilobum** (L.) Borkh., from S
Europe, was formerly natd in Cambs. **FERULA** L. is an erect
perennial to 3m with yellow flowers and finely divided
leaves; it resembles <u>Foeniculum</u> but differs in its lack of
distinctive smell and in its linear, not filiform, leaf-
segments. A few plants of **F. communis** L. (<u>Giant Fennel</u>),
from Mediterranean, were established in Northants from 1956
to 1988 but then built over; 1 plant appeared in W Suffolk in
1988.

SUBFAMILY 1 - **HYDROCOTYLOIDEAE** (genus 1). Leaves simple, $\pm$
orbicular, not or shallowly lobed, stipulate; flowers in
simple axillary umbels, often with whorls below; fruit with a
woody inner wall, but no carpophore and no oil-bodies;
chromosome base-number 8.

1. HYDROCOTYLE L. - <u>Pennyworts</u>
Perennials with thin procumbent rooting stems; leaves on
thin, usually erect petioles as long as to much longer than
axillary peduncles; flowers very small, dull; fruits strongly

laterally compressed, ± orbicular in lateral view.

Other spp. - **H. sibthorpioides** Lam., from the tropics, resembles a small H. moschata (or Sibthorpia europaea vegetatively) but with glabrous stems, patent-pubescent petioles 3-10(30)mm, and glabrous leaves 3-12(20)mm lobed 1/3 to 1/2 way to base; it has survived between paving slabs in a school garden in Staffs since c.1970.

1 Leaves peltate, glabrous; fruits wider than long
                                                    1. H. vulgaris

1 Leaves not peltate, usually pubescent, with deep
   basal sinus; fruits longer than wide   2
     2 Fruits ≥2 x 1.8mm; leaves crenate
                                        3. H. novae-zeelandiae
     2 Fruits ≤1.6 x 1.5mm; leaves serrate   **2. H. moschata**

  1. H. vulgaris L. - Marsh Pennywort. Very variable in size; 589 stems to 30cm, but usually much less; petioles 1-25cm; leaves peltate, glabrous, 8-35mm across; fruit 1.5-1.8 x 1.8-2.3mm, glabrous, with thin low ridges. Native; in bogs, fens and marshes and at sides of lakes; locally common throughout BI.
  2. H. moschata G. Forster - Hairy Pennywort. Similar to 589 small-sized H. vulgaris; differs in pubescent stems; more deeply lobed (up to 2/5 to base) leaves 5-20mm across; fruits 1.3-1.6 x 1-1.5mm; and see key. Intrd; in lawns and on grassy banks; well natd on Valencia Island, S Kerry, less so in E Sussex and Ayrs; New Zealand.
  3. H. novae-zeelandiae DC. (H. microphylla auct. non Cunn.) 589 - New Zealand Pennywort. Differs from H. moschata in fruit 2-3.5 x 1.8-2.5mm; and leaves crenate. Intrd; on lawns and golf-courses, natd in turf; W Cornwall and Angus; New Zealand.

SUBFAMILY 2 - SANICULOIDEAE (genera 2-4). Leaves mostly simple, palmately or ternately lobed, rarely pinnately lobed and then spiny, exstipulate; flowers in simple umbels or capitula; fruit with soft inner wall and with oil-bodies, but without carpophore; chromosome base-number 7 or 8.

2. SANICULA L. - Sanicle
± glabrous perennials; most leaves in basal rosette; leaves palmately lobed almost to base; petioles long; umbels with pedicellate male and sessile bisexual flowers; bracteoles several, shorter than flowers; fruits laterally compressed, covered with long, hooked bristles; sepals persistent.

  1. S. europaea L. - Sanicle. Stems erect, to 40(60)cm; 589 leaf-lobes usually 5, shallowly lobed and toothed; flowers white or pinkish-white, in small, loosely-aggregated umbels; fruits 2.2-3mm. Native; in deciduous woods on leaf-mould; locally common throughout Br and Ir.

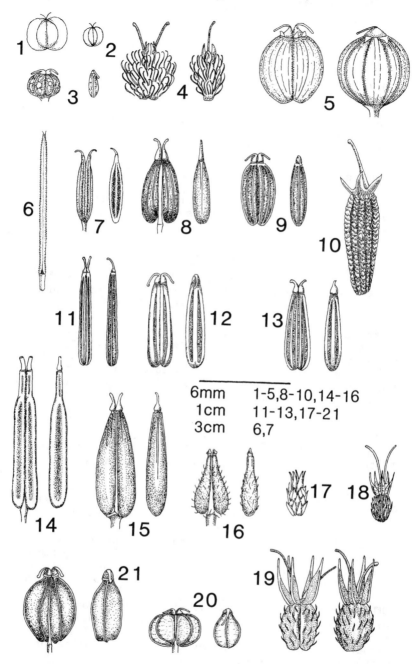

FIG 589 – Fruits (mostly with lateral and dorsal views) of Apiaceae. 1, **Hydrocotyle novae-zeelandiae**. 2, **H. moschata**. 3, **H. vulgaris**. 4, **Sanicula**. 5, **Coriandrum**. 6, **Scandix**. 7, **Myrrhis**. 8, **Conopodium**. 9, **Bunium**. 10, **Astrantia**. 11, **Chaerophyllum hirsutum**. 12, **C. aureum**. 13, **C. temulum**. 14, **Anthriscus cerefolium**. 15, **A. sylvestris**. 16, **A. caucalis**. 17, **Eryngium planum**. 18, **E. campestre**. 19, **E. maritimum**. 20, **Smyrnium perfoliatum**. 21, **S. olustrum**.

3. ASTRANTIA L. - <u>Astrantia</u>

± glabrous perennials; most leaves in basal rosette; leaves ternately or palmately lobed almost to base; petioles long; umbels with pedicellate male and bisexual flowers; bracteoles numerous, lanceolate to oblanceolate, at least as long as flowers; fruits scarcely compressed, covered with swollen scales; sepals persistent.

Other spp. - Some other spp. now supplant <u>A.</u> <u>major</u> in gardens and may occur as escapes in future; the most common is **A. maxima** Pallas, from Caucasus, with larger umbels and bracts, less deeply divided leaves, and fruits 5-6mm.

1. A. major L. - <u>Astrantia</u>. Stems erect, to 80(100)cm; **589** leaf-lobes 3-5, scarcely lobed, toothed; flowers and bracteoles whitish tinged with green and/or pink, in umbels up to 3.8cm across; fruits much longer than wide, 5-7mm. Intrd; garden escape natd in grassy and shady places; scattered in Br, mostly in N & W; Europe.

4. ERYNGIUM L. - <u>Sea-hollies</u>

± glabrous perennials; basal leaves petiolate, simple to ± pinnate; stem-leaves various, at least the upper spiny; flowers sessile, in ± globose to ovoid capitula with ± leaf-like spiny bracts at base; bracteoles 3-lobed or entire, with 1-3 spines, longer than flowers; fruits scarcely compressed, densely scaly or bristly; sepals persistent.

1   Basal leaves and lower stem-leaves pinnate or
    pinnately lobed almost to midrib   **4. E. campestre**
1   Basal leaves and lower stem-leaves not lobed or lobed
    ≤ 1/2 way to midrib   2
    2 Upper stem-leaves and bracts palmate or palmately
       lobed almost to base   **2. E. planum**
    2 Upper stem-leaves toothed, or lobed <1/2 way to
       base   3
3   Basal leaves and at least lower stem-leaves shallowly
    toothed, not spiny   **1. E. giganteum**
3   Basal leaves and stem-leaves with deep, strongly
    spiny teeth   **3. E. maritimum**

Other spp. - Several other spp. are grown for ornament and may persist for a while, but rarely or never become established. Most common is **E. amethystinum** L., from SE Europe, with capitula and bracts similar to those of <u>E.</u> <u>planum</u> but with all leaves pinnate.

1. E. giganteum M. Bieb. - <u>Tall Eryngo</u>. Stems erect, to 1.2m; basal leaves triangular-ovate, cordate at base, serrate; upper stem-leaves and bracts sharply toothed, scarcely spiny; capitula grey to bluish, oblong-ovate, 1-2.5cm across. Intrd; grown for ornament, well natd on waste ground near Otley, MW Yorks; Caucasus.

2. E. planum L. - <u>Blue</u> <u>Eryngo</u>. Stems erect, to 50(100)cm; **589** basal leaves oblong-ovate, cordate at base, serrate; upper stem-leaves deeply lobed, spiny; bracts linear-lanceolate; capitula blue, ovoid, c.1cm across; fruits densely scaly. Intrd; grown for ornament and occasionally natd in waste places; scattered in S Br, notably on sandy ground at Littlestone, E Kent since before 1965; SE Europe.

3. E. maritimum L. - <u>Sea-holly</u>. Stems erect, to 60cm; basal **589** leaves ovate, variously lobed, truncate to cordate at base, strongly spiny; stem-leaves similar but sessile; bracts ovate; capitula pale blue, ovoid, c.2cm across; fruits with hooked bristles. Native; on maritime sand and shingle; formerly around all coasts of BI, now gone from most Sc and NE En.

4. E. campestre L. - <u>Field</u> <u>Eryngo</u>. Stems erect, to 75cm; **589** basal leaves pinnate or almost so, strongly spiny; upper **RRR** stem-leaves deeply lobed, spiny; bracts linear-lanceolate; capitula green to brownish-green, ovoid, c.1cm across; fruits densely scaly. Probably native in Kent, doubtfully elsewhere; grassland or open places, especially calcareous, mostly near sea; very local in S & SW Br, scattered and mostly casual elsewhere in S & C Br and CI.

SUBFAMILY 3 - APIOIDEAE (genera 5-48).

Leaves various, often much divided, not spiny, exstipulate; flowers usually in compound umbels; fruit with soft inner wall, carpophore and usually oil-bodies; chromosome base-number various, often 11.

5. CHAEROPHYLLUM L. - <u>Chervils</u>

Biennials or perennials; stems solid; leaves 2-3-pinnate; bracts 0 or present, entire; bracteoles present; sepals $\pm$ 0; petals white or pink, actinomorphic; fruits slightly compressed laterally, >3x as long as wide, glabrous, with low, wide, rounded ridges.

1 Petals minutely ciliate; styles suberect, forming
  angle <45 degrees; petals usually pinkish
                                     1. C. hirsutum
1 Petals glabrous; styles $\pm$ divergent, forming angle
  >45 degrees; petals white   2
  2 Fruits 8-10mm; leaf-segments acute   **2. C. aureum**
  2 Fruits 4-6.5mm; leaf-segments obtuse or abruptly
    contracted to acute apex   **3. C. temulum**

1. C. hirsutum L. - <u>Hairy</u> <u>Chervil</u>. Erect, $\pm$ softly **589** pubescent perennial to 1m; basal leaves usually 3-pinnate, with rather abruptly acute segments; fruits 8-11mm, gradually tapered from about middle. Intrd; natd on grassy roadside verge since 1979, Westmorland; Europe.

2. C. aureum L. - <u>Golden</u> <u>Chervil</u>. Erect, softly to rather **589** roughly pubescent perennial to 1.2m; basal leaves 3-pinnate, with gradually acute segments; fruits 7-10mm, rather abruptly contracted near apex. Intrd; natd in grassy places; very scattered in N En and Sc, rare casual elsewhere in Br; C & S Europe.

3. C. temulum L. (_C. temulentum_ L.) - <u>Rough</u> <u>Chervil</u>. Erect, **589** roughly pubescent biennial to 1m; basal leaves 2-3-pinnate, with obtuse to abruptly acute segments; fruits 4-6.5mm, gradually tapered from about middle. Native; grassy places, hedgerows and wood-borders; common over much of BI but sparse in CI and Ir and absent from most of W & N Sc.

6. ANTHRISCUS Pers. - <u>Chervils</u>

Annuals to perennials; stems hollow; leaves 2-3-pinnate; bracts 0(-1); bracteoles present; sepals $\pm$ 0; petals white, actinomorphic; fruits slightly compressed laterally, >3x as long as wide, glabrous, ridged only near apex.

1 Fruits <5mm, with abundant hooked bristles
                                                          3. A. caucalis
1 Fruits >5mm, without bristles   2
   2 Annuals; rays pubescent; fruits with well
      differentiated beak 1-4mm   **2. A. cerefolium**
   2 Perennials; rays $\pm$ glabrous; fruits with scarcely
      differentiated beak $\leq$1mm   **1. A. sylvestris**

1. A. sylvestris (L.) Hoffm. - <u>Cow</u> <u>Parsley</u>. Erect, **589** pubescent perennial to 1.5m; basal leaves 3-pinnate, with acute segments; rays glabrous; fruits 6-10mm, glabrous, smooth, with scarcely differentiated beak $\leq$1mm. Native; grassy places, hedgerows and wood-margins; abundant throughout most of BI.

2. A. cerefolium (L.) Hoffm. - <u>Garden</u> <u>Chervil</u>. Erect to **589** spreading, sparsely pubescent annual to 70cm; leaves 2-3-pinnate, with subacute segments; rays pubescent; fruits 6-10mm, glabrous, smooth, with strongly differentiated beak 1-4mm. Intrd; formerly cultivated as herb, now rarely so, decreasing casual in waste and marginal places; very scattered in S & C Br, Guernsey; SE Europe.

3. A. caucalis M. Bieb. - <u>Bur</u> <u>Parsley</u>. Erect to decumbent, **589** sparsely pubescent annual to 70cm; leaves 2-3-pinnate, with subacute segments; rays glabrous; fruits 2.9-3.2mm, with very strongly differentiated beak <1mm. Native; waste places, open ground and open hedgerows on sandy or shingly soils, especially near sea; common in parts of E En, scattered over most of rest of BI except N & NW Sc.

7. SCANDIX L. - <u>Shepherd's-needle</u>

Annuals; stems hollow at fruiting; leaves 2-4-pinnate; bracts 0 or umbels simple; bracteoles present; sepals very small (<0.5mm) but persistent; petals white, actinomorphic; fruit $\pm$ not compressed, many times longer than wide, with beak 3-5x as long as seed-bearing part, glabrous, with wide low rounded ridges.

1. S. pecten-veneris L. - <u>Shepherd's-needle</u>. Very sparsely **589**
pubescent; stems usually erect, to 50cm; leaf-segments **RR**
linear; fruits 3-7cm, scabrid with tiny apically directed
bristles, with seed-bearing part c. as wide as beak, c.1cm.
Possibly native; weed of arable land and waste places;
formerly common in En, scattered over rest of BI, now rare
and decreasing.

8. MYRRHIS Miller - <u>Sweet Cicely</u>

Perennials smelling of aniseed when crushed; stems hollow;
leaves 2-4-pinnate; bracts 0; bracteoles present; sepals $\pm$ 0;
petals white, slightly zygomorphic; fruit slightly laterally
compressed, >3x as long as wide, glabrous or with minute
bristles, with acute well developed ridges.

1. M. odorata (L.) Scop. - <u>Sweet Cicely</u>. Softly pubescent; **589**
stems erect, to 1.8m; leaves with whitish patches, with acute
segments; fruits 15-25mm, contracted to short beak. Intrd;
banks, pathsides, waste and grassy places; common in much of
Br N from Derbys and Monts, rare in S Wa, S En and N & C Ir.

9. CORIANDRUM L. - <u>Coriander</u>

Annuals; stems solid; leaves simple to 3-pinnate; bracts
0-1(2), entire; bracteoles present; sepals conspicuous,
persistent; petals white to purplish, zygomorphic; fruits $\pm$
globose, glabrous, the mericarps not separating at maturity,
with low $\pm$ rounded ridges.

1. C. sativum L. - <u>Coriander</u>. Glabrous; stems erect, to **589**
50cm, $\pm$ unridged; basal leaves simple with 3 or more wide
lobes, or variously ternate to pinnate; cauline leaves 2-3-
pinnate, with linear lobes; fruits 2-6mm. Intrd; casual of
tips and waste places, mostly as bird-seed alien, now also a
small-scale crop for fruits or as green salad; scattered in
Br; E Mediterranean.

10. SMYRNIUM L. - <u>Alexanders</u>

Glabrous biennials (to perennials?); stems solid; leaves
simple to 3-pinnate or -ternate; bracts and bracteoles 0-few,
entire; sepals minute; petals yellow, actinomorphic; fruits
laterally compressed, not or scarcely longer than wide,
glabrous, with prominent sharp ridges.

1. S. olusatrum L. - <u>Alexanders</u>. Stems erect, to 1.5m; **589**
leaves 2-3-pinnate or -ternate, dark glossy green, with wide,
shallowly toothed lobes, all petiolate; stem-leaves with
expanded sheathing petioles; fruit 6.5-8mm, c. as long as
wide or slightly longer, blackish when ripe. Intrd; fully
natd on cliffs and banks, by roads and ditches, in waste
places, mostly near sea; common on coasts of BI N to C Sc,

very scattered inland; Europe.

2. S. perfoliatum L. - <u>Perfoliate</u> <u>Alexanders</u>. Stems erect, **589**
to 60(100)cm; leaves yellowish-green, the basal ones 2-3-
pinnate or -ternate; stem-leaves simple, deeply cordate,
sessile, strongly clasping stem, shallowly toothed; fruits
2-3.5mm, wider than long, dark brown when ripe. Intrd; natd
in grassy places and flower-borders; few places in SE En, 1
place in Co Durham; S Europe.

11. BUNIUM L. - <u>Great</u> <u>Pignut</u>
Glabrous perennials; stems solid, arising from subterranean ±
globose tuber; leaves 2-3-pinnate, with linear lobes; bracts
and bracteoles several, entire; sepals ± 0; petals white,
actinomorphic; fruits slightly laterally compressed, <2x as
long as wide, glabrous, with low, rounded ridges.

1. B. bulbocastanum L. - <u>Great</u> <u>Pignut</u>. Stems erect, to **589**
50(80)cm; leaves mostly withered by flowering time; fruits **RR**
3-4.5mm. Native; chalk grassland and banks; very local in
Herts, Bucks, Beds and Cambs.

12. CONOPODIUM Koch - <u>Pignut</u>
Glabrous perennials; stems hollow after flowering, arising
from subterranean ± globose tuber; leaves 2-3-pinnate, with
linear lobes; bracts 0(-2); bracteoles several; sepals 0;
petals white, actinomorphic; fruits slightly laterally
compressed, <2x as long as wide, glabrous, with low, rounded
ridges.

1. C. majus (Gouan) Loret - <u>Pignut</u>. Stems erect, to **589**
40(75)cm; basal leaves mostly withered by fruiting time;
fruits 3-4.5mm. Native; grassland, hedgerows, woods; common
± throughout BI.

13. PIMPINELLA L. - <u>Burnet-saxifrages</u>
Perennials; stems hollow or solid; leaves 1-2(3)-pinnate;
bracts and bracteoles 0; sepals 0; petals white or pinkish-
white, actinomorphic; fruits slightly laterally compressed,
slightly longer than wide, glabrous, with narrow low ridges.

1. P. major (L.) Hudson - <u>Greater</u> <u>Burnet-saxifrage</u>. Stems **595**
erect, to 1(2)m, hollow, ± glabrous; basal leaves simply
pinnate, with ovate, serrate lobes; fruits 3-4mm. Native;
grassland, hedgerows, wood-borders; locally common in C, E,
SE & SW En and S & WC Ir, rare casual in Wa and Sc, very rare
in Guernsey.
2. P. saxifraga L. - <u>Burnet-saxifrage</u>. Stems erect, to **595**
70(100)cm, solid, glabrous to densely puberulent; basal
leaves 1-2-pinnate, with linear to ovate-serrate lobes;
fruits 2-3mm. Native; grassland and open rocky places;
common throughout most of BI except parts of N & W Sc. Very
variable in dissection of basal leaves.

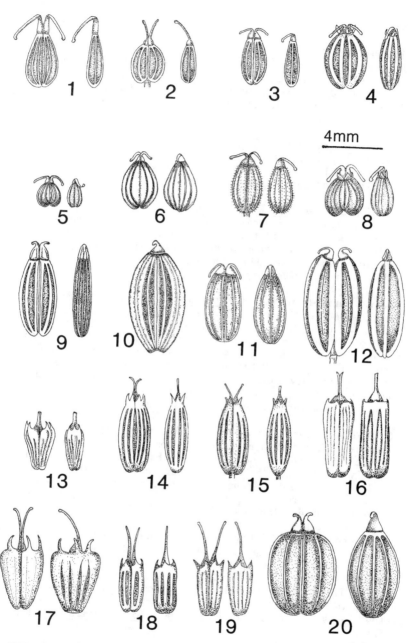

FIG 595 – Fruits (lateral and dorsal views) of **Apiaceae**. 1, **Aegopodium**. 2, **Pimpinella saxifraga**. 3, **P. major**. 4, **Sium**. 5, **Berula**. 6, **Aethusa**. 7, **Seseli**. 8, **Physospermum**. 9, **Foeniculum**. 10, **Anethum**. 11, **Silaum**. 12, **Meum**. 13, **Oenanthe lachenalii**. 14, **O. fluviatilis**. 15, **O. aquatica**. 16, **O. crocata**. 17, **O. fistulosa**. 18, **O. pimpinelloides**. 19, **O. silaifolia**. 20, **Crithmum**.

14. AEGOPODIUM L. - <u>Ground-elder</u>
Rhizomatous glabrous perennials; stems hollow; leaves 1-2-ternate with large, ovate, serrate lobes; bracts and bracteoles 0; sepals 0; petals white, ± actinomorphic; fruits laterally compressed, <2x as long as wide, glabrous, with narrow ridges.

1. A. podagraria L. - <u>Ground-elder</u>. Stems erect, to 1m; **595** rhizomes slender, far-creeping; fruits 3-4mm. Probably always intrd; waste places and cultivated and other open ground; common throughout BI; Europe.

15. SIUM L. - <u>Greater</u> <u>Water-parsnip</u>
Glabrous perennials; stems hollow; aerial leaves pinnate with ovate, serrate leaflets; bracts and bracteoles several, entire; sepals conspicuous, persistent; petals white, actinomorphic; fruits laterally compressed, distinctly longer than wide, glabrous, with thick, prominent ridges.

1. S. latifolium L. - <u>Greater</u> <u>Water-parsnip</u>. Stems erect, **595** to 2m; submerged leaves 2-3-pinnate in Spring; lower aerial **R** leaves with 3-8 pairs of leaflets each up to 15cm; fruits 2.5-4mm. Native; in ditches and fens; very local and decreasing in En and C Ir, now mostly in CE En, 1 recent record in Jersey. Often over-recorded for <u>Berula</u> <u>erecta</u>, but note strongly ridged fruits distinctly longer than wide and with persistent sepals; finely divided subaquatic leaves in Spring; and larger, more finely serrate leaflets.

16. BERULA Besser ex Koch - <u>Lesser</u> <u>Water-parsnip</u>
Glabrous stoloniferous perennials; stems hollow; lower leaves pinnate with ovate, serrate leaflets, if submerged then scarcely different; bracts and bracteoles several, often lobed; sepals distinct but not persistent; petals white, actinomorphic; fruits slightly laterally compressed, scarcely longer than wide, glabrous, with low, slender ridges.

1. B. erecta (Hudson) Cov. - <u>Lesser</u> <u>Water-parsnip</u>. Stems **595** decumbent to erect, to 1m; lower leaves with 5-10 pairs of leaflets each up to 6cm; fruits 1.3-2mm. Native; in and by water in ditches, marshes, lakes and rivers; frequent over much of Ir and Br N to C Sc. Told from <u>Apium</u> <u>nodiflorum</u> in vegetative state by presence of ring-mark on petiole some way below lowest pair of leaflets; see <u>Sium</u> <u>latifolium</u> for differences.

17. CRITHMUM L. - <u>Rock</u> <u>Samphire</u>
Glabrous perennials; stems solid, ± woody near base; leaves 2-3-pinnate, with linear, succulent lobes; bracts and bracteoles several, entire; sepals minute; petals yellowish-green, actinomorphic; fruits not compressed, each mericarp ± triangular in section, slightly longer than wide, spongy when

fresh, glabrous, with thick, prominent ridges.

1. C. maritimum L. - <u>Rock</u> <u>Samphire</u>. Stems erect to **595**
decumbent, to 45cm; fresh leaves smelling of furniture-polish
when crushed; fruits 3.5-5mm. Native; cliffs, rocks and less
often sand and shingle by sea; coasts of BI but N only to E
Suffolk on E coast.

18. SESELI L. - <u>Moon</u> <u>Carrot</u>
Puberulent biennial to monocarpic perennial; stems solid;
leaves 2-3-pinnate, with rather narrow lobes; bracts and
bracteoles numerous, entire; sepals small, sometimes ±
persistent; petals white, actinomorphic; fruits not
compressed, slightly longer than wide, puberulent, with
thick, prominent ridges.

1. S. libanotis (L.) Koch - <u>Moon</u> <u>Carrot</u>. Stems erect, to **595**
60cm, with dense sheath of fibres at base; fruits 2.5-3.5mm. **RR**
Native; grassland or rough ground on chalk; very local in E
Sussex, Herts, Cambs and Beds, and decreasing.

19. OENANTHE L. - <u>Water-dropworts</u>
Glabrous annuals to perennials, often with tuberous roots;
stems hollow or solid; leaves 1-4-pinnate; bracts 0-several,
entire; bracteoles usually numerous; sepals conspicuous,
persistent; petals white, ± actinomorphic to slightly zygomorphic; fruits not or very slightly dorsally or laterally
compressed, up to c.2.5x as long as wide, glabrous, with
obscure to rather prominent ridges.

1 Some umbels leaf-opposed, with peduncles shorter than
  rays; styles ≤1/4 as long as mature fruit   2
1 All umbels terminal, with peduncles longer than rays;
  styles >1/4 as long as mature fruit   3
  2 Fruit ≤4.5mm; stems ascending to erect, often
    terrestrial   **7. O. aquatica**
  2 Fruit ≥5mm; stems usually floating at least
    at base   **6. O. fluviatilis**
3 Ultimate clusters of ripe fruits globose; all fruits
  sessile; all leaves usually with petioles longer than
  divided part   **1. O. fistulosa**
3 Ultimate clusters of ripe fruits not globose; some
  fruits stalked; all leaves usually with petioles
  shorter than divided part   4
  4 Fruits ≥4mm; segments of mid stem-leaves ovate
    to ± orbicular, <2x as long as wide   **5. O. crocata**
  4 Fruits ≤3.5mm; segments of mid stem-leaves ±
    linear, >3x as long as wide   5
5 Bracts 0 at least on most umbels; rays ≥1mm thick
  at fruiting; stems at maturity hollow, straw-like,
  with walls c.0.5mm thick   **2. O. silaifolia**
5 Bracts (0)1-c.5; rays <1mm thick at fruiting; stems

at maturity solid to hollow with walls >0.5mm thick   6
6 Rays and pedicels thickened at fruiting, the
   pedicels >0.5mm thick; root-tubers ellipsoid,
   the proximal part of the root not thickened
                                  3. O. pimpinelloides
6 Rays and pedicels scarcely thickened at fruiting,
   the pedicels <0.5mm thick; root-tubers cylindrical
   to fusiform, gradually widening ± from base of
   root   **4. O. lachenalii**

1. O. fistulosa L. - <u>Tubular</u> <u>Water-dropwort</u>. Erect, stolon- 595
iferous perennial to 80cm; root-tubers fusiform; stems
hollow, with thin walls; leaves 1-3-pinnate, with ± linear
segments; rays and pedicels thickened in fruit; bracts 0;
fruits 3-3.5mm, with styles c. as long. Native; marshes,
ditches and other wet places; locally frequent throughout BI
N to C Sc, common in much of E En.

2. O. silaifolia M. Bieb. - <u>Narrow-leaved</u> <u>Water-dropwort</u>. 595
Erect perennial to 1m; root-tubers fusiform; stems hollow,   R
with thin walls; leaves 1-4-pinnate, with linear to narrowly
elliptic segments; rays and pedicels thickened in fruit;
bracts usually 0; fruits 2.5-3.5mm, with styles nearly as
long. Native; marshes, dykes and ditches; scattered and
decreasing in C & S En.

3. O. pimpinelloides L. - <u>Corky-fruited</u> <u>Water-dropwort</u>. 595
Erect perennial to 1m; root-tubers ellipsoid; stems solid   R
(with pith) to ± hollow, with thick walls; leaves 1-3-
pinnate, with linear to ovate segments; rays and pedicels
thickened in fruit; bracts 1-5; fruits 3-3.5mm, with styles
c. as long. Native; marshes, ditches and other wet ground;
local in En S from E Gloucs and E Kent, Mons, Co Clare,
formerly E Cork.

4. O. lachenalii C. Gmelin - <u>Parsley</u> <u>Water-dropwort</u>. Erect 595
perennial to 1m; root-tubers cylindrical to fusiform; stems
solid (with pith) to ± hollow, with thick walls; leaves 1-3-
pinnate, with linear (rarely to ovate) segments; rays and
pedicels not thickened in fruit; bracts (0)1-c.5; fruits 2.5-
3mm, with shorter styles. Native; ditches, marshes and
dykes, mostly near sea and often brackish; coasts of BI
except N & E Sc, scattered inland in En.

5. O. crocata L. - <u>Hemlock</u> <u>Water-dropwort</u>. Erect perennial 595
to 1.5m; root-tubers fusiform; stems hollow; leaves 1-3(4)-
pinnate, with mostly ovate segments; bracts 3-6; fruits
4-5.5mm, with styles c.1/2 as long. Native; ditches, pond-
sides and other wet places; locally common in BI, but absent
from much of C Ir and E Br.

6. O. fluviatilis (Bab.) Coleman - <u>River</u> <u>Water-dropwort</u>. 595
Erect to ascending or floating perennial to 1m; root-tubers 0   R
at maturity; stems hollow; leaves 1-3-pinnate, with linear to
ovate segments; bracts usually 0; fruits 5-6.5mm, with styles
<1/4 as long. Native; in slow rivers; scattered in S & E En,
C & E Ir, Denbs.

7. O. aquatica (L.) Poiret - <u>Fine-leaved Water-dropwort</u>. 595
Erect to ascending annual to biennial to 1.5m; root-tubers 0
at maturity; stems hollow; leaves 2-4-pinnate, with linear to
ovate segments; bracts usually 0; fruits 3-4.5mm, with styles
≤1/4 as long. Native; ditches and ponds, often drying up
in summer; scattered in Br and Ir N to SE Sc.

20. AETHUSA L. - <u>Fool's Parsley</u>
Glabrous annuals; stems hollow; leaves 2-3-pinnate; bracts
usually 0; bracteoles usually 3-4; sepals 0; petals white, ±
actinomorphic; fruits slightly dorsally compressed, slightly
longer than wide, glabrous, with prominent, wide, keeled
ridges.

1. A. cynapium L. - <u>Fool's Parsley</u>. Stems erect, to 595
1(1.5)m; bracteoles all on outer side of each flower-cluster,
strongly reflexed; fruits (2.5)3-4mm. Native; cultivated and
waste ground; throughout BI except much of C & N Sc.
  a. Ssp. cynapium. Stems to 1m; longest pedicels mostly <1/2
as long as bracteoles, mostly c.2x as long as fruits. The
common plant of gardens and waste places, also arable land.
  b. Ssp. agrestis (Wallr.) Dostal. Stems to 20cm; longest
pedicels c. as long as bracteoles, mostly shorter than
fruits. Arable land; distribution uncertain but mainly or
only in S Br and CI; probably intrd.
  Ssp. **cynapioides** (M. Bieb.) Nyman, a tall plant (to 1.5m)
with shorter bracteoles than ssp. <u>cynapium</u>, and terete (not
grooved) stems and narrower leaf-segments, has been recorded
from Jersey and W Kent, and might become established.

21. FOENICULUM Miller - <u>Fennel</u>
Glabrous perennials smelling strongly of aniseed; stems solid
at first, becoming ± hollow; leaves 3-4-pinnate, with long
filiform segments; bracts 0; bracteoles 0; sepals 0; petals
yellow, actinomorphic; fruits scarcely compressed, c.2-4x as
long as wide, glabrous, with prominent thick ribs.

1. F. vulgare Miller - <u>Fennel</u>. Stems erect, ± glaucous, to 595
2.5m; fruits often not developing or developing very late in
season, 4-5mm. Probably intrd; open ground and waste places,
especially near the coast; well natd in BI N to N Ir, Man and
Yorks, scarce and mainly casual further N; Europe. Nowadays
rarely grown on field-scale.

22. ANETHUM L. - <u>Dill</u>
Glabrous annuals smelling strongly of aniseed; stems hollow;
leaves as in <u>Foeniculum</u>; bracts 0; bracteoles 0; sepals 0;
petals yellow, actinomorphic; fruits strongly dorsally
compressed, c.2x as long as wide, glabrous, with prominent
slender dorsal ridges and conspicuously winged lateral ones.

1. A. graveolens L. - <u>Dill</u>. Stems erect, to 60cm; differs 595 from Foeniculum in annual habit; completely hollow stems; and totally different fruits 3-5mm. Intrd; rather frequent casual from bird-seed or grain in waste and cultivated ground; scattered in Br; W & C Asia.

23. SILAUM Miller - Pepper-saxifrage
Glabrous perennials; stems solid; leaves 1-4-pinnate; bracts 0-3, entire; bracteoles numerous; sepals 0; petals yellowish, actinomorphic; fruits scarcely compressed, c.2x as long as wide, glabrous, with prominent slender ridges.

1. S. silaus (L.) Schinz & Thell. - Pepper-saxifrage. Stems 595 erect, to 1m; leaf-segments linear to lanceolate; fruits 4-5mm. Native; grassy places; locally frequent in Br N to C Sc, mainly in E.

24. MEUM Miller - Spignel
Glabrous perennials; stems hollow; leaves 3-4-pinnate; bracts 0-few, entire; bracteoles several; sepals 0; petals white to pinkish, actinomorphic; fruits scarcely compressed, <2x as long as wide, glabrous, with prominent, rather narrow ridges.

1. M. athamanticum Jacq. - Spignel. Plant sweetly aromatic; 595 stems erect, to 60cm, with dense sheath of fibres at base; **R** most leaves basal, with filiform segments; fruits (4)5-7(10)mm. Native; mountain grassland; local in N En, N Wa, C & S Sc.

25. PHYSOSPERMUM Cusson ex A.L. Juss. - Bladderseed
Almost glabrous perennials; stems solid; basal leaves 2-ternate; stem-leaves simple to 1-ternate; bracts and bracteoles several, entire; sepals conspicuous, persistent; petals white, actinomorphic; fruits ± inflated, wider than long, glabrous, with narrow low ridges.

1. P. cornubiense (L.) DC. - Bladderseed. Stems erect, to 595 1.2m; leaves mostly basal, with ± ovate, deeply serrate **RR** segments; fruits 2.5-4mm. Native; arable fields, hedgebanks, scrub and woods; very local in E Cornwall and S Devon, natd in Bucks.

26. CONIUM L. - Hemlock
Glabrous biennials; stems hollow; leaves 2-4-pinnate; bracts and bracteoles several, entire; sepals 0; petals white, actinomorphic; fruits scarcely compressed, c. as wide as long, glabrous, with very prominent, narrow, ± undulate ridges.

1. C. maculatum L. - Hemlock. Stems erect, to 2m, usually 603 purple-spotted; leaves with ovate, deeply serrate segments; fruits 2-3.5mm. Native; damp ground, roadside banks, ditches, waste ground; common over most of BI except W & C

Sc.

27. BUPLEURUM L. - Hare's-ears

Annual to perennial herbs, rarely shrubs; stems hollow or solid; leaves simple, entire; bracts 0 to several, entire; bracteoles several; sepals 0; petals yellow, actinomorphic; fruits slightly laterally compressed, 1-1.5x as long as wide, glabrous, sometimes papillose, strongly to scarcely ridged.

```
1  Upper leaves fused right around stem; bracts 0              2
1  Leaves not fused around stem; bracts present                3
   2  Umbels with 4-8 rays; fruit smooth between
      ridges; leaves mostly <2x as long as wide
                                       5. B. rotundifolium
   2  Umbels with 2-3 rays; fruit strongly papillose
      between ridges; leaves mostly >2x as long as
      wide                              6. B. subovatum
3  Firmly rooted perennials; at least some leaves often
   >1cm wide                                                   4
3  Easily uprooted annuals; leaves all <1cm wide              5
   4  Shrubs; leaves with strong midrib and many lateral
      veins                             1. B. fruticosum
   4  Herbs, often woody at extreme base; leaves with >3
      equally strong ± parallel main veins  2. B. falcatum
5  Bracteoles linear, not concealing flowers or fruits;
   fruit tuberculate                     3. B. tenuissimum
5  Bracteoles lanceolate, concealing flowers and fruits;
   fruits smooth                         4. B. baldense
```

Other spp. - **B. fontanesii** Guss. ex Caruel, from Mediterranean region, is an annual somewhat resembling B. baldense but with much larger umbels, 5-7 rays, and bracteoles with conspicuous, abruptly recurved cross-veins; it is a rare bird-seed alien.

1. B. fruticosum L. - _Shrubby Hare's-ear_. Shrub to 2.5m, **603** but often much less due to annual frost-damage; leaves oblong to narrowly obovate, with 1 main vein, most >4 x 1.5cm; bracts several but often fallen before fruiting, ovate to obovate, shorter than rays; rays 5-many; fruits 4.5-6mm, with prominent, slender ridges, smooth. Intrd; grown in gardens, natd on roadside and railway banks since before 1909; E Suffolk, W Kent, S Devon and Worcs; S Europe.

2. B. falcatum L. - _Sickle-leaved Hare's-ear_. Herbaceous ± **603** erect perennial to 1m; leaves oblanceolate to linear-falcate, **RRR** with 5 main veins, the lower usually >5 x 1cm; bracts several, linear to lanceolate, shorter than rays; rays 5-c.11; fruits 2.5-3.5mm, with slender, prominent ridges, smooth. Possibly native; damp roadsides, ditches, hedgebanks and field-borders; 1 locality in S Essex, discovered 1831, ± eradicated 1962 but refound 1979.

3. B. tenuissimum L. - _Slender Hare's-ear_. Erect to **603**

procumbent slender annual to 50cm; leaves linear to very   R
narrowly oblanceolate, with 3 main veins, the lower mostly
>2cm x <5mm; bracts 3-5, often longer than shortest rays, ±
linear; rays 1-3, unequal; fruit 1.5-2mm, with slender
ridges, tuberculate. Native; grassy or barish brackish
ground; coasts of S Br from Glam to Durham, formerly rare
inland.

4. B. baldense Turra - <u>Small Hare's-ear</u>. Erect annual to **603**
25cm, usually much <10cm; leaves linear to very narrowly **RRR**
oblanceolate, with 3-5 main veins, the lower mostly >2cm x
<0.3mm; bracts c.4, longer than shortest rays, lanceolate;
rays 1-4, unequal; fruits 1.5-2.5mm, scarcely ridged, smooth.
Native; barish ground on fixed dunes and cliffs by sea; very
local in CI, S Devon and E Sussex, rare casual elsewhere.
Often very dwarfed and superficially resembling <u>Euphorbia</u>
<u>exigua</u>, with which it sometimes grows.

5. B. rotundifolium L. - <u>Thorow-wax</u>. Erect annual to 30cm; **603**
leaves elliptic to suborbicular, 2-6 x 1.5-4cm, with >5 main
veins; bracts 0; bracteoles ovate, longer than flowers and
fruits; fruits 3-3.5mm, with slender prominent ridges,
smooth. Intrd; formerly common in cornfields in most of En,
especially C & S, extinct since 1960s; Europe.

6. B. subovatum Link ex Sprengel (<u>B. lancifolium</u> auct. non **603**
Hornem., <u>B. intermedium</u> (Lois. ex DC.) Steudel) - <u>False</u>
<u>Thorow-wax</u>. Differs from <u>B. rotundifolium</u> in key charac-
ters; leaves ovate to narrowly so, 2.5-6 x 1.5-2.8cm; fruits
3.5-5mm. Intrd; common bird-seed alien over much of BI;
Mediterranean. Very commonly mistaken for <u>B. rotundifolium</u>.

28. TRINIA Hoffm. - <u>Honewort</u>
Glabrous, dioecious biennials to monocarpic perennials; stems
solid; leaves 1-3-pinnate; bracts 0-1, 3-lobed; bracteoles 0-
several, entire to 3-lobed; sepals 0; petals white, actinomorphic; fruits scarcely compressed, <1.5x as long as wide,
glabrous, with prominent, wide ridges.

1. T. glauca (L.) Dumort. - <u>Honewort</u>. Stems erect, to 20cm, **603**
with dense sheath of fibres at base; female plants with **RR**
longer, more unequal rays and fewer, longer-pedicelled
flowers than male plants; fruits 2.3-3mm. Native; dry limestone turf; very local in S Devon, N Somerset and W Gloucs.

29. CUMINUM L. - <u>Cumin</u>
Glabrous annuals; stems solid; leaves 2-ternate; bracts 2-4,
entire to 3-lobed with long filiform lobes; bracteoles
usually 3; sepals conspicuous, persistent; petals white to
pinkish, actinomorphic; fruits slightly dorsally compressed,
c.2x as long as wide, glabrous or more usually bristly-
pubescent, with prominent, narrow ridges.

1. C. cyminum L. - <u>Cumin</u>. Stems erect, to 50cm; leaf-lobes **603**
filiform, up to 5cm; rays 1-5; fruits 4-6mm. Intrd;

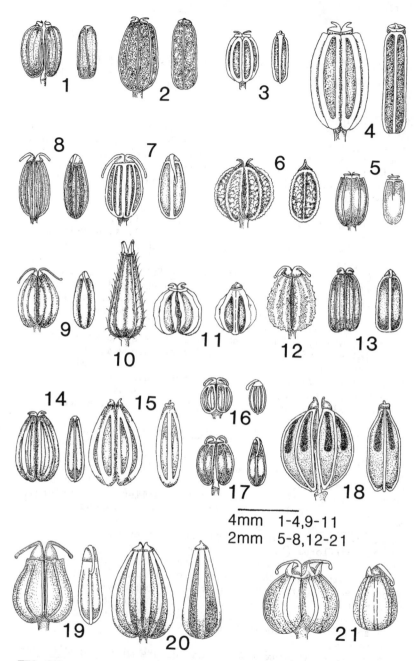

FIG 603 – Fruits (mostly with lateral and dorsal views) of Apiaceae. 1, **Bupleurum rotundifolium**. 2, **B. subovatum**. 3, **B. falcatum**. 4, **B. fruticosum**. 5, **B. baldense**. 6, **B. tenuissimum**. 7, **Ammi visnaga**. 8, **A. majus**. 9, **Trinia**. 10, **Cuminum**. 11, **Conium**. 12, **Trachyspermum**. 13, **Ridolfia**. 14, **Apium nodiflorum**. 15, **A. inundatum**. 16, **A. repens**. 17, **A. graveolens**. 18, **Sison**. 19, **Petroselinum crispum**. 20, **P. segetum**. 21, **Cicuta**.

increasingly frequent casual from bird-seed and from use as spice, on tips and in waste places; scattered in S En; N Africa, SW Asia.

30. APIUM L. - <u>Marshworts</u>
Glabrous biennials to perennials; stems hollow or solid; leaves pinnate or lower ones 2-3-pinnate; bracts and bracteoles 0-several, entire; sepals 0; petals white, actino-morphic; fruits laterally compressed, wider than long to longer than wide, glabrous, with prominent, slender to thick ridges.

1   Bracts and bracteoles 0; fresh plant smelling of
    celery   **1. A. graveolens**
1   Bracts 0-7; bracteoles 3-7; fresh plant not smelling
    of celery   2
    2   Lower leaves 2-3-pinnate, with ± filiform segments
        if submerged; styles shorter than stylopodium in
        fruit   **4. A. inundatum**
    2   All leaves 1-pinnate, even if submerged; styles
        longer than stylopodium in fruit   3
3   Bracts 0(-2); peduncles shorter than rays and adjacent
    petioles; leaflets distinctly longer than wide; fruits
    longer than wide   **2. A. nodiflorum**
3   Bracts (1)3-7; peduncles longer than rays and
    adjacent petioles; leaflets c. as long as wide;
    fruits slightly wider than long   **3. A. repens**

1. A. graveolens L. - <u>Wild Celery</u>. Usually erect biennial 603 to 1m; stems solid; leaves 1-pinnate, rarely 1-ternate or 2-pinnate, the basal ones with stalked leaflets; fruits 1-1.5mm, with slender ridges. Native; damp barish usually brackish places usually near sea; coasts of BI N to S Sc, very scattered inland. Var. **dulce** (Miller) DC. (<u>Celery</u>), with fat, ± succulent petioles, and var. **rapaceum** (Miller) DC. (<u>Celeriac</u>), with greatly swollen stem-bases, are both cultivated as vegetables but rarely or never escape.
   2. A. nodiflorum (L.) Lag. - <u>Fool's Water-cress</u>. Stems to 603 1m, suberect to procumbent, rooting at lower nodes, hollow; leaves 1-pinnate, with sessile lobes distinctly longer than wide; bracts 0(2); peduncles shorter than rays; rays 3-15; fruits 1.5-2.5mm, longer than wide, with thick ridges. Native; ditches, marshes and by lakes and rivers; common in BI N to S Sc, very local in C & N Sc. See <u>Berula erecta</u> for differences.
   2 x 3. A. nodiflorum x A. repens occurs with <u>A. nodiflorum</u> in some of the Oxon localities of <u>A. repens</u>, and in Cambs, SE Yorks, Fife and perhaps elsewhere where <u>A. repens</u> has become extinct; it usually has a more procumbent and extensively rooting stem than <u>A. nodiflorum</u>, leaflets longer than wide, peduncles shorter than to c. as long as rays, and 1-4 bracts; sterile; ?endemic.

2 x 4. A. nodiflorum x A. inundatum = A. x moorei (Syme)
Druce occurs with the parents scattered over most of Ir and
locally in CE En; all leaves 1-pinnate with obovate leaflets;
rays usually 2-3; sterile; endemic.
3. A. repens (Jacq.) Lag. - Creeping Marshwort. Differs **603**
from dwarfest plants of A. nodiflorum in stems procumbent, **RRR**
rooting at ± all nodes; fruits 0.7-1mm, with narrow ridges;
and see key. Native; open wet places; only in 4 localities
in Oxon in recent years, and even there possibly no longer in
pure state, formerly scattered in E En and C & S Sc.
4. A. inundatum (L.) H.G. Reichb. - Lesser Marshwort. Stems **603**
decumbent to procumbent, to 50cm, hollow, often largely submerged, rooting at lower nodes; lower leaves 2-3-pinnate,
with narrow segments; bracts 0; bracteoles 3-6; rays 2(-4);
fruits 2.5-3mm, with thick ridges. Native; in still, usually
shallow water and on bare mud nearby; scattered over most BI.

31. TRACHYSPERMUM Link - Ajowan
Almost glabrous annuals; stems hollow; leaves mostly 2-
pinnate; bracts several, entire or lobed; bracteoles several;
sepals small, not persistent; petals white, actinomorphic;
fruits laterally compressed, slightly longer than wide,
densely tuberculate, with rather obscure, wide ridges.

1. T. ammi (L.) Sprague - Ajowan. Stems erect, to 30cm; **603**
leaves with filiform segments to 2cm; fruits 1.5-2mm, very
characteristically tuberculate. Intrd; casual on tips and
waste ground from bird-seed and use as spice, perhaps
increasing; scattered in S En; E Mediterranean.

32. PETROSELINUM Hill - Parsleys
Glabrous annuals or biennials; stems solid; leaves 1-3-
pinnate, with wide lobes; bracts 1-several, entire or lobed;
bracteoles several; sepals 0 or very small; petals white or
yellow, actinomorphic; fruits laterally compressed, somewhat
longer than wide, glabrous, with prominent thick to narrow
ridges.

1. P. crispum (Miller) Nyman ex A.W. Hill - Garden Parsley. **603**
Erect biennials to 75cm, with characteristic parsley smell;
lower leaves 3-pinnate, shiny; bracts 1-3, often lobed or
leaf-like; petals yellow; fruits 2-2.5mm, with styles ± as
long as stylopodium. Intrd; commonly grown on small scale
and frequent escape on tips and in waste places; scattered in
BI except most of C & N Sc; E Mediterranean.
2. P. segetum (L.) Koch - Corn Parsley. Erect annuals or **603**
biennials to 1m; lower leaves 1-pinnate, with 4-12 pairs of
leaflets, matt; bracts 2-5, sometimes lobed; petals white;
fruits 2.3-3mm, with styles much shorter than stylopodium.
Native; barish or grassy places in arable fields, pastures
and hedgerows and on banks; local in S Br N to SE Yorks and
Cards, Jersey.   Easily mistaken for Sison, but with

extremely unequal lengthed rays, more leaflets, thicker
ridges on fruits, and without the characteristic smell.

33. RIDOLFIA Moris - <u>False</u> <u>Fennel</u>
Glabrous annuals; stems solid; leaves 3-4-pinnate, with
filiform segments; bracts 0; bracteoles 0; sepals 0; petals
yellow, actinomorphic; fruits laterally compressed, longer
than wide, glabrous, with slender, rather low ridges.

1. R. segetum (Guss.) Moris - <u>False</u> <u>Fennel</u>. Stems erect, to **603**
1m; leaves with filiform segments, resembling those of
<u>Anethum</u> but lacking aniseed smell; fruits 1.5-2.5mm. Intrd;
rather infrequent bird-seed alien on tips and waste ground;
very scattered in S En; Mediterranean.

34. SISON L. - <u>Stone Parsley</u>
Glabrous biennials, stems solid; leaves 1-2-pinnate, the
lower with wide lobes; bracts 2-4 and bracteoles 2-4, entire;
sepals 0; petals white, actinomorphic; fruits laterally
compressed, somewhat longer than wide, glabrous, with narrow
prominent ridges.

1. S. amomum L. - <u>Stone Parsley</u>. Fresh plant smelling **603**
rather like petrol when crushed; stems erect, to 1m; lower
leaves 1-pinnate, with 2-5 pairs of leaflets; most rays of ±
same length, often 1 much shorter; fruits 1.5-3mm. Native;
hedgebanks, grassland, roadsides; locally frequent in Br N to
Cheshire and SE Yorks, rare casual further N.

35. CICUTA L. - <u>Cowbane</u>
Glabrous perennials; stems hollow; leaves 2-3-pinnate; bracts
0; bracteoles numerous; sepals conspicuous, persistent;
petals white, actinomorphic; fruits laterally compressed but
each mericarp ± globose, wider than long, glabrous, with wide
inconspicuous ridges.

1. C. virosa L. - <u>Cowbane</u>. Stems erect, to 1.5m; leaves **603**
with narrowly elliptic to linear-lanceolate segments, up to   **R**
5(9)cm; fruits 1.2-2mm. Native; ditches, marshy fields,
pond-sides; very local in N & C Ir, E Anglia, WC En, very
rare (formerly less so) elsewhere in BI.

36. AMMI L. - <u>Bullworts</u>
Glabrous annuals to biennials; stems solid; leaves 1-3(4)-
pinnate; bracts several, mostly pinnately divided with linear
to filiform lobes; bracteoles numerous; sepals 0; petals
white, very slightly zygomorphic; fruits slightly laterally
compressed; somewhat longer than wide, glabrous, with slender
prominent ridges.

1. A. majus L. - <u>Bullwort</u>. Stems erect, to 1m; lower leaves **603**
1-2-pinnate, with elliptic to narrowly elliptic, serrate

lobes; upper leaves 2(3)-pinnate; rays remaining slender and bracts not becoming strongly reflexed in fruit; fruits 1.5-2mm. Intrd; rather infrequent casual mainly from bird-seed and wool on tips and waste ground and in fields; very scattered in CI, C & S Br; S Europe.

2. **A. visnaga** (L.) Lam. - <u>Toothpick-plant</u>. Stems erect, to **603** 1m; lower leaves 1-2-pinnate; upper leaves 2-3(4)-pinnate; all leaves with linear to filiform segments; rays becoming rigid, thick and erect, and bracts becoming strongly reflexed, in fruit; fruits 2-2.8mm. Intrd; casual in S En from same sources as <u>A. majus</u>, but rarer; Mediterranean.

37. FALCARIA Fabr. - <u>Longleaf</u>

Glabrous, glaucous, perennial; stems solid; leaves 1-2-ternate; bracts and bracteoles numerous, entire; sepals conspicuous, persistent; petals white, actinomorphic; fruits laterally compressed, 2-4x as long as wide, glabrous, with wide, low ridges.

1. **F. vulgaris** Bernh. - <u>Longleaf</u>. Stems erect, to 60cm; **611** leaf-segments c.10-30cm, linear-lanceolate, serrate; fruits 2.5-4(5)mm. Intrd; natd in grassy and waste places and scrub; very local in CI and S & C Br, especially E Anglia; Europe.

38. CARUM L. - <u>Caraways</u>

Glabrous biennials or perennials; stems hollow; leaves 1-3-pinnate; bracts and bracteoles 0-numerous; sepals 0 or minute, not persistent; petals white, actinomorphic or nearly so; fruits laterally compressed, 1.2-1.7x as long as wide, glabrous, with narrow, ± prominent ridges.

1. **C. carvi** L. - <u>Caraway</u>. Erect biennial to 60cm, without **611** fibres at base; leaves 2-3-pinnate, the lower with linear to lanceolate segments; bracts and bracteoles 0-few, the former sometimes ± leaf-like; fruits 3-4mm, with distinctive smell when crushed. Intrd; natd in fields, roadsides and waste places, sparsely scattered throughout BI, cultivated as flavouring less than formerly; Europe.

2. **C. verticillatum** (L.) Koch - <u>Whorled Caraway</u>. Erect **611** perennial to 60cm, with dense sheath of fibres at base; leaves 1-pinnate, each leaflet deeply palmately divided to base into filiform segments; bracts and bracteoles numerous, entire; fruits 2-3mm. Native; marshes, damp meadows and stream-sides; locally frequent in SW & N Ir, W parts of Br from Cornwall to Westerness, Jersey, very rare elsewhere.

39. SELINUM L. - <u>Cambridge Milk-parsley</u>

Glabrous perennials; stems solid; leaves 2-3-pinnate; bracts 0-few, entire, soon falling; bracteoles numerous; sepals ± 0; petals white, actinomorphic; fruits dorsally compressed, c. as long as wide, glabrous, with conspicuous winged ridges.

1. S. carvifolia (L.) L. - <u>Cambridge</u> <u>Milk-parsley</u>. Stems **611** erect, to 1m; lower leaves with lanceolate to ovate, deeply **RRR** lobed segments <10mm; fruits 3-4mm. Native; fens and damp meadows; 3 localities in Cambs, formerly Notts and N Lincs. See <u>Peucedanum palustre</u> for differences.

40. LIGUSTICUM L. - <u>Scots</u> <u>Lovage</u>
Glabrous perennials; stems hollow; leaves 1-2-ternate, with stalked, wide leaflets; bracts 1-several, entire; bracteoles several; sepals conspicuous, persistent; petals greenish-white, actinomorphic; fruits strongly dorsally compressed, c.2x as long as wide, glabrous, with prominent, ± winged ridges.

1. L. scoticum L. - <u>Scots</u> <u>Lovage</u>. Stems erect, to 60(90)cm; **611** leaves bright green, with ovate-trullate, serrate segments >2 x 1.5cm; fruits 4-7mm. Native; cliffs and rocky places near sea; frequent around whole coast of Sc, local in N & W Ir, formerly Cheviot.

41. ANGELICA L. - <u>Angelicas</u>
Puberulent to ± glabrous perennials; stems hollow; leaves 2-3-pinnate, the upper with very strongly inflated petioles; bracts 0-few, entire, soon falling; bracteoles numerous; sepals very small; petals white, greenish-white or pinkish-white, actinomorphic; fruits strongly dorsally compressed, slightly longer than wide, glabrous, with low dorsal and conspicuously winged lateral ridges.

1. A. sylvestris L. - <u>Wild</u> <u>Angelica</u>. Stems erect, to 2.5m, **611** often much less, usually somewhat purplish; leaflets ovate, abruptly acute, closely and finely serrate; peduncles and rays puberulent; petals white to pinkish-white; fruits 4-5mm. Native; damp grassy places, fens, marshes, by streams, ditches and ponds, in damp open woods; common throughout BI.
2. A. archangelica L. - <u>Garden</u> <u>Angelica</u>. Stems erect, to **611** 2m, usually green; leaflets narrowly ovate, acute-acuminate, coarsely but sharply serrate; peduncles glabrous; rays puberulent; petals greenish-white; fruits 5-6mm. Intrd; natd on river-banks and waste places, now cultivated less than formerly; scattered in Br, ± frequent in London area; N & E Europe.

42. LEVISTICUM Hill - <u>Lovage</u>
Almost glabrous perennials smelling of celery when crushed; stems hollow; leaves 2-3-pinnate; bracts and bracteoles numerous, entire, the latter often fused at base; sepals ± 0; petals yellow, actinomorphic; fruits strongly dorsally compressed, <2x as long as wide, glabrous; with very prominent dorsal and winged lateral ridges.

1. L. officinale Koch - <u>Lovage</u>. Stems erect, to 2.5m; **611**

leaves with trullate to rhombic, sparsely but deeply serrate leaflets; fruits 4-7mm. Intrd; natd, usually not permanently and sometimes only a relic, in rough ground, by walls and paths; very scattered in Br, mostly in Sc and N En but S to Surrey; Iran.

43. PEUCEDANUM L. - Hog's Fennels
Erect, glabrous to nearly glabrous biennials to perennials; stems hollow or solid; leaves 1-6-ternate or 2-4-pinnate; bracts 0-several, entire; bracteoles few to several; sepals ± 0 to conspicuous, persistent or not; petals white or yellow, actinomorphic; fruits strongly dorsally compressed, somewhat longer than wide, glabrous, with low dorsal and winged lateral ridge.

The 3 spp. in this genus seem at least as different as several other genera; probably it should be united with a number of related genera.

1 Petals yellow; stems solid, with dense sheath of
   fibres at base; ultimate leaf-segments >3cm, linear
   P. officinale
1 Petals white; stems hollow, without basal fibres;
   ultimate leaf-segments not linear, or if so <2cm   2
   2 Leaves 2-4-pinnate, with ultimate segments
     <1.5 x 0.5cm; bracts >3, reflexed   **2. P. palustre**
   2 Leaves 1-2-ternate, with ultimate segments
     >2 x 1cm; bracts 0(-2)   **3. P. ostruthium**

1. P. officinale L. - Hog's Fennel. Stems to 2m; leaves **611** 3-6-ternate, with ultimate segments c.4-10cm, linear; petals **RR** yellow; fruits 5-8mm. Native; rough brackish grassland, banks of creeks and path-sides near sea; extremely local in E Kent and N Essex.

2. P. palustre (L.) Moench - Milk-parsley. Stems to 1.5m; **611** leaves 2-4-pinnate, with ultimate segments linear to narrowly **R** oblong-lanceolate, c.0.5-1.5cm; petals white; fruits 3-5mm. Native; fens and marshes; local in En N to Yorks, very rare outside fens of E Anglia. Often with Selinum which differs in solid stems, patent bracts, and dorsally winged fruits.

3. P. ostruthium (L.) Koch - Masterwort. Stems to 1m; **611** leaves 1-2-ternate, with ultimate segments ovate, serrate, c.5-12cm; petals white; fruits 3-5mm. Intrd; natd in grassy places, marshy fields and river-sides, nowadays rarely cultivated; scattered and decreasing in N Ir and Br N from S Lincs and Staffs, formerly S Ir and Wa; C & S Europe.

44. PASTINACA L. - Parsnips
Somewhat pubescent biennials with strong characteristic smell; stems hollow or solid; leaves 1-pinnate, with large ovate leaflets; bracts and bracteoles 0-2, entire, soon falling; sepals 0; petals yellow, actinomorphic; fruits strongly dorsally compressed, somewhat longer than wide,

glabrous, with low dorsal and winged lateral ridges.

1. P. sativa L. - see vars. for English names. Stems erect, **611** to 1.8m; fruits 4-7mm. Both our vars. belong to ssp. <u>sativa</u>.
   a. Var. hortensis Gaudin - <u>Parsnip</u>. Rather sparsely pubescent; hairs on stems and petioles straight; root often swollen-conical. Intrd; frequent escape from cultivation, when established often losing its swollen root feature; scattered over BI; cultivated origin.
   b. Var. sylvestris (Miller) DC. (ssp. <u>sylvestris</u> (Miller) Rouy & Camus) - <u>Wild</u> Parsnip. Pubescent; hairs on stems and petioles longer and wavy; root not swollen. Native; grassland, roadsides, rough ground, especially on chalk and limestone; common in En SE of line from Humber to Severn, very local elsewhere in En and Wa.

45. HERACLEUM L. - <u>Hogweeds</u>
Erect, pubescent biennials to often monocarpic perennials; stems hollow; leaves simple and pinnately or ternately divided, or 1(-2)-pinnate, or ternate; bracts 0-several, entire; bracteoles several; sepals minute or conspicuous and persistent; petals white to purplish or greenish-white, zygomorphic to scarcely so; fruits strongly dorsally compressed, slightly to somewhat longer than wide, glabrous or pubescent, with very low dorsal and winged lateral ridges.

Other spp. - Plants named H. mantegazzianum are variable and might represent >1 sp. **H. persicum** Desf. ex Fischer, from Turkey to Iran, is 1 such taxon; it is smaller, to 2.5m, with more divided leaves with less sharp serrations.

1. H. sphondylium L. - <u>Hogweed</u>. Stems to 2(3)m, hispid; **611** leaves 1(-2)-pinnate, with large, hispid, usually lobed leaflets; bracts 0-few; fruits glabrous, without persistent sepals, with linear oil-bodies scarcely widened (<0.4mm wide) at proximal end. Native; grassy places, rough ground, roadsides and banks.
   a. Ssp. sphondylium. Petals white or pinkish-white to purplish, the outer ones on outermost flowers of umbel bilobed and c.2x inner unlobed ones; fruits 6-10mm. Common throughout BI.
   b. Ssp. sibiricum (L.) Simonkai. Petals greenish-white, scarcely zygomorphic; fruits 4-6mm (?always). NE parts of E Norfolk (?intrd).
   1 x 2. H. sphondylium x H. mantegazzianum is scattered in En, Ir and Sc, especially SE Sc and the London area; it is intermediate in size, pubescence, leaf-shape and fruit characters, and has very low fertility; endemic.
   2. H. mantegazzianum Sommier & Levier - <u>Giant Hogweed</u>. **611** Stems to 5.5m, rather softly pubescent; leaves pinnate to ternate, or simple and ternately to pinnately lobed, the lowest up to 2.5m; bracts several; petals white, zygomorphic;

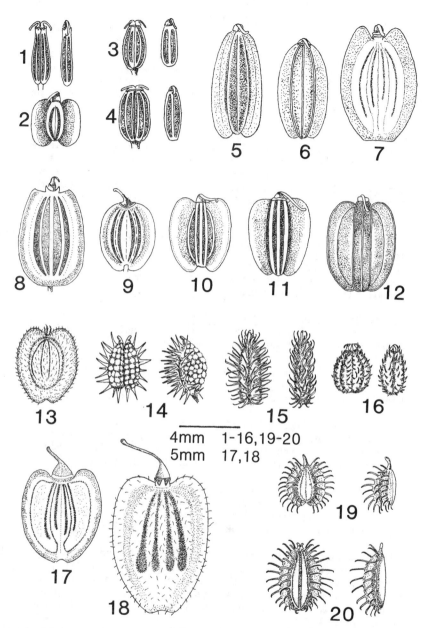

FIG 611 – Fruits (some with lateral and dorsal views) of Apiaceae. 1, **Falcaria**. 2, **Selinum**. 3, **Carum verticillatum**. 4, **C. carvi**. 5, **Ligusticum**. 6, **Levisticum**. 7, **Pastinaca**. 8, **Peucedanum officinale**. 9, **P. palustre**. 10, **P. ostruthium**. 11, **Angelica sylvestris**. 12, **A. archangelica**. 13, **Tordylium**. 14, **Torilis nodosa**. 15, **T. arvensis**. 16, **T. japonica**. 17, **Heracleum sphondylium**. 18, **H. mantegazzianum**. 19–20, **Daucus**, lateral view shows 1 mericarp only. 19, **D. carota**. 20, **D. glochidiatus**.

fruits 9–14mm, glabrous or pubescent, with persistent sepals, with conspicuous oil-bodies swollen to 0.6–1mm wide at proximal end. Intrd; natd on waste ground, roadside and riverside banks, rough grassland; scattered throughout BI and locally frequent; SW Asia.

46. TORDYLIUM L. - Hartwort
Hispid annuals or biennials; stems hollow or ± solid; leaves 1-pinnate (basal ± simple but gone by flowering); bracts and bracteoles several, entire; sepals conspicuous, persistent; petals white, zygomorphic; fruits strongly dorsally compressed, scarcely longer than wide, hispid, with very low dorsal and broadly whitish-winged lateral ridges.

1. T. maximum L. - Hartwort. Stems erect, to 1m; lower **611** leaves with 2–5 pairs of lanceolate, coarsely serrate leaflets; fruits 4.5–6(8)mm. Intrd; rough scrubby grassland; near Thames in S Essex since 1875, formerly elsewhere as casual in SE En; S Europe.

47. TORILIS Adans. - Hedge-parsleys
Hispid annuals (rarely biennials); stems solid; leaves 1–3-pinnate; bracts 0–numerous, entire; bracteoles several; sepals persistent but inconspicuous at fruiting due to spines; petals white to purplish-white, not or slightly zygomorphic; fruits ± not compressed, somewhat longer than wide, variously furnished with curved or hooked spines.

1 Fruits with dimorphic mericarps, 1 with spines, 1
    tuberculate; peduncles ≤1cm; rays <5mm, ± hidden
    by flowers or fruits   **3. T. nodosa**
1 Both mericarps with spines; peduncles >1cm; rays
    >5mm, conspicuous   2
  2 Bracts 0–1; fruits 3–4mm (excl. spines), with ±
      straight spines minutely hooked at end **2. T. arvensis**
  2 Bracts >2; fruits 2–2.5mm (excl. spines), with
      curved spines not hooked at end   **1. T. japonica**

1. T. japonica (Houtt.) DC. - Upright Hedge-parsley. Stems **611** erect, to 1.2m; leaves 1–3-pinnate; bracts 4–6(12); petals scarcely zygomorphic; fruits 2–2.5mm; both mericarps with stout, tapering, curved spines. Native; grassy places, hedgerows, wood-borders and -clearings; frequent throughout BI except N & NW Sc.
2. T. arvensis (Hudson) Link - Spreading Hedge-parsley. **611** Stems erect, with wide-spreading branches, to 50cm; leaves **RR** 1–2-pinnate; bracts 0–1; petals somewhat zygomorphic; fruits 3–4mm; both mericarps with slender, ± straight spines minutely hooked at apex. Probably intrd; weed of arable land; formerly frequent in S & C Br, now a rare casual of SE En; Europe.
3. T. nodosa (L.) Gaertner - Knotted Hedge-parsley. Stems **611**

procumbent to ascending, to 50cm; leaves 1-2-pinnate; bracts
0; petals actinomorphic; fruits 2.5-3.5mm, the outer mericarp
with minutely hooked, straight, stout spines, the inner
tuberculate. Native; arable and barish ground, especially
near sea; rather scattered in BI N to SE Sc, mostly in E En.

48. DAUCUS L. - Carrots

Glabrous annuals to biennials with strong characteristic
smell (especially in crushed root); stems solid; leaves (1)2-
3-pinnate; bracts numerous, usually longer than rays,
pinnately divided into filiform lobes; bracteoles numerous;
sepals small, scarcely visible in fruit; umbel often with 1
dark purple central flower; other petals white, slightly
zygomorphic; fruits strongly dorsally compressed, somewhat
longer than wide; mericarps each with 2 lateral and 2
secondary dorsal ridges each with row of terminally barbed
spines, the 3 primary dorsal ridges alternating with the
secondaries and bearing only short weak bristles.

1. D. carota L. - see sspp. for English names. Usually **611**
biennials; stems erect to procumbent, scarcely branched to
strongly branched with widely spreading branches, to 1m; rays
>10; fruits 2-3mm, with stout, ± straight spines.
1 Umbels convex to slightly concave in fruit
                                c. ssp. gummifer
1 Umbels very contracted in fruit, very concave   2
  2 Root swollen in 1st year, usually orange; leaves
    usually bright green   **b. ssp. sativus**
  2 Root not swollen, whitish; leaves usually grey-
    green   **a. ssp. carota**
a. Ssp. carota - Wild Carrot. Root not swollen, whitish;
stems erect, usually narrowly branching; leaves usually dark
to rather grey-green, usually thin, hispid-pubescent; umbels
very contracted and concave in fruit, with sparsely hispid to
subglabrous rays. Native; grassy and rough ground, mostly on
chalky soils and near sea (there often very stunted);
throughout most of BI but mainly coastal in N & W Br.
b. Ssp. sativus (Hoffm.) Arcang. - Carrot. Root swollen in
1st year, usually orange; leaves usually bright green, thin,
usually rather sparsely pubescent; stems and umbels as in
ssp. carota. Intrd; casual in waste places and tips and a
relic where planted; scattered over BI; garden origin.
c. Ssp. gummifer (Syme) Hook. f. - Sea Carrot. Root not
swollen, whitish; stems erect, rarely >25cm, usually widely
branching; leaves dark green, usually thick and ± succulent,
hispid-pubescent; umbels not contracted in fruit, hence
convex to slightly concave, with hispid rays. Native;
cliffs, dunes and rocky places by sea; coasts of CI, S & SE
Ir, S & W Br from Anglesey to E Kent.
2. D. glochidiatus (Labill.) Fischer, C. Meyer & Ave-Lall. - **611**
Australian Carrot. Slender annuals; stems erect, ± glabrous,
to 40cm, little branched; leaves sparsely pubescent; rays <8,

very uneven lengthed, slender, each with ≤6 flowers, or
sometimes umbels simple; fruits 3-5mm, with dense, rather
slender spines. Intrd; rather frequent wool-alien; very
scattered in En: Australia.

110. GENTIANACEAE - Gentian family

Glabrous annuals to herbaceous perennials; leaves opposite,
simple, entire, exstipulate, sessile or ± so. Flowers in
terminal dichasial cymes, sometimes monochasial cymes or
solitary, actinomorphic, bisexual, hypogynous; sepals 4-5(8),
fused at least at base; petals 4-5(8), variously coloured,
fused into corolla-tube proximally, persistent in fruit;
stamens as many as petals, borne on corolla-tube; ovary
1-celled, each cell with many ovules on 2 parietal placentas;
styles 1-2, if 1 usually divided into 2 distally, sometimes ±
0; stigmas (1-)2, capitate to variously expanded; fruit a
capsule dehiscing into 2 valves.
Easily recognized by the opposite, entire, glabrous leaves,
4-5(8) fused petals with as many stamens borne on corolla-
tube, and 1-celled ovary with many ovules forming 2-celled
capsule.

1 Petals purple to blue, rarely white, very rarely
  pinkish; style and stigmas persistent in fruit   2
1 Petals pink or yellow, rarely white; stigmas and
  sometimes style falling before fruiting   3
  2 Corolla with small inner lobes alternating with
    main ones; distal part of calyx-tube membranous
    between the calyx-lobe origins   6. GENTIANA
  2 Corolla without small inner lobes but with long
    fringes; calyx-tube without membranous part
                                            5. GENTIANELLA
3 Pairs of stem-leaves fused at base; flowers 6-8-
  merous   4. BLACKSTONIA
3 Stem-leaves not fused in pairs; flowers 4-5-merous   4
  4 Calyx-lobes shorter than calyx-tube; corolla
    yellow   1. CICENDIA
  4 Calyx-lobes longer than calyx-tube; corolla pink,
    rarely white   5
5 Calyx-lobes (4-)5, keeled; corolla-lobes (4-)5,
  ≥2mm; anthers twisting after flowering   3. CENTAURIUM
5 Calyx-lobes 4, not keeled; corolla-lobes 4, <2mm;
  anthers not becoming twisted   2. EXACULUM

1. CICENDIA Adans. - Yellow Centaury
Annuals; flowers 4-merous; calyx-lobes triangular, shorter
than calyx-tube; corolla yellow; anthers not becoming
twisted; style simple; stigma 1, peltate.

1. C. filiformis (L.) Delarbre - Yellow Centaury. Stems   R erect, very slender, to 10(18)cm; leaves few, linear, <6mm; corolla 3-7mm, opening only in the sun. Native; damp sandy and peaty barish ground mostly near coast; very local in W Ir, W Wa, S & SW En, formerly (to 1928) W Norfolk and Lincs.

2. EXACULUM Caruel - Guernsey Centaury

Annuals; flowers 4-merous; calyx-lobes linear, flat, longer than calyx-tube; corolla pale pink; anthers not becoming twisted; style 1, divided near apex; stigmas 2.

1. E. pusillum (Lam.) Caruel - Guernsey Centaury. Stems   **RR** procumbent to ascending, very slender, to 4(10)cm; leaves linear, <7mm; corolla 3-6mm, with lobes c.1.5mm. Native; with Cicendia in short ± open turf in dune-slacks; few places in Guernsey.

3. CENTAURIUM Hill - Centauries

Annuals, biennials or perennials; flowers (4-)5-merous; calyx-lobes linear, keeled, longer than calyx-tube; corolla pink, rarely white; anthers becoming twisted at fruiting; style 1, divided near apex; stigmas 2.

1 Perennials with procumbent to decumbent non-flowering
   stems; corolla-lobes >7mm   **1. C. scilloides**
1 Annuals to biennials without procumbent to decumbent
   non-flowering stems; corolla-lobes <7mm   2
  2 Usually biennials with basal leaf-rosette at
    flowering; flowers with 1-2 bracts at base of
    calyx, with stalk between 0-1mm   3
  2 Usually annuals without basal leaf-rosette at
    flowering; flowers with stalks 1-4mm between base
    of calyx and bracts   4
3 Stem-leaves narrowly oblong-elliptic, almost parallel-
   sided, obtuse to rounded at apex; calyx usually >3/4
   as long as corolla-tube; stigmas broadly rounded to
   nearly flat at apex   **3. C. littorale**
3 Stem-leaves ovate to elliptic, acute to subacute at
   apex; calyx usually <3/4 as long as corolla-tube;
   stigmas narrowly rounded to nearly conical at apex
                      2. C. erythraea
  4 Main stem with 2-4 internodes; all branches
    arising at c.30-45 degrees and forming rather open
    inflorescence   **4. C. pulchellum**
  4 Main stem with 5-9 internodes; upper branches
    arising at c.20-30 degrees and forming rather
    dense inflorescence   **5. C. tenuiflorum**

1. C. scilloides (L.f.) Samp. (C. portense (Brot.) Butcher)   **RR** - Perennial Centaury. Perennial with procumbent to decumbent sterile stems and ascending flowering stems to 30cm; leaves on sterile stems shortly petiolate, broadly elliptic; flowers in groups of 1-6; corolla-lobes 6-9mm. Native; grassy cliff-

tops and dunes by sea in Pembs and W Cornwall, formerly E
Cornwall, natd as lawn-weed from garden plants in W Kent and
E Sussex.

2. **C. erythraea** Rafn (_C. latifolium_ (Smith) Druce, _C.
capitatum_ (Willd. ex Cham.) Borbas, _C. minus_ auct. non
Moench) - Common Centaury. Erect biennial (rarely annual) to
50cm; flowers numerous in rather dense to tightly congested
inflorescences; corolla-lobes 4.5-6mm. Native; grassy and
rather open ground on well-drained soils; frequent throughout
BI except local and mostly coastal in Sc. Very variable,
often dwarfed in exposed places. Very dwarf plants with very
dense inflorescences and stamens inserted at base (not apex)
of corolla-tube (var. **capitatum** (Willd. ex Cham.) Meld.)
occur on the coast locally in En and Wa. **C. latifolium** was
known from the coast of S Lancs from 1804 to 1872 (endemic).
It differed in its corolla-lobes only 3-4mm and its broadly
elliptic to suborbicular basal leaves; like var. _capitatum_,
it was probably a mutant of _C. erythraea_.

2 x 3. **C. erythraea x C. littorale = C. x intermedium**
(Wheldon) Druce occurs with the parents on coastal dunes in S
& W Lancs, Anglesey and Merioneth. F1 plants are intermediate in all 3 key characters and like both parents have
2n=40; they have low fertility. In Lancs backcrossing to
both parents has occurred, producing fertile plants with
2n=40 appearing as introgressed _C. erythraea_ and, in other
populations, fertile plants with 2n=60 closer to _C.
littorale_. The latter could be treated as a distinct new
sp., but the parents themselves are so close that this is not
feasible. **C. intermedium** strictly refers to this taxon.

2 x 4. **C. erythraea x C. pulchellum** occurs with the parents
on the coast in N Somerset, S Essex and W Lancs; it is
intermediate in all characters, notably pedicel and corolla-
lobe lengths, and highly fertile

3. **C. littorale** (Turner ex Sm.) Gilmour (_C. minus_ Moench) -   **R**
Seaside Centaury. Erect biennial to 26cm; flowers in rather
dense inflorescences; corolla-lobes 5-6.5mm. Native; coastal
dunes and sandy turf; local in Br N from S Wa and Cheviot, S
Hants, Co Londonderry. Leaf-shape is diagnostic and differs
from that of even the most narrow-leaved variants of _C.
erythraea_.

4. **C. pulchellum** (Sw.) Druce - Lesser Centaury. Erect
annual to 20cm but often <6cm; smallest plants unbranched
with 1-few flowers, larger ones with wide-spreading branches
and diffuse inflorescence; corolla-lobes 2-4mm. Native; damp
grassy or ± open ground, especially near sea; local in Br N
to Cumberland and NE Yorks, CI, S & E Ir.

5. **C. tenuiflorum** (Hoffsgg. & Link) Fritsch - Slender   **RR**
Centaury. Erect annual to 35cm; branches usually many, the
upper usually rather narrowly divergent, forming rather dense
inflorescence; corolla-lobes 2-4mm. Native; damp grassy
places near sea; very rare in Dorset, formerly Wight, CI
pre-1840. Very similar to _C. pulchellum_ and specific

distinctness needs confirming.

4. BLACKSTONIA Hudson – Yellow-wort

Annuals; flowers 6–8-merous; calyx divided almost to base into linear, flat lobes; corolla yellow; anthers not becoming twisted; style 1, divided near apex; stigmas 2.

1. B. perfoliata (L.) Hudson – Yellow-wort. Stems erect, to 50cm, glaucous; stem-leaves triangular-ovate, pairs fused round stem at base, glaucous; corolla-lobes longer than tube, 5–10mm. Native; calcareous grassland, bare chalk and dunes; locally frequent in BI N to Co Sligo, Westmorland and S Northumb.

5. GENTIANELLA Moench – Gentians

Annuals to biennials; flowers 4–5-merous; calyx-lobes shorter or longer than tube; corolla blue or dark- to whitish-purple, rarely pink, the lobes with long fringes at margins or at base on inner face; anthers not becoming twisted; style scarcely distinct; stigmas 2.

Spp. 3–6 are variable in size and habit. The existence of hybrids and of diminutive annuals (with smaller than normal floral parts) in normally biennial taxa can make determination difficult.

1 Corolla-lobe with long narrow fringes along sides, not at base on inner side   **1. G. ciliata**
1 Corolla-lobes with long narrow fringes at base on inner side, not along sides   2
  2 Flowers 4-merous; calyx with 2 lobes several times wider than other 2 lobes   **2. G. campestris**
  2 Flowers 4–5-merous (often both on same plant); calyx with 4–5 lobes, the widest $\leq$2x as wide as others   3
3 Corolla (15)25–35mm, $\geq$2x as long as calyx; plant with 9–15 internodes   **3. G. germanica**
3 Corolla 12–22mm, $\leq$2x as long as calyx; plant with 2–11 internodes   4
  4 Internodes (2)4–9(11); apical pedicel <1/4 total height to pedicel apex   **4. G. amarella**
  4 Internodes 0–3(5); apical pedicel usually $\geq$1/2 total height to pedicel apex (except in Cornwall)   5
5 Stem-leaves lanceolate to ovate- or oblong-lanceolate; calyx-lobes $\pm$ equal, appressed to corolla   **5. G. anglica**
5 Stem-leaves ovate to ovate-lanceolate; some calyx-lobes usually distinctly longer and wider than others, somewhat divergent from corolla   **6. G. uliginosa**

1. G. ciliata (L.) Borkh. (Gentiana ciliata L.) – Fringed **618** Gentian. Erect biennial to 30cm; corolla 25–40mm, blue, with **RRR** ovate- to rhombic-oblong lobes 10–18mm; corolla-tube widest at apex. Possibly native; chalk grassland; 1 place in Bucks,

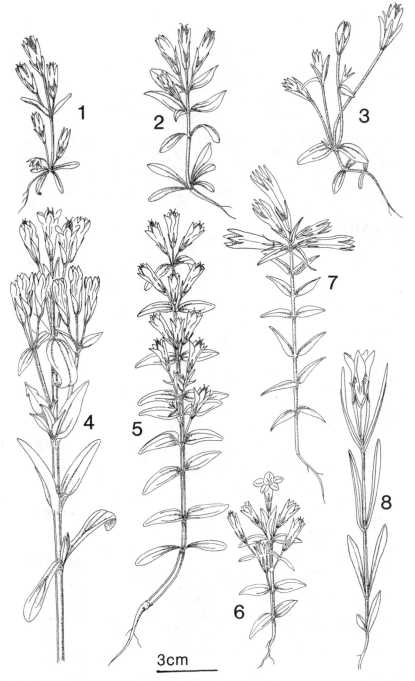

FIG 618 – Gentianella. 1–2, G. anglica. 1, ssp. anglica. 2,
ssp. cornubiensis. 3, G. uliginosa. 4, G. campestris. 5–6, G.
amarella. 5, ssp. amarella. 6, ssp. septentrionalis. 7, G.
germanica. 8, G. ciliata.

reported in 1875, disregarded, refound 1982; Europe.

2. G. campestris (L.) Boerner (<u>Gentiana</u> <u>campestris</u> L.) – **618**
<u>Field</u> <u>Gentian</u>. Erect annual or biennial to 30cm; corolla 15–
25(30)mm, bluish-purple to ± white, with narrowly oblong-
ovate lobes 6–11mm; corolla-tube ± cylindrical.   Native;
grassland and dunes; scattered over Br and Ir, locally common
in N, absent from most of S Ir and S & C Br.

3. G. germanica (Willd.) Boerner (<u>Gentiana</u> <u>germanica</u> Willd.) **618**
– <u>Chiltern</u> <u>Gentian</u>. Erect (annual or) biennial to 40cm;   **R**
corolla (15)25–35mm, bright bluish-purple, with narrowly
triangular-ovate lobes 6–11mm; corolla-tube widest at apex.
Native; chalk grassland, mostly sheltered or among scrub;
very local in SC En from N Hants and N Wilts to Herts and
Beds, formerly more widespread.

3 x 4. G. germanica x G. amarella = G. x pamplinii (Druce)
E. Warb. is intermediate in corolla-shape, -size and -colour
and is ≥50 per cent fertile; it occurs near most populations
of <u>G.</u> <u>germanica</u>.   Back-crossing to <u>G.</u> <u>amarella</u> occurs, and
introgressed <u>G.</u> <u>amarella</u> colonies exist from E Kent to Oxon,
partly outside the present area of <u>G.</u> <u>germanica</u>.

4. G. amarella (L.) Boerner (<u>Gentiana</u> <u>amarella</u> L.) – <u>Autumn</u>
<u>Gentian</u>.   Erect (annual or) biennial to 30cm; corolla 12–
22mm, with narrowly oblong-ovate lobes 4–7mm; corolla-tube ±
cylindrical.   Native; basic pastures and dunes.   Flowering
Jul-Oct.

1 Corolla creamy-white, suffused purplish-red on
  outside, (12)14–17mm   **c. ssp. septentrionalis**
1 Corolla usually dull purple, rarely pale blue, pink
  or whitish, (14)16–22mm   2
    2 Corolla (14)16–18(20)mm; Britain   **a. ssp. amarella**
    2 Corolla (17)19–22mm; Ireland   **b. ssp. hibernica**

a. Ssp. amarella. Stems with 4–9(11) internodes. Locally **618**
frequent in Br S from Angus and S Ebudes.

b. Ssp. hibernica N. Pritch. Stems with 7–11 internodes;
leaves usually narrower and less tapering than in ssp.
<u>amarella</u>. Frequent over most of Ir; endemic.

c. Ssp. septentrionalis (Druce) N. Pritch. (ssp. <u>druceana</u> N. **618**
Pritch., <u>G.</u> <u>septentrionalis</u> (Druce) E. Warb.). Stem with 2–7
internodes. Locally frequent in Sc N from N Aberdeen and M
Perths; probably endemic.   Ssp. **druceana** usually has 2–5(6)
(not 6–7) internodes and narrower leaves; it is more eastern
in distribution but is doubtfully worth ssp. recognition.

4 x 5. G. amarella x G. anglica occurs with <u>G.</u> <u>anglica</u> ssp.
<u>cornubiensis</u> in W Cornwall and N Devon and is fertile,
showing all grades of intermediacy; endemic.

4 x 6. G. amarella x G. uliginosa occurs with most colonies
of <u>G.</u> <u>uliginosa</u> in S Wa and is fertile, showing all grades of
intermediacy, sometimes pure <u>G.</u> <u>uliginosa</u> becoming rare or
absent.

5. G. anglica (Pugsley) E. Warb. – <u>Early</u> <u>Gentian</u>.   Erect **R**
(annual or) biennial to 20cm, usually much less; corolla
13–20mm, dull purple to whitish, with narrowly oblong-ovate

lobes 5-8mm; corolla-tube ± cylindrical. Native; endemic.
Flowering Mar-Jul(Aug).
   a. Ssp. anglica. Stem with 2-3(4) internodes, the uppermost **618**
usually longer than others; terminal pedicel usually ≥1/2
total height to pedicel apex; corolla 13-16mm. Chalk and
limestone grassland and fixed dunes; local in C & S En from
Devon to S Lincs and E Kent.
   b. Ssp. cornubiensis N. Pritch. Stem with 3-5 ± equal **618**
internodes; terminal pedicel <1/3 total height to pedicel
apex; corolla (15)17-20mm. Cliff-top turf and fixed dunes;
N W Cornwall and N Devon.
   6. G. uliginosa (Willd.) Boerner (Gentiana uliginosa Willd.) **618**
- Dune Gentian. Erect annual or biennial to 15cm, with **RR**
0-2(3) internodes; terminal pedical >1/2 total height to
pedicel apex; corolla 9-22mm, dull purple, with narrowly
oblong-ovate lobes 3-7mm; corolla-tube ± cylindrical.
Native; coastal dunes and dune-slacks; Glam, Carms and Pembs.

6. GENTIANA L. - Gentians
Annuals to perennials; flowers 5-merous; calyx-lobes shorter
than to c. as long as tube, with small membranous connexion
at base of sinuses; corolla blue, rarely white or pink, the
lobes not fringed, with 5 small lobes alternating with the 5
large ones; anthers, styles and stigmas as in Gentianella.

1 Corolla-tube >1cm wide, widening distally; leaves all
   or most >15(20)mm   2
1 Corolla-tube <8mm wide, ± cylindrical; leaves all
   <15mm   4
   2 Leaves crowded in basal rosette, few reduced ones
      up stem   **3. G. clusii**
   2 Leaves spread ± evenly up stem   3
3 Leaves linear, <1cm wide, with 1 vein
                                              2. G. pneumonanthe
3 Leaves lanceolate (to ovate), >1cm wide, with 3-5
   veins   **1. G. asclepiadea**
   4 Rhizomatous perennial with several rosettes of
      leaves; corolla lobes >8mm   **4. G. verna**
   4 Annual, with or without 1 basal leaf-rosette;
      corolla-lobes <6mm   **5. G. nivalis**

   1. G. asclepiadea L. - Willow Gentian. Perennial; stems
erect, to 60cm, with 1-30 flowers; leaves lanceolate to
ovate, 2-12cm; corolla blue, usually with purple spots
inside, 35-50mm. Intrd; grown in gardens and self-sowing
freely, natd by streams and in shady places; 1 place each in
E & W Sussex; C Europe.
   2. G. pneumonanthe L. - Marsh Gentian. Perennial; stems **R**
erect, to 40cm, with 1-10(28) flowers; leaves linear,
15-40mm; corolla blue with 5 green lines outside, 25-50mm.
Native; wet heathland; very local in Br from Dorset and E
Sussex to NE Yorks and Cumberland, decreasing.

3. G. clusii Perrier & Song. - <u>Trumpet Gentian</u>. Perennial; stems erect, to 6cm, each with 1 flower; basal leaves lanceolate to oblanceolate, 15-40mm; corolla blue, 40-70mm. Intrd; planted and natd since 1960 in chalk grassland in 3 nearby places in Surrey; Alps.

4. G. verna L. - <u>Spring Gentian</u>. Perennial; stems erect, **RRR** 0.5-7cm, each with 1 flower; basal leaves ovate to elliptic, 8-15(20)mm; corolla-tube 15-25mm; corolla-lobes patent at anthesis, forming brilliant blue limb 17-31mm across. Native; grassland on limestone, calcareous glacial drift and fixed dunes; extremely local in N En and W Ir.

5. G. nivalis L. - <u>Alpine Gentian</u>. Annual; stems erect, to **RRR** 15cm, with 1-10 flowers; leaves 2-10mm, the basal ones ovate to obovate, the upper ones narrower; corolla-tube 10-15mm; corolla-lobes patent at anthesis, forming brilliant blue limb 7-10mm across. Native; rock-ledges above 730m; very local and rare in M Perth and Angus.

111. APOCYNACEAE - <u>Periwinkle family</u>

Slightly woody perennials; leaves evergreen, opposite, simple, entire, shortly petiolate, exstipulate. Flowers solitary in leaf-axils, actinomorphic, bisexual, hypogynous; sepals 5, fused at base; petals 5, blue, rarely white, fused into corolla-tube proximally; stamens 5, inserted on corolla-tube proximally; stamens 5, inserted on corolla-tube, not exserted; ovaries 2, free, each with many ovules, united by common single style; stigma 1, capitate-peltate, complexly ornamented; fruit rarely produced, of 2 follicles.

The opposite, evergreen entire leaves and blue flowers of characteristic shape are diagnostic.

1. VINCA L. - <u>Periwinkles</u>

1 Margin of leaves and calyx-lobes minutely pubescent
                                         3. V. major
1 Margin of leaves and calyx-lobes glabrous   2
  2 Corolla-tube 9-11mm; corolla-limb 25-30mm across;
    calyx-lobes 3-4(5)mm   **1. V. minor**
  2 Corolla-tube 12-18mm; corolla-limb 30-45mm across;
    calyx-lobes 5-14mm   **2. V. difformis**

Other spp. - Records of the entirely herbaceous **V. herbacea** Waldst. & Kit., from E Europe, are errors for variants of <u>V. major</u>, which often is cut to the ground by frosts, or of <u>V. minor</u>.

1. V. minor L. - <u>Lesser Periwinkle</u>. Vegetative stems procumbent to arching, rooting at tips, to 1m; leaves ovate to elliptic or narrowly so, 15-45mm; flowering stems erect to ascending, to 20cm; flower dimensions as in key; corolla-

lobes usually sky-blue, obliquely truncate. Probably intrd;
well natd in woods, hedgebanks and other shady places;
scattered throughout most of BI; Europe.

2. V. difformis Pourret - <u>Intermediate Periwinkle</u>. Similar
to <u>V. minor</u> in habit but more robust, with vegetative stems
to 2m and flowering stems to 30cm; leaves lanceolate to
ovate, 25-70mm; flower dimensions as in key; corolla-lobes
pale blue, very obliquely truncate to acute. Intrd; natd on
bank in W Kent; SW Europe.

3. V. major L. (<u>V. herbacea</u> auct. non Waldst. & Kit.) -
<u>Greater Periwinkle</u>. Vegetative stems ascending-arching then
often procumbent, to 1.5m; leaves ovate, 25-90mm; flowering
stems as in <u>V. difformis</u>; calyx-lobes 7-17mm; corolla-tube
12-15mm; corolla-limb 30-50mm across; corolla-lobes
purplish-blue, obliquely truncate (violet-blue, narrower and
acute in var. **oxyloba** Stearn (ssp. <u>hirsuta</u> auct. non (Boiss.)
Stearn)). Intrd; natd in hedgebanks, shrubberies and rough
ground; over most of BI, frequent in S, rare in N; Mediterranean. Var. <u>oxyloba</u> rarely natd in S Br.

112. SOLANACEAE - <u>Nightshade family</u>

Annuals to herbaceous perennials or shrubs; leaves alternate,
rarely opposite, simple (entire to deeply lobed) or pinnate,
exstipulate, usually petiolate. Flowers solitary or in
axillary or terminal cymes or racemes, actinomorphic or
slightly zygomorphic, bisexual, hypogynous; sepals 5, fused
into tube proximally, sometimes 2-lipped; petals 5, fused
into tube proximally, variously coloured; stamens 5, borne on
corolla-tube; ovary 2(-5)-celled, each cell with many ovules
on axile placentas; style 1; stigmas 1-2, $\pm$ capitate; fruit a
berry (sometimes $\pm$ dry) or 2(-4)-valved capsule.
    Distinguished from all but the very distinctive <u>Verbascum</u> in
its $\pm$ actinomorphic 5-merous flowers with fused calyx and
corolla and usually 2-celled ovary with many ovules.

1 Open flowers with anthers touching laterally, forming
   cone-shaped group around style   2
1 Open flowers with anthers separated laterally, not
   forming cone-shaped group around style   4
   2 Stamens opening by apical pores; fruit succulent;
      if corolla yellow then plant spiny   **9. SOLANUM**
   2 Stamens opening by longitudinal slits; if fruit
      very succulent then corolla yellow and plant not
      spiny   3
3 Leaves simple; corolla white to purple; fruit rather
   dry   **7. CAPSICUM**
3 Leaves pinnate; corolla yellow; fruit very succulent
                                                8. LYCOPERSICON
   4 Woody shrubs, usually spiny; corolla purple;
      fruits red   **2. LYCIUM**

4 Stems herbaceous, not spiny, or if woody towards
  base then flowers and fruits both whitish   5
5 Calyx toothed, the teeth <1/4 total length at
  flowering   6
5 Calyx lobed, >1 lobe >1/3 total length at flowering
  (often not so at fruiting)   7
6 Calyx in fruit funnel-shaped, persistent, c.2x as
  long as fruit; capsule opening by lid; corolla
  with network of dark veins   **4. HYOSCYAMUS**
6 Calyx in fruit tubular to campanulate, withering,
  <2x as long as fruit; capsule opening by
  longitudinal valves; corolla without network of
  dark veins   **10. DATURA**
7 Fruit a capsule; corolla tubular to trumpet-shaped   8
7 Fruit a berry; corolla cup-, bowl-, bell- or star-
  shaped   9
8 Calyx-teeth >3/4 total length of calyx; flowers
  solitary in leaf-axils   **12. PETUNIA**
8 Most or all calyx teeth <2/3 total length of
  calyx; flowers opposed to or in axils of much
  reduced bracts   **11. NICOTIANA**
9 Fruit + completely enclosed in enlarged calyx   10
9 Fruit well exposed from calyx   11
10 Fruiting calyx with lobes much longer than tube;
   ovary and fruit 3-5-celled   **1. NICANDRA**
10 Fruiting calyx with tube much longer than lobes;
   ovary and fruit 2-celled   **6. PHYSALIS**
11 Corolla brownish- to greenish-purple, >20mm; largest
   leaves >5cm   **3. ATROPA**
11 Corolla whitish, <12mm; largest leaves <5cm
                                                   5. SALPICHROA

Other genera - **SCOPOLIA** Jacq. resembles a dwarf (to 60cm)
<u>Atropa</u> with reddish-brown pendent flowers, but has a
shallowly lobed calyx and a capsule with a lid. **S.
carniolica** Jacq., from C Europe, is grown in gardens and can
exist after neglect as does 1 small clump in Surrey; it has
erect, + unbranched stems to only 60cm. The genus is often
placed in Scrophulariaceae. **VESTIA** Willd. is an evergreen
shrub with pendent, tubular, pale yellow flowers. **V. foetida**
(Ruiz Lopez & Pavon) Hoffsgg. (<u>V. lycioides</u> Willd.) (<u>Huevil</u>),
from Chile, is grown in Cornwall and may persist in hedges.

1. NICANDRA Adans. - <u>Apple-of-Peru</u>
Glabrous annuals; leaves simple, toothed to + lobed; flowers
solitary, axillary; calyx deeply 5-lobed, later enlarging and
enclosing fruit; corolla campanulate, shallowly lobed; ovary
3-5-celled; fruit a rather dry berry.

1. N. physalodes (L.) Gaertner - <u>Apple-of-Peru</u>. Stems **628**
erect, to 80cm; corolla 25-40mm long and across, blue to
mauve; fruit globose, brownish, 12-20mm; fruiting calyx

25-35mm. Intrd; frequent casual in waste and cultivated ground and on tips, also alien from wool and bird-seed, sometimes persistent; scattered in CI and Br, mainly S; Peru.

2. LYCIUM L. - Teaplants
Almost glabrous, usually spiny shrubs with arching branches; leaves simple, entire, deciduous; flowers axillary, 1-few together; calyx irregularly 2-lipped, not enclosing fruit; corolla funnel-shaped, rather deeply lobed; ovary 2-celled; fruit a succulent berry.

1. **L. barbarum** L. (L. halimifolium Miller) - Duke of Argyll's Teaplant. Shrub to 2.5m; leaves narrowly elliptic, 2-10cm; calyx c.4mm; corolla purplish, 7-12mm, with narrow proximal tube 2-3mm, with lobes 4-5mm; berry bright red, ellipsoid, 12-20mm. Intrd; grown as hedging and natd in rough ground, hedges and on walls; frequent in CI, En and Wa, very scattered in Ir and Sc; China.

2. **L. chinense** Miller - Chinese Teaplant. Differs from L. barbarum in leaves lanceolate to narrowly ovate; calyx c.3mm; corolla 10-15mm, with narrow proximal tube 1.3-2mm, with lobes 5-8mm. Intrd; similar habitats to L. barbarum but probably much rarer; China. Much confused with L. barbarum and relative abundances uncertain.

3. ATROPA L. - Deadly Nightshade
Nearly glabrous to glandular-pubescent perennials; leaves simple, entire; flowers solitary, axillary; calyx rather deeply 5-lobed, slightly enlarging later but not enclosing fruit; corolla campanulate, shallowly lobed; ovary 2-celled; fruit a berry.

1. **A. belladonna** L. - Deadly Nightshade. Stems erect, to 2m; leaves ovate to elliptic, 8-20cm; corolla 24-30mm, greenish- or brownish-purple; fruit globose to depressed-globose, 15-20mm across, shiny black. Native; woods, scrub, rough and cultivated ground; locally frequent in C & S En, scattered in Ir and elsewhere in Br, probably native only in C & S Br on chalk and limestone.

4. HYOSCYAMUS L. - Henbane
Glandular-pubescent stinking annuals to biennials; leaves simple, toothed to ± lobed; flowers solitary, axillary, forming 2 rows on 1 side of stem; calyx funnel-shaped, enlarging later and becoming swollen at base to accommodate fruit, broadly 5-toothed; corolla broadly funnel-shaped, rather deeply lobed; ovary 2-celled; fruit a capsule dehiscing by lid.

1. **H. niger** L. - Henbane. Stems erect, to 80cm; leaves ovate-oblong, 6-20cm; corolla 2-3cm, yellowish with strong purple reticulate venation, 2-3cm across; capsule enclosed by

R

calyx, c.1cm, broadly ovoid. Native; maritime sand and shingle, inland rough and waste ground, especially manured by rabbits or cattle; scattered in Br and Ir, mainly C & S.

5. SALPICHROA Miers - Cock's-eggs
Pubescent perennials, somewhat woody below; leaves simple, entire, with petiole c. as long; flowers solitary, axillary; calyx cup-shaped, divided nearly to base, not enlarging; corolla bowl-shaped, with rather short lobes; ovary 2-celled; fruit an ovoid berry.

1. S. origanifolia (Lam.) Thell. - Cock's-eggs. Stems **628** much-branched, sprawling, to 1.5m; leaves suborbicular to ovate-rhombic or -trullate, 15-25mm; corolla 6-10mm, whitish, with reflexed lobes; berry 10-15mm, whitish. Intrd; grown for ornament, natd in rough ground and open places; few places on S coast of En since 1927, Guernsey since 1946, rare and impermanent elsewhere; S America.

6. PHYSALIS L. - Japanese-lanterns
Subglabrous to pubescent annuals to perennials; leaves simple, entire to coarsely dentate; flowers solitary, axillary; calyx campanulate, 5-lobed, the calyx-tube later enlarging to enclose fruit; corolla broadly campanulate to funnel-shaped, shallowly to rather deeply lobed; ovary 2-celled; fruit a globose berry.

1 Fruiting calyx and berry red to orange; corolla
  whitish   **1. P. alkekengi**
1 Fruiting calyx green to yellowish-green; berry yellow,
  green or purple; corolla yellowish   2
  2 Sparsely pubescent; leaves cuneate to ± rounded
    at base; annual   **3. P. ixocarpa**
  2 Densely pubescent; leaves cordate at base;
    perennial   **2. P. peruviana**

1. P. alkekengi L. (P. franchetii Masters) - Japanese- **635**
lantern. Rather sparsely pubescent, erect, rhizomatous perennial to 60cm; leaves broadly cuneate to subcordate at base, entire to coarsely dentate; fruiting calyx 2.5-5cm; fruit 12-17mm, not filling calyx. Intrd; grown for ornament, natd on waste land, road-sides and in shrubberies; scattered in En and Wa, mainly S; Europe.
2. P. peruviana L. - Cape-gooseberry. Densely pubescent, **635**
erect, rhizomatous perennial to 1m; leaves cordate at base, entire to obscurely dentate; fruiting calyx 3-5cm; fruit 12-20mm, yellow, not filling calyx. Intrd; imported as minor fruit and casual on tips; occasional in S Br; S America.
3. P. ixocarpa Brot. ex Hornem. - Tomatillo. Sparsely **635**
pubescent, erect annual to 60cm; leaves cuneate to rounded at base, entire to coarsely dentate; fruiting calyx 3-5cm, often dark-veined; fruit 13-40mm, green to purple, completely

filling calyx. Intrd; casual mostly as wool-alien, perhaps
sometimes from use as minor fruit; scattered in En and Wa; N
& S America. The closely related **P. philadelphica** Lam., with
larger flowers and fruit and curved (not straight) anthers,
is the sp. more usually used for fruit, but is only a rare
casual.

7. CAPSICUM L. - Sweet Pepper
Glabrous annuals; leaves simple, entire; flowers solitary,
axillary; calyx campanulate, shallowly toothed, slightly
enlarging; corolla star-shaped, deeply lobed; ovary 2-3(5)-
celled; fruit a rather dry ovoid berry, with large cavities
when mature.

1. C. annuum L. - Sweet Pepper. Stems erect, to 60cm;
leaves ovate to narrowly so, 8-15cm; corolla white to
purplish, 2-3cm across; fruit green, yellow or red, (1)3-
15(25)cm. Intrd; imported as bird-seed, as fruit for cooking
and, increasingly, as green salad, now occasional casual on
tips and sewerage works; scattered in S En; tropical America.

8. LYCOPERSICON Miller - Tomato
Glandular-pubescent annuals; leaves pinnate with mixed large
and small leaflets; flowers in leaf-opposed cymes; calyx
star-shaped, lobed nearly to base, slightly enlarged and
lobes reflexed at fruiting; corolla star-shaped, deeply
lobed; ovary 2-3(5)-celled; fruit a succulent, depressed-
globose to globose berry.

1. L. esculentum Miller (L. lycopersicum (L.) Karsten) -
Tomato. Stems erect to decumbent or scrambling, to 2m;
leaves 20-40cm, with toothed to lobed leaflets; corolla
yellow, 18-25mm across; fruit usually red, rarely yellow to
orange, 2-10cm across. Intrd; much used as salad-fruit and
vegetable and common on tips and in sewerage works and waste
places, grown on field scale in S En and CI; frequent
throughout BI; C & S America.

9. SOLANUM L. - Nightshades
Annual to perennial herbs or shrubs; leaves simple and entire
to pinnate; flowers in axillary or leaf-opposed cymes or
solitary; calyx star- to cup-shaped, usually deeply lobed;
corolla star-shaped, deeply to scarcely lobed; ovary 2(-4)-
celled; fruit a succulent to dry berry.

1 Stems and leaves with strong spines   2
1 Spines 0   4
   2 Corolla yellow; one anther longer than 4 others
                      11. S. rostratum
   2 Corolla whitish to bluish-purple; 5 anthers of
      equal length   3
3 Annual; berry red; leaves lobed mostly >1/2 way to

midrib, with toothed or lobed lobes

            10. S. sisymbriifolium

3 Rhizomatous perennial; berry yellow; leaves lobed
 <1/2 way to midrib, with ± entire lobes

            9. S. carolinense

 4 Perennials with stems ± woody below   5
 4 Annuals to perennials with entirely herbaceous
  stems   7

5 Stems scrambling to procumbent; many inflorescences
 >10-flowered; at least some leaves with 2 small
 leaflets at base   **6. S. dulcamara**

5 Stems erect to spreading; 0 or few inflorescences
 >10-flowered; leaves simple, entire to laciniate 6

 6 Corolla purple; fruits yellow to orange; plant
  subglabrous   **8. S. laciniatum**

 6 Corolla white; fruits purplish-black; plant
  appressed pubescent   **2. S. chenopodioides**

7 Perennial with subterranean stem-tubers; leaves
 pinnate   **7. S. tuberosum**

7 Annual; leaves entire to deeply pinnately lobed 8

 8 Leaves pinnately lobed >3/4 to base **5. S. triflorum**

 8 Leaves entire to toothed <1/2 to base   9

9 Plant without gland-tipped hairs

        1a. S. nigrum ssp. nigrum

9 Plant with many gland-tipped hairs   10

 10 Calyx not enlarging in fruit, with obtuse teeth;
  berries usually black, sometimes green, without
  groups of stone-cells in flesh

       1b. S. nigrum ssp. schultesii

 10 Calyx enlarging in fruit, with acute teeth; berries
  green to purplish-brown, with ≥2 groups of
  stone-cells in flesh   11

11 Calyx-lobes ≥3mm in flower, >5mm in fruit, usually
 at least as long as berry; petals 5-7mm wide; fruits
 with >50 seeds   **4. S. sarachoides**

11 Calyx-lobes ≤2mm in flower, <4mm in fruit, usually
 shorter than berry; petals 2-4mm wide; fruits with
 <30 seeds   **3. S. physalifolium**

Other spp. - **S. villosum** Miller (_S. luteum_ Miller) differs
from _S. nigrum_ in having yellow to red berries and only 3-5
flowers per cyme, and like it has 2 sspp. differing in
pubescence: ssp. **villosum** with many patent glandular hairs
and ssp. **miniatum** (Bernh. ex Willd.) Edmonds (ssp. _alatum_
(Moench) Edmonds, ssp. _puniceum_ (Kirschl.) Edmonds, _S. luteum_
ssp. _alatum_ (Moench) Dostal, _S. miniatum_ Bernh. ex Willd.)
without them. Both are rare casuals from S Europe. **S.**
pygmaeum Cav., from S America, was formerly natd on canal
path in Middlesex; it resembles some variants of _S. nigrum_
ssp. _nigrum_ but has woody stems to 20cm arising from a
rhizome. The record of **S. pseudocapsicum** L. (_Jerusalem-_
cherry) was an error based on this occurrence. **S.**

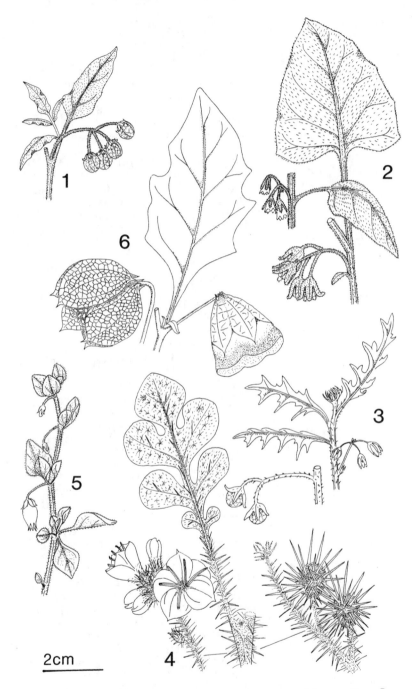

FIG 628 – Solanaceae. 1, S. chenopodioides. 2, S. physalifolium. 3, S. triflorum. 4, S. rostratum. 5, Salpichroa origanifolia. 6, Nicandra physalodes.

capsicastrum Link ex Schauer (Winter-cherry), is the common pot-plant with orange, cherry-like fruits c.2cm across; it has been found self-sown in pavements in London.

1. S. nigrum L. - Black Nightshade. Sparsely to densely pubescent, erect to decumbent annual to 70cm; leaves entire to coarsely dentate; flowers (3)5-10; petals white; fruit black, rarely green, slightly depressed-globose, 6-10mm across.

a. Ssp. nigrum. Hairs eglandular, mostly appressed, often sparse. Native; waste and cultivated ground; common in CI, most of En and S Wa, local in rest of En and Wa and Man, very scattered casual in Ir and Sc.

b. Ssp. schultesii (Opiz) Wessely. Many glandular, patent hairs present. Intrd; waste and cultivated ground; sporadic in S & E En, mainly SE; S Europe.

1 x 3. S. nigrum x S. physalifolium = S. x procurrens Leslie occurs with the parents in cultivated ground in W Suffolk, Cambs and Beds; it is intermediate in calyx characters and has black berries with 0-few seeds; endemic.

2. S. chenopodioides Lam. (S. sublobatum Willd. ex Roemer & **628** Schultes) - Tall Nightshade. Erect perennial to 1.6m, woody at least below, with ± appressed eglandular hairs; leaves entire to sparsely and shallowly dentate; flowers (1)3-8; petals / white; fruit as in S. nigrum but blackish-purple. Intrd;/ natd on rough ground in Guernsey since 1958, perhaps now in/ Surrey, very rare casual elsewhere; S America.

3. S. physalifolium Rusby (S. nitidibaccatum Bitter, S. **628** sarachoides auct. non Sendtner) - Green Nightshade. Annual with decumbent stems to 1(2)m, to 40cm tall, with many patent glandular hairs; leaves as in S. nigrum; flowers (3)4-8(10); petals white; fruit as in S. nigrum but greeen to purplish-brown, partly concealed by sepals. Intrd; natd in cultivated and waste ground in C & S En and CI, casual from wool and other sources elsewhere in Br; S America. Our plant is var. **nitidibaccatum** (Bitter) Edmonds.

4. S. sarachoides Sendtner - Leafy-fruited Nightshade. Annual with erect to decumbent stems to 2m, to 60cm tall; differs from S. physalifolium in fruit green, ± fully enveloped by sepals; and see key. Intrd; casual on tips and waste ground in S En, natd on tip at Dagenham, S Essex; S America.

5. S. triflorum Nutt. - Small Nightshade. Diffusely **628** branched, sparsely pubescent annual to 60cm; leaves deeply pinnately lobed; flowers (1)2-3; petals white; fruit marbled green and white, globose, 10-15mm. Intrd; casual in cultivated and rough ground scattered in En and Wa, natd in W Norfolk and Cheviot; W N America.

6. S. dulcamara L. - Bittersweet. Scrambling (procumbent in var. **marinum** Bab.) perennial to 3(7)m, woody below, glabrous to pubescent; leaves ovate to lanceolate, often simple and entire, ± succulent in var. marinum, at least some with 2 small lobes or leaflets near base; flowers usually >10; petals usually purple; fruit bright red, ovoid-ellipsoid, 8-12mm. Native; walls, hedges, woods, ditches, fens, pondsides, rough ground and shingle beaches; common throughout lowland BI except N Sc.

7. S. tuberosum L. - <u>Potato</u>. Sparsely pubescent herbaceous perennial with erect stems to 1m; subterranean stems bearing large terminal tubers (the only overwintering part); leaves pinnate, with mixed large and small leaflets; flowers few to many; petals white to purple or mauve; fruit greenish to purplish, depressed-globose, 2-4cm across. Intrd; much cultivated vegetable, casual and often persistent on tips, waste ground and in fields as a relic; scattered throughout BI; S America.

8. S. laciniatum Aiton (<u>S. aviculare</u> auct. non G. Forster) - <u>Kangaroo-apple</u>. Glabrous (or + so) shrub to 1(2)m; leaves entire and lanceolate, and deeply lobed and ovate (the latter commoner) on same plant; flowers 3-12; corolla purple; fruits yellow to orange, ellipsoid, 14-25mm. Intrd; rough ground, tips and maritime sand, mostly casual but sometimes natd; W Cornwall and CI; Australia.

9. S. carolinense L. - <u>Horse-nettle</u>. Erect perennial to 80(100)cm, with branched hairs; rather sparse strong spines present on stems and leaves; leaves ovate, shallowly lobed; flowers 3-8; petals white to purple; fruit yellow, globose, 10-15mm. Intrd; tips and waste places, or frequent soya-bean alien; S En; N America.

10. S. sisymbriifolium Lam. - <u>Red Buffalo-bur</u>. Erect annual to 1m, with branched and glandular simple hairs; many spines present on stems, leaves and calyx; leaves deeply pinnately lobed; flowers (1)2-10; petals white or purplish; fruit red, globose, 12-20mm, enclosed but not completely obscured by calyx. Intrd; waste places and cultivated ground, casual from wool and other sources; scattered in S Br; S America.

11. S. rostratum Dunal (<u>S. cornutum</u> auct. non Lam) - 628 <u>Buffalo-bur</u>. Similar in appearance to <u>S. sisymbriifolium</u> but stems to 60cm; glandular hairs 0; petals yellow; fruit entirely concealed by calyx; and see key. Intrd; rather frequent casual from wool, bird-seed and other sources in arable fields, tips and waste places; scattered in En, Wa and CI; N America.

10. DATURA L. - <u>Thorn-apples</u>
Glabrous to sparsely pubescent annuals; leaves simple, coarsely toothed to lobed; flowers solitary, axillary; calyx tubular, with 5 teeth; corolla trumpet-shaped, very shallowly lobed; ovary 2-celled towards apex, 4-celled towards base; fruit a usually spiny capsule dehiscing by 4 valves.

1. D. stramonium L. (<u>D. tatula</u> L., <u>D. inermis</u> Juss. ex Jacq.) - <u>Thorn-apple</u>. Stems erect, to 1(1.5)m; leaves 5-18cm, elliptic-ovate; calyx 3-5cm, with teeth (3)5-10mm;

corolla 5-10cm, white (purple in var. **tatula** (L.) Torrey);
capsule (2.5)3.5-7cm incl. spines, with slender spines 2-15mm
(spineless in var. **inermis** (Juss. ex Jacq.) Schinz & Thell.).
Intrd; casual on tips and waste and cultivated ground,
especially manured places, from several sources incl. bird-
seed, wool and soya-bean; sporadic ± throughout BI; America.
2. **D. ferox** L. - Angels'-trumpets. Differs from D.
stramonium in calyx 2.5-4cm, with teeth 3-5mm; corolla 4-6cm,
white; capsule 5-8cm incl. spines, with stout spines 1-3cm,
some broad-based and >2cm. Intrd; mainly wool-alien on tips
and in fields; occasional casual in En and Wa (mainly S); E
Asia.

11. NICOTIANA L. - Tobaccos
Glandular pubescent, ± sticky annuals (at least in BI);
leaves simple, ± entire; flowers in terminal panicle- or
raceme-like cymes; calyx tubular, with 5 unequal lobes c.1/4
to 2/3 total calyx length; corolla tubular with expanded limb
to trumpet-shaped; ovary 2-celled; fruit a capsule with 2
short valves, each 2-lobed at apex.

1 Petioles not winged **1. N. rustica**
1 Petioles broadly winged, the wing clasping stem 2
2 Inflorescence a cymose panicle; corolla-tube with
distal wider part c.1/4 total length **2. N. tabacum**
2 Inflorescence a simple raceme-like cyme; corolla-
tube with distal wider part ≤1/8 total length 3
3 Corolla-limb red to purple on upperside; filaments
inserted in basal 1/2 of corolla-tube **4. N. forgetiana**
3 Corolla-limb white on upperside; filaments inserted
in apical 1/2 of corolla-tube **3. N. alata**

Other spp. - **N. sylvestris** Speg. & Comes, from Argentina, is
also grown in gardens but less commonly and is a rare casual;
it has large white flowers in short dense panicles clustered
at stem apex.
1. **N. rustica** L. - Wild Tobacco. Erect annual to 1.5m;
leaves with unwinged petioles; inflorescence a cymose
panicle; 4 calyx-lobes shorter than -tube, 1 c. as long;
corolla-tube 12-17mm; corolla-limb 9-16mm across, scarcely
lobed, greenish-yellow. Intrd; once grown for tobacco, now
occasional casual on tips; scattered in S En; N America.
2. **N. tabacum** L. - Tobacco. Erect annual to 2(3)m; leaves
with winged petioles, the wings clasping stem; inflorescence
a cymose panicle; calyx-lobes c. as long as -tube or slightly
shorter; corolla-tube 30-55mm; corolla-limb 20-30mm across,
lobed c.1/2 way, whitish to dingy red. Intrd; rarely grown
for tobacco, now rare casual on tips and as relic, once much
commoner; scattered in S En; S & C America.
3. **N. alata** Link & Otto - Sweet Tobacco. Erect annual to
1.5m; leaves with winged petioles, the wings clasping stem;
inflorescence a simple raceme-like cyme; calyx-lobes c. as

long as -tube or somewhat longer; corolla-tube 50-100mm;
corolla-limb 35-60mm across, lobed $\geq$1/2 way, white. Intrd;
much grown for ornament and frequent casual on tips and rough
ground; scattered in S Br; S America.
   3 x 4. N. alata x N. forgetiana = N. x sanderae Will. Watson
is grown in gardens like its parents and similarly occurs as
a casual; it is intermediate in all characters, with a range
of flower colours.
   4. N. forgetiana Hemsley - Red Tobacco. Differs from N.
alata in stems to 1m; some calyx-lobes shorter than -tube;
corolla-tube 20-33mm; corolla-limb 25-40mm across, red.
Intrd; grown and found as casual as for N. alata; S America.

12. PETUNIA Juss. - Petunia
Glandular-pubescent, + sticky annuals (at least in BI);
leaves simple, + entire; flowers solitary, axillary; calyx
divided $\geq$3/4 to base into 5 narrow lobes; corolla trumpet-
shaped; ovary 2-celled; fruit a capsule with 2 valves, each
slightly notched.

   1. P. x hybrida (Hook.) P.L. Vilm. (P. axillaris (Lam.)
Britton, Sterns & Pogg. x P. integrifolia (Hook.) Schinz &
Thell.) - Petunia. Stems procumbent to erect, to 60cm;
leaves ovate-elliptic, sessile to shortly petiolate; corolla
5-12cm, often equally wide distally, white to red, mauve or
purple. Intrd; much grown in gardens and frequent on tips
and rough ground, sometimes self-sown; S Br; garden origin.

113. CONVOLVULACEAE - Bindweed family

Annuals to herbaceous perennials, with twining or procumbent
stems; leaves alternate, simple, entire, petiolate, exstipulate. Flowers 1-few in leaf axils, actinomorphic,
bisexual, hypogynous, usually with 2 bracteoles near base;
sepals 5, free; petals 5, fused almost whole way to form
funnel- to trumpet-shaped corolla, sometimes fused only to
c.1/2 way, variously coloured; stamens 5, borne on corolla-
tube; ovary 1-celled with 4 basal ovules, or 2-3-celled with
2 basal ovules per cell, or (Dichondra) the 2 cells almost as
separate ovaries; style 1, or (Dichondra) 2; stigmas 1-2 per
style, globose to linear or 2-3-lobed; fruit a capsule,
usually without proper dehiscence mechanism.
   All but the distinctive Dichondra are distinguishable by the
large funnel- to trumpet-shaped corolla, usually 2-carpellary
ovary with usually 2 stigmas and 2 ovules per cell, and often
twining stems.

1 Corolla <5mm, divided to c.1/2 way; capsule deeply
     bilobed   **1. DICHONDRA**
1 Corolla $\geq$10mm, normally not or scarcely lobed;
     capsule + not lobed   2

   2 Bracteoles ovate, often pouched, partly or wholly
     obscuring sepals   **3. CALYSTEGIA**
   2 Bracteoles linear, distant from sepals   3
  3 Stigmas linear; sepals obtuse to retuse   **2. CONVOLVULUS**
  3 Stigmas ± globose; sepals acute to acuminate
                                    4. IPOMOEA

1. DICHONDRA Forster & G. Forster - Kidneyweed
Perennials with procumbent stems rooting at nodes; leaves
orbicular to reniform; flowers 1(-few); corolla divided to
c.1/2 way, c. as long as calyx; styles 2; stigmas 1 per
style, capitate; ovary and fruit deeply bilobed.

1. D. micrantha Urban (D. repens auct. non Forster & G.
Forster) - Kidneyweed. Stems very thin, to 50cm; leaves
4-20(30)mm, appressed-pubescent; petioles 5-50mm; corolla
1.5-2.5mm, yellowish- or greenish-white, on stalk 3-20mm.
Intrd; dry fixed sand-dunes; natd near Hayle, W Cornwall,
since 1955; E Asia.

2. CONVOLVULUS L. - Field Bindweed
Rhizomatous perennials with trailing or climbing stems;
leaves triangular or ovate-oblong to linear, hastate to
sagittate at base; flowers 1-few, with 2 linear bracteoles
some way below; corolla usually scarcely divided, much longer
than calyx; style 1; stigmas 2, linear; ovary and fruit not
lobed.

1. C. arvensis L. - Field Bindweed. Stems to 1(2)m, often
much less; leaves 2-6cm, glabrous or pubescent; corolla 10-
25mm, white, pink or striped pink-and-white, rarely deeply 5-
lobed (var. **stonestreetii** Druce). Native; waste and
cultivated ground, waysides, banks and rough or short grassland; common throughout most of lowland BI except N Sc.

3. CALYSTEGIA R. Br. - Bindweeds
Rhizomatous perennials with trailing or climbing stems;
leaves triangular and sagittate, or reniform; flowers usually
1, with 2 ovate, often ± pouched bracteoles partly or wholly
concealing calyx; corolla usually scarcely divided, much
longer than calyx; style 1; stigmas 2, ellipsoid; ovary and
fruit not lobed.

1 Leaves reniform; stems not or weakly climbing
                                 1. C. soldanella
1 Leaves triangular and sagittate; stems usually
  strongly climbing   2
   2 Bracteoles 10-18mm wide when flattened, not or
     little overlapping at edges, not or little
     obscuring sepals in lateral view; ratio of midrib-
     to-midrib to edge-to-edge distances in natural
     condition 0.4-1.1   **2. C. sepium**

2 Bracteoles 18–45mm wide when flattened, strongly
   overlapping at edges, completely or nearly
   obscuring sepals in lateral view; ratio of midrib-
   to-midrib to edge-to-edge distance in natural
   condition 1.1–2.2   3
3 Corolla pink or pink-and-white striped; pedicels
   shortly pubescent (often only sparsely or partly so),
   usually with narrow wavy-edged wing near apex
                                              3. C. pulchra
3 Corolla white (sometimes narrowly pink-striped on
   outside only); pedicels glabrous, without wing
                                             4. C. silvatica

1. C. soldanella (L.) R. Br. – Sea Bindweed. Stems trailing
or weakly climbing, to 1m; bracteoles pouched, slightly overlapping laterally; corolla 3–5.5cm, trumpet-shaped, pink with
5 white stripes, yellowish in centre. Native; on sand-dunes
and sometimes shingle, by sea; coasts of BI N to C Sc.
2. C. sepium (L.) R. Br. – Hedge Bindweed. Stems strongly **635**
climbing, to 2(3)m; bracteoles flat to strongly keeled and
slightly pouched, not or slightly overlapping laterally;
corolla 3–6cm, funnel- to trumpet-shaped, rarely deeply
5-lobed (f. **schizoflora** (Druce) Stace), white, or pink with 5
white stripes. Native; hedges, ditches, fens, marshes, by
water and on rough and waste ground.
a. Ssp. sepium. Glabrous; corolla 3–5(5.5)cm, white, or
pink with 5 white stripes (f. **colorata** (Lange) Doerfler);
stamens 15–25mm. Throughout BI, but local and perhaps intrd
in N Sc.
b. Ssp. roseata Brummitt. Stems, petioles and pedicels
sparsely short-pubescent; corolla 4–5.5cm, pink with 5 white
stripes; stamens 17–25mm. Local near W coast of Br and Ir,
CI, occasionally elsewhere but perhaps intrd.
Ssp. spectabilis Brummitt, from N Eurasia, was formerly natd
at 1 place in Merioneth; it has pink corollas and differs
from the above 2 sspp. in its rounded (not acute) basal leaf-
sinus, and slightly larger parts (corolla 5–6cm; stamens
20–30mm). It may be overlooked elsewhere.
2 x 3. C. sepium x C. pulchra = C. x scanica Brummitt is
intermediate between the parents and partially fertile;
corolla 4.5–6.5cm, white to pale pink; stamens 23–28mm.
Scattered in C & S Br, CI.
2 x 4. C. sepium x C. silvatica = C. x lucana (Ten.) Don is
intermediate between the parents and highly fertile; corolla
4–6.5cm, white or very pale pink (according to C. sepium
parent); stamens 20–25mm. Scattered over C & S Br and CI,
frequent in some parts of SE En, especially Greater London.
3. C. pulchra Brummitt & Heyw. (C. sepium ssp. pulchra **635**
(Brummitt & Heyw.) Tutin nom. inval., C. dahurica auct. non
(Herbert) Don) – Hairy Bindweed. Stems strongly climbing to
3(5)m; bracteoles strongly pouched, strongly overlapping
laterally;   stems,   petioles   and   pedicels   sparsely

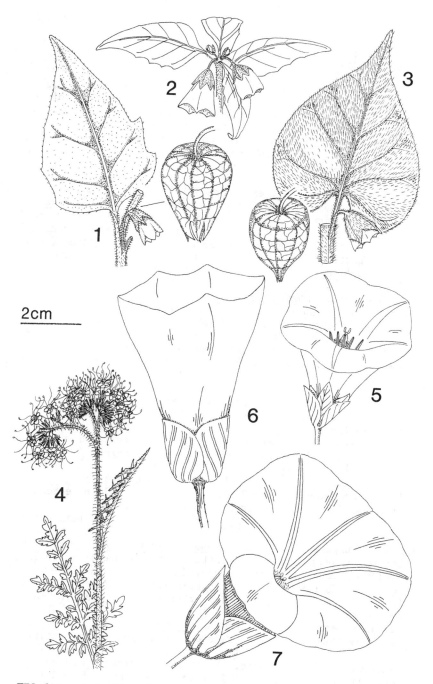

FIG 635 - 1-3, Physalis. 1, P. alkekengi. 2, P. ixocarpa. 3,
P. peruviana. 4, Phacelia tanacetifolia. 5-7, Calystegia. 5,
C. sepium. 6, C. pulchra. 7, C. silvatica.

short-pubescent; pedicels with narrow wavy wing; corolla
5-7.5cm, pink with white stripes; stamens 25-35mm. Intrd;
natd in hedges and on rough and waste ground; scattered ±
throughout BI; origin uncertain. This sp. is rarely highly
fertile and, since it shares characters (pubescence, colour)
with some sspp. of C. sepium, might be derived by
hybridization between the latter and some other sp.
    3 x 4. C. pulchra x C. silvatica = C. x howittiorum Brummitt
is intermediate between the parents and partially fertile;
corolla 5.3-6.5cm, pink; stamens 25-30mm. Very scattered in
En; endemic.
    4. C. silvatica (Kit. ex Schrader) Griseb. (C. sepium ssp. 635
silvatica (Kit. ex Schrader) Battand.) - Large Bindweed.
Glabrous; stems strongly climbing, to 3(5)m; bracteoles
strongly pouched, strongly overlapping laterally; corolla
(5)6-9cm, trumpet-shaped, white, rarely deeply 5-lobed (var.
quinquepartita Terracc.); stamens 23-35mm. Intrd; natd as
for C. pulchra, but much commoner; S Europe.

4. IPOMOEA L. - Morning-glories
Annuals with usually strongly climbing stems; leaves ovate,
cordate, entire to deeply 3-lobed; flowers 1-few, with 2
linear bracteoles some way below; corolla scarcely to
shallowly lobed, much longer than calyx, funnel- to trumpet-
shaped; style 1, stigma 1, 2-3-lobed; ovary and fruit not
lobed.

1 Flower-stalks shorter than petioles, with 0-few
    patent to forwardly directed hairs; corollas usually
    white (rarely purple)   **3. I. lacunosa**
1 Flower-stalks longer than petioles, with reflexed
    hairs; corollas usually blue, fading or drying to
    pinkish-purple (rarely white)   2
    2 Corolla <5cm; sepals abruptly long-acuminate;
        most leaves deeply 3-lobed   **2. I. hederacea**
    2 Corolla >5cm; sepals acute; most leaves entire
                                               1. I. purpurea

    1. I. purpurea (L.) Roth - Common Morning-glory. Stems to
3m; leaves mostly entire, sometimes a few 3-lobed; sepals 12-
16(20)mm; corolla 4-7cm, usually blue when fresh; ovary 3-
celled; stigma 3-lobed. Intrd; tips and waste places, casual
from soya-beans, bird-seed and other sources; sporadic in S
En; N America.
    2. I. hederacea Jacq. - Ivy-leaved Morning-glory. Stems to
2m; leaves mostly deeply 3-lobed, sometimes a few entire;
sepals 15-30mm, corolla 2-4cm, usually blue when fresh; ovary
3-celled; stigma 3-lobed. Intrd; occurrence as for I.
purpurea but less common; N America. The similar **I. nil** (L.)
Roth, with corollas 5-7.5cm, has been much confused with I.
hederacea and might occur in Br.
    3. I. lacunosa L. - White Morning-glory. Stems to 2m;

leaves entire or rather shallowly 3-lobed; sepals 6-13mm, shortly acuminate; corolla 1.5-2.5cm, usually white; ovary 2-celled; stigma 2-lobed. Intrd; a rather constant casual from soya-bean waste; very local and sporadic in S En; SE N America.

114. CUSCUTACEAE - Dodder family

Herbaceous, annual to perennial, rootless parasites without visible chlorophyll; stems twining and adherent to host plants by haustoria, very thin; leaves alternate, reduced to minute scales, exstipulate, sessile. Flowers in dense, sessile, ± globose heads, very small, actinomorphic, bisexual, hypogynous; sepals 4-5, fused at base, petals 4-5, fused into tube with 3 lobes; stamens 4-5, borne on corolla-tube, with small corolla-scale just below each; ovary 2-celled with 2 basal ovules per cell; styles 2; stigmas linear or capitate; fruit a capsule dehiscing transversely.
The very thin, chlorophyll-less, rootless stems with haustoria and small globose inflorescences are unique.

1. CUSCUTA L. - Dodders

1 Stigmas capitate (style ending in distinct knob); stems yellowish **1. C. campestris**
1 Stigmas linear (style scarcely thickened distally; stems reddish 2
  2 Styles + stigmas shorter than ovary; corolla-scales bifid, minute, or 0 **2. C. europaea**
  2 Styles + stigmas longer than ovary; corolla-scales not bifid, reaching to base of filaments **3. C. epithymum**

Other spp. - C. epilinum Weihe (Flax Dodder), from Europe, is parasitic on Linum usitatissimum and used to occur sporadically in Br and Ir as a casual in flax fields; it has been rare since 1900 and is now extinct. It differs from C. europaea in its nearly unbranched stems, 5-merous flowers and acute sepals.

1. C. campestris Yuncker - Yellow Dodder. Stems yellow; flowers pentamerous; sepals obtuse; corolla-scales not bilobed, laciniate, reaching above insertion of stamens; stamens exserted. Intrd; on a range of cultivated plants, especially carrot; scattered in En and Wa; N America.
2. C. europaea L. - Greater Dodder. Stems reddish; flowers 4-5-merous; sepals obtuse; corolla-scales deeply bilobed, each lobe entire or sparsely laciniate, ending well short of stamen insertion; stamens included. Native; on a range of hosts but usually primarily on Urtica dioica, often near water; scattered and rare in En and Wa, formerly commoner.

R

3. C. epithymum (L.) L. - <u>Dodder</u>. Stems reddish; flowers
5-merous; sepals acute; corolla-scales not bifid, laciniate,
reaching ± to stamen insertion; stamens ± exserted. Native;
on a wide range of hosts, most commonly on <u>Ulex</u> spp. and
<u>Calluna</u> on heathland; frequent in S Br and CI, scattered
elsewhere in BI N to C Sc but mostly casual.

115. MENYANTHACEAE - <u>Bogbean</u> family

Glabrous, stoloniferous, aquatic or semi-aquatic perennials;
leaves alternate, simple or ternate, entire or with ± entire
leaflets, exstipulate but with flat, sheathing petiole-bases.
Flowers showy, in axillary small clusters or elongated
racemes, actinomorphic, bisexual, hypogynous; sepals 5, fused
only at base; petals 5, fused into short tube with 5 lobes;
stamens 5, borne on corolla-tube, sometimes alternating with
short fringed scales; ovary 1-celled, with many ovules on 2
parietal placentas; style 1; stigma 2-lobed; fruit a capsule.
Easily recognized as aquatics with showy white to pink or
yellow flowers with 5 fringed petals.

1 Leaves ternate; corolla white to pink   **1. MENYANTHES**
1 Leaves simple; corolla yellow   **2. NYMPHOIDES**

1. MENYANTHES L. - <u>Bogbean</u>
Leaves ternate, all alternate, held above water level;
flowers in erect racemes; corolla white to pink, with many
long fringes on inner side of lobes; capsule dehiscing by 2
valves.

1. M. trifoliata L. - <u>Bogbean</u>. Stems procumbent or
floating, to 1.5m; racemes up to 30cm; flowers 1.5-2cm
across. Native; in shallow water, bogs and fens; throughout
Br and Ir, local in many parts of En, formerly CI.

2. NYMPHOIDES Seguier - <u>Fringed</u> <u>Water-lily</u>
Leaves simple, alternate on vegetative stems, opposite on
flowering stems, cordate, floating on water; flowers in small
axillary groups, on long pedicels; corolla yellow, with
fringes on margins of lobes; capsule dehiscing irregularly.

1. N. peltata Kuntze - <u>Fringed</u> <u>Water-lily</u>. Stems floating,   R
to 1.5m; pedicels up to 8cm; flowers 3-4cm across. Possibly
native; in ponds and slow rivers; fens of E Anglia and London
area, natd in scattered places elsewhere N to C Sc.

116. POLEMONIACEAE - <u>Jacob's-ladder</u> family

Herbaceous perennials; leaves opposite and simple or
alternate and pinnate, exstipulate, petiolate to ± sessile.

Flowers showy, in terminal corymbose racemes, actinomorphic, mostly bisexual, hypogynous; sepals 5, fused into tube proximally; petals 5, fused into short or long tube, white to purple or blue; stamens 5, borne on corolla-tube; ovary 3-celled, with several ovules per cell on axile placentas; style 1; stigmas 3; fruit a capsule opening by 3 valves.

Told from other families with 5 fused sepals and petals and actinomorphic, hypogynous flowers (except the distinctive Diapensiaceae) by the 3-celled ovary and 3 stigmas.

1 Leaves pinnate; corolla-tube much shorter than
  -lobes, with long-exserted stamens   **1. POLEMONIUM**
1 Leaves simple; corolla-tube longer than -lobes,
  with anthers at apex of corolla-tube   **2. PHLOX**

1. POLEMONIUM L. - Jacob's-ladder
Leaves pinnate, petiolate, with 6-15 pairs of entire leaflets; corolla-lobes much longer than -tube; stamens ± equal-lengthed, well exserted, with hairy base to filaments.

1. P. caeruleum L. - Jacob's-ladder. Stems erect, to 1m;   RR
flowers blue (or white), 2-3cm across. Native; limestone grassland, scree, rock-ledges, wood-borders; locally frequent in Peak District, Yorkshire Dales, 1 place in S Northumb, sporadic garden escape elsewhere in Br.

2. PHLOX L. - Phlox
Leaves simple, ± sessile, entire; corolla-lobes shorter than tube, patent; stamens with anthers at different heights, the longest at mouth of corolla-tube, glabrous.

1. P. paniculata L. - Phlox. Stems erect, woody at base, to 1.5m; flowers white to pink, purple or mauve, 2-3cm across. Intrd; much grown in gardens and natd on rough and waste ground; sporadic in En; N America.

117. HYDROPHYLLACEAE - Phacelia family

Annuals; leaves alternate, deeply pinnately lobed to ± pinnate, exstipulate, petiolate. Flowers in terminal clustered spiralled cymes, actinomorphic, bisexual, hypogynous; sepals 5, ± free; petals 5, fused into tube longer than lobes, bluish; stamens 5, borne on corolla-tube, well exserted; ovary 1-celled, with ovules borne on 2 intruded parietal placentas; style 1, divided into 2 distally; stigmas minute; fruit a 2-valved capsule.

Similar to Boraginaceae in its flowers in scorpioid cymes, but with 2-valved capsule and deeply divided style.

1. PHACELIA Juss. - Phacelia
1. P. tanacetifolia Benth. - Phacelia. Pubescent, erect to **635**

ascending annual to 70(100)cm; flowers numerous; corolla blue
or pale mauve, 6-10mm. Intrd; grown in gardens for ornament
and small-scale in fields for bees, also contaminant of crop-
and grass-seed, casual on tips, waste ground and among crops
and new grass; very scattered but increasing in En and Wa;
California.

118. BORAGINACEAE - Borage family

Annual to perennial herbs, often hispid or scabrid; leaves
alternate, simple, entire or ± so, exstipulate, sessile or
petiolate. Flowers in often spiralled cymes, actinomorphic
to weakly zygomorphic, bisexual, hypogynous; sepals 5, united
into tube with 5 lobes or teeth; petals 5, fused into tube
with distal limb, the latter with 5 lobes, mostly blue to
pink, often with a knob, scale or hair-tuft at throat of
tube; stamens 5, borne on corolla-tube; ovary 4-celled,
deeply 4-lobed, with 1 ovule per cell; style 1 (bifid at apex
in Echium), arising from base of ovary where the 4 cells
meet; stigma 1 (2 in Echium), capitate or bilobed; fruit a
cluster of 4 1-seeded nutlets (schizocarp).
Like Verbenaceae and Lamiaceae in its 4-celled ovary with 1
ovule per cell and a fruit of 4 nutlets, but differing from
both in usually alternate leaves, and spiralled cymose
inflorescence.
Much value is placed in many keys on the presence or absence
of folds, scales or bands of hairs at the throat of the
corolla-tube; these are often difficult to make out and very
little use is made of them here. When they are well
developed they may meet in the centre or around the style and
the corolla-tube appears closed.

1 Style bifid at apex; flowers distinctly zygomorphic,
   with unequal stamens and corolla-lobes   **2. ECHIUM**
1 Style simple; flowers actinomorphic to weakly zygomorphic, with equal stamens and ± equal corolla-lobes 2
   2 All anthers completely exserted   3
   2 All anthers completely included or only tips
      exserted   4
3 Annual; filaments glabrous; anthers longer than
   filaments; calyx divided nearly to base   **9. BORAGO**
3 Rhizomatous perennial; filaments pubescent; anthers
   shorter than filaments; calyx divided c.1/2 way
                                            10. TRACHYSTEMON
   4 Calyx-lobes with some small teeth between 5 main
      ones, enlarging greatly in fruit and forming 2-
      lipped covering   **14. ASPERUGO**
   4 Calyx-lobes 5, entire, not or slightly enlarging
      in fruit   5
5 Nutlets with hooked or barbed bristles   6
5 Nutlets smooth to warty, ridged or pubescent   7

6 Flowers and fruits all or mostly without bract;
    nutlets >4.5mm   **18. CYNOGLOSSUM**
6 Flowers and fruits all or mostly with bract;
    nutlets <4.5mm   **16. LAPPULA**
7 Plant glabrous, very glaucous   **11. MERTENSIA**
7 Plant bristly to (sometimes appressed-) pubescent,
    not or scarcely glaucous   8
8 At least lower leaves opposite   **13. PLAGIOBOTHRYS**
8 All leaves alternate (rarely uppermost pair
    opposite in _Myosotis_)   9
9 Open flowers pendent, with exserted stigma **4. SYMPHYTUM**
9 Open flowers erect, with stigma included or at throat
    of corolla-tube   10
10 Ripe nutlets smooth (sometimes pubescent or with
    keel round edge)   11
10 Ripe nutlets tuberculate to strongly warty and/or
    with variously branched ridges   15
11 Basal and all or most stem-leaves petiolate
                      17. OMPHALODES
11 All or most stem-leaves sessile   12
12 Corolla-tube plus -lobes >10mm   13
12 Corolla-tube plus -lobes <10mm   14
13 Calyx-lobes divided nearly to base **1. LITHOSPERMUM**
13 Calyx-lobes fused >1/2 way   **3. PULMONARIA**
14 Corolla-tube longer than -lobes; calyx-hairs
    straight; throat of corolla partially closed by
    hairy folds   **1. LITHOSPERMUM**
14 Corolla-tube usually shorter than -lobes, if
    longer then calyx-hairs hooked; throat of corolla
    closed by glabrous or papillate scales **15. MYOSOTIS**
15 Basal leaves strongly cordate at base **5. BRUNNERA**
15 Basal leaves gradually to abruptly cuneate at base 16
16 Leaves ovate to obovate, at least most basal ones
    >5cm wide   17
16 Leaves lanceolate to oblanceolate or linear-
    oblong, <5cm wide   18
17 Corolla-lobes rounded; corolla-scales closing throat
    of corolla-tube; nutlets stalked **8. PENTAGLOTTIS**
17 Corolla-lobes acute; corolla-scales not closing throat
    of corolla-tube; nutlets sessile   **9. BORAGO**
18 Nutlets tuberculate to strongly warty, not ridged
    apart from marginal keel, without collar-like
    base   19
18 Nutlets tuberculate and with strong branching
    ridges, with distinct collar-like base at point
    of attachment   20
19 Corolla yellow to orange; nutlets coarsely warty
                      12. AMSINCKIA
19 Corolla white to bluish-purple; nutlets minutely
    tuberculate   **1. LITHOSPERMUM**
20 Corolla-tube longer than -limb   **6. ANCHUSA**
20 Corolla-tube shorter than -limb   **7. CYNOGLOTTIS**

Other genera - **NONEA** Medikus resembles <u>Anchusa</u> but has very small corolla-scales and a ± tubular corolla. **N. rosea** (M. Bieb.) Link, from Caucasus, with a pinkish-purple corolla 15-18mm, is a rare casual. **N. lutea** (Desr.) DC., from Russia, with a yellow corolla 7-12mm, was formerly established in Caerns.

1. LITHOSPERMUM L. (<u>Buglossoides</u> Moench) - <u>Gromwells</u>
Pubescent to hispid annuals to perennials; leaves lanceolate to narrowly elliptic, narrowed to base; flowers solitary in leaf-axils, congested at flowering, distant at fruiting; calyx divided ± to base; corolla actinomorphic, with narrow tube at least as long as expanded limb, purplish-blue or white to yellowish; stamens equal, included; style simple, included; nutlets smooth to warty, without collar-like base.

```
1 Corolla purplish-blue, 11-16mm   1. L. purpureocaeruleum
1 Corolla usually white to yellowish, <10mm               2
    2 Leaves with lateral veins apparent on lowerside;
      nutlets white, smooth                      2. L. officinale
    2 Leaves without lateral veins apparent; nutlets
      brown, tuberculate                         3. L. arvense
```

1. L. purpureocaeruleum L. (<u>Buglossoides purpureocaerulea</u>   **RR**
(L.)   I.M. Johnston) - <u>Purple Gromwell</u>. Rhizomatous perennial with procumbent sterile and erect flowering stems to 60cm; corolla 11-16mm, purplish-blue; nutlets white, smooth, shining. Native; scrub and wood-margins on chalk and limestone; very local in SW En, S & N Wa, formerly W Kent, rare casual elsewhere.
2. L. officinale L. - <u>Common Gromwell</u>. Shortly rhizomatous perennial with erect stems to (80)100cm; corolla 3-6mm, yellowish- or greenish-white; nutlets white, smooth, shining. Native; grassy and bushy places, hedgerows and wood-borders mostly on basic soils; locally frequent in En, very local in Wa and Ir, rare casual elsewhere.
3. L. arvense L. (<u>Buglossoides</u> arvensis (L.) I.M. Johnston)   **R**
- <u>Field Gromwell</u>. Erect annual to 50(80)cm; corolla 5-9mm, whitish (rarely bluish); nutlets pale brown, tuberculate, shining on tubercles. Native; arable fields, rough ground and open grassy places; locally frequent in En, very scattered and often casual in Wa, Sc, Ir and CI.

2. ECHIUM L. - <u>Viper's-buglosses</u>
Plants hispid, monocarpic (usually biennials); leaves lanceolate to oblanceolate or the lower ovate, tapered to base; cymes terminal and lateral, forming compound narrow panicle; calyx divided nearly to base; corolla zygomorphic, pink, purple or blue with tube shorter than limb, the latter with unequal lobes; stamens unequal, at least some exserted; apex of style bifid and exserted; nutlets without collar-like base, warty to ridged.

1 Shrubs with unbranched woody stem to 75 x 3-5cm and
   with terminal panicle up to 3.5m; leaves up to 50cm,
   crowded below panicle   **4. E. pininana**
1 Stems herbaceous or ± so, <1m (incl. panicle) x 1cm;
   leaves scattered up stem, the panicle not sharply
   delimited from rest of stem   2
   2 Corolla pubescent on veins and margins only;
      usually 2 stamens exserted   **2. E. plantagineum**
   2 Corolla ± uniformly pubescent on outside; usually
      3-5 stamens exserted   3
3 Corolla blue when fully open; stems stiffly erect,
   with narrow rather dense inflorescence with bracts
   scarcely exceeding cymes   **1. E. vulgare**
3 Corolla pinkish-violet when fully open; stems
   ascending, with rather loose inflorescence with
   conspicuous leafy bracts   **3. E. rosulatum**

1. E. vulgare L. - Viper's-bugloss. Stems usually erect, to
1m; corolla 10-19mm, blue; usually 4-5 stamens exserted;
nutlets irregularly and densely ridged. Native; open grassy
places, cliffs, dunes, shingle, rough ground, usually on
light, often calcareous soils; locally frequent to common in
BI, especially S & E En.
   2. E. plantagineum L. (E. lycopsis auct. non L.) - Purple   RR
Viper's-bugloss. Stems erect to ascending, to 75cm; corolla
18-30mm, purple; usually 2 stamens exserted; nutlets warty.
Native; disturbed or open grassy, sandy ground near sea;
frequent in Jersey, very local in Scillies and W Cornwall,
rare escape or casual elsewhere in S En.
   3. E. rosulatum Lange (E. humile auct. non Desf.) - Lax
Viper's-bugloss. Stems ascending, to 75cm; corolla 11-20mm,
pinkish-violet; usually 3-4 stamens exserted; nutlets
irregularly and densely ridged. Intrd; natd on waste ground
at Barry Docks, Glam, since 1927 but perhaps now gone;
Portugal and NW Spain.
   4. E. pininana Webb & Berth. - Giant Viper's-bugloss. Stems
erect, to 4m; corolla c.13mm, blue; 5 stamens well exerted;
nutlets sharply warty. Intrd; self-sown garden escapes on
rough ground in CI, Caerns, W Cornwall and Scillies;
Canaries. Plants usually exist for several years before
flowering, followed by seeding and death. Other similar spp.
have been recorded similarly escaped in Scillies.

3. PULMONARIA L. - Lungworts
Pubescent to slightly hispid tufted perennials; leaves
lanceolate to ovate, abruptly to gradually contracted at
base; flowers in rather dense terminal clusters of cymes;
calyx divided <1/2 way to base; corolla actinomorphic, blue,
red or purple, with tube slightly shorter to slightly longer
than limb; stamens equal, included; style simple, included;
nutlets smooth, sparsely pubescent, with collar-like base.
 Leaf-characters must be observed on basal leaves that

develop during the flowering season and reach maturity during the summer. There are 5 conspicuous hair-tufts at the throat of the corolla; pubescence of inside of corolla-tube refers to region below these tufts. The flowers are heterostylous, as in _Primula_, with pin and thrum morphs.

1 Basal leaves developing at flowering cordate to
   broadly cuneate at base, abruptly contracted into
   petiole   2
1 Basal leaves developing at flowering gradually
   cuneate at base, tapered into petiole   4
   2 Corolla bright red when open; inside of corolla-
      tube pubescent below hair-tufts; basal leaves
      abruptly cuneate to rounded at base   **3. P. rubra**
   2 Corolla reddish- to bluish-violet when open;
      inside of corolla-tube glabrous below hair-tufts;
      basal leaves cordate at base   3
3 Basal leaves with large white spots, usually longer
   than petiole when mature   **1. P. officinalis**
3 Basal leaves unspotted or with faint pale green spots,
   usually shorter than petiole   **2. P. obscura**
   4 Stalked glands 0 to very sparse in inflorescence;
      leaves hispid on upperside, the basal ones white-
      spotted; flowers pin or thrum   **5. P. longifolia**
   4 Stalked glands very frequent in inflorescence;
      leaves softly pubescent on upperside, unspotted;
      flowers always thrum   **4. P. 'Mawson's Blue'**

Other spp. - **P. angustifolia** L., from Europe, is grown in gardens and might escape; it has been much confused with P. longifolia, but has unspotted basal leaves mostly >5cm wide, a less dense inflorescence, and leaf-hairs of uniform (not variable) length.

1. P. officinalis L. - Lungwort. Stems erect, to 30cm; basal leaves ovate, cordate to rounded at base, white-spotted; stalked glands fairly frequent in inflorescence; calyx-lobes subacute; corolla reddish- to bluish-violet when open. Intrd; much grown in gardens, natd on banks and in scrub, woods and rough ground; scattered throughout Br; Europe.

2. P. obscura Dumort. (P. officinalis ssp. obscura (Dumort.) Murb.) - Unspotted Lungwort. Differs from P. officinalis in darker green leaves; rather sparse stalked glands in inforescence; and see key. Probably native; woods; E & W Suffolk, rare and decreasing, perhaps natd elsewhere but confused with P. officinalis. **646** **RR**

3. P. rubra Schott - Red Lungwort. Stems erect, to 50cm; basal leaves ovate or narrowly so, rounded to very broadly cuneate at base, usually unspotted; stalked glands very frequent in inflorescence; calyx-lobes acute; corolla red when open. Intrd; grown in gardens, natd in grassy places,

hedges and scrub; scattered in N En, C & S Sc; SE Europe.
4. P. 'Mawson's Blue' - <u>Mawson's Lungwort</u>. Stems ascending **646**
to erect, to 30cm; basal leaves elliptic to narrowly so;
stalked glands very frequent in inflorescence; calyx-lobes
obtuse; corolla blue when open, glabrous on inside of tube
below hair-tufts. Intrd; grown in gardens, natd in shady
places; Surrey, S & C Sc; uncertain garden origin.
5. P. longifolia (Bast.) Boreau (<u>P.</u> angustifolia auct. non **646**
L.) - <u>Narrow-leaved Lungwort</u>. Stems erect, to 40cm; basal **R**
leaves narrowly elliptic, mostly <5cm wide; stalked glands 0
to very sparse in inflorescence; calyx-lobes acute; corolla
blue to violet when open, glabrous on inside of tube below
hair-tufts. Native; woods and scrub; extremely local in
Dorset, S Hants and Wight, perhaps rare escape elsewhere.

4. SYMPHYTUM L. - <u>Comfreys</u>
Hispid perennials; leaves ovate-elliptic, subcordate to
broadly cuneate at base, the basal long-petiolate; flowers in
rather dense cymes forming terminal panicle; calyx lobed
c.1/5 to 9/10 to base; corolla actinomorphic, various
colours, with limb little wider than tube and c.as long, the
limb with short lobes; stamens equal, included, alternating
with 5 long corolla-scales; style simple, exserted; nutlets
smooth to granulate, sometimes also ridged, with collar-like
base.

1 Plant with decumbent to procumbent leafy stolons   2
1 Plant without stolons   3
  2 Corolla pale yellow when open, often flushed
    reddish on outside; flowering stems unbranched
                              6. S. grandiflorum
  2 Corolla blue when open, often flushed reddish on
    outside; larger flowering stems branched
                          5. S. 'Hidcote Blue'
3 Nutlets ± smooth, shining; stem-leaves strongly
  decurrent, forming wings on stem extending down for
  >1 internode   **1. S. officinale**
3 Nutlets minutely tuberculate, dull; stem-leaves not
  to moderately decurrent, the wings rarely extending
  for >1 internode   4
  4 Corolla pink, purple or blue   5
  4 Corolla pale yellow to white   7
5 Calyx divided <1/2 way to base   **9. S. caucasicum**
5 Calyx divided ≥1/2 way to base   6
  6 Calyx-hairs almost all broad-based whitish
    bristles, with some much finer and smaller hairs;
    upper stem-leaves shortly petiolate, not decurrent
    or clasping stem   **3. S. asperum**
  6 Calyx-hairs a mixture of broad-based bristles and
    finer and smaller hairs and all intermediates; upper
    stem-leaves sessile, shortly decurrent or clasping
    stem   **2. S. x uplandicum**

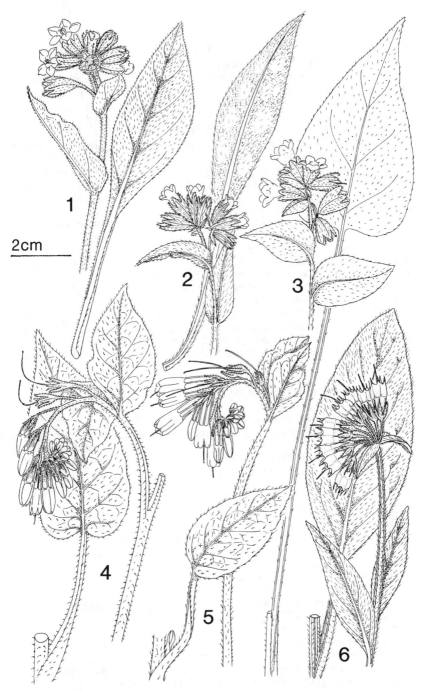

FIG 646 – Boraginaceae. 1-3, **Pulmonaria.** 1, **P.** 'Mawson's Blue'. 2, **P. longifolia.** 3, **P. obscura.** 4-6, **Symphytum.** 4, **S.** 'Hidcote Blue'. 5, **S. grandiflorum.** 6, **S. bulbosum.**

7 Corolla-scales exserted for >1mm **10. S. bulbosum**
7 Corolla-scales included 8
  8 Calyx divided <1/2 way to base; corolla pure
    white **8. S. orientale**
  8 Calyx divided >1/2 way to base; corolla yellow to
    pale yellow 9
9 Stems not or little branched; middle and upper stem-
  leaves sessile, the upper shortly decurrent; rhizomes
  with swollen tubers present **4. S. tuberosum**
9 Stems well-branched; middle stem-leaves petiolate,
  uppermost ones sessile but none decurrent; rhizomes 0
    7. S. tauricum

1. S. officinale L. - Common Comfrey. Stems erect, well-branched, to 1.5m, from thick, vertical root; stem-leaves sessile, long-decurrent; calyx divided c.2/3 to 4/5 way to base; corolla purplish or pale creamy-yellow. Native; by streams and rivers, in fens and marshy places, also roadsides and rough ground; locally frequent in BI, but less common than and over-recorded for S. x uplandicum. The flowers are often wrongly described as white; except for very rare albinos they are pale creamy-yellow (or purplish).

2. S. x uplandicum Nyman (S. officinale x S. asperum) - Russian Comfrey. Differs from S. officinale in more bristly stems, leaves and calyx; corolla blue to violet or purplish when open; and see key. Intrd originally as fodder, or possibly arisen anew in a few places; roadsides, rough and damp ground, wood-borders; frequent over most of BI. Fertile and backcrosses to S. officinale, forming a spectrum of intermediates.

2 x 4. S. x uplandicum x S. tuberosum occurs near the parents in very scattered localities in En and Sc; it is intermediate in all characters, with yellow corollas tinged with blue or purple and tuberous rhizomes, and is at least partially sterile; endemic. Some plants might be S. officinale x S. tuberosum.

3. S. asperum Lepechin - Rough Comfrey. Stems erect, well-branched, to 1.5m, from thick, vertical root; stem-leaves shortly petiolate, not decurrent; calyx divided c.2/3 to 4/5 to base, enlarging somewhat in fruit, with less sharply acute lobes than in S. officinale; corolla sky blue when open. Intrd; natd in rough and waste ground; formerly occasional, now very rare, scattered over Br, Co Sligo, much over-recorded for S. x uplandicum; SW Asia. The shortly petiolate upper stem-leaves and white, sub-spiny bristles, are diagnostic.

4. S. tuberosum L. - Tuberous Comfrey. Stems erect, little or not branched, to 60cm, from rhizomes with thick swollen regions; stem-leaves sessile, shortly decurrent; calyx divided c.3/4 to 9/10 to base; corolla pale yellow. Native; damp woods, ditches and river banks; frequent in lowland Sc, scattered in En, Sc and Ir and perhaps only intrd.

5. S. 'Hidcote Blue' (S. grandiflorum x ?S. x uplandicum) – **646**
Hidcote Comfrey. Ascending to erect flowering stems to
50(100)cm and procumbent to decumbent stolons arising from
rhizomes; stem-leaves mostly petiolate, not or scarcely
decurrent; calyx divided c.3/5 to 2/3 to base; corolla blue
when open, pink earlier. Intrd; grown in gardens, natd in
hedges and woodland; Surrey, Salop and Guernsey; garden
origin.
6. S. grandiflorum DC. (S. ibiricum Steven) – Creeping **646**
Comfrey. Differs from S. 'Hidcote Blue' in unbranched stems
to 40cm; corolla pale yellow when open, pinkish-red earlier.
Intrd; common in gardens and well natd in woods and hedges;
scattered in C & S Br, S Northumb; Caucasus.
7. S. tauricum Willd. – Crimean Comfrey. Stems erect, well-
branched, to 60cm, from thick, vertical root; stem-leaves
petiolate to sessile but not decurrent; calyx divided c.3/5
to 5/6 to base; corolla pale yellow. Intrd; natd on hedgebank; 1 place in Cambs since 1973; SE Europe.
8. S. orientale L. – White Comfrey. Stems erect, little-
branched, to 70cm, from thick branched roots; stem-leaves
petiole to sessile but not decurrent; calyx divided c.1/4 to
2/5 to base; corolla white. Intrd; natd in hedgerows and
other shady places, often self-sown; frequent in E & S En and
C Sc, very scattered elsewhere; W Russia and Turkey.
9. S. caucasicum M. Bieb. – Caucasian Comfrey. Stems erect,
to 60cm, from thick, branched roots; stem-leaves sessile,
shortly decurrent; calyx divided c.1/4 to 1/2 to base;
corolla blue. Intrd; natd in hedgerows and other shady
places; very scattered, and often not persistent, in SE En,
Flints; Caucasus.
10. S. bulbosum C. Schimper – Bulbous Comfrey. Stems **646**
simple, erect, to 50cm, from rhizome with subglobose tubers;
stem-leaves petiolate to sessile, somewhat decurrent; calyx
divided c.1/2 to 3/4 to base; corolla pale yellow. Intrd;
natd in woods and by streams; very scattered in C & S Br; SC
& SE Europe.

5. BRUNNERA Steven – Great Forget-me-not
Appressed-pubescent densely tufted perennials; basal and
lower stem-leaves ovate-cordate, petiolate; flowers in bract-
less dense cymes in terminal subcorymbose panicles; calyx
divided nearly to base; corolla actinomorphic, blue, with
tube shorter than limb, the latter with patent lobes; stamens
equal, included; style simple, included; nutlets ridged, with
collar-like base.

1. B. macrophylla (Adams) I.M. Johnston – Great Forget-me- **657**
not. Stems erect, to 50cm, the basal leaves plus petioles
often not much shorter; calyx c.1mm in flower, corolla
c.3-4mm across, resembling a Myosotis. Intrd; much grown in
gardens, very persistent throw-out, sometimes self-sown, in
rough ground and woods and on tips; scattered in Br;

Caucasus. See Omphalodes verna for differences.

6. ANCHUSA L. (Lycopsis L.) - Alkanets
Hispid annuals to perennials; leaves lanceolate to oblanceolate, sessile or narrowed to short petiole; flowers in
terminal, branched, spiralled cymes; calyx divided c.1/3 to
nearly wholly to base; corolla actinomorphic to slightly
zygomorphic, blue to purple, or yellow, with tube longer than
limb; stamens equal, included; style simple, included;
nutlets ridged and tuberculate, with collar-like base.

1 Corolla with curved tube and 5 slightly unequal
  lobes   **4. A. arvensis**
1 Corolla with straight tube and 5 equal lobes   2
  2 Corolla yellow; calyx-lobes obtuse to rounded
                                     1. A. ochroleuca
  2 Corolla blue to purple; calyx-lobes acute   3
3 Calyx divided nearly to base; nutlets >5mm in
  longest plane   **3. A. azurea**
3 Calyx divided c.1/2 way to nearly to base; nutlets
  <5mm in longest plane   **2. A. officinalis**

1. A. ochroleuca M. Bieb. - Yellow Alkanet. Rather softly 657
pubescent, erect perennial to 50cm; calyx divided c.1/3 to
1/2 to base, with obtuse to rounded lobes; corolla actinomorphic, yellow, with tube 5-10mm, with limb 7-15mm across.
Intrd; natd on rough ground on Upton Towans (W Cornwall)
since at least 1922, rare impermanent escape elsewhere; E
Europe.
  1 x 2. A. ochroleuca x A. officinalis = A. x baumgartenii
(Nyman) Gusul. occurs in W Cornwall with both parents; it is
intermediate in pubescence and calyx-lobe tips, has pale blue
and/or greyish-yellow corollas, and seems partially fertile.
  2. A. officinalis L. - Alkanet. Hispid, erect perennial to
1.5m; calyx divided c.1/2 way to nearly to base; corolla
actinomorphic, purplish-violet, with tube 5-7mm, with limb
7-15mm across. Intrd; grown in gardens, rather impermanent
escape on rough and waste ground and tips, sometimes natd;
scattered over lowland Br; Europe. White- or yellow-flowered
plants occur on the Continent, and perhaps in gardens, but
have not been reported from the wild in BI.
  3. A. azurea Miller - Garden Anchusa. Differs from A. 657
officinalis in corolla bright blue, with tube 6-10mm, with
limb 8-20mm across; and see key. Intrd; the commonest
Anchusa grown in gardens, rather infrequent escape on waste
and rough ground and tips, also bird-seed alien, very rarely
natd; scattered over lowland Br, natd in W Cornwall since at
least 1922; S Europe.
  4. A. arvensis (L.) M. Bieb. (Lycopsis arvensis L.) -
Bugloss. Hispid, procumbent to erect annual to 50cm; leaves
undulate, crenate; calyx divided nearly to base; corolla
zygomorphic (see key), blue, with tube 4-7mm, with limb 4-6mm

across. Native; weed of arable and rough ground on light,
acid or calcareous soils; locally common throughout lowland
Br, CI, E & N Ir.

7. CYNOGLOTTIS (Gusul.) Vural & Kit Tan - False Alkanet
Hispid perennials; differ from Anchusa in calyx divided >1/2
way to base; corolla actinomorphic, blue, with tube much
shorter than limb.

1. C. barrelieri (All.) Vural & Kit Tan (Anchusa barrelieri
(All.) Vitman) - False Alkanet. Plant appressed-hispid,
erect, to 70cm; calyx with dense, white, appressed, stiff
hairs, with lobes truncate to rounded or very broadly obtuse;
corolla-tube 1-2mm; corolla-limb 7-10mm across. Intrd; grown
in gardens, infrequent but pesistent escape; very scattered
in S En; SE Europe.

8. PENTAGLOTTIS Tausch - Green Alkanet
Hispid perennials with deep, thick roots; leaves ovate,
abruptly contracted at base; flowers in terminal and lateral
dense cymes; calyx divided >3/4 to base; corolla actinomorphic, blue, with tube shorter than limb; stamens equal,
included; style simple, included; nutlets ridged, with knob-
like, stalked base.

1. P. sempervirens (L.) Tausch ex L. Bailey - Green Alkanet.
Stems erect, to 1m; basal leaves up to 40cm, long-petiolate;
corolla 8-10mm across. Intrd; natd in hedges and wood-
borders and on rough ground: frequent over much of BI; SW
Europe.

9. BORAGO L. - Borages
Hispid annuals or perennials; leaves lanceolate to ovate or
obovate, the lower abruptly tapered to petiole; cymes
terminal, rather lax; calyx divided nearly to base; corolla
actinomorphic, blue (rarely white), with lobes longer than
rest of limb plus tube; stamens equal, exserted or included;
style simple, included in corolla or between anthers; nutlets
with collar-like base, ridged.

1. B. officinalis L. - Borage. Erect annual to 60cm; calyx
7-15mm at flowering, up to 20mm in fruit; corolla with patent
to reflexed lobes 7-15mm; stamens completely exposed; nutlets
7-10mm. Intrd; still grown as herb and persistent on tips,
rough ground and waysides; scattered over much of BI; S
Europe.
2. B. pygmaea (DC.) Chater & Greuter (B. laxiflora (DC.)
Fischer non Poiret) - Slender Borage. Decumbent to ascending
perennial to 60cm; calyx 4-6mm at flowering, up to 8mm in
fruit; corolla with ± erect lobes 5-8mm; stamens included;
nutlets 3-4mm. Intrd; natd on heathy ground, Jethou (CI)
since 1932, less permanent on rough ground and by paths in

very scattered places in Wa and S En; Corsica and Sardinia.

10. TRACHYSTEMON D. Don - Abraham-Isaac-Jacob
Hispid, rhizomatous perennials; leaves ovate, the basal long-
petiolate and cordate to rounded at base; cymes dense,
several in terminal panicle; calyx divided c.1/2 way to base;
corolla actinomorphic, blue, white near base, with lobes
longer than rest of limb plus tube; stamens equal, completely
exposed; style simple, included between closely appressed
stamens; nutlets with collar-like base, ridged.

1. **T. orientalis** (L.) Don - Abraham-Isaac-Jacob. Stems
erect, to 40cm; calyx 4-7mm at flowering, up to 10mm in
fruit; corolla with patent and revolute lobes 9-12mm. Intrd;
natd on shady banks and in dampish woods; scattered in En and
Wa; Caucasus and Turkey.

11. MERTENSIA Roth - Oysterplant
Glabrous, glaucous perennials; leaves papillose, obovate or
elliptic to oblanceolate, the middle and lower ones with
winged petioles, the upper sessile; flowers in terminal,
rather dense cymes; calyx divided ± to base; corolla actinomorphic, blue or blue and pink, pink in bud, with tube
shorter than limb, the latter campanulate and lobed c.1/2
way; stamens equal, included or slightly exserted; style
simple, included; nutlets slightly flattened, smooth,
succulent then papery on outside.

1. **M. maritima** (L.) Gray - Oysterplant. Stems usually   R
decumbent, to 60cm; pedicels elongating in fruit up to 2.5cm;
corolla 4-6mm, c.6mm across. Native; on bare shingle or
shingly sand by sea; local on coasts of N En, N Ir and Sc S
to Westmorland, formerly S to E Norfolk, Cards and N Kerry,
decreasing.

12. AMSINCKIA Lehm. - Fiddlenecks
Hispid annuals; leaves linear to oblong or lanceolate,
sessile; flowers in terminal, spiralled, bracteate or bract-
less cymes; calyx divided nearly to base; corolla actinomorphic, yellow to orange, with tube longer than limb;
stamens equal, included; style simple, included; nutlets
keeled, warty, without collar-like base.

1. **A. lycopsoides** (Lehm.) Lehm. - Scarce Fiddleneck. Plant
to 70cm, erect, little- to much-branched, with abundant
patent white bristles on most parts; fruiting calyx (6)8-
11(15)mm; corolla 5-8mm, with pubescent scales at apex of
tube visible from above; stamens inserted c.1/2 way up
corolla-tube or just below. Intrd; natd on rough ground in
Farne Islands (Cheviot) since 1922, becoming increasingly
frequent over much of En, especially E, on sandy soils; W N
America.

2. A. micrantha Suksd. (**A. menziesii** auct., ?(Lehm.) Nelson **657**
& J.F. Macbr., **A. intermedia** auct., ?Fischer & C. Meyer, **A.
calycina** auct., ?(Moris) Chater) – Common Fiddleneck.
Differs from **A. lycopsoides** in fruiting calyx 5-6(9)mm;
corolla paler yellow, 3-5mm, without hairs or scales inside,
with stamens borne on upper 1/2 of tube. Intrd; recently
increased casual in arable land and sandy rough ground, often
persistent; frequent in E En and E Sc N to E Ross, especially
E Anglia, rare in C & W Br; W N America.

13. PLAGIOBOTHRYS Fischer & C. Meyer – White Forget-me-not
Appressed-pubescent annuals; leaves linear-oblong, the lower
ones opposite, sessile; flowers in terminal, semi-bractless,
spiralled cymes; calyx divided >1/2 way to base; corolla
actinomorphic, white, with tube shorter than limb, the latter
divided c.1/2 way into 5 rounded lobes; stamens equal,
included; style simple, included; nutlets keeled, ridged,
minutely tuberculate, without collar-like base.

1. P. scouleri (Hook. & Arn.) I.M. Johnston – White Forget-
me-not. Stems erect to decumbent, to 20cm, often well-
branched; leaves up to 60 x 5mm; corolla 2-4mm across;
flowers resembling a white Myosotis. Intrd; casual, sometimes persistent, with grass-seed; Caithness 1974, S Hants
since 1982; W N America.

14. ASPERUGO L. – Madwort
Hispid annuals; leaves lanceolate to oblanceolate, tapered to
base; flowers 1-2 in leaf-axils; calyx with small teeth
between 5 main ones, at fruiting much enlarged and forming 2-
lipped structure around nutlets; corolla actinomorphic,
purplish-blue, with tube shorter than limb; stamens equal,
included; style simple, included; nutlets strongly compressed, without collar-like base, with dense low tubercles.

1. A. procumbens L. – Madwort. Stems procumbent to
scrambling, to 60cm; calyx growing by several times at
fruiting, very distinctive in shape; corolla 1.5-3mm,
inconspicuous. Intrd; arable fields, waste and rough ground,
occasionally natd; very scattered over Br, especially Sc,
formerly much commoner; Europe.

15. MYOSOTIS L. – Forget-me-nots
Pubescent annuals to perennials; leaves mostly narrowly
oblong to oblanceolate, the basal ones tapered to petiole-
like base, the upper ones sessile; flowers in terminal
spiralled cymes; calyx divided c.1/4 way to nearly to base;
corolla actinomorphic, blue or sometimes white or yellow,
usually pink in bud, with tube shorter to longer than limb,
with limb divided ± to base; stamens equal, included; style
simple, usually included; nutlets slightly compressed, with
distinct keel, smooth.

Maximum corolla-size is of diagnostic value, but larger-flowered spp. often produce flowers with unusually small corollas.

1 Calyx with all hairs ± straight and closely
  appressed   2
1 Calyx with some hairs patent and distally hooked or
  at least strongly curved   7
    2 Nutlets <1mm, shining olive-brown; annual without
      sterile shoots; calyx divided <1/2 way to base at
      flowering; Jersey only   **5. M. sicula**
    2 Nutlets >1.2mm, mid-brown to black, shining or
      not, or if <1.2mm then plants with axillary
      stolons; calyx often divided >1/2 way to base at
      flowering; widespread   3
3 Style longer than calyx-tube and often exceeding
  calyx-lobes at flowering; calyx divided <1/2 way to
  base at flowering, with broad teeth forming equi-
  lateral triangle   **1. M. scorpioides**
3 Style shorter than calyx-tube at flowering; calyx
  often divided >1/2 way to base at flowering, with
  narrow teeth forming isosceles triangle with base
  shorter than sides   4
    4 Lower part of stem with ± patent hairs   5
    4 Stem with only appressed hairs   6
5 Pedicels eventually 2.5-5x as long as fruiting calyx;
  nutlets <2mm   **2. M. secunda**
5 Pedicels <2x as long as fruiting calyx; nutlets
  usually >2mm   **6. M. alpestris**
    6 Stolons produced from lower nodes; leaves rarely
      >3x as long as wide   **3. M. stolonifera**
    6 Stolons 0; larger leaves >(3)4x as long as wide
                                                  4. M. laxa
7 Perennial, with sterile basal shoots at fruiting   8
7 Annual, without sterile shoots at fruiting   10
    8 Fruiting calyx narrowed and acute to subacute at
      base; nutlets obtuse to rounded at apex; only on
      hills >700m   **6. M. alpestris**
    8 Fruiting calyx rounded to broadly obtuse at base;
      nutlets acute to subacute at apex; lowland or
      upland   9
9 Corolla <8mm across, with ± flat limb; calyx-teeth
  erecto-patent, exposing ripe nutlets   **7. M. sylvatica**
9 Corolla <5mm across, with saucer-shaped limb; calyx-
  teeth erect, ± appressed and concealing ripe nutlets
  (often squashed open when pressed)   **8. M. arvensis**
10 Pedicels 1.2-2x as long as fruiting calyx
                                               8. M. arvensis
10 Pedicels shorter than to c. as long as fruiting
   calyx   11
11 Corolla cream to yellow at first (rarely white), with
   tube eventually longer than calyx; calyx-lobes

oblong-lanceolate   **10. M. discolor**
11 Corolla blue (rarely white) from start, with tube
   shorter than calyx; calyx-lobes narrowly to broadly
   triangular   **9. M. ramosissima**

Other spp. - M. **decumbens** Host has been doubtfully recorded
from mountains in Perths and needs confirming; it differs
from M. sylvatica in its corolla-tube longer (not shorter)
than calyx and nutlets c.2mm (not ≤1.8mm).

1. M. scorpioides L. - Water Forget-me-not. Erect to
ascending, appressed or sometimes partly patent-pubescent
perennial to 70cm, with rhizomes and/or stolons; calyx with
rather sparse appressed hairs, lobed <1/2 way to base at
flowering; corolla up to 8mm across. Native; by or in edges
of ponds and rivers, in damp fields; common throughout most
BI except very rare in CI (Jersey only). Distinguished by
its long style (see key).

1 x 4. M. scorpioides x M. laxa = M. x suzae Domin has been
found with the parents very scattered in En and Wa but
perhaps overlooked; it is partially fertile and intermediate,
with corolla 5-7mm across and style c. as long as calyx-tube.

2. M. secunda A. Murray - Creeping Forget-me-not. Erect to
ascending annual to perennial to 50cm, with stolons, with
hairs appressed above but patent below; calyx with appressed
hairs, lobed c.1/2 way or slightly more to base at flowering;
corolla up to 6mm across. Native; wet, often acidic places
by streams and in pools and bogs; common in many parts of BI,
especially upland and N & W, rare or absent in most of C & E
En. Distinguished by its long pedicels (see key) and patent
hairs.

3. M. stolonifera (DC.) Gay ex Leresche & Levier (M.   R
brevifolia Salmon) - Pale Forget-me-not. Erect, appressed-
pubescent perennial to 20(30)cm, with numerous leafy stolons;
calyx with appressed hairs, lobed >1/2 way to base at
flowering; corolla up to 5mm across. Native; wet flushes and
streamsides on hills; local in N En and S Sc. Distinguished
by its short leaves (see key) and abundant stolons.

4. M. laxa Lehm. (M. caespitosa Schultz) - Tufted Forget-me-
not. Erect to ascending, appressed-pubescent annual to
biennial; calyx with rather sparse appressed hairs, lobed
more or less than 1/2 way to base at flowering; corolla up to
5mm across. Native; same places as M. scorpioides and often
with it; fairly common + throughout BI. Our plant is ssp.
caespitosa (Schultz) N. Hylander ex Nordh. Best told from M.
scorpioides by its short style (see key).

5. M. sicula Guss. - Jersey Forget-me-not. Erect to   RR
decumbent, appressed-pubescent annual to 20cm; calyx with
appressed hairs, lobed <1/2 way to base at flowering; corolla
up to 3mm across. Native; damp grassland and by pond; very
local, 2 places in Jersey, discovered 1922.

6. M. alpestris F.W. Schmidt - Alpine Forget-me-not. Erect,   RR

± rhizomatous, patent-pubescent perennial to 25cm; calyx with dense appressed to erecto-patent curved hairs, with or without some patent hooked ones on tube; corolla up to 10mm across. Native; mountain slopes and ledges, 700-1200m; very local in c.11 sites in NW Yorks, Westmorland and M Perths. Often mis-recorded elsewhere for upland variants of M. sylvatica; apparently genuine but ?intrd in Angus.

7. **M. sylvatica** Hoffm. - <u>Wood</u> Forget-me-not. Erect to ascending, tufted, patent-pubescent perennial to 50cm; calyx with dense appressed to erecto-patent curved hairs and many patent hooked ones; corolla up to 8mm across. Native; woods, scree and rock-ledges; locally common in C & N Br, very local in S Br, frequent garden escape elsewhere. The common garden sp.

8. **M. arvensis** (L.) Hill - <u>Field</u> Forget-me-not. Erect to ascending, tufted, patent-pubescent annual to perennial to 40cm, more greyish-green than M. sylvatica; calyx with many patent-hooked and some suberect hairs; corolla up to 3(5)mm across. Native; open, well-drained ground in many places, including gardens; common throughout most BI. Var. **sylvestris** Schldl. (ssp. <u>umbrata</u> (Mert. & Koch) O. Schwarz, var. <u>umbrosa</u> Bab.) is said to have calyx <7mm (not <5mm) in fruit; hooked hairs <0.6mm (not <0.4mm); nutlets <2.5mm (not <2mm); and a larger corolla 3-5mm across; it is often mistaken for M. sylvatica (see key).

9. **M. ramosissima** Rochel - <u>Early</u> Forget-me-not. Erect to decumbent, patent-pubescent annual to 25cm; calyx with many patent-hooked and some suberect hairs; corolla up to 3mm across, blue (rarely white), with tube shorter than calyx. Native; dry open places on sandy or limestone soils; locally common over most of lowland BI, especially En and CI. Ssp. **globularis** (Samp.) Grau (var. <u>mittenii</u> (Baker) ined.) has calyx <2.5mm in fruit (not <4mm) with broadly (not narrowly) triangular segments, but scarcely merits var. status. Ssp. **lebelii** (Godron) Blaise is said to be intermediate between the 2.

10. **M. discolor** Pers. - <u>Changing</u> Forget-me-not. Erect, patent-pubescent annual to 25cm; differs from M. ramosissima in corolla up to 2mm across, yellow or cream at first, then pink to blue, with tube usually lengthening to longer than calyx; and see key. Native; similar places to M. ramosissima but common also in Sc, Wa and Ir, and in N Br also in damp places, marshes and dune-slacks. Subsp. **dubia** (Arrond.) Blaise differs in its small corolla (<2mm across) cream (not yellow) at first and the 2 uppermost leaves on main stem not opposite; both sorts often occur together and deserve at most varietal rank.

16. LAPPULA Gilib. - <u>Bur</u> Forget-me-not
Pubescent annuals or biennials; leaves linear-lanceolate to oblong-lanceolate, the basal ones tapered to base, the upper ones sessile; flowers in terminal, spiralled cymes; calyx

divided nearly to base; corolla actinomorphic, blue, with
tube shorter than limb, the latter 5-lobed, nearly flat;
stamens equal, included; style simple, included; nutlets
covered with hook-tipped spines.

1. L. squarrosa (Retz.) Dumort. (L. myosotis Moench) - Bur **657**
Forget-me-not. Stems erect, to 50cm, with dense appressed
and patent white hairs; corolla 2-4mm across; nutlets
2.5-4mm. Intrd; casual from wool, bird-seed, grass-seed and
grain, in waste places, rough ground and tips; very scattered
in C & S Br; Europe.

17. OMPHALODES Miller - Blue-eyed-Mary
Rhizomatous and stoloniferous, rather sparsely pubescent
perennials; leaves ovate, rounded to cordate at base, all or
nearly all long-petiolate; flowers in few-flowered terminal
cymes; calyx divided nearly to base; corolla actinomorphic,
blue, with tube shorter than limb, the latter 5-lobed, nearly
flat; stamens equal, included; style simple, included;
nutlets smooth, pubescent.

1. O. verna Moench - Blue-eyed-Mary. Stems erect to **657**
ascending, to 25cm; calyx c.4mm; corolla 10-15mm across.
Intrd; grown in gardens, persistent relic or throw-out mostly
in woods; very scattered over Br, especially in N, much rarer
than formerly; SE Europe. Over-recorded for Brunnera
macrophylla, which has much smaller, more densely congested
flowers, coarsely-pubescent leaves, no stolons and ridged,
glabrous nutlets.

18. CYNOGLOSSUM L. - Hound's-tongues
Pubescent biennials; leaves ovate to narrowly lanceolate, the
lower petiolate, the upper sessile; flowers in spiralled
terminal and lateral cymes; calyx divided nearly to base;
corolla actinomorphic, reddish-purple, with tube about as
long as funnel-shaped limb; stamens equal, included; style
simple, included; nutlets large (5-9mm in longest plane),
covered with hooked spines.

1. C. officinale L. - Hound's-tongue. Stems erect, to
60(90)cm; leaves grey-green, densely pubescent; pedicels
<5mm; corolla 6-10mm across; nutlets with distinct thickened
rim, uniformly spiny. Native; rather open ground mostly on
sand, shingle or limestone, and waste ground; locally
frequent in S, C & E Br, CI and E Ir, rare casual elsewhere.
2. C. germanicum Jacq. - Green Hound's-tongue. Differs from **RRR**
C. officinale in stems more slender, with longer, more
diffuse branches; leaves green, sparsely pubescent; pedicels
mostly >5mm; nutlets without thickened rim, but spines longer
and denser at edges. Native; woods and hedgerows; rare and
decreasing in E Gloucs, Oxon, Bucks and Surrey, formerly
widespread but very local in S & C En, very rare casual

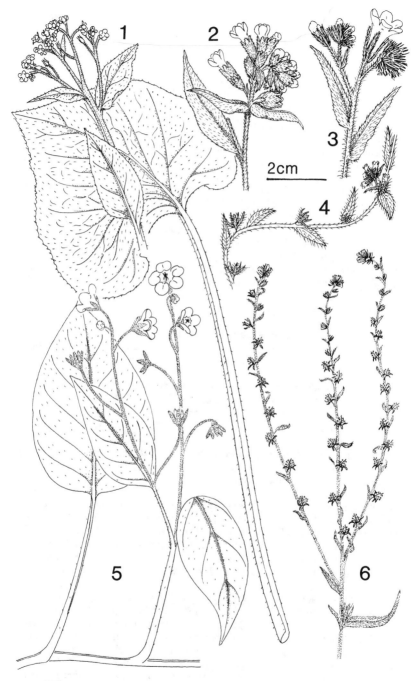

FIG 657 - Boraginaceae. 1, Brunnera. 2-3, Anchusa. 2, **A.** ochroleuca. 3, **A.** azurea. 4, **Amsinckia** **micrantha.** 5, **Omphalodes** **verna.** 6, **Lappula** **squarrosa.**

elsewhere.

119. VERBENACEAE - Vervain family

Herbaceous perennials or rarely annuals; stems square in
section; leaves opposite, simple, serrate to deeply pinnately
lobed, exstipulate, sessile to petiolate.   Flowers in
terminal elongated to corymbose spikes, zygomorphic,
bisexual, hypogynous; sepals 5, fused into tube with 5 teeth;
corolla a tube with 5 slightly unequal lobes arranged in
upper (2-lobed) and lower (3-lobed) lips, lilac to purple or
blue; stamens 4, included, 2 borne at each of 2 levels in
corolla-tube; ovary 4-celled, scarcely lobed, with 1 ovule
per cell; style 1, terminal; stigma 1, capitate; fruit a
cluster of 4 1-seeded nutlets.
Told from Boraginaceae by square-sectioned stems, opposite
leaves and 4 stamens, and from Lamiaceae by scarcely lobed
ovary, terminal style and capitate stigma.

1. VERBENA L. - Vervains

1 At least lower stem-leaves petiolate, deeply
   pinnately lobed; spikes long, slender **1. V. officinalis**
1 Stem-leaves sessile, at most sharply serrate; spikes
   short, very dense, forming sub-corymbose panicle   2
   2 Bracts shorter than to as long as calyx; corolla
      mostly <2x as long as calyx   **2. V. bonariensis**
   2 Bracts longer than calyx; corolla mostly >2x as
      long as calyx   **3. V. rigida**

Other spp. - The dwarf annual bedding plant known as Verbena
is **V. x hybrida** Voss, of uncertain parentage, sometimes found
on tips, as are a few other ornamental spp.

1. V. officinalis L. - Vervain. Stems erect, to 75cm; lower
and mid stem-leaves petiolate, deeply pinnately lobed; spikes
elongated (up to 25cm) in fruit; corolla pinkish-lilac, 3.5-
5mm, 3-5mm across. Native; barish ground and rough grassy
places, on well-drained often calcareous soils; locally
common in S Br, more scattered elsewhere in BI N to N En and
C Ir.
2. V. bonariensis L. - Argentinian Vervain. Stems erect, to
1(1.5)m; stem-leaves oblong-lanceolate, sharply serrate,
sessile; spikes rarely >2cm in fruit; calyx mostly <3mm;
corolla blue to purple. Intrd; wool-alien and garden escape
on tips and waste ground; very scattered in En; S America.
3. V. rigida Sprengel (V. venosa Gillies & Hook.) - Slender
Vervain. Stems erect, to 50cm; differs from V. bonariensis
in leaves mostly obovate-elliptic; spikes usually >2cm in
fruit, usually shorter-stalked and more crowded; calyx mostly
>3mm; and see key. Intrd; garden escape on tips; occasional

in S En; S America.

120. LAMIACEAE - <u>Deadnettle</u> <u>family</u>
(Labiatae)

Herbaceous annuals to perennials or dwarf shrubs, often
aromatic; stems usually square in section; leaves opposite,
simple, entire to serrate or rarely deeply lobed,
exstipulate, sessile to petiolate. Flowers 1-many in
contracted cymes in axils of leaf-like to much reduced
bracts, the opposite pairs of cymes often forming false whorl
of flowers, sometimes the whole forming terminal spike-like
inflorescence, zygomorphic, bisexual or male-sterile,
hypogynous; sepals 5, fused into tube with usually 5 teeth,
often separated distally into upper (usually 3-toothed) and
lower (usually 2-toothed) lips; petals 5, fused into tube
with 3-5 lobes scarcely or strongly grouped into upper
(usually 2-lobed) and lower (usually 3-lobed) lips, sometimes
upper lip scarcely lobed, or much reduced, or its 2 lobes
incorporated into lower lip, variously coloured; stamens 4, 2
long and 2 short, sometimes 2, borne on corolla-tube; ovary
4-celled, each cell with 1 ovule, deeply 4-lobed; style 1,
usually arising from base of ovary where 4 cells meet,
sometimes $\pm$ apical; stigmas usually 2, sometimes 1, linear;
fruit a cluster of 4 1-seeded nutlets.
Like Verbenaceae and Boraginaceae in its 4-celled ovary with
1 ovule per cell and fruit of 4 nutlets, but differing as
under those 2 families. Some Scrophulariaceae resemble
Lamiaceae vegetatively, but have totally different ovaries.
Several genera in both subfamilies may produce male-sterile
flowers on same or different plants as functionally bisexual
flowers; such male-sterile flowers usually have smaller
corollas and much reduced, pollen-less stamens, often
included in the corolla-tube. They are not covered in the
following keys, but usually occur with bisexual flowers or
plants. Corolla-lengths refer to length from base of calyx
to tip of longest corolla-lobe on fresh flowers; considerable
shortening occurs on drying.
8 groups of distinctive genera are first defined, but all
are included in the main keys which follow.

<u>Distinctive</u> <u>genera</u>
Plants annual   1. STACHYS, 5. LAMIUM, 6. GALEOPSIS,
            11.TEUCRIUM, 12. AJUGA, 18. CLINOPODIUM, 26.SALVIA
Shrubs >40cm high   7. PHLOMIS, 17. SATUREJA, 19. HYSSOPUS,
                  21. THYMUS, 24. LAVANDULA, 25. ROSMARINUS
Flowers with 2 fertile stamens
                  22. LYCOPUS, 25. ROSMARINUS, 26. SALVIA
Corolla yellow or yellowish
        1. STACHYS, 4. LAMIASTRUM, 6. GALEOPSIS, 7. PHLOMIS,
        11. TEUCRIUM, 12. AJUGA, 15. PRUNELLA, 16. MELISSA

Calyx-teeth spiny tipped   **1. STACHYS,**
         3. LEONURUS, 6. GALEOPSIS, 7. PHLOMIS, 22. LYCOPUS
Calyx-teeth 10, hooked at apex   **9. MARRUBIUM**
Leaves lobed >1/2 way to midrib   **3. LEONURUS,**
         11. TEUCRIUM, 12. AJUGA, 15. PRUNELLA, 22. LYCOPUS
Stamens longer than corolla (incl. lips)
                                 19. HYSSOPUS, 20. ORIGANUM,
         21. THYMUS, 22. LYCOPUS, 23. MENTHA, 25. ROSMARINUS

General key
1 Corolla with well-developed lower lip; upper lip 0 or
   represented by 1-2 short lobes   2
1 Corolla with upper and lower lips well developed, or
   ± actinomorphic (4-5-lobed)   3
   2 Corolla with ring of hairs inside tube; lower lip
      3-lobed (central lobe often ± bifid); upper lip
      of 1-2 short lobes   **12. AJUGA**
   2 Corolla without ring of hairs inside tube; lower
      lip 5-lobed (central lobe sometimes slightly bifid);
      upper lip 0   **11. TEUCRIUM**
3 Stamens 2   4
3 Stamens 4 (often very reduced in female flowers)   6
   4 Shrub; leaves entire   **25. ROSMARINUS**
   4 Herbaceous annual or perennial; leaves crenate to
      pinnately lobed, rarely some entire   5
5 Calyx and corolla both distinctly 2-lipped   **26. SALVIA**
5 Calyx with 5 equal lobes; corolla with 4 subequal
   lobes (the uppermost slightly wider and emarginate)
                                                22. LYCOPUS
   6 Calyx with 2 entire lips, the upper with a dorsal
      outgrowth   **10. SCUTELLARIA**
   6 Calyx with 5-10 lobes or teeth, often 3 forming
      upper and 2 lower lip, without dorsal outgrowth   7
7 Calyx-teeth 10, hooked at apex   **9. MARRUBIUM**
7 Calyx-teeth (4-)5, not hooked at apex   8
   8 Corolla ± actinomorphic, indistinctly 2-lipped,
      or distinctly 2-lipped with upper lip ± flat;
      stamens (except in female flowers) usually fully
      exposed from front view of flower, sometimes longer
      than corolla   Key A
   8 Corolla distinctly 2-lipped, with upper lip
      distinctly hooded and usually at least partially
      concealing stamens from front view   Key B

Key A - Calyx with 5 teeth or lobes; stamens 4, not included
        within corolla-tube (but often very reduced in female
        flowers); corolla ± actinomorphic to zygomorphic and
        2-lipped with upper lip ± flat and not concealing
        stamens from front view.
1 Corolla ± actinomorphic, with 4 ± equal lobes or
   1 lobe shortly bifid   **23. MENTHA**
1 Corolla strongly 2-lipped, or weakly zygomorphic with

upper lip of 2 and lower lip of 3 lobes   2
2 Upper calyx-tooth with widened apical appendage;
   shrub with flowers crowded in long-stalked spikes
                                                   24. LAVANDULA
2 Upper calyx-tooth without appendage; if a shrub,
   flowers not crowded in long-stalked spikes   3
3 Calyx-teeth ± equal, not forming 2 lips   4
3 Calyx-teeth unequal, the upper 3 differing markedly
   in length and/or breadth from lower 2   9
4 Evergreen shrubs up to c.60cm; leaves entire   5
4 Herbaceous perennials; leaves often crenate to
   serrate   6
5 Stamens divergent, the longer 2 longer than corolla;
   corolla bluish-violet (very rarely white)   **19. HYSSOPUS**
5 Stamens convergent, shorter than corolla; corolla
   pinkish-purple (very rarely white)   **17. SATUREJA**
6 Flowers few in axillary clusters; plant with long
   procumbent stolons   **14. GLECHOMA**
6 Flowers numerous in terminal inflorescences; plant
   without stolons   7
7 Leaves entire or nearly so   **20. ORIGANUM**
7 Leaves conspicuously crenate to serrate   8
8 Calyx 15-veined; stem-leaves many pairs   **13. NEPETA**
8 Calyx 5-10-veined; stem-leaves ≤4(5) pairs **1. STACHYS**
9 Corolla ≥25mm; calyx with upper lip 2-3-toothed, with
   lower lip with 2 much deeper but scarcely narrower
   lobes   **8. MELITTIS**
9 Corolla ≤22mm; calyx with upper lip 3-lobed, with
   lower lip with 2 much narrower lobes   10
10 Stems woody, either erect to ascending or filiform
   and procumbent to decumbent   **21. THYMUS**
10 Stems herbaceous, usually ± erect   11
11 Corolla-tube curved; stigmas ± equal; fresh plant
   lemon-scented   **16. MELISSA**
11 Corolla-tube straight; stigmas distinctly unequal;
   fresh plant variously scented but not of lemon
                                               18. CLINOPODIUM

Key B - Calyx with 5 teeth or lobes; stamens 4, not included
        within corolla-tube (but often very reduced in female
        flowers); corolla distinctly 2-lipped, with upper lip
        distinctly hooded and often at least partially
        concealing stamens from front view.
1 Plant with stolons >10cm   2
1 Plant without stolons   5
2 Corolla yellow   **4. LAMIASTRUM**
2 Corolla variously bluish, purplish or white   3
3 Flowers few in axillary clusters; bracts all leaf-
   like   **14. GLECHOMA**
3 Flowers in dense whorls forming terminal
   inflorescence; at least upper bracts much reduced   4
4 Lateral lobes of lower lip of corolla obscure or

    pointed, if rounded much <1/2 as large as terminal
    lobe; terminal lobe bifid for >1/3 length; carpels
    and mericarps truncate at apex   **5. LAMIUM**
4 Lateral lobes of lower lip of corolla conspicuous,
    rounded, usually c.1/2 as large as terminal lobe;
    terminal lobe not bifid or bifid for <1/3 length;
    carpels and mericarps rounded at apex   **1. STACHYS**
5 Calyx-teeth unequal, the upper 3 differing markedly
    in length and/or breadth from lower 2   6
5 Calyx-teeth + equal, not forming 2 lips   8
  6 Corolla >2cm; calyx >12mm   **8. MELITTIS**
  6 Corolla <2cm; calyx <12mm   7
7 Flowers forming dense terminal head; upper lip of
    calyx nearly entire, with 3 small teeth; fresh
    plant not strongly scented   **15. PRUNELLA**
7 Flowers in axillary whorls, only the uppermost of
    which merge; upper lip of calyx with 3 distinct
    lobes; fresh plant strongly lemon-scented **16. MELISSA**
  8 Lower leaves lobed >1/2 way to midrib **3. LEONURUS**
  8 Leaves entire to serrate   9
9 Calyx-teeth and associated bracteoles spine-tipped;
    lower lip of corolla with conical projection at
    base of each of 2 lateral lobes   **6. GALEOPSIS**
9 Calyx-teeth and bracteoles not spine-tipped, or
    former sometimes so; lower lip of corolla without
    2 conical projections   10
  10 Stigmas distinctly unequal; leaves whitish-
      tomentose on lowerside   **7. PHLOMIS**
  10 Stigmas + equal; leaves rarely white-tomentose
      on lowerside   11
11 Lateral lobes of lower lip of corolla obscure or
    pointed, if rounded much <1/2 as large as terminal
    lobe; terminal lobe bifid for >1/3 length;
    carpels and mericarps truncate at apex   **5. LAMIUM**
11 Lateral lobes of lower lip of corolla conspicuous,
    rounded, usually c.1/2 as large as terminal lobe;
    terminal lobe not bifid or bifid for <1/3 length;
    carpels and mericarps rounded at apex   12
  12 Upper lip of corolla scarcely hooded; calyx
      15-veined   **13. NEPETA**
  12 Upper lip of corolla strongly hooded; calyx
      5-10-veined   13
13 Calyx-teeth c. as long as wide; calyx-tube
    trumpet-shaped (cylindrical or + so for most
    part but conspicuously expanded in distal 1/2)
                        2. BALLOTA
13 Calyx-teeth usually distinctly longer than wide;
    calyx-tube cylindrical to obconical, not
    conspicuously expanded in distal 1/2   **1. STACHYS**

SUBFAMILY 1 - LAMIOIDEAE (Stachyoideae, Ajugoideae, Scutell-
arioideae) (genera 1-12).   Plants mostly not pleasantly

scented; male-sterile plants rather rare; upper lip of corolla usually conspicuously hooded (not in <u>Marrubium</u>) or 0; stamens 4, shorter than corolla; cotyledons usually at least as long as wide; seeds with endosperm; pollen with 3 furrows, 2-celled at dispersal.

1. STACHYS L. (<u>Betonica</u> L.) - <u>Woundworts</u>
Herbaceous annuals or perennials; leaves crenate to serrate; calyx with 5 ± equal acute (sometimes ± spine-tipped) lobes; corolla yellow, pink to purple, or white, with hooded upper lip, with 3-lobed lower lip; stamens 4, shorter than upper lip of corolla; whorls distant in leaf-axils, or congested in axils of reduced bracts.

```
1  Corolla cream to yellow                                    2
1  Corolla purplish, very rarely white                        3
   2  Annual; flowers <6 per node; corolla 10-16mm
                                                    9. S. annua
   2  Biennial to perennial; flowers usually >6 per
      node; corolla 15-20mm                       8. S. recta
3  Annual; corolla <10mm                         10. S. arvensis
3  Perennial; corolla >10mm                                   4
   4  Bracteoles 0 or minute                                  5
   4  Bracteoles mixed in with flowers, at least as
      long as calyx                                           7
5  Middle and upper stem-leaves sessile      7. S. palustris
5  All leaves up to 1st inflorescence node petiolate         6
   6  Petioles of middle and upper stem-leaves 1/10 to
      1/5 total leaf + petiole length; few or 0 fruits
      ripening                                 6. S. x ambigua
   6  Petioles of middle and upper stem-leaves 1/4 to
      2/5 (1/2) total leaf + petiole length; all or most
      fruits ripening                            5. S. sylvatica
7  Corolla-tube longer than calyx; calyx very sparsely
   pubescent to glabrous on outside; sterile leaf-
   rosettes present at flowering               1. S. officinalis
7  Corolla-tube shorter than calyx; calyx densely
   pubescent on outside; sterile leaf-rosettes 0 at
   flowering                                                  8
   8  Upper part of stem and calyx with stalked glands
      as well as eglandular hairs                 4. S. alpina
   8  Stem and calyx with only eglandular hairs              9
9  Leaves cuneate at base; stem and both leaf-surfaces
   densely white tomentose                      2. S. byzantina
9  At least lower leaves cordate at base; green upper
   leaf-surface showing through pubescence     3. S. germanica
```

1. S. officinalis (L.) Trev. St. Leon (<u>Betonica officinalis</u> L.) - <u>Betony</u>. Erect, sparsely pubescent perennial to 75cm, with sterile leaf-rosettes and only 2-4 pairs of stem-leaves; corolla reddish-purple, 12-18mm. Native; hedgebanks, grassland, heaths, avoiding heavy soils; common in En and Wa,

local in Jersey, extremely so in Sc and Ir. Often very dwarf
on grassy cliff-tops.
   2. S. byzantina K. Koch (S. lanata Jacq. non Crantz) -
Lamb's-ear.   Erect to ascending, densely white-woolly
perennial to 75cm, with thick surface rhizomes and many
sterile leaf-rosettes; corolla pinkish-purple, 15-25mm.
Intrd; much grown in gardens, persistent throw-out and
sometimes self-sown on tips and waste ground; scattered in S
& C Br; SW Asia.
   3. S. germanica L. - Downy Woundwort. Erect, densely white- **RRR**
pubescent to -subtomentose biennial to 80(100)cm, without
rhizomes; corolla pinkish-purple, 12-20mm. Native; rough
open grassland, hedgebanks, wood-borders on calcareous soils;
extremely local over c.20 x 7km in Oxon, formerly elsewhere
in S & C Br.
   4. S. alpina L. - Limestone Woundwort. Erect, softly **RRR**
pubescent perennial to 1m, with stalked glands in upper
parts, without rhizomes; corolla as in S. germanica. Native;
open woods; 1 site each in Denbs and W Gloucs.
   5. S. sylvatica L. - Hedge Woundwort. Erect, strong- **669**
smelling, harshly pubescent perennial to 1m, with stalked
glands in upper parts, with strong ± surface rhizomes; leaves
all petiolate, ovate, cordate at base; corolla reddish-
purple, 13-18mm. Native; woods, hedgerows, rough ground;
common over most BI.
   6. S. x ambigua Smith (S. sylvatica x S. palustris) - Hybrid **669**
Woundwort. Intermediate between parents in most characters;
leaves all petiolate, middle and upper stem-leaves with
petioles 1/10 to 1/5 total length; few or no fruits ripening.
Native; in habitats of either parent, but often in absence of
1 or both, distributed by fragmented rhizomes; scattered over
whole BI.
   7. S. palustris L. - Marsh Woundwort. Erect, slightly- **669**
smelling, pubescent perennial to 1m, with stalked glands in
upper parts, with strong ± surface rhizomes developing slight
swellings at tips late in year; upper and middle stem-leaves
sessile, narrowly oblong-ovate, cordate to rounded at base;
corolla pinkish-purple, 12-15mm. Native; damp places, by
rivers and ponds, and on rough ground; common over most BI.
   8. S. recta L. - Perennial Yellow-woundwort. Erect to
ascending, shortly pubescent perennial to 70cm, without
rhizomes, without glandular hairs; corolla pale yellow,
15-20mm. Intrd; waste ground; natd at Barry Docks, Glam,
since 1923; Europe.
   9. S. annua (L.) L. - Annual Yellow-woundwort. Erect,
shortly pubescent annual to 30cm, with or without glandular
hairs; corolla pale yellow, 10-16mm. Intrd; waste ground;
rather rare casual in C & S Br; Europe.
   10. S. arvensis (L.) L. - Field Woundwort. Erect to
ascending, pubescent annual to 25cm, without glandular hairs;
corolla dull pinkish-purple, 6-8mm. Native; arable ground on
non-calcareous soils; scattered throughout much of BI.

2. BALLOTA L. - <u>Black</u> <u>Horehound</u>
Herbaceous perennials; leaves serrate; calyx with 5 ± equal
acuminate lobes; corolla reddish-mauve, with hooded upper
lip, with 3-lobed lower lip; stamens 4, shorter than upper
lip of corolla; whorls distant, in leaf-axils.

1. B. nigra L. - <u>Black</u> <u>Horehound</u>. Erect perennial to 1m,
with unpleasant smell when bruised; leaves petiolate, ovate,
subcordate to broadly cuneate at base; corolla 9-15mm.
Native; hedgerows, waysides, rough ground; common in most of
En, Wa and CI, very local in Sc and Ir and mostly intrd. Our
plant is ssp. **foetida** (Vis.) Hayek; reports of ssp. **nigra**,
with narrower and longer calyx-teeth, are all or nearly all
errors.

3. LEONURUS L. - <u>Motherwort</u>
Herbaceous perennials; lower and middle stem-leaves and basal
leaves deeply palmately to ternately lobed; calyx with 5 ±
equal spine-tipped patent lobes; corolla pinkish-purple, with
hooded upper lip, with 3-lobed lower lip; stamens 4, shorter
than upper lip of corolla; whorls distant in leaf-axils.

1. L. cardiaca L. - <u>Motherwort</u>. Rhizomatous; stems erect,
to 1.2m, shortly pubescent; corolla 8-12mm, densely pubescent
outside on upper lip. Intrd; formerly grown medicinally,
natd in waste places and waysides; thinly scattered over much
of BI, becoming scarcer; Europe.

4. LAMIASTRUM Heister ex Fabr. (<u>Galeobdolon</u> Adans.) - <u>Yellow</u>
<u>Archangel</u>
Stoloniferous herbaceous perennials; leaves serrate; calyx
with 5 ± equal triangular-acuminate lobes; corolla yellow,
with hooded upper lip, with 3-lobed lower lip; stamens 4,
shorter than upper lip of corolla; whorls distant in leaf-
axils. Differs from <u>Lamium</u> in glabrous (not pubescent)
anthers.

1. L. galeobdolon (L.) Ehrend. & Polatschek (<u>Lamium</u>
<u>galeobdolon</u> (L.) L., <u>Galeobdolon</u> <u>luteum</u> Hudson) - <u>Yellow</u>
<u>Archangel</u>. Flowering stems erect, to 60cm; stolons leafy,
rooting at nodes, often >1m.
1 Most leaves with large conspicuous whitish blotches
   for whole year; fruiting calyx ≥(11.5)12mm; upper
   lip of corolla ≥(7.5)8mm wide   **c. ssp. argentatum**
1 Leaves without whitish blotches, or some with blotches
   for some of year; fruiting calyx ≤12(12.5)mm; upper
   lip of corolla ≤8(8.5)mm
     2 Bracts 1-2(2.2)x as long as wide, obtusely serrate;
       flowers ≤8(9) per node, at ≤4(5) nodes; flowering
       stems with hairs ± confined to 4 angles
                                          a. ssp. galeobdolon
     2 Bracts (1.5)1.7-3.6x as long as wide, acutely

serrate; flowers >(7)10 per node, at >(3)4
nodes; flowering stems with hairs on faces as well
as angles   **b. ssp. montanum**
   a. Ssp. galeobdolon (Lamium galeobdolon ssp. galeobdolon,
Galeobdolon luteum ssp. luteum). Leaves rarely whitish-
blotched; differs from ssp. montanum in less robust habit
with less extensive stolons; 2n=18; and see key. Native;
woods, wood-borders and hedgerows; very local in small area
of N Lincs, in absence of ssp. montanum.
   b. Ssp. montanum (Pers.) Ehrend. & Polatschek (Lamium
galeobdolon ssp. montanum (Pers.) Hayek, Galeobdolon luteum
ssp. montanum (Pers.) Dvorakova). Differs from ssp.
galeobdolon in more robust habit with very long stolons;
2n=36; and see key. Native; woods, wood-borders and
hedgerows; common over most of En and Wa, very local in SW
Sc, E Ir and CI.
   c. Ssp. argentatum (Smejkal) Stace (Galeobdolon argentatum
Smejkal). Differs from other 2 sspp. in large conspicuous
whitish blotches on all leaves at all seasons; 2n=36; and see
key. Intrd; much grown in gardens, natd in shrubberies and
waysides; scattered over Br and CI; origin uncertain.
Perhaps only a cultivar of ssp. montanum.

5. LAMIUM L. - Dead-nettles
Annuals or herbaceous perennials; leaves serrate to deeply
so, rarely + entire, calyx with 5 + equal narrowly
triangular-acuminate lobes; corolla white, or pink to purple
or mauve, with hooded upper lip, with + 1-lobed lower lip of
which lateral lobes are much reduced and pointed or rounded;
stamens 4, shorter than upper lip of corolla; whorls distant
in leaf-axils, or + congested in axils of modified leaves.

1 Perennials with rhizomes and/or stolons; corolla-tube
   curved   2
1 Annuals; corolla-tube straight   3
   2 Corolla white; leaves never blotched whitish; lower
     lip of corolla with 2-3 teeth each side   **1. L. album**
   2 Corolla usually pinkish-purple; leaves usually
     blotched whitish; lower lip of corolla with 1
     tooth on each side   **2. L. maculatum**
3 + all leaves petiolate   4
3 Middle and upper leaves subtending whorls sessile   5
   4 Leaves subtending whorls serrate to crenate-
     serrate, with teeth <2mm long   **3. L. purpureum**
   4 Leaves subtending whorls deeply serrate, with
     many teeth >2mm long   **4. L. hybridum**
5 Calyx 5-7mm at flowering, densely white- + patent-
   pubescent, the teeth erect to convergent at fruiting;
   lower lip of corolla <3mm   **6. L. amplexicaule**
5 Calyx 8-12mm at flowering, + appressed-pubescent,
   the teeth divergent at fruiting; lower lip of
   corolla >3mm   **5. L. confertum**

1. L. album L. - <u>White Dead-nettle</u>. Rhizomatous or sometimes stoloniferous perennial; calyx 9-15mm; corolla white, 18-25mm, with tube 9-14mm. Native; hedge-banks, waysides, rough ground; common in most of lowland Br except N & W Sc, rare in CI, local (mainly NE & CE) in Ir.

2. L. maculatum (L.) L. - <u>Spotted Dead-nettle</u>. Rhizomatous and/or stoloniferous perennial; calyx 8-15mm; corolla usually pinkish-purple, rarely white, 20-35mm, with tube 10-18mm. Intrd; much grown in gardens and natd on rough ground and tips; scattered through much of Br and CI; Europe. The garden plant, and that natd in Br, nearly always has white-blotched leaves.

3. L. purpureum L. - <u>Red Dead-nettle</u>. Annual; calyx 5-7.5mm; corolla usually pinkish-purple, 10-18(20)mm, with tube 7-12mm. Native; cultivated and waste ground; common ± throughout BI.

4. L. hybridum Villars - <u>Cut-leaved Dead-nettle</u>. Annual; calyx 5-7(10)mm; corolla usually pinkish-purple, 10-18(20)mm, with tube 7-12mm, but often much shorter and not opening. Native; cultivated and waste ground, often with <u>L. purpureum</u>; scattered over most of lowland BI.

5. L. confertum Fries (<u>L. molucellifolium</u> auct. non (Schum.) **669** Fries) - <u>Northern Dead-nettle</u>. Annual; calyx 8-12mm; corolla usually pinkish-purple, 14-20(25)mm, with tube 10-15mm. Native; cultivated and waste ground; locally frequent near coast in N, W & E Sc, Man, formerly NW En, very scattered in Ir. Records of <u>L. molucellifolium</u> from S En refer to <u>L. hybridum</u>.

6. L. amplexicaule L. - <u>Henbit Dead-nettle</u>. Annual; calyx **669** 5-7mm; corolla usually pinkish-purple, when well-developed 14-20mm with tube 10-14mm, but often much shorter and not opening. Native, open, cultivated and waste ground; over ± all BI, common in E Br and CI, scattered in W Br and Ir.

6. GALEOPSIS L. - <u>Hemp-nettles</u>

Annuals; leaves usually serrate; calyx with 5 ± equal, weakly spine-tipped lobes; corolla variously white, pinkish-purple or yellow, with hooded upper lip, with 3-lobed lower lip; stamens 4, shorter than upper lip of corolla; whorls somewhat congested, in axils of reduced leaves.

1	Stems with soft hairs, not swollen at nodes	2
1	Stems with rigid bristly hairs, swollen at nodes	3
	2 Corolla predominantly pale yellow, (20)25-30mm; leaves and calyx densely silky- or velvety-pubescent	**1. G. segetum**
	2 Corolla predominantly reddish-pink, rarely white, 14-25mm; leaves and calyx variously pubescent but not densely silky or velvety	**2. G. angustifolia**
3	Corolla (22)27-35mm, yellow with purple blotch on lower lip; corolla-tube c.2x as long as calyx (incl. teeth)	**3. G. speciosa**

3   Corolla 13-20(25)mm, variously coloured; corolla-tube
    rarely >1.5x as long as calyx (incl. teeth)   4
4   Terminal lobe of lower lip of corolla entire to very
    slightly emarginate, + flat   **4. G. tetrahit**
4   Terminal lobe of lower lip of corolla clearly
    emarginate, convex (with revolute sides)
                                                  5. G. bifida

Other spp. - **G. ladanum** L. (Broad-leaved Hemp-nettle), from
Europe, differs from G. angustifolia in its ovate to narrowly
ovate leaves, and green calyx with translucent, smooth to
finely dotted (not opaque, + densely papillose) transparent
hairs (microscope!); it has been much confused with and over-
recorded for G. angustifolia, but has never been more than an
infrequent casual and is now very sporadic.

1. G. segetum Necker - Downy Hemp-nettle. Stems erect, to **669**
50cm; leaves lanceolate to ovate; calyx 7-10mm; corolla **RR**
(20)25-30(35)mm, pale yellow.   Native in arable land in
Caerns, once constant, now sporadic; casual in arable and
waste ground, scattered and sporadic in En and Wa.
2. G. angustifolia Ehrh. ex Hoffm. - Red Hemp-nettle. Stems **669**
erect, to 50cm; leaves linear-lanceolate to narrowly ovate, **R**
rarely ovate; calyx 8-13mm; corolla 14-25mm, reddish-pink,
rarely white.   Native; arable land, open ground mostly on
calcareous soils or maritime sand or shingle; locally
frequent in S & E En, very local in Wa and Ir, ?extinct in
Sc, decreasing.
3. G. speciosa Miller - Large-flowered Hemp-nettle. Stems
erect to ascending, to 1m; leaves ovate; calyx 12-17mm;
corolla (22)27-35mm, yellow with purple blotch on lower lip
(sometimes whole lower lip purplish).   Native; arable land,
often on peaty soil with root-crops, and waste places;
locally common in C & N Br and N Ir, rare and scattered in S
Br and C & S Ir.
4. G. tetrahit L. - Common Hemp-nettle. Stems erect, to 1m; **669**
leaves lanceolate-ovate to ovate; calyx 12-14mm; corolla
13-20(25)mm, pinkish-red with darker lines and blotches on
lower lip, the darker markings falling well short of tip of
terminal lobe, often white but perhaps never yellow. Native;
arable land, rough ground, woodland clearings, damp places;
common over most of Br and Ir, rare in CI.
4 x 5. G. tetrahit x G. bifida = G. x ludwigii Hausskn. is
intermediate in corolla shape and has 20-70 per cent pollen
fertility and low seed-set; it has been recorded from Wa and
C En, but is probably frequent.
5. G. bifida Boenn. - Bifid Hemp-nettle. Differs from G. **669**
tetrahit in usually shorter corolla (13-16mm) coloured like
G. tetrahit or like G. speciosa or white; darker markings on
lower lip very extensive to near margin (sometimes whole
terminal lobe dark); and see key. Native; in similar places
to G. tetrahit and probably as common and widespread.

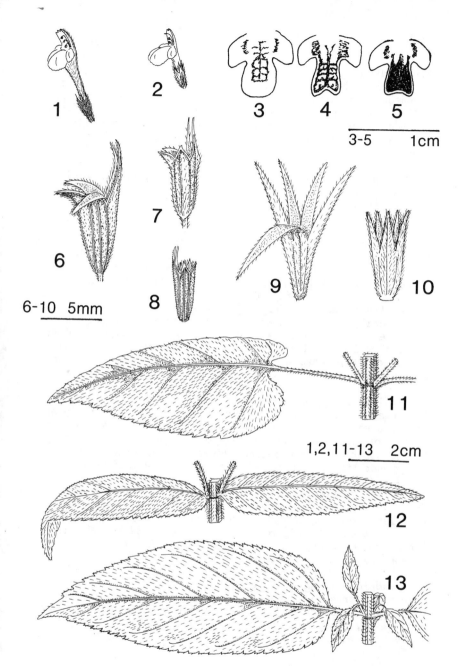

FIG 669 – Lamiaceae. 1–2, **Galeopsis** flowers. 1, G. segetum.
2, **G. angustifolia**. 3–5, **Galeopsis** lower lips of corolla. 3,
G. tetrahit. 4–5, **G. bifida**. 6–8, **Clinopodium** calyces. 6, C.
menthifolium. 7, C. ascendens. 8, C. calamintha. 9–10, **Lamium**
calyces. 9, **L. confertum**. 10, **L. amplexicaule**. 11–13, **Stachys**
leaves. 11, **S. sylvatica**. 12, **S. palustris**. 13, **S. x ambigua**.
3–5 drawn by C.A. Stace.

7. PHLOMIS L. - <u>Sages</u>
Perennial herbs or shrubs; leaves entire to crenate, white-tomentose; calyx with 5 ± equal, spine-tipped teeth; corolla yellow, rarely purplish, with strongly hooded upper lip, with 3-lobed lower lip; stamens 4, shorter than upper lip of corolla; whorls somewhat congested, in axils of reduced leaves.

Other spp. - **P. samia** L., from Greece, is grown in gardens and has been recorded as a persistent outcast in the past; it differs from P. russeliana in its purplish-mauve corolla and glandular hairs on stem and bracteoles (0 in P. russeliana).

1. P. russeliana (Sims) Benth. - <u>Turkish Sage</u>. Herb to 1m; **689** basal leaves ovate, cordate at base, 6-20cm, with longer petiole; calyx 20-25mm; corolla 30-35mm, yellow. Intrd; banks by roads and railways, rough ground; persistent or ± natd in few places in En and Sc; Turkey,

2. P. fruticosa L. - <u>Jerusalem Sage</u>. Evergreen shrub to **689** 1.3m; basal leaves 0; lower leaves elliptic-ovate, cuneate to truncate at base, 3-9cm, with shorter petiole; calyx 10-20mm; corolla 23-35mm, yellow. Intrd; grown in gardens and ± natd on sea-cliffs, banks and rough ground in S En and Ir, rare throwout elsewhere in Br; Mediterranean.

8. MELITTIS L. - <u>Bastard Balm</u>
Perennial herbs; leaves serrate; calyx with 2 lips, the upper with 2-3 short lobes, the lower with 2 deeper lobes; corolla white, or pink to mauve or purple, with flat or slightly hooded upper lip, with 3-lobed lower lip; stamens 4, shorter than upper lip of corolla; whorls distant, in leaf-axils.

1. M. melissophyllum L. - <u>Bastard Balm</u>. Stems erect, to   **R** 70cm; lower leaves ovate, cordate to rounded at base, petiolate; corolla 25-40mm, with lower lip longer than upper. Native; woods and hedgerows; very local in SW Wa, SW En, S Hants and W Sussex, formerly E Sussex, C En and C Wa, rare escape from gardens elsewhere.

9. MARRUBIUM L. - <u>White Horehound</u>
Perennial herbs; leaves serrate; calyx with 10 equal teeth, each hooked at end and patent at fruiting; corolla white, with erect, flat, bilobed upper lip, with 3-lobed lower lip; stamens 4, all included in corolla-tube; whorls distant, in leaf-axils.

1. M. vulgare L. - <u>White Horehound</u>. Plant whitish-   **R** tomentose or densely pubescent; stems erect or ascending, to 60cm; leaves ovate to suborbicular, petiolate; flowers numerous and dense in each whorl. Native; short grassland, open or rough ground and waste places; sparsely scattered in BI to Moray, becoming much rarer, probably native only near sea in S En and perhaps S & N Wa.

10. SCUTELLARIA L. - Skullcaps
Perennial herbs; leaves entire to serrate; calyx 2-lipped,
both lips entire, the upper with a dorsal outgrowth; corolla
pinkish to blue or purple, with hooded upper lip, with rather
obscurely 3-4-lobed lower lip; stamens 4, shorter than to c.
as long as upper lip of corolla; flowers 2 at each node, in
leaf-axils.

1 Flowers in axils of bracts markedly smaller than
  foliage leaves; lower leaves 5-15cm, with petioles
  >1cm   **1. S. altissima**
1 Flowers in axils of foliage leaves; leaves very
  rarely >5cm, with petioles <1cm   2
  2 Corolla 6-10mm, pale pinkish-purple, with nearly
    straight tube   **4. S. minor**
  2 Corolla 10-20mm, blue, with strongly bent tube   3
3 Calyx with glandular hairs; leaves entire at least in
  distal 1/2, hastate at base   **3. S. hastifolia**
3 Calyx with eglandular or 0 hairs; leaves crenate along
  length, cordate to rounded at base   **2. S. galericulata**

  1. S. altissima L. - Somerset Skullcap. Stems erect, to
80cm; leaves ovate-cordate, serrate, petiolate; calyx with
sparse usually glandular and eglandular hairs; corolla
12-18mm, blue with white lower lip. Intrd; natd in hedgerows
and wood borders since 1929 in N Somerset and since 1972 in
Surrey; S & E Europe.
  2. S. galericulata L. - Skullcap. Stems erect to decumbent,
to 50cm; leaves narrowly ovate-elliptic to lanceolate,
cordate to rounded at base, crenate, sessile to shortly
petiolate; calyx with rather sparse eglandular hairs; corolla
10-18mm, blue. Native; fens, wet meadows, by ponds and
rivers; locally common throughout most of BI.
  2 x 4. S. galericulata x S. minor = S. x hybrida Strail
occurs locally near the parents in S En and S Ir; it is
intermediate in corolla and leaf characters and highly
sterile, but vegetatively vigorous.
  3. S. hastifolia L. - Norfolk Skullcap. Stems erect to
decumbent, to 40cm; at least middle stem-leaves ovate-
triangular, hastate at base, with 0-2 extra serrations near
base on each side, shortly petiolate; calyx rather densely
glandular-pubescent; corolla 12-20mm, blue. Intrd; natd in
woodland at 1 site in W Norfolk; Europe.
  4. S. minor Hudson - Lesser Skullcap. Stems erect to
decumbent, to 25cm; leaves narrowly ovate, entire or with 1-2
serrations on each side near base, rounded to cordate or
subhastate at base, shortly petiolate; calyx with eglandular
hairs or glabrous; corolla 6-10mm, pale pinkish-purple.
Native; wet heaths and open woodland on acid soils; locally
frequent in S & W Br and S Ir, rare and very scattered

elsewhere in En and Jersey.

11. TEUCRIUM L. - <u>Germanders</u>
Annual or perennial herbs or very low shrubs; leaves serrate
to deeply lobed; calyx rather unequally 5-lobed, not 2-
lipped; corolla pinkish-purple or greenish-cream, with 5-
lobed lower lip and 0 upper lip; stamens 4, shorter than
lower lip; whorls distant in leaf-axils to congested in axils
of reduced bracts.

1 Leaves divided much >1/2 way to midrib; annual
                                                4. T. botrys
1 Leaves serrate much <1/2 way to midrib; perennial   2
  2 Corolla greenish-cream; upper calyx-tooth much
    wider than other 4   **1. T. scorodonia**
  2 Corolla pinkish-purple (rarely white); upper
    calyx-tooth scarcely different from other 4   3
3 Very dwarf evergreen shrub; whorls forming a terminal
              inflorescence   **2. T. chamaedrys**
3 Entirely herbaceous; whorls spaced down stem
                                              3. T. scordium

  1. T. scorodonia L. - <u>Wood</u> <u>Sage</u>. Herbaceous perennial;
stems erect, to 50cm; leaves serrate, cordate at base,
petiolate; corolla greenish-cream. Native; woods, hedgerows,
hilly areas, fixed shingle and dunes, on acidic or alkaline
usually well-drained soils; common in suitable places
throughout most of BI, but rare in much of C Ir and CE En.
  2. T. chamaedrys L. - <u>Wall</u> <u>Germander</u>. Dwarf evergreen **RR**
shrub; stems suberect to decumbent, to 40cm; leaves serrate,
cuneate at base, shortly petiolate; corolla pinkish-purple.
Native; chalk grassland at 1 site in E Sussex, discovered
1945; also grown in gardens and natd on old walls and dry
banks, very scattered in BI N to C Sc, mainly S En,
decreasing; Europe. Garden plants are often more robust and
less hairy and are probably the hybrid <u>T.</u> <u>chamaedrys</u> x <u>T.</u>
<u>lucidum</u> L.; some or most natd plants might be this also.
  3. T. scordium L. - <u>Water</u> <u>Germander</u>. Herbaceous perennial; **RRR**
stems ascending to decumbent, to 50cm; leaves crenate-
serrate, broadly crenate to rounded at base, + sessile;
corolla pinkish-mauve. Native; fens, dune-slacks and river-
banks on calcareous soils; very local, 1 site each in N Devon
and Cambs, locally frequent in WC Ir, formerly scattered in
En N to NW Yorks and in Guernsey, decreasing.
  4. T. botrys L. - <u>Cut-leaved</u> <u>Germander</u>. Erect annual to **RRR**
30cm; leaves dissected nearly to midrib, petiolate; corolla
pinkish-purple. Native; bare chalk and chalky fallow fields
on downs; c.5 places in N Hants, W Kent, Surrey and E Gloucs,
formerly elsewhere in S En, decreasing.

12. AJUGA L. - <u>Bugles</u>
Annual or perennial herbs; leaves subentire to deeply

divided; calyx ± equally 5-lobed; corolla pink, blue, white or yellow, with 3-lobed lower lip (terminal lobe often bifid), with very short 1-2-lobed upper lip; stamens 4, shorter than lower lip; whorls distant to slightly congested, in leaf-axils.

1 Leaves divided much >1/2 way to midrib; annual;
  corolla yellow   **3. A. chamaepitys**
1 Leaves subentire or serrate much <1/2 way to midrib;
  perennial with stolons and/or rhizomes; corolla blue,
  pink or white   2
  2 Plant with stolons; upper bracts shorter than
    flowers; upper part of stem hairy only on 2
    opposite sides   **1. A. reptans**
  2 Plant without stolons, with rhizomes; all bracts
    longer than flowers; stem hairy all round
                                              2. A. pyramidalis

Other spp. - A. genevensis L., from Europe, was formerly natd in chalk grassland in Berks and dunes in W Cornwall; it differs from A. reptans in absence of stolons and in stems hairy all round, and from A. pyramidalis in upper bracts shorter than flowers and without the purple or violet tint.

1. A. reptans L. - Bugle. Stolons present; flowering stems erect, to 30cm; bracts green or blue-tinged (a copper-leaved cultivar exists); corolla blue, rarely pink or white. Native; woods, shady places, damp grassland; common throughout most of BI.
1 x 2. A. reptans x A. pyramidalis = A. x pseudopyramidalis Schur (A. x hampeana A. Braun & Vatke) has occurred near the parents in Co Clare, W Sutherland, E Ross and Orkney; it is sterile and intermediate in most characters, forming stolons late in the season. Possibly now extinct.
2. A. pyramidalis L. - Pyramidal Bugle. Rhizomes present; R
flowering stems erect, to 30cm; bracts rather pale green strongly tinged with mauve or purple; corolla blue (rarely pink or white abroad). Native; rock crevices in hilly areas; very local in N & NW Sc, W C Ir, Westmorland and Co Antrim.
3. A. chamaepitys (L.) Schreber - Ground-pine. Stems erect RR
to decumbent, usually branched low down, to 20cm; leaves divided nearly to midrib in often linear segments; corolla yellow. Native; bare chalk and chalky arable fields on downs; local in S & SE En N to W Suffolk, formerly Northants, decreasing.

SUBFAMILY 2 - NEPETOIDEAE (Lavanduloideae, Rosmarinoideae) (genera 13-26). Plants mostly pleasantly scented; male-sterile plants very common; upper lip of corolla usually ± flat (hooded in Prunella, Rosmarinus and Salvia) or corolla ± actinomorphic; stamens 2 or 4, sometimes exceeding corolla; cotyledons usually wider than long; seeds without endosperm;

pollen with 6 furrows, 3-celled at dispersal.

13. NEPETA L. - Cat-mints
Perennial herbs; leaves serrate; calyx with 5 subequal teeth;
corolla white to blue, with flat to slightly hooded upper
lip, with 3-lobed lower lip; stamens 4, shorter than upper
lip of corolla.

1. **N. cataria** L. - Cat-mint. Plant softly densely grey-
pubescent; stems erect, to 1m; leaves ovate, cordate at base,
the lower usually >4cm; whorls densely crowded into terminal
inflorescences; corolla white with small purple spots, the
tube shorter than calyx. Probably native; open grassland,
waysides, rough ground on calcareous soils; rather scattered
in En and Wa, once more common, rarely natd or casual in Ir.

2. **N. x faassenii** Bergmans ex Stearn (N. mussinii auct. non
Sprengel ex Henckel; N. racemosa Lam. (N. mussinii Sprengel
ex Henckel) x N. nepetella L.) - Garden Cat-mint. Plant
softly grey-tomentose; stems ascending, to 1.2m; leaves
narrowly ovate-oblong, mostly truncate at base, rarely >3cm;
whorls in lax elongated inflorescences; corolla blue, the
tube longer than calyx. Intrd; much grown in gardens,
frequent throw-out and sometimes natd on tips and rough
ground; scattered in En, Wa and CI; garden origin. Said to
be sterile, but fertile plants exist; perhaps some plants are
N. racemosa.

14. GLECHOMA L. - Ground-ivy
Perennial herbs with long trailing stolons; leaves crenate-
serrate; calyx with 5 subequal teeth; corolla blue, rarely
pink or white, as in Nepeta; stamens as in Nepeta; flowers
only 2-4 per node, the whorls ± distant in leaf-axils.

1. **G. hederacea** L. - Ground-ivy. Stolons often >1m;
flowering stems suberect, to 30cm; leaves broadly ovate to
orbicular, cordate at base, those on stolons with long
petioles; corolla 15-20mm in bisexual flowers, smaller in
female ones. Native; woods, hedgerows, rough ground, often
on heavy soils; common throughout BI except N Sc.

15. PRUNELLA L. - Selfheals
Perennial herbs; leaves entire to divided ± to midrib; calyx
2-lobed, the upper lip ± truncate with 3 very short teeth,
the lower lip with 2 long teeth; corolla yellow, blue, pink
or white, with 3-lobed lower lip, with ± entire, strongly
hooded upper lip; stamens 4, shorter than upper lip of
corolla; whorls congested, in axils of strongly modified
bracts.

1. **P. vulgaris** L. - Selfheal. Stems erect to decumbent, to
30cm; leaves entire or shallowly toothed; corolla bluish-
violet, rarely pink or white. Native; grassland, lawns,

wood-clearings, rough ground; common throughout BI.

1 x 2. P. vulgaris x P. laciniata = P. x intermedia Link (P. x hybrida Knaf) occurs scattered in S & C En wherever P. laciniata occurs or has occurred; it is variably intermediate in leaf-shape and corolla-colour. It is said to be sterile, but intermediates of all degrees occur, suggesting backcrossing and/or segregation.

2. P. laciniata (L.) L. - Cut-leaved Selfheal. Differs from   R
P. vulgaris in leaves variably shaped, often some entire but at least the upper deeply divided + to midrib; corolla creamy-yellow or -white, rarely pale blue. Possibly native; calcareous grassland; scattered in S & C En N to N Lincs.

16. MELISSA L. - Balm
Perennial herbs; leaves serrate; calyx 2-lipped, the upper lip + truncate with 3 short teeth, the lower lip with 2 long teeth; corolla pale yellow, becoming whitish or pinkish, with 3-lobed lower lip, with 2-lobed flat or slightly hooded upper lip; stamens 4, shorter than upper lip of corolla; whorls distant, in leaf-axils.

1. M. officinalis L. - Balm. Plant lemon-scented when fresh; stems erect, to 1m; leaves ovate, the lower with cordate to truncate base and petiole >1/2 as long as lamina; corolla 8-15mm. Intrd; much grown in gardens (often variegated) and a frequent throw-out or self-sown and natd; scattered in BI N to Co Cavan, Man and Cumberland; S Europe.

17. SATUREJA L. - Winter Savory
Evergreen low shrubs; leaves entire; calyx with 5 subequal teeth; corolla pinkish-purple, rarely white, with 3-lobed lower lip, with shallowly 2-lobed + flat upper lip; stamens 4, shorter than upper lip; stigmas + equal; flowers in contracted cymes in axils of reduced leaves, the whorls slightly congested.

1. S. montana L. - Winter Savory. Stems erect to ascending,   679
to 50cm; leaves linear-lanceolate, finely acute; corolla 6-12mm. Intrd; grown in gardens and + natd on old walls in N Somerset and S Hants; S Europe.

18. CLINOPODIUM L. (Calamintha Miller, Acinos Miller) - Calamints
Herbaceous annuals or perennials; leaves subentire to serrate; calyx 2-lipped, the lower lip with 2 longer and narrower teeth than the 3 teeth of upper lip; corolla pinkish-purple or violet to pale lilac or almost white, with short 3-lobed lower lip, with shallowly 2-lobed flat upper lip; stamens 4, shorter than corolla; stigmas markedly unequal; whorls distant to congested, in axils of reduced leaves.

1 Axillary flower-clusters very dense, without common
   stalk; calyx-tube asymmetrically curved (upper side
   convex, lower side concave near apex); flowers
   pinkish-purple to violet (rarely white)   2
1 Axillary flower-clusters in contracted cymes with
   common stalk; calyx-tube straight or slightly curved
   symmetrically on upper and lower sides; corolla very
   pale lilac to mauvish-pink   3
   2 Most or all whorls with >8 flowers; calyx-tube
      not or scarcely swollen   4. C. vulgare
   2 Whorls with <6(8) flowers; calyx-tube strongly
      swollen near base on lower side, especially in
      fruit   5. C. acinos
3 Teeth of lower calyx-lobe 1-2mm, with hairs all
   0.1mm or very few longer; hairs in throat of calyx
   protruding beyond tube   3. C. calamintha
3 Teeth of lower calyx-lobe 2-4mm, with many hairs
   >0.2mm; hairs in throat of calyx usually entirely
   included   4
   4 Teeth of lower calyx-lobe 3-4mm; corolla 15-22mm;
      leaves often >4cm, with 6-10 teeth on each side
                                                    1. C. menthifolium
   4 Teeth of lower calyx-lobe 2-3(3.5)mm; corolla
      10-16mm; leaves rarely >4cm, with 3-8 teeth on
      each side   2. C. ascendens

Other spp. - **C. grandiflorum** (L.) Stace (Calamintha
grandiflora (L.) Moench, Satureja grandiflora (L.) Scheele)
(Greater Calamint), from S Europe, differs from C. sylvaticum
in its larger flowers (calyx 10-16mm, corolla 25-40mm); it is
grown in gardens and persists rarely as a throw-out.

1. C. menthifolium (Host) Stace (Calamintha sylvatica **669**
Bromf., Satureja menthifolia (Host) Fritsch) - Wood Calamint. **RRR**
Stems erect, to 60cm; stem-leaves 3-7cm; calyx 6-10mm, with
lower teeth 3-4mm; corolla 15-22mm in bisexual flowers.
Native; scrubby laneside bank on chalk; 1 site in Wight.
2. C. ascendens (Jordan) Samp. (Calamintha ascendens Jordan, **669**
C. sylvatica ssp. ascendens (Jordan) P. Ball, Satureja
ascendens (Jordan) K. Maly - Common Calamint. Stems erect,
to 60cm; stem-leaves 1.5-4(5)cm; calyx 5-8mm, with lower
teeth 2-3(3.5)mm; corolla 10-16mm in bisexual flowers.
Native; dry banks and rough grassland, usually calcareous;
local in BI N to E Donegal, Man and Durham, decreasing.
3. C. calamintha (L.) Stace (Calamintha nepeta auct. non **669**
(L.) Savi, C. nepeta ssp. glandulosa (Req.) P. Ball, Satureja **R**
calamintha (L.) Scheele) - Lesser Calamint. Stems erect, to
60cm; stem-leaves 1-2(2.5)cm; calyx 3-6mm, with lower teeth
1-2mm; corolla 10-15mm in bisexual flowers. Native; dry
banks and rough grassland, usually calcareous; local in CE
and SE En, extending W to Mons, CI, formerly Yorks,
decreasing.

4. C. vulgare L. (Satureja vulgaris (L.) Fritsch) - <u>Wild</u>
<u>Basil</u>.   Stems erect, to 75cm; stem-leaves 1.5-5cm; calyx
7-9.5mm, with lower teeth 2.5-4mm; corolla 12-22mm in
bisexual flowers, pinkish-purple. Native; hedgerows, wood-
borders and scrubby grassland on light soils; frequent in Br
N to C Sc, very scattered in Ir, Alderney.
5. C. acinos (L.) Kuntze (Acinos arvensis (Lam.) Dandy,
Satureja acinos (L.) Scheele) - <u>Basil</u> <u>Thyme</u>. Usually annual;
stems erect to decumbent, to 25cm; stem-leaves 5-15mm; calyx
4.5-7mm, with lower teeth 1.5-2.8mm; corolla 7-10mm, violet.
Native; bare or rocky ground, arable fields on dry, usually
calcareous soils; rather local in Br N to C Sc, sparse in N
and decreasing there, very local (?intrd) in C & SE Ir.

19. HYSSOPUS L. - <u>Hyssop</u>
Evergreen low shrubs; leaves entire; calyx with 5 equal
teeth; corolla blue, rarely white, with 3-lobed lower lip and
shallowly 2-lobed ± flat upper lip; stamens 4, longer than
corolla; styles ± equal; whorls distant below, fairly
congested above, in axils of reduced leaves.

1. H. officinalis L. - <u>Hyssop</u>. Stems erect to ascending, to **679**
60cm; leaves narrowly oblong-elliptic to ± linear, rounded to
obtuse or acuminate; corolla 7-12mm, the stamens <4mm longer.
Intrd; grown in gardens and ± natd on old walls in Dorset,
Berks and perhaps elsewhere (no longer S Hants), rare casual
elsewhere; S Europe.

20. ORIGANUM L. - <u>Wild</u> <u>Marjoram</u>
Herbaceous perennials; leaves entire or remotely crenate-
denticulate; calyx with 5 subequal teeth; corolla reddish-
purple, rarely white, with short 3-lobed lower lip, with
shallowly 2-lobed flat upper lip; stamens 4, longer than
corolla in bisexual flowers; inflorescence a mass of dense
cymes forming corymbose panicle, with large purple
bracteoles.

1. O. vulgare L. - <u>Wild</u> <u>Marjoram</u>. Stems erect, to 50(80)cm;
leaves ovate, 1-4cm; corolla 4-7mm. Native; dry grassland,
hedgebanks and scrub, usually on calcareous soils; locally
common in BI N to C Sc. Grown as the herb oregano; the herb
marjoram is **O. majorana** L. (Pot Majoram), a rare casual from
N Africa and SW Asia.

21. THYMUS L. - <u>Thymes</u>
Dwarf evergreen shrubs; leaves entire; calyx 2-lipped, the
upper lip with 3 short teeth, the lower lip with 2 long
teeth; corolla pinkish-purple or mauve to white, with 2 ill-
defined lips, the upper of 1 emarginate lobe, the lower of 3
lobes; stamens 4, longer than corolla in bisexual flowers;
whorls crowded into dense terminal heads or the lower ones
more distant, in axils of reduced leaves.

1 Leaf margin revolute so that leaves are linear to
   narrowly oblong-elliptic in outline; plant without
   procumbent stems rooting at nodes   **1. T. vulgaris**
1 Leaf margins not or scarcely revolute; leaves elliptic
   or elliptic-oblong to narrowly so; plant usually with
   procumbent stems rooting at nodes   2
   2 Lower internodes of flowering stems with hairs all
      or nearly all on the 4 angles   **2. T. pulegioides**
   2 Lower internodes of flowering stems with hairs
      mainly on 2 or 4 faces   3
3 Lower internodes of flowering stems with hairs on all
   faces ± evenly distributed   **4. T. serpyllum**
3 Lower internodes of flowering stems with hairs on 2
   opposite faces, the 2 other faces glabrous or nearly
   so   **3. T. polytrichus**

Other spp. - **T. x citriodorus** Pers. (T. vulgaris x T.
pulegioides) (Lemon Thyme) is grown in gardens and might
occur as an throw-out or relic; it has the upright habit of
T. vulgaris but wider, scarcely revolute leaves and a
distinct lemon scent when fresh.

1. T. vulgaris L. - Thyme. Stems erect to decumbent, not or   **679**
scarcely rooting along length, to 40cm; leaves 3-8 x 0.5-
2.5mm, grey-green with very short hairs; corolla pale purple
to very pale mauve. Intrd; grown in gardens as herb and natd
on old walls and stony banks; very scattered in CI and S En;
W Mediterranean.
  2. T. pulegioides L. - Large Thyme. Vegetative stems   **679**
procumbent to ascending, rooting but not forming dense mats;
flowering stems suberect to decumbent, to 25cm, with hairs ±
only on the 4 angles; leaves ≤12 x 6mm, usually glabrous,
green; corolla pinkish-purple. Native; short fine turf or
barish places in coarser turf on well-drained chalky or sandy
soils; locally frequent in S & C En, scattered N to SE Yorks,
very rare and scattered in Ir and Sc. More robust than next
2 spp and flowering c.1 month later in S En. The only sp. on
heaths and dunes in SE En.
  3. T. polytrichus A. Kerner ex Borbas (T. drucei Ronn., T.   **679**
praecox auct. non Opiz, T. serpyllum auct. non L.) - Wild
Thyme. Vegetative stems procumbent, abundantly rooting,
usually forming dense mats; flowering stems decumbent to
ascending, to 10cm, with hairs ± only on 2 opposite faces;
leaves ≤8 x 4mm, glabrous or pubescent, green; corolla pink
to pinkish-purple. Native; short fine turf or open sandy or
rocky places; common over BI in suitable places; more
confined than T. pulegioides to chalk in SE En, but much more
catholic in N En, Sc, Ir and CI. Our plant is ssp.
britannicus (Ronn.) Kerguelen (T. praecox ssp. britannicus
(Ronn.) Holub, ssp. arcticus (Durand) Jalas).
  4. T. serpyllum L. - Breckland Thyme. Closely resembles   **679**
small T. polytrichus but differs in flowering stems with   **RR**

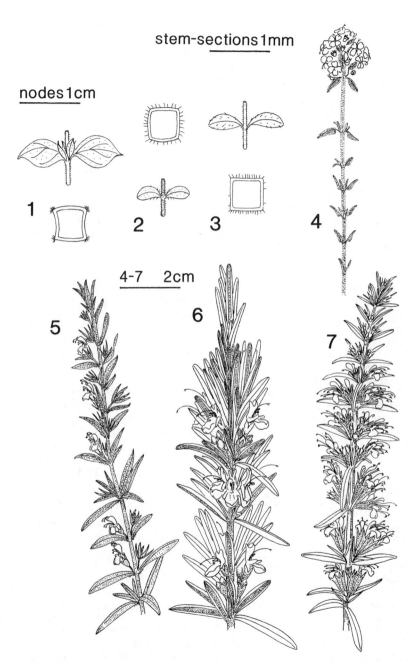

stem-sections 1mm

nodes 1cm

1

2

3

4

4-7   2cm

5

6

7

FIG 679 - **Lamiaceae**. 1–3, nodes and stem-sections of **Thymus**. 1, **T. pulegioides**. 2, **T. serpyllum**. 3, **T. polytrichus**. 4, **Thymus vulgaris**. 5, **Satureja**. 6, **Rosmarinus**. 7, **Hyssopus**.

hairs ± equally on all faces; leaves ≤5 x 1.5mm, with lateral
veins mostly ending short of margin (in T. polytrichus
lateral veins usually curve round parallel to margin and join
at leaf apex).   Native; sandy heaths in c.22 sites over
c.30km in W Suffolk and W Norfolk, formerly Cambs.

22. LYCOPUS L. - Gypsywort

Herbaceous perennials; leaves sharply serrate to deeply and
acutely lobed; calyx with 5 equal teeth; corolla white with
small purple dots, nearly actinomorphic, with 4 subequal
lobes, the uppermost usually wider and shallowly bifid;
stamens 2, longer than corolla; whorls remote, in leaf-axils.

1. L. europaeus L. - Gypsywort. Stems erect, to 1m; lower
leaves partly divided >1/2 way to midrib into very acute
lobes, tapered to short petiole; flowers densely clustered;
3-5mm. Native; fens, wet fields, by lakes and rivers; common
over most of En and Wa, much more scattered in CI, Ir and Sc.

23. MENTHA L. - Mints

Herbaceous rhizomatous and/or stoloniferous perennials with
characteristic scents when fresh; leaves entire to serrate;
calyx with 5 equal to rather unequal teeth; corolla pinkish
to bluish-mauve or white, nearly actinomorphic, with 4
subequal lobes, the uppermost usually wider and shallowly
bifid; stamens 4, longer than corolla in bisexual flowers of
fertile taxa; whorls all distant in leaf-axils, or the upper
congested in axils of reduced leaves.
  Taxonomically difficult due to well marked plasticity,
widespread hybridization, and the clonal propagation of
mutants and nothomorphs by the strongly developed rhizomes.
The stamens are typically exserted, but are included in M.
requienii, some plants of M. pulegium, female flowers of all
other spp., and most (but not all) hybrids. With practice
the scent of fresh plants is very helpful, but difficult to
describe.
  Taxa 7-11 (M. spicata group), involving M. spicata
(tetraploid), M. suaveolens (diploid) and their hybrids with
each other and with the non-British M. longifolia (diploid),
are particularly difficult, especially owing to the great
variation of M. spicata, which is itself derived from M.
longifolia x M. suaveolens. The 2 triploid hybrids are
sterile, but the diploid hybrid and the species are fertile
in bisexual plants or in female plants open to a pollen
source. Pubescent plants that are not M. suaveolens are
often impossible to name for certain. Characters of M.
suaveolens often seen in its hybrids are the broad, obtuse,
very rugose leaves with teeth partly folded under the margin
and with floccose indumentum on lowerside; of M. longifolia
are the lanceolate-oblong, acute, flat leaves with sharp
patent teeth and with felted grey indumentum; and of M.
spicata are the lanceolate to ovate, acute, not to slightly

rugose leaves with usually forward-directed teeth and relatively coarse pubescence. <u>M.</u> spicata is the most variable; very broad-leaved plants, and strongly rugose-leaved plants occur, but this is the only sp. of the 3 that can be glabrous to sparsely pubescent and the only sp. that can smell of spearmint. Hybrids (especially <u>M.</u> x <u>villosa</u>) can be very variable, showing many combinations of characters not always connected by intermediates, but they do not exactly duplicate the combinations shown by any of the spp.

1 Stems filiform, procumbent, rooting, mat-forming;
  flowers <6 per node   **13. M. requienii**
1 Stems sometimes procumbent but not rooting along
  length and mat-forming; flowers usually >6 per node   2
  2 Whorls usually all axillary, the axis terminated
    by leaves, or by a reduced whorl; bracts like the
    leaves but reduced   3
  2 Upper whorls contracted into terminal long or
    rounded head; upper bracts much reduced, unlike
    leaves   7
3 Calyx with hairs in throat; lower 2 calyx-teeth
  narrower and slightly longer than 3 upper
                               12. M. pulegium
3 Calyx without hairs in throat; calyx-teeth $\pm$ equal   4
  4 Calyx 1.5-2.5mm, incl. triangular teeth <0.5mm,
    pubescent all over; usually fertile   **1. M. arvensis**
  4 Calyx 2-4mm, incl. narrowly triangular to subulate
    teeth 0.5-1.5mm, glabrous or pubescent; usually
    sterile   5
5 Calyx 2-3.5mm, incl. teeth usually <1mm, the tube
  <2x (usually c.1.5x) as long as wide   **4. M. x gracilis**
5 Calyx 2.5-4mm, incl. teeth usually >1mm, the tube
  c.2x as long as wide   6
  6 Plant subglabrous; calyx mostly >3.5mm; stamens
    usually exserted   **3. M. x smithiana**
  6 Plant pubescent; calyx mostly <3.5mm; stamens
    usually included   **2. M. x verticillata**
7 Leaves distinctly petiolate; flower-head 12-25mm
  across   8
7 Leaves sessile or $\pm$ so; flower-head 5-15mm across   10
  8 Leaves and calyx-tube glabrous to very sparsely
    pubescent   **6. M. x piperita**
  8 Leaves and calyx-tube pubescent   9
9 Leaves ovate to ovate-lanceolate; flowers in rounded
  heads; plant normally fertile   **5. M. aquatica**
9 Leaves lanceolate to ovate-lanceolate; flowers usually
  in elongate $\pm$ pyramidal heads; plant normally
  sterile   **6. M. x piperita**
10 Leaves suborbicular to ovate, strongly rugose,
  pubescent, obtuse to $\pm$ rounded at apex, with
  teeth bent under and hence appearing as crenations
  from above; corolla whitish; fresh plant with

sickly scent   **11. M. suaveolens**
10 Leaves lanceolate to ovate-oblong, rugose or not,
   glabrous to densely pubescent, acute to subobtuse
   at apex, with teeth not bent over and hence
   appearing acute from above; if leaves close to
   those of sp. 11 then corolla pinkish and fresh
   plant with sweet scent   11
11 Plant subglabrous to densely pubescent; corolla
   white to pinkish; leaves with acute, usually
   forwardly-directed teeth unless leaves broadly ovate
   to suborbicular   12
11 Plant pubescent to densely so; corolla pinkish;
   leaves with often acuminate teeth that curve
   outwards and become patent, especially near leaf-
   base, never broadly ovate to suborbicular   13
   12 Plant normally fertile; leaves lanceolate to
   broadly ovate, rarely both rugose and pubescent
                                          7. M. spicata

   12 Plant sterile; leaves usually ovate to suborbicular, often both rugose and pubescent
                                          9. M. x villosa
13 Plant normally fertile; leaves oblong-lanceolate to
   -ovate, with nearly parallel sides and broad rounded
   base; Scotland only   **10. M. x rotundifolia**
13 Plant sterile; leaves lanceolate-elliptic, broadest
   near middle, narrowed into rounded to subcuneate
   base; widespread   **8. M. x villosonervata**

Other spp. - **M. longifolia** (L.) Hudson (<u>Horse</u> <u>Mint</u>), from
Europe, has often been misdetermined for pubescent plants of
<u>M. spicata</u> or <u>M. x villosonervata</u>, but does not occur in BI;
see characters above.

1. M. arvensis L. (?<u>M. gentilis</u> L.) - <u>Corn</u> <u>Mint</u>. Plant
pubescent, with sickly scent; stems erect to decumbent, to
60cm; leaves ovate- to lanceolate-elliptic, cuneate at base,
with shallow, blunt teeth, petiolate; whorls in leaf-axils;
calyx campanulate, with triangular teeth; corolla lilac.
Native; arable fields, damp places in wood clearings, fields
and pond-sides; fairly common throughout most BI.
  1 x 11. M. arvensis x M. suaveolens = M. x carinthiaca Host
(<u>M. x muelleriana</u> F. Schultz) occurred in S Devon and Dorset
but is now extinct; it is similar to <u>M. arvensis</u> but has
usually larger, broader leaves with deeper, sharper
serrations and is sterile.
  2. M. x verticillata L. (<u>M. arvensis</u> x <u>M. aquatica</u>) - **684**
<u>Whorled</u> <u>Mint</u>. Plant similar in habit and scent to <u>M.</u>
<u>arvensis</u> but more robust, to 90cm, and with sharper leaf-
teeth; differs in calyx characters (see key, couplets 4-6);
usually sterile. Native; similar places to both parents, but
often in absence of either; frequent throughout most BI.
  3. M. x smithiana R.A. Graham (<u>M. arvensis</u> x <u>M. aquatica</u> x **684**

M. spicata) - <u>Tall</u> <u>Mint</u>. Plant subglabrous, usually red-tinged, with spearmint scent; stems erect to scrambling, to 1.5m; leaves ovate, rounded to broadly cuneate at base, with sharp, forward-directed teeth, petiolate; whorls in leaf-axils; calyx narrowly campanulate to $\pm$ tubular, with acuminate, narrowly triangular teeth; corolla lilac; usually sterile. Probably intrd; damp places and waste ground, perhaps sometimes arising as <u>M.</u> x <u>verticillata</u> x <u>M.</u> spicata, but usually escape or throw-out; scattered $\pm$ throughout Br and Ir, mostly S & C Br; ?garden origin.

4. M. x gracilis Sole (<u>M.</u> gentilis auct. non L.; <u>M.</u> arvensis **684** x <u>M.</u> spicata) - <u>Bushy</u> <u>Mint</u>. Plant glabrous to pubescent, usually with spearmint scent; differs from <u>M.</u> x <u>smithiana</u> in stems to 90cm and leaves cuneate at base, from <u>M.</u> x <u>verticillata</u> in scent, and from both in more campanulate calyx with conspicuously pubescent, $\pm$ subulate teeth (see key, couplets 4-6); stamens usually included; usually sterile. Possibly native; damp places and waste ground, usually escape or throw-out, but possibly native in parts of Br; scattered $\pm$ throughout BI.

5. M. aquatica L. - <u>Water</u> <u>Mint</u>. Plant subglabrous to pubescent, with strong pleasant (not spearmint-like) scent; stems erect, to 90cm; leaves ovate, broadly cuneate to subcordate at base, with rather shallow, rather blunt teeth, petiolate; upper whorls in axils of small bracts, congested to form rounded head; calyx tubular, with narrowly triangular to subulate teeth, 3-4.5mm; corolla lilac. Native; marshes, ditches, wet fields and by ponds; common throughout BI.

5. x 11. M. aquatica x M. suaveolens = M. x suavis Guss. (<u>M.</u> x <u>maximilianea</u> F. Schultz) occurs near the parents in W Cornwall, N Devon and Jersey; it is similar to <u>M.</u> aquatica but has more rugose leaves with more cordate base, and much marrower (c.9-15mm) often longer terminal heads.

6. M. x piperita L. (<u>M.</u> x <u>citrata</u> Ehrh., <u>M.</u> x <u>dumetorum</u> **684** auct. non Schultes; <u>M.</u> aquatica x <u>M.</u> spicata) - <u>Peppermint</u>. Plant glabrous or pubescent, often red-tinged, variously scented, often of peppermint; stems erect, to 90cm; leaves ovate to oblong-lanceolate, cuneate to subcordate at base, with rather sharp but not deep teeth, petiolate; upper whorls in axils of small bracts, congested to form variously rounded, pyramidal or cylindrical head; calyx tubular to campanulate, 2.5-4.5mm, with subulate teeth; corolla pinkish-lilac; sterile. Native; damp ground and waste places, escape or throw-out when glabrous, usually spontaneous when pubescent; scattered throughout BI. Pubescent plants were formerly misdetermined as <u>M.</u> aquatica x <u>M.</u> longifolia = **M. x dumetorum** Schultes. **Var. citrata** (Ehrh.) Briq. (<u>Eau de Cologne</u> <u>Mint</u>) is a distinctive glabrous variant with scent of Eau de Cologne, ovate, subcordate leaves, and a rounded inflorescence; it is often grown and escapes.

7. M. spicata L. (<u>M.</u> scotica R.A. Graham, <u>M.</u> longifolia

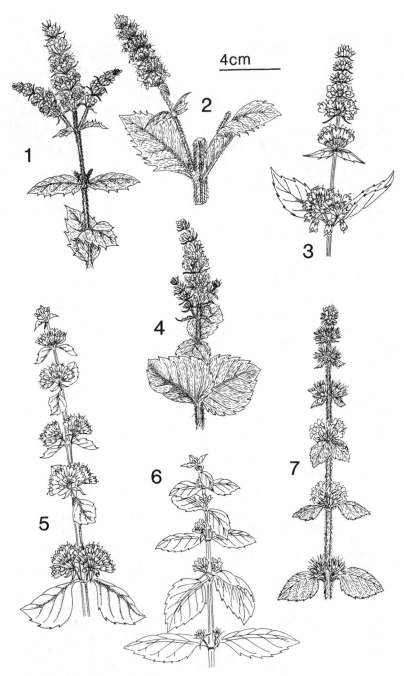

FIG 684 – Mentha. 1, **M. x rotundifolia.** 2, **M. x villosonervata.** 3, **M. x piperita.** 4, **M. x villosa.** 5, **M. x smithiana.** 6, **M. x gracilis.** 7, **M. x verticillata.**

auct. non (L.) Hudson) - <u>Spear</u> <u>Mint</u>. Plant glabrous to
tomentose, usually with characteristic spearmint scent; stems
erect, to 90cm; leaves lanceolate to ovate; sometimes broadly
ovate, usually not or slightly rugose, sometimes markedly so,
usually with sharp, forwardly-directed teeth, sessile or +
so; upper whorls in axils of small bracts, congested to form
narrow, long, not or little-branched, spike-like head; calyx
campanulate, with narrowly triangular to subulate teeth,
1-3mm; corolla white to pink or lilac. Intrd; much grown and
natd in rough and waste ground; scattered throughout most BI;
?garden origin.   Fertile   tetraploid   derived   from   <u>M.</u>
<u>longifolia</u> x <u>M.</u> <u>suaveolens</u>, variously approaching 1 or the
other; often almost impossible to distinguish from its
hybrids with these 2 spp., but the hybrids are sterile
triploids. Pubescent variants have in the past been mis-
identified as **M. longifolia** (L.) Hudson. **M. scotica** is a
pubescent variant from E Sc with lanceolate, obtuse,
shallowly serrate leaves with an unpleasant scent. A very
distinctive glabrous variant with rugose, broadly ovate
leaves and a strong spearmint scent occurs in SW Br; it was
formerly referred to as glabrous <u>M. spicata</u> x <u>M. suaveolens</u>.
   8. **M. x villosonervata** Opiz (<u>M. longifolia</u> var. horridula **684**
auct. non Briq.; (<u>M. spicata</u> x <u>M. longifolia</u>) - <u>Sharp-toothed</u>
<u>Mint</u>. Sterile triploids resembling pubescent, lanceolate-
leaved <u>M. spicata</u> but with leaf-teeth acuminate and + patent;
corolla pink; stamens included or exserted. Intrd; rough and
waste ground; sparsely scattered in Br; ?garden origin.
   9. **M. x villosa** Hudson (<u>M. x cordifolia</u> auct. ?non Opiz, <u>M.</u> **684**
x <u>niliaca</u> auct. non Juss. ex Jacq.; <u>M. spicata</u> x <u>M.</u>
<u>suaveolens</u>) - <u>Apple-mint</u>. Very variable sterile triploids,
varying in form from 1 parent to other; leaves glabrous (<u>M. x</u>
<u>cordifolia</u> auct.) to pubescent (<u>M.</u> x <u>niliaca</u> auct.),
lanceolate to suborbicular, with spearmint or various other
scent. Intrd; much grown and natd in rough and waste ground;
scattered + throughout BI; ?garden origin.   Var.
alopecuroides (Hull) Briq. is the usual Apple Mint of
gardens; it comes close to <u>M. suaveolens</u> but has spreading
leaf-teeth, not folded under, and pink corolla. The typical
var. **villosa** is more clearly intermediate.   Var.
nicholsoniana (Strail) R. Harley is natd in C to SE Wa and
adjacent En; it has lanceolate-oblong, pubescent leaves with
+ acuminate apex and sometimes + patent teeth. None of the
glabrous variants has received a varietal name.
   10. **M. x rotundifolia** (L.) Hudson (<u>M. x niliaca</u> Juss. ex **684**
Jacq.; <u>M. longifolia</u> x <u>M. suaveolens</u>) - <u>False</u> <u>Apple-mint</u>.
Close to some variants of <u>M. x villosa</u> but always pubescent;
leaves oblong-lanceolate, broadly rounded at base, rugose,
with subacute to obtuse apex and patent teeth; fertile.
Intrd; damp places and rough ground; very scattered in Br,
mainly Sc; Europe. Most of our plants are var. **webberi** (J.
Fraser) R. Harley.
   11. **M. suaveolens** Ehrh. (<u>M. rotundifolia</u> auct. non (L.)

Hudson) - <u>Round-leaved</u> <u>Mint</u>. Plant pubescent, with sickly
scent; stems erect, to 1m; leaves oblong-ovate to suborbicular, strongly rugose, with teeth bent under and hence
appearing as blunt crenations, sessile or ± so; upper whorls
in axils of bracts, congested to form very narrow often much-
branched panicle of spike-like heads; calyx campanulate,
1-2mm; corolla usually whitish. Native; ditches and other
damp places, waysides; locally frequent in W & S Wa, SW En
and CI, natd sparsely elsewhere in BI.

12. M. pulegium L. - <u>Pennyroyal</u>. Plant (often sparsely) **RRR**
pubescent, with pungent scent, resembling a small <u>M.</u>
<u>arvensis</u>; stems erect to procumbent, to 30cm; leaves
elliptic, with few very small teeth to subentire, obtuse to
rounded at apex, tapered to petiole; whorls in leaf-axils;
calyx tubular to narrowly campanulate, 2-3mm; corolla lilac.
Native; damp grassy or heathy places and by ponds; very
local, often near sea, in S En, S Wa and Ir, much decreased,
sometimes natd elsewhere.

13. M. requienii Benth. - <u>Corsican</u> <u>Mint</u>. Plant with pungent
scent, glabrous to sparsely pubescent; stems filiform,
procumbent, to 12cm; leaves <5mm, orbicular to broadly
elliptic, entire, petiolate; whorls of 2-6 flowers, in leaf-
axils; calyx 1-2mm, campanulate; corolla lilac. Intrd; natd
from gardens on damp paths and rocky places; very scattered
in S Br and 3 parts of Ir.

24. LAVANDULA L. - <u>Lavenders</u>
Evergreen shrubs; leaves entire, lanceolate to oblanceolate
or linear; calyx with 5 subequal teeth, the uppermost with
obcordate appendage at apex; corolla purple, weakly
zygomorphic, with short 3-lobed lower lip, with short
shallowly 2-lobed flat upper lip; stamens 4, included within
corolla-tube; flowers on long peduncles, congested in spikes
with much shorter bracts.

1. L. x intermedia Lois. (<u>L.</u> <u>angustifolia</u> Miller x <u>L.</u>
<u>latifolia</u> Medikus) - <u>Garden</u> <u>Lavender</u>. Erect shrub to 1m,
with characteristic scent; leaves 2-4cm, white-tomentose when
young; peduncles up to 40cm, with 0-2 pairs of reduced
leaves, simple or with 3 branches; spikes 2-8cm; corolla
9-12mm. Intrd; much grown in gardens, rarely self-sown on
walls and banks or persistent throw-out; sporadic in S En,
cultivated on field-scale in E Anglia; W Mediterranean. Most
garden, crop and natd <u>Lavandula</u> plants, which exist in many
cultivars, are probably this fertile hybrid, but certainly
some garden plants and perhaps some of those natd might
belong to 1 of the parents (particularly <u>L.</u> <u>angustifolia</u>).

25. ROSMARINUS L. - <u>Rosemary</u>
Evergreen shrubs: leaves linear, entire, with revolute
margins; calyx 2 lipped, with large ± entire upper lip, with
narrower 2-lobed lower lip; corolla pale to deep mauvish-

blue, strongly zygomorphic, with 2-lobed upper lip, with 3-lobed lower lip; stamens 2, longer than corolla; whorls ± distant in leaf-axils.

1. R. officinalis L. - Rosemary. Usually erect shrub to 2m, **679** with characteristic scent; leaves 1.5-4cm, tomentose on (mostly hidden) lowerside; corolla 10-15mm. Intrd; much grown in gardens, self-sown on walls and rough ground; sporadic in S En and CI, not fully frost-hardy; Mediterranean.

26. SALVIA L. - Claries
Herbaceous annuals to perennials or rarely small shrubs; leaves simple, serrate to crenate; calyx 2-lipped, the lower with 2 longer teeth, the upper entire or with 3 shorter teeth; corolla blue to purple or pink, or yellow, rarely red or white, strongly zygomorphic, with shortly bilobed strongly hooded upper lip and 3-lobed lower lip; stamens 2, shorter than upper lip of corolla; whorls of 1-many flowers in axils of modified bracts in terminal interrupted spikes.

1	Corolla ≥18mm	2
1	Corolla ≤18mm	4
2	Corolla yellow with reddish markings, >30mm; leaves often hastate at base	**2. S. glutinosa**
2	Corolla blue to violet, rarely pink or white, <30mm; leaves never hastate at base	3
3	Bracts green, often tinged with violet-blue, much shorter than flowers	**3. S. pratensis**
3	Bracts pink or white, at least as long as flowers	**1. S. sclarea**
4	Flowers (8)15-30 in each whorl	**7. S. verticillata**
4	Flowers 1-6(8) in each whorl	5
5	Inflorescence with conspicuous tuft of green or coloured flower-less bracts at apex	**6. S. viridis**
5	Inflorescence without terminal tuft of conspicuous flower-less bracts	6
6	Eglandular; stem-leaves usually >3 pairs, linear to lanceolate-oblong, crenate; calyx with hairs <0.2m	**5. S. reflexa**
6	Glandular above; stem-leaves ≤3 pairs, ovate to ovate-oblong, the lower doubly serrate to pinnately lobed and serrate; calyx with some hairs >0.5mm	7
7	Longest hairs on calyx white, eglandular; corolla with 0 or few glandular hairs; lower leaves often distinctly lobed	**4. S. verbenaca**
7	Longest hairs on calyx brownish, glandular; corolla with many glandular hairs; leaves at most strongly doubly serrate	**3. S. pratensis**

Other spp. - Several spp. are grown for ornament and may occasionally appear on tips or as throw-outs. **S. splendens**

Ker Gawler, from Brazil, is the familiar Scarlet Sage used as
a bedding plant; **S. officinalis** L. (Sage), from SW Europe, is
the shrubby pot-herb with purple, pink or white flowers; **S.
nemorosa** L. (S. sylvestris auct. non L.) (Balkan Clary), from
SE Europe, and its hybrids with S. pratensis (**S. x sylvestris**
L.) and S. amplexicaulis Lam. (**S. x superba** Stapf), have blue
to pink corollas 8-12mm in the axils of brightly coloured
bracts (S. nemorosa was formerly natd at docks in S Wa).

1. S. sclarea L. - Clary. Erect biennial or perennial to
1m, glandular above; leaves ovate, cordate at base, crenate-
serrate; bracts usually longer than flowers, pink or white;
corolla 18-30mm, lilac or blue. Intrd; grown in gardens,
rather frequent relic or throw-out on tips and rough ground;
scattered in S Br; S Europe.
2. S. glutinosa L. - Sticky Clary. Erect perennial to 1m, **689**
very glandular above; leaves ovate, cordate or more often
hastate at base, sharply serrate; bracts small, green;
corolla 3-4cm, yellow with reddish markings. Intrd; grown in
gardens, natd in woods, hedges and on road and river banks;
scattered in En and Sc; S Europe.
3. S. pratensis L. - Meadow Clary. Erect perennial to 80cm, **RR**
glandular above; leaves ovate or ovate-oblong, cordate at
base, doubly serrate; bracts small, green or tinged purplish;
corolla 15-30mm in bisexual flowers, down to 10mm in female
flowers, violet-blue. Native; calcareous grassland, scrub
and wood-borders; very local in c.12 places in S En from E
Kent to W Gloucs, Mons, natd elsewhere in C & S Br.
4. S. verbenaca L. (S. horminoides Pourret) - Wild Clary.
Differs from S. pratensis in key characters; corolla open and
10-15mm, or cleistogamous and 6-12mm. Native; dry grassy and
barish rough ground, roadsides, dunes; rather frequent in S &
E Br N to C Sc, N Wa, S & E Ir, CI, rare and decreasing in C
En.
5. S. reflexa Hornem. - Mintweed. Erect eglandular annual **689**
to 60cm; leaves linear- to lanceolate-oblong, cuneate at
base, entire to crenate; bracts small, green or tinged
purplish; corolla 6-12mm, pale blue; flowers only 1-4 per
whorl. Intrd; casual from bird-seed, grain, grass-seed and
wool in waste and rough ground; scattered in En and S Sc,
mainly SE En; N America.
6. S. viridis L. (S. horminum L.) - Annual Clary. Erect **689**
glandular or eglandular annual to 50cm; leaves oblong-ovate,
rounded to cordate at base, crenate; bracts conspicuous, the
upper ones and a terminal flower-less tuft green, white or
pink to purple; corolla 14-18mm, pink to purple. Intrd;
grown in gardens and casual or rarely natd on tips and waste
ground; scattered in SE En; S Europe.
7. S. verticillata L. - Whorled Clary. Erect perennial to **689**
80cm, with eglandular hairs and sessile glands; leaves ovate,
cordate at base, crenate-dentate, sometimes the lower ones
lobed; bracts small, green, brown or purplish-tinged; corolla

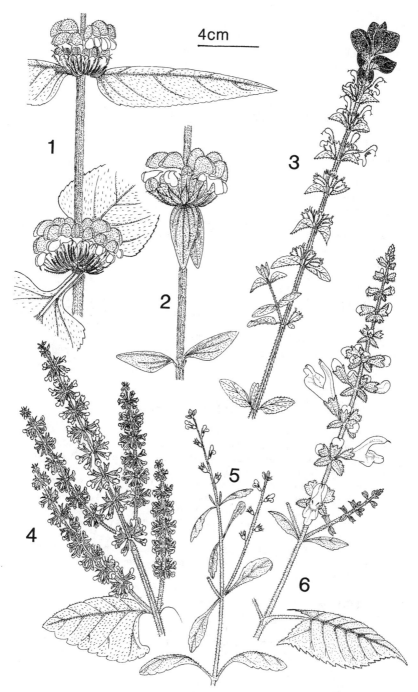

FIG 689 – Lamiaceae. 1–2, Phlomis. 1, P. russeliana. 2, P. fruticosa. 3–6, Salvia. 3, S. viridis. 4, S. verticillata. 5, S. reflexa. 6, S. glutinosa.

8-15mm, lilac-blue. Intrd; casual or sometimes natd in rough ground and by roads and railways; scattered in C & S Br; S Europe.

121. HIPPURIDACEAE - Mare's-tail family

Rhizomatous, perennial, aquatic or mud-dwelling herbs, with conspicuous air-cavities in stems; leaves in whorls of 6-12, simple, linear, sessile, exstipulate. Flowers 1 per leaf-axil, very small, often male, female and bisexual on same plant with female ones more apical, epigynous; perianth reduced to minute rim at apex of ovary; stamen 1, borne on top of ovary if present; ovary 1-celled with 1 apical ovule; style/stigma 1, filiform; fruit a 1-seeded nut.

The only aquatic or mud-plant with linear simple entire leaves >4 in a whorl.

1. HIPPURIS L. - Mare's-tail
1. H. vulgaris L. - Mare's-tail. Usually aquatic, then with stems to 1(2)m, the apical part emergent and bearing flowers, and with leaves up to 8cm; sometimes on mud, then with stems c.4-20cm and leaves c.1-2cm. Native; in ponds and slow-flowing rivers, especially base-rich; locally frequent throughout BI.

122. CALLITRICHACEAE - Water-starwort family

Annual or perennial herbs, aquatic or on mud, with filiform stems; leaves opposite, simple, linear, entire or notched at apex, often forming terminal rosette, exstipulate, ± sessile. Flowers very small, monoecious, 1 or 2 (and then 1 of each sex) per leaf-axil, each with 0 or 2 bracteoles; perianth 0; male flowers with 1 stamen; female flowers with 1 4-celled ovary, each cell with 1 apical ovule; styles/stigmas 2, filiform; fruit a group of 4 1-seeded nutlets, 2 pairs each more closely united.

Distinguished from other aquatics and mud-dwellers by the opposite, thin, narrow, often notched leaves, single stamen and distinctive fruits.

1. CALLITRICHE L. - Water-starworts
Vegetatively very variable and often shy-fruiting; difficult to identify certainly without a strong lens or microscope. The following 3 keys should be used only on fruiting material, preferably with some stamens available as well. Use of names for British plants was very confused before the work of H.D. Schotsman (1954 onwards); synonyms should be applied with great caution. Fruits should be mature. Plants normally reaching water surface but recently flooded can be told by their terminal rosettes of leaves different in shape

from those further back. Fruit-wings are whitish or translucent when viewed in transmitted light. Leaf-shape is notoriously variable and misleading.

General key

1 Fruiting plants terrestrial or on wet mud   **Key C**
1 Fruiting plants mainly or completely submerged   **2**
  2 Fruiting plants completely submerged   **Key A**
  2 Fruiting plants mainly submerged but with
    terminal leaf rosette at water surface   **Key B**

Key A - Fruiting plants completely submerged; terminal leaves not in distinct rosette; pollen grains with $\pm$ colourless unsculptured wall (beware re-submerged plants with terminal leaf-rosette).

1 Leaves transparent, all with 1 vein, without stomata; minute sessile hairs present only in leaf-axils, composed of a $\pm$ irregular cell-mass   **2**
1 Leaves opaque, often some with >1 vein, often some with stomata; minute sessile hairs present on leaves, stems and in leaf-axils, composed of cells radiating from central point   **3**
  2 Fruits $\pm$ orbicular in side view, c.1.4–2.2(3.3)mm, with conspicuous wing 0.1–0.5mm wide; leaves mostly >1cm, conspicuously notched at apex, usually pale to mid green   **1. C. hermaphroditica**
  2 Fruits wider than long, c.1–1.2 x 1.4–1.6mm, not winged nor sharply keeled; leaves usually <1cm, truncate to shallowly notched at apex; usually dark green   **2. C. truncata**
3 Leaves expanded suddenly just at apex with deep notch $\pm$ like a bicycle-spanner; fruits $\pm$ orbicular in side view, 1.2–1.5mm, with wing usually <0.1mm wide   **7. C. hamulata**
3 Leaves not expanded suddenly just at apex, with variable, often shallow or asymmetric notch; fruits often longer than wide, 1–1.4 x 1–1.2mm, with wing usually >0.1mm wide   **6. C. brutia**

Key B - Fruiting plants aquatic but with terminal rosette of leaves at water surface
1 Flowers submerged; pollen grains with $\pm$ colourless unsculptured wall; styles usually persistent, reflexed and appressed to sides of fruit   **2**
1 Flowers aerial; pollen-grains yellow, with sculptured wall; styles, if persistent, erect to patent   **3**
  2 Submerged leaves expanded at apex with deep notch $\pm$ like a bicycle-spanner; fruits $\pm$ orbicular in side view, 1.2–1.5mm, with wing usually <0.1mm wide   **7. C. hamulata**
  2 Submerged leaves not expanded at apex, with variable, often shallow or asymmetric notch;

fruits often longer than wide, 1–1.4 x 1–1.2mm,
with wing usually $\geq$0.1mm wide   **6. C. brutia**
3 Fruits without wing, rounded to obtuse at edges,
distinctly longer than wide   **5. C. obtusangula**
3 Fruits with wing 0.07–0.25mm wide often nearly
orbicular in side view   4
  4 Fruits with wing 0.12–0.25mm wide; stamens c.2mm
   at anthesis, with anthers c.0.5mm wide; pollen
   grains $\pm$ globose   **3. C. stagnalis**
  4 Fruits with wing 0.07–0.1mm wide; stamens c.4mm at
   anthesis, with anthers c.1mm wide; many pollen
   grains mis-shapen or ellipsoid   **4. C. platycarpa**

Key C – Fruiting plants terrestrial or on wet mud
1 Fruits without wing, rounded to obtuse at edges
                                    5. C. obtusangula
1 Fruits with wing 0.07–0.25mm wide   2
  2 Fruits with a stalk 2–10mm   **6. C. brutia**
  2 Fruits sessile or with stalk <2mm   3
3 Styles usually persistent, reflexed and appressed to
sides of fruit; pollen grains with $\pm$ colourless
unsculptured wall   **7. C. hamulata**
3 Styles, if persistent, erect to patent; pollen grains
yellow, with sculptured wall   4
  4 Fruits with wing 0.12–0.25mm wide; stamens c.2mm
   long, with anthers c.0.5mm wide; pollen grains $\pm$
   globose   **3. C. stagnalis**
  4 Fruits with wing 0.07–0.1mm wide; stamens c.4mm
   at anthesis, with anthers c.1mm wide; many pollen
   grains mis-shapen or ellipsoid   **4. C. platycarpa**

Other spp. – C. palustris L., a small fragile plant with
obovate fruits 0.8–1 x 0.7–0.8mm usually narrowly winged just
at apex, has been found for certain in BI once only (Surrey,
1877); other records are probable or certain errors. **C.
cophocarpa** Sendtner differs from C. platycarpa in the absence
of fruit wings and from C. obtusangula in its nearly
orbicular fruits; it has been said to occur in BI but such
statements were based on supposition, not fact.

  1. C. hermaphroditica L. – <u>Autumnal</u> <u>Water-starwort</u>. **693**
Submerged annual to 50cm; leaves 8–18mm, widest near base, **R**
tapering to emarginate apex; fruits common, $\pm$ orbicular in
side view, c.1.4–2.2(3.3)mm, with wing 0.1–0.5mm wide; 2n=6.
Native; lakes and rivers; scattered in BI N from S Lincs,
Brecs and S Kerry.
  2. C. truncata Guss. – <u>Short-leaved Water-starwort</u>. Differs **693**
from C. hermaphroditica in leaves 5–11mm, darker green when **R**
fresh; fruits rare; and see key A. Native; ponds, rivers and
canals; very local in Br, S & E of range of <u>C.
hermaphroditica</u> (S Devon and N Somerset to E Kent; Notts,
Leics and N Lincs), Co Wexford, Guernsey. Our plant is ssp.

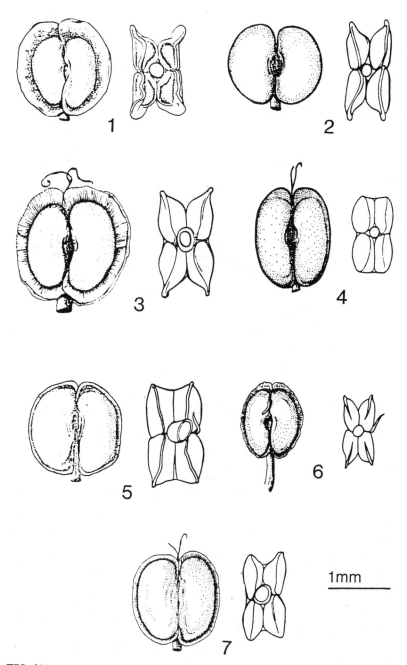

FIG 693 – Fruits of **Callitriche**, lateral and basal views. 1, **C. hermaphroditica.** 2, **C. truncata.** 3, **C. stagnalis.** 4, **C. obtusangula.** 5, **C. hamulata.** 6, **C. brutia.** 7, **C. platycarpa.** Courtesy of H.D. Schotsman.

694 CALLITRICHACEAE

occidentalis (Rouy) Braun-Blanquet.
3. C. stagnalis Scop. - <u>Common</u> <u>Water-starwort</u>. Aquatic or 693
terrestrial annual or perennial, but not fruiting if
submerged; submerged leaves narrowly elliptic; floating leaf-
blades broadly elliptic to suborbicular; terrestrial leaves
similar to last but smaller; stamens 0.5-2mm; anthers c.0.5mm
wide; pollen grains ± globose; fruits ± orbicular in side
view, 1.6-1.8mm, with wing 0.12-0.25mm wide, with styles ±
erect (or recurved when terrestrial); 2n=10. Native; in
ponds, rivers, ditches and muddy places; by far commonest sp.
throughout BI.
4. C. platycarpa Kuetz. (<u>C.</u> <u>palustris</u> auct. non L., <u>C.</u> 693
<u>polymorpha</u> auct. non Loennr.) - <u>Various-leaved</u>
<u>Water-starwort</u>. Differs from <u>C.</u> <u>stagnalis</u> in floating leaf-
blades elliptic; 2n=20; and see key B, but sometimes
extremely close to <u>C.</u> <u>stagnalis</u>. Native; more often in
flowing water than <u>C.</u> <u>stagnalis</u>; distribution uncertain due
to confusion with <u>C.</u> <u>stagnalis,</u> but probably widespread in
lowlands of Br, Ir and CI.
5. C. obtusangula Le Gall - <u>Blunt-fruited</u> <u>Water-starwort</u>. 693
Aquatic or terrestrial perennial, but not fruiting if
submerged; submerged leaves linear; floating leaf-blades
broadly rhombic; stamens c.5mm; anthers c.0.6mm wide; pollen
grains markedly ellipsoid or slightly curved; fruits c.1.5 x
1.2mm, unwinged, with ± erect styles; 2n=10. Native; in and
by ponds and streams; rather scattered in BI N to MW Yorks
and Co Antrim.
6. C. brutia Petagna (<u>C.</u> <u>pedunculata</u> DC., <u>C.</u> <u>intermedia</u> 693
Hoffm. ssp. <u>pedunculata</u> (DC.) Clapham) - <u>Pedunculate</u> <u>Water-</u>
<u>starwort</u>. Submerged, floating or terrestrial usually annual;
submerged leaves linear, variably and often shallowly
notched; floating leaf-blades narrowly elliptic; stamens
c.0.5-1mm; anthers c.0.5mm wide; pollen grains subglobose;
fruits 1-1.4 x 1-1.2mm, ± sessile when aquatic, with stalk
2-10mm when terrestrial, with wing usually ≥0.1mm wide, with
reflexed styles; 2n=28. Native; shallow water often drying
up in summer; distribution uncertain due to confusion with <u>C.</u>
<u>hamulata</u>; S & W Br N to Kintyre and E to E Kent, W & N Ir,
?CI.
7. C. hamulata Kuetz. ex Koch (<u>C.</u> <u>intermedia</u> ssp. <u>hamulata</u> 693
(Kuetz. ex Koch) Clapham) - <u>Intermediate</u> <u>Water-starwort</u>.
Submerged, floating or terrestrial annual or perennial;
differs from <u>C.</u> <u>brutia</u> in submerged leaves with bicycle-
spanner-like leaf-apex; floating leaf-blades elliptic; fruits
always sessile; 2n=38; and see key A. Native; in usually
acid ponds, rivers and ditches, less often drying out than
for <u>C.</u> <u>brutia</u>; probably frequent throughout BI, but less
common than <u>C.</u> <u>brutia</u> in W.

123. PLANTAGINACEAE - Plantain family

Annual to perennial herbs, very rarely dwarf shrubs; leaves simple, entire to deeply dissected, usually all basal, sometimes opposite on stems, exstipulate, sessile or with ill-defined petiole. Flowers solitary or in dense spikes terminal on unbranched basal or axillary stalks, actinomorphic, bisexual or unisexual, hypogynous; sepals (2-)4, papery, persistent in fruit, nearly free to fused at base; petals 2-4, fused into tube with 2-4 lobes, greenish or brownish; stamens 4, on long filaments; ovary 1-, 2- or 4-celled, each cell with 1-many axile or basal ovules; style 1, terminal; stigma 1, linear; fruit a 2- or 4-celled transversely-dehiscing capsule or indehiscent 1-seeded nut. Easily distinguished by the brownish spike-like inflorescence (or long-stalked solitary male flowers in Littorella) and leaves usually all in a basal rosette.

1 Flowers bisexual, in compact spikes; fruit a many-
 seeded dehiscent capsule   **1. PLANTAGO**
1 Flowers unisexual, the male solitary on long stalks;
 fruit a 1-seeded nut   **2. LITTORELLA**

1. PLANTAGO L. - Plantains
Stolons 0; flowers bisexual, in compact spikes, 4-merous; stamens borne on corolla-tube; ovary 2- or 4-celled, each cell with 2-many axile (and basal) ovules; capsule transversely dehiscent.

1 Spikes borne in axils of opposite stem-leaves   2
1 Spikes borne on leafless scapes   3
 2 Inflorescence and peduncles with many glandular
  hairs; spikes with all bracts similar   **7. P. afra**
 2 Inflorescence and peduncles with 0 glandular hairs;
  spikes with lower bracts strongly differing from
  upper bracts   **6. P. arenaria**
3 Scapes strongly furrowed   **5. P. lanceolata**
3 Scapes not furrowed   4
 4 Corolla-tube glabrous on outside; leaves narrowly
  elliptic to broadly ovate, at most bluntly and
  distantly toothed   5
 4 Corolla-tube pubescent on outside; linear to
  linear-elliptic, often sharply toothed or
  deeply and finely lobed   6
5 Capsule usually 4-seeded; seeds >2mm; stamens
 exserted >5mm; anthers >1.5mm   **4. P. media**
5 Capsule usually >4-seeded; seeds <2mm; stamens
 exserted <5mm; anthers <1.5mm   **3. P. major**
 6 Capsule 2-celled, with 1-2 seeds per cell; seeds
  >1.5mm; corolla-lobes with conspicuous brown midrib;
  leaves usually entire, sometimes slightly toothed
                                             2. P. maritima

6 Capsule 3(-4)-celled, with 1-2 seeds per cell;
   seeds <1.5mm; corolla-lobes with 0 or inconspicuous
   midrib; leaves usually deeply and narrowly lobed,
   sometimes toothed, rarely all entire **1. P. coronopus**

Other spp. - **P. sempervirens** Crantz, from SW Europe, was
formerly natd in Kent, but not since 1920; it superficially
resembles P. arenaria, but is a woody perennial.

1. P. coronopus L. - Buck's-horn Plantain. Usually
pubescent annual to perennial with 1-many rosettes; leaves
linear and entire to toothed, or deeply and finely divided;
scapes to 20cm; spikes up to 4(7)cm. Native; barish places
or in very short turf on sandy or gravelly soils, sometimes
on rocks, mostly near sea; common round coasts of BI, inland
in scattered lowland places in En and SW Wa.
2. P. maritima L. - Sea Plantain. Usually glabrous
perennial with 1-many rosettes; leaves linear to
linear-elliptic, entire or sparsely toothed; scapes to 30cm;
spikes up to 7(10)cm. Native; in salt-marshes, rock-crevices
and short turf near sea, and wet rocky places on mountains;
common round coasts of BI, inland on mountains in Sc, N En
and N & W Ir, rare in inland salt-marshes or by salt-treated
roads.
3. P. major L. - Greater Plantain. Glabrous to pubescent
perennial with 1 rosette; leaves ovate to broadly so,
abruptly narrowed to petiole usually ± as long, entire to
weakly toothed; scapes to 40cm; spikes up to 20cm. Native.
a. Ssp. major. Leaves mostly with 5-9 veins, usually obtuse
at apex, subcordate to rounded at base and subentire;
capsules mostly with 4-15 seeds; seeds (1)1.2-1.8(2.1)mm.
Open and rough ground, either cultivated or grassy, and on
lawns; abundant throughout BI.
b. Ssp. intermedia (Gilib.) Lange (P. intermedia Gilib., P.
uliginosa F.W. Schmidt). Plant usually smaller with much
shorter spikes; leaves mostly with 3-5 veins, usually
subacute at apex, broadly cuneate at base and ± undulate-
toothed near base; capsules mostly with (9)14-25(36) seeds;
seeds (0.6)0.8-1.2(1.5)mm. Damp, usually slightly saline
places near sea and less often inland; distribution uncertain
but probably scattered through much of BI.
4. P. media L. - Hoary Plantain. Pubescent perennial with
1-few rosettes; leaves elliptic to broadly ovate, usually
rather abruptly narrowed to usually short petiole, usually
weakly toothed; scapes to 40cm; spikes up to 6(12)cm.
Native; neutral and basic grassland; locally common in En,
very local in Ir, Wa, CI and S & E Sc.
5. P. lanceolata L. - Ribwort Plantain. Glabrous to
pubescent perennial with 1-several rosettes; leaves linear-
to ovate-lanceolate, very gradually narrowed to petiole,
entire to sparsely and weakly toothed; scapes to 50cm; spikes
up to 4(8)cm. Native; grassy places; abundant throughout BI.

6. P. arenaria Waldst. & Kit. (**P.** scabra Moench nom. illeg., **P. psyllium** L. nom. ambig., **P.** indica L. nom. illeg.) - Branched Plantain. Erect to decumbent pubescent annual to 30(50)cm, with opposite leaves and axillary peduncles; leaves linear, entire; peduncles 1-6cm; spikes 0.5-1.5cm. Intrd; open and rough ground on sandy soil, casual or sometimes natd; very scattered in S Br, especially E Anglia, less common than formerly, now often over-recorded for P. afra; S Europe.

7. P. afra L. (**P.** psyllium L. 1762 non 1753, P. indica auct. non L.) - Glandular Plantain. Differs from P. arenaria in key characters; especially note lower bracts lacking the long linear leaf-like apices of P. arenaria. Intrd; bird-seed and grain alien of tips and waste places; scattered throughout much of Br, now commoner than P. arenaria; S Europe.

2. LITTORELLA P. Bergius - Shoreweed
Stoloniferous; flowers unisexual, the male solitary on long scapes and 3-4-merous, the female 1-few sessile at base of male scape and 2-4-merous; ovary 1-celled, with 1(-2) ovules; fruit a 1-seeded nut.

1. L. uniflora (L.) Asch. - Shoreweed. Leaves in rosette, semi-cylindrical, usually subulate, 1.5-10(25)cm; male scapes up to as long as leaves; stamens 1-2cm. Native; typically in shallow water at lake edges but often on exposed shore or down to 4m, on sandy or gravelly acid soils; in suitable places through much of BI, but very local and decreasing in lowlands. Not flowering unless exposed.

124. BUDDLEJACEAE - Butterfly-bush family

Deciduous or semi-evergreen shrubs; leaves simple, entire to serrate, exstipulate, opposite or alternate, $\pm$ petiolate. Flowers in $\pm$ dense panicles, small but showy in mass, fragrant, actinomorphic, bisexual, hypogynous; sepals 4, fused into lobed tube; petals 4, fused into lobed tube, various colours; stamens 4, borne on corolla-tube; ovary 2-celled, each cell with many ovules on axile placentas; style 1; stigmas 2, linear; fruit a 2-valved capsule with many narrowly winged seeds.
Recognized by its shrubby habit and numerous small but brightly coloured flowers with 4 fused sepals and petals, 2-celled ovary and 4 stamens.

1. BUDDLEJA L. - Butterfly-bushes

1 Leaves alternate; flowers lilac, borne in small
　clusters along previous year's wood　　**1. B. alternifolia**
1 Leaves opposite; flowers lilac to orange-yellow,
　borne in panicles on current year's growth　　　2

2 Flowers in long dense pyramidal panicles, lilac to
  purple or white   **2. B. davidii**
2 Flowers in dense globose clusters arranged in
  large panicles, orange-yellow   **3. B. globosa**

Other spp. - B. davidii x B. globosa = **B. x weyeriana** Weyer
is grown in gardens and single natd plants have been recorded
in E & W Cornwall; the flowers are borne in globose heads
arranged in long slender panicles and vary from yellowish,
through greyish, to purplish.

1. B. alternifolia Maxim. - <u>Alternate-leaved Butterfly-bush</u>.
Large shrub to 8m with long thin arching branches; leaves
alternate, narrowly elliptic; flowers lilac, in ± sessile
clusters borne along last year's wood. Intrd; grown in
gardens, persistent and sometimes self-sown in woodland,
hedges or banks; very scattered in En N to MW Yorks; China.
2. B. davidii Franchet - <u>Butterfly-bush</u>. Shrub to 5m with
long ± arching branches; leaves opposite, lanceolate to
narrowly ovate; flowers usually lilac, sometimes purple or
white, in long pyramidal dense panicles borne on current
year's wood. Intrd; natd in waste ground, walls, banks and
scrub; common in S BI, decreasing northwards to C Sc; China.
3. B. globosa Hope - <u>Orange-ball-tree</u>. Shrub to 5m with
stiff erect to spreading branches; leaves opposite,
lanceolate to narrowly ovate or elliptic; flowers orange-
yellow, in dense globose clusters arranged in stiff open
panicles on current year's wood. Intrd; grown in gardens,
natd rarely on roadsides and rough ground; very scattered in
Br N to Anglesey and Yorks; Chile and Peru.

125. OLEACEAE - Ash family

Trees, shrubs or woody trailers; leaves opposite, simple to
pinnate, entire to serrate, exstipulate, petiolate. Flowers
variously arranged, often in dense clusters, small and dull
to large and showy, various colours, actinomorphic, bisexual
to dioecious or monoecious, hypogynous; sepals 0, or 4 and
united into lobed tube; petals 0, or 4 (rarely -6) and united
into lobed tube (rarely ± free); stamens 2, borne on
corolla-tube when present; ovary 2-celled, each cell with 2
ovules on axile placentas; style 1; stigmas 2, slightly
elongated; fruit a 2-valved capsule, winged achene, or
2-seeded berry.
  Distinguished by the woody habit and flowers with 4 fused
sepals and petals (0 or free in <u>Fraxinus</u>), 2-celled ovary and
2 stamens.

1 Corolla and calyx 0 (rarely both present and
  then petals free); flowers often unisexual;
  leaves usually pinnate   **2. FRAXINUS**

1 Corolla showy; calyx present; petals fused; flowers
  bisexual; leaves simple or with 1 main leaflet plus
  1 or 2 small basal ones   2
  2 Flowers yellow, appearing before leaves **1. FORSYTHIA**
  2 Flowers white, or lilac to mauve or purple,
    appearing after leaves   3
3 Fruit a capsule; leaves truncate to cordate at base;
  flowers often lilac to mauve or purple   **3. SYRINGA**
3 Fruit a berry; leaves cuneate at base; flowers always
  white   **4. LIGUSTRUM**

Other genera - JASMINUM L. (Jasmines) has pinnate or ternate
leaves, showy flowers with long corolla-tube and 4-6 lobes,
and fruit a black berry. 2 cultivated spp. sometimes persist
after neglect or give rise to stray seedlings: **J. nudiflorum**
Lindley (Winter Jasmine), from China, has yellow flowers
produced before the leaves; **J. officinale** L. (Summer
Jasmine), from SW Asia, has white fragrant flowers produced
in summer.

1. FORSYTHIA Vahl - Forsythia
Deciduous shrubs; leaves simple or with 2 small leaflets at
base; petals 4(-6), united into 4(-6)-lobed tube, yellow,
appearing before leaves; fruit a capsule.

1. F. x intermedia Zabel (F. suspensa (Thunb.) Vahl x F.
viridissima Lindley) - Forsythia. Erect to rambling or
pendent shrub to 5m; leaves ovate to ovate-lanceolate,
cuneate to rounded at base, serrate; flowers in small
clusters on old wood, 2-3.5cm. Intrd; much grown in gardens,
occasional relic or throw-out in rough ground or on road- or
river-banks; few places in SE En; garden origin. **F.
suspensa**, from China, is a stronger-growing shrub with hollow
branches and is equally grown; some escapes may be this.

2. FRAXINUS L. - Ash
Deciduous trees; leaves normally pinnate (rare form with 1
leaflet exists); petals 0 (rarely present, free, white);
fruit an achene with a single long wing.

Other spp. - F. ornus L. (Manna Ash), from S Europe, is a
more delicate tree to 22m, with flowers with white petals
appearing in large panicles with or after the leaves; it is
grown as a street tree or on road-, railway- or canal-banks
in many places.

1. F. excelsior L. - Ash. Thick-twigged tree to 37m;
leaflets serrate, lanceolate to ovate; flowers in dense
axillary or terminal panicles appearing well before leaves,
unisexual or bisexual, different sorts on same or different
trees. Native; woods, scrub and hedgerows, often the
commonest tree, especially on damp or base-rich soils; common

± throughout BI.

3. SYRINGA L. - Lilac
Deciduous shrubs; leaves simple; petals usually 4, united
into usually 4-lobed tube, white, or lilac to mauve or
purple, appearing after leaves; fruit a capsule.

1. S. vulgaris L. - Lilac. Erect suckering shrub to 7m;
leaves ovate, cordate to truncate at base, entire; flowers in
large terminal pyramidal panicles, 1.5-2cm. Intrd; much
grown in gardens, occasional relic or throw-out in hedges and
road- and railway-banks; scattered through most of BI, rarely
self-sown; SE Europe.

4. LIGUSTRUM L. - Privets
Deciduous to evergreen shrubs; leaves simple; petals 4,
united into 4-lobed tube, white, appearing after leaves;
fruit a black berry.

1. L. vulgare L. - Wild Privet. Erect, semi-deciduous shrub
to 3(5)m; 1-year-old stems and panicle branches densely
puberulous; leaves lanceolate to oblanceolate, narrowly
acute; corolla-tube c. as long as -limb. Native; hedgerows
and scrub, especially on base-rich soils; throughout most BI
except parts of N Sc.
2. L. ovalifolium Hassk. - Garden Privet. Differs from L.
vulgare in evergreen leaves usually elliptic, acute to
rounded at apex; young stems and panicle branches glabrous;
corolla-tube distinctly longer than lobes. Intrd; abundantly
planted for hedging, often yellow- or variegated-leaved,
persistent in hedges and rough ground; scattered throughout
much of BI, rarely self-sown; Japan.

126. SCROPHULARIACEAE - Figwort family

Herbaceous annuals to perennials or very rarely small shrubs
or trees; stems usually round, sometimes square, in section;
leaves alternate, opposite, whorled, or rarely all basal,
simple, entire to deeply divided, exstipulate, sessile to
petiolate. Flowers single and axillary or in terminal or
less often axillary racemes or sometimes in racemose or
cymose panicles, zygomorphic (occasionally nearly actinomorphic), bisexual, hypogynous; calyx with (2)4-5-lobes fused
just at base to most of length, never well differentiated
into upper and lower lips; corolla strongly 4-5-lobed and
sometimes nearly actinomorphic, to 2-lipped with lobes
variously or not developed, variously coloured; stamens
usually 4, sometimes with a staminode, sometimes 2 or 5 or
rarely 3, borne on corolla-tube; ovary 2-celled, each cell
with numerous ovules on axile placentas; style 1, terminal;
stigmas 1-2, short, usually ± capitate; fruit a 2-celled

SCROPHULARIACEAE 701

capsule.
Distinguished from Lamiaceae, Verbenaceae and Boraginaceae
by the totally different ovary and fruit, and from other
families with zygomorphic flowers and 2-celled ovary by the
herbaceous habit (woody in Hebe and Phygelius), 2 or 4
stamens (5 in Verbascum), and non-spiny bracts. See also
under Orobanchaceae.

1 Stamens 2   2
1 Stamens 4 or 5   4
  2 Corolla 2-lipped, the lower lip large and inflated,
    yellow to reddish-brown; stamens included
                                                    6. CALCEOLARIA
  2 Corolla 4-lobed, without inflated portion, white
    to blue or pink; stamens exserted   3
3 Stems woody; leaves evergreen, entire   17 HEBE
3 Herbaceous plants; stems not woody or woody only at
  base and then with serrate leaves   16. VERONICA
  4 Flowers 2.5-4cm, red, pendent in terminal
    panicles; stems woody   3. PHYGELIUS
  4 Flowers often <2cm, if >2cm and red then in
    racemes; stems herbaceous   5
5 Corolla with conspicuous basal spur or pouch on
  lowerside   6
5 Corolla not spurred or pouched at base   12
  6 Leaves palmately veined and lobed   7
  6 Leaves with single midrib, often also with lateral
    pinnate veins, entire to serrate   8
7 Plant glandular-pubescent; corolla >3cm, yellow
  with purple veins, pouched at base   10. ASARINA
7 Plant glabrous to puberulent, not glandular; corolla
  <3cm, mauve to purple, often with yellow centre,
  spurred at base   11. CYMBALARIA
  8 Corolla-tube with broad, rounded pouch (wider than
    long) at base   9
  8 Corolla-tube with narrow, often pointed, spur
    (longer than wide)   10
9 Calyx-lobes + equal, all shorter than corolla-tube;
  corolla >2.5cm   7. ANTIRRHINUM
9 Calyx-lobes distinctly unequal, all longer than
  corolla-tube; corolla <2cm   9. MISOPATES
  10 Leaves ovate to obovate, rounded to cordate at
     base; capsule opening by detachment of 2 oblique
     lids leaving large pores   12. KICKXIA
  10 Leaves linear to lanceolate or oblanceolate,
     narrowed to base, rarely ovate to obovate and
     rounded at base and then capsule without
     detachable lids   11
11 Mouth of corolla completely closed by boss-like
   swelling on lower lip   13. LINARIA
11 Mouth of corolla incompletely closed by small
   swelling   8. CHAENORHINUM

12 Fertile stamens 5, at least in most flowers
                                     1. VERBASCUM
12 Fertile stamens 4, sometimes with a sterile
   staminode representing fifth   13
13 Leaves all in basal rosettes, linear to spathulate
                                   5. LIMOSELLA
13 At least some leaves borne on stems   14
  14 Stems procumbent, rooting at nodes; leaves
    reniform, long-stalked; flowers solitary in leaf-
    axils   **18. SIBTHORPIA**
  14 Stems not procumbent and rooting at nodes; leaves
    not reniform; flowers in terminal inflorescences 15
15 Calyx 5-lobed or -toothed   16
15 Calyx 4-lobed or -toothed, rarely 2-5-lobed with
   toothed lobes and inflated tube   19
  16 Leaves opposite   17
  16 Leaves alternate   18
17 Calyx-tube shorter than -lobes; corolla purplish-
   brown to dull yellowish-green, with tube scarcely
   longer than wide   **2. SCROPHULARIA**
17 Calyx-tube longer than -teeth; corolla bright yellow,
   often with red spots or blotches, with tube much
   longer than wide   **4. MIMULUS**
  18 Corolla distinctly zygomorphic, with tube >2x as
    long as calyx, lobes not strongly patent
                          14. DIGITALIS
  18 Corolla scarcely zygomorphic, with tube <2x as
    long as calyx, with strongly patent lobes **15. ERINUS**
19 Calyx irregularly 2-5-lobed, with toothed lobes;
   leaves divided almost to base, the lobes toothed
                        25. PEDICULARIS
19 Calyx regularly 4-lobed, with entire lobes; leaves
   entire to simply toothed up to c.1/2 way to base 20
  20 Calyx-tube inflated, especially at fruiting; seeds
    discoid, with marginal wing   **24. RHINANTHUS**
  20 Calyx-tube not inflated; seeds not discoid,
    without marginal wing   21
21 Lower lip of corolla with 3 distinctly emarginate
   lobes   **20. EUPHRASIA**
21 Lower lip of corolla with 3 entire lobes, or sometimes middle lobe slightly emarginate   22
  22 Mouth of corolla partially closed by boss-like
    swellings on lower lip; capsules with 1-4 seeds
                         19. MELAMPYRUM
  22 Mouth of corolla open; lower lip without
    swellings; capsules with >4 seeds   23
23 Corolla yellow   24
23 Corolla pink to dark purple, rarely white   25
  24 Leaves serrate; seeds c.0.5mm, ± smooth
                      23. PARENTUCELLIA
  24 Leaves entire or ± so; seeds >1mm, ridged and
    grooved   **21. ODONTITES**

25 Perennial; corolla dark purple, >12mm   **22. BARTSIA**
25 Annual; corolla pink to reddish-purple, rarely white,
   <12mm   **21. ODONTITES**

Other genera - **NEMESIA** Vent. (Nemesias) (tribe
Hemimerideae), from S Africa, are commonly grown for ornament
and rarely occur as casuals on tips, etc. They are branched
annuals with variously coloured showy flowers with 4-lobed
upper lip, large slightly bossed lower lip, and spurred or
pouched tube. **N. strumosa** Benth., with pouched corolla-tube,
is the most grown sp. **PAULOWNIA** Siebold & Zucc. is an
ornamental tree with large ovate-cordate leaves and large
blue trumpet-shaped flowers in terminal panicles; **P.
tomentosa** (Thunb.) Steudel (Foxglove-tree), from China, has
been reported regenerating in Middlesex. For **SCOPOLIA** see
Solanaceae.

TRIBE 1 - SCROPHULARIEAE (Verbasceae, Cheloneae pro parte)
(genera 1-3). Herbaceous annuals or perennials or rarely
shrubs; leaves alternate or opposite; flowers nearly
actinomorphic to zygomorphic; corolla not to obscurely 2-
lipped, not spurred or pouched at base; stamens 4-5.

1. VERBASCUM L. - Mulleins
Herbaceous biennials, less often annuals or perennials, with
tall, erect, terete to ridged stems; leaves alternate;
corolla yellow, purplish or white, with 5 ± equal lobes and
short tube; stamens 5, rarely 4 in some flowers, all fertile,
at least the upper 3 with very hairy filaments.
Hybrids arise frequently where 2 or more spp. occur
together; they are usually highly but often not completely
sterile. Characters may be as in 1 or other parent or
intermediate. In hybrids between spp. with violet stamen-
hairs and spp. with white stamen-hairs all the stamens may
have all violet hairs, all have all pale violet hairs, all
have mixed white and violet hairs, or the upper 3 have white
and the lower 2 have violet hairs. Oblique-asymmetrical
anthers of 1 parent may or may not appear in the hybrid.
Hybrids between white-flowered V. lychnitis and yellow-
flowered spp. are yellow-flowered. Combinations other than
those listed often appear with the parents in gardens.

1 Anthers all reniform, symmetrical, placed transversely
  on filaments   2
1 Anthers of 3 upper stamens as above, of 2 lower
  stamens asymmetrical, placed obliquely or ±
  longitudinally on filaments and often ± decurrent
  on them   8
  2 All hairs on filaments yellow or white   3
  2 All or many hairs on filaments violet   5
3 Stems and leaves uniformly and persistently densely
  pubescent   **11. V. speciosum**

3 Stems and leaves unevenly mealy- or powdery-pubescent
   at first, becoming less pubescent to glabrous later   4
   4 Corolla usually white (rarely yellow); all or most
      pedicels >6mm; leaves sparsely pubescent to
      glabrous and green on upperside   **13. V. lychnitis**
   4 Corolla ± always yellow; all or most pedicels
      <6mm; leaves whitish-pubescent on upperside
                                           12. V. pulverulentum
5 Flowers 1 per node in axil of bract; bracteoles 0   6
5 Flowers 2-several per node in axil of bract; each
   pedicel with 2 small bracteoles   7
   6 Corolla violet to purple; hairs all simple, mostly
      glandular; inflorescence usually simple
                                           3. V. phoeniceum
   6 Corolla yellow; many hairs stellate; inflorescence
      much branched   **4. V. pyramidatum**
7 Basal leaves truncate to rounded at base; pedicels c.
   as long as calyx   **9. V. chaixii**
7 Basal leaves cordate at base; pedicels mostly c.2-3x
   as long as calyx   **10. V. nigrum**
   8 Hairs all simple, glandular   9
   8 At least some hairs branched (often stellate), all
      non-glandular   10
9 Flowers 1 per node; pedicels mostly longer than
   calyx; plant usually with glandular hairs only in
   upper parts   **1. V. blattaria**
9 Flowers usually >1 per node in lower parts of
   inflorescence; pedicels mostly shorter than calyx;
   plant usually with glandular hairs ± throughout
                                           2. V. virgatum
   10 Upper and middle stem-leaves distinctly decurrent   11
   10 Stem-leaves not decurrent   12
11 Stigma capitate   **8. V. thapsus**
11 Stigma elongated, spathulate   **7. V. densiflorum**
   12 Two lower stamens with filaments pubescent in
       lower 1/2, with anthers <4mm   **5. V. bombyciferum**
   12 Two lower stamens with glabrous to subglabrous
       filaments, with anthers >4mm   **6. V. phlomoides**

Other spp. - **V. sinuatum** L., from S Europe, has strongly
lobed basal leaves and stamens as in <u>V. nigrum</u>; it no longer
occurs in BI.

1. V. blattaria L. - <u>Moth Mullein</u>. Annual to biennial to
1m, ± glabrous below and on leaves, with stalked glands
above; inflorescence usually simple; flowers 1 per bract-
axil; corolla usually yellow; 3 upper anthers reniform, 2
lower decurrent; filaments all with violet hairs (upper 3
also with white hairs). Intrd; waste and rough ground;
scattered in C & S Br, formerly Ir and CI, rarely persistent;
Europe.
     1 x 10. V. blattaria x V. nigrum = V. x intermedium Rupr. ex

Bercht. & Pfund has occurred in N Wilts and N Lincs.

2. V. virgatum Stokes - <u>Twiggy</u> <u>Mullein</u>. Erect biennial to **R** 1m; differs from <u>V.</u> <u>blattaria</u> in key characters. Native; fields, waste places and dry banks; locally frequent in Devon and Cornwall, infrequent casual in waste places elsewhere in C & S Br, formerly Ir and CI.

2 x 8. V. virgatum x V. thapsus = V. x lemaitrei Boreau has occurred in Warks as a casual.

3. V. phoeniceum L. - <u>Purple</u> <u>Mullein</u>. Perennial to 1m, with simple hairs below, with stalked glands above; inflorescence usually simple; flowers 1 per bract-axil; corolla violet to purple; anthers all reniform; filaments all with violet hairs. Intrd; grown in gardens, casual on tips and waste ground and as bird-seed alien; very scattered in S & C Br; SE Europe.

4. V. pyramidatum M. Bieb. - <u>Caucasian</u> <u>Mullein</u>. Perennial to 1.5m, with branched and simple hairs throughout and stalked glands above; inflorescence well branched; flowers 1 per bract-axil; corolla usually yellow; anthers all reniform; filaments all with violet hairs. Intrd; waste and rough ground; casual, sometimes natd, in S En; Caucasus.

4 x 8. V. pyramidatum x V. thapsus has occurred in Cambs; endemic.

4 x 10. V. pyramidatum x V. nigrum has occurred in W Suffolk.

5. V. bombyciferum Boiss. - <u>Broussa</u> <u>Mullein</u>. Biennial to 2m, with dense branched hairs throughout; inflorescence usually simple; flowers several per bract-axil; corolla usually yellow; 3 upper anthers reniform, 2 lower decurrent; filaments all with whitish hairs, the 2 lower glabrous distally; stigma spathulate. Intrd; waste and rough ground; casual or rarely natd in S En and Guernsey; Turkey.

6. V. phlomoides L. - <u>Orange</u> <u>Mullein</u>. Differs from <u>V.</u> **707** <u>bombyciferum</u> in key characters (couplet 11). Intrd; waste and rough ground; frequent casual or sometimes natd in C & S Br and CI, rare in N Br; Europe.

6 x 8. V. phlomoides x V. thapsus = V. x kerneri Fritsch has occurred in Middlesex.

7. V. densiflorum Bertol. (<u>V.</u> <u>thapsiforme</u> Schrader) - <u>Dense-flowered</u> <u>Mullein</u>. Differs from <u>V.</u> <u>phlomoides</u> and <u>V.</u> <u>bombyciferum</u> in decurrent stem-leaves; from <u>V.</u> <u>bombyciferum</u> in 2 lower stamens with subglabrous filaments; and from <u>V.</u> <u>thapsus</u> in elongated spathulate stigma. Intrd; waste and rough ground; scattered casual or rarely natd in C & S En, probably over-recorded for <u>V.</u> <u>thapsus</u> and <u>V.</u> <u>phlomoides</u>; Europe.

8. V. thapsus L. - <u>Great</u> <u>Mullein</u>. Differs from <u>V.</u> **707** <u>phlomoides</u> and <u>V.</u> <u>bombyciferum</u> in decurrent stem-leaves; from <u>V.</u> <u>densiflorum</u> in capitate stigma; and from <u>V.</u> <u>bombyciferum</u> in 2 lower filaments glabrous to sparsely pubescent. Native; waste and rough ground, banks and grassy places, mostly on sandy or chalky soils; common in S & C Br and CI, locally frequent elsewhere, by far the commonest sp.

8 x 10. V. thapsus x V. nigrum = V. x semialbum Chaub. is rather frequent with its parents in much of Br; by far the commonest hybrid in the genus. Usually the upper 3 filaments have violet and the lower 2 white hairs, but all the anthers are reniform.

8 x 11. V. thapsus x V. speciosum = V. x duernsteinense Teyber has occurred in W Norfolk.

8 x 12. V. thapsus x V. pulverulentum = V. godronii Boreau has occurred in a few places in S En.

8 x 13. V. thapsus x V. lychnitis = V. x thapsi L. (V. x spurium Koch) is occasional with the parents in C & S Br.

9. V. chaixii Villars - Nettle-leaved Mullein. Biennial or perennial to 1m, with branched hairs dense below, sparse above; inflorescence usually well branched; flowers several per bract-axil; corolla usually yellow; anthers all reniform; filaments all with violet hairs. Intrd; grown in gardens, casual or natd escape on waste or rough ground; very scattered in S Br, overlooked as V. nigrum; S & C Europe.

10. V. nigrum L. - Dark Mullein. Differs from V. chaixii in stems to 1.2m; inflorescence simple to rather sparsely branched; and see key. Native; waste and rough ground, open places on banks and in grassland, mostly on soft limestone; locally common in CI and C & S Br, sparse casual elsewhere.

10 x 12. V. nigrum x V. pulverulentum = V. x mixtum Ramond ex DC. (V. x wirtgenii Franchet) occurs rarely with the parents in E Anglia and has occasionally been found elsewhere in En as a casual.

10 x 13. V. nigrum x V. lychnitis = V. x incanum Gaudin (V. x schiedeanum Koch) occurs sporadically in S En with the parents.

11. V. speciosum Schrader - Hungarian Mullein. Biennial to 2m, with dense branched persistent hairs on all parts; inflorescence well branched; flowers several per bract-axil; corolla usually yellow; anthers all reniform; filaments all with white hairs. Intrd; grown in gardens (probably the finest sp.), casual or natd escape in W Kent, W Norfolk and E Suffolk; SE Europe.

12. V. pulverulentum Villars - Hoary Mullein. Biennial to 707 1.5m; differs from V. speciosum in stems and leaves becoming **R** glabrous later, the pubescence becoming floccose as it wears off. Native; local on barish ground on chalky soils in E Anglia (mostly Norfolk) rare casual or natd escape elsewhere in Br.

12 x 13. V. pulverulentum x V. lychnitis = V. x regelianum Wirtgen has been recorded from N & S Devon as a casual; the parents are not sympatric as natives in BI.

13. V. lychnitis L. - White Mullein. Differs from V. **R** speciosum and V. pulverulentum in key characters (couplets 10 & 11); stems to 1.5m; corolla usually white (var. **album** (Miller) Druce), but yellow around Minehead, S Somerset. Native; similar places to V. pulverulentum; local in S En

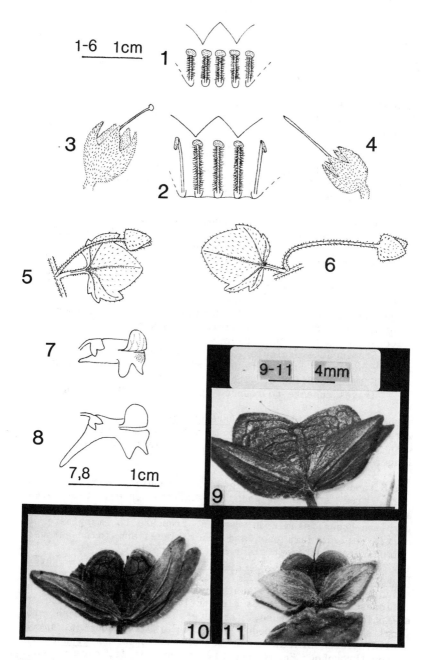

FIG 707 – Scrophulariaceae. 1–2, stamens of Verbascum. 1, **V. pulverulentum**. 2, **V. thapsus**. 3–4, calyx and style of Verbascum. 3, **V. thapsus**. 4, **V. phlomoides**. 5–6, fruiting nodes of **Veronica hederifolia**. 5, ssp. **hederifolia**. 6, ssp. **lucorum**. 7–8, corollas of **Linaria**. 7, **L. repens**. 8, **L. purpurea**. 9–11, fruits of **Veronica**. 9, **V. persica**. 10, **V. agrestis**. 11, **V. polita**.

from Devon to E Kent, infrequent casual or natd escape elsewhere in Br, and then often yellow-flowered.

2. SCROPHULARIA L. - Figworts

Herbaceous perennials, less often biennials, with erect, usually 4-angled stems usually square in section; leaves opposite; corolla dull purplish, brownish or greenish-yellow, with 5 nearly equal lobes and short tube, the lobes usually obscurely organised into 2-lobed upper and 3-lobed lower lips; fertile stamens 4, with the 5th (uppermost) usually represented by a sterile staminode.

```
1 Staminode 0; corolla-lobes equal, not organized
     into upper and lower lips                        5. S. vernalis
1 Staminode present, larger than anthers; corolla-
     lobes unequal, distinctly forming 2 weak lips            3
   2 Leaves and stems pubescent                 4. S. scorodonia
   2 Leaves and stems ± glabrous                               3
3 Stem-angles not or scarcely winged; sepals with
     narrow scarious border <0.3mm wide              1. S. nodosa
3 Stem-angles distinctly winged; sepals with
     conspicuous scarious border >0.3mm wide                  4
   4 Staminode ± orbicular, entire; leaves broadly
       cuneate to subcordate at base            2. S. auriculata
   4 Staminode bifid or with 2 divergent lobes at apex;
       leaves cuneate to rounded at base          3. S. umbrosa
```

Other spp. - **S. canina** L. (French Figwort), from S Europe, with leaves pinnately lobed ± to midrib and purplish corolla with 0 staminode, was once regular at docks in S Wa but no longer occurs.

1. S. nodosa L. - Common Figwort. Perennial to 1m, glabrous apart from stalked glands in inflorescence; stem-angles acute; leaves ovate, acute, serrate; corolla greenish to purplish-brown; staminode obovate, truncate to emarginate at apex. Native; damp open and shady places and in hedgerows; common throughout most of Br and Ir, rare in CI.

2. S. auriculata L. (S. aquatica auct. non L.) - Water Figwort. Perennial to 1.2m, usually glabrous (but sometimes slightly puberulent) apart from stalked glands in inflorescence; stem-angles distinctly winged; leaves ovate or elliptic-ovate, obtuse, crenate; corolla greenish to purplish-brown; staminode ± orbicular, rounded at apex. Native; places similar to or wetter than those of S. nodosa; common throughout En, Wa and CI, frequent in Ir, rare in Sc.

3. S. umbrosa Dumort. - Green Figwort. Differs from S.   R
auriculata in stem-angles more broadly winged; leaves acute to obtuse, serrate; bracts often leaf-like; calyx-teeth more deeply serrate; and see key. Native; damp shady places; very locally scattered in Br and Ir.

4. S. scorodonia L. - Balm-leaved Figwort. Perennial to 1m,   RR

pubescent on all parts and with stalked glands in
inflorescence; stem-angles acute to obtuse; leaves ovate,
acute to obtuse, doubly serrate; corolla dull purple;
staminode as in S. auriculata. Native; grassy field-borders
and hedgerows; very local, mostly near coast, in Devon and
Cornwall, common (the commonest sp.) in CI, natd in S Wa.
5. S. vernalis L. - Yellow Figwort. Biennial or perennial
to 50(80)cm, pubescent on all parts and with glandular hairs
in inflorescence; stem-angles obtuse; leaves broadly ovate,
acute to obtuse, strongly serrate; corolla dull yellowish;
staminode 0. Intrd; waste and rough ground, hedges and
woodland clearings; scattered throughout much of Br, locally
natd; Europe.

3. PHYGELIUS E. Meyer ex Benth. - Cape Figwort
Evergreen shrubs with 4-angled stems; leaves opposite;
corolla bright red, with 5 nearly equal lobes not organized
into 2 lips and tube much longer than lobes; stamens 4,
exserted.

1. P. capensis E. Meyer ex Benth. - Cape Figwort. Glabrous
shrub to 1.5m; leaves ovate, crenate-serrate; flowers pendent
in large terminal erect panicles, 2.5-4cm, with short, patent
lobes. Intrd; natd by rivers in Co Wicklow since at least
1970, ± natd by garden boundaries in Dunbarton; S Africa.

TRIBE 2 - GRATIOLEAE (genera 4-5). Herbaceous annuals to
perennials; leaves opposite (at least below) or all in basal
rosette; flowers nearly actinomorphic to strongly zygomorphic; corolla not to strongly 2-lipped, not spurred or
pouched at base; stamens 4.

4. MIMULUS L. - Monkeyflowers
Perennials with leafy stolons; leaves opposite; flowers
strongly zygomorphic, with well-defined upper and lower lips,
showy, yellow to red.
M. guttatus and M. luteus agg. (incl. M. variegatus and M.
cupreus) form a difficult, interfertile group; hybrids are
frequent and at least 3 occur in the absence of either
parent. Hybrids between M. guttatus and M. luteus agg. are
sterile (pollen <40 per cent full; seeds 0 or few per
capsule), but hybrids within M. luteus agg. are fertile
(pollen >50 per cent full; seeds many per capsule). In this
account M. cupreus is considered distinct from M. luteus, but
M. variegatus is included in the latter.

1 Calyx-teeth ± equal; plant glandular-pubescent ±
   all over; corolla <2.5cm   **1. M. moschatus**
1 Upper calyx-tooth distinctly longer than lower 4;
   plant glabrous below, often glandular pubescent
   above; corolla >2.5cm   2
   2 Inflorescence with abundant stalked glands; small

simple hairs present at least on keels of calyx,
often also elsewhere in inflorescence; plant
fertile or sterile   3
  2 Inflorescence glabrous or with sparse glandular
     hairs; small simple hairs 0 except inside calyx;
     plant fertile   5
3 Corolla copper-coloured, often also spotted or
   blotched red or purplish; calyx sometimes petaloid
                                          4. M. x burnetii(*1)
3 Corolla yellow, often spotted or blotched with
   orange, red or purplish; calyx not petaloid   4
     4 Throat of corolla $\pm$ closed by 2 boss-like
        swellings on lower lip; corolla wholly yellow or
        with red spots in throat, but with unmarked
        lobes; plant usually fertile   **2. M.guttatus(*2)**
     4 Throat of corolla $\pm$ open, the boss-like
        swellings low or inconspicuous; corolla lobes with
        orange, red or purplish spots or blotches; plant
        sterile   **3. M. x robertsii**
5 Leaves with even, triangular, flat teeth; corolla
   yellow, usually with coppery-orange spots or
   blotches, or mainly coppery   **6. M. x maculosus**
5 Leaves with irregular, oblong, often twisted teeth;
   corolla yellow, usually with dark red to purplish-
   brown blotches   **5. M. luteus**
(*1) M. guttatus x M. luteus x M. cupreus would also key out
here; see text.
(*2) M. guttatus is fertile; sterile plants keying out here
are hybrids of M. guttatus otherwise $\pm$ indistinguishable from
it.

Other spp. - **M. cupreus** Dombrain has been recorded from
several places, but in error for M. x burnetii.
All the spp. are natd in damp or wet places by streams and
ponds, on river shingle, in flushes and damp patches in
fields or open woods.

1. M. moschatus Douglas ex Lindley - Musk. Whole plant
glandular-pubescent; stems decumbent to ascending, to 40cm;
corolla 1-2cm, rather pale yellow. Intrd; scattered over
most of Br and Ir; W N America.
2. M. guttatus DC. - Monkeyflower. Plant glabrous below,
densely glandular-pubescent above; stems erect to ascending,
to 75cm; corolla 2.5-4.5cm, bright yellow, often with red
spots on throat, its throat $\pm$ closed by 2 bosses on lower
lip; fertile. Intrd; scattered and locally common over most
of BI, commonest lowland taxon; W N America.
2 x 6. M. guttatus x (M. luteus x M. cupreus) occurs as a
garden escape or sometimes spontaneously with its 2 parents;
it differs from M. x burnetii in its corolla-lobes having
dark blotches, and from M. luteus x M. cupreus in being
glandular-pubescent above and sterile.

3. M. x robertsii Silverside (M. luteus auct. non L.; M. guttatus x M. luteus) - Hybrid Monkeyflower. Plant glabrous below, variably glandular- pubescent above; stems erect to ascending, to 50cm; corolla 2.5-4.5cm, bright yellow with orange to red or purplish-brown spots and blotches variably developed on throat and lobes, its throat ± open; sterile. Intrd; scattered and locally common over much of Br and Ir, mostly in N & W and commonest upland taxon; garden origin.

4. M. x burnetii S. Arn. (M. cupreus auct. non Dombrain; M. guttatus x M. cupreus Dombrain) - Coppery Monkeyflower. Differs from M. guttatus x M. luteus in corolla copper-coloured with lighter throat marked with red spots but no dark blotches on corolla-lobes. Intrd; W & N Br N from Mons and Yorks; garden origin.

5. M. luteus L. (M. variegatus Lodd. nom. nud.) - Blood-drop-emlets. Plant glabrous or very sparsely glandular pubescent in inflorescence; stems decumbent to ascending, to 50cm; corolla 2.5-4.5cm, yellow, with red spots in throat and usually reddish blotches on lobes, its throat open; fertile. Intrd; rather uncommon in N Br S to Durham, much over-recorded for M. guttatus x M. luteus; Chile. Our plant is var. **rivularis** Lindley.

6. M. x maculosus T. Moore (M. luteus x M. cupreus) - Scottish Monkeyflower. Differs from M. luteus in key characters; corolla variable, usually with coppery-orange spots or blotches but sometimes more coppery and sometimes as in unblotched forms of M. luteus; fertile. Intrd; Sc N from Peebles; garden origin.

5. LIMOSELLA L. - Mudworts
Glabrous annuals (usually) with thin leafless stolons; leaves ± erect in basal rosette; flowers ± actinomorphic, small and inconspicuous, solitary on pedicels from leaf-rosettes.

1. L. aquatica L. - Mudwort. Leaves (incl. petioles) to   R
6(10)cm, mostly with narrowly elliptic blades up to 2cm; flowers 2.5-3mm, scentless; calyx longer than corolla-tube; corolla white to pale mauve. Native; wet sandy mud by ponds, often dried out in summer; very scattered and decreasing in Br and Ir N to Fife.

1 x 2. L. aquatica x L. australis occurs with the parents at Morfa Pools, Glam; it is sterile but more vigorous than either parent and often perennial, with a mixture of subulate and elliptic leaf-blades and calyx c. as long as corolla-tube; endemic.

2. L. australis R.Br. (L. subulata Ives) - Welsh Mudwort.   RR
Leaf-blades and petioles not differentiated, subulate, to 4cm; flowers 3.5-4mm, scented; calyx shorter than corolla-tube; corolla white. Native; similar places to L. aquatica; extremely local in Glam, Merioneth and Caerns (nowhere else in Europe).

TRIBE 3 - CALCEOLARIEAE (genus 6). Herbaceous annuals; leaves opposite; flowers strongly zygomorphic; corolla strongly 2-lipped, not spurred or pouched at base, with greatly incurved 'slipper-like' lower lip; stamens 2.

6. CALCEOLARIA L. - Slipperwort

Other spp. - C. integrifolia L., from Chile, is a shrub with simple serrate leaves and yellow to reddish-brown corollas; it is grown in gardens and occasional seedlings appear on walls in CI.

1. C. chelidonioides Kunth - Slipperwort. Stems well-branched, to 40cm, glandular-pubescent; leaves pinnate, with serrate leaflets; inflorescence few-flowered; corolla yellow, 10-14mm. Intrd; tips, waste places, cultivated ground; casual, sometimes natd, scattered in Br N to S Sc, especially E Anglia; C & S America.

TRIBE 4 - ANTIRRHINEAE (genera 7-13). Herbaceous annuals to perennials, rarely ± woody at base; leaves usually opposite below, often mostly alternate; flowers strongly zygomorphic; corolla strongly 2-lipped, with boss-like swelling on lower lip and conspicuous spur or pouch at base of tube; stamens 4.

7. ANTIRRHINUM L. - Snapdragon
Tufted, perennating by means of basal shoots; leaves entire, with single midrib; calyx-lobes ± equal, shorter than corolla-tube; corolla with broad, rounded pouch at base, with mouth closed by boss-like swelling on lower lip; capsule opening by 3 apical pores.

1. A. majus L. - Snapdragon. Stems erect to ascending, glabrous or glandular-pubescent above, to 1m; corolla 3-4.5cm, usually pink to purple, sometimes white, yellow, orange or combinations. Intrd; rough ground, walls, rocks, buildings; frequently natd throughout most of BI, but often killed in winter; SW Europe.

8. CHAENORHINUM (DC. ex Duby) Reichb. - Toadflaxes
Annuals, or tufted perennials with basal new shoots; leaves entire, with single midrib; calyx with slightly unequal lobes >1/2 as long to c. as long as corolla-tube; corolla with narrow conical spur at base, with mouth not completely closed by low boss-like swelling on lower lip; capsule opening by irregular large apical pores or tears.

1. C. origanifolium (L.) Kostel. - Malling Toadflax. Glandular-pubescent biennials to perennials; stems decumbent to erect, to 30cm, often several; lower leaves elliptic; corolla (incl. spur) 8-15(20)mm, bluish-mauve with pale yellow boss. Intrd; well natd on old walls at West Malling,

W Kent, since c. 1880, rare and impermanent in similar places elsewhere in En; SW Europe.

2. C. minus (L.) Lange - <u>Small Toadflax</u>. Glandular-pubescent or rarely glabrous annuals; stems erect, to 25cm, often 1; lower leaves oblong-oblanceolate; corolla (incl. spur) 6-9mm, pale purple with pale yellow boss. Native; arable land, waste places, railway tracks, in open ground; frequent over most of Br and Ir except N Sc.

9. MISOPATES Raf. - <u>Weasel's-snouts</u>
Annuals; leaves entire, with single midrib; calyx-lobes distinctly unequal, all longer than corolla-tube; corolla and capsule as in <u>Antirrhinum</u>.

1. M. orontium (L.) Raf. (<u>Antirrhinum</u> <u>orontium</u> L.) - <u>Weasel's-snout</u>. Stems erect, usually glandular-pubescent above, to 50cm; corolla 10-17mm, usually bright pink, rarely white. Probably native; weed of cultivated ground; locally frequent in S Br and CI, very scattered and decreasing in C & N Wa and C & N En, casual in Sc and Ir.

2. M. calycinum Rothm. - <u>Pale</u> <u>Weasel's-snout</u>. Differs from M. orontium in plant usually glabrous; corolla pale pink to white, 18-22mm. Intrd; bird-seed alien on tips and waste ground; rather scarce in S En; W Mediterranean.

10. ASARINA Miller - <u>Trailing Snapdragon</u>
Stoloniferous perennials; leaves crenate to shallowly lobed, with palmate main veins; calyx with slightly unequal lobes much shorter than corolla-tube; corolla as in <u>Antirrhinum</u>; capsule opening by 2 apical pores.

1. A. procumbens Miller - <u>Trailing Snapdragon</u>. Stems procumbent, glandular-pubescent, to 60cm; corolla 3-3.5cm, pale yellow with pinkish-purple veins, the boss deep yellow. Intrd; natd on dry banks, walls and cliffs; scattered in Br N to Dumfriess; mountains of S France and NE Spain.

11. CYMBALARIA Hill - <u>Ivy-leaved Toadflaxes</u>
Stoloniferous perennials; leaves subentire to shallowly and obtusely lobed, with palmate main veins; calyx with unequal lobes shorter than corolla-tube; corolla with narrow cylindrical to conical spur at base, with mouth completely closed by boss-like swelling on lower lip; capsule opening by irregular apical longitudinal slits.

1 Corolla 9-15mm, incl. spur 1.5-3mm; stems long and
  trailing, often >20cm **1. C. muralis**
1 Corolla 15-25mm, incl. spur 4-9mm; stems usually not
  trailing, <20cm 2
  2 Stems, leaves, petioles and calyx shortly and
    densely pubescent **2. C. pallida**
  2 Plant glabrous or nearly so **3. C. hepaticifolia**

714 SCHROPHULARIACEAE

1. C. muralis P. Gaertner, Meyer & Scherb. - <u>Ivy-leaved</u>
<u>Toadflax</u>. Stems long and trailing, to 60cm; corolla 9-15mm
incl. spur 1.5-3mm, mauvish-violet on lower lip. Intrd; well
natd on walls, pavements, rocky or stony banks, first
recorded 1640; frequent over most of BI; CS Europe.
  a. Ssp. muralis. Plant glabrous, or sparsely pubescent on
calyx and young parts. Well natd on walls, pavements, rocky
or stony banks, first recorded 1640; frequent over most of
BI; CS Europe.
  b. Ssp. visianii (Kuemm. ex Jav.) D. Webb. Plant pubescent
on ± all parts. Natd on waste ground since 1970 in Surrey;
Italy and Jugoslavia.
  2. C. pallida (Ten.) Wettst. - <u>Italian Toadflax</u>. Stems
<20cm, decumbent to suberect; plant shortly pubescent;
corolla 15-25mm incl. spur 6-9mm, mauvish-violet with white
boss on lower lip. Intrd; natd on walls, shingle and stony
places; scattered in Br, mainly N En and Sc, less well natd
in S; Italy.
  3. C. hepaticifolia (Poiret) Wettst. - <u>Corsican Toadflax</u>.
Differs from <u>C. pallida</u> in plant glabrous or nearly so;
corolla 15-18mm incl. spur 4-5mm. Intrd; marginally natd in
and near gardens and nurseries, perhaps overlooked for <u>C.</u>
<u>pallida</u>; very scattered in En and Sc; Corsica.

12. KICKXIA Dumort. - <u>Fluellens</u>
Annuals; leaves ovate to elliptic, ± entire to remotely
dentate, with pinnate venation; calyx with equal lobes c. as
long as corolla-tube; corolla with narrowly conical spur at
base, with mouth completely closed by boss-like swelling on
lower lip; capsule opening by 2 large oblique lids.

  1. K. elatine (L.) Dumort. - <u>Sharp-leaved Fluellen</u>. Stems
procumbent to suberect, usually well-branched at base, to
50cm; whole plant except pedicels and corolla with patent
hairs (pedicels pubescent immediately below flower); leaves
hastate; corolla 7-12mm incl. spur c.1/2 total length, yellow
with violet upper lip. Probably native; arable fields and
field-borders on light, usually calcareous soils; locally
common in Br N to N Wa and N Lincs but intrd or casual in
much of N & W, CI, intrd in S & W Ir.
  2. K. spuria (L.) Dumort. - <u>Round-leaved Fluellen</u>. Differs
from <u>K. elatine</u> in usually more robust; pedicels pubescent;
leaves rounded at base; corolla 8-15mm, with more curved spur
and purple upper lip. Probably native; similar places to <u>K.</u>
<u>elatine</u>, often with it; SE Br NW to N Lincs and S Wa.

13. LINARIA L. - <u>Toadflaxes</u>
Annuals or perennials, sometimes rhizomatous; leaves entire,
with single midrib; calyx with usually unequal lobes shorter
than corolla-tube; corolla with narrow conical spur at base,
with mouth closed (or ± so) by boss-like swelling on lower
lip; capsule opening by irregular apical longitudinal slits.

1 Spur longer than rest of corolla   **8. L. maroccana**
1 Spur much shorter than to nearly as long as rest
  of corolla   2
  2 Whole plant glandular-pubescent; corolla (incl.
  spur) 4-7mm   **6. L. arenaria**
  2 Plant glabrous below, glabrous to glandular-
  pubescent above; corolla (incl. spur) usually
  >8mm   3
3 Corolla predominantly yellow, sometimes very pale or
  with purplish tinge (if with violet veins, see L.
  vulgaris x L. repens)   4
3 Corolla predominantly mauve, violet, purple or pink,
  sometimes very pale but then with darker veins,
  sometimes with yellow to orange boss   6
  4 Annual; stems decumbent, with conspicuous region
  below inflorescence bare of leaves   **5. L. supina**
  4 Perennial; stems normally erect, with leaves +
  up to inflorescence; rhizomes often present   5
5 Seeds disc-like, with broad wing round circumference;
  plant often glandular-pubescent above; leaves linear
  to narrowly elliptic-oblanceolate, cuneate at base
                                                      1. L. vulgaris
5 Seeds angular, scarcely winged; plant always glabrous;
  at least some leaves lanceolate to ovate and subcordate at base   **2. L. dalmatica**
  6 Annual; capsule shorter than calyx; seeds disc-
  like, with broad wing round circumference
                                                 7. L. pelisseriana
  6 Perennial; capsule longer than calyx; seeds
  angular, not winged   7
7 Spur <1/2 as long as rest of corolla, straight, subacute to rounded at tip; corolla with orange patch
  on boss   **4. L. repens**
7 Spur >1/2 as long as rest of corolla, usually
  curved, acute at tip; corolla without orange (sometimes with white) patch on boss   **3. L. purpurea**

1. L. vulgaris Miller - <u>Common Toadflax</u>. Erect to ascending
perennial to 80cm, glabrous all over or glandular-pubescent
above; corolla 18-35mm incl. spur 6-13mm, yellow (sometimes
very pale) with orange boss; seeds discoid, with broad
marginal wing. Native; rough and waste ground, stony places,
banks, open grassland; common over most of BI, absent from
parts of Ir and C & N Sc.
1 x 4. L. vulgaris x L. repens = **L. x sepium** Allman is
frequent within the range of <u>L. repens</u> in BI N to Ayrs; it is
intermediate in corolla colour and fertile, often forming
hybrid swarms. The corolla is most often pale yellow with
violet veins and intermediate in size and shape, but in
hybrid swarms plants close to either parent may occur.
2. L. dalmatica (L.) Miller (<u>L. genistifolia</u> (L.) Miller
ssp. <u>dalmatica</u> (L.) Maire & Petitm.) - <u>Balkan Toadflax</u>.

Erect, glabrous perennial to 80cm; corolla 20–55mm, incl.
spur 4–25mm yellow; seeds angular, not or scarcely winged.
Intrd; natd in waste places, waysides and by railways;
scattered in S En, rare casual elsewhere; SE Europe.

3. L. purpurea (L.) Miller - <u>Purple</u> Toadflax. Erect, **707**
glabrous perennial to 1m; corolla 7–15mm incl. curved spur
3–6mm, mauve with heavy purplish-violet veins or wholly
purplish-violet, rarely pink, with concolorous or whitish
boss; seeds angular, not winged. Intrd; natd on rough
ground, walls, banks; frequent to sparse throughout much of
BI; Italy.

3 x 4. L. purpurea x L. repens = L. x dominii Druce occurs
in scattered localities usually with 1 or both parents in Br
N to Westmorland; it is intermediate in corolla characters
and fertile, sometimes segregating to give hybrid swarms.

4. L. repens (L.) Miller - <u>Pale</u> Toadflax. Glabrous, **707**
decumbent to erect perennial to 80cm; corolla 8–15mm incl.
spur 1–4mm, whitish to pale mauve with violet thin veins,
with orange spot on boss; seeds angular, not winged. Native;
stony places, rough ground, banks and walls; scattered over
much of Br, CI and E Ir, but absent from many places and
frequent only in parts of S & W Br.

4 x 5. L. repens x L. supina = L. x cornubiensis Druce was
collected from Par, E Cornwall, with both parents in 1925 and
1930; it was said to be intermediate in corolla characters
and sterile.

5. L. supina (L.) Chaz. - <u>Prostrate</u> Toadflax. Procumbent to **RR**
decumbent annual to 20cm, glabrous below, glandular pubescent
above; corolla 15–25mm incl. spur 7–11mm, pale with yellowy-
orange boss; seeds discoid, with broad marginal wing.
Possibly native; sandy ground near Par, E Cornwall, natd in
open waste ground mainly by railways in E & W Cornwall and S
Devon, rare casual elsewhere; SW Europe.

6. L. arenaria DC. - <u>Sand</u> Toadflax. Erect, glandular-
pubescent annual to 15cm; corolla 5–7mm incl. spur 1.5–3mm,
yellowish with yellowish to violet spur; seeds discoid, with
rather narrow marginal wing. Intrd; planted at Braunton
Burrows, N Devon c. 1893, now well natd on semi-fixed dunes;
W France and NW Spain.

7. L. pelisseriana (L.) Miller - <u>Jersey</u> Toadflax. Erect, **RR**
glabrous annual to 30cm; corolla 11–20mm, incl. spur 5–9mm,
purplish-violet with whitish boss; seeds discoid, with broad
marginal wing. Native; rough ground, rocky places and hedgebanks; very rare and sporadic in Jersey, last seen 1955,
perhaps not native, rare casual in S Br.

8. L. maroccana Hook. f. - <u>Annual</u> Toadflax. Erect annual to
50cm, glabrous below, glandular-pubescent above; corolla
17–30mm, incl. spur 9–17mm, usually purplish-violet with
yellow boss but white, yellow, pink, red or variously
variegated colour-forms occur; seeds angular, not winged.
Intrd; grown in gardens, frequent casual on tips and in waste
places; scattered in Br and CI, mainly S En; Morocco.

TRIBE 5 - DIGITALEAE (Veroniceae) (genera 14-18).
Herbaceous annuals to perennials or sometimes shrubs; leaves
opposite or alternate; flowers nearly actinomorphic to
zygomorphic; corolla not to weakly 2-lipped, not spurred or
pouched at base; stamens 2 or 4 (rarely 3 or 5).

14. DIGITALIS L. - Foxgloves
Herbaceous biennials to perennials; leaves alternate, with
pinnate venation; flowers in terminal racemes; corolla showy,
weakly 2-lipped, with 5 lobes shorter than tube; stamens 4.

Other spp.. - D. grandiflora Miller (Yellow Foxglove), with
large yellow flowers, **D. lanata** Ehrh. (Grecian Foxglove),
with whitish-yellow corolla with brownish veins, and **D.
ferruginea** L. (Rusty Foxglove), with yellowish- or reddish-
brown corolla with darker veins, all from S & C Europe, have
all been reported as garden escapes but have not survived.
D. lutea x D. purpurea = **D. x fucata** Ehrh. (D. x purpurascens
Roth) is not rare as a spontaneous hybrid in gardens with the
parents, but has not been recorded outside.

1. D. purpurea L. - Foxglove. Erect usually densely
pubescent biennial to short-lived perennial to 2m; leaves
ovate to lanceolate, + rugose; corolla 40-55mm, pink to
purple with dark spots inside, sometimes white. Native; many
sorts of open places, especially woodland clearings, heaths
and mountain-sides, also waste ground, on acid soils; common
throughout BI in suitable places, garden escape elsewhere.
2. D. lutea L. - Straw Foxglove. Erect glabrous to slightly
pubescent usually perennial to 1m; leaves lanceolate- to
oblanceolate-oblong, + smooth; corolla 9-25mm, pale yellow.
Intrd; natd on waste ground, roadsides and walls; scattered
in S En; W Europe.

15. ERINUS L. - Fairy Foxglove
Herbaceous perennials; leaves alternate, with pinnate
venation; flowers in terminal raceme; corolla showy, not
2-lipped, with 5 + equal, patent, emarginate lobes little
shorter than tube; stamens 4.

1. E. alpinus L. - Fairy Foxglove. Stems to 20cm, ascending
to suberect, pubescent; leaves oblanceolate to obovate,
serrate; corolla-tube 3-7mm, corolla-limb 6-9mm across,
purple, sometimes white. Intrd; natd on walls and stony
places; scattered in BI, mainly N En to C Sc; mountains of SW
Europe.

16. VERONICA L. - Speedwells
Herbaceous annuals to perennials, sometimes woody at base;
leaves opposite, at least below, with pinnate or sometimes +
palmate venation; flowers solitary in leaf-axils or in
terminal or axillary racemes; corolla showy to inconspicuous,

not 2-lipped, with 4 subequal lobes much longer than tube;
stamens 2.

General key

1 Flowers (at least mostly) in axillary racemes (include
   plants with a single raceme in 1 of the most apical
   leaf-axils)   Key A
1 Flowers in terminal racemes or solitary in leaf-axils 2
   2 Flowers in terminal racemes; bracts all or at least
      the upper very different from the foliage leaves,
      but the lower sometimes similar to them   Key B
   2 Flowers solitary in axils of leaves closely
      resembling the foliage leaves, though the upper
      often smaller than them   Key C

Key A - Flowers in axillary racemes

1 Leaves and stems glabrous, except stems sometimes
   glandular-pubescent in inflorescence   2
1 Leaves and stems pubescent   5
   2 Racemes 1 per node; capsule dehiscing into 2
      valves   **9. V. scutellata**
   2 Racemes mostly 2 per node; capsule dehiscing into
      4 valves   3
3 Leaves all shortly petiolate; flowering stems procumbent or decumbent (to ascending)   **10. V. beccabunga**
3 Upper leaves sessile; flowering stems usually erect
   (to ascending)   4
   4 Corolla pale blue; pedicels erecto-patent in fruit;
      capsule 2.5-4mm, $\pm$ orbicular
                               11. V. anagallis-aquatica
   4 Corolla pinkish; pedicels patent in fruit; capsule
      2-3mm, wider than long   **12. V. catenata**
5 Stems pubescent along 2 opposite lines only;
   capsule shorter than calyx   **7. V. chamaedrys**
5 Stems pubescent all round; capsule longer than calyx 6
   6 Petioles >6mm; capsules >6mm wide   **8. V. montana**
   6 Petioles <6mm; capsules <6mm wide   7
7 Leaves linear-oblong to -lanceolate; pedicels >6mm in
   fruit, longer than bracts   **9. V. scutellata**
7 Leaves lanceolate-oblong to ovate or elliptic;
   pedicels <6mm in fruit, up to as long as bracts   8
   8 Calyx-lobes usually 5, 1 much shorter than other 4;
      leaves sessile   **5. V. austriaca**
   8 Calyx-lobes 4; at least lower leaves petiolate
                                           6. V. officinalis

Key B - Flowers in terminal racemes

1 Annuals with 1 root system, easily uprooted   2
1 Perennials, with non-flowering shoots and/or stems
   rooted more than just at base   8
   2 Plant glabrous   **18. V. peregrina**
   2 Plant pubescent or puberulent, at least on capsules

and inflorescence-axis   3
3 Bracts much longer than fruiting pedicels   4
3 Bracts shorter than to ± as long as fruiting pedicels 6
  4 At least some upper leaves lobed >1/2 way to
    midrib   **17. V. verna**
  4 All leaves entire to crenate-serrate, toothed
    <1/2 way to midrib   5
5 All hairs glandular; leaves oblanceolate to narrowly
  oblong   **18. V. peregrina**
5 Many hairs, at least below, non-glandular; leaves
  ovate   **16. V. arvensis**
  6 Lower bracts and upper leaves lobed much >1/2 way
    to base   **15. V. triphyllos**
  6 Bracts and leaves toothed <1/2 way to midrib   7
7 Capsule notched to c.1/2 way; pedicels >2x as long as
  calyx; seeds flat   **13. V. acinifolia**
7 Capsule notched to <1/4 way; pedicels <2x as long
  as calyx; seeds cup-shaped   **14. V. praecox**
  8 Corolla-tube usually >2mm, longer than wide;
    racemes dense, long and many-flowered   9
  8 Corolla-tube usually <2mm, wider than long; racemes
    lax, short and/or few-flowered   10
9 Leaves usually widest in middle 1/3, crenate to
  serrate with usually obtuse teeth, pubescent on both
  surfaces   **26. V. spicata**
9 Leaves usually widest in basal 1/3, serrate to
  biserrate with acute to subacuminate teeth, glabrous
  to sparsely pubescent or puberulent on both surfaces
                       25. V. longifolia
  10 Corolla pink; style c.2x as long as capsule
                             2. V. reptans
  10 Corolla white to blue; style shorter than capsule 11
11 Capsule wider than long   **1. V. serpyllifolia**
11 Capsule longer than wide   12
  12 Corolla >10mm across; stems woody at base; style
    >2mm; racemes with eglandular hairs   **4. V. fruticans**
  12 Corolla <10mm across; stems herbaceous; style
    <2mm; racemes with glandular hairs   **3. V. alpina**

Key C - Flowers solitary in leaf-axils
1 Calyx-lobes apparently 2, each bilobed at apex
                       22. V. crista-galli
1 Calyx-lobes 4, each acute to rounded at apex   2
  2 Perennial; stems rooting at nodes along length;
    pedicels >2x as long as leaves + petioles
                     23. V. filiformis
  2 Annual; stems not rooting at nodes or doing so
    only near base; pedicels <2x as long as leaves +
    petioles   3
3 Calyx-lobes cordate at base; leaves with 3-7 shallow
  lobes or teeth   **24. V. hederifolia**
3 Calyx-lobes cuneate to rounded at base; leaves

crenate-serrate, most with >7 teeth   4
  4 Lobes of capsule with apices diverging at c.90
      degrees; corolla mostly ≥8mm across   **21. V. persica**
  4 Lobes of capsule with apices ± parallel or
      diverging at narrow angle; corolla ≤8mm across   5
5 Capsule with patent glandular hairs only
                                             19. V. agrestis
5 Capsule with many short eglandular arched hairs and
      some patent glandular hairs   **20. V. polita**

Other spp. - Records of **V. paniculata** L. (V. spuria auct.
non L.) as garden escapes seem to be all or mostly errors for
V. longifolia. **V. cymbalaria** Bod. (Pale Speedwell), from S
Europe, differs from V. hederifolia in its 5-9-lobed leaves,
obtuse to rounded (not acute to subacute) calyx-lobes and
pubescent (not glabrous) capsules; it occurred in W Cornwall
in 1985 and might become more frequent.

Section 1 - VERONICASTRUM Koch (sect. Berula Dum.) (spp.
1-4). Perennials; inflorescence a terminal raceme; corolla-
tube wider than long; capsule dehiscing into 2 valves,
usually only apically; seeds flat.

1. V. serpyllifolia L. - Thyme-leaved Speedwell. Stems to
30cm, herbaceous, rooting at nodes; leaves shortly petiolate,
glabrous; corolla 5-10mm across; capsule wider than long,
shorter than calyx, with style c. as long. Native.
  a. Ssp. serpyllifolia. At least 1/2 of flowering stem
upturned and erect; leaves ovate-elliptic; racemes ± glabrous
or with eglandular hairs and usually >12 flowers; pedicels c.
as long as calyx; corolla 6-8mm across, whitish to pale blue
with darker veins. Waste and cultivated ground, paths,
lawns, open grassland, woodland rides and on mountains;
common throughout BI.
  b. Ssp. humifusa (Dickson) Syme. Most of flowering stem
procumbent; leaves ovate-orbicular; racemes with glandular
hairs and often <12 flowers; pedicels longer than calyx;
corolla 7-10mm across, bright blue. Rock-ledges, flushes and
wet gravel in mountains; Scottish Highlands, less extreme
plants in N En and N & S Wa.
  Mountain plants of ssp. serpyllifolia often approach ssp.
humifusa in their bluer corollas, more procumbent stems,
fewer flowers and presence of glandular hairs, and need
investigating.
  2. V. reptans Kent (V. repens Clarion ex DC. non Gilib.) -
Corsican Speedwell. Differs from V. serpyllifolia ssp.
humifusa in pedicels usually longer than bracts; corolla
pink; style much longer than calyx. Intrd; natd weed in
lawns in very few places in Sc and N En, less common than
formerly; Corsica and S Spain.
  3. V. alpina L. - Alpine Speedwell. Stems to 15cm,   R
herbaceous, erect to ascending from short rooting portion;

leaves ovate-elliptic, subsessile, ± glabrous; racemes
glandular-pubescent; corolla 5-10mm across, dull blue;
capsule longer than wide, longer than calyx, with much
shorter style. Native; damp alpine rocks above 500m; local
in C Sc.
   4. V. fruticans Jacq. - Rock Speedwell. Stems woody below, **RR**
to 20cm, erect to ascending; leaves obovate-elliptic, subsessile, glabrous; racemes puberulent; corolla 10-15mm
across, bright blue; capsule longer than wide, longer than
calyx, with slightly shorter style. Native; alpine rocks
above 500m; very local in C Sc.

Section 2 - VERONICA (spp. 5-9). Perennials; inflorescences
opposite or alternate axillary racemes; corolla-tube wider
than long; capsule dehiscing into 2 valves, usually only
apically; seeds flat.

   5. V. austriaca L. (V. teucrium L.) - Large Speedwell.
Stems decumbent to erect, to 50cm, pubescent all round;
leaves ovate-oblong, serrate, sessile, pubescent; corolla
10-15mm across, bright blue; capsule elliptic, longer than
wide. Intrd; grown in gardens and natd in open and rough
ground, dunes; scattered in Br N to Angus; Europe. Our plant
is ssp. **teucrium** (L.) D. Webb.
   6. V. officinalis L. - Heath Speedwell. Stems procumbent to
ascending, to 40cm, pubescent all round; leaves obovate-
elliptic, serrate, petiolate, pubescent; corolla 5-9mm
across, lilac; capsule obtriangular-obovate, c. as long as
wide. Native; banks, open woods, grassland and heathland on
well-drained soils; common throughout most of BI.
   7. V. chamaedrys L. - Germander Speedwell. Stems erect to
ascending, to 50cm, pubescent only along 2 opposite lines;
leaves triangular-ovate, serrate, sessile or with petioles up
to 5mm, pubescent; corolla 8-12mm across, bright blue;
capsule obtriangular-obovate, wider than long. Native;
woods, hedgerows, grassland in damper areas; common throughout BI.
   8. V. montana L. - Wood Speedwell. Stems procumbent to
ascending, to 40cm, pubescent all round; leaves ovate to
broadly so, serrate, with petioles 5-15mm, pubescent; corolla
8-10mm across, pale lilac-blue; capsule broadly transversely
elliptic. Native; dampish woods; scattered and locally
frequent throughout Br and Ir except N Sc.
   9. V. scutellata L. - Marsh Speedwell. Stems decumbent to
scrambling-erect, to 60cm, glabrous or pubescent (var.
villosa Schum.) all round; leaves linear-oblong to
-lanceolate, entire to distantly serrate, sessile, glabrous
or pubescent; corolla 5-8mm across, whitish to pale pinkish
or lilac; capsule reniform. Native; bogs, marshes, wet
meadows, by ponds and lakes, on bare ground or among tall
vegetation; locally frequent throughout BI.

Section 3 - BECCABUNGA (Hill) Dumort. (spp. 10-12). Perennials (sometimes annual); inflorescences opposite, axillary racemes; corolla-tube wider than long; capsule dehiscing into 4 valves; seeds flat on 1 side, convex on other.

10. V. beccabunga L. - Brooklime. Stems procumbent to ascending, to 60cm; leaves ovate-oblong to elliptic or broadly so, obtuse to rounded, petiolate; pedicels ± patent in fruit, shorter to longer than bracts; corolla bright blue; capsule 2-4mm, c. as long as wide. Native; streams, ditches, marshes, pond-sides and river-banks; common throughout BI except rare in N Sc.

11. V. anagallis-aquatica L. - Blue Water-Speedwell. Stems **741** erect, to 50cm; leaves often narrowly ovate and shortly petiolate below, always lanceolate and sessile above, acute or subacute; pedicels erecto-patent in fruit, at least as long as bracts at flowering; corolla pale blue with darker veins. Native; by ponds and streams, in marshes and wet meadows; scattered and locally common throughout most BI.

11 x 12. V. anagallis-aquatica x V. catenata = V. x lackschewitzii J. Keller is frequent with the parents throughout their range but commonest in En; it is intermediate, but often more robust and with longer racemes than either parent, and partially fertile.

12. V. catenata Pennell - Pink Water-Speedwell. Differs **741** from V. anagallis-aquatica in leaves all narrow and sessile; pedicels usually shorter than bracts at flowering; and see key. Native; mostly in open muddy places with little or no flowing water; locally frequent in En, Wa and Ir, rare in Sc and CI.

Section 4 - POCILLA Dumort. (spp. 13-24). Annuals (except V. filiformis); flowers solitary in leaf-axils or in terminal racemes; corolla-tube wider than long; capsule dehiscing into 2 or 4 valves; seeds fairly flat, flat to convex on 1 side, flat to concave on other.

13. V. acinifolia L. - French Speedwell. Stems erect, to 15cm, glandular-pubescent throughout or above only and glabrous below; leaves ovate, obscurely serrate; flowers in racemes; corolla 2-3mm across, blue; capsule with patent glandular hairs. Intrd; casual or persistent weed of gardens, nurseries and public flower-beds; scattered in S En; S Europe.

14. V. praecox All. - Breckland Speedwell. Stems erect to **RR** ascending, to 20cm, pubescent, glandular-pubescent above; leaves ovate, deeply serrate; flowers in racemes; corolla 2.5-4mm across, blue; capsule with patent glandular hairs. Perhaps native; sandy arable fields; very local in W Suffolk and W Norfolk, 1st recorded 1933, rare casual elsewhere, decreasing.

15. V. triphyllos L. - Fingered Speedwell. Stems erect to **RRR**

ascending, to 20cm, glandular-pubescent; at least upper leaves and lower bracts lobed palmately almost to base with 3-7 lobes; flowers in racemes; corolla 3-4mm across, blue; capsule with patent glandular hairs. Native; sandy arable fields; very local in W Norfolk and E & W Suffolk, formerly scattered Surrey to MW Yorks, decreasing, perhaps extinct.

16. V. arvensis L. - <u>Wall Speedwell</u>. Stems erect to decumbent, to 30cm, pubescent or glandular-pubescent; leaves ovate, serrate to crenate-serrate; flowers in racemes; corolla 2-3mm across, blue; capsule glandular-pubescent. Native; walls, banks, open acid or calcareous ground and cultivated land; common throughout BI.

17. V. verna L. - <u>Spring Speedwell</u>. Stems erect, to 15cm; **RR** differs from <u>V. arvensis</u> in stems more glandular-pubescent above; capsule wider than long (not c. as long as wide); and see key. Native; open places in poor grassland on dry sandy soils, often with <u>V. arvensis</u>; very local in W Suffolk, formerly E Suffolk and E & W Norfolk.

18. V. peregrina L. - <u>American Speedwell</u>. Stems erect, to 25cm, usually glabrous; leaves oblanceolate to narrowly oblong, entire to distantly crenate; flowers in racemes; corolla 2-3mm across, blue; capsule glabrous. Intrd; casual or persistent weed of gardens, nurseries and public flower-beds; scattered in BI N to C Sc; N & S America. A glandular-pubescent variant has been recorded once as a casual in N Hants.

19. V. agrestis L. - <u>Green Field-speedwell</u>. Stems **707** procumbent to ascending, to 30cm, pubescent; leaves ovate, serrate; flowers solitary in leaf-axils; corolla 3-8mm across, whitish to pale blue or pale lilac; capsules with patent glandular hairs only. Native; cultivated ground; frequent throughout most of BI.

20. V. polita Fries - <u>Grey Field-speedwell</u>. Differs from <u>V.</u> **707** <u>agrestis</u> in lower leaves usually wider than long (not longer than wide), dull to greyish-green (not light to mid-green); calyx-lobes acute to subacute (not obtuse to subacute); corolla bright blue; and see key. Native; cultivated ground; frequent in most of BI but rare in C & N Sc.

21. V. persica Poiret - <u>Common Field-speedwell</u>. Stems **707** procumbent to decumbent, to 50cm, pubescent; leaves ovate, serrate; flowers solitary in leaf-axils; corolla 8-12mm across, bright blue; capsules with patent glandular hairs only. Intrd; well natd in cultivated and waste ground throughout most of BI, 1st recorded 1825; SW Asia.

22. V. crista-galli Steven - <u>Crested Field-speedwell</u>. Differs from <u>V. persica</u> in corolla 5-7mm across; calyx-segments fused in pairs forming conspicuous bilobed structures completely concealing much more shallowly notched capsules with non-divergent lobes. Intrd; occasional casual in cultivated and rough ground and waste places; very scattered in En and Wa, natd in N Somerset and formerly in W Sussex; Caucasus.

23. V. filiformis Smith - <u>Slender Speedwell</u>. Perennial with procumbent, puberulent stems to 50cm; leaves orbicular to reniform, crenate; flowers solitary in leaf-axils on erect long pedicels; corolla 8-15mm across, blue; capsule (seldom produced) with patent glandular hairs only, the lobes only slightly divergent. Intrd; well natd on stream-sides, lawns, grassy paths and roadsides throughout most of BI, 1st recorded 1838, then 1927 onwards; Turkey and Caucasus.

24. V. hederifolia L. - <u>Ivy-leaved Speedwell</u>. Stems procumbent to scrambling or ascending, to 60cm, pubescent, rather fleshy; leaves orbicular to reniform, with 1-3 large teeth or lobes each side; flowers solitary in leaf-axils; corolla 4-9mm across; capsule glabrous. Native; cultivated and waste ground, open woods, hedgerows, walls and banks; common throughout BI except parts of Sc and Ir.

a. Ssp. hederifolia. Apical leaf-lobe usually wider than **707** long; fruiting pedicels mostly 2-4x as long as calyx; calyx enlarging strongly after flowering, with marginal hairs mostly $\geq$0.9mm; corolla mostly $\geq$6mm across, whitish to blue; anthers blue, 0.7-1.2mm; 2n=54. Commoner in open places and cultivated ground.

b. Ssp. lucorum (Klett & H. Richter) Hartl (<u>V. sublobata</u> M. **707** Fischer). Apical leaf-lobe usually longer than wide; fruiting pedicels mostly 3.5-7x as long as calyx; calyx enlarging slightly after flowering, with marginal hairs mostly $\leq$0.9mm; corolla mostly $\leq$6mm across, whitish to pale lilac-blue; anthers whitish to pale blue, 0.4-0.8mm; 2n=36. Commoner in shady places.

Determination of the ssp. is best using as many characters as possible; even so, up to 25% of plants may be difficult to name.

Section 5 - PSEUDOLYSIMACHIUM Koch (spp. 25-26). Perennials; flowers in dense, long terminal racemes; corolla-tube longer than wide; capsule dehiscing into 2 valves usually only apically; seeds flat on 1 side, convex on other.

25. V. longifolia L. - <u>Garden Speedwell</u>. Stems erect, woody near base, to 1.2m, glabrous to puberulent; leaves lanceolate, acute to acuminate, serrate to biserrate with acute to subacuminate teeth, glabrous or sparsely pubescent, 2-4 per node; calyx-lobes acute. Intrd; much grown in gardens and natd on waste and rough ground, banks, roadsides; scattered in Br, mainly S & C; N & C Europe.

25 x 26. V. longifolia x V. spicata is probably the parentage of many garden and some natd plants; the hybrid is intermediate and fertile.

26. V. spicata L. - <u>Spiked Speedwell</u>. Stems erect to ascending, woody near base, to 60(80)cm, pubescent; leaves oblong-elliptic to narrowly so, acute to obtuse, serrate or crenate with obtuse teeth, pubescent, 2 per node; calyx-lobes obtuse. Native; rocks and short grassland on limestone

and other basic soils.
a. Ssp. spicata. Stems to 30cm; lowest leaves 15–30 x **RRR**
8–12mm, usually crenate-serrate only at widest part (near
middle), gradually narrowed into petiole. 4 sites in Cambs,
W Suffolk and W Norfolk.
b. Ssp. hybrida (L.) Gaudin. Stems to 60cm; lowest leaves   **R**
20–40 x 10–20mmm usually crenate-serrate along most of
margin, usually widest below middle, abruptly narrowed into
petiole. Very local in W Br fro N Somerset to Westmorland.

17. HEBE Comm. ex A.L. Juss. – Hedge Veronicas
Evergreen shrubs; leaves opposite, with pinnate venation but
usually with obscured lateral veins; flowers in axillary
racemes; corolla showy, as in Veronica; stamens 2.
The terminal leaf-bud is composed of successively smaller
developing leaves, without any modified bud-scales. The
outermost 2 leaves enclose all the inner ones and their
margins meet along 2 sides, but towards their base might (or
might not) leave a gap (leaf-bud sinus) on either side due to
the presence (or absence) of a distinct petiole.

1 Leaf-buds with distinct sinuses; leaves + petiolate   2
1 Leaf-buds without sinuses; leaves + sessile   4
  2 Leaves <2.5 x 0.8cm; racemes mostly <3cm
                                          4. H. brachysiphon
  2 Leaves > (2.5)3 x 0.7cm; racemes mostly >3cm   3
3 Leaves linear-lanceolate, c.8–12x as long as wide,
  narrowly acute to acuminate at apex; racemes mostly
  >10cm; corolla usually white   **1. H. salicifolia**
3 Leaves oblanceolate to obovate or oblong-obovate,
  c.2–4x as long as wide, subacute to rounded at apex;
  racemes <10cm; corolla usually blue or pink to purple
                                          3. H. x franciscana
  4 Leaves obovate to oblong-obovate, often shortly
    acuminate at apex, 2–3x as long as wide
                                          2. H. x lewisii
  4 Leaves lanceolate to oblanceolate, subacute to
    obtuse at apex, 3–5x as long as wide   5
5 Leaves 4.5–8cm, mid-green; stem glabrous to finely
  puberulent, green   **5. H. dieffenbachii**
5 Leaves 3–5cm, grey-green; stem glabrous, becoming
  purple   **6. H. barkeri**

Other spp. – Apart from some dwarf montane spp. not natd in
BI, no taxa are fully frost-hardy, and except in the extreme
SW and in CI they are killed off in hard winters; however,
viable seed can survive these winters in the ground. Many
spp., hybrids and cultivars are grown, especially by the sea
in S En, but none is as well natd as the following 6.
Records of **H. elliptica** and **H. speciosa** apparently all refer
to H. x franciscana.

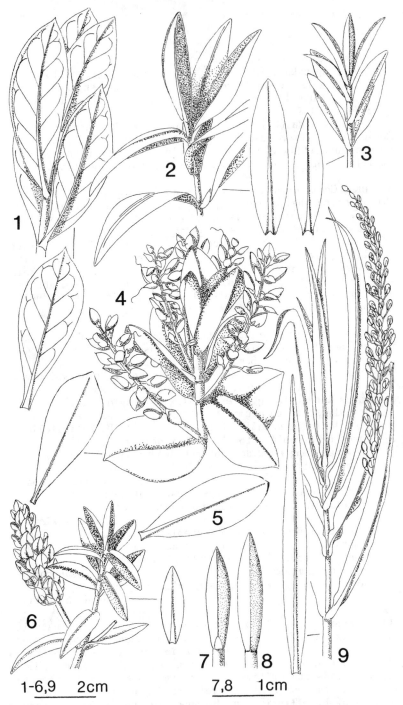

FIG 726 – Hebe. 1, **H. x lewisii.** 2, **H. dieffenbachii.** 3, **H. barkeri.** 4, **H. x franciscana.** 5, leaf of **H. x franciscana** 'Blue Gem'. 6, **H. brachysiphon.** 7, shoot apex of **H. brachysiphon.** 8, shoot apex of **H. barkeri.** 9, **H. salicifolia.**

1. **H. salicifolia** (G. Forster) Pennell - <u>Koromiko</u>. Shrub to **726** 2m; leaf-bud sinuses present; leaves 7-12 x 0.7-1.5cm, linear-lanceolate, acuminate, subglabrous to ciliate; racemes 10-20cm; corolla white or tinged pale lilac. Intrd; natd by sea in Devon and Cornwall, marginally so elsewhere in S En, W Ir and CI; New Zealand and Chile.

2. **H. x lewisii** (J. Armstr.) Wall (<u>H.</u> <u>salicifolia</u> x <u>H.</u> **726** <u>elliptica</u> (G. Forster) Pennell) - <u>Lewis's Hebe</u>. Shrub to 1.5m; leaf-bud sinuses 0: leaves 4.5-7 x 1.8-2.5cm, obovate to oblong-obovate, subacute to acute or shortly acuminate, subglabrous to ciliate; racemes 4-6.5cm, corolla violet with white tube. Intrd; natd in Devon, Cornwall and Guernsey; New Zealand.

3. **H. x franciscana** (Eastw.) Souster (<u>H.</u> x <u>lewisii</u> auct. non **726** (J. Armstr.) Wall, <u>H.</u> <u>elliptica</u> auct. non (G. Forster) Pennell, <u>H.</u> <u>speciosa</u> auct. non (R. Cunn. ex Cunn.) Cockayne; <u>H.</u> <u>elliptica</u> x <u>H.</u> <u>speciosa</u> (R. Cunn. ex Cunn.) Cockayne) - <u>Hedge Veronica</u>. Shrub to 1.5m; leaf-bud sinuses present; leaves 3-9 x 1-2.5cm, oblanceolate to obovate or oblong-obovate, subacute to rounded at apex, glabrous; racemes 5-7.5cm; corolla usually violet-blue, sometimes pinkish-purple. Intrd; well natd by sea in Devon, Cornwall, S & N Wa, S & W Ir and CI, less so elsewhere in BI N to WC Sc; garden origin. By far the most grown and commonest natd taxon, and often used for seaside hedging in SW En. Cultivar 'Blue Gem', with obtuse to rounded leaf-apex and violet-blue **726** flowers, is the mostly widely grown.

4. **H. brachysiphon** Summerh. - <u>Hooker's Hebe</u>. Shrub to 2m; **726** leaf-bud sinuses present, narrow; leaves 1.2-2.5 x 0.4-0.8cm, obovate to narrowly elliptic-oblong, obtuse to acute, ± glabrous; racemes 2-4cm; corolla white. Intrd; natd in Scillies; New Zealand.

5. **H. dieffenbachii** (Benth.) Cockayne & Allan - **726** <u>Dieffenbach's Hebe</u>. Spreading shrub to 1m; leaf-bud sinuses 0; leaves 4.5-8 x 1.2cm, narrowly oblong-elliptic, obtuse, sub-glabrous to ciliate; racemes 6-9cm; corolla white to purplish-lilac. Intrd; natd in Devon and Cornwall; Chatham Island.

6. **H. barkeri** (Cockayne) Wall - <u>Barker's Hebe</u>. Erect shrub **726** to 2.5m; leaf-bud sinuses 0; leaves 3-5 x 0.6-1.2cm, lanceolate to oblanceolate, subacute, puberulent at first, then glabrous; racemes 5-7cm; corolla white or tinged purple. Intrd; natd in Devon and Cornwall; Chatham Island.

18. SIBTHORPIA L. - <u>Cornish Moneywort</u>
Stoloniferous perennials; leaves alternate, with palmate venation; flowers solitary in leaf-axils; corolla small and inconspicuous, ± actinomorphic, with (4-)5 subequal lobes slightly longer than tube; stamens (3)4(-5).

1. **S. europaea** L. - <u>Cornish Moneywort</u>. Stems procumbent, **R** rooting at nodes, to 40cm, pubescent; leaves reniform to

orbicular, 5-9-lobed <1/2 way to base, pubescent, on long petioles; corolla 1-2.5mm across, whitish to yellowish or pinkish. Native; damp shady places; very locally frequent in SW En, S Wa, N Kerry, E Sussex and CI, also natd in Outer Hebrides and rarely on damp lawns elsewhere.

TRIBE 6 - PEDICULARIEAE (Rhinantheae) (genera 19-25). Herbaceous annuals or sometimes perennials semi-parasitic on roots of many groups of angiosperms; leaves opposite or alternate; flowers strongly zygomorphic; corolla strongly 2-lipped, not spurred or pouched at base; stamens 4.

19. MELAMPYRUM L. - Cow-wheats
Annuals; leaves opposite, mostly entire; calyx not inflated, with 4 entire lobes; corolla mostly yellowish, with opening partially closed by boss-like swellings on lower lip, with 3 entire lobes on lower lip; capsules with 1-4 seeds; seeds smooth, with oil-body.

1   Bracts densely overlapping, concealing inflorescence
    axis at least in upper part, pink or purple at least
    near base, usually with >3 teeth on either side at
    base   2
1   Bracts not or scarcely overlapping, with inflorescence
    axis well exposed, green, usually with <3 teeth on
    either side at base   3
    2   Bracts cordate at base, strongly recurved, folded
        inwards along midrib proximally   **1. M. cristatum**
    2   Bracts rounded to cuneate at base, not recurved,
        not folded inwards along midrib   **2. M. arvense**
3   Lower lip of corolla not reflexed, its underside
    forming a straight line with lower edge of tube;
    lower 2 calyx-lobes appressed to corolla and upswept;
    fruit with four seeds   **3. M. pratense**
3   Lower lip of corolla strongly reflexed (turned down);
    lower 2 calyx-lobes patent, not upswept; fruit with
    two seeds   **4. M. sylvaticum**

1. M. cristatum L. - Crested Cow-wheat. Stems erect, to   R
50cm; inflorescence dense, 4-sided; bracts strongly recurved,
the lower part infolded along midrib and with many fine teeth
on either side, bright purple towards base; corolla 12-16mm,
pale yellow with purple and darker yellow areas. Native;
wood-borders and scrub; very local in E Anglia and adjacent C
En, formerly elsewhere, decreasing.
2. M. arvense L. - Field Cow-wheat. Stems erect, to 60cm; RRR
inflorescence rather dense, cylindrical; bracts not recurved
or infolded, with >3 strong teeth on either side, pink at
first; corolla 20-25mm, pink and yellow. Possibly native;
cornfields and grassy field margins; very local in Wight, N
Essex and Beds, formerly more widespread, decreasing.
3. M. pratense L. - Common Cow-wheat. Stems erect, to 60cm; 741

inflorescence very lax; bracts entire or with 1-2(3) pairs of teeth near base, green, not recurved or infolded; corolla 10-18mm, pale to golden yellow, often with purple marks near mouth. Native; woods, scrub, heathland.

a. Ssp. pratense. Uppermost leaves (below bracts) (1)2-8(11)cm x (1)2-10(20)mm, mostly 7-15x as long as wide. On acid soils in suitable places throughout most of Br and Ir.

b. Ssp. commutatum (Tausch ex A. Kerner) C. Britton. Uppermost leaves (below bracts) (3)4-7(10)cm x (4)8-20(27)mm, mostly 3-8x as long as wide. On calcareous soils in S En and SE Wa N to Herefs.

4. M. sylvaticum L. - <u>Small</u> <u>Cow-wheat</u>. Stems erect, to 35cm; inflorescence very lax; bracts entire or the upper ones with <u><</u>2 pairs of small teeth, green, not recurved or infolded; corolla 8-12mm, usually deep, often brownish-yellow. Native; upland woods and moorland; local in Sc N to E Ross, MW Yorks, Co Durham, Co Antrim.   **741** **R**

20. EUPHRASIA L. - <u>Eyebrights</u>
Annuals; leaves mostly opposite, conspicuously toothed; calyx not inflated, with 4 entire lobes; corolla white to purple, usually with darker veins and yellow blotch on lower lip, rarely yellow all over, with open mouth, with lower lip with 3 emarginate lobes; capsules with many seeds; seeds furrowed longitudinally, without oil-body.

A highly critical genus with >60 wild hybrids, for which the key does not allow. For a good chance of correct determination at least 5 or 6 well-grown (not stunted or spindly) and undamaged plants bearing some fruits as well as open flowers should be examined from a population. Ranges, rather than means, from these should be used. Single plants, or plants not agreeing with the above definition, are not allowed for in the key. The following key and accounts are based upon the classification of P.F. Yeo; other specialists, such as T. Karlsson, hold different views with a broader sp. limit that might eventually prove more valuable. Nodes are numbered from the base upwards, excluding the cotyledonary node. Corolla length is from base of tube to tip of upper lip in fresh state; dried specimens may have shrivelled or stretched (<u><</u>1mm) corollas. All spp. are known as <u>Eyebright</u>.

1 Middle and upper leaves with glandular hairs with
  stalk (6)10-12x as long as head   2
1 Middle and upper leaves without glandular hairs, or
  with glandular hairs with stalk <u><</u>6x as long as head   9
  2 Capsule >2x as long as wide   **5. E. arctica**
  2 Capsule <u><</u>2x as long as wide   3
3 Corolla <u><</u>7mm   4
3 Corolla >7mm   5
  4 Lowest flower at node 5-8; lower bracts 5-12mm,
    often longer than flowers; plant usually with 1-4
    pairs of strong branches   **3. E. anglica**

4 Lowest flower at node (2)3-5(6); lower bracts
  3-6(7)mm, shorter than flowers; plant not branched
  or with 1-2 pairs of short branches **2. E. rivularis**
5 Lowest flower at node 2-5(6)   6
5 Lowest flower at node 5 or higher   7
  6 Corolla 9-12.5mm; lower bracts 5-12(20)mm
                     1. E. rostkoviana
  6 Corolla <u><</u>9mm; lower bracts 3-6(7)mm
                     2. E. rivularis
7 Leaves dull greyish-green, often strongly suffused
  with violet or black; corolla usually lilac to purple
                     4. E. vigursii
7 Leaves light or dark green, usually with little violet
  suffusion; corolla usually with at least lower lip
  white   8
  8 Stem usually erect, with erect or divergent
    branches; lower internodes of inflorescence mostly
    1.5-3x as long as bracts; corolla 8-12mm
                     1. E. rostkoviana
  8 Stem usually flexuous with flexuous or arched
    branches; lower internodes of inflorescence mostly
    <1.5x as long as bracts; corolla usually 6.5-8mm
                     3. E. anglica
9 Capsule glabrous or with a few short hairs; at least
  2 distal pairs of leaf-teeth (and sometimes all) not
  contiguous at base   **21. E. salisburgensis**
9 Capsule with long ± numerous hairs in distal part;
  usually all leaf-teeth contiguous at base   10
  10 Corolla >7.5mm   11
  10 Corolla <u><</u>7.5mm   16
11 Basal pair of teeth of lower bracts directed apically 12
11 Basal pair of teeth of lower bracts patent   13
  12 Stem and branches slender, flexuous; capsule
    usually c. as long as calyx; lower bracts mostly
    alternate   **9. E. confusa**
  12 Stem stout, erect, with straight or evenly curved
    branches; capsule usually much shorter than calyx;
    lower bracts mostly opposite   **10. E. stricta**
13 Lowest flower at node 8 or lower; capsule usually
  elliptic to obovate   **5. E. arctica**
13 Lowest flower at node 9 or higher; capsule oblong to
  elliptic-oblong   14
  14 Stem and branches flexuous; leaves near base of
    branches usually very small   **9. E. confusa**
  14 Stem and branches usually straight or gradually
    curved; leaves near base of branches not much
    smaller than others   15
15 Teeth of bracts acute to acuminate; capsule usually
  slightly shorter than calyx   **7. E. nemorosa**
15 Teeth of bracts mostly aristate; capsule much shorter
  than calyx   **8. E. pseudokerneri**
  16 Calyx-tube whitish and membranous, with prominent

     green to blackish veins   **17. E. campbelliae**
16 Calyx-tube green, not membranous   17
17 Lowest flower at node 6 or higher   18
17 Lowest flower at node 5(-6) or lower   38
  18 Stem internodes mostly 2-6x as long as leaves 19
  18 Stem internodes mostly $\leq$2x as long as leaves 32
19 Basal pair of teeth of lower bracts directed apically 20
19 Basal pair of teeth of lower bracts patent   23
  20 Teeth of lower bracts obtuse to acute; corolla
    $\leq$6.5mm   21
  20 Teeth of lower bracts acute to aristate; corolla
    $\geq$7mm   22
21 Leaves strongly purple-tinged, not darker on lowerside; corolla usually lilac to purple; capsule
   shorter than calyx   **18. E. micrantha**
21 Leaves weakly or moderately purple-tinged, often
   darker on lowerside; corolla usually white; capsule
   at least as long as calyx   **19. E. scottica**
  22 Capsule $\leq$3x as long as wide   **5. E. arctica**
  22 Capsule $\geq$3x as long as wide   **10. E. stricta**
23 Corolla $\geq$6.5mm   24
23 Corolla $\leq$6.5mm   25
  24 Lowest flower at node 9 or higher; leaves usually
    without glandular hairs; lower bracts smaller than
    upper leaves   **7. E. nemorosa**
  24 Lowest flower at node 8 or lower; leaves usually
    with glandular hairs; lower bracts larger than
    upper leaves   **5. E. arctica**
25 Leaves subglabrous to sparsely pubescent 26
25 Leaves densely pubescent   28
  26 Stem and branches very slender, blackish; leaves
    strongly purple-tinged, not darker on lowerside;
    corolla usually lilac to purple **18. E. micrantha**
  26 Stem and branches either stout or lightly
    pigmented; leaves weakly or moderately purple-
    tinged; corolla usually white   27
27 Lowest flower at node 8 or higher; stem stout;
   leaves not darker on lowerside; capsule usually
   shorter than calyx   **7. E. nemorosa**
27 Lowest flower at node 7 or lower; stem slender;
   leaves usually light green on upperside and purplish
   on lowerside; capsule usually longer than calyx
             19. E. scottica
  28 Lowest flower at node 9 or higher; stem $\leq$40cm;
    lower bracts often longer than wide **7. E. nemorosa**
  28 Lowest flower at node 8 or lower; stem $\leq$15cm;
    lower bracts c. as long as wide   29
29 Leaves pubescent mainly near apex, obovate to
   narrowly ovate to elliptic   **17. E. campbelliae**
29 Leaves $\pm$ uniformly pubescent, usually suborbicular,
   ovate or ovate-oblong   30
  30 Teeth of lower bracts mostly wider than long;

branches <u><</u>3 pairs   **16. E. rotundifolia**
30 Teeth of lower bracts mostly as long as wide;
   branches <u><</u>5 pairs   31
31 Corolla 5.5-7mm; capsule usually >2x as long as wide
                                                 15. E. marshallii
31 Corolla 4.5-6mm; capsule <u><</u>2x as long as wide
                                                 14. E. ostenfeldii
   32 Basal pair of teeth of lower bracts directed
      apically   33
   32 Basal pair of teeth of lower bracts patent   34
33 Teeth of lower bracts not much longer than wide
                                                 Return to 29
33 Teeth of lower bracts much longer than wide
                                                 Return to 12
   34 Lowest flower at node 10 or higher   35
   34 Lowest flower at node 9 or lower   36
35 Stem erect, stout, with stout ascending branches;
   lower bracts mostly opposite   **7. E. nemorosa**
35 Stem and branches slender and flexuous; lower bracts
   mostly alternate   **9. E. confusa**
   36 Leaves with numerous eglandular hairs   Return to 29
   36 Leaves with few eglandular hairs   37
37 Capsule 5.5-7mm, often slightly curved, as long as or
   longer than calyx   **20. E. heslop-harrisonii**
37 Capsule usually <u><</u>5.5mm, straight, usually shorter
   than calyx   **6. E. tetraquetra**
   38 Stem internodes mostly <u>></u>2.5x as long as leaves   39
   38 Stem internodes mostly <2.5x as long as leaves   44
39 Capsule broadly elliptic to obovate-elliptic   40
39 Capsule oblong to narrowly elliptic   41
   40 Teeth of lower bracts mostly subacute, not longer
      than wide; corolla 4.5-7mm; lowest flower at node
      2-4(5)   **11. E. frigida**
   40 Teeth of lower bracts usually acute or acuminate,
      longer than wide; corolla <u>></u>6.5mm; lowest flower
      usually at node 4 or higher   **5. E. arctica**
41 Lower bracts deeply serrate, with basal pair of teeth
   directed apically   **10. E. stricta**
41 Lower bracts crenate to shallowly serrate, with basal
   pair of teeth usually patent   42
   42 Upper leaves elliptic-ovate to narrowly obovate
                                                 19. E. scottica
   42 Upper leaves suborbicular to broadly ovate or
      broadly obovate   43
43 Lowest flower at node 4 or lower; lower bracts often
   considerably larger than upper leaves   **11. E. frigida**
43 Lowest flower at node 4 or higher; lower bracts
   scarcely larger than upper leaves   Return to 29
   44 Corolla <u>></u>6mm   45
   44 Corolla <6mm   47
45 Teeth of lower bracts usually very acute, all
   direct apically   Return to 12

45 Teeth of lower bracts acute to subacute, the basal
   pair patent   46
46 Capsule at least as long as calyx, usually
   emarginate   **11. E. frigida**
46 Capsule shorter than calyx, truncate to slightly
   emarginate   **6. E. tetraquetra**
47 Lower bracts ovate to rhombic, with acute to aristate
   teeth, the basal pair directed forwards   Return to 12
47 Lower bracts broadly ovate or rhombic to suborbicular, with obtuse to subacute teeth, the basal
   pair patent   48
48 Leaves with numerous hairs, all eglandular   49
48 Leaves with few eglandular hairs, sometimes with
   short glandular hairs   50
49 Lower bracts scarcely larger than upper leaves
                                                        Return to 29
49 Lower bracts considerably larger than upper leaves
                                                        11. E. frigida
   50 Capsule elliptic to obovate, emarginate
                                                      13. E. cambrica
   50 Capsule oblong to elliptic-oblong, usually
      truncate   51
51 Capsule usually shorter than calyx; distal teeth of
   lower bracts not incurved   **6. E. tetraquetra**
51 Capsule as long as or longer than calyx; distal teeth
   of lower bracts ± incurved   52
   52 Capsule 4.5–5.5(7)mm, c.2x as long as wide,
      straight; upper leaves only obscurely petiolate,
      with margins of teeth not wavy   **12. E. foulaensis**
   52 Capsule (4.5)5.5–7mm, 2–3x as long as wide, often
      slightly curved; upper leaves ± distinctly
      petiolate, with margins of teeth wavy
                                           20. E. heslop-harrisonii

Other spp. – Records of **E. hirtella** Jordan ex Reuter refer
to E. anglica or E. rostkoviana ssp. rostkoviana, and those
of **E. brevipila** Burnat & Gremli ex Gremli (=E. stricta) to E.
arctica ssp. borealis.   **E. eurycarpa** Pugsley is either a
variant of E. ostenfeldii or one of its hybrid segregates; **E.
rhumica** Pugsley is either E. micrantha or a hybrid of it.

Hybrids – Crosses between spp. of Group 2 (tetraploids of
subsection Ciliatae) are highly fertile and often common
where 2 or more spp. occur together; c.50 binary combinations
have been recorded in BI, as well as some triple hybrids.
The following 11 hybrids are frequent and may occur in the
absence of 1 or both parents, locally replacing them: E.
arctica x E. nemorosa, x E. confusa and x E. micrantha; E.
tetraquetra x E. confusa; E. nemorosa x E. pseudokerneri,   x
E. confusa and x E. micrantha; E. confusa x E. micrantha and
x E. scottica; E. frigida x E. scotica; and E. micrantha x E.
scottica.   3 spp. of this group (E. arctica, E. nemorosa and

E. micrantha) have been found to hybridise with the tetraploid E. salisburgensis (subsection Angustifoliae) in Ir; these hybrids are highly but not totally sterile.
Hybrids within Group 1 (diploids of subsection Ciliatae) are much less common, mainly because the spp. tend to be ± allopatric. 4 combinations are known, resulting in inter-gradation between the parent spp., mostly in Wa and W Br.
Hybrids between Groups 1 and 2 also occur; 7 combinations have been identified. First generation hybrids are probably highly but not totally sterile triploids, but most examined in the field are ± fertile diploid introgressants.
In this genus hybrids are not listed; for full details see P.F. Yeo in C.A. Stace (1975). Hybridization and the flora of the British Isles, pp. 373–381. Academic Press.

Group 1 – Subsection CILIATAE Joerg.: diploids (2n=22) (spp. 1–4). Middle and upper leaves with long glandular hairs with stalk (6)10–12x as long as head; usually all leaf-teeth contiguous at base; capsule with long, ± numerous hairs in distal part.

1. **E. rostkoviana** Hayne. Stems stout, erect, to 35cm;   **R** branches 0–5(more) pairs, often again branched; corolla (6.5)8–12.5mm, with usually white lower and lilac upper lip. Native; grassland in hilly areas.
   a. **Ssp. rostkoviana.** Branches 1–5(more) pairs, ascending, 736 divergent or erect; internodes shorter than to 3(4)x as long as leaves; lowest flower usually at node 6–10; corolla (6.5)8–12mm. Often in damper places and by rivers, sometimes lowland; locally frequent in Ir, Wa, N En and C & S Sc.
   b. **Ssp. montana** (Jordan) Wettst. (E. montana Jordan). 736 Branches 0–3(4) pairs, erect; internodes 2–6(10)x as long as leaves; lowest flower usually at node 2–6; corolla (7)9–12.5mm. Usually in drier upland places; very local in N En and S Sc, probably errors in Wa and Ir. Flowers earlier than and perhaps an aestival variant of the autumnal ssp. rostkoviana.
2. **E. rivularis** Pugsley. Stems flexuous–erect, to 15cm; 736 branches 0–2 pairs, short; lowest flower usually at node **RR** (2)3–5(6); corolla 6.5–9mm, with white or lilac lower and lilac upper lip. Native; damp mountain pastures and stream-sides; very local in NW Wa, Lake District; endemic. Probably a stabilized hybrid segregate of E. rostkoviana x E. micrantha.
3. **E. anglica** Pugsley. Stems flexuous–erect, to 20(30)cm; 736 branches (0)1–4(6) pairs, flexuous or arcuate, usually again branched; lowest flower at node 5–8; corolla (5)6.5–8(10)mm, with white or lilac lower and lilac upper lip. Native; grassy places on often damp soils, heathland; rather local in C & S Br N to SW Sc, Ir; ?endemic.
4. **E. vigursii** Davey. Stems erect, to 20(25)cm; branches 736 0–5(7) pairs, erect, often again branched; lowest flower at **RR**

node 7-10(12); corolla (6)7-8.5mm, with usually lilac to deep purple (occasionally white) lower and lilac to deep purple upper lip. Native; <u>Ulex gallii</u> - <u>Agrostis curtisii</u> heathland in Devon and Cornwall; endemic. Probably a stabilized segregate of E. anglica x E. micrantha.

Group 2 - Subsection CILIATAE Joerg.: tetraploids (2n=44) (spp. 5-20). Middle and upper leaves without glandular hairs, or with glandular hairs with stalk <6x as long as head; usually all leaf-teeth contiguous at base; capsule with long, ± numerous hairs in distal part.

5. E. arctica Lange ex Rostrup. Stems erect at least above, to 30(35)cm, with 0-5(6) pairs of branches sometimes again branched; lowest flower at node (3)4-8(10); corolla 6-11(13)mm, with usually white (sometimes lilac) lower and lilac to purple (sometimes white) upper lip. Native; meadows and pastures.

a. Ssp. arctica (E. borealis auct. non (F. Towns.) Wettst.). 736 Stem procumbent or flexuous at base, then erect; lower bracts suborbicular to broadly ovate; corolla 7-11(13)mm; capsule (5.5)6-7.5(8)mm. Orkney and Shetland.

b. Ssp. borealis (F. Towns.) Yeo (E. borealis (F. Towns.) 736 Wettst., E. brevipila auct. non Burnat & Gremli ex Gremli). Stem erect from base; lower bracts narrowly to broadly ovate, rhombic or trullate; corolla 6-9(10)mm; capsule (4)4.5-6.5(7)mm. Ir, N & W Br except Shetland, very scattered in S & E Br. Flowers later than ssp. arctica in Orkney.

6. E. tetraquetra (Breb.) Arrond. (E. occidentalis Wettst.). 736 Stems erect, stout, to 15(20)cm; branches 0-5(8) pairs, usually rather short and erect or ascending, but sometimes branched again; lowest flower at node (3)5-7(9); corolla (4)5-7(8)mm, with usually white (sometimes lilac) lower and white or lilac upper lip. Native; short turf on cliffs and dunes by sea, limestone pasture inland; coasts of BI except most of E En and N Sc, inland in parts of SW En.

7. E. nemorosa (Pers.) Wallr. (E. curta (Fries) Wettst. pro 736 parte). Stems erect, to 35(40)cm; branches 1-9 pairs, ascending, often again branched; lowest flower at node (5)10-14; corolla 5-7.5(8.5)mm, coloured as in E. tetraquetra. Native; pastures, scrub, marginal areas, heathland, dunes in Sc; throughout BI, the commonest lowland sp.

8. E. pseudokerneri Pugsley. Stems erect or flexuous, to 736 20(30)cm; branches (0)3-8(10) pairs, ascending to patent, R often again branched; lowest flower at node (5)10-16(18); corolla (6)7-9(11)mm, white to pale (rarely deep) lilac. Native; dry limestone (usually chalk) grassland, fens in E Anglia; S & E En N to N Lincs and E Gloucs, W Ir; ?endemic.

9. E. confusa Pugsley. Stems flexuous or procumbent at 736 base, to 20(45)cm; branches (0)2-8(10) pairs, usually long, flexuous and ascending, usually branched again; lowest flower at node (2)5-12(14); corolla 5-9mm, variously white to lilac

FIG 736 – **Euphrasia**. 1–2, **E. rostkoviana**. 1, ssp. **rostkoviana**. 2, ssp. **montana**. 3, **E. rivularis**. 4, **E. anglica**. 5, **E. vigursii**. 6, **E. nemorosa**. 7, **E. tetraquetra**. 8–9, **E. arctica**. 8, ssp. **borealis**. 9, ssp. **arctica**. 10, **E. confusa**. 11, **E. stricta**. 12, **E. pseudokerneri**.

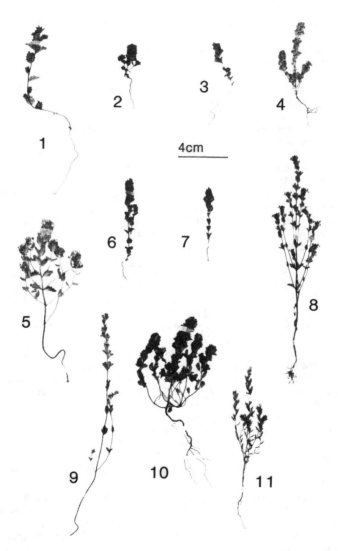

FIG 737 - **Euphrasia.** 1, **E.** frigida. 2, **E.** foulaensis. 3, **E.** cambrica. 4, **E.** ostenfeldii. 5, **E.** marshallii. 6, **E.** rotundifolia. 7, **E.** campbelliae. 8, **E.** micrantha. 9, **E.** scottica. 10, **E.** heslop-harrisonii. 11, **E.** salisburgensis.

or mixed, sometimes both lips reddish-purple or rarely
yellow. Native; short turf on moorland, heaths and dunes;
throughout most of BI but very local in CI, Ir and C & S En.
   10. E. stricta D. Wolff ex J. Lehm. Stems erect, stout, to **736**
35cm; branches (0)2-6 pairs, usually long, erect, often **RR**
branched again; lowest flower at node (3)7-14; corolla
(6)6.5-7.5(10)mm, lilac or white. ?Native; meadows and
pastures; Guernsey, also Bucks and C Sc but needing confirmation and perhaps not native.
   11. E. frigida Pugsley. Stems erect to flexuous, to **737**
20(30)cm; branches 0-2 pairs, erect, only occasionally again **R**
branched; lowest flower at node 2-4(5); corolla 4.5-7(8)mm,
with white (rarely lilac) lower and white or lilac (rarely
purple) upper lip. Native; grassland on rock-ledges on
mountains, mostly over 600m; Sc, Lake District, W Ir.
   12. E. foulaensis F. Towns. ex Wettst. Stems erect, rather **737**
stout, to 6(9)cm; branches 1-3(4) pairs, short, ascending, **R**
occasionally again branched; lowest flower at node (2)4-6;
corolla 4-6mm, white to purple. Native; Outer Hebrides,
Shetland, Orkney, N & W mainland Sc.
   13. E. cambrica Pugsley. Stem flexuous, to 8cm; branches **737**
0-2 pairs, flexuous; lowest flower at node 2-4; corolla **RR**
4-5.5mm, with white or yellowish white lower and white or
lilac upper lip. Native; mountain grassland and rock-ledges;
Caerns and Merioneth; endemic.
   14. E. ostenfeldii (Pugsley) Yeo (E. curta auct. non (Fries) **737**
Wettst.). Stems erect or flexuous below, to 12(15)cm; **R**
branches 0-4(6) pairs, erect or ascending, sometimes again
branched; lowest flower at node (3)4-7(9); corolla (3.5)4.5-
6mm, with white lower and white or lilac upper lip. Native;
grassy, stony and sandy places, mostly near sea; N & NW Br
from N Wa to Shetland.
   15. E. marshallii Pugsley. Stems erect, to 12cm; branches **737**
(0)1-5 pairs, rather long, erect, sometimes again branched; **R**
lowest flower at node (5)7-9; corolla 5.5-7mm, white or
lilac. Native; turf on sea-cliffs or dunes; extreme N Sc
from W Ross to Shetland; endemic.
   16. E. rotundifolia Pugsley. Stems erect, to 10cm; **737**
branches 0-3 pairs, short, erect; lowest flower at node **RR**
6-8(9); corolla 5-6mm, white or lilac or the upper lip
purplish, rarely pale yellow. Native; turf on sea-cliffs or
dunes; similar distribution to E. marshallii and sometimes
with it; endemic.
   17. E. campbelliae Pugsley. Stems erect, to 10cm; branches **737**
0-2 pairs, short, erect; lowest flower at node 5-7; corolla **RR**
5.5-7mm, usually with white lower and lilac upper lip.
Native; heathland near sea; Isle of Lewis (Outer Hebrides),
?Shetland; endemic.
   18. E. micrantha Reichb. Stems erect, slender, to 25cm; **737**
branches (0)2-7(10) pairs, slender, erect, usually again
branched; lowest flower at node (4)6-14(16); corolla 4.5-
6.5mm, lilac to purple or with white lower lip (rarely white

all over). Native; heathland, usually with <u>Calluna</u>,
sometimes in damp places; throughout most of BI but absent
from most of C & E En.
19. **E. scottica** Wettst. Stems erect, to 25cm; branches 0-4 **737**
pairs, long, erect to ascending; lowest flower at node (2)3-
6(8); corolla (3.5)4.5-6.5mm, white or with lilac upper lip.
Native; wet moorland; Wa, N W En, Sc and Ir. Close to <u>E.</u>
<u>micrantha</u> and perhaps not distinct; usually flowers earlier
and with less anthocyanin on corollas and vegetative parts.
20. **E. heslop-harrisonii** Pugsley. Stems erect from usually **737**
flexuous base, to 15cm; branches 0-4(5) pairs, erect or **RR**
patent, sometimes again branched; lowest flower at node
4-7(8); corolla 4.5-6(6.5)mm, white or occasionally lilac.
Native; turf in salt-marshes or drier grassy places; extreme
N & NW Sc; endemic. Close to the last 2 spp. and perhaps not
distinct.

Group <u>3</u> - Subsection <u>ANGUSTIFOLIAE</u> (Wettst.) Joerg.:
tetraploids (2n=44) (sp. 21). Middle and upper leaves
without glandular hairs; at least 2 distal pairs of leaf-
teeth (and sometimes all) not contiguous at base; capsule
glabrous or with a few short hairs.

21. **E. salisburgensis** Funck. Stems erect or flexuous, to **737**
12cm; branches (0)1-7 pairs, slender, erect or patent, often **R**
again branched; lowest flower at node 5-13; corolla 4.5-
6.5mm, white or upper lip sometimes lilac. Native; among
limestone rocks and on dunes; W Ir from Co Limerick to E
Donegal, probably mis-localized specimens from MW Yorks in
1885/6. Our plant is var. **hibernica** Pugsley.

21. ODONTITES Ludwig - <u>Bartsias</u>
Annuals; leaves opposite, entire to toothed; calyx not
inflated, with 4 entire lobes; corolla yellow or pinkish-
purple, rarely white, with open mouth, with lower lip with 3
entire to slightly notched lobes; capsules with rather few
seeds; seeds furrowed longitudinally, without oil-body.
  Intercalary leaves are those at nodes on the main stem
between the topmost branches and the lowest bract.

1. **O. jaubertianus** (Boreau) D. Dietr. ex Walp. - <u>French</u>
<u>Bartsia</u>. Stems erect, to 50cm; leaves linear, <4mm wide,
entire to slightly toothed; calyx 4-5mm; corolla 7-9mm,
yellow, often tinged pinkish. Intrd; natd on gravelly rough
ground near Aldermaston, Berks, since 1965; France.
Originally misdetermined as **O. luteus** (L.) Clairv.
2. **O. vernus** (Bellardi) Dumort. - <u>Red Bartsia</u>. Stems erect,
to 50cm; leaves lanceolate or linear-lanceolate to oblong-
lanceolate, serrate or crenate-serrate; calyx 5-8mm; corolla
8-10mm, pinkish-purple, rarely white. Native. Unbranched
(starved) plants occur in most populations and are best
ignored unless the prevalent sort; as for <u>Euphrasia</u>, a range

of individuals should be measured for determination. Ssp.
pumilus in the past has been used to cover ssp. litoralis and
maritime variants of ssp. serotinus.

1 Branches <8 pairs, held at >50 degrees to main
  stem; intercalary leaves 2-7 pairs; lowest flower at
  node 8-14   **b. ssp. serotinus**
1 Branches <4 pairs, held at <50 degrees to main
  stem; intercalary leaves 0-1 pairs; lowest flower at
  node 4-9   2
    2 Calyx-teeth narrowly triangular, acute, as long as
      tube; style exserted from corolla at full anthesis
                                                     a. ssp. vernus
    2 Calyx-teeth triangular, subacute to obtuse,
      shorter than tube; style included in upper lip of
      corolla of full anthesis   **c. ssp. litoralis**
  a. Ssp. vernus. Stems to 25cm, with (0)1-4 pairs of 741
branches; lowest flower at node 6-9; 2n=40. Grassy places,
arable and waste ground, waysides; recorded from most of BI
but many errors in S, scattered in Ir, W & N Br, replacing
ssp. serotinus in most of C & N Sc. Aestival.
  b. Ssp. serotinus (Syme) Corbiere (O. vulgaris Moench, O. 741
vernus ssp. pumilus (Nordst.) A. Pedersen). Stems to 50cm,
with (0)2-8 pairs of branches; calyx-teeth usually as in ssp.
vernus but as in ssp. litoralis in some maritime populations;
style as in ssp. vernus; 2n=18. In habitats of both sspp.
serotinus and litoralis; frequent over most BI except C & N
Sc, the only ssp. in most of C & S Br. Autumnal.
  c. Ssp. litoralis (Fries) Nyman (O. litoralis Fries, O. 741
vernus ssp. pumilus auct. non (Nordst.) A. Pedersen). Stems
to 20cm, with 0-3 pairs of branches; lowest flower at node 4-
8; 2n=18. Gravelly and rocky sea-shores and in salt-marshes;
coasts of N & W Sc from Arran to Shetland, ?Merioneth, the
only ssp. in Shetland. Aestival.

22. BARTSIA L. - Alpine Bartsia
Perennials; leaves opposite, toothed; calyx not inflated,
with 4 entire lobes; corolla dull purple, with open mouth,
with lower lip with 3 entire lobes; capsules with few seeds;
seeds with rather narrow longitudinal wings, without
oil-body.

  1. B. alpina L. - Alpine Bartsia. Stems erect, to 25cm; **RR**
leaves and bracts glandular-pubescent, ovate; bracts all
wholly or partly dark purple; corolla 15-20mm. Native;
grassy places and rock-ledges on basic, often damp, soils in
mountains; very local in hills of N En and C Sc.

23. PARENTUCELLIA Viv. - Yellow Bartsia
Annuals; leaves opposite, toothed; calyx not inflated, with 4
entire lobes; corolla yellow, very rarely white, with open
mouth, with lower lip with 3 entire lobes; capsules with
numerous seeds; seeds ± smooth, without oil-body.

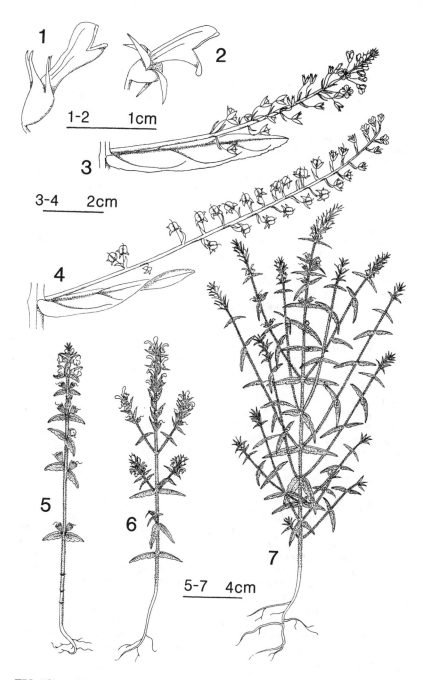

FIG 741 – Scrophulariaceae. 1–2, flowers of **Melampyrum**. 1, **M. pratense**. 2, **M. sylvaticum**. 3–4, racemes of **Veronica**. 3, **V. anagallis–aquatica**. 4, **V. catenata**. 5–7, Odontites vernus. 5, ssp. **litoralis**. 6, ssp. **vernus**. 7, ssp. **serotinus**.

1. P. viscosa (L.) Caruel - <u>Yellow Bartsia</u>. Stems erect, to   **R**
50cm; leaves and bracts glandular-pubescent, ovate to
lanceolate; corolla 16-24mm. Native; damp grassy places
mostly near coast; locally frequent in S & W Br from E Kent
to W Cornwall and N to Dunbarton, NW & SW Ir, CI.

24. RHINANTHUS L. - <u>Yellow-rattles</u>
Annuals; leaves opposite, toothed; calyx inflated, especially
at fruiting, with 4 entire lobes; corolla basically yellow to
brownish-yellow, with semi-closed to ± open mouth, with upper
lip with 1 subterminal white or violet tooth either side of
tip, with lower lip with 3 ± entire lobes; capsule with
numerous seeds; seeds discoid, usually with marginal wing,
without oil-body.
Intercalary leaves are those at nodes on the main stem
between the topmost branches and the lowest bract. Unbranched (starved) plants occur in most populations and are
best ignored unless the prevalent sort; as for <u>Euphrasia</u>, a
range of individuals should be used for determination.
Corolla colour ignores the white or violet teeth on the upper
lip.

1. R. angustifolius C. Gmelin (<u>R. serotinus</u> (Schoenheit) **744**
Oborny, incl. ssp. <u>apterus</u> (Fries) N. Hylander) - <u>Greater</u> **RRR**
<u>Yellow-rattle</u>. Stems erect, to 60cm; leaves lanceolate to
linear-lanceolate, 25-70 x 3-8mm; intercalary leaves 0-2
pairs; calyx pale green, pubescent only on margin; corolla
(15)17-20mm, yellow, with lower lip held horizontal ±
adjacent to upper lip, with teeth of upper lip mostly >1mm
and longer than wide; stigma clearly exserted; seeds winged
or not. Native; arable and grassy fields, rough ground,
sandy open places on heathland or near sea; extremely local
in Surrey, N Lincs and Angus, formerly widely scattered over
Br but over-recorded.
2. R. minor L. - <u>Yellow-rattle</u>. Stems erect, to 50cm; **744**
leaves linear to lanceolate or narrowly oblong-lanceolate;
intercalary leaves 0-6 pairs; calyx usually mid-green or
reddish-tinged, hairy only on margins or all over; corolla
12-15(17)mm, yellow to brownish-yellow, with lower lip turned
down away from upper lip, with teeth of upper lip mostly <1mm
and shorter than wide; stigma included or slightly exserted,
seeds winged. Native.
6 sspp. may be recognized in BI, but some populations do not
fit into any of them and the pattern of variation on the
Continent is more complex; the sspp. may be better abandoned.
Some of the taxa are aestival and others autumnal, as in
<u>Euphrasia</u> and <u>Odontites</u> <u>vernus</u>.

1 Calyx pubescent all over   2
1 Calyx pubescent only on margins   3
   2 Branches 0-2 pairs; intercalary leaves 0-3 pairs;
      lowest flower at node 7-10; leaves linear-

lanceolate   **e. ssp. lintonii**
2 Branches 0(-1) pairs; intercalary leaves 0; lowest
  flower at node 5-7(8); leaves linear-oblong
                                          f. ssp. borealis
3 Intercalary leaves mostly (2)3-6 pairs; lowest flower
  usually at node 14-19; leaves mostly linear
                                          d. ssp. calcareus
3 Intercalary leaves mostly 0-2(4) pairs; lowest flower
  usually at node 6-13(15); leaves mostly linear-
  narrowly oblong to linear-lanceolate   4
4 Intercalary leaves mostly 0(1) pairs; lowest
  flower usually at node 6-9; leaves mostly
  parallel-sided for most of length   **a. ssp. minor**
4 Intercalary leaves mostly (0)1-2(4) pairs; lowest
  flower usually at node (7)8-13(15); leaves mostly
  ± tapering from near base   5
5 Stems ≤50cm, usually with several pairs of long
  flowering branches from basal and middle parts; leaves
  of main stems 1.5-4.5cm; corolla usually yellow
                                          b. ssp. stenophyllus
5 Stems ≤25cm, with 0-3 pairs of long flowering
  branches from near base; leaves of main stems 1-
  2.5cm; corolla usually dull- or brownish-yellow
                                          c. ssp. monticola
  a. Ssp. minor. Stems to 40cm, usually with 0 or few **744**
flowering branches from middle or upper parts, usually with
internodes (except lowest) ± equal; leaves mostly (10)20-
40(50) x (3)5-7mm, linear-oblong to narrowly oblong. Grassy
places, especially on well-drained basic soils; ± throughout
BI, especially in lowland C & S Br and the only ssp. in most
of that area. Aestival.
  b. Ssp. stenophyllus (Schur) O. Schwarz (R. stenophyllus **744**
(Schur) Druce). Stems to 50cm, usually with several long
flowering branches from middle and lower parts, usually with
lower internodes much shorter than upper; leaves mostly 15-45
x 2-5(7)mm, linear-lanceolate. Damp grassland and fens;
throughout most of Br and Ir, common in N and often replacing
ssp. minor, local and largely replaced by ssp. minor in S.
Autumnal.
  c. Ssp. monticola (Stern.) O. Schwarz (R. monticola (Stern.) **744**
Druce, R. spadiceus Wilm.). Stems to 20(25)cm, usually with
0 or few flowering branches from near base, usually with
lower internodes much shorter than upper; leaves mostly
10-20(25) x 2-4mm, linear-lanceolate. Grassy places in hilly
areas; local in Br N from MW Yorks, N Kerry and Co
Londonderry. Autumnal.
  d. Ssp. calcareus (Wilm.) E. Warb. (R. calcareus Wilm.). **744**
Stems to 50cm, usually with flowering branches from near
middle, usually with lower internodes much shorter than
upper; leaves mostly 10-25 x 1.5-3mm, linear. Dry grassy
places on chalk and limestone; from Dorset and W Gloucs to E
Kent, ?Northants, probably errors for ssp. stenophyllus in

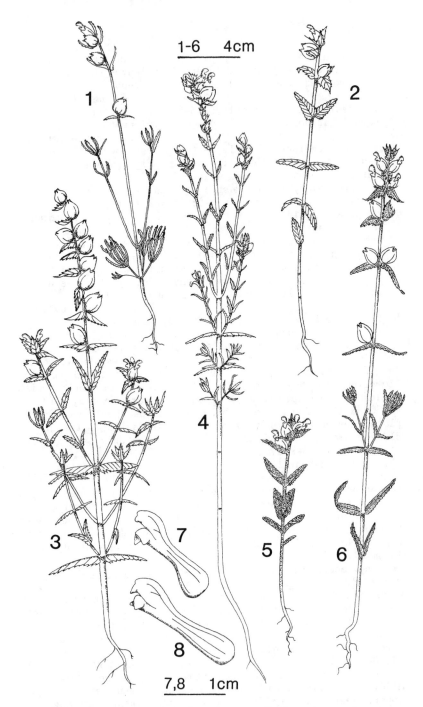

FIG 744 – **Rhinanthus.** 1–6, plants of **R. minor.** 1, ssp.
monticola. 2, ssp. **lintonii.** 3, ssp. **stenophyllus.** 4, ssp.
calcareus. 5, ssp. **borealis.** 6, ssp. **minor.** 7–8, corollas. 7,
R. minor. 8, **R. angustifolius.**

1-6   4cm

7,8   1cm

Ir. ?Autumnal.
  e. Ssp. lintonii (Wilm.) Sell (<u>R. lintonii</u> Wilm., <u>R.</u> **744**
<u>lochabrensis</u> Wilm., <u>R. gardineri</u> Druce). Stems to 30cm;
similar to ssp. <u>borealis</u> in pubescent calyx but differs in
key characters. Grassy places on mountains; C & N Sc.
Derived from hybridization between ssp. <u>borealis</u> and sspp.
<u>monticola</u> and <u>stenophyllus</u>, but often occupying exclusive
areas. Autumnal.
  f. Ssp. borealis (Stern.) Sell (<u>R. borealis</u> (Stern.) Druce). **744**
Stems to 20(28)cm, usually with 0 flowering branches, with
internodes (except lowest) + equal; leaves mostly 10-30 x
3-7mm, linear-oblong. Grassy places on mountains; Sc, mostly
C & N, ?Caerns, N & S Kerry. Autumnal.

25. PEDICULARIS L. - Louseworts.
Annuals to perennials; leaves alternate or mostly so, deeply
pinnately lobed with crenate to lobed lobes; calyx becoming
inflated, irregularly 2-5 lobed with toothed lobes; corolla
pinkish-purple, rarely white, with open mouth, with lower lip
with 3 + entire lobes; capsules with rather few seeds; seeds
+ smooth, without oil-body, winged or not.

  1. P. palustris L. - <u>Marsh</u> Lousewort. Annual to biennial;
stems usually single, usually erect, to 60cm; calyx
pubescent, with 2 short, broad, variously dissected lobes;
corolla 2-2.5cm, with upper lip with 1 terminal tooth, 1
lateral tooth on either side near apex, and 1 lateral tooth
on either side further back; capsule longer than calyx; seeds
not winged. Native; wet heaths and bogs; throughout Br and
Ir, common in N & W, rare in C & E En.
  2. P. sylvatica L. - Lousewort. (Biennial to) perennial;
stems several, procumbent to ascending or some suberect, to
25cm; calyx with 4 short dissected lobes; corolla 2-2.5cm, as
in <u>P.</u> palustris but lacking the second pair of lateral teeth;
capsule shorter than or equalling calyx; seeds usually
partially winged. Native; similar places to <u>P.</u> palustris and
sometimes with it, but often in drier habitats.
  a. Ssp. sylvatica. Calyx and pedicels glabrous. Throughout
BI except parts of C En and where replaced by ssp. <u>hibernica</u>.
  b. Ssp. hibernica D. Webb. Calyx and pedicels pubescent.
Prevalent in W Ir, very local in extreme W Sc, E & S Ir.
Intermediates with ssp. <u>sylvatica</u> occur, especially in E Ir.

127. OROBANCHACEAE - Broomrape family

Brown to whitish, reddish or bluish, herbaceous, erect,
perennial root-parasites; chlorophyll + lacking; leaves +
scale-like, alternate, simple, entire, exstipulate, sessile.
Flowers in terminal racemes or spikes, in axils of bracts,
zygomorphic, bisexual, hypogynous; calyx with 4 + equal lobes
fused to c.1/2 way, or with 2 lateral lips each not to +

deeply bifid; corolla tubular, with upper and lower lips, the
upper lip ± entire to 2-lobed, the lower lip 3-lobed,
variously coloured; stamens 4, borne on corolla-tube; ovary
1-celled, with numerous ovules borne on 4 (or 2 2-lobed)
inwards-thrusted parietal placentas; style 1, terminal;
stigmas 2, ± capitate; fruit a 2-celled capsule.
Distinguished from Scrophulariaceae by the chlorophyll-less
aerial parts and 1-celled ovary. Either the whole family or
just Lathraea is often placed in the Scrophulariaceae, the
evidence for which is equivocal.

1 Plant rhizomatous; flowers pedicellate; calyx with
   4 equal lobes   1. LATHRAEA
1 Plant not rhizomatous; flowers sessile except rarely
   some near base of inflorescence; calyx with 2-4(5)
   teeth arranged in 2 lateral lips   2. OROBANCHE

1. LATHRAEA L. - Toothworts
Plants with rhizome with ± succulent scales; flowers
pedicellate; calyx with 4 equal lobes; 2 lips of corolla held
nearly parallel to one another.

1. L. squamaria L. - Toothwort. Aerial stems whitish to
cream or pale pink, to 30cm; pedicels shorter than calyx;
calyx glandular-pubescent; corolla <2x as long as calyx,
whitish-cream usually tinged with purple or pink, 14-20mm;
capsule with numerous seeds. Native; in woods and hedgerows,
usually on moist rich soils, on a range of woody plants
especially Ulmus and Corylus; locally frequent in Br and Ir N
to C Sc.
2. L. clandestina L. - Purple Toothwort. Aerial stems 0;
flowers arising from axils of scale-leaves near apex of
rhizome; pedicels c. as long as to longer than calyx; calyx
glabrous; corolla ≥2x as long as calyx, purple-violet, 40-
50mm; capsule with 4-5 seeds. Intrd; natd in damp places on
Salix and Populus; scattered in En, S Sc and E Ir; W & SW
Europe.

2. OROBANCHE L. - Broomrapes
Rhizomes 0; flowers sessile, rarely lower ones pedicellate;
calyx with 2-4(5) teeth arranged in 2 lateral lips, the lips
usually open to the base on upper or both sides; 2 lips of
corolla held apart, the lower turned down.
   Pressed plants are difficult to determine because the
corolla shape and colour and the stigma colour are lost or
obscured. At collection some corollas should be opened out
by slitting up 1 side and pressing, the shape of the corolla
in side view and of the lower lip in front view should be
recorded, and the colour of the stem, corolla and stigmas
noted. The host sp. is a useful character, but often
difficult to ascertain with certainty. Corolla-lengths are
from base to tip of upper lip in a straight line.

1 Each flower with 2 bracteoles ± similar to the 4
 calyx-teeth (1 bracteole and 2 calyx teeth each side
 of flower) in axil of each bract; stigmas white;
 capsule-valves free **1. O. purpurea**
1 Bracteoles 0; each flower with 2–4-toothed calyx
 (1–2 calyx-teeth on each side of flower); stigmas
 yellow, red or purplish, rarely white; capsule-valves
 coherent distally 2
  2 Lower lip of corolla with minute glandular hairs
   at margins 3
  2 Margins of lower lip of corolla glabrous or with
   very few glandular hairs, but latter often frequent
   elsewhere on corolla 5
3 Stigma-lobes yellow at flower opening; filaments
 glabrous in basal 1/3 **2. O. rapum–genistae**
3 Stigma-lobes red to purple at flower opening;
 filaments pubescent at least at base 4
  4 Stigma-lobes separate; corolla not suffused dark
   red, mostly >20mm; each calyx-lip with 1–2 teeth,
   shorter than corolla-tube **3. O. caryophyllacea**
  4 Stigma-lobes partly fused; corolla suffused dark
   red, mostly <20mm; each calyx-lip with 1 tooth,
   c. as long as corolla-tube **5. O. alba**
5 Calyx with 2 lateral lips partially fused on lowerside; stamens inserted (3)4–6mm above base of
 corolla-tube **4. O. elatior**
5 Calyx with 2 lateral lips free on both upperside and
 lowerside; stamens inserted 2–4(5)mm above base of
 corolla-tube 6
  6 Corolla with sparse dark glands mostly distally,
   with very strongly curved back so that mouth is
   nearly at right angles to base **6. O. reticulata**
  6 Corolla without dark glands (often with pale ones),
   with nearly straight to slightly curved back 7
7 Corolla 20–30mm, with 2 upper and 3 lower lobes with
 conspicuous and patent margins; filaments sparsely
 pubescent along length **7. O. crenata**
7 Corolla 10–22mm, with 0 or only 3 lower lobes with
 conspicuous and patent margins; filaments glabrous or
 pubescent only proximally 8
  8 Corolla-tube constricted just behind mouth; lower
   lip of corolla with acute to subacute lobes;
   stigmas usually yellow, rarely purplish
    8. O. hederae
  8 Corolla-tube not constricted; lower lip of corolla
   with obtuse to rounded lobes; stigmas usually
   purplish, rarely yellow 9
9 Filaments with long white hairs at base, usually
 inserted >3mm above base of corolla-tube; bract
 equal to or longer than corolla; all 4 calyx-teeth
 long and filiform **9. O. artemisiae–campestris**
9 Filaments glabrous to sparsely pubescent at base,

usually inserted <3mm above base of corolla-tube;
bract shorter than or equal to corolla; calyx-teeth
various but not all 4 long and filiform   **10. O. minor**

Other spp. - O. ramosa L. (Hemp Broomrape), from S Europe,
was formerly natd, especially in CI, but not since 1928; it
resembles a small, often branched, O. purpurea but the
corolla is only 10-18(22)mm. Records of **O. cernua** Loefl. and
O. amethystea Thuill. are errors.

1. O. purpurea Jacq. - Yarrow Broomrape. Stems to 45cm, **749**
tinged bluish; corolla 18-26mm, bluish-violet at least **RR**
distally; filaments glabrous or + so, inserted >5-8mm above
base of corolla; stigma white or very pale blue. Native; on
Achillea millefolium and perhaps other Asteraceae; very local
in Dorset, Wight, N Hants, E Norfolk, N Lincs and CI,
formerly more widespread in S En and S Wa.
2. O. rapum-genistae Thuill. - Greater Broomrape. Stems to **749**
90cm, yellowish, tinged reddish-brown; corolla 20-25mm, **R**
yellow usually tinged with reddish-purple; filaments glabrous
proximally, glandular-pubescent distally, inserted <2mm above
base of corolla; stigmas yellow. Native; on various woody
Fabaceae; local in Br N to N Wa and N Lincs, CI, S & SE Ir.
Flowers with strong unpleasant scent.
3. O. caryophyllacea Smith - Bedstraw Broomrape. Stems to **749**
40cm, yellow tinged with purplish-brown; corolla 20-32mm, **RRR**
yellow tinged with reddish- or purplish-brown; filaments
pubescent usually to apex, inserted 1-3(5)mm above base of
corolla; stigma purple. Native; on Galium mollugo; very
local in E Kent, few unconfirmed records elsewhere. Flowers
pleasantly clove-scented.
4. O. elatior Sutton - Knapweed Broomrape. Stems to 75cm, **749**
yellow to orange-brown; corolla 18-25mm, yellow usually
tinged with purple; filaments pubescent usually to apex,
inserted (3)4-6mm above base of corolla; stigma yellow.
Native; on Centaurea scabiosa; on chalk and limestone in S &
E En N to NE Yorks, Glam.
5. O. alba Stephan ex Willd. - Thyme Broomrape. Stems to **749**
25(35)cm, purplish-red; corolla 15-20(25)mm, yellow usually **R**
tinged with reddish-purple; filaments pubescent usually to
apex, inserted 1-3mm above base of corolla; stigma red to
purple. Native; on Thymus, perhaps other Lamiaceae; local in
W Sc, N & W Ir, N & SW En. Flowers fragrant.
6. O. reticulata Wallr. - Thistle Broomrape. Stems to 70cm, **749**
yellowish to purplish; corolla 12-22mm, yellow with purplish **RRR**
tinge; filaments glabrous to sparsely pubescent up to apex,
inserted 2-4mm above base of corolla; stigma purple. Native;
on Carduus and Cirsium; very local in SE, MW and NW Yorks.
7. O. crenata Forsskaol - Bean Broomrape. Stems to 80cm, **749**
yellowish to purplish; corolla 20-30mm, white with mauve
veins; filaments sparsely pubescent up to apex, inserted
2-4mm above base of corolla; stigma white, yellow or pinkish.

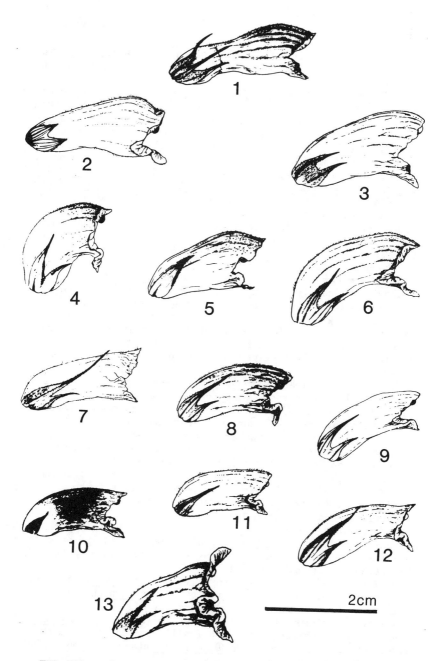

FIG 749 – Corollas of **Orobanche**. 1, **O. purpurea**. 2, **O. caryophyllacea**. 3, **O. rapum–genistae**. 4, **O. reticulata**. 5, **O. alba**. 6, **O. elatior**. 7, **O. artemisiae–campestris**. 8–11, **O. minor**. 8, var. **minor**. 9, var. **compositarum**. 10, var. **maritima**. 11, var. **flava**. 12, **O. hederae**. 13, **O. crenata**. Drawings by F.J. Rumsey.

Intrd; on herbaceous Fabaceae, often crop spp.; natd and casual in 1 part of S Essex since 1950, perhaps now extinct; S Europe.

8. O. hederae Duby - <u>Ivy</u> <u>Broomrape</u>. Stems to 60cm, **749** brownish-purple, rarely yellowish; corolla 10-22mm, cream **R** tinged and veined with reddish-purple; filaments subglabrous to sparsely pubescent, inserted 3-4mm above base of corolla; stigma usually yellow, sometimes purplish. Native; on <u>Hedera</u>; S & W Br N to S Lancs, Ir, CI, rare casual elsewhere.

9. O. artemisiae-campestris Vaucher ex Gaudin (<u>O. picridis</u> **749** F. Schultz, <u>O. loricata</u> Reichb.) - <u>Oxtongue</u> <u>Broomrape</u>. Stems **RRR** to 60cm, yellowish tinged with purple; corolla 14-22mm, white to pale yellow tinged and veined with purple; filaments pubescent at base, glabrous distally, inserted (2)3-5mm above base of corolla; stigma purple. Native; on <u>Picris</u> and <u>Crepis</u>; very local in Wight, W Sussex and E Kent. Records for N Somerset and elsewhere are errors.

10. O. minor Smith - <u>Common</u> <u>Broomrape</u> Stems to 60cm, yellowish, usually strongly tinged with red or purple; corolla 10-18mm, yellow usually strongly tinged with purple; filaments subglabrous or sparsely pubescent at base, inserted 2-3.5mm above base of corolla; stigma usually purple, sometimes yellow. Lower flowers may be pedicellate. Native.

1 Stems, corolla and stigma yellow; spike very dense
  and short   **b. var. flava**
1 Plant with at least some purple pigmentation   **2**
  2 Corollas sub-erect, + appressed to stem, sub-
    glabrous, 3.5-5mm wide   **d. var. compositarum**
  2 Corollas erecto-patent, not appressed to stem,
    subglabrous to glandular-pubescent, 5-8mm wide   **3**
3 Lower lip of corolla with large yellow bosses, with
  middle lobe the largest; lips of calyx each 1-lobed;
  stigma-lobes partly fused   **c. var. maritima**
3 Lower lip of corolla without prominent bosses, with
  3 + equal lobes; lips of calyx each usually 2-lobed;
  stigma-lobes separate   **a. var. minor**

a. Var. minor. Stems and corollas usually strongly purple- **749** tinged. On a very wide range of dicotyledons, including most specific for other spp. and vars; throughout En, Wa, Ir and CI but absent from many areas.

b. Var. flava Regel. Differs from var. <u>minor</u> in key **749** characters and 1-lobed calyx-lips. On <u>Hypochaeris</u> and relatives; Newport Docks (Mons), formerly CI.

c. Var. maritima (Pugsley) Rumsey & Jury (<u>O. maritima</u> **749** Pugsley, <u>O. amethystea</u> auct. non Thuill.). Differs from var. <u>minor</u> in key characters. On <u>Daucus</u>, <u>Plantago coronopus</u> and <u>Ononis repens</u>; coast of S En, S Wa and CI.

d. Var. compositarum Pugsley. Differs from <u>O. minor</u> in key **749** characters. Mostly on Asteraceae: Lactuceae but sometimes other families; scattered throughout much of range of var. <u>minor</u>, but mostly in E Anglia.

128. GESNERIACEAE - Pyrenean-violet family

Herbaceous perennials; stem ± absent; leaves in basal rosette, simple, dentate, exstipulate, petiolate. Flowers 1-few on long peduncles arising from rosette, nearly actinomorphic, bisexual, hypogynous; sepals 5, fused into tube proximally; petals 5, fused into tube proximally; stamens 5, borne on corolla-tube, exserted; ovary 1-celled, with many ovules on 2 parietal placentas; style 1; stigma 1, capitate; fruit a capsule dehiscing into 2 valves.

Easily recognized by the growth-form, the leaves with long brown hairs, and the attractive violet and yellow flowers distinguished from those of Scrophulariaceae by the 1-celled ovary.

1. RAMONDA Rich. - Pyrenean-violet

1. R. myconi (L.) Reichb. - Pyrenean-violet. Leaves elliptic to broadly so, with long brown hairs; flowers 3-4cm across, violet with yellow centre and yellow exserted anthers. Intrd; planted on rock-face in Cwm Glas, Caerns in 1921 and still there, often not flowering; Pyrenees.

129. ACANTHACEAE - Bear's-breech family

Herbaceous perennials; leaves in basal rosette, few opposite or alternate on stems, ± pinnate to simple and deeply pinnately lobed, exstipulate, petiolate. Flowers large, in robust terminal spikes, zygomorphic, bisexual, hypogynous, each in axils of spiny-toothed bract and 2 long linear bracteoles; sepals 4, 2 small lateral and 2 large upper and lower, fused at base, persistent to fruiting; corolla with short tube and 3-lobed lower lip (upper lip 0); stamens 4, borne on corolla-tube, the filaments free but the 1-celled anthers fused in pairs; ovary 2-celled, each cell with many ovules on axile placentas; style 1; stigmas 2, slightly unequal, linear; fruit a 2-celled capsule with persistent style.

Easily distinguished by the robust growth-habit, large pinnately lobed leaves, spiny bracts and unique flower structure.

1. ACANTHUS L. - Bear's-breeches

1. A. mollis L. - Bear's-breech. Stems to 1m, erect, glabrous at least above; leaves deeply pinnately lobed with acutely toothed but not spiny lobes, glabrous; corolla 3.5-5cm, white with purple veins; bracts glabrous, often purplish. Intrd; grown in gardens and long natd in waste places, roadside and railway banks, and scrub; scattered in S En, S Wa and CI, especially SW En; W & C Mediterranean.

2. A. spinosus L. - Spiny Bear's-breech. Differs from A. mollis in stems usually shorter (to 80cm), often pubescent;

leaves usually pubescent, with spiny teeth on more dissected
lobes; bracts usually pubescent. Intrd; similar places to A.
mollis but much less common; very scattered in S En; C
Mediterranean.

130. LENTIBULARIACEAE - Bladderwort family

Rootless aquatic, insectivorous, perennials, or rooted,
stemless, rosette-forming insectivorous perennials; leaves
simple and with many slime-oozing glands or alternate and
divided into filiform to linear segments some of which are
modified as small bladder-traps; flowers on erect stalks, in
racemes or solitary, yellow, bluish or white, zygomorphic,
bisexual, hypogynous; calyx of 5 subequal lobes or of 2
obscurely lobed lips, fused at base; corolla of 2 lips, the
upper 2-lobed, the lower 3-lobed, spurred at base, yellow,
violet or white; stamens 2, borne at base of corolla; ovary
1-celled, with many ovules on free-central placenta; style 0
or very short; stigma variously expanded; fruit a capsule.
Easily recognized by the very different insectivorous habit
of the 2 genera, both of which have a 2-lipped, spurred
corolla, 2 stamens and free-central placentation.

1 Plant rooted, with basal rosette of simple entire
   leaves; corolla white to violet   **1. PINGUICULA**
1 Plant rootless, aquatic; leaves divided into linear
   to filiform segments; corolla yellow   **2. UTRICULARIA**

1. PINGUICULA L. - Butterworts

Roots present; leaves simple, entire, in basal rosette,
covered in slime; flowers solitary on erect pedicels; calyx
of 5 subequal lobes; corolla white to violet, with open
mouth; capsule opening by 2 valves.

1 Corolla white, with 1-2 yellow spots on lower lip;
   ?extinct   **2. P. alpina**
1 Corolla pale lilac or pinkish to violet, very rarely
   white, sometimes with yellowish throat but never
   with yellow spots on lower lip   2
   2 Corolla 7-11mm incl. cylindrical spur 2-4mm,
      usually pale lilac   **1. P. lusitanica**
   2 Corolla 14-35mm incl. tapering spur 4-14mm,
      usually violet with whitish throat   3
3 Corolla 14-22(25)mm incl. spur 4-7(10)mm; lobes of
   lower lip separated laterally   **3. P. vulgaris**
3 Corolla 25-35mm incl. spur 10-14mm; lobes of lower lip
   overlapping laterally   **4. P. grandiflora**

1. P. lusitanica L. - Pale Butterwort. Overwintering as a
rosette; leaves oblong, 1-2.5cm; pedicels 3-15cm; corolla
7-11mm incl. spur 2-4mm, pale lilac with darker suffusion

proximally, often with yellow in throat. Native; bogs and wet heaths; locally frequent in Ir, W & N Sc, Man, SW Wa and SW & SC En.

2. P. alpina L. - <u>Alpine</u> <u>Butterwort</u>. Overwintering as a **RR** bud; leaves elliptic-oblong, 2-5cm; pedicels 5-11cm; corolla 10-16mm incl. tapering spur 2-3mm, white with 1-2 yellow spots on lower lip. Formerly native; boggy places in E Ross from 1831 to c.1900.

3. P. vulgaris L. - <u>Common</u> <u>Butterwort</u>. Overwintering as a bud; leaves ovate-oblong, 2-8cm; pedicels 5-18cm; corolla size as in key, violet with whitish throat. Native; bogs, wet heathland and limestone flushes; locally common over much of Br and Ir, especially N & W, absent from most of C & S En.

3 x 4. P. vulgaris x P. grandiflora = P. x scullyi Druce occurs rarely with the parents in S Kerry and Clare, perhaps elsewhere; it is intermediate in flower morphology and largely, though probably not entirely, sterile.

4. P. grandiflora Lam. - <u>Large-flowered</u> <u>Butterwort</u>. Over- **R** wintering as a bud, but leaves often persisting or precociously developing; leaves as in <u>P. vulgaris</u>; flowers as in <u>P. vulgaris</u> except as in key. Native; bogs and damp moorland; locally common in SW Ir to Clare and E Cork, planted and persistent in scattered places in SW En, Merioneth and ?Carlow.

2. UTRICULARIA L. - <u>Bladderworts</u>
Roots 0; plants free-floating or with lower stems in sub-stratum; leaves divided into linear to filiform segments, some or all bearing tiny animal-catching bladders; flowers on erect racemes emerging from water; calyx of 2 obscurely-lobed lips; corolla yellow, with mouth ± closed by swollen upfolding of lower lip; capsule opening irregularly.

The leaf-segments usually have small marginal teeth (often extremely short) which bear 1 or more long bristles. The bladders are 1-4mm long and bear hairs on their inner and outer surfaces; those on the inner surface are 2- or 4-armed. Small circular or elliptic glands are present on the stems, leaves, outside of bladders and inside of corolla-spur. The presence of bristle-bearing teeth on the leaf-segment margins, the shape of the 4-armed bladder hairs ('quadrifids'), and the distribution of the glands on the inside of the corolla-spur are of diagnostic importance. The first can be seen with a strong lens, the second 2 need a microscope. Pressed, dried material is ideal for the first 2, but pressing distorts the morphology of the quadrifids. At least 5-10 quadrifids should be examined and the range noted. Since the basal corolla-spur is forward-directed, its abaxial side is the side nearer the lower lip.

1 Margins of leaf-segments without teeth with bristles;
  lower lip of corolla <8mm; spur 1-2mm   **6. U. minor**
1 Margins of leaf-segments with teeth with bristles;

lower lip of corolla $\geq$8mm; spur 3-10mm   2
2 Stems of 1 sort, all bearing green leaves and
  bladders, free-floating; leaf-segments filiform;
  quadrifids with 2 long arms 1.2-2x as long as 2
  short arms   3
2 Stems of 2 sorts - free-floating ones bearing
  green leaves and 0 or few bladders, and ones
  often anchored in substratum and bearing very
  reduced non-green leaves and many bladders; leaf-
  segments linear; quadrifids with 2 long arms 1.8-
  2.8x as long as 2 short arms   4
3 Lower lip of corolla with reflexed margins (fresh
  material only); pedicel 8-15mm, recurved but not
  elongating after flowering; glands present on inside
  of only abaxial side of spur   **1. U. vulgaris**
3 Lower lip of corolla with flat or slightly upturned
  margins; pedicel 8-15mm at flowering, becoming
  sinuous and 10-30mm after; glands present on inside of
  both abaxial and adaxial sides of spur   **2. U. australis**
  4 Green leaves totally without traps; apex of leaf-
    segments usually obtuse; spur 8-10mm, c. as long
    as lower lip; quadrifids with 2 shorter arms $\pm$
    parallel or diverging at $\leq$21(37) degrees
                         3. U. intermedia
  4 Green leaves usually with some traps; apex of
    leaf-segments subulate; spur 3-5mm, c.1/2 as
    long as lower lip; quadrifids with 2 shorter arms
    diverging at $>$(30)52 degrees   5
5 Margin of leaf-segments with 2-7 teeth with bristles;
  lower lip of corolla with flat or slightly upturned
  margins, 9-11 x 12-15mm; quadrifids with 2 shorter
  arms diverging at (30)52-97(140) degrees   **4. U. stygia**
5 Margin of leaf-segments with 0-5 teeth with bristles;
  lower lip of corolla with flat margins at first,
  later with reflexed margins, c.8 x 9mm; quadrifids
  with 2 shorter arms diverging at (117)146-197(228)
  degrees   **5. U. ochroleuca**

1. U. vulgaris L. - <u>Greater</u> Bladderwort. Stems of 1 sort, 755
free-floating, to 1m; quadrifids with long/short arm ratio
1.8-2.8, with 2 short arms diverging at mostly 85-130
degrees; glands in corolla-spur on abaxial side only; corolla
yellow to bright yellow; upper lip c.11 x 10mm; lower lip
12-15 x 14mm; spur 7-8mm. Native; usually base-rich still or
slow water; scattered in Br and Ir, commonest in E En,
perhaps absent from N Sc.
    2. U. australis R.Br. (<u>U.</u> neglecta Lehm.) - <u>Bladderwort.</u> 755
Differs from <u>V.</u> vulgaris in stems to 60cm; quadrifids with 2
short arms diverging at mostly 100-157 degrees; corolla pale
yellow to yellow; and see key. Native; usually acidic still
or slow water; scattered throughout Br and Ir, commonest in W
& N Br.

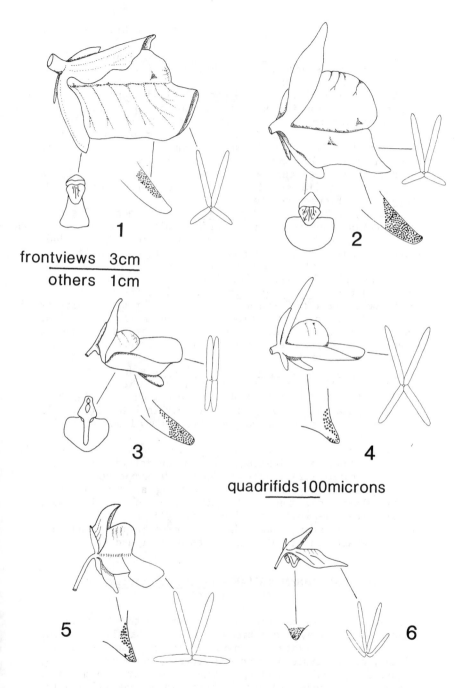

frontviews 3cm
others 1cm

quadrifids 100 microns

FIG 755 - Flowers (lateral and front views, spur showing gland distribution) and bladder quadrifids of **Utricularia**. 1, **U. vulgaris**. 2, **U. australis**. 3, **U. intermedia**. 4, **U. stygia**. 5, **U. ochroleuca**. 6, **U. minor**. Courtesy of G. Thor.

Distributions of U. vulgaris and U. australis are uncertain
as distinction is uncertain in absence of flowers, which are
rarely produced in N Br.

3. U. intermedia Hayne - Intermediate Bladderwort. Stems of   755
2 sorts, the floating ones to 20(40)cm; quadrifids with   **R**
long/short arm ratio 1.2-2, with 2 short arms diverging at
(2)6-21(37) degrees; glands in corolla spur on both abaxial
and adaxial sides; corolla yellow; upper lip 7-8 x c.7mm;
lower lip 8-10 x 12-13mm; spur 8-10mm. Native; still shallow
water in peaty bogs and marshes; very scattered in Ir, Sc, N
& SC En, E Anglia and Caerns, much over-recorded for U.
ochroleuca. Flowers not found in BI.

4. U. stygia Thor - Nordic Bladderwort. Stems of 2 sorts,   755
the floating ones to 20cm; quadrifids with long/short arm   **RR**
ratio 1.2-2, with 2 short arms diverging at (30)52-97(140)
degrees; glands in corolla-spur on both abaxial and adaxial
sides; corolla yellow with reddish tinge; upper lip c.8 x
6mm; lower lip (9)10-11 x (12)13-15mm; spur 4-5mm. Native;
similar places to U. intermedia; near Salen, Westerness.

5. U. ochroleuca R. Hartman - Pale Bladderwort. Differs   755
from U. stygia in corolla pale yellow; upper lip c. 7 x 5mm;
spur c.3mm; and see key. Native; similar places to U.
intermedia; locally frequent in Sc, very scattered in Ir, N &
SC En. Fruits not formed.

U. intermedia, U. stygia and U. ochroleuca have been much
confused in the past and their distributions are very
unclear. O. ochroleuca is probably the commonest at least in
Sc. U. stygia has so far been found only in 1 locality. All
3 flower very rarely but can be distinguished on vegetative
characters.

6. U. minor L. - Lesser Bladderwort. Stems usually of 2   755
sorts, sometimes all floating, to 40cm; quadrifids with long/
short arm ratio 1.2-2, with 2 short arms swept up to point
same way as 2 long arms hence diverging at >180 degrees;
glands almost covering whole of inside of corolla-spur;
corolla greenish- to light-yellow; upper lip c.4 x 3mm; lower
lip c.7 x 6mm; spur 1-2mm. Native; in boggy pools and fen-
ditches; scattered in suitable places over Br and Ir.

131. CAMPANULACEAE - Bellflower family
(Lobeliaceae)

Herbaceous annuals to perennials, often with white latex;
leaves alternate, sometimes mostly basal, simple,
exstipulate, petiolate or not. Flowers usually showy,
solitary or in simple or branched racemes, corymbose racemes,
congested spikes or heads, actinomorphic or zygomorphic,
bisexual, epigynous; sepals 5, fused into tube proximally;
petals 5, fused into tube proximally, the 5 lobes ± equal or
organized into 2-lobed upper and 3-lobed lower lips, often
blue but also other colours; stamens 5, borne around style-

base (not on corolla), usually closely appressed round style
and sometimes anthers or anthers and filaments fused
laterally into a ring; ovary (1)2-5-celled, each cell with
many ovules on axile placentas; style 1; stigmas usually as
many as ovary-cells, capitate to filiform; fruit a 2-5-celled
capsule opening variously or a berry.
Differ from other families with fused petals and inferior
ovaries by the 5 stamens borne on the receptacle (not on
corolla), and numerous ovules on axile placentae.

1 Flowers densely packed into flattish heads or globose
  to elongated spikes; corolla divided nearly to base,
  with linear lobes   2
1 Flowers not all packed into dense heads, or if so
  then corolla divided <2/3 way to base   3
  2 Flowers in flattish heads with conspicuous region
    of flowerless bracts at base; each flower with 0
    bract; flower buds straight; stigmas ± globose;
    stems pubescent   6. JASIONE
  2 Flowers in globose to elongated spikes, without
    flowerless bracts at base; each flower with 1
    bract; flower buds curved; stigmas linear; stems
    ± glabrous   5. PHYTEUMA
3 Corolla distinctly zygomorphic; filaments fused
  laterally at least distally to form tube round style   4
3 Corolla actinomorphic; filaments free though often
  close together round style (anthers sometimes fused
  laterally)   6
  4 Stems procumbent to decumbent, rooting at nodes;
    leaves suborbicular; fruit a berry   8. PRATIA
  4 Stems erect to ascending, not rooting at nodes;
    leaves linear to obovate; fruit a capsule   5
5 Flowers and capsules pedicellate; ovary and capsules
  <1.5cm, widening distally, 2-celled   7. LOBELIA
5 Flowers and capsules sessile; ovary and capsules >2cm,
  cylindrical, 1-celled   9. DOWNINGIA
  6 Ovary and fruits >3x as long as wide; corolla
    shorter than calyx; annual   2. LEGOUSIA
  6 Ovary and fruits <2(3)x as long as wide; corolla
    longer than (rarely ± as long as) calyx; biennial
    or perennial   7
7 Corolla-tube <2mm wide; style >1.5x as long as
  corolla (tube + lobes)   4. TRACHELIUM
7 Corolla-tube >3mm wide; style not or scarcely longer
  than corolla   8
  8 Stems filiform, procumbent, with solitary axillary
    flowers on erect stalks much longer than corolla;
    all leaves petiolate; capsule opening apically
    (i.e. within calyx)   3. WAHLENBERGIA
  8 Usually at least flowering stems erect to ascending,
    if all procumbent then not all pedicels longer than
    corolla; usually at least apical leaves sessile or

± so; capsule opening laterally or basally (i.e.
outside calyx)   **1. CAMPANULA**

1. CAMPANULA L. – <u>Bellflowers</u>
Biennials to perennials; flowers in racemes or panicles,
sometimes in ± compact heads; corolla usually blue,
actinomorphic, divided up to 1/2(2/3) way to base; filaments
and anthers free; ovary 3–5-celled; style shorter than to
slightly longer than corolla; stigmas 3–5, linear; capsule
dehiscing by lateral or basal pores.
White-flowered variants of most spp. are not rare but have
not been mentioned under each sp. Corolla-lengths given are
those in the fresh state; considerable shrinking often occurs
on drying.

1 Calyx with 5 sepal-like reflexed appendages
    alternating with 5 calyx-lobes   2
1 Calyx with 5 calyx-lobes but no extra appendages   3
    2 Biennial; all leaves cuneate at base; ovary and
        fruit 5-celled; stigmas 5   **5. C. medium**
    2 Perennial; basal and lower stem-leaves cordate at
        base; ovary and fruit 3-celled; stigmas 3
                                        6. C. alliariifolia
3 Capsule with pores in apical 1/2   4
3 Capsule with pores at or near base   7
    4 Calyx-lobes lanceolate to ovate, serrate; lower
        stem-leaves ovate to ovate-oblong   **3. C. lactiflora**
    4 Calyx-lobes linear to lanceolate, entire or with
        1–2 basal small teeth; stem-leaves linear to
        obovate   5
5 Perennial, with non-flowering rosettes arising from
    rhizomes; corollas mostly >3cm; stigmas >1/2 as long
    as styles   **4. C. persicifolia**
5 Usually biennial, without non-flowering rosettes;
    corollas mostly <3cm; stigmas <1/2 as long as style   6
    6 Tap-root thickened; inflorescence narrowly
        pyramidal; basal leaves abruptly narrowed to
        distinct petiole   **2. C. rapunculus**
    6 Tap-root thin; inflorescence widely spreading;
        basal leaves gradually narrowed to indistinct
        petiole   **1. C. patula**
7 Flowers sessile   **7. C. glomerata**
7 Flowers with distinct pedicels   8
    8 Calyx-teeth linear to filiform, <1mm wide at base   9
    8 Calyx-teeth lanceolate or narrowly triangular to
        ovate-oblong or -triangular, >1mm wide at base   10
9 Mid stem-leaves ovate-oblong or narrowly so, rounded
    at base, serrate   **14. C. rhomboidalis**
9 Mid stem-leaves linear to linear-elliptic, very
    gradually tapered to base, ± entire **15. C. rotundifolia**
10 Stems decumbent to ascending, <30(50)cm   11
10 Stems erect, usually >50cm   12

11 Leaves 1-serrate; corolla funnel-shaped (diameter at
apex much < length), lobed 1/4 to 2/5 way to base
9. C. portenschlagiana
11 Leaves 2-serrate; corolla ± star-shaped from above
(diameter at apex c. equalling length), lobed 1/2 to
3/4 way to base   **10. C. poscharskyana**
  12 Capsules erect; plant glabrous; inflorescence
  dense, pyramidal or cylindrical   **8. C. pyramidalis**
  12 Capsules pendent; plant pubescent; inflorescence
  racemose   13
13 Plant patch-forming, with shoots arising from
rhizomes and/or root-buds; calyx-teeth patent to
reflexed   **13. C. rapunculoides**
13 Plant tufted, without rhizomes or root-buds; calyx-
teeth erect to erecto-patent   14
  14 Middle and lower stem-leaves sessile, cuneate at
  base; stem bluntly ridged, softly pubescent to
  subglabrous   **11. C. latifolia**
  14 Middle and lower stem-leaves petiolate, cordate
  at base; stem sharply angled, sparsely hispid
  12. C. trachelium

Other spp. - **C. carpatica** Jacq., from the Carpathians, has
subapical pores in the capsule, ovate-cordate glabrous leaves
and large flowers on rather short weak stems. **C. fragilis**
Cirillo, from Italy, resembles C. poscharskyana but is
glabrous and has 1-serrate leaves. Both are grown in gardens
and may persist or self-sow on nearby walls and paths in S.

1. C. patula L. - Spreading Bellflower. Stems scabrid-   R
pubescent, erect, to 60cm; calyx-lobes linear, erecto-patent;
corolla 15-25mm, pale to purplish-blue, broadly funnel-
shaped, lobed c.1/2 way to base. Native; open woods, wood-
borders, hedgebanks; very local in S Br N to Salop, formerly
to SW Yorks, decreasing, often also natd garden escape.
2. C. rapunculus L. - Rampion Bellflower. Stems usually
scabrid-pubescent, erect, to 80cm; calyx-lobes linear, erect
to erecto-patent; corolla 10-22mm, pale blue, funnel-shaped;
lobed c.1/3 way to base. Intrd; grown (now rarely) as
ornament and salad-vegetable, natd in rough grassy fields and
banks; SE En, now rare, formerly scattered over much of Br N
to C Sc; Europe.
3. C. lactiflora M. Bieb. - Milky Bellflower. Stems **760**
sparsely scabrid-pubescent, erect, to 1m; calyx-lobes
lanceolate to ovate, erecto-patent; corolla 15-25mm, very
pale to bright blue, broadly campanulate, lobed 1/2 to 2/3
way to base. Intrd; grown as ornament, natd in waste and
rough ground, often in damp places; scattered in Br,
especially N En and Sc; Turkey to Iran.
4. C. persicifolia L. - Peach-leaved Bellflower. Stems
glabrous, erect, to 80cm; calyx-lobes lanceolate to linear-
lanceolate, patent to erecto-patent; corolla 25-50mm, blue,

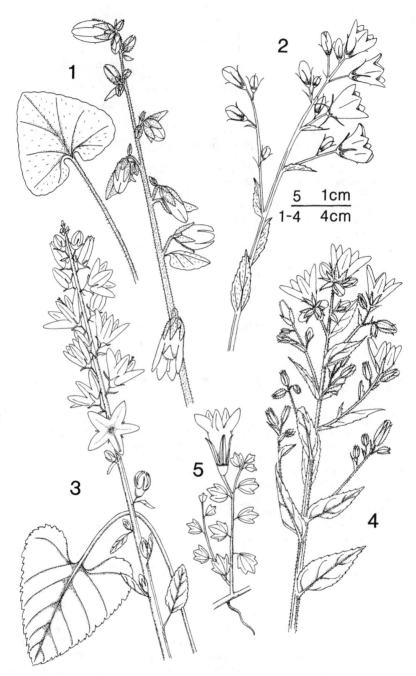

FIG 760 – Campanulaceae. 1–4, Campanula. 1, C. alliariifolia.
2, C. rhomboidalis. 3, C. pyramidalis. 4, C. lactiflora. 5,
Pratia angulata.

broadly campanulate, divided <1/4 way to base. Intrd; waste and rough ground, grassy places and banks, often natd; scattered through most of Br; Europe.

5. C. medium L. - <u>Canterbury-bells</u>. Stems hispid-pubescent, erect, to 60cm; calyx-lobes lanceolate to narrowly ovate, alternating with reflexed appendages, erect; corolla 40-55mm, bright to violet-blue, campanulate, lobed <1/4 way to base. Intrd; much grown in gardens, casual and ± natd on waste and rough ground, grassy places and banks; scattered in Br, mainly C & S; Italy and SE France. Some cultivars are to varying degrees <u>flore pleno</u>.

6. C. alliariifolia Willd. - <u>Cornish Bellflower</u>. Stems **760** pubescent, erect or ± so, to 70cm; calyx-lobes lanceolate to narrowly ovate, erect, alternating with reflexed appendages; corolla 20-40mm, white, campanulate, lobed 1/4 to 1/3 way to base. Intrd; grown in gardens, natd on banks and rough ground; S En, especially by railways in S & SW; Turkey and Caucasus.

7. C. glomerata L. - <u>Clustered Bellflower</u>. Stems pubescent, erect, to 80cm but often <20cm; calyx-lobes lanceolate, erect; corolla 12-25mm, violet- or purplish-blue, campanulate, lobed 1/4 to 1/2 way to base. Native; chalk and limestone grassland, scrub and open woodland, cliffs and dunes by sea, also casual or natd escape on rough ground; mainly S & E Br N to CE Sc, scattered escape elsewhere. Garden escapes are very variable (usually more robust) and might belong to other sspp.

8. C. pyramidalis L. - <u>Chimney Bellflower</u>. Stems glabrous, **760** erect, to 1m; calyx-lobes lanceolate, patent; corolla 10-30mm, broadly campanulate, purplish-blue, lobed c.1/2 way to base. Intrd; grown in gardens, natd on walls; Guernsey and W Kent, rare casual elsewhere; Italy and Jugoslavia.

9. C. portenschlagiana Schultes - <u>Adria Bellflower</u>. Stems glabrous to sparsely pubescent, decumbent to ascending, to 30(50)cm; calyx-lobes lanceolate, erect to erecto-patent; corolla funnel-shaped, 15-25mm, violet-blue, lobed 1/4 to 2/5 way to base. Intrd; much grown on walls and rockeries in gardens; natd on walls and rocky banks; scattered in Br, mostly C & S, CI, probably over-recorded for <u>C.</u> <u>poscharskyana</u>; Jugoslavia.

10. C. poscharskyana Degen - <u>Trailing Bellflower</u>. Differs from <u>C. portenschlagiana</u> in calyx-lobes lanceolate to narrowly ovate; corolla slatey-blue, broadly funnel-shaped with ± patent lobes; and see key. Intrd; grown and natd as for <u>C. portenschlagiana</u>; Jugoslavia.

11. C. latifolia L. - <u>Giant Bellflower</u>. Stems softly pubescent to subglabrous, erect, to 1.2m; calyx-lobes lanceolate to narrowly triangular, erect to erecto-patent; corolla campanulate, 35-55mm, pale to purplish-blue, lobed 1/3 to 1/2 way to base. Native; rich, often damp, mainly calcareous woods; most of Br but very rare to absent in S En and N Sc, NE Ir (?intrd).

12. C. trachelium L. - <u>Nettle-leaved Bellflower</u>. Differs from C. <u>latifolia</u> in stems to 80(100)cm; corolla 25-35mm, lobed 1/4 to 2/5 to base; and see key. Native; mainly base-rich woods and hedgebanks; frequent in Br N to N Lincs and N Wa, SE Ir, well natd from gardens elsewhere in Br and Ir.
13. C. rapunculoides L. - <u>Creeping Bellflower</u>. Stems sparsely pubescent, erect, to 80cm; calyx-lobes lanceolate to triangular-ovate, patent to reflexed; corolla campanulate to funnel-shaped, 20-30mm, violet-blue, lobed 1/3 to 1/2 way to base. Intrd; grown in gardens, natd in fields, woods, banks and rough ground, very persistent; widely scattered in Br and Ir; Europe.
14. C. rhomboidalis L. - <u>Broad-leaved Harebell</u>. Stems **760** sparsely pubescent, erect, to 60cm; calyx-lobes linear, patent; corolla campanulate, 15-22cm, pale to bright blue, lobed 1/4 to 1/3 way to base. Intrd; grown in gardens, natd on shady river-bank in Dumfriess and roadside bank in Westmorland; W Alps.
15. C. rotundifolia L. (C. <u>giesekiana</u> auct., ?Vest) - <u>Harebell</u>. Stems glabrous to sparsely pubescent, erect to decumbent, to 50cm; calyx-lobes linear, patent; corolla campanulate, 12-20mm, pale to bright blue, lobed 1/4 to 1/3 way to base. Native; grassy places, fixed dunes, rock-ledges, usually on acid often sandy soils; in suitable places throughout BI, but absent from CI and most S & E Ir. A difficult European complex; tetraploids and hexaploids are found in BI, the latter mainly in Ir, W Sc, Man and extreme SW En. The hexaploids often more closely resemble the Scandinavian C. <u>giesekiana</u> than tetraploid C. <u>rotundifolia</u> (often solitary, larger flowers, squat capsules and wider, blunter tipped upper stem-leaves), but C. <u>giesekiana</u> is diploid in Scandinavia and British hexaploids are not consistently separable from tetraploids.

2. LEGOUSIA Durande - <u>Venus's-looking-glass</u>
Annuals; flowers few in terminal cymes; corolla lilac to purple, actinomorphic, divided c.1/2 way to base; filaments and anthers free; ovary 3-celled; style shorter than corolla; stigmas 3, linear; capsule dehiscing by sub-apical lateral pores.

1. L. hybrida (L.) Delarbre - <u>Venus's-looking-glass</u>. Stems hispid-pubescent, erect to decumbent, to 30cm; leaves sessile, narrowly oblong-obovate; flowers sessile; corolla 4-7mm, widely funnel-shaped, much exceeded by calyx-lobes; fruit 15-30mm, with persistent calyx. Native; arable fields; scattered in S, C & E En, mostly on calcareous soils, decreasing.

3. WAHLENBERGIA Schrader ex Roth - <u>Ivy-leaved Bellflower</u>
Perennials; flowers solitary, axillary; corolla blue, actino-morphic, divided 1/3 to 1/2 way to base; filaments and

anthers free; ovary 3-celled; style shorter than corolla;
stigmas 3, linear; capsule dehiscing by apical pores.

1. W. hederacea (L.) Reichb. - <u>Ivy-leaved Bellflower</u>.
Stems filiform, procumbent, to 30cm; leaves petiolate, broadly
ovate to orbicular-reniform, angled or shortly lobed; flowers
on long, filiform, erect stalks; corolla 6-10mm, campanulate,
pendent; fruit c.3mm, with persistent calyx. Native; damp
acid places on heaths and moors, in woods, by streams; W & S
Br N to Main Argyll, S & SE Ir, formerly CI, common only in
Wa and SW En, natd rarely elsewhere in wet lawns.

4. TRACHELIUM L. - <u>Throatwort</u>
Perennials; flowers numerous in terminal corymbose compound
cymes; corolla blue, actinomorphic, with narrow tube and
small limb divided >1/2 way to base; filaments and anthers
free; ovary 2-3-celled; style longer than corolla; stigmas
2-3, capitate; capsule dehiscing by 2-3 sub-basal pores.

1. T. caeruleum L. - <u>Throatwort</u>. Stems glabrous, erect, to
1m; lower leaves petiolate, elliptic-ovate; flowers with tube
4-7 x c.0.5-1mm, with limb 2-3mm across. Intrd; natd on
walls since 1892 in Guernsey, also Jersey and W Kent; W
Mediterranean.

5. PHYTEUMA L. - <u>Rampions</u>
Perennials; flowers numerous in terminal congested heads;
corolla blue or yellow, slightly zygomorphic, tubular and
usually curved in bud but split nearly to base when open;
filaments and anthers free but appressed around style; ovary
2-3-celled; style slightly shorter than corolla; stigmas 2-3,
linear; capsule dehiscing by 2-3 lateral pores.

1 Corolla usually pale yellow, rarely blue; inflorescence oblong to cylindrical in flower   **1. P. spicatum**
1 Corolla violet-blue; inflorescence globose to very
  shortly ovoid in flower   2
  2 Corolla nearly straight in bud; lowest bracts
    longer than inflorescence (though often reflexed);
    lowest leaves strongly cordate at base
                              3. P. scheuchzeri
  2 Corolla strongly curved in bud; bracts shorter
    than inflorescence; lowest leaves rounded to
    rarely sub-cordate at base   **2. P. orbiculare**

1. P. spicatum L. - <u>Spiked Rampion</u>. Stems glabrous, erect,   **RR**
to 80cm; lowest leaves ovate-cordate, with long petiole;
inflorescence oblong to cylindrical, 3-8cm; corolla pale
yellow in native plants, 7-10mm; stigmas usually 2. Native;
woods, scrub and hedgerows on acid soils in area <20 x 10km
in E Sussex, rare escape (usually blue-flowered) elsewhere in
Br.

2. P. orbiculare L. (**P. tenerum** R. Schulz) - <u>Round-headed</u>   **R**
<u>Rampion</u>. Stems glabrous to sparsely pubescent, erect, to
50cm (but often <15cm); lowest leaves ovate to narrowly so,
rounded to rarely subcordate at base, with long petiole;
inflorescence globose, 1-2cm; corolla violet-blue, 5-8mm;
stigmas 2-3. Native; open chalk grassland; local in S En
from N Wilts to E Sussex, formerly to E Kent. Our plant
belongs to the segregate **P. tenerum**, sometimes considered a
separate sp.

3. P. scheuchzeri All. - <u>Oxford</u> <u>Rampion</u>. Stems glabrous,
erect to decumbent or pendent, to 40cm; lowest leaves ovate-
cordate, with long petiole; inflorescence globose to very
shortly ovoid, 2-2.5cm; corolla blue, 8-12mm; stigmas 3.
Intrd; natd on walls and pavements in Oxford (Oxon) since
c.1951; S Alps.

6. JASIONE L. - <u>Sheep's-bit</u>
Annuals to perennials; flowers numerous in terminal,
congested heads; corolla blue, actinomorphic, tubular and
straight in bud but split nearly to base when open; filaments
free, anthers slightly laterally fused; ovary 2-celled; style
longer than corolla; stigmas 2, subcapitate; capsule
dehiscing by 2 apical short valves.
  Superficially closely resembles a Scabious (Dipsacaceae),
but in the latter the stamens are well separated, the corolla
is split $\leq$ 1/2 way to the base, and the fruit is indehiscent
and 1-seeded.

1. J. montana L. - <u>Sheep's-bit</u>. Stems pubescent, suberect
to decumbent, to 50cm; lower leaves narrowly oblong to
oblanceolate, shortly petiolate; inflorescence depressed-
globose, 0.5-3.5cm across. Native; grassy or sandy or rocky
places on acid soils, walls, cliffs, banks; locally common in
BI, mainly in W, absent from much of C & E Br and C Ir.

7. LOBELIA L. - <u>Lobelias</u>
Annuals or perennials; flowers in terminal racemes; corolla
pale lilac to blue, zygomorphic, with tube (with deep dorsal
split) and expanded limb, the latter with 5 lobes arranged 2
in upper and 3 in lower lip; filaments and anthers fused
laterally around style; ovary 2-celled; style shorter than
corolla; stigma capitate, 2-lobed; capsule dehiscing by 2
apical valves.

1 Leaves all basal, linear, entire; plant submerged
    (except inflorescence) or at lakeside   **3. L. dortmanna**
1 Stem-leaves present, the lower ones obovate, serrate;
    plant terrestrial   2
    2 Corolla-lobes <2mm wide; pedicels <1cm   **1. L. urens**
    2 Lower 3 corolla-lobes >2mm wide; at least lower
        pedicels >1cm   **2. L. erinus**

LOBELIA 765

Other spp. - **L. siphilitica** L., from N America, is grown in gardens and is a very rare escape; it is a strong perennial to 1m, with dense raceme of blue flowers with corollas 2-3.5cm.

1. L. urens L. - Heath Lobelia. Perennials; stems sparsely **RR** and shortly pubescent, erect, to 80cm; flowers many, in long raceme with bracts much narrower than leaves, 10-15mm, purplish-blue; corolla-lobes 2-4 x <1mm, with recurved, acute apex. Native; acid heathy grassland, open woods, woodborders; very local in S En from E Cornwall to W Kent, formerly E Kent and Herefs.

2. L. erinus L. - Garden Lobelia. Annuals (in BI); stems glabrous to sparsely pubescent, erect to asending, to 30cm; flowers rather few, in racemes with lower bracts leaf-like, 8-20mm, usually purplish-blue, sometimes white or pinkish; lower corolla-lobes c.3-8 x 2-6mm, with rounded, apiculate apex; upper corolla-lobes narrower. Intrd; much grown in gardens, frequent escape, sometimes self-sown, on tips and in pavement-cracks and rough ground; scattered in Br, CI; S Africa.

3. L. dortmanna L. - Water Lobelia. Perennials; stems glabrous, erect, to 70(120)cm; flowers (1)3-10 in very lax raceme with very reduced bracts, 12-20mm, pale lilac; lower corolla-lobes 5-10 x 1-3mm, acute, upper corolla-lobes slightly smaller. Native; in stony, acid mainly montane lakes, rarely on wet ground adjacent; locally common in N & W Br S to S Wa, W, N & E Ir.

8. PRATIA Gaudich. - Lawn Lobelia
Perennials; flowers solitary in leaf-axils; corolla white; differs from Lobelia in fruit a berry.

1. P. angulata (G. Forster) Hook. f. - Lawn Lobelia. Stems **760** procumbent, rooting at nodes, glabrous, to 15cm; leaves <12mm, suborbicular, shallowly lobed, shortly petiolate; flowers 7-20mm. Intrd; grown and becoming established on damp lawns; scattered in Sc, Surrey and W Kent; New Zealand.

9. DOWNINGIA Torrey - Californian Lobelia
Annuals; corolla blue with white centre; differs from Lobelia in corolla with lower lip only shallowly lobed, flowers sessile with long, pedicel-like ovary, ovary 1-celled, and capsule dehiscing by 3-5 longitudinal slits.

1. D. elegans (Douglas ex Lindley) Torrey - Californian Lobelia. Stems erect, glabrous, to 25cm; leaves linear-lanceolate, sessile; flowers 8-18mm. Intrd; casual or perhaps natd in grassy places, probably intrd with grass-seed, first found 1978, apparently increasing; E Sussex and Bucks; W N America.

132. RUBIACEAE - <u>Bedstraw</u> <u>family</u>

Annual to perennial herbs, evergreen climbers or rarely
shrubs; leaves opposite with 1-several stipules per leaf, the
stipules usually leaf-like and as large as leaves, hence
leaves apparently in whorls of 4 or more, simple, + entire,
usually sessile and narrow. Flowers small, in usually
compound terminal and/or axillary cymes, often aggregated
into terminal panicle, actinomorphic, bisexual or bisexual
and male mixed (<u>Cruciata</u>) or + dioecious (<u>Coprosma</u>),
epigynous; sepals 0 or minute, 4-5, fused below; petals 4-5,
fused into long or short tube below, various colours; stamens
4-5, borne at apex of corolla-tube (near base in <u>Coprosma</u>);
ovary 2-celled, each cell with usually 1 ovule on axile
placenta; styles 1-2, if 1 often branched into 2; stigmas 1
per style or style-branch, capitate; fruit mostly of 2 fused
(later separating) 1-seeded nutlets, or succulent with 1-2
seeds.

Easily recognized by the small flowers with 4-5 petals fused
into an (often very short) tube, 0 or minute calyx and
inferior 2-celled ovary with 1 ovule per cell. Most spp.
have apparently whorled leaves, 4 corolla-lobes and
distinctive paired nutlets.

```
1  Leaves opposite, usually with stipules or smaller
     leaves also at same node                                2
1  Leaves in whorls of >4, + all same size in 1 whorl       4
     2  Evergreen shrub                             1. COPROSMA
     2  Procumbent to ascending herb                         3
3  Leaves linear or nearly so; fruit of 2 nutlets
                                                  5. ASPERULA
3  Leaves ovate to suborbicular; fruit succulent
                                                  2. NERTERA
     4  Most or all flowers with 5 corolla-lobes            5
     4  Most or all flowers with 4 corolla-lobes            6
5  Procumbent to ascending annual; leaves >6 in a
     whorl; corolla pink; fruit dry              4. PHUOPSIS
5  Evergreen climber or scrambler; leaves 4-6 in a
     whorl; corolla yellowish-green; fruit fleshy  8. RUBIA
     6  Calyx distinct, c.0.5-1mm at first, slightly
        enlarging in fruit                        3. SHERARDIA
     6  Calyx absent or vestigial                           7
7  Corolla-tube >1mm                                         8
7  Corolla-tube <1mm                                         9
     8  Ovary and fruit smooth to papillose       5. ASPERULA
     8  Ovary and fruit covered with hooked bristles
                                                  6. GALIUM
9  At least some whorls with >4 leaves             6. GALIUM
9  All whorls with 4 leaves                                 10
     10 Flowers in dense axillary whorls; ovary and fruit
        smooth                                    7. CRUCIATA
     10 Flowers in terminal panicles; ovary and fruit
```

covered with hooked bristles   **6. GALIUM**

1. COPROSMA Forster & G. Forster - Tree Bedstraw
Evergreen, usually dioecious shrubs; leaves opposite, with
small stipules, petiolate; flowers inconspicuous, <1cm, in
axillary clusters with 2 partly fused bracts below; calyx
minute; corolla greenish, 4-5-lobed, with long tube; fruit
succulent, with 2 nuts.

1. C. repens A. Rich. (C. baueri auct. non Endl.) - Tree
Bedstraw. Shrub to 3m; leaves broadly ovate-oblong,
(2)5-8cm, very shiny on upperside, with recurved margins;
fruit depressed-obovoid, c.10 x 8mm, orange-red.   Intrd;
planted as windbreak in Scillies and sometimes self-sown; New
Zealand.

2. NERTERA Banks & Sol. ex Gaertner - Beadplant
Herbaceous perennials; leaves opposite, with small stipules,
shortly petiolate; flowers solitary, axillary and terminal,
<5mm; calyx minute; corolla greenish, 4-lobed; fruit
succulent, with 2 nuts.

1. N. granadensis (Mutis ex L.f.) Druce (N. depressa Banks &
Sol. ex Gaertner) - Beadplant. Stems procumbent, to 15cm,
glabrous; leaves ovate to suborbicular, (3)5-8(15)mm, with
recurved margins, glabrous; fruit globose, c.4mm, bright
reddish- orange. Intrd; natd on damp lawns; few places in WC
Sc; Australia, New Zealand, S America.

3. SHERARDIA L. - Field Madder
Annuals; leaves in whorls of 4-6, sessile; flowers 4-10 in
dense terminal and axillary clusters with whorl of 8-10 leaf-
like bracts at base, the clusters stalked; calyx 0.5-1mm at
first, slightly enlarging in fruit, with 4-6 deeply toothed
lobes; corolla pale to deep mauvish-pink, 4-lobed; fruit a
pair of scabrid nutlets with persistent calyx on top.

1. S. arvensis L. - Field Madder. Stems procumbent to
ascending, to 40cm, glabrous to pubescent; lower leaves
obovate; upper leaves narrowly elliptic to oblanceolate,
5-18mm; corolla 4-5mm, with tube longer than lobes. Native;
arable fields, waste places, thin grassland and lawns;
frequent almost throughout BI but local in Sc.

4. PHUOPSIS (Griseb.) Hook. f. - Caucasian Crosswort
Annuals to perennials; leaves in whorls of 6-9, sessile;
flowers many in dense terminal clusters with whorl of many
leaf-like bracts at base; calyx minute; corolla deep pink,
5-lobed; fruit a pair of glabrous, papillose nutlets.

1. P. stylosa (Trin.) Benth. & Hook. f. ex B. D. Jackson -
Caucasian Crosswort.   Stems procumbent to sprawling, to

30(70)cm, sparsely scabrid-pubescent; leaves linear to
narrowly elliptic, 12-30mm; corolla 12-15mm, with narrow tube
much longer than lobes; style long-exserted. Intrd; grown in
gardens and sometimes persistent outside on waste and rough
ground and tips; scattered in C & S Br; Caucasus and Iran.

5. ASPERULA L. - Woodruffs
Annuals to herbaceous perennials; leaves in whorls of 4-8 and
all equal, or in whorls of 4 with 2 long and 2 short,
sessile; flowers in dense terminal clusters with whorl of
leaf-like bracts at base or in loose terminal panicles; calyx
minute; corolla various colours, 4-lobed; fruit a pair of
smooth to papillose nutlets.

1 Leaves in whorls of 6-8; corolla blue   **3. A. arvensis**
1 Leaves in whorls of 4, equal or 2 short and 2 long;
   corolla white to pink   2
   2 Leaves equal in all whorls, narrowly ovate to
     narrowly elliptic; inflorescence a compact cluster
     with whorl of leaf-like bracts below   **2. A. taurina**
   2 Leaves at upper whorls 2 long and 2 short, linear
     oblanceolate; inflorescence a diffuse panicle
                                              1. A. cynanchica

1. A. cynanchica L. - Squinancywort. Subglabrous
perennials; stems procumbent to ascending, to 50cm; at least
upper whorls with 2 long and 2 short leaves; inflorescence a
diffuse panicle; corolla white to pink, 2.5-5mm. Native.
   a. Ssp. cynanchica. Rhizomes 0 or brown; leaves mostly
linear, mostly 10-20 x 0.5-1mm; pedicels 0-1mm; corolla-lobes
usually distinctly shorter than -tube. Limestone and chalk
grassland, calcareous dunes; locally common in S Br and W Ir,
scattered N to Westmorland and SE Yorks, formerly Jersey.
   b. Ssp. occidentalis (Rouy) Stace (A. occidentalis Rouy).
Rhizomes orange; leaves linear to oblanceolate, often <10mm
and >1mm wide; pedicels usually 0; corolla-lobes usually c.
as long as -tube. Calcareous dunes; S Wa, W Ir.
Leaves of ssp. cynanchica in dune habitats resemble those of
ssp. occidentalis rather closely, but the other characters
are more constant. The sp. occurs in Alderney and needs
checking for ssp. determination.
   2. A. taurina L. - Pink Woodruff. Pubescent perennials;
stems erect, to 50cm; all whorls with 4 equal leaves;
inflorescence a dense cluster with whorl of leaf-like bracts
at base; corolla white to yellowish-pink, 10-14mm, with tube
much longer than lobes. Intrd; grown in gardens, natd in
damp woods; locally natd in C Sc, rare and impermanent
further S; S Europe.
   3. A. arvensis L. - Blue Woodruff. Annuals, glabrous apart
from long-ciliate bracts; stems erect, to 50cm; all whorls
with 6-8 equal linear leaves; inflorescence a dense cluster
with whorl of leaf-like bracts at base; corolla blue, 5-

6.5mm, with tube much longer than lobes. Intrd; casual on tips and in waste places mainly (?only) from bird-seed; scattered in most of Br and CI, mainly S; Europe. Often confused with the garden plant **A. orientalis** Boiss. & Hohen. (A. azurea Jaub. & Spach), from S W Asia, which is a much rarer casual with a longer corolla (7-12(14)mm).

6. GALIUM L. - Bedstraws
Annuals to herbaceous perennials; leaves in whorls of 4-12 and all equal, sessile; flowers in terminal panicles or axillary cymes; calyx minute; corolla white to yellow, 4-lobed; fruit a pair of smooth to bristly nutlets, the bristles sometimes hooked.

```
1  Ovaries and fruits with hooked bristles                        2
1  Ovaries and fruits smooth to rugose or papillose               5
   2  All whorls with 4 leaves                      1. G. boreale
   2  Most or all whorls with >5 leaves                           3
3  Rhizomatous perennial; flowers in terminal panicles;
   corolla-tube >1mm                                2. G. odoratum
3  Annual; flowers in axillary cymes; corolla-tube <1mm    4
   4  Fruit >3mm (excl. bristles), its bristles with
      bulbous base; corolla >1.4mm across          11. G. aparine
   4  Fruit <3mm (excl. bristles), its bristles wider
      towards but not bulbous at base; corolla <1.4mm
      across                                        12. G. spurium
5  Corolla bright yellow                             6. G. verum
5  Corolla white to pale cream (N.B. G. x pomeranicum),
   sometimes tinged pink                                          6
   6  Annuals of open ground, usually easily uprooted,
      with sparse rooting system                                 7
   6  Perennials, firmly rooted, often in grassland or
      wet ground                                                 9
7  Plant much slenderer than G. aparine; leaves with
   forward-directed marginal prickles             14. G. parisiense
7  Plant coarse, with habit of G. aparine; leaves with
   backward-directed marginal prickles                           8
   8  Fruit smooth, <3mm; peduncles and pedicels
      divaricate at various angles at fruiting, but
      straight                                      12. G. spurium
   8  Fruit papillose, >3mm; peduncles and/or pedicels
      strongly recurved                            13. G. tricornutum
9  Leaves obtuse to acute, never apiculate or mucronate  10
9  Leaves apiculate to mucronate at apex                          11
   10 Leaves linear; pedicels scarcely divaricate at
      fruiting; inflorescence obconical, widest near
      top; rare                                     4. G. constrictum
   10 Leaves linear-oblong to oblanceolate or narrowly
      elliptic; pedicels strongly divaricate at fruiting;
      inflorescence usually conical to cylindrical,
      widest well away from apex; common            5. G. palustre
11 Stems rough, with projecting minute papillae or
```

pricklets   **3. G. uliginosum**
11 Stems perfectly smooth   12
  12 Corolla-lobes apiculate to strongly mucronate at
    apex, with point $\geq$0.2mm; fruit minutely wrinkled
                         7. G. mollugo
  12 Corolla-lobes acute to minutely apiculate at apex,
    with point <0.2mm; fruit minutely tuberculate   13
13 Leaf-margins with forward-directed prickles; leaves
  on flowering shoots oblanceolate   **10. G. saxatile**
13 Leaf-margins with at least some backward-directed
  prickles; leaves on flowering shoots linear to
  linear-elliptic or -oblanceolate   14
  14 Fruit with minute high-domed subacute tubercles;
    NW of line from Severn to Humber   **9. G. sterneri**
  14 Fruit with minute low-domed to rounded tubercles;
    almost entirely SE of line from Severn to Humber
                         8. G. pumilum

Other spp. - **G. verrucosum** Hudson (G. valantia G. Weber, G. saccharatum All.), from S Europe, has been found as a casual but is very rare; it resembles G. tricornutum in its papillose fruits >3mm but has leaves with forward-directed marginal prickles.

1. G. boreale L. - Northern Bedstraw. Erect perennial to 45cm; leaves all 4 per whorl, with 3 main veins, lanceolate to narrowly ovate or narrowly elliptic; corolla white, with very short tube; fruit with hooked bristles. Native; damp grassy, rocky and gravelly places usually on hills and often by streams, also on sand-dunes; locally frequent in Br N from MW Yorks, scattered in W, C & N Ir, very local in N & S Wa.
2. G. odoratum (L.) Scop. - Woodruff. Erect perennial to 45cm; leaves 6-8(9) per whorl, with 1 main vein and distinct laterals, elliptic to oblanceolate; corolla white, with tube c. as long as lobes; fruit with hooked bristles. Native; damp, base-rich woods and hedgerows; frequent throughout most of BI except CI and Outer Isles of Sc, garden escape in Jersey.
3. G. uliginosum L. - Fen Bedstraw. Decumbent to ascending 772 or scrambling, scabrid perennial to 60cm; leaves (4)5-8 per whorl, 1-veined, linear-oblanceolate to -elliptic, strongly mucronate at apex; corolla white, with very short tube; fruit with low-domed tubercles; 2n=22, 44. Native; fens and base-rich marshy places, scattered in BI except N Sc, N Ir and CI.
4. G. constrictum Chaub. (G. debile Desv. non Hoffsgg. & 772 Link) - Slender Marsh-bedstraw. Decumbent to ascending or RR scrambling, smooth or slightly scabrid perennial to 40cm; leaves 4-6 per whorl, 1-veined, linear, obtuse to subacute (but often revolute and appearing acute) at apex; corolla white, with very short tube; fruit with high-domed tubercles; 2n=24. Native; marshy places, ditches and pond-sides; very local in SE Yorks, S Hants, S Wilts, S Devon, and CI.

5. G. palustre L. - <u>Common</u> <u>Marsh-bedstraw</u>. Decumbent to ascending or scrambling, smooth or more often scabrid perennial to 1m; leaves 4-6 per whorl, 1-veined, linear-oblong to oblanceolate or narrowly elliptic, rounded to subacute at apex; corolla white, with very short tube; fruit slightly wrinkled. Native; damp meadows, pond-sides, ditches, marshes and fens; common throughout BI.

a. Ssp. palustre (ssp. <u>tetraploideum</u> Clapham). Most leaves 772 <20mm; inflorescence ± cylindrical; pedicels mostly <4mm at flowering; corolla mostly 2-3.5mm across; fruit c.1.6mm; 2n=24, 48.

b. Ssp. elongatum (C. Presl) Arcang. (<u>G. elongatum</u> C. 772 Presl). Most leaves >20mm; inflorescence ± conical; pedicels mostly >4mm at flowering; corolla mostly 3-4.5mm across; fruit c.1.9mm; 2n=96, 144. More robust than ssp. <u>palustre</u> and commoner.

6. G. verum L. - <u>Lady's</u> <u>Bedstraw</u>. Procumbent to erect, 772 smooth perennial to 1m; leaves (6)8-12 per whorl, 1-veined, linear, apiculate to shortly mucronate at apex; corolla bright yellow, with very short tube, with acute to apiculate lobes; fruit smooth to minutely wrinkled. Native; dry grassy places especially on calcareous soils, often by sea; common throughout BI. Dwarf, procumbent plants with internodes shorter than leaves (var. **maritimum** DC.) are common on maritime dunes and cliff-tops.

6 x 7. G. verum x G. mollugo = G. x pomeranicum Retz is 772 frequent with the parents in Br N to Midlothian and in CI, and rare in S & E Ir; it is intermediate in all characters, notably petal- and leaf-shape and flower-colour, and somewhat variable, suggesting back-crossing. Often <u>G. verum</u> var. <u>maritimum</u> is involved.

7. G. mollugo L. - <u>Hedge</u> <u>Bedstraw</u>. Decumbent to erect, smooth perennial to 1.5m; leaves 5-8 per whorl, 1-veined, oblong to oblanceolate, apiculate to strongly mucronate at apex; corolla white, with very short tube, with apiculate to strongly mucronate lobes; fruit smooth to wrinkled. Native; all sorts of grassy places, hedgerows, mainly on well-drained base-rich soils; throughout most of BI but rare in Sc, Wa and Ir, common in S En and CI.

a. Ssp. mollugo. Leaves mostly oblanceolate to narrowly 772 obovate; inflorescence broad, with branches mostly at >45 degrees, rather lax; corolla 2-3mm across; pedicels strongly divaricate at fruiting.

b. Ssp. erectum Syme (<u>G. album</u> Miller). Leaves mostly 772 linear-oblanceolate to oblanceolate; inflorescence rather narrow, with branches mostly at <45 degrees, rather dense; corolla 2.5-5mm across; pedicels weakly divaricate at fruiting. Mostly on drier, more calcareous soils.

<u>G. mollugo</u> is very variable and the 2 sspp. are of doubtful value. On the Continent diploids (2n=22) and tetraploids (2n=44) occur and are usually referred to <u>G. mollugo</u> and <u>G. album</u> respectively. The correlation of these with the 2

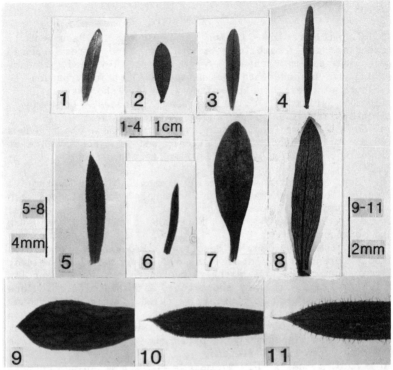

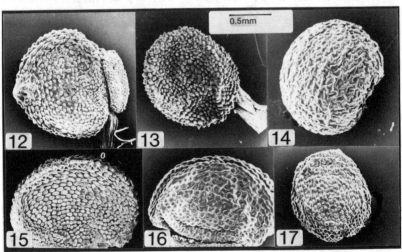

FIG 772 – Galium. 1–11, leaves. 1, G. mollugo ssp. erectum. 2, G. mollugo ssp. mollugo. 3, G. x pomeranicum. 4, G. verum. 5, G. uliginosum. 6, G. constrictum. 7, G. palustre ssp. palustre. 8, G. palustre ssp. elongatum. 9, G. saxatile. 10, G. sterneri. 11, G. pumilum. 12–17, one nutlet. 12, G. uliginosum. 13, G. constrictum. 14, G. palustre. 15, G. saxatile. 16, G. pumilum. 17, G. sterneri.

sspp. is highly uncertain; in BI only the tetraploid has been recorded but no detailed survey has been carried out.

8. G. pumilum Murray (G. fleurotii auct., ?Jordan) - Slender 772 Bedstraw. Decumbent to erect, smooth perennial to 40cm; **RR** leaves 5-9 per whorl, 1-veined, linear-oblanceolate to oblanceolate, strongly mucronate at apex, with mainly backward-directed prickles on margins; corolla white, with very short tube; fruit with low-domed tubercles; 2n=88. Native; dry chalk and limestone grassland; scattered and local in En SE of line from Severn to Humber. The plants at Cheddar, N Somerset, have been referred to **G. fleurotii**, but they have the same chromosome number as other populations and do not consistently differ by the characters said to distinguish Continental G. fleurotii.

9. G. sterneri Ehrend. - Limestone Bedstraw. Differs from 772 G. pumilum in usually many more non-flowering shoots, forming **R** a mat; flowering stems to 30cm; fruit with high-domed subacute tubercles; 2n=22, 44. Native; limestone or other base-rich grassland or rocks; local in BI NW of line from Severn to Humber. Populations in NW Wa, N Sc and W Ir are diploid (2n=22); others counted are tetraploid, but scarcely any morphological differences exist.

9 x 10. G. sterneri x G. saxatile has been cytologically confirmed (2n=33 or 55 according to G. sterneri parent 2n=22 or 44) in Caerns, Mid Perth, Easterness and W Sutherland; it is intermediate in leaf-shape and marginal prickles and highly sterile.

10. G. saxatile L. - Heath Bedstraw. Decumbent to 772 ascending, smooth perennial to 30cm; non-flowering shoots numerous, mat-forming; leaves 5-8 per whorl, 1-veined, oblanceolate to obovate, apiculate to shortly mucronate at apex, with forward-directed prickles on margins; corolla white, with very short tube; fruit with high-domed subacute tubercles; 2n=44. Native; dry grassland, rocky places and open woods on acid soils; common throughout most of BI.

11. G. aparine L. - Cleavers. Procumbent to scrambling-erect annual to 3m, with strongly recurved prickles on stems; leaves 6-8 per whorl, 1-veined, oblanceolate, strongly mucronate at apex; corolla white, with very short tube; fruit 3-5mm (excl. bristles), covered with hooked bristles. Native; cultivated and arable land, hedgerows and scrub, other open ground; common throughout BI.

12. G. spurium L. - False Cleavers. Differs from G. aparine **RR** in slightly more slender habit; stems to 1m; leaves linear-lanceolate to lanceolate; corolla greenish-cream; fruit 1.5-3mm (excl. bristles), smooth or with hooked bristles; and see key (couplet 4). Probably intrd; well natd as arable weed around Saffron Waldron, N Essex since 1844, elsewhere as casual or temporarily natd in Br N to S Sc; Europe. The N Essex plant and most others are var. **vaillantii** (DC.) Gren, with bristly fruits; the smooth-fruited var. **spurium** has occurred only as a very rare casual.

13. G. tricornutum Dandy - <u>Corn</u> <u>Cleavers</u>. Differs from <u>G.</u> **R**
<u>aparine</u> in stems to 60cm; flowers and fruits mainly in
clusters of 3 (not 2-5); peduncles and/or pedicels strongly
recurved in fruit; fruit acutely papillose. Probably intrd;
once common in arable and waste places in S C & E En, rare
elsewhere in Br, now rare and sporadic in C & SE En; Europe.
14. G. parisiense L. - <u>Wall</u> <u>Bedstraw</u>. Slender, procumbent **R**
to ascending annual to 30cm, with small recurved prickles on
stems; leaves 5-7 per whorl, 1-veined, linear-oblanceolate to
oblanceolate, mucronate at apex; corolla greenish-white
tinged reddish, with very short tube; fruit 0.8-1.2mm, very
finely papillose. Native; walls and sandy banks; scattered
in E Anglia and SE En, rare casual elsewhere, decreasing.

7. CRUCIATA Miller - <u>Crosswort</u>
Herbaceous perennials; leaves in whorls of 4, all equal, $\pm$
sessile; flowers in dense axillary whorls of cymes, the
terminal flower bisexual and the laterals male in each cyme;
calyx minute; corolla yellow, 4-lobed, with very short tube;
fruit 1 or a pair of smooth nutlets.

1. C. laevipes Opiz (<u>C.</u> <u>chersonensis</u> auct. non (Willd.)
Ehrend., <u>Galium</u> <u>cruciata</u> (L.) Scop.) - <u>Crosswort</u>. Stems
erect, to 60cm, conspicuously pubescent; leaves elliptic to
oblong- or ovate-elliptic, 10-20mm, yellowish-green. Native;
grassy places, hedgerows, scrub and rough ground, mostly on
calcareous soils; common in Br N to C Sc, intrd in 2 places
in E Ir.

8. RUBIA L. - <u>Wild</u> <u>Madder</u>
Evergreen scrambling perennials; leaves 4-6 per whorl, all
equal, narrowed to $\pm$ sessile base; flowers in diffuse
axillary and terminal panicles; calyx minute; corolla pale
yellowish-green, 5-lobed, with very short tube; fruit
succulent, with 1 seed.

Other spp. - <u>R.</u> <u>tinctorum</u> L. (<u>Madder</u>), from Asia, was
formerly grown for its dye and used to occur as a casual and
escape, but no longer; it differs in its light green leaves
2-10cm and its anthers 5-6x (not 1.3-2x) as long as wide.

1. R. peregrina L. - <u>Wild</u> <u>Madder</u>. Stems trailing to
scrambling, to 1.5m, glabrous, with strong recurved prickles;
leaves 1-5cm, elliptic to narrowly so, leathery, 1-veined,
glabrous, with strong recurved prickles on margins. Native;
hedges, scrub, rocky places; locally common in S & W Br from
E Kent to N Wa, S, E & C Ir, CI, mainly coastal.

133. CAPRIFOLIACEAE - <u>Honeysuckle</u> <u>family</u>

Deciduous or evergreen shrubs (small and procumbent in

Linnaea), small trees or woody climbers, rarely herbaceous
perennials; leaves opposite, simple (lobed or not) or
pinnate, exstipulate, petiolate. Flowers variously arranged,
axillary or terminal, zygomorphic or actinomorphic, bisexual
or some sterile, epigynous; sepals 5, fused into tube
proximally; petals 5, fused into tube proximally, sometimes
2-lipped with 4-lobed upper and 1-lobed lower lip, usually
white to yellow, sometimes pink to reddish; stamens 4-5,
borne on corolla-tube; ovary 1-5-celled, sometimes including
2 sterile cells, each cell with 1 apical ovule or many ovules
on axile placenta; style 0 or 1; stigmas 1-5, + capitate;
fruit succulent, 1-several seeded (an achene in Linnaea).
The only woody plants with fused petals, inferior ovary and
stamens only 4 or 5.

1 Leaves pinnate   **1. SAMBUCUS**
1 Leaves simple (sometimes deeply lobed)   2
   2 Flowers numerous in corymbose compound cymes;
      style + 0   **2. VIBURNUM**
   2 Flowers 2-few, not corymbose; style conspicuous   3
3 Main stems procumbent, the flowers in pairs terminal
   on erect lateral stems   **4. LINNAEA**
3 Main stems + erect or climbing   4
   4 Bracts >15mm, leaf-like, purple, or green
      strongly tinged purple   **5. LEYCESTERIA**
   4 Bracts not purple, <15mm   5
5 Ovary 4-celled, with 2 fertile and 2 sterile
   cells, the former each with 1 ovule; corolla
   actinomorphic; berry white to pink   **3. SYMPHORICARPOS**
5 Ovary 2-3-celled, all cells fertile, and with >1 ovule;
   corolla zygomorphic to + actinomorphic; berry red,
   blue, or purple to black   **6. LONICERA**

Other genera - Several other genera are grown as garden
ornamentals. Of these **WEIGELA** Thunb. and **KOLKWITZIA** Graebner
are similar to shrubby Lonicera spp. but have fruit a capsule
and an achene respectively. **W. florida** (Bunge) A.DC.
(Weigelia), with pink to deep red flowers, and **K. amabilis**
Graebner (Beauty-bush), with pale pink flowers with a yellow
throat, both from China, both occasionally produce seedlings
in waste places in S En or persist as isolated bushes.

1. SAMBUCUS L. - Elders
Deciduous shrubs or herbaceous perennials; leaves pinnate;
flowers numerous in corymbose or paniculate compound cymes,
actinomorphic; stamens 5; ovary 3-5-celled, each cell with 1
ovule; style 0; stigmas as many as carpels; fruit a drupe
with 3-5 seeds.

1 Inflorescence an ovoid to + globose panicle; ripe
   fruits red; stipules represented by stalked glands
                                              1. S. racemosa

1 Inflorescence a flat or slightly convex corymb; ripe
  fruits black to purplish-black, rarely red or
  greenish-yellow or -white; stipules 0 or subulate
  to ovate   2
2 Rhizomatous herbaceous perennial; stipules
  conspicuous, ovate or narrowly so; anthers purple
                                              4. S. ebulus
2 Erect shrub; stipules small and subulate; anthers
  cream   3
3 Fruits black, rarely greenish-yellow or -white; 2nd
  year twigs with numerous lenticels; leaflets (3)5(-7);
  not rhizomatous   2. S. nigra
3 Fruits purplish-black, rarely red; 2nd year twigs
  with few lenticels; leaflets (5)7(-11); rhizomatous
                                           3. S. canadensis

1. S. racemosa L. (_S. pubens_ Michaux, _S. sieboldiana_ (Miq.)
Graebner) - _Red-berried_ Elder. Shrub to 4m; leaves with
(3)5-7 leaflets; stipules represented by stalked glands;
flowers cream; fruits bright red; flowering Apr-May. Intrd;
well natd in hedges, woods and shrubberies; frequent in Br N
from Derbys and Cheshire, very scattered further S; Europe.
Plants with more diffuse panicles and a pubescent rhachis and
sometimes leaf lowerside come from N America (var. **pubens**
(Michaux) Koehne); those with more diffuse panicles, more
finely serrate leaflets and smaller fruits (c. 3mm, not
4-5mm) come from E Asia (var. **sieboldiana** Miq.). Both are
natd in Sc, but all 3 vars hybridise and are difficult to
delimit. Cultivars with greenish-yellow fruits, with
dissected leaflets, or with variegated leaves exist.
2. S. nigra L. - _Elder_. Shrub or small tree to 10m; leaves
with (3)5(-7) leaflets; stipules 0 or small and subulate;
flowers creamish-white; fruits black, sometimes greenish-
yellow; flowering Jun-Jul. Native; hedges, woods,
shrubberies, waste and rough ground, especially on manured
soils; common throughout BI but only intrd in Northern Isles.
Cultivars exist with dissected leaflets or with variegated
leaves; both are natd.
3. S. canadensis L. - _American_ Elder. Suckering shrub to
4m; leaves with (5)7(-11) leaflets; stipules 0 or small and
subulate; flowers white; fruits purplish-black, rarely red;
flowering Jul-Sep. Intrd; natd in scrub, rough ground and on
railway banks; very scattered in Sc, N En and Surrey.
Cultivars exist with greenish-yellow fruits, with dissected
leaflets, or with variegated leaves; the last is natd and has
red fruits.
4. S. ebulus L. - _Dwarf_ Elder. Rhizomatous perennial; stems
herbaceous, erect, to 1.5m; leaves with (5)7-13 leaflets;
stipules conspicuous, ovate to narrowly so; flowers white,
sometimes pink-tinged; fruits black; flowering Jul-Aug.
Possibly native; waysides, rough and waste ground; scattered
over most BI.

2. VIBURNUM L. - <u>Viburnums</u>

Deciduous or evergreen shrubs; leaves simple, sometimes lobed; flowers numerous in corymbose compound cymes, actinomorphic, sometimes some sterile; stamens 5; ovary 3-celled, but appearing 1-celled due to abortion of 2 cells, with 1 ovule; style 0; stigmas 3; fruit a drupe with 1 seed.

1 Leaves deciduous, lobed or serrate   2
1 Leaves evergreen, entire or obscurely denticulate   3
   2 Leaves lobed; outer flowers sterile much larger
     than inner; fruits red, subglobose   **1. V. opulus**
   2 Leaves serrate; all flowers fertile, uniform
     in size; fruits red, then black, compressed
                              2. V. lantana
3 Leaves smooth; first-year twigs glabrous or sparsely
   pubescent   **3. V. tinus**
3 Leaves strongly wrinkled; first-year twigs
   tomentose   **4. V. rhytidophyllum**

Other spp. - Several other spp., evergreen and deciduous, winter- and summer-flowering, are mass-planted on roadsides etc. and many produce occasional seedlings.

1. V. opulus L. - <u>Guelder-rose</u>. Deciduous shrub to 4m; leaves lobed, the lobes irregularly dentate; corymbs with large outer sterile flowers surrounding small fertile ones; corolla white; fruits globose, bright red. Native; woods, scrub and hedges; frequent throughout Br and Ir except N Sc.

2. V. lantana L. - <u>Wayfaring-tree</u>. Deciduous shrub to 6m; leaves regularly serrate; flowers all fertile, cream; fruits compressed, becoming red then black. Native; woods, scrub and hedges, especially on base-rich soils; common in Br SE of line from Glam to S Lincs, scattered elsewhere in BI probably always natd, much used as stock for cultivated spp. and often persisting when latter die or when suckering away from them.

2 x 4. V. lantana x V. rhytidophyllum = V. x rhytidophylloides Valcken. is grown in gardens and 2 plants have been found + in the wild in W Kent; it is intermediate in leaf-shape, -wrinkledness and -pubescence; garden origin.

3. V. tinus L. - <u>Laurustinus</u>. Evergreen shrub to 6m; leaves entire; flowers all fertile, white to pink; fruits subglobose, blue-black. Intrd; much grown in shrubberies and natd on cliffs, banks and rough ground; widespread in S En and S Wa; S Europe.

4. V. rhytidophyllum Hemsley ex Forbes & Hemsley - <u>Wrinkled Viburnum</u>. Evergreen shrub to 6m; leaves entire to nearly so; flowers all fertile, yellowish-white; fruits + globose, becoming red then black. Intrd; much grown in shrubberies and natd or a relic in old woodland or parkland; very scattered in En N to Leics; China.

3. SYMPHORICARPOS Duhamel - <u>Snowberries</u>
Deciduous shrubs; leaves simple, entire, sometimes deeply
lobed; flowers in dense terminal spikes, actinomorphic;
stamens (4-)5; ovary 4-celled, with 2 fertile cells each with
1 ovule and 2 sterile cells; style present; stigma 1,
capitate; fruit a drupe with 2 seeds.

Other spp. - Modern cultivars are mostly the so-called
'Doorenbos Hybrids', most of which are <u>S. albus</u> x <u>S.</u> x
<u>chenaultii</u> and are variously intermediate or close to one or
other parent; they might escape and surely will in future.

1. S. albus (L.) S.F. Blake (<u>S. rivularis</u> Suksd., <u>S.</u>
<u>racemosus</u> Michaux) - <u>Snowberry</u>. Strongly suckering, ± erect
then arching shrub to 2m; leaves glabrous, rounded to very
broadly obtuse or apiculate at apex, those on strong sterile
shoots often deeply lobed; corolla 5-8mm, pink; style
glabrous; fruit white, 8-15mm. Intrd; natd in woods, scrub,
rough ground; frequent throughout BI; our plant is var.
laevigatus (Fern.) S.F. Blake, from W N America.
2. S. x chenaultii Rehder (<u>S. microphyllus</u> Kunth x <u>S.</u>
<u>orbiculatus</u> Moench) - <u>Pink Snowberry</u>. Arching shrub to 1.5m,
with procumbent or arching then procumbent non-flowering
stems rooting at tips; leaves pubescent on lowerside at least
along midrib, acute to obtuse at apex, not lobed; corolla
3-5mm, pinkish-white; style pubescent; fruit pink, 6-10mm.
Intrd; natd as for <u>S. albus</u> but much rarer; scattered in Br N
to C Sc; garden origin. The parents are rare in cultivation
and records of them need checking.

4. LINNAEA L. - <u>Twinflower</u>
Procumbent, evergreen, dwarf shrubs; leaves simple, crenate;
flowers 2, each with 1 bract at base and 2 bracteoles at apex
of pedicel, actinomorphic; stamens 4; ovary with 1 fertile
cell with 1 ovule and 2 sterile cells; style present; stigma
1, bilobed; fruit an achene.

1. L. borealis L. - <u>Twinflower</u>. Stems procumbent, to 40cm;   R
leaves broadly ovate to suborbicular, 4-16mm; flowers in
pairs borne on erect leafless stems to 8cm; corolla 5-10mm,
pink; fruit, calyx, bracteoles and pedicels glandular-
pubescent. Native; on barish ground under shade of rocks or
trees, mostly in woods, especially of <u>Pinus</u>; very local in E
Sc N to E Sutherland, formerly S to NE Yorks, decreasing.
Cultivated plants are often ssp. **americana** (Forbes) Hulten
(var. <u>americana</u> (Forbes) Rehder), with corolla 10-16mm with
distinct narrow tube at base, and might escape.

5. LEYCESTERIA Wallich - <u>Himalayan Honeysuckle</u>
Deciduous shrub, often semi-herbaceous; leaves simple, entire
to serrate; flowers in crowded terminal spikes with large
purple or purple-green bracts, ± actinomorphic; stamens 5;

ovary 5-celled, each cell with several ovules; style present;
stigma 1, capitate; fruit a several-seeded berry.

1. L. formosa Wallich - <u>Himalayan Honeysuckle</u>. Stems erect,
to 2m; leaves 5-18cm, ovate, acuminate at apex; bracts
12-35mm; corolla 10-20mm, pinkish-purple; fruit purple,
subglobose, c.1cm. Intrd; natd in woods, shrubberies and
rough ground, often appearing unexpectedly from bird-sown
seed; scattered ± throughout BI; Himalayas.

6. LONICERA L. - <u>Honeysuckles</u>
Deciduous or evergreen shrubs or climbers; leaves simple,
sometimes lobed, entire; flowers sessile, in pedunculate
axillary pairs or in terminal heads; corolla zygomorphic with
4-lobed upper and 1-lobed lower lip, or ± actinomorphic with
5 lobes; stamens 5; ovary 2-3-celled with several ovules per
cell; style long; stigma capitate or slightly lobed; fruit a
several-seeded berry.

1 Flowers and fruit sessile in terminal and subterminal
   whorls; climbers   2
1 Flowers and fruit in pairs, sessile at apex of common
   axillary stalk, sometimes crowded near branch ends   4
   2 All leaves separate, not fused in pairs; berry red
                                                        7. L. periclymenum
   2 At least most apical pair of leaves on each branch
      fused around stem at base; berry orange   3
3 Bracteoles at base of each flower 0 or minute
                                                        8. L. caprifolium
3 Bractoles c.1mm, obscuring base of ovary
                                                        9. L. x italica
   4 Stems twining   5
   4 Stems not twining   6
5 Flowering nodes often clustered into terminal spikes;
   corolla glabrous on outside; 2 bracts at base of each
   flower-pair subulate   **5. L. henryi**
5 Flowering nodes not clustered near branch apex;
   corolla pubescent on outside; 2 bracts at base of each
   flower pair leaf-like   **6. L. japonica**
   6 Two bracts at base of each flower-pair (and
      bracteoles within) ovate, obscuring base of flower,
      purple and enlarging in fruit   **3. L. involucrata**
   6 Two bracts at base of each flower-pair subulate to
      linear-lanceolate, not obscuring ovaries, scarcely
      enlarging in fruit   7
7 Corolla distinctly 2-lipped; leaves deciduous,
   20(30)-60mm; first-year twigs densely pubescent
                                                        4. L. xylosteum
7 Corolla ± actinomorphic; leaves evergreen, (4)6-32mm;
   first-year twigs glabrous to sparsely pubescent   8
   8 Often >1m; leaves (4)6-16mm, mostly ovate, rounded
      to subcordate at base   **2. L. nitida**

8 Rarely >1m; leaves (6)12-32mm, mostly oblong-
elliptic to narrowly so, cuneate at base
1. L. pileata

Other spp. - **L. trichosantha** Bureau & Franchet, from China,
differs from L. xylosteum in its leaves mostly widest below
middle and corollas bright yellow; a few bushes occur as
relics of planting in Clapham Woods, MW Yorks.

1. **L. pileata** Oliver - Box-leaved Honeysuckle. Evergreen **781**
shrub to 1m, with spreading branches; leaves (6)-32mm,
oblong-elliptic to narrowly so; flowers in pairs in
leaf-axils, actinomorphic, 6-8mm, cream; berry purplish.
Intrd; much grown in shrubberies and road-borders, sometimes
self-sown; very scattered in C & S Br; China.
2. **L. nitida** E. Wilson - Wilson's Honeysuckle. Evergreen **781**
shrub to 1.8m, with erect to arching branches; leaves (4)6-
16mm, ovate; flowers in pairs in leaf-axils, actinomorphic,
5-7mm, cream; berry violet. Intrd; much grown for hedging
and then very rarely flowering, but not rarely self-sown in
scrub, hedges, woodland, banks and rough ground; scattered ±
throughout BI; China.
3. **L. involucrata** (Richardson) Banks ex Sprengel (L. **781**
ledebourii Eschsch.) - Californian Honeysuckle. Deciduous
shrub to 2m, with spreading or arching branches; leaves
4-12cm, elliptic-oblong; flowers in pairs in leaf-axils,
actinomorphic, 10-15mm, pale yellow often tinged with red;
berry shining black, subtended by usually purple bracts.
Intrd; frequent in gardens, sometimes bird-sown in rough and
marginal ground; scattered in En (mostly N) and Ir; W N
America.
4. **L. xylosteum** L. - Fly Honeysuckle. Deciduous shrub to **RR**
2m; leaves 3-7cm, obovate to elliptic or broadly so; flowers
in pairs in leaf-axils, zygomorphic, 8-15mm, pale yellow or
cream; berry red. Possibly native; woods and scrub on chalk
in W Sussex, also widely natd (bird-sown) in hedges, woods
and scrub throughout much of Br and Ir.
5. **L. henryi** Hemsley - Henry's Honeysuckle. Evergreen **781**
climber to 5(10)m; leaves 4-13cm, lanceolate to oblong-
lanceolate; flowers in axillary pairs but clustered into
terminal groups, zygomorphic, 15-25mm, yellow tinged with
red; berry black. Intrd; grown in gardens, natd from throw-
outs or bird-sown; few places in Surrey; China.
6. **L. japonica** Thunb. ex Murray - Japanese Honeysuckle. **781**
Semi-evergreen climber to 5(10)m; leaves 3-8cm, ovate to
oblong; flowers in pairs in leaf-axils, zygomorphic, 30-50mm,
pale yellow tinged purple; berry black. Intrd; often grown
in gardens, natd in hedges, scrub, banks and rough ground;
scattered in C & S Br (1 place in S Devon since 1930s), CI; E
Asia.
7. **L. periclymenum** L. - Honeysuckle. Deciduous climber to
6(10)m; leaves 3-7cm, ovate, elliptic or oblong; bracteoles

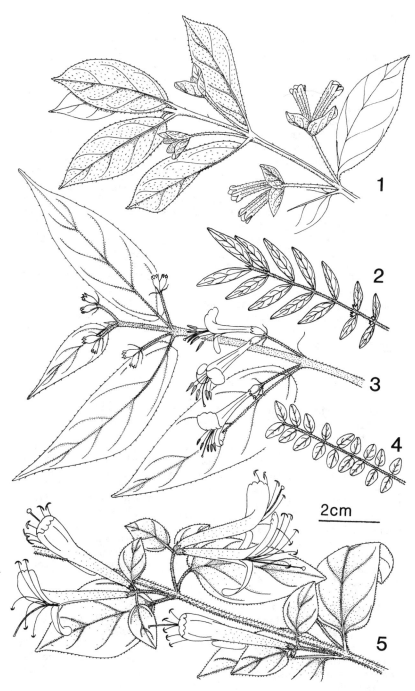

FIG 781 - Lonicera. 1, L. involucrata. 2, L. pileata. 3, L. henryi. 4, L. nitida. 5, L. japonica.

1-2mm, partly obscuring ovary, densely glandular; flowers in terminal whorls, zygomorphic, 40-50mm, pale yellow to yellow, often tinged purplish, glandular pubescent to densely so; berry red. Native; woods, scrub and hedges (often not flowering in shade); common throughout BI. Cultivars are often more robust and with more deeply purple-tinged flowers, and are often natd.

   8. L. caprifolium L. - Perfoliate Honeysuckle. Deciduous climber to 6(10)m; leaves 4-10cm, ovate to obovate, the uppermost 1-few pairs fused round stem at base; bracteoles 0 or ± so; flowers in terminal whorl and often some whorls below, zygomorphic, 40-50mm, colour as in L. periclymenum, glabrous to sparsely glandular; berry orange. Intrd; natd in hedges and rough ground; scattered in Br N to C Sc; S Europe.

   9. L. x italica Schmidt ex Tausch (L. x americana auct. non (Miller) K. Koch; L. caprifolium x L. etrusca Santi) - Garden Honeysuckle. Differs from L. caprifolium in its glabrous bracteoles c.1mm; inflorescences often large, with whorls in axils of small bracts beyond node with uppermost fused leaf-like bracts. Intrd; 1 of the commonest garden honeysuckles today, having mostly replaced L. caprifolium, natd in marginal and rough places in Surrey and several places in E Anglia and probably elsewhere, overlooked for L. caprifolium; garden origin.

134. ADOXACEAE - Moschatel family

Perennial, rhizomatous herbs; leaves 1-3-ternate, those on flowering stems opposite, exstipulate, petiolate. Flowers in compact cubical terminal head, 4 lateral and 1 terminal, actinomorphic, bisexual, 1/2-epigynous; sepals fused with 2 lobes in terminal and 3 lobes in lateral flowers; petals fused, with 4 lobes in terminal and 5 lobes in lateral flowers, yellowish-green; stamens 4 in terminal and 5 in lateral flowers, but appearing 8 and 10 due to longitudinal division into 1/2-stamens, borne at apex of corolla tube; ovary 2-5-celled (mostly 4 in terminal and 5 in lateral flowers), with 1 ovule per cell; styles as many as ovary-cells; stigmas capitate; fruit a rather dry drupe.
   Instantly recognizable by the small yellowish-green 'town-hall clock' flower-head.

1. ADOXA L. - Moschatel
   1. A. moschatellina L. - Moschatel. Long-petioled 2-3-ternate leaves, and erect flowering stems to 15cm with 2 1-ternate opposite leaves, arising from short, white, scaly rhizome; inflorescence 6-10mm long and wide. Native; woods, hedges, shady rocky places on mountains, mostly on damp, humus-rich soil; frequent throughout Br except N Sc, Co Antrim, intrd in Co Dublin.

135. **VALERIANACEAE** – <u>Valerian family</u>

Annual to perennial herbs; leaves simple or pinnate, opposite, exstipulate, petiolate or sessile. Flowers numerous in terminal paniculate cymes, often ± corymbose, ± actinomorphic to zygomorphic, bisexual or dioecious, epigynous; calyx represented by 0-many teeth, very small in flower, similar in fruit or developing long feathery appendages; petals 5, equal or slightly longer on abaxial side, fused into tube proximally, the tube straight or slightly pouched or with a long backward-directed spur at base; stamens 1 or 3, borne on corolla-tube; ovary 3-celled, 1 adaxial cell fertile with 1 ovule, 2 abaxial cells sterile and equally large to vestigial; style 1; stigma 1 and capitate or 3 and linear- oblong; fruit a 1-seeded nut.
Distinguished by the inferior ovary with 1 ovule but 2 other sterile cells (these often obscure), 0 or minute calyx at flowering, and 5-lobed tubular corolla often pouched or spurred at base.

1 Stems forked into 2 at each node; calyx remaining
  minute at fruiting   **1. VALERIANELLA**
1 Main stem simple or with lateral branches; calyx
  developing long feathery projections at fruiting   2
  2 Stamen 1   **3. CENTRANTHUS**
  2 Stamens 3   **2. VALERIANA**

1. **VALERIANELLA** Miller – <u>Cornsalads</u>
Annuals with stems repeatedly forked; leaves simple, entire to serrate or sparsely lobed; flowers in rather lax to dense compound cymes, bisexual; calyx ± 0 or small, persistent but remaining small on top of fruit, usually unequal; corolla-tube not pouched or spurred; stamens 3; stigmas 3; sterile cells of ovary small to large.
Ripe fruits are essential for determination; the key does not require fruit sections to be cut, but their appearance in section is important and is mentioned in the diagnoses. Fruit lengths exclude the calyx.

1 Calyx in fruit absent or vestigial, <1/10 as long
  as rest of fruit   2
1 Calyx in fruit distinct, c.1/4 to nearly as long as
  rest of fruit   3
  2 Fruit c. as wide as thick, much longer than wide
    or thick, with a very deep groove on abaxial face
                  2. V. carinata
  2 Fruit c.2x as thick as wide, scarcely longer than
    thick, shallowly grooved on abaxial face
                  1. V. locusta
3 Calyx in fruit with short tube, with usually 6 teeth,
  >2/3 as long as rest of fruit, nearly as wide as
  fruit   **5. V. eriocarpa**

3 Calyx in fruit with very short or 0 tube, with <6
   (often 1) teeth, <1/2(2/3) as long as rest of fruit,
   <1/2 as wide as fruit   4
   4 Main tooth of calyx in fruit scarcely or not
      toothed; fruit ± smooth on all faces, with 2-6
      fine grooves and/or longitudinal ridges, with
      easily broken walls   3. V. rimosa
   4 Main tooth of calyx in fruit usually with 2 or
      more distinct teeth; fruit with 2 distinct ribs
      on abaxial face delimiting ovate ± flat area,
      with hard walls   4. V. dentata

1. V. locusta (L.) Laterr. - Common Cornsalad. Stems erect,   785
to 15(40)cm; inflorescences compact; calyx 0 or vestigial;
fruits glabrous, 1.8-2.5mm, 1.8-2.5mm thick, 1-1.5mm wide;
fertile cell with outer (adaxial) wall spongy, ± as thick as
rest of cell; sterile cells each c. as large as fertile cell
(excl. spongy layer) scarcely grooved between them. Native;
arable and rough ground, bare places in grassland, on banks,
walls, rocky outcrops and dunes; frequent throughout BI.
Very dwarf, ± stemless plants on dunes in W Br are best
recognized as var. dunensis D. Allen (ssp. dunensis (D.
Allen) Sell).
2. V. carinata Lois. - Keeled-fruited Cornsalad. Differs   785
from V. locusta in fruits 2-2.7mm, 0.8-1.4mm thick and wide;
fertile cell without spongy wall; sterile cells slightly
smaller than fertile one, with deep groove between them.
Native; similar places to V. locusta, but (?)not dunes;
scattered in En, CI and (intrd) S & E Ir.
3. V. rimosa Bast. - Broad-fruited Cornsalad. Stems erect,   785
to 15(40)cm; inflorescences lax; calyx in fruit composed   RR
largely of 1 nearly entire tooth 0.5-1mm; fruits glabrous
(?always), 1.5-2.5mm, 1-1.5mm thick, 1.5-2.5mm wide; fertile
cell without spongy wall; sterile cells slightly larger than
fertile cell, scarcely grooved between them. Native; cornfields and rough ground; very local in S En and (?still) Ir,
formerly widespread in Ir and Br N to C Sc.
4. V. dentata (L.) Pollich - Narrow-fruited Cornsalad.   785
Stems erect, to 15(40)cm; inflorescences rather lax to dense;   R
calyx in fruit like that of V. rimosa but main tooth often
toothed; fruits glabrous or pubescent, 1.5-2mm, 0.6-0.8mm
thick, 0.7-1.2mm wide; fertile cell without spongy wall;
sterile cells reduced to 2 abaxial ridges. Native; cornfields and rough ground; scattered in Br and Ir N to C Sc,
decreasing.
5. V. eriocarpa Desv. - Hairy-fruited Cornsalad. Stems   785
erect, to 15(40)cm; inflorescences compact; calyx 0.7-1.3mm,   RR
with a very short tube and usually 6 unequal teeth; fruit
similar to that of V. dentata. Intrd; banks, walls and rough
ground; very scattered in S En and CI, decreasing; S Europe.

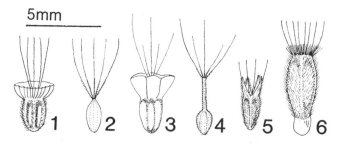

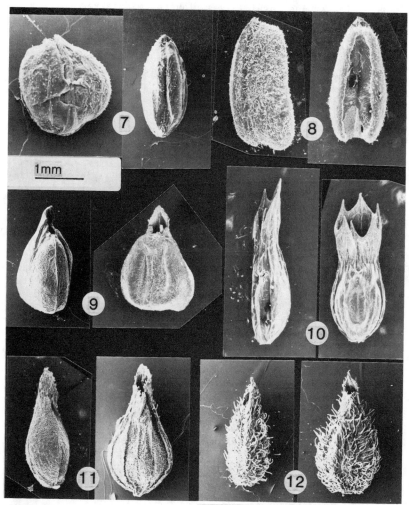

FIG 785 – 1–6, Fruits of **Dipsacaceae**. 1–2, **Scabiosa columbaria**, with and without epicalyx. 3–4, **S. atropurpurea**, with and without epicalyx. 5, **Succisa pratensis**, with epicalyx. 6, **Knautia arvensis**, with epicalyx. 7–12, Fruits of **Valerianella**, lateral and abaxial views. 7, **V. locusta**. 8, **V. carinata**. 9, **V. rimosa**. 10, **V. eriocarpa**. 11, **V. dentata** (glabrous). 12, **V. dentata** (pubescent).

2. VALERIANA L. – Valerians

Perennials with main and lateral stems; leaves pinnate and/or simple on each plant; flowers in rather dense compound cymes, bisexual or dioecious; calyx developing long feathery projections at fruiting; corolla-tube not or slightly pouched at base; stamens 3; stigmas 3; sterile cells of ovary scarcely discernible.

1 Stem-leaves and basal leaves pinnate, with several
  lateral leaflets as large as terminal one
                                              1. V. officinalis
1 At least basal leaves simple   2
   2 Basal leaves entire, cuneate to rounded at base;
     stem-leaves with several lateral leaflets as large
     as terminal one   **3. V. dioica**
   2 Basal leaves dentate, cordate at base; stem-leaves
     simple or with 1 large terminal and 1-2 much
     smaller pairs of lateral leaflets   **2. V. pyrenaica**

1. V. officinalis L. (V. sambucifolia J.C. Mikan ex Pohl) – Common Valerian. Stems erect, to 2m, sometimes with short stolons; all leaves (except very first-formed) pinnate, with linear-lanceolate to narrowly ovate, entire to serrate leaflets; flowers pink. Native; dry or damp grassy places and rough ground; frequent + throughout Br and Ir. Very variable; short tetraploid plants with very narrow, + entire leaflets (ssp. **collina** Nyman) occur on dry calcareous soils in S & C Br, while taller octoploid plants with broader leaflets (ssp. **sambucifolia** (J.C. Mikan ex Pohl) Hayw.) occur in damper places throughout Br and Ir, but every intermediate is found. The diploid (ssp. **officinalis**) has not been confirmed from BI.

2. V. pyrenaica L. – Pyrenean Valerian. Stems erect, to 1.2m; stolons 0; basal leaves simple, ovate to suborbicular, cordate, dentate; stem-leaves similar but often with 1-2 pairs of small lateral leaflets also; flowers pink. Intrd; damp woods and shady hedgebanks; natd in N & W Br from E Cornwall to N Aberdeen, frequent in S & C Sc, rare in W & NE Ir; Pyrenees.

3. V. dioica L. – Marsh Valerian. Stems erect, to 40cm; stolons well developed; basal leaves and those on stolons simple, entire; stem-leaves pinnate, with entire to slightly dentate leaflets; flowers dioecious, white to pink. Native; marshes, fens and bogs, often in shade; frequent in Br N to S Sc.

3. CENTRANTHUS DC. – Red Valerian

Annuals or perennials with main and usually lateral stems; leaves simple and entire to deeply pinnately lobed or + pinnate; flowers in dense compound cymes, bisexual; calyx developing long feathery projections at fruiting; corolla-tube with backward-directed spur; stamen 1; stigma 1; sterile

cells of ovary scarcely discernible.
1. **C. ruber** (L.) DC. - Red Valerian. Erect perennial to
80cm; leaves simple, somewhat glaucous, ovate to narrowly-
elliptic or lanceolate, entire or remotely dentate, those on
stems mostly ± sessile; flowers red, pink or white; corolla-
spur (2)5-12mm. Intrd; natd on walls, dry rocky or sandy
places, cliffs and banks; common in CI and C Br, extending to
N Ir and C Sc; Mediterranean region.
2. **C. calcitrapae** (L.) Dufr. - Annual Valerian. Erect
annual to 30cm; leaves deeply pinnately lobed to ± pinnate;
flowers pink; corolla-spur <2mm. Intrd; natd since 1982 on
open ground in churchyard, Surrey; S Europe.

136. DIPSACACEAE - Teasel family

Biennial to perennial herbs; leaves simple or pinnate,
opposite, exstipulate, petiolate or sessile. Flowers
numerous, borne on common receptacle in dense terminal heads
(capitula), weakly zygomorphic, bisexual or gynodioecious,
epigynous; calyx small, cup-shaped or divided into 4-8 teeth
or bristles; petals 4-5, fused into tube proximally, the
lobes usually larger on abaxial side, especially in outer
flowers; stamens 4, free, ± exserted, borne on corolla-tube;
ovary 1-celled with 1 ovule; style 1; stigma 2-lobed, or
simple and capitate or oblique; fruit an achene.
The calyx, which arises from the top of the ovary outside
the corolla, is often less conspicuous than the epicalyx,
which arises at the base of the ovary but encloses the latter
in a tubular structure and often expands into lobes around
the calyx. The ripe fruit remains enclosed in the epicalyx.
The receptacle bears bracts at the base of the capitulum and
usually a bract associated with each flower within the
capitulum.
The flowers with 4 stamens and borne in a capitulum, and the
ovary and fruit enclosed in a tubular epicalyx, are unique.
Before using the key a capitulum should be dissected carefully to distinguish between calyx and epicalyx.

1 Stems, and usually midribs on lowerside of
  leaves, prickly   **1. DIPSACUS**
1 Stems and leaves glabrous to pubescent, not prickly   2
  2 Corolla 5-lobed; epicalyx expanded at apex into
    membranous, veined funnel   **5. SCABIOSA**
  2 Corolla 4-lobed; epicalyx variously expanded at
    apex but not membranous   3
3 Corolla cream; calyx without teeth or bristles
                                                         2. CEPHALARIA
3 Corolla blue to purple or violet, rarely white or
  pinkish; calyx with 4-8 teeth or bristles   4
  4 Flowers ± all equal-sized; calyx with 4-5
    bristles; receptacle bearing bracts, 1 subtending

each flower as well as some at base   **4. SUCCISA**
4 Outer flowers much longer than inner ones; calyx
   with 8 bristles; receptacle bearing bracts at
   base but not subtending each flower   **3. KNAUTIA**

1. DIPSACUS L. - <u>Teasels</u>
Biennials; stems prickly; leaves simple or pinnate;
receptacle with spine-tipped bracts subtending each flower
and very long spiny ones at base; flowers all $\pm$ 1 size;
epicalyx 4-angled, scarcely toothed at apex; calyx cup-
shaped, scarcely toothed; corolla 4-lobed.

1 Capitula ovoid-cylindrical; upper stem-leaves sessile,
   fused in pairs round stem at base   2
1 Capitula globose; upper stem-leaves petiolate, not
   fused in opposite pairs   3
   2 Bracts on receptacle with stiff but flexible,
      straight apical spine   **1. D. fullonum**
   2 Bracts on receptacle with stiff, rigid, recurved
      apical spine   **2. D. sativus**
3 Capitula 15-28mm across (incl. bracts 7-13mm) in full
   flower or fruit   **3. D. pilosus**
3 Capitula 30-40mm across (incl. bracts 14-20mm) in
   full flower or fruit   **4. D. strigosus**

1. D. fullonum L. (<u>D. sylvestris</u> Hudson, <u>D. fullonum</u> ssp.
<u>sylvestris</u> (Hudson) P. Fourn.) - <u>Wild</u> <u>Teasel</u>. Stems erect,
to 2m; leaves simple, entire to dentate or serrate, prickly
on midrib on lowerside; capitula ovoid-cylindrical, 4-8cm,
with basal bracts curving upwards, often as long as
capitulum; bracts on receptacle longer than flowers, with
stiff but flexible, straight apical spine; corolla pinkish-
purple to lilac. Native; marginal habitats and rough ground
by roads, railways, streams, woods and fields; frequent in CI
and Br N to C Sc, local in Ir.
2. D. sativus (L.) Honck. (<u>D. fullosum</u> ssp. <u>sativus</u> (L.)
Thell.) - <u>Fuller's</u> <u>Teasel</u>. Differs from <u>D. fullonum</u> in
capitula with basal bracts $\pm$ patent; bracts on receptacle c.
as long as flowers, with stiff, rigid, recurved, apical
spine. Intrd; still grown for fulling in Somerset, formerly
more widely so, now a frequent casual from bird-seed on tips
and waste ground; scattered in CI and Br; uncertain origin.
3. D. pilosus L. - <u>Small</u> <u>Teasel</u>. Stems erect, to 1.5m;
lower leaves simple, upper ones often pinnate with 1 pair of
small lateral leaflets, serrate, not or slightly prickly on
midrib on lowerside; capitula globose, with basal bracts c.
as long as those on receptacle; corolla whitish. Native;
damp places in open woods, hedgerows and by streams; rather
scattered in En and Wa.
4. D. strigosus Willd. - <u>Yellow-flowered</u> <u>Teasel</u>. Differs
from <u>D. pilosus</u> in larger capitula (see key); corolla pale
yellow. Intrd; natd in rough ground and waste places; Cambs,

rare casual elsewhere in En; Russia.

2. CEPHALARIA Schrader - Giant Scabious
Bisexual perennials; stems sparsely pubescent; leaves simple
and deeply pinnately lobed to pinnate; receptacle with rather
leathery bracts subtending each flower and similar ones at
base; outer flowers longer than inner; epicalyx 8-ridged,
with 8 apical teeth; calyx cup-shaped, scarcely toothed;
corolla 4-lobed.

1. C. gigantea (Ledeb.) Bobrov - Giant Scabious. Stems
erect, to 2m, with long branches; leaves all deeply pinnately
lobed, the uppermost usually pinnate; capitula 4-10cm across;
corolla pale yellow. Intrd; grown in gardens, natd or
persistent in rough grassy places and waste ground; scattered
in C & S En, rarely further N; Caucasus.

3. KNAUTIA L. - Field Scabious
Gynodioecious perennials; stems pubescent; leaves simple and
crenate to pinnate; receptacle with herbaceous bracts at base
but none subtending each flower; outer flowers longer than
inner; epicalyx 4-ridged, scarcely toothed but with dense
hairs at apex; calyx with 8 long bristles; corolla 4-lobed.

1. K. arvensis (L.) Coulter - Field Scabious. Stems erect **785**
to ascending, to 1m; lowest leaves simple and crenate, the
upper ones pinnate, with all intermediates between; bisexual
capitula 2.5-4cm and female ones 1.5-3cm across; corolla
bluish-lilac. Native; dry grassy places on light soils;
frequent over most BI but very local in N Ir and N & W Sc.

4. SUCCISA Haller - Devil's-bit Scabious
Gynodioecious perennials; stems sparsely pubescent; leaves
all simple, entire or distantly toothed; receptacle with
herbaceous bracts at base and smaller ones subtending each
flower; flowers all ± same size; epicalyx 4-angled, with 4
teeth; calyx with 4-5 bristle-tipped teeth; corolla 4-lobed.

1. S. pratensis Moench - Devil's-bit Scabious. Stems erect **785**
to ascending, to 1m; bisexual capitula 2-3cm and female ones
1.5-2.5cm across; corolla bluish-violet, sometimes pinkish-
lilac. Native; many sorts of grassy places, wet or dry, acid
or calcareous, in open or shade; common over most BI.

5. SCABIOSA L. (Sixalix Raf.) - Scabiouses
Bisexual perennials; stems sparsely pubescent; leaves simple
and serrate or lobed, to pinnate; receptacle with herbaceous
bracts at base and narrower ones subtending each flower;
outer flowers longer than inner; epicalyx 8-ridged, expanded
at apex into membranous funnel; calyx with 5 long bristles at
top of stalk (hypanthium); corolla 5-lobed.

1. S. columbaria L. - <u>Small</u> <u>Scabious</u>. Stems erect, to 70cm; **785**
capitula 1.5-3.5cm across; corolla bluish-lilac; fruiting
capitula up to 1.5(2)cm long; epicalyx-tube 2-3mm, with
16-24-veined membranous funnel 0.8-1.5mm; hypanthium shorter
than fruit and falling short of apex of epicalyx funnel.
Native; dry calcareous grassland and rocky places; locally
common in Br N to S Sc. Often very dwarf on exposed maritime
cliff-tops.
2. S. atropurpurea L. (<u>Sixalix</u> <u>atropurpurea</u> (L.) Greuter & **785**
Burdet) - <u>Sweet</u> <u>Scabious</u>. Differs from <u>S.</u> <u>columbaria</u> in
corolla dark purple to pale lilac; fruiting capitula up to
2.5(3)cm long; epicalyx-tube 1.5-2.5mm, with 8-ribbed funnel
c. as long (shorter or longer) and inrolled at apex;
hypanthium longer than fruit and reaching at least to apex of
epicalyx funnel. Intrd; grown in gardens, natd on rough
ground by sea in W Cornwall and E Kent (at Folkestone since
1862), rare casual elsewhere; S Europe.

137. ASTERACEAE - <u>Daisy</u> <u>family</u>
(Compositae)

Annual to perennial herbs, often woody near base, rarely
shrubs; leaves simple and entire to pinnately or palmately
lobed or variously pinnately compound, alternate or sometimes
opposite, exstipulate, petiolate or sessile. Flowers usually
numerous, small, borne on common receptacle in dense terminal
heads (<u>capitula</u>), zygomorphic or actinomorphic (often both in
same capitulum), bisexual to variously monoecious or
dioecious, epigynous; sepals 0 or represented by <u>pappus</u> of
scales, teeth, bristles, hairs or a membranous ring, often
enlarging in fruit; petals 5 (rarely 0 or 4), fused into tube
with distal limb of either (a) 5 (rarely 4) actinomorphic (or
nearly so) lobes or teeth (<u>tubular</u> flowers), or (b)
unilateral strap-like <u>ligule</u> often with 3 or 5 apical teeth
(<u>ligulate</u> flowers), the tube often extremely short in latter
type; stamens 5, borne on corolla-tube, the anthers fused
laterally into cylinder around style (not in <u>Xanthium</u>); ovary
1-celled with 1 ovule; style 1, usually branched, each branch
with linear stigmatic surface, fruit an achene, often with
persistent pappus.
The capitula bear around the outside of the flower-bearing
area a series of often sepal-like bracts or <u>phyllaries</u>. They
may also bear mixed in with the flowers (often 1 per flower)
small <u>receptacular</u> <u>scales</u> or <u>bristles</u>. Each achene is often
inserted on the receptacle into a minute <u>achene-pit</u>. Each
capitulum may be <u>discoid</u>, with tubular flowers only,
<u>ligulate</u>, with ligulate flowers (usually with 5-toothed
ligules) only, or <u>radiate</u>, with a central region of tubular
flowers (<u>disc</u> <u>flowers</u>) and an outer region of ligulate
flowers (<u>ray</u> <u>flowers</u>) (usually with 3-toothed ligules).
Sometimes each capitulum has very few flowers (very rarely

only 1, e.g. <u>Olearia</u> <u>paniculata</u>). In <u>Echinops</u> the capitula
have only 1 flower but are aggregated into large spherical
compound heads. In <u>Ambrosia</u> and <u>Xanthium</u> the male and female
capitula are separate on the same plant; the females contain
only 1 or 2 flowers respectively. Otherwise, except in
dioecious spp., the disc flowers are bisexual, but ray
flowers may be bisexual, female or sterile.
The flowers borne in capitula, with 5 stamens with laterally
fused anthers (except in <u>Xanthium</u>), and the fruit an achene,
are a unique combination. The largest dicotyledon family.
Before using the key a capitulum should be dissected to
identify all its component parts. Rare variants without ray
flowers, or with all the flowers ligulate (<u>flore</u> <u>pleno</u>), are
not covered in the key, though more frequent variations in
the presence or absence of ray flowers (e.g. in <u>Aster</u>,
<u>Bidens</u>, <u>Senecio</u>) are.

<u>General key</u>
1 Shrubs, with new growth arising each year from older
   woody stems   <u>Key A</u>
1 Herbs; stems sometimes woody at base   2
   2 Flowers 1 per capitulum, the capitula aggregated
      into tight globose heads   **1. ECHINOPS**
   2 Flowers >1 per capitulum at least in some capitula,
      the capitula not in tight globose heads   3
3 Male and female capitula separate on same plant, the
   male more apical and with several flowers, the
   female lower down and with 1-2 flowers   4
3 If male and female capitula separate then on different
   plants, both with several-many flowers   5
   4 Leaves alternate; fruiting heads with 2 prominent
      terminal processes, covered with stiff hooked
      bristles   **82. XANTHIUM**
   4 Leaves mostly opposite; fruiting heads without
      terminal processes, with short straight spines or
      with tubercles   **80. AMBROSIA**
5 Flowers all ligulate, the ligules usually 5-toothed
   at ends; milky latex usually present   <u>Key B</u>
5 Flowers all tubular or tubular and ligulate in same
   head, the ligules usually 3-toothed at ends; milky
   latex 0   6
   6 Style with ring of minute hairs, or with glabrous
      thickened ring, just below the branches; anthers
      with long 'tails' at base; leaves and/or
      phyllaries often spiny; flowers mostly pink or blue
      to purple, all tubular   <u>Key C</u>
   6 Style without ring of hairs or thickened zone below
      branches; anthers mostly without basal 'tails';
      plant rarely with spines; flowers various colours,
      often yellow, the marginal ones often ligulate   7
7 At least the lower leaves opposite   <u>Key D</u>
7 All leaves basal or alternate   8

8 Capitula discoid, with ligules 0 or inconspicuous
   and not exceeding inner phyllaries   Key E
8 Capitula radiate, with >1 obvious ligules
   exceeding inner phyllaries   9
9 Pappus of at least inner flowers of hairs (sometimes
   also of scales)   Key F
9 Pappus 0, or of scales and/or few bristles   10
   10 Receptacular scales or abundant bristles present
                                                   Key G

   10 Receptacular scales and bristles 0, but some-
      times short fringes round achene-pits present Key H

Key A - Shrubs (all aliens)
1 Marginal flowers conspicuously ligulate   2
1 Flowers 1, or few to many and all tubular, the outer
   sometimes with very short ligule-like lobes not or
   scarcely exceeding phyllaries (but beware ligule-like
   inner phyllaries in Helichrysum)   4
   2 Leaves pinnate or divided nearly to midrib
                                          68. SENECIO
   2 Leaves entire to toothed <1/2 way to midrib   3
3 Ligules white; leaves evergreen, not sticky;
   glandular hairs 0   51. OLEARIA
3 Ligules yellow; leaves dying in winter, very sticky;
   glandular hairs abundant   40. DITTRICHIA
   4 Leaves divided >1/2 way to midrib, very fragrant
      when bruised   5
   4 Leaves entire to shallowly lobed or toothed,
      usually not fragrant   6
5 Leaves all <5mm wide, with crowded incurved lobes;
   capitula >5mm wide, solitary on long stalks
                                        57. SANTOLINA
5 At least lower leaves >1cm wide, with spreading
   lobes; capitula <5mm wide, clustered   56. ARTEMISIA
   6 Stems procumbent, rooting, <60cm; inner phyllaries
      patent in flower, white, ligule-like 38. HELICHRYSUM
   6 Stems erect to scrambling, often >60cm; inner
      phyllaries not patent and ligule-like   7
7 Young stems and leaf lowersides densely white- to
   buff-tomentose   8
7 Stems and leaves glabrous or nearly so   9
   8 Leaves entire   51. OLEARIA
   8 At least larger leaves remotely sinuate-lobed
                                      72. BRACHYGLOTTIS
9 Leaves linear, <5mm wide   47. CHRYSOCOMA
9 Larger leaves lanceolate to orbicular, >1cm wide   10
   10 Erect shrub; leaves rhombic to obovate, cuneate at
      base; florets white, dioecious   52. BACCHARIS
   10 Scrambler; leaves + ivy-shaped, cordate to hastate
      at base; florets yellow, bisexual   71. DELAIREA

Key B - Flowers all ligulate (tribe Lactuceae)
1 Pappus 0 or of scales   2
1 Pappus, at least in central flowers, of hairs   6
  2 Leaves spiny; plant thistle-like   13. SCOLYMUS
  2 Leaves not spiny; plant not thistle-like   3
3 Pappus of scales   4
3 Pappus 0; achene often terminating in minute collar   5
  4 Capitula >2cm across; ligules blue, rarely white;
    achenes <4mm   14. CICHORIUM
  4 Capitula <2cm across; ligules yellow; achenes
    >4mm   17. HEDYPNOIS
5 Leaves all basal; peduncles strongly dilated below
  capitula   15. ARNOSERIS
5 Stems leafy; peduncles not or scarcely dilated below
  capitula   16. LAPSANA
  6 Pappus-hairs feathery, with slender lateral
    branches visible to naked eye, at least on some
    achenes   7
  6 Pappus-hairs all simple, smooth or shortly toothed
    (lens)   11
7 Aerial stems with 0 or extremely reduced leaves   8
7 Aerial stems with well-developed leaves   9
  8 Receptacular scales present among flowers
                                             18. HYPOCHAERIS
  8 Receptacular scales 0   19. LEONTODON
9 Plant hispid, with harsh, hooked bristles   20. PICRIS
9 Plant glabrous or with very soft hairs   10
10 Phyllaries c.8, all in 1 row and of 1 length
                                            22. TRAGOPOGON
10 Phyllaries >10, the outermost c.1/2 as long as
   innermost   21. SCORZONERA
11 Leaves with strong spines; receptacular scales
   present, wrapped round achenes; pappus-hairs 2-4
                                               13. SCOLYMUS
11 Leaves not or weakly spiny; receptacular scales
   usually 0; pappus-hairs numerous   12
  12 Achenes distinctly flattened   13
  12 Achenes not or scarcely flattened   16
13 Achenes with distinct narrow beak at apex, or at
   least markedly narrowed distally   14
13 Achenes without beak and scarcely narrowed distally   15
  14 Pappus-hairs in 2 equal rows; phyllaries in
     several rows   25. LACTUCA
  14 Pappus-hairs in 2 unequal rows; phyllaries in 2
     distinct unequal rows   27. MYCELIS
15 Ligules yellow   24. SONCHUS
15 Ligules blue to mauve   26. CICERBITA
  16 Capitulum 1 per stem; stems without leaves or
     scales; rhizomes and stolons 0   28. TARAXACUM
  16 Capitula >1 per stem; or stems with leaves or
     scales; or rhizomes and/or stolons present   17
17 Pappus-hairs pure white   18

17 Pappus-hairs yellowish-white to pale brown   19
   18 Capitulum 1 per stem; long thin rhizomes present
                                            23. AETHEORHIZA
   18 Capitula normally >1 per stem; rhizomes O **29. CREPIS**
19 Stolons usually present; achenes $\leq 2(2.5)$mm, each rib
   ending in a small point at apex of achene **30. PILOSELLA**
19 Stolons O; achenes >(1.5)2.5mm, each rib ending in a
   smooth ring at apex of achene   20
   20 Plant glabrous except for phyllaries; phyllaries
      in distinct inner and outer rows   **29. CREPIS**
   20 Plant with hairs (often dense) on some (often
      most) parts; phyllaries graduated between innermost
      and outermost   **31. HIERACIUM**

Key C - Style with ring of minute hairs or with glabrous
        thickened ring just below branches (tribe Cardueae)
1 Phyllaries strongly hooked at tip, stiff, subulate
                                            3. ARCTIUM
1 Phyllaries not hooked at tip   2
   2 Corolla pale yellow to reddish-orange (beware
      inner phyllaries of <u>Carlina</u>)   3
   2 Corolla pink or blue to purple, rarely white   5
3 Pappus-hairs feathery, with slender lateral branches
   visible to naked eye   **6. CIRSIUM**
3 Pappus O, or of narrow scales, or of simple to
   toothed hairs   4
   4 Outer phyllaries large and leaf-like   **12. CARTHAMUS**
   4 Outer (and inner) phyllaries scale-like, with spiny
      apical portion   **11. CENTAUREA**
5 Pappus-hairs feathery, with slender lateral branches
   visible to naked eye   6
5 Pappus O, of narrow scales, or of simple to toothed
   hairs   8
   6 Phyllaries all obtuse to rounded at apex; leaves
      ± entire to distantly toothed, the teeth not
      spinose or bristle-like   **4. SAUSSUREA**
   6 At least outer phyllaries spinose, mucronate or
      acuminate at apex; leaves spiny or at least with
      fine bristle-like teeth   7
7 Outer phyllaries leaf-like; inner ones scarious,
   patent in dry weather, pale yellow on upperside,
   appearing like ligules   **2. CARLINA**
7 All phyllaries scale-like to ± subulate   **6. CIRSIUM**
   8 Leaves with sharp spines   9
   8 Leaves without spines   11
9 Receptacle glabrous, but achene-pits fringed with
   teeth   **7. ONOPORDUM**
9 Receptacle densely pubescent or bristly   10
   10 Stem-leaves not decurrent down stem; stems not
      spiny; outer phyllaries with spine-tipped lateral
      lobes or teeth   **8. SILYBUM**
   10 Stem-leaves decurrent down stem in a spiny wing;

phyllaries all entire, with terminal spine
                                              5. CARDUUS
11 Phyllaries simple, entire   **9. SERRATULA**
11 At least inner phyllaries with distinct apical
   portion which is scarious, toothed, or spiny   12
   12 All flowers bisexual and of same size; apical
      portion of phyllaries scarious, not separated from
      main part by constriction   **10. ACROPTILON**
   12 Marginal flowers functionally female though some-
      times with sterile stamens, often longer than inner
      flowers; apical portion usually toothed or spiny,
      if merely scarious then separated from main
      part by constriction   **11. CENTAUREA**

Key D - Plant herbaceous; at least lower leaves opposite
1 Capitula discoid, with only tubular flowers   2
1 Capitula radiate, with >1 marginal ligulate flower   6
   2 Terminal capitula male only, in elongated bract-
     less racemes   **80. AMBROSIA**
   2 All capitula bisexual, variously arranged   3
3 Pappus of hairs   **96. EUPATORIUM**
3 Pappus 0 or of scales or stout bristles   4
   4 Leaves pinnate, or simple with narrowly cuneate
     base; pappus of barbed bristles   **88. BIDENS**
   4 Leaves simple, broadly cuneate to cordate at base;
     pappus 0 or of scales   5
5 Flowers blue; receptacular scales 0; pappus of scales
                                               97. AGERATUM
5 Flowers greenish-white; receptacular scales present;
  pappus 0   **81. IVA**
   6 Capitula with 1 ligulate flower   **93. SCHKUHRIA**
   6 Capitula with 3-numerous ligulate flowers   7
7 Plant with large underground tubers   8
7 Plant without underground tubers   9
   8 Leaves (bi-)pinnate or (bi-)ternate; ligules often
     red to pink or purple, or white   **91. DAHLIA**
   8 Leaves simple; ligules yellow   **86. HELIANTHUS**
9 Ligules pink or white   10
9 Ligules yellow to greenish- or brownish-yellow   12
   10 Leaf-lobes linear to filiform; ligules usually
      pink, rarely white   **90. COSMOS**
   10 Leaf-lobes lanceolate to ovate; ligules white,
      rarely purplish   11
11 Capitula <7mm across excl. ligules; pappus of scales
                                               87. GALINSOGA
11 Capitula >7mm across excl. ligules; pappus of barbed
   strong bristles   **88. BIDENS**
   12 Pappus of barbed, strong, persistent bristles
                                               88. BIDENS
   12 Pappus 0 or of weak deciduous bristles and/or of
      scales   13
13 Most leaves divided ± to base or to midrib   14

13 Leaves simple, subentire to toothed or shallowly
   lobed   15
   14 Pappus of conspicuous scales; receptacular scales
      0   **92. TAGETES**
   14 Pappus 0, or minute, or of 2 small scales;
      receptacular scales present   **89. COREOPSIS**
15 Ligules >1cm; outer phyllaries lanceolate to ovate,
   not or scarcely glandular   16
15 Ligules <1cm; outer phyllaries linear, with dense
   glandular hairs   **84. SIGESBECKIA**
   16 Annual or perennial; leaves petiolate, or sessile
      but not clasping stem; achenes flattened in
      radial plane   **86. HELIANTHUS**
   16 Annual; leaves sessile, clasping stem at base;
      achenes flattened in tangential plane   **83. GUIZOTIA**

Key E - Plant herbaceous; leaves alternate or all basal;
        capitula discoid or with very inconspicuous ligules
        (excl. Cardueae)
1 Pappus 0, or of small scales or a few bristles   2
1 Pappus of hairs (somewhat expanded apically in male
  Antennaria and Anaphalis)   15
  2 Leaves divided <1/2 way to midrib   3
  2 Leaves divided >1/2 way to midrib   7
3 Receptacular scales present; plant densely white-
  woolly   **58. OTANTHUS**
3 Receptacular scales 0; plant not densely woolly   4
  4 Pappus of 1-8 stiff barbed bristles and sometimes
    also some small scales   **44. CALOTIS**
  4 Pappus 0 or a small membranous ring   5
5 Capitula very numerous, in racemes or panicles
                                              56. ARTEMISIA
5 Capitula 1-many, if many then in corymbs   6
  6 Achenes ± compressed; stems procumbent to weakly
    erect, <30cm   **67. COTULA**
  6 Achenes scarcely compressed; stems strong, erect,
    usually >50cm   **54. TANACETUM**
7 Capitula very numerous, in racemes or panicles;
  corolla brownish-, reddish- or greenish-yellow   8
7 Capitula 1-few, or if ± numerous then in corymbs
  and corolla bright yellow   9
  8 Flowers all identical, with functional male and
    female parts   **55. SERIPHIDIUM**
  8 Outer flowers functionally female, with filiform
    corolla; inner ones bisexual and tubular
                                             56. ARTEMISIA
9 Corolla pubescent; stem and leaves densely pubescent
                                             56. ARTEMISIA
9 Corolla glabrous; stem and leaves glabrous to
  pubescent   10
  10 Receptacular scales present   11
  10 Receptacular scales 0   12

11 Corolla with small pouch at base, obscuring top of
   ovary in 1 plane   **60. CHAMAEMELUM**
11 Corolla not pouched at base, not obscuring top of
   ovary   **61. ANTHEMIS**
  12 Achenes compressed; stems procumbent to weakly
     erect, <30cm   **67. COTULA**
  12 Achenes not or scarcely compressed; stems usually
     stiffly erect to ascending   13
13 Capitula flat, corymbose; perennial   **54. TANACETUM**
13 Capitula conical to convex, not corymbose; annual   14
  14 Achenes strongly 3-ribbed, with usually 2 resin-
     glands near apex   **66. TRIPLEUROSPERMUM**
  14 Achenes with 4-5 ribs, without resin-glands
          65. MATRICARIA
15 Leaves broadly ovate to orbicular, cordate at base,
   petiolate   16
15 Leaves linear to ovate, cuneate at base, petiolate
   or sessile   18
  16 Plant with scrambling stems bearing ivy-shaped
     leaves   **71. DELAIREA**
  16 Plants with aerial stems bearing only reduced,
     bract-like leaves, the main leaves all basal   17
17 Capitula solitary   **78. HOMOGYNE**
17 Capitula in inflorescences   **77. PETASITES**
  18 Plant densely glandular, very sticky   **40. DITTRICHIA**
  18 Plant glabrous to woolly, not obviously glandular,
     not sticky   19
19 Plant glabrous to pubescent, not woolly   20
19 Plant woolly at least in part, especially near tops
   of stems   25
  20 Leaves conspicuously serrate to deeply lobed or
     ± pinnate; phyllaries in 1 main row with much
     shorter supplementary ones near base of capitulum
          68. SENECIO
  20 Leaves entire to slightly or remotely serrate;
     phyllaries in several rows   21
21 Phyllaries totally scarious, coloured or white
          38. HELICHRYSUM
21 Phyllaries at least partly herbaceous and green   22
  22 Stems glabrous, sometimes weakly scabrid   23
  22 Stems pubescent   24
23 Pappus straw-coloured to pale reddish or brownish; at
   least lower leaves >2.5cm, either fleshy or flat
          46. ASTER
23 Pappus pure white; all leaves <2.5cm, with margins
   rolled under   **47. CHRYSOCOMA**
  24 Flowers yellow; outer phyllaries with patent to
     recurved tips   **39. INULA**
  24 Flowers white to cream or pinkish; phyllaries
     appressed to flowers   **49. CONYZA**
25 Annual with simple root system   26
25 Perennial with rhizomes or stolons (often short)   27

26 Marginal florets with receptacular scales; outer
   phyllaries herbaceous, woolly beyond half-way to
   apex   **34. FILAGO**
26 Receptacular scales 0; all phyllaries scarious,
   glabrous or woolly in lower half   **37. GNAPHALIUM**
27 Capitula in elongate panicles; capitula bisexual
                                            37. GNAPHALIUM
27 Capitula in ± crowded terminal subcorymbose
   clusters; plants dioecious or subdioecious   28
   28 Largest leaves forming basal rosettes, the stem-
      leaves much narrower; stems rarely >20cm;
      stoloniferous   **35. ANTENNARIA**
   28 Leaves not forming basal rosette, the largest
      ones being on the stems; stems rarely <25cm;
      rhizomatous   **36. ANAPHALIS**

Key F - Plant herbaceous; leaves alternate or all basal;
        capitula radiate; pappus of hairs
1 Ligules white, or pink or blue to purple or mauve   2
1 Ligules yellow to orange   9
  2 Main leaves all basal, broadly ovate to orbicular,
    cordate at base; flowering stems with only reduced
    bract-like leaves   **77. PETASITES**
  2 Leaves on flowering stems, or if ± all basal
    then not cordate at base   3
3 Outer phyllaries broad, green, ± leafy **50. CALLISTEPHUS**
3 Outer phyllaries similar to or smaller than inner
  ones   4
  4 At least the larger leaves truncate to cordate at
    base   5
  4 Leaves all cuneate at base   7
5 Phyllaries in a graded series of rows   **46. ASTER**
5 Phyllaries in 1 main row, with smaller supplementary
  ones near base of capitulum   6
  6 Leaves pinnately veined; phyllaries in 1 main row
    with smaller supplementary ones at base of
    capitulum; ligules white   **68. SENECIO**
  6 Leaves palmately veined; capitula without
    supplementary phyllaries; ligules white or
    coloured   **69. PERICALLIS**
7 Ligules linear to narrowly elliptic, >1mm wide;
  phyllaries all green or more green in apical than
  basal 1/2   **46. ASTER**
7 Ligules filiform to linear, usually <0.6 but sometimes
  up to 1.5mm wide; phyllaries more green in basal than
  in apical 1/2   8
  8 Ligules <1mm; central tubular flowers fewer than
    peripheral filiform ones   **49. CONYZA**
  8 Ligules >1mm; central tubular flowers more
    numerous than peripheral filiform ones (or the
    latter 0)   **48. ERIGERON**
9 Main leaves all basal; flowering stems with 1

capitulum and only reduced bract-like leaves
  76. TUSSILAGO
9 Leaves on flowering stems, or if ± all basal then
   capitula >1 per stem   10
   10 Phyllaries in 1 main row, often with smaller
      supplementary ones near base of capitulum   11
   10 Phyllaries in 2 or more (often indistinct) rows   15
11 Capitula with only 2-4 disc and 2-4 ray flowers,
   numerous in cylindrical or pyramidal panicles   12
11 Capitula with >5 disc and >4 ray flowers, not in
   dense pyramidal panicles   13
   12 Lower leaves deeply palmately lobed; petioles
      sheathing stem; ray flowers usually 2   **74. LIGULARIA**
   12 Lower leaves deeply pinnately lobed; petioles not
      sheathing stem; ray flowers usually 3   **73.SINACALIA**
13 Stem-leaves and basal leaves with petioles with
   broad sheathing bases   **74. LIGULARIA**
13 Stem-leaves and basal leaves without sheathing
   petioles   14
   14 Phyllaries in 1 main row with supplementary
      smaller ones near base of capitulum   **68. SENECIO**
   14 Phyllaries in 1 row, with no supplementary small
      ones near base of capitulum   **70. TEPHROSERIS**
15 Phyllaries in 2 distinct rows of ± equal length
  75. DORONICUM
15 Phyllaries in 2 or more indistinct rows, progressively
   longer towards the inside   16
   16 Plant densely glandular, very sticky   **40. DITTRICHIA**
   16 Plant not or slightly glandular, not sticky   17
17 Pappus of inner row of hairs and outer row of small
   (often laterally fused) scales   **41. PULICARIA**
17 Pappus entirely of hairs   18
   18 Capitula >(1.5)2cm across (incl. ligules); ligules
      >10mm; anthers with long filiform basal appendages
  39. INULA
   18 Capitula <1.5(2)cm across (incl. ligules); ligules
      <10mm; anthers without basal appendages **45. SOLIDAGO**

<u>Key G</u> - Plant herbaceous; leaves alternate or all basal;
      capitula radiate; pappus 0, or of scales or bristles;
      receptacular scales present
1 Receptacle with abundant bristles only; pappus of
  5-10 scales with long apical bristles   **94. GAILLARDIA**
1 Receptacular scales present; pappus 0, or of small
  scales not bristle-tipped   2
   2 Achenes strongly compressed, >2x as wide as thick,
     with lateral strong ribs or narrow wings   3
   2 Achenes angular, or not or slightly compressed, <2x
     as wide as thick, without lateral ribs or wings   4
3 Capitula <2cm across (incl. ligules); ligules <1cm
  59. ACHILLEA

3 Capitula >2cm across (incl. ligules); ligules >1cm
                                               89. COREOPSIS
   4 Ligules white   5
   4 Ligules yellow to orange   6
5 Corolla of tubular flowers with small pouch at base,
   obscuring top of ovary in 1 plane   **60. CHAMAEMELUM**
5 Corolla of tubular flowers not pouched, not obscuring
   top of ovary   **61. ANTHEMIS**
   6 Phyllaries with broad scarious margins and tips
                                                  61. ANTHEMIS
   6 Phyllaries entirely herbaceous   7
7 Ligules <2mm wide; receptacle slightly convex; lower
   leaves cordate at base   **42. TELEKIA**
7 Ligules >2mm wide; receptacle conical; leaves all
   cuneate at base   **85. RUDBECKIA**

Key H - Plant herbaceous; leaves alternate or all basal;
        capitula radiate; pappus 0, or of scales or bristles;
        receptacular scales 0
1 Capitula with 1 ligulate flower   **93. SCHKUHRIA**
1 Capitula with outer row of ligulate flowers   2
   2 Ligules yellow to orange or brownish-red, at least
      in part; pappus often of distinct scales or
      bristles   3
   2 Ligules white to pink; pappus 0 or a minute rim   9
3 At least some achenes very strongly curved, very
   warty on outer face   **79. CALENDULA**
3 Achenes not or slightly curved, if warty then not just
   on outer face   4
   4 Pappus of 1-8 bristles, sometimes with minute
      scales as well   5
   4 Pappus 0, or a minute rim, or of scales   6
5 Pappus bristles barbed, persistent; fruiting capitula
   <1cm across, not resinous   **44. CALOTIS**
5 Pappus bristles smooth or forwardly serrated,
   deciduous; fruiting capitula >1cm across, very
   resinous   **43. GRINDELIA**
   6 Pappus 0   **62. CHRYSANTHEMUM**
   6 Pappus of distinct scales   7
7 Leaves subglabrous to sparsely pubescent, decurrent on
   stems as wings; receptacle markedly convex **95. HELENIUM**
7 Leaves densely white-tomentose on lowerside, not
   decurrent; receptacle + flat to slightly convex   8
   8 Ligules pale yellow on upperside, purplish on
      lowerside; phyllaries free, glabrous to pubescent
                                              32. ARCTOTHECA
   8 Ligules orange-yellow, with basal black blotch
      bearing central white spot, or rarely plain yellow;
      outer phyllaries fused into cup-like structure,
      white-tomentose   **33. GAZANIA**
9 Rosette-plant; capitula solitary on leafless stems
                                                 53. BELLIS

9   Flowering stems bearing leaves   10
   10 Stem-leaves simple, shallowly to deeply toothed
      but not to midrib, the teeth simple   11
   10 Stem-leaves pinnately lobed to midrib or nearly
      so, the lobes further lobed   13
11   Ligules <10mm   **54. TANACETUM**
11   Ligules >10mm   12
   12 Lowest (tubular) part of corolla of ray flowers
      with 2 narrow transparent wings   **64. LEUCANTHEMUM**
   12 Lowest (tubular) part of corolla of ray flowers
      not winged   **63. LEUCANTHEMELLA**
13   Ultimate leaf-segments lanceolate to ovate, flat
                                               54. TANACETUM
13   Ultimate leaf-segments linear to filiform, not or
      scarcely flattened   14
   14 Achenes strongly 3-ribbed, with usually 2 resin-
      glands near apex   **66. TRIPLEUROSPERMUM**
   14 Achenes with 4-5 ribs, without resin-glands
                                               65. MATRICARIA

Other genera - Many other genera occur as casuals or as
non-persistent escapes or throw-outs. The following are
those most often claimed, mentioned under the key in which
they would appear.
Key B (Lactuceae) - **TOLPIS** Adans. is distinguished by its
very long, very narrow, curved, semi-patent outer phyllaries;
RHAGADIOLUS Scop. by its few, widely patent, long, narrow,
pappus-less achenes. **T. barbata** (L.) Gaertner and **R.
stellatus** (L.) Gaertner are rare bird-seed aliens from
Europe.
Key C (Cardueae) - **CNICUS benedictus** L. (Blessed Thistle),
from Europe, is a subspinose annual with single yellow
capitula partly concealed by leaves and large dimorphic
phyllaries; it used to occur as an infrequent casual. **CYNARA**
L. includes large ± thistle-like mauve-flowered vegetables
that persist for a while after neglect; **C. cardunculus** L.,
with spiny leaves, is the Cardoon, and **C. scolymus** L., with
spineless leaves, is the Globe Artichoke.
Key D (Heliantheae) - **ZINNIA** L., **HELIOPSIS** Pers. and
SANVITALIA Lam. are related ornamental American genera with
ligules persistent on the fruiting capitula. **Z. elegans**
Jacq. is an annual with red or orange ligules (often flore
pleno) and sessile upper stem-leaves; **H. scabra** Dunal is a
Helianthus-like perennial; and **S. procumbens** Lam. is a ±
procumbent annual with yellow ligules and purple disc
flowers. **SPILANTHES** Jacq. resembles a small Helianthus but
has a narrowly conical receptacle; **S. oleracea** L. is grown in
warm countries as a salad plant (Para Cress).
Key E (Anthemideae) - **SOLIVA** Ruiz Lopez & Pavon, from S
America, would key out as Cotula but has winged achenes and
sessile capitula; **S. anthemifolia** (Juss.) Sweet was formerly
an occasional wool-alien. **ONCOSIPHON** Kallersjo resembles a

ligule-less <u>Matricaria</u> with tomentose phyllaries and large
orange capitula; **O. grandiflorum** (Thunb.) Kallersjo (<u>Pentzia</u>
<u>grandiflora</u> (Thunb.) Hutch., <u>Matricaria</u> <u>grandiflora</u> (Thunb.)
Fenzl ex Harvey), from Africa, is a scarce wool-alien.
Key <u>G</u> - **BUPHTHALMUM** L. (Inuleae) resembles <u>Telekia</u> but has
dimorphic achenes and is a much smaller plant; **B.
salicifolium** L., from Europe, is an impermanent garden
escape.   **ANACYCLUS** L. (Anthemideae), from Mediterranean,
resembles <u>Anthemis</u> but the outer achenes are winged; 7 spp.
have been recorded as casuals. **MADIA** Molina and **HEMIZONIA**
DC. (<u>Spikeweed</u>) (Heliantheae), from N & C America, are coarse
unattractive annuals with receptacular scales subtending only
the outermost row of disc flowers.   In <u>Madia</u> the achenes of
ray flowers are completely enclosed by the inner phyllaries,
while in <u>Hemizonia</u> they are only 1/2-enclosed; **M. capitata**
Nutt., **M. glomerata** Hook. and **H. pungens** (Hook. & Arn.)
Torrey & A. Gray are rare casuals.
Key <u>H</u> (Anthemideae) - **ISMELIA** Cass., from Africa, resembles
<u>Chrysanthemum</u> but has flattened (and winged) disc-achenes and
keeled phyllaries; **I. carinata** (Schousboe) Schultz-Bip.
(<u>Chrysanthemum</u> <u>carinatum</u> Schousboe) is similar to <u>C.</u>
<u>coronarium</u>, with yellow ligules darker or lighter near base,
but is a rare casual. **DENDRANTHEMA** (DC.) Des Moul., from the
Orient, is the familiar <u>Florist's</u> <u>Chrysanthemum</u>, occasionally
found on tips; several spp. and hybrids are involved.

SUBFAMILY 1 - LACTUCOIDEAE (Cichorioideae, Liguliflorae)
(tribes 1-3).   Plants often producing white latex; stem-
leaves usually spiral ('alternate'), sometimes 0; capitula
ligulate, discoid or rarely (Arctotideae) radiate; tubular
flowers (if present) usually with long narrow lobes, usually
blue to red, rarely yellow; filaments joining anthers on
back; 2 style branches usually each with 1 broad stigmatic
surface on inner face; pollen grains ridged, spiny or both.

TRIBE 1 - CARDUEAE (Cynareae, Echinopeae, Carlineae) (genera
1-12).   Plant not producing white latex, often spiny;
capitula discoid, with the outermost florets often longer and
with larger lobes and hence pseudo-radiate, usually red to
blue (or white), rarely yellow.

1. ECHINOPS L. - <u>Globe-thistles</u>
Biennials to perennials; leaves deeply pinnately lobed,
white- to grey-tomentose on lowerside, with teeth ending in
rather weak spines; capitula each with 1 flower, many
arranged in compact, globose heads $\geq$2.5cm; phyllaries in c.3
rows, herbaceous, long-pointed and with long fine teeth, with
a zone of bristles outside them; corolla white to blue;
pappus of partially fused scale-like bristles.

1 Phyllaries strongly recurved at tip; glandular

```
hairs 0                                      2. E. exaltatus
1 Phyllaries erect or very slightly curved at tip;
  glandular hairs present at least on leaf upperside      2
  2 Phyllaries without glandular hairs; corolla
    bluish; plant rarely >1.25m              3. E. bannaticus
  2 Phyllaries with abundant glandular hairs; corolla
    greyish; plant often >1.5m           1. E. sphaerocephalus
```

Other spp. - Records of **E. ritro** L., from S Europe, are errors for E. bannaticus.

1. E. sphaerocephalus L. - <u>Glandular</u> Globe-thistle. Stems erect, to 2.5m, glandular-pubescent above; leaves glandular-pubescent on upperside; phyllaries with dense stalked glands; corolla white to greyish. Intrd; grown in gardens, natd in waste places and rough ground, on road- and railway-banks; scattered throughout much of Br, probably over-recorded for other 2 spp.; S Europe.

2. E. exaltatus Schrader (<u>E.</u> commutatus Juratzka) - <u>Globe-thistle</u>. Stems erect, to 2.5m, with eglandular hairs; leaves with only eglandular hairs; phyllaries without stalked glands, strongly rcurved at apex; corolla white to greyish. Intrd; grown and natd as for <u>E. sphaerocephalus</u>, but commoner; E & SE Europe.

3. E. bannaticus Rochel ex Schrader - <u>Blue</u> Globe-thistle. Stems erect, to 1.25m, with eglandular hairs; leaves glandular-pubescent on upperside; phyllaries without stalked glands; corolla bluish. Intrd; grown and natd as for <u>E. sphaerocephalus</u>, the commonest sp.; SE Europe.

2. CARLINA L. - <u>Carline</u> Thistle
Biennials; leaves pinnately lobed, very spiny; outer phyllaries ± leaf-like, innermost linear, entire, scarious, patent and ligule-like in dry weather, straw-yellow on upperside; corolla purple; pappus of feathery hairs often variously united proximally.

1. C. vulgaris L. - <u>Carline</u> Thistle. Stems erect, to 60cm, not spiny; leaves very spiny, ± glabrous to cottony-pubescent; capitula 1-6 in terminal corymb, c.1.5-3cm across when phyllaries patent. Native; open grassland, on usually calcareous but sometimes sandy soils, fixed dunes and cliff-tops; frequent in suitable places throughout most BI, only coastal in N Ir and Sc and absent from Outer Isles.

3. ARCTIUM L. - <u>Burdocks</u>
Biennials; leaves simple, entire to remotely denticulate, the lower ovate-cordate and petiolate, not spiny; phyllaries very numerous, stiff, subulate, strongly hooked at apex, spreading to form subglobose head; corolla purple, rarely white; pappus of rough yellowish free hairs.

Other spp. – **A. tomentosum** Miller, from Europe, is a rare casual; it differs from A. lappa in its smaller capitula with dense web-like pubescence.

1. A. lappa L. – <u>Greater</u> <u>Burdock</u>. Stems erect, well branched, to 1.5m; petioles solid; capitula of main branches in subcorymbose clusters, with peduncles 3-10cm, 35-42mm across (apex to apex of phyllaries) and open at apex in fruit; phyllaries usually glabrous. Native; waysides, field-borders, wood-clearings, waste places; rather scattered in S & C Br, very scattered in Ir.

1 x 2. A. lappa x A. minus = **A. x nothum** (Ruhmer) J. Weiss (A. x debrayi Senay) occurs infrequently in S & C Br with the parents; it is intermediate and fertile. A. minus ssp. pubens might be derived from this cross.

2. A. minus (Hill) Bernh. – <u>Lesser</u> <u>Burdock</u>. Differs from A. lappa in petioles of at least lower leaves hollow; capitula in racemose or spikiform clusters, with peduncles 0-4(10)cm, 15-35mm across in fruit; phyllaries frequently pubescent. Native. 3 sspp. may be recognized, but all intermediates occur.

1 Capitula 15-20 x 15-25mm (apex to apex of phyllaries) in fruit; florets exceeding phyllaries **c. ssp. minus**
1 Capitula 20-25 x 30-35mm in fruit; florets c. as long as phyllaries **2**
  2 Peduncles 1-4(10)cm; capitula open at apex (phyllaries parted) in fruit **a. ssp. pubens**
  2 Peduncles 0-1cm; capitula closed at apex (phyllaries erect) in fruit **b. ssp. nemorosum**

a. Ssp. pubens (Bab.) Arenes (A. pubens Bab.). Capitula 20-25 x 30-35mm and open at apex in fruit; peduncles 1-4(10)cm. Similar places to A. lappa; scattered in BI but distribution uncertain.

b. Ssp. nemorosum (Lej.) Syme (A. nemorosum Lej.). Capitula 20-25 x 30-35mm and closed at apex in fruit; peduncles 0-1cm. Similar places to A. lappa; commonest taxon in N Br and N Ir, apparently absent from SW Br.

c. Ssp. minus. Capitula 15-20 x 15-25mm and closed at apex in fruit; peduncles 0-1cm. Similar places to A. lappa but also in woods and shady places where other taxa seldom occur; commonest taxon in C & S Br, C & S Ir and CI, apparently absent from Sc, N En and N Ir.

4. SAUSSUREA DC. – <u>Alpine</u> <u>Saw-wort</u>
Perennials; leaves simple, subentire to denticulate, not spiny; phyllaries in many rows, simple, entire, rounded to obtuse at apex; corolla purple; pappus of feathery hairs in 1 row, often ± united proximally, with an outer row of simple shorter hairs, the former often deciduous.

1. S. alpina (L.) DC. – <u>Alpine</u> <u>Saw-wort</u>. Stems erect, to **813** 45cm; leaves ovate or elliptic to narrowly so, the basal and lower ones petiolate, densely white-pubescent on lowerside; capitula 15-20mm, 1-several in ± dense terminal cluster. Native; mountain cliffs and scree, maritime in N Sc; local in N En, N Wa, N & W Sc, scattered in Ir.

5. CARDUUS L. - Thistles

Annuals or biennials; stems with spiny wings at least in part; leaves variously lobed, sharply and densely spiny; phyllaries simple, linear-subulate to narrowly ovate, spine-tipped, in many rows; corolla purple (or white); pappus of many rows of simple hairs united proximally.

1 Capitula subcylindrical, <14mm across (excl. flowers); corolla with 5 ± equal lobes   2
1 Capitula globose to campanulate, >14mm across (excl. flowers); corolla with 1 lobe distinctly more deeply delimited than other 4   3
   2 Capitula in clusters of 3-10; stems with spiny wings right up to base of capitula; phyllaries thin and transparent on margins, without strongly thickened midrib except sometimes near apex
                       1. C. tenuiflorus
   2 Capitula in clusters of 1-3; stems with discontinuous spiny wings, with at least some peduncles unwinged distally; phyllaries with strongly thickened margins and midrib for at least distal 1/2   **2. C. pycnocephalus**
3 Capitula 15-25(30)mm across (excl. flowers), in clusters of (1)2-4(5), ± erect; phyllaries linear-subulate, not narrowed just above base; corolla 12-15mm   **3. C. crispus**
3 Capitula (20)30-60mm across (excl. flowers), usually solitary, pendent; phyllaries lanceolate, narrowed just above base; corolla 15-25mm   **4. C. nutans**

Other spp. - **C. acanthoides** L., from C Europe, is the name that has often been given to C. crispus ssp. multiflorus; it differs in its longer-spined (c.5mm, not c.2-3mm) stem-wings, web-like pubescent (not subglabrous) stems, and usually larger (25-35mm across) solitary capitula. It might occur as a rare casual, but most plants that resemble it are extreme variants of C. crispus. **C. macrocephalus** Desf. (Giant Thistle), from C & E Mediterranean, is grown in gardens and was formerly natd in S En; it differs from C. nutans in its more densely pubescent stems to 2m, longer-stalked capitula, and shorter phyllaries with prominent (not obscure) midrib.

1. C. tenuiflorus Curtis - Slender Thistle. Stems erect, to **813** 60(80)cm, continuously spiny-winged up to capitula; leaves grey- to white-cottony on lowerside; capitula ± sessile, in clusters of (1)3-10, 12-18 x 5-10mm (excl. flowers), ± erect; corollas mostly shorter than phyllaries. Native; waysides, rough and open ground; locally frequent by coasts in BI except most of N & W Sc, very scattered inland.

2. C. pycnocephalus L. - <u>Plymouth Thistle</u>. Differs from <u>C.</u> 813 <u>tenuiflorus</u> in stems and leaves more densely white-cottony; capitula subsessile to stalked, in clusters of 1-3, 14-20 x 7-12mm (excl. flowers); corollas mostly longer than phyllaries; and see key. Intrd; natd on open limestone cliff at Plymouth (S Devon) since 1868, infrequent casual elsewhere from wool and birdseed; S Europe.

3. C. crispus L. (<u>C.</u> <u>acanthoides</u> auct. non L.) - <u>Welted</u> <u>Thistle</u>. Stems erect, to 1.5m, continuously spiny-winged up to or nearly up to capitula; leaves subglabrous on lowerside; capitula mostly stalked, in clusters of (1)2-4(5), 9-15 x 15-20(25)mm (excl. flowers), campanulate, ± erect. Native; hedgerows, ditch- and steam-sides, rough ground, especially on rich or basic soils; frequent in Br N to C Sc, scattered in Ir and W & N Sc. Our plant is ssp. **multiflorus** (Gaudin) Franco.

3 x 4. C. crispus x C. nutans = C. x dubius Balbis (<u>C.</u> x <u>polyacanthus</u> Schleicher non L., <u>C.</u> x <u>orthocephalus</u> auct. non Wallr.) is found scattered in Br with the parents N to Yorks; it is partially fertile and intermediate in capitulum characters.

4. C. nutans L. - <u>Musk Thistle</u>. Stems erect, to 1m, discontinuously spiny-winged with peduncles unwinged distally; leaves pubescent on lowerside; capitula stalked, mostly solitary, 16-30 x 20-60mm (excl. flowers), depressed-globose, pendent, with strongly reflexed outer phyllaries. Native; grassy and bare places, waysides and rough ground mostly on calcareous soils; locally frequent in CI and Br N to S Sc, rare elsewhere in Sc and in Ir, sometimes casual.

6. CIRSIUM Miller - <u>Thistles</u>

Biennials to perennials; stems with or without spiny wings; leaves denticulate to deeply lobed, spiny or at least with bristle-pointed teeth; phyllaries simple, ovate to linear-subulate, spine-tipped to mucronate or acuminate at apex, in many rows; corolla purple (or white), rarely yellow; pappus of many rows of feathery hairs united proximally.

```
1 Leaves with rigid bristles on upperside                        2
1 Leaves glabrous or with soft hairs on upperside                3
    2 Stem with discontinuous spiny wings        2. C. vulgare
    2 Stem not winged                            1. C. eriophorum
3 Flowers yellow                                                 4
3 Flowers purple, rarely white                                   5
    4 Upper part of stem with only very reduced, distant
      leaves; capitula pendent                    5. C. erisithales
    4 Upper part of stem with large yellowish-green
      leaves exceeding the erect capitula
                                                  7. C. oleraceum
5 Stems continuously spiny-winged; biennial with tap-
```

root   **9. C. palustre**
5 Stems not winged or with very short wings below each
  leaf; perennial with at least short rhizomes   6
  6 Distal broad part of corolla c.1/2 as long as
    proximal narrow part, lobed >3/4 way to base; stem
    usually well branched   **10. C. arvense**
  6 Distal broad part of corolla c. as long as proximal
    narrow part, lobed c.1/2 way to base; stem usually
    unbranched   7
7 Stem 0-10cm, or if up to 30cm then with >1 well-
  developed leaf near top   **8. C. acaule**
7 Stem >10cm, with only distant much reduced leaves
  in distal 1/4   8
  8 Stem-leaves widened proximally to broad base
    clasping stem; capitula mostly >20mm to tip of
    uppermost phyllaries; mostly N En and Sc
                                            6. C. heterophyllum
  8 Stem-leaves narrowed to base, not or scarcely
    clasping stem; capitula mostly <20mm to tip of
    uppermost phyllaries; mostly S En, Wa and Ir   9
9 Lower stem-leaves deeply lobed, the lobes deeply
  lobed, green on lowerside; some roots swollen into
  tubers   **4. C. tuberosum**
9 Lower stem-leaves not or rather shallowly lobed, the
  lobes scarcely or not lobed, white on lowerside;
  roots not swollen   **3. C. dissectum**

1. C. eriophorum (L.) Scop. - <u>Woolly</u> <u>Thistle</u>.   Erect
biennial to 1.5m; stems leafy to top, not winged; leaves
deeply lobed, very strongly spiny; capitula 3-5 x 4-7cm
(excl. flowers); flowers purple.   Native; dry grassland,
scrub and banks on calcareous soil; locally frequent in Br N
to Co Durham.
  1 x 2. C. eriophorum x C. vulgare = C. x grandiflorum Kittel
(<u>C.</u> x <u>gerhardtii</u> Schultz-Bip.) has occurred rarely in En N to
SE Yorks; it is intermediate in stem wingedness and capitulum
characters and partially fertile.
  2. C. vulgare (Savi) Ten. - <u>Spear</u> <u>Thistle</u>.   Erect biennial
to 1.5m; stems leafy to top, with spiny wings; leaves deeply
lobed, very strongly spiny; capitula 2.5-4 x 2-5cm (excl.
flowers); flowers purple.   Native; grassland, waysides,
cultivated, rough and waste ground; common throughout BI.
  2 x 8. C. vulgare x C. acaule = C. x sabaudum Loehr has
occurred rarely in En N to N Lincs; it has stems to c.30cm
with short spiny wings below each leaf, a few bristles on the
leaf uppersides, intermediate capitula and is slightly
fertile.
  2 x 9. C. vulgare x C. palustre = C. x subspinuligerum
Peterm. was found in Merioneth in 1986.
  3. C. dissectum (L.) Hill - <u>Meadow</u> <u>Thistle</u>.   Erect perennial
to 80cm; stems with only very reduced leaves in upper part,
not winged; leaves scarcely to rather deeply lobed, softly

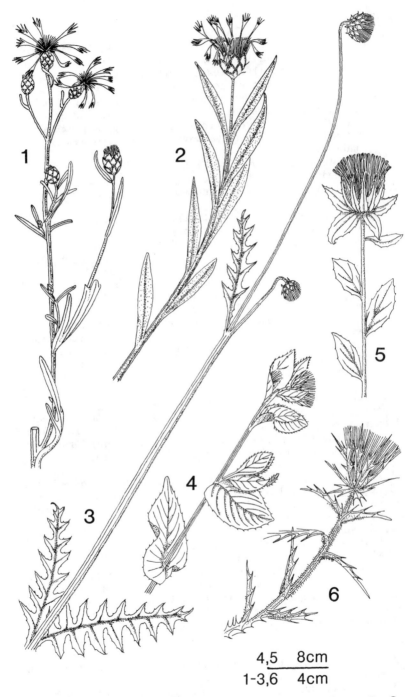

FIG 808 – Asteraceae: Cardueae. 1, Centaurea diluta. 2, C. montana. 3, Cirsium erisithales. 4, C. oleraceum. 5, Carthamus tinctorius. 6, C. lanatus.

4,5   8cm
1-3,6   4cm

spiny, white-pubescent on lowerside; capitula 1.3-2.5 x 1.2-3cm (excl. flowers); flowers purple. Native; fens, bogs, wet fields on peaty soil; local in En and Wa N to NE Yorks and Merioneth, throughout Ir, very local in SW Sc.

3 x 8. C. dissectum x C. acaule = C. woodwardii (H. Watson) Nyman once occurred near Swindon, N Wilts.

3 x 9. C. dissectum x C. palustre = C. x forsteri (Smith) Loudon is not uncommon with the parents throughout the range of C. dissectum and is the commonest hybrid thistle; it has discontinuous spiny-winged, cottony-pubescent stems, and intermediate leaves and capitula.

4. C. tuberosum (L.) All. - <u>Tuberous</u> Thistle. Erect **RR** perennial to 60cm; stems with only very reduced leaves in upper part, not winged; leaves very deeply lobed, softly spiny; capitula ± as in C. dissectum. Native; dry calcareous grassland; very local in N & S Wilts, Glam and Cambs (extinct in 1974, re-introduced from local stock 1987).

4 x 8. C. tuberosum x C. acaule = C. x medium All. (C. x zizianum Koch) occurs in the Wilts and Glam areas of C. tuberosum; it has branched stems to 60cm with both jointed (from C. acaule) and web-like (from C. tuberosum) hairs. It is female-fertile and backcrosses to C. tuberosum in Wilts.

4 x 9. C. tuberosum x C. palustre = C. x semidecurrens H. Richter occurs with C. tuberosum in Glam and formerly in S Wilts; it is intermediate in leaf, stem and capitulum characters and sterile.

5. C. erisithales (Jacq.) Scop. - <u>Yellow</u> Thistle. Erect **808** perennial to 1m; stems with only very reduced leaves in upper part, not winged; leaves very deeply lobed, softly spiny; capitula pendent, 1.3-2 x 1.5-3cm (excl. flowers); flowers yellow. Intrd; natd in disused quarry in N Somerset since 1980; S Europe.

6. C. heterophyllum (L.) Hill (C. helenioides auct. non (L.) Hill) - <u>Melancholy</u> Thistle. Erect perennial to 1.2m; stems with only reduced leaves in uppermost part, not winged; leaves not to quite deeply lobed, softly spiny, white-pubescent on lowerside; capitula 2-3 x 2-3.5cm (excl. flowers); flowers purple. Native; grassland, scrub, open woodland and stream-sides in hilly country; locally common in Br N from Derbys and Rads, Fermanagh and Co Leitrim.

6 x 9. C. heterophyllum x C. palustre = C. x wankelii Reichardt occurs with the parents in C & N Sc; it is intermediate in stem, leaf and capitulum characters.

7. C. oleraceum (L.) Scop. - <u>Cabbage</u> Thistle. Erect **808** perennial to 1.5m; stems leafy throughout, not winged; leaves slightly to deeply lobed, softly spiny, the uppermost yellowish; capitula 1.2-2.5 x 1.5-3cm (excl. flowers); flowers yellow. Intrd; natd since at least 1912 in marshes by R. Tay, E Perth, and by stream in S Lancs since c.1978, rare and impermanent elsewhere; Europe.

8. C. acaule (L.) Scop. - <u>Dwarf</u> Thistle. Perennial with basal leaf-rosette and 1-few capitula sessile in centre or on

wingless leafy stems to 10(30)cm; leaves rather deeply lobed,
strongly spiny; capitula 2-3 x 1.2-2.5cm (excl. flowers);
flowers purple. Native; short, base-rich grassland; locally
frequent in Br N to NE Yorks, Alderney, formerly Jersey.
  8 x 9. C. acaule x C. palustre = C. x kirschlegeri Schultz-
Bip. has been found with the parents in S En; it has shortly
spiny-winged stems to 40cm and intermediate leaves and
capitula.
  8 x 10. C. acaule x C. arvense = C. x boulayi Camus has
occurred for certain only in N Essex but other records exist
in En and Wa; it has branched stems to 60cm with leaves like
those of C. arvense, but is intermediate in capitulum and
pubescence characters.
  9. C. palustre (L.) Scop. - Marsh Thistle. Erect biennial
to 2m; stems leafy to top, with very spiny wings; leaves
rather deeply lobed, strongly spiny; capitula often crowded,
0.8-1.5 x 0.7-1.2cm (excl. flowers); flowers purple. Native;
marshes, damp grassland and open woods, ditch-sides; common
throughout BI.
  9 x 10. C. palustre x C. arvense = C. x celakovskianum Knaf
is very scattered in Br and Ir; it has stems winged below but
scarcely so above, and intermediate leaves, capitula and
corollas.
  10. C. arvense (L.) Scop. - Creeping Thistle. Perennial
with long rhizomes and erect stems to 1.2m; stems leafy to
top, not winged; leaves scarcely to rather deeply lobed,
usually strongly spiny; capitula 1-2.2 x 0.8-2cm (excl.
flowers); flowers purple. Native; grassland, hedgerows,
arable, waste and rough ground; common throughout BI. Var.
incanum (Fischer) Ledeb., from S Europe, has leaves
greyish-white pubescent on lowerside, and occurs as a casual.

7. ONOPORDUM L. - Cotton Thistle
Biennials; stems with spiny wings; leaves dentate to
shallowly lobed, strongly spiny; phyllaries simple,
linear-lanceolate, strongly spine-tipped, in many rows;
corolla purple (rarely white); pappus of many rows of simple
hairs united proximally.

  1. O. acanthium L. - Cotton Thistle. Stems erect, to 2.5m;
stems, leaves and phyllaries greyish-white with cottony
hairs; capitula 2-5 x 2.5-6cm (excl. flowers). Intrd; natd
or casual since at least 16th century in fields, marginal
habitats, waste and rough ground; locally frequent in En
(especially SE and E Anglia, possibly native in latter), very
scattered in Wa, Sc and CI; Europe.

8. SILYBUM Adans. - Milk Thistle
Annuals to biennials; stems not spiny; leaves shallowly to
rather deeply lobed, strongly spiny; outer phyllaries with
spine-tipped lateral lobes or teeth and strong apical spine,
in many rows; corolla purple (rarely white); pappus of many

rows of simple hairs united proximally.

1. S. marianum (L.) Gaertner - <u>Milk</u> <u>Thistle</u>. Stems erect, to 1m; leaves bright green, usually veined or marbled with white; capitula 2.5-4 x 5-14cm (excl. flowers). Intrd; frequent casual, sometimes natd, in waste and rough ground since at least 17th century, now often a wool- or bird-seed-alien; scattered throughout BI N to E Ross.

9. SERRATULA L. - <u>Saw-wort</u>
Perennials; stems and leaves not spiny; lower leaves usually lobed almost to midrib; phyllaries simple, acute to acuminate, not spiny, in many rows; corolla purple (rarely white); pappus of many rows of free, simple hairs, the outermost much shorter than the inner ones.

1. S. tinctoria L. - <u>Saw-wort</u>. Stems erect, to 70(100)cm, **813** but often very short (<10cm) in very exposed places, usually well branched above; leaves dark green, subglabrous, with bristle-tipped teeth. Native; grassland, scrub, open woodland, cliff-tops and rocky stream-sides on well-drained soils; local in Br N to SW Sc, Jersey, formerly Co Wexford.

10. ACROPTILON Cass. - <u>Russian</u> <u>Knapweed</u>
Perennials; stems and leaves not spiny; leaves simple, entire to narrowly lobed; phyllaries with conspicuous broad apical scarious border, in many rows; corolla pink, the outermost no longer than the inner; pappus of hairs, soon falling.

1. A. repens (L.) DC. (<u>Centaurea</u> <u>repens</u> L.) - <u>Russian</u> **813** <u>Knapweed</u>. Stems erect, well branched, to 70cm; leaves lanceolate to oblanceolate, densely grey-pubescent; phyllaries pale brown with paler, puberulous, scarious border at apex, becoming jagged later. Intrd; natd on waste ground at Hereford, Herefs, since 1959; Russia.

11. CENTAUREA L. - <u>Knapweeds</u>
Annuals to perennials; stems and leaves not spiny; leaves simple and entire to ± pinnate; phyllaries with distinct apical portion which is scarious, toothed or spiny, in many rows; corolla purple to pink or blue, white or yellow, that of outermost flowers often much longer than that of inner flowers (pseudo-radiate); pappus 0, or of many rows of simple to toothed free hairs, sometimes also with some scales.

1	Apical portion of phyllaries with >1 sharp spines	2
1	Apical portion of phyllaries merely scarious or variously toothed, not spiny	6
	2 Flowers yellow; stems strongly winged	3
	2 Flowers purple, rarely white; stems not winged	4
3	Apical phyllary spines <10mm, with lateral spines arranged pinnately along its proximal 1/2; corolla	

    with minute sessile glands   **7. C. melitensis**
3 At least some apical phyllary spines >10mm, with
    lateral spines arranged palmately at its base;
    corolla not glandular   **6. C. solstitialis**
    4 Apical portion of phyllaries scarious, variously
      toothed, with 1(-few) terminal spines   **8. C. diluta**
    4 Whole of apical portion of phyllaries modified
      into spines   5
5 Apical phyllary spine >10mm, >3x as long as longest
    laterals   **4. C. calcitrapa**
5 Phyllary spines 3-5, subequal, >5mm, the apical one
    <1.5x as long as laterals   **5. C. aspera**
    6 Apical portion of phyllaries with 1(-few) distinct
      terminal subspinose teeth distinctly different
      from lateral teeth   **8. C. diluta**
    6 Apical portion of phyllaries similarly toothed at
      apex and sides   7
7 Apical portion of phyllaries strongly delimited from
    basal portion, with slight constriction between   8
7 Apical portion of phyllaries ill-delimited from basal
    portion, the former decurrent down sides of latter   9
    8 Apical portion of outer phyllaries dark brown to
      black, deeply and very regularly toothed; pappus
      present or absent   **10. C. nigra**
    8 Apical portion of outer phyllaries pale to dark
      brown, irregularly toothed or deeply jagged;
      pappus absent or ± so   **9. C. x moncktonii**
9 Easily uprooted annual; basal leaves 0 at flowering
    time   **3. C. cyanus**
9 Deeply rooted perennial; basal leaves or non-flowering
    shoots present at flowering time   10
    10 Leaves simple, entire, strongly decurrent and
       forming distinct wings on stem; strongly
       rhizomatous; flowers usually blue   **2. C. montana**
    10 Leaves usually deeply lobed, rarely simple and
       entire, not decurrent on stem; scarcely
       rhizomatous; flowers usually reddish-purple
                    1. C. scabiosa

Other spp. - 2 European aliens formerly natd now seem to be
extinct. **C. jacea** L. (<u>Brown</u> <u>Knapweed</u>) differs from <u>C. nigra</u> 813
in its pale brown irregularly jagged apical portion of the
phyllaries; it formerly occurred in grassy places in S En and
CI, but now only exists as its hybrid <u>C</u> x <u>moncktonii</u> (q.v.).
C. paniculata L. (<u>Jersey</u> <u>Knapweed</u>) was natd in Jersey from 813
1851 to 1981; it differs from <u>C. jacea</u> and <u>C. nigra</u> in its
much smaller capitula (3-8mm across at level of phyllaries,
not >10mm across), phyllaries with pale brown regularly
toothed apical portion, and lower leaves very deeply divided
into very narrow lobes.

1. C. scabiosa L. - <u>Greater</u> <u>Knapweed</u>.   Erect perennial to 813

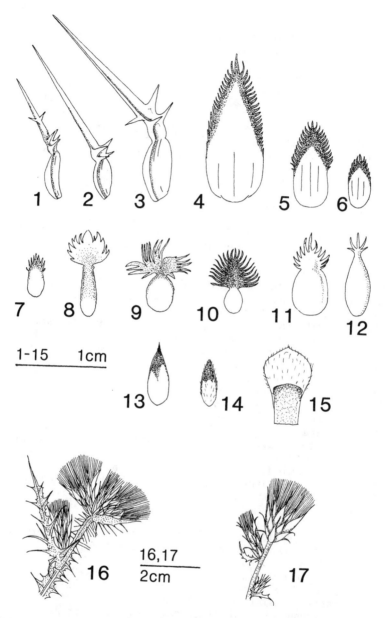

FIG 813 — Asteraceae: Cardueae. 1–15, phyllaries. 1, Centaurea melitensis. 2, C. solstitialis. 3, C. calcitrapa. 4, C. montana. 5, C. scabiosa. 6, C. cyanus. 7, C. paniculata. 8, C. jacea. 9, C. x moncktonii. 10, C. nigra. 11, C. diluta. 12, C. aspera. 13, Serratula. 14, Saussurea. 15, Acroptilon. 16–17, terminal capitula of Carduus. 16, C. tenuiflorus. 17, C. pycnocephalus.

1.2m; leaves mostly very deeply lobed, rarely all or most
simple (var. **succisiifolia** E. Marshall); phyllaries with
blackish-brown, decurrent, deeply toothed apical portion;
capitula pseudo-radiate; corolla usually reddish-purple.
Native; grassland, rough ground, cliffs and waysides mainly
on calcareous soils; locally common in BI N to N En and C Ir,
very local in mainland Sc.

2. **C. montana** L. - Perennial Cornflower. Erect rhizomatous **808**
perennial to 80cm; leaves simple, entire; phyllaries with **813**
blackish-brown, decurrent, deeply toothed apical portion;
capitula pseudo-radiate; corolla usually blue; flowers with
strong sweet scent. Intrd; much grown in gardens, natd in
grassy places and rough ground throughout most of Br,
especially C & N; mountains of C & S Europe.

3. **C. cyanus** L. - Cornflower. Erect annual to 80cm; leaves **813**
linear-lanceolate, the lower with few pinnate lobes, the **R**
upper entire; phyllaries with blackish-brown, decurrent,
deeply toothed apical portion; capitula pseudo-radiate;
corolla usually blue, sometimes pink or white. Native;
traditionally natd in cornfields, now mostly casual bird-
seed-alien or garden escape in waste places; scattered
throughout BI, much scarcer than formerly.

4. **C. calcitrapa** L. - Red Star-thistle. Erect to spreading **813**
biennial to 60cm; leaves shallowly to very deeply lobed;
phyllaries with apical portion of 1 terminal strong spine
>1cm and several much shorter laterals near its base;
capitula ± discoid; corolla reddish-purple. Intrd; waste and
rough ground, waysides, on well-drained soils; natd in very
scattered places in S En, decreasing, casual from wool,
bird-seed and other sources elsewhere; Europe.

5. **C. aspera** L. - Rough Star-thistle. Erect to spreading **813**
perennial to 80cm; leaves pinnately lobed to dentate;
phyllaries with apical portion of 3-5 subequal spines <5mm;
capitula ± discoid; corolla reddish-purple. Intrd; natd on
maritime dunes in Jersey since at least 1839, Guernsey since
1788, casual or ± natd in few places in S Br; Europe.

6. **C. solstitialis** L. - Yellow Star-thistle. Erect to **813**
spreading annual to biennial to 60cm; leaves pinnately lobed
to dentate; phyllaries with apical portion of 1 terminal
strong spine >1cm and several much shorter laterals at its
base; capitula ± discoid; corolla yellow. Intrd; formerly
natd in arable fields (especially of lucerne), now more
usually casual of waste ground mostly from wool and bird-
seed; scattered mainly in C & S Br; S Europe.

7. **C. melitensis** L. - Maltese Star-thistle. Differs from C. **813**
solstitialis in key characters. Intrd; casual of tips and
waste ground mainly from bird-seed and wool; scattered mainly
in S & C Br; S Europe.

8. **C. diluta** Aiton - Lesser Star-thistle. Erect perennial **808**
(often behaving as annual) to 80cm; leaves shallowly lobed to **813**
dentate; phyllaries with slightly decurrent pale brown apical
portion with small lateral teeth and 1(-few) weak terminal

spines; capitula pseudo-radiate; corolla pinkish-purple.
Intrd; usually casual on waste ground and tips, mainly from
bird-seed; scattered mainly in S & C Br; W Mediterranean.

9. C. x moncktonii C. Britton (C. x drucei C. Britton; C. 813
jacea x C. nigra) - Hybrid Knapweed. Erect perennial to
60cm; leaves entire to medium-lobed; phyllaries with pale to
dark brown irregularly toothed or deeply jagged apical
portion; capitula usually pseudo-radiate; corolla reddish-
purple. Native, from crosses between C. nigra and formerly
natd C. jacea; grassy places; very scattered in S En and CI,
formerly more widespread. Evidence of C. jacea genes in
plants of C. nigra: apical portion of phyllaries pale brown,
shiny, irregularly toothed; outer flowers ray-like, pale in
colour; pappus irregular or 0.

10. C. nigra L. (C. nemoralis Jordan, C. nigra ssp. 813
nemoralis (Jordan) Gremli, C. debeauxii Gren. & Godron ssp.
nemoralis (Jordan) Dostal) - Common Knapweed. Erect
perennial to 1m; leaves entire to deeply lobed; phyllaries
with dark brown to black finely and very regularly toothed
apical portion; capitula usually (?always) discoid; corolla
reddish-purple. Native; grassy places, rough ground,
waysides; common throughout BI. British botanists have been
unable to maintain a distinction between C. nigra and **C.
nemoralis**, which has narrower leaves, slenderer branches,
smaller capitula, phyllaries with paler apical portions, and
shorter pappus.

12. CARTHAMUS L. - Safflowers
Annuals; leaves entire or with acute lobes and apex tipped
with spines or bristles; phyllaries in many rows, the outer
leaf-like, the inner usually spine-tipped; corolla yellow to
orange; pappus 0 or of narrow pointed scales.

1. C. tinctorius L. - Safflower. Stems erect, glabrous, to 808
60cm; leaves and outer phyllaries entire to shallowly lobed
with bristly or softly spine-tipped lobes; corolla yellow to
reddish-orange; pappus 0. Intrd; tips, waste and rough
ground, casual from bird-seed; rather frequent in S Br, very
scattered further N; SW Asia.

2. C. lanatus L. - Downy Safflower. Stems erect, densely 808
pubescent, to 60cm; leaves and outer phyllaries with apex and
usually deep lobes with strong terminal spines; corolla
yellow; pappus c. as long as achene. Intrd; similar habitats
and distribution to C. tinctorius but mostly a wool-alien; S
Europe.

TRIBE 2 - LACTUCEAE (Cichorieae) (genera 13-31). Plants
producing white latex, rarely spiny; capitula ligulate, the
flowers all bisexual and ligulate, usually with 5-toothed
ligules, usually yellow.

13. SCOLYMUS L. - <u>Golden</u> <u>Thistle</u>
Thistle-like (annuals,) biennials or perennials; rhizomes and
stolons 0; stems leafy, with several capitula; phyllaries in
several rows; receptacular scales present, wrapped round
achenes; pappus (0 or) of a few rigid hairs; ligules yellow;
achenes flattened, not beaked.

Other spp. - **S. maculatus** L. is an annual with white-
bordered leaves and pappus 0; it is a rare casual.

1. S. hispanicus L. - <u>Golden</u> <u>Thistle</u>. Stems erect, to 80cm, **820**
pubescent; leaves linear to ovate, lobed, with strong rigid
spines; capitula 2.5-4cm across. Intrd; casual in waste
places as grain-, bird-seed- or shoddy-alien; very scattered
in S En and Wa; S Europe. Resembles a thistle but flowers
all ligulate and stems with latex.

14. CICHORIUM L. - <u>Chicory</u>
Perennials; rhizomes and stolons 0; stems leafy, with many
capitula; phyllaries in 2 rows; receptacular scales present,
small; pappus of short scales; ligules blue, rarely white;
achenes not flattened, somewhat angular, not beaked.

1. C. intybus L. - <u>Chicory</u>. Stems stiff, procumbent to
erect, glabrous to pubescent, to 1m; lower leaves oblanceolate, deeply lobed to toothed; capitula 2.5-4cm across.
Possibly native; roadsides, rough grassland, waste places;
locally common, especially on calcareous soils, in C & S Br,
scattered elsewhere in BI. Most, if not all, plants are of
the cultivated variant, sometimes separated as ssp. **sativum**
(Bisch.) Janchen.

15. ARNOSERIS Gaertner - <u>Lamb's</u> <u>Succory</u>
Annuals; stems leafless, with small scale-like bracts,
branched or not; phyllaries in 2 rows, the outer small and
incomplete; receptacular scales 0; pappus 0; ligules yellow;
achenes somewhat flattened, c.5-ribbed, not beaked.

1. A. minima (L.) Schweigger & Koerte - <u>Lamb's</u> <u>Succory</u>. **RR**
Stems erect, to 30cm, ± glabrous, conspicuously dilated some
way below capitula; leaves all basal, narrowly obovate to
oblanceolate, lobed to toothed; capitula 7-12mm across.
Probably native; sandy arable fields; formerly very local in
E En from Dorset and E Kent to SE Yorks, and in E Sc, extinct
since 1971.

16. LAPSANA L. - <u>Nipplewort</u>
Annuals to perennials; rhizomes and stolons 0; stems leafy,
with many capitula; phyllaries in 2 rows, the outer small and
incomplete; receptacular scales 0; pappus 0; ligules yellow;
achenes slightly flattened, c.20-ribbed, not beaked.

1. L. communis L. - <u>Nipplewort</u>. Stems erect, to 1m, pubescent at least below; lower leaves narrowly obovate to oblanceolate, ± pinnate with lateral leaflets much smaller than terminal one.

a. Ssp. communis. Annual; upper stem-leaves lanceolate to ovate or rhombic, usually well toothed; capitula 1.5-2cm across; ligules c.1.5x as long as inner phyllaries; ripe achenes c.1/2 as long as phyllaries. Native; open woods, hedgerows, waste and rough ground; common throughout BI.

b. Ssp. intermedia (M. Bieb.) Hayek (<u>L. intermedia</u> M. Bieb.). Usually perennial; upper stem-leaves linear to linear-lanceolate, entire or slightly toothed; capitula 2.5-3cm across; ligules 2-2.5x as long as inner phyllaries; ripe achenes <1/2 as long as phyllaries. Intrd; natd on chalk bank in Beds since 1945, in limestone grassland in Flints since c.1970 and Caerns since 1977, and rough grassland in Middlesex since 1982; SE Europe.

17. HEDYPNOIS Miller- <u>Scaly Hawkbit</u>
Annuals; stems leafy, with several capitula; phyllaries in 2 rows, the outer very small; receptacular scales 0; pappus of scales, those of inner achenes long and narrow, those of outer achenes short and ± fused laterally to form ring; ligules yellow; achenes not flattened, ribbed, not beaked.

1. H. cretica (L.) Dum.-Cours. - <u>Scaly Hawkbit</u>. Stems erect **820** to decumbent, pubescent, to 45cm, conspicuously dilated below capitula at fruiting; leaves narrowly obovate to oblanceolate, entire to lobed; capitula 5-15mm across. Intrd; casual in waste places, from bird-seed; occasional in En; S Europe.

18. HYPOCHAERIS L. - <u>Cat's-ears</u>
Annuals to perennials; rhizomes and stolons 0; stems usually leafless, usually with small scale-like bracts, mostly with >1 capitula; phyllaries in several rows; receptacular scales present; pappus of 1-2 rows of dirty-white to pale brown hairs, the single or inner row feathery; ligules yellow; achenes not flattened, finely ribbed, beaked or not.

1 Pappus-hairs all feathery, in 1 row; leaves usually spotted or streaked with purple; outer phyllaries usually uniformly pubescent   **3. H. maculata**
1 Pappus-hairs in 2 rows, outer usually simple, inner feathery; leaves rarely with purple markings; outer phyllaries glabrous to very sparsely or patchily pubescent   2
  2 Central achenes 8-17mm (incl. beak), beaked; marginal achenes usually beaked; capitula 2-4cm across, opening every day; ligules c.4x as long as wide   **1. H. radicata**
  2 Central achenes 6-9(13.5)mm (incl. beak), beaked or not; marginal achenes not beaked; capitula

1-1.5cm across, opening only in bright sunshine;
ligules c.2x as long as wide   **2. H. glabra**

1. H. radicata L. - Cat's-ear. Perennial; stems to 60cm, **832**
erect or ascending, glabrous or pubescent near base; leaves
oblanceolate to narrowly obovate, sinuate-lobed or -toothed,
usually pubescent; capitula 2.5-4cm across; achenes usually
all beaked, sometimes some outer ones unbeaked. Native;
grassy places in many situations; common throughout BI.
1 x 2. H. radicata x H. glabra = H. x intermedia H. Richter
occurs rarely in SE En and Merioneth, but may be overlooked;
it is intermediate in capitulum characters and <5% fertile.
The capitula open in dull and sunny weather.
2. H. glabra L. - Smooth Cat's-ear. Differs from H. **832**
radicata in stems to 40cm, glabrous; leaves usually glabrous; **R**
outer achenes beakless, inner usually beaked but sometimes
beakless; and see key. Native; grassy or open ground on
sandy soils; scattered in Br N to C Sc, frequent in E En and
CI, scattered elsewhere in Br N to C Sc but over-recorded for
small H. radicata, decreasing.
3. H. maculata L. - Spotted Cat's-ear. Perennial; stems to **832**
60cm, erect, often simple, usually pubescent + throughout; **RR**
leaves as in H. radicata but usually more shallowly lobed and
purple-marked; capitula 2.5-4.5cm across, paler yellow than
in H. radicata; achenes all beaked. Native; grassy or open
ground mostly on calcareous or sandy soils and on maritime
cliffs; very local in Br N to Westmorland, Jersey.

19. LEONTODON L. - Hawkbits
Perennials; rhizomes and stolons 0; stems leafless, usually
with small scale-like bracts, simple or branched; phyllaries
in several rows; receptacular scales 0; pappus of 1-2 rows of
pale brown to dirty-white hairs, the single or inner row
feathery, the outer row simple, or outer achenes with pappus
a scaly ring; ligules yellow; achenes not flattened, finely
ribbed, not or indistinctly beaked.

1 Leaves glabrous or with simple hairs; pappus of 1
  row of feathery hairs; stems usually branched
                                      1. L. autumnalis
1 Leaves with at least some hairs forked; pappus of 2
  rows of hairs, the inner feathery, the outer simple;
  stems simple   2
  2 All achenes with pappus of hairs; phyllaries
    (9)11-13(15)mm; whole of stem and phyllaries
    usually conspicuously pubescent   **2. L. hispidus**
  2 Outer achenes with pappus a scaly ring; phyllaries
    7-11mm; upper part of stem and phyllaries usually
    glabrous to sparsely pubescent   **3. L. saxatilis**

1. L. autumnalis L. - Autumn Hawkbit. Stems to 60cm, **832**
usually at least some branched, glabrous, or sparsely

pubescent below; leaves oblanceolate, deeply lobed to (rarely) ± entire; phyllaries 7-12mm, glabrous to pubescent; outermost ligules usually reddish on lowerside. Native; grassy places in very many situations; very common throughout BI. Very variable; distinctive variants occur in salt-marshes (var. **salina** (Aspegren) Lange: stems mostly unbranched; phyllaries glabrous; leaves ± entire) and on mountains (ssp. **pratensis** (Hornem.) Gremli: stems mostly unbranched; phyllaries with dense long dark hairs), but their distinctness and status need study.

2. **L. hispidus** L. - Rough Hawkbit. Stems to 60cm, simple, **832** usually conspicuously pubescent; leaves oblanceolate, sinuate-dentate to deeply lobed; phyllaries usually conspicuously pubescent; outermost ligules usually reddish (rarely greyish-violet) on lowerside. Native; basic, often calcareous, grassland; common in Br N to C Sc, locally frequent in Ir; very scattered in Sc, not in CI or Man. Nearly glabrous plants (var. **glabratus** (Koch) Bisch.) occur but are rare.

2 x 3. **L. hispidus x L. saxatilis** occurs rarely in En but may be overlooked; it is intermediate in habit and pubescence, but resembles L. saxatilis in pappus-type and is <1% fertile.

3. **L. saxatilis** Lam. (L. taraxacoides (Villars) Merat) - **832** Lesser Hawkbit. Differs from L. hispidus in stems to 40cm, glabrous or sparsely pubescent; outermost ligules usually greyish-violet on lowerside; and see key. Native; similar places and distribution to L. hispidus but frequent in CI and Man and less local in Ir.

20. **PICRIS** L. - Oxtongues
Annuals to perennials; rhizomes and stolons 0; stems leafy, with numerous stiff bristles hooked at apex, with several capitula; phyllaries in several rows; receptacular scales 0; pappus of 2 rows of white to off-white hairs, the inner or both rows feathery; ligules yellow; achenes somewhat flattened, weakly ribbed, transversely wrinkled, beaked or not.

1. **P. echioides** L. - Bristly Oxtongue. Annual or biennial; stems erect, to 80cm; leaves narrowly elliptic-oblanceolate, bristly, entire to dentate; inner phyllaries 12-20mm, the outer 3-5 ovate-cordate, >2x as wide as inner; achenes with beak ± as long as body; pappus white. Probably intrd; well natd in marginal, disturbed and rough ground and waste places; frequent in S & E En, more scattered elsewhere in Br N to N En and in CI, very local in S & E Ir; S Europe.

2. **P. hieracioides** L. - Hawkweed Oxtongue. Biennial to perennial; stems erect, to 1m; leaves narrowly elliptic to oblanceolate, bristly, sinuate-toothed to -lobed; inner phyllaries 11-13mm, the outer ones similar to inner but shorter and often patent; achenes with beak 0 or <1/2 as long

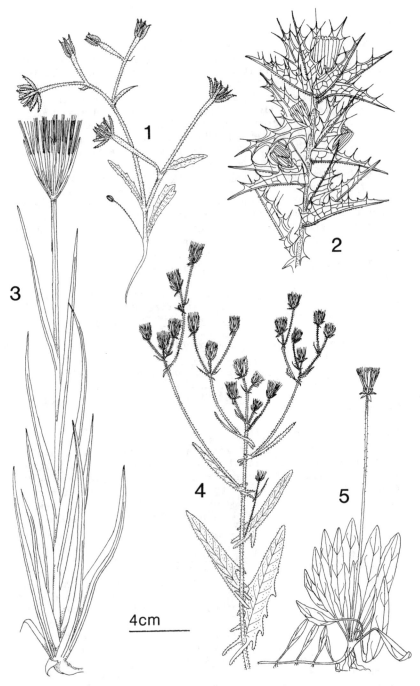

FIG 820 – Asteraceae: Lactuceae. 1, Hedypnois cretica. 2, Scolymus hispanicus. 3, Tragopogon hybridus. 4, Crepis setosa. 5, Aetheorhiza bulbosa.

as body; pappus off-white. Native; grassland and open or rough ground mostly on calcareous soils; similar distribution to P. echioides. Ssp. **spinulosa** (Bertol. ex Guss.) Arcang. (P. spinulosa Bertol. ex Guss.), from S Europe, was recorded from rough ground in N W Kent in 1935 and 1936, but the plants might have been variants of ssp. hieracioides; the true plant has subsessile crowded capitula with ± glabrous phyllaries.

21. SCORZONERA L. - Viper's-grass

Perennials; rhizomes and stolons 0; stems leafy, usually simple, sometimes with 1 or 2 branches; phyllaries in several rows; receptacular scales 0; pappus of several rows of dirty-white feathery hairs; ligules yellow, the outermost flushed crimson on lowerside; achenes slightly flattened, ribbed, not beaked.

Other spp. - **S. hispanica** L., from C & S Europe, provides the garden vegetable Scorzonera and sometimes persists briefly on tips and roadsides; it differs in its larger achenes 10-15mm with rugose (not smooth) ridges.

1. S. humilis L. - Viper's-grass. Stems erect, to 50cm, **RRR** woolly when young, becoming glabrous; leaves entire, narrowly elliptic to linear-lanceolate; capitula 2-3cm across; achenes 7-11mm. Native; marshy fields, now confined to 1 site in Dorset, formerly 1 other site in Dorset and 1 site in Warks.

22. TRAGOPOGON L. (Geropogon L.) - Goat's-beards

Annuals to perennials; rhizomes and stolons 0; stems leafy, simple or branched; phyllaries in 1 row; receptacular scales 0; pappus of 1 row of mainly feathery but some simple hairs, or the outer achenes with entirely simple hairs, dirty-white to pale brown; ligules yellow or purple; achenes not flattened, ribbed, beaked. The capitula open in the morning but close regularly about noon.

1 Ligules yellow; outer achenes often <3cm

1. T. pratensis

1 Ligules pink to purple; outer achenes >3cm 2
  2 Marginal achenes with pappus of simple hairs; achenes very gradually narrowed into beak

3. T.hybridus

  2 All achenes with pappus with some feathery hairs; achenes abruptly narrowed into beak

2. T. porrifolius

1. T. pratensis L. - Goat's-beard. Annual to perennial; stems erect, to 75cm, glabrous or woolly when young, slightly dilated below capitula; leaves linear to linear-lanceolate, entire, glabrous; ligules yellow.
a. Ssp. pratensis. Ligules as long as or longer than

phyllaries. Intrd; grassy places and open or rough ground; very scattered in S & C Br, perhaps over-recorded; Europe.

b. Ssp. minor (Miller) Wahlenb. Ligules c.1/2 to 3/4 as long as phyllaries. Native; grassy places, roadsides, rough and cultivated ground; common in most of Br N to C Sc, scattered in N Sc and Ir, Alderney.

Ssp. orientalis (L.) Celak., from E Europe, has been recorded but is only a rare casual; it has ligules as in ssp. pratensis but deeper yellow and achenes with a beak shorter than the body (about as long as body in other 2 sspp.).

1 x 2. T. pratensis x T. porrifolius = T. x mirabilis Rouy occurs rarely in C & S En near T. porrifolius but rarely persists; it has yellow ligules suffused purple distally, so that the capitulum appears purple with a yellow centre, and has a low level of fertility.

2. T. porrifolius L. - Salsify. Biennial; stems erect, to 1m, glabrous, strongly dilated below capitula; leaves as in T. pratensis but more widened at base; ligules purple, c.1/2 as long to as long as phyllaries. Intrd; grown as vegetable, casual or rarely natd in waste and rough ground, waysides; very scattered in BI, mainly S En and CI; Mediterranean.

3. T. hybridus L. (Geropogon glaber L.) - Slender Salsify. 820 Annual; stems erect, to 50cm, glabrous, strongly dilated below capitula; leaves linear, entire, glabrous; ligules pink to pinkish-purple, <1/2 as long as phyllaries. Intrd; occasional bird-seed alien in gardens, parks, tips and waste ground; scattered in S Br; S Europe.

23. AETHEORHIZA Cass. - Tuberous Hawk's-beard
Perennials with long thin rhizomes bearing large tubers; stems simple, with 1 capitulum and 0(-2) leaves; phyllaries in several rows, often weakly 2-rowed; receptacular scales 0; pappus of several rows of white, simple hairs; ligules yellow; achenes not flattened, with 4 deep grooves, not beaked.

1. A. bulbosa (L.) Cass. - Tuberous Hawk's-beard. Stems to 820 30cm, erect, glabrous; leaves obovate to narrowly so, sinuate-dentate; phyllaries (13)14-15(16)mm; achenes 3-4.5mm. Intrd; grown in gardens and very persistent when neglected; very local in E Ir; Mediterranean region.

24. SONCHUS L. - Sow-thistles
Annuals or perennials, sometimes with rhizomes; stems leafy, usually branched; phyllaries in several rows; receptacular scales 0; pappus of 2 or more rows of white, simple, hairs; ligules yellow; achenes distinctly flattened, distinctly ribbed, not beaked.

1 Plant annual or biennial, with main root and laterals 2
1 Plant perennial, rhizomatous or with thick ± erect underground portion 3

> 2 Auricles of stem-leaves pointed; achenes transversely rugose **3. S. oleraceus**
> 2 Auricles of stem-leaves rounded; achenes not rugose **4. S. asper**
> 3 Auricles of stem-leaves pointed; stems arising from thick underground root-like organ; achenes straw-coloured **1. S. palustris**
> 3 Auricles of stem-leaves rounded; plant strongly rhizomatous; achenes bright brown **2. S. arvensis**

1. S. palustris L. – <u>Marsh</u> <u>Sow-thistle</u>. Stems erect, to **R** 2.5m, glabrous below, glandular-pubescent above with usually dark glands; lower leaves deeply pinnately lobed, upper ones unlobed except for auricles; achenes 3.5–4mm, weakly transversely rugose, usually with 6–10 ribs. Native; marshes, fens and riversides; SE En from W Kent to E Norfolk, formerly N to N Lincs and W to Leics and Oxon, decreasing, intrd in Leics.

2. S. arvensis L. – <u>Perennial</u> <u>Sow-thistle</u>. Stems erect, to 1.5m; glabrous to densely glandular-pubescent above with usually yellowish glands; most leaves variably pinnately lobed, very rarely as deeply as lower ones or as little as upper ones of <u>S. palustris</u>; achenes 2.5–3.5mm, transversely rugose, usually with 12 ribs. Native; arable and waste land, waysides, dunes and shingle by sea, ditches and river-banks; common throughout lowland BI. Glabrous plants (ssp. **uliginosus** (M. Bieb.) Nyman) are said to have shorter phyllaries (mostly 12–16mm, not mostly 15–18mm) and have 2n=36 (not 54), but their status needs checking.

3. S. oleraceus L. – <u>Smooth</u> <u>Sow-thistle</u>. Stems erect, to 1.5m, usually glabrous except glandular-pubescent below capitula; leaves very variously divided, the middle ones usually deeply pinnately lobed; phyllaries usually glandular-pubescent, often also woolly at first; achenes 2.5–3.75mm, variably c.8–12-ribbed; 2n=32. Native; waste and cultivated ground, roadsides; abundant throughout lowland BI.

3 x 4. S. oleraceus x S. asper has occurred very rarely in C & S En, with more records doubtful or erroneous; it has leaf-auricles rounded and dentate as in <u>S. asper</u> but with one long pointed tooth, and is sterile (2n=25).

4. S. asper (L.) Hill – <u>Prickly</u> <u>Sow-thistle</u>. Habit and pubescence as in <u>S. oleraceus</u>; leaves often less lobed than in <u>S. oleraceus</u> but with more sharply dentate margins, but sometimes very deeply lobed; achenes 2–3mm, usually with 3 thin ribs on each face and 2 wide wing-like marginal ones; 2n=18. Native; similar places and distribution to <u>S. oleraceus</u>, and often with it. Ssp. **glaucescens** (Jordan) Ball differs in being biennial (often with a basal leaf rosette) and in having better developed recurved spinules on the achene ribs; its status needs checking.

25. LACTUCA L. (<u>Mulgedium</u> Cass.) - <u>Lettuces</u>

Annuals or biennials, or perennials with rhizomes; stems leafy, branched at least above, with many capitula; phyllaries in several rows; receptacular scales 0; pappus of 2 rows of white, simple hairs; ligules yellow or blue; achenes distinctly flattened, ribbed, beaked.

1 Achenes with beak <1/2 as long as body and of same
  colour; ligules blue; perennial   **5. L. tatarica**
1 Achenes with beak >1/2 as long as body and much
  lighter in colour; ligules yellow; annual to biennial 2
  2 Plant <1m; leaf midrib on lowerside glabrous to
    sparsely hispid; achenes without bristles   3
  2 Plant often 1-2m; leaf midrib on lowerside with
    strong prickles; achenes with minute bristles
    just below beak   4
3 Stem-leaves oblong-ovate, cordate and clasping stem
  at base; inflorescence subcorymbose   **2. L. sativa**
3 Middle and upper stem-leaves linear-oblong, sagittate
  and clasping stem at base; inflorescence very narrow
                         4. L. saligna
  4 Ripe achenes (excl. beak) (4)4.2-4.8(5.2)mm,
    maroon to blackish; stems and midribs strongly
    tinged maroon   **3. L. virosa**
  4 Ripe achenes (excl. beak) (2.8)3-4(4.2)mm, olive-
    grey; stems and midribs greenish-white
                    1. L. serriola

1. L. serriola L. - <u>Prickly</u> <u>Lettuce</u>. Stems erect, to 2m; 825 leaves deeply pinnately lobed to (more often) unlobed, sharply dentate; achenes olive-grey, mostly 3-4mm, with white beak c. as long. Probably native; waysides and waste and rough ground; frequent in En SE of line from Severn to Humber, especially in SE and E Anglia, very scattered elsewhere in En, Wa and CI.

2. L. sativa L. - <u>Garden</u> <u>Lettuce</u>. Stems erect, to 75cm; 825 leaves unlobed or the basal pinnately lobed, entire to remotely denticulate; achenes brownish-grey, 3-4mm, with white beak 1/2 to 1x as long. Intrd; casual on tips, waste ground and abandoned arable land, also bird-seed alien; scattered in Br, mainly S; probably arose from <u>L. serriola</u> in E Mediterranean.

3. L. virosa L. - <u>Great</u> <u>Lettuce</u>. Stems erect, to 2m; leaves 825 as in <u>L. serriola</u> but more often well lobed, often undulate, and usually maroon-tinged; achenes maroon to blackish, mostly 4.2-4.8mm, with white beak c. as long or shorter. Probably native; similar distribution to <u>L. serriola</u> but scattered N to C Sc and less common in SE En and E Anglia.

4. L. saligna L. - <u>Least</u> <u>Lettuce</u>. Stems erect, to 825 75(100)cm; lower leaves rather deeply pinnately lobed; middle **RRR** and upper stem-leaves linear-oblong, entire except for sagittate base; achenes olive-grey, 2.8-3.5mm, with white

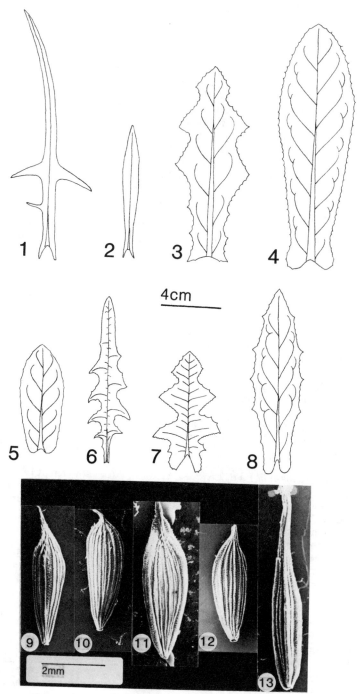

FIG 825 – Lactuca. 1–8, leaves. 1–2, **L. saligna.** 3–4, **L. virosa.** 5, **L. sativa.** 6, **L. tatarica.** 7–8, **L. serriola.** 9–13, achenes, pappus removed. 9, **L. serriola.** 10, **L. sativa.** 11, **L. virosa.** 12, **L. saligna.** 13, **L. tatarica.**

beak 1/2 to 1x as long. Native; salt-marshes, shingle, waste
places and sea-walls near sea; very local in S Essex, W Kent
and E Sussex, formerly in S Br from E Cornwall to W Norfolk.

5. L. tatarica (L.) C. Meyer (<u>Mulgedium</u> <u>tataricum</u> (L.) DC.) **825**
- <u>Blue Lettuce</u>. Stems erect, to 80cm; leaves linear-
lanceolate to narrowly elliptic, irregularly dentate, the
lowest sometimes ± deeply lobed; achenes dark brown to black,
3.5-5.5mm, with concolourous beak <1/2 as long. Intrd; natd
in rough and waste ground mostly by sea; very scattered in
En, Wa, Man, Guernsey, NE Galway; E Europe & W Asia.

26. CICERBITA Wallr. - <u>Blue-sow-thistles</u>
Perennials, often with rhizomes; stems leafy, branched above,
with many capitula; phyllaries in several rows; receptacular
scales 0; pappus of 2 rows of simple, dirty- to yellowish-
white hairs; ligules blue to mauve; achenes flattened,
ribbed, not beaked.

1 Plant glabrous   **3. C. plumieri**
1 Peduncles and/or upper parts of stems with simple or
   glandular hairs   2
   2 Upper parts of stems and sometimes peduncles with
      simple hairs; glandular hairs 0   **4. C. bourgaei**
   2 Peduncles and usually upper parts of stems with
      glandular hairs   3
3 Inflorescence narrowly pyramidal; lower leaves
   glabrous, with sharply triangular apical lobe
                                                        1. C. alpina
3 Inflorescence subcorymbose; lower leaves pubescent
   on veins on lowerside, with ovate-subcordate apical
   lobe   **2. C. macrophylla**

1. C. alpina (L.) Wallr. - <u>Alpine Blue-sow-thistle</u>. Not **RRR**
rhizomatous; stems erect, to 1.3m, bristly below, glandular-
pubescent near top; leaves with sharply triangular terminal
lobe and a few smaller laterals; ligules blue. Native; moist
mountain rock-ledges; extremely local in 4 sites in Angus and
S Aberdeen, discovered 1801.

2. C. macrophylla (Willd.) Wallr. - <u>Common Blue-sow-thistle</u>.
Strongly rhizomatous; stems erect, to 2m, glabrous below,
glandular-pubescent on peduncles and sometimes upper parts of
stems; leaves with ovate-subcordate terminal lobe and 0-1
pairs of laterals; ligules pale mauve. Intrd; well natd on
rough and waste ground and roadsides, first recorded 1915;
frequent throughout most of Br, scattered in Ir; Urals. Our
plant is ssp. **uralensis** (Rouy) Sell.

3. C. plumieri (L.) Kirschl. - <u>Hairless Blue-sow-thistle</u>.
Not rhizomatous; stems erect, to 1.3m, glabrous; leaves with
triangular terminal lobe and several pairs of laterals;
ligules blue. Intrd; natd as for <u>C. macrophylla</u>, much rarer
but often misrecorded for it; very scattered in En and Sc; WC
Europe.

4. C. bourgaei (Boiss.) Beauverd - <u>Pontic Blue-sow-thistle</u>.
Not rhizomatous; stems erect, to 2m, with rather sparse
simple hairs; leaves with ovate-rhombic terminal lobe and
1-few pairs of laterals; ligules pale mauve. Intrd; natd as
for <u>C. macrophylla</u> but much rarer; very scattered in En, Sc
and Man; Georgia & NE Turkey.

27. MYCELIS Cass. - <u>Wall Lettuce</u>
Perennials, often short-lived; stems leafy, branched above,
with many capitula; phyllaries in 2 very unequal rows;
receptacular scales 0; pappus of 2 unequal rows of white
simple hairs; ligules yellow; achenes distinctly flattened,
ribbed, very shortly beaked.

1. M. muralis (L.) Dumort. - <u>Wall Lettuce</u>. Usually suffused
maroon; stems erect, to 1m, glabrous; leaves deeply pinnately
lobed, sharply dentate; capitula 1-1.5cm across, with c.5
flowers. Native; shady places in woods, on walls and rocks
and in hedgerows; locally common in En and Wa, scattered in
Sc, Ir and Man.

28. TARAXACUM Wigg. - <u>Dandelions</u>
Perennials with tap-roots; stems usually leafless, with 1
capitulum; phyllaries in 2 often very different rows;
receptacular scales 0; pappus of several rows of white,
simple hairs; ligules yellow, usually with coloured stripe(s)
on lowerside; achenes not flattened, finely ribbed, usually
spinulose near apex, beaked.
   A very critical genus in which apomixis is the rule; 226
microspp. are currently recognized in BI, of which 39 are
thought to be endemic and only c.76 others native.
Triploids, tetraploids, pentaploids (rare) and hexaploids
(rare) occur, of which the latter 3 are probably all
obligately apomictic. In BI almost all triploids are also
obligately apomictic, but in a few sexuality occurs very
rarely, producing non-persistent sexual diploids that can
hybridize with each other and with pollen from apomictic
plants to produce diploid to triploid hybrids.
   In this work the microspp. are not treated in full but are
aggregated into 9 rather ill-defined sections, determination
of which is often not easy even after much experience.
   In most spp. the achene is spinulose near its apex, but
between that region and the beak there is a short, usually
pyramidal region known as the <u>cone</u>. Descriptions apply only
to fully ripe achenes. Leaves produced in summer do not
maintain all diagnostic characters, so determination should
be attempted only with specimens collected during the first
main flush of flowering (usually Apr to early May in the
lowlands). Plants from shaded, heavily trodden or grazed, or
mown areas should be avoided.
   This account follows the sectional and microsp. delimitation
recognized by A.J. Richards and C.C. Haworth.

1   Plants delicate, usually with strongly dissected
(often nearly pinnate) leaves; outer row of phyllaries
rarely >7mm; capitula rarely >3cm across   2

1   Plants usually medium to robust, rarely with nearly
pinnate leaves; outer row of phyllaries usually
>7mm; capitulum usually >3cm   3

  2   Achenes grey-brown, with pyramidal cone $\leq$0.5mm;
leaves often with $\geq$6 pairs of lateral lobes
      2. T. sect. Obliqua

  2   Achenes usually reddish to yellowish-brown, with
cylindrical cone c.1mm; leaves rarely with >6
pairs of lateral lobes   **1. T. sect. Erythrosperma**

3   Outer row of phyllaries appressed, ovate, with broad
scarious border; leaves very narrow, usually scarcely
lobed   **3. T. sect. Palustria**

3   Outer row of phyllaries appressed to recurved, linear
to narrowly ovate, with narrow to very narrow scarious
border; leaves broader, usually distinctly lobed   4

  4   Leaves and petioles green; rare plants of a few
mountain cliffs in Sc   **6. T. sect. Taraxacum**

  4   If on mountain cliffs then leaves usually dark
or blotched or spotted with purple and petiole
usually purple   5

5   Achene (incl. cone, excl. beak) >4.5mm, nearly
cylindrical; outer row of phyllaries erect to
appressed; ligules usually with dark red stripes on
lowerside; pollen 0   **4. T. sect. Spectabilia**

5   Achene <4mm, narrowly top-shaped; outer row of
phyllaries never appressed; ligule stripes rarely
dark red; pollen present or 0   6

  6   Leaves with large dark spots covering >10% of
surface   **5. T. sect. Naevosa**

  6   Leaves unspotted or with spots covering <10%
of blade (beware damaged or attacked leaves)   7

7   Petiole and midrib uppersides green or solid red or
purple; outer row of phyllaries usually >10mm, often
recurved, not dark on lowerside; leaves often
complexly lobed and folded in 3 dimensions
    9. T. sect. Ruderalia

7   Petiole and midrib uppersides usually minutely (lens)
stiped red or purple; outer row of phyllaries rarely
>10mm, usually patent to erect and dark on lowerside;
leaves $\pm$ flat, relatively simply lobed   8

  8   Lateral leaf-lobes broad-based, with convex front
and concave rear edge, commonly 4 pairs; outer
row of phyllaries usually arched to various
degrees, often subobtuse   **8. T. sect. Hamata**

  8   Lateral leaf-lobes rarely as above, often 5-6
pairs; outer row of phyllaries erect to recurved
all $\pm$ to same degree, often acute
    7. T. sect. Celtica

1. T. Sect. Erythrosperma (Lindb. f.) Dahlst. (T. laevigatum (Willd.) DC. group. T. simile Raunk. group, T. fulvum Raunk. group). 30 microspp. currently placed here; triploids or tetraploids. Mostly native; dry exposed places, usually on well-drained soils, short grassland, heathland, dunes; throughout BI, commonest section after Hamata and Ruderalia, mostly maritime in N; 7 endemics.

2. T. sect Obliqua Dahlst. (T. obliquum (Fries) Dahlst. R group). 2 microspp. currently placed here; triploids. Native; open sandy turf by sea; local on coasts of BI, commonest in Sc, but S to CI.

3. T. sect. Palustria (Dahlst.) Dahlst. (T. palustre (Lyons) R Symons group). 4 microspp. currently placed here; tetraploids and pentaploids. Native; wet usually base-rich meadows and fen grassland; local, scattered in BI; 2 endemics.

4. T. sect. Spectabilia (Dahlst.) Dahlst. (T. spectabile Dahlst. group). 3 microspp. currently placed here; pentaploids. Native; damp or wet acidic grassy places, often in upland areas, also roadsides etc.; throughout BI, commonest in N; 2 endemics.

5. T. sect. Naevosa M. Christiansen (T. naevosum Dahlst. group, T. praestans Lindb. f. group pro parte). 11 microspp. currently placed here; tetraploids. Mostly native; habitat and distribution as for sect. Spectabilia but uncommon in SE En; 3 endemics.

6. T. sect. Taraxacum (sect. Crocea M. Christiansen, T. R croceum Dahlst. group). 6 microspp. currently placed here; tetraploids. Native; mostly base-rich mountain rock-ledges and flushes; very local in highlands of Sc; 1 endemic.

7. T. sect. Celtica A. Richards (T. celticum A. Richards group, T. unguilobum Dahlst. group, T. adamii Claire group, T. nordstedtii Dahlst. group, T. praestans group pro parte). 30 microspp. currently placed here; triploids, tetraploids and hexaploids. Mostly native; mostly wet places in lowland grassland and in upland flushes and on rock-ledges; throughout BI, but few microspp. in lowlands; 17 endemics.

8. T. sect. Hamata Oellgaard (T. hamatum Raunk. group). 18 microspp. currently placed here; triploids. Native (c. 6 microspp.) and intrd; grassland, both damp and dry, and roadsides and rough ground; throughout BI, usually weedy.

9. T. sect. Ruderalia Kirschner, Oellgaard & Stepanek (sect. Vulgaria Dahlst. nom. illeg., T. officinale Wigg. group). 122 microspp. currently placed here; triploids, occasionally diploids. Native (c.25 microspp.) and intrd; habitat and distribution as for sect. Hamata. By far the commonest section, especially as weeds in lowland areas. c.50 of the microspp. are sporadic non-persistent casuals; 7 endemics.

29. CREPIS L. - Hawk's-beards
Annuals to perennials, sometimes shortly rhizomatous; stems branched, leafy or (C. praemorsa) all basal; phyllaries in 2

rows; receptacular scales 0 but receptacle often pubescent, sometimes each achene-pit with membranous fringe; pappus of several rows of white or (C. paludosa) yellowish-white, simple hairs; ligules yellow; achenes not flattened, ribbed, beaked or not (if not, usually slightly tapered distally). Doubtfully distinct from Hieracium.

1 Flowering stems leafless   10. C. praemorsa
1 Flowering stems bearing leaves   2
   2 Outer achenes with short or 0 beak, distinctly
      different from inner slender-beaked ones
                                                     9. C. foetida
   2 Achenes all the same, or inner and outer slightly
      different but grading into one another   3
3 Achenes distinctly beaked, the beak usually $\geq 1/2$ as
   long as body of achene   4
3 Achenes not beaked, but often narrowed at apex   6
   4 Basal lobes of upper stem-leaves not clasping
      stem; achenes with beak scarcely 1/2 as long as
      body   4. C. tectorum
   4 Basal lobes of upper stem-leaves clasping stem;
      achenes with beak c. as long as body   5
5 Upper parts of plant nearly always with many patent
   stiff bristles; achenes (excl. beak) 3-5.5mm
                                                      8. C. setosa
5 Upper parts of plant without patent stiff bristles;
   achenes (excl. beak) (5)6-8(9)mm   7. C. vesicaria
   6 Phyllaries pubescent on their inner faces   7
   6 Phyllaries glabrous on their inner faces   8
7 Achenes (2.5)3-4(4.5)mm, with 10 ribs distinctly
   rough towards apex   4. C. tectorum
7 Achenes 4-7.5mm, with 13-20 $\pm$ smooth ribs   3. C. biennis
   8 Pappus-hairs yellowish-white, brittle   1. C. paludosa
   8 Pappus-hairs pure white, $\pm$ flexible   9
9 Achenes c.20-ribbed; perennial arising from short
   rhizome; rare in N En and Sc   2. C. mollis
9 Achenes 10-ribbed; common annual or biennial with
   tap-root   10
   10 Achenes 2.5-3.8mm; outer phyllaries patent to
      erecto-patent; receptacle with laciniate membranous
      fringes around achene-pits   5. C. nicaeensis
   10 Achenes 1.4-2.5mm; outer phyllaries appressed to
      inner; receptacle with a few hairs around achene-
      pits   6. C. capillaris

1. C. paludosa (L.) Moench – Marsh Hawk's-beard. Erect 832 subglabrous perennial to 80cm; stem-leaves toothed to shallowly lobed, clasping stem at base; phyllaries with woolly eglandular and straight glandular hairs; achenes 4-5.5mm, not beaked. Native; wet places in open woodland, grassland, fens; BI N from S Ir, S Wa and Leics, frequent in Sc and N En.

2. C. mollis (Jacq.) Asch. - <u>Northern</u> <u>Hawk's-beard</u>. Erect 832 subglabrous to sparsely pubescent perennial to 60cm; stem- R leaves entire to sinuate-toothed, tapering to broad base; phyllaries with glandular hairs; achenes 3-4.5mm, not beaked. Native; grassy, often damp slopes or hills; very local from MW Yorks to Banffs.

3. C. biennis L. - <u>Rough</u> <u>Hawk's-beard</u>. Erect pubescent 832 biennial to 1.2m; stem-leaves irregularly and sharply lobed, ± clasping stem at base; phyllaries with eglandular and often glandular hairs; achenes 4-7.5mm, not beaked. Probably native; rough grassy places, waysides; scattered in Br and Ir N to C Sc, frequent in SE En.

4. C. tectorum L. - <u>Narrow-leaved</u> <u>Hawk's-beard</u>. Erect to 832 ascending subglabrous annual to 75cm; stem-leaves entire to sinuate-toothed, tapered to base; phyllaries with eglandular and glandular hairs, achenes 2.5-4(4.5)mm (incl. beak), strongly tapered at apex or shortly beaked. Intrd; occasional casual on roadsides, disturbed soil and re-seeded verges, a grain- and grass-seed-alien; scattered in En and Sc; Europe.

5. C. nicaeensis Balbis - <u>French</u> <u>Hawk's-beard</u>. Erect to 832 ascending rather sparsely pubescent annual or biennial to 80cm; leaf-shape and capitula as in <u>C. capillaris</u> but latter usually larger; phyllaries with glandular and eglandular hairs; achenes 2.5-3.8mm, not beaked. Intrd; habitat, source and distribution as for <u>C. tectorum</u>, becoming rare; W & C Mediterranean.

6. C. capillaris (L.) Wallr. - <u>Smooth Hawk's-beard</u>. Erect 832 to decumbent glabrous to sparsely pubescent annual or biennial to 75cm; stem-leaves subentire to deeply and sharply lobed, not clasping stem; phyllaries glabrous or pubescent, with (var. **glandulosa** Druce) or without glandular hairs; achenes 1.4-2.5mm, not beaked. Native; grassy places, rough and waste ground; common throughout BI, var. <u>glandulosa</u> commonest in the N, especially Sc.

7. C. vesicaria L. - <u>Beaked Hawk's-beard</u>. Erect, pubescent 832 perennial to 80cm; stem-leaves deeply sharply lobed, clasping stem at base; phyllaries usually with eglandular and often with glandular hairs; achenes 5-9mm incl. beak c. as long as body. Intrd; grassy places, waysides, walls and rough ground; common in S and CE En, scattered N to N En and in CI and Ir; Europe. Our plant is ssp. **taraxacifolia** (Thuill.) Thell. ex Schinz & R. Keller (ssp. <u>haenseleri</u> (Boiss. ex DC.) Sell).

8. C. setosa Haller f. - <u>Bristly Hawk's-beard</u>. Erect, 820 usually hispid (very rarely subglabrous) annual or biennial 832 to 75cm; stem-leaves usually toothed to sharply lobed, clasping stem at base; phyllaries with patent stiff eglandular hairs; achenes 3-5.5mm incl. beak c. as long as body. Intrd; occasional casual with crops or grass or in rough ground; scattered in Br, mostly C & S; S Europe.

9. C. foetida L. - <u>Stinking Hawk's-beard</u>. Erect, pubescent 832

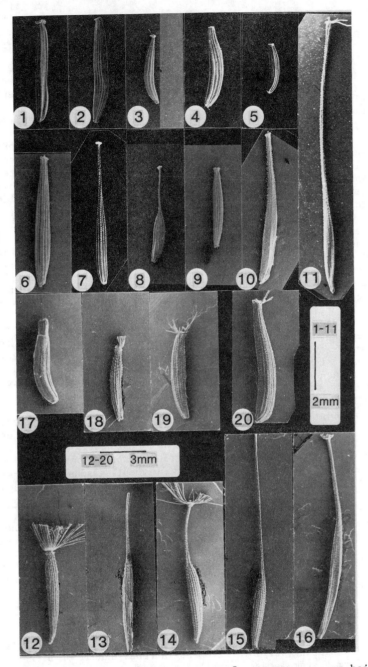

FIG 832 — Achenes of **Asteraceae**: Lactuceae, pappus-hairs removed. 1–11, **Crepis**. 1, C. paludosa, 2, C. mollis. 3, C. tectorum. 4, C. nicaeensis. 5, C. capillaris. 6, C. biennis. 7, C. vesicaria. 8, C. setosa. 9, C. praemorsa. 10–11, C. foetida (outer, inner). 12–16, **Hypochaeris**. 12–13, **H. glabra** (outer, inner). 14–15, **H. radicata** (outer, inner). 16, **H. maculata**. 17–20, **Leontodon**. 17–18, **L. saxatilis** (outer, inner). 19, **L. autumnalis**. 20, **L. hispidus**.

annual or biennial to 60cm, stinking when fresh; stem-leaves **RRR**
few, usually deeply sharply lobed, ± clasping stem at base;
phyllaries with eglandular and glandular hairs; inner achenes
10–15mm incl. beak c. as long as body; outer achenes 6–9mm
incl. beak much shorter than body. Native; waysides and
rough ground; now probably only on shingle at Dungeness (E
Kent), formerly scattered in SE En and intrd NW to Worcs.

10. C. praemorsa (L.) F. Walther - <u>Leafless</u> Hawk's-beard. 832
Erect, sparsely pubescent perennial to 60cm; stem-leaves 0;
basal leaves narrowly obovate, not or shallowly toothed;
phyllaries sparsely pubescent; achenes 3–4mm, not beaked.
Possibly native; on natural calcareous grassy bank, Westmorland, discovered 1988.

30. PILOSELLA Hill (<u>Hieracium</u> subg. <u>Pilosella</u> (Hill) Gray -
Mouse-ear-hawkweeds
Perennials, usually stoloniferous; stems usually leafless,
sometimes with few leaves, with 1-many capitula; basal leaves
oblanceolate to narrowly obovate or narrowly elliptic,
pubescent, subentire; phyllaries in several rows; receptacular scales 0 but achene-pits often variously fringed;
pappus of 1 row of dirty-white to pale brown, simple hairs;
ligules yellow to orange; achenes not flattened, 10-ribbed,
not beaked, scarcely tapered towards apex. Doubtfully
distinct from <u>Hieracium</u>.

Both apomictic and sexual plants occur, often within 1 sp;
hybridization is frequent wherever 2 spp. occur together. Of
the native spp., <u>P. peleteriana</u> is always diploid and sexual,
whereas <u>P. officinarum</u> may be sexual (tetraploid) or
apomictic (pentaploid or hexaploid); the sexual plants are
commoner in the S and rare or 0 in Sc.

1 All flowering stems with only 1 capitulum   2
1 At least some flowering stems with >1 capitulum   3
  2 Stolons elongated, slender, with spaced out small
    leaves, 0 or few ending in leaf-rosette
                                            2. P. officinarum
  2 Stolons few or 0, short (<5cm), stout, with full-
    sized ± crowded leaves, often ending in leaf-
    rosette   **1. P. peleteriana**
3 Ligules orange-brown to brick red, often turning
  purplish when dried   **6. P. aurantiaca**
3 Ligules yellow, sometimes red-striped on lowerside   4
  4 Capitula (1)2-6 per stem, not crowded; phyllaries
    (8)9-12mm   **3. P. flagellaris**
  4 Capitula (3)7-50 per stem, many closely crowded;
    phyllaries 5-9mm   5
5 Stem and leaves with or without stellate hairs, with
  0 to sparse simple (glandular or eglandular) hairs;
  largest leaves <15(20)mm wide   **4. P. praealta**
5 Stem and leaves with stellate hairs, with numerous
  simple (glandular or eglandular) hairs; largest

leaves 15-25(30)mm wide   **5. P. caespitosa**

Other spp. - P. lactucella (Wallr.) Sell & C. West (H.
lactucella Wallr.), from Europe, has 1-7 capitula per stem
and glaucous, glabrous to very sparsely pubescent leaves; it
was formerly natd in S Wilts. P. lactucella x P. caespitosa
= **P. x floribunda** (Wimmer & Grab.) Arv.-Touv. (H. x
floribundum Wimmer & Grab., P. lactucella ssp. helveola
(Dahlst.) Sell & C. West, H. helveolum (Dahlst.) Pugsley) was
formerly natd in Co Antrim.

1. P. peleteriana (Merat) F. Schultz & Schultz Bip. (H. **RR**
peleterianum Merat) - Shaggy Mouse-ear-hawkweed. Like a
robust, large-headed P. officinarum with stolons 0 or short,
thick and often ending in a leaf-rosette; phyllaries with
dense long eglandular, few or 0 glandular, and usually few
stellate hairs.   Native; short grassland on well-drained
soils, dunes; very local, but commoner in CI than P.
officinarum. The sspp. are probably better as vars.
1 Scapes up to 12(18)cm; rosette-leaves 9-20mm wide,
  not or scarcely tapered at base   **a. ssp. peleteriana**
1 Scapes (6)10-30cm; rosette-leaves 4-12(18)mm wide,
  long-tapered at base   2
  2 Phyllaries 11-15mm, lanceolate; capitula 12-17mm
    across excl. ligules   **b. ssp. subpeleteriana**
  2 Phyllaries 10-12(13)mm, linear-lanceolate;
    capitula (9)10-12(14)mm across excl. ligules
                                      c. ssp. tenuiscapa
a. Ssp. peleteriana.   Phyllaries 11-15mm, lanceolate;
capitula 12-20mm across excl. ligules. CI (all islands),
Dorset, Wight and E Kent (extinct in last).
  b. Ssp. subpeleteriana (Naeg. & Peter) Sell (H. peleterianum
ssp. subpeleterianum Naeg. & Peter). Monts (Craig Breidden).
  c. Ssp. tenuiscapa (Pugsley) Sell & C. West (H. peleterianum
ssp. tenuiscapum (Pugsley) Sell). Jersey, S Devon, Staffs,
Derbys and MW Yorks.
1 x 2. P. peleteriana x P. officinarum = P. x longisquama
(Peter) Holub (H. x longisquamum Peter, P. x pachylodes
(Naeg. & Peter) Sojak nom. illeg.) has occurred with the
parents in Jersey, Guernsey, Staffs and E Kent; it is
intermediate.
  2. P. officinarum F. Schultz & Schultz-Bip. (H. pilosella
L.) - Mouse-ear-hawkweed. Stolons long, slender, with ±
distant reduced leaves; scapes to 30(50)cm, with 1 capitulum,
densely pubescent; ligules yellow; capitula 7-12mm across
excl. ligules.   Native; short grassland on well-drained
soils, banks, rocky places; locally common throughout BI
except Shetland.   Very variable in scape height and
robustness and in pubescence. Based mainly on colour, length
and relative distribution of simple eglandular, simple
glandular and stellate hairs on phyllaries, 7 taxa can be
recognized. These are not or only partially geographically

separated and are no more than vars.

2 x 6. P. officinarum x P. aurantiaca = P. x stoloniflora (Waldst. & Kit.) F. Schultz & Schultz-Bip. (<u>H.</u> x <u>stoloniflorum</u> Waldst. & Kit.) occurs with the parents in scattered places from Guernsey to N Sc; it is intermediate in ligule colour and capitulum number per scape.

3. P. flagellaris (Willd.) Sell & C. West (<u>H.</u> <u>flagellare</u> **RR** Willd.) - <u>Shetland</u> Mouse-ear-hawkweed. Stolons long, stout, leafy; scapes to 40cm, with (1)2-4(7) capitula, with glandular and eglandular hairs; ligules yellow; phyllaries with numerous simple glandular and eglandular hairs and sparse stellate hairs.

a. Ssp. flagellaris. Scapes to 40cm; capitula 2-4(7); peduncles with simple eglandular hairs 2-3mm; phyllaries with few to numerous simple eglandular hairs <1.5mm. Intrd; natd on grassy roadsides and railway banks as garden escape; scattered in C & CS En and CE SC, first recorded 1869; C & E Europe.

b. Ssp. bicapitata Sell & C. West (<u>H.</u> <u>flagellare</u> ssp. <u>bicapitatum</u> (Sell & C. West) Sell). Scapes to 18cm; capitula (1)2(-4); peduncles with simple eglandular hairs <7.5mm; phyllaries with dense simple eglandular hairs <2.5mm. Native; dry rocky pastures, rocky slopes and outcrops; 3 localities in Shetland; endemic, discovered 1962.

4. P. praealta (Villars ex Gochnat) F. Schultz & Schultz-Bip. (<u>H.</u> <u>praealtum</u> Villars ex Gochnat) - <u>Tall</u> Mouse-ear-hawkweed. Stolons 0 to long; scapes to 65cm, with numerous capitula, pubescent above but glabrous to subglabrous below; ligules yellow. Intrd; natd garden escape on grassy road-sides, walls and railway banks; Europe.

a. Ssp. praealta. Stolons 0 or very short; phyllaries with numerous glandular and 0 or few eglandular hairs. Scattered localities in En and Wa N to Cheshire, first recorded 1899.

b. Ssp. thaumasia (Peter) Sell (ssp. <u>arvorum</u> (Naeg. & Peter) Sell & C. West, ssp. <u>spraguei</u> (Pugsley) Sell & C. West, <u>H.</u> <u>pilosella</u> ssp. <u>thaumasium</u> (Peter) Sell). Stolons long and slender; phyllaries with numerous glandular and 0 to numerous eglandular hairs. Scattered localities in SC En and W Lothian, first recorded 1918.

5. P. caespitosa (Dumort.) Sell & C. West (<u>H.</u> <u>caespitosum</u> Dumort.) - <u>Yellow</u> Fox-and-cubs. Stolons strong, with large leaves; scapes to 50(80)cm, with numerous crowded capitula, pubescent; ligules yellow. Intrd; natd garden escape on rough ground, walls and railway banks; scattered in Br and S Ir; N & E Europe. Our plant is ssp. **colliniformis** (Peter) Sell & C. West (<u>H.</u> <u>pilosella</u> ssp. <u>colliniforme</u> (Peter) Sell).

6. P. aurantiaca (L.) F. Schultz & Schultz-Bip. (<u>H.</u> <u>aurantiacum</u> L.) - <u>Fox-and-cubs</u>. Stolons strong, above and below ground, with large leaves; scapes to 40(65)cm, with numerous crowded capitula, pubescent; ligules orange-brown to brick red. Intrd; natd garden escape (setting abundant seed) on rough ground, walls, roadsides and railway banks; N & C

Europe. The 2 sspp. are of doubtful value.
   a. Ssp. aurantiaca. Stolons mostly underground; basal leaves mostly 10-20 x 2-6cm; phyllaries 8-11mm. Very scattered in Br.
   b. Ssp. carpathicola (Naeg. & Peter) Sojak (ssp. brunneocrocea (Pugsley) Sell & C. West, H. aurantiacum ssp. carpathicola Naeg. & Peter, H. brunneocroceum Pugsley). Stolons mostly above ground; basal leaves mostly 6-10(16) x 1.2-2(3)cm; phyllaries 5-8mm. Frequent throughout BI.

31. HIERACIUM L. - Hawkweeds

Perennials, without stolons or rhizomes; stems leafy or sometimes not, with (1-)few-several capitula, with or without basal rosette of leaves at flowering; phyllaries in several rows; receptacular scales 0 but achene-pits often variously fringed; pappus of 1 row of dirty-white to pale brown, simple hairs; ligules yellow; achenes not flattened, 10-ribbed, not beaked, scarcely tapered towards apex.
   All the taxa are obligate apomicts and are triploids or tetraploids so far as is known, except for the single sp. of section Umbellata (H. umbellatum L.), which exists as diploid sexual and triploid apomictic plants. About 250 microspp. are currently recognized in BI, of which many are endemic and probably a considerable no. aliens. In this work they are not treated in full, but are aggregated into sections that are recognizable after a little practice.
   Plants often exhibit a second phase of flowering on new growth, either naturally or if the first growth is damaged. Only the first growth provides reliable diagnostic characters. As a rule of thumb, identification should not be attempted on plants with 0-1 stem-leaves after mid-Jun, on plants with 2-8 stem-leaves after mid-Jul, and on others after mid-Aug.

1   Stem-leaves 8-many except in dwarfed plants; rosette
       of leaves usually 0 at flowering   2
1   Stem-leaves 0-8(12); rosette of leaves usually present
       at flowering   7
   2   Middle stem-leaves not or scarcely clasping stem
           at base   3
   2   Middle stem-leaves distinctly clasping stem at
           base, though often very narrowly so   5
3   Leaves all sessile, often linear-lanceolate, ± all
       of similar shape, with recurved margins; phyllaries
       (except innermost) with recurved tips; styles yellow
       when fresh   **1. H. sect. Umbellata**
3   Lower leaves petiolate, usually broader, middle and
       upper ones sessile or nearly so, not with recurved
       margins; phyllaries very rarely with recurved tips;
       styles usually dark when fresh   4
   4   Stem-leaves rarely <15, often crowded, upper ones
           with broad rounded bases   **2. H. sect. Sabauda**

4 Stem-leaves usually <15, rarely crowded, upper
   ones narrowed to base   **4. H. sect. Tridentata**
5 Middle stem-leaves slightly constricted just above the
   broad clasping base; peduncles with dense glandular
   hairs; achenes pale brown   **5. H. sect. Prenanthoidea**
5 Middle stem-leaves not constricted, with narrow
   clasping base; peduncles with 0-few glandular hairs;
   achenes purplish- or blackish-brown   6
6 Stem-leaves c.10-30, the lower ones clasping stem
   at base to merely sessile; phyllaries sparsely
   pubescent and glandular; ligules glabrous at tip
                                  3. H. sect. Foliosa
6 Stem-leaves c.2-10(15), the lower ones sub-
   petiolate; phyllaries moderately pubescent and
   glandular; ligules glabrous or pubescent at tip
                                  6. H. sect. Alpestria
7 Stem-leaves 1-7(12), clasping stem at base   8
7 Stem-leaves 0-8(12), not clasping stem at base   9
8 Stem-leaves yellowish-green; plant viscid-
   glandular   **9. H. sect. Amplexicaulia**
8 Stem-leaves glaucous-green; plant not viscid-
   glandular   **10. H. sect. Cerinthoidea**
9 Leaves with small glandular hairs on margins and
   sometimes on surface; phyllaries with shaggy hairs;
   almost confined to Sc, N Wa and Lake District   10
9 Leaves without stalked glands; phyllaries without
   shaggy hairs; widespread   11
10 Stem-leaves 0-4, narrow and bract-like; capitula
   1(-5); plants to 15(30)cm   **12. H. sect. Alpina**
10 Stem-leaves (0)1-4, usually at least one leaf-
   like; capitula (1)2-5; plants to 45cm
                                  11. H. sect. Subalpina
11 Leaves usually bristly at least along margins;
   phyllaries erect in bud, without dense white stellate
   hairs   **8. H. sect. Oreadea**
11 Leaves variously pubescent but not bristly; phyllaries
   incurved in bud, with dense white stellate hairs at
   least on margins   **7. H. sect. Vulgata**

1. H. sect. Umbellata F. Williams. 1 microsp. currently
placed here (**H. umbellatum** L.). Native; sandy heathland,
dunes and dry rocky places, often near the coast; scattered
throughout BI. Ssp. **bichlorophyllum** (Druce & Zahn) Sell & C.
West has broader leaves (the lower ovate-lanceolate to
oblong, not linear to linear-lanceolate) and is confined to
Wa, SW En, W Ir and CI.
2. H. sect. Sabauda F. Williams. 5 microspp. currently
placed here. Native; common in rough ground, grassy and
marginal places and on roadside and railway banks; En and Wa,
rather local in Sc, very local in E Ir. **H. sabaudum** L. (H.
perpropinquum (Zahn) Druce) is the commonest many-leaved
Hieracium of S & C En.

3. H. sect. Foliosa Pugsley. 10 microspp. currently placed here. Native; grassy and rocky places; locally common in N Br S to S Wa and Peak District, very local in E, N & W Ir.

4. H. sect. Tridentata F. Williams. 21 microspp. currently placed here. Native; grassy, rocky and marginal habitats; frequent in Br, rare in W & N Ir.

5. H. sect. Prenanthoidea Koch. 2 microspp. currently placed here. Native; grassy and rocky places, often on limestone; local in N Br S to S Wa and Peak District, Co Antrim.

6. H. sect. Alpestria F. Williams. 18 microspp. currently **RR** placed here. Native; rocky places, cliffs, hillsides; 13 microspp. endemic to Shetland, others very local in Sc and N En.

7. H. sect. Vulgata F. Williams (sects. <u>Bifida</u> (Pugsley) Clapham, <u>Caesia</u> (W.R. Linton) Clapham, <u>Glandulosa</u> (Pugsley) Clapham, <u>Sagittata</u> (Pugsley) Clapham). c.83 microspp. currently placed here. Native and intrd; rough ground, woodland, marginal habitats, cliffs and rocky places; throughout BI. The commonest microspp. of lowland Br belong here.

8. H. sect. Oreadea Zahn (sect. <u>Suboreadea</u> Pugsley). c.44 microspp. currently placed here. Native; cliffs, rocky and grassy banks, often on limestone; scattered in Ir, Wa, Sc and W, C & N En.

9. H. sect. Amplexicaulia Zahn. 3 microspp. currently placed here. Intrd; walls and rough ground; very scattered in En and Sc; C Europe to Pyrenees.

10. H. sect. Cerinthoidea Koch. 10 microspp. currently placed here. Native; cliffs and rocky streamsides; coastal and upland areas of Sc, N En, N & W Ir, 1 site in Wa. The commonest microsp. of Ir (**H. anglicum** Fries) belongs here.

11. H. sect. Subalpina Pugsley. c.31 microspp. belong here. Native; rock-ledges and rocky stream-sides usually above 450m; local in mainland Sc and N En.

12. H. sect. Alpina F. Williams. 18 microspp. belong here. Native; rock-ledges, barish slopes and scree, grassy banks, usually above 650m; local in mainland Sc, Lake District and N Wa.

TRIBE 3 – ARCTOTIDEAE (genera 32-33). Plants not producing white latex, not spiny; capitula radiate, with sterile ligulate flowers with yellow to orange, usually 3-toothed ligules.

32. ARCTOTHECA Wendl. (<u>Cryptostemma</u> R. Br.) – <u>Plain Treasure-flower</u>

Annuals or perennials (annuals in BI); lower leaves deeply pinnately lobed, white-tomentose on lowerside; phyllaries in several rows, free, glabrous to sparsely pubescent, with conspicuous scarious, rounded to obtuse tips; receptacular scales 0; achenes densely pubescent; pappus of distinct scales.

1. A. calendula (L.) Levyns (<u>Cryptostemma</u> <u>calendulacea</u> (Hill) R.Br.) – <u>Plain Treasureflower</u>. Stems decumbent, leafy only near base, to 40cm, white-pubescent; all leaves usually deeply lobed; capitula on long peduncles; ligules pale yellow on upperside, purplish on lowerside. Intrd; rather frequent wool-alien in arable fields and waste places; scattered in Br; S Africa.

33. GAZANIA Gaertner – <u>Treasureflower</u>
Differs from <u>Arctotheca</u> in outer phyllaries fused into cup-like structure, white-tomentose on lowerside, with acute to acuminate scarious tips; leaves sometimes all simple.

1. G. rigens (L.) Gaertner (<u>G.</u> <u>uniflora</u> (L.f.) Sims, <u>G.</u> <u>splendens</u> hort.) – <u>Treasureflower</u>. Stems decumbent to ascending, to 50cm, often ± woody at base, white-tomentose below; lower leaves deeply pinnately lobed, upper ones ± entire; capitula on long peduncles; ligules orange-yellow with basal black blotch bearing central white spot, or rarely plain yellow (var. **uniflora** (L.f.) Roessler). Intrd; grown in gardens, ± natd on walls, rocks and cliffs near sea; Scillies and CI; S Africa.

SUBFAMILY 2 – ASTEROIDEAE (Tubuliflorae) (tribes 4-10). Plant not producing white latex; stem-leaves usually spiral ('alternate'), sometimes opposite or 0; capitula mostly radiate, sometimes discoid; tubular flowers most commonly yellow to orange, usually with short lobes or teeth; filaments joining anthers at base; 2 style branches each with 2 stigmatic zones, 1 near each margin of inner face; pollen grains spiny.

TRIBE 4 – INULEAE (genera 34-42). Annual to perennial herbs, rarely shrubs; leaves alternate, simple; capitula discoid or radiate; phyllaries in several rows, herbaceous to scarious; receptacular scales present or (usually) 0; pappus usually of hairs, sometimes of short scales or of scales and hairs; flowers usually yellow or brown to whitish.

34. FILAGO L. (<u>Logfia</u> Cass.) – <u>Cudweeds</u>
Annuals with stems and leaves ± covered with woolly hairs; capitula small, brownish, borne in clusters of 2-c.40; phyllaries in few ill-defined rows, the outer herbaceous, the inner scarious; receptacle conical, with scales associated with outer (female) florets only; flowers all tubular, the inner bisexual, with wider corollas than the outer female; pappus of bisexual flowers of simple hairs, of female flowers of simple hairs or 0.

1 Capitula 2-7(14) in each cluster; outer phyllaries
   obtuse to subacute, patent in fruit   2
1 Capitula (5)10-c.40 in each cluster; outer phyllaries

acuminate, erect in fruit   3
2 Leaves 4-10mm, the most apical ones not over-
   topping clusters of capitula   **4. F. minima**
2 Leaves (8)12-20(25)mm, the most apical ones overtopping clusters of capitula   **5. F. gallica**
3 Leaves widest in basal 1/2, the most apical ones not
   overtopping clusters of capitula; capitula in
   clusters of (15)20-40   **1. F. vulgaris**
3 Leaves widest in apical 1/2, the most apical ones
   usually overtopping clusters of capitula; capitula in
   clusters of (5)10-20(25)   4
   4 Clusters of capitula each overtopped by (0)1-2
      leaves; outer phyllaries with erect red-tinged
      points; plant usually yellowish-woolly
                                             2. F. lutescens
   4 Clusters of capitula each overtopped by 2-4(5)
      leaves; outer phyllaries with recurved yellowish
      points; plant white-woolly   **3. F. pyramidata**

Other spp. - **F. arvensis** L. (**Logfia** arvensis (L.) Holub),
from Europe, once occurred regularly as a casual but is now
rare; it differs from **F.** minima in its racemosely branching,
not bifurcating stems and outer phyllaries woolly to the apex
(not glabrous near apex).

1. F. vulgaris Lam. (**F.** germanica L. non Hudson) - Common
Cudweed. Stems erect, to 40cm, branching below each cluster
of capitula; clusters of capitula ± globose, c.10-12mm
across. Native; barish places on sandy soils, e.g. heaths,
waysides, sand-pits; throughout most of BI, but absent from
most of N & W Sc.
2. F. lutescens Jordan (**F.** apiculata G.E. Smith ex Bab.) - RRR
Red-tipped Cudweed. Differs from **F.** vulgaris in ±
irregularly branched stems; yellowish (not whitish) hairs;
apiculate (not acute) leaves; and see key (couplet 3).
Native; similar places to **F.** vulgaris; very local in S & E En
from S Hants to SE Yorks, formerly W to Worcs.
3. F. pyramidata L. (**F.** spathulata auct. non C. Presl) - RR
Broad-leaved Cudweed. Differs from **F.** vulgaris in leaves
often apiculate (not acute); outer phyllaries with recurved
(not erect) tips; and see key (couplet 3). Native; similar
places to **F.** vulgaris; very local in S En, formerly N to N
Lincs and in Jersey.
4. F. minima (Sm.) Pers. (**Logfia** minima (Sm.) Dumort.) -
Small Cudweed. Stems erect, to 25cm, irregularly branching;
clusters of capitula ovoid, c.2-5mm across. Native; similar
habitats and distribution to **F.** vulgaris.
5. F. gallica L. (**Logfia** gallica (L.) Cosson & Germ.) -
Narrow-leaved Cudweed. Differs from **F.** minima in key
characters; inflorescence appearing very leafy. Intrd; natd
in sandy and gravelly ground; Sark since 1902, formerly S
Hants to N Essex from 1696 to 1955; S & W Europe.

35. ANTENNARIA Gaertner – <u>Mountain Everlasting</u>
Dioecious whitish-woolly perennials; capitula pale to deep pink, sometimes whitish, borne in terminal umbel-like clusters of 2-8; phyllaries in several rows, scarious, the outer ones of male capitula patent and perianth-like in flower, white to pink; receptacle flat, without scales; flowers all tubular, males with wider corollas than females; pappus of female flowers of simple hairs, of male flowers of simple hairs widened distally.

1. A. dioica (L.) Gaertner – <u>Mountain Everlasting</u>. Stems erect, to 20cm, with basal leaf-rosette; surface-creeping leafy stolons present; leaves green on upperside, white-woolly on lowerside. Native; heaths, moors, mountain slopes; common in much of N 1/2 of BI, scattered S to S Ir and W Cornwall, much reduced in S & E Br, no longer in CE or SE En except 1 place in Northants.

36. ANAPHALIS DC. – <u>Pearly Everlasting</u>
Dioecious to variably sexed white-woolly perennials; capitula white, with yellow flowers, borne in large terminal corymbose inflorescences; phyllaries in several rows, scarious, pearly white; receptacle convex, without scales; flowers all tubular, male and female variously arranged, the males with wider corollas; pappus of 1 row of hairs.

1. A. margaritacea (L.) Benth. – <u>Pearly Everlasting</u>. Stems **844** erect to ascending, to 1m, without basal leaf-rosette; rhizomes present; leaves green on upperside, white-woolly on lowerside. Intrd; common in gardens and well natd as relic or throw-out by rivers and in grassland, marginal and rough ground; scattered in BI, locally frequent in W: N America.

37. GNAPHALIUM L. (<u>Omalotheca</u> Cass., <u>Filaginella</u> Opiz, <u>Gamochaeta</u> Wedd., <u>Pseudognaphalium</u> Kirpiczn.) – <u>Cudweeds</u>
Annuals or perennials ± covered with whitish woolly hairs; capitula small, yellowish to brown, variously arranged; phyllaries whitish to yellowish or brown, subherbaceous to scarious, in several rows; receptacle flat, without scales; flowers all tubular, the inner bisexual, with wider corollas than the outer female: pappus of simple hairs.

1 Capitula in terminal, subglobose to subcorymbose
  clusters; annual   **2**
1 Capitula in elongated, racemose clusters, sometimes
  few or rarely 1; annual to perennial   **4**
  2 Phyllaries brown; clusters of capitula
    conspicuously leafy   **5. G. uliginosum**
  2 Phyllaries uniformly white to yellowish: clusters
    of capitula not leafy   **3**
3 Leaves white-woolly on both sides, not decurrent down
  stem   **6. G. luteoalbum**

3 Leaves green on upperside, white-woolly on lowerside,
   decurrent down stem   **7. G. undulatum**
   4 Annual to biennial without rhizome; achenes <1mm,
      glabrous; phyllaries acute to acuminate
                                                 4. G. purpureum
   4 Perennial with short ± surface rhizome; achenes
      >1mm, pubescent; phyllaries obtuse to rounded or
      retuse   5
5 Capitula <10 per stem; pappus-hairs free, falling
   separately   **3. G. supinum**
5 Capitula normally >10 per stem; pappus-hairs united
   at base, falling as a unit   6
   6 Stem-leaves 1-(or indistinctly 3-)veined, steadily
      diminishing in size up the stem   **2. G. sylvaticum**
   6 Stem-leaves 3(5)-veined, scarcely diminishing in
      size until above 1/2 way up stem   **1. G. norvegicum**

1. G. norvegicum Gunnerus (<u>Omalotheca</u> <u>norvegica</u> (Gunnerus)   **RR**
Schultz-Bip. & F. Schultz) - <u>Highland</u> <u>Cudweed</u>. Stems erect,
to 30cm; leaves grey-green on upperside; phyllaries grey-
green in centre, brownish-scarious around edges. Native;
mountain rocks and gravel; very local in C Sc.
   2. G. sylvaticum L. (<u>Omalotheca</u> <u>sylvatica</u> (L.) Schultz-Bip.   **R**
& F. Schultz) - <u>Heath</u> <u>Cudweed</u>. Stems erect, to 60cm; leaves
green on upperside; phyllaries green in centre, brownish-
scarious around edges. Native; rather open ground on heaths,
banks, woodland rides; locally frequent in most of Br and Ir.
   3. G. supinum L. (<u>Omalotheca</u> <u>supina</u> (L.) DC.) - <u>Dwarf</u>
<u>Cudweed</u>. Stems erect, to 12(20)cm; leaves grey-woolly on
upperside; phyllaries grey-green in centre, brownish-scarious
around edges. Native; mountain rocks and gravel; local in C,
N & W mainland Sc and Skye (N Ebudes).
   4. G. purpureum L. (<u>G.</u> <u>pensylvanicum</u> Willd. nom. illeg.) -   **844**
<u>American</u> <u>Cudweed</u>. Stems decumbent to erect, to 40cm; leaves
woolly on upperside; phyllaries green, with brown tips.
Intrd; natd in churchyard, Surrey, since at least 1978; USA.
   5. G. uliginosum L. (<u>Filaginella</u> <u>uliginosa</u> (L.) Opiz) -
<u>Marsh</u> <u>Cudweed</u>. Stems decumbent to erect, to 25cm; leaves
woolly on upperside; phyllaries pale brown with dark brown
tips. Native; damp places in fields and arable land and by
ponds and paths; common throughout BI.
   6. G. luteoalbum L. (<u>Pseudognaphalium</u> <u>luteoalbum</u> (L.)   **RRR**
Hilliard & B.L. Burtt) - <u>Jersey</u> <u>Cudweed</u>. Stems erect, to
50cm; leaves woolly on upperside; phyllaries scarious, straw-
coloured. Native; sandy fields, waste places and sand-
dunes; very local in Jersey, Guernsey and W Norfolk, formerly
E Norfolk, W Suffolk and Cambs.
   7. G. undulatum L. (<u>Pseudognaphalium</u> <u>undulatum</u> (L.) Hilliard   **844**
& B.L. Burtt) - <u>Cape</u> <u>Cudweed</u>. Stems erect, to 80cm; leaves
green on upperside; phyllaries whitish-scarious. Intrd;
rough ground, cliffs, marginal habitats; natd in CI (all main
islands) since 1888; S Africa. Resembles a small <u>Anaphalis</u>.

38. HELICHRYSUM Miller - Everlastingflowers
Mat-forming woody perennials, or annuals; capitula conspicuous, terminal, usually solitary; phyllaries in several rows, scarious, white or coloured; receptacle slightly convex, without scales; flowers all tubular, the outer female, the inner bisexual and with wider corollas; pappus of 1 row of hairs.

Other spp. - **H. bracteatum** (Vent.) Andrews, the common annual everlastingflower of florists, has capitula >20mm across with yellow, red, white or mixed-coloured phyllaries; it is sometimes found on tips in SE En.

1. H. bellidioides (G. Forster) Willd. - New Zealand Ever- 844
lastingflower. Mat-forming dwarf evergreen shrub with decumbent stems to 60cm; leaves c.5-8mm, Thymus-like, silvery on lowerside; capitula on erect stems up to 10cm, c.2-3cm across; flowers yellow, surrounded by patent white inner phyllaries up to 1cm. Intrd; natd in rocky turf by stream; 1 place in Shetland, since 1975; New Zealand.

39. INULA L. - Fleabanes
Perennial herbs, sometimes woody at base; capitula 1-many, terminal, usually subcorymbose, usually showy, yellow, usually radiate, less often discoid; phyllaries in several rows, herbaceous; receptacle flat or slightly convex, without scales; pappus of 1 row of hairs.

1 Stem and leaves succulent, glabrous   **4. I. crithmoides**
1 Stem and leaves not succulent, very sparsely to
  densely pubescent   2
  2 Ligules 0 or <1mm; capitula numerous on each stem
            3. I. conyzae
  2 Ligules conspicuous, >1cm; capitula 1-c.5 on each
    stem   3
3 Outer phyllaries ovate; capitula >5cm across (incl.
  ligules); stems rarely <1m   **1. I. helenium**
3 Outer phyllaries lanceolate; capitula <5cm across
  (incl. ligules); stems <1m   **2. I. salicina**

Other spp. - **I. britannica** L., from Europe, was natd by a reservoir in Leics from 1894 to at least 1932; it resembles I. salicina in stature but has broader leaves pubescent (not ± glabrous) on lowerside, and linear pubescent (not lanceolate glabrous) outer phyllaries.

1. I. helenium L. - Elecampane. Stems erect, to 2.5m, pubescent; stem-leaves ovate, cordate, sessile, pubescent; capitula few, c.6-9cm across (incl. ligules). Intrd; natd in fields, waysides, marginal habitats, rough ground; scattered throughout BI, less common than formerly since grown less; W & C Asia. See Telekia for differences.

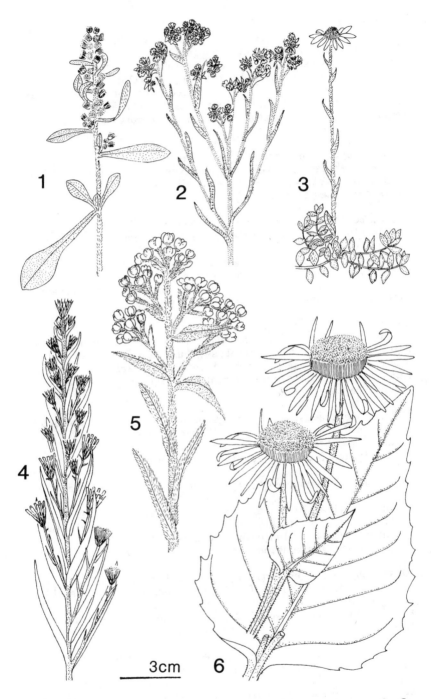

FIG 844 – Asteraceae: Inuleae. 1, *Gnaphalium purpureum*. 2, G. *undulatum*. 3, **Helichrysum bellidioides**. 4, **Dittrichia viscosa**. 5, **Anaphalis margaritacea**. 6, *Telekia speciosa*.

2. I. salicina L. - Irish Fleabane. Stems erect, to 70cm, ± RRR glabrous; leaves narrowly elliptic to narrowly ovate, subglabrous; capitula few, c.2.5-4.5cm across (incl. ligules). Native; stony limestone shores of Lough Derg, N Tipperary (and formerly SE Galway).

3. I. conyzae (Griess.) Meikle (I. conyza DC.) - Ploughman's-spikenard. Stems erect, to 1.25m, pubescent; leaves narrowly ovate to narrowly obovate, pubescent; capitula numerous, c.0.7-1.2cm across (incl. ligules). Native; scrub, grassland and barish places on calcareous soils; locally common throughout En, Wa and CI.

4. I. crithmoides L. - Golden-samphire. Stems erect to R decumbent, to 1m, glabrous; leaves linear to oblanceolate, succulent, glabrous; capitula rather few, c.1.5-2.5cm across (incl. ligules). Native; salt-marshes, shingle, cliffs, rocks and ditchsides by sea; local on coasts of BI N to Co Louth and E Suffolk.

40. DITTRICHIA Greuter - Fleabanes
Low shrubs or annual herbs with glandular-sticky stems and leaves; capitula several in racemose inflorescences, rather showy, yellow, radiate but ligules often very short; phyllaries in several rows, herbaceous; receptacle flat or slightly convex, without scales; pappus of 1 row of hairs fused at base.

1. D. viscosa (L.) Greuter (Inula viscosa (L.) Aiton) - 844 Woody Fleabane. Perennial with resinous smell when crushed; stems ascending to erect, woody, to 1m;; capitula 1-1.5(2)cm across; ligules ≤10mm, much longer than phyllaries. Intrd; ± natd in rough ground in E Suffolk, casual elsewhere in S En; S Europe.

2. D. graveolens (L.) Greuter (Inula graveolens (L.) Desf.) - Stinking Fleabane. Annual with strong camphorous smell when crushed; stems erect, to 50cm; capitula 0.6-1.2cm across; ligules ≤3mm, not or scarcely exceeding phyllaries. Intrd; a rather frequent wool-alien in fields etc.; scattered in En; S Europe.

41. PULICARIA Gaertner - Fleabanes
Annuals or perennials; capitula several to many, terminal, usually subcorymbose, usually showy, yellow, radiate; phyllaries in several rows, herbaceous; receptacle flat, without scales; pappus of 1 row of hairs plus an outer row of free or fused scales.

1. P. dysenterica (L.) Bernh. - Common Fleabane. Densely pubescent perennial with extensive rhizomes; stems erect, to 1m; stem-leaves sessile, cordate at base; capitula 1.5-3cm across; ligules usually c.1.5x as long as phyllaries. Native; marshes, ditches, wet fields and hedgebanks; common in lowland BI N to N En, rare in S Sc and N Ir.

2. P. vulgaris Gaertner - <u>Small</u> <u>Fleabane</u>. Pubescent annual; **RRR**
stems erect, to 45cm; stem-leaves sessile, rounded to cuneate
at base; capitula 0.6-1.2cm across; ligules c. as long as
phyllaries. Native; sandy places flooded in winter, often by
ponds; formerly widespread in CI and En N to Leics, now very
local in Surrey, S Hants and S Wilts.

42. TELEKIA Baumg. - <u>Yellow</u> <u>Oxeye</u>
Herbaceous perennials; capitula 1 to several, terminal, subcorymbose, showy, yellow, radiate; phyllaries in several
rows, herbaceous; receptacle convex, with scales; pappus of
fused scales. Probably best united with **Buphthalmum** L.

1. T. speciosa (Schreber) Baumg. - <u>Yellow</u> <u>Oxeye</u>. Stems **844**
erect, pubescent, to 2m; lower stem-leaves petiolate and
cordate at base, upper ones sessile and rounded to broadly
cuneate at base; capitula 5-8cm across; ligules 1-2.5cm.
Intrd; grown in gardens and natd in rough ground and by lakes
and rivers; scattered throughout most of Br, mostly in N; C &
SE Europe. Resembles <u>Inula</u> <u>helenium</u> in habit, but basal
leaves deeply cordate (not narrowly cuneate), leaf-margins
sharply dentate (not crenate-dentate), and note pappus.

TRIBE 5 - ASTEREAE (genera 43-53)
Annual to (usually) perennial herbs, rarely shrubs; leaves
alternate or all basal, simple; capitula discoid or radiate;
phyllaries in 2-several rows, usually herbaceous; receptacular scales 0; pappus usually of hairs, sometimes 0 or of
strong bristles; flowers various colours.

43. GRINDELIA Willd. - <u>Gumplants</u>
Herbaceous perennials; stem-leaves sessile, clasping stem,
serrate; capitula radiate, with yellow ray and disc flowers;
phyllaries sticky, in several rows, herbaceous, with recurved
tips; pappus of 2-8 stiff bristles.

Other spp. - Records of **G. squarrosa** (Pursh) Dunel and **G.**
rubricaulis DC., from USA, are either old or errors.

1. G. stricta DC. - <u>Coastal</u> <u>Gumplant</u>. Stems erect, to
75cm, sparsely shaggy-pubescent; capitula 3-5cm across;
ligules 8-15mm. Intrd; natd on sea-cliffs at Whitby, NE
Yorks, since 1977; W coast of N America.

44. CALOTIS R.Br. - <u>Bur Daisy</u>
Annuals or perennials; stem-leaves various; capitula radiate
or ± discoid, with yellow disc and yellow, white or mauve ray
flowers; phyllaries in ±2 rows, herbaceous; pappus of
(1-)3(-6) rigid barbed bristles and usually some extra
shorter bristles or scales; fruiting capitula forming a
globose bur 5-9mm across.

Other spp. - Several other spp. occur as wool-aliens, of which **C. lappulacea** Benth. and **C. hispidula** (F. Muell.) F. Muell., from Australia, both with short yellow ligules, are most common. The former is perennial with linear stem-leaves, glabrous bristles and ligules exceeding phyllaries; the latter is annual with oblanceolate stem-leaves, pubescent bristles and ligules shorter than phyllaries.

1. C. cuneifolia R. Br. - <u>Bur</u> <u>Daisy</u>. Perennial with 856 branching, erect to procumbent stems to 30(60)cm; leaves obtriangular, narrowed to petiole, dentate at distal end; capitula 1-2cm across in flower, with white or mauve ligules 3-9mm, much longer than phyllaries. Intrd; rather characteristic wool-alien, scattered in En; Australia.

45. SOLIDAGO L. (<u>Euthamia</u> Nutt.) - <u>Goldenrods</u>
Perennials; stem-leaves narrowly elliptic to oblanceolate or obovate, serrate, narrowed to base; capitula small, numerous, ± crowded, radiate, yellow; phyllaries in many rows, herbaceous; pappus of 1-2 rows of hairs.

1 Capitula sessile, in small clusters forming corymbose inflorescence; leaves gland-spotted, ligules 1-1.5mm
5. S. graminifolia
1 At least most capitula stalked, forming pyramidal to ± cylindrical inflorescence; leaves not gland-spotted; ligules >1.5mm   2
  2 Leaves with many pairs of short lateral veins (though often inconspicuous)   3
  2 Leaves with 1(-2) pairs of main lateral veins from near base, running parallel with midrib for most of length   4
3 Most stem-leaves rounded to acute at apex; inner phyllaries >4.5mm; disc flowers >10; ligules 4-9mm
1. S. virgaurea
3 Most stem-leaves acute to acuminate at apex; inner phyllaries <4.5mm; disc flowers <8; ligules 1.5-4mm
2. S. rugosa
  4 Leaves scabrid-pubescent on surfaces, stems pubescent at least in top 1/2   **3. S. canadensis**
  4 Leaves glabrous on surfaces or pubescent only on lowerside veins; stems ± glabrous   **4. S. gigantea**

1. S. virgaurea L. - <u>Goldenrod</u>. Stems erect, to 70(100)cm, but often much less, glabrous to pubescent; capitula in a raceme, or in a panicle with straight erect branches; disc flowers 10-30; ligules 6-12. Native; open woodland, grassland, hedgerows, rocky places, cliffs; frequent over most of BI except parts of C En, C Ir and CI. Very variable.
1 x 3. S. virgaurea x S. canadensis = S. niederederi Khek was found on a railway bank in W Kent in 1979; it is closer to <u>S. canadensis</u> in inflorescence shape and capitulum size

and to S. virgaurea in leaf venation, and is sterile.
2. S. rugosa Miller (S. altissima Aiton non L.) - <u>Rough-</u>
<u>stemmed</u> <u>Goldenrod</u>. Stems erect, to 1.5m, roughly pubescent;
capitula on erecto-patent to patent curved branches forming
terminal pyramidal inflorescence; disc flowers 3-8; ligules
6-10. Intrd; natd as for S. canadensis; 2 places in
Dunbarton, formerly Main Argyll; N America.
3. S. canadensis L. (S. altissima L.) - <u>Canadian</u> <u>Goldenrod</u>.
Stems erect, to 2.5m, pubescent; capitula on erecto-patent to
patent curved branches forming terminal pyramidal inflorescence; disc flowers 2-8;: ligules 7-15. Intrd; much grown in
gardens, fully natd on waste land, banks, waysides and rough
grassland; frequent throughout C & S Br and CI, scattered in
N Br and Ir; N America. **S. altissima** is sometimes recognized
as a distinct sp., as ssp. altissima (L.) O. Bolos & Vigo, or
as var. scabra Torrey & A. Gray; it has stems pubescent
throughout (not just in upper 1/2), phyllaries 2.5-4.5mm (not
2.1-3mm), ligules 2.4-4.1mm (not 1.5-2.8mm) and more sharply
serrate leaves, and probably occurs with us, as might hybrids
of S. canadensis with S. rugosa and S. gigantea. The precise
identity of British material of S. canadensis in the sense
used here is uncertain.
4. S. gigantea Aiton (S. serotina Aiton non Retz.) - <u>Early</u>
<u>Goldenrod</u>. Differs from S. canadensis in disc flowers
6-10(12); and see key. Intrd; natd as for S. canadensis,
often with it but less common; N America. Our plant is ssp.
serotina (O. Kuntze) McNeill (var. serotina (O. Kuntze)
Cronq., var. leiophylla Fern.).
5. S. graminifolia (L.) Salisb. (Euthamia graminifolia (L.) **856**
Elliott) - <u>Grass-leaved</u> <u>Goldenrod</u>. Stems erect, to 1.5m,
glabrous to pubescent; leaves linear to linear-lanceolate,
appearing 3(-5)-veined; inflorescence flat-topped; disc
flowers (4)5-10(13); ligules 12-15. Intrd; natd as for S.
canadensis; very scattered in En, older (?extant) records in
C & S Sc; N America.

46. ASTER L. (Crinitaria Cass.) - <u>Michaelmas-daisies</u>
Perennials; stem-leaves ovate to linear, entire, with various
bases; capitula conspicuous, radiate or discoid, with yellow
disc flowers and white to blue, pink or purple ligules;
phyllaries in 2-several rows, herbaceous or partly
membranous; pappus of 1-2 rows of hairs.
   The cultivated Michaelmas-daisies that are found in the wild
are difficult to determine due to hybridization between A.
novi-belgii and 2 other spp. These 2 hybrids and A.
lanceolatus appear to be the commonest taxa, and show every
grade of variation from 1 parent to the other.

1   Basal and lower stem-leaves with long petiole and
       cordate base   **1. A. schreberi**
1   Basal leaves cuneate at base; stem-leaves cuneate at
       base and/or sessile   2

2 Leaves all 1-veined, linear to very narrowly
elliptic, gland-spotted, not succulent;
maritime cliffs   **9. A. linosyris**
2 Leaves mostly with well-developed lateral veins,
if all 1-veined then succulent, not gland-spotted
but sometimes with stalked glands; widespread   3
3 Leaves succulent, with 0-few lateral veins mostly
running parallel with midrib; mostly maritime
                                         8. A. tripolium
3 Leaves not succulent, usually with normally developed
lateral veins, widespread   4
4 Upper part of plant with abundant long patent hairs
and shorter stalked glands   **2. A. novae-angliae**
4 Plant with 0 or rather sparse long patent hairs;
stalked glands 0   5
5 Upper leaves tapering to base, not clasping stem;
leaves rarely >1cm wide; phyllaries <5mm; ligules
usually white (see also A. x salignus)
                                         7. A. lanceolatus
5 Upper leaves tapering to base or not, but distinctly
(though often narrowly) clasping stem at base; some
leaves >1cm wide; phyllaries >5mm; ligules usually
coloured   6
6 Phyllaries with wide white borders in basal 1/2
and narrow ones in apical 1/2, leaving elliptic
to trullate green patch in centre near apex; outer
phyllaries reaching <1/2 as high as inner ones   7
6 Phyllaries wholly or mainly green in apical 1/2,
hence appearing leafy near apex; outer phyllaries
usually reaching >1/2 as high as inner ones   8
7 Leaves distinctly glaucous on upperside; outer
phyllaries usually reaching distinctly <1/2 as high
as inner ones; plant usually <1m   **3. A. laevis**
7 Leaves not glaucous; outer phyllaries often reaching
c.1/2 as high as inner ones; plant usually 1-2m
                                         4. A. x versicolor
8 Middle stem-leaves mostly 2.5-5x as long as wide,
conspicuously clasping stem; outer phyllaries
usually c.1/2-3/4x as high as inner ones
                                         4. A. x versicolor
8 Middle stem-leaves mostly 4-10x as long as wide,
usually very narrowly clasping stem; outer
phyllaries nearly as long as inner ones   9
9 Outer phyllaries widest below middle, rather neatly
appressed to capitulum   **6. A. x salignus**
9 Outer phyllaries widest at or just above middle,
with conspicuous leafy apical 1/2 loosely or unevenly
appressed to capitulum   **5. A. novi-belgii**

Other spp. - Many other spp. are grown in gardens and some
may escape, but most records are suspect due to misidentification. Other spp. that might occur are **A. puniceus** L., with

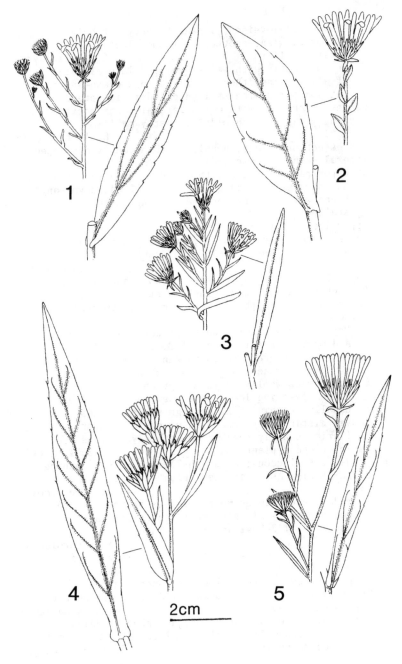

FIG 850 - Apex of inflorescence and mid stem-leaf of **Aster**.
1, **A. laevis**. 2, **A. x versicolor**. 3, **A. lanceolatus**. 4, **A.
novi-belgii**. 5, **A. x salignus**.

hispid stem and leaves and very leafy ± recurved tips to phyllaries; **A. dumosus** L., with a much-branched inflorescence, phyllaries ≤6mm and narrow non-clasping leaves (some of the dwarf Michaelmas-daisies may belong here or to hybrids of it); **A. concinnus** Willd., differing from last in wider (>1cm) leaves; and **A. foliaceus** Lindley, similar in leaf-shape to <u>A. laevis</u> but with leafy phyllaries, few, large capitula, and non-glaucous leaves. More distinct are the European spp. **A. amellus** L., with pubescent stems to 60cm, 2-6 capitula in corymb, and pubescent ± petiolate lower stem-leaves, and **A. sedifolius** L., with 1-3-veined very narrow gland-spotted leaves, numerous capitula crowded into a corymb, and scabrid stems to 60cm; both have been found to persist for a few years.

1. A. schreberi Nees (<u>A. macrophyllus</u> auct. non L.) - <u>Nettle-leaved</u> <u>Michaelmas-daisy</u>. Stems erect, to 1m, puberulent; phyllaries with narrow green midline and wide scarious margins, outer reaching <1/2 as high as inner; ligules whitish-grey tinged mauve. Intrd; natd on railway bank in Renfrews since 1931; N America.

2. A. novae-angliae L. - <u>Hairy</u> <u>Michaelmas-daisy</u>. Stems erect, to 2m, with long patent hairs; phyllaries all green and/or purple, very glandular, ± equal; ligules usually bright pinkish-purple. Intrd; natd on waste and rough ground; very scattered in Br, mainly S & C; N America. The fresh inflorescence smells strongly of <u>Calendula</u> when crushed.

3. A. laevis L. - <u>Glaucous</u> <u>Michaelmas-daisy</u>. Stems erect, 850 to 1m, ± glabrous; phyllaries as in key (couplet 6); ligules usually bluish-purple. Intrd; natd on waste and rough ground; rather rare in S & C En; N America.

4. A. x versicolor Willd. (<u>A. novi-belgii</u> ssp. <u>laevigatus</u> 850 (Lam.) Thell.; <u>A. laevis</u> x <u>A. novi-belgii</u>) - <u>Late Michaelmas-daisy</u>. Stems erect, to 2m, ± glabrous; phyllaries similar to those of <u>A. laevis</u> but more leafy at apex and less unequal; ligules usually bluish-purple. Intrd; natd on waste and rough ground; scattered in Br, probably under-recorded; garden origin. Most of the taller, larger flowered, late-flowering cultivars, often with dark red stems, belong here.

5. A. novi-belgii L. (<u>A. longifolius</u> auct. non Lam.) - 850 <u>Confused Michaelmas-daisy</u>. Stems erect, to 1.5m, glabrous to sparsely pubescent; phyllaries as in key (couplets 8 & 9), with apical 1/2 ± entirely green; ligules usually mauve, but white to purple in cultivars. Intrd; natd on waste and rough ground; scattered over Br, greatly over-recorded for <u>A. x salignus</u>; N America.

6. A. x salignus Willd. (<u>A. longifolius</u> auct. non Lam.; <u>A.</u> 850 <u>novi-belgii</u> x <u>A. lanceolatus</u>) - <u>Common Michaelmas-daisy</u>. Differs from <u>A. novi-belgii</u> in leaves narrower and scarcely clasping stem; and see key (couplet 9). Intrd; natd on waste and rough, often damp ground; easily the commonest natd

Michaelmas-daisy in BI, reproducing from seed and often weedy
in appearance; garden origin.
   7. A. lanceolatus Willd. – <u>Narrow-leaved</u> <u>Michaelmas-daisy</u>. **850**
Stems erect, to 1.2m, glabrous or sparsely pubescent;
phyllaries with conspicuous membranous borders, not leafy at
apex, the outer reaching c.1/2 as high as inner ones; ligules
white or pale mauve. Intrd; natd on waste and rough ground;
frequent throughout much of BI, under-recorded, less common
than only <u>A.</u> x <u>salignus</u>; N America.
   8. A. tripolium L. – <u>Sea</u> <u>Aster</u>. Sometimes annual or
biennial; stems erect, to 1m, glabrous; phyllaries fewer and
much blunter than in all other spp., unequal; ligules bluish-
mauve or 0. Native; salt-marshes, less often cliffs and
rocks on coasts around whole BI, rare in inland saline areas.
   9. A. linosyris (L.) Bernh. (<u>Crinitaria</u> <u>linosyris</u> (L.) **RR**
Less.) – <u>Goldilocks</u> <u>Aster</u>. Stems erect to decumbent, to
50cm, glabrous; phyllaries unequal, the outer + wholly green
and loosely appressed to capitulum; ligules 0. Native;
limestone sea-cliffs; very local in W Br from S Devon to
Westmorland.

47. CHRYSOCOMA L. – <u>Shrub</u> <u>Goldilocks</u>
Glabrous shrublets; stem-leaves linear to filiform, the edges
rolled under, entire, sessile; capitula yellow, discoid, 1 on
end of each branch; phyllaries in several rows, narrow,
herbaceous with membranous margins; pappus of 1 row of hairs.

   1. C. coma-aurea L. – <u>Shrub</u> <u>Goldilocks</u>. Stems thin, erect
to ascending, to 60cm; leaves linear, absent from apical
1-3cm of stem below capitula; capitula c.10-15mm across.
Intrd; grown in gardens, natd on walls and open ground;
Scillies; S Africa.
   2. C. tenuifolia P. Bergius – <u>Fine-leaved</u> <u>Goldilocks</u>.
Differs from <u>C.</u> <u>coma-aurea</u> in leaves filiform, ascending to
within <1cm of capitula; capitula c.5-10mm across. Intrd;
fairly frequent wool-alien; scattered in En; S Africa.

48. ERIGERON L. – <u>Fleabanes</u>
Annuals to perennials; stem-leaves linear to obovate, entire
or toothed, sessile or shortly petiolate; capitula 1-many per
stem, radiate, with whitish to yellow disc flowers and
whitish to pink or mauve ligules; central flowers tubular,
bisexual, more numerous than peripheral female filiform
flowers or the latter 0, the outermost female flowers with
obvious ligules at least as long as tubular part; phyllaries
in several rows, narrow, herbaceous with + membranous
margins; pappus of 1 row of hairs, sometimes with an outer
row of very short hairs, the ray flowers sometimes with only
very short hairs or narrow scales.
   Merges into <u>Aster</u> at 1 extreme and <u>Conyza</u> at the other.

1 Ligules ≤4mm, not or scarcely exceeding pappus
                                                    6. E. acer
1 Ligules ≥4mm, exceeding pappus by ≥2mm   2
  2 Capitula very showy, 3-5cm across, with bluish-
   mauve or pale mauve ligules 9-20mm; leaves
   succulent   **1. E. glaucus**
  2 Capitula less showy, 1.5-3cm across, with white to
   pinkish (mauve in 1 alpine sp.) ligules 4-10mm;
   leaves not succulent   3
3 Stem procumbent to ascending; lower leaves with 1 pair
 of lateral lobes or teeth   **4. E. karvinskianus**
3 Stem erect; lower leaves entire or toothed, but not
 regularly toothed as in last   4
  4 Stems rarely >20cm, with 1(-3) capitula; ligules
   mauve; rare alpine; leaves entire   **3. E. borealis**
  4 Stems rarely <20cm, with numerous capitula;
   ligules white or pale pink or blue; lowland aliens;
   leaves often serrate   5
5 Leaves clasping stem; pappus of ray and of disc
 flowers of long hairs only   **2. E. philadelphicus**
5 Leaves not clasping stem; pappus of disc flowers of
 long hairs and outer short scales, of ray flowers of
 short scales only   **5. E. annuus**

Other spp. - The arctic **E. uniflorus** L., close to E. borealis but without filiform disc flowers, has been erroneously reported from Sc. **E. speciosus** (Lindley) DC., from N America, is the familiar garden Erigeron with capitula 3-6cm across, mauve ligules 9-24mm, and glabrous, thin leaves; it is occasionally found as a relic or throw-out.

1. E. glaucus Ker Gawler - Seaside Daisy. Stems procumbent to ascending, pubescent, to 50cm; stem-leaves much smaller than basal ones, succulent, pubescent; ligules 9-15mm, mauve. Intrd; natd in rocky places and on cliffs in S En (especially Wight) and CI; N America.
2. E. philadelphicus L. - Robin's-plantain. Stems erect, pubescent, to 75cm; stem-leaves oblanceolate to narrowly obovate, irregularly dentate, pubescent; ligules 5-10mm, white to pink. Intrd; natd on walls and in rough ground; very scattered in Br; N America.
3. E. borealis (Vierh.) Simmons - Alpine Fleabane. Stems **RRR** erect, pubescent, to 20cm; stem-leaves linear to lanceolate or oblanceolate, entire, pubescent; ligules 4-6mm, mauve. Native; mountain rock-ledges above 800m; very rare in M & E Perths, Angus and S Aberdeen.
4. E. karvinskianus DC. (E. mucronatus DC.) - Mexican Fleabane. Stems procumbent to ascending; sparsely pubescent, to 50cm; stem-leaves oblanceolate to obovate, entire or with 1 pair of lobes or large teeth, sparsely pubescent; ligules 5-8mm, white or pale mauve on upperside, pink to purple on lowerside. Intrd; natd on walls, banks and stony ground;

scattered in BI N to Co Armagh and E Norfolk, especially SW
En and CI; Mexico.
5. E. annuus (L.) Pers. (_E. strigosus_ Muhlenb. ex Willd.) - 856
Tall Fleabane. Stems erect, rather sparsely pubescent, to
70(100)cm; stem-leaves lanceolate to oblanceolate, entire to
irregularly dentate, subglabrous; ligules 4-10mm, white to
pale mauve. Intrd; natd in sandy places and rough ground;
very scattered in SW En, rare casual elsewhere; N America.
Ssp. **strigosus** (Muhlenb. ex Willd.) Wagenitz, with shorter
stems, fewer narrower stem-leaves, and shorter ligules,
occurred recently in N Hants; it is doubtfully distinct.
6. E. acer L. - Blue Fleabane. Stems erect, pubescent, to
60cm; stem-leaves lanceolate to oblanceolate, entire,
pubescent; ligules 2-4mm, purplish-mauve. Native; barish
sandy or calcareous soils, banks, walls and dunes; locally
frequent in En, Wa and CI, rare in Ir, casual in Sc.

48 x 49. ERIGERON X CONYZA = X CONYZIGERON Rauschert
48/6 x 49/1. X C. huelsenii (Vatke) Rauschert (_Erigeron_ x
huelsenii Vatke; _E. acer_ x _C. canadensis_) occurs sporadically
with the parents in disturbed sandy places in S En; it is
intermediate in pubescence and capitulum size (ligules pale
mauve, 1-2mm), and sterile, somewhat resembling a small, weak
C. bonariensis (q.v.).

49. CONYZA Less. - Fleabanes
Annuals; stem-leaves linear to narrowly elliptic or oblanceolate, entire or toothed, sessile or shortly petiolate;
capitula numerous, discoid or very inconspicuously radiate,
with white to cream or pinkish flowers; central flowers
tubular, bisexual; peripheral flowers female, more numerous,
the outermost often with very short ligules (shorter than
tubular part); phyllaries as in _Erigeron_; pappus of 1 row of
hairs.

1 Phyllaries yellowish-green, glabrous to sparsely
   pubescent; disc flowers with 4-lobed corolla;
   inflorescence ± cylindrical   **1. C. canadensis**
1 Phyllaries greyish-green, pubescent to densely so;
   disc flowers with 5-lobed corolla; inflorescence
   pyramidal to corymbose   2
   2 Inflorescence pyramidal; capitula mostly 5-7mm
     wide; phyllaries not or minutely red-tipped
                                          2. C. sumatrensis
   2 Inflorescence with long lateral branches, often
     subcorymbose; capitula mostly 6-10mm wide;
     phyllaries usually conspicuously red-tipped
                                          3. C. bonariensis

1. C. canadensis (L.) Cronq. (_Erigeron canadensis_ L.) -
Canadian Fleabane. Stems erect, to 1m, green, rather
sparsely pubescent; leaves green, linear to oblanceolate, the

margins with well-spaced patent to erecto-patent hairs often
>1mm; capitula 3-5mm wide; ligules c.0.5-1mm; pappus cream.
Intrd; natd in waste and rough ground, walls, waysides and
dunes on well-drained soils; common in SE En and CI, becoming
sparser to N & W and ± absent from Sc and Ir; N America.
2. C. sumatrensis (Retz.) E. Walker (C. albida Willd. ex
Sprengel, C. floribunda Kunth, Erigeron sumatrensis Retz.) -
Guernsey Fleabane. Stems erect, to 1(1.5)m, greyish,
pubescent; leaves greyish-green, linear-oblong to narrowly
elliptic or narrowly obovate, the margins with many hooked
hairs much <0.5mm; ligules <0.5mm; pappus usually cream to
grey. Intrd; natd in waste and rough ground in protected
sunny spots; London area and CI, increasing; S America.
3. C. bonariensis (L.) Cronq. (Erigeron bonariensis L.) -
Argentine Fleabane. Stems erect, to 60cm, brownish- to
reddish-grey, pubescent; leaves greyish, linear to oblanceolate, the margins with many hooked hairs much <0.5mm;
ligules 0; pappus usually dirty white to brownish- or
reddish-grey. Intrd; rather frequent casual in waste and
cultivated ground, often as wool-alien; scattered in CI, En
and Sc; S America. The presence of glands or stickiness is
unreliable for separating this and the last sp.

50. CALLISTEPHUS Cass. - China Aster
Annuals; stem-leaves ovate, deeply toothed or lobed, the
lower petiolate; capitula very conspicuous, radiate but often
flore pleno, with yellow disc flowers when present, with
white or blue to pink or purple ligules; phyllaries in
several rows, the outer herbaceous and very leafy, the inner
membranous-bordered; pappus of 2 rows of hairs.

1. C. chinensis (L.) Nees - China Aster. Stems erect, to
75cm, stiffly pubescent; capitula few, up to 12cm across,
very showy. Intrd; garden throw-out on tips and waste
ground; scattered in En; China.

51. OLEARIA Moench - Daisybushes
Strong shrubs to small trees, evergreen; leaves simple,
alternate or opposite, white-tomentose on lowerside, entire
to sharply toothed, petiolate; capitula numerous, ± crowded
in lateral or terminal corymbose panicles, radiate or
discoid, with yellow to reddish disc flowers and (if present)
white ligules; phyllaries in several rows, rather scarious;
pappus of 1 row of hairs.

1 Leaves with conspicuous acute teeth on margin
                                             3. O. macrodonta
1 Leaves entire   2
  2 Leaves opposite   **4. O. traversii**
  2 Leaves alternate   3
3 Leaves undulate at margin, mostly >3cm; inflorescences

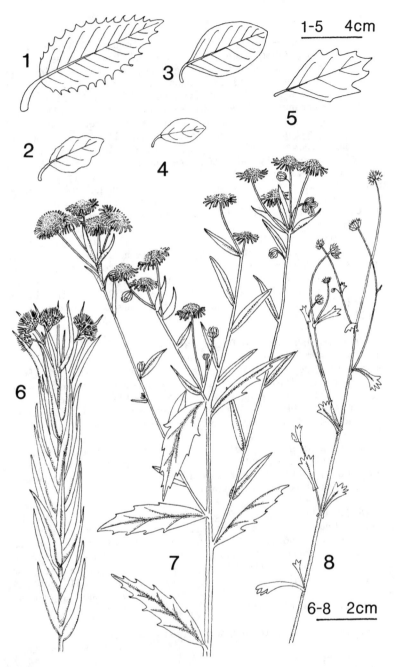

FIG 856 – Asteraceae: Astereae. 1–4, leaves of **Olearia**. 1, **O. macrodonta**. 2, **O. traversii**. 3, **O. paniculata**. 4, **O. x haastii**. 5, leaf of **Baccharis halimifolia**. 6, **Solidago graminifolia**. 7, **Erigeron annuus**. 8, **Calotis cuneifolia**.

axillary, produced Oct-Nov; capitula discoid **1. O. paniculata**
3 Leaves flat, mostly <3cm; inflorescences terminal, produced Jul-Aug; capitula radiate **2. O. x haastii**

Other spp. - **O. solandri** Hook. f., with opposite linear-obovate leaves <10(15)mm, is grown as hedging in SW En and sometimes appears semi-wild.

1. O. paniculata (Forster & G. Forster) Druce - <u>Akiraho</u>. 856 Shrub to 3(6)m; leaves alternate, 2.5-10cm, entire, undulate-margined; inflorescences axillary, produced in Autumn, pyramidal; capitula discoid, white, with only 1 flower. Intrd; grown in Guernsey for hedging, frequent relic, rarely self-sown; New Zealand. See <u>Pittosporum</u> <u>tenuifolium</u> (Pittosporaceae) for differences.

2. O. x haastii Hook. f. (?<u>O. avicenniifolia</u> (Raoul) Hook. 856 f. x <u>O. moschata</u> Hook. f.) - <u>Daisybush</u>. Shrub to 2(3)m; leaves alternate, 1-3cm, entire, flat; inflorescences terminal, produced in Summer, corymbose; capitula radiate, with white ligules and yellow disc flowers. Intrd; the hardiest sp. and much grown in gardens and shrubberies, often well established and rarely self-sown on walls and in open ground; very scattered in SW and WC Br; New Zealand.

3. O. macrodonta Baker - <u>New Zealand Holly</u>. Shrub to 3(6)m; 856 leaves alternate, 5-12cm, deeply and sharply dentate, undulate-margined; inflorescences terminal, produced in Summer, corymbose; capitula radiate, with white ligules and reddish disc flowers. Intrd; grown by sea, the best natd sp., in hedges, scrub, on banks and rough ground; scattered in Ir and W Br N to Wigtowns, Guernsey; New Zealand.

4. O. traversii (F. Muell.) Hook. f. - <u>Ake-ake</u>. Shrub or 856 tree to 10m; leaves opposite, 2.5-7cm, entire, flat; inflorescences axillary, produced in Summer, pyramidal; capitula discoid, greyish. Intrd; grown as hedging in Guernsey and Scillies, often well established but not self-sown; Chatham Island. See <u>Pittosporum</u> <u>crassifolium</u> (Pittosporaceae) for differences.

52. BACCHARIS L. - <u>Tree</u> <u>Groundsel</u>
Dioecious, deciduous shrubs; leaves simple, alternate, roughly toothed in distal 1/2, tapered to petiole; capitula ± numerous, in terminal, loose leafy panicles, small, discoid, whitish; phyllaries in several rows, herbaceous with scarious borders; pappus of 1 row of hairs, shorter in male plants.

1. B. halimifolia L. - <u>Tree</u> <u>Groundsel</u>. Erect, ± sticky 856 shrub to 4m; leaves obovate, glabrous; capitula in wide, terminal panicles, c.2mm across, white, produced in Oct; pappus conspicuous on female plants, white. Intrd; grown by sea in S due to salt-tolerance, natd in S Hants since at least 1942; N America.

53. BELLIS L. - <u>Daisy</u>
Herbaceous perennials; leaves all basal, in rosette, simple,
toothed, petiolate; capitula single on leafless stalks,
radiate, with white to pink or red ligules and yellow disc
flowers, sometimes <u>flore</u> <u>pleno</u>; phyllaries in 2 rows,
herbaceous; pappus 0.

1. B. perennis L. - <u>Daisy</u>. Leaves obovate, irregularly
serrate; stems procumbent to erect, to 12(20)cm, leafless,
with 1 capitulum; capitula 12-25mm across, up to 80mm across
and often <u>flore</u> <u>pleno</u> in cultivars. Native; abundant
throughout BI mostly in short grassland.

TRIBE 6 - ANTHEMIDEAE (genera 54-67). Annual to perennial
herbs, rarely shrubs; leaves alternate, simple to pinnate,
often finely and deeply divided; capitula discoid or radiate;
phyllaries in 2-several rows, herbaceous with scarious
margins and apex; receptacular scales 0 or present; pappus
usually 0, sometimes a short rim; usually with yellow disc
flowers and white ligules but exceptions not rare.

54. TANACETUM L. (<u>Balsamita</u> Miller) - <u>Tansies</u>
Strongly aromatic perennial herbs; leaves simple and toothed
to deeply pinnately lobed or pinnate; capitula radiate or
discoid, rarely <u>flore</u> <u>pleno</u>; disc flowers yellow; ligules
white or 0; receptacular scales 0; pappus a very short rim.

1 Leaves toothed, divided much <1/2 way to midrib
                                          4. T. balsamita
1 All or most leaves pinnate, or pinnately lobed
   much >1/2 way to midrib   2
   2 Rhizomes 0; ultimate leaf-lobes obtuse to subacute,
      sometimes apiculate   **1. T. parthenium**
   2 Rhizomatous; ultimate leaf-lobes acute to
      acuminate   3
3 Capitula discoid, >5mm across   **3. T. vulgare**
3 Capitula radiate, <5mm across excl. ligules;
   ligules white   **2. T. macrophyllum**

1. T. parthenium (L.) Schultz-Bip. (<u>Chrysanthemum</u> <u>parthenium</u>
(L.) Bernh.) - <u>Feverfew</u>. Stems erect, to 70cm; capitula
usually radiate, rarely discoid or <u>flore</u> <u>pleno</u>, in lax
corymbs, 15-23mm across (6-9mm excl. ligules). Intrd; natd
on walls, waste ground and waysides; frequent throughout BI;
Balkans.
2. T. macrophyllum (Waldst. & Kit.) Schultz-Bip. - <u>Rayed</u> **862**
<u>Tansy</u>. Stems erect, to 1.2m; capitula radiate, in dense
corymbs, 7-13mm across (3-5mm excl. ligules). Intrd; grown
for ornament, natd in grassy places and waysides since
c.1912; very scattered in En and Sc; SE Europe.
3. T. vulgare L. (<u>Chrysanthemum</u> <u>vulgare</u> (L.) Bernh.) -
<u>Tansy</u>. Stems erect, to 1.2m; capitula discoid, in dense

corymbs, 6-10mm across. Native; grassy places, waysides, rough ground; frequent throughout BI.

4. T. balsamita L. (Chrysanthemum balsamita (L.) Baillon non L., Balsamita major Desf.) - Costmary. Stems erect, to 1.2m; capitula discoid, in tight corymbs, 4-8mm across, rarely with few ligules 4-6mm. Intrd; grown for ornament and cooking, rarely natd as outcast; very scattered in C & S Br; Caucasus.

55. SERIPHIDIUM (Besser ex Hook.) Fourr. - Sea Wormwood
Aromatic perennials; differ from Artemisia in flowers all similar and functionally bisexual.

1. S. maritimum (L.) Polj. (Artemisia maritima L.) - Sea Wormwood. Stems decumbent to erect, woody below; leaves white-woolly, 1-2-pinnate with linear ultimate segments; capitula yellowish- to reddish-brown, numerous in terminal panicle, 1.5-3.5mm across. Native; dry parts of salt-marshes, sea-walls and rough ground by sea; local on coasts of Br N to C Sc, especially E En, E & W Ir.

56. ARTEMISIA L. - Mugworts
Annual to perennial herbs or small shrubs, often aromatic; leaves entire to finely divided; capitula discoid, small, brownish overall; flowers usually yellowish, the outer female, with filiform corolla, the inner bisexual, with tubular corolla; receptacular scales 0; pappus 0.

1	Leaves most or all entire	**9. A. dracunculus**
1	Most or all leaves deeply divided	2
	2 Stems woody ± to top	**6. A. abrotanum**
	2 Stems herbaceous, or woody only near base	3
3	Capitula 1-2(5); stems <10cm	**5. A. norvegica**
3	Capitula normally >10; stems >10cm	4
	4 Annual or biennial with simple root system and 0 non-flowering shoots	5
	4 Perennial with strong underground portion and non-flowering shoots	6
5	Leaves in inflorescence projecting laterally well beyond capitula, with many primary divisions >(1.5)2cm x c.1-3mm	**7. A. biennis**
5	Leaves in inflorescence extending laterally less far than capitula, with primary divisions <1(1.5)cm x c.0.5-1mm	**8. A. annua**
	6 Mature leaves densely (often whitish-) pubescent on upperside	7
	6 Mature leaves glabrous or subglabrous on upperside (beware mildew)	8
7	Plant not aromatic, rhizomatous; receptacle glabrous; capitula 6-10 x 5-9mm excl. flowers	**4. A. stelleriana**
7	Plant aromatic, at least when fresh, not rhizomatous; receptacle pubescent; capitula 1.5-3.5 x 3-5mm excl. flowers	**3. A. absinthium**

8   All leaf-lobes <2mm wide; plant not aromatic;
    achenes usually produced only by marginal flowers
                                              10. A. campestris
8   Most or all leaf-lobes >2mm wide; plant aromatic,
    at least when fresh; achenes produced by all
    flowers   9
9   Plant not or scarcely rhizomatous; terminal untoothed
    portion of middle stem-leaves usually <3cm; stem with
    central (white) pith region occupying c.4/5 of total
    (white + green) pith diameter; flowers Jul-Sep
                                                    1. A. vulgaris
9   Plant strongly rhizomatous; terminal untoothed portion
    of middle stem-leaves usually >3cm; stem with central
    (white) pith region occupying c.1/3 of total (white +
    green) pith diameter; flowers Oct-Dec **2. A. verlotiorum**

Other spp. - 11 other spp. recorded as casuals, incl. **A.
afra** Jacq., **A. pontica** L. and **A. scoparia** Waldst. & Kit., are
much rarer than the 10 spp. treated here.

1. A. vulgaris L. - Mugwort. Aromatic, tufted perennial to
1.5m; leaves glabrous on upperside, whitish-tomentose on
lowerside, with lobes c.2.5-8mm wide; capitula numerous, 1.5-
3.5mm across. Native; rough ground, waste places, waysides;
common throughout lowland BI.
    1 x 2. A. vulgaris x A. verlotiorum was discovered in
Middlesex and S Essex in 1987 and Surrey in 1989; it is
intermediate in all characters (white part of pith c.3/5
total pith width) and completely sterile (flowers appear
Oct-Dec but have abortive stamens); endemic.
    2. A. verlotiorum Lamotte - Chinese Mugwort. Rhizomatous
perennial to 1.5m; differs from A. vulgaris in leaves darker
on upperside with closer network of veins visible in fresh
state; infloresence more leafy; and see key. Intrd; natd
since 1908 in similar places to A. vulgaris; frequent in
London area, especially near R Thames, very scattered
elsewhere in S & C En, Caerns, Guernsey; China.
    3. A. absinthium L. - Wormwood. Aromatic, tufted perennial
to 1m; leaves greyish-pubescent on both surfaces, with lobes
c.2-4mm wide; capitula numerous, 3-5mm across. Native;
similar places to A. vulgaris; frequent in En, Wa and CI,
very scattered in Sc and Ir.
    4. A. stelleriana Besser - Hoary Mugwort. Non-aromatic,
rhizomatous perennial to 60cm; leaves whitish-tomentose on
both surfaces, with lobes c.3-8mm wide; capitula ± numerous,
5-9mm across. Intrd; natd on maritime dunes in Kirkcudbrights since 1979 and in Clyde Is since 1976, formerly W
Cornwall and Co Dublin; NE Asia.
    5. A. norvegica Fries - Norwegian Mugwort. Aromatic, **RR**
rosette-perennial to 8cm; leaves pubescent on both surfaces,
few and reduced on stems, with lobes 0.5-1.5mm wide; capitula
1-2(5), 8-13mm across. Native; at 3 sites at c.800m on

barish mountain-tops in E & W Ross, discovered 1950. Our
plant has been named var. **scotica** Hulten.

6. A. abrotanum L. - <u>Southernwood</u>. Very aromatic shrub to
1.2m; leaves pubescent on both surfaces, with lobes c.0.6-1mm
wide; capitula numerous, 3-4mm across. Intrd; much grown in
gardens, rarely persistent on tips and waste ground; sporadic
in S & C Br; origin unknown. Rarely or never flowers.

7. A biennis Willd. - <u>Slender Mugwort</u>. Very aromatic erect 862
annual (to biennial) to 1.5m; leaves glabrous, with primary
lobes c.1-3mm wide excl. teeth; capitula numerous, 1.5-4mm
across. Intrd; casual from grain and wool on waste ground
and reservoir mud, ± natd in few sites in S Br; Asia and N
America.

8. A. annua L. - <u>Annual Mugwort</u>. Differs from <u>A. biennis</u> in 862
key characters, but often confused with it. Intrd; casual
from same sources as <u>A. biennis</u> but rarer and not natd; very
scattered in S Br; SE Europe and Asia.

9. A. dracunculus L. - <u>Tarragon</u>. Aromatic perennial to
1.2m; leaves glabrous, mostly linear to narrowly elliptic and
entire, c.2-10mm wide; capitula numerous, 2-3mm across.
Intrd; grown for flavouring, rarely persistent on tips and
waste ground; very scattered in S En; Russia.

10. A. campestris L. - <u>Field Wormwood</u>. Non-aromatic RRR
perennial to 75cm; leaves ± pubescent when young, ± glabrous
when mature, with lobes c.0.3-1mm wide; capitula numerous,
2-4mm across. Native; grassy places by roads and on heathland; very local in W Suffolk and W Norfolk, formerly E
Norfolk and Cambs, natd in Glam.

57. SANTOLINA L. - <u>Lavender-cotton</u>
Evergreen shrubs; leaves neatly and closely pinnately lobed;
capitula discoid, yellow; receptacular scales present; pappus
0.

1. S. chamaecyparissus L. - <u>Lavender-cotton</u>. Stems
decumbent to suberect, to 60cm; whole plant white- to grey-
tomentose; capitula 6-10mm across, solitary on erect stems.
Intrd; much grown in gardens, persistent on tips, rough
ground, old gardens and rockeries, scattered in S & C Br,
especially SW En; Mediterranean.

58. OTANTHUS Hoffsgg. & Link - <u>Cottonweed</u>
Perennial, densely white-woolly herbs; leaves simple,
crenate; capitula discoid with yellow flowers; phyllaries
obscured by dense hairs; receptacular scales present; pappus
0.

1. O. maritimus (L.) Hoffsgg. & Link - <u>Cottonweed</u>. Stems RRR
erect to ascending, to 30cm; leaves oblong-obovate; capitula
few, subcorymbose, 6-9mm across. Native; maritime fixed sand
and shingle; now in 1 place in Co Wexford, formerly scattered
in BI N to E Suffolk, Anglesey and Co Wicklow.

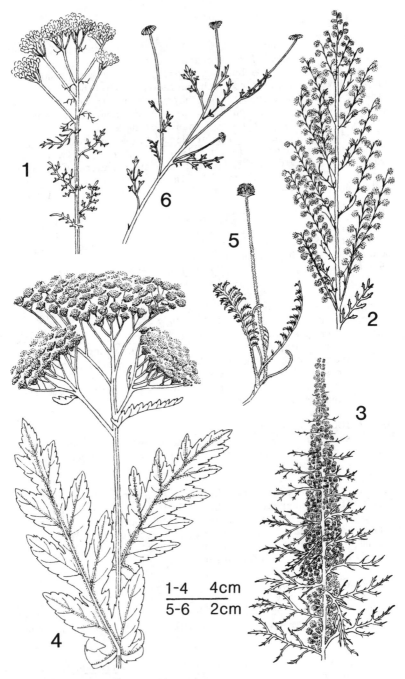

FIG 862 - Asteraceae: Anthemideae. 1, Achillea ligustica. 2, Artemisia annua. 3, Artemisia biennis. 4, Tanacetum macrophyllum. 5, Cotula squalida. 6, Cotula australis.

59. ACHILLEA L. - <u>Yarrows</u>
Perennial herbs; leaves simple and very shallowly toothed to deeply and finely dissected; capitula radiate, rarely <u>flore pleno</u>; disc flowers and ligules white to deep pink, rarely yellow; receptacular scales present; pappus 0.

1 Leaves simple, toothed much <1/2 way to midrib; capitula >1cm across   **1. A. ptarmica**
1 Leaves compound, or simple and divided much >1/2 way to midrib; capitula <u><</u>1cm across   2
  2 Middle stem-leaves <3x as long as wide; with <10 pairs of primary lateral lobes   **2. A. ligustica**
  2 Middle stem-leaves >3x as long as wide, with >15 pairs of primary lateral lobes   3
3 Leaves ± flat in fresh state, the primary lateral lobes ± contiguous on the rhachis; inner phyllaries >3.5mm   **4. A. distans**
3 Leaves with lobes spreading in 3 dimensions in fresh state, the primary lateral lobes separated by a length of winged rhachis; inner phyllaries <u><</u>3.5mm   **3. A. millefolium**

Other spp. - Several yellow-flowered spp. are grown in gardens and have been reported as rare escapes: **A. filipendulina** Lam., from W & C Asia, is the familiar border-plant >1m high with corymbs often >6cm across; **A. tomentosa** L., from SW Europe, is much smaller and is densely grey-pubescent. Similarly reported white- (or very pale yellow-) flowered spp. are the European **A. nobilis** L., differing from <u>A. ligustica</u> in its primary leaf-segments wider and more regularly subdivided; and **A. grandifolia** Friv., with middle stem-leaves >(3)4cm wide and undivided part of primary lobes >2mm wide.

1. A. ptarmica L. - <u>Sneezewort</u>. Stems erect, to 60cm; leaves linear to linear-lanceolate, finely and shallowly toothed; capitula <10(15), in lax corymbs, 12-20mm across, sometimes <u>flore pleno</u> in garden escapes; ligules white. Native; damp grassy places and marshy fields; frequent in Br and most Ir, casual in CI.
2. A. ligustica All. - <u>Southern Yarrow</u>. Resembles <u>A.</u> 862 <u>millefolium</u> but leaves shorter, wider and more finely divided (see key); capitula smaller (c.3mm across; inner phyllaries <u><</u>3mm) and more numerous; ligules white. Intrd; natd in waste ground at Newport Docks, Mons, since 1953; Mediterranean.
3. A. millefolium L. - <u>Yarrow</u>. Stems erect, to 80cm; leaves very deeply divided into many, deeply divided lateral lobes; capitula >(25)50 in ± dense corymbs, c.4-6mm across; inner phyllaries c.3-3.5mm; ligules white (to deep pink). Native; grassland (usually short), banks and waysides; very common throughout BI.
4. A. distans Waldst. & Kit. ex Willd. - <u>Tall Yarrow</u>.

Resembles A. millefolium but stems to 1.3m; leaves differ as in key; capitula larger (5-10mm across, inner phyllaries 3.5-5mm). Intrd; natd in grassy places; Derbys and MW Yorks; S & E Europe. Our plant is ssp. **tanacetifolia** Janchen.

60. CHAMAEMELUM Miller - Chamomile
Aromatic perennial herbs; leaves deeply and finely dissected; capitula radiate, rarely discoid; disc flowers yellow, with short pouch at base of tube; ligules white; receptacular scales present; pappus 0.

1. C. nobile (L.) All. - Chamomile. Stems procumbent to ascending, to 30cm; receptacular scales oblong to narrowly obovate, acuminate; achenes weakly ridged on 1 face. Native; short grassy places on sandy soils; locally frequent in CI, S Br and SW Ir, very scattered elsewhere and mostly intrd.

61. ANTHEMIS L. - Chamomiles
Aromatic annual to perennial herbs; leaves deeply and finely dissected; capitula radiate, rarely discoid; disc flowers yellow; ligules white or yellow; receptacular scales present; pappus 0 or a short rim.

```
1 Ligules yellow, occasionally 0          4. A. tinctoria
1 Ligules white, very rarely 0                         2
    2 Receptacular scales only on inner (upper) part of
    receptacle, linear-subulate; achenes tuberculate on
    ribs; fresh plant with unpleasant scent  3. A. cotula
    2 Receptacular scales all over receptacle, at least
    the inner ones lanceolate to oblanceolate; achenes
    ribbed or scarcely so, but not tuberculate; fresh
    plant with sweet scent                             3
3 Perennial, often woody near base and with non-
    flowering shoots; at least outer receptacular scales
    3-toothed; achenes not or slightly ribbed
                                           1. A. punctata
3 Annual or biennial, not woody at base and usually
    without non-flowering shoots; receptacular scales
    with single slender apex; achenes strongly ribbed
                                           2. A. arvensis
```

1. A. punctata Vahl - Sicilian Chamomile. Perennial to **866** 60cm; ligules white; at least outer receptacular scales 3-toothed at apex; achenes not or weakly ribbed, with apical rim 0.3-0.8mm. Intrd; grown in gardens, natd in rough and marginal ground and on cliffs, mostly near sea; very scattered in S Br; Sicily. Our plant is ssp. **cupaniana** (Tod. ex Nyman) Ros. Fernandes.

2. A. arvensis L. - Corn Chamomile. Annual to 50cm; ligules **866** white; receptacular scales oblong-lanceolate, cuspidate; achenes strongly ribbed, not tuberculate, with apical rim ≤0.5mm. Native; arable land, waste places and rough ground,

usually on calcareous soils, also a grass-seed alien; locally
frequent in S & C Br, rare and mainly casual in N Br and CI,
extinct in Ir.

3. A. cotula L. - <u>Stinking</u> Chamomile. Differs from <u>A.</u> 866
<u>arvensis</u> in more stiffly erect stems; leaf segments linear
and ± glabrous (not narrowly oblong and pubescent); ligules
becoming reflexed (not remaining patent); achenes without
apical rim; and see key.   Native; similar habitats to <u>A.</u>
<u>arvensis</u> but often on heavier soils; similar distribution to
<u>A. arvensis</u> but commoner, still in S & E Ir.

4. A. tinctoria L. - <u>Yellow</u> Chamomile. Biennial or 866
perennial to 50cm; ligules yellow, occasionally 0; receptacular scales as in <u>A. arvensis</u>; achenes scarcely ribbed, not
tuberculate, with apical rim <0.3mm.   Intrd; natd or casual
in waste places, rough and marginal land; rather frequent in
S & C Br, rare in N; Europe.   Ligule-less plants can be told
from similar variants of other spp. by combination of
receptacular scale and achene characters, larger capitula and
very characteristic leaf-lobing.   Sometimes mistaken for
<u>Chrysanthemum segetum</u>, but this differs in many details,
incl. leaf-lobing.

61 X 66. ANTHEMIS X TRIPLEUROSPERMUM = X TRIPLEUROTHEMIS
Stace

61/3 x 66/2. x T. maleolens (P. Fourn.) Stace (X <u>Anthemi-</u>
<u>matricaria celakovskyi</u> Geisenh. ex Domin nom. illeg.; <u>A.</u>
<u>cotula</u> x <u>T. inodorum</u>) is intermediate in the irregular
presence of receptacular scales that are intermediate between
those of the <u>Anthemis</u> parent and the phyllaries, and in the
sterile achenes with intermediate rib development and traces
of subapical oil-glands.   1 plant in Berks (1966) and 2 in
Salop (1969).

62. CHRYSANTHEMUM L. - <u>Crown Daisies</u>
Annual herbs; leaves simple, shallowly to deeply lobed;
capitula radiate; disc flowers yellow; ligules yellow, cream
or yellow and cream;   receptacular scales 0; pappus 0.

1. C. segetum L. - <u>Corn Marigold</u>. Stems decumbent to erect, 866
to 60cm; leaves glaucous, slightly toothed to deeply lobed,
at least the upper usually lobed <1/2 way to midrib; capitula
3-7cm across; ligules yellow; achenes 2.5-3mm, deeply ridged,
not winged.   Intrd; natd or casual weed of arable fields,
waste places and waysides; locally frequent throughout BI;
Europe.   See <u>Anthemis tinctoria</u> for differences.

2. C. coronarium L. - <u>Crown Daisy</u>. Stems ascending to 866
erect, to 80cm; leaves green, lobed >1/2 way to midrib, often
± to midrib; capitula 4-8cm across; ligules cream, yellow, or
cream and yellow; achenes 3-3.5mm, deeply ridged, the inner
with adaxial wing, the marginal with 2 lateral and 1 adaxial
wings.   Intrd; similar places to <u>C. segetum</u> but much rarer,
often as grain alien, not natd; very scattered in En and Wa;

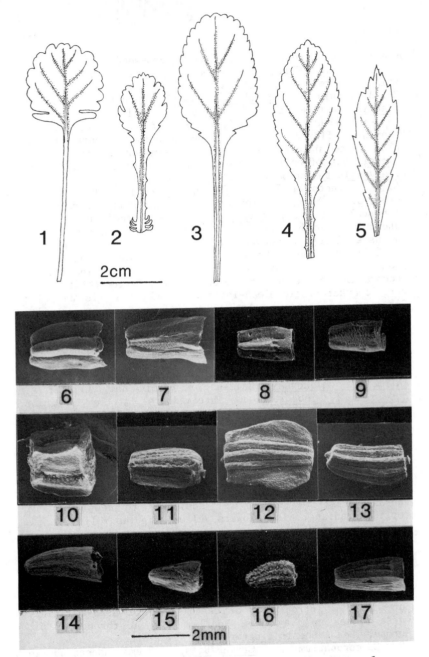

FIG 866 - Asteraceae: Anthemideae. 1–4, basal and stem-leaves of **Leucanthemum**. 1–2, **L. vulgare**. 3–4, **L. x superbum**. 5, stem-leaf of **Leucanthemella serotina**. 6–9, adaxial and abaxial faces of achenes of **Tripleurospermum**. 6–7, **T. maritimum**. 8–9, **T. inodorum**. 10–13, marginal and inner achenes of **Chrysanthemum**. 10–11, **C. coronarium**. 12–3, **C. segetum**. 14–17, achenes of **Anthemis**. 14, **A. punctata**. 15, **A. arvensis**. 16, **A. cotula**. 17, **A. tinctoria**.

Europe.

63. LEUCANTHEMELLA Tzvelev - Autumn Oxeye
Perennial herbs; leaves simple, sharply serrate; capitula radiate; disc flowers yellow; ligules white; receptacular scales 0; pappus + 0.

1. L. serotina (L.) Tzvelev (Chrysanthemum uliginosum 866 (Waldst. & Kit. ex Willd.) Pers., C. serotinum L.) - Autumn Oxeye. Stems erect, to 1.5m; resembles Leucanthemum x superbum but leaves paler green and more sharply and deeply serrate (most serrations >3mm), flowers later (Sep-Oct), and tubular part of corolla of ray flowers unwinged. Intrd; garden escape or throw-out natd on rough ground and by ditches and ponds; scattered in S En; SE Europe.

64. LEUCANTHEMUM Miller - Oxeye Daisies
Differ from Leucanthemella in tubular part of ray flowers with 2 narrow translucent wings; and achenes with translucent secretory canals.

1. L. vulgare Lam. (Chrysanthemum leucanthemum L.) - Oxeye 866 Daisy. Stems erect to ascending, to 75cm; basal and lower stem-leaves obovate-spathulate, abruptly contracted to broadly cuneate base; upper stem-leaves usually deeply serrate; capitula 2.5-6(7.5)cm across. Native; grassy places, especially on rich soils; common throughout BI.
2. L. x superbum (Bergmans ex J. Ingram) Kent (L. maximum 866 auct. non (Ramond) DC., Chrysanthemum maximum auct. non Ramond; L. lacustre (Brot.) Samp. x L. maximum (Ramond) DC.) - Shasta Daisy. Stems erect to ascending, to 1.2(1.5)m; basal and lower stem-leaves elliptic-oblong, gradually contracted to narrowly cuneate base; upper stem-leaves usually shallowly serrate to subentire; capitula (5)6-10cm across, often flore pleno. Intrd; abundant in gardens and fully fertile, well natd in waste and rough ground and grassy waysides; scattered throughout Br and CI; garden origin.

65. MATRICARIA L. (Chamomilla Gray) - Mayweeds
Annual herbs, differing from Tripleurospermum in much more conical, hollow (not solid) receptacle; ligules often 0; and achenes with 4-5 weak (not 3 strong) ribs and without (not with) oil-glands.

Other spp. - Other ligule-less spp. recorded, e.g. **M. disciformis** (C. Meyer) DC. and **M. decipiens** (Fischer & C. Meyer) K. Koch, are either rare casuals or errors for M. discoidea.

1. M. recutita L. - Scented Mayweed. Superficially much like Tripleurospermum inodorum but usually (not always) more strongly and sweetly scented when fresh; phyllaries with very

pale brown (not deep brown) scarious margins; ligules soon
very strongly reflexed; and see generic characters above.
Native; in similar places to and often with T. inodorum but
less common and more restricted to arable ground on light
soils; locally common in CI, En and Wa, very scattered in Sc,
rare casual in Ir.
2. **M. discoidea** DC. (M. matricarioides (Less.) Porter nom.
illeg., M. suaveolens (Pursh) Buchenau non L.) - Pineapple-
weed. Plant erect, to 35cm; ligules 0; differs from rare
ligule-less plants of M. recutita in sweet pineapple-like
scent; much wider, white scarious margins to phyllaries; and
disc flowers with 4-lobed (not 5-lobed) corolla. Intrd; weed
of barish places by paths and waste places; common throughout
BI, 1st recorded 1871; widespread weed.

66. **TRIPLEUROSPERMUM** Schultz-Bip. (Matricaria auct. non L.) -
Mayweeds
Annual to perennial herbs; leaves deeply and finely
dissected; capitula radiate, rarely discoid; disc flowers
yellow; ligules white; receptacular scales 0; pappus a very
short rim.

1. **T. maritimum** (L.) Koch (Matricaria maritima L.) - Sea **866**
Mayweed. Erect to procumbent (biennial to) perennial to
60cm; leaf-segments succulent, (acute) obtuse to rounded at
apex; achenes 1.8-3.5mm, with 3 strong ribs ± touching
laterally on 1 face, with 2 subapical distinctly elongated
oil-glands on opposite face. Native; sand, shingle, rocks,
walls, cliffs and waste ground near sea; locally common round
most coasts of BI. Plants from N Sc (incl. N Isles) have
dark-bordered phyllaries and have been referred to ssp.
phaeocephalum (Rupr.) Haemet-Ahti (T. maritimum var.
phaeocephalum (Rupr.) N. Hylander, M. maritima ssp.
phaeocephala (Rupr.) Rauschert), but are not as extreme as
the true Arctic taxon. Plants from S En with strong
anthocyanin development and thinner leaf-segments are
referable to var. **salinum** (Wallr.) Kay (M. maritima (Wallr.)
Clapham), which has often been misplaced under T. inodorum.
  1 x 2. **T. maritimum x T. inodorum** is intermediate in leaf
and achene characters and is ≥80 per cent fertile (with
backcrossing occurring); it is not infrequent in coastal
areas and casts doubt on the distinction of the 2 parents at
sp. level.
  2. **T. inodorum** (L.) Schultz-Bip. (T. maritimum ssp. inodorum **866**
(L.) N. Hylander ex Vaar., Matricaria perforata Merat) -
Scentless Mayweed. Erect to ascending annual to 60cm; leaf-
segments not succulent, acute and often bristle-tipped;
achenes 1.3-2.2mm, with 3 strong ribs on 1 face separated by
2 distinct granular areas, with 2 subapical orbicular to
angular oil-glands on opposite face. Native; waste, rough
and cultivated land; common throughout lowland BI. See
Matricaria recutita for differences.

67. COTULA L. (Leptinella Cass.) - Buttonweeds

Annual to perennial herbs; leaves entire to deeply pinnately divided; capitula discoid, bisexual or dioecious, yellow or white, with pedicellate flowers; in bisexual capitula outer flowers are female with 0 or minute corolla, inner ones bisexual with 4-lobed corolla; in dioecious capitula males and females both with minutely 4-lobed corolla: receptacular scales 0; pappus 0.

1 Leaves entire to very irregularly pinnately lobed with usually <6 lobes, + succulent; capitula 8-12mm across, bright yellow **1. C. coronopifolia**
1 Leaves regularly pinnately (to 2-pinnately) lobed with usually >6 lobes, not succulent; capitula 3-10mm across, white or dull yellow **2**
  2 Annual; capitula bisexual, white, the female (outer) flowers with 0 corolla; phyllaries not purple-tinged **2. C. australis**
  2 Procumbent perennial with rooting stems; capitula dioecious, dull yellow, the female flowers with corolla; phyllaries strongly purple-tinged **3**
3 Leaves with oblong-triangular abruptly apiculate teeth or shallow lobes **3. C. dioica**
3 Leaves lobed nearly to midrib, the lobes with lanceolate, acute to acuminate teeth **4. C. squalida**

1. C. coronopifolia L. - Buttonweed. Rather succulent glabrous annual to perennial with procumbent to ascending, often rooting stems to 30cm; capitula 8-12mm across, bright yellow. Intrd; wet, usually saline places; natd in Cheshire since c.1880, W Cork, rare casual elsewhere; ?S Africa.

2. C. australis (Sieber ex Sprengel) Hook. f. - Annual 862 Buttonweed. Annual with suberect to decumbent pubescent stems to 15cm; capitula 3-7mm across, white. Intrd; rather frequent wool-alien, sometimes persisting in rough or arable ground, very scattered in En, natd in S Devon; Australia, N Zealand.

3. C. dioica (Hook. f.) Hook. f. - Hairless Leptinella. Dioecious procumbent perennial with rooting stems to 20cm; capitula 4-10mm across, the male dull yellow, the female + enclosed by incurved phyllaries. Intrd; garden plant becoming natd in mown lawns, often not flowering; extremely scattered in BI; New Zealand.

4. C. squalida (Hook. f.) Hook. f. - Leptinella. Differs 862 from C. dioica as in key. Intrd; natd as for C. dioica, but commoner, scattered in Br and Ir, especially Sc; New Zealand.

TRIBE 7 - SENECIONEAE (genera 68-78).

Annual to perennial herbs, sometimes shrubs, rarely weak climbers; leaves alternate or all basal; capitula discoid or radiate; phyllaries usually in 1 or 2 rows, often in 1 main row and 1 much shorter row, herbaceous; receptacular scales 0; pappus of 1-many rows of simple hairs; corolla most often yellow in both ray and disc flowers.

68. SENECIO L. - Ragworts

Annual to perennial herbs, rarely shrubby; leaves alternate, pinnately veined; capitula discoid or radiate; phyllaries in 1 main row with short supplementary ones at base of capitulum; corolla of disc flowers yellow, of ray flowers usually yellow (rarely white or purple).

```
1   Stems woody at least towards base                          2
1   Stems entirely herbaceous                                  5
    2   Stems and leaves ± glabrous                            3
    2   At least leaf lowersides grey- or white-pubescent      4
3   Leaves linear, ± entire                         4. S. inaequidens
3   Leaves conspicuously toothed to deeply lobed
                                                    14. S. squalidus
    4   Leaves lobed ± to midrib, with obtuse to rounded
        lobes; phyllaries tomentose                 1. S. cineraria
    4   Leaves serrate, acuminate at apex; phyllaries
        glabrous                                    2. S. pterophorus
5   Ligules white or purple                                    6
5   Ligules yellow to orange, or 0                             7
    6   Leaves with linear lobes, divided >1/2 way to
        midrib; ligules usually purple              3. S. grandiflorus
    6   Leaves very shallowly and irregularly toothed;
        ligules white                               10. S. smithii
7   Leaves simple, entire to shallowly toothed                 8
7   At least some leaves lobed ≥1/2 way to midrib              14
    8   Ligules 4-8                                            9
    8   Ligules >8                                             11
9   Middle and upper stem-leaves shortly but distinctly
    petiolate                                       6. S. ovatus
9   Middle and upper stem-leaves sessile                       10
    10  Leaf-teeth divergent, often obtuse; phyllaries and
        peduncles usually glabrous                  7. S. doria
    10  Leaf-teeth with acute, ± incurved apex;
        phyllaries and peduncles pubescent          5. S. fluviatilis
11  Leaves linear, <5mm wide                        4. S. inaequidens
11  Leaves ovate or lanceolate to oblanceolate, most
    >5mm wide                                                  12
    12  Capitula 1-3(4); phyllaries 10-15mm         9. S. doronicum
    12  Capitula numerous; phyllaries 6-10mm                   13
13  Phyllaries conspicuously black-tipped; leaves ±
    glabrous                                        14. S. squalidus
13  Phyllaries not black-tipped; leaves pubescent on
    lowerside                                       8. S. paludosus
    14  Glandular hairs present at least on peduncles,
        often also on leaves and stems                        15
    14  Glandular hairs 0                                      16
15  Achenes glabrous; supplementary phyllaries at base
    of capitulum 1/3 to 1/2 as long as main ones;
```

plant sticky **19. S. viscosus**
15 Achenes minutely pubescent; supplementary
phyllaries ≤1/4(1/3) as long as main ones; plant
not or scarcely sticky **18. S. sylvaticus**
 16 Biennial to perennial, firmly rooted, usually with
  very short thick rhizome; phyllaries without
  black tips   17
 16 Annual to perennial, usually easily uprooted,
  without rhizome; at least supplementary
  phyllaries with black tips   19
17 Supplementary phyllaries at base of capitulum
c.1/2 as long as main ones; all achenes shortly
pubescent; leaves grey-pubescent on lowerside
        13. S. erucifolius
17 Supplementary phyllaries c.1/4 to 2/5 as long as
main ones; achenes of ray flowers glabrous; leaves
glabrous to sparsely pubescent on lowerside   18
 18 Achenes of disc flowers pubescent; stem-leaves
  with several pairs of lateral lobes and terminal
  lobe not much larger; corymbs dense   **11. S. jacobaea**
 18 Achenes of disc flowers glabrous to sparsely
  pubescent; stem-leaves with 1-few pairs of lateral
  lobes and terminal lobe much larger; corymbs lax
        12. S. aquaticus
19 Ligules <8mm or 0; capitula (excl. ligules)
cylindrical in flower, c.2x as long as wide   20
19 Ligules usually ≥8mm, rarely shorter or 0;
capitula (excl. ligules) campanulate in flower,
<1.5x as long as wide   21
 20 Ligules usually 0; achenes ≤2.5mm; pollen grains
  20-25 microns across, 3-pored   **16. S. vulgaris**
 20 Ligules usually present; achenes >3mm; pollen
  grains 30-36 microns across, mostly 4-pored
        15. S. cambrensis
21 Leaves ± flat, usually with lateral lobes much
longer than width of central undivided portion,
usually glabrous or nearly so   **14. S. squalidus**
21 Leaves usually undulate, with lateral lobes c. as
long as width of central undivided portion,
usually conspicuously pubescent   **17. S. vernalis**

Other spp. - Records of **S. lautus** Sol. ex Willd., from
Australia, are probably all errors for S. inaequidens.

1. S. cineraria DC. (S. bicolor (Willd.) Tod. ssp. cineraria
(DC.) Chater) - Silver Ragwort. Densely silvery-pubescent
spreading perennial with stems woody at least below; leaves
deeply pinnately lobed, all but upper ones petiolate; ligules
10-13, 3-6mm, yellow. Intrd; natd on cliffs and rough ground
mostly near sea; S & SW En, Wa, CI, Co Dublin: Mediterranean.
Not to be confusd with Cineraria (Pericallis hybrida).
1 x 11. S. cineraria x S. jacobaea = S. x albescens Burb. &

Colgan occurs $\pm$ throughout the range of S. cineraria in BI; it is intermediate in pubescence, leaf-shape and habit but has pubescent disc-achenes as in S. jacobaea. It is fertile and backcrosses, forming a range of intermediates.

1 x 13. S. cineraria x S. erucifolius = S. x thuretii Briq. & Cavill. (S. x patersonianus Burton) was found in E Kent in 1978; it differs from S. x albescens in having supplementary phyllaries c.1/3 (not $\leq$1/4) as long as the main ones and sometimes (?always) having short stolons.

2. S. pterophorus DC. - Shoddy Ragwort. Erect perennial to 1.5m (usually much less); stems woody at least below; leaves sharply and coarsely serrate, densely grey-pubescent on lowerside, subsessile; ligules c.8, 2-4mm, yellow. Intrd; occasional wool-alien in fields and waste land; very scattered in En; S Africa.

3. S. grandiflorus P. Bergius - Purple Ragwort. Erect perennial to 1.5m; leaves deeply pinnately lobed, sessile, clasping stem at base, sparsely pubescent; ligules c.8-13, 10-15mm, purple. Intrd; garden escape natd on rough ground in Guernsey; S Africa.

4. S. inaequidens DC. (S. lautus auct. non Sol. ex Willd.) - Narrow-leaved Ragwort. Subglabrous spreading perennial to 80cm, with stems often woody below; leaves linear, entire or nearly so, sessile; ligules 10-15, 5-8(10)mm, yellow. Intrd; rather frequent wool-alien in En and Sc, now natd on sandy beach in E Kent, perhaps soon to spread as in N France; S Africa.

5. S. fluviatilis Wallr. - Broad-leaved Ragwort. Erect perennial to 1.5m; leaves lanceolate to oblanceolate, shortly serrate, sessile, $\pm$ clasping stem at base; ligules mostly 6-8, 8-15mm, yellow. Intrd; natd by streams and in fens and swampy ground; scattered in Br and Ir N to C Sc, formerly more common; Europe.

6. S. ovatus (P. Gaertner, Meyer & Scherb.) Willd. (S. fuchsii C. Gmelin, S. nemorensis L. ssp. fuchsii (C. Gmelin) Celak.) - Wood Ragwort. Erect perennial to 1.5m; leaves mostly narrowly elliptic, shortly serrate, distinctly though shortly petiolate; ligules 5-6, 12-15mm, yellow. Intrd; natd in damp shady places; MW Yorks; Europe.

7. S. doria L. - Golden Ragwort. Erect perennial to 1.5m; leaves lanceolate to narrowly ovate, shortly serrate, sessile, clasping stem at base; ligules 4-6, 7-10mm, golden-yellow. Intrd; natd by streams and in wet meadows; Kirkcudbrights, Offaly, Man, perhaps elsewhere; Europe.

8. S. paludosus L. - Fen Ragwort. Erect perennial to **RRR** 1.5(2)m; leaves lanceolate to linear-lanceolate, shortly serrate, sessile, clasping stem at base; ligules 12-20, 10-14mm, yellow. Native; fenland ditches; formerly local in E En from N Lincs to Cambs, last seen 1857, refound in 1 place in Cambs in 1972.

9. S. doronicum (L.) L. - Chamois Ragwort. Erect perennial to 60cm; leaves elliptic-oblong (below) to linear-lanceolate, sessile and ± clasping stem (above), shortly to obscurely serrate; ligules 15-22, 12-20mm, golden-yellow. Intrd; natd on river banks in M Perths since 1985; mountains of Europe.

10. S. smithii DC. - <u>Magellan</u> Ragwort. Erect perennial to 1m; leaves ovate-oblong to -triangular, shortly and irregularly dentate, the lower petiolate, the upper sessile; ligules 12-20, 15-25mm, white. Intrd; natd in grassy places and by streams; extreme N Caithness, Orkney, Shetlands; S Chile and S Argentina.

11. S. jacobaea L. - <u>Common</u> Ragwort. Erect perennial to 1.5m (often much shorter), very variably pubescent; leaves deeply pinnately lobed, the lowest lobes ± clasping stem; ligules 12-15, 5-9mm (rarely shorter), rarely 0 (var. **nudus** Weston), yellow. Native; grassland, waysides, waste ground, sand-dunes; common throughout BI. Very variable; dwarf coastal plants with dense web-like pubescence and 0 ligules have been called ssp. **dunensis** (Dumort.) Kadereit & Sell, but represent only 1 line of variation.

11 x 12. S. jacobaea x S. aquaticus = S. x ostenfeldii Druce occurs throughout much of Br and Ir, more commonly in W, with the parents; it is intermediate in leaf dissection, achene pubescence and inflorescence shape. It is <15% fertile but backcrosses and sometimes forms hybrid swarms.

12. S. aquaticus Hill (<u>S.</u> erraticus auct. non Bertol.) - <u>Marsh</u> Ragwort. Erect biennial to perennial to 80cm, glabrous to sparsely pubescent; differs from <u>S. jacobaea</u> in usually laxer habit, ray flowers always present, and see key. Native; marshes, damp meadows and streamsides; frequent throughout most of BI, common in much of W.

13. S. erucifolius L. - <u>Hoary</u> Ragwort. Erect perennial to 1.2m, with short rhizomes, densely pubescent when young; leaves deeply pinnately lobed, the lowest lobes ± clasping stem; ligules 12-15, 5-9mm, yellow. Native; grassy places, banks, waysides and field-borders, usually on base-rich soils; common in most of En and Wa, very local in E Ir, rare alien in Sc.

14. S. squalidus L. - <u>Oxford</u> Ragwort. Erect to ascending, ± 874 glabrous annual to perennial to 50cm, sometimes woody below; leaves usually deeply pinnately lobed, sometimes deeply serrate, the lower petiolate, the upper sessile; ligules 12-15, (6)8-10mm, yellow, rarely 0; 2n=20. Intrd; waste ground, walls and waysides; 1st recorded 1794, now common in En and Wa, local elsewhere but still spreading; S Europe.

14 x 16. S. squalidus x S. vulgaris = S. x baxteri Druce is 874 intermediate in leaf and capitulum characters and highly sterile; it has been recorded from scattered places in Br and Ir but is probably over-recorded for radiate plants of <u>S. vulgaris</u>.

14 x 19. S. squalidus x S. viscosus = S. x subnebrodensis Simonkai (<u>S.</u> x <u>londinensis</u> Lousley) occurs sporadically on waste land in Br, mostly in S; it is intermediate in leaf-

2cm

FIG 874 – Senecio. 1, S. squalidus. 2, S. vulgaris (rayless).
3, S. vulgaris (rayed). 4, S. x baxteri. 5, S. vernalis. 6,
S. cambrensis.

shape, pubescence and capitulum characters, and highly sterile.

15. S. cambrensis Rosser - <u>Welsh Groundsel</u>. Erect glabrous **874** to sparsely pubescent annual to 30(50)cm; leaves deeply **RR** pinnately lobed, the lower petiolate, the upper sessile; ligules 8-15, 4-7mm, yellow; 2n=60. Native; waste ground and waysides; 1st found 1948 in Flints, now also in Denbs and Salop, found in Midlothian in 1982; originated as amphi-diploid of <u>S. squalidus</u> x <u>S. vulgaris</u>; endemic.

16. S. vulgaris L. - <u>Groundsel</u>. Usually erect, glabrous to **874** webby-pubescent annual to 30(45)cm; leaves shallowly to deeply pinnately lobed, the lower petiolate, the upper sessile; ligules usually 0, sometimes 7-11, <5mm, yellow; 2n=40. Native; open and rough ground in all sorts of habitats; common throughout BI. Ligulate plants are of 2 sorts: those otherwise like var. **vulgaris** and sporadic with it (var. **hibernicus** Syme); and those with ± simple stems, few **874** capitula, extensive web-like pubescence, and less deeply lobed leaves and occurring on sand-dunes in CI and W Br N to Man (var. **denticulatus** (Mueller) N. Hylander (ssp. <u>denticulatus</u> (Mueller) Sell)). There is evidence that var. <u>hibernicus</u> has arisen by introgression from <u>S. squalidus</u>. Hybrids between ligulate and eligulate plants have very short ligules.

16 x 17. S. vulgaris x S. vernalis = S. x helwingii Beger ex Hegi (<u>S.</u> x <u>pseudovernalis</u> Zabel ex Nyman nom. inval.) has occurred with <u>S. vernalis</u> in Leics (1968-9) and Moray (1983); it is intermediate and has low fertility.

17. S. vernalis Waldst. & Kit. - <u>Eastern Groundsel</u>. Erect, **874** usually webby-pubescent annual to 50cm; leaves usually rather shallowly lobed, often very like those of <u>S. vulgaris</u>, the lower petiolate, the upper sessile; ligules 8-15, (6)8-10mm, yellow; 2n=20. Intrd; road-verges and newly landscaped areas, semi-natd grass-seed alien; sporadic in Br and Man, increasing; E Europe, natd in W Europe.

18. S. sylvaticus L. - <u>Heath Groundsel</u>. Erect, often somewhat sticky, pubescent annual to 70cm; leaves deeply pinnately lobed, the lower petiolate, the upper sessile; ligules 8-15, <6mm, very soon strongly revolute, yellow. Native; open ground on heaths, banks and sandy places; locally common throughout BI.

18 x 19. S. sylvaticus x S. viscosus = S. x viscidulus Scheele occurs with the parents in scattered places in En and Sc and in M Cork; it is intermediate and has low fertility.

19. S. viscosus L. - <u>Sticky Groundsel</u>. Erect, very sticky, pubescent annual to 60cm; leaves deeply pinnately lobed, the lower petiolate, the upper sessile; ligules 8-15, <8mm, very soon strongly revolute, yellow. Possibly native; waste and rough ground, railway tracks, roadsides, walls; frequent in most of Br, intrd and local in N & C Sc and Ir.

69. PERICALLIS D. Don - <u>Cineraria</u>
Annual to perennial herbs; leaves alternate, palmately
veined; capitula radiate; phyllaries all in 1 main row;
colour of disc flowers and ligules usually contrasting, the
former darker, blue, red or pink to purple, never yellow.

1. P. hybrida R. Nordenstam (<u>Senecio</u> <u>hybridus</u> N. Hylander
nom. inval., <u>S.</u> <u>cruentus</u> auct. non Roth nec (L'Her.) DC.) -
<u>Cineraria</u>. Stms erect, pubescent, to 80cm; leaves petiolate,
palmately lobed, pubescent; capitula in ± dense corymbose
masses, 1.5-4cm across (incl. ligules).   Intrd; a very
popular but frost-sensitive pot-plant, well natd on open
ground, walls and waysides in Scillies, rarely on mainland W
Cornwall; garden origin from **P. cruenta** (L.'Her.) Bolle and
other spp. from Canaries.

70. TEPHROSERIS (Reichb.) Reichb. - <u>Fleaworts</u>
(Biennial to) perennial herbs; leaves alternate, pinnately
veined; capitula radiate; phyllaries all in 1 main row; disc
flowers and ligules yellow.

1. T. integrifolia (L.) Holub (<u>Senecio</u> <u>integrifolius</u> (L.)   **R**
Clairv.) - <u>Field</u> <u>Fleawort</u>.   Erect, densely pubescent
perennial to 60cm; basal leaves oblong-ovate, petiolate,
entire to coarsely dentate; stem-leaves much smaller, lanceolate, sessile, rarely >10; capitula <12(15), 1.5-2.5cm
across; ligules 12-15, 5-10mm. Native; short natural grassland.
   a. Ssp. integrifolia. Stems to 30(40)cm; leaves entire to
denticulate; stem-leaves usually <6; capitula rarely >6;
phyllaries 6-8.5mm. On chalk and limestone; local in S En N
to Cambs and E Gloucs, formerly to S Lincs.
   b. Ssp. maritima (Syme) R.   Nordenstam   (<u>Senecio</u>
<u>spathulifolius</u> auct. non Griess., <u>S.</u> <u>integrifolius</u> ssp.
<u>maritimus</u> (Syme) Chater). Stems to 60(90)cm; leaves usually
dentate; stem-leaves often >6; capitula often >6; phyllaries
8-12mm. On glacial drift on sea-cliffs; extremely local on
Holyhead Island (Anglesey); endemic.
   An extinct population in Westmorland probably represented an
endemic, third ssp.
   2. T. palustris (L.) Fourr. (<u>Senecio</u> <u>palustris</u> (L.) Hook.   **RR**
non Vell. Conc., <u>S.</u> <u>congestus</u> (R. Br.) DC.) - <u>Marsh</u> Fleawort.
Erect, densely pubescent perennial to 1m; basal leaves
usually withered before flowering; stem-leaves very numerous,
lanceolate, sessile, dentate, ± clasping stem at base;
capitula often >12, 2-3cm across; ligules c.21, 7-10mm;
phyllaries 10-13mm. Native; fen ditches; formerly local in E
En from W Sussex to MW Yorks; extinct (last recorded 1899).
Our plant was ssp. **congesta** (R. Br.) Holub.

71. DELAIREA Lemaire - <u>German-ivy</u>
Trailing or climbing, ± glabrous woody perennial; leaves

alternate, palmately veined; capitula discoid; phyllaries in
1 main row with short supplementary ones at base of
capitulum; disc flowers yellow.

1. D. odorata Lemaire (Senecio mikanioides Otto ex Walp.) - 878
German-ivy. Stems climbing, woody below, rather succulent
distally, to 3m; leaves succulent, palmately lobed; capitula
numerous in dense axillary and terminal panicles; phyllaries
3-4mm. Intrd; clambering over hedges and walls; natd in CI
and Scillies, rarely so in mainland E & W Cornwall; S Africa.
Flowers fragrant, appearing in Nov.

72. BRACHYGLOTTIS Forster & G. Forster - Shrub Ragworts
Evergreen shrubs; leaves alternate, pinnately veined, densely
white-felted on lowerside; capitula discoid or radiate;
phyllaries in 1 main row with short supplementary ones at
base of capitulum; disc flowers cream or yellow; ligules
yellow.

1 Many leaves >8cm, distantly sinuate-lobed; capitula
   <5mm across, cream; ligules 0; phyllaries with
   woolly hairs only at base   **3. B. repanda**
1 Leaves <8cm, entire to denticulate or tightly
   undulate; capitula >1cm across, yellow; ligules
   conspicuous; phyllaries with woolly hairs along ±
   whole length   2
   2 Leaves <4cm, tightly crenate-undulate   **2. B. monroi**
   2 Many leaves >4cm, entire to remotely denticulate
                                              1. B. 'Sunshine'

1. B. 'Sunshine' (Senecio greyi auct. non Hook. f., ?S. 878
laxifolius J. Buch. x S. compactus Kirk) - Shrub Ragwort.
Spreading shrub to 1(2)m; leaves up to 8cm, oblong-elliptic,
entire to remotely denticulate; capitula 2-4.5cm across;
ligules c.13, 10-15mm. Intrd; much grown in gardens and
often mass planted by roads, etc., persistent on rough
ground; scattered in En and N Wa; garden origin.
2. B. monroi (Hook. f.) R. Nordenstam (Senecio monroi Hook. 878
f.) - Monro's Ragwort. Spreading shrub to 1m; leaves up to
4cm, elliptic-obovate, tightly crenate-undulate; capitula
1.5-4.5cm across; ligules c.13, 6-15mm. Intrd; persistent
and ± natd on dunes near Llandudno, Caerns; New Zealand.
3. B. repanda Forster & G. Forster - Hedge Ragwort. Shrub 878
or small tree to 6m; leaves up to 25cm, remotely sinuate-
lobed; capitula c.5mm across, in large dense panicles;
ligules 0. Intrd; used as hedging in Scillies, often long
persistent after neglect; New Zealand.

73. SINACALIA H. Robinson & Brettell - Chinese Ragwort
Rhizomatous, ± glabrous herbaceous perennials; leaves
alternate, pinnately veined; capitula radiate, with 3 or 4
disc and 3 or 4 ray flowers; phyllaries all in 1 main row,

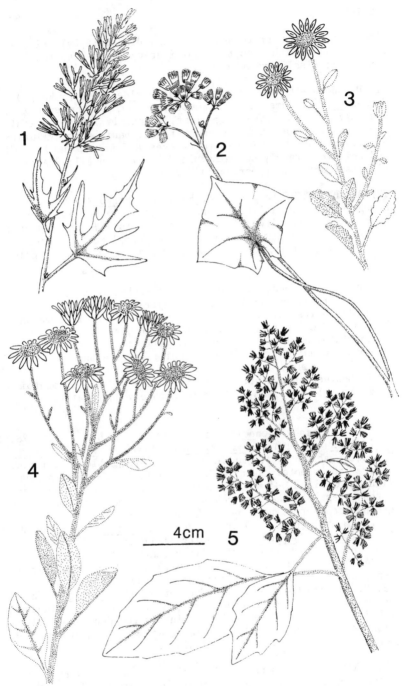

FIG 878 – Asteraceae: Senecioneae. 1, *Sinacalia tangutica*. 2,
Delairea odorata. 3, *Brachyglottis monroi*. 4, B. 'Sunshine'.
5, B. *repanda*.

but with small bracts some way below base of capitulum; disc flowers and ligules yellow.

1. S. tangutica (Maxim.) R. Nordenstam (<u>Senecio</u> <u>tanguticus</u> **878** Maxim.) - <u>Chinese</u> <u>Ragwort</u>. Stems erect, to 2m, unbranched except in inflorescence; leaves up to 20cm, ovate, deeply pinnately lobed; capitula numerous in large terminal panicles, <3mm wide excl. ligules; ligules 5-9mm. Intrd; grown in gardens, natd in damp shady places; scattered in N Wa and N En, Renfrews; China.

74. LIGULARIA Cass. - Leopard-plants

Herbaceous perennials; leaves mostly basal, those on stems alternate, cordate at base, palmately veined, with sheathing petiole bases; capitula radiate; phyllaries all in 1 row; disc flowers brownish-yellow; ligules yellow to orange.

1. L. dentata (A. Gray) H. Hara (<u>L.</u> <u>clivorum</u> Maxim., <u>Senecio</u> <u>clivorum</u> (Maxim.) Maxim. nom. nud.) - <u>Leopard-plant</u>. Stems erect, to 1.2m; basal leaves reniform, dentate, up to 50cm wide; capitula 4-10cm across, several in subcorymbose terminal cluster, with numerous disc flowers and 10-15 orange ligules 15-40mm. Intrd; grown in gardens, persistent in damp or shady places; very scattered in En and Sc, especially N; China and Japan.

2. L. przewalskii (Maxim.) Diels - <u>Przewalski's</u> <u>Leopard-plant</u>. Stems erect, to 1.8m; basal leaves deeply palmately lobed, up to 50cm wide, the lobes lobed or toothed; capitula 1.5-3cm across, numerous in long narrow terminal raceme-like panicle, with usually only 3 disc flowers and 2 yellow ligules 6-15mm. Intrd; grown in gardens, persistent by R. Tyne in S Northumb; N China.

75. DORONICUM L. - Leopard's-banes

Rhizomatous herbaceous perennials; leaves alternate, ± palmately veined, the basal ones ± withered by flowering time; capitula radiate; phyllaries in 2 rows of equal length; disc flowers and ligules yellow.

1 Basal leaves all cuneate at base   **3. D. plantagineum**
1 Most or all basal leaves cordate to rounded or
  truncate at base   2
  2 Petioles of basal leaves with many long (>1mm)
   flexuous or patent hairs; capitula usually 3-8 per
   stem   **1. D. pardalianches**
  2 Petioles of basal leaves with 0-very few long
   hairs; capitula 1-2(3) per stem   3
3 Basal leaves deeply cordate at base; all hairs on
  stems short (<1mm) and glandular   **5. D. columnae**
3 Basal leaves shallowly cordate to truncate or rounded
  at base; stems usually with a few long (>1mm)
  eglandular as well as short glandular hairs   4

4 Basal leaves acute, mainly shallowly cordate at
   base, with prominent teeth >2mm   **4. D. x excelsum**
4 Basal leaves obtuse, mainly rounded to truncate at
   base, with less prominent teeth <2mm
                                          2. D. x willdenowii

1. D. pardalianches L. - <u>Leopard's-bane</u>. Stems erect, to
80cm, with long eglandular and short glandular hairs; basal
leaves deeply cordate with narrow sinus, obscurely crenate-
dentate; capitula c.3-8, mostly 3-4.5cm across. Intrd; well
natd in woods and shady places; frequent throughout Br; W
Europe.
2. D. x willdenowii Rouy (<u>D. plantagineum</u> var. <u>willdenowii</u>
(Rouy) A.B. Jackson; ?<u>D. pardalianches</u> x <u>D. plantagineum</u>) -
<u>Willdenow's Leopard's-bane</u>. Stems erect, to 1m, with short
glandular and usually a few long eglandular hairs; basal
leaves mostly rounded to truncate at base, obscurely crenate-
dentate; capitula 1-2(3) per stem, mostly 4.5-8cm across.
Intrd; natd in woods and shady places; scattered in Br; W
Europe or garden origin. Under-recorded for <u>D. plantagineum</u>.
3. D. plantagineum L. - <u>Plantain-leaved Leopard's-bane</u>.
Stems erect, to 1m, with only short glandular hairs; basal
leaves cuneate at base, obscurely crenate-dentate; capitula
as in <u>D. x willdenowii</u>. Intrd; natd in woods and shady
places; scattered in Br but over-recorded for last and next;
W Europe.
4. D. x excelsum (N.E. Br.) Stace (<u>D. plantagineum</u> var.
<u>excelsum</u> N.E. Br.; ?<u>D. pardalianches</u> x <u>D. plantagineum</u> x <u>D.</u>
<u>columnae</u>) - <u>Harpur-Crewe's Leopard's-bane</u>. Stems erect, to
1m, with short glandular and usually a few long eglandular
hairs; basal leaves mostly shallowly cordate at base, rather
conspicuously dentate; capitula as in <u>D. x willdenowii</u>.
Intrd; habitat and distribution as for <u>D. x willdenowii</u>;
garden origin.
5. D. columnae Ten. (<u>D. cordatum</u> auct. non Lam.) - <u>Eastern</u>
<u>Leopard's-bane</u>. Stems erect, to 60cm, with usually few short
glandular hairs only; basal leaves deeply cordate with wide
sinus, conspicuously dentate; capitulum usually 1, mostly
2.5-5cm across. Intrd; natd on bank of reservoir in Surrey,
perhaps overlooked; SE Europe.

76. TUSSILAGO L. - <u>Colt's-foot</u>
Rhizomatous, herbaceous perennials; leaves all basal, cordate
at base, ± palmately veined, cottony-pubescent on lowerside;
flowering stems bearing many bracts and 1 terminal capitulum,
cottony-pubescent; capitula radiate; phyllaries all in 1 row;
disc flowers and ligules yellow.

1. T. farfara L. - <u>Colt's-foot</u>. Stems erect, to 15cm,
appearing before leaves; leaves broadly ovate, 20-30cm
across, shallowly palmately lobed, the lobes dentate to
denticulate; capitula 1.5-3.5cm across. Native; open or

semi-open or disturbed ground in many habitats, including arable land and maritime sand and shingle; common throughout BI. Leaves differ from those of most spp. of Petasites in their distinct lobes and from P. albus in their black-tipped teeth.

77. PETASITES Miller - Butterburs

Dioecious, rhizomatous, herbaceous perennials; leaves all basal, cordate at base, ± palmately veined, cottony-pubescent on lowerside; flowering stems bearing few to many bracts and a terminal raceme or panicle of capitula, cottony-pubescent; male capitula composed of male flowers with clavate sterile stigmas and sometimes some female-like sterile discoid or radiate flowers; female capitula composed of discoid female flowers and 1 or few central male-like sterile flowers; flowers white to purple or cream.

1 Marginal flowers ligulate (ligules <1cm);
  inflorescences appearing Nov-Feb, with basal leaves
  present, always male; leaves with small teeth all of
  1 size   **4. P. fragrans**
1 Marginal flowers tubular; inflorescences appearing
  Feb-Apr, before basal leaves, male or female; leaves
  unevenly dentate, with large teeth or short lobes
  dispersed among small teeth   2
    2 Basal leaf sinus with parallel or divergent sides,
      bordered by 0-1 veins on each side; flowers pure
      white; leaves <30cm across   **3. P. albus**
    2 Basal leaf sinus with convergent sides, bordered
      by ≥2 veins on each side; flowers usually cream
      or with purplish tinge; leaves often >30cm across   3
3 Leaves distinctly but very shallowly lobed; mature
  inflorescences ± cylindrical; upper part of stem
  below inflorescence with bracts <1cm wide; phyllaries
  and/or florets usually with anthocyanin; corollas
  white to purple-tinged   **1. P. hybridus**
3 Leaves scarcely or not lobed, unevenly dentate;
  mature inflorescences ± hemisperical; upper part
  of stem below inflorescence with bracts >1cm wide;
  anthocyanin absent; corollas cream in male capitula,
  whitish in female ones   **2. P. japonicus**

1. P. hybridus (L.) P. Gaertner, Meyer & Scherb. - Butterbur. Leaves suborbicular, with convergent sides to basal sinus, up to 90cm across, obscurely lobed; petioles up to 1.5m; flowering stems erect, to 30cm (to 1m in fruit), with many narrow bracts; flowers white with purplish tinge. Native; by rivers and ditches, in damp fields and waysides, often in shade; male plant frequent throughout most of BI; female plant frequent in N & C En, very sporadic elsewhere.
2. P. japonicus (Siebold & Zucc.) Maxim. - Giant Butterbur. Leaves as in P. hybridus but see key; flowering stems erect,

to 30cm, with many broad bracts; flowers cream. Intrd; natd
by rivers and in damp places in open or shade; scattered
throughout most of Br; Japan. The female plant is rarely or
perhaps never natd here.
   3. P. albus (L.) Gaertner – <u>White Butterbur</u>. Leaves
suborbicular, with divergent or parallel sides to basal
sinus, up to 30cm across, with well-developed acute lobes;
petioles up to 30cm; flowering stems erect, to 30cm (to 70cm
in fruit), with few to rather many narrow to medium-broad
bracts; flowers white. Intrd; natd in rough ground, waysides
and woods; throughout Br, rare in S, common in N, N Ir;
Europe. Female plant much less common than male here.
   4. P. fragrans (Villars) C. Presl (?*P. pyrenaicus* (L.) G.
Lopez) – <u>Winter Heliotrope</u>. Leaves suborbicular, with
divergent sides to basal sinus, up to 20cm across, not lobed;
petioles up to 30cm; flowering stems erect, to 30cm, with few
medium-broad bracts; flowers white tinged purple, strongly
vanilla-scented. Intrd; natd on waste and rough ground and
waysides; throughout BI, common in S, local in N; N Africa.
Female plant unknown in BI.

78. HOMOGYNE Cass. – <u>Purple Colt's-foot</u>
Rhizomatous, herbaceous perennials; leaves all basal, cordate
at base, palmately veined, rather sparsely pubescent on
lowerside; flowering stems bearing few bracts and 1 terminal
capitulum, cottony-pubescent; capitula discoid, the central
flowers bisexual with 5-lobed corolla, the outermost row
female with obliquely truncate tubular corolla; phyllaries ±
in 1 row; flowers purple.

   1. H. alpina (L.) Cass. – <u>Purple Colt's-foot</u>. Stems erect, **RRR**
to 35cm, appearing with leaves; leaves reniform-orbicular, up
to 4cm across, shallowly crenate-dentate; capitula 10-15mm
across. Probably intrd; 1 locality in Angus at c.600m, known
c.1813, refound 1951, perhaps originally planted; C Europe.

TRIBE 8 – CALENDULEAE (genus 79). Annual to perennial herbs;
leaves alternate, simple, ± sessile; capitula radiate;
phyllaries in 1-2 rows, herbaceous with scarious margin;
receptacular scales 0; achenes varying in one capitulum in
size, degree of curving and wartiness, all with pappus 0;
flowers yellow to orange, sometimes <u>flore pleno</u>.

79. CALENDULA L. – <u>Marigolds</u>
   1. C. officinalis L. – <u>Pot Marigold</u>. Perennial often
behaving as annual, with distinctive scent; stems erect to
procumbent, to 80cm; capitula 4-7cm across, pale yellow to
deep orange; ligules 2x as long as phyllaries. Intrd; much
grown in gardens, frequent escape or throw-out on tips and
waste ground; scattered in Br and CI, sometimes ± natd in S;
?garden origin.
   2. C. arvensis L. – <u>Field Marigold</u>. Erect to procumbent

casual elsewhere; scattered in En; tropical America.

85. RUDBECKIA L. - Coneflowers
Perennials; leaves alternate, simple to deeply lobed;
phyllaries in 2 or more rows, herbaceous; capitula radiate;
receptacle conical, with scales partly enclosing achenes;
pappus 0 or a short rim; ligules numerous, yellow to orange.

1. R. hirta L. (R. serotina Nutt.) - Black-eyed-Susan. 893
Stems erect, to 80cm; leaves simple, roughly pubescent,
entire or nearly so; capitula 5-10cm across; disc flowers
brownish-purple; ligules yellow or orange. Intrd; grown in
gardens, occasionally persistent in rough ground and waste
places; scattered in S Br; N America.
2. R. laciniata L. - Coneflower. Stems erect, to 3m; leaves
deeply divided, the lowest ± pinnate, glabrous or nearly so;
capitula 7-14cm across; disc flowers greenish-yellow; ligules
yellow. Intrd; grown in gardens, natd on waste and rough
ground, much commoner than R. hirta; scattered in Br, mainly
S & C; N America.

86. HELIANTHUS L. - Sunflowers
Annuals to perennials; leaves opposite below, alternate
above, simple; phyllaries in 2 or more rows, herbaceous;
capitula radiate; receptacle flat or slightly convex, with
scales partly enclosing achenes; pappus of 2 narrow scales
soon falling off, sometimes with some shorter extra scales;
ligules numerous, yellow (often flore pleno).

```
1  Plant annual, with simple tap-root                           2
1  Plant perennial, clump-forming, with (often very
   short) rhizomes                                              3
  2  Phyllaries ovate, abruptly contracted to acuminate
     tip; central receptacular scales inconspicuously
     pubescent                                        1. H. annuus
  2  Phyllaries lanceolate to narrowly ovate, gradually
     tapered to apex; central receptacular scales with
     conspicuous long white hairs at apex
                                                  2. H. petiolaris
3  Stems ± glabrous in lower half; at least some
   phyllaries much exceeding edge of receptacle
                                                3. H. x multiflorus
3  Stems roughly pubescent almost to base; phyllaries
   not or scarcely exceeding edge of receptacle                 4
  4  Rhizomes with swollen tubers; stems often >2m;
     phyllaries not or loosely appressed to receptacle
                                                    5. H. tuberosus
  4  Rhizomes without swollen tubers; stems rarely
     >2m; phyllaries closely appressed to receptacle
                                                 4. H. x laetiflorus
```

1. H. annuus L. - Sunflower. Annual; stems erect, to 3m, usually unbranched; capitula usually 1, with receptacle 2-30cm across. Intrd; grown as ornamental or on field scale as oil-seed, common casual on tips and in waste places, also as bird-seed alien; throughout most of BI except most rural areas; N America.

2. H. petiolaris Nutt. - <u>Lesser Sunflower</u>. Resembles a small <u>H. annuus</u>; differs in stems usually <u><</u>1m; receptacle 1-2.5cm across; and see key. Intrd; rather infrequent casual, especially from soya-bean waste, perhaps overlooked; frequent in London area; N America.

3. H. x multiflorus L. (<u>H. annuus</u> x <u>H. decapetalus</u> L.) - <u>Thin-leaved Sunflower</u>. Perennial; stems erect, to 1.5(2)m, smooth and <u>+</u> glabrous below; capitula usually >1, with receptacle 1-2cm across; often <u>flore pleno</u>. Intrd; grown in gardens, rather infrequent escape or throw-out; scattered in Br, mainly S & C, but confused with <u>H. x laetiflorus</u>; garden origin. Possibly some records refer to <u>H. decapetalus</u>.

4. H. x laetiflorus Pers. (<u>H. pauciflorus</u> Nutt. (<u>H. rigidus</u> (Cass.) Desf.) x <u>H. tuberosus</u>) - <u>Perennial Sunflower</u>. Perennial; stems erect, to 1.5(2)m, scabrid-pubescent <u>+</u> throughout; capitula usually >1, with receptacle 1-2.5cm across; often <u>flore pleno</u>. Intrd; much grown in gardens, frequent escape or throw-out; throughout most of BI; probably garden origin. The commonest garden perennial sunflower. Some records may refer to <u>H. pauciflorus</u>, which has purplish-brown (not yellow) disc-flowers.

5. H. tuberosus L. - <u>Jerusalem Artichoke</u>. Perennial, with irregular tubers developing on rhizomes; stems erect, to 3m, scabrid-pubescent <u>+</u> throughout; capitula usually several, with receptacle 1.5-2.5cm across, but stems usually frosted down before flowering. Intrd; grown as minor root-crop and very persistent in waste places; scattered throughout much of Br and CI; N America.

87. GALINSOGA Ruiz Lopez & Pavon - <u>Gallant-soldiers</u>
Annuals; leaves all opposite, simple, ovate, petiolate; phyllaries in 2 rows, largely herbaceous, the outer much shorter, the inner with membranous margins; capitula <1cm across, radiate; receptacle conical, with scales; pappus of scales; ligules few (usually 5), white or pinkish; disc flowers yellow.

1. G. parviflora Cav. - <u>Gallant-soldier</u>. Stems erect to **885** ascending, to 80cm, glabrous or sparsely pubescent; peduncles rather sparsely pubescent with glandular and eglandular hairs c.0.2mm; receptacular scales mostly distinctly 3-lobed, the central lobe the largest; pappus-scales fringed with hairs, without a terminal projection. Intrd; well natd weed of cultivated and waste ground; locally frequent in CI and Br N to C Sc, especially in London and other large cities, first record 1860; S America.

2. G. quadriradiata Ruiz Lopez & Pavon (<u>G. ciliata</u> (Raf.) **885**

S.F. Blake) - <u>Shaggy-soldier</u>. Differs from <u>G. parviflora</u> in more densely pubescent stems; peduncles with many glandular and eglandular hairs c.0.5mm; receptacular scales mostly simple, some with 1 or 2 weak lateral lobes; pappus-scales fringed with hairs and with a fine terminal projection. Intrd; similar habitats and distribution to <u>G. parviflora</u>, often with it, first record 1909; S America. Often very distinct from <u>G. parviflora</u> in pubescence, but sometimes very close to it; pappus-scale apex is best character.

88. BIDENS L. - <u>Bur-marigolds</u>

Annuals; leaves all opposite, simple and toothed to pinnate; phyllaries in 2 dissimilar rows, the outer herbaceous, the inner ± membranous with a usually scarious border; capitula usually discoid, rarely radiate; receptacle flat or slightly convex, with scales; pappus of 2-5 barbed (forwardly or backwardly), strong bristles; ligules 0, rarely yellow, very rarely white.

The achenes provide important diagnostic characters, but some of those traditionally used, e.g. bristle number and direction of barbs on bristles, are sometimes unreliable. Throughout the account 'achenes' refers to the central ones in the capitulum; the outer ones may differ considerably. All spp. are usually eligulate, but <u>B. cernua</u> sometimes has conspicuous yellow rays and <u>B. frondosa</u> and <u>B. pilosa</u> white ones.

1 Leaves not lobed, or lobed <1/2way to midrib   2
1 At least lower leaves pinnate, or lobed nearly to
  midrib   4
  2 Achenes scarcely flattened (<2x as wide as thick),
    the faces between the 4 ridges warty   **3. B. connata**
  2 Achenes strongly flattened (>2x as wide as thick),
    the faces between the 2-4 ridges smooth   3
3 At least lower leaves distinctly petiolate, with
  1(-2) pairs of distinct lobes   **2. B. tripartita**
3 All leaves tapered to base but sessile, unlobed
  (but strongly serrate)   **1. B. cernua**
  4 Leaflets lobed again to midrib or ± so
                            6. B. bipinnata
  4 Leaflets unlobed   5
5 Petioles winged to base; apical (main) lobe of leaf
  scarcely stalked or with winged stalk; barbs on edge
  of achenes (?always) backward-directed   **2. B. tripartita**
5 At least lower leaves with ± unwinged petioles and
  with apical (main) lobe with distinct ± wingless
  stalk; barbs on edge of achenes (but not on apical
  bristles) (?always) forward-directed   6
  6 Leaflet-teeth mostly wider than long; achenes
    ± parallel-sided, slightly tapered at each end
                                  5. B. pilosa
  6 Leaflet-teeth mostly longer than wide; achenes

tapered + from apex to base   **4. B. frondosa**

Other ssp. - B. vulgata E. Greene, from N America, would key
out as <u>B. frondosa</u> but has (10-)13(-16) (not (5-)8(-10))
outer phyllaries and rather pale yellow (not orange-yellow)
disc flowers; it has been recorded as a rare casual and might
be overlooked.

1. B. cernua L. - <u>Nodding Bur-marigold</u>. Stems erect, to **885**
75cm; leaves not lobed, sessile; achenes tapered from apex to
base, with 3-4 bristles, the achene and bristles with
backward-directed barbs. Native; by ponds and streams and in
ditches and marshy fields; locally common in C & S Br and Ir,
rare and very scattered in N. Radiate plants (var. **radiata**
DC.) are occasional in NW En, very rare elsewhere.
2. B. tripartita L. - <u>Trifid Bur-marigold</u>. Stems erect, to **885**
75cm; many leaves usually with 1-2 pairs of deep lateral
lobes, rarely all unlobed (var. **integra** Koch), with winged
petiole; achenes tapered from apex to base, with 2-4
bristles, the achene and bristles with backward-directed
barbs. Native; similar habitat and distribution to <u>B.</u>
<u>cernua</u>, and often with it.
3. B. connata Muhlenb. ex Willd. - <u>London Bur-marigold</u>. **885**
Differs from <u>B. tripartita</u> in leaves not lobed; achenes
(incl. bristles) with usually forward-directed (rarely
backward-directed) barbs; and see key (couplet 2). Intrd;
natd by canals; Middlesex, Herts, Bucks and W Kent; N
America.
4. B. frondosa L. - <u>Beggarticks</u>. Stems erect, to 1m; many **885**
leaves pinnate, with 1-2 pairs of lateral leaflets; petioles
and leaflet-stalks + unwinged; achenes tapered from apex to
base, with forward-directed barbs, the usually 2 bristles
with backward-directed hairs. Intrd; natd by canals and
rivers and on damp ground and waste places; scattered in En
and Wa, frequent around Birmingham and London and in S Wa; N
& S America. Sometimes with white ligules.
5. B. pilosa L. - <u>Black-jack</u>. Differs from <u>B. frondosa</u> in **885**
leaflets usually broader and less deeply serrate; bristles
usually 2-3; and see key. Intrd; a rather characteristic
wool-alien; scattered in En; S America.
6. B. bipinnata L. - <u>Spanish-needles</u>. Similar to <u>B. pilosa</u> **885**
in capitulum size and achene morphology except bristles 2-4;
leaves 1-2 pinnate, the primary leaflets divided again to
midrib or nearly so. Intrd; a characteristic wool-alien;
scattered in En and Sc; S America.

89. COREOPSIS L. - <u>Tickseeds</u>
Annuals to perennials; leaves (usually all) opposite, all or
most pinnately or ternately divided to midrib or + so, the
primary divisions often divided again; phyllaries in 2
dissimilar rows, the outer narrower and shorter and +
herbaceous, the inner partially membranous; capitula radiate;

receptacle flat or slightly convex, with scales; pappus 0 or of few very short teeth or bristles; ligules c.8, yellow, sometimes with dark basal blotch.

Other spp. - C. tinctoria Nutt., from USA, is an annual with disc flowers with 4-toothed (not 5-toothed), dark purple corollas and ligules with a brownish-purple blotch at base; it is grown in gardens and a rare escape. **C. verticillata** L., from USA, has leaves finely divided from near base, so appearing whorled, and yellow capitula; it was established for a short while in N Hants.

1. C. grandiflora Hogg ex Sweet - Large-flowered Tickseed. Tufted perennial; stems erect, to 1m, well-branched above; capitula on long slender peduncles, uniformly yellow, 3-5cm across. Intrd; grown in gardens, infrequently natd as escape or throw-out; SE En; USA.

90. COSMOS Cav. - Mexican Aster
Annuals; leaves all opposite, 2-3-pinnate with linear to filiform segments; phyllaries in 2 dissimilar rows, the outer narrower, herbaceous with membranous border, the inner membranous; capitula radiate; receptacle flat, with scales; pappus of (0)2(3) bristles with usually backward-directed barbs; ligules numerous, pinkish-purple, rarely white; disc flowers yellow.

1. C. bipinnatus Cav. - Mexican Aster. Stems erect, to 2m; capitula 4-9cm across, incl. ligules 1.5-4cm. Intrd; much grown in gardens, frequent persistent casual on tips and in waste places; scattered in Br, mostly C & S; Mexico, S USA.

91. DAHLIA Cav. - Dahlia
Perennials (but killed by first frosts); leaves all opposite, petiolate, (bi-)pinnate or (bi-)ternate, the uppermost simple, phyllaries in 2 dissimilar rows, the inner membranous, the outer ± herbaceous; capitula radiate (but most often flore pleno); receptacle flat or slightly convex, with scales; pappus 0 or of 2 obscure teeth; ligules numerous, yellow, white, pink or purple; disc flowers yellow.

1. D. pinnata Cav. (D. variabilis (Willd.) Desf.) - Dahlia. Stems erect, to 2m, rather succulent; capitula extremely variable (in different cultivars) in size (up to 30cm across but often <10cm), colour, shape of ligules, and degree to which they are flore pleno. Intrd; much grown for ornament, frequent as throw-out on tips and waste ground, not natd; scattered in En, mainly SE; Mexico. The origin of cultivated Dahlias is uncertain; other spp., especially **D. coccinea** Cav., have probably shared in their parentage.

92. TAGETES L. - <u>Marigolds</u>
Aromatic annuals; leaves opposite below, alternate above, pinnate; phyllaries in 1 row, fused for most of length to form sheath round capitulum; capitula radiate; receptacle flat, without scales; pappus of unequal scales; ligules c.3-8, yellow to orange or brownish, often <u>flore pleno</u>.

1 Ligules <3mm; phyllary-sheath <4mm wide   **3. T. minuta**
1 Ligules >(5)10mm; phyllary-sheath >5mm wide   2
  2 Peduncles conspicuously swollen below capitula;
    phyllaries mostly >1.5cm; ligules mostly 1-3cm
                                   1. T. erecta
  2 Peduncles not or scarcely swollen below capitula;
    phyllaries mostly <1.5cm; ligules mostly
    (0.5)1-1.5cm   **2. T. patula**

Other spp. - A few other spp. are grown in gardens and rarely reported as escapes or throw-outs.

1. T. erecta L. - <u>African Marigold</u>. Stems erect, to **885** 50(100)cm; capitula very showy, yellow to orange, 3-7cm across. Intrd; much grown in gardens, frequent casual as escape or throw-out; scattered in En; Mexico.
2. T. patula L. - <u>French Marigold</u>. Stems erect, to 40cm; **885** capitula showy, yellow to orange or reddish-brown, often bicoloured, 2-4cm across. Intrd; occurrence as for <u>T. erecta</u>; Mexico.
3. T. minuta L. - <u>Southern Marigold</u>. Stems erect, to 1.2m; **885** capitula inconspicuous, <1cm across, with usually 3 yellowish-green ligules c.1-2mm. Intrd; rather characteristic wool-alien; scattered in En; S America.

93. SCHKUHRIA Roth - <u>Dwarf Marigold</u>
Annuals; leaves usually alternate, rarely some opposite, pinnate with linear to filiform leaflets with many minute sunken glands; phyllaries few in 1 overlapping row, herbaceous with membranous tips; capitula <1cm across, radiate with only 1 short yellow ligule; receptacle concave, without scales; pappus of scales.

1. S. pinnata (Lam.) Kuntze - <u>Dwarf Marigold</u>. Stems erect, **893** to 60cm; capitula numerous in subcorymbose panicle, obconical, 6-10 x 2-6mm (excl. ligule). Intrd; rather characteristic wool-alien; scattered in En; S & C America.

94. GAILLARDIA Foug. - <u>Blanket-flower</u>
Annuals to perennials; leaves all alternate, simple, coarsely dentate, tapered to base but ± sessile; phyllaries in 2-3 rows, ± herbaceous, becoming reflexed before fruiting; capitula radiate; receptacle strongly convex to subglobose, with bristle-like scales; pappus of scales with apical bristle; ligules numerous, yellow or more often purple in

FIG 893 - **Asteraceae: Heliantheae.** 1, **Rudbeckia hirta.** 2,
Sigesbeckia **serrata.** 3, **Schkuhria** **pinnata.** 4, **Iva**
xanthifolia. 5, **Guizotia** **abyssinica.** 6, **Gaillardia** **x**
grandiflora.

proximal 1/4 to 3/4.

1. G. x grandiflora Van Houtte (G. aristata Pursh x G. **893** pulchella Foug.) - Blanket-flower. Annual or short-lived perennial; stems erect, to 70cm; capitula 3.5-10cm across, very showy. Intrd; much grown in gardens, occasional casual on tips and rough ground, sometimes ± natd especially on sand and shingle by sea; local in SE En; garden origin. Records for the 2 parents (both N American) are mostly (or all) errors for the hybrid.

95. HELENIUM L. - Sneezeweed
Tufted perennials; leaves all alternate, simple, subentire to shallowly dentate, tapered to base, decurrent on stem; phyllaries in 2-3 rows, herbaceous, becoming reflexed before fruiting; capitula radiate; receptacle convex, without scales; pappus of scales with apical bristle; ligules numerous, yellow to brownish-purple.

1. H. autumnale L. - Sneezeweed. Stems erect, to 1m; capitula 4-6.5cm across, the ligules soon becoming reflexed; disc and ray flowers yellow to brownish-purple in various combinations. Intrd; much grown in gardens, impermanent relic or throw-out; sporadic in SE En; N America.

TRIBE 10 - EUPATORIEAE (genera 96-97). Annual to perennial herbs; leaves opposite; capitula discoid; phyllaries in several rows, herbaceous; receptacular scales 0; pappus of 1 row of hairs or pointed scales; corolla blue or pinkish-purple, sometimes white.

96. EUPATORIUM L. - Hemp-agrimony
Perennials; at least some leaves very deeply 3-(5-)lobed or palmate, cuneate at base; pappus of hairs; corolla pinkish-purple, rarely white.

1. E. cannabinum L. - Hemp-agrimony. Stems erect, to 1.5m; capitula 2-5mm across, very numerous in compound subcorymbose panicles. Native; all sorts of damp places and by water, in shade or open, sometimes in dry grassland or rough ground; common in most of En and Wa, frequent in Ir and CI, local and mostly coastal in Sc.

97. AGERATUM L. - Flossflower
Annuals; leaves simple, ovate, cordate; pappus of narrow scales; corolla usually blue, rarely pink or white.

1. A. houstonianum Miller - Flossflower. Stems erect, to 60cm; capitula 6-10mm across, rather numerous in compound subcorymbose panicles. Intrd; grown in gardens and escaping in waste ground or on tips; occasional in SE En; Mexico.

L I L I I D A E - MONOCOTYLEDONS
(Monocotyledonidae)

Very rarely trees, rarely shrubs; rarely with secondary thickening and never from a permanent vascular cambium; vascular bundles usually scattered through stem; primary root usually short-lived; leaves usually with parallel major venation, and minor venation scarcely or not reticulate; flower parts mostly in threes; pollen grains mostly bilaterally symmetrical, commonly with 1 pore and/or furrow; cotyledon normally 1; endosperm typically helobial. Numerous exceptions to all the above occur.

138. BUTOMACEAE - Flowering-rush family

Glabrous, aquatic perennials rooted in mud, emergent through water; leaves all basal, simple, linear, entire, sessile, exstipulate. Flowers in terminal umbel with several bracts at base, bisexual, hypogynous, actinomorphic; sepals 3, free, purplish, tinged green; petals 3, free, pink; stamens 9; carpels 6, free except at extreme base, with numerous ovules all over inner side of wall; style 1; stigma slightly bilobed; fruit a group of 6 follicles.
Distinctive among petaloid monocots in its 6 ± free, dehiscent follicles, and similar sepals and petals.

1. BUTOMUS L. - Flowering-rush
1. B. umbellatus L. - Flowering-rush. Stems erect, to 1.5m; leaves usually slightly shorter; pedicels up to c.10cm; petals and sepals 1-1.5cm. Native; in ponds, canals, ditches and river-edges; rather scattered in Br N to C Sc and in Ir, but often intrd.

139. ALISMATACEAE - Water-plantain family

Glabrous, aquatic annuals or perennials rooted in mud, often emergent through water; leaves usually all basal, simple, entire, sessile or petiolate, exstipulate, sometimes produced in tufts on rooting stolons. Flowers in simple or compound umbels or in whorls, sometimes solitary, with bracts at base of umbel or whorl, bisexual or monoecious, hypogynous, actinomorphic; sepals 3, free, green; petals 3, free, white to pink; stamens 6 to numerous; carpels >6, free, with 1(-several) ovules; style 0 or very short; stigma not lobed,

usually slightly elongated; fruit a group of achenes or
few-seeded follicles.
Distinguished from Butomaceae in its very different petals
and sepals, and 1-2(few)-seeded indehiscent fruits.
Submerged leaves are often ribbon-like, often very different
from the diagnostically-shaped aerial leaves, and should be
ignored.

1 Flowers monoecious, male and female in same
  inflorescence; stamens >6   **1. SAGITTARIA**
1 Flowers bisexual; stamens 6   2
  2 Stems procumbent or floating, rooting and
    producing tufts of leaves and inflorescences   3
  2 Stems erect, leafless, all leaves basal   4
3 Floating and aerial leaves obtuse; carpels in
  irregular whorl or flattish mass   **3. LURONIUM**
3 Floating and aerial leaves acute; carpels spiral in
  ± globose (Ranunculus-like) head   **2. BALDELLIA**
  4 Carpels spiral in ± globose (Ranunculus-like)
    head   **2. BALDELLIA**
  4 Carpels in a single (often irregular) whorl   5
5 Fruits curved inwards, ± unbeaked, 1-seeded   **4. ALISMA**
5 Fruits divergent outwards, beaked, usually
  2(-few)-seeded   **5. DAMASONIUM**

1. SAGITTARIA L. - Arrowheads
Leaves all basal, linear and/or long-petiolate and sagittate;
flowers conspicuous, usually in whorls, monoecious, the male
flowers in upper and the female in lower whorls; petals
white; stamens 7-numerous; carpels spiral in ± globose head,
each with 1 ovule. In Autumn stolons tipped by small bud-
like propagules are formed.
Leaf-shape is notoriously variable in this genus and needs
to be used with great caution.

1 Most or all emergent leaves strongly sagittate, with
  two long basal lobes   2
1 Most or all leaves linear to elliptic, rarely a few
  with short basal lobes   3
  2 Achenes 4-6mm, with apical beak <1mm; petals
    usually with purple blotch at base; anthers
    purple   **1. S. sagittifolia**
  2 Achenes 2.5-4mm, with subapical beak >1mm; petals
    without basal purple blotch; anthers yellow
                                           2. S. latifolia
3 Flowers and leaves floating; many leaves linear,
  usually some or all floating ones elliptic;
  filaments glabrous   **4. S. subulata**
3 Flowers and leaves emergent; emergent leaves elliptic
  or rarely some with short basal lobes; filaments
  with scale-like hairs   **3. S. rigida**

1. S. sagittifolia L. - <u>Arrowhead</u>. Emergent leaves strongly sagittate; floating leaves often elliptic; submerged leaves linear; stems emergent, to 1m; flowers 2-3cm across; achenes 4-6mm, with beak <1mm. Native; in ponds, canals and slow rivers; frequent in En, very scattered in Wa and Ir.

2. S. latifolia Willd. - <u>Duck-potato</u>. Differs from <u>S. sagittifolia</u> in key characters. Intrd; natd in ponds and stream-side in Surrey since 1941, Jersey since 1961; N America.

3. S. rigida Pursh - <u>Canadian</u> <u>Arrowhead</u>. Emergent and floating leaves elliptic, rarely with 2 short basal lobes; submerged leaves linear; stems emergent to 75cm; achenes 2.5-4mm, with beak >1mm. Intrd; 2 places in canal in S Devon since 1898; N America.

4. S. subulata (L.) Buchenau - <u>Narrow-leaved</u> <u>Arrowhead</u>. Most leaves linear and submerged, none emergent; floating leaves elliptic; stems to 30cm, producing flowers on water surface; achenes 1.5-2.5mm, with beak <1mm. Intrd; natd in acid pond in N Hants since 1962; N America.

2. BALDELLIA Parl. - <u>Lesser</u> <u>Water-plantain</u>
Varying in vegetative and inflorescence habit from <u>Alisma</u>-like to <u>Luronium</u>-like; petals pale mauve to ± white with yellow basal blotch; stamens 6; carpels spiral in ± globose head, each with 1 ovule.

1. B. ranunculoides (L.) Parl. - <u>Lesser</u> <u>Water-plantain</u>. Rosette-plant with erect stem to 20cm bearing inflorescence in simple whorl or with 1(-2) whorls below, or plant with trailing stems producing tufts of leaves and reduced (often 1-flowered) inflorescences; leaves linear or petiolate and narrowly elliptic; flowers 10-16mm across; achenes 2-3.5mm, tapered to acute apex. Native; wet places or shallow water in ditches, stream-sides and pond-sides; scattered over most of BI.

3. LURONIUM Raf. - <u>Floating Water-plantain</u>
Stems procumbent or floating, rooting at intervals and producing tufts of leaves and inflorescences; submerged leaves linear, floating ones petiolate, elliptic, obtuse; flowers solitary or 2-5 in simple umbels, bisexual; petals pale mauve to ± white with yellow basal blotch; stamens 6; carpels in irregular whorl or flattish mass, each with 1 ovule.

1. L. natans (L.) Raf. - <u>Floating Water-plantain</u>. Stems to 75cm, often much less; floating leaves 1-2.5(4)cm; flowers 12-18mm across; achenes c.2.5mm, with apical, laterally pointed beak <1mm. Native; in acid ponds and canals; local in Wa and N & C En.

R

4. ALISMA L. - <u>Water-plantains</u>
Leaves all basal, linear and/or long-petiolate and narrowly
elliptic to elliptic-ovate; flowers in whorled panicles, or
(in small plants) in simple whorls or umbels, bisexual;
petals pale mauve to ± white with basal yellow blotch;
stamens 6; carpels in single whorl with lateral or
subterminal style, each with 1 ovule.

1 Leaves elliptic to ovate-elliptic, rounded to subcordate at base; style arising c.1/2 way up fruit
                                     1. A. plantago-aquatica
1 Leaves linear to narrowly elliptic or lanceolate-
   elliptic, cuneate at base; style arising in upper
   1/2 of fruit   2
     2 Fruits widest near middle, with ± straight,
        erect style   **2. A. lanceolatum**
     2 Fruits widest in upper 1/2, with strongly recurved
        style   **3. A. gramineum**

1. A. plantago-aquatica L. - <u>Water-plantain</u>. Stems erect,
to 1m; leaves elliptic to ovate-elliptic, rounded to
subcordate at base; flowers 7-12mm across; achenes with
straight, ± erect style; flowers open in afternoon. Native;
in or by ponds, ditches, canals and slow rivers; common
throughout BI except rare in N Sc.
  **1 x 2. A. plantago-aquatica x A. lanceolatum = A. x
rhicnocarpum** Schotsman is intermediate in all characters and
sterile; there are records from Ir, Man, C Sc, London area
and E Anglia, but all need confirming.
  2. A. lanceolatum With. - <u>Narrow-leaved Water-plantain</u>.
Differs from <u>A. plantago-aquatica</u> as in key (couplet 1);
flowers open in morning (some overlap). Native; similar
places to <u>A. plantago-aquatica</u>; scattered in Ir, in Br N to
Yorks, and in Man.
  3. A. gramineum Lej. - <u>Ribbon-leaved Water-plantain</u>. **RRR**
Differs from <u>A. lanceolatum</u> in leaves linear or very narrowly
elliptic; and see key. Possibly native, more likely
sporadically intrd; in shallow ponds; 1 place in Worcs since
1920, for short periods in S Lincs, W Norfolk and Cambs. W
Norfolk plants are said to belong to ssp. **wahlenbergii** O.
Holmb., with shorter achenes with thinner walls, and the
others to ssp. **gramineum**, but the taxa are doubtfully
distinct sspp.

5. DAMASONIUM Miller - <u>Starfruit</u>
Leaves all basal, long-petiolate, submerged, floating or
sometimes emergent, ovate-oblong with cordate base; flowers
in whorls, bisexual; petals white, with basal yellow blotch;
stamens 6; carpels 6-10 in 1 whorl, with terminal style, each
with 2-several ovules.

  1. D. alisma Miller - <u>Starfuit</u>. Stems erect, to 30(60)cm; **RRR**

leaves 3-6(8)cm; flowers 5-9mm across; follicles 5-14mm, with
long beak. Native; muddy margins of acid ponds; 1 place in
Surrey, 2 in Bucks, formerly elsewhere in S & C En,
apparently approaching extinction.

140. HYDROCHARITACEAE - Frogbit family

Glabrous, aquatic perennials, free floating or rooted in mud;
leaves submerged or floating, sometimes emergent, all basal
or along stems,. entire or serrate, sessile or petiolate,
stipulate or not. Flowers 1-few in axillary inflorescences
subtended by spathe, usually dioecious, epigynous,
actinomorphic, conspicuous or inconspicuous; sepals 3, free;
petals 3 or vestigial, white to purplish; stamens 2-12; ovary
1-celled, with 3-6 parietal placentas bearing many ovules;
styles 3 or 6; stigmas usually linear, often bifid; fruit an
irregularly opening capsule.
Distinctive among monocots in having an inferior ovary
combined with 3 sepals, usually 3 petals and dioecious
flowers.

1 Leaves with long petioles and suborbicular cordate
  blades   **1. HYDROCHARIS**
1 Leaves sessile, tapering apically or at both ends   2
  2 Leaves all in basal rosette   3
  2 Leaves borne along stems   4
3 Leaves sharply serrate all along margins, rigid;
  petals conspicuous, white   **2. STRATIOTES**
3 Leaves denticulate only near apex, flaccid; petals
  vestigial   **7. VALLISNERIA**
  4 Leaves variously whorled to spiral   **6. LAGAROSIPHON**
  4 Leaves all whorled or opposite   5
5 Middle and upper leaves in whorls of 3-6(8), with 2
  minute (c.0.5mm) fringed scales at base   **5. HYDRILLA**
5 Middle and upper leaves in whorls of 3-4(5), with or
  without 2 entire scales at base <0.5mm   6
  6 Leaves in whorls of (3)4-5; petals >5mm, white; in
    industrially warmed water only   **3. EGERIA**
  6 Leaves in whorls of (2)3-4(5); petals <5mm,
    inconspicuous; widespread   **4. ELODEA**

1. HYDROCHARIS L. - Frogbit
Plants usually floating, the roots hanging in water; leaves
with long petioles and floating suborbicular cordate blades,
all in basal rosette, entire, with large stipules; flowers
dioecious, arising on pedicels from a stalked spathe (1 in
female, 1-3(4) in male), conspicuous, with petals much larger
than sepals; stamens 9-12, some usually as sterile staminodes; female flowers with 6 staminodes only; styles 6,
bifid.

1. H. morsus-ranae L. - <u>Frogbit</u>. Main stems are floating   **R**
stolons to 50(100)cm forming terminal buds in Autumn; leaves
1.6-5cm across; flowers 2-3cm across; petals white with basal
yellow blotch. Native; in ponds, canals and ditches; locally
frequent in En, very scattered in Wa and Ir, formerly CI.

2. STRATIOTES L. - <u>Water-soldier</u>
Plant usually floating, mostly submerged, rising to surface
at flowering; leaves sessile, linear-lanceolate, tapering
from base, all in basal rosette, spinous-serrate,
exstipulate; flowers mostly dioecious, arising from a stalked
spathe (1 ± sessile in female, several pedicelled in male),
conspicuous, with petals much larger than sepals; stamens 12,
with sterile staminodes surrounding them; female flowers with
staminodes only; styles 6, bifid.

1. S. aloides L. - <u>Water-soldier</u>. Rosettes large, sturdy;   **R**
leaves up to 50 x 2cm; flowers 3-4cm across; petals white.
Native; ponds, dykes and canals, usually calcareous; very
locally frequent in NE Wa, NW & NE En and E Anglia, intrd in
scattered places in Br N to C Sc. Only female plants occur
in Br.

3. EGERIA Planchon - <u>Large-flowered</u> <u>Waterweed</u>
Stems long, branched, rooted in mud, submerged; leaves
sessile, in whorls of (3)4-5, narrowly oblong-linear,
minutely serrate, exstipulate; flowers dioecious,
pedicellate, arising from sessile axillary spathe (1 in
female, 2-4 in male), conspicuous; petals white, much larger
than sepals; stamens 9; styles 3, bifid.

1. E. densa Planchon - <u>Large-flowered</u> <u>Waterweed</u>. Stems to
2m(?more); leaves 10-30 x 1.5-4mm, 0.5-1mm wide 0.5mm behind
apex, with acute apex; flowers 1.2-2cm across. Intrd; in
warmed water of canals and mill-lodges; very local in S
Lancs, 1st recognized 1953; S America. Only male plants in
Br, flowering rarely and only in well-warmed water. Closely
resembles a robust <u>Elodea</u> when not in flower; leaves similar
in width to those of <u>E. canadensis</u>, but longer and more
acute.

4. ELODEA Michaux - <u>Waterweeds</u>
Stems long, branched, rooted in mud, submerged; leaves
sessile, the lower opposite, the upper in whorls of 3-4(5),
minutely serrate, with 2 minute entire basal scales; flowers
dioecious, solitary from sessile axillary spathe, inconspicuous; petals whitish to reddish, c. as large as sepals;
stamens 9; styles 3, bifid. Only female plants occur in BI.

1 Leaf apices obtuse to subacute; leaves (0.7)0.8-
  2.3mm wide 0.5mm behind apex   **1. E. canadensis**
1 Leaf apices acute to acuminate; leaves 0.2-

0.7(0.8)mm wide 0.5mm behind apex   2
2 Usually some leaves strongly recurved and/or
   twisted, with marginal teeth 0.05-0.1mm; root-
   tips white to greyish-green when fresh; sepals
   1.6-2.5mm   **2. E. nuttallii**
2 Usually no leaves strongly recurved or twisted,
   with marginal teeth usually 0.1-0.15mm; root-tips
   red when fresh; sepals 3-4.3mm **3. E. callitrichoides**

1. E. canadensis Michaux - Canadian Waterweed. Stems to 3m;
leaves in whorls of (2)3(-4), 4.5-17 x 1.4-5.6mm, (0.7)0.8-
2.3mm wide 0.5mm behind apex. Intrd; natd in ponds, lakes,
canals, slow rivers; common throughout BI except extreme N &
NW Sc; N America. 1st recorded 1836.
   2. E. nuttallii (Planchon) H. St. John - Nuttall's Waterweed. Stems to 3m; leaves in whorls of (2)3-4(5), 5.5-35 x
0.8-3mm, 0.2-0.7(0.8)mm wide 0.5mm behind apex. Intrd;
habitat as for E. canadensis; locally common in En, very
scattered in Wa, Sc, Ir and Jersey, rapidly spreading; N
America. Leaves longer, narrower and more acute than in E.
canadensis. 1st recorded 1966.
   3. E. callitrichoides (Rich.) Caspary (E. ernstiae H. St.
John) - South American Waterweed. Stems to 3m; leaves in
whorls of 3, 9-25 x 0.7-2.2mm, 0.2-0.6mm wide 0.5mm behind
apex. Intrd; habitat as for E. canadensis but probably not
fully natd; very local in S En and S Wa; S America. Differs
from E. nuttallii as in key. 1st recorded 1948.

5. HYDRILLA Rich. - Esthwaite Waterweed
Stems long, branched, rooted in mud, submerged; leaves
sessile, the lower opposite, the upper in whorls of 3-6(8),
minutely serrate, with 2 minute fringed scales at base;
flowers dioecious, solitary from sessile axillary spathe,
inconspicuous; petals transparent with red streaks, c. as
large as sepals; stamens 3; styles 3(-5), simple.

   1. H. verticillata (L. f.) Royle (Elodea nuttallii auct. non **RRR**
(Planchon) H. St. John) - Esthwaite Waterweed. Stems to 1m
(?more); leaves 5-20 x 0.7-2mm, 0.2-0.7mm wide 0.5mm behind
apex, with narrowly acute to acuminate apex. Native; lakes;
Esthwaite Water (Westmorland) 1914-c.1945, and Rusheenduff
Lough (W Galway) from 1935 (no flowers found in former,
female flowers once only in latter place). Closely resembles
Elodea nuttallii but less robust and leaves more per node and
with minute teeth ± to base (not just in distal 1/2).

6. LAGAROSIPHON Harvey - Curly Waterweed
Stems long, branched, rooted in mud, submerged; leaves
variously whorled to spiral, the lowest always spiral (not
opposite), subentire to minutely denticulate, with 2 minute
entire basal scales; flowers dioecious, inconspicuous,
arising from sessile axillary spathe (1 in female, several in

male); petals reddish, c. as large as sepals; stamens 3;
styles 3, bifid.

1. L. major (Ridley) Moss - <u>Curly Waterweed</u>. Stems to 3m;
leaves 6-30 x 1.3mm, usually strongly recurved, 0.2-0.5mm
wide 0.5mm behind apex, with narrowly acute to acuminate
apex. Intrd; ponds, lakes, canals, slow rivers; locally
frequent in CI and Br N to C Sc, M Cork, 1st recorded 1944; S
Africa. Only female plants occur in BI.

7. VALLISNERIA L. - <u>Tapegrass</u>
Plant rooted in mud, submerged, with stems as stolons; leaves
sessile, all in basal rosette, linear, denticulate near apex,
exstipulate; flowers dioecious (or monoecious?), solitary
(female) or many (male) in stalked spathes, inconspicuous;
petals 0 or vestigial; stamens (1)2(-3); styles 3, bifid.

1. V. spiralis L. - <u>Tapegrass</u>. Leaves 2-80cm x 1-10mm,
ribbon-like; stalks of female spathes very long, spiralling
after flowering. Intrd; natd (often not permanently) in slow
rivers and canals, usually where water is heated; very local
in S En, S Lancs, SW Yorks; warm regions of world.

141. APONOGETONACEAE - <u>Cape-pondweed family</u>

Glabrous, aquatic perennials with elongated stems rooted in
mud at tuberous base; leaves floating, all basal, entire,
long-petiolate with sheathing base. Flowers in a forked
spike at water surface on long stalk, with deciduous spathe
at base of spike, bisexual, hypogynous, actinomorphic except
for perianth, conspicuous; tepals 1(-2), white; stamens 6-18;
carpels 3, free, each with several ovules, short style, and
linear stigma; fruit a group of follicles.
The forked, white inflorescence borne just above the water
surface is unique.

1. APONOGETON L. f. - <u>Cape-pondweed</u>.
1. A. distachyos L. f. - <u>Cape-pondweed</u>. Leaves oblong-
elliptic, up to 25 x 7cm; spikes up to 6cm, each with up to
10 flowers; tepals 10-20mm. Intrd; persistent or ± natd
where planted in ponds; scattered in Br; S Africa.

142. SCHEUCHZERIACEAE - <u>Rannoch-rush family</u>

Glabrous, <u>Juncus</u>-like perennials with rhizomes clothed with
old leaf-bases and erect leafy flowering stems; leaves
alternate, linear, with sheathing base with ligule at top on
adaxial side, entire, sessile. Flowers few in terminal
raceme with bract at base of each, bisexual, hypogynous,
actinomorphic, inconspicuous; tepals 6, sepaloid; stamens 6;

carpels 3(-6), free, each with usually 2 ovules and sessile
capitate stigma; fruit a group of follicles.
Juncus-like vegetatively but with free (usually 3) carpels
each with usually only 2 seeds. The leaves are very
distinctive in having a large apical pore.

1. SCHEUCHZERIA L. - Rannoch-rush
1. S. palustris L. - Rannoch-rush. Stems to 25(40)cm; upper   **RR**
stem-leaves usually overtopping inflorescence; tepals up to
3mm; follicles up to 7mm. Native; in pools or wet Sphagnum
bogs; very rare in 2 places in M Perth, formerly scattered in
Sc and N En and 1 place in Ir.

143. JUNCAGINACEAE - Arrowgrass family

Glabrous perennials with rhizomes and erect, + leafless,
flowering stems; leaves + all basal, in rosette or a few
alternate near stem base, linear, with sheathing base with
ligule at top on adaxial side, entire, sessile. Flowers many
in bract-less simple terminal raceme, bisexual, hypogynous,
inconspicuous; tepals 6, sepaloid; stamens 6; ovary 6-celled,
all cells with 1 ovule or alternate cells small and sterile;
styles +0; stigmas papillate to long-fringed; fruit 3 or 6
1-seeded units breaking apart at maturity.
Juncus-like or Plantago-like superficially, but the 3 or 6
1-seeded fruit segments are diagnostic.

1. TRIGLOCHIN L. - Arrowgrasses
1. T. palustre L. - Marsh Arrowgrass. Stems erect, to 60cm;
leaves usually deeply furrowed on upperside near base; fruit
with 3 fertile cells, 7-10mm, extremely narrowed at base;
stigmas long-fringed. Native; marshy places and wet fields,
sometimes at back of salt-marshes; throughout BI, commonest
in N.
2. T. maritimum L. - Sea Arrowgrass. Stems erect, to 60cm;
leaves usually flat on upperside near base; fruit with 6
fertile cells, 3-5mm, + rounded at base; stigmas papillate.
Native; in salt-marshes and salt-sprayed grassland; round
coasts of whole BI, rare in inland salty areas. Ligule
usually longer than wide (not wider than long) and more
pointed than in T. palustre, but this and leaf character
given above are not completely reliable.

144. POTAMOGETONACEAE - Pondweed family

Glabrous aquatic (or + so) perennials, often with rhizomes,
with leafy submerged flowering stems; leaves alternate or
opposite, floating and/or submerged (or some + aerial),
linear to elliptic, entire or nearly so, petiolate or
sessile, at least some with sheathing base which is either

free and stipule-like or fused to leaf-base for most of
length, forming sheath and free distal ligule. Flowers in
short, axillary, stalked, bractless spikes, bisexual,
hypogynous, actinomorphic, inconspicuous, submerged or
aerial; tepals 4, sepaloid; stamens 4; carpels (1-)4, free,
each with 1 ovule and ± sessile capitate stigma; fruits in a
group of 1-4 achenes or drupes.
Distinguished from other pondweeds by the 4 tepals and 4
stamens.

1 All or most leaves alternate, all with membranous
   sheath or stipule   **1. POTAMOGETON**
1 All leaves opposite (or some in whorls of 3), only
   the uppermost with membranous stipules   **2. GROENLANDIA**

1. POTAMOGETON L. - <u>Pondweeds</u>
Leaves all alternate, or just those subtending inflorescences
opposite, all with membranous sheath or stipules; fruits with
thick pericarp, soft on outside but with bony inner layer.
   For accurate identification it is essential first to examine
thoroughly the leaf morphology, including range of leaf-
shape, leaf-blade venation, and morphology of basal sheath/
stipules.   The key covers only mature plants that have
reached flowering or are about to flower; beware recent
flooding or drying out of habitat when distinguishing
floating and submerged leaves.   In taxa that can have
floating leaves (whether actually present or not) the upper
submerged leaves often approach the former in certain
respects and may be quite different from the middle and lower
submerged leaves, which are the diagnostic ones.   In the keys
and descriptions 'veins' refers to the midrib plus its
laterals that run ± parallel to it for nearly the whole leaf
length.   Fresh material is best for determination; the
stipule characters are diffcult to see in dried material, but
are sometimes necessary for certain identification.
   Although 25 hybrids have been found in BI, hybridization is
not common and most hybrids are very rare.   8 that occur in
the absence of both parents are treated fully.   All hybrids
except <u>P.</u> x <u>zizii</u> are sterile, with no fruit developed; the
fruits of <u>P.</u> x <u>zizii</u> are of unproven viability.   Hybrids are
variously intermediate between the parents.

<u>General key</u>
1 Floating leaves present   <u>Key A</u>
1 All leaves submerged   2
   2 Leaves slightly to strongly convex-sided, >6mm
      (usually >10mm) wide, often with >7 main veins   <u>Key B</u>
   2 Leaves grass-like, parallel-sided for almost
      whole length, <5(6)mm wide, with 3-5(7) main
      veins   <u>Key C</u>

Key A - Floating leaves present
1 Floating leaves with distinct hinge-like joint
  at junction with petiole; submerged leaves (if
  present) <3mm wide   **1. P. natans**
1 Floating leaves merging gradually or abruptly into
  petiole, but without distinct joint; at least some
  submerged leaves (if present) >3mm wide   2
  2 Submerged leaves always present, sessile or with
    petiole <1cm   3
  2 Submerged leaves usually present, if so then
    distinctly petiolate with petiole >1cm   8
3 All submerged leaves strictly parallel-sided +
  throughout length, <8mm wide   **16. P. epihydrus**
3 At least some submerged leaves convex-sided for at
  least part of length, often >8mm wide   4
  4 At least lower submerged leaves broadly rounded at
    base, + clasping stem   **10. P. x nitens**
  4 All submerged leaves narrowed to base, not clasping
    stem   5
5 Submerged leaves obtuse to rounded at apex   6
5 Submerged leaves acute, cuspidate or acuminate at
  apex   7
  6 Stem terete; at least some submerged leaves >1cm
    wide, with >7 veins; fertile   **11. P. alpinus**
  6 Stem compressed; submerged leaves <1cm wide, all
    or most 7-veined; sterile   **12. P. x olivaceus**
7 Submerged leaves <1cm wide, gradually acuminate
  to acute; stipules <2cm   **9. P. gramineus**
7 Submerged leaves >1.5cm wide, rather abruptly
  acuminate to cuspidate; stipules >2cm   **7. P. x zizii**
  8 Submerged leaves always present, <1cm wide
                                        2. P. x sparganiifolius
  8 Submerged leaves usually present, if so all or
    most >1cm wide   9
9 Floating leaves translucent, the vein network clearly
  visible (fresh or dried); fruits <2mm   **4. P. coloratus**
9 Floating leaves opaque, the vein network difficult
  to see; fruits >2mm   10
  10 Most leaves floating (or + aerial), at least some
     with rounded to cordate base; fruits < 2.5mm
                                          3. P. polygonifolius
  10 Most leaves submerged; floating leaves with
     cuneate base; fruits >2.5mm   11
11 Submerged leaves with petiole >1/2 as long as blade;
   peduncles not widened distally   **5. P. nodosus**
11 Submerged leaves with petiole <1/2 as long as blade;
   peduncles distinctly widened distally   **7. P. x zizii**

Key B - All leaves submerged, convex-sided, >6mm wide, often
          with >7 veins
1 Leaves minutely serrate along + whole margin to
  naked eye, usually regularly undulate; fruits tapered

    to beak ± as long as body   **26. P. crispus**
1 Leaves entire even with x10 lens, or obscurely
    serrulate just near apex, not regularly undulate;
    fruits with beak much shorter than body   2
   2 Leaves with acute, acuminate or cuspidate apex   3
   2 Leaves with obtuse to rounded, sometimes hooded,
      apex   7
3 At least some leaves rounded at base and ± clasping
    stem; sterile hybrids   4
3 All leaves tapered to base, not clasping stem   5
   4 Stipules ≥2.5cm; leaves mostly >2cm wide
                     8. P. x salicifolius
   4 Stipules <2.5cm; leaves mostly <2cm wide
                     10. x P. nitens
5 Leaves ≤1cm wide, gradually acuminate to acute;
    stipules <2cm   **9. P. gramineus**
5 Leaves ≥1.5cm wide, rather abruptly acuminate to
    cuspidate; stipules >2cm   6
   6 Leaves all or mostly <12cm; only some leaves
      petiolate, the petiole usually narrowly winged
                     7. P. x zizii
   6 Leaves all or mostly >12cm; all leaves petiolate,
      the petiole unwinged near base   **6. P. lucens**
7 At least some leaves rounded at base, usually ±
    clasping stem   8
7 Leaves narrowed to base, not clasping stem   10
   8 Stipules conspicuous, at least some >1cm (often
      much more so), persistent; leaves mostly >10cm,
      with ≥3 faint lateral veins between midrib and
      nearest strong lateral vein   **13. P. praelongus**
   8 Stipules inconspicuous, <1cm, soon disappearing,
      leaves <10cm, with <3 faint lateral veins between
      midrib and nearest strong lateral vein   9
9 Stem compressed; leaves oblong-lanceolate, most or
    all <2cm wide; sterile   **15. P. x cooperi**
9 Stem terete; leaves oblong-ovate, most or all >2cm
    wide; fertile   **14. P. perfoliatus**
   10 Stem terete; at least some leaves >1cm wide and
      with >7 veins; fertile   **11. P. alpinus**
   10 Stem compressed; leaves <1cm wide, 3–7-veined;
      sterile   11
11 Leaves ≤7mm wide, with (3–)5 veins **18. P. x lintonii**
11 At least some leaves >7mm wide, with 7 veins
                     12. P. x olivaceus

<u>Key C</u> - All leaves submerged, parallel-sided for almost whole
      length, <5(6)mm wide, with 3–5(7) veins
1 Stipules mostly >5cm; leaves actually blade-less
    petioles, opaque, without obvious midrib **1. P. natans**
1 Stipules <5cm; leaves translucent, with obvious
    midrib   2
   2 Stipules fused to base of leaf, forming sheathing

leaf-base, free distally to form ligule   3
2 Stipules free from leaf, forming stipule-like
  outgrowth from node   5
3 Plant either with leaf apex as in P. pectinatus and
  leaf-sheath as in P. filiformis or vice versa;
  fruit not formed   28. P. x suecicus
3 Plant with both leaf apex and leaf-sheath as in P.
  pectinatus or P. filiformis; fruit well developed   4
  4 Leaves acute to acuminate at apex; sheathing leaf-
    base not fused in tube round stem, but with margins
    overlapping; fruit 3-5mm   29. P. pectinatus
  4 Leaves obtuse to rounded at apex; sheathing leaf-
    base fused in tube round stem proximally when
    young; fruit 2-3mm   27. P. filiformis
5 Most leaves <2mm wide, with 3(-5) veins   6
5 Most leaves >2mm wide, with 3-5(many) veins   9
  6 Stipules fused in tube round stem proximally when
    young; fruits 1.5-2.25mm   7
  6 Stipules not fused in tube round stem, but with
    margins overlapping; fruits 2-3mm   8
7 Leaves very gradually and finely acute to long-
  acuminate, rigid; C & N Sc only   19. P. rutilus
7 Leaves acute to obtuse and abruptly mucronate, not
  rigid; widespread   20. P. pusillus
  8 Leaves acute to finely acute, mostly <1mm wide;
    fruits 1(-3) per flower, often warty or + toothed
    near base   23. P. trichoides
  8 Leaves obtuse to subacute, often mucronate, mostly
    >1mm wide; fruits (1-)4 per flower, not warty
                                               22. P. berchtoldii
9 Leaves with 3 or 5 main veins and many finer ones
  between them   10
9 Leaves with 3 or 5(-7) main veins only   11
  10 Leaves with 3 main veins, usually acuminate;
     fruit usually with basal wart or tooth, with erect
     beak   25. P. acutifolius
  10 Leaves with 5 main veins (2 submarginal), usually
     mucronate; fruit not warted, with asymmetric,
     curved beak   24. P. compressus
11 Leaves mostly 3-veined; stems with many lateral
   branches closely placed, forming fan-like sprays;
   stipules not fused in tube round stem
                                               21. P. obtusifolius
11 Leaves mostly 5-veined; stems without many closely
   placed lateral branches; stipules fused in tube round
   stem proximally when young   12
  12 Leaves <7mm wide, usually serrulate distally,
     rounded at apex; sterile   18. P. x lintonii
  12 Leaves <3.5mm wide, entire, mucronate at apex;
     fruits 2-3mm   17. P. friesii

Subgenus 1 - POTAMOGETON (spp. 1-26). Leaves submerged

and/or floating, sessile or petiolate, with basal sheaths free or nearly free from leaf-base, forming stipule-like outgrowth; flowers pollinated above water surface by wind.

1. P. natans L. - <u>Broad-leaved</u> <u>Pondweed</u>. Floating leaves **910** opaque, elliptic to ovate-elliptic, up to 8(12.5) x 5(7)cm, **911** very rarely 0; submerged leaves opaque, linear, <3mm wide, often very long, rounded at apex, sessile; fruit 3-4mm. Native; lakes, ponds, rivers, ditches, usually over rich soils; frequent throughout BI.
  1 x 3. P. natans x P. polygonifolius = P. x gessnacensis G. Fischer has been found in Caerns and E Ross; other records are errors.
  1 x 6. P. natans x P. lucens = P. x fluitans Roth has been found scattered in S En.
  1 x 22. P. natans x P. berchtoldii = P. x variifolius Thore is known in W Mayo.
2. P. x sparganiifolius Laest. ex Fries (<u>P. natans</u> x <u>P.</u> **910** <u>gramineus</u>) - <u>Ribbon-leaved</u> <u>Pondweed</u>. Floating leaves opaque, **911** elliptic to narrowly so, up to 10 x 3cm; submerged leaves various, some linear to very narrowly elliptic and sessile, others with narrowly elliptic blade and long petiole, acute at apex, up to 30 x <1cm. Native; lakes, ponds, canals, streams; scattered in Br and Ir.
3. P. polygonifolius Pourret - <u>Bog</u> <u>Pondweed</u>. Floating (or **910** emergent) leaves opaque, similar in shape to those of <u>P.</u> **911** <u>natans</u>, up to 9 x 4cm; submerged leaves narrowly elliptic to oblanceolate, up to 20 x 3cm, petiolate, obtuse, sometimes 0; fruit 2-2.5mm. Native; shallow ponds, bogs, ditches, small streams, on acid soil; common throughout most of BI, but absent from parts of C & E En.
  3 x 9. P. polygonifolius x P. gramineus = P. x lanceolati-folius (Tisel.) Preston has been found in Wigtowns and Easterness.
4. P. coloratus Hornem. - <u>Fen</u> <u>Pondweed</u>. Floating leaves **910** ovate, up to 10 x 5cm, translucent; submerged leaves similar **911** to those of <u>P. polygonifolius</u>; fruit 1.5-2mm. Native; ponds **R** and pools, on base-rich peat; local throughout BI except N & E Sc, mostly in C Ir and CE En.
  4 x 9. P. coloratus x P. gramineus = P. x billupsii Fryer has been found in Cambs and Outer Hebrides.
  4 x 22. P. coloratus x P. berchtoldii = P. x lanceolatus Smith occurs in Anglesey, Co Clare and NE Galway, formerly in Cambs; endemic.
5. P. nodosus Poiret - <u>Loddon</u> <u>Pondweed</u>. Floating leaves **910** elliptic to narrowly so, translucent, up to 15 x 6cm; **911** submerged leaves narrowly elliptic to oblanceolate, up to 20 **RR** x 4mm, acute to obtuse at apex, petiolate; fruit 3-4mm. Native; slow-flowing base-rich rivers; very local in S En, ?decreasing.
6. P. lucens L. - <u>Shining</u> <u>Pondweed</u>. Leaves all submerged, **911** elliptic or obovate to narrowly elliptic or oblanceolate, up

to 20 x 6cm, acuminate to cuspidate at apex, petiolate; fruit 3.5-4mm. Native; lakes, ponds, canals, slow rivers on base-rich soil; locally common in En, Wa and Ir, rare in Sc and CI.

6 x 11. P. lucens x P. alpinus = P. x nerviger Wolfg. is known from Co Clare.

6 x 26. P. lucens x P. crispus = P. x cadburyae Dandy & G. Taylor was found in 1948 as a single plant in Warks; endemic. Possibly extinct.

7. P. x zizii Koch ex Roth (P. lucens x P. gramineus) - **910** Long-leaved Pondweed. Floating leaves ± opaque, elliptic or **911** narrowly so, up to 10 x 4cm, sometimes 0; submerged leaves narrowly elliptic to oblanceolate, up to 15 x 3cm, sessile or shortly petiolate (often both on 1 plant), cuspidate to mucronate at apex; fruit c.3-3.5mm. Native; lakes, ponds, streams, canals; scattered throughout most of Br and Ir, often without parents.

8. P. x salicifolius Wolfg. (P. lucens x P. perfoliatus) - **911** Willow-leaved Pondweed. Leaves all submerged, elliptic-oblong to narrowly so, up to 20 x 5cm, sessile and some clasping stem, acute to cuspidate at apex. Native; ponds, canals, rivers; scattered in C & S En, very local in E Sc, NE En and C Ir.

9. P. gramineus L. - Various-leaved Pondweed. Floating **910** leaves opaque, elliptic, up to 7 x 3cm, sometimes 0; **911** submerged leaves narrowly elliptic to oblanceolate, up to 8 x 3cm, acute to acuminate or cuspidate at apex, sessile; fruit 2-3mm. Native; lakes, ponds, canals, streams, usually on acid soils; scattered throughout most of Br and Ir, but not in most of SW Br.

9 x 11. P. gramineus x P. alpinus = P. x nericius Hagstr. is known from S Aberdeen.

10. P. x nitens G. Weber (P. gramineus x P. perfoliatus) - **910** Bright-leaved Pondweed. Floating leaves ± opaque, elliptic **911** or narrowly so, up to 6 x 1.5cm, often 0; submerged leaves narrowly elliptic to lanceolate, up to 8 x 1.5cm, acute or acuminate at apex, at least lower ones sessile and clasping stem. Native; lakes, ponds, canals, streams; scattered through most of BI, but absent from CI and most S Br.

11. P. alpinus Balbis - Red Pondweed. Plant often tinged **910** reddish; floating leaves ± translucent, elliptic to rather **911** narrowly so, up to 8 x 2cm, sometimes 0; submerged leaves lanceolate to narrowly elliptic, up to 15 x 2.6cm, obtuse at apex, all except the upper ons sessile; fruit 2.5-3mm. Native; lakes, canals, streams, especially on peaty soil; fairly frequent ± throughout BI.

11 x 13 P. alpinus x P. praelongus = P. x griffithii A. Bennett has been found in Caerns, Westerness and W Donegal; endemic.

11 x 14. P. alpinus x P. perfoliatus = P. x prussicus Hagstr. has been found in Colonsay (S Ebudes) and Benbecula (Outer Hebrides).

FIG 910 – Floating leaves of **Potamogeton**. 1, **P. natans**. 2, **P. x sparganiifolius**. 3, **P. polygonifolius**. 4, **P. coloratus**. 5, **P. nodosus**. 6, **P. alpinus**. 7, **P. gramineus**. 8, **P. x nitens**. 9, **P. x zizii**. 10, **P. epihydrus**.

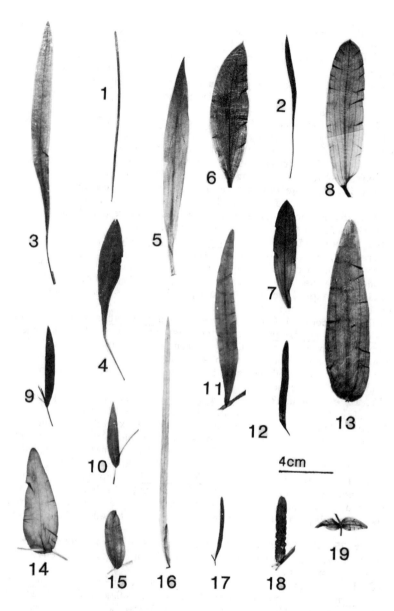

FIG 911 - 1-18, Submerged leaves of **Potamogeton**. 1, P.
natans. 2, **P. x sparganiifolius**. 3, P. polygonifolius. 4, P.
coloratus. 5, P. nodosus. 6, P. lucens. 7, **P. x zizii**. 8, **P.
x salicifolius**. 9, P. gramineus. 10, **P. x nitens**. 11, P.
alpinus. 12, **P. x olivaceus**. 13, P. praelongus. 14, P.
perfoliatus. 15, **P. x cooperi**. 16, P. epihydrus. 17, **P. x
lintonii**. 18, P. crispus. 19, Leaves of **Groenlandia**.

12. P. x olivaceus Baagoee ex G. Fischer (P. alpinus x P. 911
crispus) - Graceful Pondweed. Stem compressed; floating
leaves reported sometimes, not seen by me; submerged leaves
narrowly oblong-linear, up to 9cm, 6-10mm wide, rounded at
apex, not or scarcely undulate, narrowed to sessile or
shortly petiolate base. Native; rivers; scattered in N & W
Br S to S Wa.

13. P. praelongus Wulfen - Long-stalked Pondweed. Leaves 911
all submerged, narrowly oblong-ovate, up to 18 x 4.5cm,   R
sessile and clasping stem, blunt and hooded at apex (bifid
when pressed); fruit 4-5mm. Native; lakes, rivers, canals,
streams; scattered throughout Br and Ir except S En.

13 x 14. P. praelongus x perfoliatus = P. x cognatus Asch. &
Graebner has been found in N Lincs and W Sutherland.

13 x 26. P. praelongus x P. crispus = P. x undulatus Wolfg.
has been found in Rads and Co Antrim.

14. P. perfoliatus L. - Perfoliate Pondweed. Leaves all 911
submerged, broadly ovate to narrowly oblong-ovate, up to 10 x
6cm (often much shorter), obtuse to rounded and often ±
hooded at apex, cordate and clasping stem; fruit 3-3.5mm.
Native; lakes, ponds, canals, rivers, streams; frequent
throughout Br and Ir.

15. P. x cooperi (Fryer) Fryer (P. perfoliatus x P. crispus) 911
- Cooper's Pondweed. Stem compressed; leaves all submerged,
narrowly oblong to narrowly oblong-ovate, up to 6 x 2.5cm,
rounded at apex, mostly rounded and clasping stem at base but
the lower narrowed to sessile base. Native; lakes, ponds,
canals, streams; scattered in Br and Ir.

16. P. epihydrus Raf. - American Pondweed. Stem compressed; 910
floating leaves opaque, narrowly elliptic-oblong, up to 7 x 911
2.5cm; submerged leaves linear, up to 20 x 0.8cm, rounded to RR
subacute at apex, tapered to sessile base; fruit 2.5-3.5mm.
Native; 3 lakes in S Uist (Outer Hebrides), discovered 1943,
also natd in canals in S Lancs and SW Yorks since 1907.

17. P. friesii Rupr. - Flat-stalked Pondweed. Stems 917
compressed; leaves all submerged, linear, (1.5)2-3.5mm wide, R
mucronate at apex, (3)5(-7)-veined; fruit 2-2.5mm. Native;
lakes, ponds, canals; frequent in En, scattered in Wa, Sc and
Ir.

17 x 25. P. friesii x P. acutifolius = P. x pseudofriesii
Dandy & G. Taylor was found in 1952 in E Norfolk; endemic.
Possibly extinct.

18. P. x lintonii Fryer (P. friesii x P. crispus) - Linton's 911
Pondweed. Stem compressed; leaves all submerged, linear or
very narrowly oblong-elliptic, (2)3-7mm wide, rounded at
apex, (3-)5-veined. Native; rivers, canals, streams; very
locally frequent in En, Monts, Kirkcudbrights, Co Armagh.

19. P. rutilus Wolfg. - Shetland Pondweed. Stem compressed; 917
leaves all submerged, linear, 0.5-1(1.5)mm wide, narrowly   RR
acute to acuminate at apex, 3(-5)-veined; fruit 1.5-2mm.
Native; lakes; local in N & NW Sc, discovered 1890. Over-
recorded for other narrow-leaved spp.

20. P. pusillus L. (P. panormitanus Biv.) - <u>Lesser Pond-</u> 917
<u>weed</u>. Stem only slightly compressed; leaves all submerged,
linear, 0.5-1.5(3)mm wide, acute or obtuse and mucronate at
apex, 3(-5)-veined; fruit 2-2.5mm. Native; lakes, ponds,
canals, streams, mostly in base-rich water; frequent +
throughout BI, but over-recorded for P. berchtoldii.
20 x 23. P. pusillus x P. trichoides = P. x grovesii Dandy &
G. Taylor was found in 1 place in E Norfolk in 1897 and 1900,
but not since; endemic.

21. P. obtusifolius Mert. & Koch - <u>Blunt-leaved Pondweed</u>. 917
Stem only slightly compressed; leaves all submerged, linear,
2-4mm wide, usually <10cm, obtuse or rounded and apiculate at
apex, 3(-5)-veined; fruit 3-4mm. Native; lakes, ponds,
canals, streams; locally frequent + throughout Br & Ir.

22. P. berchtoldii Fieber (P. pusillus auct. non L.) - <u>Small</u> 917
<u>Pondweed</u>. Stem scarcely compressed; leaves all submerged,
linear, 0.5-2(2.5)mm wide, subacute to obtuse and often
mucronate at apex, 3(-5)-veined; fruit 2-2.5mm. Native;
lakes, ponds, canals, rivers, streams; fairly common
throughout most of BI, the commonest narrow-leaved sp.
**22 x 25. P. berchtoldii x P. acutifolius = P. x suder-
mannicus** Hagstr. has been found in 1 place in Dorset.

23. P. trichoides Cham. & Schldl. - <u>Hairlike Pondweed</u>. 917
Stems not compressed; leaves all submerged, linear, 0.5-   R
1(1.5)mm wide, acute at apex, 3-veined; fruit 2.5-3mm.
Native; ponds, canals, streams; rather local in C & S Br,
very local in C Sc.
23 x 26. P. trichoides x P. crispus = P. x bennettii Fryer
is extremely local in C Sc; endemic.

24. P. compressus L. - <u>Grass-wrack Pondweed</u>. Stems strongly 917
compressed; leaves all submerged, linear, 2-4mm wide, usually   R
>10cm, acute to acuminate but mostly mucronate at apex, with
5 main veins and many fine ones between them; fruit 3-4.5mm.
Native; lakes, ponds, canals, streams; locally frequent in C
& N En and N Wa, formerly Angus.

25. P. acutifolius Link - <u>Sharp-leaved Pondweed</u>. Stems 917
strongly compressed; leaves all submerged, linear, 2-4mm
wide, c.5-13cm, acute to acuminate (mostly acuminate) at
apex, with 3 main veins and many fine ones between them;
fruit 3-4mm. Native; ponds, canals, streams, mostly on
calcareous soils; local in S & E En W to Dorset and W Gloucs,
but extinct in several places.

26. P. crispus L. - <u>Curled Pondweed</u>. Stem compressed; 911
leaves all submerged, narrowly oblong-linear, (3)6-15mm wide,
3-10.5cm, acute to rounded at apex, usually closely undulate
and minutely serrate along whole length, 3-5-veined; fruit
4-5mm. Native; lakes, ponds, canals, rivers, streams;
frequent to common + throughout BI.

<u>Subgenus 2</u> - <u>COLEOGETON</u> Reichb. (spp. 27-29). Leaves all
submerged, sessile, with basal sheaths fused to leaf except
for ligule- like apex; flowers pollinated under water.

27. P. filiformis Pers. – <u>Slender-leaved</u> <u>Pondweed</u>. Stem not **917** compressed; leaves all submerged, linear, 0.3–1(1.5)mm wide, **R** obtuse to rounded at apex, 3-veined but 2 laterals very faint and submarginal; fruit 2–2.8mm, with sessile stigma. Native; lakes, rivers, streams, dykes, sometimes brackish; scattered in Sc and Ir, formerly Anglesey, mostly near coast.

28. P. x suecicus K. Richter (<u>P. filiformis</u> x <u>P. pectinatus</u>) – <u>Swedish Pondweed</u>. Plants with leaf-apex of <u>P. pectinatus</u> and leaf-sheath and stigma of <u>P. filiformis</u>, or other combinations. Native; local in MW & NW Yorks (outside range of <u>P. filiformis</u>) and E & W coasts of Sc.

29. P. pectinatus L. – <u>Fennel</u> <u>Pondweed</u>. Stem not **917** compressed; leaves all submerged, linear, 0.3–2(3)mm wide, narrowly acute to subacute and mucronate at apex, 3–5-veined but laterals very faint; fruit 3–5mm, with short neck-like style. Native; similar habitats to <u>P. filiformis</u>; frequent over most of BI.

2. GROENLANDIA Gay – <u>Opposite-leaved</u> <u>Pondweed</u>
Leaves all opposite (or rarely some in whorls of 3), only the uppermost with 2 membranous stipules fused to edges of leafbase; fruits with thin, papery pericarp.

1. G. densa (L.) Fourr. – <u>Opposite-leaved</u> <u>Pondweed</u>. Leaves **911** all submerged, ovate to lanceolate, up to 4 x 1.5cm, acute to obtuse at apex, clasping stem; fruit 3–3.5mm. Native; ponds, ditches, streams (often fast-flowing); locally frequent in En, scattered in Wa and Ir, very rare and intrd in Sc.

145. RUPPIACEAE – <u>Tasselweed</u> <u>family</u>

Glabrous aquatic perennials rooted in substratum, with leafy submerged stems; leaves mostly alternate but the upper opposite, linear, sessile, with sheathing base, minutely denticulate near apex. Flowers (1)2-few in short sub-umbellate raceme on terminal peduncle, bisexual, hypogynous, inconspicuous, submerged; perianth 0; stamens 2 but each anther with 2 widely separate lobes, so appearing 4; carpels 4, free, each with 1 ovule and ± sessile peltate stigma; fruits in a group of up to 4 drupelets each becoming very long-stalked.
Distinguished from <u>Potamogeton</u> in its terminal inflorescence, lack of perianth, fruits becoming long-stalked and leaves with only a midrib.

1. RUPPIA L. – <u>Tasselweeds</u>
1. R. maritima L. – <u>Beaked</u> <u>Tasselweed</u>. Leaves 0.15–0.5mm **917** wide, usually acute, with slightly inflated sheath; peduncles <4(6)cm, straight to curved or flexuous in fruit; anthers <1mm; drupelets 2–2.5(3)mm, with beak usually 0.5–1mm, on stalks <4cm, asymmetrically pear-shaped. Native; in brackish

ditches and pools; local round most coasts of Br and Ir, formerly Guernsey.

2. R. cirrhosa (Petagna) Grande (R. spiralis L. ex Dumort.) 917 - Spiral Tasselweed. Leaves 0.2-1mm wide, obtuse to rounded, **R** with strongly inflated sheath; peduncles >(4.5)8cm, usually spiral in fruit; anthers >1mm; drupelets (2)2.5-3mm, with beak usually <0.5mm, on stalks <4cm, pear-shaped or slightly asymmetrically so. Native; similar places to R. maritima, sometimes with it but more local, frequent only in E & SE En and Shetland, but S to CI.

146. NAJADACEAE - Naiad family

Glabrous, aquatic annuals or perennials rooted in substratum, with leafy submerged stems; leaves opposite or whorled, linear, sessile, with sheathing base, minutely denticulate to conspicuously dentate or + lobed. Flowers very inconspicuous, 1-3 sessile in each leaf-axil, monoecious or dioecious, hypogynous, submerged; male flowers each surrounded by 2 variously interpreted scales, with 1 sessile anther; female flowers without scales, with 1 carpel with 1 ovule and 2-4 elongated stigmas; fruit a sessile drupe. Distinguished from Potamogetonaceae in its linear opposite (or whorled) leaves, with only a midrib, and unisexual sessile axillary flowers with 0 perianth and 1 carpel.

1. NAJAS L. - Naiads

Other spp. - **N. graminea** Del., from the Tropics, occurred in a canal with warmed water in Manchester, S Lancs, between 1883 and 1947; it differs from N. flexilis in having long narrow auricles on its leaf-sheath, and most leaves densely packed on lateral branches.

1. N. flexilis (Willd.) Rostkov & W. Schmidt - Slender **RR** Naiad. Leaves 1-2.5(4)cm, <1mm wide incl. teeth, + entire or minutely denticulate; fruit 2.5-3.5 x 1-1.5mm, plus long filiform style. Native; clean lakes; local in W Br from Westmorland to Inner Hebrides and in W Ir. Monoecious.

2. N. marina L. - Holly-leaved Naiad. Leaves 1-4.5cm, 1-6mm **RRR** wide incl. teeth, conspicuously spinose-dentate; fruit (3)4-6(8) x 1.5-3mm, with short thick style. Native; slightly brackish waterways; extremely local in NE E Norfolk. Apparently dioecious; fruits are common but male plants have not been recorded. Our plant is spp. **intermedia** (Wolfg. ex Gorski) Casper.

147. ZANNICHELLIACEAE - Horned Pondweed family

Glabrous, aquatic rhizomatous annuals or perennials rooted in substratum, with leafy submerged stems; leaves mostly opposite (sometimes alternate on sterile shoots), linear, sessile, with sheathing base ± free from leaf, entire. Flowers very inconspicuous, solitary in leaf-axils, monoecious, hypogynous, submerged; male flowers with 1-2 stamens, naked, long-stalked; female flowers with (2)4(-8) carpels with 1 ovule and peltate stigma on distinct style, in cup-shaped scale; fruits in a group of 1-several (often 4) achenes, each with persistent style.

Distinguished from <u>Potamogetonaceae</u> in its linear mostly opposite leaves and unisexual axillary flowers with 0 perianth; and from <u>Naias</u> in its several carpels and entire leaves.

1. ZANNICHELLIA L. - <u>Horned Pondweed</u>

1. Z. palustris L. - <u>Horned Pondweed</u>. Leaves 2-10cm x **917** 0.4-1(2)mm, entire, acute to obtuse at apex, 1(-3)-veined; fruits 3-6mm incl. style, variably stalked, variably winged and toothed on dorsal and ventral edges. Native; rivers, streams, ditches and ponds, fresh or brackish; frequent throughout most BI.

148. ZOSTERACEAE - <u>Eelgrass family</u>

Glabrous, marine perennials rooted in substratum, with leafy submerged stems often exposed at low tide; leaves alternate, linear, sessile, with sheathing base with ligule at top, entire. Flowers inconspicuous, in congested compound cymes ± enclosed in leaf-sheath, monoecious, hypogynous; perianth 0; male flowers with 1 stamen; female flowers with 1 1-celled ovary with 1 ovule and 2 filiform stigmas on common style; fruit a sessile drupe.

Distinguished from other marine or brackish pondweeds in the complex, congested inflorescence ± enclosed in leaf-sheath.

1. ZOSTERA L. - <u>Eelgrasses</u>

1 Flowering stems lateral, unbranched or with few branches near base; leaves of sterile shoots 0.5-1.5mm wide, with sheaths clasping stems but not fused into tube   **3. Z. noltii**
1 Flowering stems terminal, branched; leaves of sterile shoots 1-10mm wide, with sheaths fused into tube round stem   2
  2 Leaves (2)4-10mm wide; stigmas c.2x as long as style   **1. Z. marina**
  2 Leaves 1-2(3)mm wide; stigmas c. as long as style   **2. Z. angustifolia**

1. Z. marina L. - <u>Eelgrass</u>. Leaves (2)4-10mm wide, rounded **917** or rounded-mucronate at apex, with 5-11 veins, up to **R**

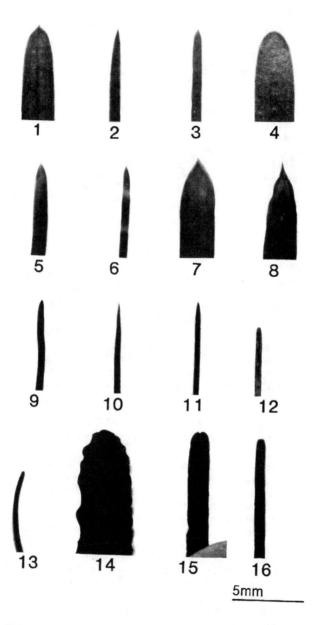

FIG 917 - Leaf-apices of narrow-leaved pondweeds. 1-10, Potamogeton. 1, *P. friesii*. 2, *P. rutilus*. 3, *P. pusillus*. 4, *P. obtusifolius*. 5, *P. berchtoldii*. 6, *P. trichoides*. 7, *P. compressus*. 8, *P. acutifolius*. 9, *P. filiformis*. 10, *P. pectinatus*. 11-12, **Ruppia**. 11, **R. maritima**. 12, **R. cirrhosa**. 13, **Zannichellia palustris**. 14-16, **Zostera**. 14, **Z. marina**. 15, **Z. angustifolia**. 16, **Z. noltii**.

50(120)cm; fruit 3–3.5mm excl. style; seed ribbed. Native;
sea-coasts, c.0–4(9)m below low-water mark; scattered round
coasts of BI.

2. Z. angustifolia (Hornem.) Reichb. – <u>Narrow-leaved</u> **917**
<u>Eelgrass</u>. Leaves 1–2(3)mm wide, rounded often becoming **R**
emarginate at apex, with 3–5 veins, up to 30cm; fruit 2.5–3mm
excl. style; seed ribbed. Native; sea-coasts and estuaries,
from half-tide to low-tide mark or rarely down to 4m below;
very scattered round coasts of BI. Perhaps only a var. (var.
angustifolia Hornem.) of Z. marina.

3. Z. noltii Hornem. – <u>Dwarf Eelgrass</u>. Leaves 0.5–1.5mm **917**
wide, emarginate at apex, with 3 veins, up to 22cm; fruit **R**
1.5–2mm excl. style; seed smooth. Native; similar habitats
and distribution to Z. angustifolia and often with it, but ±
never below low-water mark.

148A. ARECACEAE (Palmae)

PHOENIX dactylifera L. (<u>Date Palm</u>), from SW Asia and N
Africa, is frequent on rubbish-tips as seedlings killed by
first frosts; the 1-few, leathery, narrowly elliptic leaves
arise from the ground and have several parallel veins raised
alternately on either leaf surface. The familiar pinnate-
leaved palm much planted in the Mediterranean and in small
numbers in SW En is **P. canariensis** Chabaud (<u>Canary Palm</u>),
from Canaries. The only common palmate-leaved sp. planted
here is **TRACHYCARPUS fortunei** (Hook.) Wendl. f. (<u>Chusan
Palm</u>), from China.

149. ARACEAE – <u>Lords-and-Ladies</u> family

Glabrous, herbaceous perennials with rhizomes or underground
tubers giving rise to aerial leaves and flowering stems with
0-few leaves; leaves alternate, simple, linear to ovate and
then often cordate to sagittate at base, entire or rarely
deeply lobed, sessile or petiolate, exstipulate but usually
sheathing at base. Flowers closely packed on sterile axis
(<u>spadix</u>) which often extends distally as succulent <u>appendix</u>,
usually subtended or partially enclosed by leaf-like but
often coloured <u>spathe</u>, bisexual or monoecious, if latter
usually the upper male and lower female, hypogynous, actinomorphic; perianth 0 or of 4 or 6 tepals; stamens 1–6; ovary
1–3-celled, with 1-many ovules; style ±0; stigma capitate;
fruit a berry with 1-several seeds.
  The minute numerous flowers packed on to a spadix, subtended
or partially enclosed by a spathe except in the very
distinctive <u>Acorus</u>, are diagnostic.

1 Leaves linear, <u>Iris</u>-like; spadix apparently lateral
    and lacking spathe   **1. ACORUS**

1   Leaves lanceolate to ovate, narrowed at base and
    usually petiolate; spadix terminal, with spathe   2
    2   Flowers covering spadix to its apex   3
    2   Spadix with succulent sterile region (appendix)
        distal to flowers   5
3   Spathe ± flat, not enclosing spadix even at extreme
    base   3. CALLA
3   Spathe wrapped round basal part of spadix   4
    4   Leaves truncate to cuneate at base, with petioles
        shorter than blade; tepals 4   2. LYSICHITON
    4   Leaves cordate at base, with petiole longer than
        blade; tepals 0   4. ZANTEDESCHIA
5   Leaves palmately divided   6. DRACUNCULUS
5   Leaves simple   6
    6   Spathe fused into tube proximally, with distal
        filiform projection ≥5cm   7. ARISARUM
    6   Spathe overlapping at base, not fused into tube,
        no more than acuminate at apex   5. ARUM

Other genera - **COLOCASIA** Schott, from tropical Asia, has
large, entire, peltate long-petiolate leaves; **C. antiquorum**
(L.) Schott (Taro, Dasheen or Coco-yam) is a rare ephemeral
casual of rubbish-tips.

1. ACORUS L. - Sweet-flags
Rhizomatous; leaves linear, Iris-like, entire, sessile;
spadix aparently lateral, without appendix; spathe apparently
0; flowers bisexual; tepals 6; stamens 6; ovary 2-3-celled;
fruit not forming. Fresh leaves have strong spicy scent when
bruised.

1. A. calamus L. - Sweet-flag. Leaves 50-125 x 0.7-2.5cm,
with well-defined midrib, usually transversely wrinkled in
places; spadix 5-9 x 0.6-1.2cm, yellowish-green. Intrd; in
shallow water at edges of lakes, ponds, rivers and canals;
scattered over most of BI, but frequent only in En; Asia and
N America. Often not flowering.
2. A. gramineus Aiton - Slender Sweet-flag. Leaves 8-50 x
0.2-0.8cm, without obvious midrib; spadix 5-10 x 0.3-0.5cm.
Intrd; natd by lake in Surrey since 1986; E Asia.

2. LYSICHITON Schott - Skunk-cabbages
Rhizomatous; leaves ovate-oblong, entire, truncate to cuneate
at base, shortly petiolate; spadix terminal, without
appendix; spathe wrapped round and concealing spadix at base,
falling off after flowering; flowers bisexual; tepals 4;
stamens 4; ovary (1-)2-celled; fruit a green berry with 2
seeds.

1. L. americanus Hulten & H. St. John - American Skunk-
cabbage.   Leaf-blades 30-150 x 25-70cm; flowers foul-
smelling; spathe 10-35cm, yellow; spadix 3.5-12cm, greenish;

tepals 3-4mm; anthers 0.9-2mm. Intrd; grown for ornament, persistent and spreading, in swampy ground; scattered throughout Ir and S & W Br (but see next sp.); W N America.

2. L. camtschatcensis (L.) Schott - <u>Asian</u> <u>Skunk-cabbage</u>. Differs from <u>L. americanus</u> in ± scentless flowers; slightly smaller spathe and spadix; white spathe; tepals 2-3mm; and anthers 0.6-0.8mm. Intrd; similar places to <u>L. americanus</u>; distribution uncertain due to confusion with latter, but perhaps the commoner in Ir; extreme E Asia. Hybrids occur and are perhaps natd; perhaps the 2 are better as sspp.

3. CALLA L. - <u>Bog Arum</u>
Rhizomatous; leaves ovate to broadly so, cordate, entire, with long petiole; spadix terminal, without appendix; spathe open, ± flat, not concealing spadix; flowers mostly bisexual but uppermost usually male; tepals 0; stamens 6; ovary 1-celled; fruit a red berry with several seeds.

1. C. palustris L. - <u>Bog Arum</u>. Leaf-blades 5-12 x 4-10cm; spathe 3-8 x 3-6cm, white or greenish-white; spadix 1-3 x 0.7-2cm. Intrd; grown for ornament, persistent and spreading in marshy ground and shallow ponds, often in shade; very scattered in Br from SE En to C Sc; Europe.

4. ZANTEDESCHIA Sprengel - <u>Altar-lily</u>
Rhizomes short, tuberous; leaves ovate or broadly so, cordate, entire, with long petiole; spadix terminal, without appendix; spathe wrapped round and concealing spadix at base; flowers unisexual; tepals 0; stamens 2-3; ovary (1-)3-celled; fruit a yellow berry with several seeds; but very seldom (?never) produced.

1. Z. aethiopica (L.) Sprengel - <u>Altar-lily</u>. Leaf-blades 10-45 x 10-25cm; petiole up to 50(75)cm; spathe 10-25cm, pure white; spadix up to 15cm, bright yellow. Intrd; grown for ornament, persistent and spreading in ditches, damp hedgerows and scrub, and neglected fields; CI, SW En, S Wa and S Ir; S Africa.

5. ARUM L. - <u>Lords-and-Ladies</u>
Rhizomes short, tuberous; leaves triangular-ovate, hastate to sagittate, entire, with long petiole; spadix terminal, with long appendix; spathe wrapped round and concealing spadix at base, pale greenish-yellow; flowers unisexual; tepals 0; stamens 3-4; ovary 1-celled; fruit a red berry with 1-several seeds.

1. A. maculatum L. - <u>Lords-and-Ladies</u>. Leaf-blades appearing in early Spring, 7-20cm, often blackish-purple-spotted, with concolourous midrib; spadix appendage purple or yellow, usually reaching c.1/2 way up expanded part of spathe; spathe 10-25cm, sometimes with dark spots; fruiting spike 3-5cm. Native; woods and hedgerows, usually on base-rich soils; frequent throughout BI N to C Sc, scattered and natd further N.

1 x 2. A. maculatum x A. italicum occurs rarely in S & SW En, S Wa and CI; it is intermediate and probably sterile, but has leaves appearing in early Winter and often spotted.

2. A. italicum Miller - <u>Italian</u> <u>Lords-and-Ladies</u>. Leaf-   **R** blades appearing in early Winter, 15-35cm, with pale midrib; spadix appendage yellow, usually reaching c.1/3 way up expanded part of spathe; spathe 15-40cm, never spotted; fruiting spike 10-15cm.

a. Ssp. neglectum (F. Towns.) Prime. Leaves sometimes dark-spotted, with veins slightly paler than rest of leaf, with basal lobes somewhat convergent and sometimes overlapping; fruits with 1-2 seeds. Native; hedgerows, scrub and stony field-borders; extreme S & SW En, S Wa and CI, very rarely natd elsewhere, common in CI and Scillies and the only sp. in latter place.

b. Ssp. italicum. Leaves never dark-spotted, with whitish veins, with basal lobes divergent; fruits with 2-4 seeds. Intrd; persistent garden throw-out natd in similar places to ssp. <u>neglectum</u>; scattered in CI, Br N to Man, Dunbarton and S Lincs, E Ir, frequent and perhaps native CI.

6. DRACUNCULUS Miller - <u>Dragon</u> <u>Arum</u>
Rhizomes short, tuberous; leaves deeply ± palmately lobed, cordate at base, with entire lobes and long petiole; spadix terminal, with long appendix; spathe wrapped round and concealing spadix at base; flowers unisexual; tepals 0; stamens 2-4; ovary 1-celled; fruit a red berry with several seeds.

1. D. vulgaris Schott - <u>Dragon</u> <u>Arum</u>. Leaf-lobes up to 20cm; spadix appendage dark purple, nearly as long as spathe; spathe 25-40cm, dark purple. Intrd; garden throw-out natd in hedges, rough ground and old gardens; scattered in S & SE En and CI; E & C Mediterranean.

7. ARISARUM Miller - <u>Mousetailplant</u>
Rhizomatous; leaves triangular-ovate, sagittate, entire, with long petiole; spadix terminal, with long appendix; spathe fused in tube round spadix and concealing most of it, extended apically into filiform projection; flowers unisexual; tepals 0; stamen 1; ovary 1-celled; fruit green, with several seeds.

1. A. proboscideum (L.) Savi - <u>Mousetailplant</u>. Leaf-blades 6-15cm; spadix appendage whitish, concealed within spathe; spathe 2-4cm excl. filiform projection 5-15cm, dark or greenish-brown. Intrd; garden throw-out natd in hedges, rough ground and old gardens; very scattered in S En; Spain and Italy.

150. LEMNACEAE - Duckweed family

Aquatic perennial plants reduced to ± undifferentiated pad-like frond to 15mm (but often much less) floating on or under water surface (sometimes stranded on mud), not or variously adhering together, with 0-21 roots per frond. Flowers rather rarely produced, very reduced, borne in (1-)2 hollows on frond, each hollow with 1-2 stamens and 1 ovary with 1-2 ovules and funnel-shaped stigma (variously interpreted as 1 flower or 1-2 male and 1 female flower) subtended or not by minute spathe; perianth 0; fruit of 1-2 seeds in thin sac. The floating pad-like plants are unique.

Only well-grown Spring or Summer fronds should be used; poorly grown ones or those produced in Autumn and over-wintering are often atypical, being smaller and often with fewer veins and fewer or 0 roots.

1 Fronds rootless and veinless, spherical to ellipsoid
                                                      3. WOLFFIA
1 Each frond with (0)1-16(21) roots and 1-16(21) veins,
   ± flattened at least on upperside   2
   2 Each frond with (0)1 root and 1-5(7) veins   **2. LEMNA**
   2 Each frond with 7-16(21) roots and veins
                                                   1. SPIRODELA

1. SPIRODELA Schleiden - Greater Duckweed
Fronds with 7-16(21) roots, with 7-16(21) veins, floating on water surface.

1. S. polyrhiza (L.) Schleiden (Lemna polyrhiza L.) - Greater Duckweed. Fronds 1.5-10 x 1.5-8mm, ± flattened on both surfaces. Native; canals, ditches and ponds; rather local in C & S Br, very scattered in N Br and Ir, formerly CI.

2. LEMNA L. - Duckweeds
Fronds with (0)1 root, with 1-5(7) veins, floating on or below water surface.

1 Fronds narrowed to a stalk-like portion at 1 end,
   usually submerged, usually cohering in branched
   chains of 3-50   **3. L. trisulca**
1 Fronds orbicular to ellipsoid, without stalk-like
   portion, usually on water surface, cohering in small
   groups (not chains)   2
   2 Fronds usually strongly swollen on lowerside, ±
      hemispherical, with (3)4-5(7) veins originating
      from 1 point   **1. L. gibba**
   2 Fronds ± flattened on both surfaces, with 1-3(5)
      veins, if 4 or 5 then 3 originating from 1 point
      and the outermost 1 or 2 extras branching from
      near base of inner laterals   3

3 Fronds 0.8-3(4)mm long, usually elliptic, with 1 vein
                                                    4. L. minuta
3 Fronds (1)2-5(8)mm long, usually ovate, with 3(-5)
  veins
                                                    2. L. minor

1. L. gibba L. - <u>Fat</u> <u>Duckweed</u>. Fronds 1-8 x 0.8-6mm,
usually strongly swollen on lowerside (not in Autumn-produced
fronds, nor in starved plants), with (3)4-5(7) veins, with
larger air-spaces (visible as reticulum on frond upperside)
>0.3mm across. Native; ponds, ditches and canals, usually in
rich, often brackish water; frequent in C & S Br, very
scattered in CI and Ir.

2. L. minor L. - <u>Common</u> <u>Duckweed</u>. Fronds 1-8 x 0.6-5mm, $\pm$
flattened on both surfaces, with 3(-5) veins, with larger
air-spaces $\leq$0.3mm across. Native; ponds, ditches, canals and
slow parts of rivers and streams; common throughout BI except
rare in N Sc.

3. L. trisulca L. - <u>Ivy-leaved</u> <u>Duckweed</u>. Fronds 3-15 (plus
stalk 2-20) x 1-5mm, $\pm$ flattened on both surfaces, with (1-)3
veins, with 0-1 root. Native; ponds, ditches and canals;
frequent in most of BI, but very scattered in Sc.

4. L. minuta Kunth (<u>L. minuscula</u> Herter nom. illeg.) - <u>Least</u>
<u>Duckweed</u>. Fronds 0.8-4 x 0.5-2.5mm, $\pm$ flattened on both
surfaces, with 1 vein, with larger air-spaces $\leq$0.25mm across.
Intrd; same habitats as <u>L. minor</u>; scattered in CI and Br N to
E Norfolk and Flints, 1st recorded 1977, probably overlooked;
N & S America.

3. WOLFFIA Horkel ex Schleiden - <u>Rootless</u> <u>Duckweed</u>
Fronds with 0 roots, with 0 veins, usually floating on water
surface.

1. W. arrhiza (L.) Horkel ex Wimmer - <u>Rootless</u> <u>Duckweed</u>.   R
Fronds 0.5-1.5 x 0.4-1.2mm, strongly swollen on both sides
(thicker than wide). Native; ponds and ditches; very local
in extreme S En, formerly Glam.

151. COMMELINACEAE - <u>Spiderwort family</u>

Herbaceous perennials; leaves alternate, entire, with
sheathing base, glabrous or pubescent just on sheath, sessile
or nearly so, exstipulate. Flowers few in terminal paired
cymes with large leaf-like bract at base of each, bisexual,
hypogynous, actinomorphic, showy; sepals 3, green, free;
petals 3, white or coloured, free; stamens 6, with pubescent
filaments; ovary 3-celled with 2 ovules per cell; style 1;
stigma capitate; fruit a capsule (?never formed).
  Easily recognized by the flowers with 3 green sepals, 3
white or coloured petals, and 6 stamens with conspicuous long
hairs on filaments.

1. TRADESCANTIA L. (Zebrina Schnitzl.) - Spiderworts

Other spp. - T. zebrina Loudon (Zebrina pendula Schnitzl.), from Mexico, resembles T. fluminensis but has silver-striped leaves and pinkish-purple petals; it occasionally persists for a short while where thrown out.

1. T. virginiana L. - Spiderwort. Tufted, shortly rhizomatous perennial; stems erect, to 60cm; leaves linear, 15-35cm; flowers 2.5-3.5cm across; petals usually violet, sometimes white, pink or purple. Intrd; grown in gardens, occasionally persisting on tips and waste ground where thrown out; rare in SE En; N America. The commonest garden plant is probably a hybrid of T. virginiana with 1 or more other spp.; it has no valid name (T. x andersoniana W. Ludwig & Rohw. nom. inval.).

2. T. fluminensis Vell. Conc. - Wandering-jew. Stems trailing, rooting at nodes, to 1m or more; leaves ovate-oblong, 1.5-5cm; flowers 1-1.5cm across, rarely produced; petals white. Intrd; much grown as pot-plant, sometimes persisting in shrubberies and frost-free rough ground where thrown out; rare in S En and CI; S America.

152. ERIOCAULACEAE - Pipewort family

Aquatic (or ± so) herbaceous perennials rooted in substratum; leaves appearing all basal, simple, subulate, entire, sessile, exstipulate. Flowers in terminal, whitish, capitate mass on long leafless stem from basal rosette, unisexual with male flowers in centre of inflorescence and female around them, hypogynous, actinomorphic, with bracteoles below each; tepals 4, 2 outer fused (male) or ± free (female), 2 inner free, membranous and pubescent or fringed; stamens 4; ovary 2-celled, each cell with 1 ovule; style 1; stigmas 2, filiform; fruit a capsule with 2 seeds.

Unique in the subulate basal leaf-rosette and erect leafless stems bearing small capitate inflorescences of many unisexual flowers.

1. ERIOCAULON L. - Pipewort
1. E. aquaticum (Hill) Druce (E. septangulare With.) -   **R**
Pipewort. Leaves clearly transversely septate, very finely pointed, up to 10cm; stems erect, usually emergent and varying in height according to water level, up to 20(150)cm; inflorescence 5-12(20)mm across. Native; in shallow lakes and pools or in bare wet peaty ground; extremely local in W Sc, local in W Ir.

153. JUNCACEAE - Rush family

Annuals or herbaceous perennials, often ± aquatic; leaves
alternate or all basal, grass-like (bifacial) to rush-like
(cylindrical to flattened but unifacial), with sheathing base
with membranous ligule at top of sheath, the blade simple,
linear and entire or 0, exstipulate. Flowers in various
simple to complex, often congested cymes that are terminal
but often appear lateral, bisexual, hypogynous,
actinomorphic; tepals 6, 3 inner amd 3 outer, greenish,
brownish or membranous, free; stamens (3-)6; ovary 1-celled
with 3 ovules, 1-celled with many ovules on parietal
placentas, or 3-celled with many ovules per cell on axile
placentas; style 0 or 1; stigmas 3, linear; fruit a capsule
with 3 or numerous seeds.
Distinguished from other rush-, sedge- or grass-like plants
by the flowers with 6 tepals, (3-)6 stamens and a single
1-3-celled ovary with 3-many ovules.

1 Leaves bifacial to unifacial, glabrous; ovary with
   many ovules; capsule with many seeds   **1. JUNCUS**
1 Leaves bifacial, usually pubescent at least near
   base when young; ovary with 3 ovules; capsule with
   3 seeds   **2. LUZULA**

1. JUNCUS L. - Rushes

Annuals to perennials; leaves various, bifacial to unifacial,
glabrous; ovary 1-3-celled, with many ovules; capsule with
many seeds.
'Leaves' refers to stem-leaves and/or basal leaves, but
excludes leaf-like bracts immediately below or within the
inflorescence. In subg. Genuini the lowest inflorescence
bract is cylindrical and stem-like, making the inflorescence
appear lateral. A sharp scalpel or razor-blade is needed to
cut longitudinal and transverse sections of stem and leaves
to see the internal structure; dried material usually needs
resuscitation by boiling in water.

General key
1 Leaves distinctly bifacial, flat and ± grass-like
   with 2 opposite surfaces but sometimes inrolled, or
   subcylindrical and ± rush-like but with a distinct
   deep channel on upperside for most or all of length
                                                        Key A
1 Leaves unifacial or ± so, or apparently absent,
   cylindrical to flattened-cylindrical and rush-like,
   not deeply channelled or with deep channel only near
   ligule and extending <1/2 way to leaf apex, sometimes
   with shallow grooves   Key B

Key A - Leaves bifacial, flat, or subcylindrical but with a
        deep channel on upperside

1 Easily uprooted annual, with simple fibrous root-
   system   2
1 Rhizomatous perennial, usually firmly rooted
   (rhizomes often very short and plant densely tufted)   5
   2 Stems unbranched, with basal leaves and leaf-like
      bracts at top but bare between   **10. J. capitatus**
   2 Stems branched, leafy (but leaves often short and
      very narrow)   3
3 Leaves usually >1.5mm wide; tepals usually with dark
   line either side of midrib; anthers 1.2–5x as long as
   filaments; seeds with longitudinal ridges (10–15 in
   side view) clearly visible with x20 lens   **6. J. foliosus**
3 Leaves rarely >1.5mm wide; tepals rarely with
   dark line on either side of midrib; anthers usually
   0.3–1.1x as long as filaments; seeds without
   longitudinal ridges visible with x20 lens (beware
   shrivelled seeds)   4
   4 Inner tepals rounded to emarginate and mucronate
      at apex; capsule truncate at apex, at least as
      long as inner tepals   **8. J. ambiguus**
   4 Inner tepals acute to subacute at apex; capsule
      acute to obtuse at apex, rarely ± truncate,
      usually shorter than inner tepals   **7. J. bufonius**
5 Outer 3 tepals obtuse to rounded at apex   6
5 Outer 3 tepals acute to acuminate at apex   7
   6 Anthers 0.5–1mm, 1–2x as long as filaments; seeds
      0.35–0.5mm   **3. J. compressus**
   6 Anthers 1–2mm, 2–3x as long as filaments; seeds
      0.5–0.7mm   **4. J. gerardii**
7 Lowest 2 bracts of inflorescence leaf-like, usually
   far exceeding inflorescence; stem with (1)2–4(5)
   well-developed leaves (usually near base)   8
7 All bracts of inflorescence mostly scarious, usually
   much shorter than inflorescence; stem with 0(–1)
   well-developed leaves   9
   8 Inflorescence of 1–3(4) flowers in tight cluster;
      anthers longer than filaments; seeds 0.9–1.6mm,
      with long appendage at each end   **5. J. trifidus**
   8 Inflorescence of 5–40 flowers, usually ± diffuse;
      anthers shorter than filaments; seeds 0.3–0.4mm,
      with short appendages   **2. J. tenuis**
9 Leaves flat or ± inrolled; stamens 3   **9. J. planifolius**
9 Leaves rounded on lowerside, with deep channel on
   upperside; stamens 6   **1. J. squarrosus**

Key B – Leaves unifacial or ± so, cylindrical to flattened-
      cylindrical
1 Leaves on stems represented only by blade-less
   scarious sheaths near base   2
1 At least 1 stem-leaf with well-developed green blade   7
   2 Stems strongly glaucous, with pith conspicuously
      and regularly interrupted at least in region just

```
            below inflorescence                25. J. inflexus
  2 Stems not glaucous, with pith well formed and
            conspicuous or ill-formed and irregular           3
  3 Rhizomes extended, forming straight lines or diffuse
    patches of aerial stems; inflorescence usually <20-
    flowered; stems rarely >50cm                              4
  3 Rhizomes short, forming dense clumps of aerial stems
    (or large dense patches in very old plants);
    inflorescence usually >20-flowered; stems usually
    >50cm                                                     5
      4 Inflorescence in lower 2/3(3/4) of apparent stem
        ± globose; stem with subepidermal sclerenchyma
        girders, with fine longitudinal ridges when dry
                                         24. J. filiformis
      4 Inflorescence in upper 1/4 of apparent stem,
        usually elongated; stem without subepidermal
        sclerenchyma girders, not ridged when dry
                                          23. J. balticus
  5 Fresh stems dull, ridged, with usually <35 ridges;
    main (stem-like) bract opened out and ± flat
    adjacent to inflorescence, causing it to hinge over
    backwards at end of season         27. J. conglomeratus
  5 Fresh stems glossy, smooth, becoming finely ridged
    with usually >35 ridges when dry; main bract scarcely
    opened out adjacent to inflorescence, not hingeing
    over backwards at end of season                           6
      6 Capsule as long as or longer than tepals,
        2.5-3.5mm when fully fertile; stamens 6; larger
        stems usually >1.2m                    28. J. pallidus
      6 Capsule shorter than tepals, 2-2.5mm when fully
        fertile; stamens 3(-6); stems usually <1.2m
                                           26. J. effusus
  7 Leaves and main bract with very sharp apex, with
    subepidermal sclerenchyma girders, with vascular
    bundles scattered through pith                            8
  7 Leaves and main bract with soft apex, without sub-
    epidermal sclerenchyma girders, without vascular
    bundles in pith                                           9
      8 Capsule 2.5-3.5mm, not or slightly longer than
        tepals; inner tepals obtuse, the extreme apex not
        exceeded by membranous margins    20. J. maritimus
      8 Capsule 4-6mm, much longer than tepals; inner
        tepals retuse, with membranous margins extended
        into lobes on each side of extreme apex and
        exceeding it                        21. J. acutus
  9 Leaves cylindrical, with continuous pith within
    vascular cylinder; each flower with 2 small bracteoles
    immediately beneath tepals            22. J. subulatus
  9 Leaves cylindrical to flattened cylindrical, with
    pith usually interrupted by transverse septa and
    often with large cavities; flowers without
    bracteoles at base                                       10
```

10 Easily uprooted annual, with simple fibrous root
   system (W Cornwall)   **16. J. pygmaeus**
10 Perennial, usually firmly rooted, either
   rhizomatous or with ± swollen stem-bases
   (widespread)   11
11 Anthers <1/3 as long as filaments; seeds with
   conspicuous whitish appendages at each end each c.
   as long as actual seed (alpine)   12
11 Anthers >1/3 (up to 2x) as long as filaments; seeds
   with at most minute points at each end (lowland or
   alpine)   14
   12 Outer tepals acute; capsule ≥6mm; stems usually
      solitary from rhizome system   **19. J. castaneus**
   12 Outer tepals obtuse to rounded; capsule <6mm;
      stems usually in small tufts   13
13 Capsule 3-4mm, retuse at apex; lowest bract usually
   exceeding inflorescence; flowers mostly 2 per
   inflorescence   **17. J. biglumis**
13 Capsule 4-5mm, obtuse at apex; lowest bract usually
   shorter than inflorescence; flowers mostly 3 per
   inflorescence   **18. J. triglumis**
   14 Leaves with >2 empty or loosely pith-filled
      longitudinal cavities separated by thin walls
      bearing a few vascular bundles   15
   14 Leaves with 1 empty or loosely pith-filled
      longitudinal cavity   16
15 Rhizomatous; outer tepals obtuse, incurved at apex;
   flowers very rarely vegetatively proliferating
                                        11. J. subnodulosus
15 Not rhizomatous; stem-base usually swollen; outer
   tepals acute, not incurved; flowers commonly
   vegetatively proliferating   **15. J. bulbosus**
   16 Tepals acuminate, the outer with recurved
      apical points   **14. J. acutiflorus**
   16 Outer tepals obtuse to shortly acuminate with
      erect apical points; inner tepals acute to
      rounded   17
17 Outer tepals subacute to obtuse; inner tepals obtuse
   to rounded; capsule obtuse (ignore beak)
                                    12. J. alpinoarticulatus
17 Outer tepals acute; inner tepals acute to subacute;
   capsule acute (ignore beak)   **13. J. articulatus**

Other spp. - Several other spp. of subg. Genuini occur as
wool-aliens and have sometimes persisted in the past, but
none so much as J. pallidus; their determination is often
extremely difficult.

Subgenus 1 - PSEUDOTENAGEIA V. Krecz. & Gontch. (spp. 1-5).
Rhizomatous (but sometimes densely tufted) perennials; leaves
flat or strongly channelled, basal and along stem, not
sharply pointed, with 1 subepidermal sclerenchyma girder at

each margin (O in J. trifidus), with cavities developing
between vascular bundles but without pith; inflorescence
terminal, very or rather compact, exceeded or not by 1-2
leaf-like or scale-like bracts; each flower with 2 small
bracteoles; seeds with or without terminal appendages.

1. J. squarrosus L. - Heath Rush. Densely tufted; stems **930**
erect, to 50cm; leaves nearly all basal, rounded on
lowerside, deeply channelled on upperside; lowest bract much
shorter than inflorescence. Native; bogs, wet moors and
heaths, on acid soil; common throughout Br and Ir where acid
soils exist, absent from much of C Ir and C En.

2. J. tenuis Willd. (J. dudleyi Wieg.) - Slender Rush. **930**
Densely tufted; stems erect, to 80cm; leaves nearly all **938**
basal, flattened; lowest 1(-2) bracts usually much longer
than inflorescence. Intrd; damp barish ground on roadsides,
tracks and paths; locally frequent throughout most BI, 1st
recorded 1795/6; N & S America. Var. **dudleyi** (Wieg.) F.J.
Herm. differs in its leaf-sheaths ending in brown auricles
much wider than long (not scarious and much longer than
wide); it is natd in M Perths.

3. J. compressus Jacq. - Round-fruited Rush. Loosely tufted
to extensive patches; stems erect, to 50cm; leaves mostly
basal, flattened, often ± inrolled; lowest bract usually
exceeding inflorescence. Native; marshes and wet meadows,
often near sea; scattered in En and Wa, mostly C & E En.

4. J. gerardii Loisel. - Saltmarsh Rush. Differs from J.
compressus constantly in seeds and stamens (see Key A,
couplet 6); usually more extensively rhizomatous; lowest
bract usually shorter than inflorescence; tepals usually dark
(not light) brown; capsule usually subacute to obtuse (not
obtuse to rounded) and scarcely (not greatly) exceeding
tepals. Native; salt-marshes and inland saline areas;
abundant round coasts of BI, very scattered inland. Stamens
persist well after seed dispersal, flattened between tepals
and capsule.

5. J. trifidus L. - Three-leaved Rush. Densely tufted; **930**
stems erect, to 30cm; leaves few, usually 0-1 at base and 1-2 **938**
plus leaf-like lowest bract near apex of stem, flattened and
inrolled; lowest bract and 1-2 upper stem-leaves exceeding
inflorescence. Native; barish places on mountains; locally
frequent in C & W Sc, Shetland.

Subgenus 2 - POIOPHYLLI Buchenau (spp. 6-8). Annuals; leaves
flat or inrolled, all on stem, those at base often withered
by flowering time, not sharply pointed, with 1 subepidermal
sclerenhyma girder at each margin, with cavities developing
between vascular bundles but without pith; inflorescence
terminal, usually very diffuse and occupying most of plant,
interspersed with ± leaf-like bracts; each flower with 2
small bracteoles; seeds without appendages.

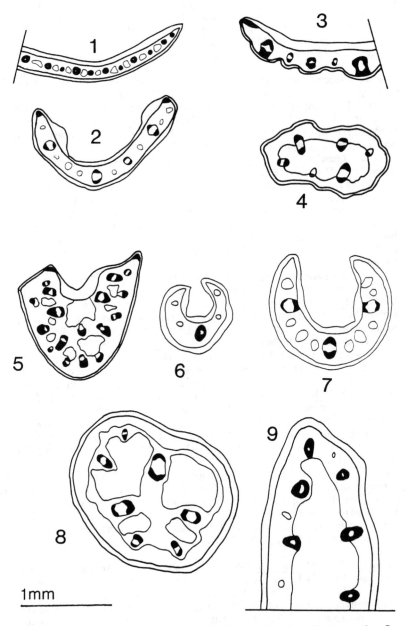

FIG 930 - Leaf-sections of **Juncus**. 1, **J. planifolius**. 2, **J. bufonius**. 3. **J. tenuis**. 4, **J. biglumis**. 5, **J. squarrosus**. 6, **J. trifidus**. 7, **J. capitatus**. 8, **J. bulbosus**. 9, **J. articulatus**. Sclerenchyma in black. Drawings by C.A. Stace.

1mm

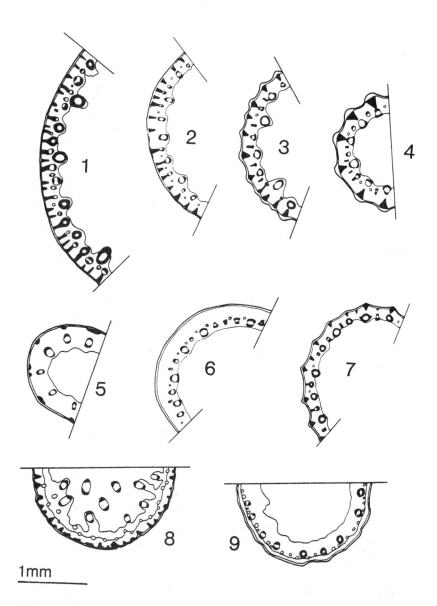

FIG 931 – Leaf-sections of **Juncus** (main bract in case of 1-7), 1, **J. pallidus**. 2, J. effusus. 3, **J. x diffusus**. 4, **J. inflexus**. 5, **J. filiformis**. 6, J. **balticus**. 7, **J. conglomeratus**. 8, **J. maritimus**. 9, **J. subulatus**. Sclerenchyma in black. Drawings by C.A. Stace.

6. J. foliosus Desf. - Leafy Rush. Differs from J. bufonius **938**
in stems erect to ascending; flowers well spaced; tepals
usually with dark line on each side of midrib; and see Key A
(couplet 3). Native; muddy margins of areas of fresh water,
wet fields, marshes and ditches; scattered through W & S BI.

7. J. bufonius L. (J. minutulus V. Krecz. & Gontch.) - Toad **930**
Rush. Stems erect to procumbent, to 35(50)cm, variable in **938**
branching but often very diffuse; inflorescence variable,
with flowers well spaced to tightly bunched at branchlet
ends; tepals rarely with dark lines; inner tepals acute to
subacute, c. as long to longer than acute to subacute or
rarely truncate capsule. Native; all kinds of damp habitats,
both fresh-water and brackish, natural and artificial; common
throughout BI.

8. J. ambiguus Guss. (J. ranarius Nees ex Song. & Perrier) - **938**
Frog Rush. Differs from J. bufonius in stems to 17cm;
flowers 2-4(5) on each ultimate branch, with usually 2-3
bunched together at tip; inner tepals rounded to emarginate
and mucronate at apex, shorter than to as long as truncate
capsule. Native; damp brackish habitats near coast and
inland, and on damp lime-waste; scattered in suitable
habitats throughout BI.

Subgenus 3 - GRAMINIFOLII Buchenau (sp. 9). Shortly
rhizomatous tufted perennials; leaves flat, basal, not
sharply pointed, without subepidermal sclerenchyma girders,
with cavities developing between vascular bundles but without
pith; inflorescence terminal, rather compact, with short
reduced main bract; flowers without bracteoles; seeds without
appendages.

9. J. planifolius R. Br. - Broad-leaved Rush. Stems erect, **930**
to 30cm; leaves light green, 2-8mm wide; inflorescence with **938**
shape of that of Luzula multiflora ssp. multiflora. Intrd;
damp pathsides, lake shores and wet meadows; over c.40 square
km in W Galway, discovered 1971; Australia, New Zealand, S
America.

Subgenus 4 - JUNCINELLA V. Krecz. & Gontch. (sp. 10).
Annuals; leaves flat to strongly channelled, all basal, not
sharply pointed, without subepidermal sclerenchyma girders,
without cavities or pith region; inflorescence terminal, very
compact, exceeded or not by 1-2 main bracts; flowers without
bracteoles; seeds ± without appendages.

10. J. capitatus Weigel - Dwarf Rush. Stems erect, to 5cm **930**
and often much less; leaves <1mm wide. Native; barish ground **938**
on heaths, usually where water stands in winter; extremely **RR**
local in W Cornwall and CI, formerly Anglesey.

Subgenus 5 - SEPTATI Buchenau (spp. 11-16). Annuals or
rhizomatous or non-rhizomatous perennials; leaves terete or

flattened-terete, all on stems or some basal, not sharply
pointed, with 1-several central cavities, divided by
transverse septa (often visible externally), without
subepidermal sclerenchyma girders; inflorescence terminal,
very compact to very diffuse, with ± leaf-like main bract
usually shorter than it; flowers without bracteoles; seeds
without appendages.

11. J. subnodulosus Schrank - <u>Blunt-flowered</u> Rush.
Rhizomatous; stems erect, to 1.2m; leaves with very distinct
transverse septa; inflorescence diffuse, with very widely
spreading to ± reflexed branchlets; tepals 1.8-2.3mm, pale
brown, with obtuse, incurved tips. Native; fens, marshes and
duneslacks on peaty base-rich soil; locally frequent in En,
Wa and Ir, very local in Jersey, SW & CE Sc.
12. J. alpinoarticulatus Chaix (<u>J. alpinus</u> Villars nom.   R
illeg., <u>J. nodulosus</u> auct. non Wahlenb.) - <u>Alpine</u> Rush.
Rhizomatous; stems erect to ascending, to 40cm; leaves with
very distinct transverse septa; inflorescence diffuse but
often rather sparse, with suberect to erecto-patent
branchlets; tepals 1.8-2.5mm, dark brown to blackish, the
outer subacute to obtuse, the inner obtuse to rounded.
Native; marshes, flushes and streamsides on mountains; local
in mainland Br N from MW Yorks. Plants from E Ross and S
Aberdeen have been referred to **J. nodulosus** Wahlenb.
(probably better as a ssp. of <u>J. alpinoarticulatus</u>), but this
is doubtful.
**12 x 13. J. alpinoarticulatus x J. articulatus = J. x
buchenaui** Doerfler has been found in E Ross, W Sutherland and
Co Durham; it is intermediate in habit and tepal shape and
has low, if any, fertility.
13. J. articulatus L. - <u>Jointed</u> Rush. Rhizomatous; very 930
variable in habit; stems erect to decumbent, to 80cm but
often much less; leaves with very distinct transverse septa;
inflorescence diffuse, with suberect to erecto-patent
branchlets; tepals 2.3-3.5mm, dark brown to blackish, the
outer acute, the inner subacute to acute. Native; damp
grassland, heaths, moors, marshes, dune-slacks, margins of
rivers and ponds; common throughout BI.
13 x 14. J. articulatus x J. acutiflorus = J. x surrejanus
Druce ex Stace & Lambinon occurs with the parents throughout
BI and is commoner than either in some places; it is
intermediate in tepal shape and size and has low fertility.
14. J. acutiflorus Ehrh. ex Hoffm. - <u>Sharp-flowered Rush</u>. 938
Stems erect, to 1.1m; differs from <u>J. articulatus</u> in longer
rhizomes; larger, more branched inflorescence; tepals 1.5-
2.7mm, mid to dark brown; and see Key B (couplet 16).
Native; marshes, bogs, damp grassland, margins of rivers and
ponds; common throughout BI.
15. J. bulbosus L. (<u>J. kochii</u> F. Schultz) - <u>Bulbous</u> Rush. 930
Not rhizomatous; extremely variable in habit; stems erect to
procumbent or floating, often corm-like at base, to 30cm,

often rooting at nodes; leaves with very indistinct transverse septa; inflorescence diffuse but often very sparse with suberect to patent branchlets, very commonly proliferating; tepals 1.5-3.5mm, green to dark brown, the outer acute, the inner obtuse. Native; in all kinds of wet and damp places, often submerged; abundant throughout BI. Often misdetermined for a range of plants from J. articulatus to Eleocharis or Isoetes. Plants known as J. kochii have been separated in different ways using various characters that are not correlated.

16. J. pygmaeus Rich. (J. mutabilis auct. non Lam.) - Pigmy   **RR** Rush. Dwarf annual; stems erect to ascending, to 8cm; leaves with rather indistinct transverse septa; inflorescence very compact; tepals 4.5-6mm, acute to obtuse, greenish- or purplish-brown. Native; damp hollows and rutted tracks on heathland; Lizard area of W Cornwall.

Subgenus 6 - ALPINI Buchenau (spp. 17-19). Dwarf, alpine, rhizomatous perennials; leaves terete or flattened-terete, all basal or some on stems, not sharply pointed, with 1-several central cavities, divided by transverse septa (often not visible externally), without subepidermal sclerenchyma girders; inflorescence terminal, very compact, exceeded or not by leaf-like or scale-like main bract; flowers without bracteoles; seeds with terminal appendages.

17. J. biglumis L. - Two-flowered Rush. Stems tufted, to **930** 12cm, erect; leaves all basal, <6cm; inflorescence 1, with   **R** (1)2(-4) flowers in a + vertical row, usually exceeded by lowest bract; capsule 3.5-4.5mm, retuse. Native; barish rocky places on mountains; very local in C, W & NW Sc.

18. J. triglumis L. - Three-flowered Rush. Stems tufted, to 20cm, erect; leaves all basal, <10cm; inflorescence 1, with (2)3(-5) flowers in + horizontal row, usually not exceeded by lowest bract; capsule 4-6mm, obtuse to rounded. Native; boggy and rocky places on mountains; local in C, W & NW Sc, very local in N En and N Wa.

19. J. castaneus Smith - Chestnut Rush. Stems not tufted, **938** to 30cm; leaves basal and 1-3 on stems, <20cm; inflorescences   **R** 1-3, each with 3-8(10) flowers in + horizontal row, usually exceeded by lowest bract; capsule 6-7.5mm, obtuse to rounded. Native; boggy places and flushes on mountains; very local in C, W & NW Sc.

Subgenus 7 - JUNCUS (Thalassii Buchenau) (spp. 20-21). Robust maritime rhizomatous (but sometimes densely tufted) perennials; leaves cylindrical, basal and on stems, very sharply pointed, with central compact pith bearing scattered vascular bundles, not transversely septate, with very numerous subepidermal sclerenchyma girders; inflorescence terminal, with sharply pointed leaf-like main bract shorter or longer than it, very to somewhat compact; flowers without bracteoles; seeds with terminal appendages.

20. J. maritimus Lam. - Sea Rush. Stems densely to scarcely **931**
tufted, erect, very stiff, to 1m; inflorescence usually
forming an interrupted panicle with erect to erecto-patent
branches, usually exceeded by lowest bract. Native; salt-
marshes; common round coasts of BI except extreme N Sc.
21. J. acutus L. - Sharp Rush. Stems very densely tufted, **938**
erect, extremely stiff, to 1.5m; inflorescence usually a   **R**
dense ± rounded head with erect to reflexed branches, usually
exceeded by lowest bract. Native; sandy sea-shores and drier
parts of salt-marshes; very local in BI N to W Norfolk,
Caerns and Co Dublin, formerly NE Yorks.

Subgenus 8 - SUBULATI Buchenau (sp. 22). Rhizomatous
maritime perennials; leaves cylindrical, basal and on stems,
not sharply pointed, with central soft pith, not transversely
septate, without subepidermal sclerenchyma girders; inflorescence terminal, with short much reduced main bract,
diffuse; each flower with 2 small bracteoles; seeds with
short terminal appendages.

22. J. subulatus Forsskaol - Somerset Rush. Stems rather **931**
weak, ± erect, to 1m; inflorescence a rather diffuse panicle **938**
with suberect branches, much exceeding lowest bract. Intrd;
saltmarsh in N Somerset, discovered 1957, and wet reclaimed
land by docks in Stirlings, discovered 1983; Mediterranean.

Subgenus 9 - GENUINI Buchenau (spp. 23-28). Rhizomatous (but
often densely tufted) perennials; leaves reduced to brown
sheaths at base of stem and the lowest bract, which is stem-
like and much exceeds inflorescence which appears lateral;
main bract cylindrical, not or slightly sharply pointed, with
central soft pith sometimes regularly interrupted, not
transversely septate, with or sometimes without subepidermal
sclerenchyma girders; inflorescence very compact to rather
diffuse; each flower with 2 small bracteoles; seeds ± without
appendages.

23. J. balticus Willd. - Baltic Rush. Strongly rhizomatous; **931**
stems erect, to 75cm, smooth and glossy when fresh, with   **R**
continuous pith, without subepidermal sclerenchyma girders;
inflorescence rather lax, with suberect branches; tepals dark
brown. Native; maritime dune-slacks, rarely on upland river
terraces, on bare or grassy ground; local in N Sc S to M
Ebudes and Fife, S Lancs, formerly W Lancs.
23 x 25. J. balticus x J. inflexus occurs as 3 large patches
of strongly rhizomatous completely sterile clones in S & W
Lancs; 2 of them (S Lancs) are very tall (to 2m) and have an
interrupted pith as in J. inflexus, the other (W Lancs) is
close to J. inflexus in height but has a continuous pith;
endemic.

23 x 26. J. balticus x J. effusus = J. x obotritorum Rothm.
occurred as 3 patches, 1 large, of strongly rhizomatous
completely sterile clones in S Lancs between 1933 and 1980
(transplanted portions still exist); they differ from <u>J.</u>
<u>balticus</u> x <u>J. inflexus</u> in being much more slender and in
minor anatomical characters.

24. J. filiformis L. - <u>Thread</u> <u>Rush</u>. Rather weakly **931**
rhizomatous; stems erect, to 45cm, very slightly ridged when **RR**
fresh, with continuous pith, with subepidermal sclerenchyma
girders; inflorescence compact or very compact, with <u><</u>10
flowers; tepals pale brown. Native; on stony, silty edges of
lakes and reservoirs; local in Br from Leics to Easterness,
increasing.

25. J. inflexus L. - <u>Hard</u> <u>Rush</u>. Densely tufted; stems **931**
erect, to 1.2m, glaucous, very strongly ridged when fresh,
with interrupted pith, with very strong subepidermal
sclerenchyma girders; inflorescence rather lax, with suberect
branches; tepals dark brown. Native; marshes, dune-slacks,
wet meadows, ditches, by lakes and rivers, usually on neutral
or base-rich soils; common throughout most of BI N to C Sc.

25 x 26. J. inflexus x J. effusus = J. x diffusus Hoppe **931**
occurs sporadically with the parents, usually as isolated
plants, within the range of <u>J. inflexus</u>. The stems are not
glaucous and have continuous pith, intermediate anatomy, and
inflorescence shape + as in <u>J. inflexus</u>; fertility low.

26. J. effusus L. - <u>Soft-rush</u>. Densely tufted or sometimes **931**
forming larger patches; stems erect, to 1.2m, smooth and **938**
glossy when fresh, with continuous pith, with subepidermal
sclerenchyma girders; inflorescence lax, with suberect to
widely divergent branches, or compact (var. **subglomeratus** DC.
(var. <u>compactus</u> Lej. & Courtois)); tepals pale brown.
Native; marshes, ditches, bogs, wet meadows, by rivers and
lakes, damp woods, mostly on acid soils; abundant through BI.

**26 x 27. J. effusus x J. conglomeratus = J. x kern-
reichgeltii** Jansen & Wachter ex Reichg. occurs sporadically
with the parents in N & W Br, and there are many other
records at least some of which are erroneous; it is
intermediate in diagnostic characters, but due to its high
fertility it is difficult to determine other than in the
field with its parents.

27. J. conglomeratus L. (<u>J. subuliflorus</u> Drejer) - <u>Compact</u> **931**
<u>Rush</u>. Densely tufted; stems erect, to 1m, distinctly ridged
and dull when fresh, with continuous pith, with subepidermal
sclerenchyma girders; inflorescence usually very compact,
sometimes of several stalked heads (var. <u>subuliflorus</u>
(Drejer) Asch. & Graebner); tepals pale brown. Native;
similar places to <u>J. effusus</u>; common throughout BI.

28. J. pallidus R. Br. - <u>Great</u> <u>Soft-rush</u>. Densely tufted, **931**
like a very large <u>J. effusus</u>, with stems to 2m; see Key B
(couplet 6) for differences. Intrd; wool-alien formerly natd
in Middlesex and Beds, now infrequently recurring in En;
Australia and New Zealand.

J. **pallidus** x J. **effusus** and J. **pallidus** x J. **inflexus**
occurred in Middlesex and Beds in the 1950s by hybridization
between the native spp. and the wool-alien J. pallidus.

2. LUZULA DC. - Wood-rushes
Perennials, vegetatively grass-like; leaves bifacial,
variously pubescent but rarely without hairs near base of
leaf on margins; ovary 1-celled, with 3 ovules; capsules with
3 seeds.

1 Flowers all or most borne singly in inflorescence
    each on distinct pedicels >3mm, rarely some in pairs   2
1 Flowers mostly borne in groups of 2 or more, each one
    in a group sessile or with pedicels <2mm, often a
    few solitary   3
    2 Basal leaves rarely >4mm wide; inflorescence
      branches erect to widely erecto-patent in fruit;
      seeds with terminal appendage ≤1/2 as long as rest
      of seed, + straight   **1. L. forsteri**
    2 Some basal leaves usually >4mm wide; lower
      inflorescence branches reflexed in fruit; seeds
      with terminal appendage >1/2 as long as (often
      longer than) rest of seed, often curved or
      hooked   **2. L. pilosa**
3 Tepals white to pale straw-coloured   **4. L. luzuloides**
3 Tepals yellowish- to dark-brown   4
    4 All or most basal leaves >8mm wide   **3. L. sylvatica**
    4 All leaves <8mm wide   5
5 Inflorescence drooping, spike-like, with the flower
    groups subsessile along main axis, or the lower them-
    selves forming lateral spikes   **9. L. spicata**
5 Inflorescence without single main axis; either all
    flower clusters congested in dense head or some or
    all with distinct stalks arising from short main
    axis near base of inflorescence   6
    6 Leaves deeply channelled, glabrous or sparsely
      pubescent just near base; seeds with inconspicuous
      appendage ≤1/10 as long as rest of seed; Scottish
      mountains   **8. L. arcuata**
    6 Leaves + flat, conspicuously pubescent; seeds
      with conspicuous whitish terminal appendage
      c.1/4 to 1/2 as long as rest of seed; widespread   7
7 Rhizomatous or stoloniferous; anthers >1.5x as long
    as filaments   **5. L. campestris**
7 Rhizomes and stolons 0; anthers <1.5x as long as
    filaments   8
    8 Tepals 1.5-2.5mm, pale yellowish-brown; Hunts & Co
      Antrim   **7. L. pallescens**
    8 Tepals 2.5-3.5mm, reddish-brown to dark brown;
      widespread   **6. L. multiflora**

Other spp. - **L. nivea** (L.) DC. (Snow-white Wood-rush), from

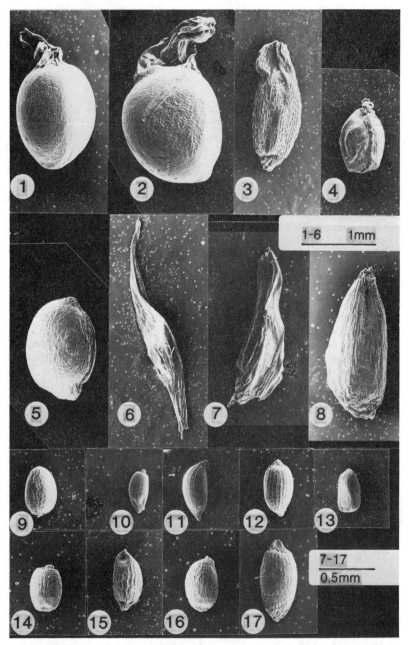

FIG 938 – 1-5, Seeds of **Luzula**. 1, **L. forsteri**. 2, **L. pilosa**. 3, **L. multiflora**. 4, **L. arcuata**. 5, **L. sylvatica**. 6–17, Seeds of **Juncus**. 6, **J. castaneus**. 7, **J. acutus**. 8, **J. trifidus**. 9, **J. planifolius**. 10, **J. capitatus**. 11, **J. effusus**. 12, **J. acutiflorus**. 13, **J. tenuis**. 14, **J. ambiguus**. 15, **J. foliosus**. 16, **J. bufonius**. 17, **J. subulatus**.

Europe, differs from L. luzuloides in its pure white tepals
4.5-5.5mm, c.2x as long as capsule (not 2.5-3.5mm, c. as long
as capsule); it has been recorded in several places, but
either in error or as non-persistent escapes.

1. L. forsteri (Smith) DC. - Southern Wood-rush. Tufted, **938**
with very short rhizomes; stems ± erect, to 35cm;
inflorescence usually slightly 1-sided, with erect to widely
erecto-patent branches bearing flowers singly (rarely in
pairs). Native; woods and hedgerows; locally common in CI
and Br N to Beds and Herefs.
1 x 2. L. forsteri x L. pilosa = L. x borreri Bromf. ex Bab.
occurs frequently within the range of L. forsteri in Br, and
formerly outside it in Co Wicklow; it is intermediate in
inflorescence shape and leaf width, with very low fertility.
2. L. pilosa (L.) Willd. - Hairy Wood-rush. Differs from L. **938**
forsteri in leaves obtuse to truncate at apex (with minute
point in L. forsteri); and see key (couplet 2). Native;
woods, hedgerows, among heather on moors; common throughout
most of Br, scattered in Ir.
3. L. sylvatica (Hudson) Gaudin - Great Wood-rush. Densely **938**
tufted and with long rhizomes; stems erect, to 80cm; inflorescence with erect to reflexed subumbellate branches bearing
flowers in groups of (2)3-5. Native; woods, moorland, shady
stream-sides; locally common through BI except parts of E En.
4. L. luzuloides (Lam.) Dandy & Wilm. - White Wood-rush.
Tufted, with short rhizomes; stems ± erect, to 70cm;
inflorescence of many erect to erecto-patent corymbose
branches bearing flowers in groups of (1)2-5(8). Intrd;
grown for ornament, natd in woods and by shady streams;
scattered throughout most of Br; Europe.
5. L. campestris (L.) DC. - Field Wood-rush. Tufted, with
short rhizomes; stems erect, to 15(25)cm; inflorescence of 1
sessile and 2-several stalked corymbose clusters of 3-12
flowers. Native; short grassland and similar places; very
common throughout BI.
6. L. multiflora (Ehrh.) Lej. - Heath Wood-rush. Tufted, **938**
usually with 0 rhizomes; stems erect, to 60cm. Native;
grassland, heaths, moors, woods on acid soil; common throughout BI except parts of C & E En.
a. Ssp. multiflora. Inflorescence of several stalked
corymbose clusters of 8-18 flowers; tepals c. as long as or
slightly shorter than capsule; 2n=36.
b. Ssp. congesta (Thuill.) Arcang. Inflorescence with all
flower clusters subsessile in compact lobed head; tepals
longer than capsule; 2n=48.
Both sspp. are widespread; partially fertile intermediates
(2n=42) occur, especially in Sc.
7. L. pallidula Kirschner (L. pallescens auct. non Sw.) - **RR**
Fen Wood-rush. Stems to 30cm; differs from L. multiflora
ssp. multiflora in paler flowers in small clusters; and
shorter tepals (see key, couplet 8). Native; open grassy

places in dry parts of fens; Hunts and Co Antrim.

8. L. arcuata Sw. - Curved Wood-rush. Tufted, with short **938** rhizomes; stems erect, to 10cm; inflorescence of several **R** variably-stalked clusters of 2-5 flowers, the longer stalks curved downwards. Native; open stony ground on high mountains; very local in C & N mainland Sc.

9. L. spicata (L.) DC. - Spiked Wood-rush. Tufted, with short rhizomes; stems erect but pendent at apex, to 30cm; inflorescence with long main axis and many subsessile many-flowered clusters. Native; open stony ground on mountains; local in C & N Sc, formerly Lake District.

154. CYPERACEAE - Sedge family

Herbaceous, usually rhizomatous, perennials, rarely annuals, with usually solid, often 3-angled stems, mostly aquatic or in wet places; leaves alternate, grass-like to ± rush-like (flattened to subcylindrical), with sheathing base often with membranous ligule at top of sheath on adaxial side, the blade simple, linear and entire or 0, exstipulate. Flowers much reduced, 1 each in axil of bract-like glume, 1-many in discrete units (spikelets), the spikelets terminal and solitary or in terminal spikes, racemes or panicles, often with extra sterile glumes, bisexual or unisexual (monoecious or sometimes dioecious), hypogynous, ± actinomorphic; perianth 0 or represented by bristles, sometimes elongating in fruit; stamens (1)2-3; ovary 1-celled with 1 ovule; style 0 or short; stigmas 2 or 3, elongated; fruit a nut.

Easily told from other grass- or rush-like plants except Poaceae by the (often unisexual) very reduced flowers with perianth 0 or in the form of bristles, and the 1-celled, 1-ovuled ovary, and from Poaceae by the absence of a bract (palea) above each flower and usually solid, often trigonous or triquetrous stems. Some Poaceae lack a palea but all these have hollow stems; the only member of the Cyperaceae with this combination is Cladium.

1 Stems hollow; leaves usually with fierce saw-edged
  margins and lowerside of midrib, easily cutting the
  skin   **14. CLADIUM**
1 Stems solid (centre often occupied by very soft
  pith); leaves not saw-edged or very mildly so   2
    2 Perianth represented by bristles which elongate
      and greatly exceed glumes at fruiting, forming a
      whitish cottony head   3
    2 Perianth 0 or represented by inconspicuous
      bristles shorter than glumes   4
3 Perianth-bristles 4-6 per flower; spikelet 1,
  terminal, <1cm excl. bristles (extinct) **2. TRICHOPHORUM**
3 Perianth-bristles numerous per flower; spikelets 1-
  several, >1cm excl. bristles   **1. ERIOPHORUM**

4 Flowers all unisexual, the male and female in
   different spikes or different parts of the same
   spike, or rarely on different plants; ovary and
   fruit enclosed or closely enfolded in membranous
   innermost glume   5
4 Flowers all bisexual; ovary and fruit not enclosed
   or closely enfolded in innermost glume   6
5 Ovary and fruit entirely enclosed in fused membranous
   glume usually ending in a short or long beak; male and
   female flowers in same or different spikes or on
   different plants, the spikes variously crowded or
   distant, and stalked or sessile; stigmas and stamens
   2 or 3   **16. CAREX**
5 Ovary and fruit closely enfolded in innermost glume
   which is not fused, leaving fruit exposed at top; male
   and female flowers in same spikes, the spikes crowded
   and ± sessile; stigmas and stamens 3   **15. KOBRESIA**
6 Inflorescence of 1 terminal spikelet; lowest
   bract not leaf-like or stem-like, shorter than
   spikelet   7
6 Inflorescence of ≥2 spikelets, or sometimes of
   1 but then lowest bract leaf-like or stem-like and
   exceeding spikelet   9
7 Most or all leaf-sheaths on stems with leafy blades
                                                              9. ELEOGITON
7 Most or all leaf-sheaths on stems without blades   8
8 Uppermost leaf-sheath on stem with short blade
                                                        2. TRICHOPHORUM
8 Uppermost leaf-sheath on stem (and all or most
   below it) without a blade   **3. ELEOCHARIS**
9 Inflorescence with ≥2 bifacial leaf-like bracts
   very close together at base   10
9 Inflorescence with basal bracts stem-like or leaf-
   like, if leaf-like then either 1 or ≥2 and well
   spaced out   12
10 Spikelets flattened, with glumes on 2 opposite
   sides of axis   **11. CYPERUS**
10 Spikelets ± terete, with glumes spirally arranged 11
11 Inflorescence dense; spikelets >8mm   **4. BOLBOSCHOENUS**
11 Inflorescence diffuse; spikelets <5mm   **5. SCIRPUS**
12 Inflorescence a flattened compact terminal head,
   with spikelets only on 2 opposite sides of main
   axis   **10. BLYSMUS**
12 Inflorescence various, if a compact terminal head
   then spikelets not only on 2 opposite sides of
   axis   13
13 Spikelets flattened, with glumes on 2 opposite sides
   of axis   **12. SCHOENUS**
13 Spikelets terete, with glumes spirally arranged   14
14 Inflorescence obviously terminal with leaf-like
   main bract; stems with several well-developed
   leaf-blades   **13. RHYNCHOSPORA**

14 Inflorescence usually apparently lateral, with
    main bract ± stem-like and continuing stem
    apically; stems with 0-1(2) reduced leaf-blades   15
15 Stems very slender, <1mm wide, rarely >20cm **8. ISOLEPIS**
15 Stems stouter, >1.5mm wide, rarely <30cm   16
    16 Inflorescence composed of (1-)several sessile to
        stalked globose apparent spikelets   **6. SCIRPOIDES**
    16 Inflorescence composed of (1-)several sessile to
        stalked ovoid spikelets   **7. SCHOENOPLECTUS**

1. ERIOPHORUM L. - Cottongrasses
Perennials with long or short rhizomes; stems terete to ±
triquetrous, leafy; leaves variously shaped in section;
inflorescence of 1-several large spikelets in a terminal
umbel; lowest bract leaf-like or glume-like; flowers
bisexual; perianth of numerous (>6) bristles elongating to
form conspicuous white, cottony head in fruit; stamens 3;
ovary not enfolded or enclosed by glume; stigmas 3.

1 Spikelet 1, erect, without leaf-like bract at base;
    leaf-blades ± triangular in section, 0 or very
    reduced on uppermost stem leaf-sheath   **4. E. vaginatum**
1 Spikelets >(1)2, ± pendent in fruit, with 1-3 ±
    leaf-like bracts at base; leaf-blades flat to
    V-shaped in section, well developed on uppermost stem
    leaf-sheath   2
    2 Stalks of spikelets smooth; stems ± terete to very
        bluntly 3-angled; anthers >2mm   **1. E. angustifolium**
    2 Stalks of spikelets with numerous minute forward-
        pointed bristles; stems distinctly 3-angled;
        anthers <2mm   3
3 Leaf-blades 0.5-2mm wide; glumes with midrib plus
    several shorter parallel veins on either side; plant
    with long rhizomes, with solitary stems   **3. E. gracile**
3 Leaf-blades 3-8mm wide; glumes with only midrib;
    plant loosely tufted   **2. E. latifolium**

1. E. angustifolium Honck. - Common Cottongrass. Rhizomes
long; stems scattered, erect, to 60cm; leaves 2-6mm wide,
V-shaped in section; inflorescence of (1)3-7 pendent
spikelets; perianth-bristles 2.5-5cm. Native; wet usually
acid bogs; common in suitable places throughout BI.
    2. E. latifolium Hoppe - Broad-leaved Cottongrass. Differs
from E. angustifolium in stems loosely tufted; leaves 3-8mm
wide, ± flat; perianth-bristles 1.5-3cm, minutely toothed
(not entire) at apex (microscope); and see key. Native; wet
base-rich marshes and flushes; scattered throughout Br and Ir
in suitable places, much less common than E. angustifolium
and extinct in much of C En.
    3. E. gracile Koch ex Roth - Slender Cottongrass. Differs **RRR**
from E. angustifolium in leaves 0.5-2mm wide; perianth-
bristles 1-2.5cm; and see key. Native; similar places to E.

<u>angustifolium</u>; very local in S Br from Surrey to Caerns, C &
W Ir, formerly NW Yorks, Northants and E Norfolk.
4. E. vaginatum L. - <u>Hare's-tail</u> <u>Cottongrass</u>. Rhizomes very
short; stems densely tufted, often tussock-forming, erect, to
50cm; leaves 0.5-1mm wide, triangular in section;
inflorescence of 1 erect spikelet; perianth-bristles 2-3cm.
Native; wet peaty places, especially on moorland bogs; common
in Ir and W, C & N Br, very local in C, E & S En.

2. TRICHOPHORUM Pers. (<u>Scirpus</u> sect. Baeothryon auct. non
(Ehrh. ex A. Dietr.) Benth. & Hook. f.) - <u>Deergrasses</u>
Tufted perennials; stems terete to trigonous, with only the
uppermost leaf-sheath with a blade; leaves thick-crescent-
shaped in section, very narrow; inflorescence of 1 terminal
spikelet; lowest bract glume-like; flowers bisexual; perianth
of 4-6 bristles, elongating or not in fruit; stamens 3; ovary
not enfolded or enclosed by glume; stigmas 3.

1. T. alpinum (L.) Pers., (<u>Scirpus</u> hudsonianus (Michaux) **945**
Fern.) - <u>Cotton</u> <u>Deergrass</u>. Rather diffusely tufted; stems **RR**
erect, to 20(30)cm, very slender, trigonous, slightly scabrid
near apex; perianth-bristles elongating to 10-25mm, forming
white cottony head, at fruiting. Native; bog in Angus from
1791 to c.1813.
2. T. cespitosum (L.) Hartman (<u>Scirpus</u> cespitosus L.) - **945**
<u>Deergrass</u>. Densely tufted; stems erect, to 35cm, very
slender, ± terete, smooth; perianth-bristles remaining
shorter than glumes, pale brown. Native; bogs, wet moors and
heaths; common throughout Br and Ir in suitables places, but
absent from most of C & E En. Our plant is ssp. **germanicum**
(Palla) Hegi; records of ssp. **cespitosum** have not been
substantiated.

3. ELEOCHARIS R. Br. - <u>Spike-rushes</u>
Perennials with long or short stout and/or slender rhizomes;
stems terete to ridged, with blade-less leaf-sheaths; leaf-
blades 0 (some spikelet-less stems may resemble basal
leaves); inflorescence of 1 terminal spikelet; lowest bract
glume-like; flowers bisexual; perianth of 0-6 bristles, not
elongating in fruit; stamens 3; ovary not enfolded or
enclosed by glume; stigmas 2 or 3.

1 Lowest glume >(2/5)1/2 as long as spikelet; spikelets
  3-12-flowered 2
1 Lowest glume <2/5(1/2) as long as spikelet; spikelets
  10-many-flowered 4
  2 Glumes greenish; uppermost stem leaf-sheath
    delicate, inconspicuous; very slender whitish
    rhizomes ending in small whitish tubers (c.2-5mm)
    present **7. E. parvula**
  2 Glumes brown (often with green midrib); uppermost
    stem leaf-sheath conspicuous, brownish; rhizomes

brownish, not bearing tubers   3
3   Stems <0.5mm wide, usually 4-ridged; spikelets 2-5mm;
    lowest glume 1.5-2.5mm   **6. E. acicularis**
3   Stems >0.5mm wide, ± terete; spikelets 4-10mm;
    lowest glume 2.5-5mm   **5. E. quinqueflora**
4   Lowest glume ± completely encircling spikelet
    at base   5
4   Lowest glume <1/2(3/4) encircling spikelet at base   6
5   Uppermost stem leaf-sheath oblique (c.45 degrees) at
    apex; stigmas 3; nuts 3-angled   **4. E. multicaulis**
5   Uppermost stem leaf-sheath ± truncate at apex;
    stigmas 2; nuts biconvex   **3. E. uniglumis**
6   Stems with 10-16 vascular bundles, showing up as
    fine ridges in dried state; bristles (4)5(-6);
    spikelets conical; swollen style-base on top of
    fruit c.1-1.5x as long as wide, slightly
    constricted at base   **2. E. austriaca**
6   Stems with >20 vascular bundles, showing up as
    fine ridges in dried state; bristles (0-)4;
    spikelets cylindrical- to ellipsoid-conical;
    swollen style-base on top of fruit wider than
    long, strongly constricted at base   **1. E. palustris**

1. E. palustris (L.) Roemer & Schultes - Common Spike-rush. 945
Stems loosely to rather densely tufted, to 75cm, often much
less; uppermost stem leaf-sheath ± truncate at apex;
spikelets 5-30mm, the lowest glume not fully encircling base;
stigmas 2.   Native; in or by ponds, marshes, ditches,
riversides.
  a. Ssp. vulgaris Walters.   Spikelets usually with <40
flowers; glumes from middle of spikelet 3.5-4.5mm; nut
(1.3)1.5-2mm; 2n=38.   Frequent throughout BI.
  b. Ssp. palustris (ssp. microcarpa Walters).   Spikelets
usually with >40 flowers; glumes from middle of spikelet
2.7-3.5mm; nut 1.2-1.4(1.5)mm; 2n=16.   S & C En from E Kent
to S Hants, Worcs and Notts, but even there much rarer than
ssp. vulgaris.
Partially fertile hybrids between the 2 sspp. (2n=27) have
been found in Oxon.
1 x 3. E. palustris x E. uniglumis occurs within populations
of E. uniglumis near to those of E. palustris on the W coasts

FIG 945 - Cyperaceae. 1-12, inflorescences. 1, **Eleocharis
palustris.** 2, **E. austriaca.** 3, **E. uniglumis.** 4, **E.
multicaulis.** 5, **E. quinqueflora.** 6, **E. acicularis.** 7, **E.
parvula.** 8, **Trichophorum cespitosum.** 9, **T. alpinum.** 10,
Eleogiton fluitans. 11, **Isolepis setacea.** 12, **I. cernua.**
13-16, glumes of **Schoenoplectus.** 13, **S. lacustris.** 14, **S.
tabernaemontani.** 15, **S. triqueter.** 16, **S. pungens.** 17-22,
inflorescences. 17, **Rhynchospora alba.** 18, **R. fusca.** 19,
Blysmus compressus. 20, **B. rufus.** 21, **Schoenus nigricans.** 22,
S. ferrugineus.

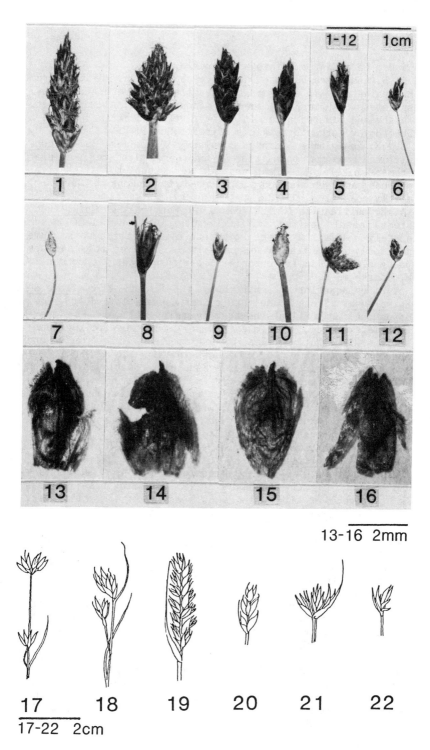

1-12   1cm

13-16   2mm

17-22   2cm

FIG 945 – see caption opposite

of Br; it is intermediate and fertile.

2. E. austriaca Hayek – <u>Northern</u> <u>Spike-rush</u>. Differs from **945**
E. palustris in stems rather fragile (not very flexible); **RR**
spikelets 8-20mm; and see key (couplet 6). Native; wet,
marshy and flushed areas in or by rivers; local in N En and S
Sc from MW Yorks to Selkirks, first found 1947.

3. E. uniglumis (Link) Schultes – <u>Slender</u> <u>Spike-rush</u>. **945**
Differs from E. palustris in shorter, slenderer, shinier
stems; spikelets 5-12mm; and see key (couplet 4). Native;
marshes and dune-slacks; scattered throughout Br and Ir,
mostly coastal.

4. E. multicaulis (Smith) Desv. – <u>Many-stalked</u> <u>Spike-rush</u>. **945**
Stems densely tufted, to 40cm; uppermost stem leaf-sheath
acutely oblique at apex; spikelets 5-15mm, the lowest glume
not fully encircling base; stigmas 3. Native; bogs and wet
peaty places, usually on acid soils; throughout BI, common in
W, sparse in E.

5. E. quinqueflora (F. Hartmann) O. Schwarz – <u>Few-flowered</u> **945**
<u>Spike-rush</u>. Stems loosely to rather densely tufted, to 30cm;
uppermost stem leaf-sheath obtusely oblique at apex;
spikelets 4-10mm, the lowest glume ± fully encircling base;
stigmas 3. Native; wet places in fens, dune-slacks and
moorland; throughout BI, commonest in NW.

6. E. acicularis (L.) Roemer & Schultes – <u>Needle</u> <u>Spike-rush</u>. **945**
Stems sparsely tufted, to 10cm (or more if submerged and **R**
sterile); uppermost stem leaf-sheath acutely to obtusely
oblique at apex; spikelets 2-5mm, the lowest glume ± fully
encircling base; stigmas 3. Native; in and by pond and lake
margins; scattered throughout Br and Ir.

7. E. parvula (Roemer & Schultes) Link ex Bluff, Nees & **945**
Schauer – <u>Dwarf</u> <u>Spike-rush</u>. Stems sparsely tufted, to 8cm; **RR**
uppermost stem leaf-sheath obtusely oblique at apex; spike-
lets 2-4mm, the lowest glume ± fully encircling base; stigmas
3. Native; wet muddy places by sea and in estuaries; very
local in SW Br from S Hants to Caerns, formerly (?still) Ir.

4. BOLBOSCHOENUS (Asch.) Palla (<u>Scirpus</u> sect. <u>Bolboschoenus</u>
(Asch.) Beetle) – <u>Sea</u> <u>Club-rush</u>
Strongly rhizomatous perennials; stems triquetrous, leafy;
leaves flattened, widely V-shaped in section; inflorescence
of (1)3-many spikelets either sessile or variously clustered
on 1-several stalks; lowest bract leaf-like; flowers
bisexual; perianth-bristles 1-6, not elongating in fruit;
stamens 3; ovary not enfolded or enclosed by glume; stigmas
2-3.

1. B. maritimus (L.) Palla (<u>Scirpus</u> <u>maritimus</u> L.) – <u>Sea</u>
<u>Club-rush</u>. Stems strong, erect, to 1m; leaves long, 2-10mm
wide; spikelets 10-30mm, dark brown. Native; wet muddy
places in estuaries or by sea; common round coasts of BI
except extreme N Sc, rarely inland.

5. SCIRPUS L. - <u>Wood Club-rush</u>
Strongly rhizomatous perennials; stems trigonous, leafy; leaves flat; inflorescence of very numerous spikelets 1-several on ends of diffusely branching panicle; lowest bract leaf-like; flowers bisexual; perianth-bristles 6, not elongating in fruit; stamens 3; ovary not enfolded or enclosed by glume; stigmas 3.

1. S. sylvaticus L. - <u>Wood Club-rush</u>. Stems strong, erect, to 1.2m; leaves long, 5-20mm wide; spikelets 3-4mm, greenish-brown. Native; by streams and in marshes and damp spots in woods or shady places; locally frequent over Br and Ir N to C Sc.

6. SCIRPOIDES Seguier (<u>Holoschoenus</u> Link, <u>Scirpus</u> sect. <u>Holoschoenus</u> (Link) Koch) - <u>Round-headed Club-rush</u>
Strongly rhizomatous perennials; stems terete; leaf-sheaths mostly blade-less but uppermost 1(-2) with well-developed blade; leaves semi-circular in section; inflorescence of (1)5-many variously stalked or sessile globular heads (apparent spikelets) each actually of numerous tightly packed spikelets; lowest bract + stem-like, making inflorescence appear lateral; flowers bisexual; perianth-bristles 0; stamens 3; ovary not enfolded or enclosed by glume; stigmas 3.

1. S. holoschoenus (L.) Sojak (<u>Scirpus</u> holoschoenus L., **RR** <u>Holoschoenus</u> vulgaris Link) - <u>Round-headed Club-rush</u>. Stems erect, to 1.5m; leaves terete but + flat on upperside, rather stem-like, 0.5-2mm wide; heads 3-10mm across, with spikelets 2.5-4mm, brown. Native; damp sandy places near sea; very rare in N Devon and N Somerset, occasional elsewhere in S Br as introduction.

7. SCHOENOPLECTUS (Reichb.) Palla (<u>Scirpus</u> sects. <u>Schoeno-plectus</u> (Reichb.) Benth. & Hook. f. & <u>Pterolepis</u> (Schrader) Asch. & Graebner) - <u>Club-rushes</u>
Strongly rhizomatous perennials; stems terete to triquetrous; leaf-sheaths mostly blade-less but uppermost 1(-3) with rather short blade; leaves crescent-shaped in section; inflorescence of (1-)few to numerous variously stalked or sessile ovoid spikelets 5-8mm; lowest bract + stem-like, making inflorescence appear lateral; flowers bisexual; perianth-bristles 0-6, not elongating in fruit; stamens 3; ovary not enfolded or enclosed by glume; stigmas 2-3.

1 Stems terete   2
1 Stems triquetrous   3
  2 Glumes (except apical projection) smooth; stigmas mostly 3; nut 2.5-3mm, mostly trigonous
                                         1. S. lacustris
  2 Glumes minutely (x20 lens) but densely papillose at

least near midrib and apex; stigmas 2; nut 2-2.5mm,
biconvex or planoconvex   **2. S. tabernaemontani**
3 Glumes with rounded to obtuse lobe on either side of
apical projection; stems with uppermost 1(-2) leaf-
sheaths with blades; perianth-bristles 6, >1/2 as
long as nut   **3. S. triqueter**
3 Glumes with acute to subacute lobe on either side of
apical projection; stems with uppermost 2-3 leaf-
sheaths with well-developed blades; perianth-bristles
0-6, <1/2 as long as nut   **4. S. pungens**

1. S. lacustris (L.) Palla (Scirpus lacustris L.) - Common 945
Club-rush. Stems erect, to 3m, green, sometimes slightly
glaucous, usually >1cm wide at mid-point; spikelets usually
numerous, variously clustered. Native; in shallow water of
lakes, ponds, slow rivers, canals and dykes; frequent over
most of Br and Ir.
1 x 3. S. lacustris x S. triqueter = S. x carinatus (Smith)
Palla (Scirpus x carinatus Smith) occurred with the parents
in R Tamar (E Cornwall and S Devon) and R Thames (W Kent,
Surrey and Middlesex), but is now very rare or extinct; it
has stems triquetrous above, terete below, and smooth glumes
with rather rounded apical lobes, and is + sterile.
2. S. tabernaemontani (C. Gmelin) Palla (Scirpus tabernae- 945
montani C. Gmelin, S. lacustris ssp. glaucus Hartman, ssp.
tabernaemontani (C. Gmelin) Syme) - Grey Club-rush. Differs
from S. lacustris in stems to 1.5m, glaucous, usually 3-8mm
wide at mid-point; and see key. Native; in similar places to
S. lacustris but also in marshes, dune-slacks and wet peaty
places, mostly near sea; frequent throughout most of BI,
except very scattered inland.
**2 x 3. S. tabernaemontani x S. triqueter = S. x kueken-
thalianus** (Junge) Kent (Scirpus x kuekenthalianus Junge, S. x
scheuchzeri Bruegger) occurs with the parents in the R Tamar
(E Cornwall and S Devon), R Arun (W Sussex) and R Medway (E &
W Kent); it differs from S. x carinatus in its papillose
glumes.
3. S. triqueter (L.) Palla (Scirpus triqueter L.) - 945
Triangular Club-rush. Stems erect, to 1(1.5)m, usually RRR
green, usually 3-8mm wide at mid-point; spikelets few to
numerous, variously clustered. Native; in tidal mud of
rivers; very local in S En and W Ir.
4. S. pungens (Vahl) Palla (S. americanus auct. non (Pers.) 945
Volkart, Scirpus pungens Vahl, S. americanus auct. non Pers.) RR
- Sharp Club-rush. Stems erect, to 60cm, green, usually
2-5mm wide at mid-point; spikelets few (rarely >6), in tight
cluster. Native; pond-margin in Jersey (not seen since early
1970s) and wet dune-slacks in S Lancs (discovered 1928, lost
c.1980, now re-introduced from same stock).

8. ISOLEPIS R. Br. (<u>Scirpus</u> sect. <u>Isolepis</u> (R. Br.) Griseb.)
- <u>Club-rushes</u>
Densely tufted annuals (to perennials); stems terete; leaf-
sheaths confined to near base of stem, the upper 1(-2) with
short blades; leaves crescent-shaped in section; inflorescence of 1-4 sessile spikelets 2-5mm; lowest bract usually
± stem-like, making inflorescence appear lateral, sometimes
very short and ± glume-like; flowers bisexual; perianth-
bristles 0; stamens 1-2; ovary not enfolded or enclosed by
glume; stigmas (2-)3.

1. I. setacea (L.) R. Br. (<u>Scirpus</u> <u>setaceus</u> L.) - <u>Bristle</u> **945**
<u>Club-rush</u>. Stems erect to ascending, <0.5mm wide, to
15(30)cm but usually <10cm; main bract usually distinctly
longer than inflorescence, ± stem-like; spikelets 1-4; glumes
reddish-brown with green midrib; nut shiny, with longitudinal
ridges.   Native; on wet open or semi-closed ground in
ditches, fens, marshes and dune-slacks, on heaths, and by
ponds and lakes; frequent throughout BI.
2. I. cernua (Vahl) Roemer & Schultes (<u>Scirpus</u> <u>cernuus</u> Vahl) **945**
- <u>Slender</u> <u>Club-rush</u>. Differs from <u>I.</u> <u>setacea</u> in main bract
at most only slightly longer than inflorescence, often ±
glume-like; spikelets 1(-3); glumes with brown area usually
only a blotch either side of midrib; nut matt, smooth.
Native; similar places to <u>I.</u> <u>setacea</u> but mostly near sea;
frequent in Ir and extreme W Br E to S Hants, very local in E
Norfolk.

9. ELEOGITON Link (<u>Scirpus</u> sect. <u>Eleogiton</u> (Link) Pax -
<u>Floating</u> <u>Club-rush</u>
Stoloniferous perennials, usually in water; stems terete,
leafy; leaves ± flat; inflorescence of 1 terminal spikelet
2-5mm; lowest bract ± glume-like; flowers bisexual; perianth-
bristles 0; stamens 3; ovary not enfolded or enclosed by
glume; stigmas 2-3.

1. E. fluitans (L.) Link (<u>Scirpus</u> <u>fluitans</u> L.) - <u>Floating</u> **945**
<u>Club-rush</u>. Stems usually floating, rooting at nodes, to
50cm, sometimes on mud or gravel and much shorter, very
leafy; spikelets green to pale brown. Native; in or by peaty
ponds, lakes and ditches; fairly frequent throughout BI,
commoner in W.

10. BLYSMUS Panzer ex Schultes - <u>Flat-sedges</u>
Rhizomatous perennials; stems subterete, leafy; leaves flat
to strongly inrolled; inflorescence a flattened ± compact
terminal head with spikelets on 2 opposite sides of axis, the
spikelets 4-10mm; lowest bract leaf-like to ± glume-like;
flowers bisexual; perianth-bristles 0-6; stamens 3; ovary not
enfolded or enclosed by glume; stigmas 2.

1. B. compressus (L.) Panzer ex Link - <u>Flat-sedge</u>. Stems **945**

erect, to 40cm; leaves flat to slightly keeled, rough, grass-
like; spikelets usually 10-20, reddish-brown; perianth-
bristles 3-6, longer than nut; nut 1.5-2mm. Native; marshy,
rather open ground; locally frequent in En but extinct in
many places and in Wa, very local in Sc.
   2. B. rufus (Hudson) Link - Saltmarsh Flat-sedge. Differs **945**
from **B. compressus** in leaves strongly inrolled, smooth, rush-
like; spikelets usually 3-8, dark brown; perianth-bristles
3-6, shorter than nut, or 0; nut 3-4mm. Native; salt-marshes
and dune-slacks in turf; locally frequent on coasts of Br and
Ir S to N Lincs and Pembs, common in W Sc.

11. CYPERUS L. - Galingales
Rhizomatous perennials or tufted annuals or perennials; stems
triquetrous to trigonous, leafy at base; leaves flat to
keeled, grass-like; inflorescence a simple or more often
compound umbel or umbel-like raceme, with grass-like many-
flowered spikelets usually clustered on ultimate branches or
all clustered in ± dense head; lowest 2-10 bracts leaf-like,
often much exceeding inflorescence; flowers bisexual;
perianth-bristles 0; stamens 1-3; ovary not enfolded or
enclosed by glume; stigmas (2-)3.

1 Tufted annual; leaves <3(5)mm wide; glumes <1.5mm
                                                     3. C. fuscus
1 Tufted or rhizomatous perennial; widest leaves
   >(2)4mm wide; glumes >1.5mm   2
   2 Spikelets reddish-brown, <2mm wide; inflorescence
      ± diffuse; rhizomes long; stamens 3   **1. C. longus**
   2 Spikelets greenish- to yellowish-brown, >2mm wide;
      inflorescence ± compact; stamen 1   **2. C. eragrostis**

Other spp. - Several spp. occur as wool-aliens, but all are
much rarer than C. eragrostis.

   1. C. longus L. - Galingale. Shortly rhizomatous perennial;   **R**
stems erect, to 1m; inflorescence diffuse, with very numerous
linear-oblong reddish-brown spikelets 4-25mm. Native;
marshes, pond-sides and ditches; very local near coast in CI
and SW Br E to E Kent and E Suffolk and N to Caerns, intrd in
scattered places elsewhere in S & C En.
   2. C. eragrostis Lam. (C. vegetus Willd.) - Pale Galingale.
Very shortly rhizomatous perennial; stems erect, to 60cm;
inflorescence rather compact, with numerous elliptic-oblong
greenish- to yellowish-brown spikelets 8-13mm. Intrd; grown
for ornament and escaping to roadsides, rough ground and by
water, also a frequent wool-alien and grass-seed alien;
scattered in CI and S Br, well natd in Guernsey; tropical
America.
   3. C. fuscus L. - Brown Galingale. Tufted annual; stems **RRR**
erect, to 20cm; inflorescence usually very compact, with
numerous narrowly oblong dark brown spikelets 3-6mm. Native;

damp barish ground by ponds and in ditches; very rare in S
Hants, N Somerset and Middlesex, formerly elsewhere in S En
and CI, refound Jersey 1989.

12. SCHOENUS L. - Bog-rushes
Densely tufted perennials; stems terete, with leaf-sheaths
only at or near base and bearing short or long blades; leaves
very thickly crescent-shaped in section to subterete;
inflorescence a compact head of 1-4-flowered flattened
spikelets; lowest bract leaf-like to + glume-like; flowers
bisexual; perianth-bristles 0 or up to 6; stamens 3; ovary
not enfolded or enclosed by glume; stigmas 3.

1. S. nigricans L. - Black Bog-rush. Stems erect, to 75cm; **945**
leaves shorter than to c. as long as stems; inflorescence of
(2)5-10 spikelets, with lowest bract usually conspicuously
exceeding it; glumes minutely rough (x20 lens) on keel.
Native; damp peaty places, serpentine heathland, bogs, salt-
marshes, fens, flushes; locally frequent in BI, especially
near W coasts and in E Anglia, absent from most of En, Wa and
E Sc.
2. S. ferrugineus L. - Brown Bog-rush. Stems erect, to **945**
40cm; leaves shorter than (often <1/2 as long as) stems;   **RR**
inflorescence of 1-3 spikelets, with lowest bract shorter
than to c. as long as it; glumes smooth on keel. Native;
semi-open ground in base-rich flushes, formerly also by lake;
2 places (formerly 3) in M Perths, first found 1884.

13. RHYNCHOSPORA Vahl - Beak-sedges
Tufted to creeping rhizomatous perennials; stems terete to
trigonous, leafy; leaves channelled; inflorescence of 1-few
rather compact heads each of several 1-3-flowered spikelets;
lowest bract leaf-like to + glume-like; flowers bisexual;
perianth-bristles 5-13; stamens 2-3; ovary not enfolded or
enclosed by glume; stigmas 2, the common style-base
persistent and forming beak to fruit.

1. R. alba (L.) Vahl - White Beak-sedge. Stems + tufted, **945**
erect, to 40cm, often forming bulbil-like buds towards base;
inflorescence whitish at flowering; lowest bract of terminal
head not or sometimes slightly longer than head. Native; wet
acid peaty places, locally common in Br and Ir, but absent
from most of En, E & S Wa and E Sc.
2. R. fusca (L.) Aiton f. - Brown Beak-sedge. Stems + **945**
scattered, erect, to 30cm, without bulbils; inflorescence   **R**
brown at flowering; lowest bract of terminal head >1cm longer
than head. Native; similar places to R. alba and usually
with it; very local in W & C Ir and S & W Br.

14. CLADIUM P. Browne - Great Fen-sedge
Rhizomatous vigorous perennials; stems terete to trigonous,
leafy; leaves channelled, usually with fiercely serrate edges

and keel; inflorescence much-branched, with many rather
compact heads each of several 1-3-flowered spikelets: lowest
bract of each head leaf-like to glume-like, but inflorescence
branches with leaf-like long bracts at base; flowers
bisexual: perianth-bristles 0; stamens 2(-3); ovary not
enfolded or enclosed by glume; stigmas (2-)3.

1. C. mariscus (L.) Pohl - Great Fen-sedge. Stems erect, to
3m; leaves up to 2m x 2cm, grey-green: inflorescence up to 70
x 10cm. Native: wet, base-rich areas in fens and by streams
and ponds; locally common but very scattered in BI. A rare
variant in parts of Ir has apple-green less stiff leaves with
only mildly serrate edges.

15. KOBRESIA Willd. - False Sedge
Rather densely tufted perennials: stems trigonous, leafy at
extreme base: leaves channelled: inflorescence of 1-flowered
spikelets arranged in terminal cluster of 3-10 spikes: lowest
bract a sheath with short usually brown blade: flowers
unisexual, the upper spikelets male and lower female in each
spike; perianth-bristles 0; stamens 3; female flowers with an
extra inner glume folded (but not fused) around ovary;
stigmas 3.

1. K. simpliciuscula (Wahlenb.) Mackenzie - False Sedge. **RR**
Stems erect, to 20cm; leaves 0.5-1.5mm wide; inflorescence
1-2.5cm x 2-6mm. Native; flushed grassy or barish areas on
mountains; very local in Upper Teesdale (NW Yorks and Durham)
and M Perths, formerly in adjacent vice-counties.

16. CAREX L. - Sedges
Extensively rhizomatous to densely tufted perennials; stems
erect, usually leafy but often only at extreme base,
triquetrous, trigonous or terete; leaves flat to channelled
or inrolled: inflorescence of 1-flowered spikelets grouped in
variously arranged spikes, all except the terminal subtended
by a bract; lowest bract leaf-like to glume-like; flowers
unisexual, each in axil of 1 glume, the sexes variously
arranged from mixed in 1 spike to dioecious, but commonly the
upper spikes entirely male and the lower entirely female;
perianth-bristles 0; stamens 2 or 3; female flowers with an
extra inner glume (utricle) completely fused around ovary,
forming a false fruit enclosing nut and usually with long or
short distal beak; stigmas 2 or 3.
Ripe fruits are essential for keying down Carex spp.; the
length of the utricle includes any beak. Two important
diagnostic characters might present some difficulties.
Genuinely 1-spiked inflorescences should not be confused with
those with several congested spikes forming a single ± lobed
head. In the former case there is 1 simple axis bearing
flowers or fruits directly upon it; in the latter case the
axis has lateral (often very short) branches, usually with a

bract at the base of each. Depauperate stems of several spp.
may rarely possess only 1 spike, but more normal stems should
also be available. The number of stigmas (2 or 3) is
important. In material with ripe fruits the stigmas might
have disappeared, but the shape of the nut and often that of
the utricle then provides the clue (see General key, couplet
4). The beak of the utricle is often bifid, and these
projections must not be mistaken for the stigmas. The glumes
subtending the male and female flowers are called 'male
glumes' and 'female glumes' respectively. The length of the
sheath of the lowest bract refers to only the portion fused
round the stem.
Many hybrids have been recorded but none except C. hostiana
x C. viridula is common; all are variously intermediate and
highly sterile, with empty though often well-developed
utricles, except for partial fertility exhibited in hybrids
between spp. 64-70.

General key
1 Spike 1, terminal   Key A
1 Spikes >1 (sometimes very close together)   2
    2 Spikes all ± similar in appearance, often very
      close together and forming single lobed head, the
      terminal spike usually at least partly female   Key B
    2 Spikes dissimilar in appearance, the upper >1 all
      or mostly male, the lower >1 all or mostly female,
      usually clearly separate and sometimes remote   3
3 Utricles pubescent on part or whole of main body
  (excl. beak and edges) (x10 lens)   Key C
3 Utricles glabrous on main body (sometimes pubescent
  on beak or along edges, sometimes papillose on
  main body)   4
    4 Stigmas 2; utricles usually biconvex or plano-
      convex; nuts biconvex   Key D
    4 Stigmas 3; utricles usually trigonous to terete;
      nuts trigonous   5
5 At least lowest spike pendent, stalked   Key E
5 All spikes erect to patent, often sessile   Key F

Key A - Spike 1, terminal
1 Spike all male   2
1 Spike female at least at base   3
    2 Plant densely tufted; stems usually scabrid and
      ± trigonous above (extinct)   19. C. davalliana
    2 Plant rhizomatous; stems usually smooth and
      terete   18. C. dioica
3 Stigmas 2; utricles usually biconvex or plano-convex;
  nuts biconvex   4
3 Stigmas 3; utricles usually trigonous to terete; nuts
  trigonous   6
    4 Utricles 4-6mm, not ribbed, strongly reflexed
      at maturity; plants monoecious, with spikes male at

```
      apex, female at base                    74. C. pulicaris
  4  Utricles 2.5-4.5mm, distinctly ribbed, not or
     weakly reflexed at maturity; plants usually
     dioecious, sometimes monoecious                           5
  5  Utricles 2.5-3.5mm, abruptly contracted to scabrid
     beak; see also couplet 2                       18. C. dioica
  5  Utricles 3.5-4.5mm, gradually contracted to smooth
     beak; see also couplet 2 (extinct)        19. C. davalliana
     6  Utricles erecto-patent to erect when ripe, obovoid,
        usually <3.5mm; leaves curved or curly
                                                   73. C. rupestris
     6  Utricles patent to reflexed when ripe, narrowly
        ovoid to narrowly ellipsoid, 3.5-7.5mm; leaves
        + straight                                             7
  7  Utricles 3.5-5(6)mm, with fine bristle (as well as
     style base) protruding 1-2mm from beak; female
     glumes c.2mm                             71. C. microglochin
  7  Utricles 5-7.5mm, with only style-base protruding
     from beak; female glumes c.4mm             72. C. pauciflora


Key B - Spikes >1, all + similar in appearance
  1  Stigmas 3; utricles usually trigonous to terete;
     nut trigonous                                             2
  1  Stigmas 2; utricles usually biconvex or plano-convex;
     nuts biconvex                                             5
     2  Lowest spike erect on short rigid stalk               3
     2  Lowest spike pendent to patent on distinct
        flexible stalk                                         4
  3  All spikes clustered and greatly overlapping; utricles
     1.8-2.5mm, greenish-brown, minutely papillose, longer
     than acute female glumes                     63. C. norvegica
  3  Spikes not or scarcely overlapping, the lowest arising
     >1cm below next; utricles 3-4.5mm, pale green, smooth,
     shorter than acuminate female glumes        62. C. buxbaumii
     4  Lowest bract with sheath 0-3mm; terminal spike
        female at top, male below                   61. C. atrata
     4  Lowest bract with sheath >5mm; terminal spike
        male at top, female below                 57. C. atrofusca
  5  Rhizomes long; stems very loosely tufted or scattered   6
  5  Rhizomes short; stems densely tufted                    13
     6  Terminal spike (not necessarily that extending
        highest) female at least at apex                       7
     6  Terminal spike male at least at apex                   8
  7  Utricles 4-5.5(7)mm, reddish-brown, narrowly winged
                                                  11. C. disticha
  7  Utricles 2-3mm, yellowish-green, not winged
                                                     22. C. curta
     8  Utricles narrowly winged on body                       9
     8  Utricles not winged on body                           10
  9  Terminal spike all male; glumes (male and female)
     >5mm; leaf-sheaths hyaline on side opposite blade
                                                  10. C. arenaria
```

9 Terminal spike male only at apex; glumes (male and
   female) <5mm; leaf-sheaths herbaceous on side opposite
   blade, except for apical hyaline rim   **11. C. disticha**
   10 Stems ± terete, smooth   11
   10 Stems trigonous or triquetrous, rough on angles
      near top   12
11 Stems rarely >15cm; leaves usually curved, reaching
   or nearly reaching inflorescence, crescent-shaped in
   section when fresh; coastal   **14. C. maritima**
11 Stems rarely <15cm; leaves usually straight, falling
   well short of inflorescence, flat or V-shaped in
   section when fresh; not coastal   **12. C. chordorrhiza**
   12 Lowest bract leaf-like, mostly at least as long
      as whole inflorescence; beak of utricle <1/2 as
      long as body; utricles pale brown   **13. C. divisa**
   12 Lowest bract not leaf-like, much shorter than
      inflorescence; beak of utricle >1/2 as long as
      body; utricles blackish-brown   **3. C. diandra**
13 All spikes female at least at top   14
13 At least uppermost or lowermost spike male at least
   at top   19
   14 Lowest bract easily exceeding inflorescence,
      leaf-like   **15. C. remota**
   14 Lowest bract shorter than inflorescence, usually
      not leaf-like   15
15 Spikes not longer than wide, each with <10 utricles;
   utricles patent   **17. C. echinata**
15 Spikes longer than wide, each usually with >10
   utricles; utricles erect to erecto-patent   16
   16 Utricles winged in upper 1/2   **16. C. ovalis**
   16 Utricles not winged   17
17 Utricles pale green to yellowish- or pale brownish-
   green   **22. C. curta**
17 Utricles reddish- to dark-brown   18
   18 Spikes (5)8-12(18); utricles divaricate, without
      slit in beak; lowland wet places   **20. C. elongata**
   18 Spikes (2)3-4(6); utricles appressed, with slit
      down back of beak; high mountains   **21. C. lachenalii**
19 Utricles 2-2.6mm   **6. C. vulpinoidea**
19 Utricles >2.7mm   20
   20 Utricles biconvex (weakly to strongly convex on
      adaxial and strongly convex on abaxial side)   21
   20 Utricles plano-convex (flat on adaxial and weakly
      convex on abaxial side)   23
21 Utricles conspicuously winged in upper 1/2
                                            1. C. paniculata
21 Utricles not or scarcely winged   22
   22 Lowest leaf-sheaths remaining whole; all spikes
      usually sessile   **3. C. diandra**
   22 Lowest leaf-sheaths decaying with fibres; lowest
      spikes usually stalked   **2. C. appropinquata**
23 Stems >2mm wide; leaves mostly >4mm wide; utricles

with distinct, often prominent veins   24
23 Stems <2mm wide; leaves mostly <4mm wide; utricles
   with obscure veins   25
   24 Ligule truncate; leaf-sheaths transversely
      wrinkled on side opposite blade; utricles matt,
      papillose, with slit down back of beak **4. C. vulpina**
   24 Ligule acute; leaf-sheaths not wrinkled on side
      opposite blade; utricles shiny, smooth, without
      slit in beak   **5. C. otrubae**
25 Roots and often base of plant purple-tinged;
   ligule acute to obtuse, distinctly longer than wide;
   utricles thickened and corky at base   **7. C. spicata**
25 Plant not purple-tinged; ligule rounded at apex, c.
   as long as wide; utricles not thickened at base   26
   26 Lowest 2-4 spikes or clusters of spikes separated
      by a gap >2x their own length; ripe utricles
      appressed to axis   **9a. C. divulsa ssp. divulsa**
   26 Lowest spikes separated by a gap <2x their own
      length; ripe utricles divaricate from axis   27
27 Utricles 4.5-5mm, cuneate at base; inflorescence
   3-5(8)cm   **9b. C. divulsa ssp. leersii**
27 Utricles 2.6-4.5mm, truncate to rounded at base;
   inflorescence (1)2-3(4)cm   28
   28 Utricles (3.5)4-4.5mm; female glumes shorter and
      darker than the utricles
                                8a. C. muricata ssp. muricata
   28 Utricles 2.6-3.5(4)mm; female glumes nearly as
      long as and similar in colour to or paler than
      the utricles   **8b. C. muricata ssp. lamprocarpa**

Key C - Spikes >1, the upper (male) different in appearance
        from the lower (female); utricles pubescent
1 Utricles with conspicuously bifid beak >0.5mm   2
1 Utricles with truncate to notched beak 0-0.5mm   3
   2 Lower leaf-sheaths pubescent; utricles 4.5-7mm,
     with beak 1.5-2.5mm   **23. C. hirta**
   2 Leaf-sheaths glabrous; utricles 3.5-5mm, with beak
     0.5-1mm   **24. C. lasiocarpa**
3 Rhizomes extended; stems not or loosely tufted, often
  borne singly   4
3 Rhizomes very short; stems densely tufted   7
   4 Male spikes (1)2-3; lowest female spike clearly
     stalked, pendent to erecto-patent   **36. C. flacca**
   4 Male spike 1; lowest female spike sessile or with
     concealed stalk, erect   5
5 Leaves ± glaucous, erect; lowest living leaf-sheaths
  reddish-brown; stems usually >20cm   **53. C. filiformis**
5 Leaves not glaucous, usually ± recurved; lowest living
  leaf-sheaths mid- to dark-brown; stems usually <20cm   6
   6 Lowest bract with sheath 3-5mm; female glumes
     acute, green to brown, with 0 or narrow scarious
     border   **52. C. caryophyllea**

6 Lowest bract with sheath 0-2mm; female glumes
obtuse to rounded, purplish-black, with wide
scarious border   **54. C. ericetorum**
7 Inflorescence occupying >1/2 of stem length; stems
much shorter than leaves; female spikes with 2-4
flowers   **51. C. humilis**
7 Inflorescence occupying <1/4 of stem length; stems
usually longer than leaves; female spikes usually
with >4 flowers   8
  8 Flowering stems arising laterally, from leaf axils,
  leafless; female spikes <3mm wide, overtopping
  male   9
  8 Flowering stems terminal, leafy at base; female
  spikes >4mm wide, falling short of top of male   10
9 Utricles 3-4mm; female glumes purplish-brown, c. as
long as utricles; female spikes arising 1 above the
other; basal leaf-sheaths crimson   **49. C. digitata**
9 Utricles 2-3mm; female glumes pale brown, much shorter
than utricles; female spikes all arising at + same
point; basal leaf-sheaths brown   **50. C. ornithopoda**
  10 Lowest bract with sheath 3-5mm   **52. C. caryophyllea**
  10 Lowest bract with sheath 0-2mm   11
11 Lowest bract usually green, leaf-like; female glumes
brown or reddish-brown; beak of utricle 0.3-0.5mm
                                           56. C. pilulifera
11 Lowest bract brown, glume-like or bristle-like;
female glumes purplish-black, beak of utricle <0.3mm   12
  12 Female glumes obtuse, with scarious ciliate margin;
  utricles 2-3mm; leaves mostly >2mm wide, rigid,
  recurved   **54. C. ericetorum**
  12 Female glumes subacute (to obtuse) and mucronate,
  with hyaline (but not scarious) glabrous margin;
  utricles 3-4.5mm; leaves mostly <2mm wide, soft,
  + erect   **55. C. montana**

<u>Key D</u> - Spikes >1, the upper (male) different in appearance
from the lower (female); utricles glabrous; stigmas 2
1 Utricles with distinct forked or notched beak >0.3mm   2
1 Utricles with 0 or indistinct truncate or minutely
notched beak <0.3mm   4
  2 Utricles not inflated; female glumes 3-4mm, acute
  to acuminate; female spikes up to 5cm
                                           25. C. acutiformis
  2 Utricles inflated; female glumes 2-3mm, subacute;
  female spikes up to 3cm   3
3 Utricles 3-3.5mm, containing nut, + not ribbed
                                           31. C. saxatilis
3 Utricles 4-5mm, empty, distinctly ribbed
                                           30. C. x grahamii
  4 At least some female glumes with apical points
  >1/2 as long as rest of glume; glumes >2x as
  long as utricles   **64. C. recta**

4 Female glumes rounded or obtuse to acuminate
   with 0 or short apical point (except sometimes the
   lowest 1(-3)); glumes <1.5x as long as utricles   5
5 Leaf-sheaths breaking into conspicuous ladder-like
   fibres on side opposite blade; stems densely tufted,
   often forming large tussocks   **69. C. elata**
5 Leaf-sheaths not breaking into fibres; stems usually
   scattered, sometimes tufted but not tussock-forming   6
   6 Utricles 3.5-5mm, prominently veined; female
      glumes 3-veined (extinct)   **67. C. trinervis**
   6 Utricles 2-3.5mm; female glumes 1-veined, or
      3-veined and then utricles veinless   7
7 Utricles without visible veins   8
7 Utricles with distinct faint to ± prominent veins   9
   8 Lowest bract exceeding inflorescence; stems
      usually >25cm, trigonous, brittle; female glumes
      often 3-veined   **65. C. aquatilis**
   8 Lowest bract shorter than inflorescence; stems
      usually <25cm, triquetrous, not brittle; female
      glumes 1-veined   **70. C. bigelowii**
9 Leaves 1-3(5)mm wide, the margins rolling inwards
   on drying; lowest bract rarely as long as
   inflorescence; male spike usually 1   **68. C. nigra**
9 Leaves 3-10mm wide, the margins rolling outwards on
   drying; lowest bract usually exceeding inflorescence;
   male spikes usually 2-4   **66. C. acuta**

Key E - Spikes >1, the upper (male) different in appearance
        from the lower (female); utricles glabrous; stigmas
        3; lowest spike pendent
1 Lower leaf-sheaths and lowerside of blades pubescent
                                                48. C. pallescens
1 Leaf-sheaths and blades glabrous   2
   2 Utricles with distinct forked or notched beak
      usually >1mm (<1mm in C. atrofusca and
      C. acutiformis)   3
   2 Utricles with beak 0 or <1mm and with truncate,
      oblique or very slightly notched apex   12
3 Male spikes >2   4
3 Male spike 1   8
   4 Rhizomes extended; stems not or loosely tufted,
      often borne singly   5
   4 Rhizomes very short; stems densely tufted   7
5 Female glumes 6-10mm, exceeding utricles **26. C. riparia**
5 Female glumes 4-6mm, mostly shorter than utricles   6
   6 Utricles 3.5-5mm, with beak <1mm   **25. C. acutiformis**
   6 Utricles (4)5-8mm, with beak 1.5-2.5mm
                                             29. C. vesicaria
7 Female spikes 6-8mm wide, on peduncles with exposed
   portion usually shorter than spike   **40. C. laevigata**
7 Female spikes 3-5mm wide, on peduncles with exposed
   portion usually much longer than spike **33. C. sylvatica**

8 Female glumes (except midrib) and utricles both
   purplish-black   **57. C. atrofusca**
8 Female glumes and/or utricles brownish or greenish 9
9 Utricles with smooth beaks; lower female spikes with
   peduncles >1/2 exposed   10
9 Utricles with scabrid beaks; lower female spikes with
   peduncles >1/2 ensheathed   11
   10 Female spikes 3-5mm wide; female glumes 3-5mm;
      ligules <5mm   **33. C. sylvatica**
   10 Female spikes 6-10mm wide; female glumes 5-10mm;
      ligules >5mm   **27. C. pseudocyperus**
11 Leaves 5-12mm wide; female glumes acuminate; ligules
   7-15mm   **40. C. laevigata**
11 Leaves 2-5(7)mm wide; female glumes obtuse and
   mucronate; ligules 1-2mm   **41. C. binervis**
   12 Male spikes (1)2-3; utricles papillose **36. C. flacca**
   12 Male spike usually 1; utricles not papillose   13
13 Rhizomes very short; stems densely tufted;   14
13 Rhizomes extended; stems not or loosely tufted,
   often borne singly   16
   14 Female spikes <2.5cm, arising very close
      together; all leaves <3mm wide; plant rarely
      >30cm   **34. C. capillaris**
   14 Female spikes >2.5cm, well spaced out along stem;
      largest leaves >4mm wide; plant rarely <30cm   15
15 Female spikes <3mm wide, with peduncle c.1/2 exposed
                                          35. C. strigosa
15 Female spikes >3mm wide, with peduncle + entirely
   ensheathed   **32. C. pendula**
   16 Female glumes distinctly narrower than utricles,
      acuminate at apex, 5-6.5mm, >1.5x as long as
      utricles; lowest spike with 1-2 male flowers
      at base   **60. C. magellanica**
   16 Female glumes at least as wide as utricles, acute
      to obtuse (sometimes mucronate) at apex, 3-4.5mm,
      <1.5x as long as utricles; lowest spike entirely
      female   17
17 Female spikes 3-4mm wide, with 5-8 flowers; utricles
   + beakless; stems usually smooth   **59. C. rariflora**
17 Female spikes 5-7mm wide, with 7-20 flowers;
   utricles with distinct beak 0.1-0.5mm; stems
   usually rough distally   **58. C. limosa**

Key F - Spikes >1, the upper (male) different in appearance
        from the lower (female); utricles glabrous; stigmas
        3; lowest spike erect to patent
1 Lower leaf-sheaths and lowerside of blades pubescent
                                          48. C. pallescens
1 Leaf-sheaths and blades glabrous   2
   2 Utricles papillose   **36. C. flacca**
   2 Utricles not papillose   3
3 Lowest bract not sheathing at base, or with sheath

```
                                                      4
     <2mm
  3  Lowest bract with distinct sheathing base >3mm   9
     4  Male glumes 7–9mm, acuminate; female glumes
        longer than utricles                 26. C. riparia
     4  Male glumes 3–6mm, obtuse to acute; female glumes
        shorter than utricles                          5
  5  Beak >1mm                                          6
  5  Beak <1mm                                          7
     6  Utricles 3.5–6.5mm, usually patent, abruptly
        contracted into beak 1–1.5mm; female glumes
        acute                                28. C. rostrata
     6  Utricles (4)5–8mm, usually erecto–patent,
        gradually contracted into beak 1.5–2.5mm; female
        glumes acuminate                     29. C. vesicaria
  7  Stems solid, triquetrous with concave faces; all or
     most leaves >5mm wide; female spikes 2–5cm
                                             25. C. acutiformis
  7  Stems hollow, trigonous with flat to convex faces;
     all or most leaves <5mm wide; female spikes 1–3cm  8
     8  Utricles 3–3.5mm, containing nut, + not ribbed
                                             31. C. saxatilis
     8  Utricles 4–5mm, empty, distinctly ribbed
                                             30. C. x grahamii
  9  Rhizomes extended; stems not or loosely tufted     10
  9  Rhizomes very short; stems densely tufted          14
     10 Utricles >5mm, with beak >1.5mm                 11
     10 Utricles <5mm, with beak <1.5mm                 12
  11 Female spikes <10–flowered; utricles with beak
     >2.5mm; male spike 1                     39. C. depauperata
  11 Female spikes >10–flowered; utricles with beak <2.5mm;
     male spikes 2–4                          29. C. vesicaria
     12 Utricles with scabrid, clearly bifid beak
                                             45. C. hostiana
     12 Utricles with smooth, truncate to shallowly
        notched, often very short beak                  13
  13 Sheaths of lowest bract inflated, loose from stem;
     utricles with beak 0.5–1mm; leaves green or
     yellowish–green                          38. C. vaginata
  13 Sheaths of lowest bract not inflated, close to stem;
     utricles with beak <0.5mm; leaves glaucous
                                             37. C. panicea
     14 At least 1/2 of female spikes close–set to
        terminal male spike; 1–several bracts far
        exceeding inflorescence                         15
     14 At most 1 female spike close–set to terminal male
        spike; bracts usually shorter than inflorescence,
        sometimes just exceeding it                     19
  15 Utricles erecto–patent, greyish–green often purple–
     blotched; leaves dark– or greyish–green, deeply
     channelled and/or with inrolled margins  44. C. extensa
  15 At least lower utricles in each spike patent to
     reflexed, bright– or yellowish–green; leaves bright–
```

or yellowish-green, flat or V-shaped   16
16 Beaks of utricles curved or bent, usually $\geq$1/2
   as long as the usually $\pm$ curved body   17
16 Beaks of utricles straight, usually <1/2 as long
   as and continuing line of $\pm$ straight body at
   least when fresh   18
17 Utricles 5.5-6.5mm; male spike usually subsessile;
   leaves $\leq$7mm wide, $\geq$2/3 as long as stems (rare)
                                                       46. C. flava
17 Utricles 3.5-5mm; male spike usually clearly
   stalked; leaves $\leq$4mm wide, <2/3 as long as stems
   (common)   **47a. C. viridula ssp. brachyrrhyncha**
18 Utricles 3-4mm, with beak 0.8-1.3mm; male spike
   usually clearly stalked; lowest female spike
   usually distant from others
                                 47b. C. viridula ssp. oedocarpa
18 Utricles 1.75-3.5mm, with beak 0.25-1mm; male spike
   usually sessile; lowest female spike usually
   bunched with others   **47c. C. viridula ssp. viridula**
19 Female spikes 2-3mm wide, lax-flowered; apex of sheath
   of uppermost stem-leaves (not bracts) truncate on
   side opposite blade   **35. C. strigosa**
19 Female spikes $\geq$4mm wide, dense-flowered; apex of
   sheath of uppermost stem-leaves (not bracts) either
   concave on side opposite blade or with an apical
   projection   20
20 Utricles patent   21
20 Utricles erecto-patent to erect   22
21 Utricles yellowish-green, not shiny, contrasting
   with female glumes (dark brown with broad scarious
   margins)   **45. C. hostiana**
21 Utricles pale green, often minutely dark-dotted,
   shiny, scarcely contrasting with female glumes (pale
   brown with narrow scarious margins)   **43. C. punctata**
22 Leaves 5-12mm wide; female glumes acuminate;
   ligules 7-15mm   **40. C. laevigata**
22 Leaves 2-5(7)mm wide; female glumes acute to
   obtuse, often mucronate or apiculate; ligules
   1-3mm   23
23 Female spikes 1.5-4.5cm; female glumes dark reddish-
   or blackish-brown; utricles with 2 conspicuous green
   lateral ribs distinct from others   **41. C. binervis**
23 Female spikes 1-2cm; female glumes either pale
   reddish-brown or dark brown with broad scarious
   margins; utricles with several $\pm$ equally
   prominent ribs   24
24 Leaf-blades and lowest bract rather abruptly
   contracted to narrow parallel-sided point;
   utricles with beak 0.8-1.2mm; female glumes dark
   brown with broad scarious margins   **45. C. hostiana**
24 Leaf-blades and lowest bract gradually contracted
   to apex; utricles with beak 0.7-1mm; female glumes

pale- to mid-brown with narrow scarious margins
                                        42. C. distans

Other spp. - **C. crawfordii** Fern., from N America, is a rare
wool-alien formerly + natd in SE En; it resembles <u>C. ovalis</u>
but differs in more numerous (7-15) spikes, shorter female
glumes (2.5-3mm), and longer leaves (+ equalling stems). 3
arctic spp. were recorded from W Isles of Sc in the 1940s but
have not been seen recently and were probably planted: **C.
capitata** L., **C. bicolor** All. and **C. glacialis** Mackenzie.

<u>Subgenus 1</u> - <u>VIGNEA</u> (P. Beauv. ex Lestib.) Kuek. (spp. 1-22).
Spikes >2, all similar in appearance though distribution of
male and female flowers varies, and some spikes may be all
male or all female; lowest bract glume-like or bristle-like
and shorter than inflorescence unless otherwise stated,
without sheath; spikes sessile or nearly so, often forming
compact, lobed inflorescence, the lateral ones without scale
between bract and lowest glume; stigmas 2; nut biconvex. In
<u>C. dioica</u> and <u>C. davalliana</u> there is a single dioecious
spike; these 2 are often placed in subg. <u>Primocarex</u>.

1. **C. paniculata** L. - <u>Greater Tussock-sedge</u>. Densely **966**
tufted, often forming large tussocks; stems to 1.5m,
triquetrous, rough; utricles 3-4mm, greenish- to blackish-
brown, winged towards apex, with beak 1-1.5mm. Native; by
lakes and streams, in marshes and fens and wet woods, on
usually base-rich soils; frequent throughout most of BI.
    1 x 2. C. paniculata x C. appropinquata = C. x rotae De
Notaris (<u>C.</u> x <u>solstitialis</u> Figert) occurs in fens with the
parents in W Suffolk, E Norfolk, Westmeath and formerly
Offaly.
    1 x 3. C. paniculata x C. diandra = C. x beckmannii Keck ex
F. Schultz occurs in SE Yorks, Ayrs, Main Argyll and M Cork.
    1 x 15. C. paniculata x C. remota = C. x boenninghausiana
Weihe is scattered throughout Br and Ir.
    1 x 22. C. paniculata x C. curta = C. x ludibunda Gay has
been found with the parents in E Sussex, Pembs, Caerns and
Brecs.
2. **C. appropinquata** Schum. - <u>Fibrous Tussock-sedge</u>. Differs **966**
from <u>C. paniculata</u> in stems to 1m; leaves 1-3mm (not 4-7mm) **R**
wide; old leaf-sheaths becoming very fibrous; and utricles
2.7-3.7mm, greyish-brown, unwinged, with beak 0.7-1.2mm.
Native; similar places to <u>C. paniculata</u>; very local in E
Anglia, N (formerly S) En, S Sc and C Ir.
3. **C. diandra** Schrank - <u>Lesser Tussock-sedge</u>. Usually with **966**
rhizomes extended, sometimes tufted but rarely forming
tussocks; differs from <u>C. appropinquata</u> in stems to 60cm,
leaves 1-2mm wide; old leaf-sheaths not becoming fibrous;
utricles 2.7-3.5mm, blackish-brown, with beak 1-1.5mm.
Native; wet peaty and acid places in ditches, meadows,
marshes and scrub; scattered throughout most of Br and Ir.

4. C. vulpina L. - True Fox-sedge. Densely tufted; stems to **966**
1m, stout, triquetrous with ± winged rough angles; utricles   **R**
4-5mm, with beak 1-1.5mm, with epidermal cells ± isodiametric
(microscope). Native; wet places on heavy soils, in ditches
and marshes and by streams; local in En N to NE Yorks, mainly
in SE.
5. C. otrubae Podp. (C. cuprina (Sandor ex Heuffel) Nendtv. **966**
ex A. Kerner nom. inval.) - False Fox-sedge. Differs from C.
vulpina in utricles 4.5-6mm with beak 1-2mm and ± winged
(making utricle more gradually tapered), with epidermal cells
markedly elongated (microscope); stems with unwinged angles;
and see Key B (couplet 24). Native; wet places on heavy
soils in a range of habitats; frequent throughout most of BI
but ± entirely coastal in Sc and very rare in N Sc.
5 x 9. C. otrubae x C. divulsa was found in W Sussex and
Worcs last century and in E Kent and Oxon in 1980s.
5 x 15. C. otrubae x C. remota = C. x pseudoaxillaris K.
Richter occurs with the parents scattered in BI N to
Midlothian.
6. C. vulpinoidea Michaux - American Fox-sedge. Densely **966**
tufted; stems to 1m, triquetrous, slightly rough; utricles
2-2.6mm, with beak 0.7-1.2mm. Intrd; wool-alien sometimes
becoming natd in rough ground; scattered in Br N to C Sc; N
America.
7. C. spicata Hudson (C. contigua Hoppe) - Spiked Sedge. **966**
Densely tufted; stems to 80cm, triquetrous, rough; utricles
4-5mm, with beak 1-2mm. Native; damp grassy places in
fields, on banks and waysides and by rivers and ponds;
frequent throughout most of En, very scattered in Wa, C & S
Sc, Ir and Guernsey.
8. C. muricata L. - Prickly Sedge. Differs from C. spicata
in utricles 2.6-4.5mm, with beak 0.7-1.3mm; and see Key B
(couplet 25). Native.
a. Ssp. muricata. Utricles (3.5)4-4.5mm, greyish-green then **966**
dark reddish-brown; female glumes 2.5-3.5mm, markedly shorter **RR**
than utricles, dark- or reddish-brown. Steep, dry, limestone
slopes; very rare, 5 places in Br from W Gloucs to Berwicks.
b. Ssp. lamprocarpa Celak. (C. pairii F. Schultz). Utricles **966**
2.6-3.5(4)mm, yellowish-green then dark brown; female glumes
3-4.5mm, somewhat shorter than utricles, pale brown later
fading. Open grassy places usually on dry acid soils;
frequent throughout Br N to C Sc, CI, S & E Ir.
8 x 9. C. muricata x C. divulsa was found in E Cork in 1987;
?endemic.
9. C. divulsa Stokes - Grey Sedge. Densely tufted; stems
trigonous, rough. Native.
a. Ssp. divulsa. Leaves usually greyish- to dark-green; **966**
inflorescence 5-18cm, with lower spikes well spaced out;
utricles 3-4(4.5)mm, usually greenish-brown when ripe.
Hedgerows, wood borders, grassy rough ground; frequent in S
Br, scattered N to N En, scattered in Ir and CI.
b. Ssp. leersii (Kneucker) W Koch (C. polyphylla Karelin & **966**

Kir.). Leaves usually yellowish-green; inflorescence 4-8cm, with lower spikes <2cm apart; utricles 4-4.5(4.8)mm, usually reddish-brown when ripe. Similar places to ssp. divulsa but only on chalk and limestone; local in Br N to Midlothian.

9 x 15. C. divulsa x C. remota = C. x emmae L. Gross has been found in E Sussex and M Cork.

10. C. arenaria L. - Sand Sedge. Very extensively **966** rhizomatous, with stems borne singly, to 90cm (often much less), triquetrous, slightly rough, utricles 4-5.5mm, winged, with beak 1-2mm. Native; bare or grassy maritime dunes or inland sandy places; round coasts of whole BI, very local inland mostly in En.

11. C. disticha Hudson - Brown Sedge. Rhizomatous, with **966** stems borne singly or in pairs, to 1m, triquetrous, rough; lowest bract usually bristle-like, sometimes leaf-like and exceeding inflorescence; utricles 4-5.5(7)mm with beak 1-1.5(3)mm. Native; marshes, fens and wet meadows; frequent throughout most of BI.

12. C. chordorrhiza L. f. - String Sedge. Rhizomatous, with **966** stems borne singly, to 40cm, + terete, smooth; utricles **RR** 3.5-4.5mm, with beak 0.5-1mm. Native; very wet acid bogs; very rare in W Sutherland and Easterness.

13. C. divisa Hudson - Divided Sedge. Rhizomatous but stems **966** often clustered, to 80cm, trigonous, rough distally; lowest **R** bract bristle- or leaf-like, usually just exceeding inflorescence; utricles 3.2-4mm, with beak 0.5-0.8mm. Native; damp, usually brackish grassy places, in marshes, pastures and ditches; locally frequent around coasts of S & C Br, scattered N to Cheviot and S Lancs, Jersey, formerly very rare in Sc and Ir.

14. C. maritima Gunnerus - Curved Sedge. Long-rhizomatous, **966** with stems single or loosely grouped, to 18cm, terete, **R** smooth; utricles 3.5-4.5mm, with beak 0.5-1mm. Native; sand-dunes and damp sandy places; local on coasts of N Br S to Cheviot, formerly S Lancs, mainly N Sc.

15. C. remota L. - Remote Sedge. Densely tufted; stems to **966** 75cm, triquetrous (or compressed distally), rough distally; lowest bract leaf-like, far exceeding inflorescence; utricles 2.5-3.8mm, with beak 0.5-0.8mm. Native; woods, hedgerows, shady banks and ditch-sides; frequent to common throughout most of BI.

16. C. ovalis Gooden. - Oval Sedge. Densely tufted; stems **966** to (40)90cm, triquetrous, rough distally; lowest bract bristle-like, sometimes as long as inflorescence; utricles 3.8-5mm, with beak 1-1.5mm, winged. Native; damp or dry grassy places; common throughout BI.

17. C. echinata Murray - Star Sedge. Densely tufted; stems **966** to 40cm, trigonous to subterete, rough distally; lowest bract glume-like or bristle-like, rarely as long as inflorescence; utricles 2.8-4mm, with beak 1-1.5mm. Native; wide range of acid to basic bogs and marshes; throughout BI, common in W & N, very scattered in C & E En.

CAREX   965

17 x 18. **C. echinata x C. dioica** = **C. x gaudiniana** Guthnick
has been found in Denbs and W Mayo.
17 x 22. **C. echinata x C. curta** = **C. x biharica** Simonkai
occurs on mountains in C Sc.
18. C. dioica L. - <u>Dioecious</u> <u>Sedge</u>. Shortly rhizomatous, **966**
with stems loosely tufted, to (20)30cm, terete, smooth;
lowest bract 0; usually dioecious but sometimes with both
sexes in the spike; utricles 2.5-3.5mm, with beak 0.7-1.3mm.
Native; base-rich bogs and flushes; throughout most of Br,
common in N, very scattered in C and absent from most of S,
scattered in Ir. .
19. C. davalliana Smith - <u>Davall's</u> <u>Sedge</u>. Densely tufted; **966**
stems to 40cm, ± trigonous, rough distally; usually **RR**
dioecious, but sometimes with both sexes in the spike; lowest
bract 0; utricles 3.5-4.5mm, with beak 1-1.7mm. Possibly
formerly native, now extinct; calcareous fen in N Somerset
until c.1845.
20. C. elongata L. - <u>Elongated</u> <u>Sedge</u>. Densely tufted; stems **966**
to 80cm, triquetrous, rough; utricles 2.5-4mm, with beak **R**
0.5-0.8mm. Native; damp places in wet meadows and boggy
woods, by ditches and streams; scattered in Br N to Dunbarton
and in N Ir.
21. C. lachenalii Schk. - <u>Hare's-foot</u> <u>Sedge</u>. Shortly **966**
rhizomatous, with stems tufted, to 20(30)cm, trigonous, rough **RR**
distally; utricles 2.5-3.3mm, with beak 0.5-0.8mm. Native;
wet acid places on mountains >750m, usually where snow lies
late; very local in C Sc.
21 x 22. **C. lachenalii x C. curta** = **C. x helvola** Blytt ex
Fries occurs with the parents in 2 places in S Aberdeen.
22. C. curta Gooden. - <u>White</u> <u>Sedge</u>. Shortly rhizomatous, **966**
with stems loosely tufted, to 50cm, triquetrous, rough
distally; utricles 2-3mm, with beak 0.5-0.7mm. Native; wet
acid places on heaths, in bogs and boggy woods and mountain-
sides; throughout most of Br and Ir, common in N 1/2 of Br, N
Ir and Wa, very scattered elsewhere.

<u>Subgenus</u> <u>2</u> - <u>CAREX</u> (spp. 23-70). Spikes >2, usually the
apical >1 male and the basal >1 female and differing
obviously in appearance, sometimes with some male flowers on
female spikes and vice versa; lowest bract leaf-like unless
otherwise stated; spikes sessile to long-stalked, sometimes
the lower ones pendent and remote from upper ones, the
lateral ones usually with a scale between bract and lowest
glume (often close to and ± hidden by former); stigmas
usually 3 and nut trigonous but often 2 and nut biconvex. In
<u>C. atrata</u>, <u>C. norvegica</u> and <u>C. buxbaumii</u>, and sometimes <u>C.</u>
<u>atrofusca</u>, the terminal spike is bisexual and all spikes are
rather similar in appearance, but all these have 3 stigmas,
trigonous nuts and a scale between bract and lowest glume.

23. C. hirta L. - <u>Hairy</u> <u>Sedge</u>. Rhizomatous or shortly so, **967**
with stems loosely tufted, to 70cm, trigonous, ± smooth;

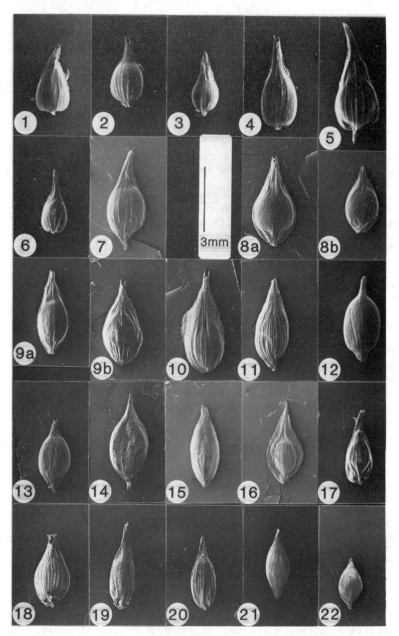

FIG 966 – Utricles of **Carex** subgenus **Vignea**. 1, **C. paniculata**. 2, **C. appropinquata**. 3, **C. diandra**. 4, **C. vulpina**. 5, **C. otrubae**. 6, **C. vulpinoidea**. 7, **C. spicata**. 8a, **C. muricata** ssp. **muricata**. 8b, ssp. **lamprocarpa**. 9a, **C. divulsa** ssp. **divulsa**. 9b, ssp. **leersii**. 10, **C. arenaria**. 11, **C. disticha**. 12, **C. chordorrhiza**. 13, **C. divisa**. 14, **C. maritima**. 15, **C. remota**. 16, **C. ovalis**. 17, **C. echinata**. 18, **C. dioica**. 19, **C. davalliana**. 20, **C. elongata**. 21, **C. lachenalii**. 22, **C. curta**.

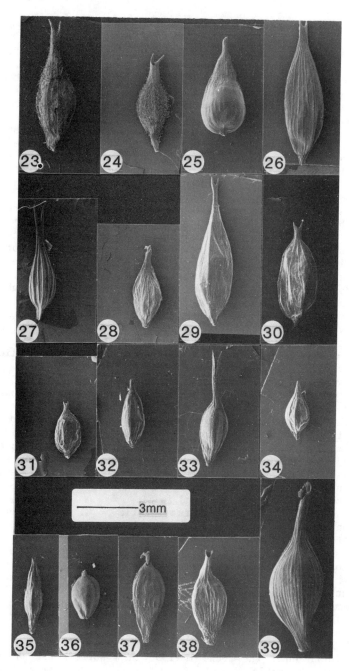

FIG 967 – Utricles of **Carex** subgenus **Carex**. 23, **C. hirta**. 24, **C. lasiocarpa**. 25, **C. acutiformis**. 26, **C. riparia**. 27, **C. pseudocyperus**. 28, **C. rostrata**. 29, **C. vesicaria**. 39, **C. x grahamii**. 31, **C. saxatilis**. 32, **C. pendula**. 33, **C. sylvatica**. 34, **C. capillaris**. 35, **C. strigosa**. 36, **C. flacca**. 37, **C. panicea**. 38, **C. vaginata**. 39, **C. depauperata**.

lowest bract rarely exceeding inflorescence, with long
pubescent sheath; utricles 4.5-7mm, with beak 1.5-2.5mm,
pubescent. Native; damp grassy places in many habitats;
common throughout BI except N Sc.
    23 x 29. C. hirta x C. vesicaria = C. x grossii Fiek occurs
in a dune-slack in Co Wicklow.
    24. C. lasiocarpa Ehrh. - <u>Slender</u> <u>Sedge</u>. Rhizomatous; stems **967**
to 1.2m, trigonous, smooth; lowest bract often exceeding
inflorescence, with short sheath; utricles 3-5mm, with beak
0.5-1mm, pubescent. Native; bogs and fens; scattered
throughout Br and Ir but absent from most of S & C En except
E Anglia. The very long narrow leaves are distinctive.
    24 x 26. C. lasiocarpa x C. riparia = C. x evoluta Hartman
has been found in N Somerset and Cambs.
    25. C. acutiformis Ehrh. - <u>Lesser</u> <u>Pond-sedge</u>. Rhizomatous; **967**
stems tufted, to 1.5m, triquetrous, rough; lowest bract
usually exceeding inflorescence, with 0 or short sheath;
utricles 3.5-5mm, with beak 0.3-0.8mm. Native; marshes, wet
meadows and swamps, by ponds and streams; common throughout
lowland Br and Ir, rare in N & C Sc, S & W Ir and SW En.
Stigmas usually 3; fertile plants with 2 stigmas might be
derivatives of C. acutiformis x C. acuta. C. acuta has
shorter, wider utricles with a much shorter beak.
    25 x 29. C. acutiformis x C. vesicaria = C. x ducellieri
Beauverd was found in S Hants in 1986.
    25 x 66. C. acutiformis x C. acuta = C. x subgracilis Druce
has been found in scattered localities in En and Wa; flowers
with 2 and 3 stigmas occur in the same spike.
    26. C. riparia Curtis - <u>Greater</u> <u>Pond-sedge</u>. Differs from C. **967**
acutiformis in larger spikes, glumes and utricles; utricles
5-8mm, with beak 1-2mm; female glumes 7-10mm (not 4-5mm); and
male spikes 3-6 (not 2-3) with acuminate (not subacute to
obtuse) glumes. Native; similar places to C. acutiformis and
often with (but not hybridizing with) it; common in C & S Br,
scattered N to C Sc, in Ir and in CI.
    26 x 29. C. riparia x C. vesicaria = C. x csomadensis
Simonkai has been found in Co Wicklow, and there are more
doubtful records in S En.
    27. C. pseudocyperus L. - <u>Cyperus</u> <u>Sedge</u>. Shortly **967**
rhizomatous, with stems loosely tufted, to 90cm, triquetrous,
rough; lowest bract far exceeding inflorescence, with short
sheath; utricles 4-5mm, with stalk-like base, with beak 1.5-
2.5mm. Native; in marshes and swamps, by ponds, rivers and
canals; locally common in C & S Br, very scattered in N Br,
Ir and CI.
    **27 x 28. C. pseudocyperus x C. rostrata = C. x justi-
schmidtii** Junge was found in 1955 in W Norfolk but has not
been seen recently.
    28. C. rostrata Stokes - <u>Bottle</u> <u>Sedge</u>. Rhizomatous; stems **967**
slightly tufted or not, to 1m, trigonous, rough distally;
lowest bract often exceeding inflorescence, with 0 or short
sheath; utricles 3.5-6.5mm, with beak 1-1.5mm. Native; acid

swamps, lake-margins and reed-beds; throughout Br and Ir, common in N & W, absent from parts of C & S En.
28 x 29. **C. rostrata x C. vesicaria = C. x involuta** (Bab.) Syme occurs in scattered places over much of Br and Ir; 1 of the commoner hybrids.
29. **C. vesicaria** L. - Bladder-sedge. Shortly rhizomatous; **967** stems slightly tufted, to 1.2m, trigonous, rough distally; lowest bract exceeding inflorescence, with short to long, rarely 0, sheath; utricles (4)5-8mm, with beak 1.5-2.5mm. Native; swamps, marshes, lake margins; frequent through BI. Often confused with C. rostrata, but has yellowish-green (not glaucous) leaves and a non-swollen (not a swollen) stem-base.
30. **C. x grahamii** Boott (C. stenolepis auct. non Less., C. **967** ewingii E. Marshall; C. saxatilis x ?C. vesicaria) - Mountain Bladder-sedge. Rhizomatous, with stems in small tufts, to 50cm, trigonous, ± rough distally; lowest bract exceeding inflorescence or not, without sheath; utricles 4-5mm, with beak 0.5-0.9mm. Native; mountain flushes >750m, often with C. saxatilis but not with C. vesicaria; very local in C Sc.
31. **C. saxatilis** L. - Russet Sedge. Differs from C x **967** grahamii in stems to 40cm; female spikes <2cm (not <3cm); and **R** see Key F (couplet 8). Native; wet places, especially where snow lies late on mountains above 750m; local in N & W Sc. Stigmas usually 3; fertile plants with 2 stigmas might be derivatives of hybrids with C. vesicaria or C. rostrata.
32. **C. pendula** Hudson - Pendulous Sedge. Densely tufted; **967** stems to 1.8m, trigonous, smooth; lowest bract usually shorter than inflorescence, with long sheath; utricles 3-3.5mm, with beak 0.3-0.6mm. Native; rich heavy soils in woods and damp copses; common in S Br, decreasing to N, scattered in Ir, Sc and CI, absent from highlands and N Sc, also grown for ornament and sometimes natd.
33. **C. sylvatica** Hudson - Wood-sedge. Densely tufted; stems **967** to 70cm, trigonous, smooth; lowest bract sometimes longer than inflorescence, with long sheath; utricles 3-5mm, with beak 1-2.5mm. Native; on damp usually heavy soils in woods, hedgerows and scrub; frequent throughout most of BI, common in S Br, rare in N Sc, not in Outer Isles.
34. **C. capillaris** L. - Hair Sedge. Densely tufted; stems to **967** (20)40cm, trigonous, smooth; lowest bract leaf-like but **R** narrow, sometimes exceeding inflorescence, with long sheath; utricles 2.5-3mm, with beak 0.5-1mm. Native; base-rich or calcareous flushes, mineral-rich bogs; local in N En and Sc, rare in Caerns.
35. **C. strigosa** Hudson - Thin-spiked Wood-sedge. Densely **967** tufted; stems to 75cm, trigonous to subterete, smooth; lowest bract usually shorter than inflorescence, with long sheath; utricles 3-4mm, with beak 0.3-0.5mm. Native; damp base-rich soils in woods or woodland clearings; locally frequent in Br N to NE Yorks, scattered in N & C Ir.
36. **C. flacca** Schreber - Glaucous Sedge. Glaucous, **967** rhizomatous; stems loosely tufted, to 60cm, trigonous to

subterete, smooth; lowest bract sometimes exceeding
inflorescence, with short to long sheath; utricles 2-3mm,
with beak <0.3mm, papillose. Native; wet or dry grassland on
chalk and limestone, sand-dunes and base-rich clay, mountain
flushes; common throughout BI (the most widespread and one of
the most variable spp.).

37. C. panicea L. - Carnation Sedge. Glaucous, shortly **967**
rhizomatous; stems loosely tufted, to 60cm, trigonous to
subterete, smooth; lowest bract shorter than inflorescence,
with long to medium sheath; utricles 3-4mm, with beak
<0.3(0.5)mm. Native; wet, usually acid, heaths and moors,
bogs, mountain flushes; throughout BI, common in N & W, local
in SE. Vegetatively distinguished from C. flacca by the
trigonous (not flat) leaf-tips.

38. C. vaginata Tausch - Sheathed Sedge. Rhizomatous; stems **967**
tufted, to 40cm, trigonous and subterete, smooth; lowest   **R**
bract much shorter than inflorescence, with long loose
sheath; utricles 4-4.5mm, with beak 0.5-1mm. Native; wet
rocky places, damp slopes and flushes above 600m; local in S,
C & N mainland Sc.

39. C. depauperata Curtis ex With. - Starved Wood-sedge. **967**
Shortly rhizomatous; stems to 1m, loosely tufted, trigonous **RRR**
to subterete, smooth; lowest bract often exceeding inflorescence, with long sheath; utricles 7-9mm (longest of our
spp.), with beak 2.5-3mm. Native; dry woods and hedge-banks
on chalk or limestone; very rare in N Somerset and M Cork,
probably extinct in Surrey, Anglesey and W Kent.

40. C. laevigata Smith - Smooth-stalked Sedge. Densely **974**
tufted; stems to 1.2m, trigonous, smooth; lowest bract
usually shorter than inflorescence, with long sheath;
utricles 4-6mm, with beak 1-2mm. Native; damp shady places,
especially woods on heavy soils; scattered throughout most of
BI, common in S & W Br, absent in most of C & E En and N Sc.

40 x 41. C. laevigata x C. binervis = C. x deserta Merino
was found in 1961 in Caerns.

40 x 47. C. laevigata x C. viridula was found in 1970 in
Merioneth; endemic. C. viridula ssp. oedocarpa was involved.

40 x 48. C. laevigata x C. pallescens was found in 1973 in
Westerness; endemic.

41. C. binervis Smith - Green-ribbed Sedge. Densely tufted; **974**
stems to 1.2m, subterete, smooth; lowest bract shorter than
inflorescence, with long sheath; utricles 3.5-4.5mm, with
beak 1-1.5mm. Native; damp heaths, moors, rocky places and
mountain-sides; frequent throughout most of BI, common in N &
W Br.

41 x 43. C. binervis x C. punctata was found in 1954 in
Merioneth.

41 x 47. C. binervis x C. viridula = C. x corstorphinei
Druce was found in 1915 in Angus; probably endemic. C.
viridula ssp. oedocarpa was involved.

42. C. distans L. - Distant Sedge. Densely tufted; stems to **974**
1m, subterete, smooth; lowest bract much shorter than

inflorescence, with long sheath; utricles 3-4.5mm, with beak
0.7-1mm. Native; brackish and fresh-water marshes, wet rocky
places, mostly near sea; round most coasts of BI, frequent
inland in S Br, rarely so elsewhere.
42 x 44. C. distans x C. extensa = C. tornabenii Chiov. has
been found in Merioneth and W Cornwall.
42 x 45. C. distans x C. hostiana = C. x muelleriana F.
Schultz has been found in N Hants and Co Dublin.
42 x 47. C. distans x C. viridula = C. x binderi Podp. has
been found in Kintyre and doubtful records exist from
elsewhere. C. viridula ssp. brachyrrhyncha was involved.
43. C. punctata Gaudin - Dotted Sedge. Differs from C.   **974**
distans in stems trigonous; lowest bract sometimes just   **R**
exceeding inflorescence; utricles patent (not erecto-patent),
usually more shiny and more strongly ribbed, often purple-
dotted. Native; similar places to C. distans and often with
it, but strictly coastal; local in S, W & N Ir, CI and S & W
Br N to Wigtowns, formerly Berwicks.
44. C. extensa Gooden. - Long-bracted Sedge. Glaucous or   **974**
bluish-greyish-green; densely tufted; stems to 40cm,
trigonous, smooth; lowest bract much longer than
inflorescence but patent to reflexed, with short sheath;
utricles 3-4mm, with beak 0.5-1mm. Native; muddy or sandy
brackish places in estuaries and by sea; frequent round
coasts of most of BI except parts of E coast of Br.
45. C. hostiana DC. - Tawny Sedge. Shortly rhizomatous;   **974**
stems loosely tufted, to 65cm, trigonous, smooth; lowest
bract shorter than inflorescence, with long sheath; utricles
3.5-5mm, with beak 0.8-1.2mm (slightly longer and more
abruptly contracted than in C. distans). Native; marshes,
flushes and fens, inland and maritime; common in W & N Br,
locally frequent in Ir and rest of Br.
45 x 47. C. hostiana x C. viridula = C. x fulva Gooden. (C.
x appeliana Zahn) occurs frequently throughout most of Br and
Ir wherever C. hostiana meets any of the 3 sspp. of C.
viridula; probably the commonest hybrid.
46. C. flava L. - Large Yellow-sedge. Densely tufted; stems **974**
to 70cm, trigonous, smooth; lowest bract much longer than   **RR**
inflorescence, often patent or reflexed, with short sheath;
utricles 5.5-6.5mm, with reflexed beak 1.5-2.5mm. Native;
base-rich fen by lake; now only in Roudsea Wood, Westmorland;
formerly occurred in localities now supporting its hybrids
and in Cumberland.
46 x 47. C. flava x C. viridula = C. x alsatica Zahn (C. x
pieperiana Junge, C. x ruedtii Kreucker) occurs (with C.
viridula ssp. oedocarpa as the parent) in the Roudsea Wood
locality of C. flava, and (with C. viridula ssp. brachyrrhyncha as the parent) in MW Yorks, Hants and NE Galway.
47. C. viridula Michaux - Yellow-sedge. Differs from C.
flava as in Key F (couplets 16 & 17). Native. Fertile
intermediates occur in all 3 combinations between the sspp.
a. Ssp. brachyrrhyncha (Celak.) B. Schmid (C. lepidocarpa   **974**

Tausch, C. marshallii A. Bennett). Stems to 70cm, usually
straight; leaves usually c.1/2 as long as stems; utricles
3.5-5mm, with very abruptly contracted reflexed beak 1.3-2mm.
Base-rich fens, flushes, and lake-sides; frequent throughout
most of Br and Ir except in acid areas and much of S & C En
and Wa.
  b. Ssp. oedocarpa (Andersson) B. Schmid (C. demissa **974**
Hornem.). Stems to 50cm, usually curved; leaves usually
nearly as long as stems; utricles 3-4mm, with abruptly
contracted not or slightly reflexed beak 0.8-1.3mm (but lower
utricles are wholly reflexed). Acid or medium base-rich
fens, bogs, flushes, wet fields and by lakes; common over
most of BI except parts of C & E En. Commoner than ssp.
brachyrrhyncha everywhere except on limestone. In dried
material unequal drying of utricles can cause beaks to become
reflexed.
  c. Ssp. viridula (C. serotina Merat, C. scandinavica E. **974**
Davies, C. bergrothii Palmgren). Stems to 25cm (often
<10cm), usually straight; leaves usually c. as long as stems;
utricles 1.75-3.5mm, with abruptly contracted ± straight beak
0.2-1mm. Acid (mostly) or basic wet places in bogs, marshes
and dune-slacks and by lakes; scattered through most of BI,
frequent by coast in W & N. Plants with utricles 1.75-2mm
with beaks 0.2-0.3mm, mainly in N & W Sc, have been separated
as var. **pulchella** (Loennr.) B. Schmid (C. scandinavica).
  48. C. pallescens L. - Pale Sedge. Densely tufted; stems to **974**
60cm, triquetrous, rough distally; lowest bract exceeding
inflorescence, ± without sheath, crimped at base; utricles
2.5-3.5mm, with beak <0.1mm. Native; damp grassland,
woodland clearings and stream-banks; frequent over most of
Br, scattered in Ir.
  49. C. digitata L. - Fingered Sedge. Densely tufted; stems **974**
to 25cm, subterete, smooth; lowest bract glume-like, **R**
sheathing; utricles 3-4mm, pubescent, with beak 0.2-0.5mm.
Native; open woodland, scrub and grassy rocky slopes on chalk
and limestone; very local in En from N Somerset (formerly
Dorset) and Mons to Westmorland and NE Yorks.
  50. C. ornithopoda Willd. - Bird's-foot Sedge. Differs from **974**
C. digitata in stems to 15(20)cm; utricle beak 0.1-0.3mm; and **RR**
see Key C (couplet 9). Native; short limestone grassland;
very local in Derbys, Westmorland, Cumberland and NW Yorks.
  51. C. humilis Leysser - Dwarf Sedge. Densely tufted; stems **974**
to 10(15)cm, subterete, smooth; lowest bract glume-like, **R**
sheathing and nearly enclosing lowest female spike; utricles
2-3mm, pubescent, with beak <0.1mm. Native; short limestone
grassland; very locally common in SW En from Dorset to
Herefs.
  52. C. caryophyllea Latour. - Spring Sedge. Shortly **974**
rhizomatous; stems loosely tufted, to 30cm, trigonous to
triquetrous, smooth; lowest bract leaf- to bristle-like, with
sheath 3-5mm; utricles 2-3mm, pubescent, with beak <0.3mm.
Native; acid or basic dry to damp short grassland; frequent

throughout BI but rare in N Sc.

53. C. filiformis L. (C. tomentosa L.) - Downy-fruited **974**
Sedge. Rhizomatous; stems not or loosely tufted, to 50cm, **RR**
trigonous, smooth to slightly rough distally; lowest bract
shorter than to ± as long as inflorescence, with 0 or very
short sheath; utricles 2-3mm, pubescent, with beak <0.4mm.
Native; damp grassy places in fields and on waysides and in
woodland rides; very local in N Wilts, Oxon and E Gloucs,
formerly W Gloucs, Middlesex and Surrey.

54. C. ericetorum Pollich - Rare Spring-sedge. Shortly **974**
rhizomatous; stems loosely mat-forming, to 20cm, trigonous, **R**
smooth; lowest bract glume- to bristle-like, with 0 or very
short sheath; utricles 2-3mm, pubescent, with beak <0.3mm.
Native; dry short calcareous grassland; very local in E & N
En from Cambs and W Suffolk to Co Durham and Westmorland.

55. C. montana L. - Soft-leaved Sedge. Rhizomatous; stems **974**
loosely tufted, to 40cm, trigonous to triquetrous, rough **R**
distally; lowest bract glume- to bristle-like, with 0 or very
short sheath; utricles 3-4mm, pubescent, with beak <0.3mm.
Native; wet or dry grassy places in open or light shade on
acid or basic soils, very local in S C Br N to Derbys.

56. C. pilulifera L. - Pill Sedge. Densely tufted; stems to **974**
40cm, triquetrous, rough distally; lowest bract bristle- to
leaf-like, shorter than to rarely slightly longer than
inflorescence, with 0 or very short sheath; utricles 2-3.5mm,
pubescent, with beak 0.3-0.5mm. Native; grassy or barish
places in open or woodland on base-poor sandy or peaty soils;
frequent ± throughout BI.

57. C. atrofusca Schkuhr - Scorched Alpine-sedge. Shortly **975**
rhizomatous; stems loosely tufted, to 35cm, trigonous to **RR**
triquetrous, smooth; lowest bract leaf- to bristle-like,
shorter than inflorescence, with short to long sheath;
utricles 4-4.5mm, with beak 0.3-0.7mm. Native; mountain
flushes 540-1050m; very rare in M Perth, Westerness and Main
Argyll.

58. C. limosa L. - Bog-sedge. Rhizomatous; stems not or **975**
loosely tufted, to 40cm, trigonous, usually rough distally;
lowest bract c. as long as spike, with 0 or very short
sheath; utricles 3-4mm, with beak <0.2mm. Native; very wet
blanket- or valley-bogs; locally frequent in N & C Ir and N &
W Br, very scattered in and absent from most of S Ir and En.

59. C. rariflora (Wahlenb.) Smith - Mountain Bog-sedge. **975**
Shortly rhizomatous; stems loosely tufted or carpet-forming, **RR**
to 20cm, trigonous, usually smooth; lowest bract shorter than
inflorescence, with short sheath; utricles 3-4mm, with beak
<0.2mm. Native; wet peaty mountain slopes or in flushes at
750-1050m, where snow lies late; very local in CE Sc.

60. C. magellanica Lam. (C. paupercula Michaux) - Tall **975**
Bog-sedge. Shortly rhizomatous; stems to 40cm, trigonous, **R**
smooth; lowest bract usually exceeding inflorescence, with 0
or very short sheath; utricles 3-4mm, with beak <0.1mm.
Native; wet bogs with some water movement, often among

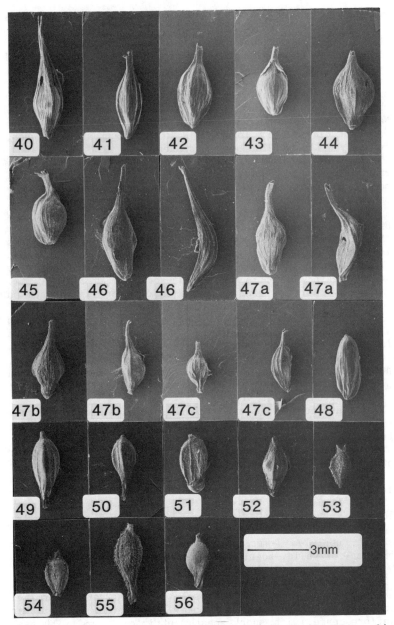

FIG 967 – Utricles of **Carex** subgenus **Carex**. 23, **C. hirta**. 24,
C. lasiocarpa. 25, **C. acutiformis**. 26, **C. riparia**. 27, **C.
pseudocyperus**. 28, **C. rostrata**. 29, **C. vesicaria**. 30, **C. x
grahamii**. 31, **C. saxatilis**. 32, **C. pendula**. 33, **C. sylvatica**.
34, **C. capillaris**. 35, **C. strigosa**. 36, **C. flacca**. 37, **C.
panicea**. 38, **C. vaginata**. 39, **C. depauperata**.

FIG 975 – Utricles of **Carex** subgenera **Carex** and **Primocarex**.
57, **C. atrofusca**. 58, **C. limosa**. 59, **C. rariflora**. 60, **C.
magellanica**. 61, **C. atrata**. 62, **C. buxbaumii**. 63, **C.
norvegica**. 64, **C. recta**. 65, **C. aquatilis**. 66, **C. acuta**. 67,
C. trinervis. 68, **C. nigra**. 69, **C. elata**. 70, **C. bigelowii**.
71, **C. microglochin**. 72, **C. pauciflora**. 73, **C. rupestris**. 74,
C. pulicaris.

Sphagnum; scattered in Br from Lake District to NW Sc, N Wa, Co Antrim. Our plant is ssp. **irrigua** (Wahlenb.) Hiit.

61. C. atrata L. - Black Alpine-sedge. Shortly rhizomatous; **975** stems loosely tufted, to 55cm, triquetrous, smooth or rough **R** distally; lowest bract exceeding inflorescence, with 0 or very short sheath; utricles 3-4mm, with beak 0.3-0.5mm. Native; wet rock-ledges above 720m; locally frequent in C & NW Sc, very rare in S Sc, Lake District and N Wa.

62. C. buxbaumii Wahlenb. - Club Sedge. Shortly rhizoma- **975** tous; stems not or loosely tufted, to 70cm, triquetrous, **RR** smooth; lowest bract exceeding inflorescence or not, with 0 or very short sheath; utricles 3-4.5mm, with beak 0-0.2mm. Native; wet fens; very rare in Main Argyll, Westerness and Easterness, formerly Co Antrim.

63. C. norvegica Retz. - Close-headed Alpine-sedge. Densely **975** tufted; stems to 30cm, trigonous, smooth or rough distally; **RR** lowest bract exceeding inflorescence, with 0 or very short sheath; utricles 1.8-2.5mm, minutely papillose, with beak 0.1-0.3mm. Native; N-facing damp rock-ledges and rocky slopes at 690-990m, where snow lies late; 5 places in C Sc.

64. C. recta Boott - Estuarine Sedge. Rhizomatous; stems **975** tufted, to 1.1m, triquetrous to trigonous, smooth; lowest **RR** bract exceeding inflorescence, with 0 or very short sheath; utricles 2.5-3mm, with beak $\leq$0.3mm. Native; forming extensive patches in wet estuarine areas; very local in NE Sc S to Easterness.

64 x 65. C. recta x C. aquatilis = C. x grantii A. Bennett occurs on banks of the Wick River, Caithness, where it forms a hybrid swarm.

65. C. aquatilis Wahlenb. - Water Sedge. Differs from C. **975** recta in leaves green (not glaucous) on upperside; stems **R** trigonous to subterete, brittle (not flexible); utricles 2-2.5mm; and female glumes subacute to obtuse and shorter than utricles (not long-pointed and much longer than utricles). Native; swampy areas by lakes and rivers and in marshes; locally frequent in Br N from Lake District, scattered in Wa and Ir.

65 x 68. C. aquatilis x C. nigra = C. x hibernica A. Bennett occurs in scattered places in Br N from C Sc, W Ir.

65 x 70. C. aquatilis x C. bigelowii = C. x limula Fries occurs at c.800-900m in E highlands of C Sc.

66. C. acuta L. - Slender Tufted-sedge. Differs from C. **975** recta in stems triquetrous, rough distally; utricles 2-3.5mm, obviously ribbed (not $\pm$ smooth); and acuminate female glumes usually longer than utricles. Native; by ponds, ditches, canals and rivers and in marshes; locally frequent throughout BI N to C Sc.

66 x 68. C. acuta x C. nigra (? = C x elytroides Fries) occurs in very scattered places in Br, including some sites in N Br without C. acuta.

66 x 69. C. acuta x C. elata = C. x prolixa Fries has been found in a few places in S En and S Wa.

67. C. trinervis Degl. - Three-nerved Sedge. Rhizomatous; **975** stems scarcely tufted, to 40cm, trigonous, smooth; lowest **RR** bract exceeding inflorescence, with 0 or very short sheath; utricles 3.5-5mm, distinctly veined, with beak <0.1mm. Possibly formerly native, now extinct; found in inland area of E Norfolk in 1869 (dune-slack sp. on Continent), probably with or entirely C. trinervis x C. nigra.

68. C. nigra (L.) Reichard - Common Sedge. Rhizomatous to **975** very shortly so; stems tufted to single, to 70cm, trigonous, smooth; lowest bract shorter than to c. as long as inflorescence, without sheath; utricles 2.5-3.5mm, distinctly ribbed, with beak <0.2mm. Native; wide range of wet acid to basic places, especially marshes and flushes; common throughout BI. Probably our commonest and most variable sp.

68 x 69. C. nigra x C. elata = C. x turfosa Fries occurs in scattered places in Ir and C & S Br.

68 x 70. C. nigra x C. bigelowii = C. x decolorans Wimmer occurs in the same area as C. aquatilis x C. bigelowii, often near C. aquatilis x C. nigra as well.

69. C. elata All. - Tufted-sedge. Densely tufted, often **975** tussock-forming; stems to 1m, triquetrous, rough distally; **R** lowest bract shorter than inflorescence, with 0 or very short sheath; utricles 2.5-4mm, distinctly ribbed, with beak <0.3mm. Native; bogs, fens, reedswamps and by rivers and lakes; locally frequent in Ir and C Br, very scattered in S En and N to C Sc.

70. C. bigelowii Torrey ex Schwein. - Stiff Sedge. Shortly **975** rhizomatous; stems scarcely tufted but often close, to 30cm, triquetrous, rough; lowest bract shorter than inflorescence, without sheath; utricles 2-3mm, smooth, with beak <0.2mm. Native; stony and heathy areas, often where snow lies late, and in flushed gullies, above 600m; common in highlands of Sc, scatterd S to N En and N Wa, very scattered in N & W (formerly E) Ir. Usually distinguishable from mountain variants of C. nigra by its reddish-brown (not blackish-brown) basal leaf-sheaths.

Subgenus 3 - PRIMOCAREX Kuek. (spp. 71-74). Spike 1, terminal, with male flowers at top and female below; bract 0; stigmas 2 and nut biconvex (C. pulicaris) or stigmas 3 and nut trigonous (other 3 spp.). An unsatisfactory subgenus, whose members would be better placed in 1 of the other 2 subgenera if only their relationships were clear.

71. C. microglochin Wahlenb. - Bristle Sedge. Rhizomatous; **975** stems usually single, to 12cm, terete to trigonous, smooth; **RR** utricles usually strongly reflexed, 3.5-5(6)mm tapered to beak 1-1.5mm, plus bristle exserted 1-2mm; stigmas 3. Native; base-rich flushes on open stony slopes at 600-900m, 1 locality in M Perth, discovered 1923.

72. C. pauciflora Light. - Few-flowered Sedge. Shortly **975** rhizomatous; stems not or very loosely tufted, to 25cm,

trigonous, smooth; utricles usually strongly reflexed,
5-7.5mm tapered to beak 1-2mm; stigmas 3. Native; acid
blanket bogs; frequent in C & N Sc, scattered S to NE Yorks
and Caerns, Co Antrim.

73. C. rupestris All. - <u>Rock Sedge</u>. Shortly rhizomatous; **975**
stems loosely tufted, to 20cm, triquetrous, ± smooth; **R**
utricles ± erect, 2-3.5mm, with beak 0.2-0.3mm; stigmas 3.
Native; on rock ledges or stony ground on limestone or
calcareous-flushed sandstone usually above 600m but in N Sc ±
down to sea-level; locally frequent in Sc from M Perth and
Angus to W Sutherland.

74. C. pulicaris L. - <u>Flea Sedge</u>. Densely tufted to shortly **975**
rhizomatous; stems to 30cm, terete, smooth; utricles usually
strongly reflexed, 3.5-6mm, with beak 0.2-0.5mm; stigmas 2.
Native; bogs, fens and flushes, usually base-rich; frequent
throughout most of BI, but absent from much of C & E En.

155. POACEAE - <u>Grass family</u>
(Gramineae)

Annuals or herbaceous perennials (rarely woody perennials -
bamboos), often with rhizomes or stolons, with usually
hollow, cylindrical (rarely flattened or other shapes but not
3-angled) stems; leaves alternate, with long, usually linear, **979**
entire, thin (but often rolled up or folded along long axis)
blade (<u>leaf</u>), with long, stem-sheathing, often cylindrical
lower part (<u>sheath</u>), usually with <u>ligule</u> (a membrane, a
fringe of hairs, or a membrane with a distal fringe of hairs)
at top of sheath on adaxial side, sometimes with small wing-
like extension (<u>auricle</u>) on either side at top of sheath.
Flowers much reduced, 1-many in discrete units (<u>spikelets</u>)
very variously arranged in terminal inflorescences, mostly
bisexual but often unisexual and bisexual mixed in same
spikelet, rarely male and female in different spikelets or
parts of plants (dioecious in <u>Cortaderia</u>), hypogynous;
perianth represented by 2 minute scales (<u>lodicules</u>) at base
of ovary (rarely fused or 0, 1 or 3); stamens usually 3,
rarely 2, 4 or 6; ovary 1-celled, with 1 ovule; styles 2,
rarely 1 or 3; stigmas elongated, feathery; fruit a typical
<u>caryopsis</u>, rarely the wall not fused to the seed inside.
Spikelets consisting of a series of bracts; usually 2 **979**
(sterile) <u>glumes</u> (<u>lower</u> and <u>upper</u>) at base, rarely 1 or 0,
with empty axils; 1-many <u>florets</u> above consisting of the
bisexual or unisexual flower proper plus 2 (fertile) glumes
on either side - <u>lemma</u> on abaxial and <u>palea</u> on adaxial side
(latter sometimes 0); the florets borne on slender axis
(<u>rhachilla</u>), often 1 or more sterile or even reduced to
vestigial scales; lemmas often with horny region (<u>callus</u>) at
base, this often vestigial but sometimes well developed
(often pointed and bristly); lemmas and/or glumes often with
short to long dorsal to terminal bristles (<u>awns</u>).

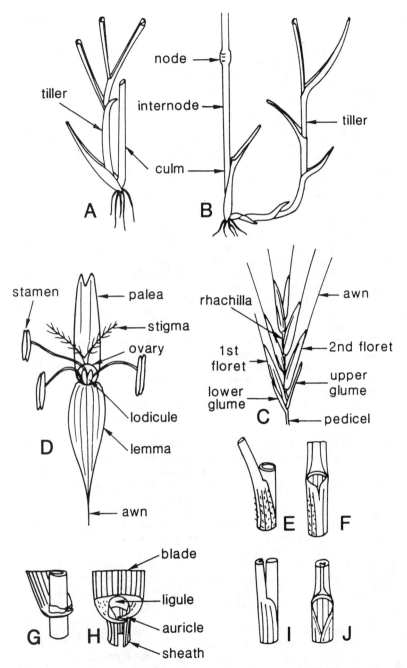

FIG 979 – **Poaceae** terminology. A, intravaginal innovation shoot. B, extravaginal innovation shoot. C, spikelet. D, floret with lemma pulled back. E–F, innovation leaf-sheath of **Festuca rubra**. G–H, innovation leaf-sheath of **F. pratensis**. I–J, innovation leaf-sheath of **F. ovina**. Drawn by S. Ogden.

Easily distinguished from all other grass-like plants by the distinctive inflorescence and flower structure, and usually from the Cyperaceae by the hollow stems. Before attempting use of the keys it is essential to dissect a spikelet and to understand thoroughly its structure, the detailed characteristics of all its parts, and the arrangement of spikelets in the inflorescence. The growth-form of the plant is also important; perennials can be distinguished by the presence of sterile leafy shoots (tillers) as well as flowering stems (culms), and they might have rhizomes and/or stolons as well. Spikelet, glume and lemma lengths exclude awns unless otherwise stated.

General key

1   Ligule a dense fringe of hairs, or membranous but
    breaking into dense fringe of hairs distally   Key A
1   Ligule membranous, sometimes jagged or pubescent but
    not densely fringed with hairs distally, sometimes 0   2
    2   Bamboos – stems woody; leaves with distinct short
        petiole between blade and sheath   Key B
    2   Stems not woody; leaves without petiole between
        blade and sheath   3
3   Maize – female spikelets in simple raceme (cob) low
    down on plant; male spikelets in terminal panicle or
    umbel of racemes (tassel)   **91. ZEA**
3   Male and female spikelets not in separate
    inflorescences   4
    4   Spikelets arising in groups of 2-7, one fatter
        and bisexual, the other 1-6 thinner and sterile
        or male   Key C
    4   Spiklets all bisexual and similar   5
5   Inflorescence a simple spike, or a simple raceme
    whose spikelets have pedicels <2mm   Key D
5   Inflorescence more complex than a simple spike or
    raceme of spikelets (but often very condensed)   6
    6   Inflorescence an umbel or raceme of spikes, or
        of racemes whose spikelets have pedicels <2mm   Key E
    6   Inflorescence a panicle, or a raceme whose
        spikelets have pedicels >3mm   7
7   Spikelets regularly proliferating to form small
    leafy plantlets (sexual spikelets present or not)   8
7   Spikelets not or only irregularly proliferating (if
    so, diseased, or because of sterility, or very late
    in season)   10
    8   Lemmas (whether proliferating or not) distinctly
        keeled on back along midrib   **20. POA**
    8   Lemmas (whether proliferating or not) rounded
        on back   9
9   Lemmas with awn arising from dorsal surface, with
    tuft of hairs arising just below base   **36. DESCHAMPSIA**
9   Lemmas with awn 0 or terminal, without tuft of hairs
    arising just below base   **13. FESTUCA**

10 Spikelets with only 1 floret (bisexual, not
   accompanied by vestigial florets or scales)   Key F
10 Spikelets with >2 bisexual florets, or with
   only 1 fertile floret but also >1 male or
   sterile florets or scales   11
11 Spikelets with 1 bisexual terminal floret and 1 or
   more male or sterile florets or scales below it   Key G
11 Spikelets with >2 bisexual florets, or with
   only 1 fertile floret but also >1 male or sterile
   florets or scales above it   12
   12 Ovary with pubescent terminal appendage extending
      beyond base of styles   Key H
   12 Ovary glabrous or pubescent, but style bases on
      apex of ovary (sometimes wide apart) and not
      exceeded by ovary appendage   13
13 Lemmas with dorsal or sub-terminal, usually bent
   awn, often bifid at apex, sometimes awnless and
   then clearly bifid at apex   Key I
13 Lemmas awnless and entire at apex, or with terminal
   straight or curved awn and then sometimes bifid or
   several-toothed at apex   Key J

Key A - Ligule, at least distally, a dense fringe of hairs
1 Maize - female spikelets in simple raceme (cob) low
   down on plant; male spikelets in terminal panicle
   or umbel of racemes (tassel)   **91. ZEA**
1 Male and female spikelets not in separate
   inflorescences   2
   2 Spikelets arising in pairs, one fatter and
     bisexual (sessile, often awned), the other thinner
     and male or sterile (awnless, often stalked)
                                             90. SORGHUM
   2 Spikelets all bisexual and similar   3
3 Spikelets with stout hooked spines on dorsal surface
   of glume   **80. TRAGUS**
3 Spikelets without hooked spines (but sometimes
   with barbed awns)   4
   4 Inflorescence a spike or contracted spike-like
     panicle; spikelets or groups of spikelets with
     1-several (sometimes proximally fused) barbed
     bristles at base   5
   4 Inflorescence of several spikes or racemes or
     obviously a panicle; spikelets without barbed
     bristles at base (though often with soft hairs) 6
5 Bristles fused proximally to form small cupule
   around spikelets, falling with spikelets to form a
   bur   **89. CENCHRUS**
5 Bristles remaining on axis when spikelets or florets
   fall, not fused   **87. SETARIA**
   6 Inflorescence an umbel or raceme of spikes, or
     of racemes whose spikelets have pedicels <2mm   7
   6 Inflorescence a panicle   16

7 Spikes all arising from same point at tip of stem   8
7 At least some spikes arising at different (though
   often very close) points along apical part of stem   11
   8 Spikes terminating in bare prolongation of
      axis   **75. DACTYLOCTENIUM**
   8 Spikes terminating in spikelet   9
9 Strongly stoloniferous; spikelets with 1 (bisexual)
   floret only   **78. CYNODON**
9 Plant tufted, not stoloniferous; spikelets with
   1-several bisexual florets, if only 1 then with
   sterile florets or scales distal to it   10
   10 Spikelets with >3 bisexual florets; lemmas
      not awned   **74. ELEUSINE**
   10 Spikelets with 1 bisexual floret plus 1-few
      sterile ones more distal; lemmas awned   **77. CHLORIS**
11 Spikelets with >3 florets   **74. ELEUSINE**
11 Spikelets with 1-2 florets   12
   12 Spikelets >8mm   **79. SPARTINA**
   12 Spikelets <6mm   13
13 Spikelets with small globose swelling at base,
   immediately below glumes   **85. ERIOCHLOA**
13 Spikelets without globose swelling at base   14
   14 Spikelets with 2 scales (upper and lower glume)
      at base <2/3 as long as spikelet, with 1
      (bisexual) floret   **76. SPOROBOLUS**
   14 Spikelets with 0 or 1 scale (lower glume) at base
      <2/3 as long as spikelet, with 2 florets (the
      upper bisexual, the lower male or sterile)   15
15 Upper lemma with awn 0.3-1mm (hidden between upper
   glume and lower lemma); lower lemma and upper glume
   shortly acuminate   **84. UROCHLOA**
15 Upper lemma awnless; lower lemma and upper glume
   obtuse   **83. BRACHIARIA**
   16 Most leaves >1m long, with very rough, cutting
      edges; plants dioecious   **69. CORTADERIA**
   16 Leaves <1m, without cutting edges; plants
      bisexual   17
17 Glumes minute, distant from lowest lemmas and
   separated from them by conspicuously pubescent
   callus   **8. EHRHARTA**
17 Glumes overlapping lowest lemmas   18
   18 Spikelets with basal tuft of long silky hairs
      becoming very conspicuous in fruit   **71. PHRAGMITES**
   18 Spikelets glabrous to pubescent, but not with
      basal tuft of long silky hairs   19
19 Spikelets with 1 floret only   20
19 Spikelets with >2 florets, sometimes only 1
   bisexual and >1 male or sterile   22
   20 Lemmas with conspicuous, bent awn   **10. STIPA**
   20 Lemmas awnless   21
21 Lemmas with 1 vein; spikelets all with 1 floret
                                             76. SPOROBOLUS

21 Lemmas with 3-5 veins; usually many spikelets with
    >1 floret **70. MOLINIA**
  22 Spikelets with 2 florets, the distal bisexual,
    the proximal male or sterile **81. PANICUM**
  22 Spikelets with >2 florets, at least the lowest
    bisexual 23
23 Lemma entire at apex, not awned, (1-)3(5)-veined;
    glumes much shorter than rest of spikelet 24
23 Lemma notched or 2-3-lobed at apex, sometimes awned,
    5-9-veined; glumes as long as spikelet (excl. awns)
    or nearly so 25
  24 Perennial of wet peaty areas; lemmas >3mm
                            70. MOLINIA
  24 Alien annuals or sometimes perennials; lemmas
    <3mm **73. ERAGROSTIS**
25 Lemmas with bent awn >5mm **67. RYTIDOSPERMA**
25 Lemmas awnless or with straight awn <1mm 26
  26 Lemmas deeply 2-lobed; spikelets falling whole;
    lemmas <3mm, with very broad hyaline margins;
    annual **68. SCHISMUS**
  26 Lemmas minutely 2-3-toothed; florets falling
    separately from glumes; lemmas >4mm, with very
    narrow hyaline margins; perennial **66. DANTHONIA**

Key B - Bamboos (stems woody; leaf-blades with short petiole)
1 Main stems ± square in section **6. CHIMONOBAMBUSA**
1 Main stems cylindrical, or flattened or grooved on
  one side 2
  2 Main stems flattened or grooved on 1 side at least
    at upper internodes 3
  2 Main stems ± cylindrical throughout except
    sometimes just above each node 4
3 Nodes of mid-region of main stems mostly with 2
  unequal branches and often a very small 3rd one;
  stems flattened or grooved on 1 side throughout
                            (PHYLLOSTACHYS)
3 Nodes of mid-region of main stems mostly with 3-5
  branches; stems often flattened or grooved only at
  upper internodes **(SEMIARUNDINARIA)**
  4 Nodes of mid-region of main stems mostly with
    1(-2) lateral branches 5
  4 Nodes of mid-region of main stems mostly with
    >3 lateral branches 8
5 Leaves pubescent on lowerside **4. SASAELLA**
5 Leaves glabrous (except sometimes on margin) 6
  6 Leaves 4-5x as long as wide, with 5-14 veins on
    either side of midrib **3. SASA**
  6 Leaves >6x as long as wide, with 2-9 veins on
    either side of midrib 7
7 Stems 2.5-5m; leaves 15-30cm, 2-4cm wide, with 5-9
  veins on either side of midrib **5. PSEUDOSASA**
7 Stems 0.2-2m; leaves 2.5-20cm, 0.3-2.5cm wide, with

	2-7 veins on either side of midrib	**2. PLEIOBLASTUS**
8	Stems $\leq$2.5m	**2. PLEIOBLASTUS**
8	Stems $>$2.5m	9
9	Leaves 15-25mm wide, mostly with 4-7 veins on either side of midrib	**2. PLEIOBLASTUS**
9	Leaves $<$12(15)mm wide, mostly with 2-4 veins on either side of midrib	10

10 Stems and/or sheaths strongly tinged brownish or purple; leaves acute to acuminate; inflorescence an open panicle, with narrow sheaths at base   **1. SINARUNDINARIA**

10 Stems bright to yellowish-green; sheaths straw-coloured; leaves drawn out into long fine point $\geq$1cm; inflorescence short, partly enclosed by large, wide sheaths   **(THAMNOCALAMUS)**

Key C - Spikelets arising in groups of 2-7, one fatter and bisexual, the other 1-6 thinner and male or sterile (N.B. Tragus (Keys A & D) may have 2-4 fertile and 1-3 sterile spikelets in groups)

1 Inflorescence a spike with 3 spikelets (each with 1 floret) per node, the central one bisexual, the 2 laterals male or sterile   **63. HORDEUM**

1 Inflorescence a panicle (sometimes strongly contracted and spike-like), not with regularly 3 spikelets per node and not all spikelets with 1 floret   2

2 Sterile or male spikelets with $\leq$3 florets, with glumes at least nearly as long as spikelet   3

2 Sterile or male spikelets clearly with $>$5 florets, with glumes much shorter than spikelet   4

3 Spikelets in pairs, 1 bisexual, 1 male or sterile, not falling as a unit   **90. SORGHUM**

3 Spikelets in groups of 3-7, 1 bisexual, the rest male or sterile, all falling as a unit   **42. PHALARIS**

4 Bisexual spikelets with 1 floret plus a sterile vestige, accompanied by 2-4 sterile spikelets, all falling as a unit   **17. LAMARCKIA**

4 Bisexual spikelets with (1)2-5 florets, accompanied by 1-few sterile spikelets, the latter and the glumes of fertile spikelets not falling   **16. CYNOSURUS**

Key D - Spikelets all bisexual and similar; inflorescence a simple spike or raceme with pedicels $\leq$2mm

1 Spikelets with stout hooked spines on dorsal surface of glume   **80. TRAGUS**

1 Spikelets without hooked spines (but sometimes with barbed awns)   2

2 Spikelets or groups of spikelets with 1-several (sometimes proximally fused) stiff bristles at base (NB do not confuse with bristle-like glumes)   3

   2 Spikelets without stiff bristles at base   4
3 Bristles fused proximally to form small cupule
  around spikelets, falling with spikelets to form
  a bur   **89. CENCHRUS**
3 Bristles remaining on axis when spikelets or grains
  fall, not fused   **87. SETARIA**
   4 One 1-floreted spikelet at each node   5
   4 Each node with >1 spikelet or with 1 spikelet
     with >1 floret   8
5 Densely tufted perennial; lemma awned; glumes 1-2,
  much shorter than floret   **9. NARDUS**
5 Annual; lemma not awned; glumes 1-2, at least as
  long as floret   6
   6 Spikelets not sunk in hollows in axis; glumes
     obtuse; lemma pubescent   **49. MIBORA**
   6 Spikelets sunk in hollows in axis; glumes acute
     to acuminate; lemma glabrous   7
7 Glumes 2 on all spikelets   **25. PARAPHOLIS**
7 Glumes 2 on terminal spikelet, 1 on lateral
  spikelets   **26. HAINARDIA**
   8 Spikelets 2-3 at each node   9
   8 Spikelets 1 at each node   12
9 Spikelets with 1(-2) florets; lemmas long-awned   10
9 Spikelets with (2)3-6 florets; lemmas awned or not   11
10 Two glumes of each spikelet free to base **63. HORDEUM**
10 Two glumes of each spikelet fused at base
                        62. HORDELYMUS
11 Lowest lemma >14mm, awnless; leaves very glaucous;
  all or ± all nodes with 2 spikelets   **61. LEYMUS**
11 Lowest lemma <14mm, awned; leaves not glaucous;
  upper nodes with 1 spikelet **60 x 63. X ELYTRORDEUM**
  12 Lower glume (except in terminal spikelet) 0 or
    ≤3/4 as long as upper   13
  12 Lower glume ≥3/4 as long as upper   16
13 Lemma with a bent awn from dorsal surface **32. GAUDINIA**
13 Lemma awnless or with straight to curved terminal
  or subterminal awn   14
   14 Upper glume with 1-3 veins; lemma very gradually
    narrowed to long terminal awn   **15. VULPIA**
   14 Upper glume with 5(-9) veins; lemma awnless or
    rather abruptly narrowed to long or short
    terminal or subterminal awn   15
15 Lower glume 0 except in terminal spikelet **14. LOLIUM**
15 Lower glume present on all or most spikelets
                13 x 14. X FESTULOLIUM
  16 Perennial; sterile shoots and often rhizomes
    present   17
  16 Annual; sterile shoots and rhizomes 0   21
17 Spikelets scarcely flattened, on distinct
  pedicels 0.5-2mm   **58. BRACHYPODIUM**
17 Spikelets flattened, sessile or with vestigial
  pedicel <0.5mm   18

18 Inflorescence axis not breaking up at maturity   19
18 Inflorescence axis breaking up at maturity,
    1 segment falling off with each spikelet   20
19 Plant densely tufted, without rhizomes; lemmas
    usually with awns >7mm; spikelets breaking up below
    each lemma at maturity, leaving glumes on rhachis;
    anthers <3mm   **59. ELYMUS**
19 Plant with long rhizomes, not densely tufted; lemmas
    rarely with awns >7mm; spikelets eventually falling
    whole, not leaving 2 glumes alone on rhachis;
    anthers >3.5mm   **60. ELYTRIGIA**
20 Lemmas awned   **60 x 63. X ELYTRORDEUM**
20 Lemmas not awned   **60. ELYTRIGIA**
21 Glumes truncate at apex (awned or not)   22
21 Glumes acuminate to obtuse, sometimes shouldered or
    notched, at apex (awned or not)   23
    22 Glumes rounded on back (except near apex),
        sometimes awned   **(AEGILOPS)**
    22 Glumes 1-2-keeled, not awned   **65. TRITICUM**
23 Glumes and lemmas <4mm, not awned   **23. CATAPODIUM**
23 Glumes and lemmas >4mm, the lemmas long-awned   24
    24 Glumes linear-lanceolate; spikelets with 2(3)
        florets (all bisexual)   **64. SECALE**
    24 Glumes lanceolate to ovate; spikelets with 2-3
        bisexual florets plus 1-few distal sterile ones,
        or with >4 bisexual florets   25
25 Spikelets with >4 bisexual florets, usually
    flattened narrow-side to inflorescence axis
                                          58. BRACHYPODIUM

25 Spikelets with 2-3 bisexual plus 1-few distal sterile
    florets, flattened broad-side to inflorescence axis
                                         (X TRITICOSECALE)

Key E - Spikelets all bisexual and similar; inflorescence an
        umbel or raceme of >2 spikes or racemes with pedicels
        <2mm
1 Spikes all arising from same point at tip of stem   2
1 At least some spikes arising at different (though
    often very close) points along apical part of stem 7
    2 Lemmas conspicuously awned   **77. CHLORIS**
    2 Lemmas not awned   3
3 Spikelets with 2-c.10 florets, if with 2 then
    lower one bisexual   4
3 Spikelets with 1-2 florets, if with 2 then lower
    one male or sterile   5
    4 Spikes terminating in bare prolongation of axis
                                        75. DACTYLOCTENIUM
    4 Spikes terminating in a spikelet   **74. ELEUSINE**
5 Fertile floret with 3-4 scales (1-2 glumes, lemma
    and palea of lower sterile floret) below it;
    annual   **88. DIGITARIA**
5 Fertile floret with 2 scales (2 glumes, or upper

glume and lemma representing lower sterile floret)
below it; stoloniferous perennial 6
6 Spikelet with 2 lower scales (glumes) 1-veined,
shorter than fertile lemma **78. CYNODON**
6 Spikelet with 2 lower scales (glume and sterile
lemma) 3-veined, longer than fertile lemma
86. PASPALUM
7 Spikelets with 2 florets, the upper bisexual, the
lower male or sterile 8
7 Spikelets 1-many flowered, at least the lower
bisexual 12
8 Ligule 0 **82. ECHINOCHLOA**
8 Ligule present (membranous or a fringe of hairs 9
9 Ligule membranous Return to 5
9 Ligule a fringe of hairs or membranous with a
fringe of hairs distally 10
10 Spikelets with small bead-like swelling at base;
lower glume $\pm$ 0 **85. ERIOCHLOA**
10 Spikelets without bead-like swelling at base;
lower glume present 11
11 Upper lemma with awn 0.3-1mm (hidden between upper
glume and lower lemma) **84. UROCHLOA**
11 Upper (and lower) lemma awnless **83. BRACHIARIA**
12 Spikelets with 1-2 florets 13
12 Spikelets with $\geq$3 florets 17
13 Spikelets $\geq$8mm **79. SPARTINA**
13 Spikelets $\leq$6mm 14
14 Ligule membranous 15
14 Ligule a fringe of hairs 16
15 Spikes 2(-4), 2-7cm; spikelets appressed-pubescent
on 1 side, glabrous on other **86. PASPALUM**
15 Spikes usually >4, 0.5-2cm; spikelets glabrous on
both sides **52. BECKMANNIA**
16 Two basal scales of spikelet much shorter than
rest of spikelet, glabrous **76. SPOROBOLUS**
16 Two basal scales of spikelet longer than rest
of spikelet, densly silky-pubescent **85. ERIOCHLOA**
17 All except terminal spikelet in each branch with
1 glume **14. LOLIUM**
17 All spikelets with 2 glumes 18
18 Lemmas 5-veined, >5.5mm **13 x 14. X FESTULOLIUM**
18 Lemmas 1-3-veined (sometimes with 1-3 extra
veins very close to midrib, forming thickened
keel), <5.5mm, keeled 19
19 Axis of inflorescence much longer than longest
spike; tip of lemma bifid (often awned from notch)
72. LEPTOCHLOA
19 Axis of inflorescence usually shorter (rarely
slightly longer) than longest spike; tip of lemma
acute to obtuse (often awned from tip) **74. ELEUSINE**

Key F - Spikelets all bisexual and similar; inflorescence a

panicle; spikelets with 1 (bisexual) floret and no
other scales or sterile florets

1 Spikelets with small bead-like swelling at base
                                        85. ERIOCHLOA

1 Spikelets without bead-like swelling at base   2
    2 Glumes ± 0 (reduced to rims below floret) **7. LEERSIA**
    2 Both glumes well developed   3

3 Panicle a soft woolly ovoid dense head; lemmas
    tapered to 2 apical bristles, with a longer awn
                                        47. LAGURUS

3 Panicle rarely a soft woolly ovoid dense head,
    if so then lemmas blunt and awnless; lemmas awned
    or not but without 2 long apical bristles   4
    4 Both glumes notched at apex and with awn from
        sinus; lemma with dorsal awn   5
    4 Glumes usually not awned, often tapered to fine
        apex but not notched at apex, if awned then lemma
        awnless   6

5 Awns of glumes (not lemma) >3mm; lemma awned from
    apex; fertile annual with deciduous spikelets
                                        50. POLYPOGON

5 Awns of glumes <3mm; lemma awned from below apex;
    sterile perennial with ± persistent spikelets
                                    43 x 50. X AGROPOGON

    6 Lemmas with long terminal awn with much twisted,
        stout proximal part   **10. STIPA**
    6 Lemmas with terminal, dorsal or 0 awn but
        without twisted proximal part   7

7 Floret with tuft of white hairs at base, the hairs
    >1/4 as long as lemma   8
7 Floret variously pubescent or glabrous, but without
    basal tuft of white hairs >1/4 as long as lemma 12
    8 Spikelets <9(10)mm; anthers <3mm   9
    8 Spikelets >(9)10mm; anthers >3mm   11

9 Lemma keeled, acute to obtuse, awnless; palea
    almost as long as lemma   **20. POA**
9 Lemma rounded on back, truncate or variously
    toothed at apex, often awned; palea c.2/3 to 3/4
    as long as lemma   10
    10 Tuft of hairs at base of lemma c.0.3–0.6mm
                                        43. AGROSTIS

    10 Tuft of hairs at base of lemma >1mm
                                        44. CALAMAGROSTIS

11 Panicle very pale, spike-like; spikelets 10–16mm;
    lemmas with basal hairs <1/2 as long as lemma, with
    awn 0–1mm; anthers 4–7mm, shedding pollen **45. AMMOPHILA**
11 Panicle usually purplish-green, usually ± lobed;
    spikelets 9–12mm; lemmas with basal hairs >1/2 as
    long as lemma, with awn 1–2mm; anthers 3–4.5mm,
    not opening   **44 x 45. X CALAMMOPHILA**
    12 Both glumes <3/4 as long as spikelet, usually
        1 or both c.1/2 as long   13

12 Both glumes >3/4 as long as spikelet, usually
    1 or both as long or longer   16
13 Lemmas with awns >3mm; annual   **48. APERA**
13 Lemmas awnless; normally perennial   14
    14 Lemmas ± truncate; ligule membranous   **22. CATABROSA**
    14 Lemmas obtuse to acute; ligule a fringe of hairs   15
15 Lemmas with 1 vein, <3mm; spikelets all with 1
    floret   **76. SPOROBOLUS**
15 Lemmas with 3-5 veins, >3mm; usually some spikelets
    with >1 floret   **70. MOLINIA**
    16 Glumes with swollen ± hemispherical base and
        long very tapering distal part   **46. GASTRIDIUM**
    16 Glumes without swollen base   17
17 Lemma shiny and much harder and tougher than glumes
    when mature   18
17 Lemma about same texture as glumes or more flimsy
    and delicate   19
    18 Lemmas awnless; ligules 2-10mm   **12. MILIUM**
    18 Lemmas with deciduous awn 2-5mm; ligules c.1mm
                11. ORYZOPSIS
19 Lemmas with subterminal awn 4-10mm   **48. APERA**
19 Lemmas with awn 0 or <4mm, or with awn ≥4mm and
    arising from lower 1/2 of back of lemma   20
    20 Panicle very compact, ovoid or oblong to
        cylindrical, spike-like, with very short branches   21
    20 Panicle diffuse or contracted but with obvious
        branches   22
21 Lemmas awnless; palea present; florets falling at
    maturity leaving glumes on panicle   **53. PHLEUM**
21 Lemmas usually awned (often shortly); palea absent;
    spikelets falling as a whole at maturity   **51. ALOPECURUS**
    22 Lemmas acute to obtuse, strongly keeled   **20. POA**
    22 Lemmas truncate or variously toothed at apex,
        rounded on back   23
23 Glumes with short pricklets or at least rough all
    over back; palea nearly as long as lemma; lemma
    c.1/2 as long as glumes   **50. POLYPOGON**
23 Glumes often with short pricklets on midrib but
    ± smooth otherwise; palea ≤3/4 as long as lemma;
    lemma c.2/3 to 3/4 as long as glumes   24
    24 Panicle branches bearing spikelets ± to base,
        disarticulating (if at all) near base of
        pedicels; very rare   **43 x 50. X AGROPOGON**
    24 Panicle branches with clear bare region at base,
        disarticulating (sometimes not) at base of
        lemmas; very common   **43. AGROSTIS**

<u>Key G</u> - Spikelets all bisexual and similar; inflorescence a
        panicle; spikelets with 1 bisexual floret plus 1 or
        more male or sterile florets or scales below it
1 Spikelets or groups of spikelets with 1-several
    (proximally fused or free) stiff bristles at base   2

1 Spikelets without stiff bristles at base   3
  2 Bristles fused proximally to form small cupule
    around spikelets, falling with spikelets to form
    bur   **89. CENCHRUS**
  2 Bristles not fused, retained on axis when
    spikelets or florets fall   **87. SETARIA**
3 Glumes minute, well separated from lowest lemmas
  by conspicuously pubescent callus   **8. EHRHARTA**
3 Glumes overlapping lowest lemmas   4
  4 Ligule a fringe of hairs, or membranous with a
    distal fringe of hairs   **81. PANICUM**
  4 Ligule membranous or 0   5
5 Bisexual floret with 1 male or sterile floret below
  it   6
5 Bisexual floret with 2 male or sterile florets below
  it   8
  6 Lower lemma awned from proximal 1/2 of back;
    perennial   **30. ARRHENATHERUM**
  6 Lower lemma awnless or with terminal awn; annual   7
7 Ligule membranous; sterile floret a minute scale;
  glumes equal, keeled   **42. PHALARIS**
7 Ligule 0; sterile floret as large as fertile one;
  glumes very unequal, not keeled   **82. ECHINOCHLOA**
  8 Lower glume c.1/2 as long as upper; lemmas of
    lower 2 florets with awns $\geq$1.5mm   **41. ANTHOXANTHUM**
  8 Glumes equal or subequal; lemmas of lower 2
    florets awnless or with awn <1mm   9
9 Panicle $\pm$ diffuse; lower 2 florets male, longer
  than bisexual floret   **40. HIEROCHLOE**
9 Panicle compact; lower 2 florets sterile, shorter
  than bisexual floret   **42. PHALARIS**

Key H - Spikelets all bisexual and similar; inflorescence a
    panicle or a raceme with pedicels >3mm; spikelets
    with $\geq$2 bisexual florets; ovary with pubescent
    terminal appendage extending beyond base of styles
    (NB Brachypodium and Gaudinia have similar ovaries,
    but the inflorescence is a spike or raceme, not a
    panicle - Key D).
1 At least lowest lemma with conspicuously bent awn
  arising dorsally from lower 2/3 of its length, or
  awnless with glumes longer than rest of spikelet   2
1 Lemmas with straight or slightly bent awns arising
  from apex or subapically, or awnless with glumes
  much shorter than rest of spikelet   4
  2 Easily uprooted annual without non-flowering
    shoots; upper glume >(15)20mm   **31. AVENA**
  2 Firmly rooted perennial with non-flowering
    shoots; upper glume <15(20)mm   3
3 Spikelets usually <11mm (excl. awns); awn of lowest
  lemma arising c.1/3 from base; upper glume 7-11mm;

lowest floret usually male (sometimes bisexual)
30. ARRHENATHERUM

3 Spikelets usually >11mm (excl. awns); all lemmas with
awn arising near middle or above it; upper glume
10-20mm; lowest floret bisexual   **29. HELICTOTRICHON**
  4 Lemmas strongly keeled on back   **57. CERATOCHLOA**
  4 Lemmas rounded on back or keeled near apex   5
5 Perennial, with sterile shoots at flowering time,
often with rhizomes   6
5 Annual (or biennial), without sterile shoots at
flowering time, without rhizomes   7
  6 Spikelets <15mm (excl. awns), narrowed to apex;
    lemmas <8mm (excl. awns)   **13. FESTUCA**
  6 Spikelets >15mm (excl. awns), ± parallel-sided
    almost to apex; lemmas >8mm (excl. awns)
55. BROMOPSIS
7 Spikelets ovate to lanceolate, slightly compressed,
markedly narrowed towards top; lemmas 5-11mm, with
awn c. as long or shorter; lower glume 3-7-veined;
upper glume 5-9-veined   **54. BROMUS**
7 Spikelets ± straight-sided, widening distally;
lemmas 9-36mm, with awn c. as long or longer, lower
glume 1(-3)-veined; upper glume 3(-5)-veined
56. ANISANTHA

Key I - Spikelets all bisexual and similar; inflorescence a
    panicle or a raceme with pedicels >3mm; spikelets
    with >2 bisexual florets, or with only 1 and that
    basal and with male or sterile florets distal to it;
    ovary glabrous or pubescent but without pubescent
    terminal appendage extending beyond base of styles;
    lemmas with subterminal to dorsal, usually bent awn,
    often bifid at apex, sometimes unawned and then
    always bifid at apex
1 Lemmas >6mm, usually with awn >10mm   2
1 Lemmas <6mm, never with awn >10mm   4
  2 Easily uprooted annual without non-flowering
    shoots; upper glume >(15)20mm   **31. AVENA**
  2 Firmly rooted perennial with non-flowering
    shoots; upper glume <15(20)mm   3
3 Spikelets usually <11mm (excl. awns); awn of lowest
lemma arising c.1/3 from base; upper glume 7-11mm;
upper lemma(s) often ± unawned   **30. ARRHENATHERUM**
3 Spikelets usually >11mm (excl. awns); all lemmas
with awns arising near middle or above it; upper
glume 10-20mm   **29. HELICTOTRICHON**
  4 Easily uprooted annual without non-flowering
    shoots   5
  4 Firmly rooted (unless in sand) perennial with
    non-flowering shoots   6
5 Spikelets with 2 florets; lemmas with bent awn
arising from below 1/2-way   **39. AIRA**

5 Spikelets with 3-5 florets; lemmas with usually ±
   straight subterminal awn   **35. ROSTRARIA**
   6 Awns slightly but distinctly widened towards
      apex (club-shaped)   **38. CORYNEPHORUS**
   6 Awns parallel-sided or tapered to apex   7
7 Glumes pubescent at least along midrib, falling with
   florets at maturity; lower lemma unawned; upper
   floret usually male   **37. HOLCUS**
7 Glumes glabrous, remaining on plant when florets
   fall; lowest lemma awned; all florets usually
   bisexual   8
   8 Spikelets with 2 florets; upper glume as long as
      spikelet or almost so; lemmas truncate to very
      blunt and usually jagged at apex   **36. DESCHAMPSIA**
   8 Spikelets with 2-4 florets; upper glume distinctly
      shorter than spikelet; lemmas acute and finely
      bifid at apex   **33. TRISETUM**

Key J - Spikelets all bisexual and similar; inflorescence a
        panicle or a raceme with pedicels >3mm; spikelets
        with >2 bisexual florets, or with only 1 and that
        basal and with male or sterile florets distal to it;
        ovary glabrous or pubescent but without pubescent
        terminal appendage extending beyond base of styles;
        lemmas unawned and entire at apex, or with terminal
        staight to curved awn and then sometimes bifid or
        several-toothed at apex
1 Leaf-sheaths fused almost to apex to form tube
   round stem   2
1 Leaf-sheaths with free, usually overlapping margins   4
   2 Lemmas finely pointed or awned, 5-veined **13. FESTUCA**
   2 Lemmas acute to rounded or ± 3-lobed at apex,
      7(-9)-veined   3
3 Fertile florets 1-3, with a club-shaped group of
   sterile florets beyond   **28. MELICA**
3 Fertile florets 4-c.16, with 0 or 1 reduced sterile
   floret beyond   **27. GLYCERIA**
   4 Lemmas with 3-5 short very pointed or shortly
      awned teeth at ± truncate apex   5
   4 Lemmas 1-pointed to rounded or 2-lobed at apex,
      awned or not   6
5 Ligule a fringe of hairs; glumes 3-5-veined (lateral
   veins often short); lemmas 7-9-veined, the veins
   ending short of the apical points   **66. DANTHONIA**
5 Ligule membranous; glumes 1-veined; lemmas
   3-5-veined, the veins running into the apical
   points   **24. SESLERIA**
   6 Lemmas 2-lobed or -toothed at apex, awned or
      not from the sinus   7
   6 Lemmas 1-pointed (sometimes very minutely notched)
      to rounded at apex, awned or not from tip   9
7 Lemmas with bent awn >5mm   **67. RYTIDOSPERMA**

7 Lemmas awnless or with straight awn <1mm   8
  8 Lemmas deeply 2-lobed; spikelets falling whole;
    annual   **68. SCHISMUS**
  8 Lemmas minutely 2-toothed; florets falling
    separately from glumes; perennial   **66. DANTHONIA**
9 Ligule a fringe of hairs   10
9 Ligule membranous   11
  10 Perennial of wet peaty areas; lemmas >3mm
                                                         70. MOLINIA
  10 Alien annuals or sometimes perennials; lemmas
    <3mm   **73. ERAGROSTIS**
11 Lemma strongly keeled throughout length   12
11 Lemma rounded on back or keeled only distally   14
  12 Lemmas 3-veined, with wide membranous ± shiny
    margins, rendering panicle silvery   **34. KOELERIA**
  12 Lemmas 5-veined, often with membranous margins
    but scarcely shiny and panicle not silvery   13
13 Spikelets borne in dense 1-sided clusters; lemmas
  long-acuminate to shortly awned   **21. DACTYLIS**
13 Spikelets not aggregated into dense 1-sided
  clusters; lemmas obtuse to acute   **20. POA**
  14 Lemmas awned   15
  14 Lemmas not awned   18
15 Easily uprooted annual without non-flowering shoots
                                                        15. VULPIA
15 Firmly rooted perennial with non-flowering shoots   16
  16 Lower glume mostly >3/4 as long as upper;
    anthers dehiscent   **13. FESTUCA**
  16 Lower glume mostly <3/4 as long as upper;
    anthers indehiscent   17
17 Pointed auricles present at junction of leaf-sheath
  and -blade; lemmas rather abruptly narrowed to awn;
  upper glume with 5-9 veins   **13 x 14. X FESTULOLIUM**
17 Auricles 0; lemmas very gradually narrowed to awn;
  upper glume with 1-3 veins   **13 x 15. X FESTULPIA**
  18 Lemmas acute to acuminate at apex   19
  18 Lemmas subacute to rounded at apex   20
19 Both glumes >1/2 as long as spikelet; lemmas
  3-veined   **34. KOELERIA**
19 At least lower glume, and usually both glumes, <1/2
  as long as spikelet; lemmas 5-veined   **13. FESTUCA**
  20 Lemmas strongly cordate at base, wider than long;
    spikelets pendent at maturity   **19. BRIZA**
  20 Lemmas not or scarcely cordate at base, longer
    than wide; spikelets not pendent   21
21 Lemmas 3-veined; spikelets with 2-3 florets; leaf-
  sheaths compressed, keeled on the back   **22. CATABROSA**
21 Lemmas 5-veined; spikelets usually with >4 florets;
  leaf-sheaths rounded on the back   22
  22 Pointed auricles present at junction of leaf-
    sheath and -blade; lower glume mostly <3/5 as
    long as upper   **13 x 14. X FESTULOLIUM**

22 Auricles 0; lower glume $\geq 3/5$ as long as upper   23
23 Lemmas minutely pubescent at base, with conspicuous
   membranous tips and margins; usually perennial
                                                      18. PUCCINELLIA
23 Lemmas glabrous, with very narrow membranous margins;
   annual   **23. CATAPODIUM**

Other genera - The following are all covered in the generic
keys.
Of the many extra genera occurring as rare casuals,
especially as bird-seed aliens or wool-aliens, mention should
be made of the wild-wheat **AEGILOPS** L. (Triticeae, Key D),
from Mediterranean, of which several annual spp. have been
recorded.
A new grain-crop, **X TRITICOSECALE** Wittm. ex A. Camus
(Triticale) (Triticeae, Key D), derived by artificial
hybridization of Triticum and Secale, is being grown on a
small scale and might increase in abundance. It has a
pendent spike and long awns; several widely differing
cultivars are grown.
2 densely clump-forming bamboos (Bambuseae, Key B) differ
from those treated in having culm internodes flattened or
grooved on 1 side (alternating in successive internodes) at
least at upper internodes. **PHYLLOSTACHYS** Siebold & Zucc. is
represented mainly by **P. bambusoides** Siebold & Zucc., from
China; it has stems to 12m and 5-12cm thick. **SEMIARUNDINARIA**
Nakai is represented by **S. fastuosa** (Latour-Marl. ex Mitf.)
Makino ex Nakai (Arundinaria fastuosa (Latour-Marl. ex Mitf.)
Houz.) (Narihira Bamboo), from Japan; it has stems to 8m and
2-4cm thick. These are occasionally found in neglected parks
and estates, as is **THAMNOCALAMUS** spathaceus, covered under
Sinarundinaria (genus 1).

SUBFAMILY 1 - BAMBUSOIDEAE (tribes 1-3). Woody bamboos or
herbaceous perennials; leaves often with short false-petiole
separating them from sheath, often with conspicuous cross-
veins as well as longitudinal veins; ligule membranous,
rarely a membrane fringed with hairs; sheaths not fused, the
lower ones often without or with very reduced leaves;
inflorescence a panicle; spikelets with ≥ 3-florets, with all
florets (except often the distal ones) bisexual, or with 1
bisexual floret with or without 2 large sterile ones below
it; glumes (0-)2(3); lemmas firm, 5-many-veined, awnless or
with terminal awn; paleas with 1-many veins, 1 on mid-line;
stamens 3-6; stigmas 2 or 3; lodicules 2 or 3; embryo
bambusoid or oryzoid; photosynthesis C3-type alone, with
non-Kranz leaf anatomy; fusoid cells often present; microhairs present; chromosome base-number 12.

TRIBE 1 - BAMBUSEAE (Arundinarieae) (genera 1-6). Woody
bamboos; leaves with short false-petiole and with cross-veins
as well as longitudinal veins; ligule membranous; lower

sheaths usually without or with very reduced leaves; spikelets with >3 florets, with all florets (except often the distal ones) bisexual; glumes (0-)2(3); lemmas 5-many-veined, awnless; paleas with >4 veins; stamens 3-6; stigmas 2 or 3; lodicules 2 or 3.

Flowering in all bamboos is very spasmodic and some have never been known to flower. Many of our spp. flowered for the first time in living memory in 1960s to 1980s, and some still are producing flowers, but flower production is too irregular and uncommon to be of much use in determination. Our spp. are grown in parks, wild gardens and estates, and become natd after neglect because of their persistence as thickets suppressing other vegetation. The 7 spp. treated here all have a strong rhizome system; other spp. forming compact clumps are no less well established but persist rather than spread.

1. SINARUNDINARIA Nakai - Fountain-bamboos
Stems mostly >3m, terete except just above nodes; nodes mostly with >3 lateral branches; leaves 5-12mm wide, glabrous, mostly with 2-4 veins on either side of midrib; stamens 3; stigmas 2-3.

Other spp. - S. nitida (Stapf) Nakai (Arundinaria nitida Stapf) (Chinese Fountain-bamboo), from W China, differs from S. anceps in stems brownish-green; leaves 3.5-8cm, mostly with 3-4 veins on either side of midrib; and sheaths pubescent. The related genus **THAMNOCALAMUS** Munro (Fargesia Franchet) contains **T. spathaceus** (Franchet) Soederstrom (Arundinaria spathacea (Franchet) D. McClint., A. murielae Gamble) (Umbrella Bamboo), from W China. This differs from S. anceps and S. nitida in its leaves up to 16 x 1.5cm and drawn out at apex into a long fine point; yellowish-green stems; and contracted inflorescence (rarely produced) enclosed in large sheaths. S. nitida and T. spathaceus are sometimes found in neglected parks and estates.

1. S. anceps (Mitf.) C.S. Chao & Renvoize (Arundinaria anceps Mitf., A. jaunsarensis Gamble) - Indian Fountain-bamboo. Stems 3-6m, 1-2cm thick, purplish-green; leaves 6-12 x 0.5-1.2cm, + glabrous, mostly with 2-3 veins on either side of midrib; sheaths glabrous. Intrd; natd in W & S Br, Ir and CI; India.

2. PLEIOBLASTUS Nakai - Bamboos
Stems 0.2-5m, terete except just above nodes; nodes with 1-many lateral branches; leaves 0.3-3.5cm wide, glabrous, mostly with 2-7 veins on either side of midrib; stamens 3; stigmas 3.

Other spp. - 2 other low-growing and 1 tall sp. from Japan are sometimes found in neglected parks and estates. **P.**

simonii (Carriere) Nakai (<u>Arundinaria</u> <u>simonii</u> (Carriere) Riv.
& M. Riv.) is >2.5m tall with stems 2-3cm thick, and has
leaves 15-25 x 1-3.5cm and mostly >3 branches per node. **P.**
chino (Franchet & Savat.) Nakai (<u>Arundinaria</u> <u>chino</u> (Franchet
& Savat.) Makino) is 1.2-2(3)m tall with stems 1-1.5cm thick,
and has leaves 8-20 x 0.8-3cm. **P. humilis** (Mitf.) Nakai
(<u>Arundinaria</u> <u>humilis</u> Mitf.) is 0.6-2m tall with stems 0.3-1cm
thick, and has leaves 8-20 x 1-2.5cm. The latter 2 have 1-3
branches per node. <u>P.</u> <u>chino</u> has leaves the same colour on
both surfaces, whereas all 3 other spp. have leaves paler on
1 or both sides of midrib of lowerside than on upperside.

  1. P. pygmaeus (Miq.) Nakai (<u>Arundinaria</u> <u>pygmaea</u> (Miq.)
Asch. & Graebner) - <u>Dwarf</u> <u>Bamboo</u>. Stems (0.2)0.4-0.75
(1.2)m, 1-3mm thick; leaves 2.5-7 x 0.3-1.5cm, with 2-3 veins
on either side of midrib; sheaths glabrous except on margins.
Intrd; natd rarely but widely in BI; origin unknown. Our
plant is var. **distichus** (Mitf.) Nakai.

3. SASA Makino & Shib. - <u>Bamboos</u>
Stems 0.5-3m, terete except just above nodes; nodes mostly
with 1 lateral branch; leaves 2.5-9cm wide, glabrous except
often with marginal hairs or bristles, mostly with 5-13 veins
on either side of midrib; stamens 6; stigmas 3.

  1. S. palmata (Burb.) Camus - <u>Broad-leaved</u> <u>Bamboo</u>. Stems
2-3m, 7-10mm thick; leaves 12-30(40) x 3.5-9cm, with 8-14
veins on either side of midrib; false-petioles usually green;
sheaths glabrous. Intrd; natd widely in BI; Japan. The
widest-leaved of our bamboos; flowered very abundantly in the
1960s.
  2. S. veitchii (Carriere) Rehder - <u>Veitch's</u> <u>Bamboo</u>. Stems
0.5-1.5m, 5-7mm thick; leaves 10-25 x 2.5-6cm, with 5-9 veins
on either side of midrib; false-petioles often purplish;
sheaths pubescent especially near base when young. Intrd;
natd widely but rarely in BI; Japan.

4. SASAELLA Makino - <u>Hairy</u> <u>Bamboo</u>
Stems 0.5-1.5m, terete except just above nodes; nodes mostly
with 1 lateral branch; leaves 1-3cm wide, sparsely pubescent
on upperside, pubescent to densely so on lowerside, mostly
with 3-5 veins either side of midrib; stamens 6; stigmas 3.

  1. S. ramosa (Makino) Makino & Shib. (<u>Arundinaria</u> <u>vagans</u>
Gamble) - <u>Hairy</u> <u>Bamboo</u>. Stems 0.5-1.5m, 3-8mm thick; leaves
8-20 x 1-3cm. Intrd; natd widely but rarely in BI,
especially in woods; Japan.

5. PSEUDOSASA Makino ex Nakai - <u>Arrow</u> <u>Bamboo</u>
Stems to 5m, terete except just above nodes; nodes mostly
with 1 lateral branch; leaves 2-5cm wide, glabrous, mostly
with 5-9 veins either side of midrib; stamens 3; stigmas 3.

1. P. japonica (Siebold & Zucc. ex Steudel) Makino ex Nakai **998** (Arundinaria japonica Siebold & Zucc. ex Steudel) - Arrow Bamboo. Stems 2.5-5m, 1-2cm thick; leaves 15-30 x 2-4cm. Intrd; frequently natd throughout much of BI, easily the commonest grown and natd bamboo, frequently flowers; Japan and Korea.

6. CHIMONOBAMBUSA Makino - Square-stemmed Bamboo
Stems to 8m, square in section with rough edges and faces; nodes mostly with 3 lateral branches; leaves 1-3cm wide, minutely pubescent when young, mostly with 8-14 veins on either side of midrib; stamens 3; stigmas 2.

1. C. quadrangularis (Fenzi) Makino (Arundinaria quadrangularis (Fenzi) Makino) - Square-stemmed Bamboo. Stems 5-8m, 2-4cm thick; leaves 10-20 x 1-3cm. Intrd; natd in SW En and SW Ir; China.

TRIBE 2 - ORYZEAE (genus 7). Herbaceous rhizomatous perennials; leaves without false-petiole, without cross-veins; ligule membranous; spikelets with 1 floret; glumes 0 or vestigial; lemmas 5-veined, awnless; palea 3-veined; stamens 3; lodicules 2.

7. LEERSIA Sw. - Cut-grass
1. L. oryzoides (L.) Sw. - Cut-grass. Culms to 1.2m, the **RR** sheaths and leaves rough with downward-pointed bristles; ligule 0.5-1.5mm; panicle diffuse or remaining 1/2-enclosed in sheaths (and then cleistogamous); florets 4-6mm. Native; wet meadows, ditches, canal- and river-sides; very locally frequent from S Somerset to Surrey and W Sussex, decreasing.

TRIBE 3 - EHRHARTEAE (genus 8). Herbaceous rhizomatous perennials; leaves without false-petiole, without cross-veins; ligule a membrane fringed with hairs; spikelets with 1 small bisexual floret with awnless 3-veined lemma and 1-veined palea, and two much larger sterile florets below each consisting of 1 long-awned 5-veined lemma only, the former obscured by the latter two; glumes 2, very small and separated from florets by stalk-like callus; stamens 4; lodicules 2.

8. EHRHARTA Thunb. (Microlaena R.Br.) - Weeping-grass
1. E. stipoides Labill. (Microlaena stipoides (Labill.) R. **998** Br.) - Weeping-grass. Culms erect to procumbent, to 70cm, the leaves rough with upward-directed bristles; ligule c.0.5mm; panicle sparsely branched, narrow; glumes <1mm; sterile lemmas 8-16mm plus awn up to 20mm. Intrd; rather infrequent wool-alien; scattered over En; Australia and New Zealand.

SUBFAMILY 2 - POOIDEAE (Festucoideae) (tribes 4-14). Annual

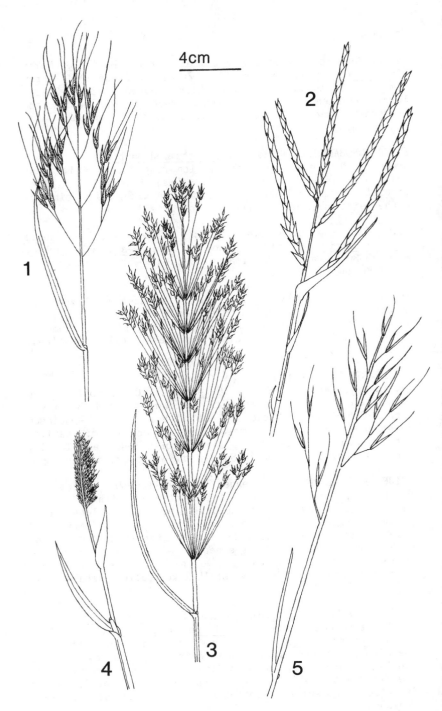

4cm

FIG 998 – Poaceae. 1, Stipa neesiana. 2, Pseudosasa japonica. 3, Oryzopsis miliacea. 4, Lamarckia aurea. 5, Ehrharta stipoides.

to perennial herbs; leaves without false-petiole or cross-veins; ligule membranous, sometimes a membrane fringed with hairs; sheaths usually not fused, sometimes fused; inflorescence a spike, raceme or panicle; spikelets with 1-many florets, with all florets (except often the distal ones) bisexual or some variously male or sterile; glumes (0-)2; lemmas firm to membranous, 3-9-veined, awnless or with terminal or dorsal awn; palea normally 2(-4)-veined without vein on mid-line, rarely with midrib or 0; stamens 1-3; stigmas 1-2; lodicules 0 or 2(-3), or 2 fused laterally; embryo pooid; photosynthesis C3-type alone, with non-Kranz leaf anatomy; fusoid cells 0; microhairs usually 0, present in <u>Nardus</u>; chromosome base-number usually 7, rarely higher or lower.

TRIBE 4 - NARDEAE (genus 9). Perennials; ligule membranous; sheaths not fused; inflorescence a ± 1-sided spike with 1 spikelet at each node; spikelets with 1 bisexual floret; glumes very short, upper often 0; lemma with 2-3 keels, 3-veined, with terminal awn; stamens 3; stigma 1; lodicules 0; ovary glabrous.

9. NARDUS L. - <u>Mat-grass</u>
  1. N. stricta L. - <u>Mat-grass</u>. Densely tufted; culms to 40(60)cm, wiry; leaves tightly inrolled; ligules 0.5-2mm; lemmas 6-9mm plus awn 1-3mm. Native; acid heaths, moors and mountain slopes, often dominant over large areas; throughout BI in suitable places, but not in much of C & E En or C Ir.

TRIBE 5 - STIPEAE (Milieae) (genera 10-12). Annuals or perennials; ligule membranous or sometimes a membrane with a fringe of hairs; sheaths not fused; inflorescence a ± diffuse panicle; spikelets with 1 bisexual floret; glumes equal or ± so, longer than body of lemma; lemma rounded on back, 3-5-veined, awnless or with long terminal awn, becoming hard and tightly wrapped round caryopsis; stamens 3; stigmas 2; lodicules 2-3; ovary glabrous.

10. STIPA L. - <u>Needle-grasses</u>
Annuals or perennials; ligule a membrane with a fringe of hairs; lemma with long basal callus pointed and with apically directed bristles, with long terminal awn with stouter spirally twisted proximal part separated from narrow distal part by conspicuous bend; lodicules 2-3.

Other spp. - >25 other spp. have been recorded as casuals, mainly wool-aliens due to the very penetrating propagules (grain tightly enclosed by lemma with its pointed callus and long awn). The 2 least rare are probably **S. aristiglumis** F. Muell., from Australia, and **S. formicarum** Del., from S America. Both are mainly wool-aliens with slenderer and shorter awns (<4cm) than the 2 spp. treated fully. <u>S.</u>

aristiglumis has lemmas c.6mm with densely appressed hairs, and awns <2cm; S. formicarum has lemmas c.5mm and glabrous apart from basal hair-tuft, and awns 2-4cm.

1. S. neesiana Trin. & Rupr. - American Needle-grass. **998** Perennial; culms to 60cm; glumes c.2cm incl. long fine 3-pronged apex, hyaline at margins and apex; lemma c.10mm incl. callus, glabrous except for dense tuft of hairs at base and on callus, with distinct membranous cupule at apex; awn 5-9cm. Intrd; wool-alien, sometimes ± natd in SE En; scattered in En; S America.

2. S. capensis Thunb. - Mediterranean Needle-grass. Annual; culms to 60cm; glumes c.1.5-2cm incl. long fine 1-pointed apex, wholly hyaline; lemma 6-8mm incl. callus, pubescent, with cupule of hairs at apex; awn 7-10cm. Intrd; wool-alien; scattered in En; Mediterranean.

11. ORYZOPSIS Michaux (Piptatherum P. Beauv.) - Smilo-grass
Perennials; ligule membranous; lemma shiny, with short smooth basal callus, with long straight deciduous terminal awn; lodicules ?2.

1. O. miliacea (L.) Benth. & Hook. ex Asch. & Schweinf. **998** (Piptatherum miliaceum (L.) Cosson) - Smilo-grass. Stems to **1052** 1.5m, densely tufted; panicle with long central axis bearing at each node many slender branches with spikelets clustered at ends; glumes 3-4mm; lemma 2-2.5mm plus awn 3-5mm. Intrd; casual wool-alien and ornamental; scattered in En and S Wa, natd in Jersey since 1931 and in W Kent; S Europe.

12. MILIUM L. - Millets
Annuals or perennials; ligule membranous; lemma shiny, with minute callus, awnless; lodicules 2.

1. M. effusum L. - Wood Millet. Tufted perennial; stems erect, to 1.5m; leaves 5-15mm wide; sheaths smooth; ligules 3-10mm; panicle 10-40cm, with patent to reflexed slender whorled branches bearing spikelets near ends; glumes 2.5-4mm; lemma nearly as long as glumes. Native; moist shady woods, usually on humus-rich soil; locally frequent throughout En, scattered in Wa, Ir and lowland Sc but absent from Outer Isles. Flowers May-Jul.

2. M. vernale M. Bieb. (M. scabrum Rich.) - Early Millet. **RR** Annual; stems procumbent to decumbent, to 10(15)cm; leaves 1-3mm wide; sheaths scabrid; ligules 2-3mm; panicle 1-4cm, with appressed branches; glumes 2-3mm; lemma 1.5-2mm. Native; short turf on sand-dunes and cliffs by sea; 2 places in Guernsey, 1st found 1899. Our plant has been segregated as ssp. **sarniense** D. McClint. Flowers Apr.

TRIBE 6 - POEAE (Festuceae, Seslerieae) (genera 13-24). Annuals or perennials; ligule membranous; sheaths not fused

or sometimes fused; inflorescence a panicle, or a raceme or
spike with normally 1 spikelet at each node; spikelets with
(1)3-many florets all (except the most apical) bisexual or
with a group of apical male or sterile florets, sometimes
spikelets in pairs, one of each pair entirely sterile: glumes
equal to very unequal, sometimes the lower vestigial or 0;
lemma rounded or keeled on back, 3-9-veined, awnless or with
terminal awn; stamens 1-3; stigmas 2; lodicules 2; ovary
glabrous or pubescent at apex.

13. FESTUCA L. - Fescues
Perennials with or without rhizomes, without stolons; sheaths
not fused or fused; inflorescence a panicle; spikelets with
(2)3-many florets all (except the most apical) bisexual,
sometimes proliferating; glumes subequal; lemmas rounded on
back, 3-5-veined, acuminate to subacute at apex, awned or
not; stamens 3.
   The F. rubra (spp. 6-7) and F. ovina (spp. 8-15) aggs are
extremely critical, and accurate identification requires
considerable experience and often the aid of leaf-sections,
leaf epidermis characters and chromosome numbers.   Shaded,
droughted, crowded-out or otherwise starved plants should be
avoided.   The classifications of these aggs adopted here
follow the results of intensive studies by A.-K.K.A. Al-
Bermani and M.J. Wilkinson respectively.
   Leaf, sheath and ligule characters refer to leaves on
tillers unless otherwise stated.   Spikelet lengths are not
total spikelet lengths, but the length from the base of the
lower glume to the apex (excl. awn) of the fourth lemma;
lengths of spikelets with only 3 florets are obtained by
addition to the total length of an increment equal to the
distance between the tips of the second and third lemmas.
Lemma lengths exclude awns, and refer to only the lowest 2
per spikelet.   Awn lengths are calculated by averaging the
lengths of all the awns of one spikelet, and then finding the
mean of this value for 5-10 spikelets.   All measurements
given are a range of means of 5-10 measurements per plant
(not a range of extremes).   The distinction between
extravaginal and intravaginal tillers is fundamental.   The **979**
latter arise parallel to the parent shoot and remain enclosed
within the parental leaf-sheath for some distance.   The
former arise ± at right angles to the parent shoot and break
through the parental leaf-sheath at its base.   Rhizomes
always start off extravaginally; hence the presence of
rhizomes indicates the existence of extravaginal shoots, but
the absence of rhizomes does not necessarily mean that all
tillers are intravaginal.   Leaf-sheath fusion should be
observed on the next to most apical sheath of a tiller by
stripping off all the more mature sheaths below it.

1 Base of leaf on each side extended into pointed
   auricles clasping stem at level of and on opposite

side to ligule   2
1 Leaves without auricles, or with short rounded
  auricles not clasping stem   4
  2 Lemmas with awns longer than body; exposed nodes
    of culms dark violet-purple   **3. F. gigantea**
  2 Lemmas with awns 0 or much shorter than body;
    exposed nodes of culms green, sometimes tinged
    purplish   3
3 Auricles glabrous; lowest 2 panicle nodes with
  two unequal branches, the shorter with 1-2(3)
  spikelets   **1. F. pratensis**
3 Auricles usually fringed with minute hairs (often few,
  wearing off with age); lowest 2 panicle nodes with
  two subequal branches, the shorter with (3)4-many
  spikelets   **2. F. arundinacea**
  4 Leaves of culms and tillers flat (or folded
    longitudinally when dry), >4mm wide (or >2mm from
    midrib to edge)   5
  4 Leaves of tillers and usually culms folded
    longitudinally or the edges also inrolled, <4mm
    wide (or <2mm from midrib to edge)   6
5 Leaves 4-14mm wide; ligules >1mm; lemmas 3-veined,
  awnless; rhizomes 0; ovary with pubescent apex
                                                   4. F. altissima
5 Leaves <5mm wide; ligules <1mm; lemmas 5-veined,
  awned; rhizomes present; ovary with glabrous apex
                                                      7. F. rubra
  6 Young leaves with sheaths fused almost up to top;
    some or all tillers extravaginal   7
  6 Young leaves with sheaths not fused but with
    overlapping margins; all tillers intravaginal
    (F. ovina agg.)   9
7 Ovary (and caryopsis) with pubescent apex; leaves
  with 3(-5) veins; leaves of culms and tillers
  markedly different, the former flat and 2-4mm wide,
  the latter folded and <0.6mm from midrib to edge
                                                 5. F. heterophylla
7 Ovary (and caryopsis) glabrous; leaves with 5-9(11)
  veins; leaves of culms and tillers similar to
  obviously different   8
  8 Leaves with densely pubescent adaxial ribs,
    rounded on midrib abaxially, with abaxial
    sclerenchyma usually continuous or semi-
    continuous, with distinct sclerenchyma bundles in
    adaxial ribs; always coastal   **6. F. arenaria**
  8 Leaves with scabrid or sparsely pubescent adaxial
    ribs, obtuse to keeled on midrib abaxially, with
    abaxial sclerenchyma in discrete bundles, often
    without or with very sparse sclerenchyma in
    adaxial ribs; coastal or inland   **7. F. rubra**
9 Leaves with 5-9 veins, with 4(-6) adaxial grooves;
  lemmas with awns usually >1.2mm and often >1.6mm   10

9 Leaves with 5-7 veins, with 2(-4) adaxial grooves;
lemmas with awns <1.6mm and often <1.2mm   11
  10 Pedicels 1.2-2.8mm; spikelets 6.1-8.5mm; lemmas
    with awns 1.2-2.6mm; sheaths often sparsely
    pubescent; leaves not or slightly glaucous, with
    abaxial sclerenchyma in 3 main islets (at midrib
    and edges)   **15. F. brevipila**
  10 Pedicels 0.5-1.8mm; spikelets 5.4-7mm; lemmas
    with awns 0.5-1.5mm; sheaths glabrous; leaves
    usually very glaucous, with abaxial sclerenchyma
    in >5 main islets (at midrib, edges and
    variously in between)   **14. F. longifolia**
11 Spikelets 4.7-7mm; lemmas with awns usually 0-1mm;
panicles mostly <8cm, with <26 spikelets and lowest
2 nodes <2cm apart; leaves usually <0.57mm midrib
to edge   12
11 Spikelets often >7mm, rarely <6mm; lemmas with awns
usually >1mm; panicles <13cm, with <40 spikelets
and lowest 2 nodes <3.7cm apart; leaves >0.57mm
midrib to edge   14
  12 Spikelets all or mostly proliferating; sexual
    florets (if present) with lemmas 3.4-4.2mm and
    awns 0-0.2mm   **9. F. vivipara**
  12 Spikelets not or rarely some proliferating; sexual
    florets with lemmas 2.5-4.9mm and awns 0-1mm   13
13 Spikelets <5.5mm; lemmas <3.5mm, with awns 0-0.6mm;
leaves glabrous, 0.3-0.45(0.53)mm from midrib to
edge   **10. F. filiformis**
13 Spikelets >5.2mm; lemmas >3.2mm, with awns 0-1.6mm;
leaves often pubescent at base, 0.3-0.75mm from
midrib to edge   **8. F. ovina**
  14 Culms 19-66cm; panicles 3.7-8.6cm; pedicels 0.6-
    2.5mm; widespread   **13. F. lemanii**
  14 Culms 11-35cm; panicles 2.3-5.9cm; pedicels 0.8-
    1.5mm; coastal in CI only   15
15 Panicles well exserted from sheath at anthesis;
leaves usually not glaucous; culms erect; usually on
dunes   **11. F. armoricana**
15 Panicles not completely or only just exserted from
sheath at anthesis; leaves often slightly glaucous;
culms erect to procumbent; usually on cliffs
                              12. F. huonii

Other spp. - Much mis-application of names has occurred in
this genus, and records of other spp. are mainly attributable
to that. **F. glauca** Vill. is often grown for ornament but is
not natd; records of it refer to F. longifolia, most of whose
records refer in turn to F. brevipila and/or F. lemanii.
Recent records of **F. guestfalica** Boenn. ex Reichb. refer to
F. brevipila, not to F. ovina ssp. ophioliticola. Records of
F. arvernensis Auq., Kerguelen & Markgr.-Dann. and **F.
indigesta** Boiss. are also errors.

1. F. pratensis Hudson (F. elatior auct. non L.) - Meadow **979**
Fescue. Culms to 80(120)cm; rhizomes 0; sheaths not fused;
leaves flat, 3-8mm wide, with glabrous pointed auricles;
ligules <2mm; panicle erect to pendent; spikelets 9-11mm;
lemmas 6-7mm, awnless; 2n=14. Native; meadows, hedgerows,
waysides, by ditches and rivers, usually on rich moist soil;
frequent throughout most of BI.
   1 x 2. F. pratensis x F. arundinacea = F. x aschersoniana
Doerfler resembles F. arundinacea but shows some influence of
F. pratensis, has 2n=28, and is sterile; it has been reported
only from Main Argyll but is probably widespread.
   1 x 3. F. pratensis x F. gigantea = F. x schlickumii
Grantzow resembles F. gigantea but shows some influence of F.
pratensis, has 2n=28, and is sterile; it has been reported
only from W Norfolk.
   2. F. arundinacea Schreber (F. elatior L.) - Tall Fescue.
Culms to 120(200)cm; rhizomes 0; sheaths not fused; leaves
usually flat, (1)3-12mm wide, with usually ciliate pointed
auricles; ligules <2mm; panicle erect or pendent; spikelets
(8)9-12mm; lemmas 6-7.5(9)mm with awn 0-4mm; 2n=42. Native;
grassy places, rough and marginal ground on wide range of
soils; common throughout BI, generally under-recorded. Very
variable, probably represented by several ecotypes of
different habitats, but the situation is obscured by use of
the sp. as a hay-grass and its frequent naturalization from
this source.
   2 x 3. F. arundinacea x F. gigantea = F. x fleischeri
Rohlena (F. x gigas O. Holmb.) resembles F. x aschersoniana
but has hairs on the auricles; 2n=42 but sterile; scattered
in Br N from Hunts, probably elsewhere.
   3. F. gigantea (L.) Villars - Giant Fescue. Culms to
100(150)cm; rhizomes 0; sheaths not fused; leaves flat,
6-18mm wide, with glabrous pointed auricles; ligules <2.5mm;
panicle pendent; spikelets 8-13mm; lemmas 6-9mm with usually
wavy awn 10-18mm; 2n=42. Native; woods, hedgerows and other
shady places; common in Br and Ir except N & NW Sc. Sometimes confused with Bromopsis ramosa, with which it very
often grows, but the latter has pubescent (not glabrous)
leaves and sheaths and awns shorter than body of lemma.
   4. F. altissima All. - Wood Fescue. Culms to 120(150)cm;   **R**
rhizomes 0; sheaths not fused; leaves flat, 4-14mm wide,
without auricles; ligules 1-5mm; panicle ± pendent; spikelets
5-8mm; lemmas 4-6mm, awnless; 2n=14. Native; moist stony
slopes and ravines in woods and copses; very scattered in Br
and Ir, especially N & W.
   5. F. heterophylla Lam. - Various-leaved Fescue. Culms to **1008**
100(120)cm, densely tufted; rhizomes 0; sheaths fused ± to
apex; leaves folded, 0.3-0.6mm from midrib to edge, without
auricles, with 5 small abaxial sclerenchyma islets; ligules
<1mm; spikelets 8-11.5mm; lemmas (4.7)5-6.5(8)mm, with awn to
4.5(6)mm; 2n=28. Intrd; grown for ornament and appearing as
contaminant of other grass-seed, natd in woods and wood-

borders on light soils; scattered in Br, mainly S En; Europe.
6. F. arenaria Osbeck (F. juncifolia Chaub., F. rubra ssp. **1008**
arenaria (Osbeck) F. Aresch.) - Rush-leaved Fescue. Culms to **R**
75(90)cm, scattered; rhizomes very long, sheaths fused ± to
apex; leaves folded, 0.5-1.9mm from midrib to edge, without
auricles, usually with continuous or subcontinuous abaxial
sclerenchyma, often with islets with 'tails', sometimes with
5-13 discrete islets; ligules <1mm; spikelets 8-14.2mm; upper
glume 3.5-9.1mm; lemmas 6-10.1mm, with awns 0.5-2.6mm; 2n=56.
Native; mobile sand-dunes and sandy shingle by sea; frequent
on coasts of Br N to E Ross and S Wa, rare to W Ross.
7. F. rubra L. - Red Fescue. Culms to 75(100)cm, often much **979**
less; sheaths fused ± to apex; leaves flat or folded,
0.5-1.4(2.5)mm from midrib to edge, without auricles, with
abaxial sclerenchyma in 5-9 discrete islets; ligules <1mm.
Extremely variable; just worthy of division into 7 (perhaps
more) sspp., some of which are often recognized as spp. 6 of
the sspp. represent ± distinct combinations of characters,
but all are closely approached by variants of ssp. rubra.
Several of the sspp. are sold as grass-seed and become natd
on road-verges, etc.
1 Rhizomes 0 or very few and very short **d. ssp. commutata**
1 Rhizomes well developed (plants densely tufted or
　not)   2
　2 Leaves 0.8-1.4(2.5)mm from midrib to edge, with
　　distinct islets of sclerenchyma in adaxial ribs;
　　lemmas usually 6-8mm   3
　2 Leaves 0.5-1.2mm from midrib to edge, usually
　　without or with very sparse sclerenchyma in
　　adaxial ribs (except in ssp. juncea); lemmas
　　4.2-6.5mm   4
3 Leaves of culms and often tillers flat; culms up to
1m; panicle diffuse; widespread   **g. ssp. megastachys**
3 Leaves of culms and tillers folded; culms up to
70cm; panicle compact with ± appressed branches;
Sc only   **f. ssp. scotica**
　4 Lemmas 4.2-6mm, with awns 0.1-1.1mm, glaucous,
　　usually with dense white hairs (sometimes glabrous
　　but usually some pubescent plants nearby); Sc and
　　Caerns only   **e. ssp. arctica**
　4 Lemmas 4.4-8mm, with awns 0.5-2.8mm, glaucous
　　or not, the hairs (if present) not dense and
　　white; widespread   5
5 Spikelets 8.7-11.2mm; lemmas 5.7-8 x >2mm, with
awns 1.1-2.8mm; saline sand or mud, often forming
dense mats   **c. ssp. litoralis**
5 Spikelets 6.8-10.2mm; lemmas 4.4-6.7 x 1.5-2.4mm,
with awns 0.5-2.2mm; rarely in saline soil   6
　6 Rhizomes short, forming dense tufts, often some
　　plants in population with very glaucous leaves
　　　　　　　　　　　　　　　　　　　　　　　 b. ssp. juncea
　6 Rhizomes medium to long, forming loose patches;

plants rarely glaucous   **a. ssp. rubra**
a. Ssp. rubra. Rhizomes well developed, usually forming **1008**
loose patches; culms to 75cm; leaves folded, rarely flat,
0.6–1.3mm from midrib to edge; spikelets 6.9–10.2mm; upper
glume 2.6–5.3mm; lemmas 4.6–6.7mm with awns 0.8–2.1mm; 2n=42,
56. Native; all kinds of grassy places; common through BI.
b. Ssp. juncea (Hackel) K. Richter (ssp. pruinosa (Hackel) **1008**
Piper). Rhizomes short, forming dense tufts; culms to 75cm;
leaves folded, 0.5–1.2mm from midrib to edge; spikelets
6.8–9.9mm; upper glume 3.2–5.3mm; lemmas 4.4–6.5mm, with awns
0.5–2.2mm; 2n=42. Native; maritime cliffs and inland grassy
rocky places; round coasts of BI and in hilly areas of N Br,
rarely inland elsewhere. Separation of ssp. pruinosa from
ssp. juncea is often impossible in practice. The former is
nearly always coastal and often has markedly pruinose leaves,
but even on the coast wholly non-pruinose populations occur.
It is also usually a shorter plant, but this might be a
dwarfing effect of the coastal environment.
c. Ssp. litoralis (G. Meyer) Auq. Rhizomes rather short, **1008**
often forming dense mats; culms to 55cm; leaves folded,
0.6–1mm from midrib to edge; spikelets 8.7–11.2mm; upper
glume 4.3–6.2mm; lemmas 5.7–8mm, with awns 1.1–2.8mm; 2n=42.
Native; salt-marshes and other muddy or sandy saline areas;
probably in suitable places round coasts of BI. Has
characteristically short culms, large lemmas and narrow,
short leaves, but the distinctive habitat is diagnostic.
d. Ssp. commutata Gaudin (ssp. caespitosa Hackel, var. **1008**
fallax sensu Tutin, F. nigrescens Lam.). Rhizomes 0 or very
few and short, forming dense tufts; culms to 75cm; leaves
folded, 0.6–1.1mm from midrib to edge; spikelets 7.2–9.1mm;
upper glume 3.2–5mm; lemmas 4.4–6.3mm, with awns 1.1–2.6mm;
2n=42. Native; grassy places and rough ground, usually in
well-drained soils; probably throughout Br due to extensive
use as grass-seed (Chewing's Fescue).
e. Ssp. arctica (Hackel) Govoruchin (F. richardsonii Hook.). **1008**
Rhizomes well developed, forming loose patches; culms to
50cm; leaves folded, 0.5–1.2mm from midrib to edge; spikelets
6.6–8.8mm; upper glume 3.1–5.5mm; lemmas 4.2–6mm, with awns
0.1–1.1mm; 2n=42. Native; wet mountain slopes and gulleys,
rock-crevices and flushes down to sea-level, often on
serpentine; scattered from C Sc to Shetland, Caerns, probably
under-recorded. The densely white-pubescent, glaucous
spikelets with short awns are diagnostic, but in many
populations some plants are glabrous.
f. Ssp. scotica S. Cunn. ex Al-Bermani. Rhizomes well **1008**
developed, forming loose patches; culms to 70cm; leaves
folded, 0.8–1.4mm from midrib to edge; spikelets 9.4–11.8mm;
upper glume 4.6–6.5mm; lemmas 6.1–7.9mm, with awns 1.2–2.3mm;
2n=56, 70. Native; grassy and rocky places from near sea-
level to >800m; Sc from Main Argyll to Shetland and Outer
Hebrides, probably under-recorded. The long spikelets,
lemmas and awns and erect panicle-branches are diagnostic.

g. Ssp. megastachys Gaudin (ssp. fallax (Thuill.) Nyman, **1008**
ssp. multiflora Piper, var. planifolia Hackel, F. diffusa
Dumort., F. heteromalla Pourret, non F. rubra var. fallax
sensu Tutin). Rhizomes well developed, forming diffuse
patches; culms to 100cm; leaves often flat, sometimes folded,
1-1.4(2.5)mm from midrib to edge; spikelets 7-11.2mm; upper
glume 3.5-6.2mm; lemmas 4.7-7.9mm, with awns 0.5-3mm; 2n=56,
?70. Probably intrd; grassy places, especially on waysides
where it is much planted; scattered throughout Br; Europe.
8-15. F. ovina L. agg. Plants densely tufted, without
rhizomes; sheaths not fused; tiller and culm leaves folded,
0.3-0.95mm from midrib to edge, with very small rounded
auricles, with continuous or discontinuous abaxial
sclerenchyma; ligules <1mm.
8. F. ovina L. - Sheep's-fescue. Culms to 50cm; leaves not **979**
glaucous or sometimes slightly so, 0.4-0.7mm from midrib to
edge, with 5-7 veins and 2(-4) adaxial grooves, with abaxial
sclerenchyma in thin broken or sometimes continuous band;
panicles 1.5-7.8cm; pedicels 0.8-2.8mm; spikelets 5.3-7.2mm;
lemmas 3.3-4.8mm, with awns 0-1mm.
1 Spikelets 5.5-7.5mm; lemmas 3.6-4.9mm; leaves with
  (5-)7 veins   **c. ssp. ophioliticola**
1 Spikelets 5.3-6.3mm; lemmas 3.1-4.2mm; leaves with
  5-7 veins   **2**
  2 Awns 0-0.8mm; leaves usually pubescent at base;
    lemmas usually pubescent; stomata mostly >31.5
    microns   **b. ssp. hirtula**
  2 Awns 0-1.2mm; leaves and lemmas usually scabrid;
    stomata mostly <31.5 microns   **a. ssp. ovina**
a. Ssp. ovina. Culms 10-35cm; leaves 0.33-0.67mm from **1009**
midrib to edge; panicles 2.2-7.3cm; pedicels 0.8-2.2mm;
lemmas with awns 0-1.1mm; 2n=14. Native; grassy places on
well-drained, usually acid soils; common in N, C & SW Br,
very sparse in SC & SE En, ?Ir.
b. Ssp. hirtula (Hackel ex Travis) M. Wilkinson (F. **1009**
tenuifolia var. hirtula (Hackel ex Travis) Howarth, F.
ophioliticola ssp. hirtula (Hackel ex Travis) Auq.). Culms
6-45cm; leaves 0.35-0.67mm from midrib to edge; panicles
1.5-6.6cm; pedicels 0.9-2.7mm; lemmas with awns 0-0.8mm;
2n=28. Native; grassy places on well-drained, usually acid
soils; common throughout BI.
c. Ssp. ophioliticola (Kerguelen) M. Wilkinson (?ssp. **1009**
guestfalica (Boenn. ex Reichb.) K. Richter, F. ophioliticola
Kerguelen, ?F. guestfalica Boenn. ex Reichb.). Culms
20-50cm; leaves 0.39-0.75mm from midrib to edge; panicles
2.8-8cm; pedicels 1.3-3.6mm; lemmas with awns 0-1.6mm; 2n=28.
Native; grassy places on well-drained, often calcareous or
serpentine soils; locally common throughout Br and Ir.
9. F. vivipara (L.) Smith. - Viviparous Sheep's-fescue. **1009**
Culms to 44(50)cm; differs from F. ovina in all or most
spikelets proliferating; completely fertile spikelets (if
present) 5.5-6.2mm, with lemmas 3.4-4.2mm with awns 0-0.2mm;

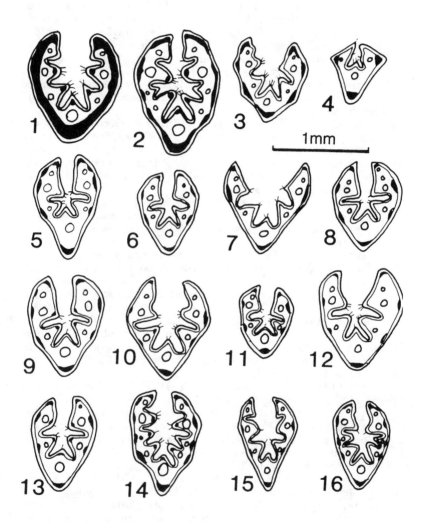

FIG 1008 – Transverse sections of innovation leaves of *Festuca rubra* agg. 1–3, **F. arenaria.** 4, **F. heterophylla.** 5–16, **F. rubra.** 5, ssp. **rubra.** 6, ssp. **littoralis.** 7–8, ssp. **arctica.** 9–10, ssp. **juncea.** 11–12, ssp. **scotica.** 13–14, ssp. **megastachys.** 15–16, ssp. **commutata.**
Black areas = sclerenchyma. Redrawn by C.A. Stace from photographs by A.-K. Al-Bermani.

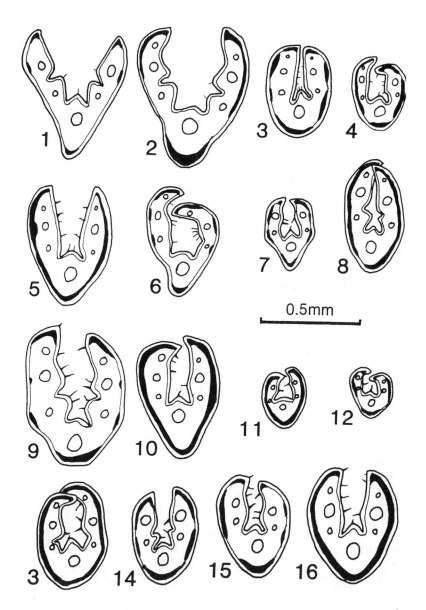

FIG 1009 – Transverse sections of innovation leaves of
Festuca ovina agg. 1–2, **F. brevipila.** 3, **F. ovina** ssp. **ovina.**
4, **F. ovina** ssp. **hirtula.** 5–6, **F. lemanii.** 7, **F. vivipara.** 8,
F. ovina ssp. **ophioliticola.** 9–10, **F. longifolia.** 11–12, **F.
filiformis.** 13–14, **F. armoricana.** 15–16, **F. huonii.**
Black areas = sclerenchyma. Redrawn by C.A. Stace from
drawings by M.J. Wilkinson.

2n=28. Native; grassy places in hilly districts, usually on rocky ground; common in C & N Sc, local in S Sc, N En, Wa and Ir.
Triploid (often proliferating) Sheep's-fescues (2n=21) have been claimed as hybrids between F. ovina, F. filiformis and F. vivipara in various combinations; possibly all 3 hybrids occur, but they need careful study.

10. F. filiformis Pourret (F. tenuifolia Sibth., F. ovina **1009** ssp. tenuifolia (Sibth.) Dumort.) - Fine-leaved Sheep's-fescue. Culms to 35cm; leaves not glaucous, 0.3-0.53mm from midrib to edge, with (4)5(-7) veins and 2(-3) adaxial grooves, with abaxial sclerenchyma in thin broken or sometimes continuous band; panicles 2.1-7.2cm; pedicels 0.6-2.3mm; spikelets 4.5-5.5mm; lemmas 2.5-3.5mm, with awns 0-0.6mm; 2n=14. Native; grassy places on usually acid sandy soils; frequent throughout BI.

11. F. armoricana Kerguelen (F. ophioliticola ssp. **1009** armoricana (Kerguelen) Auq.) - Breton Fescue. Culms to 40cm; **RR** leaves not glaucous, 0.5-0.88mm from midrib to edge, with 7 veins and 2-4 adaxial grooves, with abaxial sclerenchyma in thin broken or sometimes continuous band; panicles 3.3-5.9cm; pedicels 0.8-1.5mm; spikelets 6-8mm; lemmas 3.9-5.1mm, with awns 0.8-1.6mm; 2n=28. Native; fixed dunes on W coast of Jersey.

12. F. huonii Auq. - Huon's Fescue. Culms to 25cm, often **1009** procumbent; leaves sometimes ± glaucous, often strongly **RR** curved, 0.4-0.75mm from midrib to edge, with 5-7 veins and 2-4 adaxial grooves, with abaxial sclerenchyma in thin broken or sometimes continuous band; panicles 2.3-5.4cm; pedicels 0.7-1.8mm; spikelets 5.5-7.5mm; lemmas 3.5-4.9mm, with awns 0.9-1.8mm; 2n=42. Native; grassy cliff-tops and -bases; Jersey, Guernsey and other islands in CI.

13. F. lemanii Bast. (F. bastardii Kerguelen & Plonka, F. **1009** longifolia pro parte sensu C.E. Hubb. et al. non Thuill.) - **R** Confused Fescue. Culms to 66cm; leaves often ± glaucous, 0.43-0.95mm from midrib to edge, with (5-)7 veins and 2-4 adaxial grooves, with abaxial sclerenchyma in continuous or subcontinuous (sometimes broken) band; panicles 3.7-8.6cm; pedicels 0.6-2.5mm; spikelets 6.1-8.5mm; lemmas 4-5.3mm, with awns 0.3-1.9mm; 2n=42. Probably native; grassy places on well-drained, acid or calcareous soils, often with F. ovina; very scattered in Br, much under-recorded.

14. F. longifolia Thuill. (F. caesia Smith, F. glauca var. **1009** caesia (Smith) Howarth, F. glauca auct. non Vill.) - Blue **RR** Fescue. Culms to 40cm; leaves usually strongly glaucous, 0.55-0.9mm from midrib to edge, with 7(-9) veins and 4(-6) adaxial grooves, with abaxial sclerenchyma in thick broken or sometimes continuous band; panicles 2.4-5.6cm; pedicels 0.5-1.8mm; spikelets 5.4-7mm; lemmas 3.5-4.8mm, with awns 0.5-1.5mm; 2n=14. Native; very dry acid heaths and maritime cliff-tops; on the Breckland of W Suffolk, possibly in similar sites in N Lincs, on Jersey, Guernsey and other

islands in CI.
15. F. brevipila Tracey (**F.** <u>trachyphylla</u> (Hackel) Kraj. non **1009**
Hackel ex Druce, **F.** <u>longifolia</u> <u>pro</u> <u>parte</u> sensu C.E. Hubb. et **R**
al. non Thuill.) - <u>Hard Fescue</u>. Culms to 70cm; leaves
usually somewhat glaucous, 0.6-1mm from midrib to edge, with
(5)7-9 veins and 4(-6) adaxial grooves, with abaxial
sclerenchyma usually as 3 islets with tails at midrib and
edges, often with smaller islets opposite other veins, rarely
sub-continuous; panicles 3.5-9.5cm; pedicels 1.2-2.8mm;
spikelets 6.1-8.5mm; lemmas 3.9-5.5mm, with awns 1.2-2.6mm;
2n=42. Intrd; much grown from grass-seed mixtures, natd on
roadsides, commons and rough ground especially on acid well-
drained soils; frequent in SE En and E Anglia, scattered N to
SW Yorks, Jersey, under-recorded; Europe.

13 x 14. FESTUCA x LOLIUM = X FESTULOLIUM Asch. & Graebner
Inflorescences variously intermediate, at 1 extreme a simple
raceme (rarely a spike as in <u>Lolium</u>), at the other a
branching panicle, often a raceme with a few racemose
branches near base; most spikelets with 2 glumes, but lower
much shorter than upper; anthers ± indehiscent with ± empty
pollen grains, but some degree of fertility exists and some
back-crossing may occur.
<u>L.</u> <u>perenne</u> and <u>L.</u> <u>multiflorum</u> hybridise with <u>F.</u> <u>pratensis</u>,
<u>F.</u> <u>arundinacea</u> and <u>F.</u> <u>gigantea</u>, and 5 of the 6 possible
combinations have been found. A single report of <u>F.</u> <u>rubra</u> x
<u>L.</u> <u>perenne</u> = **X F. fredericii** Cugnac & A. Camus was an error,
but all combinations are probably under-recorded.

13/1 x 14/1. F. pratensis x L. perenne = X F. loliaceum
(Hudson) P. Fourn. (<u>Hybrid Fescue</u>) has glabrous auricles and
awnless lemmas; inflorescence usually a simple or little-
branched raceme. Native; pastures, meadows, river-sides and
roadsides, often on damp rich soils; throughout most of BI,
commonest in S En.
13/1 x 14/2. F. pratensis x L. multiflorum = X F. braunii
(K. Richter) A. Camus differs from X <u>F.</u> <u>loliaceum</u> in its
awned lemmas. Native and probably intrd in grass-seed;
grassy places, rough ground, waysides; scattered in En,
Denbs.
13/2 x 14/1. F. arundinacea x L. perenne = X F. holmbergii
(Doerfler) P. Fourn. has a well-branched inflorescence but
often with subsessile spikelets; auricles ciliate. Native;
similar places to X <u>F.</u> <u>loliaceum</u>; very scattered in S En N to
Warks.
13/2 x 14/2. F. arundinacea x L. multiflorum differs from X
<u>F.</u> <u>holmbergii</u> in its awned lemmas. Native and probably intrd
in grass-seed; similar places to X <u>F.</u> <u>braunii</u>; very scattered
in S En N to Warks.
13/3 x 14/1. F. gigantea x L. perenne = X F. brinkmannii
resembles either <u>F.</u> <u>pratensis</u> x <u>L.</u> <u>multiflorum</u> or <u>F.</u>
<u>arundinacea</u> x <u>L.</u> <u>multiflorum</u> in inflorescence shape; differs

from both in its longer lemmas and wider leaves. Native;
grassy places near the parents; Cambs, Pembs and Merioneth.
F. gigantea x L. multiflorum = X F. nilssonii Cugnac & A.
Camus might occur but would be very difficult to distinguish
from X F. brinkmannii.

13 x 15. FESTUCA X VULPIA = X FESTULPIA Meld. ex Stace & R.
Cotton
Plants perennial and vegetatively close to the Festuca
parent, but with fewer and shorter (or 0) rhizomes and some
overlapping sheaths; panicles narrower and less branched than
in Festuca, with markedly longer awns; lower glume c.1/2 as
long as upper; anthers indehiscent with $\pm$ empty pollen
grains, but some degree of fertility may exist.
F. arenaria and F. rubra hybridise with V. fasciculata, V.
bromoides and V. myuros, and 4 of the 6 possible combinations
have been found.

13/6 x 15/1. F. arenaria x V. fasciculata = X F. melderisii
Stace & R. Cotton has lower glume 5.2-8mm; upper glume
8-11.5mm (incl. awn); lemmas 9.5-10.5mm plus awn 3.5-5mm;
anthers 3, 1.5-2mm; 2n=42. Native; on open sand-dunes with
parents; very local in S En and S Wa.
   13/7 x 15/1. F. rubra x V. fasciculata = X F. hubbardii 1023
Stace & R. Cotton has lower glume 2.4-4.4mm; upper glume
(incl. awn) 3.5-7.2mm; lemmas 6-9.5mm plus awn 2-5.5mm;
anthers 3, 1.5-2mm; 2n=35. Native; on open sand-dunes with
parents; probably frequent in CI and Br N to S Lancs.
   13/7 x 15/2. F. rubra x V. bromoides has lower glume
2-3.4mm; upper glume 3.4-5.9mm; lemmas 4.5-7mm plus awn
3.2-6mm; anthers 3, 0.8-1.7mm; 2n=28. Differs from X F.
hubbardii in not having markedly distally thickened pedicels,
in having only 1 (not 2-several) ovary-less floret at apex of
spikelets, and in the awnless upper glumes. Native; waste
and rough ground with the parents; scattered in Br N to SE
Yorks. Probably both ssp. rubra and ssp. commutata of F.
rubra have produced this hybrid.
   13/7 x 15/3. F. rubra x V. myuros has lower glume 1.5-3.3mm;
upper glume 3.2-5mm; lemmas 4.5-6.2mm plus awn 3-6mm; anthers
3, 0.6-1.5mm; 2n=42. Differs from F. rubra x V. bromoides in
its long narrow panicle, showing the influence of V. myuros.
Native; waste and rough ground with the parents; scattered in
Br N to S Lancs. Probably ssp. rubra, ssp. commutata and
ssp. litoralis of F. rubra have produced this hybrid.

14. LOLIUM L. - Rye-grasses
Annuals to perennials without rhizomes or stolons; sheaths
not fused; inflorescence normally a spike with spikelets
compressed and orientated with backs of lemmas adjacent to
spike axes; spikelets partly lying in concavities in spike
axis, with 2-many florets all (except the most apical)
bisexual; glumes 2 in terminal spikelet, 1 (the upper,

abaxial) in lateral spikelets; lemmas rounded on back,
5-9-veined, obtuse to subacute or minutely bifid at apex,
with or without subterminal awn; stamens 3. Abnormal plants
with branched inflorescences are not rare. Sp. limits are
very unsatisfactory; all spp. may have awned or unawned
lemmas, and all except L. perenne have rolled (not folded)
young leaves. All the spp. are diploids and produce fertile
hybrids. Apart from the 1 hybrid covered, L. multiflorum x
L. temulentum, L. multiflorum x L. rigidum and L. rigidum x
L. temulentum have all been recorded as rare casuals with
grain, wool and distillery refuse, imported as hybrid
caryopses. Their determination is very difficult and must be
considered tentative.

1 Lemmas ovate to elliptic, <3x as long as wide,
  becoming thick and hard at base in fruit; caryopsis
  <3x as long as wide   2
1 Lemmas narrowly oblong-ovate, >3x as long as wide,
  not becoming thick or hard; caryopsis >3x as long
  as wide   3
    2 Lowest 2 lemmas (4.6)5-8.5mm; glume (7)10-30mm
                                                            4. L. temulentum
    2 Lowest 2 lemmas 3.5-5(5.5)mm; glume 5-12(15)mm
                                                               5. L. remotum
3 Perennial, with tillers at flowering and fruiting
  time; leaves folded along midrib when young; lemmas
  usually unawned   **1. L. perenne**
3 Annual or biennial, without tillers at flowering
  and fruiting time; leaves rolled along long axis
  when young; lemmas awned or not   4
    4 Lemmas nearly always awned; spikelets usually
      with >11 florets   **2. L. multiflorum**
    4 Lemmas usually unawned; spikelets usually with
      <11 florets   **3. L. rigidum**

1. L. perenne L. - Perennial Rye-grass. Perennial to
50(90)cm; spikelets with 4-14 florets; glume from 1/3 to >1x
as long as rest of spikelet; lowest 2 lemmas 3.5-9mm; awns
very rare (<8mm). Native; grassy places, waste and rough
ground, also a common escape from lawns, roadside plantings
and pastures; abundant throughout BI.
  1 x 2. L. perenne x L. multiflorum = L. x boucheanum Kunth
(L. x hybridum Hausskn.) occurs occasionally in Br and Ir as
a natural product and now commonly in lowland Br and CI as an
escape from its cultivation as a valuable pasture or meadow
grass; it is annual to perennial, and has rolled young leaves
and intermediate spikelet structure (always shortly awned).
  2. L. multiflorum Lam. - Italian Rye-grass. Annual or
biennial to 100(120)cm; spikelets with (5)11-22 florets;
glume from 1/4 to 3/4 as long as rest of spikelet; lowest 2
lemmas 4-8mm; awns nearly always present (<15mm). Intrd;
rough and waste ground, field borders, waysides; scattered

throughout BI, common in lowland Br and CI; S Europe.
3. L. rigidum Gaudin (L. loliaceum (Bory & Chaub.) **1038**
Hand.-Mazz.) - Mediterranean Rye-grass. Annual to 70cm;
spikelets with 4-8(11) florets; glume from 3/4 to >1x as long
as rest of spikelet; lowest 2 lemmas 3.5-8.5mm; awns very
rare (<10mm). Intrd; rather frequent alien from grain, wool
and other sources in waste ground and on tips; scattered in
Br; Mediterranean. Spikelets usually very narrow, well sunk
in concavities of rhachis and largely concealed by glume;
easily confused with poorly-grown L. perenne if vegetative
characters are ignored.
4. L. temulentum L. - Darnel. Annual to 75(100)cm;
spikelets with 4-10(15) florets; glume from 3/4 to 1.5x as
long as rest of spikelet; lowest 2 lemmas (4.6)5-8.5mm; awns
present (often >10mm) or less often 0. Intrd; formerly
common in cornfields, now casual on tips and waste places
from many sources especially grain; scattered in BI;
Mediterranean.
5. L. remotum Schrank - Flaxfield Rye-grass. Annual to
75cm; spikelets with 4-10 florets; glume from 2/3 to 1.5x as
long as rest of spikelet; lowest 2 lemmas 3.5-5(5.5)mm; awn
present (<10mm) or more often 0. Intrd; formerly a typical
flaxfield alien, now a very occasional alien from grain and
other sources; scattered and sporadic in En; E Europe.

15. VULPIA C. Gmelin (Nardurus Reichb.) - Fescues
Annuals; sheaths not fused; inflorescence a sparsely branched
narrow panicle or a raceme; spikelets with 2-many florets all
(except the most apical) bisexual or with a group of sterile
florets at apex; glumes 2, very unequal, the lower at most
3/4 as long as upper; lemmas rounded on back, 3-5-veined,
acute or acuminate with long terminal awn; stamens 1-3.
Glume lengths and ratios must be measured in spikelets that
are not apical on the inflorescence or its branches; in
apical spikelets the lower glume is often much longer than
normal.

1 Lemmas with basal pointed minutely scabrid callus;
   ovary and caryopsis with minute apical hairy
   appendage; lemmas 8-18mm excl. awn; upper glume
   10-30mm incl. awn   **1. V. fasciculata**
1 Lemma with basal rounded glabrous callus; ovary
   and caryopsis glabrous; lemmas 3-7.5mm excl. awn;
   upper glume 1.5-9mm incl. awn if present   2
   2 Anthers 3, 0.7-1.3(1.9)mm, well exserted at
     anthesis; lemmas 3-5mm excl. awn; inflorescence
     ± always a raceme   **5. V. unilateralis**
   2 Anthers 1(-3), 0.4-0.8(1.8)mm, usually not
     exserted at anthesis; lemmas 4-7.5mm;
     inflorescence a panicle except in starved plants   3
3 Spikelets with 1-3 bisexual and 3-7 distal sterile
   (but not smaller) florets; lemma of fertile florets

3(-5)-veined   **4. V. ciliata**
3 Spikelets with 2-5 bisexual and 1-2 distal much
  reduced sterile florets; lemma of fertile florets
  5-veined   4
  4 Lower glume 2.5-5mm, 1/2 to 3/4 as long as upper;
    lemmas usually >1.3mm wide when flattened;
    inflorescence normally well exserted from upper-
    most sheath at maturity   **2. V. bromoides**
  4 Lower glume 0.4-2.5mm, 1/10 to 2/5 as long as
    upper; lemmas usually <1.3mm wide when flattened;
    inflorescence normally not fully exserted from
    from uppermost sheath at maturity   **3. V. myuros**

Other spp. - Several spp. are found as rare casuals,
especially as wool-aliens. The least rare is **V. muralis**
(Kunth) Nees (**V. broteri** Boiss. & Reuter), from Europe, which
resembles **V. bromoides** in its inflorescence shape but has
narrow lemmas as in **V. myuros**, and lower glume 1-3mm and 1/4
to 1/2 as long as upper; perhaps overlooked.

1. V. fasciculata (Forsskaol) Fritsch (**V. membranacea** auct.   R
non (L.) Dumort.) - **Dune Fescue.** Culms to 50cm;
inflorescence a panicle or raceme 3-11cm; spikelets 10-18mm
excl. awns, with distal group of 3-6 reduced sterile florets;
lower glume 0.1-2.6mm, <1/6 as long as upper; upper glume
10-30mm incl. awn 3-12mm; fertile lemmas 8-18mm excl. awn.
Native; open parts of sand-dunes; locally frequent on coasts
of BI N to W Norfolk, Cumberland and Co Louth.
2. V. bromoides (L.) Gray - **Squirreltail Fescue.** Culms to
50cm; inflorescence a panicle or sometimes a raceme 1-11cm;
spikelets 6.5-11.5mm excl. awns, with 1-2 distal sterile
reduced florets; lower glume 2.5-5mm, 1/2 to 3/4 as long as
upper; upper glume 4.5-9mm incl. awn 0-2mm; fertile lemmas
4.5-7.5mm excl. awn. Native; open grassy places on well
drained soils, rough and waste ground; frequent over most of
BI, common in S Br.
3. V. myuros (L.) C. Gmelin (**V. megalura** (Nutt.) Rydb.) -
Rat's-tail Fescue. Culms to 65cm; inflorescence a panicle or
raceme 5-35cm; spikelets 6-10.5mm excl. awns, with 1-2 distal
sterile reduced florets; lower glume 0.4-2.5mm, 1/10 to 2/5
as long as upper; upper glume 2.5-6.5mm incl. awn 0-1mm;
fertile lemmas 4.5-7.5mm excl. awn. Probably native; open
ground, walls, rough or waste ground, roadsides and by
railways, also frequent grain- and wool-alien; increased in
recent years, now throughout BI except C & N Sc and most of N
Ir, common in S Br. Lemmas are usually scabrid, but may be
ciliate (f. **megalura** (Nutt.) Stace & R. Cotton) or pubescent
dorsally (f. **hirsuta** (Hackel) Blom) in intrd material.
4. V. ciliata Dumort. - **Bearded Fescue.** Culms to 45cm;   R
inflorescence a panicle or raceme 3-20cm; spikelets with
distal group of 3-7 sterile (but not smaller) florets; lower
glume 0.1-1mm, <1/4 as long as upper; upper glume 1.5-4mm,

never awned.

a. Ssp. ciliata. Spikelets mostly 7-10.5mm excl. awns; fertile lemmas 5-6.5mm excl. awn, pubescent on dorsal midline and margins; sterile lemmas <8mm excl. awn, densely ciliate. Intrd; occasional casual from wool and grain in S En, natd in railway sidings at Ardingly, E Sussex, since 1967; Europe.

b. Ssp. ambigua (Le Gall) Stace & Auq. (V. ambigua (Le Gall) More). Spikelets mostly 5-7mm excl. awns; fertile lemmas 4-5mm excl. awn, minutely scabrid; sterile lemmas <6mm excl. awn, minutely scabrid. Native; on maritime or submaritime sand or shingle; local in S Br N to W Norfolk and N Somerset, CI.

5. V. unilateralis (L.) Stace (Nardurus maritimus (L.) R
Murb.) - Mat-grass Fescue. Culms to 40cm; inflorescence usually a raceme 1-10cm, rarely slightly branched below; spikelets 4-7(8)mm excl. awns, with 1-2 distal sterile reduced florets; lower glume 1.5-3.5mm, 1/2 to 3/4 as long as upper; upper glume 3-5mm, unawned; fertile lemmas (2)3-4(5)mm excl. awn. Native; open grassy places on chalk, also in waste places and waysides; very scattered in Br N to E Gloucs and W Norfolk, formerly to Derbys.

16. CYNOSURUS L. - Dog's-tails
Annuals or perennials without rhizomes or stolons; sheaths not fused; inflorescence a compact spike-like panicle; spikelets of 2 kinds - fertile with (1)2-5 florets all (except the most apical) bisexual, and sterile consisting of numerous very narrow acuminate lemmas in herring-bone arrangement, normally 1 fertile and 1 sterile together; glumes 2, subequal; fertile lemmas rounded on back, 5-veined, acute to obtuse or minutely bifid, awnless or with long terminal or subterminal awn; stamens 3.

1. C. cristatus L. - Crested Dog's-tail. Tufted perennial to 75cm; leaves 1-4mm wide; ligules 0.5-1.5mm, + truncate; panicle linear-oblong, 1-10(14) x 0.4-1cm; fertile lemmas 3-4mm plus awn 0-1mm. Native; grassy places on a great range of soils; common throughout BI.

2. C. echinatus L. - Rough Dog's-tail. Tufted annual to 75(100)cm; leaves 2-10mm wide; ligules 2-10mm, acute to obtuse; panicle asymmetrically ovoid, 1-4(8) x 0.7-2cm; fertile lemmas 4.5-7mm plus awn 6-16mm. Intrd; casual on waste and rough open ground scattered in BI N to C Sc, natd in sunny places on sandy or rocky ground on coasts of S En and CI; Europe.

17. LAMARCKIA Moench - Golden Dog's-tail
Annuals; sheaths not fused; inflorescence a rather compact panicle; spikelets of 2 kinds - fertile with 1 bisexual and 1 vestigial floret, and sterile consisting of numerous flat, overlapping obtuse lemmas, normally 3 sterile and 2 fertile together, but 1 of the 2 fertile often reduced and not

producing a caryopsis; glumes 2, subequal; lemmas rounded on back, 5-veined, minutely bifid, with long awn from sinus; stamens 3.

1. L. aurea (L.) Moench - <u>Golden</u> <u>Dog's-tail</u>. Culms tufted, **998** to 20(30)cm; leaves 2-6mm wide; ligules 5-10mm, obtuse to **1052** jagged at apex; panicle 3-9 x 2-3cm, oblong; fertile lemmas 2-3.5mm, with awn 5-10mm; vestigial lemma also long-awned but lemmas of sterile spikelets unawned. Intrd; casual occasionally persisting for few years, in waste or rough ground and on tips mainly as wool-alien; very scattered in Br, mainly S & C En and S Wa; Mediterranean.

18. PUCCINELLIA Parl. - <u>Saltmarsh-grasses</u>
Annuals to perennials without rhizomes, with or without stolons; sheaths not fused; inflorescence a panicle; spikelets with 2-many florets all (except the most apical) bisexual; glumes 2, slightly unequal; lemmas rounded on back, 5-veined, subacute to rounded at apex, unawned; stamens 3.

1 Lemmas 2.8-4.6mm   2
1 Lemmas 1.8-2.5(2.8)mm   3
  2 Perennial with many tillers and usually rooting
    stolons at flowering; anthers 1.3-2.5mm
                                          1. P. maritima
  2 Annual or biennial with 0 or few tillers and 0
    stolons at flowering; anthers 0.75-1mm
                                          4. P. rupestris
3 At least some of panicle branches at lower nodes
  bearing spikelets ± to base; lemmas subacute to
  obtuse, with midrib reaching apex; anthers mostly
  <0.75mm   **3. P. fasciculata**
3 Panicle branches at lower nodes ± all with
  conspicuous basal region bare of spikelets; lemmas
  broadly obtuse to rounded, with midrib falling short
  of apex; anthers mostly >0.75mm   **2. P. distans**

1. P. maritima (Hudson) Parl. - <u>Common</u> <u>Saltmarsh-grass</u>. Perennial, usually with long stolons forming large patches; culms to 80cm; spikelets 5-13mm, with 3-10 florets; lemmas 2.8-4.6mm, subacute to obtuse, with midrib usually just reaching apex; anthers 1.3-2.5mm; 2n=56 (and 63?, 70?). Native; bare or semi-bare mud in salt-marshes and estuaries, rarely saline areas and by salted roads inland; common round coasts of BI, few places in C En.
1 x 2. P. maritima x P. distans = P. x hybrida O. Holmb. occurs rarely on coasts of En from W Sussex to Durham, and has been found by a salted road in S Northumb and (involving <u>P. distans</u> ssp. <u>borealis</u>) in Outer Hebrides; it often has stolons, has intermediate panicle-shape and lemma size, and is sterile; 2n=49, 51.
1 x 4. P. maritima x P. rupestris = P. x krusemaniana Jansen

& Wachter was recorded from W Sussex in 1920 and S Hants in 1977; it is said to be intermediate and sterile.

2. P. distans (Jacq.) Parl. - <u>Reflexed</u> <u>Saltmarsh-grass</u>. Tufted perennial; culms to 60cm; spikelets 3-9mm, with 2-9 florets; lemmas broadly obtuse to rounded, the midrib not reaching apex; anthers 0.5-1.2mm; 2n=42. Native.

a. Ssp. distans. Culms to 60cm; leaves usually flat; lower panicle branches strongly reflexed at maturity; lemmas 2-2.5mm. Semi-bare mud, rough and waste ground in estuaries, salt-marshes, inland saline areas and by salted main roads; round coasts of BI N to CE Sc, rare in W, common in E, now common inland in C & E En.

b. Ssp. borealis (O. Holmb.) W.E. Hughes (<u>P. capillaris</u> (Lilj.) Jansen). Culms to 40cm; leaves usually folded along midrib; lower panicle branches patent (sometimes weakly reflexed) to suberect at maturity; lemmas (1.8)2.2-2.8mm. Stony or rocky, sometimes sandy places and on sea-walls; coasts of N & E Sc (incl. N Isles) S to Fife.

2 x 3. P. distans x P. fasciculata occurs rarely on coasts of En from W Sussex to E Norfolk; it has intermediate panicle- and lemma-shape and is sterile; 2n=35.

2 x 4. P. distans x P. rupestris = P. x pannonica (Hackel) O. Holmb. occurs rarely on coasts of En from S Devon to E Norfolk; it has intermediate panicle-shape and lemma-length and is sterile; 2n=42.

3. P. fasciculata (Torrey) E. Bickn. (<u>P. pseudodistans</u>   **R** (Crepin) Jansen & Wachter) - <u>Borrer's</u> <u>Saltmarsh-grass</u>. Tufted perennial; culms to 50cm; spikelets 3-6mm, with 3-8 florets; lemmas 1.8-2.3mm, the midrib reaching apex; anthers 0.4-0.8mm; 2n=28. Native; in barish places, on sea-walls and banks and by dykes; locally frequent on coasts of S Br from Carms to W Norfolk, by salted roads in E & W Kent. **P. pseudodistans** appears to some extent intermediate between <u>P. distans</u> and <u>P. fasciculata</u>, especially in its panicle shape, but some of its branches bear spikelets to the base, its lemma midrib reaches the apex, and it has 2n=28; it is best considered a var. of <u>P. fasciculata</u> and has a similar distribution in Br.

4. P. rupestris (With.) Fern. & Weath. - <u>Stiff</u> <u>Saltmarsh-</u>   **R** <u>grass</u>. Tufted annual or biennial; culms to 40cm; spikelets 5-9mm, with 3-6 florets; lemmas 2.8-4mm, the midrib usually reaching apex; anthers 0.75-1mm; 2n=42. Native; in bare places on mud and clay and among rocks and stones; coasts of Br from Pembs to Yorks (formerly Cheviot), by salted road in W Kent.

19. BRIZA L. - Quaking-grasses

Annuals or perennials without rhizomes or stolons; sheaths not fused; inflorescence a panicle or sometimes a raceme with pedicels >5mm; spikelets with 4-many florets all (except most apical) bisexual, characteristically flattened, broadly ovate and pendent; glumes 2, subequal; lemmas rounded on back,

cordate at base, rounded to very obtuse at apex, 7-9-veined,
the veins not reaching apex, unawned; stamens 3.

1 Perennial, with tillers at flowering; ligule 0.5-
  1.5mm; leaves 2-4mm wide   **1. B. media**
1 Annual, without tillers at flowering; ligule 2-6mm;
  leaves 2-10mm wide   2
  2 Spikelets 2.5-5mm, >20 per panicle   **2. B. minor**
  2 Spikelets 8-25mm, <15 per panicle   **3. B. maxima**

1. B. media L. - <u>Quaking-grass</u>. Perennial; culms to 75cm;
spikelets 4-7mm, >20 per panicle. Native; grassland on light
to heavy, acid to calcareous, very dry to damp, but usually
base-rich soils; locally common through BI except N & NW Sc.
  2. B. minor L. - <u>Lesser</u> Quaking-grass. Annual; culms to
60cm; spikelets 2.5-5mm, >20 per panicle. Intrd; natd in
arable fields, bulb-fields and waste places; locally frequent
in SW & CS En and CI, rare casual elsewhere; Mediterranean.
  3. B. maxima L. - <u>Greater</u> Quaking-grass. Annual; culms to
75cm; spikelets 8-25mm, <15 per panicle. Intrd; natd in dry
open places and on banks and field-margins; distribution as
for <u>B. minor</u>, usually as garden escape; Mediterranean.

20. POA L. (<u>Parodiochloa</u> C.E. Hubb.) - <u>Meadow-grasses</u>
Annuals or perennials with or without stolons or rhizomes;
sheaths not fused; inflorescence a panicle; spikelets with
1-many florets all (or all except the most apical) bisexual,
sometimes proliferating; glumes 2, subequal; lemmas keeled on
back, 5-veined, awnless or rarely with short terminal awn,
its callus often with tuft of cottony hairs; stamens 3.

1 At least some spikelets proliferating   2
1 Spikelets not proliferating   4
  2 Base of culms swollen, bulb-like   **14. P. bulbosa**
  2 Base of culms not swollen   3
3 Leaves 2-4.5mm wide when flattened, parallel-sided
  and ± abruptly narrowed to apex, the uppermost
  usually arising below 1/2 way up culm   **15. P. alpina**
3 Leaves 1-2mm wide when flattened, gradually tapered
  to apex, the uppermost usually arising above
  1/2 way up culm   **9. P. x jemtlandica**
  4 Plant with distinct, often far-creeping rhizomes   5
  4 Plant without rhizomes, sometimes with stolons   11
5 Culms strongly compressed, 4-6(9)-noded, usually
  slightly bent at each node   **10. P. compressa**
5 Culms terete to somewhat compressed, 2-4-noded,
  usually straight except near base   6
  6 Glumes subequal, both usually 3-veined and
    distinctly acuminate; culms usually all solitary
                                        4. P. humilis
  6 Glumes distinctly unequal, the lower often
    1-veined, acute; culms usually in small or dense

    clusters   7

7 Tiller leaves 0.5-2mm wide; lemmas 2-3mm; culms
  usually in dense clusters   **6. P. angustifolia**

7 Tiller leaves 2-4(5)mm wide; lowest lemmas 3-4mm;
  culms usually in small clusters   **5. P. pratensis**
    8 All or some leaves >5mm wide   9
    8 Leaves <5mm wide   11

9 Panicle-branches erect; lowest lemma usually >4.5mm,
  with awn 1.2-2.4mm   **16. P. flabellata**

9 Panicle-branches patent; lowest lemma <4.5mm,
  awnless   10
    10 Leaves <6mm wide; lowest lemma <3mm; ligules
      often >2mm; uppermost culm-leaf usually as long
      or longer than its sheath   **11. P. palustris**
    10 Leaves often >6mm wide; lowest lemma >3mm;
      ligules <2mm; uppermost culm-leaf much shorter
      than its sheath   **7. P. chaixii**

11 Base of culms swollen, bulb-like   **14. P. bulbosa**
11 Base of culms not swollen   12
    12 Base of culms surrounded by dense mass of dead
      leaf-sheaths; mountains   13
    12 Base of culms with few or no persistent dead
      leaf-sheaths; widespread   14

13 Leaves 2-4.5mm wide when flattened, parallel-sided
  and + abruptly narrowed to apex, the uppermost
  usually arising below 1/2 way up culm   **15. P. alpina**

13 Leaves 1-2mm wide when flattened, gradually tapered
  to apex, the uppermost usually arising above 1/2
  way up culm   **8. P. flexuosa**
    14 Annual, or perennial due to procumbent stems
      rooting; culms usually procumbent to ascending;
      anthers <1(1.3)mm; usually some leaves
      transversely wrinkled   15
    14 Perennial; culms usually erect; anthers >1.3mm;
      leaves not transversely wrinkled   16

15 Anthers 0.6-0.8(1.3)mm, 2-3x as long as wide;
  panicle-branches usually patent to reflexed at
  fruiting   **2. P. annua**

15 Anthers 0.2-0.5mm, 1-1.5x as long as wide; panicle-
  branches usually erect to suberect at fruiting
                               1. P. infirma
    16 Ligule of uppermost culm-leaf 4-10mm, acute;
      sheaths rough   **3. P. trivialis**
    16 Ligule of uppermost culm-leaf <5mm, obtuse to
      rounded; sheaths smooth   17

17 Ligule of uppermost culm-leaf 0.2-0.5mm, usually
  truncate   **13. P. nemoralis**

17 Ligule of uppermost culm-leaf 0.8-4(5)mm, usually
  obtuse   18
    18 Lowland plant c.30-150cm; lowest panicle node
      with (3)4-6(8) branches; ligule of uppermost
      culm-leaf 2-4(5)mm   **11. P. palustris**

18 Mountain plant 10–40cm; lowest panicle node with
   1–2(4) branches; ligule of uppermost culm–leaf
   1–2.5(3)mm   **12. P. glauca**

1. P. infirma Kunth – <u>Early</u> <u>Meadow–grass</u>. Annual; culms **RR**
erect to procumbent, to 10(25)cm; leaves 1–3(4)mm wide;
ligules 1–3.5mm; lowest panicle–node with 1–2 branches.
Native; rough ground, waysides and on paths, usually near
sea; CI, Scillies, E & W Cornwall, S Devon, common in
Scillies and CI.
2. P. annua L. – <u>Annual</u> <u>Meadow–grass</u>. Annual, or perennial
due to procumbent stems rooting; culms erect to procumbent,
to 20(30)cm; leaves 1–4(5)mm wide; ligules 1–5mm; lowest
panicle–node with 1–2(3) branches. Native; rough, waste and
cultivated ground, waysides, on paths, in lawns and other
close–cut turf; abundant throughout BI.
3. P. trivialis L. – <u>Rough</u> <u>Meadow–grass</u>. Perennial with
many, often procumbent tillers, some becoming stolons; culms
erect, to 70(90)cm; leaves 1.5–5mm wide; ligules 4–10mm;
lowest panicle–node with 3–7 branches. Native; open woods,
marshes, ditches, river–sides, damp grassland, by ponds and
lakes, cultivated and rough ground; very common through BI.
4. P. humilis Ehrh. ex Hoffm. (<u>P.</u> <u>subcaerulea</u> Smith, <u>P.</u>
<u>pratensis</u> ssp. <u>irrigata</u> (Lindman) Lindb. f.) – <u>Spreading</u>
<u>Meadow–grass</u>. Perennial with extensive rhizomes; culms
erect, to 30(40)cm; leaves 1.5–4mm wide; ligules 0.5–2mm;
lowest panicle–node with 2–5 brnches. Native; grassland,
roadsides, on old walls, usually on sandy soil but often near
water; throughout BI, but greatly under–recorded.
5. P. pratensis L. – <u>Smooth</u> <u>Meadow–grass</u>. Perennial with
strong but short rhizomes; culms erect, to 75(100)cm; leaves
2–4(5)mm wide; ligules 1–3mm; lowest panicle–node with 3–5
branches. Native; meadows, pastures, waysides, rough and
waste ground; very common throughout BI.
6. P. angustifolia L. (<u>P.</u> <u>pratensis</u> ssp. <u>angustifolia</u> (L.)
Dumort. – <u>Narrow–leaved</u> <u>Meadow–grass</u>. Perennial with
rhizomes; culms erect, to 70cm; leaves 0.5–2(3)mm wide;
ligules 0.5–2mm; lowest panicle–node with 3–5 branches.
Native; grassy places, rough ground, on and by walls, banks,
on well–drained soil; probably frequent throughout BI but
greatly under–recorded.
7. P. chaixii Villars – <u>Broad–leaved</u> <u>Meadow–grass</u>. Densely
tufted perennial; culms erect, to 1.3m; leaves (4)6–10mm
wide; ligules 0.5–2mm; lowest panicle–node with 4–7 branches.
Intrd; grown for ornament, natd in woods and copses;
scattered throughout Br, Co Down; Europe.
8. P. flexuosa Smith – <u>Wavy</u> <u>Meadow–grass</u>. Tufted perennial; **RR**
culms erect, to 25cm; leaves 1–2mm wide; ligules 1–3.5mm;
lowest panicle–node with 2–5 branches. Native; mountain
screes and ledges at 800–1100m; very local in highlands of C
Sc.
9. P. x jemtlandica (S. Almq.) K. Richter (<u>P.</u> <u>flexuosa</u> x <u>P.</u>

alpina) - <u>Swedish</u> <u>Meadow-grass</u>. Differs from <u>P.</u> <u>flexuosa</u> in spikelets all or nearly all proliferating, with very little external suggestion of <u>P.</u> <u>alpina</u>. Native; similar habitat and range to <u>P.</u> <u>flexuosa</u>, often (?always) with it and <u>P.</u> <u>alpina</u>.

10. P. compressa L. - <u>Flattened</u> <u>Meadow-grass</u>. Rhizomatous perennial; culms erect to ascending, to 60cm; leaves 1-3(4)mm wide; ligules 0.5-2mm; lowest panicle-node with 2-3 short branches. Native; walls, paths, waysides, stony ground and banks on well-drained soils; rather scattered throughout BI except N Sc, perhaps under-recorded.

11. P. palustris L. - <u>Swamp</u> <u>Meadow-grass</u>. Tufted perennial; culms erect, to 1(1.5)m; leaves 2-6mm wide; ligules 2-4(5)mm; lowest panicle-node width 3-8 branches. Intrd; marshes, fens, ditches and damp grassland, natd from previous use as fodder-grass; scattered in BI N to C Sc (mainly S & C En), often sporadic; Europe. No evidence of its being native as sometimes claimed.

12. P. glauca Vahl (<u>P.</u> <u>balfourii</u> Parnell) - <u>Glaucous</u> **R** <u>Meadow-grass</u>. Tufted perennial; culms erect, to 40cm; leaves usually glaucous, 2-4mm wide; ligules 1-2.5(3)mm; lowest panicle-node with 1-2(4) branches. Native; damp mountain rock-ledges and -crevices and rocky slopes at 610-910m; very local in C & N Sc (not Outer Isles), Lake District, Caerns. <u>P. balfourii</u> appears to be a shade variant of <u>P. glauca</u> with laxer habit and ± non-glaucous leaves.

13. P. nemoralis L. - <u>Wood</u> <u>Meadow-grass</u>. Tufted perennial; culms erect, to 75(90)cm; leaves 1-3mm wide; ligules 0.2-0.5mm; lowest panicle-node with 3-6 branches. Native; woods, hedge-banks, walls and other shady places; frequent to common in most of BI, but probably intrd in much of Ir and NW Br.

14. P. bulbosa L. - <u>Bulbous</u> <u>Meadow-grass</u>. Tufted perennial; **R** culms erect to decumbent, to 40cm, with swollen bulb-like base; leaves 0.5-2mm wide; ligules 1-3(4)mm; lowest panicle-node with 2-3 branches. Native; barish places in short grassland and open ground on sandy soil, shingle or limestone near sea, very rare inland; coasts of S Br from Glam to N Lincs, CI. Spikelets rarely (?never) proliferating except in Glam (all plants) and Jersey (both sorts present).

15. P. alpina L. - <u>Alpine</u> <u>Meadow-grass</u>. Tufted perennial; **R** culms erect to ± pendent, to 40cm; leaves 2-4.5mm wide; ligules 2-5mm; lowest panicle-node with 1-3 branches. Native; damp mountain rock-ledges and -crevices and rocky slopes at 300-1200m; local in C & N Sc, NW En, Caerns, S Kerry, Co Sligo. Spikelets usually proliferating; sexual plants (sometimes alone) confined to a few places in Sc and NW En.

16. P. flabellata (Lam.) Rasp. (<u>Parodiochloa</u> <u>flabellata</u> **1023** (Lam.) C.E. Hubb.) - <u>Tussac-grass</u>. Very densely tufted perennial forming large tussocks up to 1m x 1m (excl. culms); culms erect, to 80cm; leaves (3)5-10mm wide; ligules 4-10mm; lowest panicle-node with 2-4 branches. Intrd; very

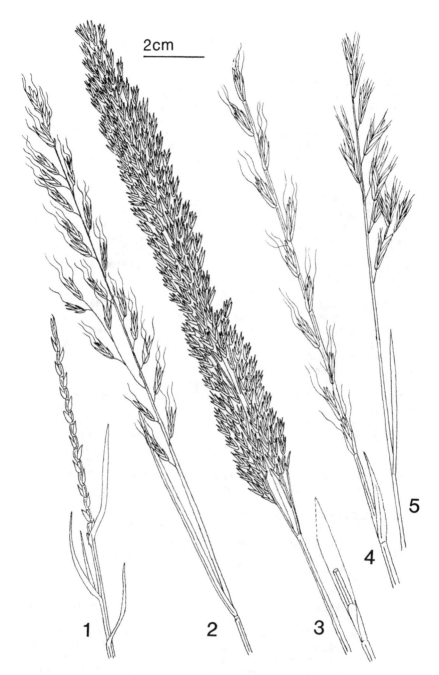

2cm

1   2   3   4   5

FIG 1023 – Poaceae. 1, **Hainardia cylindrica.** 2,
Helictotrichon neesii. 3, **Poa flabellata.** 4, **Gaudinia**
fragilis. 5, **X Festulpia hubbardii.**

persistent where planted in yards and on walls in Shetland; extreme S S America.

21. DACTYLIS L. - Cock's-foot

Perennials without stolons or rhizomes, with strongly compressed tillers; sheaths not fused; inflorescence a ± 1-sided panicle simply lobed or formed of stalked dense clusters of spikelets; spikelets with 2-5 florets all (except the most apical) bisexual; glumes 2, unequal; lemma keeled on back, 5-veined, without or with short terminal awn; stamens 3.

Other spp. - D. hispanica Roth (D. glomerata ssp. hispanica (Roth) Nyman), from SW Europe, was reported from W Cork in 1963; it is a short, densely tillering tetraploid with compact lobed panicles, and differs from similar coastal variants of D. glomerata in its split lemma. It was possibly intrd, perhaps casual, in W Cork, but its presence in SW Ir as a native needs investigating.

1. **D. glomerata** L. - Cock's-foot. Culms ± densely tufted, to 1.4m; leaves and sheaths very rough, ± glaucous; lemmas with hairs or prickles on keel, with awn 1.5-2mm; 2n=28. Native; grassland, open woodland and rough, waste and cultivated ground; common throughout BI. Formerly much grown for hay and pasture; the more robust plants of artifical habitats are probably of intrd stock.
2. **D. polygama** Horvatovszky (D. glomerata ssp. aschersoniana **1038** (Graebner) Thell., ssp. lobata (Drejer) Lindb. f.) - Slender Cock's-foot. Differs from D. glomerata in leaves and sheaths not or only slightly rough, green; inflorescence slenderer, with smaller clusters of spikelets; lemmas glabrous to obscurely prickly on keel, with awn 0 or <0.5mm; 2n=14. Intrd; grown for ornament and natd in woods; scattered in S En from Bucks and Surrey to Dorset; Europe.

22. CATABROSA P. Beauv. - Whorl-grass

Perennials with stolons, without rhizomes; sheaths not fused; inflorescence a diffuse panicle; spikelets with 1-3 florets all or all except the most apical bisexual; glumes 2, unequal; lemma rounded on back, 3-veined, truncate, awnless; stamens 3.

1. **C. aquatica** (L.) P. Beauv. - Whorl-grass. Culms erect to decumbent, to 75cm, often purple-tinged, glabrous or ± so. Native; wet meadows, marshes, ditches, by ponds and streams, often on barish mud; scattered throughout most of lowland BI, rare and decreasing in S. Plants from coastal parts of NW Br are short and ± consistently have spikelets with 1-2 florets; they are best recognized as var. **uniflora** Gray (ssp. minor (Bab.) F. Perring & Sell).

23. CATAPODIUM Link - Fern-grasses
Annuals; sheaths not fused; inflorescence a stiff raceme or
little-branched panicle; spikelets with 3(5)-14 florets all
(except the most apical) bisexual; glumes 2, subequal; lemma
rounded on back or keeled distally, 5-veined, acute to obtuse
or emarginate at apex, awnless; stamens 3.

1. C. rigidum (L.) C.E. Hubb. (Desmazeria rigida (L.) Tutin)
- Fern-grass. Culms erect to procumbent, to 15(60)cm;
inflorescence a spike-like raceme to little-branched panicle;
lower glume (1)1.3-2mm; upper glume (1.4)1.5-2.3(2.5)mm;
lemmas 2-2.6(3)mm. Native; dry, barish places on banks,
walls, sand, shingle, chalk and stony ground, especially near
sea; locally common in BI N to C Sc. Some coastal plants in
SW Br, S Ir and CI have branched panicles with the branches
spreading in 3 dimensions, not racemes or panicles spreading
only in 1 plane; they are best recognized as var. **majus** (C.
Presl) Lainz (ssp. majus (C. Presl) F. Perring & Sell).
1 x 2. C. rigidum x C. marinum (1 sterile plant) was found
in 1960 in Merioneth; endemic.
2. C. marinum (L.) C.E. Hubb. (Desmazeria marina (L.) Druce)
- Sea Fern-grass. Culms erect to procumbent, to 25cm;
inflorescence a spike-like raceme, rarely with few simple
branches below; lower glume 2-3mm; upper glume 2.3-3.3mm;
lemmas 2.2-3mm. Native; dry barish places by sea on walls,
banks, sand and shingle, often with C. rigidum. Native;
locally common round coasts of BI except parts of E & N coast
of Br.

24. SESLERIA Scop. - Blue Moor-grass
Tufted perennials with short rhizomes, without stolons;
sheaths not fused; inflorescence a small, very compact
panicle; spikelets with 2(-3) florets all bisexual; glumes 2,
subequal; lemma rounded on back, 3-5-veined, 3-5-toothed at
apex, each tooth with awn 0-1.5mm; stamens 3.

1. S. caerulea (L.) Ard. (S. albicans Kit. ex Schultes) -   R
Blue Moor-grass. Culms to 45cm, usually bare of leaves in
distal 1/2; ligule extremely short; panicle 1-3cm, ovoid,
usually bluish-violet tinged. Native; barish grassland,
rock-crevices and -ledges, screes and 'pavement' on limestone
and (in Sc) calcareous mica-schists; locally common in N En S
to MW Yorks and in C & NW Ir, very local in C Sc, very rare
in N Wa.

TRIBE 7 - HAINARDIEAE (Monermeae auct. non C.E. Hubb.)
(genera 25-26). Annuals; ligule membranous; sheaths not
fused; inflorescence a very slender cylindrical spike with
alternating spikelets partly sunk in cavities in rhachis
which breaks up into 1-spikeleted segments at fruiting;
spikelets with 1 (bisexual) floret; glumes 1-2, strongly
veined and + horny; lemma delicate, 3-veined with very short

lateral veins, acute, unawned; stamens 3; stigmas 2; ovary
with rounded glabrous appendage beyond style-bases; lodicules
2; ovary glabrous.

25. PARAPHOLIS C.E. Hubb. - Hard-grasses
Glumes 2, inserted side-by-side and together covering
rhachis-cavity except at anthesis; lemma with its side
towards rhachis.

1. P. strigosa (Dumort.) C.E. Hubb. - Hard-grass. Culms
usually erect, sometimes ascending or curved, rarely
procumbent, to 25(40)cm, very slender; spike fully exserted
from uppermost leaf-sheath or not, 2-20cm; anthers
1.5-3(3.5)mm. Native; sparsely-grassed ground on salt-
affected soil by salt-marshes and creeks and on rough and
waste ground; frequent on coasts of BI N to C Sc.
2. P. incurva (L.) C.E. Hubb. - Curved Hard-grass. Culms   R
decumbent to ascending, to 10(20)cm; spike usually not fully
exserted from uppermost leaf-sheath, 1-8(15)cm; anthers
0.5-1(1.5)mm. Native; similar places to P. strigosa but also
drier spots on cliff-tops and banks; local on coasts of Br N
to S Lincs and Merioneth, CI, very local in SE Ir.

26. HAINARDIA Greuter (Monerma auct. non P. Beauv.) - One-
glumed Hard-grass
Glumes 2 in terminal spikelet, 1 (the upper) in all others
and inserted so as to cover the rhachis-cavity except at
anthesis; lemma with its back towards rhachis.

1. H. cylindrica (Willd.) Greuter (Monerma cylindrica **1023**
(Willd.) Cosson & Durieu) - One-glumed Hard-grass. Culms
erect to ascending, straight or curved, to 30(45)cm; spike
fully exserted from uppermost leaf-sheath or not, 2-25cm;
anthers 1.5-3.5mm. Intrd; fairly frequent casual from
bird-seed on tips and waste ground; scattered in S En; S
Europe.

TRIBE 8 - MELICEAE (Glycerieae) (genera 27-28). Perennials
with rhizomes or stolons; ligule membranous; sheaths fused
into tube (often splitting later); inflorescence a little- or
much-branched panicle or a raceme with pedicels >3mm;
spikelets with 1-many bisexual florets, if with <4 florets
then with group of sterile ones beyond; glumes 2, subequal or
equal; lemmas rounded on back, 7-9-veined, subacute to
rounded, awnless; stamens 3; stigmas 2; lodicules fused
laterally into single scale shorter than wide; ovary
glabrous.

27. GLYCERIA R. Br. - Sweet-grasses
Aquatic or marsh grasses with rhizomes and/or stolons;
panicles much- to rather little-branched; ligules acute or
acuminate to rounded, 2-10(15)mm; spikelets with 4-16 florets

all (except the most apical) bisexual and each falling
separately when fruit ripe; glumes 1(-3)-veined; lemmas
7-veined.

1 Spikelets 5-12mm, with 4-10 florets; paleas not
  winged on keels; culms erect and self-supporting,
  usually >1m   **1. G. maxima**
1 Spikelets 10-35mm, with 6-17 florets; paleas winged
  on keels distally; culms decumbent to ascending,
  if erect not self-supporting, rarely >1m   2
  2 Anthers remaining indehiscent; pollen grains all
    or mostly empty and shrunken; spikelets remaining
    intact after not flowering, not forming fruits
                                                   3. G. x pedicellata
  2 Anthers dehiscent; pollen grains full and turgid;
    spikelets breaking up between florets when fruit
    ripe   3
3 Lemmas 5.5-6.5(7.5)mm; anthers 1.5-2.5(3)mm
                                                      2. G. fluitans
3 Lemmas 3.5-5mm; anthers 0.6-1.3mm   4
  4 Lemmas distinctly 3(-5)-toothed at apex, exceeded
    by 2 sharply pointed teeth of palea   **4. G. declinata**
  4 Lemmas not or scarcely toothed at apex, not
    exceeded by 2 (very short) teeth of palea
                                                      5. G. notata

1. G. maxima (Hartman) O. Holmb. - Reed Sweet-grass. Culms
erect, to 2.5m; panicles much-branched; lemmas 3-4mm, entire;
palea scarcely toothed at apex; anthers (1)1.2-1.8(2)mm;
2n=60. Native; in and by rivers, canals, ponds and lakes,
usually in deeper water than other spp.; common in most of En
except N, scattered in Wa, Ir and Sc, 1 record in Guernsey,
not in N or NW Sc.
2. G. fluitans (L.) R. Br. - Floating Sweet-grass. Culms
decumbent to ascending (or erect), to 1m; panicles sparsely
branched, narrow; lemmas 5.5-7(7.5)mm, entire; palea sharply
toothed at apex but shorter than lemma; anthers 1.5-2.5(3)mm;
2n=40. Native; on mud or in shallow water by ponds, rivers
and canals and in marshes, ditches and wet meadows; common
throughout BI.
2 x 4. G. fluitans x G. declinata is a sterile hybrid
differing from G. x pedicellata in its obscurely 3-toothed
lemmas and 2n=30; it occurs rarely with the parents very
scattered in En.
3. G. x pedicellata F. Towns. (G. fluitans x G. notata) -
Hybrid Sweet-grass. Differs from G. fluitans in lemmas
(4)5-5.5(6)mm; anthers 1-1.8mm; and see key (couplet 2).
Native; similar places to G. fluitans, with 1, both or
neither parents, often forming large patches; scattered over
most of BI, frequent in En, sometimes in areas (e.g. Jersey,
NW Sc) where G. notata does not occur.
4. G. declinata Breb. - Small Sweet-grass. Differs from G.

fluitans in culms to 60cm; lemmas 4-5mm, distinctly 3(-5)-
toothed at apex; palea sharply pointed, the points exceeding
lemma; anthers 0.6-1.3mm; 2n=20.   Native; similar places to
G. fluitans; scattered throughout BI, probably under-
recorded.
    5. G. notata Chevall. (G. plicata (Fries) Fries) - Plicate
Sweet-grass.   Differs from G. fluitans in panicles more
branched and less narrow; lemmas 3.5-5mm, much blunter at
apex; palea with very short teeth; anthers 0.7-1.3mm.
Native; similar places to G. fluitans; frequent throughout
most of BI except most of N & NW Sc and most of CI, perhaps
under-recorded.

28. MELICA L. - Melicks
Woodland or mountain grasses with short rhizomes;
inflorescence a raceme or a sparsely branched panicle;
ligules truncate, <2mm; spikelets with 1-3 bisexual florets
plus distal ± club-shaped cluster of sterile vestiges, all
the florets falling as a unit when fruit ripe; glumes
3-5-veined; lemmas 7-9-veined.

    1. M. nutans L. - Mountain Melick.   Culms erect or pendent
distally, to 60cm; sheaths without apical bristles;
inflorescence usually simple raceme; spikelets pendent, with
2-3 fertile florets; lower glume 5-veined.   Native; woods,
scrub and shady rock-crevices on limestone; scattered in N &
W Br S to Mons and Northants.
    2. M. uniflora Retz. - Wood Melick.   Culms erect or pendent
distally, to 60cm; sheaths with 2 long bristles at apex;
inflorescence a sparsely branched panicle; spikelets erect,
with 1 fertile floret; lower glume 3-veined.   Native; woods
and shady hedgebanks; scattered and locally common throughout
BI except N Sc and CI.

TRIBE 9 - AVENEAE (genera 29-39).   Annuals or perennials with
or without rhizomes, without stolons; ligule membranous;
sheaths not fused or rarely fused to form tube; inflorescence
a panicle, sometimes densely contracted, rarely a spike;
spikelets with 2-6(11) florets, sometimes 1 or more floret
(basal or apical) male, sometimes proliferating; glumes 2,
equal to unequal, often nearly as long to longer than rest of
spikelet, often with wide ± shiny hyaline margins; lemmas
5-9-veined, rounded on back, acute or obtuse to bifid or
variously toothed at apex, awnless or with short terminal awn
or more often with dorsal awn with conspicuous bend; stamens
3; stigmas 2; lodicules 2; ovary glabrous or pubescent at
apex or all over.

29. HELICTOTRICHON Besser ex Schultes & Schultes f. (Avenula
(Dumort.) Dumort., Avenochloa Holub, Amphibromus Nees) -
Oat-grasses
Tufted perennials with short or 0 rhizomes; inflorescence a

rather sparsely branched, + diffuse panicle; spikelets with 2–7 florets all except the most apical bisexual; glumes clearly unequal, the lower 1–3-veined, the upper 3–5-veined; lemmas 5–7-veined, variously shortly toothed at apex, with long, bent, dorsal awn; rhachilla-segments pubescent, the hairs in a longer tuft at apex of each segment around lemma-base; ovary glabrous or pubescent at apex.

1 Ovary and caryopsis glabrous; lower glume 4–5mm; upper glume 5–7mm; lemmas 5–8mm (excl. awn), papillose   **3. H. neesii**
1 Ovary and caryopsis with pubescent apex; lower glume 7–15mm; upper glume 10–20mm; lemmas 9–17mm (excl. awn), smooth to scabrid   2
  2 Lower culm-sheaths softly pubescent; spikelets with 2–3(4) florets; rhachilla hair-tuft 3–6(7)mm; palea with smooth keels   **1. H. pubescens**
  2 Culm-sheaths glabrous; spikelets with 3–6(8) florets; rhachilla hair-tuft 1–3mm; palea with scabrid keels   **2. H. pratense**

1. H. pubescens (Hudson) Pilger (<u>Avenula</u> <u>pubescens</u> (Hudson) Dumort., <u>Avenochloa</u> <u>pubescens</u> (Hudson) Holub) – <u>Downy</u> <u>Oat-grass</u>. Culms erect, to 1m; lower culm-sheaths and usually leaves softly pubescent; spikelets 10–17mm, with 2–3(4) florets; lower glume 7–13mm; upper glume 10–18mm; lemmas 9–14mm with awn to 22mm. Native; grassland usually on base-rich soils; + throughout BI, common on chalk and limestone in En.

2. H. pratense (L.) Besser ex Pilger (<u>Avenula</u> <u>pratensis</u> (L.) Dumort., <u>Avenochloa</u> <u>pratensis</u> (L.) Holub) – <u>Meadow</u> <u>Oat-grass</u>. Culms usually erect, to 80cm; leaves and sheaths glabrous; spikelets 11–28mm, with 3–6(8) florets; lower glume 10–15mm; upper glume 12–20mm; lemmas 10–17mm with awn to 22(27)mm. Native; similar places to <u>H.</u> <u>pubescens</u> and often with it, but usually in shorter turf and commoner in mountains; + throughout Br except in Outer Isles of Sc, C Wa and extreme SW En, common on chalk and limestone.

3. H. neesii (Steudel) Stace (<u>Amphibromus</u> <u>neesii</u> Steudel) – **1023** <u>Swamp</u> <u>Wallaby-grass</u>. Culms erect, to 1m; leaves and sheaths glabrous; spikelets 9–15mm, with (2)4–7 florets; awn up to 18mm; rhachilla hair-tuft 1–2mm. Intrd; rather characteristic wool-alien on tips, waste ground and fields; scattered in En; Australia.

30. ARRHENATHERUM P. Beauv. – <u>False</u> <u>Oat-grass</u>
Loosely tufted perennials; inflorecence a fairly well-branched + diffuse panicle; spikelets with 2(–5) florets, the lower (lowest) male, the upper bisexual, rarely both bisexual; glumes unequal, the lower 1-veined, the upper 3-veined; lemmas 7-veined, bifid at apex; lemma of male floret with long bent dorsal awn, that of bisexual floret(s)

awnless or with short terminal awn or rarely with dorsal bent awn; rhachilla-segments with apical hair-tuft 1-2mm; ovary pubescent at apex.

1. A. elatius (L.) P. Beauv. ex J.S. & C. Presl – <u>False Oat-grass</u>. Culms usually erect, to 1.8m; leaves and sheaths glabrous to sparsely pubescent; spikelets 7-11mm; lower glume 4-6mm; upper glume 7-10mm; lowest lemma 7-10mm with awn up to 20mm. Native; coarse grassy places, waysides, hedgerows, maritime sand and shingle, rough and waste ground; abundant throughout BI. Var. **bulbosum** (Willd.) St Amans (ssp. <u>bulbosum</u> (Willd.) N. Hylander, <u>A. tuberosum</u> (Gilib.) F. Schultz) has the basal, very short culm internodes swollen and corm-like; they are effective propagules in arable land (<u>Onion Couch</u>). Variants with >1 bisexual floret or >1 floret with a long awn are easily confused with <u>Helictotrichon</u>.

31. AVENA L. – <u>Oats</u>
Annuals; inflorescence a diffuse panicle; spikelets with 2-3 florets all bisexual or the distal 1 or 2 reduced and male or sterile; glumes subequal, 7-11-veined; lemmas 7-9-veined, bifid or with 2 bristles at apex, with or without long, bent, dorsal awn; rhachilla-segments with or without hair-tuft; ovary pubescent at apex or all over.
A difficult genus, in which general appearance and size of parts are often of little value. For accurate determination spikelets with fully ripe fruits are needed.

1 Lemmas bifid, the 2 apical points (1)3-9mm and each
   with 1 or more veins entering from main body of
   lemma and reaching apex   2
1 Lemmas bifid, the 2 apical points 0.5-2mm and with-
   out veins or vein(s) not reaching apex   3
     2 Rhachilla disarticulating between florets at
        maturity, releasing 1-fruited disseminules each
        with elliptic basal scar; lemmas with dense long
        hairs on lower 1/2   **1. A. barbata**
     2 Rhachilla not disarticulating at maturity, whole
        spikelets acting as disseminules, or the florets
        breaking away irregularly without basal scar;
        lemmas glabrous or sparsely pubescent on lower
        1/2   **2. A. strigosa**
3 Rhachilla not disarticulating at maturity, whole
   spikelets acting as disseminules, or the florets
   breaking away irregularly without basal scar;
   lemmas usually unawned, if awned then awn nearly
   straight, usually glabrous   **5. A. sativa**
3 Rhachilla disarticulating at maturity at least above
   glumes, often also between florets, hence at least
   lowest floret with basal scar; lemmas with long,
   strongly bent awns, usually pubescent   4
     4 Rhachilla disarticulating at maturity above

glumes only, releasing 2-3-fruited disseminules,
hence only lowest floret with (ovate) basal scar;
longer glume 24-30mm   **4. A. sterilis**
4 Rhachilla disarticulating at maturity between
  florets, releasing 1-fruited disseminules each with
  ovate basal scar; longer glume 18-25mm   **3. A. fatua**

Other spp. - A. byzantina K. Koch (Algerian Oat) is a minor
Mediterranean crop occasionally occurring as a grain
contaminant; it differs from A. sativa (in which it is often
included) in that the rhachilla eventually breaks just above
(not just below) each floret and hence remains attached to
the next floret above (not to the next floret below).

1. A. barbata Pott ex Link - Slender Oat. Stems to 1m;
spikelets with 2-3 florets, each with basal scar; longer
glume 15-30mm; lowest lemma 12-18mm, pubescent in lower 1/2,
plus 2 apical points 3-5mm. Intrd; rare grain-alien, natd in
Guernsey since 1970, perhaps overlooked for A. strigosa;
Mediterranean.
2. A. strigosa Schreber - Bristle Oat. Stems to 1.2m;
spikelets with 2-3 florets, none with basal scar; longer
glume 15-26mm; lowest lemma 10-17mm, usually glabrous in
lower 1/2, plus 2 apical points 3-9mm and 2 smaller fine
bristles alongside. Intrd; formerly (rarely still) grown as
minor crop in Wa, Sc and Ir and then a frequent (now very
local) cornfield weed sometimes natd, also infrequent casual
grain-alien; Spain.
3. A. fatua L. - Wild-oat. Stems to 1.5m; spikelets with
2-3 florets, each with basal scar; longer glume 15-25mm;
lowest lemma 14-20mm, pubescent in lower 1/2, plus 2 apical
points <0.5mm. Intrd; weed of arable, waste and rough
ground; common in most of En, very scattered elsewhere in BI;
Europe.
3 x 5. A. fatua x A. sativa occurs rarely in Br in and
around fields of A. sativa infested by A. fatua; it resembles
A. sativa but shows the influence of A. fatua in its longer
awns and tardily disarticulating spikelets, and has low
fertility.
4. A. sterilis L. (A. ludoviciana Durieu) - Winter Wild-oat.
Stems to 1.5m; spikelets with 2-3 florets, of which only the
lowest has basal scar; longer glume 20-32mm; lowest lemma
16-25mm, pubescent on lower 1/2, plus 2 apical points <1.5mm.
Intrd; similar places to A. fatua but usually on heavy soils
and replacing it there; formerly frequent, now scattered in
CS & SE En, rare grain-casual elsewhere; S Europe. Our plant
is ssp. **ludoviciana** (Durieu) Gillet & Magne. Ssp. **sterilis**
is a rare grain-alien; it differs in its larger parts
(spikelets with 3-5 florets; longer glume 32-45mm; lemmas
25-33mm; ligule >5mm (not <5mm)).
5. A. sativa L. - Oat. Stems to 1(-5)m; spikelets with 2-3
florets, none with basal scar; longer glume 17-30mm; lowest

lemma 12-20mm, usually glabrous, plus 2 apical points <1mm.
Intrd; locally common crop, frequent on tips and waysides and
as grain-alien; throughout BI; W Mediterranean.

32. GAUDINIA P. Beauv. - French Oat-grass
Annuals sometimes lasting a few years; inflorescence a spike
whose axis breaks giving 1-spikeleted segments at fruiting;
spikelets with 4-11 florets all (except most apical)
bisexual; lower glume c.1/2 as long as upper, 3-5- and
5-11-veined respectively; lemmas 5-9-veined, minutely bifid
at apex, with long, bent, dorsal awn; rhachilla segments +
glabrous; ovary with distinct hairy apex remaining
conspicuous as projection on fruit.

1. G. fragilis (L.) P. Beauv. - French Oat-grass. Stems to **1023**
45(100)cm; leaves and sheaths pubescent; spikelets 9-20mm; **1052**
lowest lemma 7-11mm with awn 5-13mm. Intrd; natd in grassy
fields, rough ground and waysides on a wide range of soils;
natd locaNly in SC En, SW Ir and CI, infrequent casual grain-
alien elsewhere; S Europe.

33. TRISETUM Pers. - Yellow Oat-grass
Perennials without rhizomes or stolons; inflorescence a well-
branched panicle; spikelets with 2-4 florets all (except the
most apical) bisexual; glumes unequal, the lower 1-, the
upper 3-veined; lemmas 5-veined, bifid with 2 short bristle-
points at apex, with long, bent dorsal awn; rhachilla-
segments pubescent, the hairs <1mm at apex; ovary glabrous.

1. T. flavescens (L.) P. Beauv. - Yellow Oat-grass. Stems
loosely tufted, to 80cm; lower leaves and sheaths pubescent;
spikelets 5-7.5mm (excl. awns); lowest lemma 4-5.5mm with awn
4.5-9mm. Native; meadows, pastures, grassy waysides,
especially on base-rich soil, throughout most of lowland BI,
common in En and SE Sc, rather scattered elsewhere.

34. KOELERIA Pers. - Hair-grasses
Tufted perennials without rhizomes or stolons; inflorescence
a spike-like panicle with very short branches; spikelets with
2-3(5) florets all (except most apical) bisexual; glumes
unequal, the lower 1-, the upper 3-veined; lemmas 3-veined,
acute or with extremely short apical awn, otherwise awnless;
rhachilla-segments shortly pubescent; ovary glabrous.

1. K. vallesiana (Honck.) Gaudin - Somerset Hair-grass. **RR**
Stems erect to procumbent, to 40cm; stems and sheaths very
shortly pubescent; lower sheaths very persistent and rotting
to form reticulated network of fibres round swollen culm-
bases; panicles 1.5-7 x 0.6-1.2cm; spikelets 4-6mm, shortly
appressed-pubescent; 2n=42. Native; short limestone grassland; 7 sites in Mendip Hills, N Somerset.
    1 x 2. K. vallesiana x K. macrantha occurs in most of the K.

vallesiana populations; it is sterile and has intermediate leaf-sheath characters, but chromosome number (2n=35) is the main diagnostic character; endemic.

2. **K. macrantha** (Ledeb.) Schultes (K. cristata auct. non (L.) Pers., K. gracilis Pers. nom. illeg., K. albescens auct. non DC., K. britannica (Domin ex Druce) Ujh., K. glauca auct. non (Schrader) DC.) - Crested Hair-grass. Stems erect to ascending, to 60cm; stems and sheaths variously pubescent to glabrous; lower sheaths not very persistent, not forming reticulated fibres and culm-base not swollen; panicles 1-10 x 0.5-2cm; spikelets 3.5-6mm, glabrous to pubescent; 2n=28. Native; short limestone or sandy base-rich grassland, dunes, less often on inland sandy soils; throughout most of BI, mostly on calcareous soils in S, mostly coastal in N. Very variable in stature, pubescence, leaf rigidity and inrolling, and colour, shape and denseness of panicle. Several segregates have been recognized, variously said to occur throughout Br or be confined to coasts of S & SW, but no thorough studies of our plants have been made. Until they have, and until the identity of our plants with the types of names used for them is demonstrated, acceptance of any of these segregates is premature.

35. **ROSTRARIA** Trin. (Lophochloa Reichb.) - Mediterranean Hair-grass
Annuals; inflorescence a spike-like panicle with very short branches; spikelets with 3-5(11) florets all (except apical 1-few) bisexual; glumes unequal, the lower 1-, the upper 3-veined; lemmas 5-veined, shortly bifid, with short subterminal awn; rhachilla-segments pubescent; ovary glabrous.

1. **R. cristata** (L.) Tzvelev (Lophochloa cristata (L.) N. **1038** Hylander, Koeleria phleoides (Villars) Pers.) - Mediterranean **1052** Hair-grass. Stems erect, to 20(60)cm; stems and sheaths glabrous to pubescent; panicles 1-10 x 0.4-1cm; spikelets 3-8mm; awns 1-3mm. Intrd; fairly characteristic wool-alien; scattered in Br; Mediterranean.

36. **DESCHAMPSIA** P. Beauv. - Hair-grasses
Densely tufted perennials usually without rhizomes or stolons; inflorescence a very diffuse panicle with fine branches; spikelets with 2 florets both bisexual or sometimes proliferating; glumes unequal, the lower 1-, the upper (1-)3-veined; lemmas 4-5-veined, rounded, obtuse or jagged-toothed at apex, with dorsal (rarely subterminal) straight or bent awn; rhachilla-segments pubescent with longer hair-tuft at base of each; ovary glabrous.

1 Spikelets proliferating   **1. D. cespitosa**
1 Spikelets not proliferating   2
  2 Leaves >1mm even if rolled up; awns not or

scarcely exceeding glumes   **1. D. cespitosa**
2 Leaves <1mm even if opened out; awns
 conspicuously exceeding glumes   3
3 Lemmas 2-3mm, toothed at apex with marginal teeth
the longest; ligule 2-8mm, very acute; palea bifid
             2. D. setacea
3 Lemmas 3-5.5mm, subacute to minutely toothed at
apex with marginal teeth not longer than inner ones;
ligule 0.5-3mm, obtuse; palea entire at apex
             3. D. flexuosa

1. D. cespitosa (L.) P. Beauv. - <u>Tufted</u> <u>Hair-grass</u>. Stems
to 1.5(2)m; leaves >2mm wide when flattened out, with very
rough edges cutting flesh; spikelets proliferating, or sexual
and 2.5-6mm. Native.
1 At least some spikelets proliferating   2
1 Spikelets not proliferating   3
 2 Leaves distinctly hooded at apex; panicle-
  branches and pedicels smooth, the main branches
  usually reflexed; awn arising from middle or
  upper 1/2 of lemma   **c. ssp. alpina**
 2 Leaves scarcely or not hooded at apex; panicle-
  branches and pedicels with minute (sometimes
  very sparse) pricklets, the main branches rarely
  reflexed; awn arising from lower 1/2 of lemma
          a. ssp. cespitosa
3 Spikelets 2-3(3.5)mm; hair-tuft at base of lower
lemma not reaching apex of rhachilla-segment above;
lowland woodland   **b. ssp. parviflora**
3 Spikelets (3)3.5-5(6)mm; hair-tuft at base of
lower lemma reaching apex of rhachilla-segment
above; lowland meadows and uplands **a. ssp. cespitosa**
 a. Ssp. cespitosa. Leaves scarcely or not hooded at apex;
panicle-branches and pedicels with minute pricklets; awn
arising from lower 1/2 of lemma; spikelets (3)3.5-5(6)mm, or
proliferating. Common throuhgout BI, a tetraploid (2n=52) in
lowland damp meadows, waysides and ditches and a diploid
(2n=26), triploid (2n=39) or tetraploid in similar habitats
and hilly country in N.
 b. Ssp. parviflora (Thuill.) Dumort. Differs from ssp.
<u>cespitosa</u> as in key; diploid (2n=26). Woods and shady
hedgerows; common in lowland Br, ?Ir.
 c. Ssp. alpina (L.) Hook. f. (<u>D. alpina</u> (L.) Roemer &
Schultes). Leaves distinctly hooded at apex; panicle-
branches and pedicels smooth; awn arising from middle or
upper 1/2 of lemma; spikelets usually all proliferating;
triploid (2n=39) or tetraploid (2n=52). Damp grassy places
on mountains; frequent in W & C Highlands of Sc, very local
in Caerns, S (formerly N) Kerry and W Mayo.
 2. D. setacea (Hudson) Hackel - <u>Bog</u> <u>Hair-grass</u>. Stems to R
70cm; leaves <1mm wide, usually bluish-green, slightly
scabrid; culm-sheaths smooth; spikelets 3-5mm; awn arising

from lower 1/2 of lemma. Native; bogs and boggy pools and ditches; very local and scattered in Br and Ir, mostly near coasts of W & N Sc and SC En.

3. D. flexuosa (L.) Trin. - <u>Wavy Hair-grass</u>. Stems to 60(100)cm; leaves <1mm wide, usually mid to dark green, slightly scabrid; culm-sheaths slightly scabrid; spikelets 4-6(7)mm; awn arising from lower 1/2 of lemma. Native; acid heaths, moors and open woods, drier parts of bogs; throughout BI, but absent from much of C En and Ir without suitable soils.

37. HOLCUS L. - <u>Soft-grasses</u>

Densely tufted or rhizomatous perennials; inflorescence a rather compact panicle; spikelets with 2 florets, the lower bisexual, the upper male; glumes subequal in length but unequal in width, the lower 1-, the upper 3-veined; lemmas 5-veined, ± rounded at apex, the lower awnless, the upper with dorsal awn arising from upper 1/2; rhachilla-segments ± glabrous but lemmas with basal tuft of hairs; ovary glabrous.

1. H. lanatus L. - <u>Yorkshire-fog</u>. Stems densely tufted, to 1m; rhizomes 0; leaves and sheaths softly and (sparsely to) densely patent-pubescent; glumes obtuse to subacute at apex; awn of upper lemma recurved to backwardly-hooked, included in glumes. Native; rough grassland, lawns, arable, rough and waste ground, open woods; common throughout BI.

1 x 2. H. lanatus x H. mollis = H. x hybridus Wein is very scattered in Br and Ir but probably under-recorded; it is sterile and resembles <u>H. mollis</u>, but has blunter glumes, less exserted awns and more pubescent culms.

2. H. mollis L. - <u>Creeping soft-grass</u>. Stems loosely tufted, to 1m; plant rhizomatous; leaves and sheaths variably pubescent (tiller sheaths often ± densely and softly pubescent) but uppermost culm-sheath glabrous with conspicuously patent-pubescent node below it; glumes acute to acuminate at apex; awn of upper lemma slightly bent, well exserted from glumes. Native; woods, hedgerows, less often open grassland, mostly on acid soils; common throughout most of BI but absent from areas of calcareous or base-rich soil.

38. CORYNEPHORUS P. Beauv. - <u>Grey Hair-grass</u>

Densely tufted perennial without rhizomes; inflorescence a rather compact panicle; spikelets with 2 florets, both bisexual; glumes subequal, 1-veined or the upper also with 2 very short laterals; lemmas with 1 central and often 2 pairs of very short lateral veins; obtuse to very shortly bifid at apex, with dorsal bent awn with club-shaped apex; rhachilla-segments pubescent and with tuft of hairs at base of each lemma; ovary glabrous.

1. C. canescens (L.) P. Beauv. - <u>Grey Hair-grass</u>. Stems R erect to decumbent, to 35cm; leaves very glaucous, <1mm wide

and tightly inrolled; ligules 2-4mm, very acute; spikelets
3-4mm; awns not exceeding glumes.   Native; secondarily open
sand on leached fixed dunes and inland sandy heathland on
acid soils; very local on and near coasts of E Suffolk and E
& W Norfolk, rare inland in E Suffolk, probably intrd on
coasts of Moray and formerly S Lancs and Glam, natd in
Staffs.

39. AIRA L. - Hair-grasses
Annuals; inflorescence a compact to diffuse panicle;
spikelets with 2 florets, both bisexual; glumes subequal,
1-3-veined; lemmas with 5 short veins, shortly bifid, with
dorsal, slightly bent awn; rhachilla-segments extremely
short; lemma with short hair-tuft at base; ovary glabrous.

    1. A. caryophyllea L. (A. multiculmis Dumort., A. armoricana
Albers) - Silver Hair-grass. Stems erect to decumbent, to
25(50)cm; sheaths slightly rough; panicle diffuse, with
conspicuous suberect to erecto-patent branches. Native; dry
sandy, gravelly or rocky ground, on walls, heaths and dunes;
frequent throughout BI.   3 segregates are sometimes
recognised but are not consistently separable on external
characters and are doubtfully worth maintaining even as sspp.
Ssp. **caryophyllea** has spikelets 2.5-3mm; anthers 0.3-0.45mm;
caryopsis 1.2-1.6mm; and some pedicels >5mm.   Ssp.
multiculmis (Dumort.) Bonnier & Layens has spikelets
2.2-2.6mm; anthers 0.3-0.5mm; caryopsis 1.1-1.5mm; and all
pedicels usually <5mm. Ssp. **armoricana** (Albers) Kerguelen,
recorded only from W Cornwall, has spikelets 3-3.5mm; anthers
0.3-0.5mm; caryopsis 1.5-1.9mm; and all pedicels usually
<5mm. The relative distributions of sspp. caryophyllea and
multiculmis are disputed.
    2. A. praecox L. - Early Hair-grass. Stems erect to
procumbent, to 10(15)cm; sheaths smooth; panicle compact,
with short erect branches largely obscured; spikelets
2.5-3.5mm; anthers 0.3-0.3mm; caryopsis 1.4-1.9mm. Native;
similar places to A. caryophyllea; common throughout BI.

TRIBE 10 - PHALARIDEAE (genera 40-42). Annuals or perennials
with or without rhizomes, without stolons; ligule membranous;
sheaths not fused; inflorescence a panicle, usually
contracted; spikelets with (2-)3 florets, the lower (1-)2
male or sterile and often much reduced, the upper 1 bisexual,
rarely the spikelets in groups consisting of 1 central
bisexual and 4-6 surrounding sterile (or male) spikelets;
glumes 2, equal or unequal, 1 or both nearly as long as to
longer than rest of spikelet; lemmas of bisexual florets
3-7-veined, keeled or rounded on back, awnless; palea with 1
vein; lower 2 lemmas minute to longer than upper 1, 0- or
4-5-veined, awnless or with dorsal, bent awm; stamens 2 or 3;
stigmas 2; lodicules 0 or 2; ovary glabrous.
Anthoxanthum and Hierochloe may be better placed in Aveneae.

40. HIEROCHLOE R. Br. - <u>Holy-grass</u>
Rhizomatous perennials; inflorescence a diffuse panicle;
lower 2 florets with 3 stamens, with 5-veined lemma slightly
longer than lemma of bisexual floret; terminal floret with 2
stamens, with 5-veined lemma, with 2 lodicules; glumes
subequal, keeled, slightly shorter than rest of spikelet,
1-3-veined. Crushed or dried plant smells strongly of hay.

1. H. odorata (L.) P. Beauv. - <u>Holy-grass</u>. Stems to 60cm;   **RR**
panicles with patent to erecto-patent thin branches;
spikelets 3.5-5mm, greenish-purple, becoming characteristic
golden-brown. Native; banks of rivers and lakes, wet
meadows, flushed cliff-bases by sea; very local on and near
coast in Sc, Co Antrim.

41. ANTHOXANTHUM L. - <u>Vernal-grasses</u>
Annuals or tufted perennials; inflorescence a contracted
panicle; lower 2 florets sterile, with 4-5-veined lemma
slightly longer than lemma of bisexual floret and with long
dorsal awn; terminal floret with 2 stamens, 5-veined awnless
lemma and 1-veined palea; lodicules 0; glumes very unequal,
lower 1-veined, upper 3-veined and longer than rest of
spikelet. Crushed or dried plant smells strongly of hay.

1. A. odoratum L. (<u>A.</u> alpinum auct. non A. & D. Loeve) -
<u>Sweet</u> <u>Vernal</u> <u>Grass</u>. Tufted perennials; culms unbranched, to
50(100)cm; ligules 1-5mm; spikelets 6-10mm; glumes usually
pubescent; awns not or only slightly exceeding glumes; awn of
lower lemma 2-4mm, of 2nd lemma 6-9mm; anthers 3-4.5mm.
Native; in all kinds of grassy places, acid and calcareous,
heavy or light soils, lowland or montane; abundant throughout
BI.
2. A. aristatum Boiss. (<u>A.</u> puelii Lecoq & Lamotte) - <u>Annual</u>
<u>Vernal-grass</u>. Tufted annuals; culms usually well branched,
to 40cm; ligules 0.6-2mm; spikelets 5-7.5mm; glumes glabrous;
awns much exceeding glumes; awn of lower lemma 4-5mm, of 2nd
lemma 7-10mm; anthers 2.5-3.5mm. Intrd; formerly natd in
sandy cultivated or rough ground in Surrey, E Suffolk and
perhaps elsewhere but not seen since 1970, now only a rare
casual; scattered records in En, S Wa and CI; S Europe.

42. PHALARIS L. - <u>Canary-grasses</u>
Annuals or rhizomatous perennials; inflorescence a contracted
(often spike-like) panicle; lower 2 florets reduced to scales
or rarely only 1 present; terminal floret with 3 stamens,
with 5-veined awnless lemma, with 2 lodicules; glumes equal,
sharply keeled, longer than rest of spikelet.

1 Perennial with very short to long rhizomes and
   tillers at flowering time   2
1 Annual without rhizomes or tillers   3
   2 Panicle at least distinctly lobed, usually with

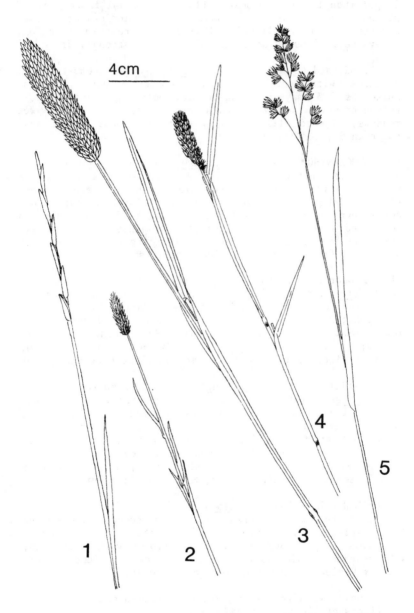

4cm

1

2

3

4

5

FIG 1038 – Poaceae. 1, Lolium rigidum. 2, Rostraria cristata.
3, Phalaris aquatica. 4, P. paradoxa. 5, Dactylis polygama.

        conspicuous branches; glumes strongly keeled
        but not winged   **1. P. arundinacea**
    2 Panicle not lobed, oblong to lanceolate in
        outline, without visible branches; glumes with
        distinct wing on keel   **2. P. aquatica**
    3 Spikelets in groups of 3-7, one bisexual the rest
        sterile, falling as a group when fruits ripe
                                                6. P. paradoxa
    3 Spikelets all bisexual, the (2)3 florets of each
        falling at maturity leaving glumes on panicle   4
        4 At least 1 glume with its keel-wing minutely
            toothed on at least some spikelets   **5. P. minor**
        4 Wings of glume-keels entire   5
    5 Sterile florets 2, >1/2 as long as fertile floret
                                                3. P. canariensis
    5 Sterile florets 1-2, <1/3 as long as fertile floret
                                                4. P. brachystachys

   1. P. arundinacea L. - <u>Reed</u> <u>Canary-grass</u>. Perennial to 2m,
with long rhizomes; panicle distinctly branched, rather
<u>Dactylis</u>-like; spikelets 4.5-6.5mm, all bisexual; glumes not
winged on keel; sterile florets 1/4 to 1/2 as long as
bisexual floret. Native; by lakes and rivers, in ditches,
wet meadows and marshes, also rough and waste ground; common
throughout most of BI.
   2. P. aquatica L. (<u>P.</u> <u>tuberosa</u> L.) - <u>Bulbous</u> <u>Canary-grass</u>. 1038
Perennial to 1.5m, with short rhizomes; panicle spike-like;
spikelets 4.5-6mm, all bisexual; glumes with entire wings on
keels; sterile florets <1/4 as long as bisexual floret.
Intrd; grown as game-cover and -food, sometimes for grazing
or silage, natd in fields and rough ground, also casual wool-
alien; increasingly frequent in C & S En, more widely
scattered as casual; S Europe.
   3. P. canariensis L. - <u>Canary-grass</u>. Annual to 1.2m;
panicle spike-like, usually ovoid; spikelets (4)6-10mm, all
bisexual; glumes with entire wings on keels; sterile florets
>1/2 as long as bisexual floret. Intrd; usually casual
bird-seed alien on tips and waste ground, sometimes + natd,
also grown as bird-seed crop in S & C En; frequent throughout
BI; NW Africa and Canaries.
   4. P. brachystachys Link - <u>Confused</u> <u>Canary-grass</u>. Annual to
60cm; differs from <u>P.</u> <u>canariensis</u> in panicles usually
slenderer (narrowly ovoid); and see key. Intrd; rather
frequent wool-alien on tips and waste ground; scattered in
En, perhaps ovelooked for <u>P.</u> <u>canariensis</u>; Mediterranean.
   5. P. minor Retz. - <u>Lesser</u> <u>Canary-grass</u>. Annual to 60cm;
differs from <u>P.</u> <u>canariensis</u> in panicles usually slenderer
(narrowly ovoid) and sterile floret(s) <1/3 as long as
fertile floret; and from <u>P.</u> <u>canariensis</u> and <u>P.</u> <u>brachystachys</u>
as in key (couplet 4). Intrd; rather frequent casual
wool-alien in Br, natd in sandy places in Guernsey since at
least 1791; SW & S Europe.

6. P. paradoxa L. - <u>Awned</u> <u>Canary-grass</u>. Annual to 1.6m; **1038**
panicle spike-like, narrowly elliptic to cylindrical;
spikelets in groups of 3-7 (1 bisexual, others sterile), the
fertile ones 3-6(8)mm with glumes narrowed to short awns and
with an apically toothed wing on the keel; the sterile ones
all or mostly with club-shaped glumes; sterile florets <1/4
as long as fertile floret. Intrd; natd weed of arable
fields, increasingly common in C & S En and S Wa, also a more
widespread casual in Br from wool, grain and other sources;
SW & S Europe.

TRIBE 11 - AGROSTIDEAE (genera 43-53). Annuals or perennials
with or without rhizomes and/or stolons; ligule membranous;
sheaths not fused; inflorescence a panicle, contracted or
diffuse, often spike-like, rarely a raceme or spike;
spikelets with 1 (bisexual) floret (in <u>Beckmannia</u> with 2
bisexual or 1 bisexual plus 1 male or sterile florets);
glumes 2, usually equal or nearly equal, usually both at
least as long as rest of spikelet; lemmas 3-7-veined, rounded
on back, awnless or with terminal, subterminal or dorsal awn;
stamens 3; stigmas 2; lodicules 0 or 2; ovary glabrous.

43. AGROSTIS L. - <u>Bents</u>
Annuals or perennials with or without rhizomes and/or
stolons; inflorescence a slightly contracted to very diffuse
panicle with obvious branches; glumes 2, equal or nearly so,
1-3-veined, longer than rest of spikelet; lemmas 3-5-veined,
awnless or with subterminal or dorsal awn, with or without
hair-tuft on callus; palea <3/4 as long as lemma, sometimes
vestigial, (usually weakly) 2-veined or veinless; disarticulation at maturity at base of lemma.
A difficult genus due to plasticity and genetic variation,
hybridization, and over-use in the past of unreliable
characters, notably presence of awns. The palea length and
presence of hair-tufts on the lemma-callus are important
(though minute) characters (x>10 lens essential). Presence
of rhizomes (as opposed to stolons) is also very reliable but
their absence is not, as they are often not developed until
late on (in fruit) and in some habitats never.

1 Palea minute, <2/5 as long as lemma   2
1 Palea >2/5 as long as lemma (c.1/2 as long to
  nearly as long)   6
    2 Anthers 0.2-0.6mm; main panicle-branches bare
      of spikelets for proximal >2/3 of length; alien
      (often casual) of waste ground   3
    2 Anthers 1-2mm; main panicle-branches bare of
      spikelets for proximal <1/2 of length; natives   4
3 Spikelets >2mm; lemma 1.5-1.7mm, distinctly
  exceeding caryopsis; leaves 1-3mm wide   **10. A. scabra**
3 Spikelets <2mm; lemma 1-1.2mm, not or scarcely
  exceeding caryopsis; leaves <1mm wide   **11. A. hyemalis**

  4 Tiller leaves ≤0.3mm wide, bristle-like;
     panicle always with ± erect branches; rhizomes
     and stolons 0   **7. A. curtisii**
  4 Tiller leaves flat or inrolled, >(0.6)1mm wide
     even when inrolled; panicle often with patent to
     erecto-patent branches at or after flowering;
     rhizomes or stolons usually present   5
5 Stolons 0; rhizomes usually present; ligule
  usually ≤1.5x as long as wide, acute to subobtuse
                                                 9. A. vinealis
5 Rhizomes 0; stolons usually present, bearing tufts
  of leaves or shoots at nodes; ligule usually ≥1.5x
  as long as wide, acute to acuminate   **8. A. canina**
  6 Anthers 0.2-0.8mm; lemma-callus with tuft
     of hairs ≥0.3mm   7
  6 Anthers 1-1.5mm; lemma-callus glabrous or
     pubescent   8
7 Lemmas with awn ≥2mm, well exserted from glumes;
  main panicle-branches bare of spikelets for proximal
  ≥2/3 of length, patent to erecto-patent after
  flowering; rhachilla extended above lemma-base,
  reaching ≥1/2 way up lemma   **6. A. avenacea**
7 Lemmas with awn 0-0.5mm, not exserted from glumes;
  main panicle-branches bare of spikelets for
  proximal ≤1/2 of length, erect to erecto-patent
  after flowering; rhachilla not extended above
  lemma-base   **5. A. lachnantha**
  8 Lemma-callus with tuft of hairs >0.3mm; awn 0
     or, if present, arising from basal 1/3 of lemma,
     often exceeding glumes   **3. A. castellana**
  8 Lemma-callus glabrous or with hairs <0.2mm; awn
     0 or, if present, arising from apical 1/2 of
     lemma, rarely exceeding glumes   9
9 Panicle contracted after flowering; rhizomes 0 or
  short and with ≤3 scale-leaves; ligules of culm-
  leaves subacute to rounded; stolons usually well
  developed   **4. A. stolonifera**
9 Panicle-branches patent or nearly so after
  flowering; rhizomes usually present, with >3 scale-
  leaves; ligules of culm-leaves truncate; stolons
  0 or poorly developed   10
  10 Ligules of tillers shorter than wide; fruiting
     panicle-branches with spikelets all well
     separated; leaves rarely >5mm wide   **1. A. capillaris**
  10 Ligules of tillers longer than wide; fruiting
     panicle-branches bearing spikelets in small ±
     dense clusters at tips; leaves often >5mm wide
                                           2. A. gigantea

Other spp. - A. exarata Trin. (Spike Bent), from N America,
has been found once in Bucks as a weed of grass-seed but may
be commoner and overlooked as a contaminant of Highland Bent.

It has a minute palea and anthers <0.6mm, but panicle-branches bearing spikelets ± to the base; the lemma-callus is glabrous and the lemmas awned or unawned. The panicle-branches are ± erect forming a narrow inflorescence super-ficially resembling that of _Apera_ _interrupta_.

1. **A. capillaris** L. (_A. tenuis_ Sibth.) – Common Bent. Rhizomatous perennial to 75cm; ligules truncate or rounded, <2mm, those of tillers shorter than wide; panicle diffuse at fruiting, with spikelets all separated; awns 0 or short; 2n=28. Native; all kinds of grassy places and rough ground, usually on acid soils; abundant throughout BI.

1 x 2. A. capillaris x A. gigantea = A. x bjoerkmanii Widen is a vigorous highly sterile pentaploid (2n=35) with intermediate ligules and panicle; very few scattered records in Br but probably overlooked.

1 x 3. A. capillaris x A. castellana = A. x fouilladei P. Fourn. is partially fertile and variably intermediate between the parents and hence difficult to determine; it has scattered records in En arising from intrd seed-mixture and perhaps in situ, but is likely to become commoner.

1 x 4. A. capillaris x A. stolonifera = A. x murbeckii Fouill. ex P. Fourn. is a vigorous highly sterile tetraploid (2n=28) with usually both rhizomes and stolons and inter-mediate ligule- and panicle-shape; it has been recorded from scattered localities in Br but is probably common throughout BI.

1 x 9. A. capillaris x A. vinealis (?=_A. x sanionis_ Asch. & Graebner) is a highly sterile tetraploid (2n=28) with awns 0 to long and basal, palea c.1/3 as long as lemma, and ligule of tillers c. as long as wide; there are scattered records in Br on poor sandy soils.

2. **A. gigantea** Roth – Black Bent. Rhizomatous perennial to 1(1.2)m; ligules truncate or rounded, <6mm, those of tillers longer than wide; panicle diffuse at fruiting, with spikelets in small clusters at branch-tips; awns 0 or short; 2n=42. Native; grassy places, rough, cultivated and waste ground, mostly on disturbed sandy soils; throughout BI, common in S & C En, scattered elsewhere but probably overlooked.

2 x 4. A. gigantea x A. stolonifera is a vigorous highly sterile pentaploid (2n=35) with intermediate ligule- and panicle-shape and often strong rhizomes and stolons; there are very scattered records in Br.

3. **A. castellana** Boiss. & Reuter – Highland Bent. Rhizom-atous perennial to 60cm; ligules subacute to rounded, <3mm, those of tillers c. as long to slightly longer than wide; panicles contracted at fruiting; awns 0, short or long; 2n=28. Intrd; lawns, roadsides, amenity and sports areas where sown and escaped; throughout Br, becoming commoner, much under-recorded.

4. **A. stolonifera** L. – Creeping Bent. Stoloniferous perennial to 75cm, with rhizomes 0 or very short; ligules

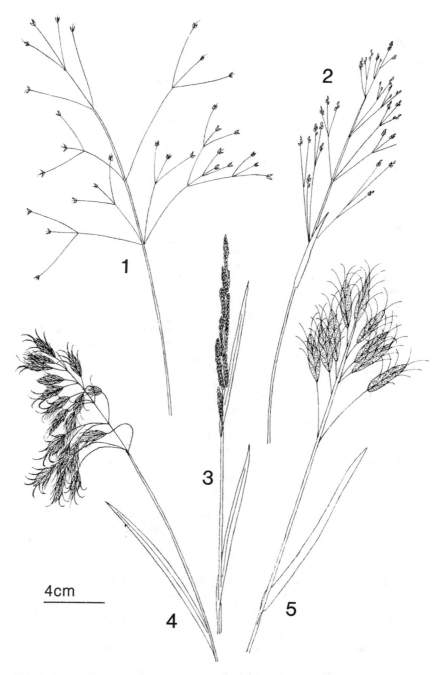

FIG 1043 – Poaceae. 1, **Agrostis avenacea**. 2, **A. hyemalis**. 3, **Beckmannia** **syzigachne**. 4, **Bromus** **japonicus**. 5, **B.** **lanceolatus**.

subacute to rounded, $\leq$7mm, those of tillers longer than wide;
panicles contracted at fruiting; awns 0 or short; 2n=28.
Native; damp meadows, ditches, marshes, by lakes, ponds,
canals and rivers, damp arable and rough ground, dune-slacks;
abundant throughout BI.
4 x 9. A. stolonifera x A. vinealis is a highly sterile
tetraploid (2n=28) with awns 0 to long and basal, palea c.1/2
as long as lemma, ligule of tillers longer than wide, and
often stolons and rhizomes; it has been recorded only from W
Cornwall, but, although rare, is probably overlooked.
5. A. lachnantha Nees - <u>African Bent</u>. Loosely tufted annual
(with us) or perennial; ligules acute to obtuse, $\leq$7mm, those
of tillers longer than wide; panicles contracted at fruiting;
awns 0 or $\pm$ so. Intrd; a rather infrequent wool-alien, but
perhaps overlooked; scattered in En; C & S Africa.
6. A. avenacea J. Gmelin - <u>Blown-grass</u>. Tufted perennial **1043**
(often annual with us) to 60cm; ligules $\leq$10mm, acute to
rounded, those of tillers longer than wide; panicles very
widely spreading at fruiting; awns long. Intrd; frequent
wool-alien now natd in waste and rough ground and by roads
and railways; scattered in Br; Australia and New Zealand.
7. A. curtisii Kerguelen (<u>A. setacea</u> Curtis non Villars) -
<u>Bristle Bent</u>. Densely tufted perennial to 60cm; ligules
$\leq$4mm, acute, those of tillers longer than wide; panicles
contracted at fruiting; awns long; 2n=14. Native; dry sandy
or peaty heaths; locally common in SW En, extending to S Wa
and Surrey, formerly E Sussex.
8. A. canina L. - <u>Velvet Bent</u>. Stoloniferous perennial to
75cm, without rhizomes; ligules $\leq$4mm, acute to obtuse, those
of tillers longer than wide; panicles loosely contracted at
fruiting; awns 0 to long; 2n=14. Native; damp or wet
meadows, marshes, ditches, pond-sides, on acid soils;
frequent to common throughout most of BI, but records
confused with those of <u>A. vinealis</u>.
9. A. vinealis Schreber (<u>A. canina</u> ssp. <u>montana</u> (Hartman)
Hartman, var. <u>arida</u> Schldl.) - <u>Brown Bent</u>. Rhizomatous
perennial to 60cm; ligules $\leq$5mm, acute to obtuse, those of
tillers longer than wide; panicles strongly contracted at
fruiting; awns 0 to long; 2n=28. Native; dry sandy or peaty
heaths, moors and hillsides; frequent to common throughout
most of BI, but see under <u>A. canina</u>.
10. A. scabra Willd. - <u>Rough Bent</u>. Tufted perennial (often
annual with us) to 60cm; ligules $\leq$5mm, acute to obtuse, those
of tillers longer than wide; panicles very widely spreading
at fruiting, resembling those of <u>A. avenacea</u> in shape; awns 0
or very short. Intrd; frequent grain-alien now natd in waste
and rough ground and by roads and railways; scattered in Br;
N America.
11. A. hyemalis (Walter) Britton, Sterns & Pogg. - <u>Small</u> **1043**
<u>Bent</u>. Tufted perennial (often annual with us) to 40cm;
differs from <u>A. scabra</u> as in key. Intrd; wool- and
grain-alien perhaps natd as for <u>A. scabra</u>; scattered in Br,

but distribution uncertain due to confusion with A. scabra; N
America.

43 x 50. AGROSTIS x POLYPOGON = X AGROPOGON P. Fourn.
Variously intermediate between A. stolonifera and Polypogon
and sterile; differs from A. stolonifera in more compact
panicle with shorter pedicels and disarticulation (if any) at
maturity near base of pedicel; differs from P. monspeliensis
and P. viridis as under each hybrid.

**43/4 x 50/1. A. stolonifera x P. monspeliensis = X A.
littoralis** (Smith) C.E. Hubb. (Perennial Beard-grass) is a
perennial vegetatively resembling A. stolonifera; differs
from A. stolonifera additionally in bifid glumes rough on
back with long apical awn and long-awned lemmas; and from P.
monspeliensis in clearly branched panicle, and lemmas with
subterminal (not terminal) awn. Native; sporadic with the
parents, usually on maritime damp sand or mud, and as
rubbish-tip casual from bird-seed and other sources; coasts
of En from Dorset to W Norfolk, casual elsewhere in Br.
Rather closely resembles an awned variant of P. viridis, and
sometimes occurs with it on tips.
43/4 x 50/2. A. stolonifera x P. viridis = X A. robinsonii
(Druce) Meld. & D. McClint. (Agrostis x robinsonii Druce) is
a short-lived perennial; differs from A. stolonifera as
above; and from P. viridis in glumes only sparsely scabrid
apart from midrib, and palea <3/4 as long as lemma. Native;
has occurred with the parents in Guernsey in 1924 and 1953
(erroneous record in 1958).

44. CALAMAGROSTIS Adans. - Small-reeds
Rhizomatous perennials; inflorescence a + diffuse to slightly
contracted panicle; glumes 2, equal or nearly so, 1-3-veined,
longer than rest of spikelet; lemmas 3-5-veined, with apical
or dorsal awn, with conspicuous basal hair-tuft >1/2 as long
as lemma; palea c.2/3 as long as lemma, 2-veined; disarticulation at maturity at base of lemma.

1 Hairs at base of lemma not reaching lemma-apex;
   lemma minutely rough, with awn arising from
   middle or lower 1/2   2
1 Hairs at base of lemma reaching at least to apex
   of lemma; lemma smooth, with awn arising from
   middle, apex or upper 1/2   3
    2 Spikelets 3-4(4.5)mm; lower glume acute
                                              4. C. stricta
    2 Spikelets (4)4.5-6mm; lower glume acuminate
                                              5. C. scotica
3 Culms mostly with 5-8 nodes; ligules 7-10(14)mm;
   pollen 0; anthers indehiscent   **3. C. purpurea**
3 Culms mostly with 2-5 nodes; ligules (1)2-9(12)mm;
   pollen present; anthers dehiscent   4

4   Leaf uppersides (often sparsely) pubescent;
    ligules (1)2-4(5)mm; lemmas 3-5-veined, with basal
    hairs 1.1-1.5x as long as lemma   **2. C. canescens**
4   Leaf uppersides scabrid, not pubescent; ligules
    4-9(12)mm; lemmas 3-veined, with basal hairs
    >1.5x as long as lemma   **1. C. epigejos**

1. C. epigejos (L.) Roth - <u>Wood</u> <u>Small-reed</u>. Culms to 2m,
with 2-4 nodes; leaves not pubescent on upperside; ligules
4-9(12)mm; glumes 4-7mm; lemmas with basal hairs c.1.5-2x as
long as lemma; 2n=28. Native; damp woods and wood-margins,
ditches, fens, dune-slacks; scattered throughout much of Br,
common in parts of S & E En, rare in most of Sc, Wa, Ir and
CI.
2. C. canescens (Wigg.) Roth - <u>Purple</u> <u>Small-reed</u>. Culms to
1.2m, with 3-5 nodes; differs from <u>C. epigejos</u> in less dense
panicles, less finely pointed glumes; and see key; 2n=28.
Native; fens, marshes and open wet woods; scattered in En
(mostly CE), Kircudbrights and Selkirks.
2 x 4. C. canescens x C. stricta = C. x gracilescens (Blytt)
Blytt occurs for certain only by a canal in SE Yorks where
partially fertile octoploids (2n=56) intermediate between the
parents and sterile tetraploids (2n=28) closer to <u>C. stricta</u>
occur along with <u>C. stricta</u>. Other populations of <u>C. stricta</u>
in En and Sc probably have been introgressed by <u>C. canescens</u>
in the past (glumes >4mm, longer lemma-hairs, etc.).
3. C. purpurea (Trin.) Trin. - <u>Scandinavian</u> <u>Small-reed</u>.   **RR**
Culms to 1.5m, with 5-8 nodes; leaves pubescent on upperside;
ligules 7-10(14)mm; glumes 4-6(7)mm; lemmas with basal hairs
1.1-1.5x as long as lemma; 2n=56. Native; fens, marshes,
ditches and lake-sides; c.6 localities in Main Argyll, E
Perth, S Aberdeen and Angus, 1 locality in Westmorland, not
recognized until 1980. An apomictic taxon probably derived
from <u>C. epigejos</u> x <u>C. canescens</u>. Our plant is ssp.
phragmitoides (Hartman) Tzvelev.
4. C. stricta (Timm) Koeler - <u>Narrow</u> <u>Small-reed</u>. Culms to   **R**
1m, with 2-3 nodes; leaves pubescent on upperside; ligules
1-3mm; glumes 3-4(4.5)mm; lemmas with basal hairs 0.5-0.8x as
long as lemma; 2n=28. Native; marshes, fens and lake-sides;
very scattered in N BI S to W Suffolk, Cheshire and Co Antrim
(formerly Co Armagh).
5. C. scotica (Druce) Druce - <u>Scottish</u> <u>Small-reed</u>. Differs   **RR**
from <u>C. stricta</u> as in key; perhaps an introgressed variant of
it. Native; marshes and fens; 1 locality (formerly 3) in
Caithness, possibly 1 in Roxburghs (identity not certain);
endemic.

44 x 45. CALAMAGROSTIS x AMMOPHILA = X CALAMMOPHILA Brand (X
<u>Ammocalamagrostis</u> P. Fourn.)
44/1 x 45/1. X C. baltica (Fluegge ex Schrader) Brand (X
<u>Ammocalamagrostis</u> <u>baltica</u> (Fluegge ex Schrader) P. Fourn; <u>C.</u>
<u>epigejos</u> x <u>A. arenaria</u>) (<u>Purple</u> <u>Marram</u>) occurs on maritime

dunes with <u>A. arenaria</u> and near <u>C. epigejos</u> in E Suffolk, E Norfolk, Cheviot and W Sutherland, and planted for sand-binding with <u>A. arenaria</u> in E Suffolk, E & W Norfolk and S Hants. It is vigorous, sterile and more closely resembles <u>A. arenaria</u>, with long rhizomes and culms to 1.5m with dense, linear-ellipsoid, often purple-tinged panicles; spikelets intermediate, 9–12mm, with lemmas 7–9mm with 3–7 veins, subterminal awn 1–2mm and basal hair-tuft >1/2 as long as lemma; palea 2–4-veined. Our plant is var. **baltica** (2n=42); on the Continent vars more exactly intermediate and nearer <u>C. epigejos</u> respectively, with different origins, also occur.

45. AMMOPHILA Host - <u>Marram</u>
Strongly rhizomatous perennials; inflorescence a compact linear-ellipsoid panicle; glumes 2, subequal, the lower 1–3-, the upper 3-veined, slightly longer than rest of spikelet; lemmas 5–7-veined, awnless or with minute subapical awn <1mm, with basal hair-tuft <1/2 as long as lemma; palea nearly as long as lemma, 2–4-veined; disarticulation at maturity at base of lemma.

1. A. arenaria (L.) Link - <u>Marram</u>. Culms to 1.2m; leaves usually tightly inrolled with minutely densely pubescent upperside; ligules 10–30mm; glumes 10–16mm; 2n=28. Native; on mobile sand-dunes, often dominant; common round coasts of BI, rare casual inland.
2. A. breviligulata Fern. - <u>American</u> Marram. Differs from <u>A. arenaria</u> in ligules 1–3mm. Intrd; planted on sand-dunes at Newborough, Anglesey in late 1950s, now well natd in 1 small area; E N America. Perhaps only a ssp. of <u>A. arenaria</u>.

46. GASTRIDIUM P. Beauv. - <u>Nit-grasses</u>
Annuals; inflorescence a compact linear-ellipsoid panicle; glumes 2, unequal, 1-veined, much longer than rest of spikelet, linear-lanceolate, with swollen ± hemispherical base; lemmas 5-veined, with dorsal usually bent awn, without basal hair-tuft; palea ± as long as lemma, 2-veined; disarticulation at maturity at base of lemma.

1. G. ventricosum (Gouan) Schinz & Thell. - <u>Nit-grass</u>. **RR**
Stems procumbent to erect, to 50(90)cm; panicles 0.5–10(16)cm; spikelets (2)3–5mm; lemmas c.1mm, subglabrous to sparsely pubescent, with awn 0–4mm, rarely exceeding glumes. Native; barish or sparsely grassed, well-drained, calcareous ground, also a frequent wool- and grain-alien, formerly a frequent weed of arable land; very local as native mostly near coast in SW Br from S Devon and Wight to Glam (plants small and often procumbent to ascending), perhaps CI, casual scattered in En and S Wa, formerly frequent as cornfield weed in SE En (plants often robust and usually erect).
2. G. phleoides (Nees & Meyen) C.E. Hubb. - <u>Eastern Nit-grass</u>. Differs from <u>G. ventricosum</u> in stems usually erect;

spikelets (4)5-8mm; lemma pubescent to densely so, with awn
4-7(8)mm, often exceeding glumes. Intrd; rather infrequent
wool-alien; scattered in En; SW Asia. Scarcely distinct from
G. ventricosum.

47. LAGURUS L. - Hare's-tail
Annuals; inflorescence a very compact, ovoid, densely silky-
pubescent panicle; glumes 2, equal, 1-veined, linear-
lanceolate and tapered to apical awn, longer than rest of
spikelet, lemmas 5-veined, with 2 apical bristles reaching c.
as far as glume-awns, with dorsal bent awn well exserted from
glumes, pubescent but without basal hair-tuft; palea shorter
than body of lemma, 2-veined; disarticulation at maturity at
base of lemma.

1. L. ovatus L. - Hare's-tail. Stems to 60cm, erect; leaves
and sheaths softly pubescent; panicle 1-7 x 0.5-2cm;
spikelets 7-10mm; awns 8-20mm. Intrd; planted and now often
abundant on mobile sand-dunes in Jersey and Guernsey, planted
but still rare at Dawlish (S Devon), rare casual elsewhere in
En and S Wa; S Europe.

48. APERA Adans. - Silky-bents
Annuals; inflorescence a diffuse to + contracted panicle;
glumes 2, unequal, the lower 1-, the upper 3-veined, + as
long as rest of spikelet; lemmas 5-veined, with subterminal +
straight awn much longer than body, minutely pubescent at
base; palea shorter than to + as long as body of lemma,
2-veined; disarticulation at maturity at base of lemma.

1. A. spica-venti (L.) P. Beauv. - Loose Silky-bent. Stems   R
to 1m; leaves 2-10mm wide; ligules 3-10mm; panicles up to 25
x 15cm, very diffuse, the longer branches bare for proximal
1/2; spikelets 2.4-3mm; awns 5-10mm; anthers 1-2mm. Native;
locally frequent in dry sandy arable fields and marginal
habitats in E Anglia, intrd elsewhere in Br and CI, locally
frequent as weed in parts of E & SE En, elsewhere very
scattered and usually casual.
2. A. interrupta (L.) P. Beauv. - Dense Silky-bent. Stems
to 40(70)cm; leaves 0.5-4mm wide; ligules 2-5mm; panicles up
to 20 x 1.5cm, loosely contracted, the branches with
spikelets nearly to base; spikelets 1.8-2.5mm; awns 4-10mm;
anthers 0.3-0.4mm. Intrd; natd in dry sandy fields and rough
ground; locally frequent in E Anglia, rare casual elsewhere;
S Europe.

49. MIBORA Adans. - Early Sand-grass
Annuals; inflorescence a slender 1-sided raceme with pedicels
<0.5mm; glumes 2, equal, 1-veined, longer than rest of
spikelet; lemmas 5-veined, awnless, pubescent but without
basal hair-tuft; palea as long as lemma, 2-veined; disarticulation at maturity at base of lemma.

1. M. minima (L.) Desv. - <u>Early Sand-grass</u>. Culms usually **RR**
densely tufted, very slender, to 8(15)cm, with leaves usually
all on basal 1/3; racemes 0.5-2cm; spikelets 1.8-3mm; glumes
and lemmas ± truncate. Native; on loose sand of maritime
dunes and similar places near sea in Anglesey and CI,
possibly native in Glam, natd in few places on S & E coasts
of Br N to E Lothian, rare casual elsewhere.

50. POLYPOGON Desf. - <u>Beard-grasses</u>
Annuals or perennials with stolons; inflorescence a
contracted or semi-diffuse panicle; glumes 2, equal, 1-veined
or the upper 3-veined, much longer than rest of spikelet,
awned from apex or unawned; lemmas 5-veined, truncate and
finely toothed at apex, without basal hair-tuft, awnless or
with short terminal awn; palea nearly as long as lemma,
2-veined; disarticulation at maturity near base or apex of
pedicel.

Other spp. - **P. maritimus** Willd., from S & W Europe, differs
from <u>P. monspeliensis</u> in its slenderer habit, much more
deeply bifid and longer-pubescent glumes, and unawned lemma;
it is a very infrequent wool-alien but might be overlooked.

1. P. monspeliensis (L.) Desf. (<u>P. paniceus</u> (L.) Lag.) - **R**
<u>Annual Beard-grass</u>. Annual; culms to 80cm; inflorescence
densely contracted, ± cylindrical; spikelets disarticulating
near apex of pedicel; glumes 2-3mm, notched at apex, each
with apical awn 3.5-7mm; lemma with awn (0)1-2mm; anthers
0.4-0.7mm. Native; drier parts of salt-marshes and damp
places near sea; near coasts of S & SE En from Dorset to E
Norfolk, formerly W Norfolk and Guernsey, elsewhere in Br N
to C Sc as casual of tips and waste ground, sometimes natd.
2. P. viridis (Gouan) Breistr. (<u>P. semiverticillatus</u>
(Forsskaol) N. Hylander, <u>Agrostis semiverticillata</u>
(Forsskaol) C. Chr.) - <u>Water Bent</u>. Stoloniferous perennial;
culms to 60(100)cm; inflorescence semi-diffuse; spikelets
disarticulating near base of pedicel; glumes 1.6-2.3mm,
obtuse, unawned; lemma unawned; anthers 0.4-0.7mm. Intrd;
natd on roadsides and rough ground and by pools in Guernsey
and Jersey, scattered casual elsewhere in Br on tips and
waste land; S Europe.

51. ALOPECURUS L. - <u>Foxtails</u>
Annuals or perennials without rhizomes, sometimes with
stolons; inflorescence a very contracted spike-like panicle;
glumes 2, equal, 3-veined, sometimes with their margins fused
proximally round spikelet, very slightly shorter to very
slightly longer than rest of spikelet, keeled, rounded to
very acute or apiculate at apex; lemmas 4-veined, obtuse to
truncate or notched, without basal hair-tuft, sometimes with
margins fused proximally round carpel and stamens, with
dorsal long or short awn from lower 1/2; palea 0; disarticul-

ation at maturity near base of pedicel.

1 Lemmas unawned or with awn shorter than body and not
  exserted from glumes or exserted by <0.5mm   2
1 Lemmas with awn longer than body and exserted from
  glumes by >1mm   3
  2 Panicles >3x as long as wide; glumes with hairs
    <0.5mm; lemma with margins fused proximally for
    1/3 to 1/2 their length; lowlands   **4. A. aequalis**
  2 Panicles <3x as long as wide; glumes with hairs
    >(0.5)1mm; lemma with margins fused proximally
    for <1/4 their length; mountains   **5. A. borealis**
3 Margins of lemma free or fused proximally for <1/4
  their length; 2 glumes fused only at extreme base   4
3 Margins of lemma fused proximally for c.1/3 to 1/2
  their length; 2 glumes fused proximally for c.1/4
  to 1/2 their length   5
  4 Glumes acute; basal culm internode swollen,
    (1)2-4.5(6)mm wider than normal culm width
                                               3. A. bulbosus
  4 Glumes obtuse; basal culm internode 0-1(1.5)mm
    wider than normal culm width   **2. A. geniculatus**
5 Annual; glumes fused proximally for 1/3 to 1/2
  their length, subglabrous or with hairs <0.5mm
  on keel, margins and at base, with winged keel
                                             6. A. myosuroides
5 Perennial; glumes fused proximally for c.1/4
  their length, conspicuously pubescent with hairs
  >0.5mm, with keel unwinged   **1. A. pratensis**

1. A. pratensis L. - <u>Meadow</u> <u>Foxtail</u>. Perennial; culms
usually erect, to 1.2m; panicles 2-12 x 0.5-1.2cm; spikelets
4-6mm; glumes acute, fused proximally for c.1/4 their length,
conspicuously pubescent; anthers 2-3.5mm. Native; grassy
places, mostly on damp rich soils; common + throughout
lowland BI.
  1 x 2. A. pratensis x A. geniculatus = A. x brachystylus
Peterm. (<u>A.</u> x <u>hybridus</u> Wimmer) is scattered with the parents
throughout Br; it is intermediate in habit, fusion of lemma
and glumes, spikelet length (3-4.5mm), and anther length and
is usually highly sterile.
  2. A. geniculatus L. - <u>Marsh</u> <u>Foxtail</u>. Perennial; culms
usually decumbent to ascending, often rooting at lower nodes,
to 40(50)cm; panicles 1.5-7 x 0.3-0.7cm; spikelets
2-3(3.5)mm; glumes obtuse, fused only at extreme base,
conspicuously pubescent; anthers 0.8-2mm, yellow or purple.
Native; wet meadows, marshes, ditches, pond-sides; frequent
to common throughout BI.
  2 x 3. A. geniculatus x A. bulbosus = A. x plettkei Mattf.
has been found with the parents in S & E En from Dorset to N
Lincs; it is intermediate in culm shape and culm swelling and
highly sterile.

2 x 4. **A. geniculatus x A. aequalis = A. x haussknechtianus**
Asch. & Graebner has been found scattered with the parents
from Merioneth and Brecs to W Norfolk and Beds; it is
intermediate in spikelet and awn lengths and lemma margin
fusion and is highly sterile.
3. **A. bulbosus** Gouan - Bulbous Foxtail. Perennial; culms   R
usually erect to ascending, not rooting at nodes, to
30(40)cm; differs from A. geniculatus in spikelets 2.5-3.3mm;
anthers 1.2-2.2mm; and see key. Native; wet grassy places,
usually brackish and grazed, near sea or in estuaries; local
on coasts of S & E Br from Carms and E Cornwall to N Lincs,
Guernsey.
4. **A. aequalis** Sobol. - Orange Foxtail. Annual to   R
perennial; culms usually decumbent to ascending, sometimes
rooting at lower nodes, to 40cm; panicles 1-6 x 0.3-0.6cm;
spikelets 2-2.5mm; glumes obtuse, fused only at extreme base,
conspicuously shortly pubescent; anthers 0.8-1.3mm, orange.
Native; similar places to A. geniculatus; scattered in C & S
Br N to NW Yorks (mainly C & E En), Easterness.
5. **A. borealis** Trin. (A. alpinus Smith non Villars) - Alpine R
Foxtail. Perennial; culms ± erect, to 50cm; panicles 1-3 x
0.7-1.2cm; spikelets 3-4.5mm; glumes acute, fused only at
extreme base, very conspicuously long-pubescent; anthers
2-2.5mm. Native; mountain springs and flushes usually
dominated by bryophytes, at 600-1200m; very local in N Br
from Westmorland to E Ross.
6. **A. myosuroides** Hudson - Black-grass. Annual; culms
erect, to 80cm; panicles 2-12 x 0.3-0.6cm; spikelets 4.5-7mm;
glumes acute, fused proximally for 1/3 to 1/2 their length,
very inconspicuously pubescent on keels, edges and base only;
anthers 2.5-4mm. Native; weed of arable fields and waste
ground; frequent (but decreasing) in S, C & E En, very
scattered to SW En, CI, Wa and C Sc and there mostly casual.

52. BECKMANNIA Host - Slough-grasses
Annuals or perennials without stolons, rarely with rhizomes;
inflorescence a long raceme of closely packed appressed
spikes forming a long narrow inflorescence; glumes 2, equal,
3-veined with connecting veins between, enclosing rest of
spikelet except for apiculate tip of lemma(s), strongly
keeled and hooded; florets 1 or 2, the 1st bisexual, the
second bisexual, male or sterile; lemmas 5-veined, apiculate
but not otherwise awned, without basal hair-tuft; palea
nearly as long as lemma, 2-veined; disarticulation at
maturity below glumes.
Abnormal in Agrostideae in racemose inflorescence and often
2-floreted spikelets with glumes with branched veins, but
even more out of place elsewhere.

Other spp. - B. eruciformis (L.) Host (European Slough-
grass), from E Europe, might occur as more than a rare casual
but has been much confused with B. syzigachne and recent

FIG 1052 – Spikelets of **Poaceae**. 1, **Oryzopsis**. 2, **Lamarckia** (group of 3 sterile and 2 fertile florets). 3, **Schismus**. 4, **Rostraria**. 5, **Beckmannia**. 6, **Rytidosperma**. 7, **Gaudinia**. 8, **Cortaderia**. 9, **Eriochloa**. 10–11, **Brachiaria** (11, upper glume removed to show upper lemma). 12–13, **Urochloa** (13, upper glume removed to show upper lemma). 14, **Paspalum**.

records of it are errors for the latter; it differs in having 2 regularly bisexual florets per spikelet, usually pubescent lemmas, and anthers 1.5–2mm, and is a longer lived perennial with culms often swollen at base.

1. B. syzigachne (Steudel) Fern. – <u>American Slough-grass</u>. **1043** Annual or short-lived perennial to 50(100)cm; inflorescence **1052** up to 15(25)cm; spikelets flattened, closely packed on spikes 5–20mm, 2.2–3.2mm, with 1 floret or sometimes a 2nd male, bisexual or sterile one; glumes and usually lemmas glabrous; anthers 0.7–1.2mm. Intrd; casual of waste ground and tips, mainly from grain or bird-seed, scattered in S En and S Wa, natd in few sites in Bristol Channel area; N America.

53. PHLEUM L. – <u>Cat's-tails</u>
Annuals or perennials, sometimes with rhizomes or stolons; inflorescence a very contracted spike-like panicle; glumes 2, equal, 3-veined, strongly keeled with stiff hairs on keel, apiculate to shortly awned at apex; lemmas 3–7-veined, irregularly truncate to rounded at apex, without basal hair-tuft, unawned; palea as long or nearly as long as lemma, 2-veined; disarticulation at maturity below the lemma.

1 Annual without tillers at flowering; glumes acute
  to subacute at apex, gradually narrowed to awn;
  anthers <1mm   **5. P. arenarium**
1 Perennial with tillers at flowering; glumes obtuse
  or truncate at apex, abruptly or very abruptly
  narrowed to awn; anthers >1mm   2
  2 Glumes 5–8.5mm incl. awns 2–3mm; panicles
    (1)2–3(5)x as long as wide; mountains >600m
    in N Br   **3. P. alpinum**
  2 Glumes 2–5.5mm incl. awns <2mm; panicles
    (2)3–20(30)x as long as wide; widespread   3
3 Glumes obtuse (and shortly awned) at apex; culms
  not swollen at base; ligules 0.5–2mm; CE En only
                                                 4. P. phleoides
3 Glumes truncate (and shortly awned) at apex; culms
  usually swollen at base; ligules 1–9mm; widespread   4
  4 Spikelets (3.5)4–5.5mm incl. awns (0.8)1–2mm;
    panicle 6–10mm wide; leaves 3–9mm wide; ligule
    usually obtuse   **1. P. pratense**
  4 Spikelets 2–3.5mm incl. awns 0.2–1(1.2)mm;
    panicle 3–6mm wide; leaves 2–6mm wide; ligule
    usually acute   **2. P. bertolonii**

1. P. pratense L. – <u>Timothy</u>. Perennial; culms erect, to 1.5m; panicles up to 20(30) x 1cm, cylindrical; glumes (3.5)4–5.5mm incl. awns, truncate at apex with awn 0.8–2mm; 2n=42. Native; grassy places and rough ground; common throughout BI.
2. P. bertolonii DC. (<u>P. nodosum</u> auct. non L., <u>P. hubbardii</u>

D. Kovats, **P.** pratense ssp. bertolonii (DC.) Bornm., ssp.
serotinum (Jordan) Berher) - Smaller Cat's-tail. Differs
from **P.** pratense in culms to 50(100)cm; panicles up to 8 x
0.5cm; glumes 2-3.5mm incl. awns 0.2-1.2mm; 2n=14; and see
key. Native; similar places to **P.** pratense but usually
confined to grassland; probably throughout BI but uncertain
due to confusion with **P.** pratense. It is possible that **P.**
bertolonii refers to a non-British tetraploid (2n=28), in
which case **P. serotinum** Jordan would be our plant; ssp.
serotinum is the correct name at ssp. level in either case.

3. **P. alpinum** L. (**P.** commutatum Gaudin) - Alpine Cat's-tail.   **R**
Perennial; culms erect, to 50cm; panicles up to 5 x 1.2cm,
ovoid to shortly cylindrical; glumes 5-8.5mm incl. awns,
truncate at apex with awn 2-3mm. Native; grassy, rocky or
mossy wet places on mountains at 600-1200m; very local in N
Br fom Westmorland to E Ross. Similar distribution to
Alopecurus borealis and not rarely with it.

4. **P. phleoides** (L.) Karsten - Purple-stem Cat's-tail.   **RR**
Perennial; culms erect, to 60cm, the lower sheaths often
purplish-tinged; panicles up to 10 x 0.7cm, ± cylindrical but
often narrowed at each end; glumes 2.5-3mm incl. awns, obtuse
at apex with awn ≤0.5mm. Native; dry sandy and chalky
pastures and adjacent rough ground; very local in CE En from
Beds and Herts to W Norfolk.

5. **P. arenarium** L. - Sand Cat's-tail. Annual; culms erect,
to 20(30)cm; panicles up to 5 x 0.7cm, subcylindrical to
narrowly ellipsoid; glumes 3-4mm incl. awns, acute or
subacute at apex with awn ≤0.7mm. Native; maritime sand-
dunes and inland on sandy heaths; frequent on most coasts of
BI except N & W Sc, inland in E Anglia.

TRIBE 12 - BROMEAE (genera 54-57). Annuals or perennials
without stolons, with or without rhizomes; ligule membranous;
sheaths usually fused when young but soon splitting;
inflorescence a panicle, rarely slightly contracted;
spikelets with several to many bisexual florets, the apical 1
or 2 (sometimes more) often reduced and male or sterile;
glumes 2, unequal, 1-9-veined, much shorter than rest of
spikelet, unawned; lemmas 5-11-veined, rounded or keeled on
back, usually minutely bifid at apex, usually with long subterminal awn; stamens 2-3; stigmas 2; lodicules 2; ovary with
pubescent terminal appendage (the styles arising below it).

54. BROMUS L. - Bromes
Annuals; spikelets ovoid to narrowly so, terete or slightly
compressed; lower glume 3-5(7)-veined; upper glume 5-7(9)-
veined; lemmas 7-9(11)-veined, rounded on back, subacute to
obtuse or rounded-obtuse, and minutely bifid at apex; stamens
3.
All the spp. are very variable in habit, becoming very small
(often with only 1 spikelet) in dry conditions, and in
extreme cases spikelet and lemma measurements also vary

outside the normal range. Dwarf, starved plants should be avoided; more normal plants are usually nearby. Lemma measurements should be made from the middle or low part of spikelets.

1 Caryopsis thick, with inrolled margins; lemma with
  margins wrapped around caryopsis when mature,
  hence lemma margins not overlapping next higher
  lemma, but rhachilla ± revealed between florets;
  rhachilla disarticulating tardily   2
1 Caryopsis thin, flat or with weakly inrolled margins;
  lemma margins not wrapped round caryopsis, overlapping next higher lemma and obscuring rhachilla;
  rhachilla disarticulating readily   3
    2 Spikelets 12-20mm, glabrous or pubescent; lemmas
      6.5-9(10)mm; palea equalling lemma; caryopsis
      6-9mm; sheaths usually glabrous or sparsely
      pubescent   **8. B. secalinus**
    2 Spikelets 8-12mm, glabrous; lemmas 5-6mm;
      palea shorter than lemma; caryopsis 4-4.5mm;
      sheaths pubescent   **9. B. pseudosecalinus**
3 Palea divided nearly to base; panicle with mostly
  subsessile spikelets densely clustered in groups of
  3; extinct   **7. B. interruptus**
3 Palea entire to shortly bifid; panicle various but
  spikelets not subsessile in groups of 3   4
    4 Anthers 3.5-5mm, >1/2 as long as lemmas;
      panicle branches long, forming very open panicle
                                                 1. B. arvensis
    4 Anthers 0.2-3mm, <1/2 as long as lemmas;
      panicles various, often not very open   5
5 Awns curved or bent outwards at maturity, their
  apices widely diverging   6
5 Awns ± straight to slightly flexuous or curved at
  maturity, the apices of those of the more apical
  lemmas ± parallel or even convergent (if ± curved
  outwards the culms procumbent to ascending)   9
    6 Panicle-branches and pedicels much shorter than
      spikelets; pedicels <10mm; spikelets <18mm   7
    6 At least some panicle-branches and pedicels on
      well grown plants longer than spikelets;
      pedicels often >15mm; spikelets >18mm   8
7 Lemmas 6.5-8.5mm; culms usually <15cm, with
  usually <10 spikelets; maritime native
                                      4c. B. hordeaceus ssp. ferronii
7 Lemmas 8-11mm; culms usually >15cm, with usually >10
  spikelets; alien   **4b. B. hordeaceus ssp. divaricatus**
    8 Lemmas 11-18mm; panicles usually rather
      stiffly erect   **11. B. lanceolatus**
    8 Lemmas 8-10mm; panicles usually lax with patent
      pendent branches   **10. B. japonicus**
9 Lemmas (4.5)5.5-6.5mm; caryopsis longer than

palea   **6. B. lepidus**
9 Lemmas 6.5-11mm; caryopsis shorter than to as long
  as palea   10
  10 Panicle ± lax with at least some pedicels
    longer than spikelets; lemmas rather coriaceous,
    with rather obscure veins; anthers 1-3mm   11
  10 Panicle ± dense, usually with all pedicels
    shorter than spikelets; lemmas papery, with
    prominent veins; anthers 0.5-1.5(2)mm   12
11 Lemmas 6.5-8mm; anthers mostly 1.5-3mm; spikelets
  10-16mm; lowest rhachilla-segment mostly 0.7-1mm
                                               3. B. racemosus
11 Lemmas 8-11mm; anthers mostly 1-1.5mm; spikelets
  15-28mm; lowest rhachilla-segment mostly 1.3-1.7mm
                                               2. B. commutatus
  12 Lemmas 8-11mm, usually pubescent
                                 4a. B. hordeaceus ssp. hordeaceus
  12 Lemmas 6.5-8mm, usually glabrous   13
13 Culms to 8(12)cm, procumbent to ascending; caryopsis
  shorter than palea; maritime dunes
                                   4d. B. hordeaceus ssp. thominei
13 Culms usually >10cm, usually erect; caryopsis c. as
  long as palea; widespread   **5. B. x pseudothominei**

1. B. arvensis L. - Field Brome. Culms erect, to 1m;
panicle diffuse, ± pendent at maturity, with branches much
longer than spikelets; lemmas 7-9mm, ± glabrous, with
straight awn 6-10mm. Intrd; usually casual weed of arable
and grass fields and waste ground; very scattered in C & S Br
(mostly C & E En), decreasing, extremely sporadic elsewhere,
natd in E Gloucs; Europe.
  2. B. commutatus Schrader - Meadow Brome. Culms erect, to
1m; panicle ± diffuse, drooping to 1 side at maturity, with
most branches longer than spikelets; lemmas 8-11mm, usually
glabrous, with straight awn 3-10mm. Native; grassy places,
waysides and rough ground, especially in damp rich meadows;
locally frequent in C & S Br, decreasing, very scattered
casual elsewhere. Perhaps better as B. racemosus ssp.
commutatus (Schrader) Tourlet.
  2 x 3. B. commutatus x B. racemosus occurs rather frequently
with the parents in C & S Br; it is fertile and backcrosses,
forming a spectrum of intermediates between the parents and
making them difficult to separate in some areas.
  3. B. racemosus L. - Smooth Brome. Culms erect, to 1m;
panicle fairly open but rather narrow and erect, sometimes
drooping to 1 side at maturity, with many branches longer
than spikelets; lemmas 6.5-8mm, glabrous, with straight awn
5-9mm. Native; similar places and distribution to B.
commutatus.
  4. B. hordeaceus L. - Soft-brome. Culms erect to
procumbent, to 80cm; panicle rather to very compact, erect,
sometimes drooping to 1 side at maturity, with branches

shorter than spikelets; lemmas 6.5-11mm. The 4 sspp. are
separated in the key to spp.

a. Ssp. hordeaceus (B. mollis L.). Stems to 80cm (often
much less), erect; panicle up to 10(16)cm, with few to many
spikelets, often drooping to 1 side at maturity; lemmas
8-11mm, pubescent, with ± straight awn 4-11mm. Native;
grassy places, waysides, rough ground; frequent throughout
lowland BI.

b. Ssp. divaricatus (Bonnier & Layens) Kerguelen (ssp.
molliformis (Lloyd) Maire & Weiller, B. molliformis Lloyd).
Stems to 60cm, erect; panicle up to 10cm, with many
spikelets, stiffly erect, with very short branches and
pedicels; lemmas 8-11mm, pubescent, with awn 5-10mm and
curved outwards at maturity. Intrd; casual from wool, grass-
seed and other sources in waste places and waysides;
scattered in En; S Europe.

c. Ssp. ferronii (Mabille) P.M. Smith (B. ferronii Mabille).
Stems to 15(20)cm, erect to ascending; panicle up to 5cm,
with few spikelets, stiffly erect; lemmas 6.5-8.5mm,
pubescent, with awn 2-6mm and curved outwards at maturity.
Native; grassy cliff-tops and sandy or shingly ground by sea;
locally frequent on coasts of CI and S & SW Br scattered N to
Kirkcudbrights and Angus, Man.

d. Ssp. thominei (Hardouin) Braun-Blanquet (B. thominei
Hardouin). Culms to 8(12)cm, procumbent to ascending;
panicle up to 3cm, with few spikelets, erect; lemmas
6.5-7.5mm, glabrous (usually) or pubescent, with straight or
slightly curved-out awn 3-7mm. Native; sandy places by sea;
coasts of CI and Br probably N to Sc, but much over-recorded
for B. x pseudothominei and distribution uncertain.

5. B. x pseudothominei P.M. Smith (B. thominei auct. non
Hardouin; B. hordeaceus x B. lepidus) - Lesser Soft-brome.
Culms erect, to 60cm; panicle ± compact, erect, with all
branches shorter than spikelets; lemmas 6.5-8mm, usually
glabrous, with straight awn 3-7mm. Probably native; grassland, waysides, rough ground, usually with B. hordeaceus and
often with B. lepidus (and commoner than it), but often
without either close by; scattered throughout BI, frequent in
C & S Br.

6. B. lepidus O. Holmb. - Slender Soft-brome. Culms erect,
to 80cm; panicle compact, erect, with branches shorter than
spikelets; lemmas (4.5)5.5-6.5mm, usually glabrous, with
straight awn 2-5.5mm. Probably intrd; grassland, waysides
and rough ground; scattered throughout BI, frequent in C & S
Br; origin uncertain.

7. B. interruptus (Hackel) Druce - Interrupted Brome. Culms **RR**
erect, to 1m; panicle compact, often spaced out below, erect,
with very short branches and spikelets often tightly
clustered in threes; lemmas 7.5-9mm, pubescent, with straight
awn 3-8mm. Native; arable and waste land, especially as weed
in Onobrychis, Lolium or Trifolium crops; formerly scattered
in SC and SE En, from E Kent and N Somerset to S Lincs, last

seen in 1972 in Cambs but retained in cultivation; endemic.
8. B. secalinus L. - Rye Brome. Culms erect, to 1.2m;
panicle rather diffuse to ± compact, erect or drooping to 1
side at maturity, with some branches longer than spikelets;
lemmas 6.5-9(10)mm, glabrous or (usually) pubescent, with
straight awn 0-8mm. Intrd; weed of cereals, marginal and
waste ground, much decreased and now an infrequent casual;
scattered throughout most of BI except N & W Sc, formerly
frequent in C & S En; Europe.
9. B. pseudosecalinus P.M. Smith - Smith's Brome. Culms
erect, to 60cm; differs from B. secalinus in awn 2-6mm; and
see key (couplet 2). Intrd; grassy fields and waysides,
probably a grass-seed contaminant; very scattered in Br and
Ir; origin uncertain.
10. B. japonicus Thunb. ex Murray - Thunberg's Brome. Culms **1043**
erect, to 80cm; panicle diffuse, ± pendent at maturity, with
branches usually much longer than spikelets; lemmas 8-10mm,
glabrous or pubescent, with awns 4-14mm and widely divergent
at maturity. Intrd; sporadic casual in waste places from
bird-seed, wool and other sources; scattered in En; Europe.
11. B. lanceolatus Roth (B. macrostachys Desf.) - Large- **1043**
headed Brome. Culms erect, to 70cm; panicle rather dense,
erect, with most branches shorter but some usually longer
than spikelets; lemmas 11-18mm, densely pubescent, with awns
6-20mm and widely divergent at maturity. Intrd; sporadic
casual in waste places from wool, bird-seed and as garden
escape; scattered in C & S Br; S Europe.

55. BROMOPSIS (Dumort.) Fourr. (Zerna auct. non Panzer,
Bromus sect. Pnigma Dumort.) - Bromes
Perennials with long to very short rhizomes; spikelets
narrowly oblong then tapered to apex, terete or slightly
compressed; lower glume 1(-3)-veined; upper glume
3(-5)-veined; lemmas 5-7-veined, rounded or slightly keeled
on back, acute to shortly acuminate and minutely bifid at
apex; stamens 3.

1 Inflorescence very lax, the branches pendent or all
   swept to 1 side; leaf-sheaths with distinct pointed
   auricles at apex   2
1 Inflorescence dense to fairly lax, the branches
   erect to erecto-patent; leaf-sheaths without or
   with short rounded auricles at apex   3
   2 Lowest panicle-node with usually >2 branches,
     some with 1 or very few spikelets, and with
     small ± glabrous scale; panicle branches swept
     to 1 side   **2. B. benekenii**
   2 Lowest panicle-node usually with 2 branches,
     both long and with >3 spikelets, and with small
     pubescent scale; panicle branches pendent
                                                    1. B. ramosa
3 Plant densely tufted, with short rhizomes; lemmas

with awns (2)3–8mm; leaves of tillers usually
folded or inrolled along long axis   **3. B. erecta**
3 Plant not densely tufted, with long rhizomes;
  lemmas awnless or less often with awns up to
  3(6)mm; leaves of tillers usually flat   **4. B. inermis**

1. B. ramosa (Hudson) Holub (<u>Bromus</u> <u>ramosus</u> Hudson, <u>Zerna</u>
<u>ramosa</u> (Hudson) Lindman) – <u>Hairy-brome</u>. Rhizomes very short;
culms to 2m; leaf-sheaths all or sometimes all except uppermost with long patent to down-pointed hairs; panicle with
pendent branches at maturity; lemmas 10–14mm, with awns
4–8mm. Native; woods, wood-margins and hedgerows; frequent
throughout most of lowland BI except Jersey and Outer Isles
of Sc.

2. B. benekenii (Lange) Holub (<u>Bromus</u> <u>benekenii</u> (Lange)   R
Trimen, <u>Zerna</u> <u>benekenii</u> (Lange) Lindman) – <u>Lesser</u> <u>Hairy-</u>
<u>brome</u>. Differs from <u>B.</u> <u>ramosa</u> in culms to 1.2m; uppermost
leaf-sheath usually glabrous; and see key. Native; similar
places to <u>B.</u> <u>ramosa</u>, which often accompanies it; very
scattered in mainland Br, perhaps overlooked. Distinction
from <u>B.</u> <u>ramosa</u> needs investigation; ssp. status might be
preferable.

3. B. erecta (Hudson) Fourr. (<u>Bromus</u> <u>erectus</u> Hudson, <u>Zerna</u>
<u>erecta</u> (Hudson) Gray) – <u>Upright</u> <u>Brome</u>. Rhizomes very short;
culms to 1.2m; leaf-sheaths glabrous or with patent hairs;
panicle ± erect; lemmas 8–15mm, with awn (2)3–8mm. Native;
dry grassland and grassy slopes, especially on calcareous
soils; common on base-rich soils in C, S & E En, scattered N
to Fife and W to W Cornwall, Pembs and Caerns, very scattered
in Ir and CI.

4. B. inermis (Leysser) Holub – <u>Hungarian</u> Brome. Rhizomes
long; culms to 1.5m; leaf-sheaths glabrous or sometimes with
patent hairs; panicle ± erect; lemmas (7)9–13(16)mm. Intrd.

a. Ssp. inermis (<u>Bromus</u> <u>inermis</u> Leysser, <u>Zerna</u> <u>inermis</u> 1062
(Leysser) Lindman). Leaf-sheaths usually glabrous; culm-
nodes glabrous or with short hairs just below; lemmas
glabrous to scabrid or with sparse hairs on margins, with awn
0(–3)mm. Formerly sown for fodder, now mostly a seed-
contaminant, natd (and casual) in rough grassy places,
waysides and field-margins; scattered in Br; Europe.

b. Ssp. pumpelliana (Scribner) W.A. Weber (<u>B.</u> <u>pumpelliana</u>
(Scribner) Holub, <u>Bromus</u> <u>pumpellianus</u> (Scribner) Wagnon).
Leaf-sheaths usually pubescent; culm-nodes pubescent; lemmas
appressed-pubescent on margins, with awn 0–6mm. Natd in 1
site in S Essex; N America. A ± distinct ssp. in N America
(where ssp. <u>inermis</u> is well natd), but in Br the 2 merge and
probably hybrids rather than true ssp. <u>pumpelliana</u> exist.

56. ANISANTHA K. Koch (<u>Bromus</u> sect. <u>Genea</u> Dumort.) – <u>Bromes</u>
Annuals; spikelets ± parallel-sided or widening distally,
slightly compressed; lower glume 1(–3)-veined; upper glume
3(–5)-veined; lemmas 7-veined, rounded on back, acute to

acuminate and minutely bifid at apex; stamens 2 or 3.
Lemma-lengths should be measured on only the two basal
florets.

```
1  Lemmas 20-36mm                                                2
1  Lemmas 9-20mm                                                 3
   2  Panicle lax, with branches spreading laterally or
      pendent; callus-scar at base of lemma nearly
      circular                                           1. A. diandra
   2  Panicle dense, with erect branches; callus-scar
      at base of lemma elliptic                          2. A. rigida
3  Panicle lax, with branches spreading laterally or
   pendent                                                        4
3  Panicle dense, with stiffly erect branches                    5
   4  Lemmas 9-13mm; inflorescence compound, the larger
      branches with 3-8 spikelets (except in
      depauperate plants); spikelets with >3 apical
      sterile florets                                    4. A. tectorum
   4  Lemmas 13-20m; inflorescence simple or the
      larger branches slightly branched and with up
      to 3(5) spikelets; spikelets with 1-2 sterile
      apical florets                                     3. A. sterilis
5  Lemmas 12-20mm; spikelets with 1-2 sterile apical
   florets                                               5. A. madritensis
5  Lemmas 9-13(15)mm; spikelets with >3 apical sterile
   florets                                               6. A. rubens
```

1. A. diandra (Roth) Tutin ex Tzvelev (<u>A. gussonei</u> (Parl.)
Nevski, <u>Bromus diandrus</u> Roth) - <u>Great Brome</u>. Stems to 80cm;
inflorescence lax, with spikelets pendent at maturity; longer
branches longer than spikelets, bearing mostly 1-2 spikelets;
lemmas 20-36mm, with awns 25-60mm. Intrd; natd in rough and
waste ground, waysides and open grassland on warm sandy
soils; frequent and well natd in CI and E Anglia, very
scattered and usually casual (especially from wool) elsewhere
in Br N to C Sc; Europe.
2. A. rigida (Roth) N. Hylander (<u>Bromus rigidus</u> Roth) -
<u>Ripgut Brome</u>. Stems to 60cm; inflorescence stiffly erect;
all branches shorter than spikelets; lemmas 20-25mm, with
awns 25-50mm. Intrd; similar places to <u>A. diandra</u>; rather
infrequently natd in CI and S Br, rare casual (especially
from wool) elsewhere; Europe.
3. A. sterilis (L.) Nevski (<u>Bromus sterilis</u> L.) - <u>Barren
Brome</u>. Stems to 80cm; inflorescence as in <u>A. diandra</u>; lemmas
13-20mm, with awns 15-35mm. Native; rough and waste ground,
way-sides, open grassland, weed of arable land and gardens;
throughout lowland BI, common in C & S Br, scattered in Sc
and Ir.
4. A. tectorum (L.) Nevski (<u>Bromus tectorum</u> L.) - <u>Drooping
Brome</u>. Stems to 60cm; inflorescence lax, with spikelets
pendent at maturity; longer branches longer than spikelets,
bearing mostly 4-8 spikelets; spikelets with apical group of

c.4-6 sterile florets becoming dispersed with topmost fertile
floret; lemmas 9-13mm, with awns 10-18mm. Intrd; similar
places to A. diandra; natd in W Suffolk and W Norfolk,
infrequent casual (especially from wool) elsewhere; Europe.
5. A. madritensis (L.) Nevski (Bromus madritensis L.) -
Compact Brome. Stems to 60cm; inflorescence stiffly erect;
all branches shorter than spikelets; lemmas 12-20mm, with
awns 12-25mm. Intrd; natd in similar places to A. diandra;
locally natd in SW En, S Wa and CI, occasional casual
(especially from wool) elsewhere in Br N to C Sc and S Ir;
Europe.
6. A. rubens (L.) Nevski (Bromus rubens L.) - Foxtail Brome. 1062
Stems to 40cm; inflorescence stiffly erect; all branches
shorter than spikelets; spikelets with apical group of c.4-6
sterile florets becoming dispersed with topmost fertile
floret; lemma 9-13(15)mm, with awns 8-15mm. Intrd; rather
infrequent wool-alien, casual in waste places; sporadic and
very scattered in Br; Europe.

57. CERATOCHLOA DC. & P. Beauv. (Bromus sect. Ceratochloa
(DC. & P. Beauv.) Griseb.) - Bromes
Perennials, often short-lived (?sometimes annual or
biennial), usually without rhizomes; spikelets ovoid to
narrowly so, compressed; lower glume 3-5-veined; upper glume
5-7-veined; lemmas 7-11(13)-veined, strongly keeled on back,
acute to shortly acuminate and minutely bifid at apex;
stamens 3.

1 Leaves <3(4)mm wide; lemmas 8-13mm; leaves and
  sheaths densely pubescent with long patent hairs   2
1 Leaves 4-10mm wide; lemmas (10)12-18mm; leaves and
  sheaths glabrous to conspicuously long-patent-
  pubescent   3
    2 Lemmas glabrous or sparsely pubescent, awnless
      or with awn up to 1(2)mm   5. C. brevis
    2 Lemmas sparsely to densely pubescent, with awn
      (3)4-8(12)mm   4. C. staminea
3 Lemmas awnless or with awn up to 3(5)mm, with
  9-11(13) veins; palea 1/2 to 3/4 as long as body of
  lemma   3. C. cathartica
3 Lemmas with awn (3)4-10mm, with 7-9 veins; palea
  3/4 to 1x as long as body of lemma   4
    4 Lemmas glabrous to sparsely and shortly pubescent,
      with awns (4)6-10(12)mm; leaves and sheaths
      glabrous to sparsely pubescent   1. C. carinata
    4 Lemmas conspicuously pubescent, with awns
      (3)4-6(7)mm; leaves and sheaths sparsely
      pubescent to pubescent   2. C. marginata

1. C. carinata (Hook. & Arn.) Tutin (Bromus carinatus Hook. 1062
& Arn.) - California Brome. Culms to 80cm; leaves and
sheaths glabrous to sparsely pubescent; lemmas (12)14-18mm,

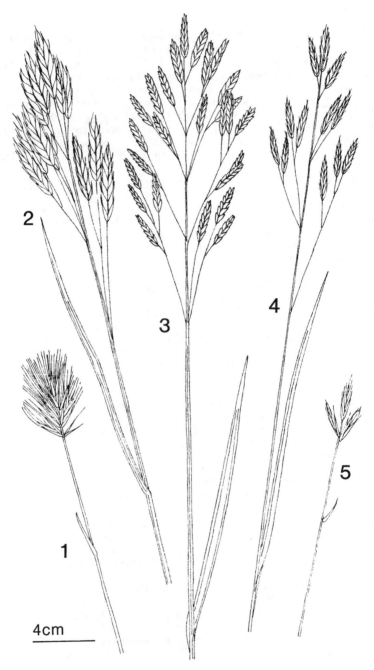

FIG 1062 - Poaceae. 1, Anisantha rubens. 2, Ceratochloa carinata. 3, Bromopsis inermis ssp. inermis. 4, Ceratochloa cathartica. 5, Brachypodium distachyon.

glabrous to shortly and sparsely pubescent, with awn
(4)6-10(12)mm. Intrd; seed contaminant and rarely grown for
fodder, natd in rough ground, field borders, waysides and on
river-banks; scattered in Br, mostly S (especially by R
Thames), Guernsey, Co Dublin; W N America.
2. C. **marginata** (Nees ex Steudel) B.D. Jackson (Bromus
marginatus Nees ex Steudel) - Western Brome. Culms to 1m;
differs from C. carinata in leaves, sheaths and lemmas more
pubescent, and awn usually shorter (see key). Intrd; casual
or sometimes natd in rough and waste ground; a few places in
SE En; W N America.
3. C. **cathartica** (Vahl) Herter (C. unioloides (Willd.) P. 1062
Beauv., Bromus catharticus Vahl, B. unioloides (Willd.)
Kunth, B. willdenowii Kunth) - Rescue Brome. Culms to 1m;
differs from C. carinata as in key (couplet 3). Intrd;
grain- and wool-alien, rarely grown for fodder, casual or
natd on rough ground, roadsides and field-borders; scattered
in C & S Br and CI; C & S America. Resembles Bromopsis
inermis in general appearance, but the longer strongly keeled
lemmas with more veins are diagnostic.
4. C. **staminea** (Desv.) Stace (Bromus stamineus Desv., B.
valdivianus Philippi) - Southern Brome. Culms to 1m; leaves
and sheaths densely long-pubescent; lemmas 8-13mm, sparsely
to densely pubescent, with awn (3)4-8(12)mm. Intrd; casual,
sometimes natd for a while, on rough ground and waysides; a
few places in S En; S America.
5. C. **brevis** (Nees ex Steudel) B.D. Jackson (Bromus brevis
Nees ex Steudel) - Patagonian Brome. Differs from C.
staminea as in key (couplet 2). Intrd; a rather characteristic wool-alien; very scattered in Br; S America. This sp.
has lemmas with curved keels and hence very closely overlapping (usually awnless) tips, producing a very neat
narrowly ovate outline to the spikelet distinct from that of
all other spp.

TRIBE 13 - BRACHYPODIEAE (genus 58). Annuals or rhizomatous
perennials; ligule membranous; sheaths not fused; inflorescence a raceme with usually 1 spikelet at each node with
pedicels <2(2.8)mm; spikelets with many bisexual florets, the
apical 1 or 2 reduced and male or sterile; glumes 2, unequal,
3-9-veined, much shorter than rest of spikelet, sometimes
shortly awned; lemmas mostly 7-veined, rounded on back, acute
to acuminate and usually with short to long awn; stamens 3;
stigmas 2; lodicules 2; ovary with pubescent terminal
appendage (the styles arising from below it).

58. BRACHYPODIUM P. Beauv. - False Bromes

1 Annual; anthers 0.3-1mm; spikelets distinctly
   compressed   **3. B. distachyon**
1 Perennials with rhizomes and tillers; anthers
   3-6(8)mm; spikelets subterete   2

2 Plant weakly rhizomatous, usually densely tufted;
  culms with 4-8 internodes; raceme usually
  pendent at apex; leaves usually >6mm wide;
  lemmas with awns 7-15mm   **2. B. sylvaticum**
2 Plant strongly rhizomatous, usually scarcely
  tufted; culms with 3-4(5) internodes; raceme
  usually erect; leaves usually <6mm wide; lemmas
  with awns 1-5mm   **1. B. pinnatum**

1. B. pinnatum (L.) P. Beauv. - <u>Tor-grass</u>. Extensively
rhizomatous; culms to 1.2m, usually erect; leaves usually
3-6mm wide; sheaths usually glabrous, sometimes pubescent;
raceme with (3)6-15 spikelets; lemmas 6-11mm, often glabrous,
with awn 1-5mm; 2n=28. Native; grassland, mainly on chalk
and limestone; common and often dominant in suitable places
in C, S & E En, very scattered in NW & SW En, Wa and Ir,
sporadic and casual in Sc and CI.
1 x 2. B. pinnatum x B. sylvaticum = B. x cugnacii A. Camus
is probably the identity of some variably sterile intermediates found very scattered in En and Ir with the parents,
but they have not been cytologically confirmed and all
intermediates examined have the chromosome number of one or
other parent. Although the 2 spp. are easily recognized once
familiar, no one character can unequivocally separate them.
The best single character is awn length, with the other key
characters in support; pubescence is of very limited value.
2. B. sylvaticum (Hudson) P. Beauv. - <u>False</u> Brome. Shortly
(or scarcely) rhizomatous; culms to 1m; usually pendent at
tip of raceme; leaves usually 6-10mm wide; sheaths usually
pubescent, sometimes glabrous; raceme with (3)5-12 spikelets;
lemmas 6-12mm, rarely glabrous, with awn 7-15mm; 2n=18.
Native; woods, scrub and shady wood-borders and hedgerows, in
open grassland mainly in the N; common throughout BI except
much of N Sc.
3. B. distachyon (L.) P. Beauv. - <u>Stiff</u> Brome. Stiffly **1062**
erect annual to 15cm; leaves mostly 2-4mm wide, usually
pubescent; sheaths usually glabrous; raceme with 1-4(6)
spikelets; lemmas 7-10mm, glabrous or pubescent, with awn
7-15mm. Intrd; casual in waste places from wool or rarely
grain; very scattered in En; Mediterranean.

TRIBE 14 - TRITICEAE (Hordeeae) (genera 59-65). Annuals or
perennials with or without rhizomes, without stolons; ligule
membranous; sheaths not fused; inflorescence a spike with 1-3
spikelets at each node; spikelets with 1-many florets, often
some florets (or some spikelets if spikelets >1 per node)
male or sterile; glumes 2, equal or slightly unequal,
1-11-veined, often with long terminal awn; lemmas mostly
5-7-veined, rounded on back, very acute to obtuse, awnless or
with terminal awn; stamens 3; stigmas 2; lodicules 2; ovary
usually pubescent or pubescent at apex, sometimes with
pubescent terminal appendage (the styles arising from below

it).

59. ELYMUS L. (<u>Agropyron</u> auct. non Gaertner, <u>Roegneria</u> K. Koch) - <u>Couches</u>
Perennials without rhizomes; spikelets 1 per node, with several to many florets with all but the apical 1 or 2 bisexual, flattened broadside on to rhachis; glumes 2-5-veined, acute or narrowly acute, often awned; lemmas 5-veined, usually long-awned, sometimes awnless or short-awned; spikelets breaking up below each lemma at maturity, leaving glumes on rhachis.

Other spp. - **E. canadensis** L., from N America, was formerly natd in Middlesex; it would key out as <u>E. caninus</u> but the glumes both have awns c. as long as the body.

1. E. caninus (L.) L. (<u>Agropyron</u> <u>caninum</u> (L.) P. Beauv., <u>A. donianum</u> Buch.-White, <u>Elymus</u> <u>trachycaulus</u> (Link) Gould ex Shinn. ssp. <u>donianus</u> (Buch.-White) A. Loeve). - <u>Bearded Couch</u>. Culms to 1.2m; leaves mostly flat, 4-12mm wide; spikelets ± contiguous; glumes very acute, finely pointed with awn <4mm, reaching >1/2 way up body of adjacent lemma; lemmas gradually narrowed to awn 7-18mm or rarely with awn 1-3(7)mm. Native; woods, hedgerows, shady river-banks and mountain gullies and cliff-ledges; scattered throughout Br and Ir but absent from large parts of C & N Sc and Ir. Var. **donianus** (Buch.-White) Meld., from mountains in C & N Sc, has awns 0-3mm but forms fertile hybrids with var. **caninus** that show every degree of intermediacy.
2. E. scabrus (Labill.) A. Loeve (<u>Agropyron</u> <u>scabrum</u> **1070** (Labill.) P. Beauv.) - <u>Australian</u> <u>Couch</u>. Culms to 1.2m; leaves mostly flat, 2-6mm wide; at least in lower part of spike spikelets ± distant, each (excl. awns) not reaching as far as next spikelet on opposite side of rhachis; glumes acute, unawned, reaching <1/2 way up body of adjacent lemma; lemmas gradually narrowed to awn 10-25mm. Intrd; occasional wool-alien of fields and waste ground; scattered in En; Australia.

60. ELYTRIGIA Desv. (<u>Agropyron</u> auct. non Gaertner, <u>Thinopyrum</u> A. Loeve) - <u>Couches</u>
Perennials with long rhizomes; spikelets 1 per node, with several to many florets with all but the apical 1-2 bisexual, flattened broadside on to rhachis; glumes 3-11-veined, acute to very obtuse, rarely awned; lemmas 5-veined, unawned or with short to rarely long awns; spikelets not breaking up easily at maturity, usually falling whole or rhachis breaking up.

1 Rhachis breaking up between each spikelet at maturity, smooth; ribs on leaf upperside densely minutely pubescent   **3. E. juncea**(*)

1   Rhachis not breaking up between each spikelet at
    maturity, scabrid; ribs on leaf upperside glabrous
    to scabrid, sometimes with sparse long hairs   2
    2   Leaf-sheaths with glabrous margin; leaves usually
        flat, their upperside ribs with rounded tops
                                              1. E. repens(*)
    2   At least middle and lower leaf-sheaths with
        minute (often sparse) fringe of hairs on exposed
        free margin; leaves usually inrolled, their
        upperside ribs ± flat-topped   **2. E. atherica**(*)
(*)Also hybrids of E. juncea with E. repens or E. atherica.

1. E. repens (L.) Desv. ex Nevski - Common Couch. Culms to
1.5m; spikelets contiguous to rather distant; glumes acute,
unawned or rarely awned, with 3-7 veins; lemmas obtuse to
narrowly acute, awnless or with awn rarely up to 15mm.
Native.
   a. Ssp. repens (Agropyron repens (L.) P. Beauv., Elymus
repens (L.) Gould). Culms erect; leaves mostly flat, 3-10mm
wide, with fine well-spaced round-topped ribs on upperside;
spikes up to 20(30)cm; spikelets with 3-8 florets; glumes
7-12mm, with (3)5-7 veins; lemmas 8-12mm, sometimes with awns
to 15mm. Cultivated, waste and rough ground; common throughout BI, incl. a range of maritime habitats.
   b. Ssp. arenosa (Spenner) A. Loeve (Elymus repens ssp.
arenosus (Spenner) Meld., Agropyron maritimum Jansen &
Wachter non (L.) P. Beauv.). Culms abruptly bent at lowest
few nodes; leaves often inrolled, 3-5mm wide, with thick,
close, round-topped ribs on upperside; spikes mostly <8cm;
spikelets with (2)3-6 florets; glumes 5-7mm, with mostly 3
veins; lemmas 6-8mm, awnless. Maritime sand-dunes; S & E
coasts of En N to ?, distribution and taxonomic status very
uncertain.
   1 x 2. E. repens x E. atherica = E. x oliveri (Druce)
Kerguelen ex Carreras Mart. (Agropyron x oliveri Druce,
Elymus x oliveri (Druce) Meld. & D. McClint.) occurs with the
parents in scattered places round the coasts of Br from S En
to N Sc and in SW Ir, but is probably not common; it has
intermediate leaf-ribs and sheath-margin pubescence, and is
sterile, with indehiscent anthers and empty pollen.
   1 x 3. E. repens x E. juncea = E. x laxa (Fries) Kerguelen
(Elymus x laxus (Fries) Meld. & D. McClint., Agropyron x
laxum (Fries) Tutin) occurs rather frequently with the
parents on coasts of Br N of Merioneth and SE Yorks, and on
the coasts of Ir and CI, but is rare in S 1/2 of Br; it is
sterile and has a tardily breaking smooth or slightly scabrid
rhachis, distant lower spikelets, and conspicuously scabrid
to ± pubescent leaf-ribs. Distinguished from E. x acuta by
the glabrous free margin of the sheaths.
   2. E. atherica (Link) Kerguelen ex Carreras Mart. (Elymus
athericus (Link) Kerguelen, E. pycnanthus (Godron) Meld.,
Agropyron pycnanthum (Godron) Gren. & Godron, Agropyron

pungens auct. non (Pers.) Roemer & Schultes) - Sea Couch.
Culms to 1.2m; leaves flat or inrolled, mostly 2-6mm wide,
with prominent, rather crowded, + flat-topped ribs on
upperside; spikelets + contiguous,; glumes acute to obtuse,
sometimes acuminate, rarely awned, with 4-7 veins; lemmas
acute to obtuse, awnless or with awn rarely up to 10mm.
Native; wet sandy, gravelly or muddy places by sea, often at
margins of dunes, creeks or salt-marshes; frequent round
coasts of BI N to Co Louth, Dumfriess and NE Yorks. The
leaves are often very glaucous, but this is equally true of
maritime variants of E. repens.

    2 x 3. E. atherica x E. juncea = E. x obtusiuscula (Lange)
N. Hylander, (Elymus x obtusiusculus (Lange) Meld. & D.
McClint., Agropyron x obtusiusculum Lange) occurs rather
frequently with the parents on coasts of Br N to Durham and
Cumberland, and in SW Ir and CI; it closely resembles E. x
laxa but has some minute hairs on free margin of leaf-
sheaths.

    3. E. juncea (L.) Nevski (Agropyron junceum (L.) P. Beauv.,
Elymus farctus (Viv.) Runem. ex Meld., non E. junceus
Fischer) - Sand Couch. Culms to 60(80)cm; leaves usually
inrolled, mostly 2-6mm wide, with prominent, crowded, densely
and minutely pubescent ribs on upperside; spikelets distant,
at least in lower part of spike; glumes subacute to obtuse,
unawned, with 7-11 veins; lemmas obtuse, awnless. Native;
maritime sand-dunes; common round coasts of BI. Our plant is
ssp. **boreoatlantica** (Simonet & Guin.) N. Hylander (Agropyron
junceiforme (A. & D. Loeve) A. & D. Loeve, Elymus farctus
ssp. boreoatlanticus (Simonet & Guin.) Meld.).

60 x 63. ELYTRIGIA x HORDEUM = X ELYTRORDEUM N. Hylander
    60/1 x 63/8. X E. langei (K. Richter) N. Hylander (X **1070**
Elyhordeum langei (K. Richter) Meld., X Agrohordeum langei
(K. Richter) Camus ex A. Camus; E. repens x H. secalinum) has
been found in wet meadows, fields and roadsides in a few
scattered localities in En from Scillies to Northumb. It
exists in two variants: 1 clearly intermediate (known only in
1 place in W Gloucs, 1945-1954), differing from E. repens in
that the rhachis disarticulates and the glumes and lemmas are
awned, and from H. secalinum in having rhizomes and mostly 1
spikelet with 2-4 florets per node; the other close to E.
repens var. aristatus, from which it differs in that the
rhachis disarticulates and the spikelets have 3-5 florets.
Both sorts are usually sterile, but the latter sometimes
produces some good pollen and caryopses.

61. LEYMUS Hochst. (Elymus auct. non L.) - Lyme-grass
Perennials with long rhizomes; spikelets 2(-3) per node, with
3-6 florets, with all but the most apical bisexual, flattened
broadside on to rhachis; glumes 3-5-veined, finely pointed
but not awned; lemmas mostly 7-veined, acute, unawned;
spikelets breaking at maturity below each lemma.

1. L. arenarius (L.) Hochst. (<u>Elymus</u> <u>arenarius</u> L.) - <u>Lyme-</u>
<u>grass</u>. Culms to 1.5(2)m, very glaucous; leaves flat,
becoming inrolled, 8-20mm wide; spike up to 35cm, dense;
spikelets 20-32mm, overlapping. Native; mobile sand on
maritime dunes, rare casual or garden escape inland; frequent
round coasts of most of BI, but absent from large parts of S
En and S Ir and from CI.

62. HORDELYMUS (Jessen) Jessen - <u>Wood Barley</u>
Perennials with very short rhizomes; spikelets (2-)3 per
node, with 1(-2) florets, all bisexual; glumes 1-3-veined,
finely pointed and with long awn, each pair fused at base;
lemmas 5-veined, with very long awn; spikelets breaking at
maturity above the glumes.

1. H. europaeus (L.) Jessen - <u>Wood Barley</u>. Culms to 1.2m;   **R**
leaves flat, 5-14mm wide; spike 5-10cm, dense; lemmas 8-10mm
with awns 15-25mm. Native; woods and copses; local in Br N
to S Northumb, formerly Co Antrim and Berwicks.

63. HORDEUM L. (<u>Critesion</u> Raf.) - <u>Barleys</u>
Annuals or less often perennials without rhizomes; spikelets
3 per node, each with 1 floret, the central spikelet
bisexual, the laterals bisexual, male or sterile; glumes very
narrow, 1-3-veined, with long awn; lemmas of bisexual florets
5-veined, with very long awn; rhachis breaking up at each
node at maturity, or (in cultivated taxa) below each bisexual
lemma.
Most of the spp. are superficially very similar. For
accurate identification a <u>triplet</u> of spikelets (the 3
spikelets at 1 rhachis node) from near the middle of the
spike should be isolated, and the 6 glumes and 3 florets
identified. The spikelets, florets and lemmas are here
referred to as central or lateral according to their position
in the triplet.

1 Rhachis not breaking up at maturity, the caryopsis-
   containing florets breaking away from the rest of
   the spikelet which remains on the rhachis; awns of
   central lemmas usually >10cm (rarely very short)   2
1 Rhachis breaking up at maturity, the triplet of
   spikelets forming the dispersal unit; all awns <10cm   3
   2 All 3 florets in each triplet producing a
      caryopsis and with a long-awned lemma   **1. H. vulgare**
   2 Only central floret of each triplet producing
      a caryopsis; lateral lemmas awnless or ± so
                                             2. H. distichon
3 Glumes of lateral spikelets >3cm, awn-like from
   base to apex; lateral florets extremely reduced,
   usually simply an awn-like outgrowth; awn of
   central lemma usually >5cm   **6. H. jubatum**
3 Glumes of lateral spikelets <3cm, if >2cm then at

least 1 of each pair distinctly widened at base;
lateral florets male or sterile but with obvious
floret construction; awns of central lemma <5cm   4
4 Perennial, with tillers at flowering   5
4 Annual, without tillers   6
5 Proximal part of glumes with very short soft hairs
   >0.1mm; anthers 1-2mm; awns strongly divergent at
   maturity; upper leaf-blades 1.5-2(3)mm wide
                                        7. H. pubiflorum
5 Proximal part of glumes with minute rough prickles
   <0.1mm; anthers 3-4mm; awns stiffly erect at
   maturity; upper leaf-blades 2-6mm wide   **8. H. secalinum**
   6 Glumes of central spikelet with conspicuous
      marginal hairs >0.5mm; leaves usually with well-
      developed pointed auricles   **3. H. murinum**
   6 Glumes of central spikelets with only pricklets
      <0.1mm; leaves usually without or with small
      rounded auricles   7
7 Lateral florets distinctly stalked, the stalk
   (above glumes) c.1mm and c. as long as stalk (below
   glumes) of lateral spikelets; longest awns of
   triplet usually <1cm   8
7 Lateral florets sessile or nearly so, the stalk
   (above glumes) 0-0.5mm and much shorter than 1-1.5mm
   stalk (below glumes) of lateral spikelets; longest
   awns of triplet usually >1cm   9
   8 Lateral lemmas obtuse to acute, 1.7-3.3mm
                                        5. H. euclaston
   8 Lateral lemmas strongly acuminate or with awn
      <2mm, 2.8-6mm incl. awn   **4. H. pusillum**
9 Glumes of lateral spikelets slightly heteromorphic,
   the inner with flattened basal part 0.3-0.7mm wide
   (c.2x as wide as basal part of outer glume); lower
   leaf-sheaths pubescent with hairs >0.5mm
                                        10. H. geniculatum
9 Glumes of lateral spikelets strongly heteromorphic,
   the inner with + winged basal part 0.7-1.2mm wide
   (c.3-4x as wide as basal part of outer glume);
   lower leaf-sheaths glabrous to pubescent with hairs
   <0.25mm   **9. H. marinum**

1. H. vulgare L. - Six-rowed Barley. Annual to 1m; all 3
spikelets in each triplet bisexual, + sessile; lemmas usually
with awns >10cm (almost awnless cultivars exist); glumes with
awn 0-2cm. Intrd; a barley now rarely cultivated, casual as
grain-alien and rare relic in waste places, waysides and
field-borders; fairly frequent throughout BI; SW Asia.
Usually the 3 fertile florets per triplet produce 6 vertical
rows of caryopses in the spike, but in some cultivars the 2
lateral florets of triplets on opposite sides of the rhachis
are superimposed, producing 4 vertical rows (Four-rowed
Barley).

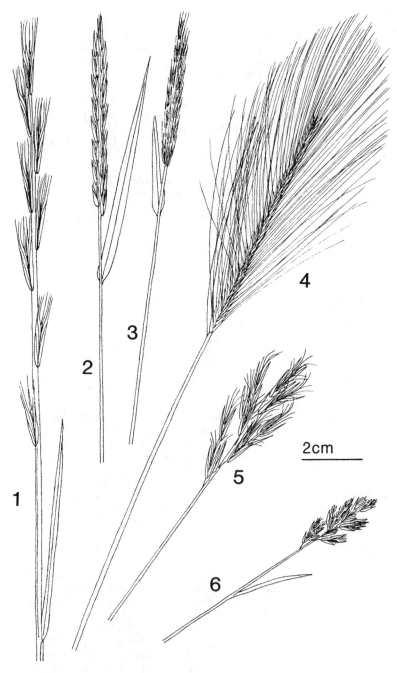

FIG 1070 — Poaceae. 1, **Elymus scabrus**. 2, **X Elytrordeum langei**. 3, **Hordeum pusillum**. 4, **H. jubatum**. 5, **Rytidosperma racemosum**. 6, **Schismus barbatus**.

2. H. distichon L. - <u>Two-rowed</u> <u>Barley</u>. Annual to 75cm; differs from **H.** <u>vulgare</u> in lateral spikelets sterile, much reduced, ± awnless, usually stalked. Intrd; the common cultivated barley, common as a relic in waste places, fields, waysides, etc. throughout BI; SW Asia. On biological grounds better amalgamated with **H.** <u>vulgare</u>, but very distinct morphologically.

3. H. murinum L. - <u>Wall</u> <u>Barley</u>. Annual to 60cm; central spikelet bisexual, sessile; lateral spikelets male or sterile but not or little reduced, stalked; lemmas with awns 1-5cm; glumes with awns 1-3cm.

1 Lemma-body and palea of central floret longer than
  those of lateral florets; central floret with stalk
  (above glumes) <0.6mm   **a. ssp. murinum**
1 Lemma-body and palea of central floret shorter than
  those of lateral florets; central floret with
  stalk (above glumes) 0.6-1.5mm   2
  2 Anthers of central florets usually blackish,
    <0.6mm, ≤1/3 as long as those of lateral florets;
    leaves usually glaucous   **c. ssp. glaucum**
  2 Anthers of central florets usually yellowish,
    >0.6mm long, 1/2 to 1x as long as those of lateral
    florets; leaves not glaucous   **b. ssp. leporinum**

a. Ssp. murinum. Central floret longer than lateral florets, with pedicel <0.6mm; anthers 0.7-1.4mm. Native; weed of waste and rough ground and barish patches in rough grassland; common in C, S & E En and CI, becoming more scattered N & W to NE & SW Sc and N Wa, local in Ir, mainly in S & E.

b. Ssp. leporinum (Link) Arcang. (**H.** <u>leporinum</u> Link). Central floret shorter than lateral florets, with pedicel ≥0.6mm; anthers 0.7-1.4mm. Intrd; rather frequent casual of waste ground, from wool and other sources, sometimes natd for short time; scattered in En; S Europe.

c. Ssp. glaucum (Steudel) Tzvelev (**H.** <u>glaucum</u> Steudel). Differs from ssp. <u>leporinum</u> as in key (anthers of central florets 0.2-0.5mm). Intrd; habitat and distribution as for ssp. <u>leporinum</u>, but recorded from Sc; Mediterranean.

4. H. pusillum Nutt. - <u>Little</u> <u>Barley</u>. Annual to 45cm; **1070**
central spikelet bisexual, sessile; lateral spikelets male or sterile, considerably reduced, stalked; central lemma with awn 0.5-1cm; glumes with awns 0.4-0.8cm. Intrd; rather frequent wool-alien on tips and waste ground and in fields; scattered in En; N America. See **H.** <u>euclaston</u> for confusion.

5. H. euclaston Steudel - <u>Argentine</u> <u>Barley</u>. Differs from **H.** <u>pusillum</u> as in key (couplet 8). Intrd; source, habitat and distribution as for **H.** <u>pusillum</u>, but relative abundance and distribution uncertain due to confusion; S America.

6. H. jubatum L. - <u>Foxtail</u> <u>Barley</u>. Tufted perennial to **1070**
60cm; central spikelet bisexual, sessile; lateral spikelets sterile, greatly reduced (floret often simply a single awn-like lemma), stalked; central lemma with awn (2)5-10cm;

glumes awn-like to base, 3-8cm. Intrd; alien from wool, bird-seed and grass-seed and garden outcast, casual in waste places, now natd especially along main roads, especially those salted in winter; locally frequent in E Br from E Kent to C Sc, scattered casual elsewhere; N America.

7. H. pubiflorum Hook. f. - <u>Antarctic Barley</u>. Tufted perennial to 40cm; central spikelet bisexual, sessile; lateral spikelets sterile, considerably reduced, stalked; central lemma with awn 1-1.6cm; glumes ± awn-like to base, 1.5-2cm. Intrd; wool-alien on tips, waste ground and in fields; scattered in En, possibly overlooked for <u>H. jubatum</u>; S S America.

8. H. secalinum Schreber - <u>Meadow Barley</u>. Tufted perennial to 80m; central spikelet bisexual, sessile; lateral spikelets male or sterile, slightly reduced, stalked; central lemma with awn 0.6-1.2cm; glumes ± awn-like to base, 1-1.6cm. Native; meadows and pastures, especially on heavy soils; common in C, S & E En, very scattered W & N to Scillies, NW Wa and Durham, formerly Fife, rare and very scattered in CI and Ir.

9. H. marinum Hudson - <u>Sea Barley</u>. Annual to 40cm; central **R** spikelet bisexual, sessile; lateral spikelets sterile, considerably reduced, stalked; central lemma with awn 1.5-2.5cm; glumes ± awn-like to base and 1-2.5cm except inner ones of lateral spikelets which have expanded base. Native; barish or sparsely grassed often salty ground near sea, by salt-marshes, on banks and walls and in rough or waste ground; locally common in S Br N to N Lincs and Carms, very scattered N to Cheviot (formerly CE Sc), rare casual elsewhere.

10. H. geniculatum All. (<u>H. hystrix</u> Roth, <u>H. marinum</u> ssp. <u>gussoneanum</u> (Parl.) Thell.) - <u>Mediterranean Barley</u>. Differs from <u>H. marinum</u> as in key (couplet 9). Intrd; wool-alien on tips, waste ground and in fields, natd in Guernsey during 19th Century; scattered in Br; S Europe.

64. SECALE L. - <u>Rye</u>
Annuals; spikelets 1 per node, each with 2(-3) bisexual florets; glumes very narrow, acute, 1-veined, awnless or shortly awned; lemmas 5-veined, keeled, acuminate, usually very long-awned; spikelets disarticulating at maturity below each caryopsis, leaving glumes, lemma and palea on rhachis.

1. S. cereale L. - <u>Rye</u>. Culms to 1.5m; spikes 5-15cm, usually pendent at maturity; lemmas with awn c.2-5cm. Intrd; cultivated now on rather small scale, casual as relic and from grain on tips and in waste places; frequent throughout most of BI; SW Asia. Easily confused at a glance with barley (but spikelet structure totally different) or awned wheat (florets >2, only lower fertile; glumes truncate, keeled).

65. TRITICUM L. - <u>Wheats</u>
Annuals; spikelets 1 per node, each with 3-7(9) florets, the

apical >2 sterile and reduced; glumes keeled, truncate to
bifid, apiculate to shortly awned; lemmas 5-veined, keeled,
truncate to bifid, apiculate to very long-awned; spikelets
disarticulating at maturity below each caryopsis, leaving
glumes, lemma and palea on rhachis.

Other spp. - T. durum Desf. (Pasta Wheat) is rare casual
but perhaps under-recorded for T. turgidum, from which it
differs in the usually black (not green to yellowish-brown)
often longer (to 20cm) awns, flinty (not floury) texture to
endosperm when chewed, and usually longer glumes (nearly as
long as lowest lemmas, not c.2/3 as long).

1. T. aestivum L. - Bread Wheat. Culms to 1.5m; spikes
4-18cm; rhachis glabrous; glumes strongly keeled in upper
1/2, scarcely so in lower 1/2; lemmas awnless or with awn up
to 16cm. Intrd; the common cultivated wheat, common as relic
in fields and waste ground and on roadsides and tips throughout BI; SW Asia.
2. T. turgidum L. - Rivet Wheat. Culms to 1.5m; spikes
7-12cm; rhachis with hair-tufts at each node; glumes strongly
keeled throughout; lemmas with awn 8-16cm. Intrd; formerly a
crop in N & W Br and Ir, now rarely grown and when so mainly
for animal feed, occasional casual as relic or grain-alien;
scattered in Br and Ir; SW Asia.

SUBFAMILY 3 - ARUNDINOIDEAE (tribe 15). Annual or perennial
herbs; leaves without false-petiole or cross-veins; ligule a
fringe of hairs; sheaths not fused; inflorescence a panicle;
spikelets with (1)2-many florets, with all florets (except
often the distal ones) bisexual or the lowest male or
sterile; glumes 2; lemmas firm, 1-9-veined, sometimes bifid
at apex, awnless or with terminal awn; stamens 3; stigmas 2;
lodicules 2; embryo arundinoid; photosynthesis C3-type alone,
with non-Kranz leaf anatomy; fusoid cells 0; microhairs
present; chromosome base-number 9 or 12.

TRIBE 15 - ARUNDINEAE (Cortaderieae, Danthonieae, Molinieae)
(genera 66-71).

66. DANTHONIA DC. (Sieglingia Bernh.) - Heath-grass
Densely tufted perennials; inflorescence a small panicle with
rarely >12 spikelets; spikelets with 4-6 florets, all or all
except most apical bisexual; glumes ovate, c. or nearly as
long as rest of spikelet, 3-7-veined; lemmas 7-9-veined,
minutely 3-toothed at apex, awnless, with tuft of short hairs
at base and fringe up each side to c.1/2 way.

1. D. decumbens (L.) DC. (Sieglingia decumbens (L.) Bernh.)
- Heath-grass. Culms decumbent to erect, to 40(60)cm;
panicle 2-7cm; spikelets 6-12mm. Native; sandy or peaty
often damp soil, usually acid but also mountain limestones,

mostly on heaths, moors and mountains; common in suitable
places throughout BI. Florets often cleistogamous, and
sometimes (often whitish) solitary spikelets occur hidden in
basal leaf-sheaths.

67. RYTIDOSPERMA Steudel - Wallaby-grass
Densely tufted perennials; inflorescence a rather compact to
elongated panicle; spikelets with 6-10 florets, lower ones
bisexual, upper ones male or sterile and reduced; glumes
lanceolate to narrowly ovate, with wide hyaline margins,
slightly longer to slightly shorter than rest of spikelet
(excl. awns), 5-7-veined; lemmas 7-veined, with 2 long
acuminate lobes at apex tipped with straight awns, with long,
bent terminal awn from sinus, with dense white silky hairs at
base and middle reaching or nearly reaching apex of body of
lemma.

1. R. racemosum (R. Br.) Connor & Edgar (Danthonia racemosa 1052
R. Br.) - Wallaby-grass. Culms erect, to 60cm; panicles 1070
3-5cm; spikelets 7-16mm excl. awns; lemmas with bent terminal
awn 5-15mm, lateral awns 2-8mm. Intrd; occasional wool-alien
in fields and waste places and on tips; scattered in En;
Australia and New Zealand. Not over-wintering with us.

68. SCHISMUS P. Beauv. - Kelch-grass
Tufted annuals; inflorescence a rather compact panicle;
spikelets with 4-10 florets, lower ones bisexual, upper ones
male or sterile and reduced; glumes lanceolate, with wide
hyaline margins, slightly longer to slightly shorter than
rest of spikelet, 5-7-veined; lemmas 9-veined, deeply and
acutely 2-lobed at apex, awnless or ± so, with short hairs at
base and long silky hairs on back not reaching apex.

Other spp. - **S. arabicus** Nees, from E Mediterranean and W
Asia, occurs similarly but is rather rare; it is very similar
to S. barbatus but has apical lobes of lemma longer than (not
c. as long as) wide, and tip of palea reaching lemma notch to
1/2-way up lobes (not from 1/2-way to near tip of lobes).

1. S. barbatus (L.) Thell. - Kelch-grass. Culms erect, to 1052
25cm; panicle 1-4cm; spikelets 5-6mm. Intrd; rather 1070
characteristic wool-alien in fields and waste land and on
tips; scattered in En; Mediterranean.

69. CORTADERIA Stapf - Pampas-grass
Densely tufted perennials; inflorescence a very large
spreading panicle; dioecious or gynodioecious (with female
and bisexual plants); spikelets with 2-7 florets; glumes
lanceolate, hyaline, slightly unequal, 1-veined, at least
upper ± as long to longer than rest of spikelet and with long
terminal awn; lemmas 3-5-veined, acuminate and long-awned at
apex, with tuft of long fine hairs at base reaching ± to apex

of body of lemma.

1. C. selloana (Schultes & Schultes f.) Asch. & Graebner - **1052**
Pampas-grass. Plant forming tussocks often >1m across, with
long (>1m) leaves with fiercely cutting serrated edges; culms
erect, to 3m; panicles up to 1m, becoming a silky-pubescent
mass at fruiting; spikelets 12-16mm incl. awns. Intrd; grown
as ornament, becoming natd where thrown-out or planted; rough
ground, waysides, old gardens, maritime cliffs and dunes;
scattered in S En, S Wa, CI and Co Wexford; S America.

70. MOLINIA Schrank - Purple Moor-grass
Densely tufted perennials; inflorescence a ± diffuse to ±
contracted panicle; spikelets with 1-4 florets, all except
most apical bisexual; glumes ovate, slightly unequal,
1-3-veined, much shorter than rest of spikelet; lemmas
3-5-veined, acute to obtuse, awnless, glabrous.

1. M. caerulea (L.) Moench - Purple Moor-grass. Plant often
forming tussocks; culms to 1.3m, erect; panicles up to 60cm,
often purplish; spikelets 3-7.5mm. Native.
a. Ssp. caerulea. Culms usually <65cm; panicle usually
<30cm, narrow, with branches mostly <5cm; spikelets 3-5.5mm;
lemmas 3-4mm. Heaths, moors, bogs, fens, mountain grassland
and cliffs and lake-shores, always on at least seasonally wet
ground; common in suitable places throughout BI.
b. Ssp. arundinacea (Schrank) K. Richter (ssp. altissima
(Link) Domin, M. litoralis Host). Culms mostly 65-
125(160)cm; panicles mostly 30-60cm, with very uneven
lengthed branches often >10cm, usually spreading at least
during flowering; spikelets (3)4-7.5mm; lemmas (3.2)3.5-
5.4(5.7)mm. Fens, fen-scrub, fen-type vegetation by rivers
and canals; scattered but frequent in suitable places in C &
S Br, very scattered N to S & W Sc and in S Ir.

71. PHRAGMITES Adans. - Common Reed
Extensively rhizomatous perennials; inflorescence a very
large spreading panicle; spikelets with 2-6(more) florets,
the lowest male or sterile, the rest bisexual; glumes
unequal, narrowly elliptic-ovate, 3-5-veined, much shorter
than rest of spikelet; lemmas lanceolate, acute to acuminate
at apex, with 1-3 veins, awnless; rhachilla-segments with
long white-silky hairs becoming very conspicuous in fruit.

1. P. australis (Cav.) Trin. ex Steudel (P. communis Trin.)
- Common Reed. Culms to 3.5m but sometimes <1m; panicles
20-60cm, usually purple; spikelets 8-16mm, with rhachilla-
hairs up to 10mm. Native; on mud or in shallow water by
lakes, rivers, canals, marshes, fens, bog-margins and edges
of salt-marshes and estuaries; common in suitable places
throughout BI.

SUBFAMILY 4 - CHLORIDOIDEAE (tribes 16-17). Annual to perennial herbs; leaves without false-petiole or cross-veins; ligules usually a fringe of hairs, sometimes membranous or a membrane fringed with hairs; sheaths not fused; inflorescence usually an umbel or a raceme of spikes or racemes, sometimes a panicle; spikelets with 1-many florets with all florets (except often the distal ones) bisexual; glumes 2; lemmas membranous to firm, 1-3(9)-veined, awnless or less often awned; stamens 3; stigmas 2; lodicules 0 or 2; embryo chloridoid (or rarely arundinoid); photosynthesis C3- plus C4-type, with Kranz leaf anatomy; fusoid cells 0; microhairs present; chromosome base-number >7.

TRIBE 16 - ERAGROSTIDEAE (Sporoboleae) (genera 72-76). Annuals or perennials with or without rhizomes; ligules membranous or a fringe of hairs; inflorescence a diffuse to contracted panicle, an umbel of spikes, or a raceme of racemes; spikelets with 3-many florets (or with 1 floret in Sporobolus), all (except 1-2 most apical) bisexual; glumes 2, much shorter than rest of spikelet; lemmas (1-)3-veined (with extra veins close to midrib in Eleusine), usually keeled, awnless (very shortly awned in Leptochloa).

72. LEPTOCHLOA P. Beauv. (Diplachne P. Beauv.) - Beetle-grasses
Rhizomatous perennials (but annual with us); ligule membranous; inflorescence a loose, long raceme of racemes, the spikelets well spaced out, the rhachis ending in a spikelet; spikelets with 6-10(14) florets; glumes unequal, 1-veined; lemmas 3-veined, slightly keeled with prominent midrib and submarginal laterals, with long silky hairs at base and on margins and base of dorsal midline, bifid or shouldered or with 1-few teeth on either side at apex, with very short apical awn; spikelets disarticulating between florets; pericarp adherent to seed.

Other spp. - **L. muelleri** (Benth.) Stace (Diplachne muelleri Benth.), from Australia, and **L. uninervia** (C. Presl) Hitchc. & Chase (D. uninervia (C. Presl) Parodi), from N & S America, resemble L. fusca closely; they have been recorded rarely but may be overlooked. L. muelleri differs in the lemmas with rounded subapical teeth and a minute awn shorter than them; L. uninervia differs in the lemmas with 1-2 minute teeth or merely a shoulder either side of the shortly apiculate apex, and anthers <0.5mm (not >1mm). The former is perhaps not distinct from L. fusca.

1. L. fusca (L.) Kunth (Diplachne fusca (L.) P. Beauv.) - **1078** Brown Beetle-grass. Culms to 1m; inflorescences up to 40cm, **1081** with long, straight erecto-patent unbranched racemes from main axis; spikelets 8-15mm; lemmas 3-6mm, with awn <1.6mm between 2 shorter subapical acute teeth. Intrd; occasional

casual wool-alien in fields and waste places and on tips;
scattered in En; Australia to tropical Africa.

73. ERAGROSTIS Wolf - Love-grasses
Annuals or tufted perennials; ligule a fringe of hairs;
inflorescence a usually diffuse panicle; spikelets with 3-
many florets, often very narrow and parallel-sided; glumes
subequal to unequal, (0)1(-3)-veined; lemmas 3-veined,
keeled, acute to obtuse, rounded or emarginate; awnless
(sometimes apiculate); spikelets disarticulating between
florets or the caryopsis falling free leaving persistent
rhachilla; pericarp adherent to seed.
Superficially often similar to <u>Poa</u>, but the 3-veined lemmas
and ligule a ring of hairs distinguish it.

1 Anthers (0.8)1-1.3mm; plant perennial   **1. E. curvula**
1 Anthers 0.2-0.6mm; plant annual   2
  2 Plant with minute, sessile, wart-like glands
    (x$\geq$10 lens) on leaf-margins, sheath-midrib,
    lemma- and glume-veins, panicle-branches and/or
    pedicels (if 0 on leaves then always $\geq$1 on
    pedicels)   3
  2 Plant without minute sessile glands   4
3 Leaves often >5mm wide; spikelets >2mm wide; lemmas
  (1.7)2-2.8mm; sessile glands prominent on lemma
  veins, usually not so on pedicels   **2. E. cilianensis**
3 Leaves <5mm wide; spikelets $\leq$2mm wide; lemmas
  1.5-2mm; 1-2 sessile glands prominent near pedicel-
  apex, usually none on lemmas   **3. E. minor**
  4 Caryopsis 1-1.3mm; upper glume 1.7-3mm; lemmas
    2-2.7mm   **6. E. tef**
  4 Caryopsis 0.5-0.8mm; upper glume 0.7-1.4mm;
    lemmas 1.2-1.8mm   5
5 Panicle-branches erecto-patent at maturity, with
  spikelets borne nearly to base; lemmas greenish-grey,
  purple-tinged at apex   **4. E. pilosa**
5 Panicle-branches patent at maturity, bare of
  spikelets for lowest c.1/4-1/3; lemmas dark grey,
  not purple-tinged   **5. E. parviflora**

Other spp. - >40 other spp. have been found on tips and
docksides, etc., as rare casuals from wool, grain and other
commodities. The following 8 are the least rare: **E. lugens**
Nees, **E. neomexicana** Vasey ex L. Dewey and **E. pectinacea**
Michaux) Nees (E. <u>virescens</u> Nees), from America; **E.
lehmanniana** Nees and **E. plana** Nees from Africa; **E. brownii**
(Kunth) Nees ex Steudel and **E. trachycarpa** (Benth.) Domin
from Australia; and **E. barrelieri** Daveau from Mediterranean.

1. E. curvula (Schrader) Nees (E. <u>chloromelas</u> Steudel) - **1078**
<u>African</u> Love-grass.   Tufted perennial but often annual with **1081**
us; culms erect, to 1.2m; leaves usually <2mm wide and

FIG 1078 – Spikelets of Poaceae. 1-6, **Eragrostis**. 1, **E. minor**. 2, **E. parviflora**. 3, **E. cilianensis**. 4, **E. tef**. 5, **E. curvula**. 6, **E. pilosa**. 7, **Leptochloa**. 8, **Eleusine indica**. 9, **Chloris truncata**. 10, **Sporobolus**. 11, **Dactyloctenium**. 12, **Cynodon dactylon**.

inrolled; panicle up to 30 x 15cm, very diffuse; spikelets linear, <2mm wide, greenish-grey; lemmas 2-3mm; caryopsis <0.8mm. Intrd; occasional wool-alien on tips and waste ground; casual scattered in En and Wa, ± natd in Southampton (S Hants); tropical Africa.

2. E. cilianensis (All.) Vign. ex Janchen (*E.* megastachya **1078** (Koeler) Link - <u>Stink-grass</u>. Annual; culms erect or **1081** ascending, to 75cm; leaves usually >5mm wide, flat; panicle up to 20 x 8cm, fairly dense; spikelets ovate to oblong-ovate, 2-3(4)mm wide, greenish-grey; lemmas (1.7)2-2.8mm; caryopsis <0.8mm. Intrd; rather frequent casual wool- and grain-alien on tips and waste ground; scattered in En and Wa; S Europe.

3. E. minor Host (*E.* pooides P. Beauv.) - <u>Small</u> <u>Love-grass</u>. **1078** Annual; culms erect or ascending, to 50cm; leaves 2-5mm wide, flat; panicle up to 20 x 8cm, fairly dense; spikelets linear to narrowly ovate-elliptic, 1.3-2mm wide, yellowish-green often purplish- or grey-tinged; lemmas 1.5-2mm; caryopsis <0.8mm. Intrd; fairly frequent casual wool- and grain-alien on tips and waste ground; scattered in C & S Br; S Europe.

4. E. pilosa (L.) P. Beauv. (*E.* multicaulis Steudel) - **1078** <u>Jersey Love-grass</u>. Annual; culms erect or ascending, to **1081** 70cm; leaves 1-3.5mm wide, flat; panicle up to 25 x 15cm, diffuse; spikelets linear to linear-lanceolate, 1-1.5mm wide, greenish-grey, tinged purple; lemmas 1.2-1.8mm; caryopsis 0.5-0.8mm. Intrd; natd in Jersey since 1961; Mediterranean.

5. E. parviflora (R.Br.) Trin. - <u>Weeping</u> Love-grass. **1078** Differs from *E.* pilosa in culms to 60cm; panicle up to 30 x 20cm, very diffuse but spikelets often lying ± appressed to panicle-branches; and see key. Intrd; fairly frequent casual wool- and grain-alien on tips and waste ground; scattered in En; Australia.

6. E. tef (Zucc.) Trotter - <u>Teff</u>. Annual; culms erect, to **1078** 1m; leaves flat, 2-4mm wide; panicle up to 30 x 15cm, diffuse or contracted; spikelets linear to linear-lanceolate, 1-1.5(2)mm wide, straw-coloured or green, often tinged reddish; lemmas 2-2.7mm; caryopsis 1-1.3mm. Intrd; minor grain-crop in warm countries, casual wool- or grain-alien on tips and waste ground; scattered in En; tropical Africa.

74. ELEUSINE Gaertner - <u>Yard-grasses</u>
Annuals or tufted perennials; ligule membranous with a sparse or dense fringe of hairs; inflorescence an umbel or very short raceme of spikes with very crowded spikelets, the rhachis ending in a spikelet; spikelets with 3-6 florets; glumes subequal to unequal, 1-7-veined; lemmas with 3 main veins and 2-4 extra veins close to midrib, keeled, acute to apiculate, not awned; spikelets disarticulating between florets; pericarp not adherent to seed, which eventually falls out separately.

1 Lemmas and glumes obtuse, with hooded tip;

perennial   **2. E. tristachya**
1 Lemmas and glumes pointed (either acute, or subacute
  to obtuse and apiculate), without hooded tip; annual   2
  2 Spikes (3.5)5-15cm, all or most in terminal
    umbel; lemmas <1mm wide from keel to edge
                                              1. E. indica
  2 Spikes 1-3cm, in very short terminal raceme;
    lemmas c.1.5mm wide from keel to edge
                                          3. E. multiflora

1. E. indica (L.) Gaertner   -   Yard-grass. Annual; culms to 1078
90cm; spikes (1)2-c.10, all or most in umbel, (3.5)5-15 x 1081
<1cm. Intrd; casual bird-seed-, grain-, wool-, cotton- or
pulse-alien, on tips and waste ground; scattered in En.
Relative abundance and distribution in En of the 2 sspp. are
unknown due to confusion, except that ssp. africana is much
the commoner (or the only) wool-alien, and ssp. indica is the
commoner grain-, pulse- and cotton-alien.
  a. Ssp. indica. Ligule sparsely and minutely pubescent at
apex; lower glume 1-veined, 1.1-2.3mm; upper glume 1.8-3mm;
lemmas 2.4-4mm; seeds 1-1.3mm, with very fine close
striations between and at right-angles to main ridges (x$\geq$20
lens). Asia, but now world-wide.
  b. Ssp. africana (Kenn.-O'Byrne) S. Phillips (E. africana
Kenn.-O'Byrne). Ligule with strong pubescent fringe at apex;
lower glume (1)2-3-veined, 2-3.5mm; upper glume 3-4.7mm;
lemmas 3.7-5mm; seeds 1.2-1.6mm, with granulations between
main ridges (x$\geq$20 lens). Africa, but now more widespread.
  2. E. tristachya (Lam.) Lam. - American Yard-grass.
Perennial (but not over-wintering with us); culms to 40cm;
spikes (1)2-4, all in umbel (rarely 1 below), 1-4 x <1cm.
Intrd; fairly frequent wool-alien on tips and waste ground;
scattered in En; S America.
  3. E. multiflora A. Rich - Fat-spiked Yard-grass. Annual;
culms to 45cm; spikes 2-8, in short raceme but congested at
stem-apex, 1-3 x >1cm. Intrd; fairly frequent wool-alien on
tips and waste ground; scattered in En; tropical Africa.

75. DACTYLOCTENIUM Willd. - Button-grass
Annuals but sometimes rooting at lower nodes; ligule
membranous, sometimes slightly fringed at apex; inflorescence
an umbel of spikes with very crowded spikelets, the rhachis
ending in a short projection; spikelets with 3-5 florets;
glumes unequal, 1-veined, the upper with a long awn-like
point; lemmas 3-veined, keeled, acuminate to acute-apiculate,
not awned; spikelets disarticulating above glumes (not
between florets); pericarp not adherent to seed, which
eventually falls out separately.

  1. D. radulans (R. Br.) P. Beauv. - Button-grass. Stems 1078
erect to decumbent, to 40cm; inflorescence of 4-10 crowded 1081
umbellate spikes each (0.5)1-2cm; spikelets 3-5mm. Intrd;

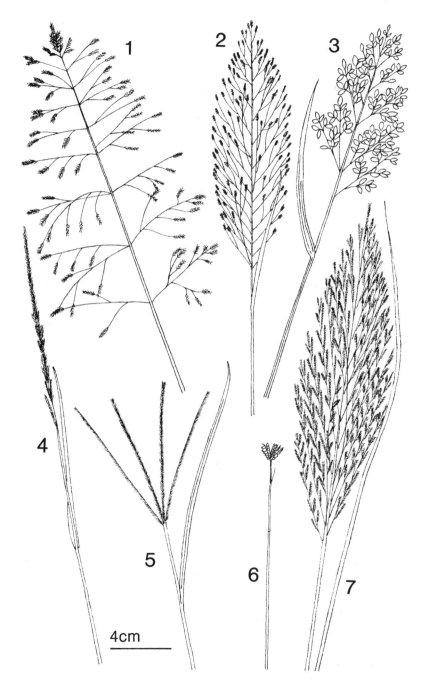

FIG 1081 – Poaceae. 1, Eragrostis curvula. 2, E. pilosa. 3, E. cilianensis. 4, Sporobolus africanus. 5, Eleusine indica. 6, Dactyloctenium radulans. 7, Leptochloa fusca.

fairly frequent wool-alien on tips and waste ground;
scattered in En; Australia.

76. SPOROBOLUS R. Br. - Dropseeds
Tufted perennials; ligule a fringe of hairs; inflorescence a
narrow panicle with short to long, closely appressed, erect
branches (the ultimate branches with closely borne small
spikelets resembling a spikelet with many florets at a casual
glance); spikelets with 1 floret; glumes unequal, 0-1-veined,
much shorter than rest of spikelet; lemma 1-3-veined, rounded
on back, acute to acuminate, awnless, inrolled and +
cylindrical but tapering at apex; spikelet disarticulating
below lemma; pericarp not adherent to seed, which eventually
falls out separately.

Other spp. - **S. indicus** (L.) R. Br., from tropical America,
and **S. elongatus** R. Br., from Australia, closely resemble S.
africanus; they have both been recorded rarely but may be
overlooked. Both differ from S. africanus in having lemmas
1.3-1.9mm; S. elongatus differs from the other 2 spp. in
having longer (>3cm) lower panicle-branches and 2 (not 3)
stamens; S. indicus has the caryopsis nearly as long as
lemma.

1. S. africanus (Poiret) Robyns & Tournay - African 1078
Dropseed. Culms erect to ascending, to 1m; panicle very 1081
narrow (<1cm), up to 35cm, with short (<2cm) erect branches;
spikelets (lemma) 2.1-2.5mm; caryopsis 1/2 to 2/3 as long as
lemma. Intrd; rather frequent wool-alien on tips and rough
ground; scattered in En; Africa.

TRIBE 17 - CYNODONTEAE (Spartineae, Chlorideae, Zoysieae)
(genera 77-80). Annuals or perennials, with or without
rhizomes or stolons; ligules membranous or a fringe of hairs;
inflorescence an umbel or raceme of spikes, or a spike-like
panicle; spikelets with 1 bisexual floret, sometimes with
1-2(3) extra sterile or male florets distal to it; glumes
(1-)2, shorter to longer than rest of spikelet; lemmas
1-3(9)-veined, usually keeled, awned or not.
Very close to Eragrostideae, but differ in spikelet with
only 1 bisexual floret (only Leptochloa of Eragrostideae has
1 floret, but this has inflorescence a panicle).

77. CHLORIS Sw. - Rhodes-grasses
Annuals or perennials, sometimes with stolons; ligule
membranous with a well-marked fringe of hairs; inflorescence
an umbel of (4)6-many slender long spikes; spikelets with
2-3(4) florets, the lowest bisexual, the others reduced and
male or sterile; glumes unequal, 1-veined, narrowly acute,
shorter to slightly longer than rest of spikelet; lowest
(bisexual) lemma 3-veined, keeled, minutely to deeply bifid
at apex with long terminal or subterminal straight awn; upper

lemma(s) variously reduced, but 2nd of similar shape and only that long-awned (hence spikelets 2-awned); spikelets disarticulating above glumes.

1 Lemma with low rounded to transversely or obliquely truncate lobes either side of awn, forming very shallow notch   **1. C. truncata**
1 Lemma with sharply acute lobes or teeth either side of awn, forming deep notch   2
  2 Fertile lemma with dense tuft of silky hairs at apex, producing feathery spikes   **3. C. virgata**
  2 Fertile lemma without apical tuft of hairs, producing ± glabrous spikes   **2. C. divaricata**

1. C. truncata R.Br. - <u>Windmill-grass</u>. Perennial, often **1078** stoloniferous; culms to 45cm; spikes 7-15cm; spikelets **1086** 2.5-3.5mm; fertile lemma truncate or slightly rounded either side of apical notch, with awn 10-15mm, ± glabrous at apex. Intrd; fairly frequent wool-alien on tips and waste ground and in fields; scattered in En; Australia.

2. C. divaricata R. Br. - <u>Australian Rhodes-grass</u>. Tufted perennial (?stoloniferous); culms to 60cm; spikes 7-15cm; spikelets 3-4.5mm; fertile lemma with very narrowly acute apical lobes forming deep notch, with awn 6-12mm, ± glabrous at apex. Intrd; habitat and distribution as for C. truncata; Australia.

3. C. virgata Sw. - <u>Feathery Rhodes-grass</u>. Annual; culms to 60(100)cm; spikes 2-8cm; spikelets 3-4mm; fertile lemma with acute apical lobes forming fairly deep notch, with awn 5-12mm, with dense tuft of silky hairs 1.5-4mm at apex. Intrd; habitat and distribution as for C. truncata; tropical Africa.

78. CYNODON Rich. - <u>Bermuda-grasses</u>
Perennials with rhizomes and/or stolons; ligule membranous or a fringe of hairs; inflorescence an umbel of 3-6 slender long spikes; spikelets with 1 (bisexual) floret; glumes subequal, shorter than rest of spikelet, 1-veined, narrowly acute; lemma 3-veined, keeled, subacute, unawned; spikelets disarticulating above glumes.

1. C. dactylon (L.) Pers. - <u>Bermuda-grass</u>. Extensively **1078** rhizomatous and stoloniferous; culms erect, to 30cm; ligule a **RR** fringe of short hairs <0.5mm with longer tuft at each edge; spikes 2-5cm; spikelets 2-3mm; rhachilla of spikelet continued beyond base of floret as fine projection between upper glume and floret >1/2 as long as floret. Probably intrd; natd in rough sandy ground, waysides and short grassland near sea; local in SW En, S Wa and CI, casual from wool and other sources scattered elsewhere in En and Wa, sometimes ± natd for short while; world-wide in warm areas.
2. C. incompletus Nees - <u>African Bermuda-grass</u>. Differs

from <u>C. dactylon</u> in ligule membranous, 0.4–1mm, with sparse
hairs at apex and longer tuft at base of each edge; rhachilla
of spikelet not extended beyond base of floret. Intrd;
fairly frequent wool-alien on tips and rough ground and in
fields; casual or ± natd for short while in scattered places
in En; S Africa.

79. SPARTINA Schreber – <u>Cord–grasses</u>
Strongly rhizomatous perennials; ligule a dense fringe of
hairs; inflorescence a raceme of (1)2–12(30) long ± erect
spikes; spikelets with 1 (bisexual) floret; glumes unequal,
the upper as long as or longer than rest of spikelet, 1–9–
veined, narrowly acute, awned or not; lemma 1–3–veined,
keeled, acute or minutely notched, unawned; spikelets falling
entire at maturity.

1 Upper glume very scabrid on keel with rigid pricklets
  ≥0.3mm, with awn 3–8mm; spikes with 2 rows of
  spikelets each crowded 4–10/cm; inland   **5. S. pectinata**
1 Upper glume glabrous or with soft hairs <0.3mm on
  keel, awnless; spikes with 2 rows of spikelets each
  spaced out 1–3/cm; coastal   2
  2 Glumes glabrous or with hairs on keel only, some-
    times very sparse on body also   **4. S. alterniflora**
  2 Glumes softly pubescent on keel and body   3
3 Ligules 1.8–3mm at longest point (beware damaged
  ones); anthers (5)7–10(13)mm, with full pollen
  >45 microns across   **3. S. anglica**
3 Ligules 0.2–1.8mm at longest point; anthers
  4–8(10)mm, if >7mm then indehiscent with empty
  pollen; pollen <45 microns across   4
  4 Ligules 0.2–0.6mm; anthers 4–6.5mm, dehiscent,
    with full pollen   **1. S. maritima**
  4 Ligules 1–1.8mm; anthers 5–7(10)mm, indehiscent,
    with empty pollen   **2. S. x townsendii**

1. S. maritima (Curtis) Fern. – <u>Small Cord–grass</u>. Culms to   R
50(80)cm; ligules 0.2–0.6mm; spikes (1)2–3(5), 3–8cm;
spikelets 11–15mm; glumes softly appressed-pubescent,
awnless; anthers 4–6.5mm, with full pollen <45 microns
across; 2n=60. Native; tidal sandy or muddy bare places by
sea or in estuaries; local in S & E En from Wight (formerly S
Devon) to N Lincs, intrd in Co Dublin.
  2. S. x townsendii Groves & J. Groves (<u>S. maritima</u> x <u>S.
alterniflora</u>) – <u>Townsend's Cord–grass</u>. Culms to 130cm;
ligules 1–1.8mm; spikes 2–8, 6–24cm; spikelets (12)14–
18(20)mm; glumes softly appressed-pubescent, awnless; anthers
5–7(10)mm, with empty pollen <45 microns across; 2n=61.
Native; tidal mud-flats; arose prior to 1870 in Southampton
Water (S Hants) and still there with the parents, now spread
Dorset to W Sussex, but also scattered elsewhere in S Br and
E Ir, either intrd with <u>S. anglica</u> or derived from it, in

absence of 1 or both parents (but not of S. anglica).
3. S. anglica C.E. Hubb. (S. x townsendii auct. non Groves &
J. Groves) - Common Cord-grass. Culms to 130cm; ligules
1.8-3mm; spikes 2-12, 7-23cm; spikelets (15)17-21(26)mm;
glumes softly appressed-pubescent, awnless; anthers
(5)7-10(13)mm, with full pollen >45 microns across; 2n=122.
Native; tidal mud-flats; arose c.1890 in Southampton Water as
amphidiploid of S. x townsendii and spread with it naturally
from Dorset to W Sussex, planted extensively elsewhere in BI
N to C Sc and becoming dominant over large areas.
4. S. alterniflora Lois. (S. glabra Muhlenb. ex Bigelow) -
Smooth Cord-grass. Culms to 120cm; ligules 1-1.8mm; spikes
3-13, (3)5-15cm; spikelets 10-18mm; glumes glabrous or
sparsely pubescent (usually on keel only), awnless; anthers
5-7mm, sometimes indehiscent, with full or sometimes empty
pollen <45 microns across; 2n=62. Intrd; planted in 3 sites
in S Hants from early 1800s but now extinct in all but 1
despite earlier spread, also planted in E Ross from 1920 and
in Dorset 1963; N America.
5. S. pectinata Bosc ex Link - Prairie Cord-grass. Culms to **1086**
180cm; ligules 0.5-3mm; spikes 5-20(30), 2-8(10)cm; spikelets
8-12mm; glumes with strongly scabrid keel, glabrous to
sparsely pubescent elsewhere, the upper with awn 3-8mm;
anthers 5-7mm, with full pollen <45 microns across; 2n=40.
Intrd; grown for ornament, persistent and spreading where
neglected by fresh-water lake in W Galway since 1967 and by
canal in N Hants since 1986; N America.

80. TRAGUS Haller - Bur-grasses
Annuals; ligule a dense fringe of hairs; inflorescence a
spike or spike-like, with 2-5 spikelets on extremely short
branch at each node; spikelets with 1 (bisexual) floret;
glumes very unequal, the lower 0 or vestigial, the upper at
least as long as floret, 5-7-veined, each vein with line of
strong hooked spines, acute, unawned; lemma 3-veined, not
keeled, acute; each nodal group of spikelets falling as a bur
at maturity, the spikelets facing the centre of the bur with
the glume-hooks outermost.
Distinctive vegetatively in the strong curved spines on
proximal part of leaf-margins.

1 Upper glume 2-3mm   **3. T. berteronianus**
1 Upper glume 3.5-4.5mm   2
  2 Spikelets 2 at all or almost all nodes; upper
    glume with 5 veins (and rows of spines)
                     2. T. australianus
  2 Spikelets 3-5 at all or almost all nodes; upper
    glume with 7 veins (and rows of spines) (one
    pair of veins sometimes thinner and with smaller
    spines)   **1. T. racemosus**

1. T. racemosus (L.) All. - European Bur-grass. Culms to

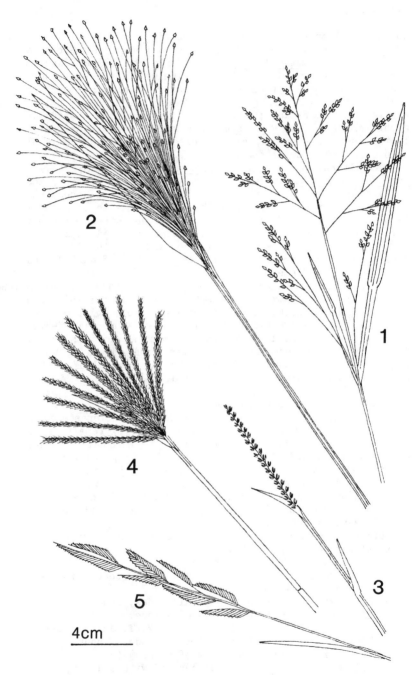

FIG 1086 – Poaceae. 1, *Panicum dichotomiflorum*. 2, *P. capillare*. 3, *Tragus australianus*. 4, *Chloris truncata*. 5, *Spartina pectinata*.

40cm; inflorescence 2-10 x 0.6-1cm; spikelets 3-5 on common
stalk at each node, 1-3 of them often reduced, each
3.5-4.5mm; upper glumes with 7 veins, each bearing row of
hooked spines. Intrd; rather frequent wool-alien on tips and
rough-ground and in fields; scattered in En; S Europe.
2. **T. australianus** S.T. Blake - <u>Australian</u> <u>Bur-grass</u>. **1086**
Differs from <u>T. racemosus</u> as in key; the 2 spikelets at each **1089**
node (both bisexual) are closely facing with the glume-hooks
outermost, the pair resembling a fruit of <u>Anthriscus</u>
<u>caucalis</u>. Intrd; habitat and distribution as for <u>T.</u>
<u>racemosus</u>; Australia.
3. **T. berteronianus** Schultes - <u>African</u> <u>Bur-grass</u>. Culms to
60cm; inflorescence 2-15 x 0.4-0.6cm; spikelets 2 at all or
almost all nodes, 2-3mm; upper glumes with 5 veins, each
bearing row of hooked spines. Intrd; habitat and
distribution as for <u>T. racemosus</u>; tropical Africa.

SUBFAMILY 5 - PANICOIDEAE (tribes 18-19). Annual to
perennial herbs; leaves without false-petiole or cross-veins;
ligule 0 or a fringe of hairs or a membrane fringed with
hairs, rarely membranous; sheaths not fused; inflorescence a
panicle, a spike, or an umbel or raceme of racemes or spikes;
spikelets with 2 florets, the upper bisexual, the lower male
or sterile (<u>Zea</u> is monoecious; <u>Sorghum</u> has paired spikelets,
1 bisexual the other male or sterile); glumes 2; lemmas firm
to thick, 5-11-veined, awnless or with terminal awn; stamens
3; stigmas 2; lodicules 2 (0 in female <u>Zea</u>); embryo panicoid;
photosynthesis C3-type alone or C3- plus C4-type, with or
without Kranz leaf anatomy; fusoid cells 0; microhairs
present; chromosome base-number 5, 9 or 10.

TRIBE 18 - PANICEAE (genera 81-89). Annuals or less often
perennials with or without rhizomes or stolons; ligule 0,
membranous, a fringe of hairs or a membrane fringed with
hairs; inflorescence a panicle, a spike, or an umbel or
raceme of racemes or spikes; spikelets all the same, all
bisexual.

81. PANICUM L. - <u>Millets</u>
Annuals; ligule a dense fringe of hairs or membranous with
distal fringe of hairs; inflorescence a diffuse panicle;
spikelets with 2 florets, the lower male or sterile with
lemma ± as long as spikelet, the upper bisexual, smaller,
concealed between upper glume and lower lemma; glumes
unequal, the lower much shorter than, the upper ± as long as
the spikelet, the upper closely resembling the lower lemma;
lower lemma 5-11-veined, awnless; upper lemma awnless;
spikelets falling whole at maturity.

1 Leaf-sheaths with long patent hairs; lower glume
  >1/3 as long as spikelet   2
1 Leaf-sheaths glabrous; lower glume ≤1/3 as long

```
                                                          3
as spikelet
2 Spikelets (4)4.5-5.5mm                    4. P. miliaceum
2 Spikelets 2-3.5mm                         3. P. capillare
3 Spikelets 2-2.8mm, subacute to obtuse at apex;
  lower floret usually male, with well developed
  palea >1/2 as long as lemma               1. P. schinzii
3 Spikelets 2.7-3.5mm, acute to acuminate at apex;
  lower floret sterile, with O or much reduced palea
                                            2. P. dichotomiflorum
```

Other spp. - Over 20 other spp. have been recorded mostly as wool- or grain-aliens, but most are rare. Records of **P. subalbidum** Kunth, from tropical Africa, are most if not all errors for P. dichotomiflorum.

1. P. schinzii Hackel (P. laevifolium Hackel) - Transvaal 1089 Millet. Culms to 1m; leaves and sheaths glabrous; panicle up to 35cm; spikelets 2-2.8mm, subacute to obtuse; lower floret usually male, with well developed palea. Intrd; frequent casual wool- and bird-seed-alien on tips and waste ground; scattered in En; tropical and S Africa.

2. P. dichotomiflorum Michaux - Autumn Millet. Culms to 1m; 1086 sheaths glabrous; leaves usually glabrous, scabrid on 1089 margins; panicle up to 40cm; spikelets 2.7-3.5mm, acute to acuminate; lower floret sterile, with O or very reduced palea. Intrd; constant casual alien from soya-bean waste, occasional from wool and other sources; scattered in S En; N America.

3. P. capillare L. - Witch-grass. Culms to 1m; sheaths and 1086 usually leaf lowerside midrib patent-pubescent; panicle up to 1089 40cm; spikelets 2-3.5mm, acuminate; lower glume 2/5 to 1/2 as long as spikelet; lower floret sterile, with O or much reduced palea. Intrd; frequent casual from bird-seed and sometimes other sources on tips and waste ground; scattered in Br and CI; N America.

4. P. miliaceum L. - Common Millet. Differs from P. 1089 capillare in spikelets (4)4.5-5.5mm; lower glume 1/2 to 2/3 as long as spikelet; lower floret sterile, with O or much reduced palea. Intrd; common casual from bird-seed and sometimes other sources on tips and waste ground; scattered in Br and CI; Asia. Pre-flowering plants can be mistaken for Zea mays or Sorghum bicolor, but in these the proximal, membranous part of the ligule is longer (not shorter) than the distal fringe and the sheaths do not have long patent hairs over their whole surface as in P. miliaceum.

82. ECHINOCHLOA P. Beauv. - Cockspurs
Annuals; ligule O; inflorescence a raceme of ± dense spikes or racemes, or the secondary racemes again racemosely branched, often with long stiff hairs especially in tufts at branch-points, the spikelets usually in >2 rows; spikelets with 2 florets, the lower male or sterile with lemma ± as

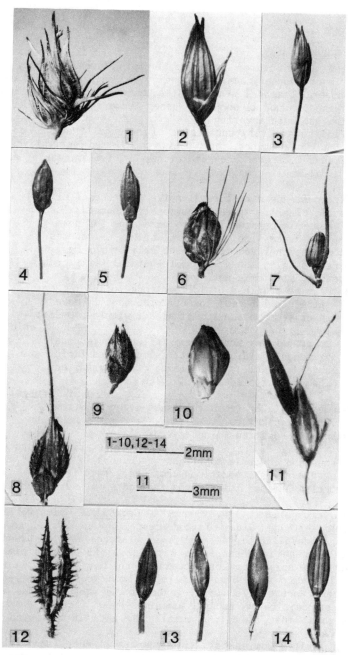

FIG 1089 – Spikelets of Poaceae. 1, **Cenchrus**. 2, **Panicum miliaceum**. 3, **P. capillare**. 4, **P. schinzii**. 5, **P. dichotomiflorum**. 6, **Setaria pumila**. 7, **S. verticillata**. 8, **Echinochloa crusgalli**. 9, **E. colona**. 10, **E. frumentacea**. 11, **Sorghum halepense** (group of 1 fertile and 1 sterile spikelets. 12, **Tragus australianus**. 13–14, **Digitaria** (2 views). 13, **D. sanguinalis**. 14, **D. ischaemum**.

long as spikelet, the upper bisexual, smaller, concealed
between upper glume and lower lemma; glumes unequal, the
lower much shorter than, the upper ± as long as the spikelet,
the upper closely resembling the lower lemma; lower lemma
5-7-veined, awned or awnless; upper lemma awnless; spikelets
falling whole at maturity.

Sp. limits are very uncertain. E. crusgalli and E. colona
are usually distinct, but their cultivated derivatives E.
utilis and E. frumentacea are extremely similar, making all 4
into a difficult complex.

1 Inflorescence with lateral spikes or racemes all or
   mostly clearly separate, obviously branched; lower
   floret male or sterile, its lemma awned or not   2
1 Inflorescence with fat lateral spikes or racemes
   close together and forming entirely or for most
   part a single, lobed, elongate head; lower floret
   sterile, its lemma not awned   3
   2 Primary branches of inflorescence simple, <3cm;
      spikelets 1.5-3mm; lower floret usually male;
      lower lemma acute to apiculate, with awn 0-2mm;
      leaf-blades <8mm wide   3. E. colona
   2 Lower primary branches of inflorescence usually
      branched again; spikelets 3-4mm; lower floret
      usually sterile; lower lemma often awned (awn
      <5cm); leaf-blades mostly >1cm wide
                                          1. E. crusgalli
3 Glumes and lower lemma yellowish-green to straw-
   coloured; spikelets 2.5-3.5mm; lower lemma acute to
   subacute, sometimes minutely apiculate
                                        4. E. frumentacea
3 Glumes and lower lemma bright green usually strongly
   tinged purplish, sometimes completely purplish;
   spikelets 3-4mm; lower lemma acuminate   2. E. utilis

   1. E. crusgalli (L.) P. Beauv. - Cockspur. Culms to 1.2m; 1089
leaves 1-3(8)cm wide; inflorescence up to 20cm, usually
strongly purplish-tinged, the lower branches usually branched
again (except in depauperate material), all or most clearly
separated; spikelets long-acuminate to long-awned.   Intrd;
casual mostly from bird-seed, also from wool and soya-bean
and other sources, on tips, waysides and waste ground, also
weed of cultivated ground sometimes natd there; scattered
throughout most of BI, especially S, natd in S Br and CI;
tropics.
   2. E. utilis Ohwi & Yab. (E. frumentacea auct. non Link) -
Japanese Millet. Differs from E. crusgalli in fat congested
inflorescence-branches forming lobed single head; spikelets
unawned.   Intrd; casual bird-seed-alien on tips and waste
ground; scattered in Br, mainly S; cultivated derivative of
E. crus-galli, originated in Japan.
   3. E. colona (L.) Link - Shama Millet. Differs from E. 1089

<u>crusgalli</u> in main branches usually borne at points separated
by >1/2 their length; and see key (couplet 2). Intrd; casual
on tips and waste ground from same sources as E. crusgalli;
occasional in S Br; tropics.
4. E. frumentacea Link - <u>White</u> <u>Millet</u>. Habit of <u>E.</u> <u>utilis</u> **1089**
but differs as in key (couplet 3). Intrd; source, habitat
and distribution as for <u>E.</u> <u>utilis</u>; cultivated derivative of
<u>E.</u> <u>colona</u>, originated in India. The vernacular <u>Japanese</u>
<u>Millet</u> is often misapplied to this sp.

83. BRACHIARIA (Trin.) Griseb. - <u>Signal-grasses</u>
Annuals; ligule a dense fringe of hairs; inflorescence a
raceme of racemes with spikelets in 2 rows on 1 side of
rhachis; spikelets with 2 florets, the lower male or sterile
with lemma + as long as spikelet, the upper bisexual,
smaller, concealed between upper glume and lower lemma;
glumes unequal, the lower <1/2 as long as upper, the upper
closely resembling the lower lemma; lower lemma 5-7-veined,
obtuse, awnless; upper lemma obtuse to rounded, awnless;
spikelets falling whole at maturity. Possibly better united
with <u>Urochloa</u>.

Other spp. - **B. eruciformis** (Smith) Griseb., from Africa, is
a rare bird-seed alien differing from <u>B.</u> <u>platyphylla</u> in its
much smaller (1.7-2.7mm) pubescent spikelets and 3-14
racemes. It resembles some plants of <u>Echinochloa</u> <u>colona</u>, but
differs in its more pubescent spikelets borne in only 2 rows,
obtuse lower lemma, well developed ligule, and often
pubescent (not glabrous) leaf-sheaths.

1. B. platyphylla (Griseb.) Nash - <u>Broad-leaved</u> <u>Signal-</u> **1052**
<u>grass</u>. Culms decumbent to erect, to 50cm; leaves up to 12 x **1093**
1.2cm, glabrous; sheaths glabrous; racemes 2-6, 3-8cm;
spikelets 3.5-4.5mm, glabrous. Intrd; casual from bird-seed
and soya-bean waste; occasional on tips and waste ground;
scattered in S En; N America.

84. UROCHLOA P. Beauv. - <u>Signal-grasses</u>
Differs from <u>Brachiaria</u> in upper glume and lower lemma
shortly acuminate, 7-veined; and upper lemma with distinct
terminal awn.

1. U. panicoides P. Beauv. - <u>Sharp-flowered</u> <u>Signal-grass</u>. **1052**
Differs from <u>Brachiaria</u> <u>platyphylla</u> in leaves, sheaths and **1093**
inflorescence-rhachis glabrous or with sparse long hairs;
spikelets 3.5-5mm, glabrous to pubescent; and in generic
characters above. Intrd; casual from bird-seed on tips and
waste ground; scattered in C & S En; Africa and W Asia.

85. ERIOCHLOA Kunth - <u>Cup-grasses</u>
Annuals or perennials; ligule a dense fringe of hairs;
inflorescence a rather irregular raceme of racemes, the main

branches ± appressed to main axis and often slightly branched
again, with spikelets scarcely in recognisable rows;
spikelets with 2 florets, the lower sterile with lemma almost
as long as spikelet and palea 0, the upper bisexual, smaller,
concealed between upper glume and lower lemma and with palea,
with small bead-like swelling at apex of pedicel; lower glume
± 0; upper glume as long as spikelet (slightly longer than
lower lemma but otherwise very similar); lower lemma
5-veined, acuminate to awned to 2mm; upper lemma obtuse to
rounded, with awn 0.3-1mm (as in Urochloa); spikelets falling
whole, disarticulating immediately below bead-like swelling.
The ± absence of a lower glume, and the lower floret being
represented by only a glume-like lemma, produces an
apparently 2-glumed spikelet with 1 floret.

Other spp. - The spp. found in Br have not been fully
investigated; **E. crebra** S.T. Blake, from Australia, and **E.
fatmensis** (Hochst. & Steudel) W. Clayton, from Africa,
closely resemble E. pseudoacrotricha and records of the last
could refer to all 3 and others too.

1. E. pseudoacrotricha (Stapf ex Thell.) S.T. Blake - 1052
Perennial Cup-grass. Perennials; culms to 60(100)cm; 1093
inflorescence up to 15 x 1cm; spikelets 3.6-6mm incl. awn
<2mm, with dense white silky hairs, acuminate. Intrd; rather
infrequent wool-alien on tips and in fields and waste places;
scattered in En; Australia.

86. PASPALUM L. - Finger-grasses
Perennials with stolons; ligule membranous; inflorescence a
raceme of 2(-4) racemes with spikelets in 2 rows on 1 side of
rhachis; spikelets with 2 florets, the lower sterile with
lemma as long as spikelet and palea very small, the upper
bisexual, smaller, concealed between upper glume and lower
lemma and with palea; lower glume ± 0; upper glume very
similar to lower lemma; lower lemma 3-5-veined, acute or
slightly apiculate; upper lemma subacute, apiculate;
spikelets falling whole.
The ± absence of a lower glume, and the lower floret being
represented by little more than a glume-like lemma, produces
an apparently 2-glumed spikelet with 1 floret.

1. P. distichum L. (P. paspalodes (Michaux) Scribner) - 1052
Water Finger-grass. Decumbent stoloniferous perennial; culms 1093
to 50cm but usually <20cm high, subglabrous; racemes 2-7cm;
spikelets 2.5-3.5mm, appressed-pubescent on upper glume,
glabrous on lemma. Intrd; natd in damp ground by sea at
Mousehole (W Cornwall) since 1971 and by canal in E London
(Middlesex) since 1984; tropics, natd in Mediterranean.

87. SETARIA P. Beauv. - Bristle-grasses
Annuals, or perennials with rhizomes; ligule a dense fringe

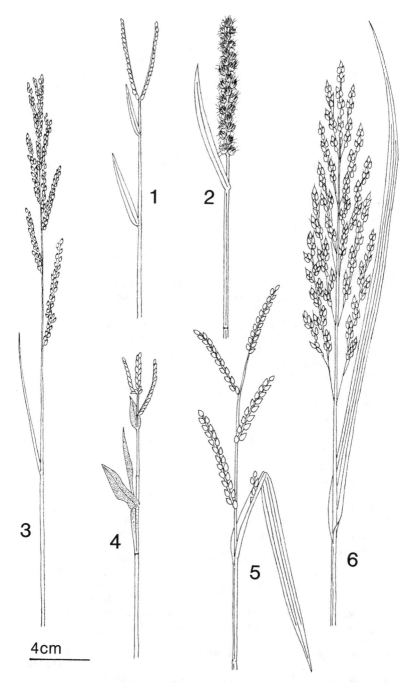

4cm

FIG 1093 – Poaceae. 1, Paspalum distichum. 2, Cenchrus echinatus. 3, Eriochloa pseudoacrotricha. 4, Urochloa panicoides. 5, Brachiaria platyphylla. 6, Sorghum halepense.

of hairs; inflorescence a dense spike-like panicle with very
short or ± vestigial crowded branches, sometimes interrupted
in lower part; spikelets with 2 florets, the lower male or
sterile with lemma as long as spikelet, the upper bisexual,
slightly smaller, concealed between or protruding from within
upper glume and lower lemma, with 1-c.12 strong bristles
borne on pedicel and usually awn-like and exceeding
spikelets; glumes unequal, the lower ±0 to c.2/3 as long as
spikelet, the upper c.2/3 to as long as spikelet; lower lemma
5-veined, obtuse, awnless; upper lemma obtuse to rounded,
awnless; spikelets falling whole at maturity, or (in S.
italica), the upper floret falling leaving the lower floret
and glumes on panicle, the bristles always remaining on
panicle.
The number of bristles (1-3, or more) borne on the pedicel
below each spikelet is a valuable character but easily over-
estimated where spikelets have aborted; in such cases the
bristles of aborted spikelets can be wrongly counted in with
those of an adjacent well-developed spikelet, but close
inspection shows the presence of spikelet-less pedicel(s).

1 Bristles (4)6-8(12) below each spikelet; upper
  glume scarcely longer than lower glume, c.1/2 to 2/3
  as long as spikelet; lower (sterile) floret with
  palea almost as long as lemma   2
1 Bristles 1-3 below each spikelet (see note above);
  upper glume much longer than lower glume, c.2/3 to
  as long as spikelet; lower (sterile) floret with
  palea ≤1/2 as long as lemma   3
  2 Annual; spikes ≥6mm wide when mature (excl.
    bristles); spikelets (2.5)3-3.3mm   **2. S. pumila**
  2 Perennial; spikes <5mm wide when mature (excl.
    bristles); spikelets 2-2.5(3)mm   **1. S. parviflora**
3 Main rhachis (often rather sparsely) hispid, with
  pricklets <0.2mm; bristles usually with backward-
  directed (rarely with forward-directed) barbs   4
3 Main rhachis densely pubescent with hairs
  >(0.2)0.5mm; bristles always with forward-
  directed barbs   5
  4 Spikelets 1.7-2mm; leaf-sheaths glabrous
                                            4. S. adhaerens
  4 Spikelets 2-2.3mm; leaf-sheaths pubescent on
    margin   **3. S. verticillata**
5 Spikelets disarticulating below upper lemma, leaving
  glumes and lower lemma on rhachis; panicle often
  >15 x 1.5cm, the bristles often not or scarcely
  longer than spikelet-clusters; upper lemma smooth
                                                7. S. italica
5 Spikelets falling whole, leaving only pedicels
  and bristles on rhachis; panicle rarely as much as
  15 x 1.5cm, the bristles always much longer than
  spikelets; upper lemma finely transversely rugose   6

6 Upper glume c.3/4 as long as spikelet; spikelets
  (2.5)2.7-3mm; leaves pubescent, often sparsely
  so   **6. S. faberi**
6 Upper glume as long or almost as long as spikelet;
  spikelets (1.8)2-2.5(2.7)mm; leaves glabrous
                                  5. S. viridis

1. S. parviflora (Poiret) Kerguelen (S. geniculata P.
Beauv., Panicum geniculatum Willd. non Lam.) - Knotroot
Bristle-grass. Shortly rhizomatous perennial; culms to 75cm;
panicles up to 10 x 0.4cm, not interrupted, with densely
shortly pubescent rhachis; bristles with forward-directed
barbs; spikelets 2-2.5(3)mm; upper lemma rather coarsely
transversely rugose. Intrd; casual from grain and bird-seed
on tips and waste ground; scattered in S En and S Wa; N
America. Annual with us and often not developing rhizomes.
2. S. pumila (Poiret) Roemer & Schultes (S. lutescens Hubb., **1089**
S. glauca auct. non (L.) P. Beauv.) - Yellow Bristle-grass.
Annual; culms to 75cm; panicles up to 15 x 1cm, not
interrupted, yellowish at maturity, with densely shortly
pubescent rhachis; bristles with forward-directed barbs;
spikelets (2.5)3-3.3mm; upper lemma coarsely rugose. Intrd;
casual weed of cultivated and waste ground and on tips, from
wool, bird-seed, soya-beans and other sources; occasional in
S & C Br and CI, very scattered in N Br, ?Ir; warm-temperate
Old World.
3. S. verticillata (L.) P. Beauv. (S. verticilliformis **1089**
Dumort., S. ambigua (Guss.) Guss.) - Rough Bristle-grass.
Annual; culms to 60cm; panicles up to 12 x 1cm, often ±
interrupted below; bristles with backward-directed or rarely
(var. **ambigua** (Guss.) Parl.) forward-directed barbs;
spikelets 2-2.3mm; upper lemma finely transversely rugose.
Intrd; casual from wool and bird-seed and other sources;
habitat and distribution as for S. glauca; warm-temperate Old
World. S. verticilliformis (S. ambigua) has been interpreted
as the hybrid S. verticillata x S. viridis, but is in fact a
var. of S. verticillata with forward-barbed bristles.
4. S. adhaerens (Forsskaol) Chiov. (S. verticillata ssp.
aparine (Steudel) T. Durand & Schinz) - Adherent Bristle-
grass. Differs from S. verticillata in bristles always with
backward-directed barbs; and see key (couplet 4). Intrd;
casual from bird-seed and perhaps other sources on tips and
waste and cultivated ground; scattered in S En and CI;
tropics of Old World. Perhaps better a ssp. or var. of S.
verticillata.
5. S. viridis (L.) P. Beauv. - Green Bristle-grass. Annual;
culms to 1m; panicles up to 12(17) x 1.2cm, not interrupted;
bristles with forward-directed barbs; spikelets (1.8)2-
2.5(2.7)mm; upper lemma finely transversely rugose. Intrd;
weed of cultivated and waste ground, sometimes natd, also
casual from wool, bird-seed and other sources (rarely soya-
beans) on tips, etc.; frequent in S & C Br and CI, sporadic

elsewhere; warm-temperate Old World.
6. S. faberi Herrm. - <u>Nodding</u> Bristle-grass. Differs from
<u>S. viridis</u> in usually rather larger with markedly curved to
pendent (not ± erect) panicles; and see key (couplet 6).
Intrd; characteristic casual from soya-bean waste, also from
grain; scattered in S En, probably overlooked; E Asia, but
arriving here from N America (natd).
7. S. italica (L.) P. Beauv. - <u>Foxtail</u> <u>Bristle-grass</u>.
Annual; culms to 1.5m; panicles up to 30 x 3cm, usually
pendent at apex, often interrrupted below; bristles with
forward-directed barbs, often shorter than spikelet-clusters;
spikelets 2-3mm; upper lemma smooth. Intrd; common casual on
tips and waste ground, from bird-seed; scattered in Br and
CI, mostly S; derived from <u>S. viridis</u> as crop probably in
China. The well-known cage-birds' 'millet-spray'.

88. DIGITARIA Haller - <u>Finger-grasses</u>
Annuals sometimes rooting at lower nodes; ligule membranous;
inflorescence an umbel of 2-many long narrow racemes or with
some racemes borne just below terminal cluster; spikelets
with 2 florets, the lower sterile with lemma as long as
spikelet, the upper bisexual, as long or slightly shorter,
concealed between or protruding from within upper glume and
lower lemma; glumes very unequal, the lower very short to ±0,
the upper 1/3 to as long as spikelet; lower lemma 5-7-veined,
obtuse, awnless; upper lemma acute to subacute, awnless;
spikelet falling whole.

1 Upper glume and lower lemma ± same length;
      spikelets 2-2.3(2.5)mm, conspicuously minutely
      pubescent; leaf-sheaths glabrous except at mouth;
      pedicel-apex slightly cup-shaped   **1. D. ischaemum**
1 Upper glume <3/4 as long as lower lemma; spikelets
      2.5-3.5(3.8)mm, sparsely appressed-pubescent;
      pedicels often slightly thicker near apex but
      simply truncate; leaf-sheaths usually pubescent,
      sometimes not; pedicel-apex truncate   2
    2 Lower lemma with smooth veins, often ciliate;
         upper glume (1/2)2/3 to 3/4 as long as spikelet,
         tapering-acute   **3. D. ciliaris**
    2 Lower lemma with minutely scabrid veins (x20
         lens), rarely ciliate; upper glume 1/3 to
         1/2(2/3) as long as spikelet, rather abruptly
         acute   **2. D. sanguinalis**

1. D. ischaemum (Schreber ex Schweigger) Muhlenb. - <u>Smooth</u> **1089**
<u>Finger-grass</u>. Culms decumbent to erect, to 35cm, glabrous
except at mouth of sheaths; racemes 2-8, 2-8cm. Intrd; weed
of cultivated ground, waste places and tips, etc., sometimes
± natd; very scattered in S En, formerly commoner; S Europe.
2. D. sanguinalis (L.) Scop. - <u>Hairy</u> <u>Finger-grass</u>. Culms **1089**
decumbent to erect, to 50cm, often pubescent on sheaths and

leaves; racemes (2)4-16, 3-16cm. Intrd; weed of similar habitats to D. ischaemum, sometimes + natd, but commoner as a bird-seed, wool or soya-bean alien and less common in cultivated ground; scattered in S Br and CI, once rare but now much commoner than D. ischaemum; S Europe.

3. D. ciliaris (Retz.) Koeler (D. adscendens (Kunth) Henrard) - Tropical Finger-grass. Differs from D. sanguinalis in leaves usually glabrous on upperside; and see key. Intrd; casual from similar sources as D. sanguinalis, not natd; scattered in S Br; tropics of Old World.

89. CENCHRUS L. - Sandburs
Annuals to tufted perennials; ligule a fringe of hairs; inflorescence spike-like, with rhachis bearing groups of 1-few spikelets on very short stalk, the group surrounded and enclosed by spiny bur composed of fused spines and bristles; spikelets with 2 florets, the lower sterile, the upper bisexual; glumes very unequal, the lower often + 0; lower lemma 5-veined, awnless; spikelets falling within bur.

Other spp. - A difficult genus; our spp. have not been worked out, and perhaps several are recorded as C. echinatus. The most often reported of the others are **C. incertus** M. Curtis and **C. pauciflorus** Benth., from America, and **C. longispinus** (Hackel) Fern., from Australia.

1. C. echinatus L. - Spiny Sandbur. Culms to 60cm; panicles **1089** to 10cm; burs + globose, 8-15mm across incl. spines; spines **1093** varying from thin bristles at base to flattened plates at apex, all with minute backward-directed barbs. Intrd; infrequent casual on tips and waste ground from wool, bird-seed and soya-bean waste; scattered in S En; originally America, now pan-tropical.

TRIBE 19 - ANDROPOGONEAE (Maydeae) (genera 90-91). Annuals, or perennials with rhizomes; ligule membranous, breaking into distal fringe of hairs; inflorescence a compact to diffuse panicle or umbel of racemes; spikelets in pairs, 1 bisexual and the other male or sterile in Sorghum, separated into male and female panicles in Zea and only the male paired.

90. SORGHUM Moench - Millets
Annuals or perennials; inflorescence a large panicle with spikelets in pairs, 1 bisexual and sessile, the other much thinner, sterile or male and stalked; bisexual spikelets with 2 florets, the upper bisexual, the lower reduced to a lemma; glumes 2, both long, the lower becoming hardened and + enclosing florets; spikelets falling whole or persistent.

1. S. halepense (L.) Pers. - Johnson-grass. Rhizomatous **1089** perennial to 1.5m; leaves usually <2cm wide; panicle + **1093** diffuse at anthesis; bisexual spikelets ovoid-ellipsoid or

narrowly so, narrowly acute, usually with bent twisted awn to
16mm, falling whole at maturity. Intrd; casual from bird-
seed, wool, grain, soya-bean waste and other sources on tips
and waste ground, occasionally natd for short while in S En;
scattered in S Br; N Africa.
2. **S. bicolor** (L.) Moench (S. vulgare Pers.) - Great Millet.
Annual to 2m; leaves usually >2cm wide; panicle compact at
anthesis; bisexual spikelets broadly ovoid-ellipsoid,
subacute, usually awnless, persistent at maturity. Intrd;
casual from bird-seed and other grain, sometimes other
sources; scattered in S Br and CI; Africa. Pre-flowering
plants resemble those of Zea mays, but leaves and sheaths are
completely glabrous and the base of the leaf-blade scarcely
clasps the stem; see also Panicum miliaceum for differences.

91. ZEA L. - Maize
Annuals; male and female inflorescences separate, the male a
large terminal panicle of spike-like racemes, the female
axillary (the familiar 'cob') forming a compact elongated
mass seated on a spongy axis; male spikelets in pairs, 1
subsessile the other stalked, with 2 florets and equal
glumes, awnless; female spikelets with 2 florets, the lower
sterile and much reduced, the upper female, with equal
glumes, awnless; spikelets persistent.

1. **Z. mays** L. - Maize. Culms to 3(-more)m; leaves 3-c.12cm
wide; male panicle up to 20cm; female inflorescence up to
c.20cm, with immense styles to 25cm. Intrd; casual from
grain and bird-seed on tips and waste ground, also grown on
large scale in S En as fodder and on small-scale as grain
crop; scattered in S & C Br and CI; C America, now world-
wide. Pre-flowering plants resemble those of Panicum
miliaceum and Sorghum bicolor, but the ligule has the
proximal membranous part longer than the distal fringe, the
base of the leaf-blade strongly clasps the stem, and the
leaf-sheaths have short hairs mainly near the top.

156. SPARGANIACEAE - Bur-reed family

Glabrous, aquatic or semi-aquatic, rhizomatous, herbaceous
perennials rooted in mud; leaves alternate, simple, linear,
entire, sessile with sheathing base, exstipulate. Flowers in
a terminal spike, raceme or panicle of globose unisexual
heads, the more distal heads male, the more proximal female,
hypogynous, actinomorphic; perianth of 1-6 inconspicuous
scales, often not easy to distinguish from bracts nor to
ascribe tepals to 1 or adjacent flowers; male flowers with
3-8 stamens; female flowers with 1-2(3) fused carpels with as
many cells, ovules, styles and stigmas; stigmas linear to
subcapitate; fruit a small, dry, spongy drupe with 1-2(3)
seeds.

Unmistakable aquatic plants with globose unisexual heads of flowers and fruits.

1. SPARGANIUM L. - Bur-reeds

1 Inflorescence branched, the male heads borne at apex
   of branches as well as of main axis; tepals dark-
   tipped   **1. S. erectum**
1 Inflorescence not branched, the male heads all at apex
   of main axis; tepals ± translucent, not dark-tipped   2
   2 Stem-leaves not inflated but strongly keeled at
     base; male heads 3-10, clearly separated
                                                  2. S. emersum
   2 Stem-leaves often inflated but not keeled at base;
     male heads 1-2(3), if >1 then close together and
     appearing ± as 1 elongated head   3
3 Bract of lowest female head >10cm, >2x as long as
  inflorescence; male heads mostly 2   **3. S. angustifolium**
3 Bract of lowest female head <10cm, barely longer than
  inflorescence; male head usually 1   **4. S. natans**

1. S. erectum L. - <u>Branched Bur-reed</u>. Stems erect, to 1.5m; leaves strongly keeled throughout; inflorescence branched, with >1 head on main branches. Native; by ponds, lakes, slow rivers and canals, in marshy fields and ditches; common throughout most of BI.

1 Fruits distinctly shouldered below beak   2
1 Fruits gradually rounded below beak   3
   2 Fruit with flat top (excl. beak) (3)4-6(7)mm
     across   **a. ssp. erectum**
   2 Fruit with rounded (domed) top (excl. beak)
     2.5-4.5mm across   **b. ssp. microcarpum**
3 Fruit ellipsoid, gradually tapered to beak,
  2-4.5mm across   **c. ssp. neglectum**
3 Fruit subglobose, abruptly contracted to beak,
  4-7mm across   **d. ssp. oocarpum**

a. Ssp. erectum. Fruits often dark brown at apex, with flat **1100** top with beak in centre. Mainly C & S Br, Man, rare in Ir.

b. Ssp. microcarpum (Neumann) Domin. Fruits often dark **1100** brown at apex, with rounded top very abruptly narrowed to beak. Throughout Br, Ir and Man.

c. Ssp. neglectum (Beeby) Schinz & Thell. Fruits pale **1100** brown, ellipsoid, gradually narrowed to beak. Throughout Br and Ir.

d. Ssp. oocarpum (Celak.) Domin. Fruits pale brown, **1100** subglobose, abruptly narrowed to beak. Ir, S & C Br, CI. Usually very few fruits form per head; possibly the hybrid ssp. <u>erectum</u> x ssp. <u>neglectum</u>.

2. S. emersum Rehmann - <u>Unbranched Bur-reed</u>. Stems erect, **1100** sometimes ± floating, to 60cm; aerial leaves strongly keeled at least in lower part, not inflated at base; inflorescence simple, the heads sessile or stalked; fruits ellipsoid,

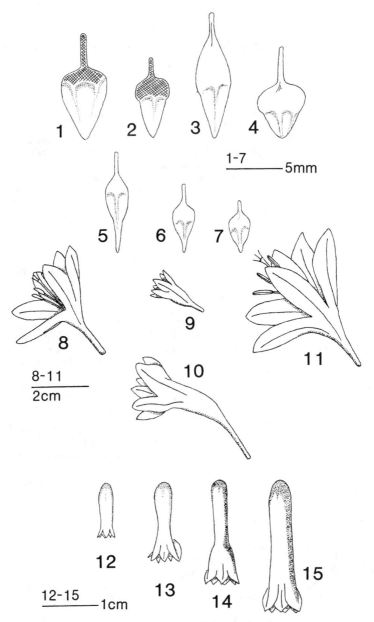

FIG 1100 – 1–7, Fruits of **Sparganium**. 1, **S. erectum** ssp.
erectum. 2, **S. erectum** ssp. **microcarpum**. 3, **S. erectum** ssp.
neglectum. 4, **S. erectum** ssp. **oocarpum**. 5, **S. emersum**. 6, **S.
angustifolium**. 7, **S. natans**. 8–11, flowers (excl. calyx) of
Crocosmia. 8, **C. x crocosmiiflora**. 9, **C. pottsii**. 10, **C.
paniculata**. 11, **C. masoniorum**. 12–15, corollas of
Polygonatum. 12, **P. verticillatum**. 13, **P. multiflorum**. 14, **P.
x hybridum**. 15, **P. odoratum**.

tapered to beak. Native; similar places to S. erectum but
rarely not in water; frequent throughout BI.
2 x 3. S. emersum x S. angustifolium = S. x diversifolium
Graebner has been found in a few places in W Sc from Wigtowns
to W Sutherland; it has the inflated leaf-bases of S.
angustifolium and the remote male heads of S. emersum, and is
fertile and backcrosses.
3. S. angustifolium Michaux - Floating Bur-reed. Stems 1100
usually floating (rarely erect), to 1m; leaves usually
floating, inflated at base, flat distally; inflorescence
simple, the heads sessile or stalked; fruits ellipsoid,
tapered to beak. Native; in peaty, acid lakes or pools;
local in W Wa, NW En, W & N Sc, Ir.
4. S. natans L. (S. minimum Wallr.) - Least Bur-reed. Stems 1100
usually floating (very rarely erect), to 50cm; leaves usually
floating, barely inflated at base, flat; inflorescence
simple, the heads sessile or stalked; fruits ellipsoid-
obovoid, tapered to beak. Native; in acid or alkaline lakes,
pools or ditches with high organic content; scattered over
most of Br and Ir but absent from most of C & S En and S Wa.

157. TYPHACEAE - Bulrush family

Glabrous, aquatic or semi-aquatic, rhizomatous, herbaceous
perennials rooted in mud; leaves alternate, simple, linear,
entire, sessile with sheathing base, exstipulate. Flowers
very numerous in dense, cylindrical, complex spike,
unisexual, the male distal, the female lower down (either
contiguous or separated by short bare axis), hypogynous,
actinomorphic; perianth of few-many bristles and/or narrow
scales; male flowers with 1-5 stamens with usually fused
filaments; female flowers with 1 1-celled, stalked ovary with
1 ovule; style 1; stigma clavate to linear; fruit a small
1-seeded capsule.
Unmistakeable aquatic plants with conspicuous cylindrical
inflorescence, male distally, female proximally.

1. TYPHA L. - Bulrushes
1. T. latifolia L. - Bulrush. Stems erect, to 3m, usually
overtopped by some leaves; leaves 8-20mm wide; male and
female parts of spike usually contiguous (rarely <2.5cm
apart); female part mostly 18-30mm wide, without scales (but
with many bristles). Native; reed-swamps, lakes, ponds, slow
rivers, ditches; frequent throughout most of BI but absent
from much of N & W Sc.
1 x 2. T. latifolia x T. angustifolia = T. x glauca Godron
occurs in scattered places throughout En and is probably
overlooked; it is variously intermediate in all characters
and highly (?completely) sterile.
2. T. angustifolia L. - Lesser Bulrush. Differs from T.
latifolia in leaves 3-6(10)mm wide; male and female parts of

inflorescence separated by (0.5)3-8(12)cm; female part mostly
13-25mm wide, with dark scales as well as bristles. Native;
similar places to T. latifolia but often on more organic
soils; scattered throughout most of BI but absent from C & N
Sc, Man and most of Ir.

158. BROMELIACEAE - Rhodostachys family

Glabrous, glaucous, dome-shaped, evergreen almost woody
plants; leaves in close spiral, simple, linear, with strong
spines on margins, sessile with sheathing base, exstipulate.
Flowers ± sessile in dense ± globose terminal heads c.5cm
across, bisexual, epigynous, ± actinomorphic; perianth of 3
outer and 3 inner free tepals, coloured; stamens 6; ovary
3-celled, each cell with numerous ovules on axile placenta;
style 1; stigmas 3, linear; fruit a berry.
Unmistakeable large dome-shaped pineapple-like plants with
spiny linear leaves and globose inflorescences.

1   Inflorescences ± sessile in terminal leaf-rosettes;
    petals blue, with basal scale-like nectaries; leaves
    flat distally, with marginal spines ± all apically
    directed   1. FASCICULARIA
1   Inflorescences arising from terminal leaf-rosettes on
    distinct stalks >10cm; petals pink, without scale-
    like nectaries; leaves concave throughout, with the
    lower marginal spines patent to recurved   2. OCHAGAVIA

1. FASCICULARIA Mez - Rhodostachys
Inflorescences sessile; petals blue, with basal scale-like
nectaries.

1. F. pitcairniifolia (Verlot) Mez - Rhodostachys. Plant
dome-shaped, to 75cm high and 2m across; leaves up to 35cm,
with sharp marginal forward-directed spines, the lower spines
patent with forward-directed tips, the most apical leaves
turning red at base at flowering. Intrd; well established
where planted on maritime dunes or shingle in Scillies and
Guernsey; Chile.

2. OCHAGAVIA Philippi - Tresco Rhodostachys
Inflorescences stalked; petals pink, without scale-like
nectaries.

1. O. carnea (Beer) Lyman B. Smith & Looser - Tresco
Rhodostachys. Very similar to F. pitcairniifolia in general
appearance but leaves shorter (<25cm), with longer, stouter
spines of which the lower are patent to recurved, and not or
scarcely turning red at base; and see key. Intrd; well
established where planted on dunes in Tresco, Scillies;
Chile.

159. PONTEDERIACEAE - Pickerelweed family

Aquatic, glabrous, rhizomatous, herbaceous perennials rooted in mud; leaves emergent, alternate, simple, entire, petiolate, with sheathing base, exstipulate. Flowers sessile in simple terminal emergent spike with bladeless leaf-sheath below it, bisexual, hypogynous, slightly zygomorphic; perianth of 6 violet-blue tepals fused at base; stamens 6; ovary 3-celled with 2 cells empty, or 1-celled; ovule 1; style 1; stigma capitate; fruit dry, 1-seeded, indehiscent. The only aquatic with ovate-cordate leaves and a spike of blue flowers.

1. PONTEDERIA L. - Pickerelweed
1. **P. cordata** L. - Pickerelweed. Semi-submerged, with creeping or floating stems to 1m; leaves triangular-ovate to narrowly so, mostly basal, 5-25cm, with long petioles; spike 3-15cm, dense; flowers 6-10mm across. Intrd; grown for ornament, natd at edges of ponds; scattered in S En; N America.

160. LILIACEAE - Lily family
(Alliaceae, Alstroemeriaceae, Amaryllidaceae, Asparagaceae, Asphodelaceae, Colchicaceae, Convallariaceae, Hemerocallidaceae, Hyacinthaceae, Melanthiaceae, Ruscaceae, Trilliaceae)

Usually erect, mostly glabrous, herbaceous perennials or occasionally small evergreen shrubs, rhizomatous or with a corm or bulb or tuberous roots; leaves all basal or alternate, occasionally whorled, simple, entire, sessile or petiolate, with or without sheathing base, exstipulate, sometimes reduced to scales and functionally replaced by leaf-like lateral stems (cladodes). Flowers solitary (terminal or axillary) or in racemes or umbels, less often in cymes or panicles, bisexual or rarely functionally dioecious, usually conspicuous, hypogynous or epigynous, actinomorphic or slightly zygomorphic; perianth usually of 6 free to fused tepals, often brightly coloured, the 3 outer not or obviously different from the 3 inner, rarely 4 or in 2 whorls of 4-6 each, sometimes with long or short funnel- or collar-shaped corona within the rows of tepals; stamens usually 6, rarely 3, 4 or 7-12; ovary 3(-5)-celled with numerous (or fewer, down to 2) ovules on axile placentas; styles 3-4(5) with capitate to linear stigma, or 1 with 1 or 3-4(5) stigmas; fruit a capsule or berry, sometimes a fleshy capsule.
A very variable family, usually recognized by conspicuous flowers with 6 petaloid tepals, 6 stamens and 3-celled ovary; exceptions are Maianthemum, Ruscus and Paris. All but Ruscus are herbaceous perennials arising from a corm, bulb, rhizome or tuberous roots. See Pontederiaceae, Iridaceae and

Agavaceae for differences.
Cronquist's circumscription of the Liliaceae is drawn very
widely; some other classifications differ considerably and up
to 13 extra families are sometimes recognized. The
subfamilial classification adopted here is more or less that
of Melchior, but numerous other versions exist.

General key
1 Leaves Iris-like, i.e. vertical, with 2 identical
   surfaces, borne on 2 opposite sides of stem, each
   with leaf-base sheathing that of next higher leaf   2
1 Leaves not Iris-like   3
   2 Styles 3; corolla creamy-white; filaments
     glabrous   1. TOFIELDIA
   2 Style 1; corolla yellow; filaments densely
     pubescent   2. NARTHECIUM
3 Leaves (3)4(-8), in 1 whorl on stem below single
   flower   17. PARIS
3 Leaves not all in 1 whorl on stem below flower;
   flowers often >1   4
   4 Flowers Crocus-like, appearing from soil in
     Autumn without leaves or stems; ovary subterranean,
     emerging with leaves to fruit in Spring   7. COLCHICUM
   4 Flowers with stems and leaves; ovary above ground
     at flowering   5
5 Leaves all reduced to small scales, with green +
   leaf-like stems (cladodes) arising from main stems
   in axils   6
5 At least some leaves green and photosynthetic   7
   6 Evergreen shrub; cladodes elliptic, very rigid,
     borne singly   35. RUSCUS
   6 Herb; cladodes linear, soft, borne in clusters
     of >4   34. ASPARAGUS
7 Flowers in an umbel with 1-few spathe-like bracts at
   base, or solitary with spathe-like bract(s) at base,
   sometimes flowers replaced by bulbils   Key A
7 Flowers in a cyme or raceme with or without bracts
   but without basal spathe-like bract(s), or rarely
   solitary or in umbel without spathe-like bract(s)
   at base   Key B

Key A - Flowers in an umbel or solitary, with 1-few spathe-
        like bracts at base (excl. Colchicum)
1 Flowers entirely replaced by bulbils   2
1 At least some flowers present   3
   2 Stems <4cm; leaves <10cm   9. GAGEA
   2 Stems >5cm; at least some leaves >15cm   24. ALLIUM
3 Funnel- or collar-like corona present inside
   perianth   33. NARCISSUS
3 Corona absent   4
   4 Ovary inferior or semi-inferior   5
   4 Ovary superior   9

5 Perianth yellow   **30. STERNBERGIA**
5 Perianth white, pink or red, often tinged green   6
  6 Flowers pink or red, often tinged green   7
  6 Flowers white with green or yellow patches   8
7 Ovary inferior; perianth >4cm, bright pink, with a
  proximal tube   **29. AMARYLLIS**
7 Ovary semi-inferior; perianth <2cm, greenish-red,
  with free tepals   **25. NECTAROSCORDUM**
  8 All 6 tepals similar   **31. LEUCOJUM**
  8 3 inner tepals much shorter and blunter than 3
    outer   **32. GALANTHUS**
9 Perianth yellow   10
9 Perianth white to various shades of red or blue   11
  10 Leaves >12mm wide; bracts (spathes) at base of
    inflorescence ovate, <15mm; flowers >5   **24. ALLIUM**
  10 Leaves <12mm wide; bracts (spathes) at base of
    inflorescence linear, >15mm; flowers 1-5   **9. GAGEA**
11 Perianth with tube >10mm, usually some shade of blue 12
11 Perianth with tepals free or fused for <5mm, very
  rarely blue or bluish   13
  12 Flowers in umbel, held horizontally, slightly
    zygomorphic, 3-5cm   **27. AGAPANTHUS**
  12 Flowers solitary, erect, actinomorphic, 2.5-3.5cm
                                             28. TRISTAGMA
13 Tepals free; style arising from base of ovary;
  plant with onion-like smell when fresh   **24. ALLIUM**
13 Tepals fused at base; style arising from top of
  ovary; plant without onion-like smell   **26. NOTHOSCORDUM**

Key B - Flowers in a cyme or raceme, rarely in an umbel or
    solitary, without spathe-like bract at base (excl.
    Tofieldia, Narthecium, Paris)
1 Ovary inferior   **36. ALSTROEMERIA**
1 Ovary superior   2
  2 Tepals and stamens 4; leaves strongly cordate at
    base   **15. MAIANTHEMUM**
  2 Tepals and stamens 6(-8); leaves cuneate to
    rounded at base   3
3 Tepals united into proximal tube >1/5 of their length 4
3 Tepals free or united just at extreme base   11
  4 Perianth yellow to orange or red, >3.5cm   5
  4 Perianth white to blue, pink or purple, very
    rarely pale yellow, <3.5cm   6
5 Flowers very numerous; perianth <5cm, tubular to
  narrowly campanulate   **6. KNIPHOFIA**
5 Flowers <c.20; perianth >5cm, funnel-shaped
                                            5. HEMEROCALLIS
  6 Flowers borne in groups of 1-c.5 in axils of main
    foliage leaves   **14. POLYGONATUM**
  6 Flowers terminal, or in terminal inflorescences,
    with bracts 0 or much reduced from leaves   7
7 Leaves linear to narrowly elliptic, narrowed at base;

```
      plant rhizomatous                                          8
  7 Leaves linear, not narrowed at base; plant with bulb        9
      8 Flowers pendent, stalked, usually white, with
         perianth-tube longer than -lobes; leaves with
         distinct petiole, up to 30 x 10cm incl. petiole
                                                    13. CONVALLARIA
      8 Flowers erect to patent, sessile, pink, with
         perianth-tube shorter than -lobes; leaves only
         slightly narrowed at base, up to 40 x 2cm
                                                    16. REINECKEA
  9 Perianth-tube >2x as long as lobes; corolla
      contracted at mouth                           23. MUSCARI
  9 Perianth-tube <2x as long as lobes; corolla spread
      open at mouth                                            10
     10 Perianth-tube much shorter than -lobes, the lobes
         bent outwards at junction with tube      22. CHIONODOXA
     10 Perianth-tube c. as long as -lobes, the lobes
         gradually curved outwards                21. HYACINTHUS
 11 Leaves all basal; inflorescence bractless or
      with bracts much reduced from leaves                     12
 11 Stems bearing at least 1 leaf, or leaves all basal
      but at least lowest bract + leaf-like                    16
     12 Inflorescence a panicle; filaments densely
         pubescent                                   4. SIMETHIS
     12 Inflorescence a raceme; filaments glabrous           13
 13 Bracts 2 per flower; tepals fused at extreme base
                                                    20. HYACINTHOIDES
 13 Bracts 0 or 1 per flower; tepals free                     14
     14 Tepals usually blue, sometimes pink or pure white
                                                    19. SCILLA
     14 Tepals white with green to reddish-brown stripe
         on abaxial surface                                   15
 15 Plant with bulb; bracts whitish          18. ORNITHOGALUM
 15 Plant without bulb, with swollen roots; bracts
      brown                                       3. ASPHODELUS
     16 Tepals <2cm                                           17
     16 Tepals >2cm                                           18
 17 Tepals white with purplish veins             8. LLOYDIA
 17 Tepals yellow with green stripes or tinge on
      abaxial side                                 9. GAGEA
     18 Stigmas sessile                           10. TULIPA
     18 Stigmas on obvious style                             19
 19 Stigmas linear; filaments + rigidly fixed to base
      of anthers                                  11. FRITILLARIA
 19 Stigmas not or scarcely longer than wide; filaments
      loosely fixed to middle of anther           12. LILIUM
```

Other genera – Several genera of garden plants are very persistent where thrown out or neglected. **ERYTHRONIUM dens-canis** L. (Dog's-tooth-violet) (Lilioideae), from S Europe, is a short Spring-flowering plant with a bulb, 2 purple-spotted leaves, and solitary pendent flowers with large purplish-pink or white reflexed tepals; it is sometimes planted in woodland gardens. **VERATRUM viride** Aiton (<u>Green</u> <u>False-helleborine</u>) (Asphodeloideae), from N America, differs from <u>Asphodelus</u> in having elliptic leaves on the stems and green flowers in dense terminal panicles; 1 clump has survived in Dunbarton for many years.

SUBFAMILY 1 - MELANTHIOIDEAE (tribe Narthecieae) (genera 1-2). Plant rhizomatous; leaves all or mostly basal, <u>Iris</u>-like (vertical, flat with 2 identical faces); inflorescence a terminal raceme; tepals yellow to greenish-white, free; styles 1 or 3; ovary superior; fruit a capsule.

1. TOFIELDIA Hudson - <u>Scottish</u> <u>Asphodel</u> Tepals greenish-white; filaments glabrous; anthers c. as long as wide, dehiscing inwards; styles 3; capsule splitting where ovary-cells meet; seeds ovoid-curved.

1. T. pusilla (Michaux) Pers. - <u>Scottish</u> <u>Asphodel</u>. Stems to **R** 20cm, with 5-10 flowers near apex; leaves up to 8cm x 3mm; tepals 1.5-2.5mm. Native; by streams and in flushes on mountains; very local in N En (Upper Teesdale), locally frequent in C & N Sc.

2. NARTHECIUM Hudson - <u>Bog Asphodel</u>
Tepals yellow; filaments densely pubescent; anthers >2x as long as wide, dehiscing outwards; style 1; capsule splitting along centre of ovary-cells; seeds with long fine projections at each end.

1. N. ossifragum (L.) Hudson - <u>Bog Asphodel</u>. Stems to 45cm, with 6-20 flowers near apex; leaves up to 30cm x 5mm; tepals 6-9mm. Native; bogs and other wet peaty acid places on heaths, moors and mountains; common in Ir and W & N Br, absent from most of C & E En.

SUBFAMILY 2 - ASPHODELOIDEAE (tribes Asphodeleae, Hemerocallideae) (genera 3-6). Plant rhizomatous or with swollen roots; leaves all or nearly all basal, linear; inflorescence a raceme or terminal compound cyme; tepals various colours (not blue); ovary superior; style 1; fruit a capsule splitting along centre of ovary-cells.

3. ASPHODELUS L. - <u>White Asphodel</u>
Plant with swollen roots; inflorescence racemose; tepals white with greenish to reddish-purple stripe on outside, erecto-patent, free; filaments glabrous; ovules 2 per cell; flowers actinomorphic.

1. A. albus Miller - <u>White Asphodel</u>. Stems to 1m, with dense ± unbranched terminal raceme; leaves up to 60 x 3cm, strongly keeled; tepals 15-20mm. Intrd; natd on grassy bank in Jersey since early 1970s; S Europe.

4. SIMETHIS Kunth - <u>Kerry Lily</u>
Rhizomatous; inflorescence cymose; tepals purplish outside, white inside, ± patent, free; filaments densely pubescent; ovules 2 per cell; flowers actinomorphic.

1. S. planifolia (L.) Gren. - <u>Kerry Lily</u>. Stems to 40cm, **RRR** with terminal ± lax panicle; leaves up to 50cm x 7.5mm; tepals 8-11mm. Native; rocky heathland near sea with <u>Ulex</u> over c.30 square km near Derrynane, S Kerry, natd on heathland near sea in Dorset and S Hants.

5. HEMEROCALLIS L. - <u>Day-lilies</u>
Rhizomatous; inflorescence cymose; tepals yellow to orange, fused to form proximal tube, funnel- to trumpet-shaped; filaments glabrous; ovules many per cell; flowers slightly zygomorphic, at least by upward curvature of stamens and style in laterally-directed flowers.

1. H. fulva (L.) L. - <u>Orange Day-lily</u>. Stems to c.1m, with up to c.20 flowers; leaves up to 90 x 2.5cm; flowers dull orange, 7-10cm, ± scentless. Intrd; much grown in gardens and very persistent when neglected or thrown out, often forming dense clumps on rough ground, banks and grassy places; scattered ± throughout Br and CI; garden origin.
2. H. lilioasphodelus L. - <u>Yellow Day-lily</u>. Stems to c.80cm, with up to c.12 flowers; leaves up to 65 x 1.5cm; flowers yellow, 7-8cm, sweetly scented. Intrd; natd as for H. fulva; scattered in Br, less common than H. fulva in En and Wa but more common in Sc; E Asia.

6. KNIPHOFIA Moench - <u>Red-hot-pokers</u>
Densely tufted, with short rhizomes; inflorescence a dense raceme; tepals red to yellow, fused to form proximal tube; cylindrical to narrowly campanulate; filaments glabrous; ovules many per cell; flowers very slightly zygomorphic due to curvature of perianth and stamens. Sometimes placed in Aloeaceae.

Other spp. - The garden plants escaping into the wild are not well understood; some may be hybrids and **K. linearifolia** Baker might also be present.

1. K. uvaria (L.) Oken - <u>Red-hot-poker</u>. Stems to 1.2m; leaves up to 80 x 1.8cm, V-shaped in section; bracts 3-9mm, ovate to oblong-ovate, rounded to subacute at apex; peduncle up to 1m; raceme up to 12cm; perianth 2.8-4cm, red at first, becoming yellow and pendent; stamens included or just exserted. Intrd; much grown in gardens, very persistent where thrown out or planted, dunes or waste ground usually near sea; scattered in extreme S & W Br from E Kent to

Flints, CI; S Africa.
2. **K. praecox** Baker - <u>Greater Red-hot-poker</u>. Differs from
<u>K. uvaria</u> in stems to 2m; leaves up to 200 x 4cm; bracts
8-12mm, lanceolate to linear-oblong, acute to acuminate at
apex; peduncle 1.2-2m; raceme 12-30cm; perianth 2.4-3.5cm;
stamens exserted 4-15mm. Intrd; grown and natd as for <u>K.
uvaria</u> and much confused with it, probably similar
distribution, certainly in E & W Cornwall and CI; S Africa.

SUBFAMILY 3 - WURMBAEOIDEAE (tribe Colchiceae) (genus 7).
Plant with a corm; leaves ± all on stem, appearing in Spring
with fruits, linear-oblong; flowers appearing in Autumn
without leaves, 1-few each arising from ground; tepals
pinkish to pale purple, united into long tube proximally;
ovary superior, but subterranean at flowering, emerging above
ground at stem apex at fruiting; styles 3; fruit a capsule
splitting where ovary-cells meet. <u>Crocus</u>-like, but with
6(not 3) stamens.

7. COLCHICUM L. - <u>Meadow Saffron</u>
1. **C. autumnale** L. - <u>Meadow Saffron</u>. Tepals with narrow
erect tube 5-20cm; capsule(s) produced on stem with sheathing
leaves; leaves up to 35 x 5cm. Native; damp meadows and open
woods on rich soils; local in C & S Br (common around Severn
estuary), very local in S Ir, natd in Jersey and Co Armagh.

SUBFAMILY 4 - LILIOIDEAE (tribes Tulipeae, Lilieae, Convall-
arieae, Polygonateae, Parideae) (genera 8-17). Plant
rhizomatous or with a bulb; leaves all basal, all on stem, or
both, linear to ovate or elliptic; inflorescence a single
flower or a spike or raceme; tepals various colours but not
blue, free or fused; style 1 or 0; ovary superior; fruit a
berry or a capsule splitting along centre of ovary-cells.

8. LLOYDIA Salisb. ex Reichb. - <u>Snowdon Lily</u>
Plant with a bulb; leaves mostly basal, some on stem, linear
or ± so; flowers 1-2(3) at stem-apex; tepals free, erecto-
patent; style 1; fruit a capsule.

1. **L. serotina** (L.) Reichb. - <u>Snowdon Lily</u>. Stems to 15cm, **RRR**
with 2-4 leaves; basal leaves up to 25cm x 2mm; tepals
9-12mm, white with purple veins. Native; cracks in basic
mountain rocks; very local in c.5 sites in Caerns.

9. GAGEA Salisb. - <u>Star-of-Bethlehems</u>
Plant with 1-2 bulbs; leaves basal (1-4) and on stem (1-4),
linear or ± so; flowers 1-5 at stem-apex; tepals free, ±
patent; style 1; fruit a capsule.

1. **G. lutea** (L.) Ker Gawler - <u>Yellow Star-of-Bethlehem</u>.   **R**
Stems to 25cm, with 1 bulb at base, with 1-5 subumbellate
flowers and 2-3 leaf-like bracts; basal leaf usually 1,

15-45cm x 7-15mm; tepals yellow, 10-18mm. Native; damp
base-rich woods, hedgerows and rough fields; scattered in Br
N to C Sc, rare except in C & N En. Flowers Mar-May.
2. G. bohemica (Zauschner) Schultes & Schultes f. - Early **RRR**
Star-of-Bethlehem. Stems to 4cm, with 2 bulbs at base, with
1(-4) subumbellate flowers and usually 4-6 leaf-like bracts;
basal leaves usually 2 per bulb, 4-9cm x c.1mm; tepals
yellow, 12-18mm. Native; cracks and ledges of basic rocks; 1
site in Rads, discovered 1965. Flowers Jan-Mar.

10. TULIPA L. - Tulips
Plant with a bulb; leaves basal and on stem, elliptic to
linear-elliptic; flowers 1-2 at stem apex, erect at maturity;
tepals free, forming cup-shaped flower; style 0; fruit a
capsule.

1   Flower rounded at base; filaments completely glabrous;
    buds erect   **3. T. gesneriana**
1   Flowers cuneate at extreme base due to narrowing of
    perianth; filaments pubescent near base; buds pendent   2
    2   Tepals + uniform yellow   **1. T. sylvestris**
    2   Tepals pink to purple with yellow blotch on inside
        at base   **2. T. saxatilis**

1. T. sylvestris L. - Wild Tulip. Stems to 50cm; leaves
linear-elliptic, up to 30 x 1.8cm; flowers 1(-2); tepals +
uniform yellow, 2-6cm. Intrd; natd in woods, meadows and
neglected estates; scattered and rare in En and S & C Sc,
formerly much more frequent; S Europe.
2. T. saxatilis Sieber ex Sprengel - Cretan Tulip. Stems to
50cm; leaves linear-elliptic to narrowly elliptic, up to 35 x
4.5cm; flowers 1-2(4); tepals pink to purple with basal
yellow blotch inside, 3.5-5.5cm. Intrd; natd on stony rough
ground, Tresco (Scillies), since 1976; Crete.
3. T. gesneriana L. - Garden Tulip. Stems to 60cm; leaves
elliptic to narrowly so, up to 35 x 8cm; flower 1; tepals
variously yellow to red, pink, purple or white, often with
basal blackish blotch inside, 4-8cm. Intrd; the common
garden tulip, persistent on tips, waysides and rough ground
where thrown out; scattered in C & S En and CI; garden
origin.

11. FRITILLARIA L. - Fritillary
Plant with a bulb; leaves all or most on stem, linear;
flowers 1(-2) at stem apex, pendent even at maturity; tepals
free, forming cup-shaped flower; style 1, with 3 linear
stigmas; fruit a capsule.

1. F. meleagris L. - Fritillary. Stems to 30(50)cm, with   **R**
3-6(8) leaves up to 20 x 1cm; tepals conspicuously chequered
light and dark purple and cream, sometimes white, 3-5cm.
Native; damp meadows and pastures; local in En N to Leics and

Staffs, much less common than formerly, now frequent only in
Suffolk and Thames Valley, planted and natd more widely in En
and in Wa.

12. LILIUM L. - <u>Lilies</u>
Plant with a bulb; leaves all or most on stem, linear to
elliptic; flowers few to many in terminal raceme, with free
tepals; flowers pendent with tepals diverging and very
strongly rolled back forming 'Turk's-cap' shape in our 2
spp.; flowers erect or facing sideways with slightly recurved
tepals in other spp.; style 1, with 3-lobed stigma; fruit a
capsule.

Other spp. - Several showy garden taxa may persist for a
short while where thrown out or neglected. Most common are
L. candidum L. (<u>Madonna Lily</u>), from Greece, with white tepals
and elliptic leaves; **L. regale** Wilson (<u>Royal Lily</u>), from
China, with tepals purple outside and white inside and linear
leaves; and **L. x hollandicum** Woodcock & Stearn (<u>L. bulbiferum</u>
L. x <u>L. maculatum</u> Thunb.) (<u>Orange Lily</u>), of garden origin,
with red, orange or yellow tepals and narrowly elliptic
leaves.   None of these has 'Turk's-cap' flowers; the first
two have erect to horizontal funnel-shaped and the third
erect cup-shaped flowers.

  1. **L. martagon** L. - <u>Martagon Lily</u>. Stems to 1.5m; leaves
oblanceolate to elliptic, at least some in whorls; flowers
c.4cm across, with tepals up to 3.5cm when straightened,
purple with darker spots, sometimes white, sickly scented.
Intrd; well natd in woods; scattered in Br N to C Sc; Europe.
  2. **L. pyrenaicum** Gouan - <u>Pyrenean Lily</u>. Stems to 1m; leaves
linear, spiral; flowers c.5cm across, with tepals up to 6.5cm
when straightened, greenish- to orange-yellow with darker
spots, strongly sickly scented.   Intrd; well natd in woods,
hedgerows and field margins; scattered in Br, mostly W & N,
Jersey; Pyrenees.

13. CONVALLARIA L. - <u>Lily-of-the-valley</u>
Plant rhizomatous; leaves all basal or ± at base of stem,
elliptic or narrowly so; flowers 6-20 in long terminal
raceme, pendent; tepals fused into tube longer than lobes,
forming campanulate flower, white; style 1; fruit a red
berry.

  1. **C. majalis** L. - <u>Lily-of-the-valley</u>. Stems to 25cm;
leaves up to 30 x 10cm; flowers 5-10 x 5-10mm, strongly
sweetly scented.   Native; dry woods, scrub and hedgebanks
usually on base-rich soil; scattered through most of Br N to
C Sc, much grown in gardens and natd in many places (often
more robust plants than native ones).

14. POLYGONATUM Miller - Solomon's-seals
Plant rhizomatous; leaves all on stem, linear-elliptic to
elliptic; flowers 1-6 in axillary, stalked, pendent clusters;
tepals fused into ± cylindrical tube longer than lobes, white
or cream with green markings; style 1; fruit a purple or
bluish-black berry.

1 Leaves linear, most in whorls of 3-8
                                                    4. P. verticillatum
1 Leaves elliptic, all alternate   2
   2 Perianths not contracted in middle; filaments
      glabrous; flowers 1 or 2 per leaf-axil
                                                    3. P. odoratum
   2 Perianths slightly contracted in middle; filaments
      sparsely pubescent; flowers 1-6 per leaf-axil   3
3 Perianths 9-15(20) x 2-4mm; stems terete
                                                    1. P. multiflorum
3 Perianths 15-22(25) x 3-6mm; stems ridged to
   slightly angled   **2. P. x hybridum**

1. P. multiflorum (L.) All. - Solomon's-seal. Stems erect 1100
then arching, to 80cm, terete; flowers 1-6 per leaf-axil,
9-15(20) x 2-4mm, contracted in middle; berry bluish-black.
Native; woods, mostly on basic soils; locally frequent in S
Br, scattered N to N En.
2. P. x hybridum Bruegger (P. multiflorum x P. odoratum) - 1100
Garden Solomon's-seal. Stems erect, then arching, to 1m,
ridged to slightly angled; flowers 1-6 per leaf-axil, 15-
22(25) x 3-6mm, slightly contracted in middle; berry rarely
produced. Intrd; much grown in gardens (by far the commonest
taxon there), natd in woods, scrub and rough ground throughout CI and Br, very scattered in Ir; garden origin.
3. P. odoratum (Miller) Druce - Angular Solomon's-seal. 1100
Stems erect or erect then arching, to 40cm, distinctly   R
angled; flowers 1-2 per leaf-axil, 15-30 x 4-9mm, not
contracted; berry bluish-black. Native; in woods on limestone; very local in NW En, Peak District, and around Severn
estuary, very scattered in Wa, rarely natd elsewhere in Br.
4. P. verticillatum (L.) All. - Whorled Solomon's-seal. 1100
Stems erect, to 80cm, distinctly angled; flowers 1-4 per RRR
leaf-axil, 5-10 x 1.5-3mm, slightly contracted in middle;
berry turning from red to purple. Native; mountain woods;
very rare in M Perth, formerly E Perth, Angus and S Northumb.

15. MAIANTHEMUM Wigg. - May Lily
Plant rhizomatous; leaves few from rhizome, 2(-3) on stem,
ovate, cordate; flowers numerous in terminal raceme; tepals
4, free, patent, white; stamens 4; style 1; fruit a red
berry.

1. M. bifolium (L.) F.W. Schmidt - May Lily. Stems to 20cm; RR
leaves up to 6 x 5cm; raceme 1-4(5)cm; flowers 2.5-6mm

across. Native; woods on acid soils; extremely local in Durham, NE Yorks, N Lincs and E Norfolk (probably intrd), formerly elsewhere in N & E En and Midlothian.

16. REINECKEA Kunth - Reineckea
Plant rhizomatous; leaves all basal, linear; flowers numerous in short terminal spike, erect to patent; tepals fused into tube slightly shorter than patent to reflexed lobes, pink; style 1; fruit a red berry.

1. R. carnea (Andrews) Kunth - Reineckea. Leaves in rosette, up to 40 x 2cm; spike 4-9cm on stalk 3-5cm; flowers rather crowded, with tube c.5mm and lobes c.7mm. Intrd; planted as ground-cover, natd in woodland near Lizard, W Cornwall; Japan and China.

17. PARIS L. - Herb Paris
Plant rhizomatous; leaves (3)4(-8), all in 1 whorl at top of stem, elliptic to obovate; inflorescence a single terminal, erect, long-stalked flower; tepals 8(-12), 4(-6) outer lanceolate, 4(-6) inner linear, all green, free, patent; stamens 8(-12); ovary superior, 4(-5)-celled; style 1, short; stigmas 4(-5), linear; fruit a dehiscent black berry.

1. P. quadrifolia L. - Herb Paris. Stems to 40cm; leaves up to 15 x 8cm; tepals 2-3.5cm. Native; in moist woods on calcareous soils; rather local in Br, absent from most of Wa, SW En and N & W Sc.

SUBFAMILY 5 - SCILLOIDEAE (tribe Scilleae) (genera 18-23). Plant with a bulb; leaves all basal, linear or nearly so; inflorescence a terminal raceme; tepals usually white or blue, sometimes pink or brownish, very rarely pale yellow, free or fused; style 1, with 1 (often 3-lobed) stigma; ovary superior; fruit a capsule splitting along centre of ovary-cells.

18. ORNITHOGALUM L. - Star-of-Bethlehems
Flowers each with 1 bract; tepals free, white with green stripe(s) on outside; stamens inserted on receptacle, with flattened filaments.

1 Bracts longer than pedicels; filaments with 1 acute
   lobe at apex on either side of anther   **3. O. nutans**
1 Bracts shorter than pedicels (at least on lower
   flowers); filaments without apical lobes   2
  2 Inflorescence corymbose; tepals >14mm
                      2. O. angustifolium
  2 Inflorescence an elongated raceme; tepals <14mm
                      1. O. pyrenaicum

1. O. pyrenaicum L. - Spiked Star-of-Bethlehem. Stems to   **R**

80(100)cm; inflorescence a narrow elongated raceme with >20
flowers; bracts shorter than pedicels; tepals 6-13mm.
Native; woods and scrub; very local in SC En N to Hunts,
rarely natd elsewhere in En.
   2. O. angustifolium Boreau (O. umbellatum auct. non L.) -
Star-of-Bethlehem. Stems to 30cm; inflorescence corymbose,
with 4-12 erect flowers; bracts shorter than pedicels; tepals
15-20mmm. Native; grassy places, rough ground and open
woods; scattered throughout Br and CI, but perhaps native
only in E En. The closely related **O. umbellatum** L., from
Europe, has longer tepals (up to 30mm), more flowers (up to
20) and a bulb producing numerous subspherical (not few
elongated) bulblets, and is hexaploid (2n=54) (not triploid,
2n=27). It might be common in gardens and some natd plants
could be this; a plant with 2n=54 was reported from Cambs in
1968. Further study of wild plants is needed.
   3. O. nutans L. - Drooping Star-of-Bethlehem. Stems to
60cm; inflorescence a 1-sided raceme of 2-12 ± pendent
flowers; bracts longer than pedicels; tepals 15-30mm. Intrd;
grown in gardens, natd in grassy places; scattered in C & S
Br; C Europe.

19. SCILLA L. - Squills
Flowers each with 0 or 1 bract; tepals free, blue, rarely
white or pink; stamens inserted on base of perianth, with
narrow or flattened filaments.

1 Bracts >4mm   2
1 Bracts 0 or <4mm   4
   2 Flowers usually >20; lower bracts ≥3cm
                                                    6. S. peruviana
   2 Flowers <15; bracts <3cm   3
3 Leaves 2-5mm wide; tepals 5-8mm   **4. S. verna**
3 Leaves 10-30mm wide; tepals 8-12mm
                                              5. S. lilio-hyacinthus
   4 Flowers pendent; tepals 12-16mm   **3. S. siberica**
   4 Flowers erect to patent; tepals 3-10mm   5
5 Flowering Jul-Sep without leaves; tepals 3-6mm
                                              7. S. autumnalis
5 Flowering Feb-Apr with leaves; tepals 5-10mm   6
   6 Pedicels <10mm; leaves usually 3-7, sheathing
     flowering stem only at base; flowers mostly >10
     per stem   **2. S. messeniaca**
   6 Lower pedicels 10-30mm; leaves usually 2-3,
     sheathing flowering stem 1/4 to 1/2 way up from
     base; flowers mostly <10 per stem   **1. S. bifolia**

Other spp. - Several other spp. are grown in gardens and
some persist locally where neglected or thrown out. S.
bithynica Boiss., from SE Europe, differs from S. messeniaca
in tepals 9-10mm and bracts 2-3mm.

1. S. bifolia L. - <u>Alpine</u> <u>Squill</u>. Stems to 20cm, with **1116**
usually <10 + erect flowers; bracts 0 or minute; tepals +
patent, 5-10mm. Intrd; grown in gardens, natd and spreading
where planted in churchyards and on banks; very scattered in
S En; Europe.
2. S. messeniaca Boiss. - <u>Greek</u> <u>Squill</u>. Differs from <u>S.</u>
<u>bifolia</u> as in key. Intrd; similar status to <u>S.</u> <u>bifolia</u> in
churchyards and open woods; N Somerset; Greece.
3. S. siberica Haw. - <u>Siberian</u> <u>Squill</u>. Stems to 20cm, with **1116**
usually <5 + pendent flowers; bracts 1-2mm; tepals forming
cup- to funnel-shaped flower, 12-16mm. Intrd; much grown in
gardens, + natd where planted and neglected; scattered in SE
En; USSR.
4. S. verna Hudson - <u>Spring</u> <u>Squill</u>. Stems to 15cm, with
usually <12 suberect flowers; bracts 5-15mm; tepals + patent,
5-8mm. Native; dry short grassland near sea, especially
cliff-tops; locally common on coasts of W Br from S Devon to
Shetland, down E coast S to Cheviot, E coast of Ir.
5. S. lilio-hyacinthus L. - <u>Pyrenean</u> <u>Squill</u>. Stems to 40cm,
with c.5-15 suberect flowers; bracts 10-25mm; tepals +
patent, 9-12mm. Intrd; natd where planted or neglected in
open woodland; Roxburghs and Berwicks; France and Spain.
6. S. peruviana L. - <u>Portuguese</u> <u>Squill</u>. Stems to 50cm, with
20-100 suberect flowers; lower bracts 30-80mm; tepals +
patent, (5)8-15mm. Intrd; natd where planted or neglected;
very scattered in Br and CI; W Mediterranean.
7. S. autumnalis L. - <u>Autumn</u> <u>Squill</u>. Stems to 25cm, with   R
4-20 suberect flowers; bracts 0; tepals + patent, 3-6mm.
Native; short grassland usually near sea; local in CI, SW En
scattered E to S Essex and Surrey, formerly Glam.

20. HYACINTHOIDES Heister ex Fabr. (<u>Endymion</u> Dumort.) -
<u>Bluebells</u>
Flowers each with 2 bracts; tepals free, blue, sometimes
white or pink; stamens inserted on or at base of perianth,
with narrow filaments.
Doubtfully distinct from <u>Scilla</u>; <u>H.</u> <u>italica</u> fits + equally
well into either genus.

1 Tepals + patent, 5-8mm; all stamens inserted at base
   of perianth   **1. H. italica**
1 Tepals erect to erecto-patent, >10mm; at least 3
   outer stamens fused from base to >1/4 way up perianth 2
   2 Racemes pendent at apex; perianth tubular;
     anthers cream   **2. H. non-scripta**
   2 Racemes erect; perianth campanulate; anthers same
     colour as tepals   **3. H. hispanica**

1. H. italica (L.) Rothm. (<u>Scilla</u> <u>italica</u> L.) - <u>Italian</u> **1116**
<u>Bluebell</u>. Stems to 40cm; leaves up to 12mm wide; racemes
pyramidal; flowers suberect, scentless; tepals + patent,
5-8mm. Intrd; natd in neglected estate, S Essex; SW Europe.

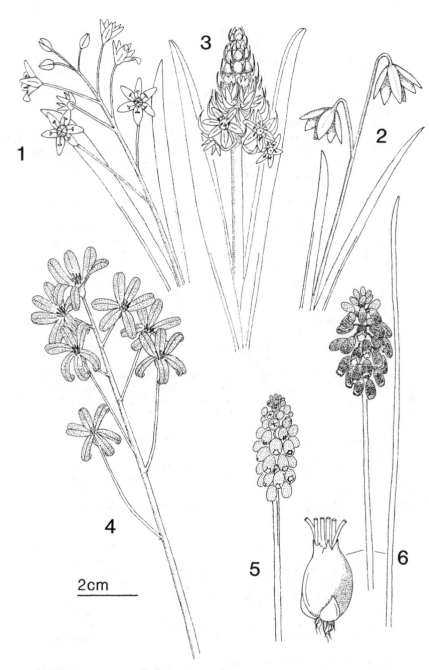

FIG 1116 – Liliaceae: Scilloideae. 1, Scilla bifolia. 2, S. siberica. 3, Hyacinthoides italica. 4, Chionodoxa forbesii. 5, Muscari armeniacum. 6, M. neglectum.

2. H. non-scripta (L.) Chouard ex Rothm. (<u>Endymion</u> <u>non-</u><u>scriptus</u> (L.) Garcke) - <u>Bluebell</u>. Stems to 50cm; leaves up to 20mm wide; racemes pendent at apex, 1-sided, with pendent strongly sweetly scented flowers; tepals 14-20mm, forming $\pm$ parallel-sided tubular perianth, strongly recurved at apex, outer 3 stamens fused to perianth for >3/4 their length. Native; woods, hedgerows, shady banks, grassland in wetter regions; frequent to abundant throughout BI.

2 x 3. H. non-scripta x H. hispanica is more commonly grown in gardens and natd than H. hispanica and arises naturally where natd H. hispanica meets native or natd H. non-scripta; frequent in Br and CI in similar places to H. hispanica. It is intermediate in all characters and fertile, forming a complete spectrum between the parents and often natd in absence of both.

3. H. hispanica (Miller) Rothm. (<u>Endymion</u> <u>hispanicus</u> (Miller) Chouard) - <u>Spanish</u> <u>Bluebell</u>. Stems to 40cm; leaves up to 35mm wide; racemes erect, not 1-sided, with erect to patent, faintly scented flowers; tepals 12-18mm, forming campanulate perianth, not recurved at apex; outer 3 stamens fused to perianth for <3/4 their length. Intrd; grown in gardens, natd in woods, copses, shady banks and field-borders; scattered in Br and CI; Spain and Portugal. Over-recorded for H. hispanica x H. non-scripta.

21. HYACINTHUS L. - <u>Hyacinth</u>
Flowers with 1 minute bract; tepals fused into tube c. as long as lobes, blue, pink or white, very rarely pale yellow; stamens inserted on perianth-tube, the filaments almost wholly fused to it.

1. H. orientalis L. - <u>Hyacinth</u>. Stems to 30cm; leaves up to 40mm wide; racemes erect, often $\pm$ parallel-sided, with few to many, pendent to suberect strongly sweetly scented flowers; perianth 10-35mm, the tube slightly constricted near middle, the lobes very strongly recurved. Intrd; much grown in gardens and long persistent where thrown out or neglected; scattered in S En; SW Asia.

22. CHIONODOXA Boiss. - <u>Glory-of-the-snows</u>
Flowers with 1 rudimentary or 0 bract; tepals fused into tube much shorter than lobes, blue, often with white central area; stamens inserted at apex of perianth-tube, fully exserted, with very flattened, white filaments.

1. C. forbesii Baker (<u>C. luciliae</u> auct. non Boiss.) - <u>Glory-</u> 1116 <u>of-the-snow</u>. Stems to 30cm (often much less), with (1)3-12 flowers; perianth-tube 3-5mm; perianth lobes $\pm$ patent, 10-15mm, white for proximal c.1/3, blue for distal c.2/3. Intrd; much grown in gardens, well natd from seeds where neglected or thrown out; scattered in Br, mainly S En; Turkey. Frequently mis-named C. luciliae.

2. C. sardensis Drude - <u>Lesser Glory-of-the-snow</u>. Differs
from <u>C.</u> <u>forbesii</u> in stems to 20cm; perianth-lobes (5)8-10mm,
wholly blue (but filaments white). Intrd; natd as for <u>C.</u>
<u>forbesii</u>, less common though perhaps overlooked; Surrey;
Turkey.

23. MUSCARI Miller - <u>Grape-hyacinths</u>
Flowers with 1 minute or 0 bract, the apical group sterile,
the lower ones fertile and often of different colour; tepals
fused for most of length, blue to blackish-blue, with white
lobes, or brownish; stamens inserted c.1/2 way up perianth-
tube, included, with narrow filaments.

1 Fertile flowers brownish-buff, on pedicels mostly
   >5mm; apical sterile flowers bright bluish-violet,
   some on pedicels >5mm   **4. M. comosum**
1 All flowers blue to blackish-blue, on pedicels <5mm   2
   2 Leaves linear with + spathulate apex; corolla +
      spherical with strongly recurved lobes
                                                       3. M. botryoides
   2 Leaves linear to oblanceolate; corolla ellipsoid-
      ovoid, distinctly longer than wide, with erecto-
      patent lobes   4
3 Perianth of fertile flowers blackish-blue
                                                       1. M. neglectum
3 Perianth of fertile flowers bright blue **2. M. armeniacum**

1. M. neglectum Guss. ex Ten. (<u>M.</u> <u>atlanticum</u> Boiss. &   1116
Reuter, <u>M.</u> <u>racemosum</u> Lam. & DC. non (L.) Miller) - <u>Grape-</u>   RR
<u>hyacinth</u>. Stems to 30cm; leaves up to 30cm, 2-8mm wide,
linear to oblanceolate; racemes dense, 1.5-5cm; fertile
flowers 3.5-7.5 x 1.5-3.5mm, dark violet-blue to blackish-
blue with white lobes; sterile flowers smaller, pale blue.
Native; dry grassland, hedgebanks and field borders; very
local in E & W Suffolk and Cambs, formerly E & W Norfolk,
rarely natd elsewhere but over-recorded for <u>M.</u> <u>armeniacum</u>.
2. M. armeniacum Leichtlin ex Baker - <u>Garden Grape-hyacinth</u>. 1116
Differs from <u>M.</u> <u>neglectum</u> in fertile flowers 3.5-5.5mm; and
see key (couplet 3). Intrd; the common garden <u>Muscari</u>, an
escape and throw-out spreading vegetatively and by seed on
rough ground, banks and grassy places, sometimes with <u>M.</u>
<u>neglectum</u>; scattered in Br, mainly S, and CI; Balkans to
Caucasus.
3. M. botryoides (L.) Miller - <u>Compact Grape-hyacinth</u>.
Stems to 25cm; leaves up to 20cm, conspicuously broadened at
apex; racemes dense, 1-6cm; fertile flowers 2.5-5 x 2.5-4mm,
bright blue with white lobes; sterile flowers smaller and
paler. Intrd; garden plant natd where thrown out or
neglected; Surrey; S Europe.
4. M. comosum (L.) Miller - <u>Tassel Hyacinth</u>. Stems to 60cm;
leaves up to 40 x 2cm, linear; racemes rather loose, 4-8cm;

fertile flowers 5-10 x 3-5mm, brownish-buff with pale lobes; sterile flowers bright bluish-violet, forming conspicuous apical tuft. Intrd; persistent weed of cultivated and rough ground, sometimes dunes and open grassland; local in SW En, S Wa and CI, rare and usually casual elsewhere; Europe.

SUBFAMILY 6 - ALLIOIDEAE (tribes Allieae, Agapantheae) (genera 24-28). Plant with a rhizome or a bulb, mostly smelling of garlic or onion when fresh; leaves usually all basal, sometimes on stem, linear to ± cylindrical, linear-oblong or elliptic; inflorescence a terminal umbel with usually scarious spathe at base, sometimes reduced to 1 flower or some or all flowers replaced by bulbils but still with spathe; tepals various colours, free or fused proximally; style 1, with capitate to 3-lobed stigma; ovary superior or semi-inferior; fruit a capsule splitting along centre of ovary-cells.

24. ALLIUM L. - Onions
Plant with bulb(s), smelling of onion or garlic when fresh; leaves linear to ± cylindric, or elliptic; flowers in umbel, some or all often replaced by bulbils; tepals free or ± so, white to greenish, pink, purple or yellow; ovary superior; ovules usually 2 per cell.

General key
1 Inflorescence consisting entirely of bulbils   Key A
1 Inflorescence with at least 1 flower   2
  2 Inflorescence with bulbil(s) and flower(s)   Key B
  2 Inflorescence with flowers only   Key C

Key A - Inflorescence consisting entirely of bulbils
1 Leaves circular to semi-circular or subcircular in
  section   2
1 Leaves obviously bifacial, flat to strongly keeled   4
  2 Stem hollow, inflated and bulging just below
    middle; leaves usually >4mm wide   **2. A. cepa**
  2 Stem solid or nearly so, not inflated; leaves
    <4mm wide   3
3 Spathe of 2 persistent valves each with apical
  attenuate part much longer than basal part
                      12. A. oleraceum
3 Spathe of 1 ± deciduous valve with apical
  attenuate part c. as long as basal part   **19. A. vineale**
  4 Stems triangular in section; leaves all basal
                      10. A. paradoxum
  4 Stems ± circular in section; at least some
    leaves borne on stem   5
5 Leaves <4mm wide   **13. A. carinatum**
5 Leaves >5mm wide   6
  6 Leaves 2-5; main bulb single, with often numerous

small bulblets outside its covering
                                    17. A. scorodoprasum
  6 Leaves 4-10; main bulb composed of several ±
    equal bulblets within common cover   **14. A. sativum**

Key B - Inflorescence consisting of both flowers and bulbils
1 Leaves circular to semi-circular or subcircular in
  section   2
1 Leaves obviously bifacial, flat to strongly keeled   5
  2 Stem hollow, inflated and bulging just below
    middle; leaves usually >4mm wide   **2. A. cepa**
  2 Stem solid or ± so, not inflated; leaves <4mm
    wide   3
3 Stamens shorter than tepals; filaments simple
                                             12. A. oleraceum
3 Stamens longer than tepals; inner 3 filaments divided
  distally into 3 points, the middle one anther-bearing   4
  4 Spathe 1-valved; lateral points of inner 3
    filaments >2x as long as central point
                                             19. A. vineale
  4 Spathe 2-valved; lateral points of inner 3
    filaments <2x as long as central point
                                          18. A. sphaerocephalon
5 Stems triangular in section   **10. A. paradoxum**
5 Stems ± circular in section   6
  6 Tepals yellow   **7. A. moly**
  6 Tepals pink to white, greenish or purplish   7
7 Filaments simple; leaves often <5mm wide   8
7 Inner 3 filaments divided distally into 3 points,
  the middle one anther-bearing; leaves >5mm wide   9
  8 Stamens shorter than tepals; spathe shorter than
    pedicels   **4. A. roseum**
  8 Stamens longer than tepals; spathe longer than
    pedicels   **13. A. carinatum**
9 Stamens longer than tepals   10
9 Stamens shorter than tepals   11
  10 Bulb scarcely swollen at base, without bulblets;
     style shorter than tepals; spathe persistent at
     least until flowering   **16. A. porrum**
  10 Bulb swollen at base, with bulblets around it
     within common cover; style longer than tepals;
     spathe usually deciduous before flowering
                                          15. A. ampeloprasum
11 Leaves 2-5; main bulb single, with often numerous
   small bulblets outside its cover; common part of
   inner 3 filaments 2-3x as long as central distal
   anther-bearing point   **17. A. scorodoprasum**
11 Leaves 4-10; main bulb composed of several ± equal
   bulblets within common cover; common part of inner
   3 filaments c. as long as central distal anther-
   bearing point   **14. A. sativum**

Key C – Inflorescence consisting entirely of flowers
1 Leaves circular to semi-circular or subcircular in
   section   2
1 Leaves obviously bifacial, flat to strongly keeled   5
   2 Stem hollow, inflated and bulging just below
      middle; leaves usually >4mm wide   **2. A. cepa**
   2 Stem solid or nearly so, not inflated; leaves
      <4mm wide   3
3 Filaments simple; stamens shorter than tepals
                                              1. A. schoenoprasum
3 Inner 3 filaments divided distally into 3 points, the
   middle one anther-bearing; stamens at least as long
   as tepals   4
   4 Spathe 1-valved; lateral points of inner 3
      filaments >2x as long as central point
                                              19. A. vineale
   4 Spathe 2-valved; lateral points of inner 3
      filaments <2x as long as central point
                                              18. A. sphaerocephalon
5 Tepals yellow   **7. A. moly**
5 Tepals white to pink, greenish or purplish   6
   6 Leaves with distinct petiole, the blade elliptic
      to narrowly so   **11. A. ursinum**
   6 Leaves without petiole, linear to filiform   7
7 Stem triangular in section   8
7 Stem ± circular in section   10
   8 Stigma simple; spathe 1-valved   **5. A. neapolitanum**
   8 Stigma 3-lobed; spathe 2-valved   9
9 Umbel 1-sided, with pendent flowers; tepals never
   opening >45 degrees   **8. A. triquetrum**
9 Umbel not 1-sided, with erect and pendent flowers;
   tepals opening >45 degrees at first, less so later
                                              9. A. pendulinum
   10 Inner 3 filaments divided distally into 3 points,
      the middle one anther-bearing   11
   10 Filaments simple   12
11 Bulb scarcely swollen at base, without bulblets;
   style shorter than tepals; spathe persistent at
   least until flowering   **16. A. porrum**
11 Bulb swollen at base, with bulblets around it within
   common cover; style longer than tepals; spathe
   usually deciduous before flowering   **15. A. ampeloprasum**
   12 Leaves conspicuously ciliate   **6. A. subhirsutum**
   12 Leaves glabrous   13
13 Stamens longer than tepals; leaves <3mm wide; spathe
   with valves much longer than pedicels **13. A. carinatum**
13 Stamens shorter than tepals; leaves most or all
   >4mm wide; spathe with valves rarely as long as
   pedicels   14
   14 Leaves >2cm wide   **20. A. nigrum**
   14 Leaves <1.5cm wide   15
15 Covering of bulb minutely pitted; spathe with 1

primary valve, often deeply >2-lobed   **4. A. roseum**
15 Covering of bulb with undulating or net-like
    markings; spathe 2-valved   **3. A. unifolium**

Other spp. - **A. fistulosum** L. (Welsh Onion) is grown in
gardens as a minor vegetable and might occur as a relic or
throw-out; it resembles variants of A. cepa with several
bunched narrow bulbs but differs in its cylindrical leaves,
stem widest at middle, and longer tepals (7-9mm).

1. A. schoenoprasum L. - Chives. Stems to 50cm, terete,   R
hollow; leaves terete, hollow, 1-5mm wide; inflorescence of
flowers only; tepals 7-14mm, pink to pale purple; stamens
shorter than tepals; filaments simple. Native; rocky ground,
usually on limestone, also grown as leaf-vegetable; local in
SW & N En and S Wa, formerly Berwicks and E Mayo, very
scattered relic or throw-out elsewhere in Br.
2. A. cepa L. - Onion. Stems to 1m, terete, hollow, bulging
just below middle; leaves subterete to hemi-cylindrical,
hollow, 2-20mm wide; inflorescence of flowers, bulbils or
both; tepals 3-4.5mm, greenish-white; stamens longer than
tepals; filaments simple except for small basal tooth each
side. Intrd; much grown as vegetable, frequent throw-out or
relic; scattered through Br; garden origin. Many variants
occur: the commonest sorts have a large single bulb and
usually no bulbils in the inflorescence; Shallot (A.
ascalonicum auct. non L.) has a number of much smaller bulbs;
Spring Onion has a cluster of small elongated bulbs with
white (not brown) covering; Tree Onion produces mostly or
only bulbils in the inflorescence.
3. A. unifolium Kellogg - American Onion. Stems to 40cm,
terete; leaves flat, scarcely keeled, 2-8mm wide; inflorescence of flowers only; tepals 10-17mm, bright pink; stamens
shorter than tepals; filaments simple. Intrd; garden plant
natd in wood in Dunbarton; W N America.
4. A. roseum L. - Rosy Garlic. Stems to 75cm, terete; 1123
leaves flat, scarcely keeled, 4-12mm wide; inflorescence of
flowers with or without bulbils; tepals 7-12mm, pink, rarely
white; stamens shorter than tepals; filaments simple. Intrd;
garden plant and weed well natd in rough or cultivated
ground, old dunes, hedgerows and waysides; frequent in SW En,
S Wa and CI, scattered elsewhere; Mediterranean. Presence or
absence of bulbils is not worth ssp. ranking.
5. A. neapolitanum Cirillo - Neapolitan Garlic. Stems to 1123
50cm, triquetrous with 2 edges much more acute than other;
leaves flat, keeled, 5-20mm wide; inflorescence of flowers
only; tepals 7-12mm, white; stamens shorter than tepals;
filaments simple. Intrd; garden plant natd in rough and
cultivated ground, hedgebanks and waysides; frequent in SW En
and CI; Mediterranean.
6. A. subhirsutum L. - Hairy Garlic. Stems to 45cm, terete; 1123
leaves flat, scarcely keeled, 2-10mm wide; inflorescence of

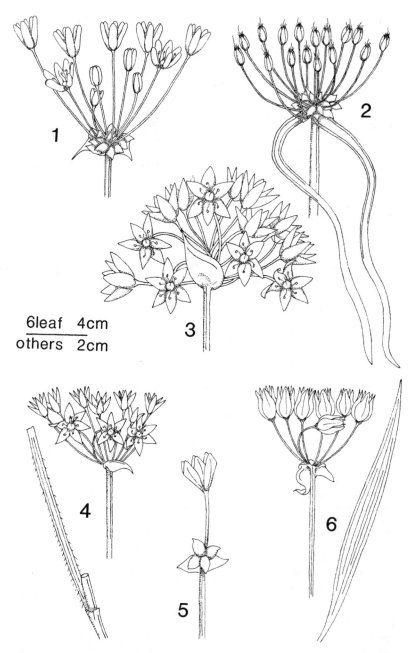

6leaf 4cm
_____
others 2cm

FIG 1123 - Allium. 1, A. roseum. 2, A. carinatum. 3, A. neapolitanum. 4, A. subhirsutum. 5, A. paradoxum. 6, A. moly.

flowers only; tepals 7-9mm, white; stamens shorter than tepals; filaments simple. Intrd; similar habitats and distribution to A. neapolitanum; Mediterranean.

7. A. moly L. - <u>Yellow Garlic</u>. Stems to 45cm, terete; 1123 leaves flat, narrowly elliptic, 8-35mm wide; inflorescence of flowers only; tepals 9-12mm, yellow; stamens shorter than tepals; filaments simple. Intrd; garden plant natd on warm banks and hedgerows; scattered in S En, Jersey, rarely elsewhere; Spain and France.

8. A. triquetrum L. - <u>Three-cornered Garlic</u>. Stems to 45cm, sharply triquetrous; leaves 2-5, flat, scarcely keeled, 4-12mm wide; inflorescence of flowers only; tepals 10-18mm, white with strong green line; stamens shorter than tepals; filaments simple. Intrd; weed of rough, waste and cultivated ground, copses, hedgerows and waysides; common in SW En and CI, scattered elsewhere in En, Wa and Ir; W Mediterranean.

9. A. pendulinum Ten. - <u>Italian Garlic</u>. Differs from <u>A. triquetrum</u> in stems to 25cm; leaves 3-8mm wide, usually 2; and see Key C (couplet 9). Intrd; natd in neglected estates, S Essex and Middlesex; Italy.

10. A. paradoxum (M. Bieb.) Don - <u>Few-flowered Garlic</u>. 1123 Stems to 40cm, triquetrous; leaves flat, scarcely keeled, 5-25mm wide; inflorescence of bulbils with or without flowers; tepals 10-12mm, white; stamens shorter than tepals; filaments simple. Intrd; natd in woods, grassy places, rough ground and waysides; scattered through much of En and Sc; Caucasus. Plants without bulbils are grown in gardens; they differ from <u>A. triquetrum</u> in having 1 leaf and tepals without green line.

11. A. ursinum L. - <u>Ramsons</u>. Stems to 45cm, variously ± terete but ridged to trigonous; leaves flat, 15-75mm wide, with narrow petiole up to as long as blade; inflorescence of flowers only; tepals 7-12mm, white; stamens shorter than tepals; filaments simple. Native; woods and other damp shady places; frequent, often abundant, over most of BI, but only 1 site in CI.

12. A. oleraceum L. - <u>Field Garlic</u>. Stems to 80cm, terete,   R slightly ridged; leaves hemi-cylindrical to rounded-channelled, 2-4mm wide; inflorescence of bulbils with or rarely without flowers; tepals 5-7mm, pinkish-, greenish- or brownish-white; stamens shorter than tepals; filaments simple. Native; dry grassy places; scattered throughout En, very scattered in Wa, Sc and Ir.

13. A. carinatum L. - <u>Keeled Garlic</u>. Stems to 60cm, terete, 1123 faintly ridged; leaves flat and keeled to crescent-shaped and channelled, 1.5-3mm wide; inflorescence of flowers, bulbils or both; tepals 4-7mm, bright pink; stamens much longer than tepals; filaments simple. Intrd; natd in rough ground, grassy places and waysides; scattered in Br and CE & NE Ir; Europe. Presence or absence of bulbils is not worth ssp. ranking.

14. A. sativum L. - <u>Garlic</u>. Stems to 1m, terete; leaves

ALLIUM

1125

flat, strongly keeled, 5-25mm wide; inflorescence of bulbils
with or without flowers; tepals 2-5mm, pinkish- to greenish-
white; stamens shorter than to c. as long as tepals; inner 3
filaments 3-pointed. Intrd; grown and imported on small-
scale as flavouring, casual where thrown out; very scattered
in En and Wa; unknown origin.
    15. A. ampeloprasum L. - (A. babingtonii Borrer) - Wild RR
Leek). Stems to 2m, terete; leaves flat, keeled, 5-40mm
wide; inflorescence of flowers with or without bulbils;
tepals 4-6mm, pale purple to pinkish-white; stamens longer
than tepals; inner 3 filaments 3-pointed. Native; rocky or
sandy places and rough ground near sea; very local in SW En,
S & NW Wa, N & CW Ir and CI. Var. ampeloprasum, with a
compact globose umbel without bulbils, occurs in SW En and
Wa; var. bulbiferum Syme, with bulbils (6-8mm) as well as
flowers in a rather compact umbel, occurs in CI; var.
babingtonii (Borrer) Syme, with bulbils (8-15mm) and flowers
in a rather loose umbel, often with some pedicels bearing
secondary heads, is endemic to SW En and Ir.
    16. A. porrum L. - Leek. Differs from A. ampeloprasum in
stems to 1m; leaves up to 100mm wide; and see Key C (couplet
11). Intrd; much grown as vegetable, casual where thrown out
or a relic; very scattered in Br; garden origin.
    17. A. scorodoprasum L. - Sand Leek. Stems to 80cm, terete; R
leaves flat, slightly keeled, 7-20mm wide; inflorescence with
bulbils with or rarely without flowers; tepals 4-8mm, deep
pink to reddish-purple; stamens shorter than tepals; inner 3
filaments 3-pointed. Native; dry grassland and scrub; local
in Br from Derbys and S Lincs N to S Aberdeen, natd in SW Ir
and rarely elsewhere.
    18. A. sphaerocephalon L. - Round-headed Leek. Stems to RRR
80cm, terete, finely ridged; leaves subcylindric, hollow,
1-3mm wide; inflorescence of flowers only; tepals 3.5-6mm,
pinkish-purple; stamens longer than tepals; inner 3 filaments
3-pointed. Native; on limestone rocks, W Gloucs, found 1847,
and on sandy waste ground by sea, Jersey, found 1836.
    19. A. vineale L. - Wild Onion. Differs from A. sphaero-
cephalon in inflorescence of flowers, bulbils or both; tepals
3-5mm, pink or greenish-white (blue-flowered plants occur in
N Kerry); and see Key C (couplet 4). Native; grassy places,
rough ground, banks and waysides; common in S En, frequent to
scattered in rest of BI except absent in N Sc. By far the
commonest narrow-leaved Allium.
    20. A. nigrum L. - Broad-leaved Onion. Stems to 80cm,
terete; leaves flat, 30-80mm wide; inflorescence of flowers
only; tepals 6-9mm, pink to white with green line; stamens
shorter than tepals; filaments simple. Intrd; garden plant
natd in rough ground in few places in S En; S Europe.

25. NECTAROSCORDUM Lindley - Honey Garlic
Plant with bulb, smelling of garlic when fresh; leaves
linear, strongly keeled; flowers in umbel, sweetly scented;

tepals free, greenish-red; ovary semi-inferior; ovules numerous per cell.

1. N. siculum (Ucria) Lindley (<u>Allium</u> <u>siculum</u> Ucria) - Honey **1144**
<u>Garlic</u>. Stems to 1.2m, terete; leaves 1-2cm wide; tepals
12-17mm; stamens shorter than tepals. Intrd; garden plant
natd in rough ground; very scattered in S En; S Europe.
Probably both ssp. **siculum** and ssp. **bulgaricum** (Janka) Stearn
(tepals redder, less acute, forming more broadly campanulate
perianth) occur, but are scarcely distinct.

26. NOTHOSCORDUM Kunth - Honey-bells
Plant with bulb, not smelling of garlic or onion; leaves
linear, scarcely or not keeled; flowers in umbel, sweetly
scented; tepals fused at base, greenish-white with pink
midrib; ovary superior; ovules numerous per cell.

1. N. gracile (Aiton) Stearn (N. inodorum auct. non (Aiton) **1144**
Nicholson, N. fragrans (Vent.) Kunth) - Honey-bells. Stems
to 60cm, terete; leaves 4-15mm wide; tepals 8-14mm; stamens
shorter than tepals. Intrd; garden plant natd in rough and
arable land and neglected estates; scattered in S & SW En and
CI. Distinguished from white-flowered spp. of <u>Allium</u> lacking
bulbils by the fused tepals and lack of garlic scent.

27. AGAPANTHUS L'Her. - African Lily
Plant with short tuber-like rhizome, not smelling of garlic
or onion; leaves oblong-linear, scarcely keeled; flowers in
umbel, slightly zygomorphic; tepals fused in lower 1/2,
bright blue, very rarely white; ovary superior; ovules
numerous per cell.

1. A. praecox Willd. - African Lily. Plant forming dense
clump; stems to 1m, terete; leaves 20-70 x 1.5-5.5cm;
perianth 30-50mm; stamens c. as long as perianth. Intrd;
well natd on sandy soil by sea in Scillies, scarcely so in
CI; S Africa. Our plant is spp. **orientalis** (F.M. Leighton)
F.M. Leighton.

28. TRISTAGMA Poeppig (<u>Ipheion</u> Raf.) - Starflowers
Plant with bulb, smelling of garlic when fresh; leaves
linear, slightly keeled; flowers solitary, terminal, with
spathe below, sweetly scented; tepals fused in lower 1/2,
pale bluish-violet with dark midrib outside; ovary superior;
ovules numerous per cell.

1. T. uniflorum (Lindley) Traub (<u>Ipheion</u> <u>uniflorum</u> (Lindley) **1128**
Raf.) - <u>Spring</u> <u>Starflower</u>. Stems to 35cm, terete; leaves
4-8mm wide; perianth 25-35mm, the tube 12-16mm, narrow, the
lobes patent, acute, forming flower 30-45mm across; stamens
as long as perianth-tube. Intrd; weed of cultivated and
waste ground; natd in W Cornwall, Scillies and CI, rare

casual elsewhere; S America.

SUBFAMILY 7 - AMARYLLIDOIDEAE (genera 29-33). Plant with a
bulb; leaves all basal, linear to narrowly elliptic or
linear-oblong; inflorescence of 1 flower or a terminal umbel
with usually scarious spathe at base; tepals white to yellow
or orange, rarely pink, free or fused proximally, sometimes
with funnel- or collar-shaped corona within the rows of
tepals; style 1, with simple or slightly 3-lobed stigma;
ovary inferior; fruit a capsule dehiscing irregularly or
along centre of ovary cells, often slightly succulent.

29. AMARYLLIS L. - Jersey Lily
Flowers in umbel, erecto-patent, trumpet-shaped, slightly
zygomorphic, without corona; tepals fused at base into short
tube, pink, all ± similar; flowers appearing in late Summer
or Autumn, before leaves.

1. A. belladonna L. - Jersey Lily. Stems very stout, to
60cm; leaves 30-45 x 1.5-3cm; perianth 5-10cm, with tube
1-1.5cm. Intrd; grown for ornament in CI (especially Jersey)
and a frequent relic in old fields, rough ground and sandy
places; S Africa. Can be confused with **Nerine sarniensis**
(L.) Herbert (Guernsey Lily), also from S Africa, grown in CI
but not natd, but this has much more open flowers with well
exposed stamens and perianth-tube <5mm.

30. STERNBERGIA Waldst. & Kit. - Winter Daffodil
Flowers solitary, erect, Crocus-like, actinomorphic, without
a corona; tepals fused into narrow proximal tube, yellow, all
± similar; flowers appearing in Autumn, ± with leaves.

1. S. lutea (L.) Ker-Gawler ex Sprengel - Winter Daffodil.
Stems 2.5-10(20)cm; leaves 7-10 x 0.4-1.5cm; perianth with
tube 0.5-2cm, the lobes 3-5.5cm. Intrd; natd on grassy
slopes by sea in Jersey since before 1919, formerly Guernsey;
Mediterranean. Differs from Crocus in its ± entire stigma
and leaves without central pale stripe.

31. LEUCOJUM L. - Snowflakes
Flowers solitary or few in umbel, pendent, campanulate or
bowl-shaped, actinomorphic, without a corona; tepals free,
white with green or yellow patches, all ± similar; flowers
appearing in late Winter or Spring, ± with leaves.

1. L. aestivum L. - Summer Snowflake. Forming clumps; stems **1128**
to 60cm, with (1)2-5(7) flowers; leaves up to 50 x 0.5-1.5cm; **RR**
tepals 10-22mm, each with green patch near apex; usually
flowers Mar-May; seeds black.
    a. Ssp. aestivum. Flowers (2)3-5(7); spathe mostly 30-50 x
7-11mm; tepals 13-22mm. Native; wet meadows and willow scrub
by rivers; very local in S En N to Oxon (mainly in Thames

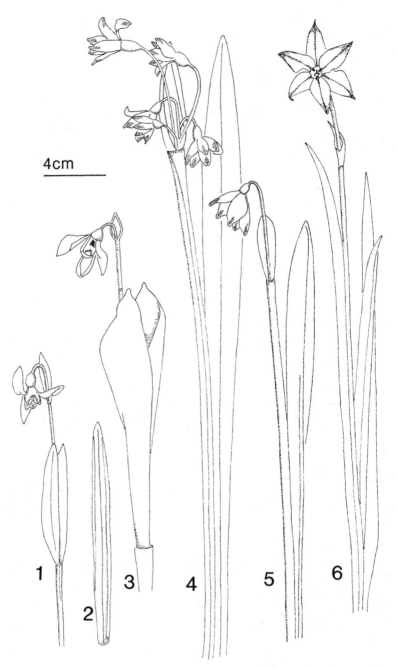

4cm

FIG 1128 – Liliaceae. 1, Galanthus nivalis. 2, G. plicatus (leaf only). 3, **G. elwesii**. 4, **Leucojum aestivum**. 5, **L. vernum**. 6, **Tristagma uniflorum**.

Valley) and S Ir, natd as garden escape elsewhere in BI but
much confused with ssp. pulchellum.
b. Ssp. pulchellum (Salisb.) Briq. Flowers (1)2-4; spathe
mostly 30-50 x 4-7mm; tepals 10-15mm. Intrd; more common in
gardens than ssp. aestivum, natd in damp places and rough
ground; scattered through BI from CI to Easterness; W
Mediterranean. Doubtfully distinct from ssp. aestivum.
2. L. vernum L. - Spring Snowflake. Differs from L. **1128**
aestivum in stems to 40cm, with 1(-2) flowers; leaves up to **RR**
30 x 0.5-2.5cm; tepals 15-25mm, with green or yellow patch
near apex; usually flowers Jan-Apr; seeds pale. Possibly
native; damp scrub and stream-banks; 2 sites, S Somerset and
Dorset, rarely natd elsewhere in En.

32. GALANTHUS L. - Snowdrops
Flowers solitary, pendent, actinomorphic, without a corona;
tepals free, the inner white with green patch(es) and forming
campanulate to bowl-shaped whorl, the outer white and
spreading when in full flower; flowers appearing in late
Winter or Spring, + with leaves, sometimes flore pleno.

1   Leaves with margins folded back or under at least
    along part of length, especially when young
                                              4. G. plicatus
1   Leaves flat, or with margins inrolled especially when
    young   2
  2   Leaves flat in bud, + linear, <1cm wide
                                              1. G. nivalis
  2   Leaves inrolled in bud, oblanceolate, at least
      one >1cm wide   3
3   Inner tepals with green patches at apex and base
                                              2. G. elwesii
3   Inner tepals with green patch at apex only
                                              3. G. caucasicus

Other spp. - A few other spp. are grown and might escape.
G. ikariae Baker, from Turkey to Caucasus and Iran, should be
sought; it has bright green leaves (not glaucous as in our
spp.) usually >1cm wide.

1. G. nivalis L. - Snowdrop. Stems to 20cm; leaves 9-20(30) **1128**
x 0.5-1cm, flat, linear; outer tepals 12-25mm; inner tepals
4-10mm, with green patch at apex only; often flore pleno.
Possibly native; woods, damp grassy places, banks and stream-
sides; scattered throughout Br and CI, usually if not always
intrd and rarely seeding, rare escape in Ir.
1 x 2. G. nivalis x G. elwesii occurs rarely in SE En where
both parents are natd together, it is intermediate in leaf
characters and tepal-patches.
1 x 4. G. nivalis x G. plicatus - see G. plicatus.
2. G. elwesii Hook. f. - Greater Snowdrop. Stems to 30cm; **1128**
leaves 6-15(25) x 0.6-3cm, inrolled, hooded at apex, oblanc-

eolate; outer tepals 15-30mm; inner tepals 8-13mm, with green
patches at base and apex. Intrd; commonly grown in gardens,
natd in woods and damp grassland; local in S En; SE Europe.
3. G. caucasicus (Baker) Grossh. - Caucasian Snowdrop. Stem
to 15cm; leaves 5-11(20) x 1-2cm, inrolled, slightly hooded
at apex, oblanceolate; outer tepals 15-20mm; inner tepals
9-12mm, with green patch at apex only. Intrd; natd as for G.
elwesii but much less frequently grown; very local in SE En;
Caucasus. Differs from G. elwesii ± only in smaller parts
and absence of basal green patch on inner tepals, and
probably better a ssp. of it.
4. G. plicatus M. Bieb. - Pleated Snowdrop. Stems to 25cm; 1128
leaves 5-15(30) x 0.4-1.5cm, with edges folded back, linear
to slightly oblanceolate; outer tepals 15-30mm; inner tepals
7-12mm, with green patch at apex or also at base. Intrd;
natd as for G. elwesii; local in S En; Romania, Turkey and
Crimea. Both ssp. **plicatus** (inner tepals with green patch at
apex only) and ssp. **byzantinus** (Baker) D. Webb (green patches
at apex and base) occur, but many plants (especially of
former) are probably G. nivalis x G. plicatus, intermediate
in leaf-folding and often occurring in absence of 1 or both
parents.

33. NARCISSUS L. - Daffodils
Flowers solitary or few in umbel, pendent to erecto-patent,
actinomorphic, with a corona; tepals and corona fused to form
hypanthial tube between base of tepals and apex of ovary;
tepals white to yellow; corona white to yellow or orange;
flowers appearing in Spring, with leaves, sometimes flore
pleno.
An extremely popular garden genus with numerous interspecific hybrids and thousands of cultivars, many with
uncertain parentage. Many occur natd in fields, waysides,
woods, rough ground, banks, etc., and are very difficult to
classify. The descriptions mainly apply to the commonest
variants of each taxon, but others (especially colour
variants) occur and are then best assigned to the recognized
Divisions of the International Daffodil Checklist:

Division 1 - Flower 1, with corona at least as long as tepals
Division 2 - Flower 1, with corona >1/3 as long as but
            shorter than tepals
Division 3 - Flower 1, with corona ≤1/3 as long as tepals
Division 4 - Flore pleno variants of garden origin of any
            affinity
Division 5 - Derivatives of N. triandrus, with character-
            istics of that species evident
Division 6 - Derivatives of N. cyclamineus, with character-
            istics of that species evident
Division 7 - Derivatives of N. jonquilla, with character-
            istics of that species evident
Division 8 - Derivatives of N. tazetta, with characteristics

of that species evident
Division 9 – Derivatives of <u>N. poeticus</u>, with characteristics of that species evident

1 Hypanthial tube parallel-sided, sometimes abruptly
   expanded at apex, >10mm; corona <10mm, wider than
   high; stamens of 2 distinct lengths   2
1 Hypanthial tube distinctly widening towards apex,
   2–25mm; corona usually >10mm, if <10mm then longer
   than tepals, c. as long as to much longer than wide;
   stamens all of same length or ± so   5
     2 Flower 1; corona yellow with sharply contrasting
        narrow red rim   **4. N. poeticus**
     2 Flowers (1)2–8(20); corona white or yellow all
        over   3
3 Flowers (1)2(–3); hypanthial tube >20mm; pollen
   sterile   **3. N. x medioluteus**
3 Flowers (2)3–8(20); hypanthial tube <20mm; pollen
   fertile   4
     4 Corona yellow; flowers <8(15)   **1. N. triandrus**
     4 Corona white; flowers <20   **2. N. papyraceus**
5 Tepals linear to very narrowly triangular or
   lanceolate, <5mm wide   6
5 Tepals ovate or triangular-ovate to suborbicular,
   >1cm wide   7
     6 Tepals reflexed back through 180 degrees;
        hypanthial tube 2–3mm; corona ± as wide at apex
        as at base; filaments straight   **9. N. cyclamineus**
     6 Tepals patent to erecto-patent; hypanthial tube 4–
        25mm; corona much wider at apex than at base;
        filaments curved   **7. N. bulbocodium**
7 Corona c. as long as tepals; hypanthial tube
   <2x as long as greatest width   **8. N. pseudonarcissus**
7 Corona distinctly shorter than tepals; hypanthial
   tube >2x as long as greatest width   8
     8 Flower 1; corona usually conspicuously deeper in
        colour than tepals; leaves ± glaucous, >8mm
        wide; stem distinctly 2-edged   **5. N. x incomparabilis**
     8 Flowers (1)2–4; tepals and corona the same colour;
        leaves green, <8mm wide; stem subterete
                                          6. N. x odorus

Other spp. – Several other spp., hybrids and groups of
cultivars are grown and may occasionally persist in the wild.
N. jonquilla L., from Spain and Portugal, differs from its
hybrid <u>N.</u> x <u>odorus</u> in its ± terete stem, cylindrical
hypanthial tube and 2–5 smaller flowers (tepals 10–15mm,
corona 3–5mm); **N. x intermedius** Lois. (<u>N. tazetta</u> x <u>N.
jonquilla</u>), garden origin, differs from <u>N.</u> x <u>odorus</u> in its
cylindrical hypanthial tube and smaller flowers (tepals
10–18mm, corona 3–4mm), and from <u>N. jonquilla</u> in its corona
darker in colour than the tepals and wider (5–8mm) leaves; **N.**

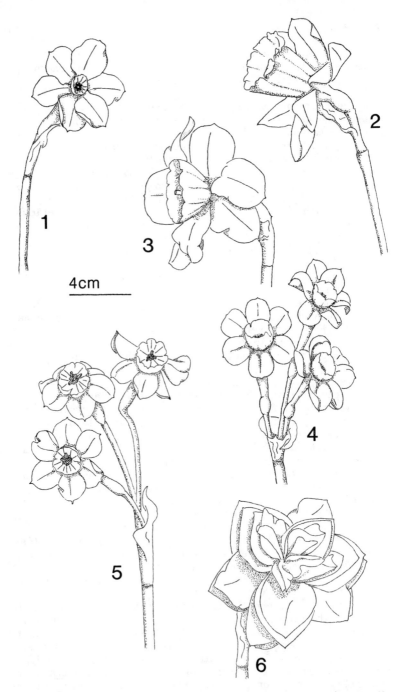

FIG 1132 – **Narcissus**. 1, **N. poeticus** (Division 9). 2, **N. pseudonarcissus** (Division 1). 3, **N. x incomparabilis** (Division 2). 4, **N. x odorus** (Division 7). 5, **N. tazetta**. 6, **N. pseudonarcissus** (Division 4).

bicolor L. (<u>N.</u> <u>abscissus</u> (Haw.) Schultes & Schultes f.) and
N. minor L., from SW Europe, are close to <u>N.</u> <u>pseudonarcissus</u>
- the former differs in its hypanthial tube 8-12mm and
slightly larger tepals and corona, the latter in its
hypanthial tube 9-15mm and slightly smaller tepals and
corona.

   1. N. tazetta L. - <u>Bunch-flowered</u> <u>Daffodil</u>. Leaves flat, 1132
5-15mm wide; flowers (2)3-8(15); hypanthial tube 12-18mm;
tepals 8-22mm, white; corona 3-6mm, yellow to deep yellow.
Intrd; rather rare relic in SW En and CI; W & C Mediterranean.
   2. N. papyraceus Ker Gawler - <u>Paper-white</u> <u>Dafodil</u>. Differs
from <u>N.</u> <u>tazetta</u> in flowers <20; hypanthial tube 10-20mm;
tepals 8-18mm; corona white. Intrd; natd as for <u>N.</u> <u>tazetta</u>;
W & C. Mediterranean.
   3. N. x medioluteus Miller (<u>N.</u> x <u>biflorus</u> Curtis; <u>N.</u> <u>tazetta</u>
x <u>N.</u> <u>poeticus</u>) - <u>Primrose-peerless</u>. Leaves flat, 7-10mm
wide; flowers (1)2(-3); hypanthial tube 20-25mm; tepals
18-22mm, usually white; corona 3-5mm, usually yellow. Intrd;
rather frequent relic in most of BI, mainly S; garden origin.
This belongs to Division 8; Primrose-peerless itself often
closely resembles <u>N.</u> <u>tazetta</u> (see key for differences), but
other cultivars may have tepals and corona the same colour
(white or yellow).
   4. N. poeticus L. - <u>Pheasant's-eye</u> <u>Daffodil</u>. Leaves flat, 1132
5-13mm wide; flower 1; hypanthial tube 20-30mm; tepals 15-
30mm, white; corona 1-3mm, yellow with sharply contrasting
narrow red rim. Intrd; S Europe. Cultivars of this belong
to Division 9.
   a. Ssp. poeticus (<u>N.</u> <u>majalis</u> Curtis). Leaves 6-10(12)mm
wide; tepals usually 20-25mm, obovate to suborbicular,
strongly overlapping; corona c.14mm across, + discoid; lower
3 stamens included. Commonly natd in BI, especially in S.
   b. Ssp. radiiflorus (Salisb.) Baker (<u>N.</u> <u>radiiflorus</u>
Salisb.). Leaves 5-8mm wide; tepals usually 22-30mm, obovate
to narrowly obovate, scarcely overlapping; corona c.8-10mm
across, shortly cylindrical; all 6 stamens partly exserted.
Frequent relic in CI, perhaps elsewhere.
   5. N. x incomparabilis Miller (<u>N.</u> <u>poeticus</u> x <u>N.</u> 1132
<u>pseudonarcissus</u>) - <u>Nonesuch</u> <u>Daffodil</u>. Leaves flat, 8-12mm
wide; flower 1; hypanthial tube 20-25mm; tepals 25-35mm,
usually yellow; corona 10-25mm, usually deeper yellow or pale
orange. Intrd; commonly natd in BI, especially S; garden
origin. Cultivars of this belong to Division 2, some of
which have a white perianth or are all white. Cultivars of
Division 3 differ in having the corona <1/3 as long as
corona, and show the same colour range. They are often
called **N. barrii** Baker, but may have the same origin (and
hence name) as <u>N.</u> x <u>incomparabilis</u>.
   6. N. x odorus L. (<u>N.</u> x <u>infundibulum</u> Poiret; <u>N.</u> <u>jonquilla</u> x 1132
<u>N.</u> <u>pseudonarcissus</u> agg.) - <u>Hybrid</u> <u>Jonquil</u>. Leaves deeply

channelled on upperside, 6-8mm wide; flowers (1)2-4;
hypanthial tube 15-25mm; tepals 15-25mm, yellow; corona 13-
18mm, yellow. Intrd; rarely natd in Scillies and Cornwall;
garden origin. Cultivars form Division 7.

7. N. bulbocodium L. - <u>Hoop-petticoat</u> <u>Daffodil</u>. Leaves
hemi-cylindrical, 1-2mm wide; flower 1; hypanthial tube
4-25mm; tepals 6-15mm, linear or nearly so, yellow; corona
7-25mm, yellow. Intrd; very distinctive sp. rarely natd; S
En; SW Europe.

8. N. pseudonarcissus L. - see sspp for English names. 1132
Leaves flat, 5-15mm wide; flower 1; hypanthial tube 15-25mm;
tepals 18-40(60)mm, usually yellow; corona ± as long as
tepals, usually yellow. Cultivars belong to Division 1;
<u>flore pleno</u> variants belong to Division 4 and are frequent.   1132

1 Tepals paler than corona   **a. ssp. pseudonarcissus**
1 Tepals and corona same colour   2
  2 Pedicels (between origin of spathe and base of
    ovary) mostly <15mm; leaves up to 30 x 1cm; tepals
    distinctly shorter than corona, not twisted
                               b. ssp. obvallaris
  2 Pedicels mostly >15mm; leaves up to 50 x 1.5cm;
    tepals as long as corona, twisted at base
                                  c. ssp. major

a. Ssp. pseudonarcissus (<u>N.</u> <u>gayi</u> (Henon) Pugsley) -
<u>Daffodil</u>. Tepals paler than corona, as long as corona,
18-40mm, twisted at base. Native; woods and grassland; local
(but often abundant) in En, Wa and Jersey, cultivars commonly
natd throughout BI. Cultivars of many colour variations
exist, but the corona is always slightly to much darker in
colour than the tepals. **N. gayi**, of unknown origin, differs
in having pedicels 10-35mm (not <10mm).

b. Ssp. obvallaris (Salisb.) Fernandes (<u>N.</u> <u>obvallaris</u>
Salisb.) - <u>Tenby</u> <u>Daffodil</u>. Tepals of same colour as corona
(yellow), 25-30mm, not twisted; corona 30-35mm. Intrd; long
natd in Pembs and Carms, probably elsewhere but overlooked;
garden origin.

c. Ssp. major (Curtis) Baker (<u>N.</u> <u>hispanicus</u> Gouan) - <u>Spanish</u>
<u>Daffodil</u>. Tepals of same colour as corona (white to yellow),
same length as corona, 18-40(60)mm, twisted at base. Intrd;
commonly natd over most of BI; SW Europe. Cultivars of many
colour variations exist.

9. N. cyclamineus DC. - <u>Cyclamen-flowered</u> <u>Daffodil</u>. Leaves
± flat or shallowly grooved on upperside, 3-6mm wide; flower
1; hypanthial tube 2-3mm; tepals 15-22mm, linear or nearly
so, yellow, reflexed through 180 degrees; corona 15-22mm,
yellow. Intrd; very distinctive sp. rarely natd; Surrey;
Spain and Portugal. Several cultivars commonly grown (incl.
'February Gold') are hybrids of <u>N.</u> <u>cyclamineus</u> with other
spp. and form Division 6; some might occur in the wild.

SUBFAMILY 8 - ASPARAGOIDEAE (tribes Asparageae, Rusceae)
(genera 34-35). Plant rhizomatous; leaves all on stems,

reduced to small scales, replaced functionally by stems
(cladodes) arising from their axils; flowers dioecious or ±
so; inflorescence an inconspicuous cluster of 1-few flowers
borne in scale-leaf axil on main stem or on cladode; tepals
greenish to yellowish-white, free or fused just at base,
without corona; style 1 or ±0, with capitate or 3-lobed
stigma; ovary superior; fruit a spherical red berry.

34. ASPARAGUS L. - Asparagus
Stems herbaceous; cladodes borne in cluster of 4-10(more),
cylindrical to slightly flattened; flowers 1-2(3) in axils of
scale-leaves on main stems; tepals fused at base; berry
5-10mm.

1. **A. officinalis** L. - see sspp. for English names.
   a. **Ssp. prostratus** (Dumort.) Corbiere - Wild Asparagus. **RR**
   Stems procumbent to decumbent, to 30(60)cm; cladodes on main
   lateral branches 4-10(15)mm, rigid, usually glaucous;
   pedicels 2-6(8)mm; seeds usually <5. Native; grassy sea-
   cliffs; very local in E & W Cornwall, Pembs, SE Ir and CI,
   formerly elsewhere in SW En and S Wa N to Anglesey,
   decreasing (many colonies now with only 1 sex).
   b. **Ssp. officinalis** - Garden Asparagus. Stems erect, to
   1.5(2)m; cladodes on main lateral branches (5)10-20(25)mm,
   flexible, usually green; pedicels 6-10(15)mm; seeds usually
   5-6. Intrd; grown as vegetable, very well natd in dry sandy
   soils among sparse grass, especially on maritime dunes and E
   Anglia heathland; scattered throughout BI N to C Sc; Europe.

35. RUSCUS L. - Butcher's-brooms
Evergreen shrubs; cladodes borne singly, flattened and leaf-
like; flowers 1-2 in axils of scale-leaves borne in centre of
adaxial side of cladode; tepals free; berry 8-13mm.

1. **R. aculeatus** L. - Butcher's-broom. Stems erect, to
   75(100)cm, much branched; cladodes 1-3(4) x 0.4-1cm, with
   sharp spine at apex; bract of inflorescence <3 x 1mm, scale-
   like. Native; woods, hedgerows, rocky places on dry soils;
   rather local in CI and Br N to Norfolk and N Wa, natd further
   N and in native areas.
2. **R. hypoglossum** L. - Spineless Butcher's-broom. Stems
   oblique, to 40cm, simple; cladodes 3-10 x 1-3.3cm, spineless;
   bract of inflorescence 10-30 x 3.5-13mm, green. Intrd; grown
   as ornament, natd in few shady places in W En and Midlothian;
   SE Europe.

SUBFAMILY 9 - ALSTROEMERIOIDEAE (genus 36). Plant with
tuberous roots; leaves all on stems, lanceolate; inflorescence a terminal simple or compound umbel without a spathe;
tepals orange with darker markings, free, without corona;
flowers slightly zygomorphic, with curved stamens; style 1,
with 3-lobed stigma; ovary inferior; fruit a capsule

splitting along centre of ovary cells.

36. ALSTROEMERIA L. - <u>Peruvian</u> <u>Lily</u>
   1. A. aurea Graham (<u>A.</u> <u>aurantiaca</u> D. Don) - <u>Peruvian</u> <u>Lily</u>.
Stems erect, to 1m; leaves 7-10cm; tepals 4-6cm, orange with
red spots and streaks; stamens shorter than tepals. Intrd;
much grown in gardens, natd in grassy places, rough ground
and old garden sites; scattered in Br N to C Sc; Chile.

161. IRIDACEAE - <u>Iris</u> family

Usually erect, mostly glabrous, herbaceous perennials,
rhizomatous or with a corm or rarely a bulb or swollen roots;
leaves all or mostly basal, those on stems alternate and
usually smaller and few, simple, entire, sessile, usually
with sheathing base, exstipulate. Flowers solitary or in
terminal spikes or panicles usually with sheathing (often
spathe-like) bracts, bisexual, conspicuous, epigynous,
actinomorphic or less often zygomorphic; perianth of 6 tepals
usually fused into tube proximally, sometimes free, petaloid,
the 3 outer not or obviously different from 3 inner; stamens
3; ovary 3-celled, with numerous ovules on axile placentas,
rarely 1-celled with 3 parietal placentas; style usually with
3 branches, the branches divided or not with stigmas at tips,
rarely simple with 3 stigmas, the branches sometimes +
petaloid; fruit a capsule splitting along centre of ovary-
cells.
Easily distinguished from Liliaceae by the 3 stamens (only
the distinctive <u>Ruscus</u> in the latter family has 3 stamens).

1 Style-branches broad and petaloid; flowers
   <u>Iris</u>-like   2
1 Style-branches not petaloid, narrow; flowers not
   <u>Iris</u>-like   3
   2 Plant with rhizome or bulb; roots not tuberous;
      ovary 3-celled   **5. IRIS**
   2 Plant without rhizome or bulb; roots tuberous;
      ovary 1-celled   **4. HERMODACTYLUS**
3 Flowers 1-few, erect, arising direct from ground or
   on very short stems, <u>Crocus</u>-like   4
3 Flowers few-many, erect or laterally-directed,
   arising from aerial green stems in spikes or
   panicles, not <u>Crocus</u>-like   5
   4 Leaves subterete, without white line; perianth-
      tube <1cm, sheathed by green bract   **7. ROMULEA**
   4 Leaves flat, channelled and with central whitish
      line on upperside; perianth-tube >1.5cm, sheathed
      by white or brown bract   **8. CROCUS**
5 Perianth actinomorphic, with radially symmetrical
   lobes and straight tube; plant with or without a corm 6
5 Perianth zygomorphic, often with bilaterally

symmetrical lobes but sometimes only due to curved
tube; plant with a corm   11
6 Perianth-lobes fused proximally into lobe >5mm   7
6 Perianth-lobes completely free or fused proximally
  into tube <5mm   8
7 Perianth-tube >3cm, the lobes shorter; bracts <2cm,
  3-toothed at apex, without dark streaks   **10. IXIA**
7 Perianth-tube <2cm, the lobes >2cm; bracts >2cm,
  deeply and jaggedly toothed at apex, with irregular
  dark longitudinal streaks   **11. SPARAXIS**
8 Inner tepals c.2x as long as outer   **1. LIBERTIA**
8 Inner and outer tepals same or nearly same length   9
9 Stem terete, arising from a corm; flowers sessile
                                                  10. IXIA
9 Stem flattened, narrowly winged, arising from rhizome
  or fibrous roots; flowers stalked   10
10 Tepals twisting spirally after flowering;
   filaments free, arising from top of short
   perianth-tube   **3. ARISTEA**
10 Tepals not twisting after flowering; filaments
   fused either just at base or for most of length,
   arising from base of perianth   **2. SISYRINCHIUM**
11 Style 3-branched, each branch bifid   12
11 Style with 3 simple branches, or unbranched with
   3-lobed stigma   13
12 Bracts >2cm; spike erect, with flowers on 2 sides
   of axis; leaves tough; seeds winged   **6. WATSONIA**
12 Bracts <1.5cm; spike bent horizontally near lowest
   flower, with flowers on 1 side; leaves soft;
   seeds not winged   **12. FREESIA**
13 Uppermost perianth-lobe >2x as long as rest
                                            14. CHASMANTHE
13 Uppermost perianth-lobe slightly longer to slightly
   shorter than rest   14
14 Perianth-lobes strongly narrowed at base; style-
   branches greatly widened distally   **9. GLADIOLUS**
14 Perianth-lobes not narrowed at base; style-
   branches filiform with minutely capitate stigmas
                                            13. CROCOSMIA

Other genera - **HOMERIA** Vent., from S Africa, would key out
as <u>Sisyrinchium</u> because of its actinomorphic flowers, free
tepals and united filaments, but has a corm; **H. collina**
(Thunb.) Salisb. (<u>H. breyniana</u> (L.) G. Lewis) (<u>Cape-tulip</u>)
has pale yellow or pink fragrant flowers 3-3.5cm and occurs
self-sown in large gardens in Scillies and Guernsey, and
might escape.

1. LIBERTIA Sprengel - <u>Chilean-irises</u>
Plants with short rhizomes; leaves <u>Iris</u>-like; inflorescence a
small terminal panicle; flowers actinomorphic; tepals white,
free, the inner c.2x as long as outer but of similar shape;

filaments slightly fused at base; style with 3 entire linear
branches.
   1. L. formosa Graham (<u>L. chilensis</u> (Molina) Klotsch ex Baker
nom. illeg.) - <u>Chilean-iris</u>. Densely tufted; stems erect, to
1.2m, unbranched; leaves dark green, up to 75 x 1.2cm;
flowers with pedicels shorter than and obscured by bracts,
c.25mm across; inner tepals 12-18mm. Intrd; garden plant
natd on rough ground, waysides and rocky lakeshore and
coasts; Scillies, W Cornwall, Man, CW Sc, S Kerry; Chile.
   2. L. elegans Poeppig - <u>Lesser Chilean-iris</u>. Differs from
<u>L. formosa</u> in more delicate habit; inflorescence branched;
flowers with pedicels exceeding bracts by c.5-10mm, c.13mm
across; inner tepals 6-9mm. Intrd; rare garden plant well
natd on railway embankment in Dunbarton; Chile.

2. SISYRINCHIUM L. - <u>Blue-eyed-grasses</u>
Plants with fibrous roots and short or 0 rhizomes; leaves
<u>Iris</u>-like; inflorescence a terminal cyme or a panicle of
terminal and lateral cymes; flowers actinomorphic; tepals
pale to bright yellow or blue, nearly free, ± equal;
filaments slightly fused at base to fused for most of length;
style with 3 entire linear branches.

```
1  Tepals blue                                                     2
1  Tepals cream to yellow for most part                           3
   2  Stem unbranched, with 1 terminal inflorescence;
      perianth 25-35mm across, violet-blue; pedicels
      erect in fruit                                  2. S. montanum
   2  At least some stems branched, each branch with
      1 terminal inflorescence; perianth 15-20mm
      across, pale blue; pedicels arched to pendent in
      fruit                                          1. S. bermudiana
3  Stem unbranched, with terminal and several lateral
   cymes; leaves >1cm wide                            5. S. striatum
3  Stem branched or unbranched, with 1 terminal cyme
   on each branch; leaves <1cm wide                                4
   4  Tepals bright yellow; stem unbranched
                                               3. S. californicum
   4  Tepals cream to pale yellow; stem branched
                                                      4. S. laxum
```

   1. S. bermudiana L. (<u>S. graminoides</u> Bickn., <u>S. hibernicum</u> A.   R
& D. Loeve) - <u>Blue-eyed-grass</u>. Stems to 50cm, usually
branched; leaves up to 5mm wide; tepals blue, 6-10mm.
Probably native; wet meadows and stony ground by lakes; very
local in W Ir from W Cork to W Donegal, known only since
1845. There is argument as to whether our plant is a native
endemic or an American introduction, and in latter case as to
its correct name.
   2. S. montanum E. Greene (<u>S. bermudiana</u> auct. non L.) -
<u>American Blue-eyed-grass</u>. Differs from <u>S. bermudiana</u> in

tepals 10-18mm; and see key. Intrd; natd in grassy places,
rough ground and waysides; scattered in Br N to Easterness; N
America. Much confused with S. bermudiana; identity of natd
plants in Ir needs checking.
  3. S. californicum (Ker Gawler) Dryander (S. boreale
(Bickn.) J.K. Henry) - Yellow-eyed-grass. Stems to 60cm,
unbranched; leaves up to 6mm wide; tepals bright yellow,
12-18mm. Intrd; grown in gardens, natd in damp grassy places
near sea; Mons, Pembs, Co Wexford, W Galway; W N America. S.
boreale may be a different sp., to which our plant belongs.
  4. S. laxum Otto ex Sims (S. iridifolium Kunth ssp.
valdiviense (Philippi) Ravenna) - Veined Yellow-eyed-grass.
Stems to 45cm, branched; leaves up to 10mm wide; tepals
whitish to pale yellow with purple veins, 12-15mm. Intrd;
garden plant natd on gravelly paths in Jersey; S America.
  5. S. striatum Smith - Pale Yellow-eyed-grass. Stems to
75cm, unbranched; leaves up to 20mm wide; tepals pale yellow,
15-18mm. Intrd; grown in gardens, natd (often short-lived)
on tips, waste ground, banks and waysides; scattered in S En
from Surrey to Scillies; S America.

3. ARISTEA Sol. ex Aiton - Blue Corn-lily
Plants with rhizomes; leaves Iris-like; inflorescence a loose
terminal panicle of few-flowered clusters; flowers actinomorphic; tepals united proximally into tube <5mm, blue, +
equal; filaments free, arising from top of perianth-tube;
style very slender, with 3-lobed stigma.

  1. A. ecklonii Baker - Blue Corn-lily. Stems to 60cm,
flattened, bearing reduced leaves; leaves up to 60 x 1.2cm,
linear; perianth 8-15mm. Intrd; natd in rough ground as
garden escape, Tresco, Scillies; S Africa. Often misdetermined as various other spp.

4. HERMODACTYLUS Miller - Snake's-head Iris
Plants with tuberous roots (no rhizomes, bulbs or corms);
leaves subterete, very long, 4-angled; flowers solitary,
terminal, actinomorphic, Iris-like; tepals united proximally
into tube; tepals, stamens and styles + like those of Iris.

  1. H. tuberosus (L.) Miller - Snake's-head Iris. Stems to
40cm; leaves longer than stems, up to 50cm x 3mm; outer
tepals yellowish-green; inner tepals yellowish-green on claw,
purplish-brown to blackish on blade. Intrd; garden plant
natd in grassy places and hedgerows; N Somerset, N Devon and
W Cornwall; Mediterranean.

5. IRIS L. - Irises
Plants with rhizomes or rarely bulbs; leaves Iris-like
(vertical, flat, with 2 identical faces) or subterete or
4-angled; inflorescence terminal, rather simple, cymose;
flowers actinomorphic; tepals united proximally into

perianth-tube; outer tepals usually longer and wider than the inner, patent, recurved or reflexed, with a narrow proximal part (claw) and expanded distal part (blade); inner tepals usually erect, less differentiated into blade and claw; filaments free, borne at base of outer tepals; style with 3 long, broad, petaloid branches each with 2 lobes at apex beyond stigma, each covering a stamen.

1 Leaves subterete to slightly flattened, angled or
  channelled; plant with a bulb   2
1 Leaves flat, not channelled or angled, vertical,
  with 2 identical faces; plant with rhizome   4
    2 Perianth-tube >10mm   **12. I. x hollandica**
    2 Perianth-tube <10mm   3
3 Leaves evergreen; claw of outer tepals <10mm wide,
  1.5-2x as long as blade   **11. I. xiphium**
3 Leaves dying down in winter; claw of outer tepals
  >20mm wide, no longer than blade   **10. I. latifolia**
    4 Outer tepals bearded, i.e. with mass of stout
      multicellular hairs on inner face   **1. I. germanica**
    4 Outer tepals not bearded, sometimes softly
      pubescent with unicellular hairs   5
5 Tepals predominantly yellow or yellow and white,
  without blue, purple, mauve or violet or only small
  spots or veining of it   6
5 Tepals predominantly of some shade of blue, purple,
  mauve or violet   8
    6 Leaves evergreen, dark green, with stinking smell
      when crushed; seeds bright orange; stems
      distinctly compressed   **9. I. foetidissima**
    6 Leaves dying in winter, mid- to pale-green, not
      stinking; seeds brownish; stems subterete   7
7 Inner tepals white; outer tepals white with large
  yellow patch on blade; petaloid style-lobes
  subentire   **8. I. orientalis**
7 Tepals yellow all over, the outer often with
  brownish or purple spots or veins; petaloid style-
  lobes deeply serrate   **3. I. pseudacorus**
    8 Leaves evergreen, dark green, with stinking smell
      when crushed; seeds bright orange **9. I. foetidissima**
    8 Leaves dying in winter, mid- to pale-green, not
      stinking; seeds brownish   9
9 Stems hollow; perianth-tube 4-7mm; bracts brown and
  papery at flowering   **2. I. sibirica**
9 Stems solid; perianth-tube 7-20mm; bracts at least
  partly green at flowering   10
    10 Upper part of ovary sterile, narrower than ovary
       below and perianth-tube above, forming
       acuminate beak on capsule >5mm   11
    10 Ovary without sterile apical part; capsule with
       0 or short beak <5mm   12
11 Capsule with beak 5-8mm, with 1 rib where 2 ovary-

cells meet; leaves mostly <10mm wide; flowers mostly
>8cm across   **6. 1. ensata**
11 Capsule with beak 8-16mm, with 2 ridges where 2
   ovary-cells meet; leaves mostly >10mm wide; flowers
   mostly <8cm across   **7. I. spuria**
   12 Outer tepals glabrous on central patch; capsules
      setting many seeds   **4. I. versicolor**
   12 Outer tepals pubescent on central patch; capsules
      setting 0-few seeds   **5. I. x robusta**

1. I. germanica L. - Bearded Iris. Rhizomatous; leaves
flat, (20)30-60mm wide; stems to 90cm, usually branched;
flowers 8-15cm across, usually blue to violet or purple with
yellow bearded region on outer tepals. Intrd; much grown in
gardens, natd on banks, rough and waste ground, waysides, old
planted areas; frequent in C & S Br and CI; garden origin.
The true I. germanica is only 1 of a group of hybrid origin,
for which no overall name exists; flower colour varies from
white or yellow to blue, violet or purple, with many
variations in size and shape.
2. I. sibirica L. - Siberian Iris. Rhizomatous; leaves
flat, (2)4-10mm wide; stems to 1.2m, usually branched;
flowers 6-7cm across, blue or bluish-violet with white patch
on outer tepals. Intrd; garden plant natd in rough, often
wet or shaded, ground; scattered throughout Br N to
Easterness; C Europe to S Asia.
3. I. pseudacorus L. - Yellow Iris. Rhizomatous; leaves
flat, 10-30mm wide; stems to 1.5m, usually branched; flowers
7-10cm across, yellow with deeper yellow patch on outer
tepals. Native; wet meadows, fens and ditches, by lakes and
rivers; common throughout BI.
4. I. versicolor L. - Purple Iris. Rhizomatous; leaves
flat, 8-25mm wide; stems to 1m, usually branched; flowers
6-8cm across, purple to violet with greenish-yellow patch
surrounded by whitish area on outer tepals. Intrd; natd by
lakes and rivers and in reed-swamps; scattered in En and N to
C Sc; E N America.
5. I. x robusta E.S. Anderson (I. versicolor x I. virginica
L.) - Windermere Iris. Differs from I. versicolor as in key
(couplet 12). Intrd; natd in reed-swamp and rough pasture by
Lake Windermere, Westmorland, without either parent; E N
America and garden origin.
6. I. ensata Thunb. (I. kaempferi Siebold ex Lemaire) -
Japanese Iris. Rhizomatous; leaves flat, 4-12mm wide; stems
to 90cm, usually unbranched; flowers 8-15cm across, purple
with yellow claws and base of blade of outer tepals. Intrd;
natd in swamp in W Kent; E Asia.
7. I. spuria L. - Blue Iris. Rhizomatous; leaves flat,
6-20mm wide; stems to 90cm, usually unbranched; flowers 6-8cm
across, bluish-violet, sometimes with small yellow or white
area at base of blade of outer tepals. Intrd; natd in Dorset
and N Somerset and by fen ditches in N Lincs since 1836;

Europe.
8. I. orientalis Miller (_I. ochroleuca_ L., _I. spuria_ ssp.
ochroleuca (L.) Dykes) - Turkish Iris. Rhizomatous; leaves
flat, 10-25mm wide; stems to 1.2m, usually little-branched;
flowers 8-10cm across, white with yellow patch on outer
tepals. Intrd; natd in limestone scrub in N Somerset since
at least 1950 and on banks in W Kent since 1984; E
Mediterranean.
9. I. foetidissima L. - Stinking Iris. Rhizomatous; leaves
flat, 10-25mm wide, evergreen; stems to 80cm, branched;
flowers 5-7cm across, dull purplish, rarely pale yellow.
Native; dry places in woods, hedges, banks and cliffs near
sea, mostly on calcareous soils; locally frequent in CI and
Br N to N Wa and Norfolk, natd elsewhere in Br and Ir.
10. I. latifolia (Miller) Voss (_I. xiphioides_ Ehrh.) -
English Iris. Bulbous; leaves channelled, whitish in
channel, 5-8mm wide; stems to 50cm, unbranched; flowers
8-12cm across, bluish-violet with yellow patch on outer
tepals. Intrd; grown in gardens, natd in grassy places in
Shetland and W Kent; Pyrenees.
11. I. xiphium L. - Spanish Iris. Differs from _I. latifolia_
in leaves often <5mm wide, not or scarcely whitish in
channel; perianth-tube 1-3mm (not 3-5mm); and see key
(couplet 3). Intrd; grown in gardens and bulb-fields, natd
as relic in old fields, rough ground and waste places in
Scillies; SW Europe.
12. I. x hollandica hort. (_I. filifolia_ Boiss. x _I._
tingitana Boiss. & Reuter) - Dutch Iris. Bulbous; leaves
0.5-3mm wide, channelled, evergreen; stems to 50cm,
unbranched; flowers 8-12cm across, white, yellow or blue to
purple. Intrd; grown and natd as for _I. xiphium_; frequent in
CI; garden origin.

6. WATSONIA Miller - Bugle-lily

Plants with corms; leaves Iris-like; inflorescence a spike;
flowers slightly zygomorphic, with curved perianth-tube;
tepals united into tube longer than the lobes, white, the
lobes ± equal; filaments free, borne in perianth-tube; style
very slender, with 3 bifid stigmas.

1. W. borbonica (Pourret) Goldblatt (_W. ardernei_ Sander) -
Bugle-lily. Stems to 1m; leaves up to 60 x 4cm; perianth-
tube 2.5-3.5cm; perianth-lobes 2.5-3.5cm. Intrd; grown in
gardens, natd in rough ground; Tresco, Scillies; S Africa.
Our plant is the white-flowered variant of ssp. **ardernei**
(Sander) Goldblatt.

7. ROMULEA Maratti - Sand Crocuses

Plants with corms; leaves subterete, 4-grooved; flowers
1-several on very short stem, Crocus-like, actinomorphic;
tepals united into tube much shorter than lobes, white or
mauve, the lobes equal; filaments free, borne in

perianth-tube; style very slender, with 3 bifid stigmas; ovary above ground at flowering.

1. R. columnae Sebast. & Mauri - <u>Sand Crocus</u>. Corm **RRR** obliquely narrowed at base; leaves 5-10cm x 0.6-1mm, recurved; perianth 7-15mm incl. tube 2.5-5.5mm, usually mauve, sometimes white, pale yellow inside at base. Native; maritime sandy turf; very local near Dawlish, S Devon, common in all isles of CI, formerly E Cornwall.

2. R. rosea (L.) Ecklon - <u>Onion-grass</u>. Corm rounded at base; leaves 15-25cm x 1-2.5mm, erect or ± so; perianth 15-45mm incl. tube 2.5-8mm, white with yellow inside at base. Intrd; natd at wall-base and in sparsely grassy area on gravel in Guernsey since at least 1969; S Africa. Our plant is var. **australis** (Ewart) De Vos.

8. CROCUS L. - <u>Crocuses</u>
Plants with corms; leaves linear, ± flattened, with central whitish channel; flowers erect, 1-few on short underground pedicels that elongate at fruiting, actinomorphic; tepals united into long narrow tube, various colours, the lobes equal; filaments free, borne at apex of perianth-tube; style slender, with 3 (or more) branches near apex, each with variously divided stigmas; ovary subterranean at flowering.
 Below the flower and sheathing the ovary and part of the perianth-tube are 1-2 bracts, borne immediately below the ovary, and 0-1 spathe, borne at the base of the pedicel.

1 Flowers appearing in Autumn (Sep-Dec), without
  leaves, never predominantly yellow   2
1 Flowers appearing in Spring (Jan-Apr), with or
  immediately before leaves, often predominantly
  yellow   5
  2 Filaments densely pubescent; throat uniformly
    deep yellow; anthers white to cream **9. C. pulchellus**
  2 Filaments glabrous to minutely pubescent; throat
    not yellow or very pale yellow or with yellow
    blotches only; anthers often yellow   3
3 Perianth not conspicuously veined; throat not yellow
                                          3. C. nudiflorus
3 Perianth conspicuously darker veined outside; throat
  often pale yellow or with yellow blotches   4
  4 Anthers white to cream; style with 3 main
    branches; corm with covering becoming fibrous,
    not splitting into rings at base **4. C. kotschyanus**
  4 Anthers yellow; style usually with many slender
    branches; corm with covering splitting into rings
    at base, not becoming fibrous   **8. C. speciosus**
5 Perianth predominantly pale to deep yellow, sometimes tinged or striped dark purple   6
5 Perianth predominantly white or pale mauve to dark
  purple, sometimes yellow on throat   7

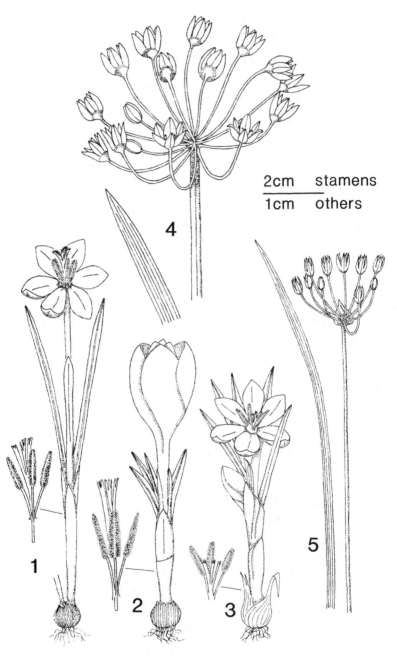

2cm stamens
1cm others

FIG 1144. - 1-3, Crocus. 1, C. tommasinianus. 2, C. vernus. 3, C. flavus. 4, Nectaroscordum siculum. 5, Nothoscordum gracile.

6 Perianth uniformly bright yellow; leaves 2.5–4mm
  wide; corm with covering splitting vertically,
  not horizontally   **7. C. flavus**
6 Perianth various shades of yellow or cream, often
  brownish– or purplish–streaked or tinged outside;
  leaves 0.5–2.5mm wide; corm with covering
  splitting horizontally into rings at base,
  scarcely vertically   **6. C. chrysanthus**
7 Perianth-lobes 2–3 x 0.7–1.1cm; throat yellow;
  spathe 0; bracts 2, white   **5. C. sieberi**
7 Perianth-lobes 2.4–5.5 x 0.8–2cm; throat white to
  mauve or purple; spathe 1, papery; bract 1, white   8
8 Leaves mostly 4–8mm wide; flowers white to deep
  purple, often with dark stripes outside; perianth-
  tube usually mauve to purple, white only if rest
  of perianth is white   **1. C. vernus**
8 Leaves mostly 2–3mm wide; flowers mauve to pale
  purple with white perianth-tube   **2. C. tommasinianus**

Other spp. – **C. sativus** L. (Saffron Crocus), of unknown
origin, was formerly grown for its styles (the spice) and was
natd in a few places; it flowers in Autumn with the leaves,
and has yellow anthers, a strongly striped perianth not
yellow inside at base, and 3 simple style–branches >1/2 as
long as perianth–lobes. **C. biflorus** Miller (Silvery Crocus),
from S Europe, was formerly natd (c.1830–1950) in E Suffolk;
it differs from C. chrysanthus in perianth very pale lilac
with very dark stripes outside, yellow throat and purple
tube.

1. **C. vernus** (L.) Hill (C. purpureus Weston) – Spring 1144
Crocus. Corm–covering fibrous; leaves mostly 4–8mm wide;
perianth white to deep purple, often strongly more darkly
striped outside, with tube not paler than lobes, with throat
white to purple, glabrous to pubescent. Intrd; the most
commonly grown sp., frequently natd in grassy places,
meadows, banks, churchyards; scattered in BI; S Europe.
2. **C. tommasinianus** Herbert – Early Crocus. Differs from C. 1144
vernus in slenderer flowers usually appearing earlier in year
and with white throat; and see key. Intrd; grown in gardens,
natd as for C. vernus; scattered in Br N to MW Yorks; SE
Europe.
3. **C. nudiflorus** Smith – Autumn Crocus. Corm–covering
becoming fibrous; leaves mostly 2–4mm wide; perianth purple
to pale mauve, not striped outside, white or pale mauve on
throat and tube; throat usually glabrous. Intrd; fields,
parks, grassy banks, the most thoroughly natd sp.; scattered
in En and Wa, especially NW En; SW Europe.
4. **C. kotschyanus** K. Koch – Kotschy's Crocus. Corm–covering
fibrous; leaves mostly 1.5–4mm wide; perianth pale mauve to
purple with conspicuous dark veining outside, with pubescent,
whitish throat with yellow blotches, with white or nearly

white tube. Intrd; natd in meadows and on grassy tracksides;
Surrey and E & W Suffolk since 1981; Turkey.
5. C. sieberi Gay - Sieber's Crocus. Corm-covering fibrous;
leaves mostly 1.5-2mm wide; perianth white to pale mauve,
with tube paler or darker, with yellow, glabrous throat.
Intrd; natd by path on common in Surrey; Balkans.
6. C. chrysanthus (Herbert) Herbert - Golden Crocus. Corm-
covering not fibrous, splitting horizontally at base; leaves
mostly 0.5-2.5mm wide; perianth deep yellow, usually with
dark streaks outside, with darker or bronzy or purplish tube,
with yellow, glabrous throat. Intrd; natd on grassy path-
sides in Surrey; Balkans and Turkey.
7. C. flavus Weston (C. aureus Smith) - Yellow Crocus. 1144
Corm-covering becoming fibrous; leaves mostly 2.5-4mm wide;
perianth bright yellow ± uniformly or tube brownish; throat
glabrous or pubescent. Intrd; much grown in gardens, natd as
for C. vernus; scattered in En; SE Europe. Many garden, and
perhaps most or all natd, plants are said to be C. flavus x
C. angustifolius Weston.
8. C. speciosus M. Bieb. - Bieberstein's Crocus. Corm-
covering not fibrous, splitting horizontally at base; leaves
mostly 3-5mm wide; perianth pale mauve to purple with
conspicuous dark veining outside, with white or nearly white
tube; throat whitish, glabrous. Intrd; natd in churchyards
and on waysides in Surrey and E Suffolk; SW Asia.
9. C. pulchellus Herbert - Hairy Crocus. Differs from C.
speciosus in throat yellow, glabrous or slightly pubescent;
anthers white to cream (not yellow); filaments pubescent (not
glabrous). Intrd; natd in churchyard in W Suffolk since
1983; Balkans and W Turkey.

9. GLADIOLUS L. - Gladioluses
Plants with corms; leaves Iris-like; inflorescence a spike;
flowers zygomorphic, with curved perianth-tube; tepals united
into tube shorter than lobes, pinkish- to purplish-red, the
lobes unequal, much narrowed at base; filaments free, borne
on perianth-tube; style slender, with 3 short terminal lobes
greatly widened distally.

Other spp. - Various garden and florists' Gladioluses rarely
occur on tips, but do not persist. **G. italicus** Miller (G.
segetum Ker Gawler), from S Europe, used to be natd in parts
of S En, but modern records seem to be errors for G. communis
from which it differs in the anthers longer than the
filaments and the unwinged (not winged) seeds.

1. G. illyricus Koch - Wild Gladiolus. Stems to 50(90)cm, **RRR**
unbranched; leaves up to 40 x 1cm; flowers 3-8(10); perianth
3.5-5cm, the lobes 2.5-4 x 0.6-1.6cm. Native; among bracken
in scrub in New Forest, S Hants, formerly Wight; records of
natd plants elsewhere may be other spp.
2. G. communis L. (G. byzantinus Miller) - Eastern

Gladiolus. Stems to 1m, often branched; leaves up to 70 x
2.5cm; flowers mostly 10–20; perianth 4–5.5cm, the lobes
3–4.5 x 1.5–2.5cm. Intrd; persistent relic of cultivation in
old bulb-fields, field-margins, roadsides and rough ground;
scattered in extreme S En, frequent in CI and Scillies;
Mediterranean. Our plant is ssp. **byzantinus** (Miller) A.P.
Ham.

10. IXIA L. - Corn-lilies
Plants with corms; leaves Iris-like; inflorescence a spike or
a raceme of spikes; flowers actinomorphic; tepals united
proximally into short or long tube, variously white or yellow
to red, + equal; filaments free, arising from top of or
within perianth-tube; style very slender, with 3-lobed
stigma.

1. I. campanulata Houtt. (I. speciosa Andrews) - Red Corn-
lily. Stems to 15(40)cm; leaves up to 10(25) x 0.5(1.5)cm;
perianth-tube 2–3mm; perianth-lobes 1.2–2.5cm, mainly red but
with white and/or yellow stripes. Intrd; grown in gardens
and bulb-fields, natd in old fields or rough ground;
Scillies; S Africa.
2. I. paniculata Delaroche - Tubular Corn-lily. Stems to
1m; leaves up to 60 x 1.2cm; perianth-tube 3–7cm, very
slender; perianth-lobes 1.2–2.5cm, cream to pale yellow
tinged with red. Intrd; grown and natd as for I.
campanulata; Scillies; S Africa.

11. SPARAXIS Ker Gawler - Harlequinflowers
Plants with corms; leaves Iris-like; inflorescence a spike;
flowers + actinomorphic, with brown, jaggedly-toothed, dark-
streaked bracts; tepals united into straight tube shorter
than lobes, mostly red and white, the lobes + equal;
filaments free, borne in perianth-tube, slightly asymmetrically arranged; style very slender, with 3 linear branches.

1. S. grandiflora (Delaroche) Ker Gawler - Plain Harlequin-
flower. Stems to 45cm; leaves up to 30 x 1.3cm; perianth-
tube 0.8–1.4cm; perianth-lobes 2–3cm, red or red-and-white
striped, often yellow near base. Intrd; grown in bulb-
fields, natd in and by old fields; Scillies; S Africa.

12. FREESIA Ecklon ex Klatt - Freesia
Plants with corms; leaves Iris-like; inflorescence a spike;
flowers slightly zygomorphic, with curved perianth-tube;
tepals united into tube longer than lobes, white, yellow,
orange, pink, purple or mauve, the lobes + equal; filaments
free, borne in perianth-tube; style very slender, with 3
bifid stigmas.

1. F. x hybrida L. Bailey (F. refracta auct. non (Jacq.)
Ecklon ex Klatt) - Freesia. Stems to 40cm; leaves up to 30 x

1cm; perianth-tube 1.5-3cm; perianth-lobes 0.8-1.5cm. Intrd;
grown in bulb-fields, natd in and by old fields; Guernsey; S
Africa.

13. CROCOSMIA Planchon (<u>Curtonus</u> N.E. Br.) - <u>Montbretias</u>
Plants with corms that produce stolons; leaves <u>Iris</u>-like;
inflorescence an often branched spike; flowers zygomorphic,
with curved perianth-tube; tepals united into tube longer or
shorter than lobes, orange to brick-red, the lobes rather
unequal; filments free, borne asymmetrically in perianth-
tube; style slender, with 3 short branches.

1 Leaves ribbed and pleated at least when young, at
  least some >3cm wide; perianth >4.5cm   2
1 Leaves ribbed but not pleated, <3cm wide; perianth
  <4(5)cm   3
  2 Perianth-lobes c.1/2 as long as -tube, erecto-
    patent; stamens shorter than perianth
                                            1. C. paniculata
  2 Perianth-lobes c. as long as -tube, widely
    spreading; stamens slightly longer than perianth
                                            2. C. masoniorum
3 Perianth-lobes c.1/2 as long as -tube, + erect;
  perianth-tube very narrow at base, abruptly widened
  distally   **3. C. pottsii**
3 Perianth-lobes c. as long as -tube, + patent;
  perianth-tube gradually expanded distally
                                        4. C. x crocosmiiflora

  1. C. paniculata (Klatt) Goldblatt (<u>Curtonus</u> <u>paniculatus</u> 1100
(Klatt) N.E. Br.) - <u>Aunt-Eliza</u>. Stems to 1.2m; leaves up to
90 x 8cm; perianth 4.5-6cm, with tube narrow at base then
rather abruptly expanded, with erecto-patent lobes 1-2cm.
Intrd; grown in gardens, natd in marginal habitats, rough and
waste ground; local in W Sc, W Ir and S Br; S Africa.
  2. C. masoniorum (L. Bolus) N.E. Br. - <u>Giant</u> Montbretia. 1100
Differs from <u>C.</u> paniculata in perianth 4.5-6cm, with tube
rather gradually widened, with spreading lobes 2.4-3cm; and
see key. Intrd; grown and natd as for <u>C.</u> paniculata; 3 sites
in Dunbarton; S Africa.
  3. C. pottsii (Macnab ex Baker) N.E. Br. - <u>Potts'</u> 1100
Montbretia. Stems to 80cm; leaves up to 80 x 2.5cm; perianth
2-3cm, with tube narrow at base then abruptly expanded, with
+ erect lobes 0.5-1cm; stamens slightly longer than perianth-
tube. Intrd; rarely grown in gardens but natd by roads and
rivers in Kirkcudbrights, Wigtowns & Clyde Is; S Africa.
  4. C. x crocosmiiflora (Lemoine ex Burb. & Dean) N.E. Br. 1100
(<u>Tritonia</u> x <u>crocosmiiflora</u> (Lemoine ex Burb. & Dean)
Nicholson; <u>C.</u> <u>pottsii</u> x <u>C.</u> <u>aurea</u> (Hook.) Planchon) -
Montbretia. Differs from <u>C.</u> pottsii in stems to 60cm;
perianth 2.5-4(5)cm, with tube gradually widened distally,
with spreading lobes 1.2-2.6cm; stamens nearly as long as

perianth. Intrd; much grown in gardens, very well natd in hedgerows, woods, by lakes and rivers, and on waste ground; scattered throughout BI, common in Ir, W Br and CI.

14. CHASMANTHE N.E. Br. - <u>Chasmanthe</u>
Plants with corms; leaves <u>Iris</u>-like; inflorescence a spike; flowers strongly zygomorphic, with curved perianth-tube; tepals united, variously red to orange, the lobes extremely unequal, the uppermost at least as long as tube and continuing its curvature, the 2 adjacent parallel to it but shorter, the 3 lower much shorter and slightly down-turned; filaments free, borne on perianth-tube; stamens conspicuously exserted; style long-exserted, very narrow, with 3 terminal thin lobes.

Other spp. - The reported presence of **C. aethiopica** (L.) N.E. Br. natd in the Scillies needs investigating; it differs in having leaves to 2.5cm wide, lateral perianth-lobes (7)10-15mm and patent or recurved (not 5-8mm and suberect), and style-branches 4-5mm (not 3-3.5mm).

1. C. bicolor (Gasp. ex Ten.) N.E. Br. - <u>Chasmanthe</u>. Stems to 1.3m; leaves to 80 x 3.5cm, with strong midrib; perianth-tube 3-3.5cm, yellow on lowerside, orange-red on upperside; uppermost perianth-lobe orange-red, 2-3.5cm. Intrd; grown for ornament, natd in damp shady places nearby; Tresco, Scillies; S Africa.

162. AGAVACEAE - <u>Centuryplant family</u>

Perennials with thick, woody, sparsely branched stems with leaves tufted at branch-ends, or with stemless giant rosette of leaves; leaves in rosettes, tough, often succulent, simple, sessile, entire or distantly toothed, with ± sheathing base, exstipulate. Flowers in large terminal panicles, bisexual, hypogynous or epigynous, actinomorphic or slightly zygomorphic; perianth of 6 lobes fused into tube proximally; stamens 6, borne on perianth-tube; ovary 3-celled, each cell with numerous ovules on axile placentas; style 1, short and thick to long and slender, with 3-lobed stigma; fruit a dehiscent or indehiscent capsule or a berry.
Flower structure as in Liliaceae, but the huge rosettes of tough leaves, on the ground or at the ends of branches, separate Agavaceae at a glance.

1 Leaf-rosettes sessile on ground or ± so; perianth
  greenish- or brownish-yellow   2
1 Leaf-rosettes at ends of woody branches; perianth
  whitish   3
  2 Leaves with extremely strong spines at margins
    and apex   **2. AGAVE**

2 Leaves spineless   **4. PHORMIUM**
3 Leaves spine-tipped, strongly recurved; perianth
   >4cm   **1. YUCCA**
3 Leaves often sharply pointed but not spine-tipped,
   not recurved; perianth <1cm   **3. CORDYLINE**

1. YUCCA L. - Spanish-daggers
Stems usually branched, with leaf-rosettes at ends of
branches; leaves with sharp spine at apex, entire or with few
inconspicuous teeth; flowers hypogynous; perianth actinomorphic, campanulate, with lobes much longer than tube; fruit
an indehiscent capsule.

1. Y. recurvifolia Salisb. - Curved-leaved Spanish-dagger.
Plant to 2(5)m; leaves up to 100 x 5cm, most strongly
recurved; perianth creamy- or greenish-white, 5-8cm. Intrd;
natd on sand-dunes (presumably where planted) in Glam since
1982; SE USA.

2. AGAVE L. - Centuryplant
Stems + 0, the leaf-rosette sessile on ground; leaves with
entremely sharp spine at apex and many more along margins;
flowers epigynous; perianth + actinomorphic, tubular, with
tube much longer than lobes; fruit a dehiscent capsule.

1. A. americana L. - Centuryplant. Rosettes mostly 2-3m
across, with massive succulent, tough very spiny leaves 1-2m
x 15-30cm; flowering stem rarely produced, to 7(12)m;
perianth greenish-yellow, 7-10cm. Intrd; very persistent
where planted, surviving from suckers when main rosette dies
after the single, rare flowering; CI; Mexico.

3. CORDYLINE Comm. ex Adr. Juss. - Cabbage-palm
Stems well-developed, simple or branched, with leaf-rosettes
at ends of branches; leaves entire, sharply pointed but
without a spine; flowers hypogynous; perianth actinomorphic,
with short tube and wide-spreading much longer lobes; fruit
a berry, becoming dry with age.

1. C. australis (G. Forster) Endl. - Cabbage-palm. Stems to
20m, becoming branched after first flowering; leaves up to
100 x 6cm; perianth white, 5-6mm, c.1cm across. Intrd; much
planted in W Br, Ir and CI, persistent in SW En, S Ir and CI,
producing seedlings in CI; New Zealand.

4. PHORMIUM Forster & G. Forster - New Zealand Flaxes
Stems + 0, the leaf-rosette sessile on ground; leaves entire,
not spiny, folded proximally, nearly flat distally; flowers
hypogynous; perianth slightly zygomorphic, with short tube
and longer lobes, + tubular; fruit a dehiscent capsule.

1. P. tenax Forster & G. Forster - New Zealand Flax.

Rosettes 1-2m across, with suberect, extremely tough, fibrous leaves up to 3m x 12cm: flowering stem to 4m; perianth 3-5cm, with outer lobes brownish-red, with inner lobes greenish-yellow with not or slightly recurved tips. Intrd: very persistent where planted on cliffs or rocky places by sea; natd in W Cornwall, Scillies, W Cork, Man and CI, self-sown mainly in Scillies; New Zealand.

2. P. cookianum Le Jolis (P. colensoi Hook. f.) - Lesser New Zealand Flax. Differs from P. tenax in smaller size; leaves up to 2m x 7cm; flowering stem to 2m; perianth 2.5-4cm, with greenish-yellow lobes tinged with red, >1 of the inner lobes strongly recurved at tip. Intrd; planted and natd as for P. tenax; Scillies, often self-sown; New Zealand.

163. DIOSCOREACEAE - Black Bryony family

Glabrous, twining, herbaceous perennials with subterranean tuber; leaves alternate, simple, entire, petiolate, stipulate. Flowers in axillary, simple or branched racemes, dioecious, inconspicuous, epigynous, actinomorphic; perianth of 6 tepals united only at base, sepaloid, all ± similar; stamens 6, vestigial in female flowers; ovary 3-celled, each cell with 2 ovules on axile placentas; style 1; stigmas 3, recurved, each 2-lobed, vestigial in male flowers; fruit a berry with <6 seeds.

The only herbaceous twiner with dioecious inconspicuous flowers and red berries except Bryonia (Cucurbitaceae), which has pubescent, palmately lobed leaves and stem-tendrils.

1. TAMUS L. - Black Bryony

1. T. communis L. - Black Bryony. Stems to 5m; leaves c.5-15 x 4-11cm, broadly ovate, strongly cordate; perianth yellowish-green, 3-6mm across; berry 10-13mm across, bright red. Native; scrambling over hedges, shrubs and wood-margins; local in CI, common in Br N to Cumberland and Durham, rarely intrd further N and in Man and Ir.

164. ORCHIDACEAE - Orchid family
(Cypripediaceae)

Erect, herbaceous perennials, sometimes ± chlorophyll-less saprophytes, with fleshy roots, subterranean tubers or rhizomes; leaves alternate, sometimes mostly basal, those on stem often reduced, simple, entire, sessile, usually clasping stem, exstipulate. Flowers in terminal raceme or spike with bract to each flower (rarely only 1 flower), very rarely single and terminal, bisexual, usually conspicuous, epigynous, zygomorphic; perianth of 6 free tepals in 2 whorls of 3, usually all petaloid (though sometimes greenish or brownish), the 3 outer ones ('sepals') similar, 2 of the

inner ones ('petals') similar and often ± similar to 3 outer,
the other (usually apparently the lowest but actually the
uppermost due to twisting of the flower through 180 degrees)
usually strongly different and forming a lip (<u>labellum</u>) which
is usually the largest and most conspicuous part of the
flower, and often is extended from its base behind the flower
into a hollow spur; stamens and stigmas borne on a special
structure (<u>column</u>) in the centre of the flower; stamens 2
(<u>Cypripedium</u>) or 1 (others), each with sessile anther
containing pollen as many single grains (<u>Cypripedium</u>) or as
2-4 masses (<u>pollinia</u>) (others); pollinia often stalked, often
provided with a sticky pad (<u>viscidium</u>) (at base of stalk when
present); ovary 1-celled, with extremely numerous ovules on 3
parietal placentae; style 0; stigmas 3, either all receptive
(<u>Cypripedium</u>) or 2 receptive and the third a sterile
protrusion (<u>rostellum</u>) or ±0 (others); fruit a capsule.

Many of the spp. are prone to produce plants with unusual
labellum-shapes or flower-colours (e.g. albinos), but these
usually occur in populations of normal plants.

The distinctive flowers could be confusd only by the very
inexperienced with a few petaloid monocotyledons or
dicotyledons (e.g. <u>Impatiens</u>, <u>Orobanche</u>, <u>Pinguicula</u>); the
inferior ovary and 1(-2) stamens on a column will dispel the
confusion.

Hybrids occur frequently within the tribe Orchideae,
especially within <u>Dactylorhiza</u>, and should be looked for
whenever ≥2 spp. of the tribe occur close together. Hybrids
between 2 spp. with the same chromosome number are usually
fertile, and even those between spp. with different
chromosome numbers are not always completely sterile. 7
intergeneric hybrid combinations also occur, 6 of them
between genera 13-18. Other intergeneric hybrid combinations
have also been recorded from BI, several of them (e.g.
<u>Dactylorhiza</u> x <u>Platanthera</u>) known from the Continent, but are
of uncertain or erroneous identity.

```
1  Plants saprophytic, without green leaves              2
1  Plants holophytic, with green leaves                  4
   2  Flowers not twisted upside down, hence labellum
      and spur directed ± upwards; spur >5mm   4. EPIPOGIUM
   2  Flowers twisted upside down (as normal in orchids),
      hence labellum directed downwards; spur 0          3
3  Labellum 8-12mm, c.2x as long as other tepals,
   brown; flowers usually >20                    5. NEOTTIA
3  Labellum ≤6mm, c. as long as other tepals, whitish-
   cream with reddish markings; flowers ≤12
                                      11. CORALLORRHIZA
   4  Spur present, sometimes very short                 5
   4  Spur 0                                            14
5  Labellum with 3 lobes, the central 3-6cm, linear,
   ribbon-like, in a loose spiral        22. HIMANTOGLOSSUM
5  Labellum without a central ribbon-like lobe >3cm      6
```

6 Spur <3mm   7
6 Spur >3mm   10
7 Flowers greenish-brown; labellum parallel-sided, with
   3 short apical lobes   **17. COELOGLOSSUM**
7 Flowers white, cream or greenish-white, sometimes
   with red or pink markings; labellum not parallel-
   sided, with 1 or more laterally protruding lobes on
   either side   8
8 Central lobe of labellum entire; flowers never
   with pink or red tinge or markings   **15. PSEUDORCHIS**
8 Central lobe of labellum conspicuously 2-3-lobed
   at apex; flowers often with pink or red tinge
   or markings   9
9 Flowers white to pinkish; labellum <5mm, scarcely
   longer than outer tepals; leaves usually spotted;
   Ir and Man only   **19. NEOTINEA**
9 Flowers white and dark purple, the latter pre-
   dominating in unopened flowers hence at top of
   spike; labellum >5mm, c.2x as long as outer tepals;
   leaves not spotted; Br only   **20. ORCHIS**
10 Labellum linear, entire; flowers always white
   (or green-tinged)   **13. PLATANTHERA**
10 Labellum not linear, lobed (if scarcely so then
   very wide); flowers pure white only in rare
   albinos   11
11 Spur >(8)11mm, filiform, >6x as long as widest point   12
11 Spur <11mm, not filiform, <6x as long as widest
   point   .   13
12 Labellum plane, not raised into plates; spike +
   cylindrical; flowers strongly scented; each
   pollinium becoming detached separately, each with
   a small stalk and basal sticky pad   **16. GYMNADENIA**
12 Labellum with 2 raised plates near its base;
   spike pyramidal; flowers not scented; each
   pollinium with its own stalk but becoming detached
   together on a common sticky pad   **14. ANACAMPTIS**
13 Lower bracts herbaceous, green, often suffused
   purplish; labellum terminating in a single pointed
   to rounded tooth or lobe much smaller than or
   rarely nearly as large as the portions on either
   side   **18. DACTYLORHIZA**
13 Bracts membranous, brown, often suffused purplish;
   labellum terminating in a forked lobe (often with a
   tooth in the notch) larger than lobes on either side,
   or in a small truncate or notched lobe smaller than
   portions on either side   **20. ORCHIS**
14 Labellum yellow, c.3cm, concavely bowl-shaped;
   other tepals maroon, 3-5cm   **1. CYPRIPEDIUM**
14 Labellum not concavely bowl-shaped; other tepals
   <2cm   15
15 Labellum velvety in texture, resembling insect's
   abdomen   **23. OPHRYS**

15 Labellum not velvety, not resembling insect's
  abdomen   16
  16 At least some flowers not twisted upside down,
     hence labellum directed ± upwards   17
  16 Flowers all twisted upside down, hence labellum
     directed downwards   18
17 Labellum entire and flat at margin; leaves <2cm,
  with minute tubercles near apex   **10. HAMMARBYA**
17 Labellum crenate or crisped at margin; leaves >2cm,
  without tubercles   **9. LIPARIS**
  18 Labellum with 2 lobes at its apex exceeding
     all others   19
  18 Labellum with 1 lobe at its apex exceeding all
     others, sometimes this slightly notched   20
19 Leaves 2, on stem; labellum with 0 or 2 small
  lateral lobes much shorter than 2 apical lobes
                                             6. LISTERA
19 Leaves usually >2, the largest ones basal; labellum
  with 2 lateral lobes longer than 2 apical lobes
                                            21. ACERAS
  20 Flowers white to greenish- or yellowish-white,
     in 1-3 distinct spirals in spike, or if forming
     a strictly 1-sided spike then leaves all in basal
     rosette   21
  20 Flowers white or whitish only in rare albinos,
     not in distinct spirals, if forming a strictly
     1-sided spike then leaves mainly on stems   22
21 Shortly rhizomatous; leaves conspicuously net-veined,
  in basal rosettes; labellum not frilly at edge
                                            8. GOODYERA
21 Tufted, with swollen roots; leaves not or
  inconspicuously net-veined, if in basal rosette then
  labellum frilly at edge   **7. SPIRANTHES**
  22 Main leaves 2(-3), basal; labellum with apical
     lobe and 2 shorter laterals, not constricted
                                          12. HERMINIUM
  22 Main leaves (2)3-many, on stems (sometimes near
     base); labellum constricted at base of apical
     lobe, the proximal part often with 2 lateral
     lobes   23
23 Apical lobe of labellum pendent, flat, entire
                                             (SERAPIAS)
23 Apical lobe of labellum variously held, variously
  crenate, frilly, undulate or toothed   24
  24 Flowers erect to erecto-patent, sessile; proximal
     part of labellum partly wrapped round column
                                        2. CEPHALANTHERA
  24 Flowers patent to pendent, stalked; proximal part
     of labellum not wrapped round column   **3. EPIPACTIS**

Other genera - **SERAPIAS** L. (Tongue-orchids) can be
identified in the above key. 2 plants of **S. parviflora**

Parl., from Mediterranean, were found in E Cornwall in 1989, probably the result of deliberate introduction; stems to 30cm; leaves linear-elliptic; labellum 15-20mm, dark brownish-red.

TRIBE 1 - CYPRIPEDIEAE (genus 1). Stamens 2 plus a large sterile projection (staminode); pollen dispersed as separate grains; receptive stigmas 3; labellum a deeply concave bowl; spur 0; flower single (very rarely 2), terminal.

1. CYPRIPEDIUM L. - Lady's-slipper
1. C. calceolus L. - Lady's-slipper. Stems to 30cm, rather **RRR** pubescent; leaves 3-4, all on stem, ovate, pubescent; labellum mainly yellow, c.3cm; other tepals maroon, 3-5cm, patent. Native; north-facing grassy slope on limestone; 1 locality in MW Yorks, formerly widespread on limestone in N En.

TRIBE 2 - NEOTTIEAE (genea 2-8). Fertile stamen 1; pollen dispersed as 2 often rather friable (or 2 each with 2 halves) pollinia; receptive stigmas 2; third stigma 0 or a sterile bulge (rostellum); stamen borne at back of column; pollinia sessile or with an apical stalk, with or without a sticky pad (pollinia with basal stalk and sticky pad in Epipogium); labellum of various shapes, often without well-marked lobes, often constricted near middle so delimiting proximal and distal parts; spur 0 (present in Epipogium).

2. CEPHALANTHERA Rich. - Helleborines
Shortly rhizomatous; leaves several, all on stem; flowers sessile or ± so, borne spirally, white or purplish-pink; spur 0; labellum constricted c.1/2 way into proximal and distal parts, neither markedly lobed, the proximal part partly wrapped round column; rostellum 0; pollinia without stalks.

1 Flowers purplish-pink; ovaries with glandular hairs; labellum acute **3. C. rubra**
1 Flowers white with yellow or orange marks on labellum; ovaries glabrous; labellum obtuse **2**
　2 Lower leaves ovate to rather narrowly so; bracts longer than ovaries; sepals obtuse **1. C. damasonium**
　2 Lower leaves lanceolate to narrowly elliptic-oblong; bracts shorter than ovaries; sepals acute **2. C. longifolia**

1. C. damasonium (Miller) Druce - White Helleborine. Stems to 60cm; lower leaves ovate to rather narrowly so; bracts longer than ovaries; flowers 3-11(16), white, 15-25mm; sepals obtuse. Native; shady woods, commonly of Fagus with little ground cover, on chalk and limestone; locally frequent in S En N to Northants and Herefs.
1 x 2. C. damasonium x C. longifolia = C. x schulzei Camus,

Bergon & A. Camus was found in 1974 and 1975 in S Hants in woodland with both parents; it is intermediate in leaf and flower characters.

2. C. longifolia (L.) Fritsch – Narrow-leaved Helleborine. **R** Stems to 60cm; lower leaves lanceolate to narrowly elliptic-oblong; bracts shorter than ovaries; flowers 3-15(20), white, 10-16mm; sepals acute. Native; woods and shady places on calcareous soils; much less common but much more widespread than C. damasonium, scattered in Br and Ir N to W Sutherland, decreasing.

3. C. rubra (L.) Rich. – Red Helleborine. Stems to 60cm; **RRR** lower leaves lanceolate to narrowly elliptic-oblong; bracts longer than ovaries; flowers 3-8(15), purplish-pink, 15-25mm; sepals acute. Native; Fagus woods on chalk or limestone, rarely with other vegetation; very rare in N Hants, Bucks and E Gloucs, formerly elsewhere in S En.

3. EPIPACTIS Zinn – Helleborines
Rhizomatous, mostly shortly so; leaves several, all on stem; flowers distinctly pedicellate, borne spirally or ± on 1 side of stem, various dull colours; spur 0; labellum usually differentiated c.1/2 way into proximal and distal parts, neither markedly lobed, the proximal part ± cup-shaped and not wrapped round column; rostellum obvious and secreting a white sticky cap (viscidium), or minute and with 0 or vestigial viscidium; pollinia without stalks.
E. phyllanthes, E. leptochila and E. youngiana form a problematical complex of self-pollinated plants in which sp. limits are uncertain and disputed. They can be distinguished from the other spp. by having a rostellum which secretes little or no viscidium that usually withers soon after the flower opens. In the cross-pollinated spp. the rostellum is obvious and the viscidium remains in the open flower until removed along with the pollinia by visiting insects. This can be effected with a match-stick or similar object; when the viscidium is touched and the object pulled away the pollinia are drawn out with it. In the self-pollinated spp. the pollinia crumble apart and cannot easily be pulled out whole.

1 Rhizome long; labellum strongly constricted separating
  proximal and distal portions, the proximal with
  erect triangular lobe on each side   **1. E. palustris**
1 Rhizome short or ± 0; labellum not or slightly
  constricted between proximal and distal portions,
  the proximal with 0 or obscure lateral lobes   2
  2 Inflorescence-axis glabrous or nearly so; flowers
    pendent as soon as they open; leaves often shorter
    than internodes   **7. E. phyllanthes**
  2 Inflorescence-axis pubescent to densely so; at
    least younger flowers usually patent to erecto-
    patent; leaves usually longer than internodes   3

3 Ovary pubescent to densely so; perianth usually
   reddish-purple all over   **2. E. atrorubens**
3 Ovary glabrous to sparsely pubescent; perianth
   usually greenish often marked or tinged with pink,
   purple or violet, but not so coloured all over   4
4 Upper leaves usually spirally arranged; rostellum
   secreting obvious, white, persistent viscidium;
   pollinia becoming detached as integral units   5
4 Upper leaves usually obviously 2-ranked; rostellum
   without or with sparse, soon disappearing
   viscidium; pollinia crumbling apart   6
5 Leaves dark green, the lowest wider than long or
   almost so; distal part of labellum wider than long,
   with 2 usually rough brownish bosses near base
                                                   4. E. helleborine
5 Leaves greyish-green, often tinged violet, the
   lowest considerably longer than wide; distal part of
   labellum at least as long as wide, with 2 smoothly
   pleated pinkish bosses near base   **3. E. purpurata**
6 Rostellum >1/2 as long as anthers; stigma with
   2 basal bosses, with the rostellum forming a 3-
   horned shape; ovary usually glabrous; petals
   pinkish; distal part of labellum wider than
   long   **5. E. youngiana**
6 Rostellum <1/2 as long as anthers; stigma
   without marked basal bosses hence not 3-horned;
   ovary usually pubescent; petals pale green;
   distal part of labellum longer than wide or
   wider than long   **6. E. leptochila**

1. **E. palustris** (L.) Crantz - <u>Marsh</u> Helleborine. Stems to
45(60)cm, pubescent; leaves spiral, mostly >2x as long as
wide; perianth predominantly white, with red and yellow
markings; viscidium well developed; ovary pubescent. Native;
fens, base-rich marshy fields, dune-slacks; locally frequent
in BI N to C Sc, extinct in many inland sites.
2. **E. atrorubens** (Hoffm.) Besser - <u>Dark-red</u> Helleborine. **R**
Stems to 30(60)cm, densely whitish-pubescent; leaves
2-ranked, mostly >2x as long as wide; perianth usually
reddish-purple; viscidium well developed; ovary pubescent.
Native; limestone scrub, grassland, scree and rocky places;
very locally frequent in CW Ir, N Sc, N Wa, N En S to Derbys,
formerly Brecs.
2 x 4. **E. atrorubens** x **E. helleborine** = **E. x schmalhausenii**
K. Richter has been reported from N Wa, N En and N Sc with
both parents, but due to its fertility is very difficult to
determine certainly; the most convincing specimens are from
Arnside Knott, W Lancs.
3. **E. purpurata** Smith - <u>Violet</u> Helleborine. Stems to
60(80)cm, often densely clumped, pubescent above; leaves
spiral, mostly >2x as long as wide; perianth predominantly
green with white, pink-tinged labellum; viscidium well

developed; ovary shortly pubescent. Native; woods (often
dense) on calcareous or sandy soil; frequent in SE & SC En N
to Salop and Leics.
3 x 4. **E. purpurata** x **E. helleborine** = **E. x schulzei** P.
Fourn. has been reported frequently from within the area of
E. purpurata, but is difficult to determine certainly due to
its fertility.
4. E. helleborine (L.) Crantz - Broad-leaved Helleborine.
Stems to 80(100)cm, pubescent above; upper leaves spiral, <2x
as long as wide, the lowest usually wider than long; perianth
predominantly green, tinged pink; viscidium well developed;
ovary usually sparsely pubescent. Native; woods, scrub and
hedgerows; frequent in Br and Ir except rare in N Sc.
5. E. youngiana A. Richards & A. Porter - Young's **RRR**
Helleborine. Stems to 60cm, pubescent above; upper leaves ±
2-ranked, distinctly longer than wide; perianth predominantly
green and pink; rostellum well developed; ovary ± glabrous.
Native; woodland on heavy, often heavy-metal-polluted, soils;
S Northumb and Lanarks; endemic. Described in 1982 but of
uncertain origin; perhaps a hybrid derivative of E.
helleborine and E. phyllanthes.
6. E. leptochila (Godfery) Godfery (E. cleistogama C.   **R**
Thomas, E. dunensis (Stephenson & T.A. Stephenson) Godfery,
E. muelleri Godfery) - Narrow-lipped Helleborine. Stems to
60cm, pubescent at least above; leaves 2-ranked, mostly >2x
as long as wide; perianth predominantly yellowish-green, the
labellum mottled or tinged with red or pink; viscidium minute
or 0; ovary pubescent. Native; woods mostly on calcareous or
heavy-metal-polluted soils, river-gravels and dunes; locally
scattered in Br N to Lanarks. Var. **dunensis** Stephenson & T.
A. Stephenson occurs on dunes in C & N Br but intermediate
plants occur inland in N En; it differs from var. **leptochila**
in distal part of the labellum wider than long and reflexed
(not longer than wide and non-reflexed). **E. cleistogama** has
green ± cleistogamous flowers; it occurs in 1 area of W
Gloucs and is probably best considered another var. of E.
leptochila. **E. muelleri** has been recorded in E Sussex but
seems scarcely different from var. dunensis.
7. E. phyllanthes G.E. Smith (E. cambrensis C. Thomas, E.   **R**
vectensis (Stephenson & T.A. Stephenson) Brooke & F. Rose, E.
pendula C. Thomas non A.A. Eaton, E. confusa D.P. Young) -
Green-flowered Helleborine. Stems to 40cm, glabrous or ± so;
leaves 2-ranked, mostly distinctly longer than wide; perianth
predominantly green; viscidium minute; ovary glabrous.
Native; woods on calcareous or sandy soils, heavy-
metal-polluted, and on dunes; scattered in Br N to Cheviot
and Westmorland, Lanarks, very scattered in Ir. Very
variable from area to area, as in many self-pollinating
plants. The above 4 synonyms and other taxa are probably
best recognized as vars.; some of these are cleistogamous.
The type var. has an undifferentiated green labellum
resembling the 2 petals.

4. EPIPOGIUM Gmelin ex Borkh. - Ghost Orchid
Saprophytic, chlorophyll-less, with coral-like rhizome
producing thin creeping rhizomes; leaves few, small and
scale-like; flowers shortly pedicellate, not twisted upside
down so spur and labellum point upwards, pale pink; spur
present, c. as long as labellum; labellum with 2 short
rounded lateral lobes at base; rostellum well developed;
pollinia with basal stalk ending in viscidium.

1. E. aphyllum Sw. - Ghost Orchid. Stems to 25cm, pinkish; **RRR**
flowers 1-2(4), c.15-20mm vertically across, patent to
slightly pendent. Native; in deep shade of Fagus or Quercus
woods on leaf-litter or rotten stumps; very rare in 1 site
each in Herefs, Oxon and Bucks, formerly Salop.

5. NEOTTIA Guett. - Bird's-nest Orchid
Saprophytic, chlorophyll-less, with very short rhizome
wrapped with succulent roots; leaves few, small and scale-
like; flowers shortly pedicellate, pale brown; spur 0;
labellum divided apically into 2 lobes and with 2 small
lateral teeth near base; other 5 tepals convergent to form
loose hood; rostellum well developed; pollinia not stalked.

1. N. nidus-avis (L.) Rich. - Bird's-nest Orchid. Stems to
50cm, brown; flowers numerous, crowded, c.15-20mm vertically
across, patent, with labellum c.10-12mm. Native; on leaf-
litter in shady woods, often of Fagus on calcareous soils;
scattered throughout most of Br and Ir, locally frequent in S
En.

6. LISTERA R. Br. - Twayblades
Shortly rhizomatous; leaves normally 2, in opposite pair on
stem (on non-flowering stems the leaves are at the apex);
flowers shortly pedicellate, yellowish-green to dull reddish;
spur 0; labellum deeply divided apically into 2 lobes,
sometimes with short tooth between them; rostellum well
developed; pollinia not stalked.

1. L. ovata (L.) R. Br. - Common Twayblade. Stems
20-60(75)cm; leaves ovate-elliptic, 5-20cm, with 3-5
prominent longitudinal veins; labellum yellowish-green,
7-15mm; other 5 tepals ± convergent to form loose hood;
flowers usually >15(<c.100). Native; woods, hedgerows,
grassy fields, dune-slacks, sometimes on Calluna-moors;
frequent throughout BI.
2. L. cordata (L.) R.Br. - Lesser Twayblade. Stems to
10(25)cm; leaves triangular-ovate, 1-2.5cm, with prominent
midrib; labellum dull reddish, 3.5-4.5mm; other 5 tepals ±
patent, 2-2.5mm; flowers c.3-15. Native; upland woods and
moors in usually wet, acid places, often among Sphagnum or
other moss and under Calluna or other moorland shrubs;
frequent in Sc, scattered in Ir and in Br S to Derbys and N

Devon, not in C, SC or SE En except 1 site in E Sussex.

7. SPIRANTHES Rich. - Lady's-tresses

With tuberous roots; leaves several, basal and on stem or +
all basal; flowers sessile, white with green markings,
usually borne spirally in tight spike; spur 0; labellum +
unlobed, with slightly to markedly frilly distal edge, +
appressed to other tepals to form tubular or trumpet-shaped
perianth; rostellum well developed; pollinia not stalked.

1 Leaves at flowering time obovate-elliptic, all in
  tight rosette adjacent to base of flowering stem
  which bears only reduced scale-leaves   **1. S. spiralis**
1 Leaves at flowering time linear-lanceolate to
  -oblanceolate, around base of flowering stem and
  short way up it   2
  2 Flowers in 1 spiral row in spike, 6-8mm excl.
    ovary; bracts 6-9mm; leaves subacute to obtuse
                                              2. S. aestivalis
  2 Flowers in 3 spiral rows in spike, 10-14mm excl.
    ovary; bracts 10-20(30)mm; leaves acute
                                           3. S. romanzoffiana

1. S. spiralis (L.) Chevall. - Autumn Lady's-tresses. Stems
to 15(20)cm, with flowers in single spiral or 1-sided spike;
leaves all in basal rosette, patent; flowers 4-6mm excl.
ovary. Native; short permanent grassland and grassy dunes;
locally frequent in BI N to NE Yorks, Man and Co Sligo,
extinct in many inland sites.
2. S. aestivalis (Poiret) Rich. - Summer Lady's-tresses.   **RR**
Stems to 40cm, with flowers in single spiral; leaves basal
and on stem, + erect; flowers 6-8mm excl. ovary. Native;
marshy ground in Hants until 1959, bog in Guernsey until
1914, by pond in Jersey until 1926; extinct.
3. S. romanzoffiana Cham. - Irish Lady's-tresses. Stems to   **R**
30cm, with flowers in 3 close spirals; leaves basal and on
stem, + erect; flowers 10-14mm excl. ovary. Native; marshy
meadows near streams, rivers or lakes; extremely local in SW,
W & NE Ir, CW & NW Sc, S Devon.

8. GOODYERA R. Br. - Creeping Lady's-tresses

With rhizomes giving rise to sterile leaf-rosettes and
flowering stems with basal leaf-rosette and reduced
scale-like stem-leaves; flowers sessile, like those of
Spiranthes but in weak spiral or 1-sided spike, and labellum
with entire distal edge.

1. G. repens (L.) R.Br. - Creeping Lady's-tresses. Stems to   **R**
20(25)cm; leaves ovate-elliptic, + patent; flowers 3-5mm
excl. ovary. Native; on barish ground under Pinus or Betula
or rarely on moist dunes; local in N Br S to Cumberland
(formerly SE Yorks), ?intrd in E & W Norfolk and E Suffolk.

TRIBE 3 - EPIDENDREAE (genera 9-11). Fertile stamen 1; pollen dispersed as 4 (or 2, each with 2 halves) pollinia; receptive stigmas 2; third stigma a minute sterile bulge (rostellum); stamen borne at apex of column; pollinia sessile, with minute sticky pads; labellum rather small, simple or with 2 short lateral lobes; spur 0 or ± so.

9. LIPARIS Rich. - Fen Orchid
Leaves usually 2, on stem, green; stem with 2 basal tubers side-by-side; labellum directed upwards, downwards or any intermediate direction, frilly on margins, scarcely lobed.

1. L. loeselii (L.) Rich. - Fen Orchid. Stems to 20cm; RRR
leaves 2.5-8cm, elliptic; flowers <20, yellowish-green, c.10mm vertically across, the labellum c. as long as other tepals. Native; wet peaty fens and dune-slacks; very local in N Devon, E & W Norfolk, Glam and Caerns, formerly elsewhere in E Anglia and E Kent, greatly decreased.

10. HAMMARBYA Kuntze - Bog Orchid
Leaves 2(-4), on stem, green; stem with 2 basal tubers, 1 above the other; labellum directed upwards, entire on margins, not lobed.

1. H. paludosa (L.) Kuntze - Bog Orchid. Stems to 8(12)cm; R
leaves 0.5-2cm, elliptic, with marginal fringe of tiny bulbils; flowers <20, yellowish-green, c.7mm vertically across, the labellum somewhat shorter than other tepals. Native; on wet Sphagnum in bogs; formerly scattered throughout most of Br and Ir except C En, now very rare except in CW & NW Sc and locally in S Hants.

11. CORALLORRHIZA Ruppius ex Gagnebin - Coralroot Orchid
Yellowish-brown or yellowish-green, saprophytic; leaves all on stem, scale-like; stem with coral-like rhizome; labellum directed downwards, distinctly 3-lobed, entire on margins.

1. C. trifida Chatel. - Coralroot Orchid. Stems to 20cm; R
leaves reduced to few sheaths on stem; flowers <12, yellowish-green tinged brown, c.6mm vertically across, the labellum somewhat shorter than other tepals. Native; damp peaty or mossy ground under trees or shrubs in woods, scrub and dune-slacks; scattered in N Br S to MW Yorks.

TRIBE 4 - ORCHIDEAE (genera 12-23). Fertile stamen 1; pollen dispersed as 2 pollinia; receptive stigmas 2; third stigma a small to large sterile bulge (rostellum); stamen borne in front of column; pollinia on long (short in Herminium) stalks each with a sticky pad or the 2 sharing 1 sticky pad; labellum often large and conspicuous, very variably lobed; spur 0 to very long.

12. HERMINIUM L. - <u>Musk</u> <u>Orchid</u>
Leaves 2(-4) near base of stem, plus usually 1 reduced leaf
higher up,; all tepals ± incurved; labellum narrow, with 2
short lateral lobes; the 2 petals rather similar but with
shorter lobes; spur 0; plant with 1 ± globose underground
tuber.

1. **H. monorchis** (L.) R. Br. - <u>Musk</u> <u>Orchid</u>. Stems to   R
15(25)cm; leaves elliptic-oblong, 2-7cm; flowers yellowish-
green, c.6-8mm vertically across, rather dense in spike.
Native; chalk and limestone grassland; local in S Br N to
Glam, E Gloucs and Beds, formerly to W Norfolk, decreasing.

13. PLATANTHERA Rich. - <u>Butterfly-orchids</u>
Leaves 2(-3), near base of stem, plus few reduced leaves
higher up; upper 3 tepals ± incurved; labellum linear-oblong,
entire; 2 lateral sepals spreading; spur long and slender;
plant with 2 ellipsoid underground tubers with tapering
apices.

1. **P. chlorantha** (Custer) Reichb. - <u>Greater</u> <u>Butterfly-</u>
<u>orchid</u>. Stems to 60cm; leaves 5-15cm; flowers pure white to
greenish-white, c.18-23mm transversely across; labellum
10-16mm; pollinia 3-4mm (conspicuously yellow against white
perianth), divergent downwards along stalks so that viscidia
are c.4mm apart; spur 19-28 x c.1mm, often strongly curved.
Native; woods and (in N) in open grassland, usually on
calcareous soils; locally frequent throughout Br and Ir
except N Isles of Sc, much commoner than <u>P. bifolia</u> in S Br.
1 x 2. **P. chlorantha x P. bifolia = P. x hybrida** Bruegger
has been reported from several places but never confirmed;
the parents sometimes occur together.
2. **P. bifolia** (L.) Rich. - <u>Lesser</u> <u>Butterfly-orchid</u>. Differs
from <u>P. chlorantha</u> in usually smaller stature and smaller in
all parts; flowers usually fewer, c.11-18mm transversely
across; labellum 6-10mm; pollinia c.2mm, parallel, the
viscidia c. 1mm apart; spur 15-20 x c.1mm, usually slightly
curved. Native; similar habitats and distribution to <u>P.</u>
<u>chlorantha</u>, commoner than it in N, hence more often in open
habitats.

13 x 15. PLATANTHERA X PSEUDORCHIS = X PSEUDANTHERA McKean
13/1 x 15/1. X P. breadalbanensis McKean (<u>Pl.</u> <u>chlorantha</u> x
<u>Ps.</u> <u>albida</u>) was described from a specimen found with the
parents in M Perth in 1980.

14. ANACAMPTIS Rich. - <u>Pyramidal</u> <u>Orchid</u>
Leaves several, decreasing in size up stem; upper 3 tepals ±
incurved; labellum deeply and nearly equally 3-lobed, with 2
raised plates near its base; 2 lateral sepals spreading; spur
long and slender; plant with 2 subglobose underground tubers.
See <u>Gymnadenia</u> for differences.

1. A. pyramidalis (L.) Rich. - <u>Pyramidal</u> <u>Orchid</u>. Stems to 60cm; leaves lanceolate, the lowest c.8-15cm; flowers in $\pm$ pyramidal dense spike, pinkish-purple (rarely white), c.10-12mm vertically across, with spur 12-14 x <1mm. Native; chalk and limestone grassland, calcareous dunes; locally frequent in BI N to S Ebudes and Fife.

14 x 16. ANACAMPTIS x GYMNADENIA = X GYMNANACAMPTIS Asch. & Graebner
14/1 x 16/1. X G. anacamptis (Wilms) Asch. & Graebner (<u>A.</u> <u>pyramidalis</u> x <u>G. conopsea</u>) has been recorded from Hants, Gloucs and Co Durham; it has the labellum plates of <u>Anacamptis</u> and the scent and cylindrical spike of <u>Gymnadenia</u>.

15. PSEUDORCHIS Seguier (<u>Leucorchis</u> E. Meyer) - <u>Small-white</u> <u>Orchid</u>
Leaves several, decreasing in size up stem; upper 5 tepals $\pm$ incurved; labellum quite deeply and nearly equally 3-lobed, without raised plates at base; spur short, wide, rounded at apex; plant with cluster of tapering underground tubers.

1. P. albida (L.) A. & D. Loeve (<u>L. albida</u> (L.) E. Meyer) - <u>Small-white</u> <u>Orchid</u>. Stems to 20(40)cm; leaves oblong-oblanceolate, the lowest 2.5-8cm; flowers in a cylindrical dense spike, creamy-white, c.2-4mm vertically across, with spur 2-3mm. Native; short grassland, usually base-rich and upland; frequent in C, W & N Sc, very scattered elsewhere in N Br, Ir and Wa, formerly S to Derbys and in Sussex, now very rare in En, extinct in many places. Superficially resembles <u>Neotinea</u>, but labellum-shape is totally different.

15 x 16. PSEUDORCHIS x GYMNADENIA = X PSEUDADENIA P. Hunt
15/1 x 16/1. X P. schweinfurthii (Hegelm. ex A. Kerner) P. Hunt (X <u>Gymleucorchis</u> <u>schweinfurthii</u> (Hegelm. ex A. Kerner) Stephenson & T.A. Stephenson; <u>P. albida</u> x <u>G. conopsea</u>) has been recorded from several places in N Br and is still frequent in NW Sc with both parents; it is intermediate in size, perianth shape (especially spur) and colour (pale pink).

15 x 18. PSEUDORCHIS x DACTYLORHIZA = x PSEUDORHIZA P. Hunt
15/1 x 18/2. x P. bruniana (Bruegger) P. Hunt (<u>P. albida</u> x <u>D. maculata</u>) was found in Orkney in 1977; it resembles <u>D. maculata</u> in stem and leaf characters, and <u>P. albida</u> in inflorescence shape, size and colour, but has intermediate floral characters.

16. GYMNADENIA R. Br. - <u>Fragrant</u> <u>Orchid</u>
Leaves several, decreasing in size up stem; upper 3 tepals $\pm$ incurved; labellum shallowly 3-lobed, without raised plates at base; 2 lateral sepals spreading; spur long and slender; plant with several divided tapering underground tubers.

Often grows with <u>Anacamptis</u>, but flowers earlier (little or no overlap); the shape of the tubers, labellum and spike, and the scented flowers distinguish it.

Other spp. - Old records of the European **G. odoratissima** (L.) Rich., with smaller flowers (lateral sepals 2.5-3mm; labellum 2.3-3mm; spur 4-5mm) have never been confirmed.

1. G. conopsea (L.) R. Br. - <u>Fragrant</u> Orchid. Stems to 40(75)cm; leaves linear-lanceolate, the lowest c.6-15cm; flowers in ± cylindrical dense spike, sweetly scented, usually lilac-purple, sometimes other shades of red (rarely white), c.8-12mm vertically across, with spur 8-17 x c.1mm. Native.

1 Lateral sepals mostly 4-5 x c.2mm; labellum
  obscurely lobed, (3)3.5-4(5)mm wide   **c. ssp. borealis**
1 Lateral sepals mostly 5-7 x c.1mm; labellum
  conspicuously lobed, (4.5)5.5-7(8)mm wide   2
    2 Labellum scarcely wider than long, mostly 5-6 x
      5.5-6.5mm; lateral sepals c.5-6mm; spur mostly
      12-14mm   **a. ssp. conopsea**
    2 Labellum much wider than long, mostly 3.5-4 x
      6.5-7mm; lateral sepals c.6-7mm; spur mostly
      14-16mm   **b. ssp. densiflora**

a. Ssp. conopsea. Flowers (7)10-11(13)mm horizontally across; labellum conspicuously lobed, (4)5-6(6.5) x (4.5)5.5-6.5(7)mm; lateral sepals bent downwards, 5-6 x c.1mm; spur (11)12-14(17)mm. Dry chalk or limestone grassland; frequent in Br N to Co Durham, N Ir.

b. Ssp. densiflora (Wahlenb.) Camus, Bergon & A. Camus. Flowers (10)11-13(14.5)mm horizontally across; labellum conspicuously lobed, (3)3.5-4(4.5) x (5.5)6.5-7(8)mm; lateral sepals held horizontally, 6-7 x c.1mm; spur (13)14-16(17)mm. Base-rich fens and usually N-facing chalk grassland; scattered and local in Br N to Westmorland, scattered through Ir (commonest ssp.).

c. Ssp. borealis (Druce) F. Rose. Flowers (7)8-10(12)mm horizontally across; labellum obscurely lobed, (3.5)4-4.5(5) x (3)3.5-4(5)mm; lateral sepals bent downwards, 4-5 x c.2mm; spur (8)11-14(15)mm. Base-rich to -poor hilly grassland in Sc and N & SW En, bogs in S Hants and E Sussex, ?Wa, ?Ir.

16 x 17. GYMNADENIA x COELOGLOSSUM = X GYMNAGLOSSUM Rolfe
16/1 x 17/1. X G. jacksonii (Quirk) Rolfe (<u>G. conopsea</u> x <u>C. viride</u>) has been recorded sporadically throughout much of Br and Ir; the inflorescences resemble those of <u>Gymnadenia</u> but are tinged green and have a much shorter spur, and labellum- and leaf-shapes are intermediate.

16 x 18. GYMNADENIA x DACTYLORHIZA = X DACTYLODENIA Garay & H. Sweet (X <u>Dactylogymnadenia</u> Soo)
Hybrids of this combination often resemble <u>Dactylorhiza</u> in

general appearance but have usually faintly spotted leaves
and scented flowers with a longer spur; the perianth is
variously intermediate in details. Precise parentage is
difficult to determine without knowledge of the sp. or spp.
of Dactylorhiza present nearby.

16/1 x 18/1. X D. st-quintinii (Godfery) J. Duvign. (? X D.
heinzeliana (Reichardt) Garay & H. Sweet, X Dactylogymnadenia
cookei (J. Heslop-Harrison) Soo; G. conopsea x D. fuchsii)
has been found in scattered localities throughout most of Br
and Ir.

16/1 x 18/2. X D. legrandiana (Camus) Peitz (X Dactylogymnadenia legrandiana (Camus) Soo; G. conopsea x D. maculata) has
been found in scattered localities throughout most of Br and
Ir.

16/1 x 18/3. X D. vollmannii (Schulze) Peitz (X Dactylogymnadenia vollmannii (Schulze) Soo; G. conopsea x D.
incarnata) was confirmed for W Cornwall in 1984, an earlier
record being doubtful.

16/1 x 18/4. X D. wintoni (Druce) Peitz (X Dactylogymnadenia
wintoni (Druce) Soo; G. conopsea x D. praetermissa) has been
found in S En; ?endemic.

16/1 x 18/5. X D. varia (Stephenson & T.A. Stephenson)
Averyanov (X Dactylogymnadenia varia (Stephenson & T.A.
Stephenson) Soo; G. conopsea x D. purpurella) has been found
in Cumberland, various parts of Sc (mostly W) and Co Down;
?endemic.

17. COELOGLOSSUM Hartman - Frog Orchid

Leaves several, decreasing in size up stem; upper 5 tepals
incurved; labellum oblong, shallowly 3-lobed near tip with
the central lobe the shortest; spur very short, rounded;
plant with 2 divided tapering underground tubers.

1. C. viride (L.) Hartman - Frog Orchid. Stems to 20(35)cm;
leaves elliptic-oblong, sometimes broadly so, the lowest
c.1.5-5cm; flowers in ± cylindrical dense spike, yellowish-
green tinged with reddish-brown, c.6-10mm vertically across;
labellum 3.5-6mm; spur c.2mm. Native; grassland, especially
on base-rich or calcareous soils; locally frequent throughout
Br and Ir, extinct in most places in S Sc and C & E En.

17 x 18. COELOGLOSSUM X DACTYLORHIZA = X DACTYLOGLOSSUM P.
Hunt & Summerh.

Hybrids of this combination usually have flowers of the
colour of the Dactylorhiza parent variously tinged or
overlaid with green; other characters of habit, leaves and
perianth shape are variously intermediate. Precise parentage
is difficult to determine without knowledge of the sp. or
spp. of Dactylorhiza present nearby. Hybrids involving D.
praetermissa, D. incarnata and D. majalis have been reported,
but the evidence is weak.

17/1 x 18/1. X D. mixtum (Asch. & Graebner) Rauschert (C. viride x D. fuchsii) has been found widely scattered in Br and in Co Down.
17/1 x 18/2. X D. dominianum (Camus, Bergon & A. Camus) Soo (x D. drucei (Camus) Soo; C. viride x D. maculata) has very scattered records in Br.
17/1 x 18/5. X D. viridella (J. Heslop-Harrison) Soo (C. viride x D. purpurella) has been found in Co Durham, M & N Ebudes and Outer Hebrides; ?endemic.

18. DACTYLORHIZA Necker ex Nevski (Dactylorchis (Klinge) Vermeulen) - Marsh-orchids
Leaves several, the lower sheathing stem, the upper transitional to bracts and not sheathing (though often clasping) stem; upper 3 tepals ± incurved; labellum usually shallowly 3-lobed, sometimes more deeply so or ± unlobed, nearly always as wide as or wider than long; 2 lateral sepals spreading, erect or bent down; spur down-pointed, usually <10mm, mostly rather wide; plant with divided tapering underground tubers.
A very difficult genus owing to ready hybridization between any of the spp., and the complex pattern of variation within most spp. whereby considerable differences between populations are often evident. There is much disagreement as to sp. limits; the views of R.H. Roberts are in the main followed here. Except with typical material it is often not possible to identify single specimens; before using the key the population should be surveyed and means of 5-10 non-extreme plants calculated.
Of the 8 spp., D. fuchsii and D. incarnata are diploid (2n=40) and the others are tetraploid (2n=80). Hybrids within a ploidy level are highly fertile, those between ploidy levels highly but not completely sterile; even in the case of triploid hybrids (2n=60) backcrossing and intro-gression can occur. Identification of hybrids involves careful examination of the characters of the putative parents in the sites concerned. The hybrids are intermediate in most characters, have low pollen fertility if triploid, and often show marked hybrid vigour.

1 Stem solid; leaves nearly always spotted; usually
   2-6 reduced non-sheathing leaves present on stem
   transitional between main (sheathing) leaves and
   bracts; lateral sepals spreading horizontally or
   bent down; spur usually <2mm wide at midpoint   2
1 Stem usually hollow, at least below; leaves often
   not spotted; usually 0-2 reduced non-sheathing
   leaves present on stem transitional between main
   (sheathing) leaves and bracts; lateral sepals ±
   erect; spur usually >2mm wide at midpoint   3
   2 Labellum lobed c.1/2 way to base with central
      lobe usually exceeding the two laterals and

$\geq 1/2$ as wide as them; leaves mostly subacute to
obtuse, with spots usually $\pm$ transversely
elongated   **1. D. fuchsii**
2 Labellum lobed much $<1/2$ way to base, with central
lobe as long as or shorter than 2 laterals and
much $<1/2$ as wide as them; leaves mostly
narrowly acute to subacute, with usually $\pm$
circular spots   **2. D. maculata**
3 Leaves unspotted or with spots on both surfaces,
yellowish-green, narrowly hooded at apex; labellum
usually with markedly reflexed sides (if slightly
reflexed then leaves with spots on both surfaces)
hence appearing very narrow from front, usually
with 2 distinct dark loop-shaped marks side by side
                                              3. D. incarnata
3 Leaves unspotted or with spots mostly on upperside,
mid-, dark- or greyish-green, not or broadly hooded
at apex; labellum usually without markedly reflexed
sides, often nearly flat, usually without 2
distinct dark loops   4
  4 Total number of leaves usually <5, the widest
  $<1.5(2)$cm wide; labellum usually distinctly
  3-lobed, with central lobe distinctly exceeding
  2 laterals and usually $>1/2(1/3)$ as long as
  unlobed basal part   5
  4 Total number of leaves usually ≥ 5 (except in
  D. majalis var. scotica), the widest $>(1.5)$2cm
  wide; labellum usually not strongly 3-lobed, if so
  then central lobe usually $<1/3$ as long as unlobed
  basal part (except D. majalis)   6
5 Leaves with strong dark spots and rings on upperside;
bracts predominantly green, suffused reddish-purple
at margins and apex, with dark spots and rings on
upperside; transitional non-sheathing leaves
(0-)2   **8. D. lapponica**
5 Leaves with 0 or faint dark spots or rings on
upperside; bracts strongly suffused reddish-purple
all over, without dark spots or rings; transitional
non-sheathing leaves 0-1(2)   **7. D. traunsteineri**
  6 Labellum with well-marked narrow central lobe
  usually 1/3-1/2 as long as unlobed basal part,
  with dark spots or lines mostly in central part but
  usually some extending almost to margins; leaves
  unmarked (parts of Ir only) or with strong spots or
  blotches mostly >2mm across   **6. D. majalis**
  6 Labellum usually obscurely lobed, with dark
  spots or lines usually confined to central part;
  leaves usually unmarked, sometimes with small
  spots <2mm across or with rings (rarely larger
  spots)   7
7 Leaves usually broadly hooded at apex, their dark
markings (if present) small spots <2mm across

or rarely larger; labellum ± rhombic, usually
reddish-purple, usually <7.5 x 9.5mm   **5. D. purpurella**
7 Leaves flat or slightly hooded at apex, their dark
markings (if present) mainly as rings; labellum
orbicular to transversely broadly elliptic, usually
pinkish- or pale mauvish-purple, usually >7.5 x
9.5mm   **4. D. praetermissa**

1. D. fuchsii (Druce) Soo (D. maculata ssp. fuchsii (Druce) **1171**
N. Hylander, D. longebracteata auct. non (F.W. Schmidt)
Holub, Dactylorchis fuchsii (Druce) Vermeulen) - Common
Spotted-orchid. Stems to 50(70)cm; largest leaves <4(5.5)cm
wide, usually with dark transversely elongated spots, flat at
apex; labellum lobed c.1/2 way to base, the central lobe the
longest, with pale pink to white ground-colour. Native; damp
woods, banks and meadows, marshes and fens, usually on
base-rich soil; ± throughout BI, the commonest orchid in En
and Ir. Some populations in Ir, Sc, Man and Cornwall are
statistically separable and have received ssp. status, but
are worth only var. rank.
  1 x 2. D. fuchsii x D. maculata = D. x transiens (Druce) Soo
has been found scattered throughout BI but is over-recorded
for plants difficult to determine.
  1 x 3. D. fuchsii x D. incarnata = D. x kerneriorum (Soo)
Soo has been found scattered throughout Br and Ir.
  1 x 4. D. fuchsii x D. praetermissa = D. x grandis (Druce)
P. Hunt occurs throughout the range of D. praetermissa in Br
(and CI?); probably the commonest hybrid orchid in S Br.
  1 x 5. D. fuchsii x purpurella = D. x venusta (Stephenson &
T.A. Stephenson) Soo occurs throughout the range of D.
purpurella in Br and Ir; it is sometimes partially fertile;
?endemic.
  1 x 6. D. fuchsii x D. majalis = D. x braunii (Hal.) Borsos
& Soo has been found in Anglesey, SE Yorks, W Sc and Co Clare
populations of D. majalis.
  1 x 7. D. fuchsii x D. traunsteineri (= D. x kelleriana P.
Hunt nom. inval.) has been found in the Irish, Anglesey and
Yorks areas of D. traunsteineri.
  2. D. maculata (L.) Soo (Dactylorchis maculata (L.) **1171**
Vermeulen) - Heath Spotted-orchid. Stems to 40(50)cm;
largest leaves <2(2.5)cm wide, usually with dark ± rounded
spots, usually flat (sometimes hooded) at apex; labellum
lobed much <1/2 way to base, the central lobe shorter than to
as long as laterals, with pale pink to ± white ground-
colour. Native; damp peaty places in bogs, marshes and
ditches; ± throughout BI, the commonest orchid in Sc and Ir.
Our plant is ssp. **ericetorum** (Linton) P. Hunt & Summerh.;
sspp. **maculata** and/or **elodes** (Griseb.) Soo have also been
reported here but the evidence is weak. Ssp. **rhoumensis** (J.
Heslop-Harrison) Soo, from Rhum, is a diploid (2n=40)
variously placed under D. fuchsii and D. maculata but is of
very doubtful ssp. status anyway.

2 x 3. D. maculata x D. incarnata = D. x carnea (Camus) Soo (D. x maculatiformis (Rouy) Borsos & Soo, D. x claudiopolitana Soo nom. nud.) has been found scattered in Br and Ir.

2 x 4. D. maculata x D. praetermissa = D. x hallii (Druce) Soo probably occurs throughout the range of D. praetermissa in Br (and CI?), but much more rarely than D. x grandis due to different habitat preferences.

2 x 5. D. maculata x D. purpurella = D. x formosa (Stephenson & T.A. Stephenson) Soo occurs throughout the range of D. purpurella in Br and Ir; probably the commonest hybrid orchid in N Br and Ir.

2 x 6. D. maculata x D. majalis = D. x dinglensis (Wilm.) Soo (D. x townsendiana auct. non (Rouy) Soo) has been found in the Irish and Cards populations of D. majalis.

2 x 7. D. maculata x D. traunsteineri = D. x jenensis (Brand) Soo has been found in the NW Wa, Yorks and Irish populations of D. traunsteineri.

3. D. incarnata (L.) Soo (Dactylorchis incarnata (L.) **1171** Vermeulen) - Early Marsh-orchid. Stems to 40(80)cm; largest leaves 1-2(3.5)cm wide, narrowly hooded at apex; labellum obscurely lobed or lobed much <1/2 way to base, ± rhombic, the sides usually strongly reflexed. Native; decreasing due to land-drainage. Very variable; some of the main variants occupy distinct habitats or geographical regions and have been recognized as sspp. Of these, the most distinct is ssp. cruenta; most of the others are possibly better considered vars. of ssp. incarnata, but the present consensus is followed here.

1 At least some plants with leaves with spots on both
  surfaces   **d. ssp. cruenta**
1 Plants with unmarked leaves, rarely some with spots
  on upperside only   2
  2 Perianth pale yellow or cream; labellum usually
    >6.5 x 8mm and with well-marked lobes, the
    lateral ones usually indented; lowest bract
    usually >30 x 20mm   **e. ssp. ochroleuca**
  2 Perianth variously pink to purple, rarely white
    or cream (if so then labellum usually <6.5 x 8mm
    and without indented lateral lobes and lowest
    bract usually <30 x 20mm)   3
3 Ground-colour of perianth pink; bracts usually
  lacking anthocyanin   **a. ssp. incarnata**
3 Ground-colour of perianth red to purple; bracts
  usually strongly suffused with anthocyanin   4
  4 Plants mostly >20cm; perianth ground-colour
    reddish-purple   **c. ssp. pulchella**
  4 Plants mostly <20cm; perianth ground-colour vivid
    ruby- or crimson-red   **b. ssp. coccinea**
a. Ssp. incarnata (ssp. gemmana (Pugsley) Sell, Dactylorchis incarnata ssp. gemmana (Pugsley) J. Heslop-Harrison, Orchis strictifolia Opiz). Plants mostly 20-40cm; leaves not

spotted or rarely with few small dots on upperside; perianth ground-colour pale pink; labellum usually <7 x 8.5mm, usually obscurely lobed. Wet meadows, fens and marshes on base-rich or neutral soils; locally frequent in En and Wa, extremely scattered in Sc and Ir. Robust plants to 50(80)cm with large labella (>7 x 8.5mm) and spurs (usually >7.5mm) are sometimes separated as ssp. **gemmana**; they have been found scattered in En and Ir but their status needs investigating.

b. Ssp. coccinea (Pugsley) Soo (Dactylorchis incarnata ssp. coccinea (Pugsley) J. Heslop-Harrison). Plants mostly <20cm; leaves not spotted; perianth ground-colour vivid ruby- or crimson-red; labellum usually <6.5 x 8mm, usually obscurely lobed. Dune-slacks and other damp base-rich sandy areas near sea, damp inland lake-shores in Ir; frequent throughout Ir, locally common by coast in W Br N to Shetland, on E coast only in N En (rare); endemic.

c. Ssp. pulchella (Druce) Soo (Dactylorchis incarnata ssp. pulchella (Druce) J. Heslop-Harrison). Plants mostly 20-40cm; leaves not spotted or rarely with few small dots on upperside; perianth ground-colour reddish-purple; labellum usually <7 x 8.5mm, usually obscurely lobed. Bogs and other neutral to acid wet peaty places; scattered throughout Br and Ir; ?endemic.

d. Ssp. cruenta (Mueller) Sell (D. cruenta (Mueller) Soo, Dactylorchis incarnata ssp. cruenta (Mueller) Vermeulen). Plants mostly 15-40cm; leaves often with heavy spots on both surfaces; perianth ground-colour pinkish-mauve; labellum usually <7.5 x 8mm, usually with obvious central lobe. Marshes on limestone by lakes in WC Ir, neutral mountain flushes in W Ross.

e. Ssp. ochroleuca (Boll) P. Hunt & Summerh. (Dactylorchis 1171 incarnata ssp. ochroleuca (Boll) J. Heslop-Harrison). Plants 20-50cm; leaves not spotted; perianth cream to pale yellow; labellum usually >6.5 x 8mm, usually obviously 3-lobed with notched lateral lobes. Calcareous fens in E Anglia, ?Carms.

3 x 4. D. incarnata x D. praetermissa = D. x wintoni (A. Camus) P. Hunt is scattered in S & C Br N to S Lancs.

3 x 5. D. incarnata x D. purpurella = D. x latirella (P. Hall) Soo is scattered in N & W Br S to MW Yorks and Cards; ?endemic.

3 x 6. D. incarnata x D. majalis = D. x aschersoniana (Hausskn.) Soo has been found in Cards, Outer Hebrides and Co Limerick populations of D. majalis.

3 x 7. D. incarnata x D. traunsteineri = D. x dufftii (Hausskn.) Peitz (D. x lehmanii (Klinge) Soo) has been found in NW Wa, Yorks and Co Wicklow populations of D. traunsteineri.

4. D. praetermissa (Druce) Soo (D. majalis ssp. praetermissa 1171 (Druce) David Moore & Soo, Dactylorchis praetermissa (Druce) Vermeulen, Orchis pardalina Pugsley) - Southern Marsh-orchid. Stems to 50(70)cm; largest leaves (1.5)2-2.5cm wide, usually unspotted, rarely with large rings, flat or slightly

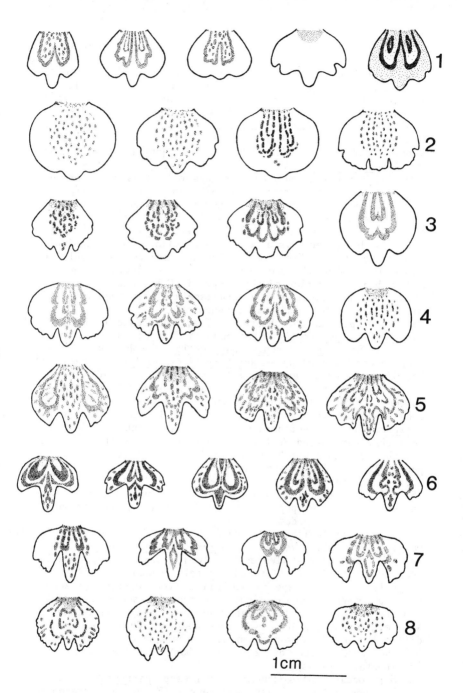

FIG 1171 – Labella of **Dactylorhiza**. 1, **D. incarnata** (4th in row, ssp. **ochroleuca**). 2, **D. praetermissa**. 3 (1st 3), **D. purpurella** (last in row, **D. praetermissa** var. **junialis**). 4, **D. majalis**. 5, **D. traunsteineri**. 6, **D. lapponica**. 7, **D. fuchsii**. 8, **D. maculata**. Drawings by R.H. Roberts.

hooded at apex; labellum scarcely or shallowly 3-lobed,
orbicular to transversely broadly elliptic, with pale to
medium pinkish-purple ground-colour.   Native; slightly acid
to calcareous damp places in fens, marshes, bogs, meadows,
gravel-pits and waste alkali-, colliery- and ash-tips;
frequent in Br N to S Northumb and W Lancs, CI, the commonest
marsh-orchid in S & C En and S Wa. Ssp. **junialis** (Vermeulen)
Soo (Orchis pardalina) is the name often given to variants
with large ring-shaped dark markings on leaves and unbroken
purple loops (rather than dots and dashes) on labellum, but
such plants usually occur with normal ones and are probably
best treated as var. **junialis** (Vermeulen) Senghas; in the 1171
past they were thought to be hybrids with D. fuchsii.
4 x 5. D. praetermissa x D. purpurella = D. x insignis
(Stephenson & T.A. Stephenson) Soo has been recorded only in
Cards and Merioneth, but probably occurs elsewhere in the
narrow band of overlap of the 2 spp. in C Br; endemic.
4 x 7. D. praetermissa x D. traunsteineri has been recorded
only from W Norfolk and Cambs; endemic.
5. D. purpurella (Stephenson & T.A. Stephenson) Soo (D. 1171
majalis ssp. purpurella (Stephenson & T.A. Stephenson) David
Moore & Soo, Dactylorchis purpurella (Stephenson & T.A.
Stephenson) Vermeulen) - Northern Marsh-orchid. Stems to
25(40)cm; largest leaves 1.5-2.5cm wide, unspotted or with
few spots <2mm across or rarely larger, moderately to broadly
hooded at apex; labellum obscurely lobed, ± rhombic, with
reddish-purple ground-colour. Native; similar places to D.
praetermissa (its northern vicariant); frequent in N Br S to
SE Yorks, Derbys and Pembs, frequent in N Ir, scattered in S
Ir, the commonest marsh-orchid in Sc, N En, N Wa and Ir. D.
purpurella often approaches D. majalis closely, especially in
Sc, and can be difficult to separate.
5 x 6. D. purpurella x D. majalis has been found in the
Anglesey and W Sc populations of D. majalis, and probably
occurs in Ir too; endemic.
6. D. majalis (Reichb.) P. Hunt & Summerh. (Dactylorchis 1171
majalis (Reichb.) Vermeulen) - Western Marsh-orchid. Stems R
to 30(35)cm; largest leaves mostly 1.5-2.8cm wide, usually
heavily spotted or blotched, rarely unspotted, flat or
slightly hooded at apex; labellum usually distinctly lobed
with prominent central lobe, subrhombic to transversely
broadly elliptic, with light reddish-purple ground-colour.
Native; marshes, fens, wet meadows and dune-slacks.   A
difficult sp. whose variation can perhaps best be summarized
by the recognition of 2 sspp., both distinct from the
Continental type ssp.
a. Ssp. occidentalis (Pugsley) Sell (ssp. kerryensis (Wilm.)
Senghas nom. illeg., var. kerryensis (Wilm.) R. Bateman &
Denholm, ssp. scotica E. Nelson, var. scotica (E. Nelson) R.
Bateman & Denholm, D. kerryensis (Wilm.) P. Hunt & Summerh.,
Dactylorchis majalis ssp. occidentalis (Pugsley) J.
Heslop-Harrison). Stems mostly <20cm; largest leaves mostly

<10cm x >2cm; spur usually <3.5mm wide at entrance. Scattered over Ir (mostly S & W), N Uist (Outer Hebrides); endemic.
  b. Ssp. cambrensis (Roberts) Roberts (var. cambrensis (Roberts) R. Bateman & Denholm, D. purpurella ssp. majaliformis E. Nelson, Dactylorchis majalis ssp. cambrensis Roberts). Stems mostly >20cm; largest leaves mostly >10cm x <2cm; spur usually >3.5mm wide at entrance. NW Wa, SE Yorks, and NW & N Sc; endemic. More work is needed on Scottish populations to see whether they are worth separating from ssp. cambrensis (as ssp. majaliformis) and whether ssp. scotica is worth recognizing in addition.
  7. D. traunsteineri (Sauter ex Reichb.) Soo (D. majalis ssp. 1171 traunsteinerioides (Pugsley) R. Bateman & Denholm, var. R eborensis (Godfery) R. Bateman & Denholm, var. francis-drucei (Wilm.) R. Bateman & Denholm, Orchis latifolia L. var. eborensis Godfery, O. francis-drucei Wilm., Dactylorchis traunsteineri (Sauter ex Reichb.) Vermeulen, D. traunsteinerioides (Pugsley) Vermeulen) - Narrow-leaved Marsh-orchid. Stems to 30(40)cm; widest leaves mostly <1.5cm wide, unspotted or rather faintly spotted, slightly hooded at apex; labellum distinctly 3-lobed but even if not then with a prominent central projection, with light reddish-purple ground-colour and usually many dark markings ± to margins. Native; calcareous fens and other damp base-rich grassy places; local in Ir, W Sc, NW Wa, N En and E Anglia, very scattered in C S En.
  8. D. lapponica (Hartman) Soo (D. traunsteineri ssp. 1171 lapponica (Hartm.) Soo, D. pseudocordigera (Neuman) Soo) - RR Lapland Marsh-orchid. Stems to 21cm; widest leaves 1.1-1.5cm wide, heavily spotted and blotched, not or scarcely hooded at apex; labellum usually distinctly lobed but even if not then with a prominent central projection, with reddish-purple ground-colour and usually heavy dark spots and lines. Native; slightly acidic to base-rich hillside flushes; very local in Westerness, Kintyre, N Ebudes and Outer Hebrides, 1st recognized in Br in 1986.

19. NEOTINEA Reichb. f. - Dense-flowered Orchid
Leaves 2-3(4) near base of stem, with reduced leaves up stem; upper 5 tepals incurved; labellum with 2 large lateral lobes and larger terminal lobe usually shallowly subdivided at apex or with small tooth between the divisions; spur short, rounded at apex; plant with 2 ovoid underground tubers.

  1. N. maculata (Desf.) Stearn (N. intacta (Link) H.G. R Reichb.) - Dense-flowered Orchid. Stems to 30(40)cm; leaves oblong-elliptic, sometimes purple-spotted (in W Ir only) but usually not, 2-6cm; flowers in dense cylindrical spike, creamy-white, sometimes pink-tinged, c.5-6mm vertically across, with labellum 4-5mm, with spur c.2mm. Native; rocky and sandy grassy places and maritime dunes; very local in CW

& SW Ir, very rare in W Donegal (found 1983) and Man (found
1966). Superficially resembles Pseudorchis, but labellum-
shape is totally different.

20. ORCHIS L. - Orchids
Leaves several, on stem, with few smaller ones above; upper 3
or upper 5 tepals incurved, the 2 lateral sepals incurved or
erect to patent; labellum with 2 lateral lobes and a terminal
lobe, the latter often larger than the laterals and usually
2-3-lobed at its apex; spur long or short, rounded to
truncate or emarginate at apex; plant with 2 ovoid underground tubers.

1 Upper 3 tepals incurved to form a 'helmet'; 2 lateral
  sepals erect to patent   2
1 All 5 upper tepals incurved to form a 'helmet'   3
  2 Labellum with terminal lobe exceeded by 2
    laterals, sometimes +0; spur shorter than ovary;
    bracts 3-veined or the lowest few 5-veined;
    leaves never spotted   **1. 0. laxiflora**
  2 Labellum with terminal lobe exceeding laterals;
    spur at least as long as ovary; bracts 1-veined
    or the lowest few 3-veined; leaves usually dark-
    spotted   **2. 0. mascula**
3 Area of central lobe of labellum from smaller than
  to slightly larger than that of each lateral lobe   4
3 Area of central lobe of labellum at least 2x that
  of each lateral lobe   5
  4 Spur horizontal or directed upwards, + as long
    as ovary; labellum mauvish-purple with paler,
    spotted, central area   **3. 0. morio**
  4 Spur directed downwards, <1/2 as long as ovary;
    labellum white with reddish-purple spots
                                                 4. 0. ustulata
5 Outside of all 3 sepals (forming 'helmet') dark
  reddish-purple, contrasting strongly with very pale
  labellum; 2 main sublobes of terminal lobe of
  labellum wider than long   **5. 0. purpurea**
5 Outside of sepals pale pinkish-purple, scarcely or
  not contrasting with labellum; 2 main sublobes of
  terminal lobe of labellum longer than wide   6
  6 Two main sublobes of terminal lobe of labellum
    oblong, >2x as wide as lateral lobes **6. 0. militaris**
  6 Two main sublobes of terminal lobe of labellum
    linear, c. as wide as lateral lobes   **7. 0. simia**

1. 0. laxiflora Lam. - Loose-flowered Orchid. Stems to **RR**
50(80)cm; leaves lanceolate to linear-oblong, unspotted;
flowers rather uniformly purple; labellum with 2 large
lateral and 0 or smaller terminal lobe. Native; marshy
meadows and by lakes; locally common in Jersey and Guernsey.
1 x 3. 0. laxiflora x 0. morio = 0. x alata Fleury occurs

sporadically with the parents in Jersey and occurred on Guernsey in 1949.

2. O. mascula (L.) L. - Early-purple Orchid. Stems to 40(60)cm; leaves elliptic-oblong to narrowly so, usually spotted; flowers uniformly pinkish-purple to purple; labellum with 3 often ± equal lobes, the terminal one shallowly bilobed. Native; neutral or base-rich grassland, scrub and woods, usually in shade in S but in open in N; frequent to common throughout BI.

2 x 3. O. mascula x O. morio = O. x morioides Brand occurs rarely and sporadically in En and Wa.

3. O. morio L. - Green-winged Orchid. Stems to 20(40)cm; **R** leaves narrowly elliptic-oblong, unspotted; flowers rather uniformly mauvish-purple but labellum with pale, darker-spotted central region and upper tepals with green veins; labellum with 3 often ± equal lobes, the terminal one often shallowly bilobed. Native; base-rich to neutral short undisturbed grassland; formerly frequent over most of En, Wa, Ir and CI, now greatly reduced and very local.

4. O. ustulata L. - Burnt Orchid. Stems to 15(30)cm; leaves **R** elliptic-oblong, unspotted; outside of sepals dark reddish-purple, contrasting strongly with white labellum with reddish-purple spots, the contrast greatest when upper flowers are still unopened; labellum with 3 ± equal lobes, the terminal one shallowly bilobed. Native; short grassland on chalk and limestone; formerly locally frequent over much of En, now greatly reduced and extremely local.

5. O. purpurea Hudson - Lady Orchid. Stems to 50(100)cm; **R** leaves elliptic-oblong, unspotted; outside of sepals dark reddish-purple, contrasting strongly with white or pale pink labellum with pink to reddish-purple spots, the contrast greatest when upper flowers are still unopened; labellum with 3 lobes, the terminal one much larger than laterals and with 2 broad sublobes with usually small tooth between. Native; woods and scrub, rarely open grassland, on chalk; locally frequent on N Downs in E & W Kent, 1 site in Oxon, formerly a few sites elsewhere in SE En.

6. O. militaris L. - Military Orchid. Stems to 45(60)cm; **RRR** leaves elliptic-oblong, unspotted; flowers pinkish- to reddish-purple; sepals paler on outside than labellum or sepals on inside; labellum with 3 lobes, the terminal one much larger than laterals with 2 oblong sublobes with small tooth between, paler or white with purplish spots in central part. Native; chalk grassland and old chalk-pit with invading trees and shrubs; 1 site each in Bucks and W Suffolk, 2 sites sporadically in Oxon, formerly more widespread in M Thames valley.

6 x 7. O. militaris x O. simia = O. x beyrichii A. Kerner occurred up to mid-19th Century in M Thames valley when the 2 parents co-existed there.

7. O. simia Lam. - Monkey Orchid. Stems to 30(40)cm; leaves **RRR** elliptic-oblong, unspotted; flowers pinkish-purple with a pale or white pinkish-purple-spotted area in centre of labellum; labellum with 3 lobes, the terminal one much larger than laterals with 2 linear sublobes with small tooth between. Native; chalk grassland and open scrub; 2 sites in E Kent, 2 in Oxon, formerly very scattered elsewhere in SE En and 1 site in SE Yorks.

20 x 21. ORCHIS x ACERAS = X ORCHIACERAS Camus
20/7 x 21/1. x O. bergonii (Nanteuil) Camus (O. simia x A. anthropophorum) was found at an O. simia site in E Kent in 1985; it has a labellum shape similar to that of O. simia but is intermediate in flower colour and sepal shape and size.

21. ACERAS R. Br. - Man Orchid
Leaves several, near stem-base, with few smaller ones above; upper 5 tepals incurved to form 'helmet'; labellum with 3 lobes, the lateral linear, the terminal larger and with 2 linear terminal sublobes; spur 0; plant with 2 ovoid underground tubers. Differs from Orchis only in absence of spur and probably better umited with it.

1. **A. anthropophorum** (L.) Aiton f. - Man Orchid. Stems to   R
40(50)cm; leaves narrowly elliptic-oblong, unspotted; flowers greenish-yellow, often tinged reddish-brown. Native; chalk and limestone grassland or scrub; local in SE En (frequent only in E & W Kent), scattered W to N Somerset and N to S Lincs and Derbys.

22. HIMANTOGLOSSUM Koch - Lizard Orchid
Leaves several, decreasing in size up stem; upper 5 tepals incurved to form 'helmet'; labellum very long and narrow, with 2 linear lateral lobes and a long terminal lobe spirally coiled at first and with 2 small sublobes at apex; spur short; plant with 2 ovoid underground tubers. Close to Orchis and perhaps not distinct.

1. **H. hircinum** (L.) Sprengel - Lizard Orchid. Stems to **RRR**
70(90)cm; leaves elliptic-oblong, purple-mottled or not; flowers greyish-yellowish-green; labellum 4-7cm, the central lobe 3-6cm. Native; on calcareous soils in rough ground, dunes, scrub and marginal places usually among tall grass; scattered places (often sporadic) in S & E En W to N Somerset (formerly N Devon) and N to W Suffolk (formerly NE Yorks), formerly S Lancs and Jersey.

23. OPHRYS L. - Spider-orchids
Leaves several, decreasing in size up stem; upper 5 tepals all patent, the 2 petals markedly different from the 3 sepals; labellum velvety in texture, subentire or with 2 small lateral lobes, the terminal lobe large and resembling an insect's abdomen; spur 0; plant with 2 ovoid underground tubers.

1 Labellum with distinct lateral lobes; the 2 petals
   filiform; labellum distinctly longer than wide
                                                1. O. insectifera
1 Labellum with 0 or obscure lateral lobes; the 2
   petals oblong to linear-oblong; labellum not or only
   just longer than wide   2
2 Sepals yellowish- to brownish-green; the 2
   petals yellowish-green, >1/2 as long as sepals
                                                2. O. sphegodes
2 Sepals pink or greenish-pink; the 2 petals pink
   or greenish-pink, <1/2 as long as sepals   3
3 Apex of labellum shortly bilobed with short simple
   projection between directed downwards and backwards
   (hence + invisible from front of flower)   **3. O. apifera**
3 Apex of labellum shortly bilobed with short
   projection between directed prominently forwards
   and often 3-toothed   **4. O. fuciflora**

Other spp. - O. bertolonii Moretti, from Mediterranean, was
found in Dorset in 1976 but was almost certainly planted and
has been removed.

1. O. insectifera L. - <u>Fly</u> Orchid. Stems to 60cm; sepals
yellowish-green; petals purplish-brown, filiform, slightly
<1/2 as long as sepals; labellum with well-marked lateral
lobes and terminal lobe conspicuously bilobed, purplish-brown
with shining blue central area. Native; woods, scrub,
grassland, spoil-heaps, fens and lakesides on calcareous
soils; scattered throughout Br N to Westmorland and NE Yorks,
C Ir, frequent in SE En.
1 x 2. O. insectifera x O. sphegodes = **O. x hybrida** Pokorny
has occurred sporadically in E Kent; it is intermediate in
flower shape.
1 x 3. O. insectifera x O. apifera (<u>O.</u> x <u>pietzschii</u> Kuempel
nom. inval.) has occurred in woodland in N Somerset since
1968; it is intermediate in sepal colour and petal and
labellum shape.
2. O. sphegodes Miller - <u>Early Spider-orchid</u>. Stems to **RRR**
20(35)cm; sepals yellowish- to brownish-green; petals
yellowish, oblong, >1/2 as long as sepals; labellum subentire
or notched (sometimes with small tooth in notch) at apex,
dark purplish-brown with variable blue markings. Native;
grassland or spoil-heaps on chalk or limestone; very local
from E Kent to Dorset and W Gloucs, intrd and natd in Herts,
formerly to W Cornwall and Northants and in Denbs and Jersey.
3. O. apifera Hudson - <u>Bee</u> Orchid. Stems to 45(60)cm;
sepals pink or greenish-pink, sometimes very pale; petals
greenish-pink, oblong, <1/2 as long as sepals; labellum with
short lateral lobes, shallowly bilobed at apex with backward-
directed projection in notch, reddish-brown with various
markings of yellow, gold and brown. Native; grassland,
scrub, spoil-heaps and sand-dunes on calcareous or base-rich

soils; locally frequent in Br N to Cumberland and Durham, CI,
scattered in Ir.
    3 x 4. O. apifera x O. fuciflora = O. x albertiana Camus has
been found in E Kent.
    4. O. fuciflora (Crantz) Moench (O. holoserica auct. non **RRR**
(Burm. f.) Greuter) - Late Spider-orchid. Stems to 35(55)cm;
differs from O. apifera in sepals and petals usually clear
pink; and see key. Native; short grassland on chalk; very
local in E Kent.

GLOSSARY

For some special terms, used only in 1 or few families, direct reference to the relevant family (families) is made; in those cases the family description and the notes immediately following it should be consulted.

abaxial - of a lateral organ, the side away from the axis, normally the **lowerside**

achene - a dry, indehiscent, 1-seeded fruit, $\pm$ hard, with papery to leathery wall; **achene–pit**, see 137. Asteraceae

acicle - a slender prickle with scarcely widened base

actinomorphic - of a flower with radial (i.e. >1 plane of) symmetry

acuminate - gradually tapering to a point; Fig 1186

acute - with point <90 degrees; Fig 1186

adaxial - of a lateral organ, the side towards the axis, normally the **upperside**

adherent - joined or fused

aerial - above-ground or above-water

alien - not **native**, introduced to a region deliberately or accidentally by man

alternate - lateral organs on an axis 1 per **node**, successive ones on opposite sides

anastomosing - dividing up and then joining again, usually applied to veins

androecium - the group of male parts of a flower; all the **stamens**; Fig 1189

andromonoecious - having male and **bisexual** flowers on the same plant

angustiseptate - fruit with the **septum** across the narrowest diameter

annual - completing its life-cycle in $\leq$12 months (but often not within 1 calendar year)

anther - pollen-bearing part of a **stamen**, usually terminal on a stalk or **filament**; Fig 1189

anthesis - flowering time; strictly pollen-shedding time

apiculus - a small, abruptly delimited point; **apiculate**, with an apiculus; Fig 1186

apomictic - producing seed wholly female in origin, without fertilization

appendage - small extra protrusion or extension, such as on a **petal**, **sepal** or seed

appendix - see 149. Araceae

appressed - lying flat against another organ

aril - **succulent** covering around a seed, outside the **testa**

(not the **pericarp**)

aristate – extended into a long bristle; Fig 1186

ascending – sloping or curving upwards

auricle – basal extension of a leaf-blade; Fig 979

awn – see 155. Poaceae

axil – angle between main and lateral axes; **axillary**, in the axil; see also **subtend**

axile – of a **placenta** formed by central axis of an **ovary** that is connected by **septa** to the wall; Fig 1187

beak – a narrow, usually apical, projection

berry – a **succulent** fruit, the seeds usually >1 and without a stony coat

biennial – completing its life-cycle in >1 but <2 years, not flowering in the first year

bifid – divided into two, usually deeply, at apex

bifurcate – dividing into two branches

biotype – a genetically fixed variant of a **taxon** particularly adapted to some (usually environmental) condition

bird-seed alien – **alien** introduced as contaminant of bird-seed

bisexual – of a plant or flower, bearing both sexes

blade – main part of a flat organ (e.g. **petal**, leaf); cf. **claw**, **petiole**; Figs 979, 1186

bloom – delicate, waxy, easily removed covering to fruit, leaves etc.; see also **pruinose**

bract – modified, often scale-like, leaf **subtending** a flower, less often a branch; **bracteate**, with bract(s); Fig 1188

bracteole – a supplementary or secondary **bract** or a bract once removed; Fig 1188

bud-scales – scales enclosing a bud before it expands

bulb – swollen underground organ consisting of condensed stem and **succulent** scale-leaves

bulbil – a small **bulb** or **tuber**, usually **axillary**, on an **aerial** part of the plant

bullate – with the surface raised into blister-like swellings

caducous – falling off early

callus – see 155. Poaceae

calyx (plural **calyces**) – the outer **whorl(s)** of the **perianth**, if different from the inner; all the **sepals**; **calyx-tube**, **calyx-lobes**, the **proximal** fused and **distal** free parts of a calyx in which the **sepals** are partly fused; Fig 1189

campanulate – bell-shaped, widest at the mouth

capillary – hair-like

capitate – head-like, such as a tight inflorescence on a stalk, a knob-like stigma on a style, or a stalked gland

capitulum – see 136. Dipsacaceae, 137. Asteraceae; Fig 1188

capsule – a dry, many-seeded dehiscent fruit formed from >1 carpel

carpel – the basic female reproductive unit of Magnoliopsida, 1-many per flower, if >1 then separate or fused; Fig 1189

carpophore – a stalk-like sterile part of a flower between the **receptacle** and **carpels**, as in some Apiaceae and

Caryophyllaceae; Fig 1189

cartilaginous - cartilage-like in consistency, hard but easily cut with a knife, not green

caryopsis - see 155. Poaceae

casual - an **alien** plant not **naturalized**

catkin - a condensed **spike** of reduced flowers on a long axis, often flexible and wind-pollinated

cauline - pertaining to the stem

cell - of an **ovary**, the chambers into which it may be divided (often each one corresponding to a **carpel**); Fig 1187

ciliate - with an edge fringed with hairs

cladode - see 160. Liliaceae

clavate - club-shaped, slender and distally thickened

claw - **proximal**, narrow part of a flat organ such as a **petal**, bearing the **blade distally**

cleistogamous - of flowers, not opening, becoming self-pollinated in the bud stage

column - a stout stalk formed by fusion of various floral parts in, e.g., Orchidaceae, Geraniaceae, Rosa; **columnar**, column-like

commissure - see 109. Apiaceae

compound - not **simple**, of a leaf divided right to the **rhachis** into **leaflets**; Fig 1187

compressed - flattened

cone - compact body composed of axis with lateral organs bearing spores or seeds, as in Lycopodiopsida, Equisetopsida, Pinopsida; **cone-scales**, the lateral organs of a cone

connective - part of **anther** connecting its 2 halves

contiguous - touching at the edges with no gap between

convergent - of ≥ 2 organs with apices closer together than their bases

cordate - of the base of a flat organ; see Fig 1186

coriaceous - of leathery texture

corm - short, usually erect, swollen underground stem

corolla - the inner **whorls** of the **perianth**, if different from the outer; all the **petals**; **corolla-tube**, **corolla-lobes**, the **proximal** fused and **distal** free parts of a **corolla** in which the **petals** are partly fused; Fig 1189

corona - see 160. Liliaceae

corymb - a **raceme** in which the lower flowers have longer **pedicels**, producing a $\pm$ flat-topped **inflorescence**; **corymbose**, corymb-like though strictly not a corymb; Fig 1188

cotyledon - the first leaves of a plant, 1 in Liliidae, usually 2 in Magnoliidae, 2-several in Pinopsida, usually quite different in appearance from all subsequent leaves

crenate - of the margin of a flat organ; see Fig 1186

culm - see 155. Poaceae

cuneate - of the base of a flat organ; see Fig 1186

cupule - see 40. Fagaceae

cuspidate - abruptly narrowed to a point; Fig 1186

cyme - an **inflorescence** in which each flower terminates the growth of a branch, more **distal** flowers being produced by longer branches lateral to it; **cymose**, in the form of a cyme; Fig 1188

decaploid - see **polyploid**

deciduous - not persistent, e.g. leaves falling in Autumn or **petals** falling after **anthesis**

decumbent - **procumbent** but with the apex turning up to become **ascending** or **erect**

decurrent - of a lateral organ, having its base prolonged down the main axis

decussate - opposite, with successive pairs at right angles to each other

dehiscent - opening naturally

dentate - of the margin of a flat organ; see Fig 1186

denticulate - minutely or finely **dentate**

depressed–globose - similar to **globose** but wider than long

dichasium - cyme with 2 lateral branches at each **node**; Fig 1188

dimorphic - occurring in 2 forms

dioecious - having the 2 sexes on different plants

diploid - having 2 matching sets of chromosomes, as in **sporophytic** tissue

disc - anything disc-shaped, e.g. top of <u>Rosa</u> or <u>Nuphar</u> fruit, nectar-secreting ring inside flower at base; **disc flowers**, see 137. Asteraceae

dissected - deeply divided up into segments

distal - at the end away from point of attachment

divaricate - dividing into widely **divergent** branches

divergent - of ≥ 2 organs with apices further apart than their bases

dorsiventral - with distinct upperside and lowerside

drupe - a **succulent** or spongy fruit, the seeds usually 1 and with a stony coat

dry - not **succulent**

e– - without, e.g. **eglandular, ebracteate**

ellipsoid - a solid shape elliptic in side view

elliptic - a flat shape widest in middle and 1.2–3x as long as wide (if less **broadly** so; if more, **narrowly** so); Fig 1186

embryo–sac - see **gametophyte, ovule**

endemic - confined to one particular area, i.e. (in this book) to BI

endosperm - In Magnoliopsida the nutritive tissue for the embryo in the developing seed; it might or might not remain as the food-store in the mature seed

entire - of the margin of a flat organ, not **toothed** or **lobed**; see Fig 1186

epicalyx - organs on the outside of a flower, calyx-like but outside and additional to the **calyx**

epicormic - of new shoots borne direct from the trunk of a tree

epigynous – of a flower with an **inferior ovary**; Fig 1189
erect – upright
erecto–patent – between **erect** and **patent**
escape – a plant growing outside a garden but having spread
  vegetatively or by seed from one
evergreen – retaining leaves throughout the year
exceeding – longer than
exserted – protruding from
exstipulate – without **stipules**; Fig 1187
falcate – sickle- or scythe-shaped
false–fruit – an apparent fruit actually formed by tissue
  (e.g. **receptacle, bracts**) in addition to the real fruit
fasciculate – in tight bundles
fastigiate – a plant with upright branches forming a narrow
  outline
fibrous roots – a root system in which there is no main axis;
  cf. **tap–root**
filament – stalk part of a **stamen**; Fig 1189
filiform – thread- or wire-like
flexuous – of a stem or hair, wavy
flore pleno – 'double' flower, with many more **petals** than
  normal, usually due to conversion of **stamens** to petals; in
  Asteraceae, a 'double' **capitulum**, with **disc flowers** all or
  many converted to **ray flowers**
floret – see 155. Poaceae; Fig 979
foliaceous – leaf-like, of an organ not normally thus
follicle – a dry, usually many-seeded fruit **dehiscent** along 1
  side, formed from 1 **carpel**
free – separate, not fused to another organ or to one another
  except at point of origin
free–central – of a **placenta** formed by central axis of any
  ovary that is not connected by **septa** to the wall; Fig 1189
fruit – the ripe, fertilized **ovary**, containing seeds
fusiform – spindle-shaped
gametophyte – the **haploid** generation of a plant that bears
  the true sex-organs (that produce the gametes), in
  pteridophytes the **prothallus,** in **spermatophytes** the pollen
  grains (male) and **embryo–sac** (female)
glabrous – hair-less; **glabrescent,** becoming glabrous
gland – a secreting structure, usually round or ± so, on the
  surface of an organ, below the surface, or raised on a
  stalk
glandular – with the functions of, or bearing, glands
glaucous – bluish-white in colour (rather than green)
globose – spherical
glume – see 154. Cyperaceae, 155. Poaceae; Fig 979
grain–alien – alien introduced as contaminant of grain
granulose – with a fine sand-like surface texture
gynodioecious – having female and **bisexual** plants
gynoecium – the group of female parts of a flower; all the
  carpels; Fig 1189
half–epigynous – of a flower with a **semi–inferior ovary**; Fig

1189
haploid - having only 1 set of chromosomes, as in
    gametophytic tissue
hastate - of the base of a flat organ; Fig 1186
heptaploid - see **polyploid**
herb - a plant dying down to ground-level each year
herbaceous - not woody, dying down each year; leaf-like as
    opposed to **woody, horny, scarious** or spongy
heterophyllous - having leaves of >2 distinct forms
heterosporous - having spores of 2 sorts (**megaspores**, female;
    and **microspores**, male), as in all **spermatophytes** and a few
    pteridophytes
heterostylous - having 2 forms (not sexes) of flower on
    different plants, the 2 sorts with different styles (and
    pollen)
hexaploid - see **polyploid**
hilum - scar on a seed where it left its point of attachment
hispid - with harsh hairs or bristles
homosporous - having spores all of 1 sort, as in most
    pteridophytes
homostylous - not **heterostylous**
hypanthium - extension of **receptacle** above base of **ovary**, in
    perigynous and **epigynous** flowers; Fig 1189
hyaline - thin and + transparent
hypogynous - of a flower with a **superior ovary**, the **calyx**,
    corolla and **stamens** inserted at the base of the ovary; Fig
    1189
imbricate - overlapping at edges
imparipinnate - pinnate with an unpaired terminal **leaflet**;
    Fig 1187
included - not exserted
incurved - curved inwards
indehiscent - not **dehiscent**
indusium - small flap or pocket of tissue covering groups of
    sporangia in many Pteropsida
inferior - of an **ovary** that is borne below the point of
    origin of the **sepals, petals** and **stamens** and is fused with
    the **receptacle** (**hypanthium**) surrounding it; Fig 1189
inflated - of an organ that is dilated, leaving a gap between
    it and its contents
inflorescence - a group of flowers with their branching
    system and associated **bracts** and **bracteoles**; Fig 1188
insertion - the position and form of the point of attachment
    of an organ
internode - the stem between adjacent **nodes**; Fig 979
introduced - a plant that owes its existence in this country
    to importation (deliberate or not) by man
introgression - the acquiring of characteristics by one
    species from another by hybridization followed by back-
    crossing
isodiametric - of any shape or organ, + the same distance
    across in any plane

isophyllous – not **heterophyllous**
keel – a longitudinal ridge on an organ, like the keel of a boat; see also 78. Fabaceae
labellum – see 164. Orchidaceae
laciniate – irregularly and deeply toothed; Fig 1186
laminar – in the form of a flat leaf
Lammas growth – extra, usually abnormal, growth put on in Summer by some trees (around Lammas Day, 1st Aug)
lanceolate – very narrowly **ovate**, c.6x as long as wide; Fig 1186
latex – milky juice
latiseptate – fruit with the **septum** across the widest diameter
lax – loose or diffuse, not dense
leaflet – a division of a **compound** leaf; Fig 1187
leaf–opposed – a lateral organ borne on the stem on opposite side from a leaf, not in a leaf-**axil** as usual
leaf–rosette – a radiating cluster of leaves, often at the base of a stem at soil level
legume – a usually dry, usually many–seeded fruit **dehiscent** along 2 sides, formed from 1 **carpel**
lemma – see 155. Poaceae; Fig 979
ligulate – see 137. Asteraceae
ligule – minute membranous flap at base of leaf of Isoetes or Selaginella; see also 137. Asteraceae, 155. Poaceae; Fig 979
limb – distal expanded part of a **calyx** or **corolla**, as distinct from the **tube** or **throat**
linear – long and narrow with $\pm$ parallel margins, i.e. extremely narrowly **oblong**; Fig 1186
lip – part of the **distal** region of a **calyx** or **corolla** sharply differentiated from the rest due to fusion or close association of its parts
lobe – a substantial division of a leaf, **calyx** or **corolla**; cf. **tooth**
lodicule – see 155. Poaceae; Fig 979
long–shoot – stem of potentially unlimited growth, especially in trees or shrubs
lowerside – the under surface of a flat organ
lunate – crescent moon–shaped
mealy – with a floury texture
megasporangium – in a **heterosporous** plant, the **sporangia** bearing **megaspores**
megaspore – in a **heterosporous** plant, the female spores that give rise to female **gametophytes**
meiosis – special form of cell–division (in **sporangia**, **pollen–sacs** or **ovules**) in which the chromosome number is halved, producing **haploid spores**
membranous – like a membrane in consistency
mericarp – a 1–seeded portion formed by splitting up of a 2–many seeded fruit, as in Geraniaceae, Apiaceae, Boraginaceae, Malvaceae; see also **schizocarp**

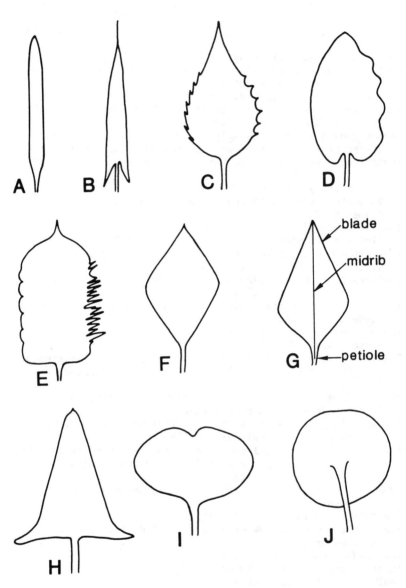

FIG 1186 – Simple leaf-shapes. A, **linear**, apex acute, base cuneate, margin entire. B, **lanceolate**, apex aristate, base sagittate, margin entire. C, **ovate**, apex acuminate, base rounded, margin serrate (left) and dentate (right). D, **elliptic**, apex obtuse, base cordate, margin entire (left) and sinuous (right). E, **oblong**, apex cuspidate, base truncate, margin crenate (left) and laciniate (right). F, **rhombic**, apex mucronate, base cuneate, margin entire. G, **trullate**, apex acute, base cuneate, margin entire. H, **triangular**, apex apiculate, base hastate, margin entire. I, **transversely elliptic**, apex retuse, base rounded, margin entire. J, **orbicular** and peltate, margin entire. Drawings by S. Ogden.

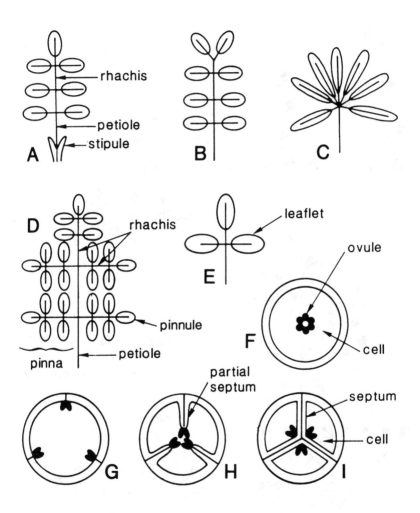

FIG 1187 – A–E, Compound leaf-types. A, **pinnate** (imparipinnate, stipulate). B, **pinnate** (paripinnate, exstipulate). C, **palmate**. D, **2-pinnate**. E, **ternate**. F–I, Ovaries in transverse section to show septa and placentation. F, 1-celled, **free-central** placentation. G, 1-celled, **parietal** placentation. H, 1-celled, **parietal** placentation. I, 3-celled, **axile** placentation. Drawings by S. Ogden.

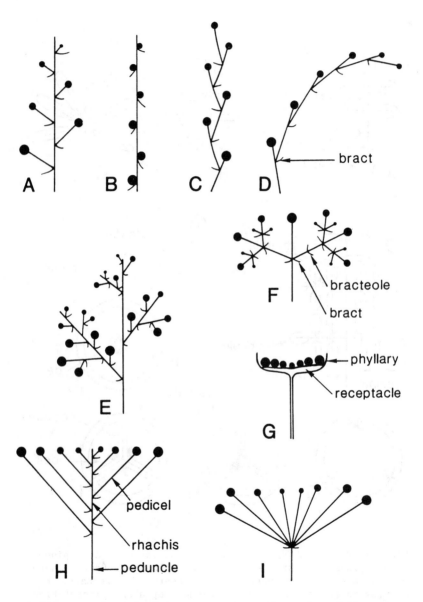

FIG 1188 – Inflorescences. A, **raceme**. B, **spike**. C–D, **cymes** (monochasial). E, **panicle**. F, **cyme** (dichasial). G, **capitulum**. H, **corymb**. I, **umbel**. Drawings by S. Ogden.

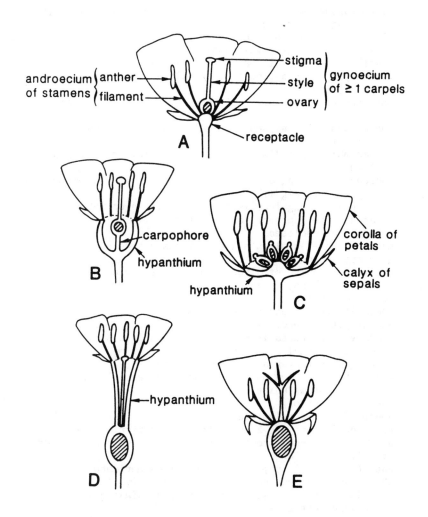

FIG 1189 – Half-flowers to show ovary position and other parts. A, **hypogynous** flower, ovary superior, carpels 1 or >1 and fused. B, **perigynous** flower with cup–shaped hypanthium, ovary superior, with carpophore, carpels 1 or >1 and fused. C, **perigynous** flower with flat hypanthium, ovary superior, carpels 4, free. D, **epigynous** flower with tubular hypanthium, ovary inferior, carpels 1 or >1 and fused. E, **epigynous** flower, ovary inferior, carpels 3, fused. Drawn by S. Ogden.

–merous – divided into particular number of parts; see
  trimerous, tetramerous, pentamerous
micron – a micrometre, i.e. one-thousandth of a millimetre or
  one–millionth of a metre
microsporangium – in a **heterosporous** plant, the **sporangia**
  bearing **microspores**
microspore – in a **heterosporous** plant, the male spores that
  give rise to male **gametophytes**
midrib – the central, main vein; Fig 1186
monocarpic – living for >1 year, flowering and fruiting, and
  then dying
monochasium – **cyme** with 1 lateral branch at each **node**; Fig
  1188
monoecious – having separate male and female flowers on the
  same plant
monomorphic – not dimorphic or trimorphic, occurring in 1
  form
mucronate – having a very short bristle-like tip; **mucro**, the
  tip itself; Fig 1186
mycorrhiza – fungal cells that live within or intimately
  around the roots of vascular plants
naked – not enclosed
naturalized – an **alien** plant that has become self-
  perpetuating in BI
native – opposite of alien, a plant that colonizied BI by
  natural means, often long ago
nec – nor, nor of
nectariferous – nectar-bearing
nectar–pit – a **nectariferous** pit
nectary – any **nectariferous** organ, usually a small knob or a
  modified **petal** or **stamen**
nodding – bent over and **pendent** at tip
node – the position on a stem where leaves, flowers or
  lateral stems arise; Fig 979
non – not, not of
nonaploid – see **polyploid**
nothomorph – one of >2 variants of a particular hybrid
nut – a dry, indehiscent 1-seeded fruit with a hard **woody**
  wall, often large
nutlet – a small nut, or a woody-walled **mericarp**
ob– – the other way up from normal, usually flattened or
  widened at the **distal** rather than **proximal** end, e.g.
  obovoid, obtrullate
oblong – a flat shape with middle part ± parallel-sided,
  1.2-3x as long as wide (if less, **broadly** so; if more,
  narrowly so); Fig 1186
obtuse – with a point >90 degrees; Fig 1186
octoploid – see **polyploid**
opposite – of 2 organs arising laterally at 1 node on
  opposite sides of the stem
orbicular – a flat shape circular in outline; Fig 1186
ovary – the basal part of the **gynoecium** containing the

ovules; Figs 979, 1187, 1189
ovate - a flat shape widest nearer the base and 1.2-3x as
  long as wide (if less, **broadly** so; if more, **narrowly** so);
  Fig 1186
ovoid - a solid shape **ovate** in side view
ovule - organ (inside the **ovary** in Magnoliopsida, naked
  in Pinopsida) that contains the **embryo-sac** (which in turn
  contains the egg, developing into seed after fertilization; Fig 1189
palea - see 155. Poaceae: Fig 979
palmate - a **compound** leaf, with >3 **leaflets** all arising at 1
  point; Fig 1187
panicle - a compound or much-branched **inflorescence**, either
  racemose or **cymose**; **paniculate**, in the form of a panicle;
  Fig 1188
papilla - small nipple-like projection; **papillose** or
  papillate, covered with papillae
pappus - see 137. Asteraceae
parasite - plant that gets all or some of its nourishment by
  attachment (often under the ground) to other plants
parietal - of a **placenta** formed by central axis of an **ovary**
  that is connected by **septa** to the wall; Fig 1187
paripinnate - **pinnate** without an unpaired terminal **leaflet**;
  Fig 1187
partial septum - a **septum** that is incomplete; Fig 1187
patent - projecting + at right-angles
pedicel - stalk of a flower; **pedicellate**, having a pedicel;
  Figs 979, 1188
peduncle - stalk of a group of >2 flowers; **pedunculate**,
  having a peduncle; Fig 1188
peltate - of a flat shape with its stalk arising from the
  plane surface, not the edge; Fig 1186
pendent - hanging down
pentaploid - see **polyploid**
perennial - living >2 years
perianth - the outer non-sexual covering layers of the flower
  (the **calyx** and **corolla** together), usually used when the
  calyx and corolla are not or little differentiated;
  perianth-lobes, perianth-tube, the **lobes** and **tube** of a
  partially fused perianth
pericarp - the wall of a fruit, originally the **ovary** wall
perigynous - of a flower with a **superior ovary** but with the
  calyx, corolla and **stamens** inserted above the base of the
  ovary on an extension of the **receptacle** (**hypanthium**) that
  is not fused with the ovary; Fig 1189
persistent - remaining attached longer than normal
petal - one of the segments of the inner **whorl(s)** of the
  perianth; **petaloid**, petal-like; Fig 1189
petiole - the stalk of a leaf; **petiolate**, with a petiole;
  Figs 1186, 1187
phyllary - see 137. Asteraceae; Fig 1188
pinna - the primary division of a >2-**pinnate** leaf; Fig 1187

pinnate - a **compound** leaf, with >3 **leaflets** arising in
   opposite pairs along the **rhachis**; 2-(etc)**pinnate**, pinnate
   with the **pinnae** pinnate again (etc.); see also
   paripinnate, imparipinnate; Fig 1187
pinnule - the ultimate division of a >2-**pinnate** leaf, usually
   applied only in ferns; Fig 1187
placenta - points of origin of **ovules** in an **ovary**; **placentation**, the arrangement of placentae; Fig 1187
plastic - varying in form according to environmental
   conditions, not according to genetic characteristics
pollen-sac - the microsporangium of a **spermatophyte**; one of
   the chambers in an **anther** in which the pollen is formed
pollinium - see 164. Orchidaceae
polyploid - having >2 sets of chromosomes, e.g. 3(**triploid**),
   4(**tetraploid**), 5(**pentaploid**), 6(**hexaploid**), 7(**heptaploid**),
   8(**octoploid**), 9(**nonaploid**), 10(**decaploid**)
prickle - spiny outgrowth with a broadened base
procumbent - trailing along the ground
proliferating - with inflorescences bearing plantlets instead
   of flowers or fruits
pro parte - partly; in part
prothallus - small **gametophyte** generation of a plant bearing
   the true sex-organs, mostly applied to the free-living
   gametophytes of **pteridophytes**
proximal - at the end near the point of attachment
pruinose - with a **bloom**
pteridophytes - ferns and fern allies, i.e. Lycopodiopsida,
   Equisetopsida, Pteropsida
puberulous - with very short hairs
pubescent - with hairs; **pubescence**, the hair-covering
punctate - marked with dots or transparent spots
raceme - an **inflorescence** with the oldest flowers (or
   spikelets in Poaceae) the most **proximal** lateral ones and a
   potentially continuously growing apex; **racemose**, in the
   form of a raceme; Fig 1188
radiate - see 137. Asteraceae
rank - a vertical file of lateral organs; 2-**ranked**, etc.,
   with 2 (etc.) ranks of lateral organs
ray - anything that radiates outwards, e.g. branches of an
   umbel, stigma-ridges in <u>Papaver</u> or <u>Nuphar</u>; **ray flowers**,
   see 137. Asteraceae
receptacle - the usually expanded, often cup-shaped or
   tubular, apical part of a **pedicel** on which the flower
   parts are inserted; **receptacular scales** or **bristles**, see
   137. Asteraceae; Figs 1188, 1189
recurved - curved down or back
reflexed - bent down or back
reniform - kidney-shaped
resiniferous - producing resin; **resinous**, resin-like; **resin-
   duct**, microscopic canal producing resin
reticulate - forming or covered with a network
retuse - notched at the apex; Fig 1186

revolute - rolled back or down
rhachilla - see 155. Poaceae; Fig 979
rhachis - the axis (not the stalk) of an **inflorescence** or
  pinnate leaf; Figs 1187, 1188
rhizome - an underground or ground-level, usually horizontal
  or down-growing stem, often ± swollen; **rhizomatous,**
  bearing or in the form of a rhizome; cf. **stolon**
rhombic - a flat shape, widest in the middle and ± angled
  (not rounded) there, 1.2-3x as long as wide (if less,
  broadly so; if more, **narrowly** so); Fig 1186
rigid - stiff, not flexible
rostellum - see 164. Orchidaceae
rounded - without a point or angle; Fig 1186
rugose - with a wrinkled surface
rugulose - finely rugose
sagittate - of the base of a flat organ; see Fig 1186
saprophyte - a plant deriving its nourishment from decaying
  organisms, usually leaf-mould
scabrid - rough to the touch, with minute prickles or
  bristly hairs
scale-leaf - a leaf reduced to a small scale
scape - a flowering stem of a plant in which all the leaves
  are basal, none on the scape
scarious - of thin, papery texture and not green
schizocarp - a fruit that breaks into 1-seeded portions or
  mericarps
sclerenchyma - woody tissue in a partly or mostly non-woody
  organ
scorpioid - a **monochasial cyme** that is coiled up like a
  scorpion's tail when young
scrambler - a plant sprawling over other plants, fences, etc.
seed - a fertilized **ovule**
self-compatible - self-fertile, able to self-fertilize
self-incompatible - self-sterile, not able to self-fertilize
semi-inferior - of an **ovary** of which the lower part is
  inferior, but the upper part is **free** and projects above
  the **sepals** etc.
sensu lato - in the broad sense
sensu stricto - in the narrow sense
sepal - one of the segments of the outer **whorl**(s) of the
  perianth; sepaloid, sepal-like; Fig 1189
septum - a wall or membrane dividing the **ovary** into **cells;**
  Fig 1187
sericeous - with silky, **appressed** straight hairs
serrate - with a row of ± apically directed **teeth; serration,**
  that sort of toothing; Fig 1186
sessile - not stalked
sheath - see 155. Poaceae; Fig 979
short-shoot - a short stem of strictly limited growth,
  usually lateral on a **long-shoot,** especially on **trees** and
  shrubs
shrub - a **woody** plant that is not a **tree**

silicula – a **dehiscent**, 2-**valved**, 2-**celled capsule** <3x as long as wide, in Brassicaceae

siliqua – a **dehiscent**, 2-**valved**, 2-**celled capsule** >3x as long as wide, in Brassicaceae

simple – not compound; Fig 1186

sinuous – wavy, either of a hair or stalk, or of the margin of a leaf and then the sinuation in the same plane as the leaf surface; Fig 1186

sinus – the space or indentation between 2 **lobes** or **teeth**; **basal sinus** the sinus at the base of a leaf, either side of the **petiole** if present

solitary – borne singly

sorus – a group of sporangia in Pteropsida

spadix – see 149. Araceae

spathe – an ensheathing **bract**, as in Lemnaceae, Araceae, Hydrocharitaceae

spathulate – paddle- or spoon-shaped

spermatophyte – a seed-plant, i.e. Pinopsida and Magnoliopsida

spike – a **racemose inflorescence** in which the flowers (or **spikelets** in Poaceae) have no stalks; Fig 1188

spikelet – see 49/1. Plumbaginaceae/<u>Limonium</u>, 154. Cyperaceae, 155. Poaceae; Fig 979

spine – a sharp, stiff, straight woody outgrowth, usually not greatly widened at base; **spinose**, spine-like; **spinulose**, diminutive of last; **spiny**, with spines

spiral – lateral organs on an axis 1 per node, successive ones not at 180 degrees to each other

sporangium – a body producing **spores** in **pteridophytes**

spore – the **haploid** product of **meiotic** division, produced on the **sporophyte** and developing into the **gametophyte**

sporophyte – the **diploid** generation of a plant that bears the **sporangia** (**ovules** and **pollen-sacs** in Magnoliopsida); the main plant body of all vascular plants

spreading – growing out **divergently**, not straight or **erect**

spur – a protrusion or tubular or pouch-like outgrowth of any part of a flower

stamen – the basic male reproductive unit of Magnoliopsida, 1-many per flower, sometimes fused; Figs 979, 1189

staminode – a sterile **stamen**, sometimes modified to perform some other function, e.g. that of a **petal** or **nectary**

standard – see 78. Fabaceae

stellate – star-shaped, with radiating arms

stem-leaves – leaves borne on the stem as opposed to basally

stigma – the apical part of a **gynoecium** that is receptive to pollen, simple to much branched; Figs 979, 1189

stipule – one of a (usually) pair of appendages at the base of a leaf or its **petiole**, often but often not **foliaceous**; **stipulate**, with stipules; Fig 1187

stolon – an **aerial** or **procumbent** stem, usually not swollen; **stoloniferous**, bearing stolons; cf. **rhizome**

style – the stalk on any **ovary** bearing the **stigma**(s),

sometimes absent; Fig 1189
stylopodium – see 109. Apiaceae
sub– – almost, as in **subacute**, **subglabrous**, **subglobose**,
   subentire, **subequal**; sometimes under, as in **subaquatic**
subshrub – a **perennial** with a short **woody** surface stem
   producing **aerial herbaceous** stems
subtend – of a lateral organ, to have another organ in its
   axil
subulate – narrowly cylindrical but drawn out to a fine point
succulent – fleshy and juicy or pulpy
sucker – a new **aerial** shoot borne (often underground) on the
   roots of a **tree** or **shrub**
superior – of any **ovary** that is borne above the **calyx**,
   corolla and **stamens** or, if below or partly below them,
   then not fused laterally to the **receptacle**; Fig 1189
suture – a seam of a union, often splitting open in later
   development
tap–root – a main descending root bearing laterals; cf.
   fibrous roots
taxon – any taxonomic grouping, such as a genus or species
tendril – a spirally coiled thread-like outgrowth from a
   stem or leaf, used by the plant for support
tepal – one of the segments of the **perianth**, used when **sepals**
   and **petals** are not differentiated
terete – rounded in section
terminal – at the very apex
ternate – a **compound** leaf with 3 **leaflets**; 2-(etc.)**ternate**,
   ternate with the 3 divisions ternate again; Fig 1187
testa – the outer coat of a seed
tetrad – a group of 4 **spores** or pollen grains formed by
   meiotic division
tetraploid – see **polyploid**
tetraquetrous – square in section
throat – the opening where the **tube** joins the **limb** of a
   corolla or **calyx**
tiller – see 155. Poaceae; Fig 979
tomentose – a very dense often $\pm$ matted hair-covering
tooth – a shallow division of a leaf, **calyx** or **corolla**, or of
   the apex of a **capsule**; cf. **lobe**, **valve**
transverse – lying cross-ways; **transversely elliptic** (etc.),
   elliptic (etc.) but with the point of attachment at the
   side, not at one end; Fig 1186
tree – a woody plant, usually >5m, with a single trunk
triangular – a flat shape widest at the base and 1.2-3x as
   long as wide (if less, **broadly** so; if more, **narrowly** so);
   Fig 1186
trifoliolate – a term used in Fabaceae in lieu of **ternate**
trigonous – **triangular** in section, with **obtuse** to **rounded**
   angles
trimorphic – occurring in 3 forms
tripartite – divided into 3 parts
triploid – see **polyploid**

triquetrous – **triangular** in section, with **acute** angles
trullate – a flat shape widest nearer the base and ± angled
(not rounded) there, 1.2–3x as long as wide (if less,
broadly so; if more, **narrowly** so); Fig 1186
truncate – of the base or apex of a flat organ, straight or
flat; Fig 1186
tube – narrow, cylindrical, **proximal** part of a **calyx** or
corolla, as distinct from the **limb**, **lobes** or **throat**
tuber – swollen roots or subterranean stems; **tuberous**, tuber-
like
tubercle – a small ± spherical or **ellipsoid** swelling;
tuberculate, with a surface texture covered in minute
tubercles
tubular – in the form of a hollow cylinder; **tubular** flowers,
see 137. Asteraceae
tufted – of elongated organs or stems that are clustered
together
twig – ultimate branch of a woody stem
umbel – an **inflorescence** in which all the **pedicels** arise from
one point; **compound umbel**, an umbel of umbels; **umbellate**,
umbel-like; Fig 1188
undulate – wavy at the edge in the plane at right-angles to
the surface
unifacial – with only 1 surface, not with a **lowerside** and
upperside
unisexual – of a plant or flower, bearing only 1 sex
upperside – the upper surface of a flat organ
valve – a deep division or **lobe** of a **capsule** apex; cf. **tooth**
vascular – pertaining to the **veins** or wood (i.e. conducting
tissue) of an organ; **vascular bundle**, one anatomically
discrete file of vascular tissue
vegetative – not reproductive
vein – a strand of **vascular** tissue consisting of >1 **vascular**
bundles; **venation**, the pattern of veins
verrucose – covered in small wart-like outgrowths
viscid – sticky
viscidium – see 164. Orchidaceae
whorl – a group of lateral organs borne >2 at each **node**;
whorled, in the form of a whorl
wing – any **membranous** or **foliaceous** extension of an organ,
e.g. a stem, seed or fruit; **winged**, with a wing; see also
78. Fabaceae
woody – hard and wood-like, not quickly dying or withering
wool-alien – an **alien** introduced as a contaminant of raw
wool imports
woolly – clothed with shaggy hairs
zygomorphic – of a flower with bilateral (i.e. 1 plane of)
symmetry

INDEX

Only Latin families (accepted ones in capitals), Latin generic names (accepted ones in bold) and English 'generic names' are indexed, with two major exceptions. Firstly, for 8 of the largest Latin genera there is a species index as well; hybrid binomials are included but not differentiated from species binomials. Secondly, English 'generic names' are often differentiated by addition of the English 'specific name' where they occur in more than one Latin genus. Reference is always given to the main text entry, not to the keys, illustrations or secondary entries, and for generic names (both English and Latin) reference is given only to the first main mention in each genus (usually the generic heading). Where the English and Latin generic names are the same in the same genus, only the Latin is referred to.